SURVEY OF SUBSAHARAN
AFRICA

SURVEY OF SUBSAHARAN AFRICA

A Regional Geography

ROY COLE
Grand Valley State University

H.J. DE BLIJ
Michigan State University

NEW YORK OXFORD
OXFORD UNIVERSITY PRESS
2007

Oxford University Press, Inc., publishes works that further Oxford University's objective of excellence in research, scholarship, and education.

Oxford New York
Auckland Cape Town Dar es Salaam Hong Kong Karachi
Kuala Lumpur Madrid Melbourne Mexico City Nairobi
New Delhi Shanghai Taipei Toronto

With offices in
Argentina Austria Brazil Chile Czech Republic France Greece
Guatemala Hungary Italy Japan Poland Portugal Singapore
South Korea Switzerland Thailand Turkey Ukraine Vietnam

Published by Oxford University Press, Inc.
198 Madison Avenue, New York, New York 10016
http://www.oup.com

Library of Congress Cataloging-in-Publication Data

Cole, Roy.
Survey of Subsaharan Africa: a regional geography / Roy Cole and Harm J. De Blij.
p. cm.
Includes bibliographical references and index.
ISBN-13: 978-0-19-517080-1
ISBN-10: 0-19-517080-6 (cloth.)
1. Africa, Subsaharan--Geography. I. De Blij, Harm J. II. Title.

DT351.9.C65 2005
916.7--dc22 2005054675

Printing number: 9 8 7 6 5 4 3 2 1

Printed in the United States of America
on acid-free paper

CONTENTS

CENTRAL AFRICA 403

EAST AFRICA, THE HORN, AND SUDAN 449

SOUTHERN AFRICA 567

FIGURES AND TABLES

FIGURES

TABLES

PREFACE

The purpose of this text is to provide an empirical, analytical, and in places even provocative text for undergraduate students on the regional geography of Africa. The book is meant to be informative as well as challenging, and is intended to stimulate discussion and debate. It covers mainly the geographic realm of Subsaharan Africa, but where appropriate it extends beyond this conceptual region to address topics in North Africa as well. Africa is vast, complex, diverse, and constantly changing, and while no book can address every relevant issue, every effort has been made to present balanced coverage.

This book constitutes a revision and expansion of a text now more than three decades out of print, *African Survey* by A. C. G. Best and H. J. de Blij (Wiley 1977). That book was unique among regional geographies of Africa because it focused on themes ranging from environmental and economic to social and cultural. All of those themes were linked to development issues.

Although the present book retains aspects of its predecessor, it also differs from it in several significant respects. Several new chapters form part of an expanded discourse on the continent as a whole. Other chapters were combined or divided; virtually every chapter was substantially rewritten and enlarged. All the cartography was updated and numerous new maps, graphs, tables, and photographs were included. In addition, readers may download all of the maps and graphs in color at the following Internet address: http://www4.gvsu.edu/coler/ssa. Regional chapters on North Africa are available for download at the same site. Readers are also invited to contact Dr. Cole with questions at coler@gvsu.edu.

Many of the themes presented in the regional parts of the original *African Survey* still are relevant in this new century and have been retained, although in some instances with a touch of irony—take, for example, the title of Chapter 26, "Zimbabwe: Failure in Partnership." In the 1970s, that failure alluded to the racist policies of Ian Smith's white minority regime in what was then still called Rhodesia. Today, after more than 25 years of majority rule, there still is a failure in partnership, but now it has a different cause. Other African states have changed so much that they are unrecognizable in the earlier book. Botswana, for example, has been evolving from a mainly pastoral economy to a banking and information-technology center. South Africa is emerging as the powerhouse engine of a wider Southern African region. Angola, once the "insurgent state," is now recovering from decades of externally infused civil war. Democratic Ghana has become a cornerstone of the new West Africa. In more general terms, socialism and the one-party state, once believed to be the best way forward for all of Africa, have been discredited. Statist models of development once thought to be the answer to the ills of African economies are now viewed as fundamental impediments to the very process itself. Current "winds of change" are those of democracy, human rights, and private enterprise. North Africa's proximity to Europe is transforming this region. Radical Islam and theocracy came and went, although some challenges are surely yet in store. Our views of human-environment relations have become much more nuanced than they were in the 1970s, when analysis tended to be informed by neo-Malthusian theory.

Survey of Subsaharan Africa addresses the physical and environmental geography of Africa as well as such topics as indigenous states, the evolution of the political-geographic framework, demographic issues, urbanization, and health conditions. The emphasis varies in the regional sections, because African countries are far more varied and diverse than they are in the public (and, we know, student) imagination. This, therefore, is a readable volume as well as a rigorous text, and as such it reveals the integrating nature of the discipline of geography, whose unifying spatial perspective prevails throughout. We hope that the book also reflects our affection for a matchless part of the world, where the saga of humanity started and where physical and cultural landscapes have meanings like nowhere else in our collective experience.

Roy Cole
Holland, Michigan
coler@gvsu.edu

H.J. de Blij
East Lansing, Michigan
deblij@c4.net

ACKNOWLEDGMENTS

The authors would like to thank the following people at Oxford University Press for their excellent and tireless work on *Survey of Subsaharan Africa:* John Challice, Vice President/Publisher for Higher Education; Lisa Grzan, Production Editor; Elyse Dubin, Director, Editing/Design/Production; Chelsea Gilmore, Assistant to the Publisher; and the superb copyeditor, Brenda Griffing. The authors would like to thank the reviewers who read the text and provided comments: Byron D. Augustin of Texas State University, Alyson L. Greiner of Oklahoma State University, Garth Myers of the University of Kansas, Earl P. Scott of the University of Minnesota—Minneapolis, and all of the other anonymous reviewers who provided such helpful comments during the preparation of this book. This work would not have been possible without the support of Chris Rogers at Oxford University Press during its early stages.

The faculty of the Department of Geography and Planning at Grand Valley State University has been especially encouraging of this work, particularly the chair, Dr. Jeroen Wagendorp and Drs. Elena Lioubimtseva and James Penn. Dr. Fred Antczak, Dean of the College of Liberal Arts and Sciences (CLAS), has supported this project from the beginning and deserves special mention. The encouragement of the African and African American Studies Center at GVSU, particularly that of Dr. Jacques Mangala, the director, has been very important, as has the advice and support of Dean Wendy Wenner of the College of Interdisciplinary Studies (CoIS), where the African and African American Studies Center is housed. The Department of Geography and Tourism at the University of Cape Coast, Ghana, where much of the manuscript editing was accomplished during the fall of 2005 was very supportive as well. Dr. Kofi Awusabo-Asare, Dean of the Faculty of Social Sciences, and Dr. Albert Abane, Head of the Department of Geography and Tourism, deserve personal thanks for their support, as does Mr. Ekow Afful-Wellington for his thoughtful comments on the legacy of Jerry Rawlings.

During my almost 10 years of (mostly) village life in Africa innumerable people contributed in small and large ways to the ideas in this book. Although I have lost touch over the years with many of the people who are named below, they each deserve an explicit word of thanks for helping me on a long journey. Profound thanks to two of my adoptive families: Mor Diakhèté and his family and the people of Dara Palmeo, the Traoré family and neighbors in Djado, and a special friend, Daouda Tikampou, who is a Bamanake at heart. Many Ngarakow have welcomed me over the years, especially the Tangara family. Thank you Mamadou Diallo, Pullo gorko lobbo, and Coumbel Barry for accepting me into your family. Thank you Pierre Hiernaux and Fanta Sangho.

Dr. J. O. McShine sparked my interest in Africa many years ago. Dr. David Wiley, Director of the African Studies Center at Michigan State University, supported my work in graduate school. Drs. Kulikoyeli Kahigi, Abd al-Ghaffar ad Damatty, and Malik Towghi, student colleagues and mentors of my graduate school days, helped me see many different points of view.

For the kind permission to use their photographs I would like to thank Steven Blakeway of Vetwork/UK; Dr. Christine Drake in the Department of Political Science and Geography at Old Dominion University; Sara Fisher, formerly of Grand Valley State University; "Gritty.org" (Michael Wadleigh and Cleo Huggins); Dr. Vera Matlova at the Research Institute of Animal Production in the Czech Republic; Steven Montgomery; David Moorehead at the United States Department of

Agriculture (USDA); the National Aeronautics and Space Administration (NASA); the United States Geological Survey (USGS); A. van Genderen Stort at the United Nations High Commission for Refugees (UNHCR); and the World Health Organization.

Without the strong support and patience of my family this book would never have been completed. But what kept me going was "ni san t'i pan, a den ka ŋuŋuma."

I also acknowledge with great appreciation the support of my coauthor, Dr. H. J. de Blij, whose sustained interest in this project has helped keep it on track and led to its completion.

Roy Cole

CHAPTER 1

The Physiography of Africa

Alone among the continents, Africa is positioned astride the equator, extending beyond latitude 35° north and nearly reaching 35° south. The continent's northernmost areas, the countries of the Maghreb, lie in the general latitude of North and South Carolina. South Africa shares its latitude with southern Brazil and Uruguay. Tunis, capital of Tunisia, lies in the approximate latitude of Richmond, Virginia, and San Francisco, California. Cape Town, South Africa's second city, lies almost exactly due east of Buenos Aires, Argentina, and Montevideo, Uruguay. Coupled with this equator-straddling situation is Africa's position at the heart of the land hemisphere. Alone among the landmasses, only Africa lacks a Pacific coastline. This means that Africa has a minimum aggregate distance to the world's other continents as well as a central location that is antipodal to the Pacific Ocean (Figure 1-1).

The African continent constitutes the second largest landmass on this planet, after Eurasia. It is 11.7 million square miles in area (30.3 million km^2) and stretches some 5,000 miles (8,000 km) from Bizerte, Tunisia, to Cape Agulhas, South Africa, and an almost equal distance from Dakar in the west to Gees Gwardafuy (Cape Gardafui) in the east. Africa's dimensions and its compact shape are two of its most distinctive physical features. Africa's coastline is remarkably straight and unbroken over distances of many hundreds of miles; there are few really large bays or estuaries, and no substantial island arcs lie off its shores. Nor is Africa flanked by continental shelves comparable to those off Eurasia or North America. In general, the ocean floor drops sharply, and quite near the coastline, to great depths. Almost equally abrupt is the sharp rise of the land, in many places as a spectacular mountain wall. The Great Escarpment, as this wall is called, exists in places as widely distributed as Sierra Leone, Namibia, Lesotho, and Ethiopia. The escarpment in many areas (such as Angola and South Africa) lies within a few dozen miles of the coast, so that Africa, per unit area, has relatively little low-lying territory and even less coastal plain.

PRINCIPAL FEATURES

Plateaus and Basins

Africa's continental physiography is that of an enormous plateau, higher in the east and south, and lower in the west and north, although the Atlas, Ahaggar (Figure 1-2), and Tibesti mountains rise above the lower northern elevations. High Africa includes the Ethiopian Massif, where elevations exceed 14,000 feet (4,250 m), the extensive East African plateau, and South Africa's Drakensberg Mountains, which reach over 11,000 feet (3,350 m). The highest point along this eastern axis of the great African plateau is Mount Kilimanjaro, a dormant volcano whose snow-capped peak exceeds 19,300 feet (5,880 m). Much of the plateau of eastern Africa, however, presents a gently undulating landscape broken only by hills and ridges sustained by more resistant rocks; elevations average between 4,000 and 5,000 feet (1,200–1,500 m). Along the plateau's eastern margin, from the Red Sea to South Africa, lies the Great Escarpment, almost everywhere a prominent obstacle to road, rail, and water transport.

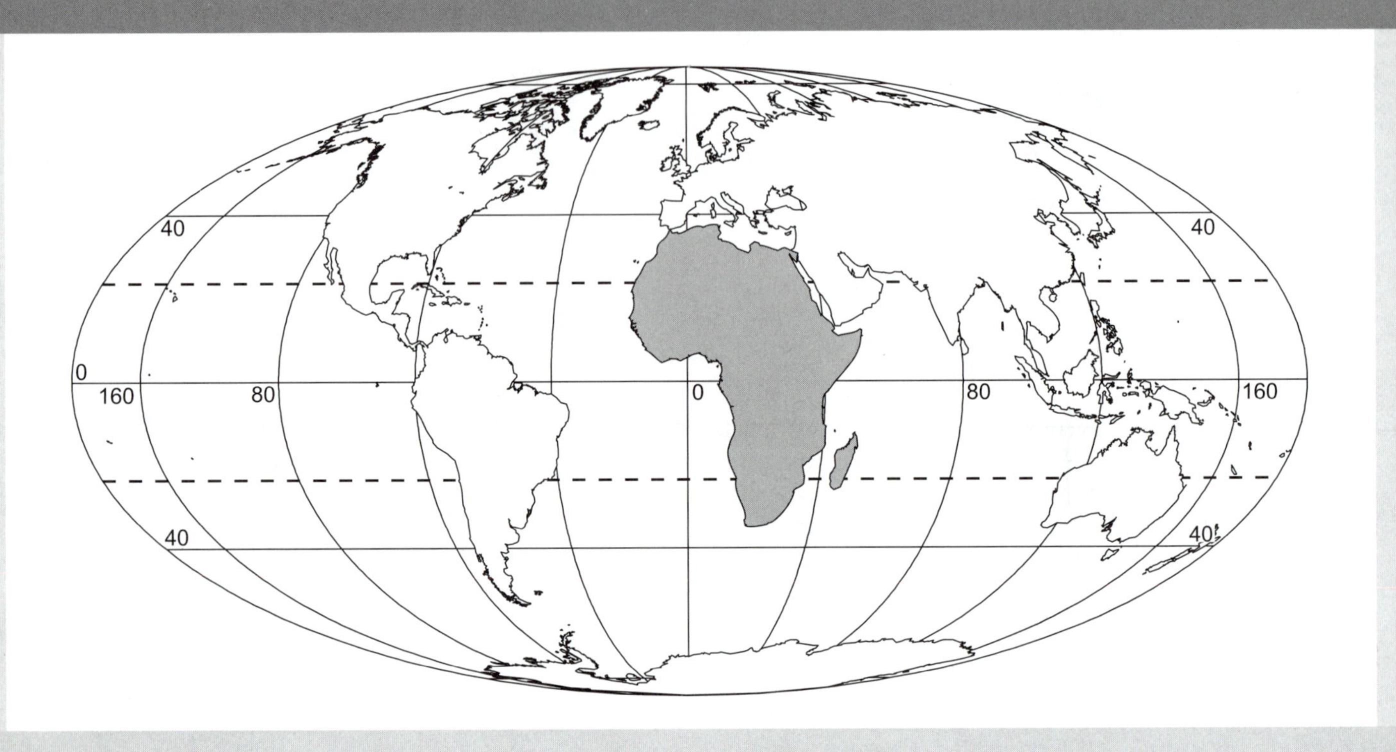

FIGURE 1-1 Location of Africa.

Madagascar, 250 miles (400 km) from the southeast African coast, mirrors the East African physiography. A high-elevation plateau axis forms the island's eastern backbone, with some mountains exceeding 8,000 feet (2,400 m) and a steep escarpment facing the Indian Ocean. Westward, elevations decline. Again the crystalline rocks support the plateau surface, volcanics sustain the greater heights, and sedimentary rocks underlie the lower west.

To the west of Africa's great eastern upland, the continental surface subsides into several major basins separated by plateau-level drainage divides (Figure 1-3). Astride the equator lies the Congo Basin, a great structural depression filled with sedimentary layers and surrounded by the crystalline rock of the plateau. To the south, the Kalahari Basin occupies the area between the South African–Zimbabwean highveld and the high coastal ranges of Namibia. Unlike the forested Congo Basin, the Kalahari Basin is a sandy desert and steppe, but the two depressions share the accumulations of thick sediments that weighed down the underlying crystalline floor. North of the Congo Basin, and beyond the North Equatorial Divide, lie three major interior basins. Easternmost is the Sudan Basin, drained by the White Nile and the Blue Nile. Much of this basin lies below 1,000 feet (300 m) in elevation, and in the south it contains the Sudd, one of the world's most extensive marshlands. Here the White Nile divides into numerous distributaries as it penetrates dense masses of floating vegetation; its gradient is reduced to under 140 feet (43 m) over a distance of nearly 800 miles (1,250 km).

The western boundary of the Sudan Basin is marked by the Plateau of Darfur and the Ennedi Plateau, and there the crystalline rocks reappear in such prominent landmarks as the Marra Mountains (Jebel Marra). Still farther to the west lies the Chad Basin, the second of the great northern depressions (Figure 1-4). The core of the Chad Basin is constituted by the Bodélé Depression and Lake Chad. Formerly much more extensive than it is today, Lake Chad has

FIGURE 1-2 Ahaggar Mountains, located in the central Sahara. *Short and Blair (1986) USGS.*

dwindled to a swampy area that is inundated after heavy rains and is being encroached on by sand from the Sahara. The Chad Basin's northern boundaries are defined quite clearly by the Tibesti Massif and the Ahaggar Mountains. The western boundary is less well expressed, although a southward extension of the Ahaggar reaches close to the plateau of northern Nigeria, the Jos Plateau. The westernmost of the great basins, the Depression of Djouf, occupies the western Sahara and reaches its fullest development in the region along the Mauritania–Mali boundary. It is separated from the coast by high ranges such as the Fouta Djallon (Futa Jallon) and forms the drainage area for the upper and middle Niger River. The heart of the Djouf Basin, an area of sedimentary accumulation like other basins, lies below 500 feet (150 m) in elevation.

Rift Valleys

A rift valley results from the spreading of crust in the interior of a continental plate (Goudie 1994) and is a form of crustal divergence that includes such phenomena as seafloor spreading. Although the prevailing scenery in Africa is that of the flat or gently undulating plateau, the surface is broken—often in spectacular fashion—by a great system of troughs that extends from the Dead Sea through the Red Sea to Ethiopia, and ultimately to South Africa. The system's total length in Africa exceeds 6,000 miles (9,600 km), beginning in the north with the Red Sea and extending through the heart of Ethiopia to Lake Turkana, then dividing into eastern and western segments in East Africa and continuing through Lake Malawi and ultimately to Swaziland and Natal in southern Africa. These are East Africa's great rift valleys, oriented north to south with subsidiary faults running north to northwest and north to northeast. In spite of their length, the rift valleys remain remarkably uniform over great distances. In general, the rifts from Lake Turkana southward are between 20 and 40 miles (30–90 km) wide, and the walls, sometimes sheer and sometimes steplike, are well defined. Almost wherever it may be observed, whether in Ethiopia, Kenya, or Swaziland, the rift is unmistakable in appearance.

From the plateau rim, the land falls suddenly to a flat lowland that often possesses climatic and vegetative characteristics quite unlike those above the fault scarp (Figure 1-5). In the far distance lies the opposite scarp, and the plateau resumes. The floor of the rift

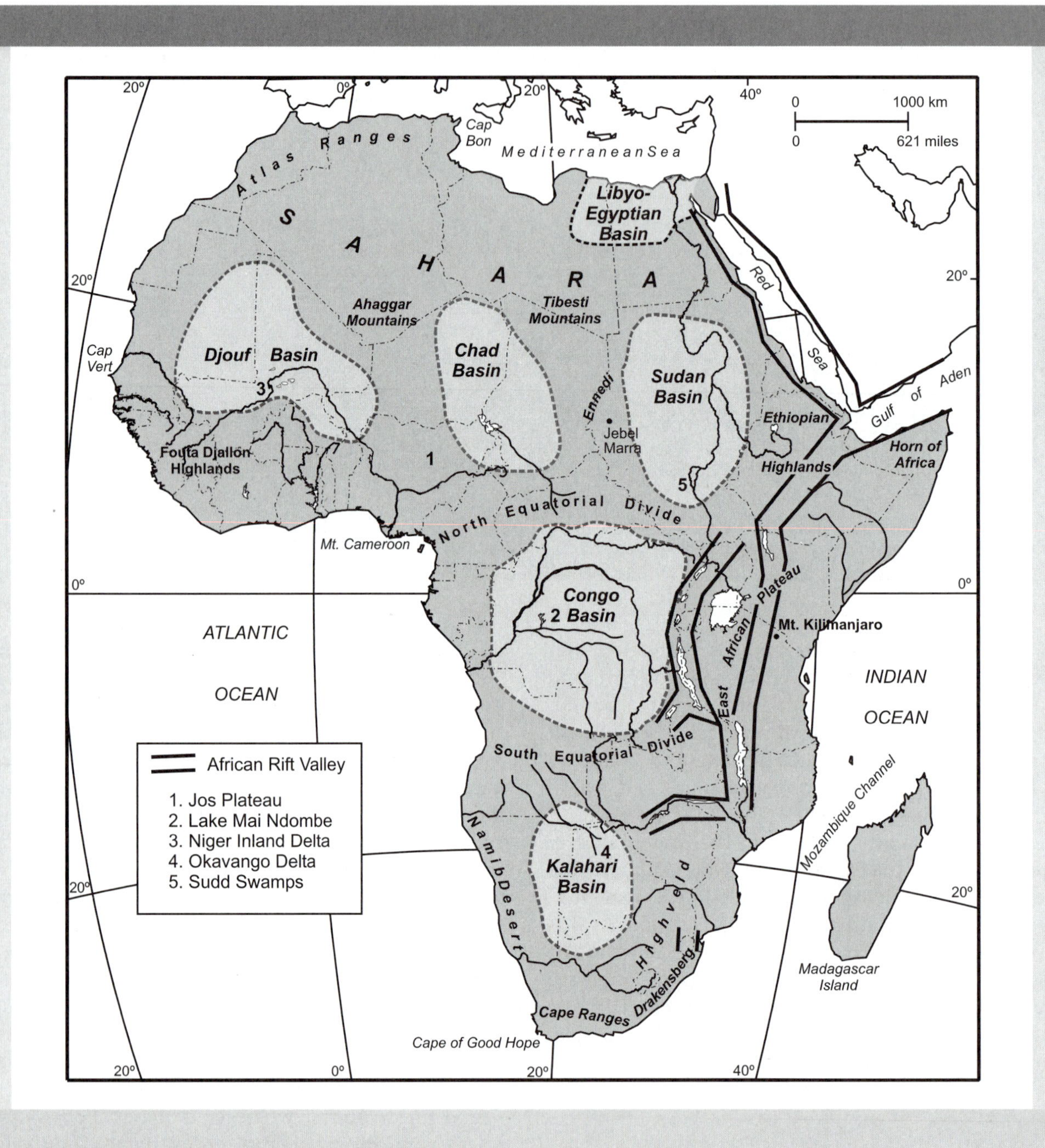

FIGURE 1-3 Basins, mountains, and drainage systems of Africa.

is not uniform in elevation. The narrow strip of land between the echelon faults lies far below sea level in some places and thousands of feet above it in others. The degree of variation is expressed especially well in the Western Rift of East Africa, where the floor of Lake Tanganyika is as much as 2,140 feet (650 m) below sea level—but some distance to the north, the rift floor lies more than 5,000 feet (1,500 m) *above* sea level.

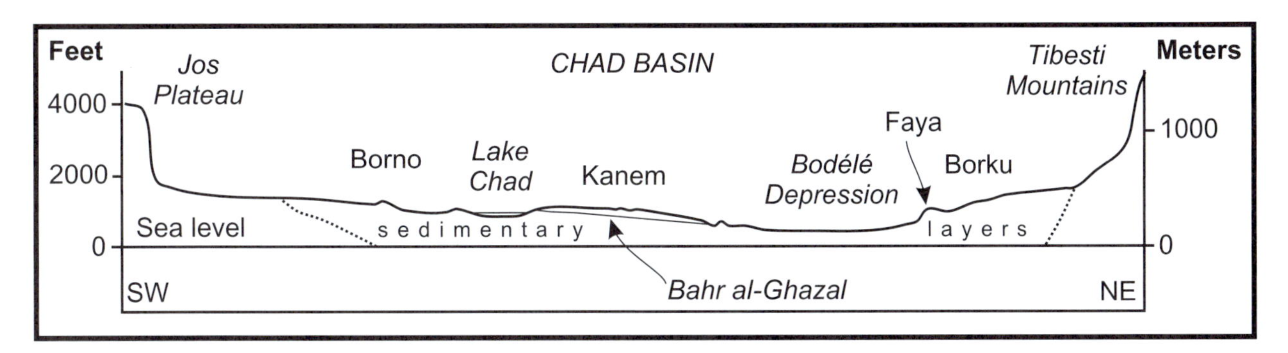

FIGURE 1-4 Cross section of the Chad Basin from the Jos Plateau to Tibesti.
Adapted from Grove (1989).

The association between the rift valleys and the lakes of East Africa is so obvious that it hardly requires emphasis. The lakes of the Western Rift are larger than those of the eastern segment, but there is evidence that lake levels have fluctuated considerably during geologic time and that the rift lakes were larger than they are at present. Raised beaches and terraces indicate higher stages during the Pleistocene (see the Table of geologic time in Appendix 1, page 16), but elsewhere there is evidence of more recent (perhaps temporary) rise of lake levels. In any case, the lake waters are not (not yet, at least) invasions of the ocean. Their saltiness relates to local source areas and rapid evaporation from the surface.

One prominent East African lake, Africa's largest, lies between, not within, the rifts: Lake Victoria, with an area of 24,300 square miles (63,000 km^2). Lake Victoria is also one of Africa's shallower lakes, under 300 feet (90 m) in depth, and it does not display the attenuated shape of the other larger East African lakes. This is because Lake Victoria lies in a broad depression rather than a rift valley. The lake's shorelines and adjacent drainage lines reveal that the area was formerly a divide, with rivers flowing eastward to the Indian Ocean and westward into the Congo Basin. Then this broad plateau was depressed, forming the basin now occupied by the lake. River flows were reversed, and

FIGURE 1-5 Ngorongoro volcano caldera in East Africa, forms part of a rift valley.
Gritty.org.

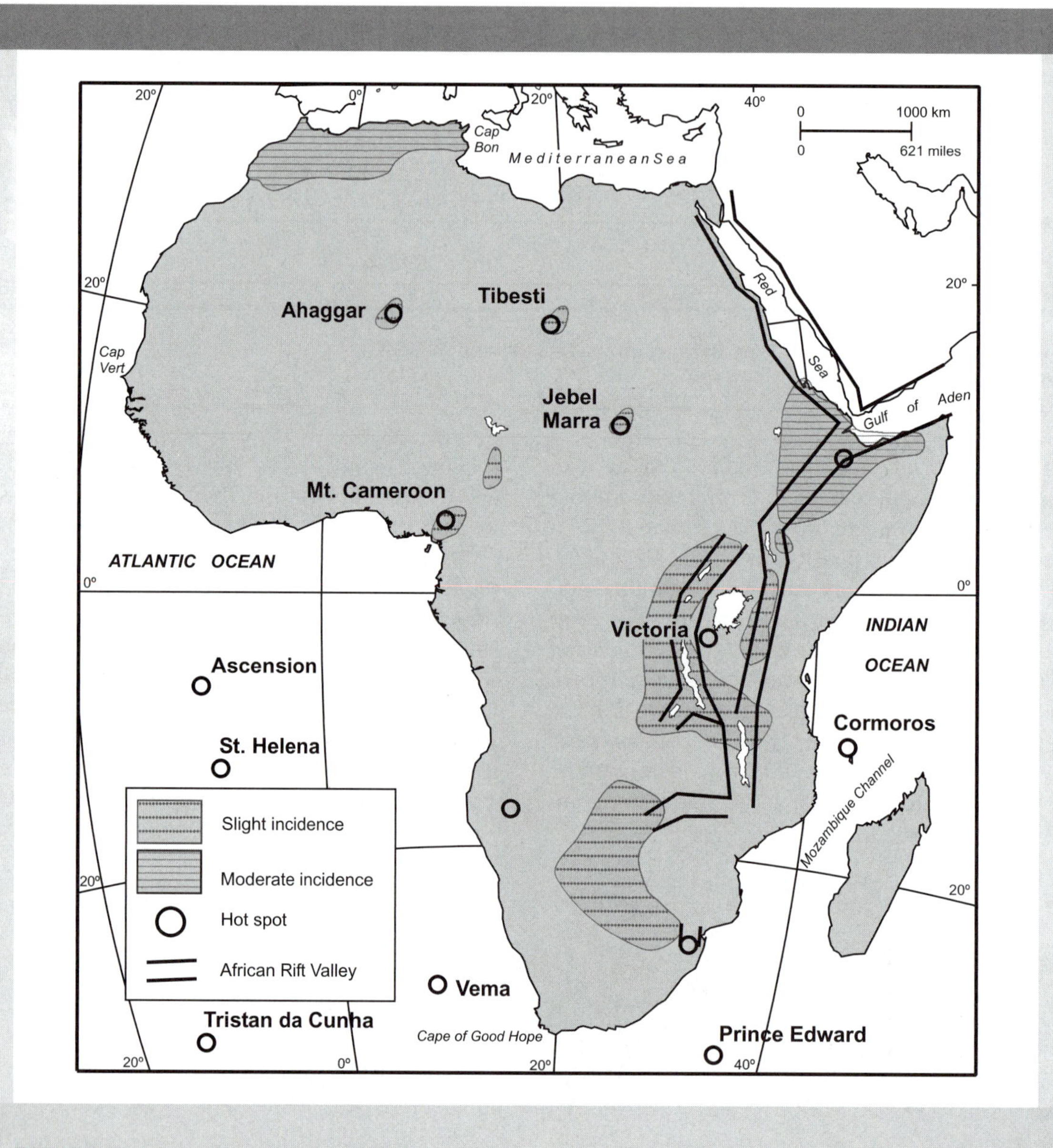

FIGURE 1-6 Seismic and volcanic activity in Africa.

Lake Victoria began to empty northward, feeding the White Nile. A relationship exists between the rift movements (which further disrupted East Africa's drainage patterns) and the formation of the Victoria Depression between the rifts.

Maps of seismic activity and the incidence of volcanoes in Africa reflect another spatial relationship with the rift valley system. East Africa, from Ethiopia to Tanzania, is the continent's great volcanic zone, and while there are extinct and apparently dormant volcanoes, volcanic activity continues in several areas (Figure 1-6). Outside the East African volcanic region, Mount Cameroon (13,350 ft; 4,000 m), the largest active volcano in Africa, erupted in 1959. Mount

Cameroon lies near the coast at the point of contact between western and equatorial Africa, and at the focus of a belt of volcanic activity that extends landward as well as seaward. Offshore, Bioko (Fernando Póo), as well as São Tomé and Principe, are volcanic islands, and inland there are areas underlain by comparatively recent volcanic rock, which, here as in East Africa, generates productive soils. Elsewhere, the Ahaggar and Tibesti areas in North Africa also have volcanic origins, and the volcanism continued into the Quaternary. More recently, water and wind erosion have carved the volcanic rocks into jagged shapes reminiscent of moon landscapes.

The "hot spots" around the continent of Africa, for example, the Ahaggar, Tibesti, and Marra mountains already mentioned, are caused by plumes of hot, molten material that rise from below the crust in the mantle. These plumes can cause surface uplift or "doming" (Summerfield 1991) and are responsible for major uplift in these areas, while the volcanics comprise the greatest heights. The African plate is unusual in having over 17 plumes of abnormally hot mantle. Seven of these plumes are located on the continental crust itself; the rest are associated with rifting of oceanic crust. All in all, this represents a high concentration of hot spots in comparison with other continents (Burke and Wilson 1972; Bond 1979; Doglioni, Carminati, and Bonatti 2003).

The major theater of volcanic activity lies in the east, where crustal instability is greatest. Volcanic mountains arose within the rifts (Mount Longonot: 9,111 ft; 2,780 m), between the rift valleys (Mount Elgon: 14,178 ft; 4,320 m), and outside the rift zone (Mount Kilimanjaro). Volcanic activity has continued into very recent times, and Quaternary volcanism produced such peaks as Mount Kenya (17,065 ft; 5,200 m) and Mount Meru (14,979 ft; 4,570 m). Pleistocene-age volcanic lavas and ashes overlie strata containing human fossils and artifacts, and while the largest volcanoes appear to be dormant or extinct, there is widespread, continuing activity. Mount Meru showed signs of life in the 1870s, as did Teleki's Volcano, near the southern end of Lake Turkana, in the 1890s. Ol Doinyo Lengai, near Lake Natron, remains an active volcano, and there has been other activity in recent years in northern Tanzania and in the Kivu region of the Democratic Republic of the Congo. In addition, local folklore is full of references to events involving volcanic activity, indicating other eruptions during the past several centuries. The East Africa sector of Africa's crustal plate is obviously under stress.

Africa's Rivers

Several great rivers drain the African continent. The White Nile, with its chief tributary, the Blue Nile, drains the northeast. The Niger, joined in Nigeria by the Benue River, traverses West Africa. The Congo River, with its large Ubangi tributary, flows across equatorial Africa in a broad arc. In southern Africa, the Zambezi River is the major artery. Africa is distinct among the continents in the high variability in the flow of its rivers (Walling 1996). This variability is determined by the annual water balance of precipitation, evapotranspiration, and runoff. Precipitation, the most important of these variables, varies from zero or near zero in the Sahara to 400 inches (10,000 mm) or more on Mount Cameroon.

African drainage lines (Figure 1-7) reveal that Africa's largest rivers share a number of properties, some of which are easily explained by the plateau nature of the continent, while others appear contrary to what would be expected. The Niger River is a good case in point. On a continental landmass that has been tilted quite strongly to the west, the Niger rises only 175 miles (280 km) from the Atlantic Ocean in the Guinea Highlands near its westernmost bulge and then heads north into the Sahara. As it penetrates the Djouf Basin, the Niger develops a network of distributaries resembling a delta (the Niger Inland Delta, shown in Figure 1-3), which eventually unite again into a single channel. The river then traverses the West African interior, turning 90 degrees to flow southeastward into relatively humid Nigeria, leaving the Djouf Basin. It plunges over a great falls, is joined by the Benue, and forms a large delta upon entering the Gulf of Guinea.

The Niger is peculiar because in point of fact it is made up of two rivers, one of which was "captured" by the other. During a wet period during the late Pleistocene, great lakes existed at the heart of each of the basins in the present Sahara, and the Niger River flowed from the Guinea Highlands as it does today to the center of the now-arid Djouf Basin to (former) Lake Arouane. A completely different river rose in the Adrar des Ifoghas Mountains, located southwest of the Ahaggar Mountains in the central Sahara, and flowed south through the Tlemsi Valley into the Gulf

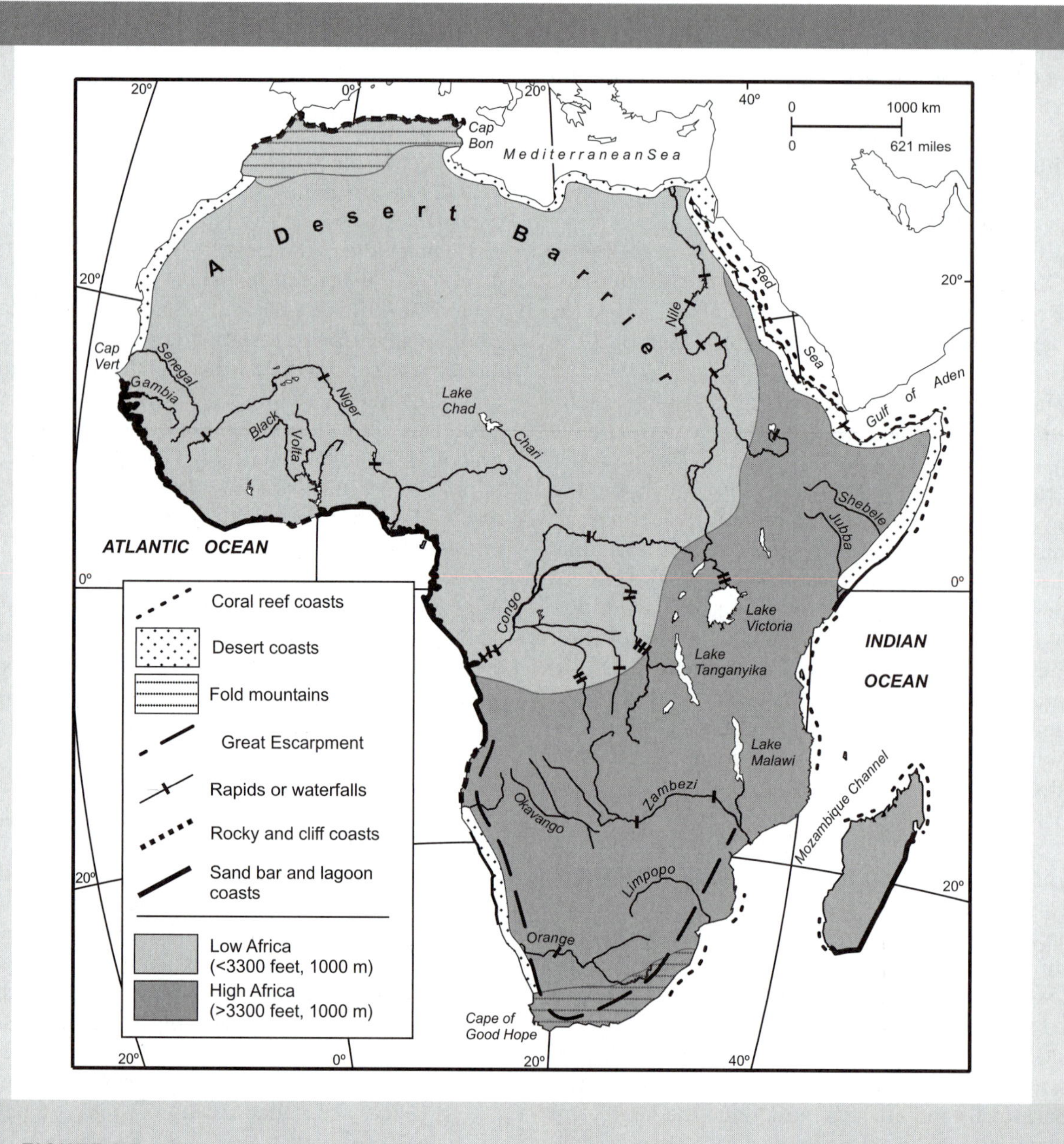

FIGURE 1-7 Coasts and rivers of Africa.

of Guinea. The climate in the Sahara became drier, and sand, borne by the dry, desiccating winds blowing from the north and northeast, pushed the waters of the lake and the flow of the Niger River south, ultimately into the bed of the river originating in the Adrar des Ifoghas (Goudie 1996), which by that time had become an intermittent stream. Occasionally even today, during the dry season, sand will cut off the flow of the Niger along its Saharan reaches until the water builds up enough to break through the sand barrier.

The course of the Nile is similarly noteworthy. The White Nile rises in the drainage entering and leaving

Lake Victoria, and on entering the Sudan Basin this river also divides into a large number of distributaries; its gradient is reduced to such an extent that, over a distance of 800 miles (1,250 km), it drops only 140 feet (43 m). After traversing its vegetation-clogged Sudd ("blockage" or "dam" in Arabic), the braided Nile unites again to form a single channel. It is joined by its largest tributary, the Blue Nile (whose source is Lake Tana in Ethiopia), and on its course through arid northeast Africa to the Mediterranean the Nile plunges over a series of cataracts (rapids). In the heart of the Sudan Basin, the Nile's northward course is interrupted as the channel turns southwest, a direction it sustains for over 150 miles (250 km) before resuming its northward orientation (Figure 1-8). On the North African coast, the Nile forms the world's most famous delta.

In equatorial Africa, the Congo River mirrors several of the features of the Nile and Niger (Figure 1-8). The most obvious of these is its course. Rising (as the Lualaba River on the South Equatorial Divide, the Congo River actually commences its course by flowing northeastward. It then proceeds northward for more than 600 miles (1,000 km) before turning westward in the vicinity of Kisangani. Eventually it turns to the southwest, is joined by its major tributary, the Ubangi River, and cuts through the Crystal Mountains to reach its mouth. Like the Nile, the Congo River drops over several cataracts and falls, including those in the Crystal Mountains.

Southern Africa's drainage pattern is dominated by the Zambezi River. The Zambezi's history involves also the Okavango and Kwando rivers, predecessors of the upper Zambezi's shifting trajectory. The Zambezi rises on the southern slope of the South Equatorial Divide in an area where drainage lines are oriented toward the Kalahari Basin rather than the Indian Ocean coast. The large delta that marked the entry of this drainage into the heart of the depression is still visible in the Okavango Swamp, as is the course that carried the water to the Zambezi's main channel above the Victoria Falls (Figure 1-9). Following its plunge over the great falls, the river traverses a deep trough now occupied by artificial Lake Kariba, crosses its rapids at Cahora Bassa in the Mozambique protrusion, and develops a substantial delta on the Indian Ocean coast north of Beira. It is especially instructive to observe the behavior of the Zambezi River's northern tributary, the Kafue. From the northern Zambia boundary for 250 miles (400 km) the Kafue River flows almost due south, headed directly for the delta in the Okavango Swamp. Then it suddenly abandons that direction and its course, which can be seen to continue as an empty valley, and turns due eastward, joining the Zambezi as it emerges from the Kariba Gorge.

The impression conveyed by these river patterns of the African continent is that the upper courses above the interior deltas are oriented toward the great interior basins, while the lower courses appear to be directed toward existing continental coastlines. This suggestion is reinforced by a study of the rivers' longitudinal profiles: they display clear evidence of a dual genesis, two approximately graded segments separated by falls or a series of rapids (Figure 1-7). Hence the African rivers at one time appear to have entered waters that stood in the great interior basins as huge lakes or seas. As they did so, they formed deltas whose remnants still are visible on the map. Nor have the lakes that once filled the basins entirely disappeared. In the swampy Sudd, in Lake Chad, in the Okavango area, and in Lake Mai Ndombe, their remnants survive. When the interior lake waters were released, they rushed out along the courses of the lower segments of the Congo, Niger, Nile, and Zambezi, and in the process they carved the falls and rapids that interrupt the continent's natural waterways. As will be seen later, Africa's rivers are keys to the interpretation of Africa's past.

Mountain Ranges

From what has been said about Africa's plateau character, it should not be surprising to find that the continent does not possess linear mountains of Andean or Himalayan dimensions. Nevertheless, the world distribution of such mountain chains and their virtual absence in Africa, a landmass comprising one-fifth of the earth's land area, raises a question in addition to those already posed by rifts, basins, rivers, and interior deltas. Every other continent is associated with a mountain backbone: South America's Andes, North America's Rocky Mountains, Europe's Alps, Asia's Himalayas, and even the Great Dividing Range of plateau-dominated Australia. No African equivalent exists; the Atlas ranges in the northwest are extensions of the trans-Eurasian Alpine system and occupy a mere

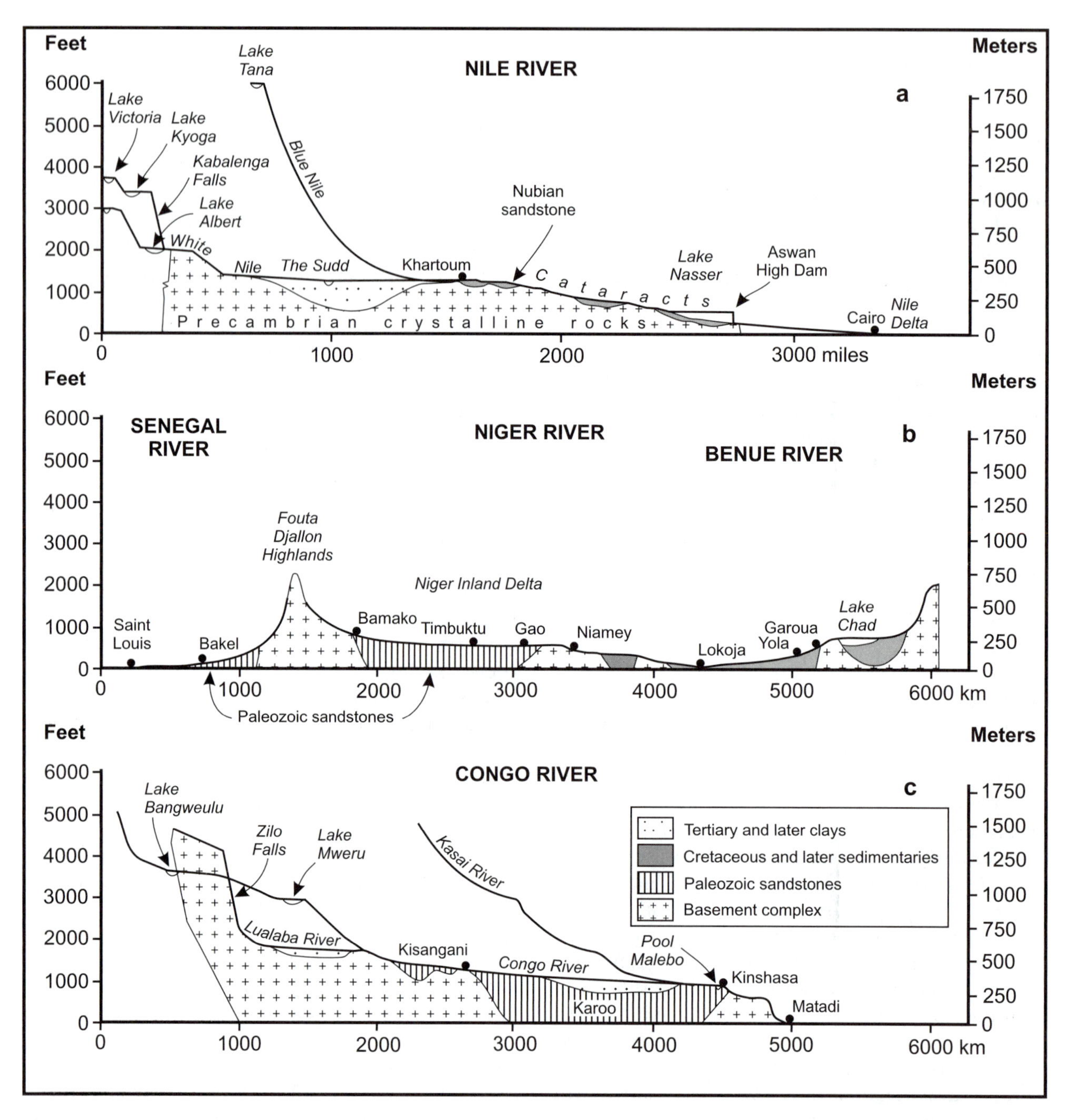

FIGURE 1-8 Cross sections of (a) the Nile River, (b) the Sahel Corridor from the Senegal River to Lake Chad, and (c) the Congo River. *Adapted from Grove (1989).*

FIGURE 1-9 Victoria Falls on the Zambezi River.
Steven Montgomery.

corner of the continent (Figure 1-10). The other, much older linear mountains in Africa lie at the Cape, in South Africa. Again, the Cape ranges are no match for the Andes or the Himalayas.

FIGURE 1-10 Aerial view of the Atlas Mountain ranges.
Short and Blair (1986) USGS.

What makes Africa different from the other continents is *not* mountainous topography, but fold mountains of structural equivalence to those of the other continents. Africa's mountainous topography (reflected in regional identities old and new: *Abyss*inia, *Sierra* Leone, Drakens*berg*, *Kilima*njaro, *Ahaggar*) owes its origins mainly to uplift and differential erosion of ancient rocks and to volcanic activity, and much less to the compression and folding of younger sedimentaries. East Africa's highlands, from Ethiopia to South Africa, are sustained by ancient crystallines, old, lava-capped sedimentaries, and volcanics.

AFRICA FORMS THE KEY

It has been almost a century since Alfred Wegener gave substance to the idea of continental drift: his book *The Origins of Continents and Oceans* was published in Germany in 1912 and first translated into English in 1924; and close to 70 years ago Alex Du Toit, following in Wegener's footsteps, published *Our Wandering Continents* (1937). Wegener focused attention on the jigsaw-like fit of opposing Atlantic coastlines, notably between Africa and South America, and Du Toit produced a mass of evidence to support a

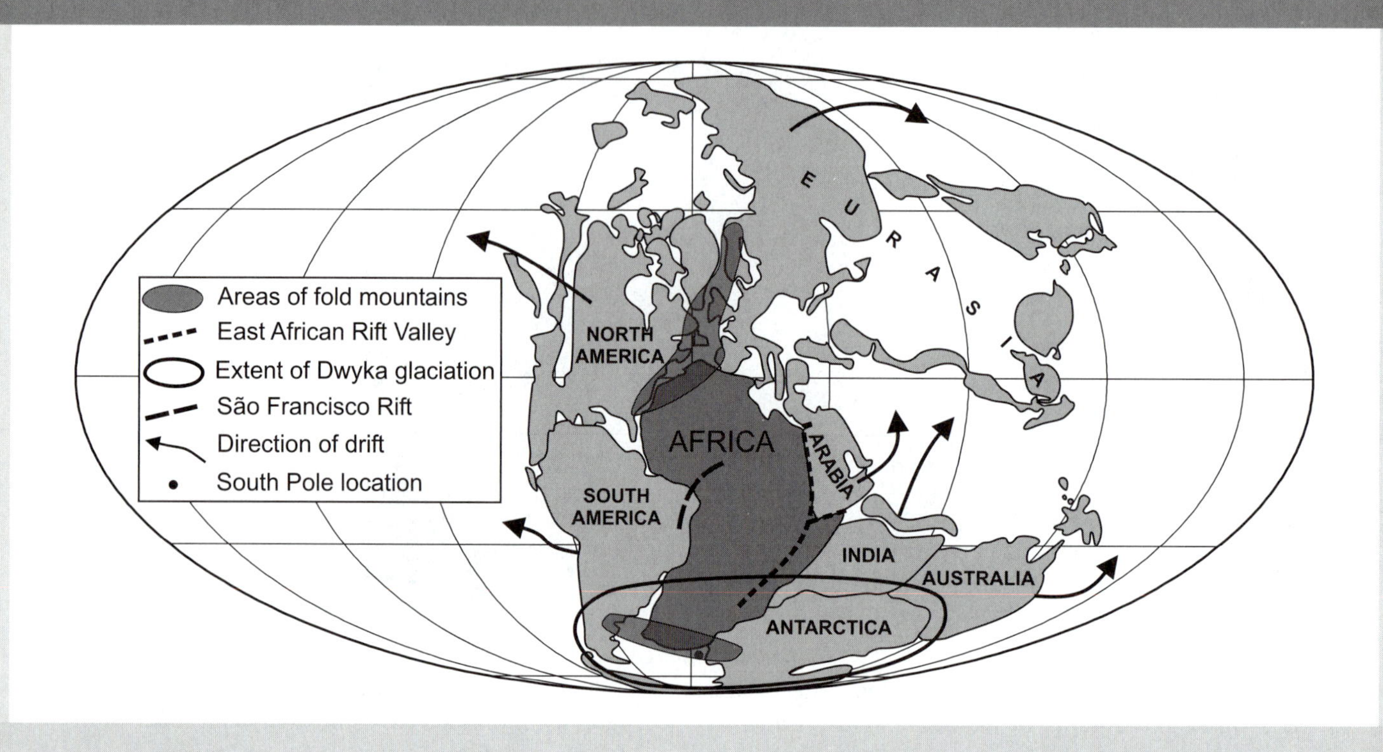

FIGURE 1-11 Africa, center of Gondwana.

former assembly not only of Africa and South America, but of the other landmasses of Gondwana as well (Figure 1-11). As Du Toit states on the frontispiece of his volume, "Africa Forms the Key."

This reference is to the surface characteristics of Africa just described, and to subsurface properties not yet enumerated. Only one hypothesis, Du Toit reasoned, could at once account for all the qualities of the African landscape: the marginal escarpment, the plateau shield with no linear mountain spine, the great interior basins, the inland deltas. That hypothesis was that Africa once lay at the heart of the supercontinent, Gondwana, and that its great interior basins were filled by sediments from great rivers flowing into growing landlocked seas. As rifts developed in Gondwana and the pieces of the jigsaw puzzle began to drift apart, Africa acquired its scarp-dominated coastlines, interior seas emptied through river channels to the newly exposed coasts, and carved gorges and falls. Eventually the process of breakup affected the crystalline rocks of the plateau itself as the rift valleys revealed the lines of weakness, first with the separation of Madagascar around 160 million years before present (mybp) and later in interior East Africa proper. Volcanic outpourings marked the original dismemberment of Gondwana and volcanism has continued into recent times (Box 1-1).

If there is indeed an essential unity to continental Africa's physiography, a unity attributable to Africa's central position in Gondwana, then the landmass should provide additional evidence for the reassembly in related contexts—for example, in subsurface geology, and the fossil record, in the distribution of mineral resources, in geomorphology and erosion surfaces, and in external structural and stratigraphic associations. And certainly the geological sequence supplies the ingredients for such a test. While Gondwana existed, a large basin of sedimentary accumulation developed, extending across southern Africa, Madagascar, areas

BOX 1-1 CONTINENTAL DRIFT, PLATE TECTONICS, AND THE SUPERCONTINENT CYCLE

Continental drift is part of the theory of plate tectonics that emerged some 30 years ago. Researchers have increasingly come to realize that Pangaea was not the first such supercontinent. Murphy and Nance (1992) have proposed a supercontinent cycle having a period of about 500 million years—from supercontinent to breakup to supercontinent again. Pangaea, which existed about 240 million years ago, was the last supercontinent. A supercontinent that existed before that roughly 700 million years ago has been called Rodinia. It is thought that the first supercontinent formed about 3 billion years ago. Clearly radioactive heat is the power behind the convection cells in the asthenosphere that move the continental plates.

What is not so clear is why the continents come together into a supercontinent every 250 million years and then break apart. One theory explains supercontinent breakup with reference to the abnormal heat that accumulates under the massive supercontinent. This heating and expansion of molten material is thought to generate hot spots, rifting, the opening up of oceans, seafloor spreading, and drift. Another possible explanation for the separation of the supercontinent is that the angular momentum of a mass that size is so high that the entity throws itself apart as the earth revolves. On the other hand, it is thought that the continents drift together as a result of the cooling and sinking of oceanic crust.

of southern South America, and parts of India. Two phases of accumulation are especially significant: first, a period of glaciation that witnessed the development of large ice sheets whose evidence exists in Africa as well as in the other fragments of Gondwana; and second, a deep sedimentary accumulation that concluded with the outpouring of basaltic lavas that still sustain not only South Africa's Drakensberg Mountains, but also India's Deccan Plateau and highlands in South America and Madagascar.

The Dwyka glaciation occurred during the beginning of Gondwana's last hundred million years of existence, and the ice sheets surrounded a polar core that appears to have been located not far off the southeast African coast (Figure 1-11). Scouring the Gondwana crystalline subsurface, these ice sheets deposited hundreds of feet of glacial debris that later became tillite, leaving widespread evidence for their distribution. This ice age provides one of the most convincing pieces of evidence for Gondwana's former unity, for its spatial properties are easily accounted for in the context of the reassembled landmass.

Following this glacial age, a lengthy period of sedimentary deposition began, and we can interpret the environmental conditions that prevailed from the color and content of the accumulated strata. Southern Africa became progressively drier, and the sediments of the great Kalahari Basin were laid down, as well as those of the even deeper Karoo Basin farther to the south and east. Sedimentary deposits into the Karoo Basin had amounted to some 25,000 feet (7,600 m), by the time the huge fissure eruptions that built the Drakensberg, the Deccan, and much of ice-covered Antarctica began, signaling the breakup of Gondwana.

Gondwana's fossil record provides ample evidence for the supercontinent's former existence, but perhaps none is more dramatic than that involving Madagascar and southern Africa. Madagascar's paleontology was similar to that of southeast Africa until the Cretaceous, but with the separation began a growing contrast that is pervasive today. Though only a few hundred miles from Africa, Madagascar today has no carnivores (lions, leopards, cheetahs) and no poisonous snakes (mambas, puff adders); it does, however, have marsupials not found on mainland Africa at all.

The distribution of prominent structural features in Gondwana still is imprinted on the continental fragments. South Africa's Cape ranges have their equivalent in Argentina, (numerous fault lines and other linear features found in Africa continue in South America), and the Karoo depositional sequence is mirrored in Madagascar and, less perfectly, in India and Antarctica. One of the most interesting continuities begins in the Bodélé Depression in the Sahara, continues through Lake Chad and the Adamawa Highlands to Mount Cameroon and to the volcanic islands in the Gulf of Guinea. These features lie in a nearly straight line—a line that can be seen to continue in South America in the São Francisco river valley, a rift feature. From the seismic incidence and volcanic activity in West Africa it could be concluded that the landmass is in the process of breaking open there; the São Francisco rift in South America may signal the first phase of a Madagascar-like breakoff as well. Figure 1-11 suggests it.

The distribution of Africa's mineral resources should also be viewed in the context of the continent's position in Gondwana. Africa's mineral-rich backbone begins in southeast Democratic Republic of the Congo, then continues through Zambia's Copper Belt and Zimbabwe's Great Dyke and includes the South African Bushveld Basin, where platinum and chromium occur, the Witwatersrand and Free State gold fields, and the diamond areas that extend to Kimberley in South Africa's Cape Province. Diamonds that are thought to be of African origin have been found in sedimentary strata in South America, and the coal fields located in and beyond South Africa's Gauteng region and in Zimbabwe form part of a belt that extends from southern Brazil to central India. Furthermore, the relatively high concentration of isolated hot spots in Africa, such as the Ahaggar Mountains and Tibesti, can be accounted for with reference to the low rate of drift of the African continent.

If the continent had drifted significantly, rather than finding isolated hot spots, we would see a series of these anomalies and their associated volcanos, one after another, marking the drift of the continent over the mantle plume. A glance at the Hawaiian Islands on the map will show the series of volcanoes that resulted from significant drifting over a hot mantle plume.

In all this it was Africa, the heart of the African tectonic plate, that was positioned in the center of Gondwana, the only continent-to-be that possessed no outer coastline but did have a vast region of internal drainage whose endorheic streams filled expanding basins of deposition, basins that became huge interior seas. As the drift, or spreading, process dismembered Gondwana, Africa began to acquire its modern physiography. Unlike South America, India, and Australia, however, Africa's lateral or radial movement was comparatively slight, a circumstance that may help explain the absence of a world-scale mountain chain.

The African physiographic map, which was Du Toit's key interpretive element in his analysis of Gondwana's former unity, suggests that the concept of a major, single African tectonic plate is incorrect. It suggests that seafloor spreading is a misnomer. The globe-girdling system of midoceanic ridges, the magma-producing rifts that form the foci of crustal divergence, commenced as fractures across the Gondwana landmass; only after the fragments began to drift away and intervening area widened did ocean water invade, whereupon the newly formed, separating crust became "sea floor."

The map of African may once again provide the crucial insight: it appears to reveal several stages in the crustal spreading process. Oldest is the stage that began along the Mid-Atlantic Ridge perhaps 180 million years ago and has caused Africa and South America to diverge by several thousand kilometers. The Madagascar separation suggests a more recent stage, and the Red Sea Rift is just the beginning of a portion of ocean that will eventually be quite wide. East Africa's rift system probably resembles the Africa–America contact just before the separation began, and this is a still younger stage than that of the Red Sea. That linear zone extending from the Bodélé Depression through Lake Chad and the Adamawa Highlands to Mount Cameroon and the islands in the Gulf of Guinea could comprise a preliminary to the whole process. We now know that "the African plate" consists of several plates.

Africa's particular position in Gondwana, central to that landmass as it is today to the land hemisphere, accounts for the essential unity of the continent's physiography. It is no accident that the course characteristics of the Niger River resemble those of the Zambezi, that rift scenery in Swaziland resembles that of Ethiopia, or that an erosion surface in South Africa is similar to one in Nigeria. Perhaps no other landmass on earth still carries so faithfully the imprint of the distant past.

CHAPTER REVIEW QUESTIONS

1. Describe Africa's location in the world as accurately as possible. Describe your continental location in similar terms.
2. What is the evidence for the existence of Gondwana?
3. Ocean crust along the margins of the Atlantic Ocean is about 180 million years old. What does supercontinent cycle theory say about the next 250 million years of plate tectonics?
4. Describe the uncanny similarity of Africa's major rivers, and explain why these rivers are the key to interpreting Africa's past.
5. How did it happen that Madagascar has animals that do not exist on mainland Africa?
6. It was stated in this chapter that Africa occupies a central position in the land hemisphere and that Africa has a "minimum aggregate distance" to the other continents. Nevertheless, a glance at the map may show relative isolation rather than proximity. What may account for this paradoxical isolation?

GLOSSARY

Antipodes Two points, located directly opposite each other on the earth's surface, through which a straight line passes through the center of the earth.

Basalt A dark-colored, relatively heavy, igneous rock, which is the primary constituent of oceanic crust.

Continental plate See **Plate tectonics.**

Endorheic Concerning a drainage system that does not flow to the sea. An internal drainage system.

Fault line A crack in rocks indicating the displacement of formerly adjacent rocks. Faults are produced by forces associated with tectonic forces that fold, warp, and crack the earth's crust.

Highveld Open land in southern Africa at an elevation from 5,000 to 6,000 meters (16,404–19,685 feet) that is generally used for grazing.

Mantle A region of the earth's interior that extends from about 25 km (16 mi) below the earth's surface to the partially molten core of the earth, 1,800 miles (3,000 km) deep. The upper part of the mantle includes the lower crust (lithosphere) and the asthenosphere, located below the crust.

Plate tectonics Fundamental theory and paradigm of geology and physical geography. Plate tectonics depicts the earth's outer layer (lithosphere or crust) as made up of eight major and eleven minor plates that are in motion relative to one another and to the asthenosphere, the hotter, more mobile, interior region of the earth located below the lithosphere.

Pleistocene In geologic time, an epoch characterized by cold climate spanning the last 2 million years up until about 10,000 years ago.

Quaternary The most recent period of geologic time, comprising the Pleistocene and the Holocene. The length of the Quaternary is 2 million years.

Steppe Semiarid grasslands.

Tectonic Describing the processes of breaking, bending, and warping of the earth's crust.

Tillite Rock made up of consolidated glacial debris (till).

BIBLIOGRAPHY

Bond, C. G. 1979. Evidence for some uplifts of large magnitude in continental platforms. *Tectonophysics*, 61: 285–305.

Burke, K., and J. T. Wilson. 1972. Is the African Plate stationary? *Nature*, 239: 387–390.

Doglioni, C., E. Carminati, and E. Bonatti. 2003. Rift asymmetry and continental uplift. *Tectonics*, 22(3): 1–13.

Du Toit, G. O. 1937. *Our Wandering Continents*. London: Oliver and Boyd.

Goudie, A. S. 1994. *The Encyclopedic Dictionary of Physical Geography*, 2nd ed. Oxford: Blackwell.

Goudie, A. S. 1996. Climate: Past and present. In *The Physical Geography of Africa*, W. M. Adams, A. S. Goudie, and A. R. Orme, eds. Oxford: Oxford University Press, pp. 34–59.

Grove, A. T. 1989. The Changing Geography of Africa. New York: Oxford University Press.

Murphy, J. B., and R. D. Nance. 1992. Mountain Belts and the Supercontinent Cycle. *Scientific American*, 266 (April): 84–91.

Short, N. M., and R. W. Blair. 1986. Geomorphology from Space: A Global Overview of Regional Landforms. Washington, D.C.: NASA.

Summerfield, M. A. 1991. *Global Geomorphology*. London: Longman.

Walling, D. E. 1996. Hydrology and Rivers. In *The Physical Geography of Africa*, W. M. Adams, A. S. Goudie, and A. R. Orme, eds. Oxford: Oxford University Press, pp. 103–121.

Wegener, A. 1967 (1912, initial publication in German). *The Origins of the Continents and the Oceans*. Translated by John Biram. London: Methuen.

APPENDIX 1 TABLE OF GEOLOGIC TIME

Era	Period	Epoch	Date of Boundary (Mybp)	Length of Time (Years)
Cenozoic	Quaternary	Holocene	Present	10,000
		Pleistocene	2	2,000,000
	Tertiary	Pliocene	5	3,000,000
		Miocene	24	19,000,000
		Oligocene	37	13,000,000
		Eocene	58	21,000,000
		Paleocene	66	8,000,000
Mesozoic	Cretaceous		144	78,000,000
	Jurassic		208	64,000,000
	Triassic		245	37,000,000
Paleozoic	Permian		286	41,000,000
	Carboniferous	Pennsylvanian	320	75,000,000
		Mississippian	360	40,000,000
	Devonian		408	48,000,000
	Silurian		438	30,000,000
	Ordovician		505	67,000,000
	Cambrian		570	65,000,000
Precambrian	Z		750	180,000,000
	Y	Proterozoic	1,600,000,000	850,000,000
	X		2,500,000,000	900,000,000
	W	Archean	4,600,000,000	2,100,000,000
	Age of earth		4,600,000,000	

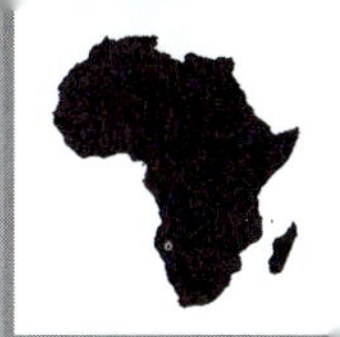

CHAPTER 2

Climate, Vegetation, and Soils of Africa

AFRICA'S CLIMATE

Africa straddles the equator, and much of it lies in the tropics. Consequently, the image most often associated with Africa is one of heat and humidity, and over a large part of the continent this impression is verified by the facts. The absence of any lengthy mountain chains or other weather or climate divides permits a free circulation of tropical air over the continent, so that, in general, changes from one place to another occur very gradually. This does not mean that there is little diversity in climate. Because of the elevation and extent of the African Plateau, the escarpment that forms its rim, the narrow coastal belt, and the ocean currents along the shores, there is variation, although perhaps not as much in temperature as in precipitation.

Figure 2-1 shows mean annual rainfall is highest in low equatorial Africa, along the Guinea Coast, in the highlands of Ethiopia and Madagascar, and on parts of the East African Plateau. In these regions the mean annual rainfall generally exceeds 56 inches (1,400 mm); however parts of the Cameroon Highlands receive over 300 inches (7,620 mm), and Robertsfield, Liberia, has an average annual rainfall of 132 inches (3,354 mm). Mean annual rainfall is lowest in the Sahara and Namib deserts, and parts of Somalia and Ethiopia. For example, the mean annual rainfall at Khartoum is only 5 inches (127 mm), and on the Namibian coast at Swakopmund it is just over half an inch (0.65 in. 16 mm). These same desert regions experience the highest variability of rainfall (Figure 2-2).

Much of Africa has distinct wet and dry seasons: the northern and southern extremities record winter maximums; the savanna areas have a pronounced summer maximum; while the rain forest regions have a more even distribution of rain through the year. Figure 2-3 shows that in January the mean monthly precipitation is highest south of the equator, especially in Madagascar, the Democratic Republic of the Congo, and Zambia, and that there is a narrow belt of rainfall along the Mediterranean coast. In July rainfall is heaviest north of the equator, especially in Ethiopia and along the Guinea Coast.

Variations also exist in temperature regimes, although in general the extremes are not as great. Interior desert regions experience the greatest temperature ranges. For example, the mean January and July temperatures of the town of In Salah, Algeria, are 57.7° F (14.3° C) and 99° F (37° C), respectively. The average daily maximum in July is 117° F (47° C), while the nighttime temperatures in winter drop below freezing. Throughout the Sahara, the mean July temperature is above 80° F (26.6° C), while in winter the mean temperature ranges from 55° F (12.7° C) along the Mediterranean coast to 70° F (21° C) along the southern border. In contrast, in low equatorial Africa the average monthly temperatures throughout the year are about 75 to 80° F (24–27° C).

Elevation has a modifying effect on temperature, and, since many eastern and southern areas are above 3,000 feet (914 m), their temperatures are not as high as might be expected given their tropical location. On the East African Plateau, for example, Nairobi, with an elevation of 5,540 feet (1,661 m) records 58.5° F (14.4° C), but the port of Lagos, which is at the same general latitude, has 78° F (25.5° C). In the Ethiopian Highlands, the Inyanga Mountains of Zimbabwe, and other highland areas within the tropics, winter frosts

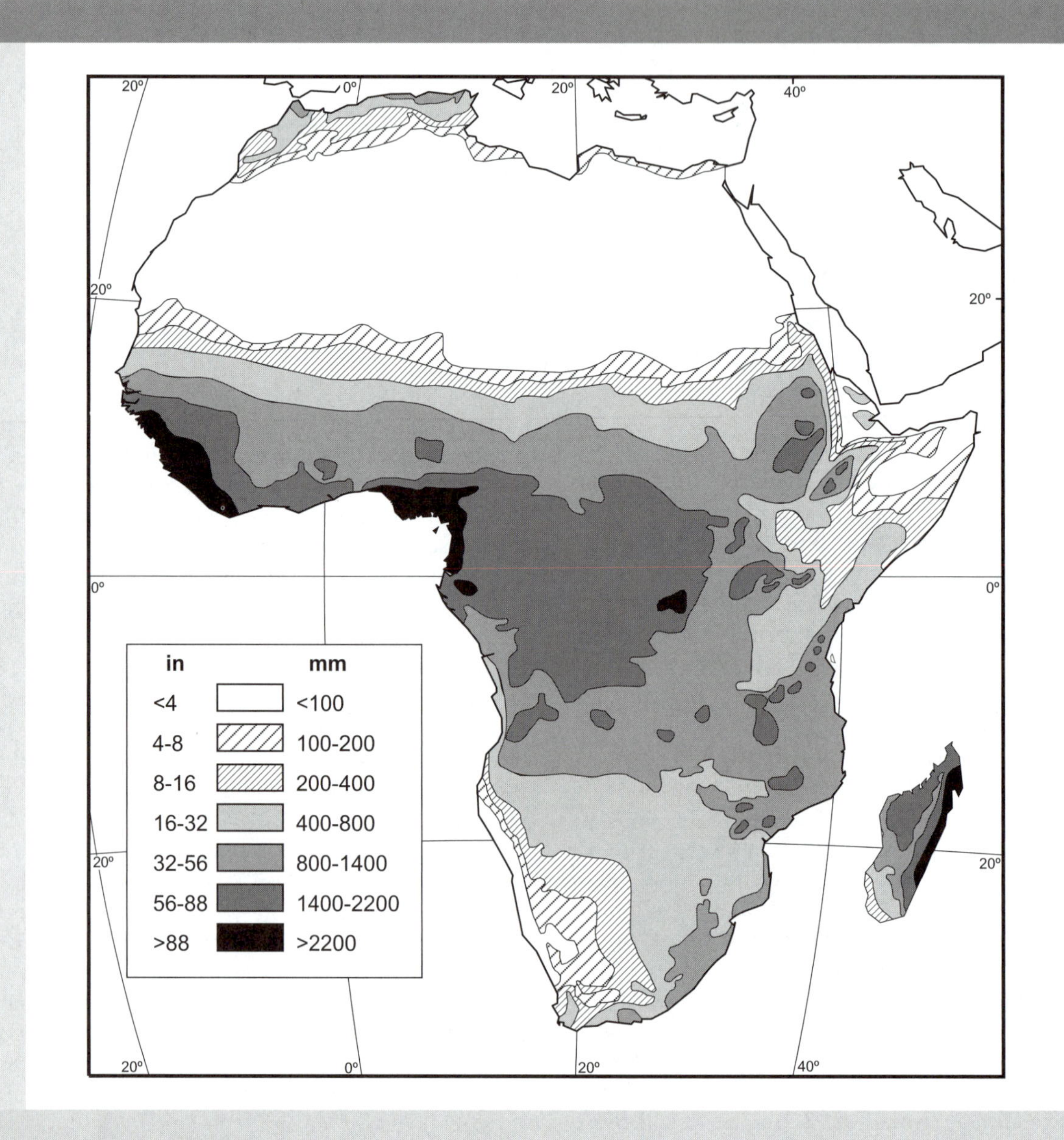

FIGURE 2-1 Mean annual precipitation in Africa.

are not uncommon. At the highest elevations in Africa, Mount Kenya (17,058 ft; 5,200 m) and Mount Kilimanjaro (19,340 ft; 5,895 m), and in the Ruwenzori Mountains (16,763 ft; 5,109 m), glaciers are found but the total area is very small, less than 4 square miles (10 km^2). There is evidence of widespread glaciations in the past.

Pressure, Air Masses, and Fronts

The basic rainfall and temperature patterns just described are affected by permanent and semipermanent pressure systems and their associated winds and air masses (Figure 2-4). Locational changes in the pressure systems bring changes in the amount and distribution

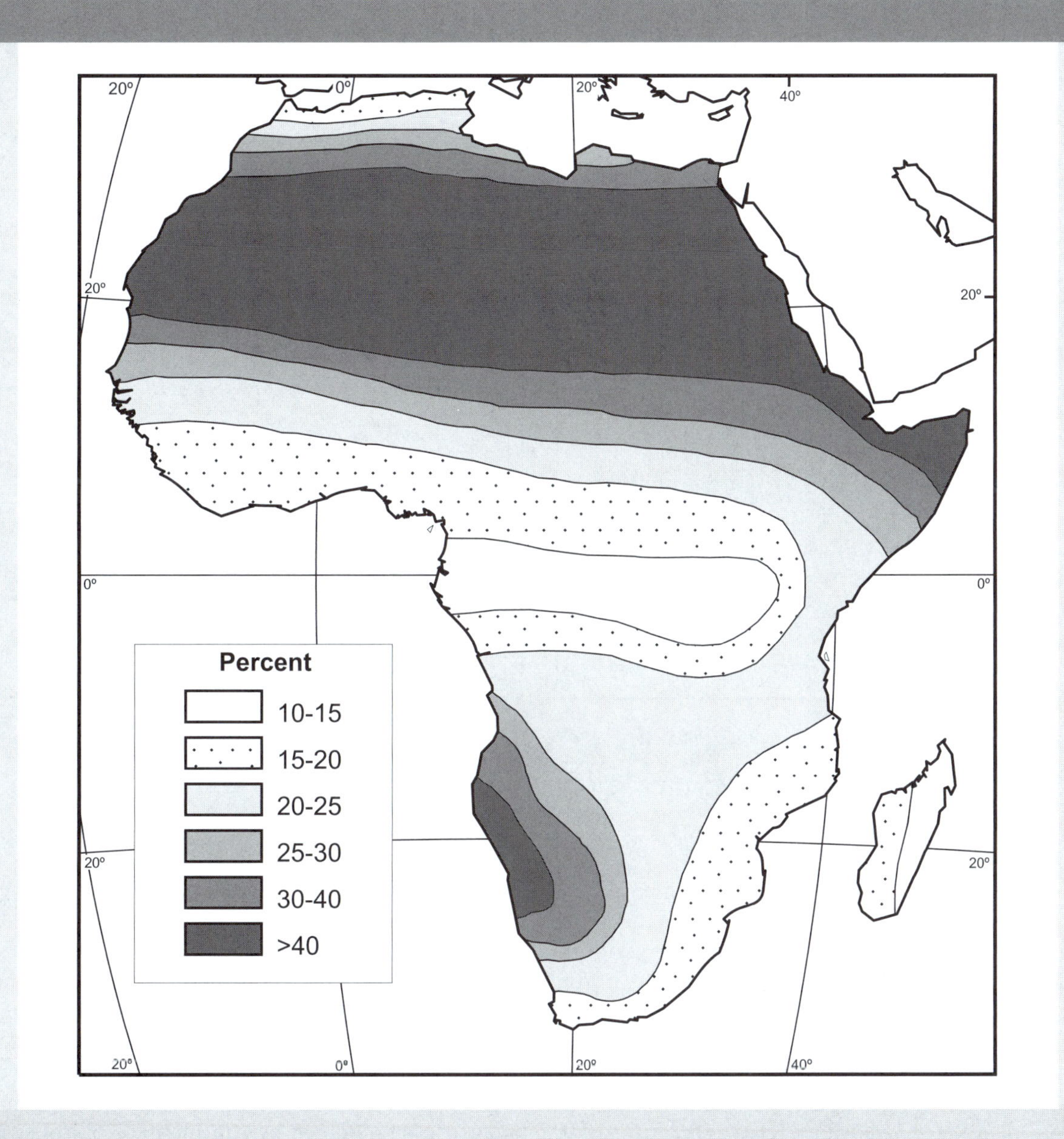

FIGURE 2-2 Precipitation variability in Africa.

of precipitation and in the intensity of heat. In January, the Southern Hemisphere tropical anticyclones are situated at about latitude 30° S in both the Atlantic and Pacific oceans, and the equatorial low-pressure belt has shifted to the south and is linked with the weak thermal heat low of the Kalahari. Thus the entire continent south of the equator is under the influence of low pressure during this high-sun period, while high pressure predominates north of the equator. In July the pattern is essentially reversed. High pressure extends across southern Africa and the adjoining Atlantic and Indian Oceans, while an intense thermal low situated over the Sahara is linked to another low-pressure system over southwest Asia. These conditions result in two

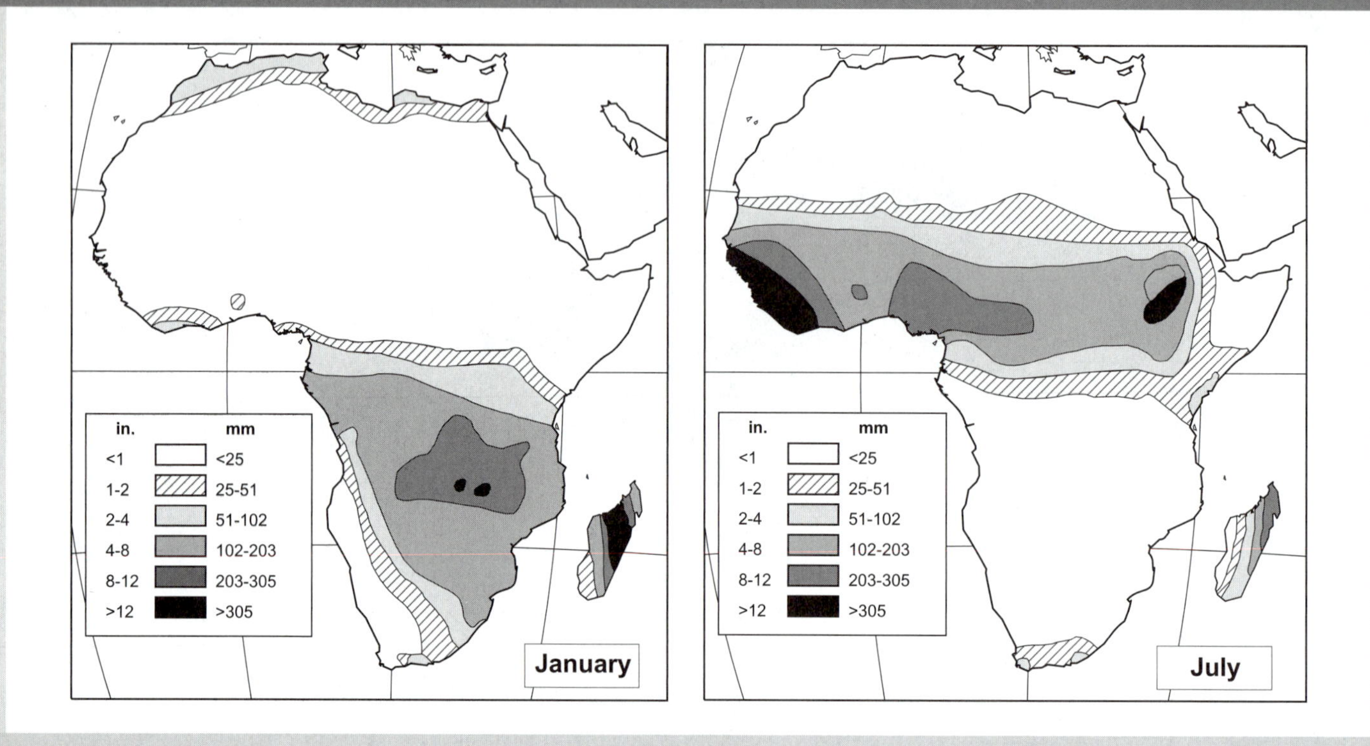

FIGURE 2-3 Mean annual precipitation January and July.

distinct seasonal air circulation patterns across Africa (Figure 2-5).

In January the circulation north of the equator is circular around the anticyclone, that is, from the north and northeast to the south, and across the equator in the eastern regions. Along the coast of the South Atlantic, however, the flow is around the South Atlantic anticyclone and from the south toward the equator. In July, the wind flow across Africa north of latitude 20° N is from the northeast, but the dominant flow across the rest of Africa is from south to north. Once the winds cross the equator, they are drawn in a northeasterly direction toward the low-pressure areas of the Arabian peninsula.

Air masses of several types are associated with these circulation systems and play important roles in the precipitation regimes and distributions. Most of the air masses are tropical. Continental tropical (cT) air originates in the upper levels of the atmosphere above the Sahara and Kalahari deserts and brings very dry and warm conditions to the surface in its descent. Subsident maritime tropical (mT) air originates in the eastern sectors of the Azores and South Atlantic anticyclone cells, and it too is dry and stable. It controls the climates of Mauritania and Morocco, and the coastal areas of Namibia and southern Angola. A similar air mass dominates the Somalian coastlands, although it may also be modified by continental tropical air from Asia. As the maritime tropical (mT) air masses pass over the oceans, they gather moisture, become unstable, and then appear as maritime equatorial (mE) air masses. The subsident maritime tropical air masses on the east side of the South Atlantic anticyclone cell, for example, become warmer, moister, and unstable as they are drawn toward the equatorial low pressure areas along the Guinea Coast. With uplift, they can give heavy rains on reaching land, such as in the Cameroon Highlands and the Fouta Djallon.

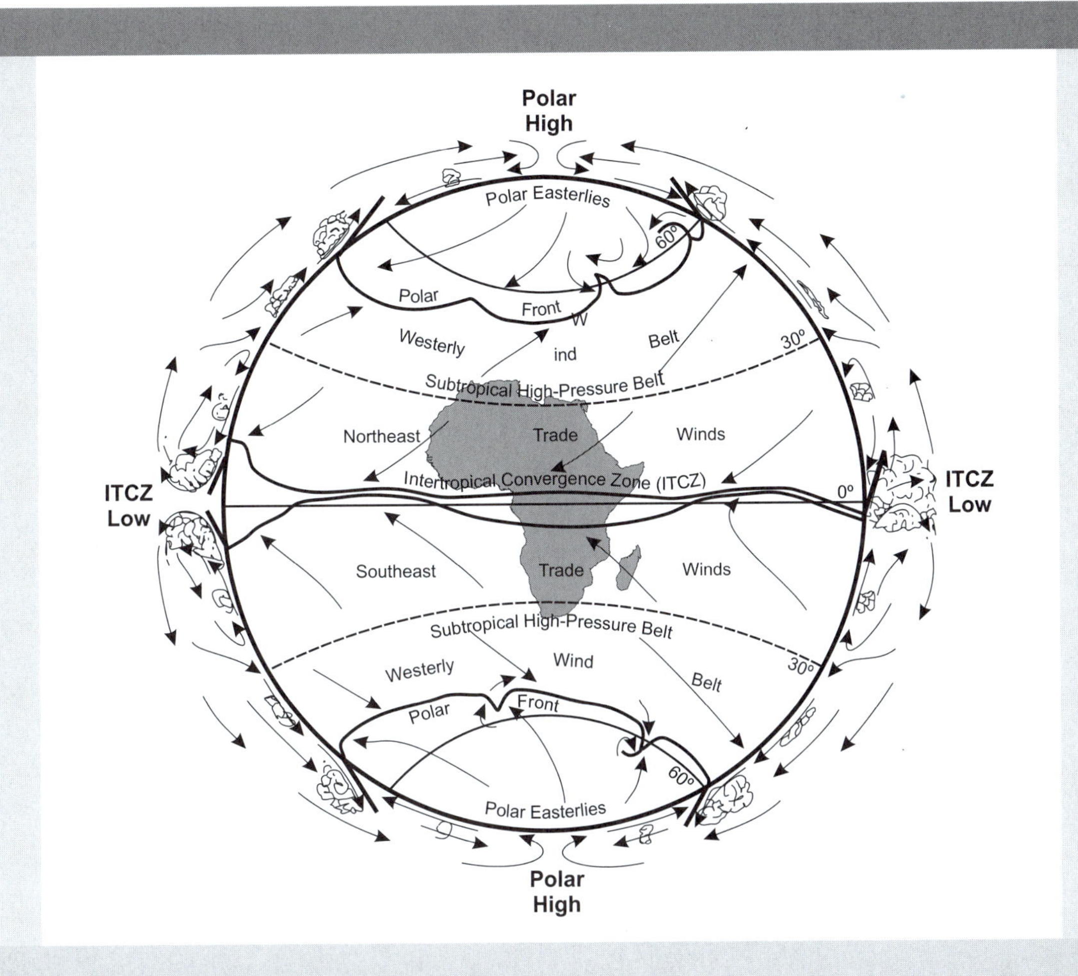

FIGURE 2-4 Global pressure belts and winds.

Similarly, the maritime tropical air masses originating in the South Indian Ocean can bring heavy rains as they ascend eastern Madagascar and the mainland's southeast coast. When two different types of air mass converge, a front is formed, along which there can be unsettled weather. In the midlatitudes, frontal activity is common where cold and warm air masses meet. In the Mediterranean regions, and along the South African coast, there are frequent fronts during the winter months as modified polar air meets warmer tropical air. Cloudy and rainy conditions prevail, followed by clearing. Most climatologists agree that the most important frontal activity in tropical Africa is associated with the Intertropical Convergence Zone (ITCZ).

The ITCZ oscillates north and south across Africa between latitudes 20° N, and 20° S. In July it extends across the continent at the southern margins of the Sahara, although its position varies considerably from year to year. In January it parallels the Guinea Coast at about 5° N, turns sharply southward to about 20° S, and then trends northeast across Zimbabwe to northern Madagascar. Its irregular movement is of critical importance, since rainfall is associated with the front. In July the ITCZ represents the boundary between the hot, dry Saharan air of anticyclonic origin and the cooler maritime equatorial air from the south (Figure 2-6). The hotter, more stable, and less dense air of the Sahara is forced to rise at a low angle of inclination

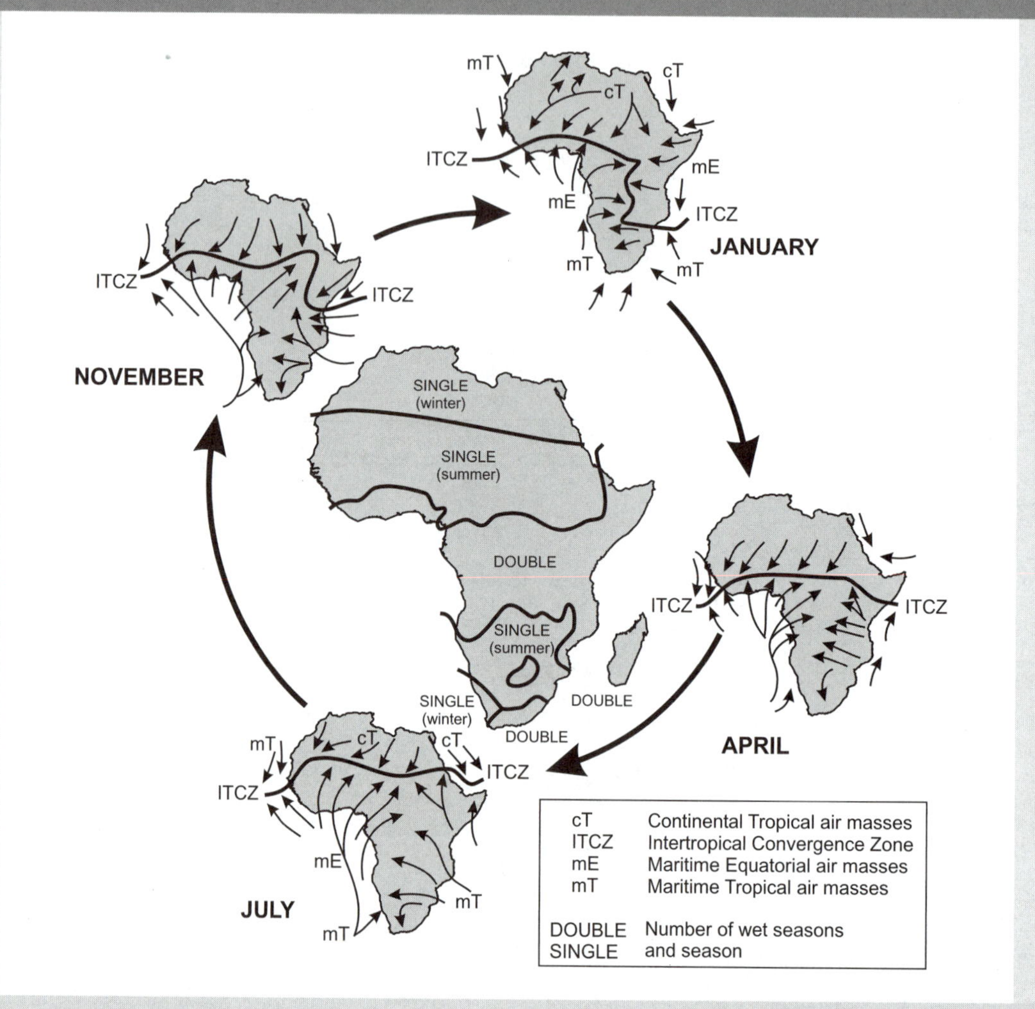

FIGURE 2-5 Air masses, winds, and the seasonal movement of the Intertropical Convergence Zone. *Adapted from Rasmusson (1978), Best and de Blij (1977).*

over the cooler, moister air from the south. But the dry and stable Saharan air precludes the development of precipitation as it rises; the underlying moister air becomes drier as it moves north. Cumulus clouds that may develop are soon desiccated as they break into the overriding Saharan air, so that little or no rain falls as the front advances. However, further to the south, where the maritime air is sufficiently deep to allow the development of cumulonimbus clouds, heavy rains may occur.

The ITCZ, therefore, does not resemble midlatitudinal warm or cold fronts that frequently bring rain. Its behavior is less predictable than the cyclonic fronts of the westerlies. Its migration is the result of contrasts in strength between the humid and dry air masses, and in West Africa its frequent failure to move northward, preventing the penetration of moist air inland, is a prime cause of unreliable precipitation only short distances from the coast. In East Africa, the ITCZ forms the zone of convergence between air masses flowing in

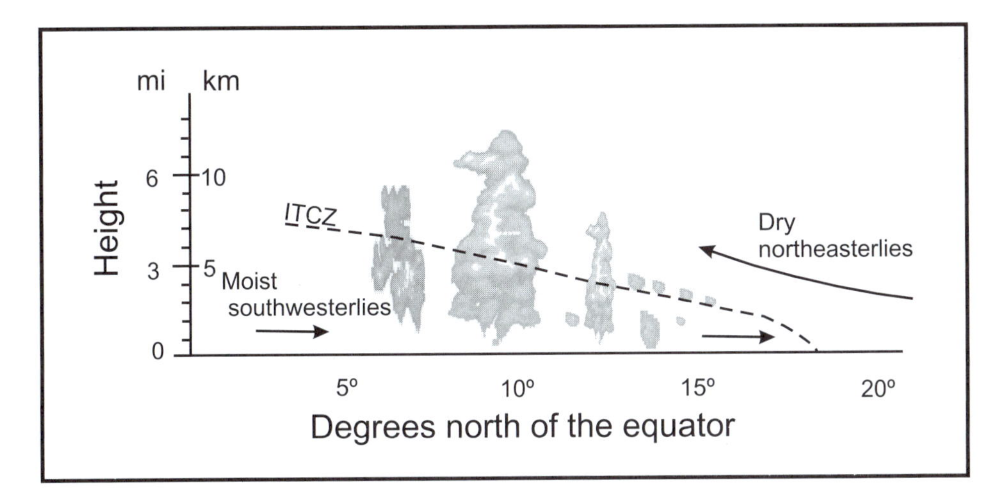

FIGURE 2-6 Cross section of the ITCZ in Africa in July.

from the northeast and southeast. Here, however, its movement is restricted and modified by the high plateaus, and it has proved a far less useful tool for the interpretation of weather phenomena and for predicting rainfall than in West Africa. Figure 2-7 illustrates the rainfall gradient from Lagos, Nigeria, in the rain forest belt, to Agades, Niger, at the southern edge of the Sahara.

Regional Climates

For the key to some abbreviations customarily used by climatologists, see Figure 2-8.

Wet Equatorial Climates (Af and Am)

A large part of equatorial Africa from the Congo Basin west along the Guinea Coast to Guinea has tropical

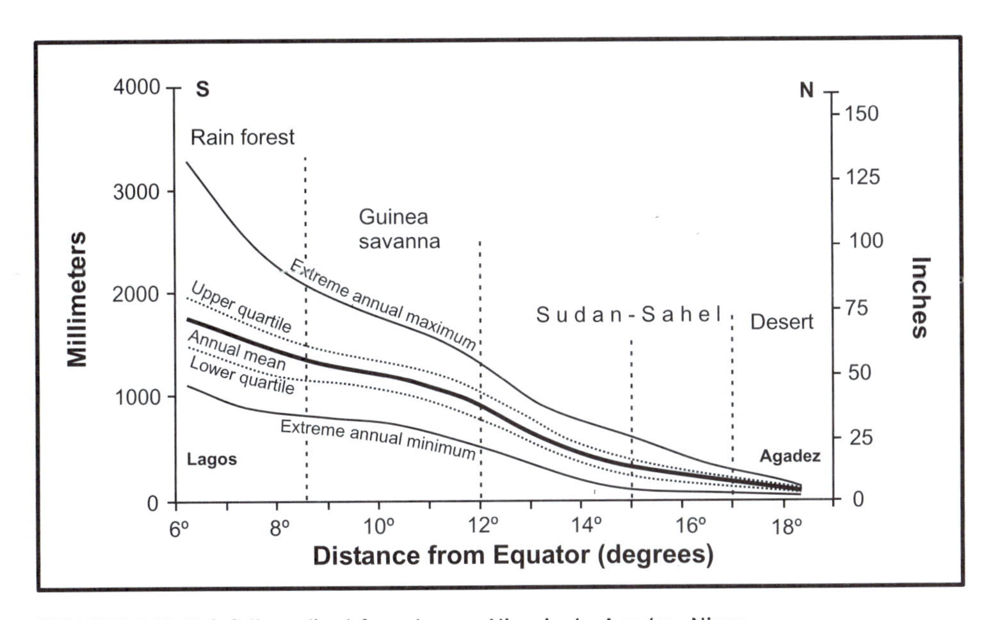

FIGURE 2-7 Rainfall gradient from Lagos, Nigeria, to Agadez, Niger.

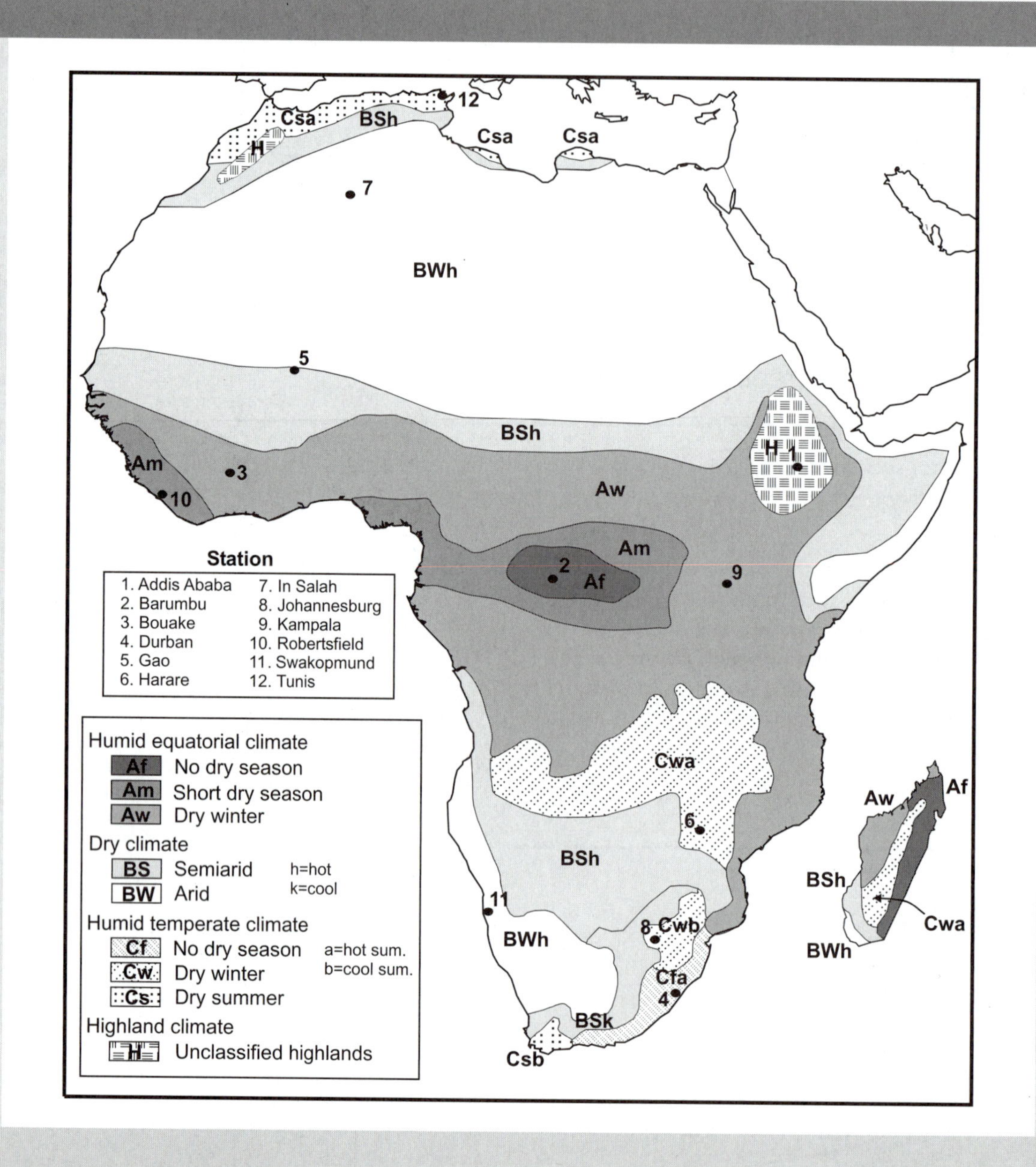

FIGURE 2-8 Climate regions of Africa according to the Köppen classification.

humid or wet equatorial climates (Figure 2-8). In the eastern sections the climate is known as tropical rainforest (Af). There the seasons pass virtually unnoticed as the unbroken monotony of hot, humid rainy days is only slightly ameliorated by the nightly drop in temperatures of about 15° F (9.4° C). Even so, since the annual temperature range (the difference between the means for the warmest and coolest months) is less than the daily range, "night is the winter in the tropics." Tropical rain forest climate is characterized by continuously high temperatures, no month below 68° F (20° C), and an annual rainfall exceeding

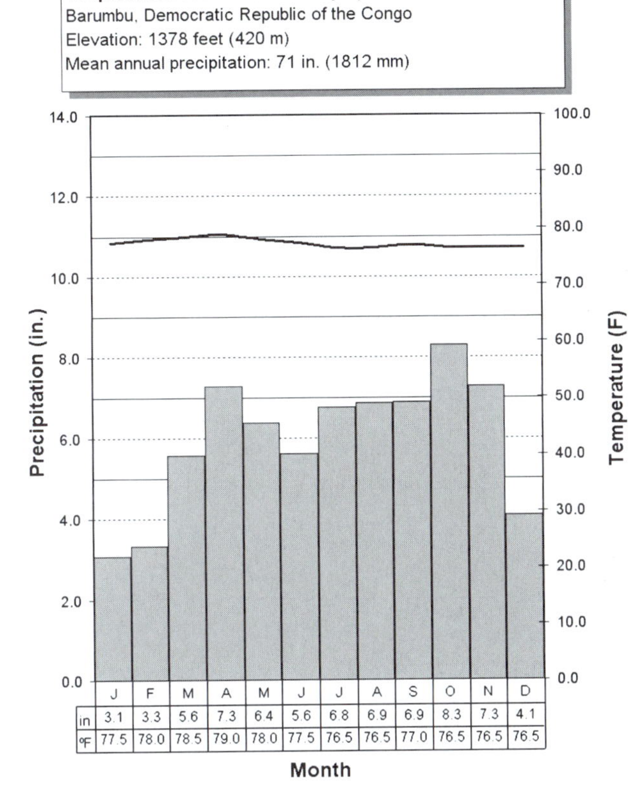

FIGURE 2-9 Barumbu, Democratic Republic of the Congo: rain forest (Af) climate.

50 inches (1,270 mm) without any dry season. Temperature and rainfall data for Barumbu, Democratic Republic of the Congo, a typical tropical rainforest station, are shown in Figure 2-9.

Like most rainforest stations, Barumbu receives heavy rains each month, although there is a distinct period of maximum precipitation. In several places, because of the migration of the heat equator and associated features (e.g., the ITCZ), there is a double maximum of rainfall (Figure 2-5). Throughout much of the rain forest regions, there are frequent, almost daily, late afternoon thundershowers derived from heat-of-day thermal convection. The heaviest falls are found where mountains force the unstable air to rise, as in the Cameroon Mountains. It is a common misconception that the hottest temperatures recorded are those of the tropics. In fact, in the tropical rain forest areas, daytime temperatures rarely rise above 100° F (37.7° C); they are usually about 84 to 86° F (30° C).

In the region of the Congo Basin, two major air circulations have been recognized as important in creating the conditions just described. There is a constant inflow of maritime equatorial (mE) air from the southwest, while the trade winds blow in from the northeast. Because of the higher elevations east of the Congo Basin, tropical rain forest conditions do not extend to the Indian Ocean, but parts of southern Uganda have a modified rain forest climate. Kampala, Uganda, for example, has the characteristically small annual range of temperature, but the warmest month averages only 71° F (21.6° C) (Figure 2-10). Like Barumbu, Kampala's rainfall is heavy and there are two peaks—one in April–May, another in November–December.

Toward the western end of the Guinea Coast, and along the Madagascar escarpment, there are narrow

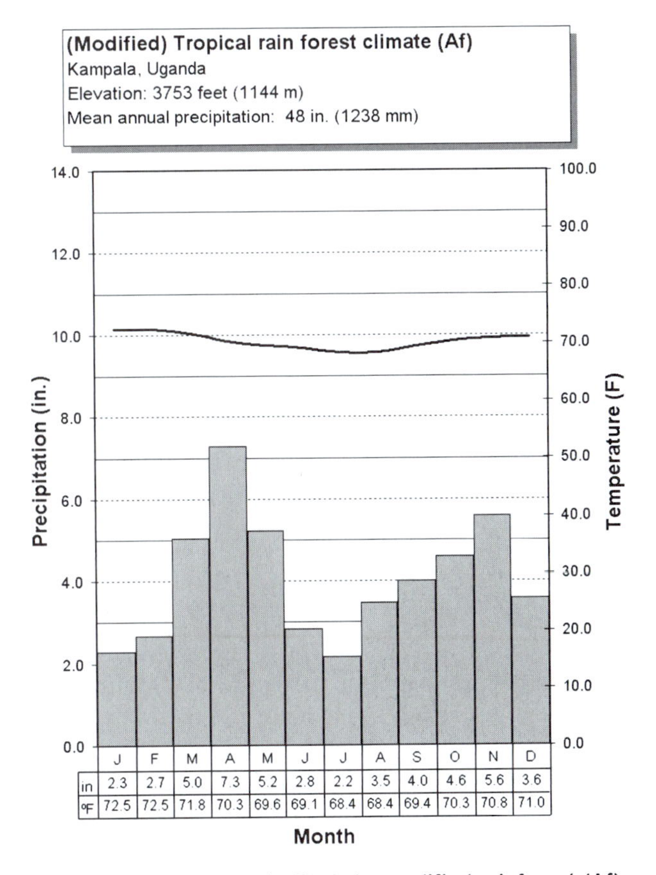

FIGURE 2-10 Kampala, Uganda: modified rainforest (Af) climate.

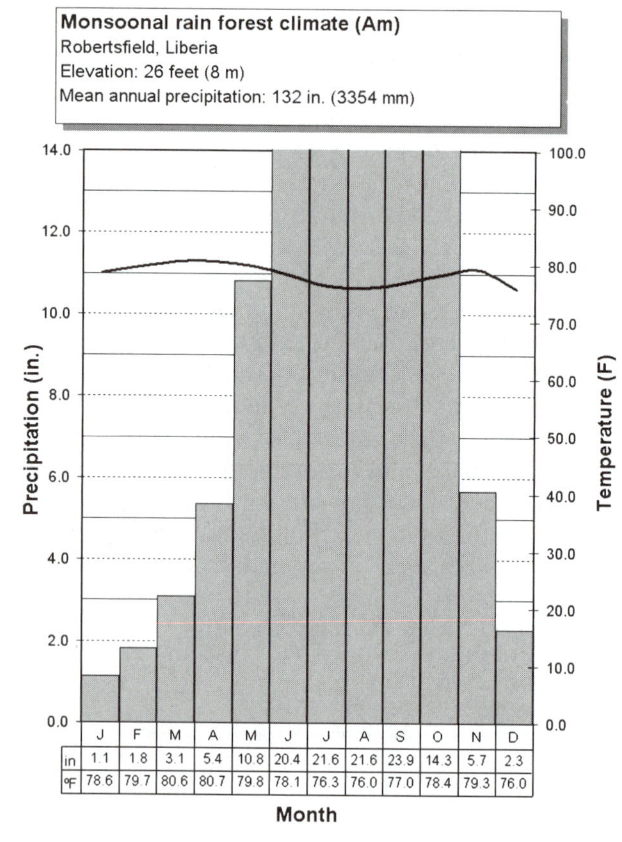

FIGURE 2-11 Robertsfield, Liberia: monsoon (Am) climate.

zones of monsoon climate (Am) (Figure 2-8). There the annual rainfall is heavy, but there is a marked dry season and a very pronounced wet season. Robertsfield, Liberia, has a monsoon climate. Its dry season extends from December through March, when less than 4 inches (100 mm) of rain are recorded, while from June through September rainfall exceeds 110 inches (2,860 mm). Monthly temperatures throughout the year are above 76° F (24.5° C), the highest occurring immediately before and after the heavy rains (Figure 2-11). At Tamatave, Malagasy, the monsoon rains begin in December and extend through March, the high-sun period in the Southern Hemisphere, but the dry season is less pronounced than in West Africa. In both regions there is orographic precipitation as the moist air masses cross from the oceans onto the plateaus, but there is also thermal convectional rainfall, as well as rainfall associated with fluctuations in the ITCZ.

Tropical Wet-and-Dry Climates (Aw)

On either side of the wet equatorial climates, lies one of the most widespread and distinctive African climates, known as the tropical wet-and-dry, or savanna climate. The major difference between rain forest and savanna conditions lies not so much in temperature as in the amount and distribution of precipitation. Temperature conditions show a slightly greater annual range that is usually less than 15° F (9° C).

At Bouaké, Côte d'Ivoire, for example, the mean annual temperature range is 13.1° F (7.3° C), and the hottest months are February and March (Figure 2-12). Daily maximums during this period can reach an average around 100° F (37.7° C). Total annual precipitation is considerably less than in the wet equatorial areas, and there are distinct dry and wet seasons.

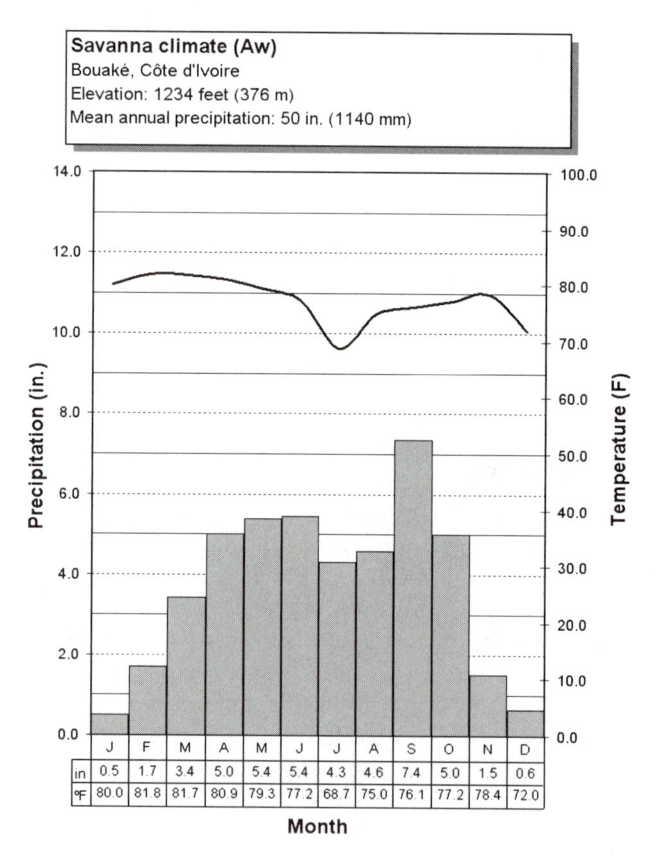

FIGURE 2-12 Bouaké, Côte d'Ivoire: savanna (Aw) climate.

BOX 2-1

THE HARMATTAN

During the winter months, the coastal regions of West Africa are influenced by northeasterly winds originating in the subtropical high-pressure region over the Sahara. The excessively moist conditions of the coasts are driven away, to be replaced by the dry, often cool air from the interior. Relative humidity may drop to between 10 and 20%; skies are clear, and the general effect is one of relief for the inhabitants. Known as the *harmattan*, this outflow of air may be strong and very persistent, not only affecting the coastal regions but continuing far out into the Gulf of Guinea.

While bringing relief to the coasts, the harmattan is often hot, dust-laden, and stifling in the Savanna and Sahel areas. Its effect on people, animals, and vegetation can be disastrous when its influence remains unbroken for long periods. Swirling dust storms and severe droughts occur, as well as failure of vegetation and consequent overgrazing results in destruction of the vulnerable grass cover and desertification.

Throughout the savanna belt of West Africa, the rainy season corresponds with the high-sun period beginning generally in May and terminating in September. Rainfall becomes lighter and more variable with distance from the equator. There is no sharp break between the wet equatorial and tropical wet-and-dry areas, so that the inner margins of the latter resemble the rain forest. The heart of the savanna, nonetheless, displays seasonally dry conditions that are absent from the wet rainforest.

There are three general areas of tropical wet-and-dry climate. The first lies north of the rain forest in West Africa and extends from Senegal to the southern Sudan, where rainfall decreases appreciably from south to north. During the dry winter months, the Harmattan winds blow from northeast to southwest, whereas in summer tropical maritime air is transported inland from the southwest (Box 2-1). The second region lies south of the Congo Basin and up to the Bié Plateau. The third zone occupies the eastern highlands and coastal areas from Kenya to the South African border.

Humid Climates

Humid, Temperate, Dry Summer (Csa and Csb) The northern and southern extremities of Africa experience dry summer subtropical climates, commonly known as Mediterranean. Both come under the influence of dry subsiding anticyclonic air in the summer and modified maritime polar air with its associated frontal disturbances in winter. At Tunis, Tunisia, (Figure 2-13), the mean temperatures for June, July, August, and September range from 72.8 to 75.2° F (22–24° C), with the total rainfall for these same months is only 2.1 inches (52.7 mm). The six months from October through March, however, see approximately 13 inches (329 mm) of rain. A small area of Mediterranean climate is found around Cape Town, where the annual range of temperature is only 15° F (9° C) which is considerably less than along coastal Tunisia. In winter, especially in July and August, the westerly wind belt migrates northward, bringing with it the cyclonic storms that produce rain and many gray, overcast days. In summer, strong southeasterly winds originating over the cool ocean may produce a "tablecloth" of cloud on Table Mountain, which dominates the Cape Town skyline.

Humid Subtropical Climate (Cfa) Along the southeast coast of South Africa lies a narrow region of humid subtropical climate, characterized by summer maximum precipitation. The area is somewhat cooler than the tropical savanna to the north, but summers are nevertheless hot, with lower temperature conditions mainly in the winter period. The region is limited by the rapid rise of the plateau slopes toward the interior. Except for some frost due to air drainage, which is confined to valley lowlands, subfreezing temperatures do not occur. In terms of precipitation, the contrast between summer and winter is not as great as in the

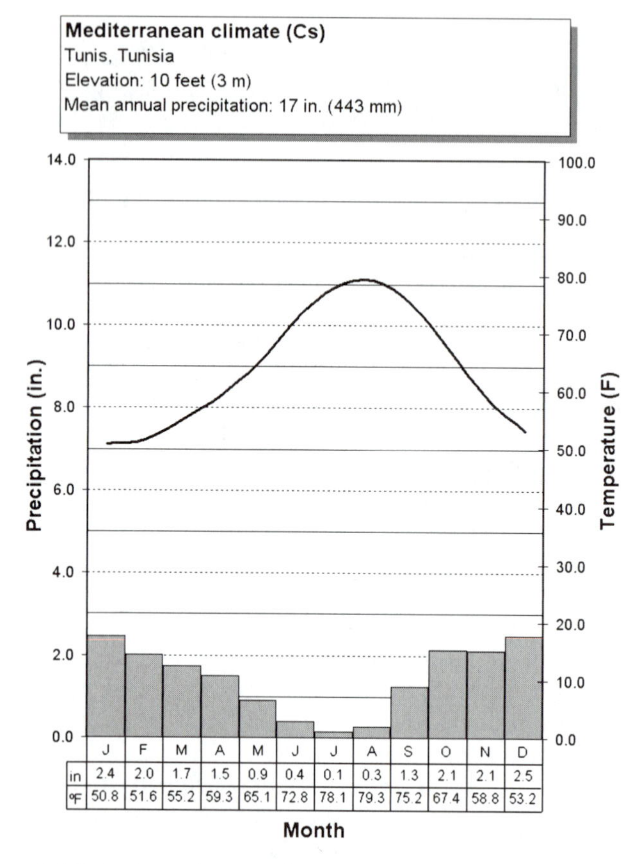

FIGURE 2-13 Tunis, Tunisia: Mediterranean (Cs) climate.

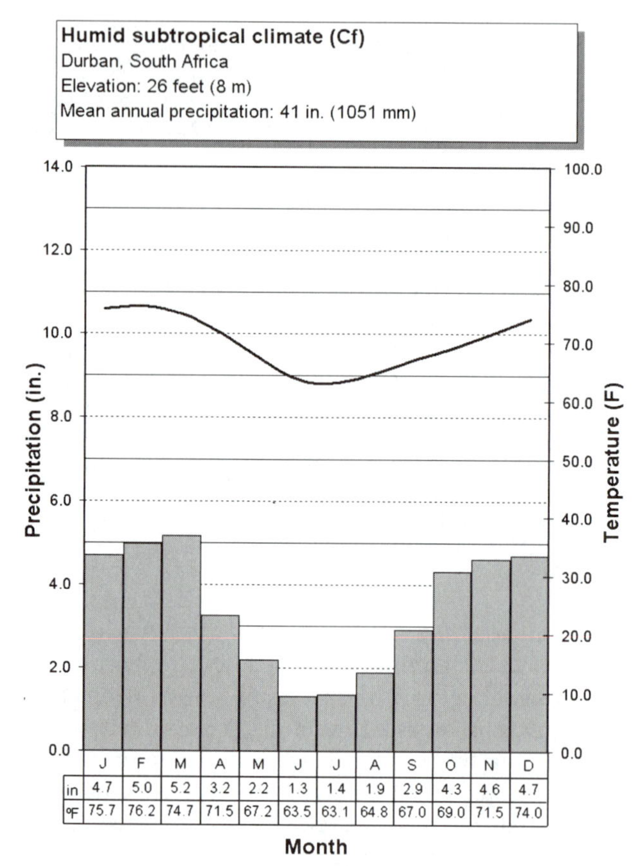

FIGURE 2-14 Durban, South Africa: humid subtropical (Cf) climate.

interior. Although most rain comes during the summer, there is no month with less than one inch of rain. Rainfall is orographic, as might be expected from the topography of the area and the prevailing winds, but, of course, convectional storms occur in the summer. The coasts are influenced by the warm Mozambique Current, and the southeast trades arrive laden with moisture from these waters. In winter, when the entire system moves to the north, it is the coasts of Mozambique and Madagascar that continue to benefit from the moisture (particularly exposed Madagascar), and the drier period sets in further south. The situation is illustrated by climatological data for Durban (Figure 2-14).

Humid, Temperate, Dry Winter, Warm (Cwa) and Cool (Cwb) Since a large portion of the East Africa savanna lies at considerable elevations, temperatures are generally lower than in the West Africa savanna belt. At Harare, Zimbabwe, for instance, situated at an elevation of 4,826 feet (1,471 m) and well within the tropics, the mean annual temperature is 65° F (18.3° C), and the warmest month averages 70.3° F (21.3° C). Harare is part of the Cwa climate zone, which extends along the Bié Plateau, the southern African highveld, and the East African Plateau. Although similar to the Aw climate, the Cw is cooler and drier; and it receives at least 10 times as much precipitation in its wettest summer month as in its driest winter month (Figure 2-15).

Humid, Temperate, Dry Winter, Cool (Cwb) and Highland (H) The plateau of South Africa is known as the *highveld,* suggesting its elevated nature and grassland vegetation. There is justification for singling out this region climatically, since its latitudinal location (well outside the tropics) and altitude combine in creating a temperate climate. Extremes are considerable on the

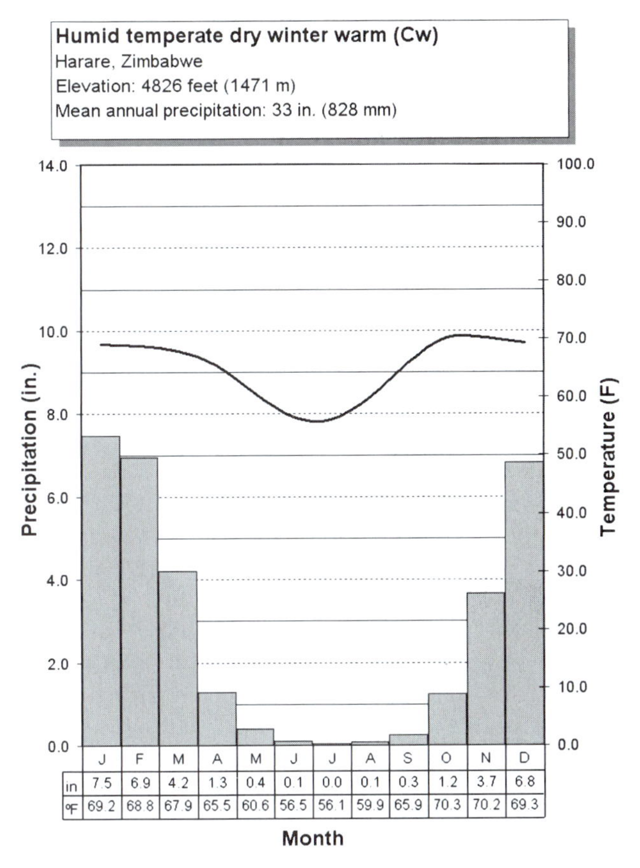

FIGURE 2-15 Harare, Zimbabwe: humid, temperate, dry winter, warm (Cw) climate.

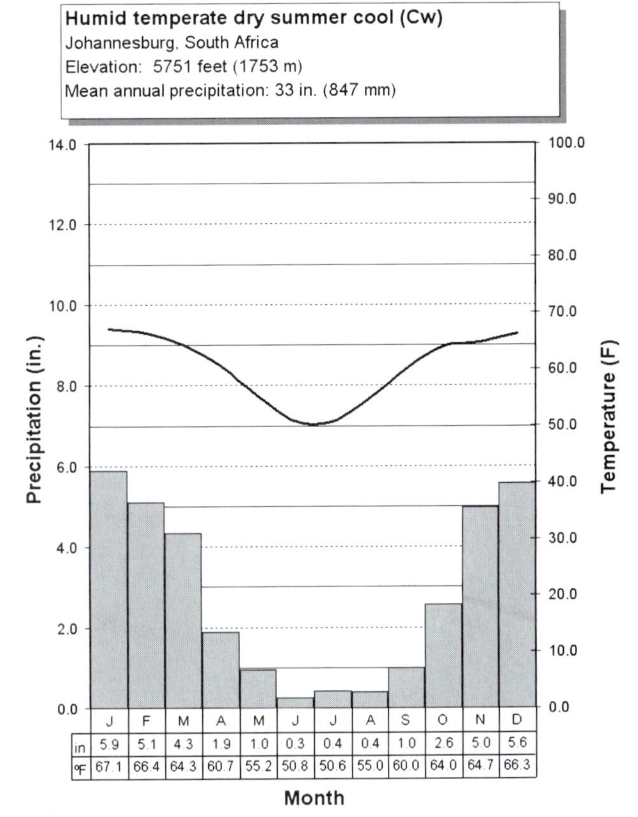

FIGURE 2-16 Johannesburg, South Africa: humid, temperate, dry summer, cool (Cw) climate.

plateau. Summer days can be oppressively hot, while frosts occur in winter. In high regions such as the Drakensberg, snowfalls are not infrequent. Winters are dry, and, after sunny days, rapid radiation losses cause inversions of temperature.

It has become clear that the circulation over the plateau is essentially the same throughout the year, namely, a rather weak anticyclone centered over the eastern margin of the subcontinent. In winter, this cell is somewhat stronger than in summer, when it is forced to move southward. During the dry winter season, the result is a high frequency of westerly and northwesterly winds. In summer, the anticyclonic circulation transfers maritime, potentially unstable air from the region of the mouth of the Limpopo River and further north onto the plateau. The actual mechanism producing the rainfall, which is particularly heavy in the east, probably lies in the convergence between the plateau anticyclone and adjacent oceanic anticyclones. In the interior, convectional storms are common, with the formation of cumulonimbus clouds, severe thunderstorms, and rapid clearing. Summer days on the plateau are hot, but they are tempered by the elevation. Conditions in Johannesburg (Figure 2-16) illustrate the situation.

A second region of warm, temperate climate in a highland environment occurs in Ethiopia, although the specific climatic characteristics and causes differ substantially from those in South Africa. Ethiopia has such great topographical and altitudinal variation that different climates exist short distances apart, and the major distinctions are caused by altitude. In the low-lying Danakil Depression and in Eritrea are found desert conditions similar to those of the Sahara; and on

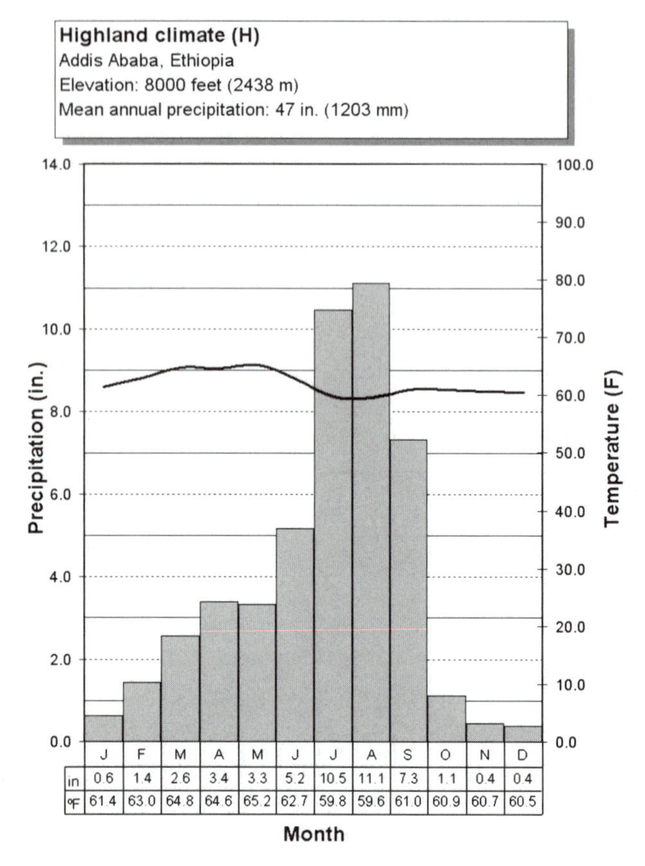

FIGURE 2-17 Addis Ababa; Ethiopia: highland (H) climate.

the lower slopes of the highlands below 5,000 feet (1,520 m), warm tropical conditions persist. But between 6,000 and 9,000 feet (1,830–2,740 m), in what is known as the *woina dega,* or wine highlands, the climate is temperate; summers are warm and rainy. At Addis Ababa (Figure 2-17), the mean monthly temperatures vary from 59 to 65° F (15–19° C), and the rainy season extends from June to September. The rains are brought by maritime equatorial air from the southwest in association with the northward movement of the ITCZ, while for the remainder of the year the Ethiopian Highlands are affected by dry continental tropical air from Arabia. Above the *woina dega* lies the *dega* (highlands), where temperatures are lower; on the peaks above 13,000 feet (3,960 m) there are occasional winter snows.

Low-Latitude Steppe (BSh) and Tropical Desert (BWh)

Africa's low-latitude steppes lie on the drier margins of the savanna, and in a narrow band north of the Sahara (Figure 2-8). The steppe zone that extends across the continent south of the Sahara is known as the *Sahel* (Arabic for "border" or "coast"). Being the southern edge of the Sahara, this region is subject to variable precipitation. It suffered a 4- to 6-year drought from the late 1960s to early 1970s when an estimated 5 million head of livestock and possibly thousands of people perished after having been weakened by lack of water and food. Drought in the early 1980s was also severe.

Recurrent drought is a feature of Africa's steppelands, and as long as the human and livestock populations grow rapidly, the potential for catastrophic consequences will exist. In the Sahel, overgrazing, poor groundwater management, and continuous cultivation of the land without rotation have produced dustbowl conditions and desertification. The steppelands are not devoid of tree growth, and particularly in the Kalahari, a thorny acacia manages to survive on the generally poor and shallow soils. Its existence is of great importance to livestock herders, for without the shade trees, the cattle would not survive.

The steppe is characterized by light and unreliable precipitation, dry, warm winters, and very hot summers. Average annual precipitation rarely exceeds 20 inches (508 mm), and because evapotranspiration rates are high, the effectiveness of even this moderate amount is reduced. The rain is almost always concentrated in the high-sun season, with the winter period being extremely dry. At Gao, Mali, for example (Figure 2-18), almost all the rain falls between June and September, and during this period the daily maximums are over 100° F (37° C), so that evaporation rates are high. Even during the cooler drier season, the daily maximums are approximately 90° F (32.2° C).

Adjoining the steppe regions are the tropical deserts (Figure 2-8). The Sahara, the largest hot desert in the world, stretches from the Atlantic to the Red Sea, and from the southern Mediterranean to the Sahel. Its aridity is due to the dry, extremely stable, continental tropical air mass that dominates the region. The air, part of the subtropical high-pressure belt, almost continuously descends from the upper levels of the atmosphere (Figure 2-4). The infrequent and isolated

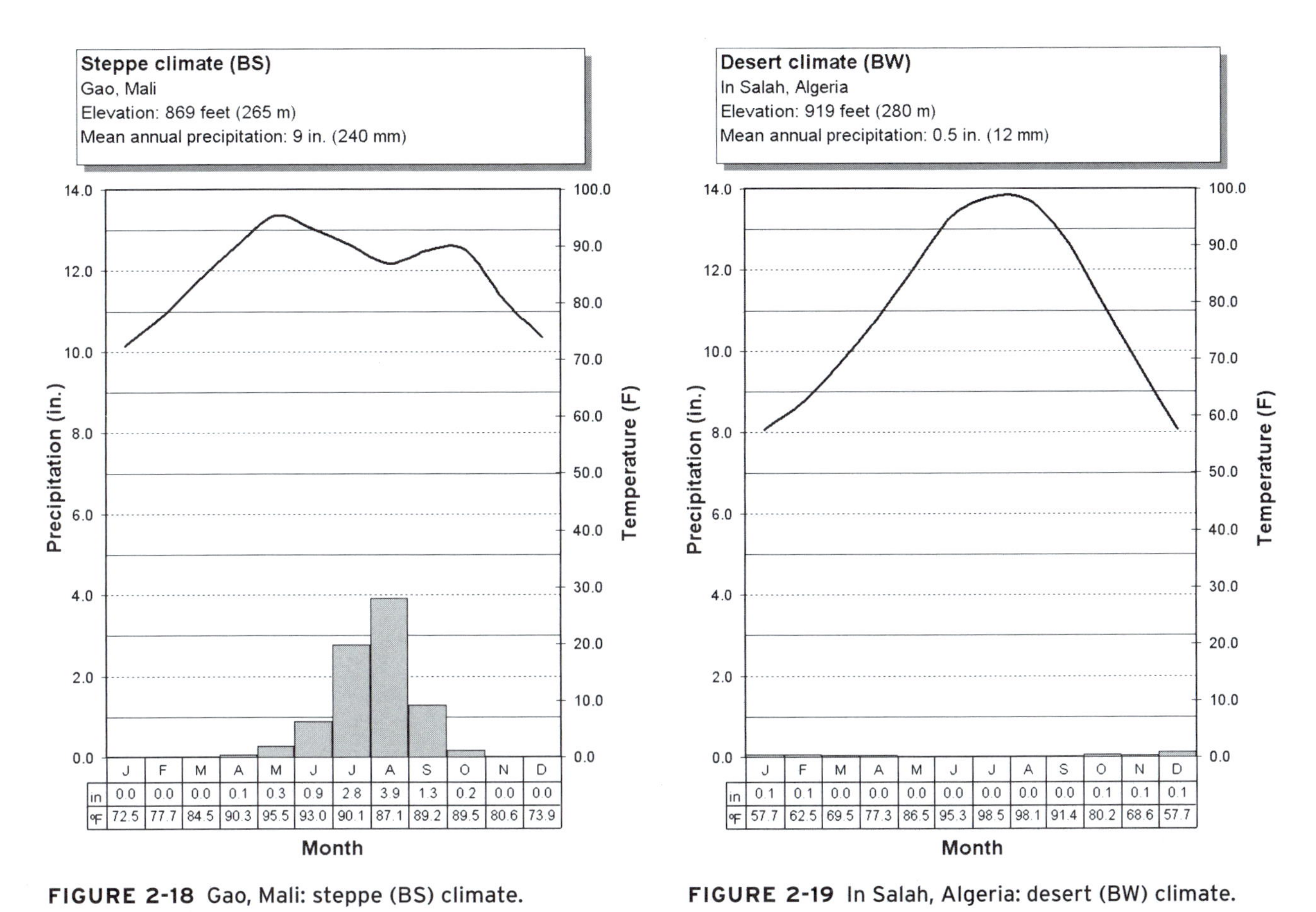

FIGURE 2-18 Gao, Mali: steppe (BS) climate.

FIGURE 2-19 In Salah, Algeria: desert (BW) climate.

rainstorms that occur, more especially in summer, come from the inflow of unstable maritime air masses from the northwest, or the occasional inflow of maritime tropical air (mT) behind the ITCZ from the south. This rain falls from convectional storms in these air masses, and since it is very localized, average rainfall figures mean very little. Few stations record more than 4 inches (102 mm), and most only a trace. For the town of In Salah, Algeria (Figure 2-19), the average annual precipitation is only 0.6 inch (15 mm); Cairo receives only 1.1 inches (28 mm) and Khartoum 5.1 inches (130 mm) on the average.

A desert characteristic is temperature extremes. Hot days may be followed by cold nights as the cloudless sky permits the loss of great amounts of long-wave radiation in short periods of time. In summer, daytime temperatures have been known to exceed 130° F (54.4° C), while in winter there are often subfreezing temperatures. Desert conditions extend south from the Red Sea to northern Somalia. These regions lie in the rain shadow of the Ethiopian Massif and thus do not receive moisture that falls from the maritime equatorial air masses that originate in the southwest in the high sun period. During the low-sun period (January), they receive dry tropical continental air from Arabia that absorbs very little moisture as it crosses the Gulf of Aden. Again, rainfall is generally less than 4 inches (102 mm), and temperatures are high. The third area of desert climate stretches along the coast of Namibia (Figure 2-8), one of the most barren and desolate regions on earth. A combination of factors produces this aridity. The Namib Desert lies in the northwestern and southwestern (dry) quadrants of the Kalahari anticyclone, so that the pressure situation prevents precipitation. In

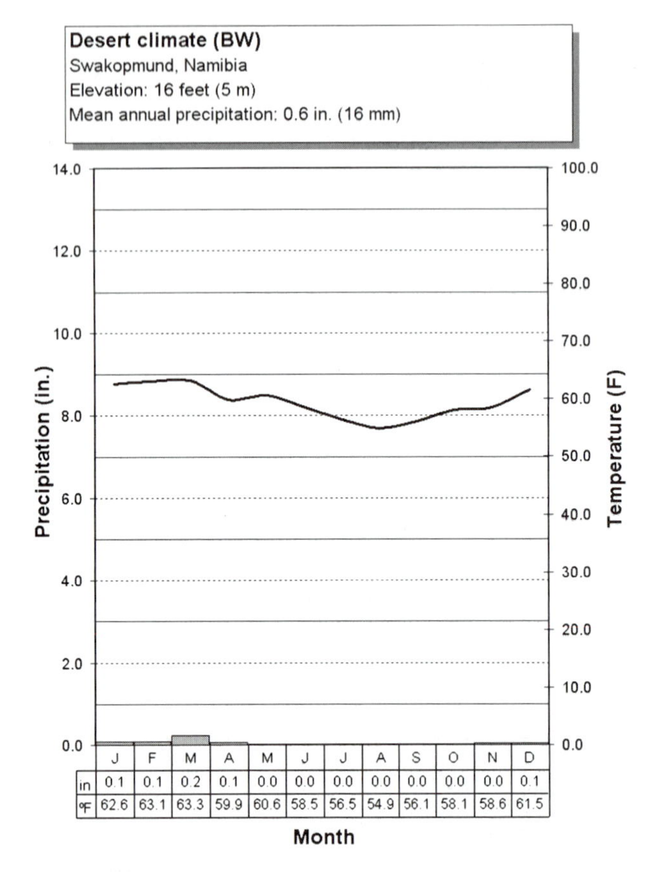

	J	F	M	A	M	J	J	A	S	O	N	D
in	0.1	0.1	0.2	0.1	0.0	0.0	0.0	0.0	0.0	0.0	0.0	0.1
ºF	62.6	63.1	63.3	59.9	60.6	58.5	56.5	54.9	56.1	58.1	58.6	61.5

FIGURE 2-20 Swakopmund, Namibia: desert (BW) climate.

addition, the coastal margin (to a width of only a few miles) is influenced by the cool Benguela Current that produces relatively cool temperatures, and, not infrequently, a fog that forms over the water and drifts slowly inland. Namib conditions are typified by Swakopmund (Figure 2-20).

Similar climatic conditions prevail along the Mauritanian coast because of the cooling effect of the Canary Current. Nouadhibou, for example, receives only an inch (25 mm) of rain on average, and in the warmest month the temperature is only 60° F (20.5° C). The interior of the Namib Desert, merging into the western Kalahari, is a little moister than the coast. Five inches (127 mm) of rain is recorded on average about 70 miles (113 km) east of Swakopmund, and the region around Windhoek, Namibia's capital, actually gets about 15 inches (380 mm). The coastal regions of the southwest are the driest, but, with the ascent of the plateau, the precipitation totals rise, to drop again toward the heart of the Kalahari.

AFRICAN VEGETATION AND SOILS

Vegetation

The natural vegetation has been modified by people and their livestock in all but a few extensive areas of Africa. For centuries pastoralists and cultivators have cleared the grasslands and forests by fire and hoe, and hunters (in places such as the Kalahari Desert) have burned the scrublands to drive out the game. Areas once covered with thick forest or tall grasses now stand open, exposed to the sun, wind, and rain. Excessive grazing in the Sahel, for example, has destroyed the natural ecology and promoted desertification. Microclimatic changes in turn have produced different vegetation covers. It is impossible, perhaps pointless, to describe and analyze Africa's "natural vegetation." What exists more properly could be termed anthropogenic vegetation.

Instead of discussing the distribution of potential vegetation, what follows is a discussion of the existing biomes of Africa, their vegetation types and their distributions, not the presumed climax types. The basic distributional pattern of vegetation types, as might be expected, corresponds quite closely with that of climate at the continental level but is modified by local soil, drainage, topographic, and bedrock conditions, and by the incidence of fire, cultivation, and pastoralism. The main biomes from the moister to the drier regions are tropical rain forest, savanna woodland and grassland, arid and semiarid vegetation, Mediterranean vegetation, and highland vegetation (Figure 2-21).

Tropical Rain Forest

Tropical rain forest covers low-lying areas of central and West Africa and eastern Madagascar including the Congo Basin, the lower altitudes of the Cameroon Highlands, the Niger Delta, and the Guinea Coast east of Sierra Leone. As indicated in Figure 2-21, the Congo Basin contains the greatest area of tropical rain forest in Africa: almost 700,000 square miles (1.8 million km^2).

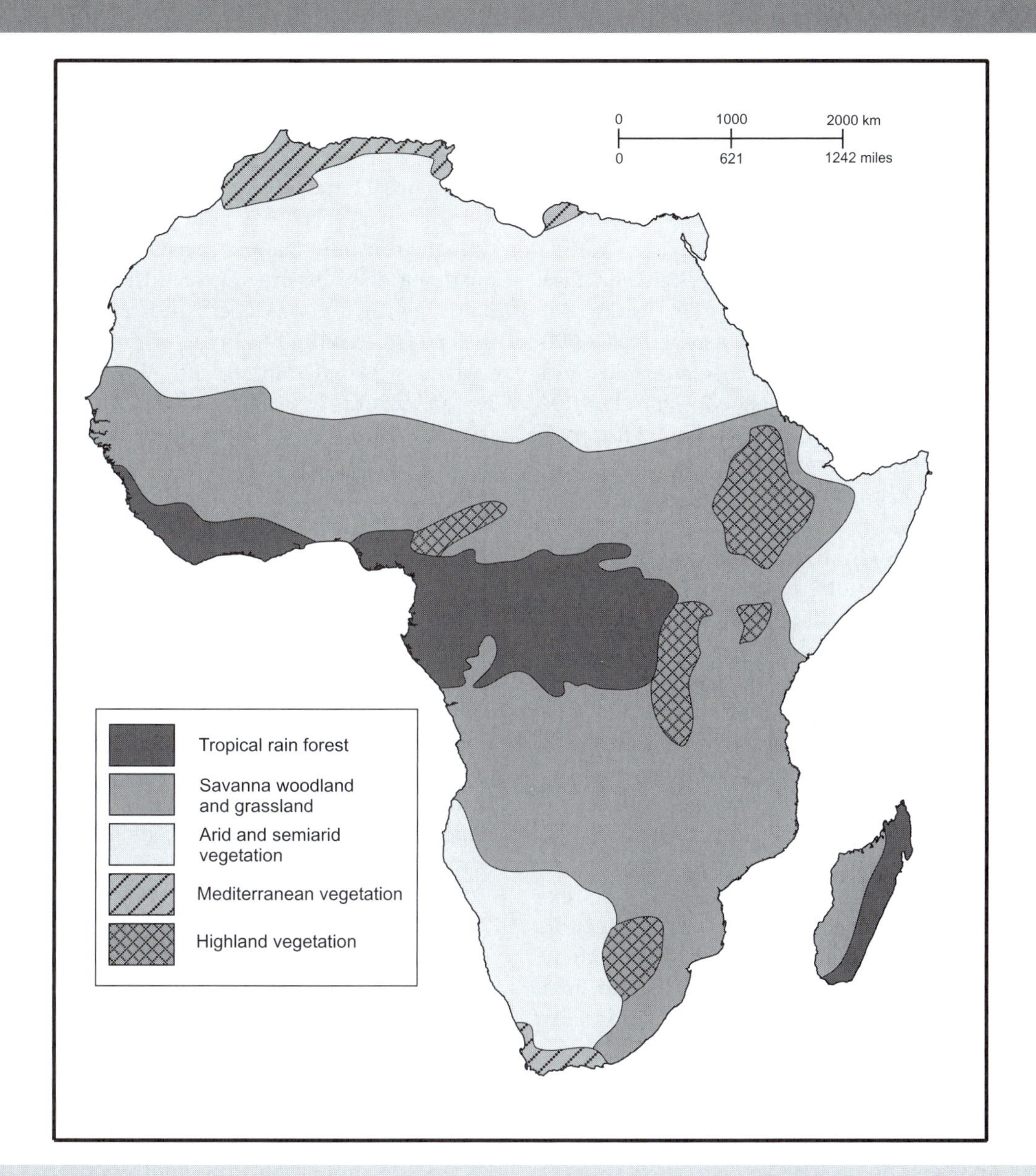

FIGURE 2-21 Biomes of Africa.
Adapted from Meadows (1996).

It is far less extensive than at one time believed, perhaps because early reports were based on observations made from rivers, the banks of which were covered in thick luxuriant forest encouraged by exposure to the sun, light, and water, whereas only short distances away the vegetation was less dense. The tropical rain forest develops in tropical climates where the rainfall exceeds 55 inches (1,400 mm) a year, and at least 2 inches (51 mm) falls in every month. In general, the warmer and wetter the climate, the more luxuriant the forest, although excessive rain, cloud cover, and poor drainage may inhibit growth.

Typical rain forest contains a great many species of trees, bushes, and ferns, and characteristically there are three layers of vegetation: a ground cover of shrubs and ferns from 6 to 10 feet (2–3 m) tall; a middle story of trees, palms, and woody climbers or lianas that reach about 60 feet (18 m) in height; and a dominant top canopy of broad-leaved evergreens of up to 150 feet (45 m) high, with spreading crowns that permit only a little light to filter through to the ground. As a result, the ground cover is not especially dense, but the lianas and epiphytes that are attached to the trunks and branches may give that appearance. As many as 3,000 species of plants may be found in a square mile, and rarely are there extensive homogeneous stands, thus making specialized lumber extraction almost impossible. Among the principal trees with considerable commercial value are mahogany, rubber, silk-cotton, and the Guinea oil palm.

Meadows (1996) indicates that older reports of the high productivity and nutrient-poor soils of tropical rain forests have been questioned in recent years. He has found that the primary productivity of African rain forests is not significantly higher than some temperate forests, where respiration losses are less, and that soils under tropical rain forests are more variable in nutrients than once thought. On certain low-lying, swampy areas such as the outer delta of the Niger, and at the mouth of the Congo, Zambezi, and Ravuma rivers, there are stretches of mangrove swamps composed of trees whose stilted roots entrap the river sediments, extending the land and hindering transport. There is no sharp outer boundary to the rain forest; instead, there is a zone of transition into the surrounding savanna grasslands. Here there are patches of forest (especially along streams) and tall broad-leaved grasses such as the "elephant grass." The reasons for this mosaic are not altogether known, but the combination of climate and the practice of burning the vegetation is undoubtedly the most important. There is evidence that forests have increased in area as a result of human activity in the forest-savanna mosaic (Fairhead and Leach 1996).

Savanna Woodland and Grassland

The most distinctive and widespread vegetation association in Africa is the savanna. The various savanna types stretch in a series of belts across West Africa to the upper reaches of the White Nile. Savanna also forms a broad zone from Angola to the Indian Ocean, and southward to the lowvelds of Zimbabwe and South Africa, including those of the Zambezi and Limpopo valleys. On the wetter margins, both the grasses and deciduous trees may be dense enough for a wooded savanna to prevail; but toward the drier limits, tree growth becomes restricted and the grasses predominate. Characteristic of the savanna belt are the flat-topped acacia trees and the large barrel-shaped, drought-resistant baobab trees, which along with bushes and shrubs are scattered throughout the grasslands, giving an open "parkland" appearance. Grasses instead of trees predominate, not only because of the marked seasonal rainfall, but also because of widespread burning, especially during the dry season, when the old "bush" is deliberately set on fire. This is to dispose of the old growth and to stimulate new, or as a means of clearing the land prior to cultivation. Only fire-resistant trees survive, and if burning were discontinued for long periods, there would be a reinvasion of woodland species. When rainfall is high, the savanna grasses may reach heights of 10 to 12 feet (4 m). Although hard and not particularly nutritious, these grasses sustain the bulk of Africa's cattle population as well as a diversity of grazing herbivores that is second to none in the world (Meadows 1996). This is also the land of the great herds of antelope, wildebeest, buffalo, elephants, and giraffe, now protected in a number of world-famous game reserves such as Serengeti (Tanzania) and Hwange (Zimbabwe).

Arid and Semiarid Vegetation

Progressing still further toward the more arid regions, the dry savannas give way to grass steppes, which give way to the subdesert steppes or thornbush. The thorny acacias and grasses become shorter and more sparse, and no longer form a continuous sod cover. In these regions of the Sahel, the East African Plateau, the Kalahari, Namibia, and the South African Karoo, the average annual rainfall varies from approximately 4 to 15 inches (100–380 mm), and the reliability of precipitation is very low, so that with prolonged and repetitive drought there can be serious implications for the increasing populations of people and livestock. In these subdesert areas, rainfall is both insufficient and too poorly distributed in time to permit a continuous vegetation cover, so that open spaces of ground separate the low shrubs and grasses that may be green only a few weeks in the year following rain. The true desert regions themselves are virtually devoid of vegetation other than widely scattered xerophytic plants such as the *stipa* grass, the tamarisk, and the date palm.

Mediterranean Vegetation

Beyond the tropics, and coinciding with the Mediterranean climate areas of South Africa and the Atlas Mountains, is a very distinctive vegetation type that is well adapted to the summer drought conditions: namely, hard-leaved evergreen shrubs and trees known as sclerophyll forest. Short, stunted junipers and pines are common, as are species of cork, oak, cedar, and olive. Tree growth is generally insufficient for extensive cutting, and the vegetation performs an important soil maintenance function on the steeper slopes, which in Morocco, Tunisia, and Algeria are all too frequently overgrazed by goats and sheep. Goudie (1990) indicates that the low, degraded, shrubland in the Mediterranean biome may be a product long overuse by people. The South African Mediterranean vegetation is characterized by diversity: over 8,500 species of plant grow in the 35,000 square miles (90,000 km^2) area.

Highland Vegetation

Forests and grasslands occupy the highland regions, where altitude rather than latitude is the dominant control factor. In the highlands there is a great variety of vegetation types, reflecting the diversities of climate, soil, drainage, and exposure. On the South African highvelds above 3,000 feet (1,100 m), there are vast expanses of temperate grasslands mixed with occasional stands of acacia. These grasslands, especially the "sweet velds," provide good grazing, although in winter the grasses are generally less palatable. In the highlands of East Africa and Ethiopia, both montane grasslands and montane forest occur in scattered locations. The latter resemble temperate deciduous forests found at lower elevations further poleward and contain many different species of plants and trees. Above 10,000 feet (3,048 m), where nightly frosts do occur, associations of many heath plants, sedges, and tussock grasses comprise the vegetation type.

Soils

The vast majority of Africans live close to the land as farmers or pastoralists, and as a consequence there is much truth in the observation that the soil is probably Africa's greatest resource (Areola 1996). The poverty of African soils and the harshness of the African environment have received an inordinate amount of attention according to Stocking (1996) and Areola (1996), but in recent years scientists have begun to reevaluate their understanding of African soils, land use, land degradation, and erosion (Adams 1996). There have been serious consequences for our ignorance in the past. We are only now beginning to understand soil characteristics, soil formation processes, and soil distributions in Africa.

It will be many years before accurate maps for the continent as a whole are constructed. Soil surveys are costly, time consuming, and most often confined to small areas. Frequently soils are used without adequate knowledge of other environmental conditions such as vegetation and climate. The consequences can be far reaching. In Tanzania, for example, about 2.4 million acres (942,000 ha) of land was temporarily destroyed in the East African Groundnut Scheme of the late 1940s. After only superficial surveys of the soil, climate, drainage, and other production factors, the land was stripped of its vegetation and planted with groundnuts. Exposed to the sun and rain, and turned over by inappropriate machinery, the soil quickly deteriorated, and the project had to be abandoned.

It can be said that in every climate region of Africa there are good soils and there are bad soils in reference to human use of the land. As is true elsewhere in the world, in Africa the alluvial soils are among the best on the continent, and in some areas extensive deposits have been laid down by the great rivers. The Niger becomes braided between Bamako and Timbuktu, and the alluvial soils form the basis of a productive dryland agriculture and extensive irrigation schemes. Similarly, alluvial soils lie between the White Nile and its tributary, the Blue Nile, and another such scheme, the Jezira, has been developed on the vast, fertile plains of the Sudan Basin. Developments on a larger or smaller scale can be found along almost every sizable river in Africa: the Gash and Baraka in Sudan, the Juba in Somalia, the Zambezi in Mozambique, and the Senegal in West Africa.

Certain nonalluvial soils also have been found to be of good quality, such as those on the South African Highveld, the Zimbabwean Plateau, and the Highlands of Kenya, as well as on the slopes of the Cameroon Highlands. These soils have resulted from particular parent materials or unusual climatic conditions, and

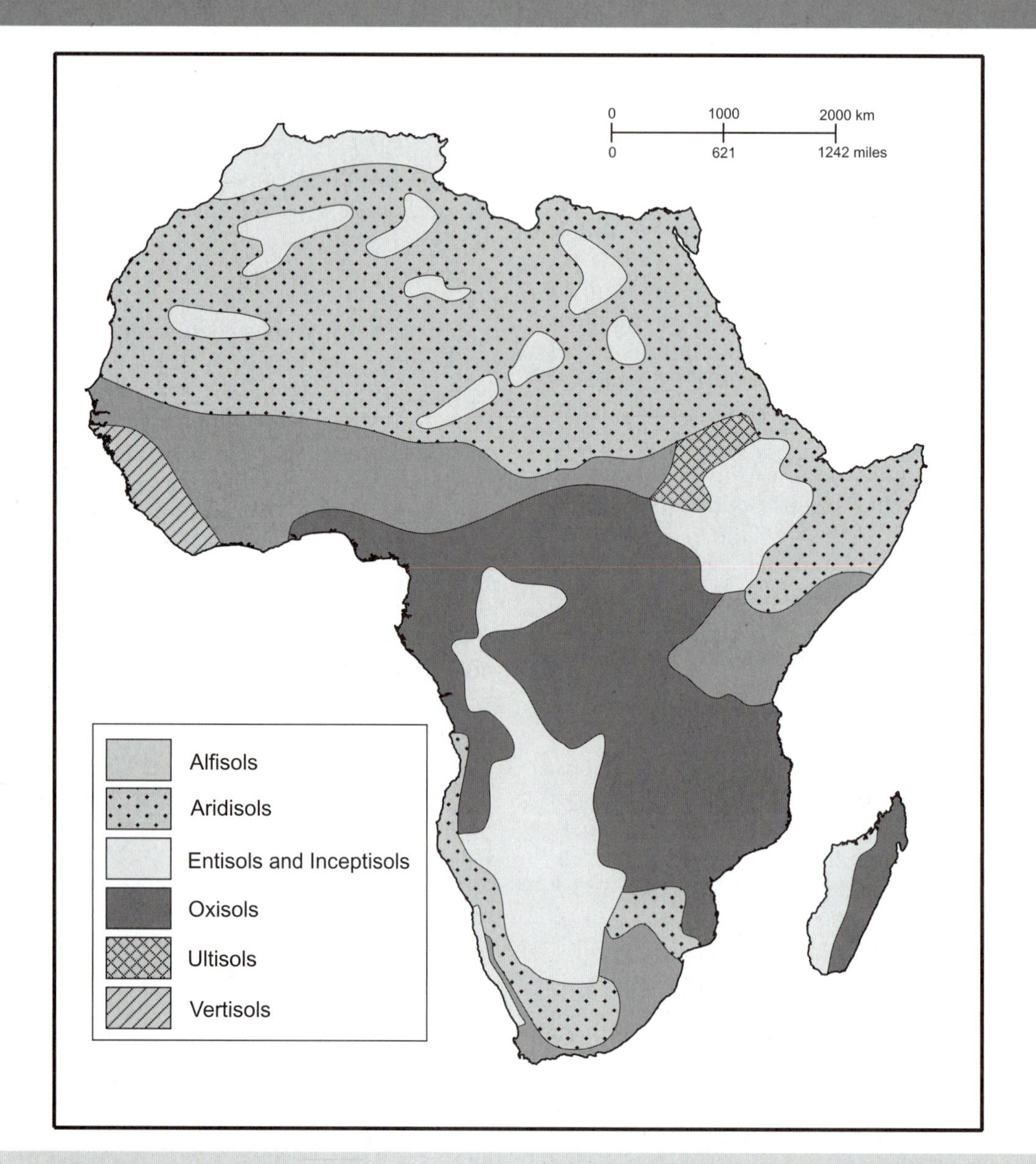

FIGURE 2-22 Major soil regions of Africa.
Adapted from Foth and Schafer (1980).

they support the few really important cash-cropping regions of the continent. But there is nothing in Africa to compare with the agricultural cores of Europe and Anglo-America in terms of size and yields per acre, a somber reality in a continent where the vast majority of the people still make their living off the land. Nevertheless, geographers are beginning to view the problems of African agriculture as an issue of development rather than an immutable reality linked to an inhospitable climate and environment. Figure 2-22 presents six broad groupings of African soils. Closer to the ground, the picture becomes more complicated and diverse.

Several continent-wide systems of soil classification have been developed in recent times, first by European soil scientists who classified African soils as part of the colonial enterprise (Kellogg 1949; Areola 1996) and later, after independence, by Africans. Indeed, most African farmers and pastoralists have a rather profound knowledge of local soils and ecology that is reflected in their use of the soil and in their soil classification systems. The terminology used in these indigenous classifications is very local, and there exist probably as many local classifications as languages in Africa. Nevertheless, the local classification is real, being rooted in experience and experimentation with many different crops, and crop varieties, and observation of natural vegetation (Cole 1991). Perhaps, then, local classifications would be more useful as a starting point for understanding African soils than sweeping generalizations made from small-scale maps of the continent.

In any event, discussion of local systems of soil classification is beyond the scope of this book, and the focus will turn to the continental system developed by the FAO and UNESCO. The classification system of the Food and Agriculture Organization of the United Nations (FAO) includes African soils (FAO-UNESCO 1988) and the observations of African soil scientists about local soils. The FAO-UNESCO classification recognizes three broad categories of soils: pedalfers, pedocals, and hydromorphic soils. We shall begin with pedalfers, the largest soil class in Africa, are found in tropical (ferrasols, nitisols, acrisols, lixisols), subtropical, and temperate regions (luvisols, planosols, arenosols); Figure 2-22 shows the distribution in Africa of some subtypes in the FAO-UNESCO system. Ferrasols, centered in the humid Congo Basin, contain iron, as evidenced by their color and name. Although chemically poor and requiring management to become productive, ferrasols are stable and not easily eroded. Nitisols are deep, agriculturally productive soils that are resistant to erosion. Acrisols are found where there is considerable precipitation, principally in Liberia, Côte d'Ivoire, and Guinea and on the East African Plateau. They are acid and relatively poor but do well with management. Acrisols are very erodible where vegetation has been removed. Lixisols are found in the savanna and dry lands of Africa. These soils have good chemical properties compared with ferrasols and acrisols, but they are erodible when vegetation is removed and they become exposed to the elements.

Soils containing laterite (plinthite or ironstone) are found in a variety of soil groupings in the tropical rain forest (ferrasols) and dryland regions but exist predominantly within the dryland soils on gently sloping terrain that has a fluctuating water table. Laterite consists of iron concretions that may be continuous at some depth in the soil as pans, outcrops, or caps on the higher points of old erosion surfaces. Areola (1996) classifies a soil as a plinthosol if 25% or more of its volume consists of a laterite horizon, or layer, that is at least 6 inches (15 cm) thick. Laterite remains soft until it is exposed to the air, after which, it becomes rock hard. Laterite when it lies close to the surface is an impediment to agriculture.

Luvisols contain clay horizons below the surface and are found principally in the Mediterranean climates of northern and southern Africa on gentle slopes near highlands. These soils are fertile but may be susceptible to erosion because of their high silt content. Planosols are seasonally waterlogged soils, used generally for livestock, and found on flat or gently rolling terrain in South Africa and in some parts of the Sahel.

The pedocals comprise vertisols, calcisols, and solonchaks. The former have a high clay fraction and are associated with river floodplains and lake beds that were created during past humid periods. The vertisols, located in the vast interior basins of Africa (Figure 2-22), are evidence of extensive lakes that once existed there. These soils are fertile and are farmed under dryland and irrigated conditions. Calcisols, soils of arid regions, are fertile, generally support pastoralism, and are most productive for farming under irrigation. The principal characteristic of calcisols is the presence of a calcium carbonate layer in their profiles. The solonchak group of soils are salty and found in waterlogged depressions in arid areas. Extreme forms of solanchak are locally called "sebkha" in Arabic and are characterized by salt at the surface and through the soil profile. Solonchaks are caused by high evaporation in comparison to precipitation or seepage rates. These soils are generally not usable by farmers, but livestock can be trained to consume some salt-loving plant species that are able to grow (Cole 1990).

There are two types of hydromorphic soil, gleysols and fluvisols. Gleysols, which developed under waterlogging, are found in low-lying land that has a high water table. Such land is found in the interior basins of Africa, for example, in the Niger Inland Delta and the

Congo Basin, and in valley bottoms and basins called *niayes* in Senegal, *fala* in central Mali, and *fadama* in northern Nigeria. Fluvisols are found on floodplains, valleys, and in estuaries. They occur along the West African coast, in the Lake Chad Basin, and in the Nile and Zambezi deltas. Both groups of soils are fertile.

Soils designated azonal in the FAO-UNESCO classification possess little or no development of horizons. They are used mainly for grazing or, when moisture is more abundant, forest.

PEOPLE AND CLIMATE

The vast majority of Africans depend on agriculture and pastoralism for their livelihoods. Compared with most Americans, Europeans, or city dwellers anywhere in the world, for that matter, Africans live close to the land. Thus the natural environment plays an important but not necessarily decisive role in influencing the distribution and density of rural activity. It is clear that people are not victims of their environment but rather make adaptations to it.

Climate, especially the precipitation factor, has clearly affected the distribution of people around the world and in Africa. The British geographer, Sir Dudley Stamp has claimed that the "key to the whole of Africa's development is control of water" (Stamp and Morgan 1972). While not all geographers accept this statement, most would agree that water supply and management are important for a great many developing states and individuals. Water issues concern all those who live on the land, and not infrequently those who live in cities. Compared with other tropical areas, however, the percentage of land in Africa that receives a high annual precipitation is relatively small, and a large part of the region that does receive much rain, the Congo Basin, is rather sparsely populated. Most Africans live in the climate zone defined as savanna, and again, the savanna is subhumid over vast areas. The bulk of East Africa's savanna lands, for instance, receive their annual average rainfall in only one year out of three. A marked dry season brings serious problems for the pastoralist; and somewhere in Africa, thousands of livestock die every year as wells and streams run dry and fodder becomes unavailable. For the farmer, much depends on the amount of moisture the preceding wet season has brought. Despite the environmental variability and uncertainty, African farmers and pastoralists are used to their environments and their variations and cope with environmental change by means of a variety of strategies based on experience and innovation (Cole 1991).

The Burton, Kates, and White (1993) model (Figure 2-23) illustrates how people who live in hazardous environments respond to environmental variation in

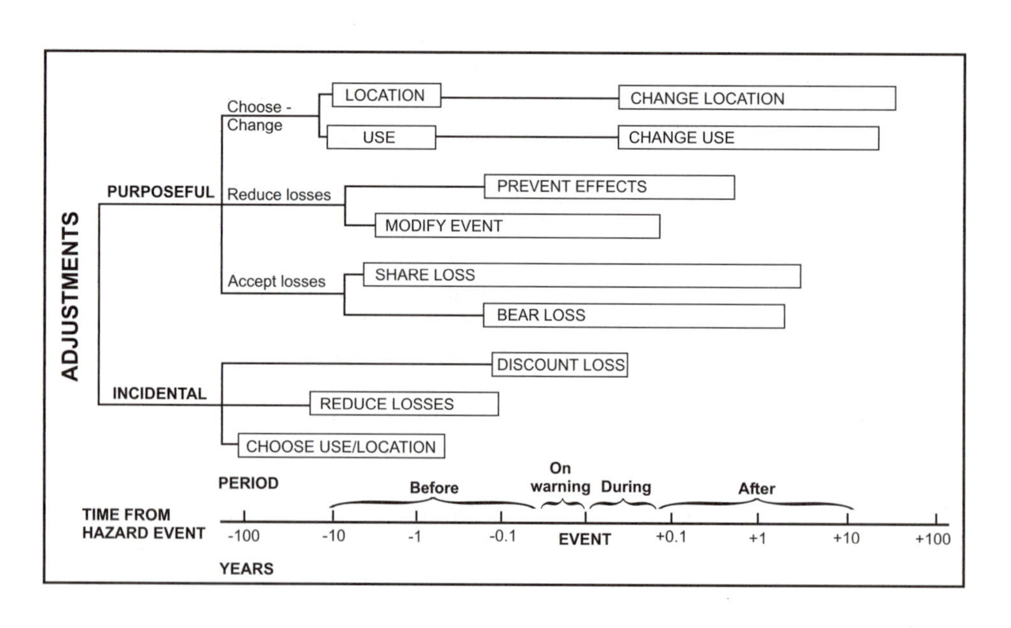

FIGURE 2-23 A choice tree of adjustment to the environment. *Adapted from Burton, Kates, and White (1993).*

positive and sometimes negative ways. The time frame for response extends in both directions from the moment—to decades before and after a natural hazard. These authors indicate that the basic choice involves choosing a particular activity (land use or economic activity) at a particular location.

> Once located and committed to a particular resource use, people use a variety of psychological, personal, and social devices to (1) discount losses by disregarding them or including them with other costs of location, or to (2) accept losses or to distribute and share them with other people. Whether people seek to reduce losses only after the occurrence of a hazard event or during the event (with a warning of possible occurrence), or much before the event, reflects increasing commitment of effort, forethought, and choice. They reduce losses by reducing the damage potential or by modifying the events themselves. As more and more substantial actions are seen to be required, the inhabitants of hazard zones may again be confronted with the choices of changing use in place or of migrating to another place. By no means are all adjustments to hazards purposeful, nor are all the theoretically available purposeful adjustments necessarily employed in a given place in relation to a given hazard. (Burton, Kates, and White 1993: 57–59)

The most disruptive adjustments to a natural hazard are changing the original land use and moving to another location.

African people are not helpless victims of their environment, but traditional systems are under strain from bearing the impacts of many processes that occur differentially over time and over social and geographic space. For those who live close to nature, coping with environmental changes occurs in a social, political, economic, and environmental context in which one or more of these factors may push individuals or groups of people to the economic margins. Watts (1983) has pointed out that response to the environment necessarily occurs under sociopolitical conditions that may impede the ability of some groups of people to successfully cope. Mortimore (1998) characterizes rural economics, especially those in the drylands, as in disequilibrium. He means that in good years farmers and lenders accumulate wealth and invest in a variety of assets. During the bad years of drought, such assets are drawn upon to sustain the family until conditions improve. But drought, for example, may become overwhelming in the face of nonenvironmental hazards such as widespread poverty, rapid population growth, a large dependent population, the breakdown of the state, economic and political instability, open war, rural unemployment, and widespread debilitating or terminal disease.

CHAPTER REVIEW QUESTIONS

1. What are some characteristics of the Mediterranean vegetation type?
2. What midlatitude climates are found in Africa? Why are they found there?
3. What makes the annual temperature of the tropical rain forest climate so uniform?
4. In which climates is annual precipitation variation greatest?
5. What climatic explanation accounts for the existence of the Sahara?
6. Why don't we know more about African soils?
7. Compare and contrast three African climate zones and their related vegetation.
8. With which climate zone would savanna vegetation be associated?
9. Explain how latitude helps determine annual temperature fluctuations.
10. Describe the environmental coping strategies of African farmers at a conceptual or theoretical level.
11. In the Köppen climate classification system, what sort of climate is designated by "A"?

GLOSSARY

Air mass A body of air that possesses relatively uniform characteristics of temperature and humidity based on the characteristics of the source area. Air masses are conventionally sorted into five categories as defined in the Köppen classification: maritime tropical (mT), warm and very moist; maritime polar (mP), cool and very moist; continental tropical (cT) hot and dry; continental polar (cP), cold and dry; and Arctic or Antarctic (A or AA), very cold and very dry.

Alluvial soils Soils deposited by moving water.

Anthropogenic vegetation Vegetation heavily influenced by humans.

Hydromorphic soils Soils that have developed under the periodic or permanent presence of excess water.

Laterite A type of soil in which iron and aluminum accumulate and form a distinct layer or horizon. Contact with air causes the layer to become hard.

Pressure systems World-girdling atmospheric belts of pressure and wind.

Subsident Describing descending or sinking air or land.

Tropical anticyclones Relatively dry and warm high-pressure cells located in the tropics.

Tropics The area on the earth's surface that lies between the Tropic of Cancer and the Tropic of Capricorn, respectively, 23.5° north and south of the equator, and defines the area of the earth's surface that receives direct, rather than oblique, rays of the sun at some time of the year.

BIBLIOGRAPHY

Adams, W. M. 1996. Irrigation, Erosion, and Famine: Visions of Environmental Change in Marakwet, Kenya. In *The Lie of the Land: Challenging Received Wisdom on the African Environment,* M. Leach and R. Mearns, eds. International African Institute. Oxford: James Curry, pp. 1155–1167.

Areola, O. 1996. Soils. In *The Physical Geography of Africa,* W. M. Adams, A. S. Goudie, and A. R. Orme, eds. Oxford: Oxford University Press, pp. 134–147.

Best, A., and H. J. de Blij. *African Survey.* New York: Wiley.

Burton, I., R. W. Kates, and G. White. 1993. *The Environment as Hazard,* 2nd ed. New York: Guilford Press.

Cole, R., ed. 1990. *Measuring Drought and Drought Impacts in Red Sea Province.* Oxfam Research Paper Number 2. Oxford: Oxfam House.

Cole, R. 1991. Changes in Drought Coping Strategies in the Ségu Region of Mali. Ph.D. dissertation, Department of Geography, Michigan State University, East Lansing.

Fairhead, J., and M. Leach. 1996. Rethinking the Forest–Savanna Mosaic: Colonial Science and its Relics in West Africa. In *The Lie of the Land: Challenging Received Wisdom on the African Environment,* M. Leach and R. Mearns, eds. International African Institute. Oxford: James Curry, pp. 105–121.

FAO-UNESCO. 1988. Soil Map of the World. New York: United Nations.

Foth, H. D. and J. W. Schafer. 1980. *Soil Geography and Land Use.* New York: Wiley.

Goudie, A. S. 1996. Climate: Past and Present. In *The Physical Geography of Africa,* W. M. Adams, A. S. Goudie, and A. R. Orme, eds. Oxford: Oxford University Press, pp. 34–59.

Goudie, A. S. 1990. *The Human Impact on the Natural Environment.* 3rd ed. Oxford: Oxford University Press.

Kellogg, C. 1949. Preliminary Suggestions for the Classification and Nomenclature of Great Soil Groups in Tropical and Equatorial Regions. *Commonwealth Bureau of Soils Technical Communication,* 46: 76–85.

Meadows, M. E. 1996. Biogeography. In *The Physical Geography of Africa,* W. M. Adams, A. S. Goudie, and A. R. Orme, eds. Oxford: Oxford University Press, pp. 161–172.

Mortimore, M. 1998. Roots in the African Dust: Sustaining the Drylands. Cambridge: Cambridge University Press.

Rasmussen. 1978.

Stamp, L. D., and W. T. W. Morgan. 1972. *Africa: A Study in Tropical Development,* 3rd ed. New York: John Wiley & Sons.

Stocking, M. 1996. Soil erosion: Breaking New Ground. In *The Lie of the Land: Challenging Received Wisdom on the African Environment,* M. Leach and R. Mearns, eds. International African Institute. Oxford: James Curry, pp. 140–154.

Watts, M. 1983. *Silent Violence: Food, Famine, and Peasantry in Northern Nigeria.* Berkeley: University of California Press.

CHAPTER 3

African Cultural and Political Hearths

A Historical Geography

Movement and migration have been so emblematic of the human experience that today they seem to be an axiomatically human characteristic. Although the pace and amplitude of movement in more developed regions of the world has picked up sharply over the last 50 years, and those of you reading this book live at an exceedingly fast pace compared with your ancestors, we rarely enlarge our time horizon to consider the ancient migrations that laid the broad lines of people and cultures around the world, what geographer James Newman calls the "peopling" of an area and the cultural, social, and political entities that ensued. Our understanding of the evolution of humans in Africa and their migration and colonization of the world, the differentiation of African languages and migration of their speakers, and the rise of states and empires is a fascinating story that is based on research in the areas of archaeology, political and cultural geography, linguistics, and history. Despite the broad disciplinary focus, Africa's prehistory and early history are known only in fragments, owing to the paucity of documentation. Over the last several decades African and non-African scholars alike have made concentrated efforts to reconstruct the African past. That they have succeeded to a remarkable degree is partly because of the unprecedented involvement of academicians in African studies, and it is also a consequence of political independence for African countries. Africans themselves desire to reestablish their links with a history full of achievements and potential, which was interrupted by the colonial episode.

Until the last decades of the twentieth century, many who studied Africa's past held that Africa and Africans were passive players in history—even victims of history—lacking in political sophistication, economic organization, and artistic achievement. This image has been shown to be false:

> . . . the emergence of African history has changed our understanding of general history, and of Europe's place in the world, in profound ways. It is no longer possible to defend the position that historical processes among non-European peoples can be seen as the consequence of all-encompassing influences emerging from a dominant European center. This shift in our understanding is uncomfortable for those who see history as the spread of civilization from a European center, and it is equally uncomfortable for those who sketch history in terms of an all-determining system of capitalist exploitation. (Feierman 1993: 182)

Indeed, it has been shown that Africans were in control of their relations with outsiders and engaged with them as equal or better partners, first in the ancient trans-Saharan trade with North Africa and Southwest Asia and later with Europeans in all aspects of the Atlantic trade (Thornton 1998). The very lateness of the European invasion, conquest, and establishment of political control in Africa is evidence for the degree to which Africans were organized and in control of their center and periphery. While sometimes at enmity with one another, they protected their inland trade routes effectively from outsiders. For example, in the hinterlands of the Gold Coast trading stations and in the gold-producing areas of Zimbabwe, the African political and merchant elites for centuries successfully repelled European encroachments and effectively controlled the African side of the Atlantic trade network.

In this chapter we will examine African prehistory and early history from the cultural and political geographic perspectives. Topics include the rise and dispersion of people, the development and diffusion of African languages, religion, early political organization, the trade in Africans and, the imposition of European control.

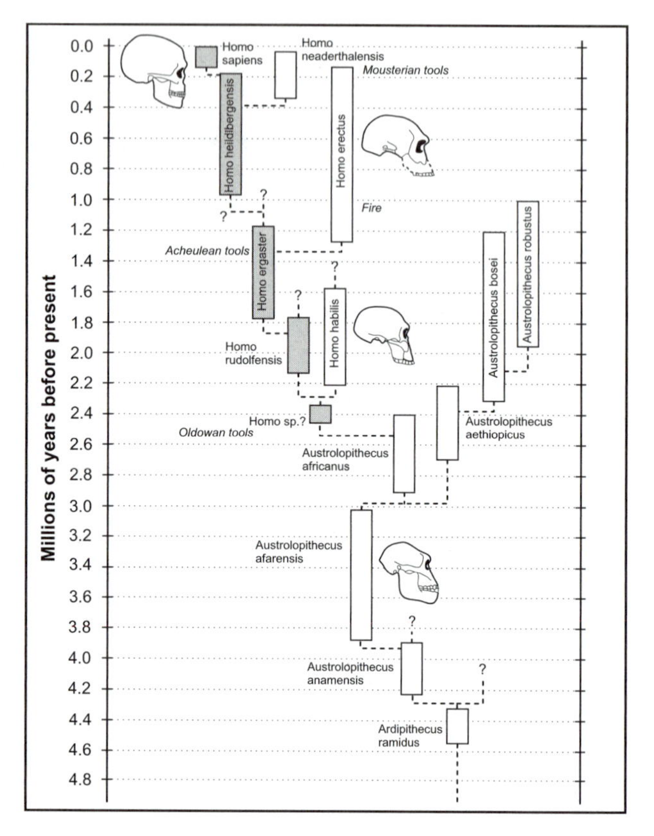

FIGURE 3-1 The descent of *Homo sapiens*.

EARLY HUMANS IN AFRICA

That humans evolved in Africa from primate populations is no longer a controversial or debated statement. Why Africa and not somewhere else? Because that is where the primates were located (Cavalli-Sforza et al. 1995). There, the evolution of *Homo sapiens* has been traced from very early beginnings, when our early ancestor, *Homo habilis* ("handy man"), commenced the making of tools. Africa is the only area on earth where evidence of *H. habilis* is found, although the later species *H. erectus* and *H. sapiens* colonized the world. Africa's Pleistocene stratigraphy has yielded an orderly sequence of objects made by our ancestors, ranking from the earliest stone pebble tools to hand axes. At Olduvai Gorge in Tanzania, a sequence covering about 2.5 million years has been found. During the later phases of the Pleistocene glaciation, which was felt in East Africa as a series of wet and dry periods and was reflected by variations in the Rift Valley lake levels, early humans discovered the use of fire. Although there is evidence of fire at Swartkrans (southern Africa) and at Koobi Fora (East Africa) at least 1.5 million years ago (mya), we do not know whether the early humans controlled the fire.

Finally, *Homo sapiens* appeared on the scene (Figure 3-1). People sought shelter in caves, began to make new and different tools, including many with wooden handles, and started to use bone and wear skins. These developments were undoubtedly related to the greater information-processing ability of *Homo sapiens*. *Homo sapiens* had a larger brain than either *H. erectus* or *H. habilis:* 1,300 to 1,400 cubic centimeters, on average, compared with 900 to 1,100 and 650 to 800 cm^3, respectively. Modern *Homo sapiens* has an average cranial capacity of 1,450 cm^3. The entire sequence of human development is recorded in Africa more perfectly than anywhere else, and "there is little doubt that throughout all but the last small fraction of [the] long development of the human form, Africa remained at the center of the inhabited world" (Oliver and Fage 1962).

Africa's first human occupants, *Homo habilis,* were meat-eating scavengers whose precarious existence was centered around a constant search of food. It is thought that they lived on remains, carcasses that had been left by predator beasts; the bone marrow, which was obtained by crushing the bone with crude stone tools, was particularly nutritious. Later evidence indicates that *H. erectus* hunted in groups using more sophisticated tools than *H. habilis,* rather than simply

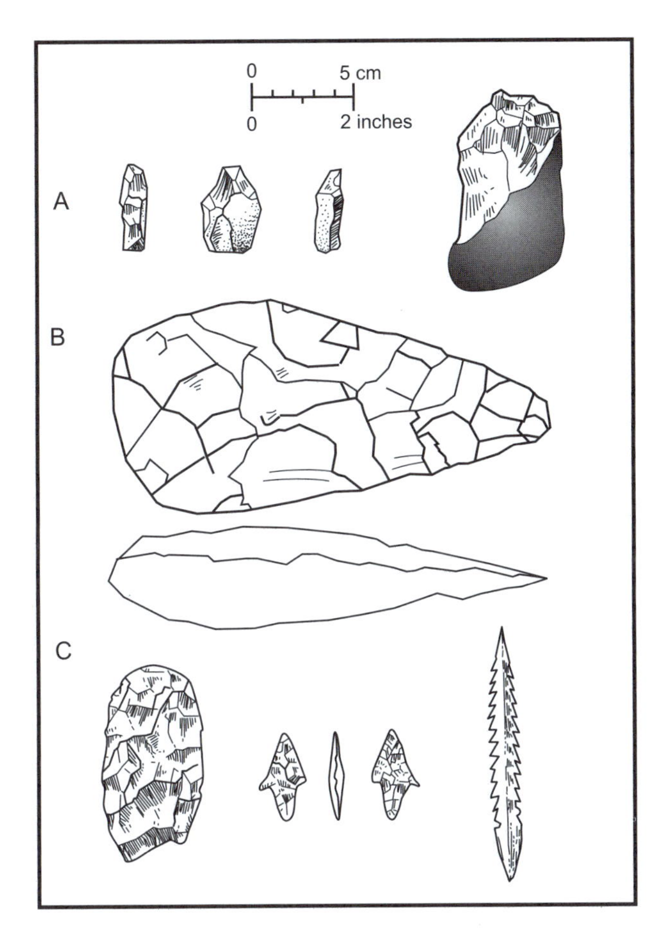

FIGURE 3-2 Tools of early humans: (A) *Homo habilis*, (B) *H. erectus*, and (C) *H. sapiens*.
Adapted from Shillington (1989).

scavenging. Early *Homo sapiens* had an even more sophisticated toolkit, including hafted axes, bone fish hooks, and later the bow. Figure 3-2 shows the evolution of the tools used by our ancestors.

Hunting and gathering, the principal occupation of these early people and one that has been our single occupation for most of human history, was an uncertain enterprise: drought forced long migrations from places of immediate privation to areas with prospects for more abundant collecting and hunting; disease and accident kept life expectancy very low. Births were controlled because of the difficulty of transporting small children, particularly infants, over long distances. In relative terms, the material level of culture was very low. Population densities also were very low, and people scattered across the continent.

Hunters and gathers still exist in isolated areas of the world today, for example, the San (or Khoisan) in the Kalahari Desert of Namibia, Botswana, and South Africa, and the Mbuti pygmies, who live in the Ituri Forest in the Democratic Republic of the Congo. The San live in small bands under the leadership of the most skillful hunter, moving from place to place in search of antelope, giraffe, ostrich, and other game that they hunt principally with nets. Hunting is also conducted using spears and the bow and arrow. The San are expert trackers and are able to locate animals by considering the most minute bits of evidence. Smaller game is taken in snares and traps, and roots and berries are gathered most frequently by women and children. Their shelters, simply made of branches and grasses gathered from the bush, are abandoned when an area becomes depleted of food. In the past, the San lived in rock shelters, and much can be learned of their culture from the rock paintings they have left behind.

THE DIFFUSION OF AFRICAN LANGUAGES

It has been well established for forty years that four ancient language families developed in Africa among the later hunter–gatherer populations (Greenberg 1963). Speakers of two other families, Indo-European and Austronesian, migrated into Africa more recently. What existed before that time, roughly 10,000 years ago, is unknown. Based on the scanty evidence available, however, we are able to construct a picture of that time. Table 3-1 shows the most recent data on African languages and their speakers.

Figure 3-3 illustrates the present distribution of these language families across Africa. Note that some of the language families occupy large areas of the continent while others are not widely distributed. Language subfamilies are equally diverse in size; some are located far from the main concentration of the subfamily and appear as islands in a sea. Notice the fragmented appearance of some families and subfamilies. Geography provides an intriguing explanation for the

TABLE 3-1 AFRICAN LANGUAGE FAMILIES, NUMBER OF SPOKEN LANGUAGES, ESTIMATED NUMBER OF SPEAKERS AND MOST PROBABLE ORIGIN.

Language Family	Number of Languages in Africa	Estimated Number of Speakers in Africa	Provenance
Afroasiatic	343	135,000,000	Probably northeast Africa
Khoisan	35	100,000	Southern Africa
Niger-Congo	1,436	400,000,000	West-Central Africa
Nilo-Saharan	194	100,000,000	West-Central Africa

Source: Grimes (2000), Heine and Nurse (2000).

present map of African languages. To understand the development of this map, we must turn to the Old Stone Age and look for evidence.

At the end of the Old Stone Age, 15,000 to 10,000 years ago, when the African population was probably around one million people, four groups of hunter–gathers were sprinkled across the continent. Newman (1995) calls them Caucasoids, Tall Negroids, Pygmoids, and Capoids. Each group appears to have distinctive material cultural remains, principally pottery. Moreover, skeletal differences between groups are distinctive; the genetic distance between the groups is great, meaning that they experienced long isolation from one another; and although there is some debate about the original language of the Pygmoids, each group gave rise to one of the four African language families spoken on the continent today. Figure 3-4 shows the geographic distribution of the four population groups at the end of the Old Stone Age, as determined on the basis of skeletal artifacts and remains.

Archaeologists, anthropologists, geographers, historians, and linguists have attempted to reconstruct the movement and migration of people as far back in time as there is evidence. Using linguistic data in present-day Africa as their guide, they suggest that the language hearths of the four African language families can be determined and that the present geographic distribution of the speakers of three of these families or their subfamilies is associated with the discovery of agriculture and, later, iron.

Blench (1993) argues that the major diaspora of people speaking early forms of Niger–Congo, Nilo-Saharan, and Afroasiatic occurred before agriculture in Africa. He suggests further that the push of increasing aridity and the concomitant pull of migrating game populations in a drier climate would increasingly fragment the human populations of hunter–gatherers. On the other hand, "[w]hen the climate improved, the expansion of both hunted and gathered resources would allow a corresponding diversification of human subsistence patterns" (Blench 1993: 137). He believes that Nilo-Saharan originated in central Africa north of the present rain forest around 13,500 years ago in an extremely arid period; then later, during a wet period when great lakes and vast grasslands existed in the Sahara, speakers of early Nilo-Saharan expanded. This would explain the deep divisions between branches of the Nilo-Saharan language family. Niger–Congo and Afroasiatic may have developed between 12,500 and 10,000 years ago west and east of the Nilo-Saharan heartland during the same wet phase, and their expansion could account for the isolation of some Nilo- Saharan branches such as Songhay.

Newman (1995) argues that the original speakers of Niger–Congo may have been pygmies, ancestors of the present-day Mbuti and Twa, rather than West Africans, who he holds may not represent an independent population but rather a result of mixing between Pygmoid peoples and Tall Negroids. Blench (1993) believes that the origin of Niger–Congo is further to the west, since evidence of the oldest divergences of Niger–Congo is found there. Additionally, since the original language of the pygmies is now lost, attributing the origin of Niger–Congo to their ancestors is conjecture.

Although there is not much evidence regarding the source area of Afroasiatic, Trigger et al. (1983) point out that the principle of least moves means that northeast Africa, probably somewhere in northeast Sudan or

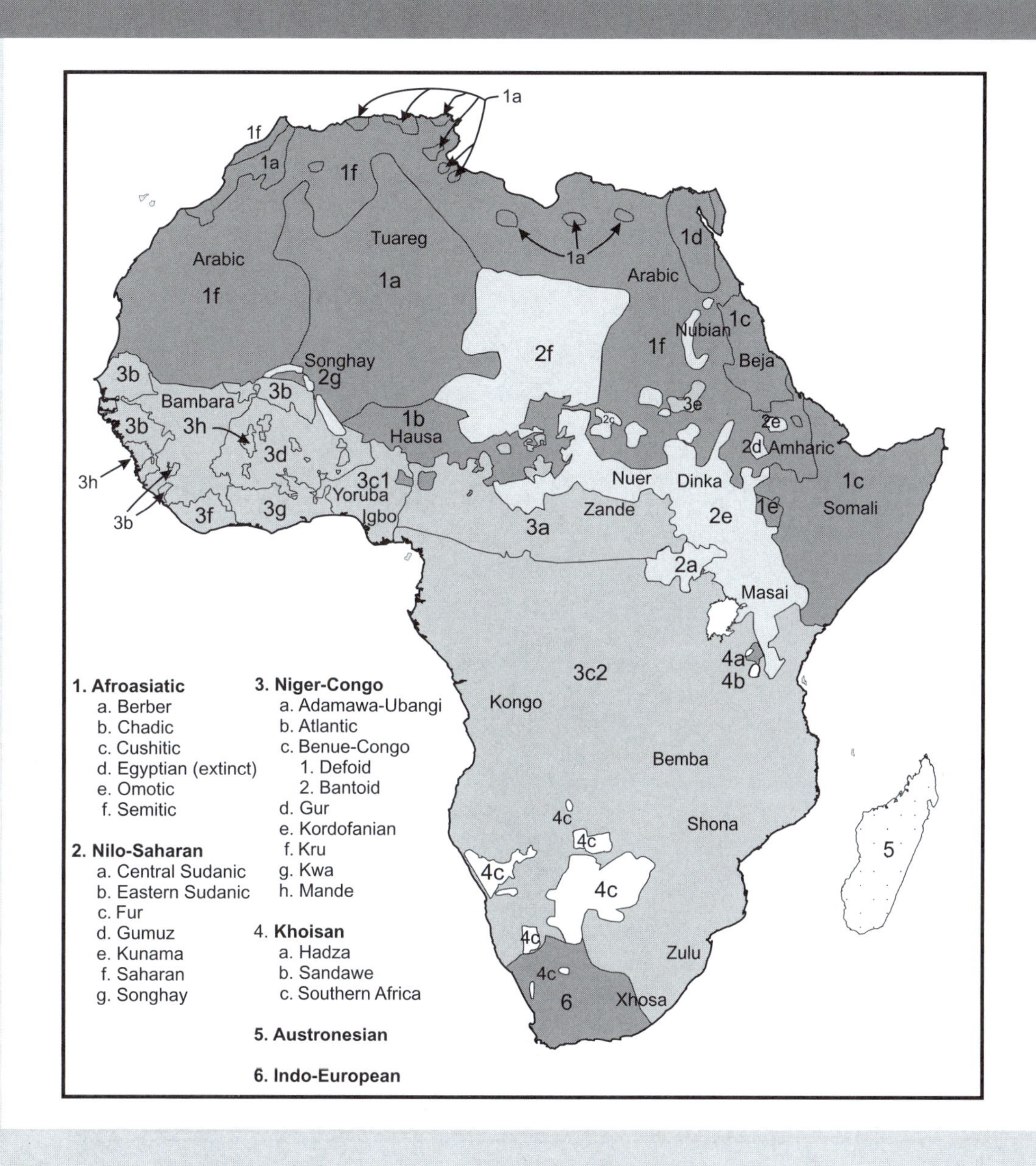

FIGURE 3-3 Present distribution of languages in Africa.

southeast Egypt, could be the source area. The antiquity of the split between Omotic, a subfamily of Afroasiatic, and Afroasiatic supports a source area somewhere in northeast Africa (Blench 1993). Ancient Egyptian began to diverge about 7,000 years ago into its branches, and Hayward (2000) suggests that the protolanguage must be dated to before 10,000 years ago. Today's Khoisan speakers comprise a small and dwindling population, and they inhabit a very small area; thus it is difficult to reconstruct the hearth of their language. Judging from skeletal remains, however, it must be in Southeast Africa.

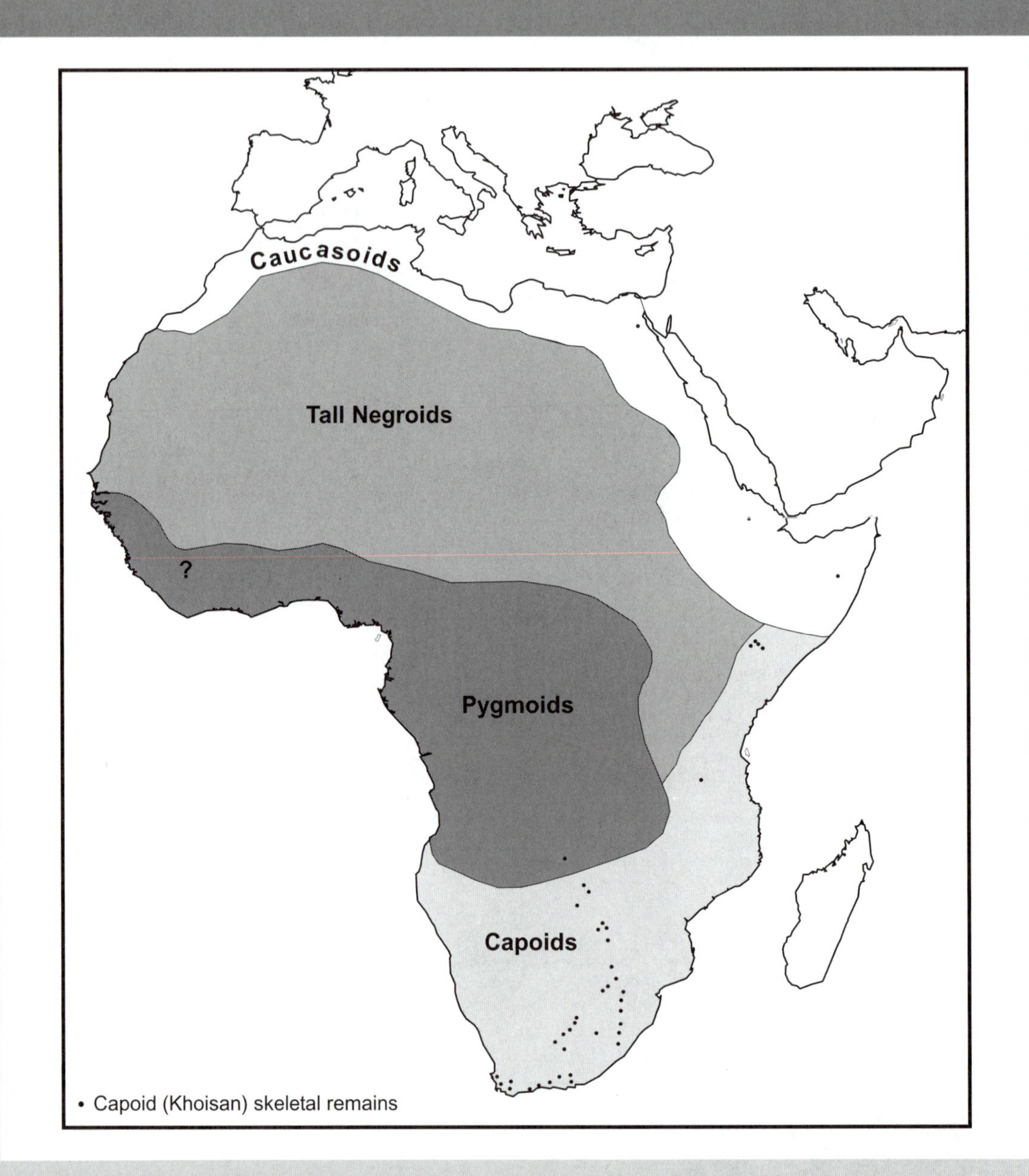

FIGURE 3-4 Four Late Stone Age regional genetic populations from which all Africans are descended. *Adapted from Newman (1995).*

THE IMPACT OF THE NEOLITHIC REVOLUTION

The invention of the pastoral way of life and the discovery of farming (called the Neolithic revolution) was to improve standards of living and change the distribution of Africans across the continent in profound ways. The agricultural revolution experienced in Africa came in several distinct phases. Agriculture was invented some 10,000 years ago in Mesopotamia, although it was later invented independently in many places in the world, including a variety of locales in Africa. It is difficult to determine a precise chronology of the diffusion or invention of agriculture in Africa. The oldest confirmed agricultural site in Africa is in Egypt, 2,000 years younger than the earliest sites in Southwest Asia. The evidence indicates that agriculture (wheat and barley cultivation) spread down the Nile Valley and west through the central Sahara 7,000 years ago and across to the Maghreb, 5,000 years ago. There is evidence that agriculture was invented independently in West Africa and in the Horn of Africa.

Why did our ancestors invent agriculture, and why would hunters and gatherers adopt agriculture from others? Speaking of the transition to food production in the central Sahara, Muzzolini (1993) notes that population pressure, particularly in relation to water, and a type of privatization of property by ethnic group or clan (the division of land into "territories") must have been major factors (Box 3-1).

A wide variety of plants were domesticated by Africans. It is thought that pearl millet was domesticated in the Sahara during a pluvial period. There is evidence from 3,100 BP (before the present) at Dar Tichett, Mauretania but cultivation is surely older. It later spread to Sahel and elsewhere in the dry lands of Africa. African rice and fonio were domesticated in West Africa, probably by people speaking one of the early Mande languages of the Niger–Congo family. Although it is not known when these two domestications occurred, they probably were contemporary with that of pearl millet at Dar Tichett. Rice was later adapted to more humid areas, from Sierra Leone to the Gambia. In the savanna lands, the growing of sorghum probably was first learned in the Sudan–Chad region and spread elsewhere (Harlan 1993).

Sorghum domestication appears to be associated with proto-Nilo-Saharan speakers who once hunted and gathered across the Sahara during a humid period but migrated south of the Sahara as the region increasingly became drier. They were distributed from the bend in the Niger River to the confluence of the Blue and White Niles. Guinea yams had been domesticated in the area of present-day Nigeria 5,000 to 4,000 years ago, probably by ancestors of peoples presently living in Nigeria who speak one of the many Kwa or Benue–Congo languages of the Niger–Congo family (Andah 1993).

In Ethiopia, locally domesticated small grains, teff, and finger millet had been domesticated by 5,000 BP. Wheat and barley were being cultivated in the Horn by 3,000 BP. We are uncertain about the dates of domestication of other plants found in the Horn, such as noog, ensete, coffee, and sesame (Phillipson 1993). Evidence of oil palm cultivation begins to appear in southern Nigeria around 2,800 years ago. Table 3-2 lists many of the most important African domesticates.

Many crops cultivated in Africa today were introduced from other continents and probably arrived less than 2,000 years ago. Bananas and yams, for instance, came from Asia, possibly during the first centuries of the common era; the present-day staples of corn (maize) and cassava (manioc) came from the Americas as late as the sixteenth century and slowly diffused across the continent. Corn is now grown in most parts of Africa, displacing sorghum over wide areas south of the equator, and the Asian yam has become widespread. Chickens were introduced to Africa by the same Southeast Asians who brought the banana and yam. The mango and citrus crops are recent introductions. Non-African vegetables such as the tomato, onions, eggplant, cabbage, lettuce, potatoes, and carrots are widely grown, especially for sale in the city. Tomatoes have been widely adopted as a necessary ingredient in the preparation of sauces to accompany the main food course at lunch or dinner. Rice varieties from China and Korea are used in the old African rice hearth, the middle Niger valley. African horticulturalists and plant geneticists have created from foreign stock plants of new types that are adapted to local ecological conditions. A notable example was the creation, in the early 1960s, of the widely successful Mali mango in the Sotuba Agricultural Experimental Station near Bamako.

There is evidence of cattle in the central Sahara around 7,000 BP. Cattle, sheep, and goats were widely herded in the Horn by 5,000 BP, while there is evidence for dwarf cattle and goats in West Africa by 3,800 BP. Longhorn cattle were dominant in Egypt by 4,000 BP. Zebu cattle, the most widespread breed in Africa today, were introduced into Africa from India, probably through the Horn. There is evidence of the zebu in semi-arid West Africa around 1,000 BP. The camel was introduced to Egypt from Arabia some 2,900 years ago, but its use was not common in northern Africa until much later, about 1,700 BP. The camel was the major means of transportation of the trans-Saharan trade, which began to develop by 1,500 BP. According to Bulliet (1990), the camel was introduced to the Horn from Arabia between 4,450 to 3,450 BP but was not widely used in transportation until 1,700 BP.

BOX 3-1

CAUSES OF THE TRANSITION TO FOOD PRODUCTION IN THE SAHARA

The transition to food production . . . is the response by a social system . . . either to an external stimulus (diffusion) or to an internal disequilibrium.

Climate is a vitally important factor which could accelerate or retard any such change in an arid zone, but it could not, alone, have set in motion the transition to food production. In an arid zone, a degree of sedentism is necessary to maintain a close link to a water supply. Complete sedentism, with permanent dwellings, appears to be a more or less obligatory condition for the cultivation of cereals, but not for the pastoral system of the Saharan Neolithic.

The density of remains and sites in certain areas of the Sahara gives clear evidence of demographic pressure; for every one Aterian or Acheulean site there are a hundred Neolithic. The suddenness with which this population explosion burst upon the scene—and the uneven way in which it then progressed in comparison with other factors—suggests that, although not the sole cause of the Neolithic upheaval, it must have been a major one.

[In an arid zone] the vital link with a water source, for animals as well as humans, rigidly determines land occupation as soon as a critical population threshold is reached and all the sources of water are in use (that is, either permanently occupied or regularly visited, according to the "rights" of each group). Neither animals nor humans then have any further possibility of migrating to another supply of water. This first "crisis," which had never occurred before the Neolithic, had the end result of creating and fixing "territories" within finite limits.

There followed an increase in sedentism, competition between groups, specialized hunting of the only species existing in the biotope and an intensification of local food collecting; all this resulted in an accelerated population increase. This continued until a second "crisis" was set off by the inevitable natural limitations and carrying capacity of the biotope, which is more fragile and, above all, less flexible in an arid zone, if it continues to be exploited in the traditional way by semi-sedentary people. Two solutions were possible. Warfare, in the modern sense of the term, emerged as a new form of relationship between organized groups. Evidence of this exists in Jebel Sahaba in Nubia, around 12,000 BP, where a burial ground yielded fifty-nine projectile points protruding into twenty-four bodies (Wendorf 1968).

Alternatively, a transition to a still more intensive and planned exploitation of the "territory" could be made. The possibility had been known; now it became a necessity. It consisted of the regulation of reserves to a far higher degree than simple storage of collected grain in pots or granaries; it involved the accumulation of protein in managed livestock or, where possible, in cultivated cereals.

Admittedly the critical threshold for the breakdown of the equilibrium of the biotope which would set off such a second "crisis" depended not only on its physical potential, but also on group values, social organization and symbolic worlds (a "sacred" animal is never hunted)—all features which exclude a hasty environmental determinism.

A. Muzzolini (1993: 239). The emergence of a food producing economy in the Sahara. In T. Shaw, P. Sinclair, B. Andah, and A. Opoko. The Archeology of Africa: Food, Metals, and Towns. London: Routledge.

The spread of agriculture took many centuries, and the groups of people who acquired agricultural technology grew faster than their hunter–gatherer neighbors. Agriculture made settled life possible and removed some of the uncertainty of the hunting and gathering way of life: population growth became an advantage, and sedentism permitted the development of more material elements of culture—the improvement of creature comforts, the accumulation of material possessions and wealth, and the use of hard-to-transport tools. The surplus produced from agriculture can be stored over several years, and the sheer volume of production can support many more people in greater densities than hunting and gathering. It is for this

TABLE 3-2 IMPORTANT CROPS OF AFRICAN ORIGIN.

Common Name	Scientific Name	Area of Domestication
Sorghum	*Sorghum vulgare*	Sahel-savanna
Bulrush millet	*Pennisetum typhoideum*	Sahel-savanna
Finger millet	*Eleusine coracana*	Ethiopian highlands
African rice	*Oryza glabberima*	Upper Niger valley
Fonio	*Digitaria exilis*	Upper Niger Valley
Teff	*Eragrostif tef*	Ethiopian highlands
Cowpea	*Vigna unguiculata*	Forest-savanna ecotone
Pigeon pea	*Cajanus cajan*	Forest-savanna ecotone
Bambara groundnut	*Voandzeia subterranea*	West African savanna
Guinea yam	*Dioscorea rotundata*	Forest-savanna ecotone
Watermelon	*Citrullus lanatus*	West African rain forest
Okra	*Hibiscus esculentus*	West African rain forest
Oil palm	*Elaeis guiniensis*	West African forest-savanna ecotone
Ensete	*Ensete ventricosa*	Ethiopian humid forest
Noog	*Guizotia abyssinica*	Ethiopia
Coffee	*Coffea arabica*	Ethiopia
Sesame	*Sesamum indica*	West Africa

Source: Adapted from Newman (1995), Andah (1993).

reason that hunting and gathering have virtually disappeared from Africa, indeed, from most of the world—these ancient practices have simply been outcompeted by one that produces more food: agriculture.

Newman (1995) and Cavalli-Sforza (1995) suggest that cultivation and the raising of livestock were the principal causes of later migrations that culminated in the present-day distribution of African peoples. Although people continued to fish, hunt, and gather, the new activities stimulated population growth and expansion. About 3,000 to 3,500 years ago, people speaking an early form of a Benue–Congo language, proto-Bantu, began expanding east and south. The Bantu farmers were in competition with early speakers of Adamawa–Ubangian languages, who were migrating east and south. The latter speakers were more successful in colonizing the moist and dry woodlands because they had acquired the knowledge of sorghum cultivation from Nilo-Saharan groups to the north, and the Bantu moved on to more southerly locations. Later, Bantu farmers learned how to cultivate sorghum.

Iron smelting and banana cultivation, two new technologies, facilitated the later spread of the Bantu-speaking farmers (Figure 3.5). Bantu-speaking farmers began to smelt iron about 2,500 years ago. With iron tools these farmers were able to clear the forest and cultivate more efficiently. Newman (1995) describes the movement of Bantu as a "hopscotch of villages" in which newcomers from far behind the frontier of Bantu settlement forged the way south and east. The banana, which they acquired on the eastern side of the rain forest about 2,000 years ago, produces more than yams with little effort. The banana was introduced to mainland Africa by Southeast Asians (Austronesian speakers) who were colonizing the uninhabited island of Madagascar. Ultimately, the Bantu speakers outcompeted the pygmies, who were hunters and gatherers in the equatorial rain forest and, further south, the Khoisan hunter–gatherers. Only relict populations of these groups exist today. Two big questions have not been answered about these two groups.

The first question concerns the original language spoken by pygmies. Since all pygmies speak Bantu languages today, it is not known what language they spoke prior to the Bantu colonization. Newman has put forward the interesting hypothesis that proto-Niger–Congo was the original language of the pygmies, postulating that they hunted and gathered across West and equatorial Africa 10,000 years ago. He suggests that people living in West Africa today represent a mix between the pygmies who once lived there and the Nilo-Saharans, who are tall and thin. The

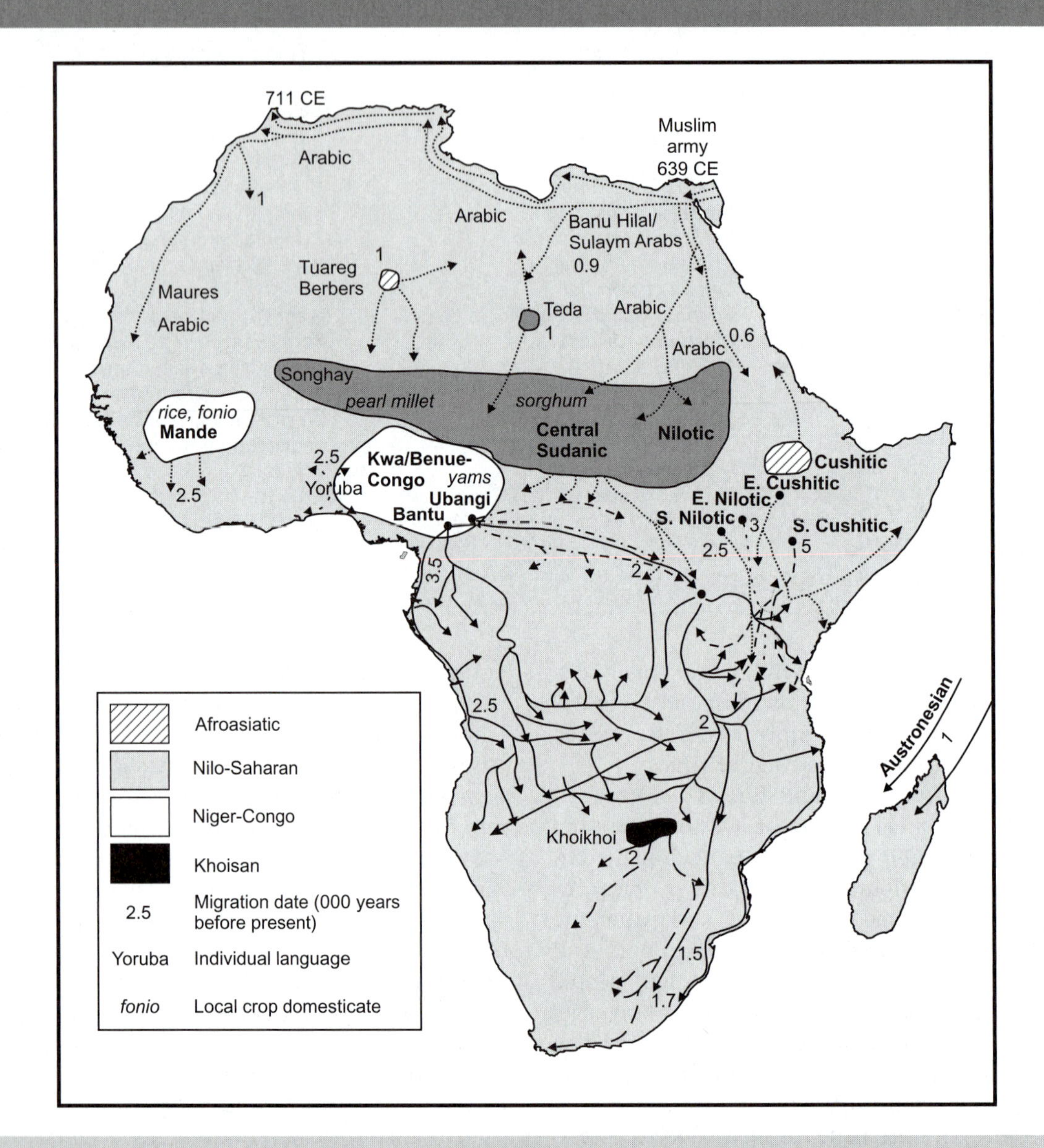

FIGURE 3-5 African migrations of farmers and/or herders over the last 5000 years. *Adapted from Cavali-Sforza et al. (1995), Newman (1995).*

geneticist Cavalli-Sforza, using 110 genetic markers, found that West Africans are as genetically distinct from pygmies and Nilo-Saharans as Europeans are from East Asians. At present, we can say that these populations existed in isolation from one another for a long time: the ancestors of the present West Africans in West Africa, the ancestors of the pygmies in central Africa, and the ancestors of Nilo-Saharan speakers just south of the Sahara. The identity of the language the Bantu originally spoke remains a mystery.

The second question concerns the maximum extent of Khoisan-speaking hunters and gatherers. Some skeletal evidence suggests that Khoisan peoples may have been distributed along eastern Africa right up to lower Egypt. Although the distribution of skeletal remains is clear, as indicated in Figure 3-4, controversy surrounds the interpretation of the Egyptian and Ethiopian remains, and experts are debating the significance of the available data.

Since the migrations of Bantu-speaking farmers, the most significant migrations have been due to colonization over the last 1,400 years of most of northern Africa and parts of East Africa by Arabic-speaking peoples and of parts of North Africa, East Africa, and southern Africa by Europeans. Although most Europeans left Africa at independence, their cultural impact was profound and continues to this day. The process of Arab relocation, diffusion, and cultural dominance was as profound as its European counterpart but began over a thousand years earlier—and the diffusion of Arabs, Islam, the Arabic language, and seventh-century Arab culture continues today south of the Sahara, from Senegal to the Sudan and from the Horn to the Cape.

EARLY STATES IN AFRICA

From ancient times to relatively recently, the state in Africa was different from the state as we know it today in Africa and elsewhere. Borders were not fixed and power declined in a distance–decay relationship from the capital town or city where the plenipotentiary resided, however he or she was called: king, queen, amir, ghana, sultan, or fama. The further away from that center, the more likely some other power could compete for the hearts and minds—and especially the taxes—of people. It was a question of political and military geography and the necessity of being in many places at one time to maintain order; a state's size depended directly on its ability to project power from its center to the unincorporated area ("lawless" frontier) located on its periphery. The geographer Brauer (1995) calls the spatiopolitical structure just described the "sovereignty-field" model (Figures 3-6 and 3-7).

> The intensity of sovereignty . . . both in terms of power and of self-identification of the subjects, is to be conceived of as being maximal and relatively constant over a region centered in its capital and the territories immediately adjacent to it. Beyond this zone, intensity of sovereignty is conceived of as radiating in all directions, diminishing with increasing distance from the capital. . . . Frontiers in such a system would correspond to regions where the intensity of sovereignty of one political unit had decreased sufficiently to be overlaid by the field of sovereignty of a neighboring unit, the intensity of which in turn would rise gradually as one approached the political center of that second entity, to peak at its capital district. (Brauer 1995: 28)

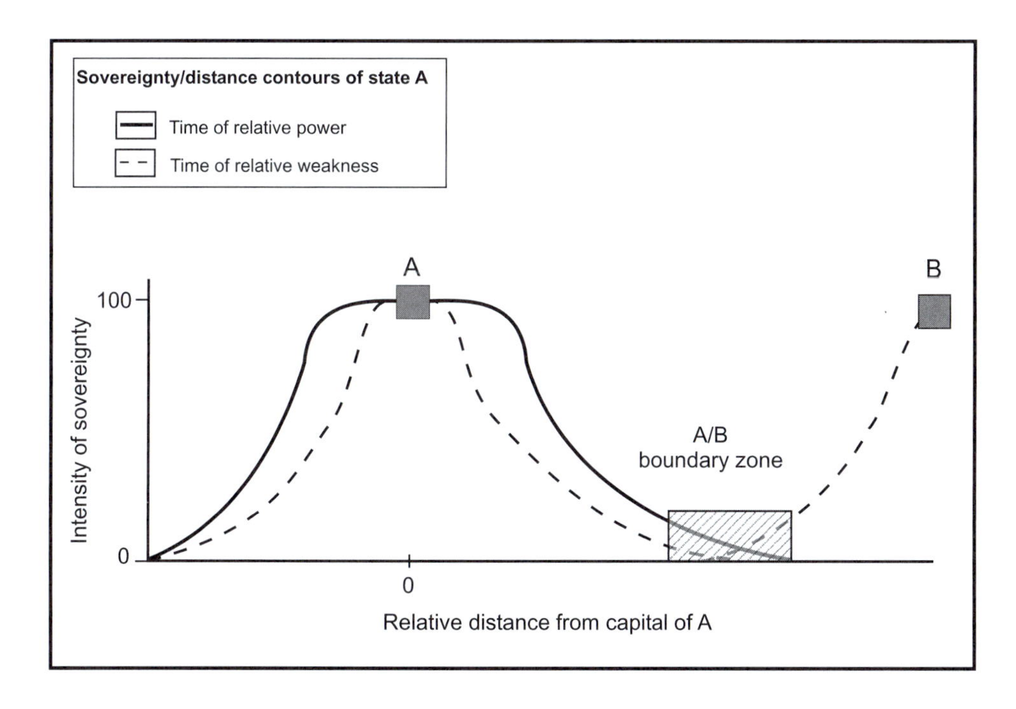

FIGURE 3-6 Hypothetical relation of intensity of sovereignty to distance from the capital cities of two states, A and B, and their relation to the boundary zone between them.
Adapted from Brauer (1995).

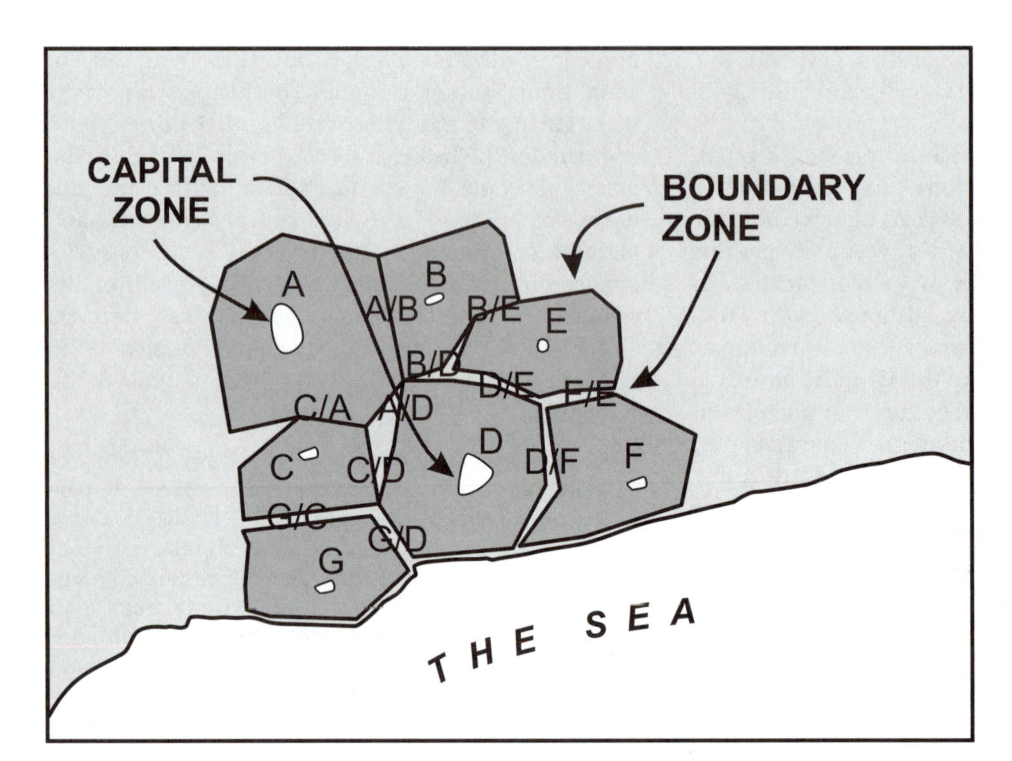

FIGURE 3-7 Map illustrating the relations between states illustrated in Figure 3-6. *Adapted from Brauer (1995).*

The size of the African state was variable too; although there were a few big states or empires, the vast majority of political entities were tiny statelets, about the size of a small county in the United States, centered on an ethnic group (Thornton 1998). If and when a state expanded beyond its core area or "homeland," new provinces and peoples would be incorporated into it, generally by conquest. Usually, the states newly incorporated into another would continue to "exercise local authority, and the rulers of a large state found his powers checked by them" (Thornton 1998: 91). As states declined, provinces would break away or perhaps join a rising competitor state.

The morphology or shape of the African state was variable too, depending on where the people lived: the state would be more linear along a river, but compact where population was rather uniformly settled. The shape of the state was also influenced by involvement in long-distance trade and the necessity to control strategic market towns or natural resources peripheral to the core area of the state. States also tended to have a more compact shape where the competitor states were close and numerous.

After the modern African states gained independence, Africanists began to endeavor to show that Africa was not a continent of benighted anarchy but instead had a similar history of political organization into empires and kingdoms just like any other region of the world. Accordingly, much postindependence scholarship focused on the great states and empires of Africa, Ghana, Mali, Songhai, Zimbabwe, and so on. Recent scholars, while acknowledging the importance of understanding the state in Africa, have looked again at the so-called acephalous (headless) societies and have found much of interest in the way these societies were governed. In the pages that follow, we will examine political organization in its larger forms up to the time of the European imposition of bounded states.

Early States of North and Northeast Africa

While the coming of cultivation and iron implements was producing major population growth and migrations in Africa, the focus of an even more important transformation lay in the northeast of the continent (Figure 3-8). For when the concept of organized agriculture had spread to Egypt, urban centers began to develop, great strides were made very rapidly in political and religious thinking, great artistic achievements occurred,

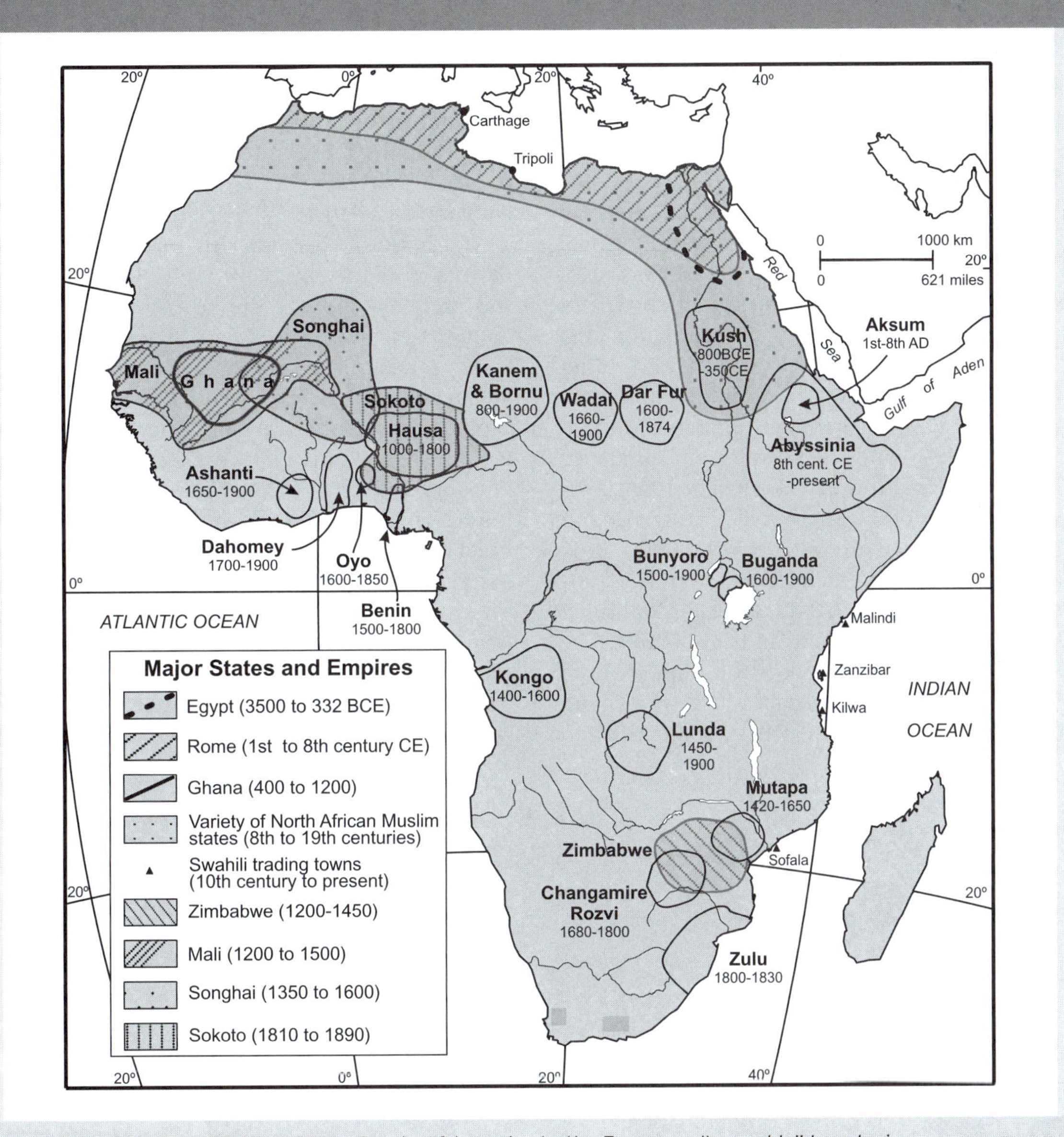

FIGURE 3-8 Some states and empires in Africa prior to the European "scramble" to colonize the continent.

and economic activity was carried farther than ever before. In an amazingly brief period, political unity was achieved in the most highly complex, most densely populated state of its time in the world. For several thousand years, dynastic Egypt was the heart of power and the source of ideas for Africa and the world. Fertile soil and an enriching flood each year permitted dense populations in comparison to the hunter–gathers located elsewhere. For a time, Egypt was the most populous state in the world, and for many centuries it contained half the population of all Africa. Egypt's population multiplied 20-fold from 7,000 to 4,000 years ago, from 100,000 to 2 million, while the rest of Africa only doubled from 500,000 to 1 million people (McEvedy 1995).

Historians conventionally agree that Egypt began about 5,000 years ago, in 3,100 BCE, when King Narmer unified Lower and Upper Egypt by conquest. This date marks the first dynasty of Egypt, but it is clear that the conditions of settled life and political organization were developing much earlier. The Old Kingdom (2,685–2,200 BCE) was the period of the great pyramids and tombs, when Egypt had a strong government that exercised control over the annual floods of the Nile and constructed a vast system of dikes and irrigation channels. The regular surplus of food supported an expanding population, stimulated commerce, and paid for the pyramids and other great monuments. During the Middle Kingdom (2,040–1,785 BCE), the Egyptian pharaohs looked westward into Libya, and their armies moved southward, up the Nile and through the Red Sea, to subjugate the inhabitants of the Middle Nile, returning with new treasures and slaves. By the year 1,900 BCE, Egyptian power had been extended to Aswan, and during the New Kingdom (1,570–1,085 BCE), Ramses II built the famous temple of Abu Simbel in Nubia. The New Kingdom pharaohs moved the capital from Memphis in Lower Egypt to Thebes in Upper Egypt, extended the empire to include Palestine and Syria in the north and to the Fourth Cataract in Nubia, and expanded the trade in gold, ivory, incense, timber, and other commodities. By the last millennium before the common era, Egypt appeared to be politically unstable in comparison to the preceding mainly tranquil centuries: it had begun to lose power, provinces began to break away, and it was eventually taken over by Nubians, Assyrians, Persians, Greeks, and Romans.

Kush, one of the states that eventually conquered Egypt, existed from about 1000 BCE to 350 CE and was centered around the towns of Napata, Meroë (Figure 3-9), and Naga, situated between the Fifth and Sixth cataracts of the Nile near the present-day Sudanese cities of Atbara and Khartoum. From their capital in Napata, the Kushites extended their sphere of influence up the rivers to the borders of modern-day Ethiopia and Uganda. By 730 BCE they had successfully embarked on the conquest of Egypt itself and ruled there for over 60 years. In 670 BCE, however, the Assyrians invaded Lower Egypt. The Kushites, unable to repel the attack primarily because of inferior weapons, retreated southward and a century later transferred their capital to Meroë. The reasons for this transfer are not entirely clear, but it is likely that increasing desert encroachment had prevented the

FIGURE 3-9 Pyramids at Meroë.
Roy Cole.

Kushites from producing sufficient food in their original stronghold. Indeed, for centuries much of the Sahara had become increasingly arid, and the human and livestock populations were being forced to retreat to the desert margins, oases, certain sections of the Nile, and south to areas of where rainfall is regular and abundant. But other reasons for the change of capital were possibly equally important. Meroë lay closer to the caravan routes along the Atbara River leading to Ethiopia and indirectly to the Indian Ocean. The move may have been made to avoid an anticipated invasion by Egyptians, which had happened in 593 BCE at Napata. For the next 800 years, Kush—and especially Meroë—was the center of a large iron-smelting industry.

There has been a debate among Africanists on the origins of iron working in Subsaharan Africa. Twenty-five years ago researchers thought that iron working had diffused from Meroë to West and southern Africa. Another theory, that iron working diffused from Meroë to West and East Africa, has been rejected by recent research (Cornevin 1993). Meroitic dates are younger than those in West Africa, and Meroitic smelters were different from West African smelters: in Meroë, the smelter was of Roman design, unlike those found in Taruga, in present-day Cameroon (Okafor 1993).

Another hypothesis was that iron working had diffused across the Sahara from Phoenician settlements in North Africa such as Carthage. The Carthage origin hypothesis is widely accepted today, but there is mounting evidence that iron smelting was invented independently elsewhere in Africa, as well. In Nigeria and Niger, French archaeologists have found three sites that date much earlier than those in North Africa or Meroë (Figure 3-10). These sites are Termit Massif located in Niger 155 miles (250 km) northwest of Lake Chad (attested dates range from 1300 to 300 BCE at various sites in this area, Taruga near the confluence of the Niger and Benue Rivers (dates range from 900 to 300 BCE), and Agades (dates range from 800 to 300 BCE). These results support the contention of Cornevin (1993) that we must "accept the possibility of one or more local inventions of iron working in West Africa." The data that are available seem to show that iron working probably was independently invented in West Africa and diffused elsewhere in Subsaharan Africa.

There appear to have been three reasons for the decline of the kingdom of Kush: an environmental crisis occasioned by agricultural overuse of the land and deforestation to produce charcoal for iron smelting, declining demand in the Roman Empire for the goods of Kush, and the shift in trade from Kush to the Red Sea port of Adulis, founded by Kush's competitor state of Aksum (modern Axum). Although Meroë was probably abandoned by the time that the Aksumite ruler, Ezana, came to occupy it in 350 CE (Munro-Hay 1991), the impact of changing patterns of trade helped accelerate the deterioration of Meroë.

The founders of Aksum were Yemeni Semites (Sabaens) who first began to colonize the African continent 2,500 years ago and later extended their authority over the northern half of present-day Ethiopia and eastern Sudan. Aksum existed for about 600 years from the first to the seventh centuries of the common era. The Aksumites intermarried with the local Cushitic-speaking people, developing a distinctive Semitic language, Ge'ez, an ancestor of modern Amharic. Trade relations were maintained with Greek and Syrian merchants who came to buy ivory, gold, and incense from the African interior, and it was because of funds generated by this commence that Aksum was able to defeat the Kushites. The importance of Aksum lies in its adherence to Christianity after the fourth century, and the spread of the Coptic Christian Church to much of Ethiopia and parts of Nubia. With the spread of Islam in the surrounding states beginning in the seventh century, Aksum became an island of Coptic Christianity that proved to be a religious, cultural, and political barrier to the passage of Islamic ideas and practices. Toward the end of the tenth century, the Kingdom of Aksum disintegrated, and there followed a period of struggle between Christian and Muslim that continues to the present.

Aksum at its height was able to conquer modern-day Yemen, but the state began to decline at the end of the sixth century, and the Aksumites were expelled from Yemen by their trading competitors the Persians, who also conquered Jerusalem and Alexandria. Red Sea trade declined and, as international trade routes shifted to the Persian Gulf, the negative impact on the Aksumite economy deepened. There is evidence of overuse of the land, overcropping, and wood cutting for charcoal production. A drying climate may have transformed overuse of the land into an environmental crisis, further weakening Aksum (Butzer (1981); plague in the sixth century and revolts in the

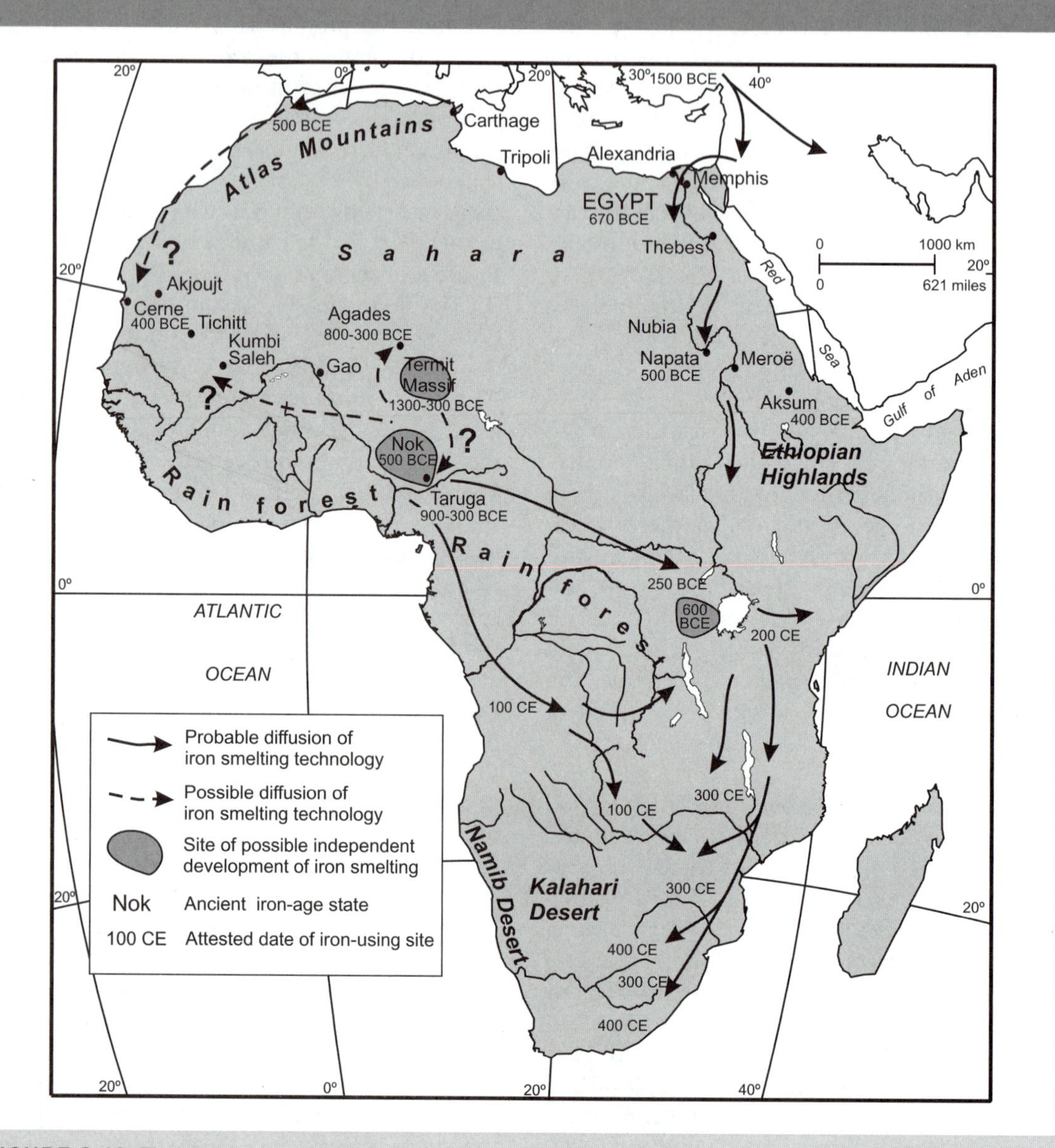

FIGURE 3-10 The probable spread of iron working in Africa.
Adapted from Kwamena-Poh et al. (1982), Shillington (1989), Cornevin (1993).

provinces contributed to the deterioration of the kingdom. Trade declined still further as Aksum was unable to keep the sea-lanes free of pirates (Munro-Hay 1991). In addition, the rise and expansion of Islam in western Asia and North Africa caused a further deterioration in trade (Box 3-2).

The history of the Saharan and North African regions is long and complex, yet a number of themes and events need to be reviewed to understand their role in the cultural and political development of Africa as a whole but, more especially, in the development of West Africa. In the second millennium before the

BOX 3-2

AKSUM

As far as the history of civilisation in Africa is concerned, the position of Aksum in international terms followed directly on to that of Pharaonic and Ptolemaic Egypt and Meroë; each was, before its eclipse, the only internationally recognised independent African monarchy of important power status in its age. Aksumite Ethiopia, however, differs from the previous two in many ways. Its economy was not based on the agricultural wealth of the Nile Valley, but on the exploitation of the Ethiopian highland environment and the Red Sea trade; unlike Egypt and Meroë, Aksumite Ethiopia depended for its communications not on the relatively easy flow along a great river, but on the maintenance of considerably more arduous routes across the highlands and steep river valleys. For its international trade, it depended on sea lanes which required vigilant policing. Most important, Aksum was sufficiently remote never to have come into open conflict with either Rome or Persia, and was neither conquered by these contemporary superpowers, nor suffered from punitive expeditions like Egypt, South Arabia or Meroë. Even the tremendous changes in the balance of power in the Red Sea and neighbouring regions caused by the rise of Islam owed something to Aksum. It was an Ethiopian ruler of late Aksumite times who gave protection and shelter to the early followers of the prophet Muhammad, allowing the new religious movement the respite it needed. Ethiopia, the kingdom of the "najashi of Habashat" as the Arabs called the ruler, survived the eclipse of the pre-Islamic political and commercial system, but one of the casualties of the upheaval was the ancient capital, Aksum, itself; various factors removed the government of the country from Aksum to other centres. The Ethiopian kingdom remained independent even though the consolidation of the Muslim empire now made it the direct neighbour of this latest militant imperial power. But eventually Ethiopia lost its hold on the coastal regions as Islam spread across the Red Sea. Nevertheless, the Aksumite kingdom's direct successors in Ethiopia, though at times in desperate straits, retained that independence, and with it even managed to preserve some of the characteristics of the ancient way of life until the present day.

Munro-Hay (1991. page 1).

common era, Phoenician traders and adventurers established a number of city-states along the Maghreb coast from the Gulf of Sirte to the Atlantic coast of Morocco. The largest and most important of these was Carthage, which may have had a population exceeding a half-million people at its zenith. Like other cities, Carthage was dependent on trade and the surrounding regions for grain. Additional coastal lands had to be cultivated as the Phoenician populations increased, but there was no need for expansion inland. Following the fall of Carthage in 146 BCE, the Roman victors, faced with the need to produce grain for their metropole, pushed southward into the drier lands. There they irrigated the land (especially in Numidia, the eastern half of modern Algeria), and established a settled agricultural way of life among the Berbers, to whom they also introduced Christianity. By the fifth century of the common era, Roman power was declining and the more nomadic Berbers undertook systematic raids on the agricultural communities, with the result that pastoralism once more expanded.

Neither Carthaginian nor Roman rule ever reached across the Sahara, and while the Carthaginian empire was never really territorial, it had, like imperial Rome, well-defined commercial links with points to the south. Carthage was linked with Gao, while the Roman city of Leptis Magna, close to modern Tripoli, was the northern terminus for the trans-Saharan caravan routes through the Fezzan, the desert region in the southwest of present-day Libya, and Marrakech was the Moroccan terminus of a trade route across the Mauritanian desert (Figure 3-11). Berber pastoralists were the agents in this contact and, quite likely, had been so even prior to the Mediterranean civilizations.

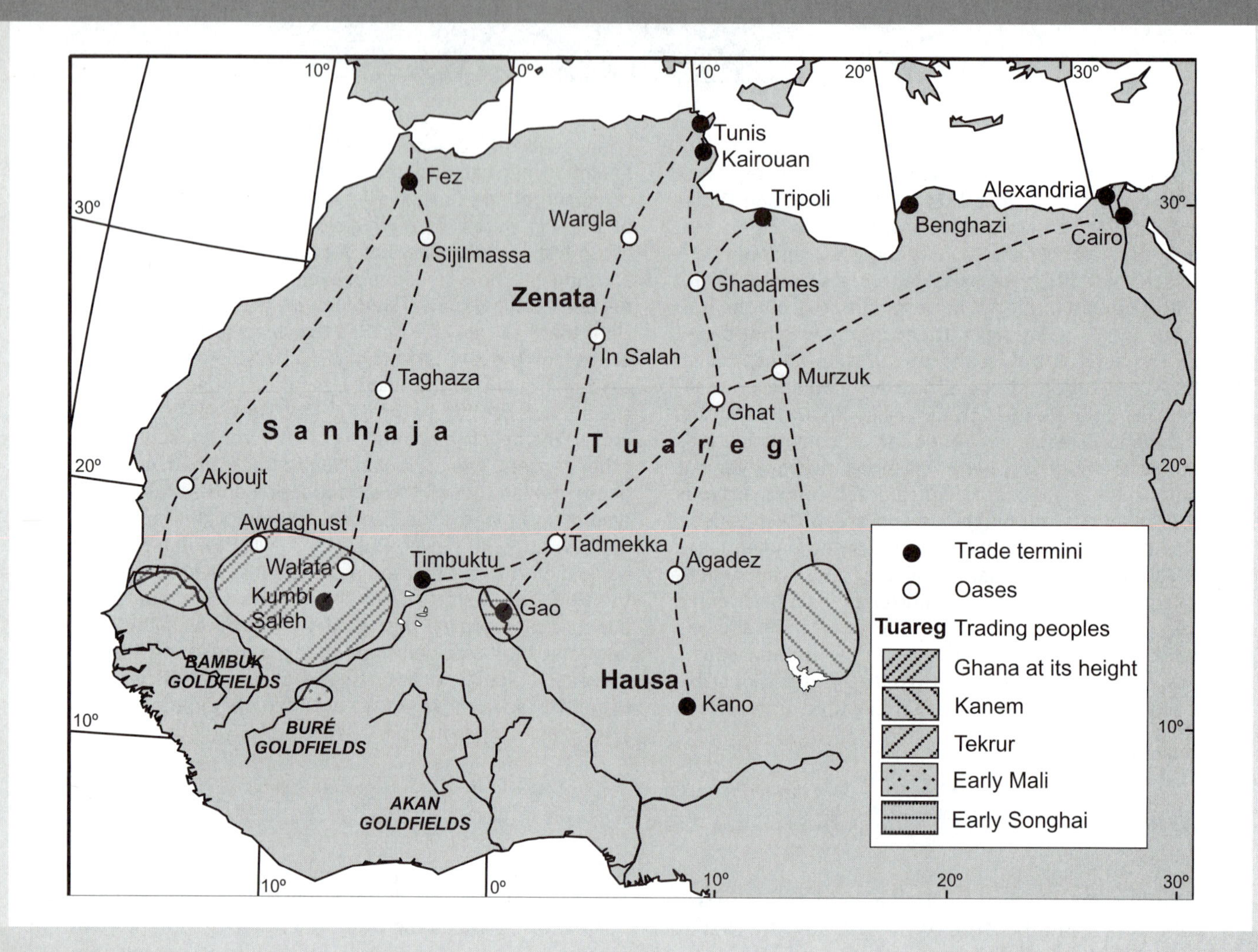

FIGURE 3-11 Trans-Saharan trade to 1200 CE.

Throughout the centuries, the main exports of the western Sudan were gold, captives, ivory, ostrich feathers, and hides, while salt, cloth, iron, and copper goods were imported from the north.

Trans-Saharan trade would have provided no extreme hardship several millennia ago when much of the Sahara was more humid than now and supported rich grasslands for the Berber pastoralist traders but, as the Sahara became more arid, the routes became more selective and the journey more hazardous. With the gradual desiccation of the landscape through climatic change, the former residents of the Sahara fled to North Africa, the Nile Valley, and south of the growing Sahara, where they were absorbed into the local populations.

Before the introduction of the camel into Africa, there is evidence of the use of pack oxen, donkeys, horses, and wheeled carts and chariots to cross the Sahara. Shillington (1989), who describes the trade as small in scale and sporadic before the introduction of the camel, says that up to about the fourth century, trans-Sahara trade was mostly local: salt from the desert for food grains. However, a hierarchy of urban places was developing that would be instrumental in the development of the trade at its peak after the eighth century (Davidson 1991). By the fifth century, the camel was a major form of transportation in the Sahara, and the nature of the trans-Sahara trade changed dramatically; the wealth generated by long-distance trade enriched and helped build states and empires, particularly just south of the Sahara in the savanna lands.

The final phase of occupation and control of the Saharan and North African regions came with the Arabs in the seventh and eighth centuries, and like their predecessors in the western regions, they initially confined themselves to the Mediterranean lands, entering the Sahara proper in significant numbers only during the twelfth century. The Arabs brought with them Islam, giving profound unity to this otherwise heterogeneous region, which, since the end of the Roman occupation, had not been a unified political or even cultural area. Arab diffusion through the Maghreb met with widespread resistance from sedentary and nomadic Berbers alike, and from Byzantine armies. Nevertheless, an Arab empire was established and Islam prevailed. By the tenth century, the empire collapsed, later to be reorganized by other Muslims, the Ottoman Turks.

Early States of the Western Sudan

The earliest of the West African states about which much information has been gathered is Ghana. It was located in the western part of present-day Mali and southeast Mauritania, to the north of the major headwaters of the Senegal and Niger rivers (Figure 3-8). It has not been possible to date the origins of the Ghana state with exactness, but it probably was in existence by about the fourth century of the common era. In any case its economic foundations are quite well understood. The river valleys to the south contained alluvial gold, much of which was mined and exported northward. The actual mining was done by people who lived outside the sphere of Ghana proper, who exchanged the gold for salt and other products they required. Also taken from the south were slaves, to be sold in northern markets. From the north, caravans brought copper, dried fruit, cowries, cloth, and other merchandise for exchange in Ghana, followed by distribution throughout the region.

The beginnings of the Ghana state are shrouded in mystery, and the identities of the earliest rulers are not even certain. Nevertheless, it is easy to see that Ghana's location was an extremely favorable one from the economic point of view and that control of its area would yield great rewards. Ghana is associated with the Soninke people of West Africa. The Soninke speak a language that belongs to the Mande subfamily of the Niger–Congo phylum. During the reign of the Soninke kings, the state thrived under strong central government, experienced many years of stability, achieved its greatest territorial expansion, extending from Timbuktu in the east to the borders of present-day Senegal in the west, and from the headwaters of the Senegal and Niger rivers in the south to the land of the Berbers in the north.

North African Arabs became increasingly involved in the trans-Saharan gold trade after the eighth century, and their involvement increased the intensity of the trade. They founded trading towns, for example Sijilmassa, to act as northern termini. The Arabs also brought Islam, which eventually reached Ghana, though indirectly: the people of the northern desert, converted to a version of Islam, launched a war against pagan Ghana in 1054 (Davidson 1991). But Ghana did not fall apart immediately. On the contrary, it appears that its rulers converted to Islam, and over the ensuing two centuries Ghana began to break up into various smaller kingdoms and tribal entities, and there was a relatively brief period of instability during which the diffusion of Islam continued.

The decline of Ghana did not end the efforts to establish states and empires in West Africa—indeed one of the reasons for the later weakness of Ghana was certainly the discovery of gold at Buré, 125 miles (200 km) southeast of the Bambuk goldfields. The control of the trans-Saharan trade was beginning to shift to the east, and a new and even more impressive empire arose during the middle of the thirteenth century, *Mali* (Figure 3-8). Like Ghana, Mali's strength depended on the caravan routes to the north; but unlike Ghana, its area incorporated the gold-producing territories, which gave the state greater security. Relations with the Islamic peoples of the desert and the north were good because Mali's leaders were Muslim. Many of the characteristics of Ghana were replicated in the empire of Mali. There was comparative stability, and a succession of rulers held the empire together. During the first quarter of the fourteenth century, the state reached its zenith. Territorially, it extended from Gao and the middle Niger in the east to the Atlantic Ocean in the west, covering the entire West African savanna belt, and from the Fouta Djallon Highlands in the south to areas deep in the Sahara. Trading caravans came from as far afield as Egypt and Morocco. The cities of Mali became urban centers of renown. Architectural improvements were made and great mosques constructed, an institution of higher learning was established at Timbuktu. The capital, Niani, was about a hundred miles below present-day Bamako on the Niger River.

But the adjutants of Mali's leaders extended the empire farther than control could be effective, and the fourteenth century saw the beginning of the end. The Tukolor of the west, the Tuaregs of the northern desert, and the Mossi of the south raided with success, and by the beginning of the fifteenth century the kings of the Songhay had also broken away. But again, the principles of state organization were not lost on those who rebelled against the weakening central authority in Mali. Other empires soon appeared.

In the region of Gao, on the middle Niger, the land and people were long subject to the rulers of Mali. When Mali fell apart another state of significance, *Songhay,* arose in this area. Songhay was located farther to the east than either Ghana or Mali, so that its influence in the far west was probably less. But the tradition of greatness of the central city continued, as Gao replaced the town of Niani, which itself had taken over from the town of Ghana. From the environs of Gao, the leaders of Songhay set about extending the domain of the state, which by the early sixteenth century encompassed the lands of the Hausa in the east almost as far as Bornu, in the north to the very margins of Morocco, in the West almost to the ocean, and in the south well into the forests. Only in the south did some of the Mossi states manage to hold their own against the Songhay armies. Meanwhile, trade and agriculture flourished, and the heart of the empire during the first quarter of the sixteenth century experienced stability and order.

Later, Songhay suffered from intrafamily rivalries for the leadership and from competition among the various generals of the successful armies. These were the seeds of weakness which, during the later 1500s, began to affect the state. In 1591 Moroccan invaders from the north defeated the Songhay army and captured Gao, while internal rebellions fragmented the empire. But Songhay had continued to spread Islam among the peoples it had subjected; it had continued to keep close ties with the Arab world of Islam, and Muslim influence was felt long after the empire fell.

Viewing the sequence from a spatial perspective, it is notable that Mali arose well to the south and somewhat east of Ghana, and that Songhay subsequently developed east of Mali. These shifts had several causes and consequences. The general easterly migration of the core areas may have been related to the discovery of new gold-producing areas east of the Bambuk fields, which had been the source of gold for ancient Ghana. The shift eastward may also have been influenced by events taking place in North Africa, where the center of power moved from Morocco to Tunis during the same period. Doubtless, the fall of Ghana and the decline of the westerly trade routes led to a search for new, necessarily more eastern crossings of the Sahara. Mali's more southerly location rendered it less vulnerable to raids of the desert peoples and, more importantly, the gold-producing area of Buré was located in Mali's core territory.

Mali was well known outside West Africa. One of the Malian kings, Mansa Musa, accompanied by a large contingent of notables, made a famous pilgrimage to Mecca in 1323–1325 and very favorably impressed Egyptians along the way. Fourteenth-century Spanish maps of Africa depict Mansa Musa on a throne holding a large gold nugget. Mali's decline began with a series of weak rulers, breakaway provinces, and raids from the Mossi and Tuareg. By 1500, little remained other than the original core area, and as Mali declined, the power of Songhay increased.

The trading empire of Songhay was founded by farmers, hunters, and fishers who established trading towns along the Niger River. Some time in the ninth century, the state of Songhay emerged and began to expand. Gao, a city on the Niger founded by Berbers or Egyptians interested in the gold trade, became Songhay's capital. By the fifteenth century, Songhay had become larger than Mali. Since Gao lay some 700 miles (1,125 km) downstream on the Niger from Mali, goods began to move along the Niger trade route to Timbuktu and Gao, and beyond, and the whole orientation of the region was more and more toward the central Sahara and the eastern parts of the north, ultimately to Cairo.

With the invasion of Songhay by the Moroccans and the ensuing chaos, the center of African power shifted southward, toward the region between the forests and the steppe in the elbow of the Niger River. There, some smaller but remarkably stable political entities, known as the Mossi–Dagomba states, survived until the Europeans entered the scene in numbers. The kingdom of Ségu, sometimes called the Ségu warrior state, rose in the seventeenth century in the power vacuum left by the fall of Songhay. Its capital was the city of Ségu-Koro, located along the Niger River in present-day of Mali. In the far east, between Songhay and Lake Chad, the kingdoms of the Hausa and the state of Kanem and Bornu also had some organization and stability. Slavery was the economic foundation of Ségu, which exported human beings north and south. Hausa and Kanem and Bornu were based on agriculture, manufacturing, and trade. Much later, these Hausa city-states were overthrown by the rising group of Fulani (Fulbe), who had come from the west to live in Hausaland, bringing with them a strict adherence to Islam. The holy war initiated by the Fulani resulted in the rise of the emirate of Sokoto in the early nineteenth century and had important political repercussions throughout the West African savanna belt.

Early States of the Guinea Coast

The sequence described in the preceding section is, of course, a very generalized one, and Figure 3-8 does not show all the states that came and went in Africa. Many states about which something is known—and probably others about which little or nothing has been learned—have not been included in the story of the complex political transitions in West Africa. It is even more difficult to comprehend the political changes going on at the same time in the Guinea area, the region between the savanna–steppe of the north and the coast. Details are even harder to obtain for the Guinea states than for the savanna states. If the peoples of the savanna did not write, the Arabs did, and much of what is known about Ghana, Mali, and Songhay has been gleaned from Arab writings. But the southerly states were beyond the orbit of the Arabs, and the reconstruction of their past is far more difficult. When Europeans arrived in the fifteenth century, they began to report on the political entities, which had been in existence for centuries, unbeknownst to the West.

Actually, the Europeans contributed considerably to the strengthening and emergence of some of these southern states. In the heyday of the trans-Saharan trade, the forests, especially the northern margins, were tributary areas from which products entered into the trade to the north. The local chiefs, who could do very little against the powerful armies of the empires, had been obliged to move deeper into the forest to avoid slave raids and the royal cavalries. The great armies of the Kings were effective on the savanna, on the open lands, where grazing was available for the horses, but in the dense undergrowth they ran the risk of defeat. Thus very little forest was actually controlled by the savanna empires. Moreover, sleeping sickness, spread by the tsetse fly, was endemic to the more humid, southerly areas and was a threat to humans as well as their horses and cattle. Armed parties did enter the forest, for raiding purposes but could maintain no power base or effective government there.

Generally, it is thought that the first state in the forest region was *Ife,* noted for its magnificent bronzes. From Ife, which may have arisen as early as the tenth century of the common era, ideas spread that led to the foundation of several states populated by the Yoruba people. Perhaps the best example of the type was *Oyo* (Figure 3-8), located in present-day western Nigeria and eastern Benin. Oyo, like other forest empires, probably had its beginnings along the northern fringes rather than within the forests, and was extended southward as time went on. Oyo was noteworthy for several reasons: it was among the first of such states to form, it survived for a long time, and it brought such stability to the Yoruba people that they withstood the European impact from the south for many years.

The core areas of these Yoruba states consisted of urban centers of considerable size, surrounded by a wall that usually enclosed some farmland. Thus, the town could withstand a siege, the protected farmland being just sufficient to supply some food to the people seeking shelter. There was much military activity in the expansion of the Yoruba states: the leaders of Oyo, for instance, sought to incorporate not only all Yoruba people, but non-Yorubas as well. The economic base of the rulers of Oyo was slave labor on royal plantations. Oyo also collected tribute from its towns, territories, and neighbors. Trade from the coast to Hausaland was taxed. Captives were acquired in the continuous wars, and those that Oyo was not able to use internally were sold to Europeans on the coast for a variety of goods: firearms, European cloth, metal goods such as knives, buckets, and fish hooks, and cowrie shells, used as a currency in some places up until the twentieth century. Eventually, this enforced tribute and strife resulting from the slave trade caused internal fragmentation. Oyo was paramount in its region from the seventeenth century until the first decades of the nineteenth, and had been growing in significance and influence long before that.

Another offshoot of Ife was the state of *Benin*. Originally a group of states, it was consolidated by the first of a strong dynasty of kings in the fifteenth century. Benin is known, among other things, for its fabulous bronzes and other works of art, products of skills introduced from Ife. The state centered on the city of Benin, which grew in importance with European contact and trade. Benin's merchants became important middlemen between Europeans and the Yoruba of the interior, and through the city passed pepper, cotton cloth, captives, and beads. The power of the merchants was increased by their procurement of firearms, and Benin's armies raided farther and farther into the interior. The kingdom reached its zenith during the late sixteenth and early seventeenth centuries, but eventually dynastic disputes, civil war, and the devastation of the hinterland proved fatal. Interestingly, although all of the states of the Bight of Benin reaped great riches from the Atlantic and trans-Saharan slave trade, Benin was one of the few African states that later chose to restrict the sale of slaves (Lovejoy 2000). Fewer products and slaves reached the capital, fewer Europeans called there, and by the end of the nineteenth century the state was in decay.

The state of *Dahomey* never reached the importance of Benin or Oyo, and its consolidation was largely the result of economic conditions created by the slave trade. Dahomey was internally divided, the Europeans encouraging this division by their support of whatever parties would deliver the slaves to the coast most efficiently. From the outside, Dahomey was threatened by Oyo and encroached upon by *Ashanti,* a state of major proportions far more powerful than Dahomey. Like the kingdoms of the Yoruba, Ashanti, located in the interior of what came to be called the Gold Coast, resulted from a union of a number of states under a powerful ruler who was succeeded by equally effective kings. In this case, a common effort on the part of Akan-speaking people to prevent the invasion of competitors led to a further combined effort to throw off the yoke imposed by a rival Akan state called Denkyira. Thus, the Ashanti kingdom was forged, with headquarters in Kumasi, under the leadership of the *Asantehene,* guardian of the Golden Stool, the symbol of national unity.

The state had its roots in the fifteenth century, when Akan farmers, who had previously settled on the forest fringes, began to clear the forest for farming. For this heavy work, they obtained slaves from Mande (Djula) merchants from the north, in exchange for gold; later, the slave labor was used to mine gold. The Portuguese arrived on the scene in the 1480s and provided Ashanti with captives obtained from neighboring Benin, cottons, metal goods, and other European commodities in exchange for gold. Cassava (manioc), a high-yielding tuber, and maize were brought to Ashanti from the Americas by the Portuguese and strengthened their food economy.

Ashanti grew during the eighteenth century, and, by the beginning of the nineteenth, the empire was encroaching on the coastal states of the Fante, where the Europeans had their forts and trading stations. Captives and gold were exchanged for arms and ammunition. Eventually, Ashanti invaded the nominal British protectorates along the coast. The termination of the slave trade had removed one of the economic mainstays of Ashanti, and relations between the Ashanti and the British had steadily deteriorated. Ashanti's attack, which came in 1863, was not repaid until several years later, and then ineffectively. It is a measure of Ashanti resilience that the state survived the 1874 punishment inflicted by Britain and required reoccupation in 1896. In 1900 when the British governor demanded the surrender of the Golden Stool, Ashanti rose for the last time in rebellion. Only as late as 1901 did Ashanti territory become part of a British Crown colony.

Early States of Equatorial and Southern Africa

Large, militarily strong, and economically diversified states also existed in equatorial and southern Africa during the period of the Guinea states (Figure 3-8). When the Portuguese arrived at the coast near the mouth of the Congo River at the end of the fifteenth century, they found the flourishing state of *Kongo,* centered on Mbanza, located near present-day São Salvador, Angola. The Bakongo king willingly entered into trade with the Portuguese, exchanged ambassadors with Lisbon, and permitted Christian missionaries to proselytize among his subjects. At the same time, he allowed the Portuguese to remove countless thousands of captives, which, together with internal wars, contributed to the kingdom's decline by the end of the seventeenth century. Slaving was lucrative and a local slave plantation economy developed in addition to the export trade. The wars were focused on capture of civilians for plantation labor in the export trade with the Portuguese. In point of fact, early sales of people to the Portuguese came from the Kongo's prisons. Later armed groups raided their neighbors and the familiar pattern of flight of refugees and depopulation set in.

East of the *Kongo* kingdom in the light woodland regions bordering on the southern rim of the Congo Basin, where conditions favored prosperous agricultural and fishing economies, were similar societies including the *Bushongo* of the Kasai, and the *Luba–Lunda* of Katanga. Skilled in weaving and in iron and copper smelting, the Bushongo also developed a sophisticated hierarchical system of chiefs and political behavior, and a division of labor. The rich mineral deposits of the Katanga, mined as early as the eighth century of the common era provided the economic base of the Luba–Lunda states. These states reached their climax in the late sixteenth century, and while their capitals were not as permanent as those in West Africa, they were considerable centers of government and trade. Court officials and skilled artisans—smiths, weavers, carvers, and others—congregated around the capitals and lived off the tribute paid in foodstuffs from the surrounding countryside. They controlled a prosperous business in ivory from Lake Bangwelu to the Benguela coast.

A series of kingdoms existed in the vicinity of the West Rift Valley between Lake Albert and Lake Tanganyika. They included *Buganda, Bunyoro, Ankole* (all within the present-day boundaries of Uganda), and *Karagwe, Rwanda,* and *Burundi,* each quite likely having at its zenith populations of half a million or more people. As with the pre-Islamic states of the Sudan, they were ruled by kings believed to be divine who governed through an elaborate hierarchy of court officials and provincial chiefs.

Throughout the centuries, the East African highlands attracted waves of foreign traders and conquerors, mainly from the Ethiopian borderlands and the southern Nile region. Slave traders periodically pushed southward up the Nile, and by the middle of the nineteenth century, Arab merchants had visited the court of Buganda to obtain ivory and captives in exchange for cloth and guns. The main trade route ran from the East African coast to Tabora, through Karagwe and on to Ankole, Buganda, and Bunyoro. The Buganda kingdom was the largest and most powerful in the region and occupied a strategic location on the north shore of Lake Victoria. Its warriors not only controlled its lakeshore stronghold, but regularly crossed Lake Victoria to control trade to the south. Like most kingdoms of its time, Buganda had succumbed to colonial domination by the turn of the twentieth century.

At the time that the states of Mali, Ghana, and Songhay flourished in West Africa, there existed between the Zambezi and Limpopo rivers a number of Iron Age states that had well-established trade connections with the east coast and points beyond to India and China. The states of the people called Shona were the most significant, and the most imposing remnants are those of Great Zimbabwe, a group of stone ruins a few miles southeast of Masvingo, Zimbabwe. These massive walls and towers, rounded gateways and strategic siting, are evidence of power and ordered settlement. The largest structure occupies an almost impregnable position atop a granite hill, while on the plains below stands a large elliptical building. Both buildings are made of local granite blocks; the larger measures some 300 by 220 feet (90 by 67 m) and its girdling walls stand 30 feet (9.1 m) high and greater than 20 feet (6.1 m) thick. The total complex gives a sense of power, skill, and permanence.

The precise chronology of the Zimbabwe complex is still unknown, but it was probably built and rebuilt by at least three successive occupants. The oldest part of the ruins date back to the fifth century, when simple structures of wood and straw predominated. By the twelfth century Zimbabwe had commercial links with Sofala and other coastal city-states through which it exchanged gold and ivory for porcelain, copper coins, glass beads, and other luxuries. By the middle of the fifteenth century, the focus of power had shifted northeast to the Mutapa state, located closer to the Zambezi and Inyanga mountains bordering modern Mozambique. There smaller stone structures were built, the hillsides were terraced, and gold and other metals were mined. But by the seventeenth century control over this vast and rich interior was threatened by Arab and Portuguese entrepreneurs, the Portuguese eventually installing a puppet king over Zimbabwe's successor state, Mutapa, in 1630. Another successor state to Great Zimbabwe and Mutapa was the Changamire Rozvi empire, which existed to the end of the eighteenth century and the arrival of Nguni-speaking peoples from the south.

Along the East African coast, city-states developed focused on the Indian Ocean trade that extended from Africa to Arabia, India, Indonesia, and China. These Swahili (Arabic, "coastal") city-states, many of which still exist as towns and cities today, having far outgrown their origins, had a golden age of trade and prosperity from 1100 to 1400. Paradoxically, the distinctive and diverse Swahili coastal culture that developed from the tenth century was at the same time the author of and the product of the interaction of diverse East African and Asian populations (Kusimba 1999) brought together through the long-distance Indian Ocean trade. Over time, the Swahili language—with significant dialectical variations due to sociogeographic causes—became the lingua franca of East Africa and even parts of the Congo Basin (Allen 1993). Mogadishu, Kismayu, Lamu, Mombasa, Zanzibar, Kilwa, and Sofala are a few of the better known Swahili city-states.

After the Portuguese rounded the Cape of Good Hope in 1498, they attempted to gain control of the Indian Ocean trade. Although it would have been impossible for them to control by force the large Indian trading companies upon whom they later came to depend for loans, the Portuguese attacked and took control of several Swahili city-states located on the southern East African coast. These ultimately formed the bridgehead for greater Portuguese colonization over 300 years later in Mozambique. The Swahili and their allies went on to extend their control of interior peoples, especially during the nineteenth century, through the slave trade, as described in the next section.

Finally, a powerful nineteenth-century military empire was situated in Zululand in southern Africa. Under Shaka and later his half-brother and assassin, Dingane, the Nguni-speaking *Zulus* controlled an extensive pastoral and game-rich region from Natal north to the Zambezi, while the Zulu core area itself focused on the Tugela (Thukela) River. Shaka Zulu organized great regiments of warriors, well disciplined and trained in hand-to-hand fighting, to extend his control over weaker groups. He was the most widely feared of the Zulu leaders and, in the course of his ten-year reign, may have been responsible for the killing of up to a million people (Hamilton 1998).

Military action by Shaka's soldiers set in motion one of the last and greatest population movements in southern Africa, the *mfecane,* or "scattering." The roots of the mfecane lie in drought, famine, and the often violent competition for scarce resources between groups of people, both indigenous and non-African. It was the better organized and powerful Zulu people who began to expand their state in the early part of the nineteenth century, displacing weaker Nguni-speaking peoples. Flight was exacerbated by Portuguese slave raiding and Dutch trekkers seeking "open" land for their cattle and settlement. Between 1816 and 1840, the displacements of weaker Nguni-speaking groups, who in turn preyed upon peripheral Sotho chiefdoms, profoundly altered the political and cultural geography of southern Africa:

> [The Zulus'] devastating raids set off a chain-reaction which spread warfare and destruction right across the highveld. As one lot were dispossessed of cattle and stores of grain, they turned in desperation on their neighbors. Villages were sacked and burnt, and thousands were killed in battle or died of starvation. (Shillington 1989: 262).

Over time, the displaced people formed new polities, many of which existed up to the time of the European conquest of Africa.

Shaka's first capital, located near Babanango, was appropriately called Kwa Bulawayo or "the Place of Killing." He later established another capital, also called Kwa Bulawayo, near Eshowe. An even larger military capital, Kraal Dukuza, was built near Stanger. Shaka Zulu had a small group of councillors, but he personally made every important decision. He was the commander in chief, the high priest, the ultimate court of appeal, and the sole source of laws. He was the wealthiest man in the kingdom, owning thousands of head of cattle and commanding the services of thousands of warriors and hundreds of women. He ruled by fear, and fear became an important nation-building factor as he amalgamated many separate tribes into the strongest single nation in southern Africa (Hamilton 1998). Following Shaka's assassination in 1828, the Zulu kingdom lost its impetus, and under another tyrannical leader, the regicide Dingane, the Zulus gradually succumbed to British power.

THE AFRICAN SLAVE TRADE

Slavery is an ancient phenomenon that has existed almost everywhere in the world. Today, although it is still practiced, it is found almost nowhere. Slavery in its most basic definition is a form of labor, like hourly labor, salaried labor, and even the "lifetime employment" that Japanese salaried workers enjoyed for decades. The difference between the latter three types of labor and slavery is that the slave is owned by his or her employer. Although slaves have always had rights, in some areas more rights than in others, the masters generally own the product of a slave's labor, including children, as well as the slave. A slave may be sold at any time, although who could actually be sold at any given time was regulated by customs, which differed from place to place.

Africa has been intimately associated with slavery as a source of slaves for other regions of the world and is an area where slavery has been common. The African slave trade has been the largest human migration in history, involving millions of captives who were subjected to such brutality and dehumanization that it seems absolutely beyond belief today that an institution sanctioning these practices could have existed.

Before the African slave trade, the Mediterranean world took its slaves from eastern Europe; in fact the word used for "slave" in most European languages today, including English, comes from the word "Slav." Philip Curtin, one of the foremost historians of slavery, has identified two simultaneous events that impelled the western slave trade to seek new slaving grounds (Curtin 1998). The first was the Turkish capture of Constantinople (today's Istanbul) in 1453. This made it very difficult for Christian slave traders to reach the slave ports located to the north of the Black Sea. The second event was the Portuguese arrival in Subsaharan Africa. The Portuguese were looking for gold, which they found in Ashanti (former Gold Coast). They also found a ready source of slaves along the Gambia River. In the early years of the Atlantic slave trade, slaves who had been captured in the region of the Gambia River were sold to *Ashanti* for gold, whereupon they were put to work in the gold mines. Ultimately the Portuguese and their successors tapped into the source areas of the trans-Saharan slave trade and diverted slaves to the Atlantic market.

The trans-Saharan trade is ancient and developed as did the trans-Saharan trade in other commodities. South of the Sahara the main participants in this trade were private traders of the empires and states such as Ghana, Mali, Oyo, and Ségu. The traders went by a variety of names: Jula, Wangara, Jellaba, Jabarti, or other local names. They were almost uniformly Muslim, and were connected to Berber and Arab traders in North Africa, along the Nile River, and on the Red Sea. The Portuguese controlled the slave trade in equatorial Africa, but in East Africa the enterprise was controlled by Swahili and Omani Arab merchants based along the east coast in the so-called Swahili trading towns and in Zanzibar. The people that they traded were soldiers captured in war or raiding, ordinary folk (farmers, herder, and fishers who were in the wrong place at the wrong time), or criminals.

From Table 3-3, which shows the number of captives taken from the major regions of Africa from 1500s to 1900, we see that the Atlantic slave trade involved by far the greatest number of captives Figure 3-12). Nevertheless, although the Atlantic trade began around the year 1500, the Saharan, Red Sea, and Indian Ocean trade had been carried on for centuries. Lovejoy (2000) estimates the trans-Saharan trade to have exported 4,820,000 persons between 640 and 1600 and 2,400,000 persons for both the Red Sea and Indian Ocean trade between 800 and 1600, making a total of 7,420,000 for the Saharan and 4,134,000 for the Red Sea and Indian Ocean trades.

Research by Africanist historians (Boahen 1986; Thornton 1998; Lovejoy 2000) has shown that the African slave trade was voluntarily engaged in and controlled by African states and private African traders. This trade, with its terrible cost in human life and suffering, was not voluntary for those who were enslaved. For those who survived capture and transportation, it entailed a dramatic change in status and life circumstances. It is estimated that as many as 9 to 15% of

TABLE 3-3 SOURCE AND NUMBERS OF CAPTIVES EXPORTED FROM THE SIXTEENTH TO THE NINETEENTH CENTURIES.

	Number of slaves in each of four centuries				
Exporting Region	**1500-1600**	**1600-1700**	**1700-1800**	**1800-1900**	**Total**
Atlantic	409,000	1,348,000	6,090,000	3,466,000	11,313,000
Saharan	550,000	700,000	700,000	1,200,000	3,150,000
Red Sea	100,000	100,000	200,000	492,000	892,000
Indian Ocean	100,000	100,000	400,000	442,000	1,042,000
Total	1,160,500	2,249,600	7,391,700	5,601,800	16,403,600

Source: Various studies summarized in Lovejoy (2000).

Africans shipped to the New World perished on the way (Lovejoy 2000). Miller (1981) estimated for the trade in Angola that 40% of captives died before they reached the African coast for shipment overseas. Contrary to popular assumptions, Europeans were rarely involved in the trade on the mainland. African leaders wished to keep them in the dark about source areas and prices. In addition, concerns about diseases unknown in Europe or the Mediterranean world motivated Europeans to remain on their island enclaves or on ships just off the African coast.

> When they arrived on the African coasts, the Europeans found a situation different from anything they encountered in the Americas or Asia. The West African disease environment was nearly as dangerous for them as their European diseases were for the American Indians. And it was equally dangerous for the North African traders, who rarely took up residence south of the port cities on the desert edge. From any point of view, tropical West Africa had a terrible disease environment for human beings of any origin. Infection rates with yaws, Guinea worm, trypanosomiasis (sleeping sickness), onchocerciasis (river blindness), and schistosomiasis (liver flukes) were extremely high. In addition, the Africans had the usual range of Old World diseases, such as smallpox, measles, and the common childhood diseases that killed the American Indians. What made the environment so dangerous for outsiders, however, was a combination of yellow fever and *Plasmodium falciparum*, the most fatal form of malaria. (Curtin 2000: 38)

Around 1600, at the time the African slave trade was really beginning, slavery was disappearing in the other long-inhabited regions of the world. Why did slavery become the dominant institution in Africa and in the Americas after that time? Manning (1990) has linked the rise of the Atlantic trade to the enormous demand for labor in the production of sugar and the mining of precious metals in the Americas, the willingness of Africans to sell captives at low prices, and the fact that selling people was a social tradition of long standing in Africa.

Why did African enslave African? Perhaps because Subsaharan Africans are culturally fragmented, divided into a constellation of ethnolinguistic, cultural, political, and interest groups which, in many cases, have had a long history of competition (and cooperation—it must be added) with their neighbors. In this sense it is probably true that African states were motivated to enslave their neighbors in the same way that European countries were impelled to carve up Africa at the end of the nineteenth century: interstate competition.

Closer to the ground, on the other hand, Lovejoy believes that people were in it for the money: that "economic factors were major determinants of the response of African suppliers to rising demand [and that] ". . . slave labour became one of the most attractive investments for those [Africans] with capital to invest" Lovejoy (2000: 53, 144). It was profitable to sell captives, and it was profitable to employ captives in agriculture. It was profitable to use slaves in a variety of occupations.

Enslaved Africans were used in Africa as domestics, as field labor in ordinary agriculture and on plantations, in the military, in industry, and for ritual sacrifice. It has been estimated that the number of slaves employed in West Africa probably equaled the number exported (Thornton 1998). In some states, for example, Ségu, located in the country of Mali, slaves became

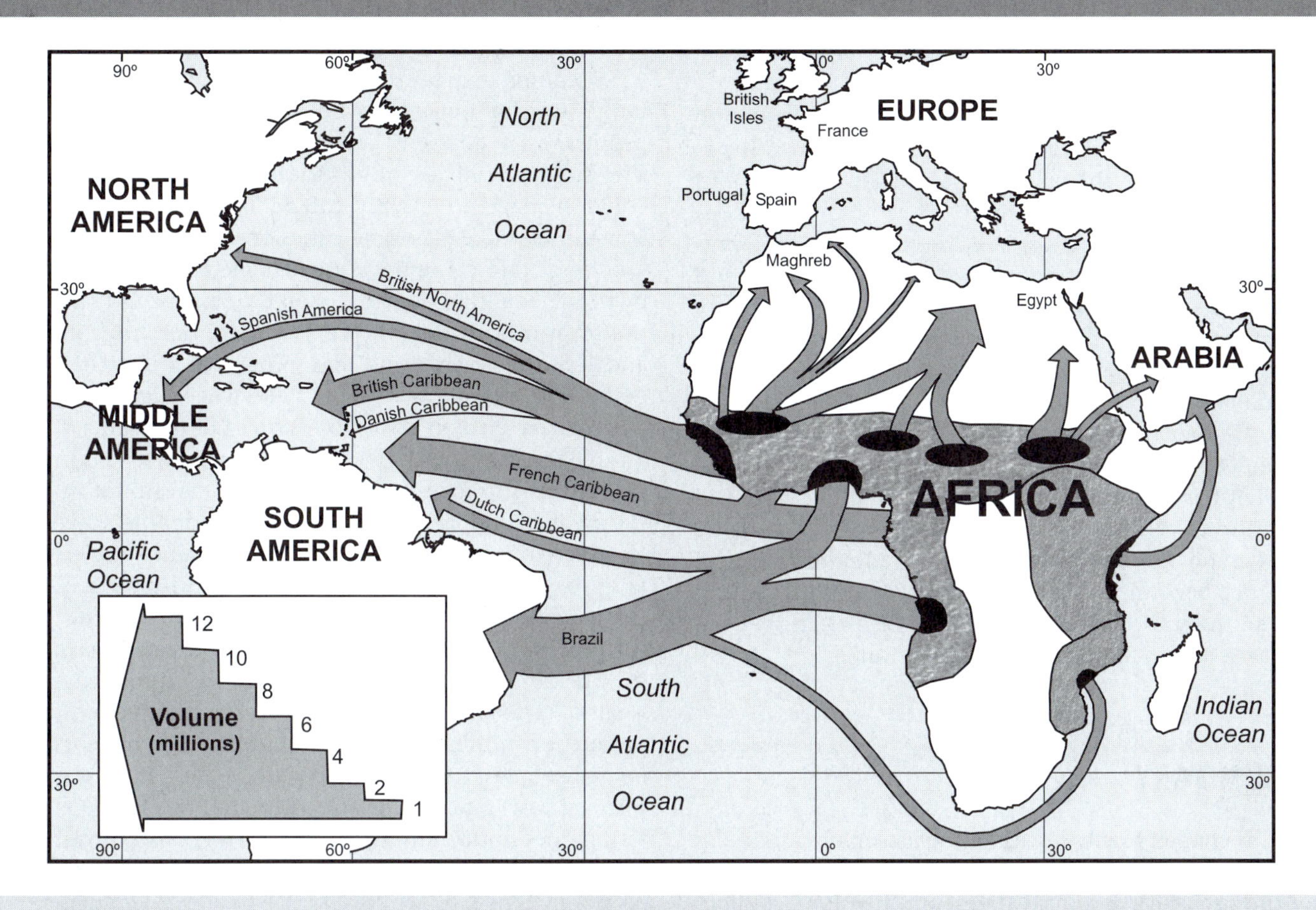

FIGURE 3-12 Source and volume of the African slave trades, 640-1867.
Adapted from de Blij and Muller (2006).

kingmakers. In others, notably Mamluk Egypt, they ran the country for hundreds of years. But the fate of the vast majority of slaves was far different from these extreme examples.

The British outlawed the slave trade in 1807 and banned this form of servitude altogether in 1833, Slavery continued as a thriving institution in Africa until the occupation of the continent by Europeans, however, and did not disappear from some areas until the early decades of the twentieth century (Manning 1990). The abolition of slavery was the result of campaigns by which Europeans and Americans intended to substitute "legitimate trade"—that is, trade in palm oil, food stuffs, and other commodities—for the illegitimate trade of slavery. Ironically, the transition from slave exports to exporting other goods caused an increase in the use of slavery in Africa in the short term. As we shall see in Chapter 4, the advent of direct European rule in the late nineteenth century terminated slavery as a legitimate use of labor in Africa, although the practice lingered for a few decades in several areas.

> . . . the transition to colonial rule and political independence occurred in the context of the collapse of slavery. Where the old elites survived, despite the turmoil of the colonial years, then the transition was less abrupt. Where the old elite was eliminated, slaves could sometimes assert themselves sooner and more fully. None the less, the question is not whether or not

> there was a dramatic break with the past, but the ways in which people were able to shape the new order, preserving the old or rising above it. (Lovejoy 2000: 288–289)

The long history of African slavery has had a profound impact on African societies to this day, just as it has deeply affected the importing countries. Africans with an ancestry, however remote, of servility have been stigmatized in many parts of the continent; for example, marriage between those with a servile background and the old nobility is not possible except under very unusual circumstances. There is some anonymity in the city for those who wish to change their identity (Gallais 1962).

The next chapter examines the "transition," mentioned by Lovejoy in the paragraph just quoted, to what has been called the "colonial interlude." We will return to some of the issues addressed earlier when we examine the impact of European administration of Africa beginning at the end of the nineteenth century and the subsequent independence of African states in the second half of the twentieth century.

SUMMARY

This chapter briefly discussed the rise of humankind in Africa, the changing distribution of African languages, the rise of early African states, and involuntary servitude. From this review of Africa's sociocultural history, it can be seen that Africa's early states were deeply rooted in the past and were the result of sometimes incremental developments in the relations between diverse peoples and in improvements in transportation and trade. Clearly a number of general conclusions may be drawn about the development of polities in Africa. It should be clear that much of the continent has enjoyed a long and rich history, a history of diverse sociopolitical organization. Like states of today, the pre-European states of Africa had towns and cities, clearly defined divisions of labor, class structures, communications networks, and spheres of influence and possession. Most of them were multicultural, involving a plethora of sociolinguistic and culture groups. Commercial linkages were widespread—the biggest trading networks being the trans-Saharan trade, centered on a succession of states located in the savanna of West Africa, and the Indian Ocean trade, which was focused on the East African Swahili trading cities.

Commonly states resulted from conquest by militarily superior invaders, who imposed their will on peoples with different customs and languages, and they survived only as long as the conquering minority could manage to extract tribute from its subjects. Sometimes there was little cohesion in the state, the only real bond being the periodic collection of tribute by the ruler and his agents. Often the state would take the form of a cluster, with a strong central kingdom and less effectively controlled provinces around the periphery. But internally, these states were not ruled in a hereditary fashion by a succession of descendants of a privileged family. They were not normally feudal in character—although Africa has seen feudalism, for instance in Ethiopia. Rather, the great administrative force serving the divine king was a sometimes endless array of viceroys, officials, chiefs, and other agents. In an economic sense, involuntary servitude in Africa was a form of private ownership of labor, which gave the African entrepreneur a "secure and reproducing wealth" (Thornton 1998). Labor was a scarce factor of production, while land was ubiquitous; entrepreneurs owned people and put them to work because there was no legal basis for owning land. Although this interpretation does not communicate the horrors of slavery, it is quite correct as a description of African slavery in an economic sense.

CHAPTER REVIEW QUESTIONS

1. In what ways is the historic relationship between the people of the kingdom of Benin and the Europeans at the Bight of Benin similar to that between the Iroquois and the British in the Great Lakes region in the United States?
2. What are some characteristics of African states?
3. How have outsiders viewed African history until relatively recent times?
4. What were the causes of African slavery?
5. The African slave trades (Atlantic, trans-Saharan, and East African) were organized and executed

by Africans with willing African sellers and willing buyers, both African and non-African. The question to you is this: who was responsible for the slave trade?

6. Describe the process of state formation in Africa. Are there some similarities between states that seem to contribute to the rise or fall of a state?
7. Name the major language phyla of Africa and describe their changing geography.
8. Where and how did agriculture originate in Africa?
9. Compare and contrast three hypotheses about the discovery of iron working in Africa?
10. Compare and contrast slavery in the United States with that practiced in Africa.

GLOSSARY

Berber A term commonly used to refer to the indigenous Caucasian peoples of North Africa to distinguish them from the Arabs, who arrived in the seventh century and later. The term is pejorative and comes from the Greek *barbar*, meaning a person who cannot be understood, a foreigner.

Distance decay The weakening of a pattern or a process with distance from a point or area of concentration or higher density.

Factor of production The items necessary in the production process to produce wealth. In an agricultural example, one would need three factors of production: land, labor, and capital. One needs capital to invest in the development of the productive resources of the farm, and one needs labor to work on the farm. In a market economy, one creates wealth on a farm by buying land and hiring labor. In the precolonial African context, land was abundant and belonged to the community rather than to individuals. Labor was the critical factor of production for the generation of wealth in Africa, and entrepreneurs sought to control it directly.

Metropole The capital city of an empire. London and Paris were the metropoles of the British and French empires, respectively.

Relocation diffusion The physical movement of people and their culture from one place to another.

BIBLIOGRAPHY

Allen, J. de V. 1993. *Swahili Origins*. London: James Curry.

Andah, B. 1993. Identifying Early Farming Traditions of West Africa. In *The Archaeology of Africa: Food, Metals, and Towns*, T. Shaw, P. Sinclair, B. Andah, and A. Okpoko, eds. London: Routledge, pp. 240–254.

Blench, R. 1993. Recent Developments in African Language Classification and Their Implications for Prehistory. In *The Archaeology of Africa: Food, Metals, and Towns*, T. Shaw, P. Sinclair, B. Andah, and A. Okpoko, eds. London: Routledge, pp. 126–138.

Boahen, A. 1986. *Topics in West African History*, 2nd ed. Harlow, Essex, U.K.: Longman.

Brauer, R. W. 1995. Borders and Frontiers in Medieval Muslim Geography. *Transactions of the American Philosophical Society*, 85(6): 1–73.

Bulliet, R. W. 1990. *The Camel and the Wheel*. New York: Columbia University Press.

Butzer, K. W. 1981. Rise and Fall of Axum, Ethiopia: A Geo-archaeological Interpretation. *American Antiquity* 46: 471–95.

Cavalli-Sforza, L., P. Menozzi, and A. Piazza. 1995. *The History and Geography of Human Genes*. Princeton, N. J.: Princeton University Press.

Cornevin, M. 1993. *Archéologie Africaine*. Paris: Maisonneuve et Larose.

Curtin, P. D. 1998. *The Rise and Fall of the Plantation Complex*, 2nd ed. Cambridge, U.K.: Cambridge University Press.

Davidson, B. 1991. *Africa in History: Themes and Outlines*. New York: Macmillan.

De Blij, H., and P. Muller. 2006. *Geography: Realms, Regions and Concepts*. New York: John Wiley & Sons.

Feierman, S. 1993. African Histories and the Dissolution of World History. In *Africa and the Disciplines: The Contributions of Research in Africa to the Social Sciences and the Humanities*, R. H. Bates, V. Y. Mudimbe, and J. O'Barr, eds. Chicago: University of Chicago Press, pp. 167–212.

Gallais, J. 1962. Signification du Groupe Ethnique au Mali. *L'Homme*, May–August: 106–129.

Greenberg, J. 1963. *The Languages of Africa*. The Hague: Mouton.

Grimes, B. 2000. *Ethnologue: Languages of the World*, 14th ed. Dallas, Tex.: SIL.

Hamilton, C. 1998. *Terrific Majesty: The Powers of Shaka Zulu and the Limits of Historical Invention*. Cambridge, Mass.: Harvard University Press.

Harlan, J. R. 1993. The Tropical African Cereals. In *The Archaeology of Africa: Food, Metals, and Towns*, T. Shaw, P. Sinclair, B. Andah, and A. Okpoko, eds. London: Routledge, pp. 53–60.

Hayward, R. 2000. Afroasiatic. In *African Languages: An Introduction*, B. Heine and D. Nurse, eds. Cambridge, U.K.: Cambridge University Press, pp. 74–98.

Heine, B., and D. Nurse, eds. 2000. *African Languages: An Introduction*. Cambridge, U.K.: Cambridge University Press.

Kusimba, C. M. 1999. *The Rise and Fall of Swahili States*. Walnut Creek, Calif.: Altamira Press.

Kwamena-Poh, M., J. Tosh, R. Waller, and M. Tidy. 1982. *African History in Maps*. Harlow, Essex, U.K.: Longman.

Lovejoy, P. 2000. *Transformations in Slavery: A History of Slavery in Africa*. 2nd ed. Cambridge, Mass.: Cambridge University Press.

Manning, P. 1990. *Slavery and African Life: Occidental, Oriental, and African Slave Trades*. Cambridge, U.K.: Cambridge University Press.

McEvedy, C. 1995. *The Penguin Atlas of African History*. London: Penguin Books.

Miller, J. 1981. Mortality in the Atlantic slave trade: Statistical evidence on causality. *Journal of Interdisciplinary History*, 11: 385–423.

Munro-Hay, S. 1991. *Aksum: An African Civilisation of Late Antiquity*. Edinburgh: University of Edinburgh Press.

Muzzolini, A. 1993. The Emergence of a Food-producing Economy in the Sahara. In *The Archaeology of Africa: Food, Metals, and Towns*, T. Shaw, P. Sinclair, B. Andah, and A. Okpoko, eds. London: Routledge, pp. 227–239.

Newman, J. 1995. *The Peopling of Africa*. New Haven, Conn.: Yale University Press.

Okafor, E. 1993. New evidence on Early Iron-smelting from Southeastern Nigeria. In *The Archaeology of Africa: Food, Metals, and Towns*, T. Shaw, P. Sinclair, B. Andah, and A. Okpoko, eds. London: Routledge, pp. 432–448.

Oliver, R., and Fage. J. D. 1962. *A Short History of Africa*. Baltimore: Penguin Books.

Phillipson, D. W. 1993. The Antiquity of Cultivation and Herding in Ethiopia. In *The Archaeology of Africa: Food, Metals, and Towns*, T. Shaw, P. Sinclair, B. Andah, and A. Okpoko, eds. London: Routledge, pp. 344–357.

Shillington, K. 1989. *History of Africa*. London: Macmillan.

Thornton, J. 1998. *Africa and Africans in the Making of the Atlantic World, 1400–1800*. Cambridge, U.K.: Cambridge University Press.

Trigger, B., B. Kemp, D. O'Connor, and A. Lloyd. 1983. *Ancient Egypt: A Social History*. Cambridge, U.K.: Cambridge University Press.

Wendorf, F. 1968. Site 117: A Nubian Final Palaeolithic Graveyard near Jebel Sahaba, Sudan. In *The Prehistory of Nubia*, F. Wendorf, ed. Dallas, Tex.: Southern Methodist University Press.

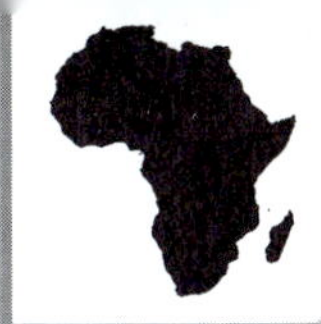

CHAPTER 4

The Colonial Interlude

The course of African history was profoundly changed by the colonial powers of Europe. Each power came for its own reasons with its own values, perceptions, and institutions; and each left its mark on the peoples and landscapes in the form of language, education, law, technology, and so on. To understand Africa's contemporary political and economic geographies, it is necessary to have an appreciation of the ways in which colonial Europe partitioned and conquered Africa, viewed its peoples and resources, and brought to Africa instruments of change. This chapter emphasizes these processes in Africa south of the Sahara, where the colonial interlude was most complex.

Although the present chapter is concerned with more recent times, it might be said that European involvement in Africa began with the Greeks and Romans over two and one-half thousand years ago. Greeks took control of Egypt after 331 BCE from the Persians and the Romans took over in 30 BCE. If one considers the Byzantine Empire as the inheritor of Rome, then Roman rule extended to the middle of the seventh century of the common era—up to the invasion of Arab Muslims from the Arabian Peninsula. Greek, Roman, and Byzantine rule involved only North Africa and the Red Sea coast, not Subsaharan Africa. Arabs and Persians have been involved in East Africa for over a thousand years. It was only in the last 500 years that Europeans arrived south of the Sahara; for over 300 years powerful African rulers and tropical disease had kept the Europeans along the coastline or on nearby islands. At the end of the eighteenth century and especially in the nineteenth century, European explorers began to crisscross Africa.

The long European involvement with Africa culminated in what is often called the scramble for Africa, the determined attempts of individual nations to gain political and economic control of huge portions of the largely unknown continent. These activities began slowly in the nineteenth century but became more and more pronounced over time. The now-familiar straight-line boundaries were drawn on a map during a conference held in Berlin from November 15, 1884, to February 27, 1885: Britain, France, Portugal, Italy, Germany, and Spain simply agreed to divide among themselves vast tracts of African land. It is remarkable that these rather hastily defined political units would prove so enduring. In the pages that follow we will explore the involvement of Omani Arabs, Turks, and Europeans in the formation of political units in Africa. Ultimately, the Omani and Turk hegemony was not to last; both groups were squeezed out during the European "scramble" for territory in Africa.

EARLY ARAB, PERSIAN, AND TURKISH INTERESTS

Arab and Persian involvement with East Africa dates back to the eighth century with Shiite settlement. These Muslims established towns along the East African coast, the Land of Zanj (Zanjibar, "Land of the Blacks," in the original Persian). A century later, well-established market towns had developed along the coast which were involved in the Indian Ocean trade. This trade was centered on states in the Persian (Arab) Gulf and extended from East Africa to India and beyond. The Arabs and Persians and the Bantu-speaking Africans who lived along the East African coast intermarried, and over the centuries a well-articulated trading diaspora developed centered on a cluster of city-states comprising three classes of people: the Arabic- and Swahili-speaking elites, who were Muslim; the Swahili Muslim townspeople (traders, artisans, etc.); and finally, the non-Muslim slaves. The ruler of each city-state was called a sultan, and his power extended into the hinterland, where raw materials such as ivory, gold, furs, and slaves were gathered for export. Luxury goods were the principal imports: cloth, beads, pottery, and so on. Over the ensuing centuries, more and more Arabs settled along the East African coast, extending their culture and religion to Africa.

During the middle of the nineteenth century Arab/Swahili slaving activities accelerated, and a host of indigenous groups of Africans were shattered by slave raids. Some groups of Africans, for example, the Yao and the Nyamwezi, worked with the Arab and Swahili traders as slave raiders. One of the most famous Swahili traders, Muhammad bin Hamid (or Tippu Tip), extended his slaving operations deep into the Congo Basin. The principal market for slaves was Zanzibar, where slaves were purchased for work on clove plantations on Zanzibar or other islands.

> For much of the nineteenth century the economic life of eastern–central Africa had been disrupted and dominated by the violence of the trade in firearms, ivory and captives. Apart from the wealth gained by individual traders and rulers, the vast majority of central African peoples gained little benefit from the trade. And the persistence of slavery and the slave trade and the dislocation of people that went with it provided potential European colonists with both opportunity and excuse to intervene in the region. (Shillington 1989: 256–257)

So profitable were the clove, slave, and ivory trades that in 1840 the sultan of Oman, Said Said (an ancestor of the present sultan Qaboos al-Said), moved the capital of the Omani Empire to Zanzibar (Figure 4-1). The slave trade continued until the British compelled the sultan to end it—although the use of slave labor continued for many more years. Turkish hegemony in Africa concerned Libya and Egypt, which were conquered in 1517 and, by the time of the "scramble," Egypt had conquered the Nile Valley south to Lake Victoria.

Muhammad Ali (c.1769–1849), the ruthless, Albanian-born, ruler of Egypt from 1805 to 1849, was an empire builder. He arrived in 1801 as part of a Turkish army sent to expel Napoleon Bonaparte, who had conquered Egypt in 1798, and reclaim Egypt for the Ottoman sultan. Later, the ambitious Muhammad Ali successfully broke away from the Turkish sultan in Istanbul and began to modernize and extend his state along European lines. He introduced the printing press, a salaried civil service, reorganized the Egyptian army along European lines, and modernized agriculture. As pasha, Muhammad Ali established a dynasty that ruled Egypt until the revolt by army officers of 1952. In 1820 he invaded the Sudan, principally to control the source and increase the supply of slaves to Cairo. His conquests were continued by his descendants right up to the European partition of Africa.

The extravagance of Muhammad Ali and his successors led to deep and chronic indebtedness. An agreement reached in 1878 between Ismail Pasha, Muhammad Ali's free-spending great-grandson, and Egypt's creditors stipulated that two Europeans, a Briton and a Frenchman, would be appointed as ministers in the Egyptian cabinet to ensure fiscal responsibility and discharge of the national debt. But by 1895, Egypt was totally bankrupt; Ismail was forced to sell Egypt's stake in the Suez Canal, which had been built with French assistance and completed just a few years before, in 1869. Egypt's creditors, principally Britain, gradually and reluctantly took control of the economy and governance and kept control for 44 years until 1922, when Egypt was declared an autonomous monarchy.

EARLY PORTUGUESE INTERESTS

Portuguese navigators made contact with West Africa in the first half of the fifteenth century. There they carried on what the Arabs had initiated in East Africa: a profitable traffic in slaves, for many centuries Africa's major export. While fewer slaves were shipped to the Iberian Peninsula than from East Africa to the Middle East and South Asia, the trade was nevertheless significant and set the stage for the extension of human exploitation to the New World.

The fifteenth and sixteenth centuries in Africa may well be called Portugal's centuries, although they were not only Portugal's. It was the Portuguese, however, who first rounded the Cape of Good Hope and sailed past the tip of Africa into the Indian Ocean, settling along the east coast and contesting possession of the lands of the east with the Arabs and Muslims of the Swahili trading cities. Although some individual explorers penetrated far inland, the Portuguese settlements were mainly peripheral. The city of Benguela, Angola, dates from this period, as does Lourenço Marques (Maputo), Mozambique, although settlement

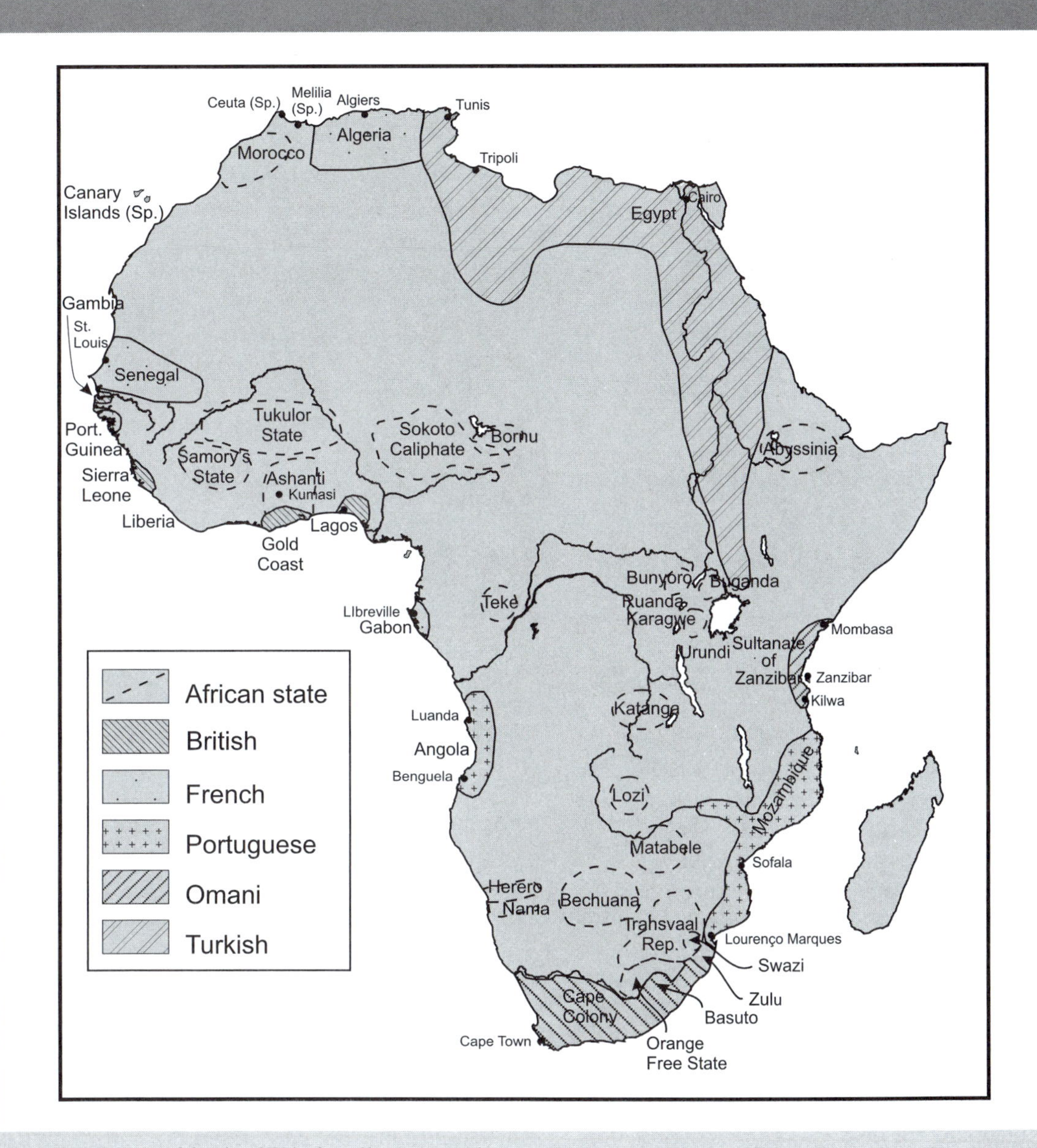

FIGURE 4-1 Africa in 1871 before the "scramble."
Adapted from Pakenham (1991).

on Delagoa Bay, near Sofala was intermittent until the eighteenth century.

Initially, the voyages of Bartholomeu Diaz and Vasco da Gama produced friendly contact with the Arabs and the Swahili, and it was an Arab guide who first led Portuguese explorers to the Indies. But the interests of the Portuguese were in conquest and control of the Indian Ocean trade, and soon Portuguese power subjugated the Arab and Swahili holdings on the East African coast and spread as far north as Arabia. By not consolidating their settlements, but maintaining real interest only in the route to the Indies, the Portuguese made the error that was eventually to lead to their defeat. As the sixteenth century drew to a close, the

FIGURE 4-2 Restored seventeenth-century Portuguese fort (Later Dutch and English), at Cape Coast, Ghana. *Steven Montgomery.*

Portuguese empire in Europe was declining, and the Arabs began systematically pushing the invaders from East Africa. Mombasa was besieged and bombed numerous times, as both sides took and lost the city repeatedly. Shortly after 1700, the Portuguese had lost all the land they once held north of Cape Delgado, the present northeastern extremity of Mozambique.

In West Africa, meanwhile, the Portuguese slave-trading stations, thriving from Senegal to Angola, also shared in the decline of Portuguese power, and Britain, the Netherlands, and France appeared on the scene. A glance at a seventeenth-century map reveals that by this time other products were being taken: in addition to the Slave Coast, a Gold Coast and an Ivory Coast are labeled. The slave trade remained the most profitable of all, stimulated by the demand in the Americas. Nevertheless, few real European settlements were established. Along the coast of West Africa, there were some forts (Figure 4-2), but they were there primarily for the protection of the trade. Slave traders were associated with Africans who helped carry out the capture of slaves, and thus the real penetration of Africa by the Europeans was delayed. In tropical Africa, it was not until late in the eighteenth century that European exploration (sometimes called the European "age of imperialism") commenced, as the slave trade began to wane. To all intents and purposes, the accumulation of knowledge concerning interior Africa began only during the nineteenth century.

DUTCH AND BRITISH AT THE CAPE

Portugal yielded its power position in Europe and on the seas to Holland, which had emerged undefeated from 80 years of war with Spain. During the period of slave trading and intermittent, peripheral European settlement in tropical Africa, the Dutch established a revictualling station for their ships at a place neglected by the Portuguese, Table Bay. Founded in 1652, the settlement at the foot of Table Mountain was the first and for many decades the only European base to possess some characteristics of permanence. Out of it grew the city of Cape Town, today the largest city in South Africa.

Having failed to oust the Portuguese permanently from Benguela, the Dutch chose the site of Cape Town mainly because it had not been taken by others, and they found themselves in possession of one of Africa's best natural harbors. In the centuries to come, it became clear that Cape Town dominated what is economically the richest part of Africa, the south. Initially, however, the Dutch were not interested in the colonization of the Cape. Having learned from the Portuguese the need for a revictualling station, they established one, but European immigration was actually discouraged, owing to fears that a large Cape colony would become an administrative liability. Thus Cape Town's growth was slow, and it has been estimated that a century and a half later, in 1800, the total European population of the city and the colony that had inevitably developed was only about 25,000.

The Dutch East India Company, engaged on behalf of the Netherlands government in the trade with the Indies, introduced a number of crops to the Cape, traded with the local pastoralists for meat, and imported a large number of slaves from Madagascar, Malaysia, and even West Africa, when local labor ran in short supply. Even so, the company was unable to contain the Dutch citizens who had fulfilled their tour of duty with the government as farmers or employees and refused to return to Europe. Many of these Europeans left the environs of Cape Town and trekked into the interior, warring with local hunter–gathers and farmers over the lands they desired. There, before the end of the eighteenth century, the Dutch were entering Africa's interior, and for some time the southern tip of the continent was better known than most other regions.

Like Portugal before it, the Netherlands declined from its position as the leading sea power of Europe. After a brief temporary wardship, Britain in 1806 took permanent possession of the Cape, finding it a stagnating, backward settlement and an asset mainly as a means of preserving sea power. Britain also showed little interest in the interior. In this hinterland of southern Africa, meanwhile, a most significant event was taking place. Among the many Bantu peoples that had migrated southward were the Zulu, who settled in the eastern region of Natal. Strong leadership by a succession of chiefs who developed a high degree of military organization thrust the Zulu empire into prominence as the most powerful on the subcontinent. Waging war on people in every possible direction, the Zulu decimated the African population on the plateau, defied only by the Basuto and the Swazi.

With the Europeans confined mainly to the Cape and the powerful Zulu concentrated in Natal, direct conflict did not seem inevitable. Events in Britain, however, resulted in a mass exodus of Dutch settlers from the Cape into the regions over which the Zulu held sway. The efforts of the humanitarian William Wilberforce, aimed at the elimination of the slave trade, were reaching the public conscience in Britain, and, in 1833, slavery was abolished as an institution throughout the British Empire, of which the Cape Colony was a part. The Dutch settlers who remained after the demise of the Dutch East India Company had shown increasing dissatisfaction with Britain's efforts to anglicize the Cape, and the termination of the practice of slavery was the last straw. So began, in 1836, the mass movement of Europeans, most of Dutch ancestry, onto the plateau of southern Africa in what has become known as the Great Trek. It was the vanguard of a series of waves of European immigration, resulting in the only really large European population accumulation in all Africa.

The Great Trek eventually brought Dutch Europeans and the Zulu into conflict, the decisive battle being fought at Blood River in Natal, where the Zulu, although numerically greatly in the majority, were defeated. It was the desire of the Dutch settlers ("Boers," as they were often called) to establish, beyond the borders of the British Cape Colony, pastoral republics where they might retain their cultural and religious heritage and not give up their slaves. Having defeated the Zulu, the Boers seemed a step closer to their ideal, and although Britain invaded Natal and removed the Dutch from that region, the Boers remained on the plateau itself. There, the Orange Free State Republic and the Republic of South Africa (Transvaal) were founded by independence-minded settlers of Dutch origin. Although there were some border skirmishes and occasional friction with Africans, the Boer republics were recognized by Britain, and it appeared that the goal of the trekkers had been achieved.

The temporary status quo was disturbed in 1867 by the discovery of diamonds on the banks of one of southern Africa's great rivers, the Orange. The richness of the fields and their relative accessibility brought fortune hunters not only from all parts of the south but also from overseas, producing a second Great Trek, this time initiated by economic forces.

Kimberley became the economic capital of southern Africa, and the building of railroads into the interior was begun. By annexing the diamondiferous region to the Cape Colony, the British positioned themselves to administer the new city and its inhabitants. Although displeased with this action, the leaders of the Boer republics did not attempt seriously to alter it, largely in order not to threaten the continued existence of their states.

Another mineral find, this time a much more significant one, occurred in 1886, when gold was discovered in the Republic of South Africa near what is today the city of Johannesburg (Lester 2000). This brought a third Great Trek, but there was no way for the British at the Cape to annex this territory without war. The republic was unequipped for the mass of foreign intruders, and administrative chaos resulted. Johannesburg began its uncontrolled growth, within a decade reaching 100,000, while the capital, Pretoria, was eclipsed. Johannesburg became the economic capital of the region, a position retains to this day. President Paul Kruger and his stubborn Afrikaner (Boer) government, resentful of the disruption created by Johannesburg's large immigrant population, continually refused to cooperate in city development and denied the allocation of funds for necessary amenities. Friction between the Afrikaners and foreigners was rife. Britain, meanwhile, partly on the advice of such people as Cecil Rhodes and Dr. Leander Starr Jameson, continued to cast covetous eyes on this rich economic prize. Troops massed on the republic's borders, and in 1899 war broke out.

In the Anglo-Boer War, European faced European on a scale unprecedented in southern Africa. Britain, after some initial setbacks, at length defeated the Boers, who maintained a guerrilla campaign after their defeat on the battlefield. Having finally subdued the obstinate Afrikaner opposition, the British made an effort to grant the Boers some participation in the affairs of what had once been their country. In 1910 the Union of South Africa came into being, joining the two British colonies (Cape and Natal) and the two defeated republics (Orange Free State and Transvaal). Although victorious, the British gave in to several of the Afrikaners' desires when the new political entity was created. One of these was the elimination from the political scene of all indigenous African and other nonwhite elements: the Union became a state ruled by an all-white electorate. Nonwhites at first retained some indirect representation, but this was subsequently discontinued. The Union survived for just over half a century, for the Afrikaners came to dominate the British and other settlers, and in 1948 an Afrikaner government was returned to power. It promised economic development, racial segregation, and a revival of the republic; by 1961, all three promises had been kept, and on May 31 of that year a new Republic of South Africa came into being. Apartheid and related issues will be explored in the chapter on South Africa.

THE EXPLORERS

The developments that took place in what is today South Africa did not mirror the sequence of events in tropical Africa. European involvement in the south came early, and numbers were comparatively large. European explorers and adventurers from several European countries traveled in Africa from the end of the eighteenth century right up to the late 1800s. Some were after discovery and fame, some were after souls, while others were helping to spread European imperialism.

Prior to the European exploration of Africa, which began in the late eighteenth century, there were few outside accounts of African life. Of note are the following: Al-Mas'udi, who visited the East African Swahili trading city states in 916; Ibn Battuta, who voyaged from North Africa to the Mali Empire from 1351 to 1352 and traveled across North Africa to Egypt (1325–1326); Ibn Khaldun (1332–1395), a North African born in Tunis, who visited Egypt and East Africa; and Leo Africanus (1485–1554), who traveled from Morocco to the Songhay state in 1510 and in 1513.

For a long time, investigation of the interior of tropical Africa was left to a few explorers. Some performed heroic deeds but have fallen into relative obscurity: James Bruce, for example, entered what is today Ethiopia and the Sudan and began, in 1768, to follow the Blue Nile to its confluence with the White Nile (Figure 4-3). Among the early explorers to achieve fame was Mungo Park, who attempted to solve the riddle of the Niger River. On many early maps, the Niger is seen to flow from east (the general vicinity of Lake Chad) to west, reaching the ocean as the Gambia. Park, in 1795, traveled up the Gambia, hoping to find an actual link with the Niger. On his first journey, he did indeed find the Niger and became aware that its flow is eastward, not west, but he was unable to follow the great river to its mouth, as he had intended. Having disproved the Gambia–Niger connection, Park in 1805 set out again in an effort to trace the Niger to its mouth which he apparently believed to be the Congo. The great explorer met his death in the waters of the river

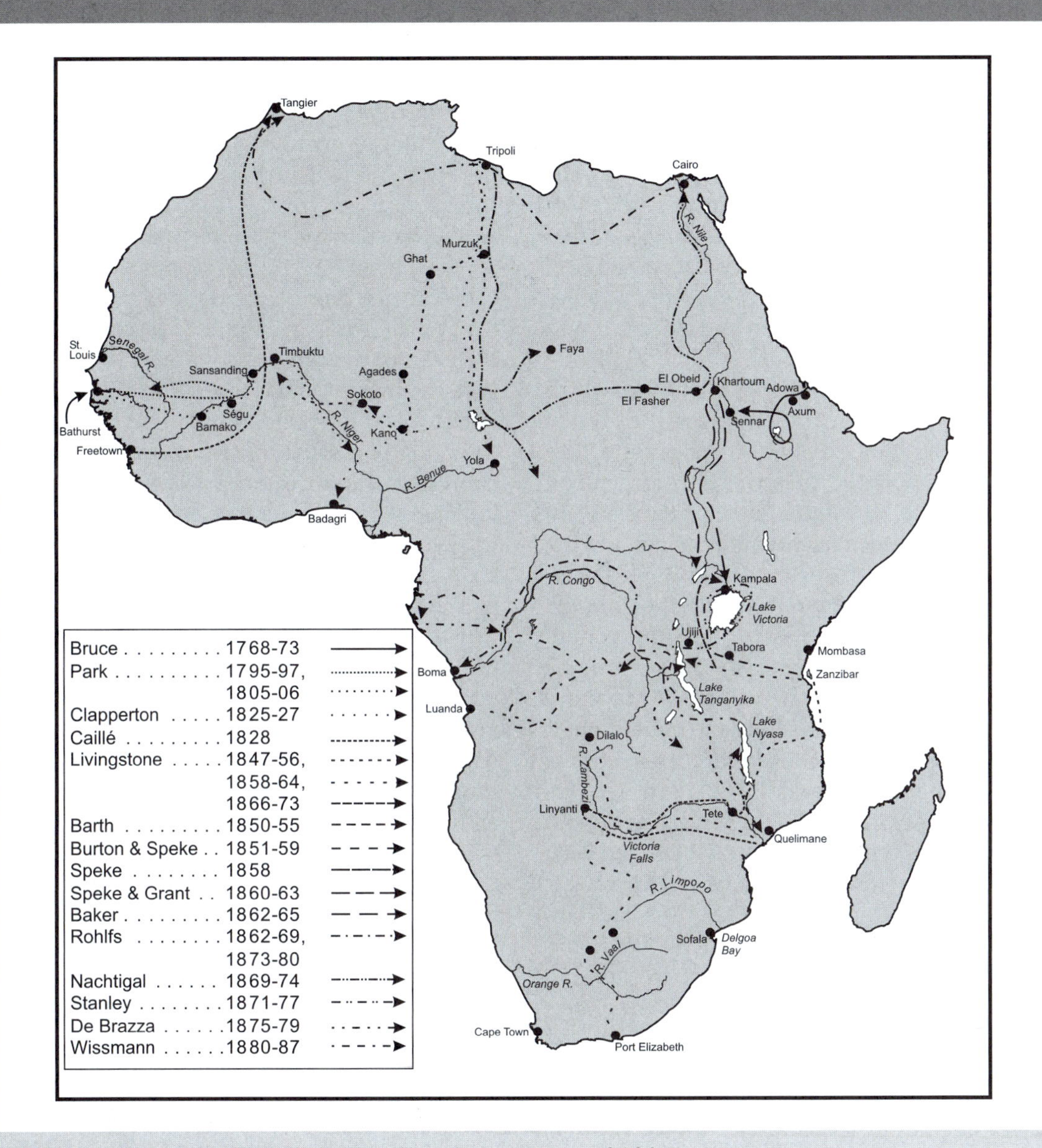

FIGURE 4-3 Eighteenth- and nineteenth-century explorers of Africa.
Barraclough, G. 1993. Times Atlas of World History, 4th Springfield, New Jersey: Hammond.

that was the object of his search: he drowned during a battle with local Nigerians. Only in 1830 did Richard Lander complete Park's task and prove that the Niger flows into its delta on the Guinea Gulf.

While explorers like Hugh Clapperton, Rene Caillé, and Richard Lander were traversing the western part of Africa, equatorial Africa remained barely touched. South of the Congo River was the kingdom of Kongo, which was entered by a number of Portuguese from the coastal settlements during the sixteenth and seventeenth centuries; the Katanga region, however, was not reached until 1798, when the Portuguese explorer Lacerda arrived there. Although there were sporadic efforts by Portuguese and Arabs to enter the interior, the real age of the explorers in equatorial Africa did not begin until the middle of the nineteenth century with the travels of the most famous of all, David Livingstone, who covered extensive areas. Livingstone was one of the emissaries of the London Missionary Society, an organization that was very active in southern Africa principally for the purposes of ending the African slave trade and spreading Christianity. He arrived in Cape Town in 1841 and traveled to Bechuanaland (now Botswana). There began his first series of traverses, and in 1849 he reached Lake Ngami, a part of the Makadikgadi–Okavango swamp and delta complex in the Kalahari Desert.

Subsequently, he crossed Angola to Luanda, which he reached in 1854. The next year, he returned to the region of the upper Zambezi River, intending to follow the stream to its mouth. In the process, he came upon the stupendous falls which he named Victoria Falls, and in 1856 he succeeded in reaching the delta of the Zambezi. The second phase of Livingstone's explorations began in 1858, as he traveled up the Zambezi from its mouth, past the rapids to the Shire confluence. He then proceeded along the Shire until he reached I.ake Nyasa (Lake Malawi). The lake may have been seen earlier by Portuguese and Arab traders, but it was Livingstone whose vivid descriptions made these areas known to the outside world.

In his third series of traverses, Livingstone set out to find the source of the (White) Nile, which had been a problem for geographers for many years. During this period the great explorer became lost, his fate unknown to the world, for half a decade. After Henry Morton Stanley succeeded in locating Livingstone in 1871, at Ujiji on Lake Tanganyika, the two explorers together sought the origins of the Nile, traversing the area around the south end of Lake Tanganyika. Stanley departed in 1873 and Livingstone, weakened by malaria and other diseases, died soon after. Meanwhile, the Nile problem was solved by John Speke, who had crossed the East African Plateau alone, and later, with Richard Burton, reached Lake Tanganyika in 1858. Subsequently, without Burton, Speke traversed the region about Lake Victoria, although he did not on that first journey discover the Nile outlet of the lake. In 1860, after having been taken prisoner by the Baganda, he did locate the outlet, realizing that the question of the Nile's origin had finally been solved, unless there should be a southerly connection between Lakes Victoria and Tanganyika. Samuel Baker in 1863 placed the Nile outlet of Lake Victoria beyond doubt by traveling up the Nile through the Sudd, meeting Speke and his companion, James Grant, who were coming down the same river.

Stanley returned to Africa in 1874 and proved that the rivers he and Livingstone had seen west of Lake Tanganyika were in the drainage basin of the Congo and could be followed to the Atlantic Ocean. It so happened that in 1876 King Leopold II of Belgium convened a meeting of geographers with knowledge of equatorial Africa, and Stanley's report on the Congo, in the heart of the continent, impressed Leopold with the potentialities of a transportation route there. Consequently, Stanley in 1879 mounted an expedition into the Congo on behalf of the king. Ruthlessly obtaining concessions and treaties, within 5 years he had accumulated for Leopold a tremendous territory that came to be known as the Congo Free State. Stanley thus changed from explorer to land hunter, and as such he played a significant role in the struggle for control of Africa's territory (Driver 2001).

Although it cannot be said that the diaries of the nineteenth-century explorers were without bias, their vivid and sensational descriptions of Africa's landscapes and peoples did much to arouse the interest of missionary societies bent on spreading the gospel and eliminating slavery and human suffering. They also stirred the imagination of the monied class, who urged their respective governments to act on their behalf in establishing control of the newly found riches and regions. But for the majority of peoples in Europe and America, the diaries meant little more than excitement and drama set on a totally foreign and exotic stage.

THE EUROPEAN SCRAMBLE FOR AFRICA

The idea of empire is about more than the control of natural resources; but as Freund (1984) has pointed out, it is undeniable that control over resources is a central part of the imperial equation. What were the causes that impelled Europeans to seek political and economic control of Africa? It did not happen overnight; they had to co-opt or destroy the old colonial powers of Africa, the Turks in North Africa, and

Omani Arabs in East Africa, as well as co-opt or defeat local African states that wished to stay in power. Ironically, the states that were able to resist European aggression were those that had been actively oppressing their people, usually through slaving, and had formidable military might.

Although many hypotheses have been put forward, it is difficult to ascribe the so-called scramble to any single cause. No one explanation seems satisfactory to all—perhaps rightly so since single-cause explanations seem grossly naïve in our complex and multifaceted world. What were the root causes of what historian Thomas Pakenham (1991) called "this undignified rush by the leaders of Europe to build empires in Africa"? A rich diversity of individuals were involved in the "scramble," each with his or her own motivations. Some Africans welcomed it, many opposed it, but almost all were transformed by it to varying degrees. The following reasons probably get to the heart of the keen European interest in occupying Africa.

1. The national rivalry between the English and the French. England and France had been competing for political and economic hegemony, having fought in North America and elsewhere on several occasions in the eighteenth and early nineteenth centuries. They continued to battle in India and elsewhere.
2. The desire to end the slave trade. Although the Atlantic slave trade was virtually dead, slaving still continued across the Sahara and in East Africa. Omani and Swahili slavers were pushing deep into central Africa, creating havoc in the process. Slavery was a central institution in African economies, and the slave trade continued.
3. The desire to extend Christianity.
4. The desire to relieve the suffering of the poor. Africans in general were impoverished by European standards of the nineteenth century, and there was agreement that trade, education, and freedom would help diminish individual poverty and generate wealth for the homeland.
5. National prestige. Belgium was a good example of the quest for overseas possessions to appear as a world power like Britain and France.
6. To gain control over and profit from natural resources.

Reasons

While exploration was still in progress in equatorial Africa, the struggle among European powers for possession of Africa's land—a struggle that was to lead to the colonial partition of virtually the entire continent—had already begun elsewhere. As early as 1857, France and Britain came to an agreement in which France recognized Britain's sovereignty over the Gambia River and its valley, while Britain consented to France's occupancy of the area around the Senegal River. Thus was the concept of "spheres of influence" born. The only territories to escape Europe's nineteenth-century invasion of Africa were coastal Sierra Leone, itself a British colony but established in the 1790s with the special purpose of providing a home for freed slaves, and Liberia, founded by American interests for a similar purpose in the 1820s. Even Ethiopia, the feudal empire that had survived the upheavals of Africa over several centuries, was eventually overrun in the twentieth century by the European state last in colonial expansion, Italy.

The French were probably the first to recognize the value of a continuous, interconnected empire in Africa; in any event, the French dream of "Africa French from Algeria to Congo" preceded the British ideal of "Africa British from the Cape to Cairo." Thus, France concentrated its efforts in West Africa and the Maghreb and made considerable headway at an early stage. Britain continued to view its African possessions as isolated stations rather than as a contiguous empire, and then France's expansionism was temporarily halted by the war with Germany. After its 1871 defeat in the Franco-Prussian War and some territorial loss in Europe, France, stimulated by the need for a revival of national pride and prestige, again focused its interest upon West Africa. By then Germany and especially Britain were also engaged in the struggle for land in this region.

Viewing with concern the spread of French power across the Sudan, the British decided to acquire the hinterland of their coastal trading stations, which, they feared, might be cut off from the interior upon which they depended for survival. Thus, Britain began to occupy the interior of Sierra Leone, the Gold Coast (now Ghana), and Nigeria. In Nigeria, the British sphere of influence was expanded by the Royal Niger Company, formed by British trading interests and supported by the British government. Its forerunner, the United Africa Company, established in 1879, was instrumental in containing the latest European power to enter with colonialist designs, Germany. Although individual land hunters had been active on behalf of the German state earlier, it was only in 1884 that Lomé on the Togoland coast was proclaimed German territory. A narrow strip of hinterland was also claimed, effectively separating the Gold Coast sphere of influence from the Nigeria region. Clearly, the major object of German claims there was the obstruction of the designs of the rival colonial powers.

In equatorial Africa, the Belgian sphere of influence was in the west, and during the first half of the nineteenth century Arab power continued to dominate the east. While Stanley was gathering treaties and concessions on behalf of the king of the Belgians, creating the Congo Free State, East Africa remained under African control in the interior and Arab hegemony in the littoral. Arab power centered on the island of Zanzibar, which had long been a focus of Arab activity (Figure 4-1). With the opening of the Suez Canal in 1869, the sultanate of Zanzibar attained unprecedented importance, and the sultan laid claim to large sections of the African coast. Although both Britain and Germany were somewhat interested in East Africa by this time, largely because of the reports of the explorers, the sultan's claims were recognized. Britain, less concerned with obtaining territory than with the elimination of the slave trade, sought good relations with the Zanzibar rulers to achieve this end. It was not until the 1880s that the British were finally successful in their efforts, and by then the sultanate had begun to crumble.

The interest shown by various European colonial powers in Africa was due to a great extent to the efforts of the explorers. The success of colonization must in large measure be attributed to individuals who could also be called explorers, though less in search of truth than of gain. Stanley, an explorer at first, became one such land hunter, seeking to establish treaties with chiefs whom he met in his travels. Some of the later "explorers" became as famous (or infamous) as their predecessors. For Germany, Robert Flegel in West Africa (especially the fringes of the Nigeria sphere of influence) and Karl Peters in East Africa obtained concessions from local chiefs. Pierre de Brazza worked for France in the region that was to become French Equatorial Africa. For Britain, Rhodes penetrated Zambezia, later to be named after him, and obtained vast concessions. It is due to these men more than any others that the spread of European influence in various parts of Africa was rapid once it began.

As the spheres of influence of the colonial powers expanded and rival claims were made to certain parts of the continent, it became clear that a discussion of Africa's colonial partition was necessary. There was real danger of open hostilities in some areas, and in others, the local chiefs had ceded their land more than once, first to the representatives of one European government, then to those of another. It was such concerns that prompted the Germans to convene the three-month conference in Berlin at which various colonial possessions were consolidated, problematic boundaries defined and delimited, and some sections of land exchanged. Rules for the "effective occupation" of the territorial claims were established, and by the end of the proceedings, in February 1885, some order had been brought out of political chaos. The story of East Africa subsequently, that is, during the last decades of the nineteenth century and the beginning of the twentieth, is one of rivalry and friction between Britain and Germany.

In the south of this region, Karl Peters, an emissary of the German Colonization Society, traveled through what is today Tanzania, obtaining treaties and concessions from African chiefs. Even chiefs who were already under the jurisdiction of the Zanzibar sultanate gave Peters concessions, so that by 1885 he had claimed for Germany most of the area of Tanganyika (now Tanzania). The sultan of Zanzibar objected to claims by Peters of land belonging to the sultanate, but Germany responded by sending a fleet to support them. Although the threat of naval bombardment forced the sultan to yield, there were uprisings by both Arabs and Africans in Tanganyika itself, and some years were to elapse before German power was undisputed. Some of the Arabs and Swahili fought because they feared that the slave trade would be ended by German occupation, others because of nationalism. Moreover, internal centrifugal forces had been building up (Glassman 1994), and later revolutions were sustained by Africans who had been affected by slavery but never by actual territorial subversion. Among the African peoples, the Hehe distinguished themselves by courage and perseverance. The last uprising was the famous "Maji-Maji" rebellion of 1905–1906, involving most of southern Tanganyika, which was put down with great bloodshed. Germany's activities in East Africa and South West Africa (now Namibia), which included a bloody campaign against the Herero, a Bantu people, are among the darkest chapters of European history in Africa, rivaling the atrocities committed in the Congo and Portuguese and Arab terror in sustaining the slave trade.

While Tanganyika was falling into German control, Britain was engaged in the colonization of other areas. In 1888 the British East Africa Company was created, with aims similar to those of the Royal Niger Company but with substantially less capital. Leaders of the company were aware that the Baganda people were likely to play a dominant role in a developing East Africa, and a railroad was begun to connect Mombasa, on the Kenya coast, with Kisumu, on the shores of Lake Victoria. In 1890 the failing sultanate of Zanzibar became a British protectorate, which included a strip on the Kenya coast but not the portion that had been claimed by Germany. Thanks to the efforts of the East Africa Company, this territory in 1893 also became a part of the British sphere of influence as a protectorate.

The British East Africa Company, unlike the British South Africa Company and the Royal Niger Company, did not thrive on rich mineral finds and agricultural development. Burdened with a multitude of administrative functions, which London was reluctant to take over, the company was frequently in financial difficulties, as when it was building the railroad to Uganda. Indeed, its charter was terminated in 1895, eight years before the railroad reached the shores of Lake Victoria. At the turn of the century, nevertheless, a British protectorate existed over Uganda, Kenya, and Zanzibar, and the company had made a major contribution in bringing this about.

In southern Africa, the events leading to the formation of the Union of South Africa had occupied center stage to such an extent that the huge territory of South West Africa was virtually neglected until German claims to it were substantiated by armed force. A war had long been in progress between the Khoisan speakers and the Herero, and the lives of European missionaries and traders were endangered. Appeals for protection from London or Cape Town were unavailing, but a number of people of German nationality were living in South West Africa, and the German rulers took a sympathetic interest in the plight of their fellow Europeans. In the 1880s, German claims in the region were expanded through the efforts of Adolph Lüderitz, and the appearance of German ships off the shores of southwest Africa removed all doubt that Germany's presence was to be permanent. At the 1884–1885 Berlin Conference, the Germans insisted upon connecting their territory with the Zambezi River, and thus the Caprivi Strip came into being, extending eastward from the northern edge of the main body of German South-West Africa.

One of the regions whose ownership was not settled was Zambezia, the area around the great Zambezi River, lying between Angola and Mozambique, north of the Transvaal and south of the Congo and Tanganyika. In 1886 France and Germany appear to have agreed that Portugal should be allowed to extend its possessions in coastal Angola and Mozambique across the entire continent in order to link these two portions of empire. This agreement was less an act of friendship toward Portugal than an effort to obstruct British imperialism in the south, but Cecil Rhodes and his supporters helped foil this plan. The Pioneer Column, a vanguard of white settlers equipped by Rhodes, penetrated Matabeleland in 1890 and began a white immigration that was to reach sizable proportions. The British South Africa Company obtained concessions from African chiefs, and Portuguese expansion into the interior was limited by agreement.

In the northeast, including the Horn of Africa and the Sudan, the center of the historicogeographical stage during the last decades of the nineteenth century was occupied by Britain and several powerful local rulers. Also interested in this area were France and Italy. France wished to extend its West African domain to the Red Sea and the Gulf of Aden, and Italy, having occupied coastal sections of Eritrea and Somaliland, desired the Abyssinian (Ethiopian) interior. Britain's protectorate over Egypt helped lead to British subjugation of a region that had long been the object of Egyptian expansionism, the Sudan.

The process of colonization in the northeast was only partially completed. Ethiopia was consolidated under a powerful leader, Menelik II, who was also a skillful negotiator and military tactician. While the colonial powers were encroaching upon this part of Africa, Menelik himself embarked on a program of Ethiopian expansionism, and the boundaries he pushed far beyond the limits of the Ethiopian Plateau were eventually recognized. He defeated the Italians in the Battle Adowa in 1896 and laid the foundations of the Ethiopian state. In the Sudan, joint British–Egyptian control was interrupted in 1881 by the Mahdist revolt, led by Muslim religious leader Muhammad Ahmad. In 1898, however, Kitchener defeated the remainder of the Dervish army. Immediately afterward, French encroachment on the Sudan was repudiated by a show of force, and joint British–Egyptian government of the region took effect once more.

COLONIAL POLICIES

The European nations partitioned Africa primarily to ensure that they would not be excluded from areas that might prove valuable in the future. Few areas were expected to produce immediate wealth, however, and pace of "development" was slow. The colonial powers believed that their primary responsibility was to maintain law and order at minimum cost to the European taxpayer. Thus the first two decades of colonial rule were characterized by the definition and delimitation of administrative boundaries (Figure 4-4), the establishment of government machinery, the immigration of administrators, missionaries, and settlers, and the building of railways and other strategic services that later facilitated the exploitation and development of Africa's resources and peoples. Each colonial power approached these and other issues from its own particular political and cultural perspective. Their policies changed through time and differed from place to place. These are highlighted in the paragraphs that follow and are developed more fully in subsequent chapters.

Germany

Germany's colonial tenure in Africa was short-lived and unsuccessful: what Germany acquired by conquest and treaty after 1885, it lost to its conquerors following World War I. In West Africa the Germans established themselves in Togoland, a narrow finger of land east of the Gold Coast, and in Cameroon, where they pushed far northward and came close to fragmenting the extensive French West African realm. In East Africa they obtained a part of what is today Tanzania, thus dealing a blow to British intentions of a Cape-to-Cairo axis, while in southern Africa, Germany acquired the desert lands of South-West Africa, today called Namibia.

Germany's economic and political interests in Africa were pursued with a fervor unmatched by other colonial powers. In South-West Africa, Germany conducted a cruel and oppressive campaign against the Herero people, while its infrastructural investments in Cameroon and German East Africa made these two territories profitable possessions by 1914. Germans were encouraged by their government and colonial societies to settle the country's African territories; land was set aside for them, particularly in areas destined to be opened up by railways. These included highland areas in southern Cameroon and in the vicinity of Mount Kilimanjaro. A railway was built from Dar es Salaam to Lake Tanganyika to compete with British interests, while in both Togoland and Cameroon powerful monopoly companies were chartered to operate in and develop each protectorate. Nevertheless, these firms suffered from insufficient capital and poor organization. Full political power, even at the local level, was in the hands of German colonial authorities, and back in Europe the Reichstag believed that the colonial governments should be self-supporting. Consequently, development grants were rare, and revenues were derived primarily from head taxes, import duties, and the sale of commercial franchises.

Following World War I, Germany's colonies were divided among the victor nations according to principles established by the League of Nations. South-West Africa was entrusted to the Union of South Africa as a class C mandate, which meant that it would be administered as an integral part of South Africa. Cameroon was similarly partitioned, the smaller British section being administered with Nigeria, and the larger share being made a mandate of France. Most of German East Africa became a class B mandate of Britain under the name Tanganyika, while Belgium took charge of a small mountainous section consisting of Ruanda and Urundi (now Rwanda and Burundi). The countries undertaking the task of administrator agreed to govern their new territories until such time as they were "able to stand on their own feet in the strenuous conditions of the modern world" (Oliver and Atmore 1967).

Belgium

Apart from the mandate and trusteeship over Ruanda–Urundi, the Belgians possessed only one colony: the Congo, once the personal property of

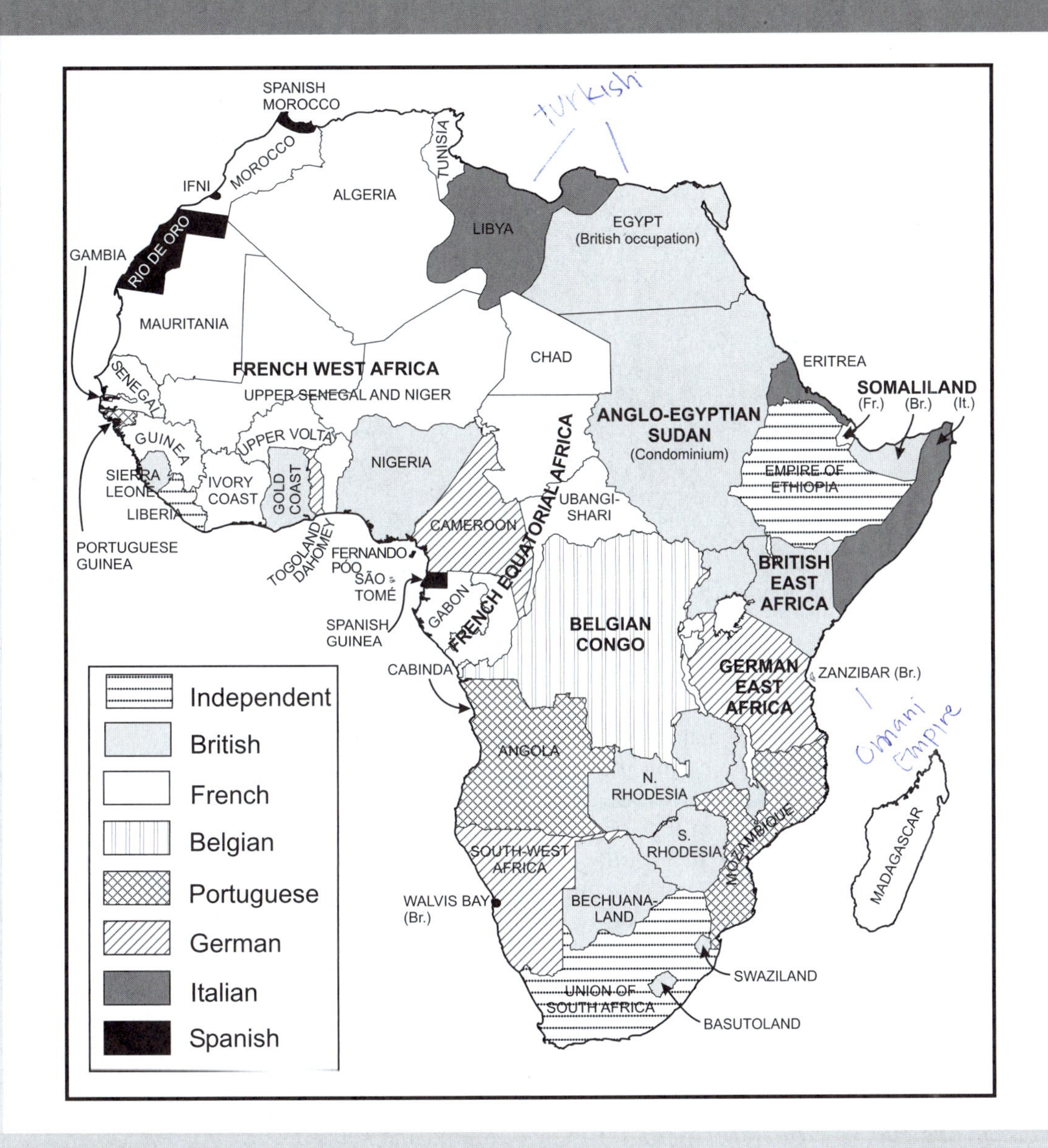

FIGURE 4-4 Africa after the "scramble," 1912.
Adapted from Barraclough (1993), Pakenham (1991).

Leopold II. In 1907, following reports of merciless exploitation of Congolese labor by Belgians and others to meet quotas of ivory and rubber (Figure 4-5), the Congo's administration was taken over by the Belgian state. Belgium saw its colonial task as a paternal one. It never viewed the Congo as an "overseas Belgium," and it never intended to make Belgians of the Congolese people.

Belgian colonial policies dictated that decisions regarding the Congo be made only in Brussels. In the Congo itself, neither the European representatives of the Belgian government nor the Belgian settlers

FIGURE 4-5 The brutality of King Leopold's state is captured in this photograph acquired from an abolitionist group.
From Anti-Slavery International in Pakenham (1991).

possessed any political rights or voice in the fate of the territory. Neither, until a matter of months before independence, did the local Congolese majority. Paternalism extended also into the economic, educational, and social spheres. Much progress was made in elementary education, but higher education for Africans was virtually nonexistent, and when independence came (1960) the new state had only 30 Congolese university graduates. A broad elementary education was preferred to one that produced a highly educated African elite, and even the small group of *évolués* (middle-class "evolved" Africans) could not aspire to Belgian nationality and equality. Rather, they were viewed as in need of paternal restrictions.

Belgium, not the Congo, benefited most from the Congo's output of industrial diamonds, copper, cobalt, and forest products. While Belgian financial interest groups thrived in the Congo, the colonial administration struggled, usually with inadequate resources, to overcome the internal diversity of the country. In the tripartite power structure of the colonial government, the Roman Catholic Church and mining–industrial concerns each had separate goals, but paternalism was the common denominator. The transport system was designed for the coastward transit of interior products. The churches and missions, while accomplishing much in primary education and medicine, strengthened the image of the European as the "father" of the African masses. And under the colonial administration, little attempt was made to produce a homogeneous people in a unified state.

To administer its policies in a land 80 times its own size, Belgium created six major provinces, each with a capital, and each subdivided into districts. The provincial capitals became places of power under lieutenant governors, and the provinces in some ways became separate colonies, with Léopoldville (now Kinshasa), the colonial capital, being both physically and psychologically removed. Following independence, the provincial capitals themselves became centers of control for secessionist leaders who failed to heed Léopoldville's dictates.

France

France possessed the most extensive colonial empire in Africa. It extended in a continuous block from Tunisia, Algeria, and Morocco across the Sahara to the Guinea Coast and south to the Congo Basin. Together with the two outliers of Madagascar and French Somalia, this empire encompassed, at the height of French power, more than 4 million square miles (10.4 million km^2)

FIGURE 4-6 French-financed floodgate on Niger River for rice irrigation.
Roy Cole.

and 65 million people. A great diversity of peoples, institutions, and environments comprised this empire, yet France imposed a remarkably uniform colonial policy. Colonial administration was to be financially self-supporting, and the colonies themselves were to provide France with raw materials and with markets for French manufactured goods. Although taxes were not new in Africa, the French imposed a head tax to finance the administration of French West Africa (Mauritania, Senegal, Upper Volta, Niger, the Ivory Coast, French Guinea, Dahomey, and French Sudan). To pay the tax, the Africans sold a variety of locally produced items. Pastoralists and other livestock owners sold small ruminants like sheep or goats; farmers sold food crops (millet, sorghum, and yams/cassava) and cash crops (groundnuts, rice, cotton) that had been introduced by the French (Figures 4-6 and 4-7).

Africans also provided labor (*la corvée*) and services required by the French. Large companies having commercial monopolies over extensive areas, and the power to use local labor as they chose, generated much of the capital used by the colonial governments in French Equatorial Africa (Chad, Ubangi-Shari, Gabon, and Moyen Congo).

France's ultimate objective in Africa was the assimilation of Africans into French culture through the adoption of the French language and education system. Unlike the Belgians, the French desired the quick development of an educated, acculturated elite, which would have French interests at heart and French culture to boast. This elite, it was reasoned, would support the French presence in the empire as a matter of self-protection, for by accepting and adopting French values and ways of life, educated Africans often separated themselves from their own countrymen and women, most of whom remained bound to local tradition. French colonial subjects, then, were to be assimilated in the greater French empire, and they could and did obtain a voice in the politics of the French realm through representation in Paris.

Important changes in French colonial policy came with the passage of the *Loi Cadre* (Outline Law) of 1956. France kept control of foreign policy, defense, and overall economic development of the French

FIGURE 4-7 New crops: onions, tomatoes, sweet potatoes, tobacco. *Roy Cole.*

territories, but all other aspects of government became the responsibility, not of the existing federal governments of French West and French Equatorial Africa, but of the 12 individual colonies of which they were composed. Two years later, the concept of a *France d'Outre Mer*, an overseas France, was largely abandoned when President Charles de Gaulle established a new French constitution and offered the colonial peoples the choice between autonomy (self-government) as separate republics within a "French community" and immediate independence, with the severance of all links with France. Only Guinea chose complete independence outside the community, while the others chose autonomy buttressed by continuing French economic aid. By 1960, the other territories, too, had received their independence while retaining close economic ties with France.

Throughout the post–World War II period, Algeria occupied a very special position in the decolonization program. Because Algeria had almost one million European settlers *(colons),* as well as rich deposits of iron ore, oil, and natural gas, and proximity to and effective communications with the metropole, France was reluctant to withdraw. Following a period of uprising, turmoil, and eventually 8 years of war, nationalist demands were met, and Algeria was granted its independence in 1962. Morocco and Tunisia were neither territories nor parts of France; they were protectorates, which meant that French policy attempted to combine French interests with those of the local people. No overt attempt was made to replace traditional institutions with those of France. Men and women were encouraged to emigrate from France and settle as *colons,* and much of the economic development that took place under French rule may be attributed to these Europeans, although their presence caused political and social difficulties. They not only occupied good land, but they competed for jobs with the native-born Muslim peoples.

United Kingdom

Britain, like France, was responsible for a multiplicity of peoples and cultures operating under an incredible diversity of environmental conditions and traditional systems; but, unlike the French, the British formulated policies that were more often adapted to the specific requirements of the individual dependencies. The basic premise was indirect rule, and flexibility was an essential ingredient. The principle of indirect rule was intended to prevent the destruction of indigenous culture and organization and to facilitate effective British control. In such fields as tribal authority, law, and education, the British often recognized local customs and permitted their perpetuation, initially outlawing only those practices that constituted, in British eyes, serious transgressions of human rights. Then the people were slowly introduced to the changes British rule inevitably brought. There was rarely any sense of urgency and need for radical change.

Mamdani (1996) views the genesis of indirect rule as solidly rooted in the British experience in South Africa as an attempt at separation of the races. On the other hand, it can be argued that the policy developed even earlier than Mamdani suggests: for example, in North America with the so-called British Proclamation Line, established in 1763 to divide the area where Europeans could settle from "Indian country," where they could not. Nevertheless, Britain professed from almost the beginning that the ultimate goal of British colonial control was the independence and self-determination of the Africans under the colonial flag. This was a major difference between British policy and the policies of all other European powers, which were either directed at converting the colonies into integral parts of the metropoles or bent on preserving the status quo indefinitely. British policy was strongly influenced not only by traditional systems, but by the size and character of the European settler populations. In Kenya and Rhodesia, where the settler populations at their maximum were large and powerful (70,000 and 225,000, respectively), the desires of London were often overridden by those of the Europeans living in Africa. But in the Sudan, Ghana, Nigeria, and elsewhere, European colonists presented fewer obstacles to either the administration from the parent country or to the path toward independence on which the Africans would embark.

British policy also differed according to the political status of the dependencies. Territories that were conquered and settled by European immigrant groups became colonies, such as Kenya, implying a considerable amount of self-determination from the settler population. Territories like Bechuanaland (Botswana), whose indigenous leaders had requested and been granted Crown "protection," became protectorates. While the principle of indirect rule in the European-controlled colonies was mainly replaced by control by local Europeans on their descendants, the principle was adhered to quite strictly in any territory that had been granted protectorate status. Southern Rhodesia, by virtue of conquest from the south, became a British colony, while Northern Rhodesia (Zambia), somewhat more isolated by the Zambezi River, became a protectorate. In Nigeria, two different kinds of administrative control prevailed simultaneously: southern Nigeria, which had a long history of contact with Britain, became a colony, while northern Nigeria remained under indirect rule. Indirect rule fostered a greater sense of individuality among the various colonies than was possible under the French regime; and thus not unexpectedly, the timing and consequences of independence differed. We shall have more to say about this in subsequent chapters.

The geographical contiguity of the three East African possessions (Kenya, Uganda, and Tanganyika) permitted closer economic, monetary, and customs links than in West Africa where British territories were geographically separated from one another. The three landlocked territories of Swaziland, Bechuanaland, and Basutoland, were joined in a Customs Union with South Africa (1910), which drew them economically into the South African sphere of influence, while their political administrations remained separate from South Africa's. Not all the possessions were economic assets (Bechuanaland and Lesotho, for example), while others, including Kenya and Ghana, yielded valuable agricultural products and minerals required in Britain.

Portugal

Portuguese rule in Africa came to an end in 1975, after more than four centuries of neglect and exploitation by Lisbon, and after two decades of struggle for liberation in the territories themselves. Portugal looked on Mozambique, Angola (including Cabinda), Portuguese Guinea, and the offshore islands of São Tomé, Principe, and the Cape Verde Islands, as reservoirs of human and natural resources to be exploited for the benefit of the metropole. Little or no concern was shown for the welfare of the peoples themselves. Angola and Mozambique in particular were viewed as richly endowed but sparsely populated lands that would take care of Portugal's excess population and provide the fragile Portuguese economy with much-needed revenues, foods, and industrial resources. By the early 1970s there were over 400,000 Portuguese settlers in Angola and 150,000 in Mozambique, and large-scale agricultural and mining projects remained firmly in European hands. The widespread poverty and the social and political backwardness that characterized these regions merely reflected the state of Portugal itself, the poorest of Africa's rulers.

Portugal adhered to the legal position that its African possessions were, in fact, "provinces" of the metropolitan Portuguese state whose residents could aspire to Portuguese citizenship and to representation in the government in Lisbon. Like the French, the Portuguese encouraged a small elitist class who could gain virtual equality with the European settlers, provided a number of educational, economic, and religious qualifications were met. Only a very small number (perhaps 70,000 out of a total population of 12 million) became *assimilados,* but those who did were accepted by the Portuguese citizenry in a manner unique in southern Africa. The vast majority of the population were *indigenas* (natives), whose main function in the eyes of the administration was to provide labor for the mines, agricultural schemes, and other government projects.

Control over the populations and local economies was maintained by dictatorial means just as at home in Portugal. The "provinces" were subdivided into a number of districts, for administrative convenience in implementing decisions made in Portugal and transmitted through governors general situated in the capital cities. The provinces were also divided into numerous economic units, or circumscriptions, controlled by administrators and assisted by "chiefs of post" who were given extraordinary powers over the essentially rural populations. The administrators had absolute jurisdiction over the amount of land to be placed under cultivation, the types of crops to be grown, and the prices to be paid for the crops; in addition, they formulated and enforced the labor laws that guaranteed the state a permanent and low-paid workforce. The system fragmented an already tribally heterogeneous population, prevented effective communication between groups and regions, seriously hindered the emergence and development of African nationalism, and left the Africans poorly prepared for an independence never foreseen in Portugal.

Italy

Italy was the last of the European states to participate in the "scramble for Africa," primarily because of the late unification of Italy itself. In 1883 Italian troops occupied part of the Eritrean coast of the Red Sea, and soon after Italy laid claim to the eastern Somali coastline, then part of the domain of the sultan of Zanzibar. In 1889 Italy signed the Treaty of Wishale, which defined the boundary between Ethiopia and Italian Eritrea, and in 1896 Italy tried unsuccessfully to occupy Ethiopia. It failed to capitalize on its position in Eritrea, and Italian commercial companies in Somaliland were bankrupt by 1904.

Italy turned next to Libya, ousting the Turks in 1912, and for the next 27 years Italian occupation became more militaristic and, under Mussolini, more fascistic. An even more aggressive attitude was taken toward what in 1936 became Italian East Africa—composed of Ethiopia, Eritrea, and Somaliland. In Ethiopia, the Italians immediately set about improving the communications system, realizing that development depended on it and that good communications would strengthen their political control. Indeed, it was Italian-built roads from Massawa (Mitsiwa) in Eritrea and Mogadishu in Somaliland, that facilitated Italy's occupation shortly before the out break of World War II. In Somaliland, the Italian interests focused on government-sponsored agricultural schemes geared to Italian markets, but little attempt was made to improve the education levels, health standards, economic well-being, and political awareness of the majority.

Spain

Of all the colonial possessions in Africa, Spain's were the smallest in area and population. They also derived little benefit from Spain despite a long history of Spanish control. For more than four centuries, parts of the northwest coast belonged to Spain, and even today there are city-sized colonies (Ceuta and Melilla), or *presidios*. The largest Spanish possession was Spanish Sahara, once a protectorate and then an overseas province of Spain until its incorporation into Morocco and Mauritania in 1976. The status of the former Spanish Sahara (today's Western Sahara) as an independent country is in dispute, Morocco claiming the entire area. Spanish Sahara and the other coastal outliers were poor and treated as integral parts of Spain. Further south, wedged between Gabon and Cameroon, was Spanish Guinea, composed of mainland Rio Muni and the island of Fernando Póo (now Bioko). It provided Spain with forestry products, cocoa, coffee, and other tropical products, and until independence (1968), its administration (like that of Spain itself) was highly centralized and autocratic.

THE "WIND OF CHANGE"

The "scramble for Africa" was completed in less than two decades after 1884–1885, the period of the division of territories among six European nations at the Berlin Conference. For the next half-century, the European powers consolidated their positions, formulated their policies, and in general reaped the rewards of their investments. In an equally short time, beginning soon after the close of World War II, the colonial powers withdrew from Africa, and their colonies and possessions became independent states (Figure 4-8). With the exception of Namibia, the French Territory of the Afars and Issas, the Spanish *presidios,* and Rhodesia, independence was completed with remarkable speed and relative calm. Few political observers foresaw this rapid transformation from colonialism to independence, least of all the colonial governments themselves. The Portuguese especially, until 1974, saw their African possessions as permanent and integral parts of European Portugal. There were many, even among the African intelligentsia, who felt that they were not yet ready for independence (Gann and Duignan 1967). This "wind of change," so named by a British prime minister, Harold Macmillan, began in northern Africa with the independence of Libya (1951) and then swept south, fanned by an increasing awareness of African nationalism and the desire for self-determination. It was ushered in by the declining strength of the European powers following World War II, and the shift in world power to two professedly anticolonial nations—the United States and the Soviet Union. Both the United States and the Soviet Union were competing for the hearts and minds of the colonized millions of the world. It was a global geopolitical struggle with profound consequences for all. In 1945 only four independent states existed in Africa: Ethiopia, Liberia, Egypt, and South Africa, and of these both Ethiopia and Egypt had seen colonial rule in the twentieth century, while South Africa was created in 1910 out of two Boer republics and two British colonies without approval of the African majority. Following the independence of Libya, the Sudan was liberated from Britain and Egypt in 1956, and later in the same year France withdrew its administrations from Morocco and Tunisia.

The second wave of the wind of change began in West Africa with the independence in 1957 of Ghana (formerly the Gold Coast), and quickly spread east and south. In 1958 Guinea became the first black African territory to gain its independence from France, and in 1960 all of former French West Africa and French Equatorial Africa made the transition. Of the 18 independent states created in the world in 1960, 17 were in Africa, and 14 were formerly French. In the same year Britain relinquished Nigeria, the Italian trusteeship over Somalia was terminated, and the Belgians withdrew from the Congo. In most cases the transfer of authority was completed without major incidents, but in the Congo separatist movements vied for control, and the economy and administration collapsed.

Sierra Leone received its independence from Britain in 1961, as did Tanganyika (now Tanzania). After almost eight years of civil war, Algeria received its independence in 1962, and during the same year, independence was granted to Ruanda–Urundi and to Uganda. The short-lived Central African Federation of Northern Rhodesia, Southern Rhodesia, and Nyasaland was dissolved, and in 1964, and Nyasaland became the independent state of Malawi. The following year Tanganyika and Zanzibar joined to form the United Republic of Tanzania, and the British protectorate of Northern Rhodesia became the independent Republic of Zambia.

Thus by 1964 the wind of change had reached the Zambezi River, there to be checked temporarily by colonial interests of Britain and Portugal. Further south lay white-ruled South Africa and its mandated territory of South-West Africa, and the three British High Commission Territories of Bechuanaland, Basutoland, and Swaziland. In 1966 Bechuanaland became Botswana and Basutoland became Lesotho; two years later Swaziland achieved independence and chose to retain its old name. The ruling white minority of the self-governing colony of Southern Rhodesia made a unilateral declaration of independence (UDI) in 1965 and renamed the territory Rhodesia. It would take civil war and fifteen long years of struggle for the white minority government in Rhodesia to see the handwriting on the wall. In 1980 Rhodesia was renamed Zimbabwe, and a black majority government took over.

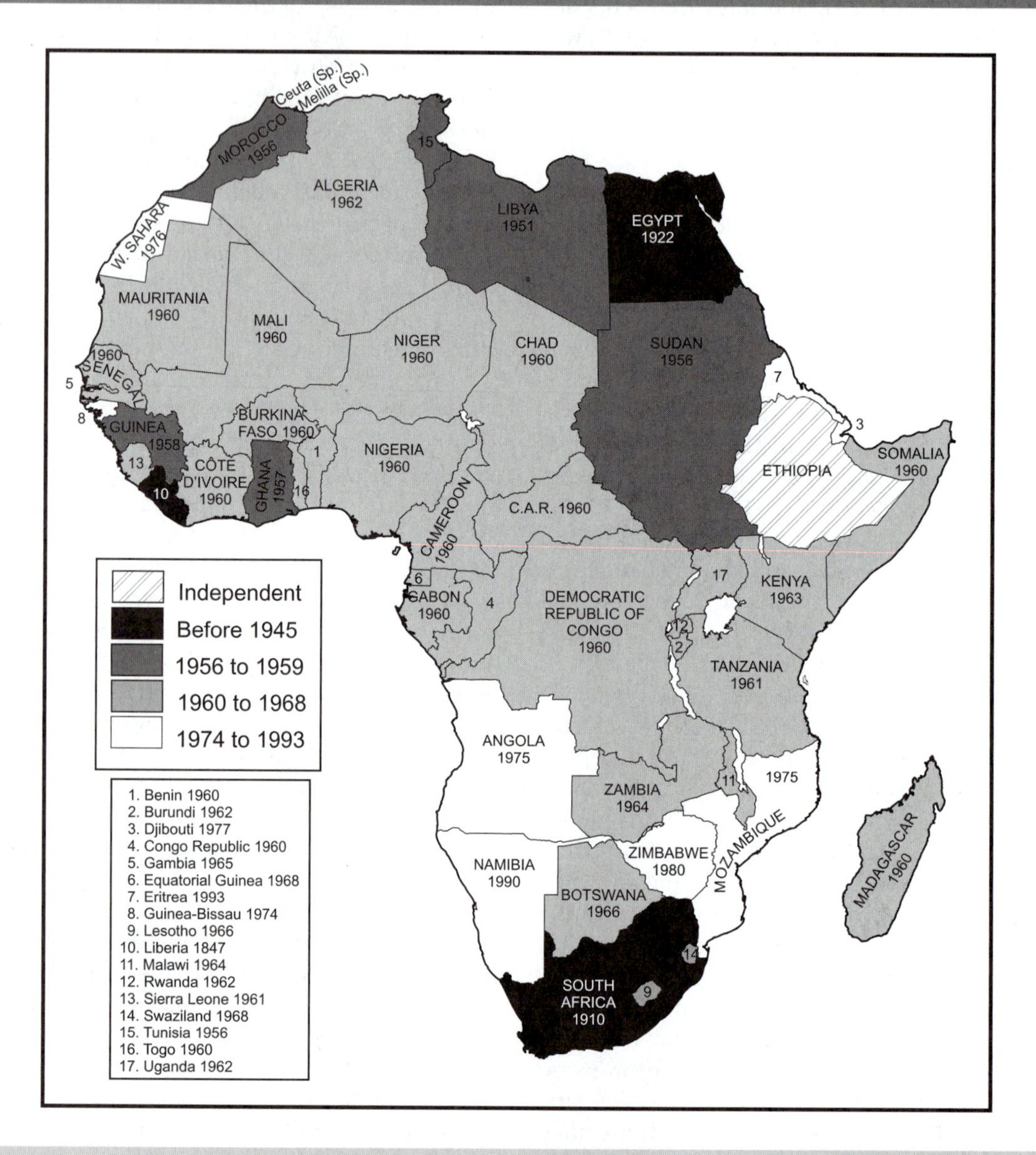

FIGURE 4-8 African states in 2004 and dates of independence.
Roy Cole.

After years of revolutionary warfare by liberation units in the Portuguese possessions, Guinea-Bissau became independent in 1974; Mozambique, São Tomé and Principe, the Cape Verde Islands, and Angola followed in 1975.

Although South Africa has been independent since 1910, it continued for decades to be ruled by a white minority government with a policy of racial separation. This policy of apartheid was institutionalized in 1948, when the racist National Party took control of the government. The exclusionary policy led to civil disobedience, rebellion, violence, and, ultimately international sanctions and a pariah status for South Africa. In 1990 the apartheid government began to make a transition to majority rule, and in 1994 elections were held that were open for the first time to the black population. Nelson Mandela, long a political prisoner on the notorious Robbin Island, was elected the first black president of South Africa.

African independence was brought about by many forces emanating from within and beyond the continent. Local nationalist groups emerged in opposition to colonial injustices and were championed by articulate leaders; in some cases, liberation movements resorted to violence. The wind of change was quickened by effective debate in the United Nations, where the newly independent states took their places to argue for the self-determination of all Africans. Supranational organizations, especially the Organization of African Unity (OAU) and the Arab League, became effective platforms for anticolonialism and pro-African ideals. Colonialism's impress on the landscape is unmistakable and not erasable: the roads and railways, mines and factories, cities and rural development schemes, schools and churches, all begun within a colonial framework, are there to stay. So too are the European languages, political institutions and boundaries, and legal and educational systems, besides many far less tangible legacies. These are the subjects of the following chapters.

CHAPTER REVIEW QUESTIONS

1. What was the "scramble for Africa"?
2. Why were European countries interested in occupying Africa?
3. What would be some differences between European colonialism and, say, Omani or Turkish colonialism in Africa?
4. What sorts of change did the Europeans impose on their African possessions?
5. List and discuss at least three negative impacts of European colonialism in Africa.
6. List and discuss at least three positive impacts of European colonialism.
7. Give several examples and discuss the African response to the Europeans' attempts to take control?
8. Compare and contrast the British and French colonial policies. In what ways were the Belgians and Portuguese different from the British and French?

GLOSSARY

Apartheid The one-time South African policy of separate development of the races.

Diamondiferous Rich in diamonds.

La corvée Forced labor.

Ruminants Animals that chew the cud, or "ruminate."

Trading diaspora A group of people (an ethnic group) geographically dispersed in another group's country or territory for the purposes of trade.

BIBLIOGRAPHY

Barraclough, G., ed. 1993. *The Times Atlas of World History,* 4th ed. Maplewood, N. J.: Hammond.

Colvin, P. 1998. Muhammad Ali Pasha, the Great Exhibition of 1851, and the School of Oriental and African Studies Library. *Libraries and Culture,* 33(3).

Driver, F. 2001. *Geography Militant.* Oxford: Blackwell.

Freund, B. 1984. *The Making of Contemporary Africa: The Development of African Society Since 1800.* Bloomington: Indiana University Press.

Gann, L. H., and P. Duignan, 1967. *Burden of Empire: An Appraisal of Western Colonialism in Africa South of the Sahara.* Stanford, Calif.: Hoover Institution Press.

Glassman, J. 1994. *Feasts and Riot: Revelry, Rebellion, and Popular Consciousness on the Swahili Coast: 1856–1888.* Portsmouth, N.H.: Heinemann.

Lester, A. 2000. Historical Geography of South Africa. In *The Geography of South Africa in a Changing World.* R. Fox and K. Rowntree, eds. Oxford: Oxford University Press, pp. 60–86.

Mamdani, M. 1996. *Citizen and Subject: Contemporary Africa and The legacy of Late Colonialism.* Princeton, N. J.: Princeton University Press.

Oliver, R., and A. Atmore. 1967. *Africa Since 1800.* Cambridge, U.K.: Cambridge University Press.

Pakenham, T. 1991. *The Scramble for Africa: White Man's Conquest of the Dark Continent.* New York: Avon Books.

Shillington, K. 1989. *History of Africa.* London: Macmillan.

CHAPTER 5

The Contemporary Map of Africa

When the states of Africa gained their independence, they inherited a number of important political and economic institutions from their former colonial powers, which have, of necessity, provided the basis of operation and development in the postcolonial era. While there have been changes in response to changing political and economic conditions both within and beyond the various state boundaries, the fundamental economic patterns and, to a lesser extent, the political institutions, have remained essentially the same. When the Europeans left, their places were taken by African elites who had been educated in the colonial schools and sometimes in the metropole. That there was little institutional change is not unexpected in view of the communications systems, administrative networks, educational systems, economic frameworks, and political structures either implanted or fostered by the colonial governments in their half-century or more of tenure. All of these contemporary systems operate within yet another set of inherited phenomena, the political boundaries established in the colonial period; and since the boundaries were drawn to satisfy colonial raisons d'être, they may not necessarily satisfy the needs and desires of the respective independent states of today. Furthermore, the nature of the colonial state in Africa left its stamp on the states that began to develop after independence.

This chapter briefly reviews several essential elements of the state, as well as some that were partly prescribed during the colonial era, and then examines the ways in which the independent states have attempted to overcome some of these limitations to better meet their own needs. The principal politicogeographic components of the state reviewed are boundaries; the size, shape, and location of states; capital cities and core regions; and political ideals, systems, and supranational organizations. During the colonial period, the map of Africa showed broad regional power blocs, which have since disappeared. Nevertheless, they have been replaced by other power blocs, some political, others cultural–ideological, and still others economic. Let us briefly look at some characteristics of a state. These will be revisited later.

ATTRIBUTES OF THE STATE

The modern state is a formally bounded geographic space or entity in which power is exercised over a population that resides within those bounds. The state possesses characteristics that, taken together, distinguish it from other types of political organization. These attributes, or imperatives, according to Young (1997), may be thought of as requirements for existence; they vary in quantity from state to state and include hegemony, autonomy, security, legitimacy, revenue, and accumulation. These requirements were just as necessary for precolonial African states, from the petty kingdom to the vast empire, as they are necessary in contemporary Africa today.

The state continually asserts its authority and control (hegemony) by controlling the mechanisms of coercion and violence: the police, the court systems, and the military. Although the state endeavors to exercise power within its entire territory, it is sometimes constrained in so doing. Brauer (1995) and Herbst (2000) describe ancient, medieval, modern and African states, in which they found a core of power, usually the state capital, in which the state is strong; outside that core, power diminishes in relation to distance from the core (see Figure 3-6). The authority of the state may be internal, exercised over the subjects of the state, or external, exercised internationally. The elaboration and application is essential to the maintenance of a state's hegemony.

> Law is crucial to the bonding of the state and civil society. Public law defines the relationships between the two and sets rules to which the state itself is subject. The criminal code establishes boundaries of acceptable behavior. Civil law brings a wide array of private transactions under state regulation, creating in the process rights in property. A host of subsidiary administrative rules complete the finely spun web of law in which civil society is enmeshed. (Young 1997: 32)

Formal law makes the relationship between civil society and the state more formal and procedural rather than personal or patrimonial. Civil society refers to groups of citizens that are organized to represent themselves and counter the power of the state. Civil society is "the capacity of social groups to come together in order to organize politically, above and beyond existing sociological, religious or other cleavages. . . . [C]ivil society [amounts] to the creation of social networks distinct from the state and capable of transcending primordial family, kin or even communal ties" (Chabal and Daloz 1999: 19). Autonomy, a second characteristic, has two faces for a state: external sovereignty and independence from claims by other states to act within its national territory. A third characteristic is security. A state requires security, both external and internal, to prosper. Legitimacy is a fourth requirement for a state to exist. A state is said to be "legitimate" when it is recognized by its population as the lawful and genuine government. A government or ruler without the support of the people is said to lack legitimacy. A state seeks to legitimize its rule in a number of ways: a unifying ideology such as religion, distributing wealth, etc. A fourth requirement of a state is revenue. Young calls this the "bedrock postulate of state behavior" (1997: 38). The state has a need and perhaps a right to demand resources from its population, but the relationship between a state's calls for revenue and its legitimacy is inverse. Likewise, the state must accumulate wealth and resources to survive and grow.

BOUNDARIES AND TERRITORIES

Many of the international boundaries of contemporary Africa, and especially Africa south of the Sahara, were defined during the Berlin Conference of 1884–1885 by land-hungry governments without real knowledge of, or concern for, the peoples and regions being divided. At the time, Europeans had only superficial knowledge gained from records of missionaries, explorers, and traders about Africa's rivers, natural divides, resources, and cultural patterns, and the territorial extent of traditional political systems. Nevertheless, boundary lines were defined by treaty, demarcated on maps, and later, in some cases, delimited in the landscape. Where practically no information existed about the region, such as the Sahara and the interior regions of southern Africa, geometrical boundaries following lines of longitude, latitude, and directional compass bearings were superimposed on the map. Most traversed sparsely populated regions, and most were drawn as a matter of expediency.

The boundaries separating Somalia from Kenya and Ethiopia (Figure 5-1) were drawn without concern for the local pastoralists, who for generations had freely migrated from highland to lowland across the region. The straight-line boundary between Angola and the Democratic Republic of the Congo was superimposed on the Bakongo peoples, making some subject to Portuguese rule, and others to Belgian. The initial consequence was local population redistribution, accompanied by social-political unrest. The pattern was repeated literally hundreds of times across Africa, and the consequences were similar. For example, the Gambia, the most artificial of all African states, is bounded by a series of arcs, drawn with the compass at regular intervals along the Gambia River; two parallel lines were extended from near the mouth of the river to connect with these arcs. Thus the state of Gambia in no way corresponds with the natural hinterland of the river, which includes much of the surrounding state of Senegal, a former French possession.

Physical features including rivers, mountain ranges, escarpments, and lakes became international boundaries primarily because they were readily identifiable and rarely needed delimiting. Rivers were seen as natural dividers between peoples and regions, whereas in fact, rivers can be integrating or binding elements of

FIGURE 5-1 Contemporary African states.

the landscape and thus very unsatisfactory boundaries. Some of Africa's large rivers, such as the Zambezi, Limpopo, and Ravuma, chosen as international boundaries, have always been effective obstacles to movement and historically have separated different culture groups. In contrast, the Congo and its tributary the Bangui, which separate the Democratic Republic of the Congo from the Congo Republic and the Central African Republic, and the Senegal River, which divides Mauritania and Senegal, are navigable for considerable distances and have tended to draw together people on both sides of the rivers. In West Africa, rivers form boundaries between Liberia and its coastal neighbors, Sierra Leone and Côte d'Ivoire, while shorter stretches of rivers partially bound Burkina Faso, Togo, Benin, Niger, Nigeria, Mali, Chad, Guinea, and Ghana. Indeed, river boundaries are common in Africa, and rarely have they been contested.

International boundary lines drawn through lakes and along escarpments and mountain ranges also have been imposed. Lakes Victoria, Turkana (Rudolf), Albert, Edward, Kivu, Tanganyika, Nyasa (Malawi), and Chad were selected for reasons of convenience and then divided by geometric lines. The boundaries between the Democratic Republic of the Congo and the Sudan, the Central African Republic and the Sudan, and Malawi and Mozambique follow natural divides (watersheds), while the Great Escarpment forms the eastern boundary of Lesotho. Since the international boundaries were defined and adopted as quickly as possible, few cultural criteria were used in the selection process. Compare, for example, the colonial boundaries of the contemporary African states with traditional ethnolinguistic boundaries (Figure 5-2). One thing is clear: this is a very complex mosaic with many contemporary countries including a bewildering number of ethnolinguistic groups. Some exceptions: the boundaries of Lesotho and Swaziland, today define ethnically homogeneous regions; but it is a case of the people having adjusted to boundaries created without much respect for the cultural patterns. Internal or provincial boundaries, such as those in the Democratic Republic of the Congo, more frequently corresponded with tribal or ethnic lines, since the colonial governments found that this facilitated their administration. Another exception is Ghana in West Africa. The core of the country is centered on the populous Ashanti region and the city of Kumasi, the seat of the powerful precolonial Ashanti state. This historical centeredness gives Ghana a politicogeographic integrity that most modern African states lack.

Independent Africa inherited not only unsatisfactory international boundaries but grossly misshapen and unfortunately positioned territories. In West Africa, boundary lines were extended inland only short distances around strategic river courses and traditional settlements, and they defined relatively small areas (other than Nigeria); for the Gambia, Togo, and Benin (Dahomey), they defined small elongated states. One result was a vast residual interior that was subsequently divided into five landlocked states (Niger, Burkina Faso, Mali, Chad, and the Central African Republic), all of which had been part of French Africa. Similarly, following the initial partition of the coasts further south and east, nine other landlocked countries were created: Uganda, Malawi, Zambia, Zimbabwe (Rhodesia), Botswana, Lesotho, and Swaziland, all of which were British, and Rwanda (Ruanda) and Burundi (Urundi), once mandates under Belgian administration. The independence of Eritrea in 1993 created another elongated state, this time along the coast, and caused Ethiopia to become landlocked. Africa today has almost as many landlocked states as the rest of the world combined (15 to 19). Of the countries that do possess a coastline, two stand out as having very little coast in relation to the size of the country: the Sudan and the Democratic Republic of the Congo.

A landlocked state is in a disadvantageous position unless it is guaranteed the right to use the high seas like the coastal states, the right of innocent passage in other states' territorial waters, a share of port facilities along suitable coasts, and a means of transit from any such port to the state territory for its external trade. Furthermore, landlocked states have no claims to the continental shelves, some of which are proving to be rich in mineral and food resources. In theory, landlocked states have access to the coast and the high seas by three means: international rivers, land corridors, and free transit rights. Any navigable river that traverses both the landlocked state and the coastal state may be declared by agreement an international river, similar to the high seas in that it is not subject to the control of any one nation. The Congo, Niger, Zambezi, and Shire were declared international rivers. In theory this provided Burkina Faso, Niger, the Democratic Republic of the Congo, Zimbabwe, Malawi, and Zambia access to the sea; but since only portions of these rivers are navigable, the measure does not truly alleviate the problems of location.

Land corridors have not been widely adopted. The Democratic Republic of the Congo has a vital corridor to the Atlantic via the Congo River. The Belgian government exchanged land with the Portuguese as late as 1927 in order to provide its then colony, the Belgian Congo, more effective access to the sea along that corridor. The Belgians acquired a little more than a square mile for the development of the ocean port of Matadi, in exchange for 480 square miles (1,243 km^2)

FIGURE 5-2 Colonial boundaries (heavy lines) superimposed on Murdock's (1949) ethnolinguistic map of Africa.

elsewhere. Similarly, the Portuguese acquired from the British a corridor up the Zambezi as far as Zumbo.

All the landlocked states of Africa have free transit rights to the coast, the terms of which have been defined by international treaties. Nevertheless, coastal states have the upper position, and can affect the amount and composition of traffic bound to and from the interior states. When the landlocked states of West Africa were under French control, they used the ports of Dakar and Abidjan without restriction. Since independence and because of local political friction, however, Mali diverted much of its traffic through the far less convenient port of Abidjan, and Niger now uses the more suitably positioned ports in Nigeria. Zambia's main links with the outside world are through Tanzania, Mozambique, and Angola, although during the postindependence Angolan civil war, the Angolan outlet via the Benguela Railway was temporarily closed. When Mozambique was Portuguese, white-minority ruled Rhodesia depended almost exclusively on Beira and Lourenço Marques (Maputo), but in the 1970s, Mozambique's revolutionary wars of independence spurred Rhodesia into building a railway to South Africa. Rhodesia's other rail link to the sea was through Botswana and South Africa.

Landlocked Botswana, Lesotho, and Swaziland faced particularly difficult problems in the past because of their opposition to apartheid in neighboring South Africa. Lesotho is completely surrounded by South Africa, and Botswana is virtually enclosed by the formerly white-minority controlled territory. Both small countries were totally dependent on South African ports for access to overseas markets, and on occasion South Africa refused to permit certain goods and individuals to cross its space. Swaziland had a slightly less vulnerable position, in that it shares a boundary with Mozambique and has rail links with Maputo. Although apartheid ended in the early 1990s, dependence on South Africa has not.

Landlocked Malawi has always depended on the ports of Mozambique, especially Beira and Naçala, and throughout Portugal's tenure in Africa it had to place its economic interests before international opinion. An independent Mozambique has removed that particular issue, but not the general problem of access.

Uganda's only practical exit is through the Kenyan port of Mombasa, and for many years Uganda has been guaranteed the right of transit across Kenya. In 1976, however, Kenya threatened to block all Ugandan transit trade following the slaughter of Kenya nationals residing in Uganda. Finally, Rwanda and Burundi have transit rights across Tanzania, while Chad has rights to use Nigerian and Cameroonian ports, and the Central African Republic uses Pointe Noire in the Congo Republic.

The "scramble for Africa" left a number of very small territories such as Lesotho, Swaziland, Rwanda, Burundi, Gambia, Togo, and Guinea-Bissau. It also left a number of prorupt states, that is, states possessing an extension of territory in the form of a peninsula or a corridor leading away from the main body of the territory. Such prorupt states often face serious internal difficulties, for the proruption is either the most important part of the state or a distant problem of administration. The Democratic Republic of the Congo has two proruptions: the western one contains the capital city (Kinshasa), the port of Matadi, the mouth of the Congo River, and the important Inga hydroelectric scheme; the southeastern proruption has the economic core based on copper and other the resources of Katanga (Figure 5-3). Between the two areas lies the vast Congo Basin, in many ways more a liability than an asset to the state at present.

Another example is found in Namibia, where the Caprivi Strip reaches deep into the interior of southern Africa and touches southeast Angola, southwest Zambia, western Zimbabwe, and northern Botswana. On the face of it, such a narrow strip of territory should have little importance in the course of African affairs. During the years of the apartheid state in South Africa, however, it became an area of guerrilla warfare directed at the overthrow of white South Africa. Since independence for Namibia and the dismantling of apartheid, the Caprivi Strip has lost its political significance.

What would the political geography of Africa look like if the "scramble" had never happened? African polities probably would have followed the nationalist trend that began transforming states and empires in other parts of the world at the end of the eighteenth and especially during the nineteenth century. Nationalism swept around the world and is still a potent force not to be underestimated—people will lay down their lives for their perceived homeland, as Eritreans did for decades until they achieved their state in 1993. Even the independence of Europe's colonies in Africa was part of the world nationalist movement. Had there been no European partition of Africa into large, ethnically diverse entities, there would have been many more culturally homogeneous states, surely hundreds, and many more landlocked states than exist today. It is difficult to say whether there would have been more or less postindependence conflict in Africa, had there been no colonialism. Fewer states mean fewer neighbors and less potential for conflict. On the other hand, large countries like Sudan and Nigeria might have naturally split into several smaller, more homogeneous entities, obviating the felt need for secession and civil war.

CORE REGIONS AND CAPITALS

In the preceding chapter we discussed certain locational, functional, and structural attributes of Africa's precolonial states, emphasizing that they contained an

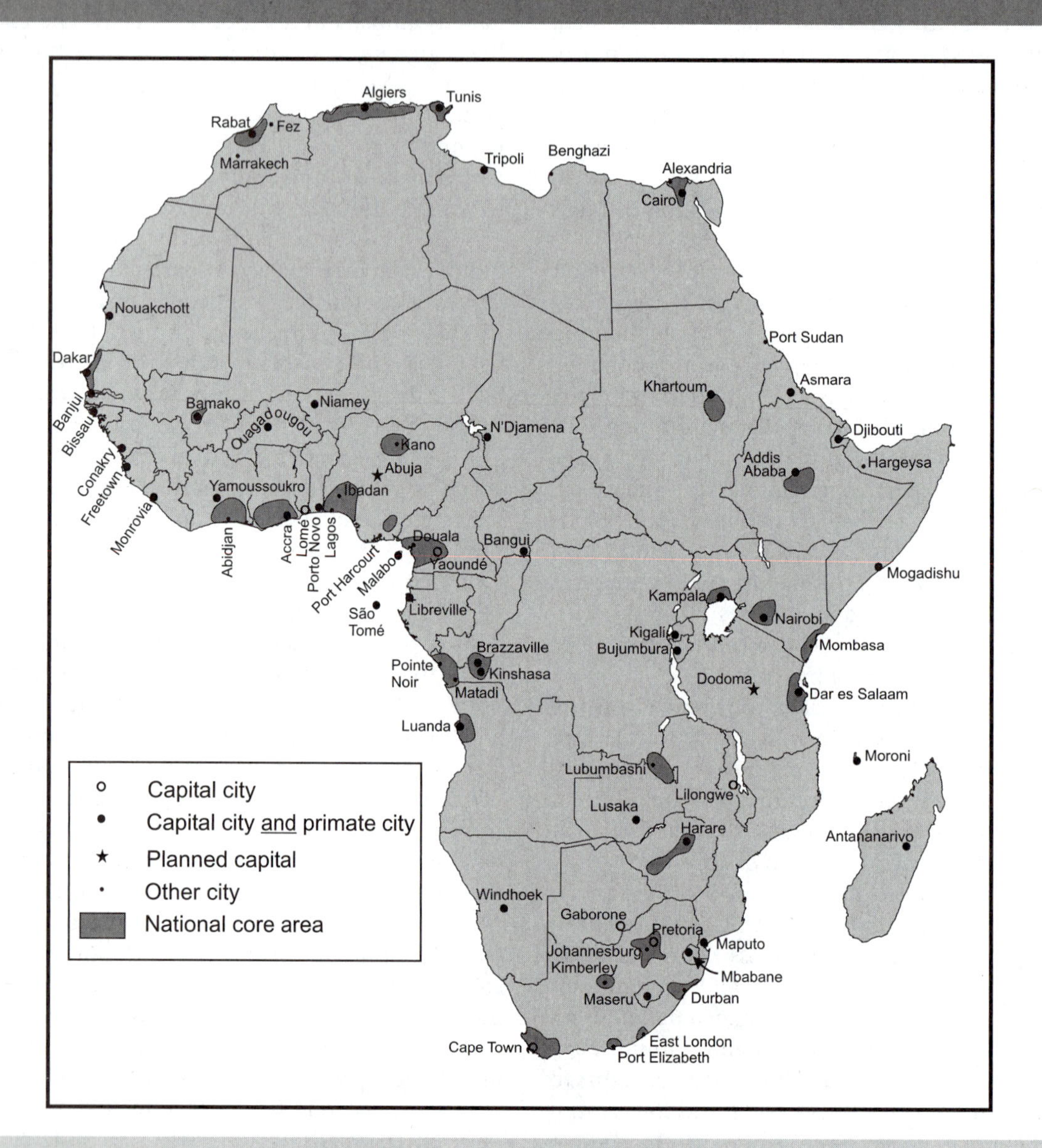

FIGURE 5-3 Contemporary national core regions and capitals.
Adapted from Best and de Blij (1977), de Blij and Muller (2006).

administrative center, an elite class, specialized labor, and a communications network that linked the capital with its periphery. Commonly the capital was the largest and most important place within the state area, and it was supported by places of lower order over which it exerted some control. There existed a pattern of core areas and peripheries: the core was the political, economic, and cultural heart of the state, as well as the locus of military power; the periphery comprised the less developed, less integrated, and less controlled surroundings. In the traditional African state people ethnically different from those in power at the core would likely be found in the peripheral areas. The Ségu Bamana of Mali, descendants of the former Ségu warrior state, celebrate the old, glorious days in song saying that the people of Ségu Town knew of the extent of the expanding state only through the sight of strangers among them ("Segukaw t'o yòrò dòn janko dunan"), many of whom were captives.

In economic geography, the term "core region" refers to a central area in a state that acts as the main organizing force and engine of economic growth for other areas of the state. Today, Africa's core regions are still not as developed as those of the industrial world, or even those of most less developed countries in Asia and Latin America, but they are growing. In view of the artificial nature of the states themselves, the relatively short history of modern economic development, and the as-yet elusive *raisons d'être* of so many states, this is not unexpected. Africa's cores differ considerably in size and composition, and in the roles they play in their respective states. In some states, such as Chad, there are no real core areas (Figure 5-3). In others they focus on the capital, which more often than not is also the primate city, that is, the largest city according to population (Table 5-1), and the most important in terms of manufacturing, services, trade, cultural, and political influence. They are also the country's prime generator, transformer, and distributor of the forces of change and development. Most are several times larger than the second city and exert greater influence on their states' political and economic systems than would be suggested by size alone. A few core regions and capitals serve to illustrate these points and to emphasize their great diversity in origin, form, and function.

The Republic of South Africa is the only modern industrial state in Africa, and the only state to possess a truly diversified national core supported by secondary development elsewhere in the state (Figure 5-3). The national core centers on the primate city (but not capital city) of Johannesburg (2 million in 2006) and extends east and west along the minerally rich Witwatersrand. Here there are gold-mining towns, iron and steel industries, several centers of higher learning, a modern transport system, the state's administrative capital, Pretoria (1.7 million), and a host of industries that account for almost half of South Africa's industrial output. It is supported by four secondary metropolitan cores, Cape Town (3.5 million), Durban (3.2 million), Port Elizabeth (1 million), and East London (400,000), each of which is supported in turn by smaller industrial centers and agricultural areas. The industrial and economic wealth of each of South Africa's four metropolitan subcores surpasses that of most primary cores elsewhere in Africa.

Another good example of a national core exhibiting a high level of urban development and industrial output is to be found in Egypt. Centering on Cairo (7.8 million), the largest city in Africa, and Alexandria (3.8 million), this core comprises the entire delta region, where almost half the Egyptian population lives. Spatially this is one of Africa's most distinct core regions, being bounded by the coast and desert. Unlike the South African core, that of Egypt dates back thousands of years, and irrigated agriculture rather than mining was the original basis for its development.

By far the most important multicore state in tropical Africa is populous Nigeria. Its core areas are not only clearly defined, but they are also individually unique in terms of ethnic composition, economic activity, historical association, and many other aspects. Nigeria has three major core areas, each of which would do justice to any single West African state. The primary core lies in the southwest (Figure 5-3), and includes a number of major urban centers such as Nigeria's most populous city of Lagos (9 million), and the Yoruba towns of Ibadan (3.7 million), Ogbomosho (900,000), Oshogbo (525,000), and Abeokuta (600,000). This is one of Africa's most highly urbanized regions, whose exchange economy is based on both agriculture (especially cocoa) and industry. Northern Nigeria is predominantly rural and Muslim, and its core focuses on the ancient cities of Kano (3.7 million) and Zaria (1 million). The third core region lies in the southeast, where population densities are among the highest in Africa, and the economy is based on petroleum, oil palm, and other products. The federal capital, Abuja, is centrally located outside any core area.

In these and other countries, the capital city plays an especially important role in the spatial organization of the state, generally having a nodal location and being the dominant center of innovation, education, political influence, and capital accumulation. In more than half the countries, the capital and the core region are virtually one and the same, and 40 capitals are, at the same time, primate cities. Table 5-1 gives some selected characteristics of Africa's capitals and some measures of primacy. Only eight capitals are not primates: Abuja (Nigeria), Rabat (Morocco), Yamoussoukro (Côte

TABLE 5-1 SELECTED CHARACTERISTICS OF AFRICAN CAPITAL CITIES.

Country	Capital	Characteristics										
		Indigenous	Former colonial capital	Planned capital	Primate	Only major city	Main seaport	Lake/river port	Main airport	Main rail port	Main industrial center	Population (millions)
North Africa												
Algeria	Algiers	X	X		X		X		X	X	X	1.7
Egypt	Cairo	X	X		X			X	X		X	7.8
Libya	Tripoli	X	X		X		X		X	X	X	1.2
Morocco	Rabat	X	X				X					1.6
Tunisia	Tunis	X	X		X	X	X		X	X	X	0.697
West Africa												
Benin	Porto Novo	X	X				X					0.227
Burkina Faso	Ouagadougou	X	X		X		X		X			0.840
Cape Verde Islands	Praia	X	X		X	X			X			0.098
Chad	N'Djamena		X		X	X			X		X	0.602
Côte d'Ivoire	Yamoussoukro	X				X			X		X	0.196
Gambia	Banjul		X		X	X	X	X	X		X	0.058
Ghana	Accra		X		X				X		X	1.6
Guinea	Conakry		X		X	X	X		X	X	X	1.6
Guinea Bissau	Bissau		X		X	X	X	X	X		X	0.289
Liberia	Monrovia		X		X	X	X		X	X	X	0.540
Mali	Bamako	X			X	X			X		X	0.907
Mauritania	Nouakchott			X	X	X	X		X			0.426
Niger	Niamey		X		X	X			X			0.723
Nigeria	Abuja		X									0.160
Sénégal	Dakar		X			X	X		X	X	X	2.4
Sierra Leone	Freetown		X			X			X	X	X	1.0
Togo	Lomé		X			X			X	X	X	0.658
Central Africa												
Cameroon	Yaoundé		X									1.3
Central African Republic	Bangui		X			X	X		X		X	0.655

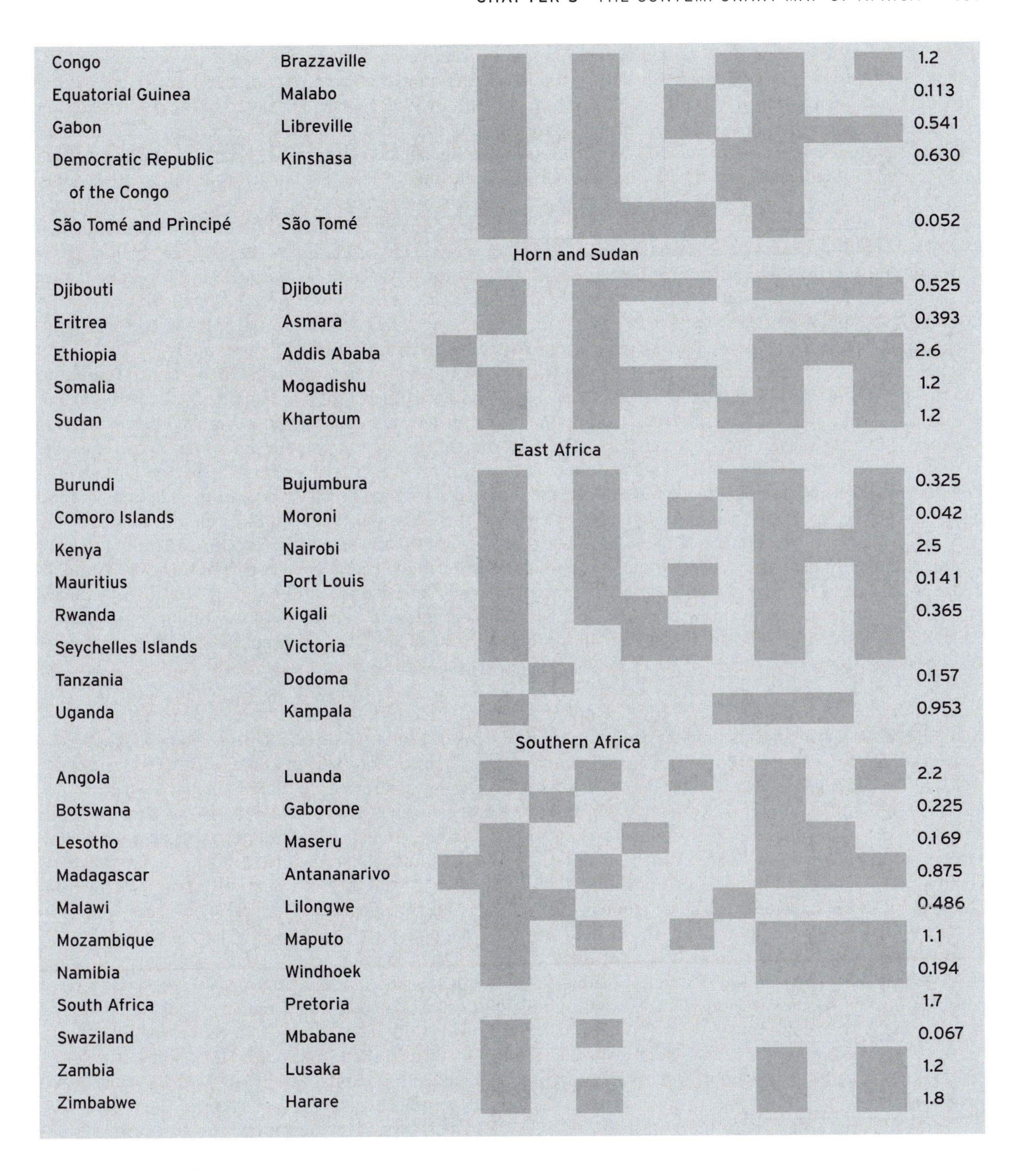

Congo	Brazzaville	1.2
Equatorial Guinea	Malabo	0.113
Gabon	Libreville	0.541
Democratic Republic of the Congo	Kinshasa	0.630
São Tomé and Prìncipé	São Tomé	0.052
Horn and Sudan		
Djibouti	Djibouti	0.525
Eritrea	Asmara	0.393
Ethiopia	Addis Ababa	2.6
Somalia	Mogadishu	1.2
Sudan	Khartoum	1.2
East Africa		
Burundi	Bujumbura	0.325
Comoro Islands	Moroni	0.042
Kenya	Nairobi	2.5
Mauritius	Port Louis	0.141
Rwanda	Kigali	0.365
Seychelles Islands	Victoria	
Tanzania	Dodoma	0.157
Uganda	Kampala	0.953
Southern Africa		
Angola	Luanda	2.2
Botswana	Gaborone	0.225
Lesotho	Maseru	0.169
Madagascar	Antananarivo	0.875
Malawi	Lilongwe	0.486
Mozambique	Maputo	1.1
Namibia	Windhoek	0.194
South Africa	Pretoria	1.7
Swaziland	Mbabane	0.067
Zambia	Lusaka	1.2
Zimbabwe	Harare	1.8

d'Ivoire), Porto Novo (Benin), Yaoundé (Cameroon), Gaborone (Botswana), Dodoma (Tanzania), and Pretoria (South Africa). Five new capitals—Nouakchott (Mauritania), Abuja (Nigeria), Gaborone, Lilongwe (Malawi), and Dodoma (Tanzania)—have been built since independence. In several other countries the capital has been moved at one time or another.

All the North African capitals predate colonial occupation and remained administrative centers under foreign rule. The ancient Arab capital of Marrakech (Morocco) was superseded by another Arab city, Rabat, which from the French point of view occupied a more favorable position on the coast. In West Africa, the colonial powers selected places that satisfied their particular economic and political interests. Along the Guinea Coast that meant ports (except Porto Novo, capital of Benin), and in the interior it meant either river locations with some nodality (Bamako, Mali, and Niamey, Niger) or some densely populated area (Ouagadougou, Burkina Faso, and N'Djamena, Chad). As the colonial systems developed, so did the ports. Today in 10 of the 16 West African states, the capital city is both the primate and the only major city. Advantages accrued to the colonial capitals have in most cases persisted to the present.

The capitals of East Africa and the Horn are markedly more central, with respect to their populations and territories, than those of West Africa. Nairobi and Kampala are central within their respective and well-defined core regions, while Bujumbura and Kigali are the only significant towns within compact, small, and very densely populated mountain states. The Ethiopian capital of Addis Ababa, which occupies a commanding position in the densely populated highlands, was established in 1889, the year Menelik II became emperor. Mogadishu is virtually the only urban center and port in Somalia, and because of the country's peculiar shape, it is peripheral and not readily accessible from the northern regions. Finally, Khartoum, which adjoins the indigenous city of Omdurman, is central within a very large area and exhibits the blend of traditional and foreign influences so typical of African capitals.

Three of Africa's new (postindependence) capitals have had insufficient time to develop into primate cities, and because of their locations and the level of development elsewhere in their respective states, they are unlikely to become the largest cities, or indeed anything more than administrative centers. Gaborone was built in 1965 to meet the needs of the newly independent state of Botswana. Under British rule, the then Bechuanaland Protectorate was administered from Mafeking, located not in the protectorate itself, but in neighboring South Africa. To avoid favoring one ethnic group over another, none of the local administrative centers was chosen to become the capital, and a new town was built. Gaborone, built for a population of 20,000, now houses over 10 times that number is merging with one of the South African core areas centered on Pretoria and Johannesburg.

Malawi's capital was shifted from Zomba to Lilongwe in 1975 to encourage economic development in the less populous and poorer areas of the center and north. To date most development has occurred in the south, around Lilongwe, the primate city. Tanzania has been slowly moving its capital from its largest port and primate city, Dar es Salaam (2.7 million), to the smaller town of Dodoma (181,000) located 250 miles (400 km) farther west. The move was part of socialist Tanzania's largely unsuccessful effort shortly after independence to decentralize development, to stimulate growth in a peripheral area, and to dispel the colonial image associated with Dar es Salaam. Nouakchott, capital of Mauritania, has grown from a small nomadic camp into a modern city of 612,000 people and, like most African capitals, it is unable to provide adequate jobs and housing for those who have drifted to it from the periphery. During the periodic Sahelian droughts since independence, many pastoralists who had lost their livestock joined the squatter settlements around Nouakchott. Abuja, the planned new capital of federal Nigeria, is centrally and neutrally located in the Middle Belt of Nigeria, far from the core areas and traditional seats of power.

Other states whose present capitals are not the original seats of administration are Niger, the Congo Republic, Zambia, the Democratic Republic of the Congo, Côte d'Ivoire, and Cameroon. In 1926 Niger's capital was shifted from Zinder to Niamey; the Congo Republic's first capital was Pointe Noire; Zambia's capital was officially moved in 1935 from the peripherally situated town of Livingstone to the more centrally situated Lusaka; Kinshasa replaced Boma as the capital of the present day Democratic Republic of the Congo in 1929. Yamoussoukro became the capital of the Côte d'Ivoire in 1983, replacing the former capital of Abidjan, the designated capital since 1934; two nearby towns, Grand-Bassam and Bingerville, had briefly enjoyed this status. Yaoundé became the capital of Cameroon under the French mandate, the colony previously being administered from the "hill station" of Buea and the present primate city of Douala. This shifting of administration from one city to another explains in part the lack of absolute primacy in the Congo Republic, Zambia, and the Democratic Republic of the Congo. South Africa differs from all the other states in that it has three capitals: Cape Town is the legislative capital, Pretoria the administrative capital, and Bloemfontein the judicial capital. Before the establishment of the state of South Africa, each of these was a locally important administrative and commercial center. Today they are all overshadowed in size and industrial importance by Johannesburg.

INSTITUTIONS AND GOVERNMENTS

The mosaic of arbitrary and illogical boundaries, the diversity of African cultures and environments, and the superimposition of several foreign value systems and institutions have combined to create a very complex political map of Africa that defies simple analysis and generalization. Nevertheless, a number of common issues and problems concerning the forms of government, nationalist movements, political stability (or instability as the case may be), and interstate relations can be identified that help explain some of the spatial patterns of development considered in subsequent chapters.

When the African territories gained their independence, each attempted to devise means to promote internal unity, foster a greater sense of nationalism, and provide the machinery for social and political modernization and economic development. This, of necessity, meant the abandonment of many traditional values and behavior patterns, and the adoption of constitutions and attitudes that were more European than African. Most states adopted constitutions that provided for democratic elections and multiparty systems of government that were based on the European experience. Both unitary and federal states were created. All the former French territories except Cameroon became unitary states, while Nigeria, Tanzania, and Uganda, once British territories, became federal states. A preindependence federal constitution was superimposed on Northern Rhodesia, Southern Rhodesia, and Nyasaland in 1953, but terminated a decade later. Ethiopia joined in a federation with Eritrea in 1952, followed by annexation in 1962 and independence for Eritrea in 1993. South Africa, while exhibiting certain federal characteristics, is a highly centralized unitary state.

In theory, unitary states have a high degree of internal homogeneity and cohesiveness among their populations and institutions: there is a uniformity of language, culture, history, and national ideals. The central authority controls all local governments and determines how much power they shall have, and under certain circumstances may exercise the functions of local governments. The ideal unitary state would be compact in shape and not unduly large in area, having an even distribution of similar peoples and only one centrally located core region, focused on the state capital and primate city. Few unitary states in the world display such characteristics, and many in Africa lack ethnic homogeneity and centrally located core regions and capitals.

During the early years of independence, many African states developed a highly centralized form of government for the purpose of reducing dissension among the different peoples, which might have disrupted the state system. One-party systems were common and more often the rule than the exception in Africa from the 1960s to the 1990s. Here the ruling party was absolute, and power was concentrated in the hands of the party, and especially the party leader, who was generally president and head of government. The one-party system was adopted to centralize the state's efforts in economic development, the argument being that opposition works against the interest of the people. Moreover, opposition fosters tribalism and secessionist tendencies, it was said, since political parties are often identified with certain regional interests. Commonly when dissatisfaction with the ruling party has intensified, the government has been toppled by a revolution or military *coup*. During the 1960s, for example, military *coups* occurred in 15 African states, because there was no other way in which the ruling clique could be removed.

Events in Ghana illustrate the shift toward increasing centralization within a unitary state. At independence in 1957, Ghana inherited a constitution and governmental apparatus that was based on the Westminster model. There were regional and tribal problems (in parts of northern Ghana the majority of people favored a delay in the granting of independence), and the opposition party in parliament was essentially a regional phenomenon. As government plans and projects were thwarted, and the power of tribal chiefs continued to influence Ghanaian politics, the ruling party sought ways to diminish the effectiveness of the opposition. The main object was to gain greater control over the recalcitrant north and to silence the chiefs who opposed the new government's programs. This eventually led to the elimination of the opposition, the proclamation of a republic, the establishment of a one-party state, and the assumption of virtual dictatorial powers by Kwame Nkrumah, who was president from 1960 to 1966. In one way or another, this basic pattern has been repeated in Uganda, Zambia, the Central African Republic, the Congo Republic, and elsewhere.

Federal states, in contrast to unitary states, permit a central government to represent the entities within the state where they happen to have common interests—defense, foreign affairs, communications, and so on—while allowing the various units to retain their identities and to have their own laws, policies, and customs in certain fields. Federation thus does not create unity out of diversity but enables the two to coexist. The federal arrangement is often the most suitable for political entities characterized by a diversity of peoples, languages, religions, cultures, and historical backgrounds, with these differences having regional expression in that various peoples see individual parts of their country as a homeland. In theory, therefore, conditions conducive to federalism are likely to exist in relatively large states, where the diverse populations are organized around several core regions separated by sparsely populated and relatively unproductive regions.

While many of the foregoing conditions exist in Africa, federal states are not common. This is due primarily to the absence of political leaders, and populations committed to federalism and the existence of broad cultural diversities of the peoples (Figure 5-2). The primary factor necessary to ensure the survival of a viable federation is the commitment of both the political leaders and the populations at large to federation for its own sake; an ideological commitment to federation as a means of achieving some secondary objective is not likely to suffice. The population as a whole must feel federal, thinking of themselves as one people with an overriding federal identity, so that the value of federation becomes the most important fact in the national community. The durability and viability of federation are determined by the degree of symmetry or level of conformity in the relations of each separate political unit of the system both to the system as a whole and to the other component units. The higher the level of symmetry at the national and regional levels (measured in terms of language, cultural heritage, economic welfare, political awareness, etc.), the greater the likelihood that federation will endure. A state encompassing enough diversities in its peoples, attitudes, values, and other essential characteristics to result in distinct regional patterns may in fact not possess sufficient uniformity or symmetry for successful federation. Strong regional interests constitute asymmetry. These principles are evident in the federation of Nigeria.

The initial federal framework for Nigeria was developed by British and Nigerian political leaders prior to independence in 1960. When independence was granted, Nigeria had an estimated population of 50 million and three major core regions that formed the nucleus for three of the dominant peoples of the country: the Yoruba in the southwest, the Ibo in the southeast, and the Hausa in the north. In point of fact, Nigeria contains literally hundreds of ethnic groups, and more than 200 languages are spoken, so that additional political divisions were deemed desirable. In 1963 the decision was reached to establish a Midwest Region, carved out of the territory's Western Region and populated mainly by Edos. Additional subdivisions were thought necessary so that no single region would be in a position to dominate the rest. The federation, however, had been so divided that the Northern Region, by far the most populous unit in the state, was able to dominate the others. Counterbalancing this was the landlocked north's dependence on southern ports for its external trade; but exacerbating the situation was the vast chasm of contrast in religion, political expertise, and economy between the northern and southern states.

Following independence, there were outbreaks of violence against the Ibo (Igbo) people living in the north, and an exodus of survivors resulted. Eventually the southeastern section of the federation (the original

Eastern Region and Igbo homeland) seceded from Nigeria and pronounced itself the sovereign state of Biafra. Nigeria was plunged into a costly and bitter civil war, and the failure of the federal framework was a factor. In 1968, while the war still raged, the Nigerian government decided to redivide the country into 12 political regions in the hope that the arrangement would preclude future Biafra-style conflicts. At the same time, a military administration came to power, so that Nigeria had moved first from a position of superimposed federation and later of compromise to centralization. Since that time there have been several redivisions of Nigeria's political space.

Seven new states were added in 1975, including the administrative district of Abuja, the new capital of Nigeria. From 1987 to 1996, 17 new states were added, for a total of 36 states. Three states had become 36 in 36 years.

Making the formation of a national identity more complicated is that Islamic law (Sharia) has been accepted in 12 northern states and secular/customary law rules elsewhere. Use of two legal systems, particularly one that is revered as holy, is problematic in a modern state. Why would this be so? In the Nigerian state, according to the laws promulgated in the Sharia states, Christians (and other non-Muslims) and Muslims who come into contact with the law work within their respective legal systems: Christians and others work within the secular legal framework inherited from the colonial period, and Muslims are judged within Islamic law. The problem arises when people of the two systems meet as adversaries. In Islamic law, a Christian (or other) cannot be a witness in an Islamic court of law against a Muslim. In Kano State, for example, if a Christian witnesses a Muslim stealing a radio from his house, the Christian cannot testify against the thief in the court of law that has jurisdiction for the Muslim. Such problems are difficult to resolve without resort to extremely cumbersome mechanisms.

The Central African Federation is a good example of an imposed federation in which the desires of a minority were imposed on the majority to perpetuate the privileged white settler position. There was much economic justification for federation: Nyasaland (Malawi) was dependent primarily on agriculture and Northern Rhodesia (Zambia) on mining, while Southern Rhodesia (Zimbabwe) had a more diversified economy. Federation, however, was imposed on 10 million Africans by 225,000 whites, most of whom lived in Southern Rhodesia. The economic benefits, promised as a political lure, failed to materialize except in the white strongholds, so that the African majorities in Nyasaland and Northern Rhodesia called for the federation's dissolution, which came about in 1963.

AFRICAN NATIONALISMS

Historically, nationalism refers to movements and ideologies that seek to unite people who share a common national identity into a sovereign state. In a sovereign state there is complete autonomy on internal matters; the state has legal equality and inviolability within the international context; and there is some precise delimitation of territory. Within the European context, those with a common national identity regard themselves as sharing a common language, religion, values, goals, and institutions, as having a common history, and as being different from other peoples, who have their own language, values, institutions, and history. Thus it is the combination of cultural variables, and the people's identity with a specific territory that defines a nation, and where this emotional nation corresponds with the legal nation, a nation-state exists.

In Africa, nationalism is a very different matter. Almost every colony incorporated many diverse and distinct ethnic groups and several languages or mutually unintelligible dialects that greatly impeded communications among the peoples, to say nothing of the development of common cultures. Interest groups were local rather than national, and there was traditionally little sharing of values and institutions in the artificially prescribed colonial boundaries. Nevertheless, African nationalisms emerged, and their movements have differed considerably in their causes and expressions. In Kenya, for instance, there was strong resentment against the European settlers who appropriated the most fertile lands and forced the Africans into less desirable regions. In Ghana, Nigeria, Guinea, and the Democratic Republic of the Congo, there was widespread frustration at not being able to control the economy, wages, employment, and living standards. In all areas, nationalism was a reaction to poverty, deprivation, and an inferior status, the consequences of colonial and racist ideologies; and everywhere, nationalism was an attempt to remove the indignities of colonial rule and a search for local (African) identity and participation in the political affairs by the mass of the people. Thus nationalisms have been directed at receding objects (colonialism and racism) for which replacements were found, some imagined, some real.

In most African states, nationalist movements first emerged during the early years of colonial conquest. The Ashanti wars (1899) and the Fulani battle at Burmi, Nigeria (1903), were resistance movements against British rule. The "Maji-Maji" (1905–1906) and Bambata (1906) rebellions were unsuccessful attempts to oust the Germans from Tanganyika and the British from Natal, respectively. As the colonial economies developed, and Africans were drawn into urban regions, a number of urban-based political organizations emerged aimed at increasing African participation in the civil service, administration, and the economy. World War II brought more Africans more directly into the European realm, which quickened the process of acculturation and reinforced rising expectations. An African elite emerged in most colonies: people who had been educated in Europe but were not immediately able to participate as equals with the colonizer in the social, economic, and political spheres.

There was much debate during this preindependence nationalist period over what should be the ultimate boundaries of the nation-states-in-the-making, some favoring smaller units respecting ethnic groupings, others preferring political groupings of several of the existing colonial units, while some others called for a "Pan-African" state encompassing all or most of the continent. The problem was that the masses of people did not identify with the state—colonial or independent—but instead had very local allegiance to tribe and clan. Pan-Africanism remained ideology. Once independence was won, there was great reluctance to give up sovereignty and become part of a larger territory. A number of Pan-African movements developed that called for continental nationhood.

President Léopold Senghor of Senegal believed the state was the expression of the nation, and was primarily a means of realizing the nation. He based his concept of African nationhood on the assumption there is a commonality of values characteristic of traditional black Africa. Known as *négritude,* this form of African nationalism asserted that all black people throughout the world share an unconscious experience that distinguishes them from all others, and that blackness in itself is a positive thing.

The racial criterion of nationhood has been rejected by some African nationalists (many of them Muslims) and denounced by Pan-Africanists who seek a continental rather than a specifically racial basis for unity. The late Ghanaian leader Kwame Nkrumah was one of the most outstanding spokesmen for Pan-Africanism. For Nkrumah, Africa comprised three major civilizations—traditional Africa, Islamic Africa, and the Euro-Christian Africa—and these were to blend and form a new and uniquely African civilization. The common denominator was Africa, not race. In Tanzania during the early years of independence, the ideology of nationalism combined traditional values of familyhood (*ujamaa*) with contemporary socialist thought. Under the leadership of President Julius Nyerere, Tanzania was committed to egalitarianism and greater self-reliance.

In South Africa, African nationalist movements, which were instrumental in exposing the inequities of the white oligarchy, were banned in 1960, and Afrikaner nationalism and apartheid prevailed up until 1991. Apartheid was expressed in the landscape in many forms, such, as the reservation of specific regions for whites and others—the Bantustans—for black Africans. Under apartheid, cities were divided into group areas that lessened interracial contact and thus perpetuated white supremacy.

Although Africa's European and Turkish colonial masters have left the continent—at least politically—many Africans oppressed by a ruling minority or majority or stigmatized by their social origin, have fought and are still fighting for self-determination from other masters. The revolt of the Igbo in southeastern Nigeria in 1967 is probably the bloodiest case of post independence struggle for self-determination. About a million people were killed in an unsuccessful bid for an independent state.

Tuareg have periodically revolted against their national rulers across the Sahelian states. In Mali, they have been brought into a national democratic dialogue, and prospects look good for long-term national reconciliation. Southern Sudanese fought for self-determination for forty years. Eritrea fought for 30 years before Ethiopia granted it independence.

North African Berbers continue to struggle, not for independence at this point, but for recognition of their language and culture by the dominant Arab colonizers. Berbers are the ancient inhabitants of North Africa; they were subjugated during the Arab Muslim conquests that began after the death of Muhammad in 632. The Berbers have been forced to learn Arabic in school; their own language, Tamazight, has been forbidden until recently. Beginning in the 1990s, Berber nationalists have marched and demostrated, sometimes being violently suppressed, against Arab oppression. Kateb Yacine (1929–1989), Berber and Algerian writer, had the following to say about French and Arab colonialism in Algeria:

> Our armed struggle ended the destructive myth of French Algeria, but we have succumbed to the power of the even more destructive myth of Arab–Islamic Algeria. French Algeria lasted 104 years. Arab–Islamic Algeria has lasted 13 centuries! The deepest form of alienation is no longer the belief that we are French, but the belief that we are Arabs. There is no Arab race and no Arab nation. There is a sacred language, that of the Koran, used by the rulers to prevent the people from discovering their own identity. (Quoted in Ibn Warraq 1996)

STATE POLITICAL PHILOSOPHY SINCE INDEPENDENCE

African states inherited more than boundaries and institutions when they became independent. The African state possessed distinctive characteristics based on the impress of colonialism. Furthermore, the states at independence, driven by nationalist ideologies, entered into a world riven by Cold War superpower rivalry. Research by Crawford Young, an Africanist political scientist, underlines the impact of the colonial experience on the postcolonial state. Young notes that the operations of the colonial state—the rapid and forcible domination of Africa, the use of sophisticated technologies of control (the punitive expedition), ruthless extraction of natural resources, the common use of forced labor, and welfare ideology as a method of gaining legitimacy from the people—provided a model of governance that postindependence governments followed or fell back upon when other models began to fail. Although postcolonial African governments must bear the burden of responsibility for their actions and surely there are other variables involved, Young makes a strong argument as to why things fell apart after independence because of the very nature of the colonial state.

One external factor that complicated the task of nation building at independence was the Cold War. When most African states became independent (Eritrea was an exception), the capitalist and socialist/communist worlds were locked in a struggle to win the hearts and minds of the peoples of the world. Powerful communist countries such as the Union of Soviet Socialist Republics (USSR) and China competed with such capitalist countries as the United States, the United Kingdom, and France for influence in Africa, and the two sides often fought each other using proxies, or stand-in armies. For example, for almost three decades, the United States fought a proxy war against the USSR in Angola through its support of Jonas Savimbi's National Union for the Total Independence of Angola (UNITA). The USSR had two proxies in Angola to fight Savimbi's rebels, who were armed by the United States: the unelected, Marxist–Leninist government of the Popular Movement for the Liberation of Angola (MPLA), which had taken control when the Portuguese left in 1975, and Cuban soldiers sent to Angola by Fidel Castro to further the spread of communism.

Complicating the Cold War struggle was the support of racist South Africa for UNITA against the MPLA, which South Africa correctly perceived as hostile to its capitalism as well as to apartheid. It was in this context that the United States found itself on the side of a country with social policies Americans would never support. In the same vein, the United States supported such brutal African dictators as Mobutu Sese Seku of Zaïre (now Congo) simply because of his anticommunist rhetoric (Bayart et al. 1999). For as long as the Cold War lasted, Cold War policies tainted and hobbled meaningful and genuine dialogue between the United States and African countries. Many African governments, however, learned quickly a cynical game, playing one Cold War power off another to obtain aid and concessions from capitalists and socialists alike.

When the Cold War ended in 1989, support for rebel movements began to ebb and the proxy wars ended shortly thereafter, although some rebels managed to continue fighting into the twenty-first century: case in point, Jonas Savimbi was finally killed by the MPLA armed forces of Angola in February 2002. Ironically, the MPLA abandoned Marxism–Leninism and embraced private property and capitalism during the 1990s (Hodges 2001).

Other ideologies that have surfaced in Africa have brought in foreign powers and sometimes complicated local politics. The governments of oil-rich Gulf states, financed Islamic centers, mosques, and Islamic groups in an effort to spread Islam in Africa. These states also gave economic assistance during times of drought and other catastrophes. Some states, Libya is an excellent example, freely interfered in the internal affairs of other African states in order to support a national agenda through the efforts of rebel groups. The civil wars in Chad and in Liberia are two well-known examples among many. In addition, Burkina Faso, Côte d'Ivoire, and Nigeria had a hand in the failure of the Liberian state (Ellis 1999).

From independence, African leaders employed a variety of ideologies that were used to organize the state and citizenry for independent development. During the heady days just after independence, African leaders optimistically believed that by this means they would be able to rapidly raise the standards of living for their peoples. All were disappointed, as the Nigerian author Chinwe Achebe noted in an influential novel, *Things Fall Apart.* Some leaders opted for extremes of socialism or communism, others opted for the market economy and capitalism in some form. Most ended up with a hybrid political organization despite the rhetoric of the state.

In the subsections that follow, we discuss five political philosophies—Marxism–Leninism, African socialism, capitalism, Islamism, and democracy—drawing examples from all over postindependence Africa. It is noted at the outset that all forms of political organization that emerged after independence were characterized by statism: that is, the state played a prominent role in planning and making decisions. In states opting for socialism and communism, state involvement was heavy; in others it was more moderate. In some states, statism turned into and facilitated personal rule by an unelected president for life (Chabal and Daloz 1999). A constant in Africa was the military government (Figure 5-4).

Marxism-Leninism

One rarely finds talk of communism or Marxism–Leninism as an organizing principle to be applied in Africa today. The hothouse years for Marxism–Leninism in Africa were the 1960s to the

FIGURE 5-4 The military commandant of Mopti Region, of Mali, surrounded by local Tuareg residents, 1978. *Roy Cole.*

1980s; with the fall of its principal patron, the USSR, communism lost its main voice and model in Africa. Marxist–Leninist states were typified by Angola under Augustino Neto, Mozambique after independence in 1974, Ghana under Kwame Nkrumah, Mali under Modibo Keita, and Guinea under Sékou Touré. Under Mozambique's leader at independence, Samora Machel, the state took control of the means of production: factories, farms, commercial enterprises, and services. No private enterprise was permitted—the state, particularly the party (the vanguard of the revolution, according to Lenin), controlled everything. African Marxist–Leninists viewed Africans as oppressed by colonialism and capitalism, thus as similar to people all over the world whom they supposed are oppressed by capitalism. Colonialism was viewed as a pernicious outgrowth of capitalism, and that the fight against colonialism must be continued against capitalism.

The main task in the Marxist–Leninist model was to organize production in accordance with communist principles: collectivization of agriculture, industrialization along Soviet lines, and indoctrination and "mobilization" of the masses for socialist development. The main issue after the revolution was deciding how to finance the transformations envisioned by the leaders. In the event, development was top-down; the party bureaucracy simply ordered things to happen without involving the citizens of Mozambique. The USSR had been rather successful in the top-down taxation of farmers to finance industrialization, but this did not work in Mozambique or other like minded countries. Instead, the country stumbled along with assistance from the USSR, Cuba, and other left-leaning states into the 1980s when it collapsed—the modestly promising colonial legacy of infrastructure destroyed by unrealistic planning, government mismanagement, and civil war. The government later abandoned Marxism–Leninism and embarked on more market-oriented endeavors.

African Socialism

African socialist countries are typified by Tanzania under Julius Nyerere, its first president after independence. Nyerere advocated a nonaligned policy for Tanzania, favoring neither the socialist nor the capitalist camp. Ideologically, the African socialists attempted to theorize a developmental path that was neither capitalist nor communist but more in keeping with the putative communalist traditions of Africa. The state took an active role in the operation of business, controlling enterprises from soap factories to flour mills. In an attempt to build a national identity and to transcend the local, ethnic, and tribal identities, the Tanzanian government stopped collecting ethnic/tribal data in the national census. Swahili, originally an East African trade language and a mixture of Bantu languages and Arabic, was adopted as the national language. The African socialists borrowed a page from the Maoists in China in their emphasis on self-reliance rather than outside assistance, and in their focus on farm rather than factory as the starting point for economic and social development.

One of the first development efforts after independence in Tanzania was the reorganization of the rural agricultural geography. Tanzanian farmers were forced to move from the scattered farmsteads where they lived with their families to government-built villages. The country's single party wanted to provide modern services to farmers (schools, clinics, clean water, etc.). The leaders also wanted to efficiently transport agricultural produce from the rural areas to the cities and thought that this flow of goods and services (and taxes presumably) would be facilitated by "villagization." Although the African socialists in Tanzania emphasized that the *ujamaa* approach to development comes from the bottom up, in actual fact the orders for villagization and the collectivization of agriculture came from the Tanzanian bureaucracy and, like it or not, the Tanzanian farmer had to comply.

The efforts to erase tribal and ethnic distinctions, the adoption of an African national language, the collectivization of the farm populations, and the indoctrination of the citizenry regarding a mythologized communal past that was supposed to have existed before colonialism represented enormous, costly efforts on the part of the state. In the end these efforts failed. In the late 1980s, amid economic crises, Tanzania abandoned socialism and embarked on a more market-oriented program of development.

African Capitalism

A few African countries chose cooperation with the world market economies and even the former colonial power as a development strategy. Neither of the countries examined here was a democracy, and in both countries the man who became leader at independence died in office. Côte d'Ivoire and Kenya embraced expatriate technical assistance and multinational corporations after independence, using international aid as a tool for development. In both countries the state became very powerful in the hands of the president. These countries based growth on tourism, and on agricultural exports of commodities such as coffee, pineapples, cocoa, and tea. Export-led growth would be the engine of development for the African capitalists. In the 1990s, however, corruption, political instability, and declining Western confidence (in Kenya's autocratic leadership) and the resulting decline in investment pushed both countries into crisis. Analysts criticized the lack of transparency in governments that were notorious for the personal rule by the respective leaders and "crony capitalism" (Reno 1999).

In a rather different vein, South Africa, a market-oriented, capitalist country by any measure, developed a peculiar economic form that critics came to call "racial capitalism." That is, the distribution of employment, the geography of residence, and the politics of international and domestic labor migration were determined by race. There was a white core and a black periphery. The whites controlled the means of production and possessed the managerial positions and skilled occupations. Blacks occupied the lowest occupational strata. Their travel within South Africa was tightly controlled by means of a network of checkpoints, where they were required to furnish internal passports on demand to the authorities. The practice of checking travel documents was not restricted to South Africa; black African authoritarian regimes commonly exerted strict control over the movements of their citizens, forcing them to register with the police when they travelled and examining identity and travel documents at checkpoints. Fortunately, these practices have greatly diminished.

The Islamic State

In some parts of Africa, Libya for example, Islam grew in resistance to European colonialism. In other areas, Senegal for example, the Muslim establishment mainly cooperated with the European colonialists. Where Muslims were the occupying, colonial power, as in southern Sudan, Islam did not grow in the same way. After independence, the influence of Islam grew almost everywhere in Africa. Islamic law (Sharia) was instituted in the Sudan in 1983 by the then president, Muhammad Jaafar al-Nimeiry, and was further reinforced after an Islamic extremist government, led by Dr. Hassan al-Turabi and General Omar Bashir, came to power after a coup d'état in 1989. Islamic law became the law of the land in 12 provinces (states) in northern Nigeria from 2000–2003. An Islamic party won the Algerian elections in 1992, and the military government annulled them with the rationale that it would be the last election Algeria would ever have if the Islamists were to take power. During the 1990s, Egypt fought a civil war against Islamic extremists who wished to take power by force. Some view the rise in the Islamic model of development as a response to the failure of capitalism and socialism, the Western models of political, economic, and social organization, to raise standards of living and increase development.

The philosophy of the Islamic state is anti-Western and mildly anticapitalist—interest on investments is forbidden, and there is the concept of the "just" price for a good. The agenda of an Islamic state is more social than economic: "Islamic" dress becomes a norm; the role of women is circumscribed; the legal system introduced by the Europeans is dismantled, and Islamic law is instituted in its place. In Islamic or Islamizing states, Friday, the Muslim day of prayer, is declared the weekly holiday rather than Sunday, the holy day of the colonialists. In extreme Islamic states, there is an attempt to purge the present of features deemed "un-Islamic," thus restoring the Muslim community to that thought to have existed in Mecca during the lifetime of Muhammad.

There are positive and negative aspects to the Islamic model. Although Islam originated in the same Middle Eastern cultural hearth as Judaism and Christianity, it may be a mode of political and social organization that is felt by Muslims to be more "African" than European models—perhaps only in opposition to the colonizer's religion. On the other hand, it may be a recipe for confessional conflict. In all provinces in Africa where Islamic law has been instituted, Christian and animist populations exist and conflicts have resulted in much bloodshed. Intolerance, ethnic cleansing and genocide are part of the spread of radical Islam, and the scramble for natural resources, in the Sudan. Lastly, because for the strictly orthodox Muslim there is no separation between religion and politics, Islamic Law is an unsuitable form of political organization for the modern state. The modern state is secular; it does not advocate or promote a religion. Governance in the secular state is carried out to serve the people not to execute a god's will.

Democracy

Whether Marxist, African socialist, capitalist, or Islamist, late twentieth-century African political philosophy, however well articulated on paper, tended to be undemocratic and repressive of the people. Military dictatorships of many political stripes have been the norm for decades of postindependence history. Bratton and van de Walle, leading analysts of democracy in Africa, had this to say about the movement to democracy over the last 20 years.

> By the 1980s, authoritarian rulers in Africa's patrimonial regimes typically faced a crisis of political legitimacy. This crisis was manifest in a loss of faith among African citizens that state elites were capable of solving basic problems of socioeconomic and political development. Leaders had damaged their own claim to rule by engaging in nepotism and corruption, which led to popular perceptions that those with access to political office were living high on the hog while ordinary people suffered. The erosion of political legitimacy built to crisis proportions because authoritarian regimes did not provide procedures for citizens to peacefully express such grievances and, especially, to turn unpopular leaders out of office. (Bratton and van de Walle 1997: 98–99)

Since the end of the Cold War, much impetus has been given to experimentation with democracy around the continent, and popular protests led to regime transitions in the early 1990s. Some multiethnic states have called national conventions so that all voices will be heard in the discussion of the transformation of the state from dictatorship to democracy. The Republic of Mali's democratically elected government is a result of such a convention. Kieh (2001) argues that African countries need such conventions to move toward accountable, representative government. Based on the disappointing track record of Africa's one-party states, Kieh sees no way forward but with democracy and accountability.

Figure 5-5 presents evaluations of African political rights and civil liberties from 1972 to 2003, calculated by the Freedom House, an independent advocacy group founded 60 years ago and based in the United States. Countries were rated on a 7-point scale for political rights and civil liberties. A score of "1" represents "most free" and a "7" is "least free." The countries are then divided into three categories: "free" (ratings of 1 to 3); partly free (ratings from 3 to 5.5); and not free (ratings above 5.5). The Freedom House research is unique because the ratings reflect whether citizens are actually able to exercise freedoms or whether the privileges described exist on paper only.

The democratic trend is unmistakable but fragile. The African trend toward democracy is part of a worldwide movement for political rights and civil liberties. In 1973, only 29% of all countries of the world were free, while in 2003, 46% were. Research by Freedom House has indicated that there are characteristics of countries that make for freedom. Significantly for Africa, ethnic homogeneity has been identified as an important factor: a state with a numerically dominant ethnic group is three times more likely to be free than a multiethnic state. It was also found that democracy was opposed by extremist Islamic movements and authoritarian regimes. Islamic regimes are disproportionately classed among the "non-free" because of the traditional relegation of women to second-class status and the lack of separation between religion and state. Authoritarian regimes are characterized by corruption, cronyism, and statist economies. It was also found that oil and gas exert a corrosive power on governments, conferring vast riches on a narrow elite who sought to preserve their position and power.

Although there have been many successes in Africa (Cape Verde, Ghana, Mali, Namibia, Botswana, South Africa, Mauritius, and São Tomé and Principe), recent setbacks in the development of freedom have occurred in Benin, the Central African Republic, Egypt, Eritrea, Liberia, Morocco, Nigeria, and Zimbabwe. Rising Islamic militancy and the resulting fear of Islamic movements in Egypt have been used to deny civil rights. In Liberia, periodic political instability has, until recently, caused an oscillation from "partly free" to "not free." Presidential elections were held in 2005 and for the first time in Liberia—and Africa—a woman was elected president. In Zimbabwe, the incumbent president, Robert Mugabe, has been using terror to intimidate the opposition and voters alike in his bid to keep himself and his party in power. Curtailing of political rights was the cause of the lower rating in 2002 and 2003 for the other countries that experienced setbacks just mentioned.

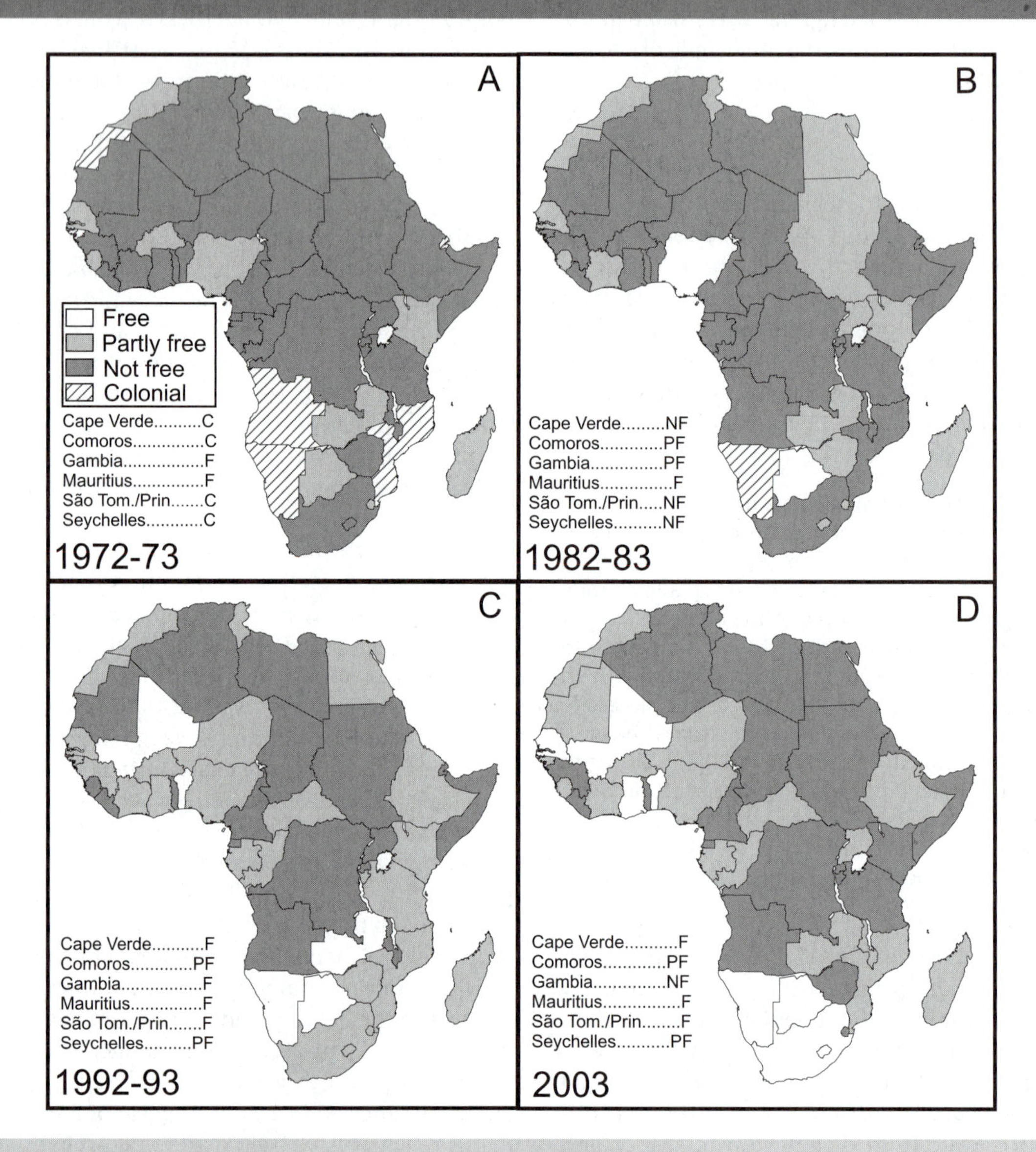

FIGURE 5-5 Political rights and civil liberties in Africa, 1972-2003.
Freedom House.

CHALLENGES TO THE AFRICAN STATE

Early after independence, African states began to experience challenges to their hegemony, autonomy, security, legitimacy, revenue, and accumulation. The inevitability of challenge seems obvious with hindsight, given the weakness of the African state, the low levels of economic and social development, the poorly developed infrastructure, the divisions along ethnic, linguistic, and religious lines, and the poor resource endowments; paradoxically rich resource endowments led to unrest in some instances. Some of the challenges to the hegemony of newly independent

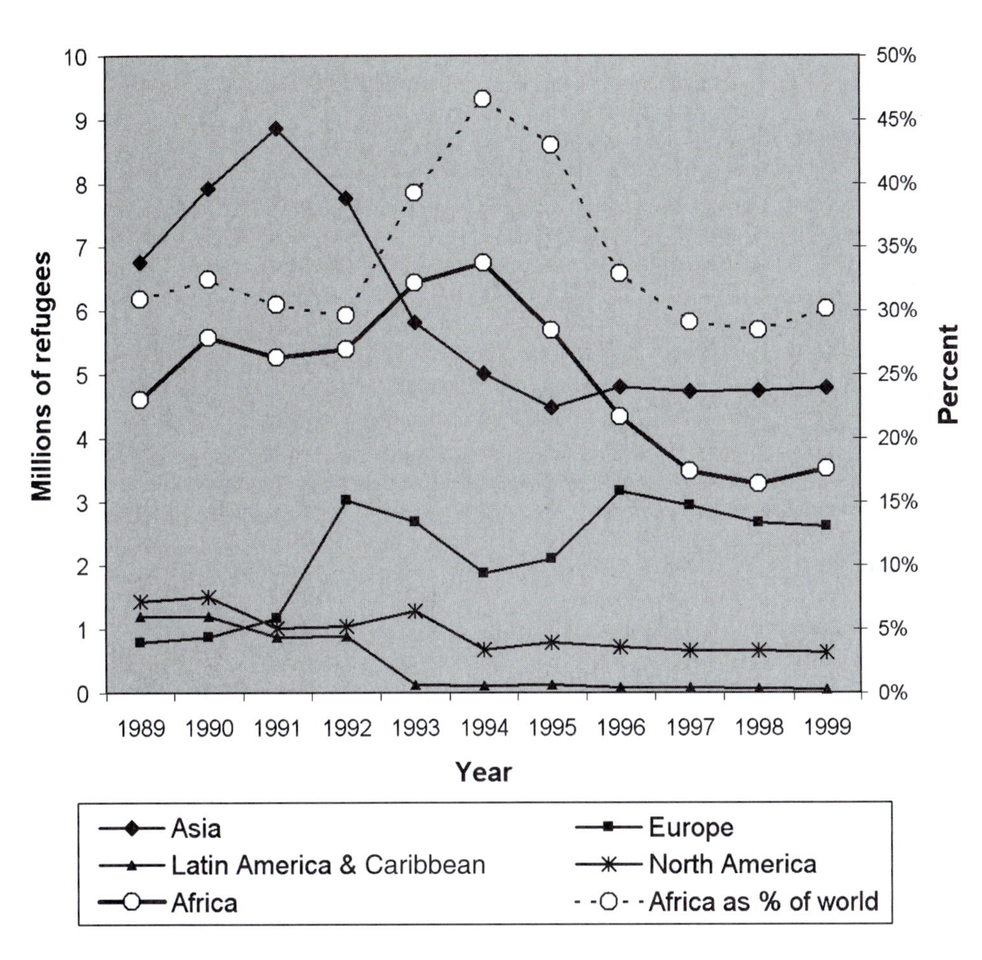

FIGURE 5-6 Refugee populations by world region, 1989-December 1999.
Source of data: UNHCR.

African states have been discussed and more detail will be given in the regional section of this book, beginning in Chapter 10.

Challenges to a state's hegemony and security have resulted in political instability in a variety of locations across the continent in the years since independence. A state that is weak and unable to project power across its national space loses political and economic control over sometimes vast swathes of its territory. This has been the case in Angola, Liberia, Sierra Leone, Liberia and the Democratic Republic of the Congo, where rebels have taken control of diamonds, iron mines, and other valuable natural resources that were once managed by the state. These actions have deprived states of revenue and the main sources of accumulation, further weakening established governments. It is difficult not to be pessimistic about the future of these African states. Their problems, along with their refugees spill across borders, create regional instability. Let us look at refugees and other displaced persons as one tragic product of the weak state.

In 1999, of the 21,793,300 people who constituted the worldwide population of concern to the UN High Commission for Refugees (UNHCR), 6,072,900, or 28%, of refugees, asylum seekers, internally displaced persons, and returnees lived in Africa. This number has been highly variable over the past decade or so (Figure 5-6). Referring only to refugees, the genocide in Rwanda occasioned a large outmigration of people in the mid-1990s. According to Figure 5-6, it is evident that Africa possesses a disproportionate share of refugees compared with the rest of the world. The population of Africa represented 14% of the world total in 1999, but the refugee population of the continent represented 30% of the world total. During the political turmoil in Rwanda during the mid-1990s, almost half (47%) of the world's

TABLE 5-2 REGIONS OF POLITICAL INSTABILITY AND FORCED MIGRATIONS, JANUARY 1, 2001.

Region	Refugees	Asylum Seekers	Returnees	Internally Displaced Persons and Others	Total
Great Lakes[1]	964,000	14,980	841,100	64,590	1,884,671
East Africa and the Horn[2]	1,117,300	8,405	23,970	0	1,149,677
West and Central Africa[3]	971,750	12,260	926,030	617,000	2,527,043
Southern Africa[4]	247,410	31,990	42,590	0	321,994
North Africa[5]	183,650	5,860	0	0	189,515
Total	3,484,110	73,495	1,833,690	681,590	6,072,900

1. Burundi, Democratic Republic of the Congo, Rwanda, Tanzania.
2. Eritrea, Ethiopia, Kenya, Somalia, Sudan, Uganda.
3. Benin, Burkina Faso, Cameroon, Central African Republic, Chad, Côte d'Ivoire, Equatorial Guinea, Gabon, Gambia, Ghana, Guinea Bissau, Guinea, Liberia, Mali, Niger, Nigeria, São Tomé and Principe, Senegal, Sierra Leone, Togo.
4. Angola, Botswana, Comoros, Lesotho, Madagascar, Malawi, Mauritius, Mozambique, Namibia, Seychelles, South Africa, Swaziland, Zambia, Zimbabwe.
5. Algeria, Egypt, Libya, Morocco, Tunisia.

refugees lived in Africa. Fortunately, political instability appears to be on the decline.

At the regional level today, there are three major areas of political instability and forced migration in Africa according to the UNHCR: West Africa, particularly Liberia and Sierra Leone; the Great Lakes region, especially Rwanda and the eastern Democratic Republic of the Congo; and the Horn. Table 5-2 presents data on populations of concern to the UNHCR in these regions.

Dr. Jacques Mangala, a Congolese scholar of international law, has observed that population displacements are have two causes: an immediate cause, which impels people to take flight, and a *cause profonde*, or root cause. The root causes, Mangala argues, "designate a complex of fundamental social dysfunctions that create a favorable context for displacement. Considered as 'causes of causes' they are often very complex and difficult to measure but certain constant themes come up time and time again that are basically of a political, economic, and ethnic nature" (2001: 1076).

Preventing population displacements means that intervention must begin before a situation becomes intolerable and flight seems like the only option. Mangala holds that what is needed to address the refugee problem is the promotion of human rights, economic and social development, conflict resolution, democratic institutions, and environmental protection. These wider developmental issues cannot be effectively addressed by weak states, acting alone. The international community has taken a leadership role in some cases, but international organizations do not ordinarily intervene in "internal" affairs. A conflict or population displacement must cross borders before international organizations will agree to get involved; and this may happen after the fact, or not at all.

Upon examining the political, social, and economic challenges to the African state, Chabal and Daloz (1999) concluded that a distinctly African response to the challenges has emerged since independence. This response, rooted in the precolonial African past, is the result of the poor institutionalization of the state (lack of a truly independent bureaucracy) and the nature of African society itself; Chabal and Daloz call it the "retraditionalization" of African society. African society is vertically rather than horizontally organized. This means that bonds or ties between kin, clan, and tribe (patrimonial ties) are "more significant than horizontal functional bonds or ties of solidarity between those who are similarly employed or professionally linked" (Chabal and Daloz 1999: 20). There is no doubt that many Africans are working sometimes at great personal risk, to develop a true civil society and independent bureaucracy and the fruits of labor are evident in the increasingly favorable Freedom House scores over the years.

SUPRANATIONAL AND REGIONAL ORGANIZATIONS

A number of supranational and regional organizations exist in Africa to further the economic and political aims of several states collectively. Some organizations, such as the Commonwealth (formerly British Commonwealth) and the European Union, have preserved and strengthened long-standing economic ties between Africa and Europe. Others, such as the African Union (AU), the Arab Maghreb Union (AMU), the Economic Community of West African States (ECOWAS), the East African Community (EAC), the Monetary and Economic Community of Central Africa (CEMAC), the Southern Africa Development Conference (SADC), and the Southern African Customs Union (SACU), whose memberships are entirely African, have addressed themselves almost exclusively to political and economic issues based in Africa (Figure 5-7).

In the immediate preindependence period, as African nationalisms intensified, Britain attempted to reorganize some of its colonies into larger political units. It tried to create an organization to be called the East African Federation of Kenya, Uganda, Tanganyika, and Zanzibar. Lack of local support prevented such a body from materializing, although the Common Services Organization survived. The Central African Federation collapsed for reasons described in the section on challenges to the African state. Most former British colonies joined the Commonwealth, which meant preferential trade agreements, financial aid, loans, and other benefits. South Africa withdrew in 1960 after member states voiced strong criticism of its racial policies, and because the precepts of Afrikaner nationalism differed so strongly from those held elsewhere in the Commonwealth.

Similarly, French-speaking Africa has economic and financial ties with Europe, especially with France and the European Union (EU). Most French-speaking states belong to the monetary union known as the Communauté Financière Africaine (CFA), which has linked members', or African franc zone (AFZ), currencies with the former French franc at a fixed rate of exchange. Whereas many Anglophone countries that created their own currencies at independence experienced inflationary conditions, the CFA system has kept currencies stable. In an economic sense, such currency linkage keeps the CFA countries in the French sphere of influence. This relationship has not changed with the advent of the euro in 2002, although ECOWAS has indicated that it intends on developing a currency for member nations over the next 5 years. France's ties with the African Franc Zone countries involve not only monetary arrangements, but also comprehensive French assistance in the forms of budget support, foreign aid, technical assistance, and subsides on commodity exports. Several French-speaking states belong to a series of regional organizations such as L'Organisation Commune Africaine et Mauricienne (OCAM) and the Central African Customs Union (UDEAC), the purposes of which are to accelerate political, economic, social, and technical development of their members.

The Economic Community of West African States (ECOWAS), founded in 1975, links 15 former British, French, and Portuguese colonies. Its mission is to promote economic integration in "all fields of economic activity, particularly industry, transport, telecommunications, energy, agriculture, natural resources, commerce, monetary and financial matters, social and cultural issues." ECOWAS member states are working toward the establishment of a single currency. Progress in economic integration has been below expectations in the history of the organization, due in large part to the low levels of development of the ECOWAS states.

> Numerous problems have been encountered by ECOWAS in the enhancement of the process of regional integration of West Africa. Among the most important of these problems are: the political instability and bad governance that have plagued many of the countries; the weakness of the national economies and their insufficient diversification; the absence of reliable road, telecommunications and energy infrastructure; the insufficient political will exhibited by some member States; the bad economic policies in certain cases; the multiplicity of organisations for regional integration with the same objectives; the irregularity in the payment of financial contributions to the budgets of the institutions; the failure to involve the civil society, the private sector and mass movements in the process of integration; the defective nature of the integrational machinery in certain cases.
> ECOWAS (2002)

Other regional groupings in Africa have had similar disappointing results for the very same reasons.

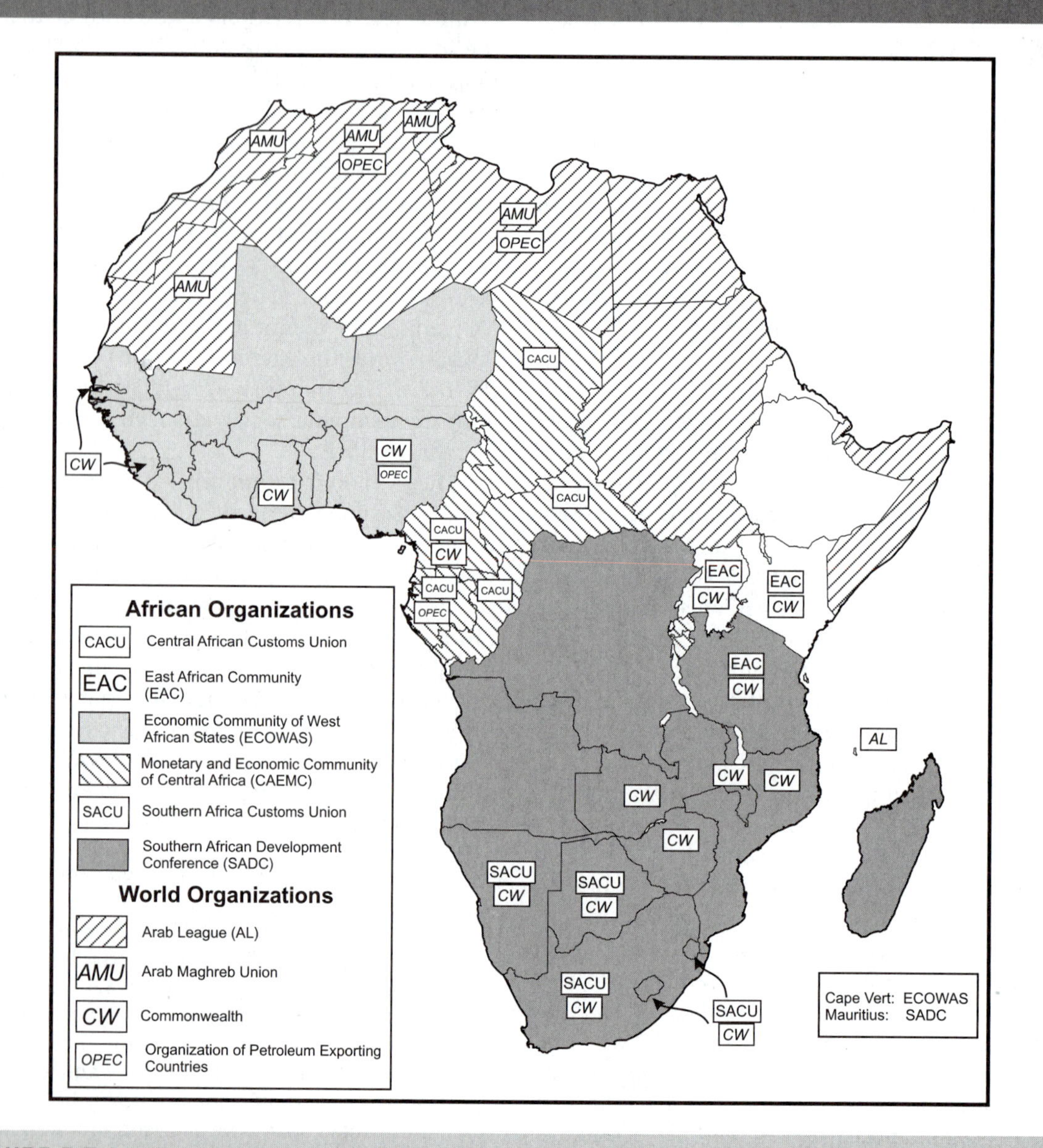

FIGURE 5-7 Memberships in selected continental African and world organizations. ***Source of data: World factbook.***

One of the oldest supranational organizations in Africa is the Arab League. Founded in 1945, the league is based on the perceived cultural unity of the states that subscribe to it. The peoples of the Arab world are united by the Arabic language, the Muslim religion (for the most part), and a history of colonial subjugation by Turks and Europeans. Ironically, Arabs in Africa were and still are colonial powers—mostly of a relatively long duration compared with the Europeans. But there are also many issues on which the Arab peoples continue to be divided: democracy, the necessity of monarchy, and the importance of Islamic law in the modern state are just three. In addition, there are questions about minority rights in every Arab country in and outside Africa.

The greatest unifying element of all 22 member states is hostility to the state of Israel and to Zionism.

Arab nationalism arose first in the Middle East, after World War I. But by the end of World War II it had spread across North Africa, and there was a move for the establishment of a great Arab state. That state was never established. Hopes were destroyed by the many divisive forces that still prevail: ethnic strife, poverty, illiteracy, and vested interests. The Arab League has been a loose and rather ineffective organization whose only apparent unifying factor remains the common opposition to Israel.

In the economic sphere, attempts have been made to coordinate the policies of oil-producing members toward the major concessionaires. Steps were taken in the 1960s to create an Arab common market, but competitive economies and conflicting policies of the members prevented that from materializing. Nevertheless, much has changed in the Arab world since the formation of the league: Egypt attained representative government; Algeria rid itself of French control; the Suez invasion was warded off; Libya charted a "third way" toward social and political organization, combining elements of socialism, capitalism, and Islam; Sudan became independent; Morocco is considering constitutional monarchy. All of these league members are facing challenges to their legitimacy by Islamist Movements. Algeria and Libya, together with Nigeria and Gabon, are also members of the powerful *Organization of Petroleum Export Countries* (OPEC) (Figure 5-7).

The largest of the exclusively African supranational organizations is the *African Union (AU),* formerly the *Organization of African Unity* (OAU). Established in 1963, and headquartered in Addis Ababa, the AU, like the Arab League, is viewed optimistically as the embryonic form of an eventual all-African government. Its charter defines its aims as seeking to promote the unity of Africa, to defend the independence and sovereignty of African states, and to coordinate and harmonize policies between member states on a strictly nonalignment basis. A United States of Africa is the ideal of a number of representatives, among whom Nkrumah of Ghana was especially vocal. While some wish to see the rapid implementation of such a federation, others desire to maintain total sovereignty and are reluctant to enter new political unions. The whole argument is not entirely dissimilar from that which attended the birth of the Council of Europe, and ultimately it goes back to the basic question of whether political unification should precede or follow collaboration in other spheres.

The AU's major practical achievements have included mediation in the variety of conflicts that have occurred in Africa since the inception of the organization. Before the last country was decolonized in Africa, the AU was in the forefront of the struggle for national liberation. It also is committed to the spread of education, the elimination of illiteracy, the improvement of health standards, the support of cultural exchange programs, and economic regionalization. It supports the equality of all African states. The AU has a policy of noninterference in internal political affairs of other African states, and it respects the boundaries established by the colonial powers.

The AU continues to help mediate conflicts in Africa, combat corruption, and work toward building civil society in Africa. In 1999 member states resolved to establish an African Union modeled after the European Union and its associated bodies and to accelerate the establishment of the stalled African Economic Community (AEC). In 1991 the AU's parent organization, the OAU, elaborated a 30-year, process for the creation of the AEC. The objectives of the community, to be achieved over 30 years, are as follows.

1. *To promote economic, social and cultural development and the integration of African economies in order to increase economic self-reliance and promote an endogenous and self-sustained development.*
2. *To establish, on a continental scale, a framework for the development, mobilization and utilization of the human and material resources of Africa in order to achieve a self-reliant development.*
3. *To promote cooperation in all fields of human endeavour in order to raise the standard of living of African peoples, and maintain and enhance economic stability, foster close and peaceful relations among Member States and contribute to the progress, development and the economic integration of the Continent.*
4. *To coordinate and harmonize policies among existing and future economic communities in order to foster the gradual establishment of the Community.* (OAU 1991)

All of Africa's independent states belong to the United Nations, and thus to several affiliated organizations. These include the United Nations Economic Commission for Africa (UNECA), the United Nations Development Program (UNDP), the International Bank for Reconstruction and Development (IBRD) of the World Bank, the International Development Association (IDA) of the World Bank, and the International Finance Corporation (IFC). All the independent states also belong to the African Development Bank (AFDB), established in 1964 under the aegis of UNECA. While this list is not complete, it illustrates the nature of supranational and regional organizations in Africa today, and these and others are discussed in the regional chapters that follow.

The contemporary map of Africa thus differs dramatically from that of the colonial era. While the state boundaries are essentially the same, the internal political organizations, *raisons d'être,* and approaches to both national and international issues are very different. Commanding a powerful position in the United Nations, today, the African bloc is a force that should not be underestimated in world affairs. Nevertheless, with notable exceptions, Africans face many challenges in their efforts to build viable states.

CHAPTER REVIEW QUESTIONS

1. Explain the role of environmental and cultural geographic factors on the siting of new capital cities in Botswana, Malawi, Mauritania, Nigeria, and Tanzania.
2. Describe the methods that were used to delimit the boundaries of African countries.
3. Evaluate the case made by former Senegalese president Léopold Sédar Senghor, that all black Africans have a commonality of values, an unconscious experience, that distinguishes them from all others. Can a group of people be set apart in this way based on skin pigmentation alone, or is something else involved?
4. Discuss the contemporary African state and Crawford Young's attributes of the state.
5. What are the possibilities and problems inherent in the boundaries of African states?
6. Both unitary and federal states exist in Africa. Discuss the positive and negative aspects of each type of state. Which one do you think would be better for African states?
7. African governments elaborated a variety of political frameworks to help them in the nation-building task after independence. Discuss the frameworks mentioned in the text and link your discussion to current events. Use a newspaper, magazine, or Internet article on African current events to show examples of the various political frameworks. For example, you might find an article on the presidential or parliamentary elections in one African country or discussion of the legacy of socialism in another country. Good places to read about current events in Africa are the *Christian Science Monitor* and the *Economist.* See this chapter's Appendix 2 for some African newspapers that can be accessed on the Internet.
8. List and discuss some challenges that African states face.

GLOSSARY

Civil society Groups of citizens that are organized to represent their interests and counter the power of the state.

Collectivization A process by which landownership is transferred from individuals or groups to the state. Collectivization is a characteristic of some socialist and all communist governments.

Core region A region that dominates production and trade, and often contains the largest urban area in a country as well as the dominant culture group.

Council of Europe The oldest institution of the European Community. It was founded in 1949 as an intergovernmental cultural and social body to protect human rights, and democracy in Europe. It also is concerned with the promotion of European cultures and the resolution of European social problems such as minorities, xenophobia, and bioethics.

Customs union An international organization to promote free trade between its members by lowering or eliminating tariffs and other restrictions on the exchange of goods.

Dysfunction A malfunction.

Federal state A method of organizing power in a country such that a central government represents a number of regional units, provinces, or states having common interests: for example in defense, foreign affairs, and communications. Each of the regional units has its own government, legislative body and budget, and policies and laws to regulate the areas not delegated to the central, or federal, government. The word "federal" is derived from the Latin word for "league."

Nation-state A state in which the citizens are culturally homogeneous.

Nationalism The desire of a particular culture or ethnic group for its own state; advocacy and support for one's country over others.

Nationalization A process occurring in socialist and communist countries in which the private property of individuals is taken over by the state, ostensibly to be administered for the benefit of the entire citizenry. Banks and businesses have been targets for nationalization, particularly foreign-owned institutions.

Negritude A philosophy of being that arose in the 1940s, propounded by former president Senghor of Senegal and others. Senghor argued that African thinking is principally marked by an emotional rationality rather than by Western analytical rationality.

Nodality Characterized by a center, usually, an urban center that is the focus of trade, industry, and other activities for a wider area.

Primate city A dominant city characterized by inordinate population size, cultural importance, and intensity of economic activities with respect to other cities in its country. Paris is the primate city of France, just as London is the primate city of the United Kingdom. Etymologically, the word "primate" stems from "first."

Prorupt state An otherwise compact state with an elongated strip of territory extending directly away from it.

Proxy war Hostility between two powers expressed in a war that is fought by others. Proxy wars were characteristic of the Cold War era.

Raison d'être Reason to be, or exist. Rationale.

Sharia Body of law based on the Koran, traditions of the prophet (hadith), and law; the guiding authority for traditional Muslims.

Statism The assumption by a national government of a disproportional amount of concern and control over social and especially economic affairs in a country.

Supranationalism Political, economic, or cultural cooperation between three or more states to promote mutually beneficial objectives.

Unitary state A method of organizing power in a country such that a central government exercises power uniformly across the levels of administration and controls local governments from the center. The central government is responsible for the development of policy, laws, governance, and budgets for all units of government.

Villagization In authoritarian countries, the process of forcing farmers who live on their farmsteads to move to a new or existing village. *Ujamaa* villages built during the 1960s and 1970s in Tanzania are characteristic of this use of governmental power.

BIBLIOGRAPHY

Bayart, J.-F., S. Ellis, and B. Hibou. 1999. *The Criminalization of the State in Africa*. Oxford: James Curry.

Best, A., and de Blij, H. J. 1977, *African Survey*. New York: Wiley.

Bratton, M., and N. van de Walle. 1997. *Democratic Experiments in Africa: Regime Transitions in Comparative Perspective*. Cambridge, U.K.: Cambridge University Press.

Brauer, R. W. 1995. Borders and Frontiers in Medieval Muslim Geography. Transactions of the American Philosophical Society, 85(6):1–73.

Chabal, P., and J.-P. Daloz. 1999. *Africa Works: Disorder as a Political Instrument*. Oxford: James Curry.

de Blij, H. J., and P. O. Muller. 2006. Geography: Realms, Regions and Concepts. New York: Wiley.

ECOWAS. 2002. *Achievements and Prospects*. Lagos, Nigeria: Economic Community of West African States. http://www.ecowas.int/.

Ellis, S. 1999. *The Mask of Anarchy: The Destruction of Liberia and the Religious Dimension of an African Civil War*. New York: New York University Press.

Freedom House. 2004. *Freedom in the World: The Annual Survey of Political Rights and Civil Liberties*. Washington, D.C.: Freedom House.

Herbst, J. 2000. States and Power in Africa: Comparative Lessons in Authority and Control. Princeton: Princeton University Press.

Hodges, T. 2001. *Angola from Afro-Stalinism to Petro-Diamond Capitalism*. Oxford: James Curry.

Ibn Warraq. 1996. *Why I Am Not a Muslim*. Amherst, N. Y.: Prometheus Books.

Kieh, G. 2001. Civil Wars in Africa. Paper presented as part of the African Studies Forum, Alumni House, Grand Valley State University, Allendale, Mich., November 19, 2001.

Mangala, J. 2001. Prévention des Déplacements Forcés de Population. *International Review of the Red Cross*, 83(844): 1067–1095.

Murdock, G. P. 1949. *Africa: Its Peoples and Their Culture History*. New York: McGraw-Hill.

OAU. 1991. The Abuja Treaty of 1991. Addis Ababa: Organization of African Unity.

Reno, W. 1999. *Warlord Politics and African States*. Boulder, Colo.: Lynne Rienner Publishers.

Young, C. 1997. The African Colonial State in Comparative Perspective. New Haven: Yale University Press.

APPENDIX 1 INTERNET ADDRESSES OF SOME AFRICAN AND INTERNATIONAL ORGANIZATIONS

Organization	Web Address
Arab League	www.leagueofarabstates.org
Commonwealth	http://www.thecommonwealth.org
Economic Community of West African States (ECOWAS)	http://www.ecowas.int/sitecedeao/english/ppt-ecowas.htm
Freedom House	http://www.freedomhouse.org/
African Union (AU)	http://www.Africa-Union .org
Organization of Petroleum-Exporting Countries (OPEC)	www.opec.org
Southern African Development Conference (SADC)	http://www.sadc.int/

APPENDIX 2 INTERNET ADDRESSES OF SOME AFRICAN AND INTERNATIONAL NEWSPAPERS

Country	Newspaper	Internet address
Algeria	*Djzair Online*	http://www.djazaironline.net/
Angola	*Jornal de Angola* (Po.)	http://www.ebonet.net/jornaldeangola/
Cameroon	*Le Messager* (Fr.)	http://wagne.com/messager/
Botswana	*Botswana Gazette*	http://www.info.bw/~gazette/
Egypt	*Cairo Times*	http://www.cairotimes.com
Ethiopia	*Addis Tribune*	http://www.addistribune.com
Ghana	*Ghanaian Chronicle*	http://www.ghanaian-chronicle.com/220320/index.html
International	*North Africa Journal*	http://www.north_africa.com/one.htm
International	PanAfrican News Agency	http://allafrica.com/
Kenya	*Daily Nation*	http://www.africaonline.com/nation/
Kenya	*East African*	http://www.nationaudio.com/News/EastAfrican/current
Morocco	*Morocco Today*	http://www.morocco-today.com/indexk.htm
Namibia	*Namibia Economist*	http://www.economist.com.na/
Namibia	*The Namibian*	http://www.namibian.com.na/
Nigeria	*Post Express*	http://www.postexpresswired.com/
South Africa	*Daily Mail and Guardian*	http://www.mg.co.za/mg/
Swaziland	*The Guardian*	http://www.theguardian.co.sz/
Tanzania	*The Guardian, Nipashe*	http://www.ippmedia.com
Zambia	*Zambia Post*	http://www.zamnet.zm/zamnet/post/post.html
Zimbabwe	*Zimbabwe Independent*	http://www.mweb.co.zw/zimin/
Zimbabwe	*Zimbabwe Standard*	http://www.samara.co.zw/standard/index.cfm

CHAPTER 6

The Population of Africa

Almost 450 million people lived in Africa in the mid 1970s. At the time, Africa's population represented one-tenth of the world's total. Over 35 years later (2005), Africa's population had more than doubled to 906 million, representing 14.1% of the world's total population. It has continued to grow.

The distribution of Africa's population is very uneven: extensive areas of the Sahara, the Kalahari, and the Horn are uninhabited, while major concentrations occur along the Gulf of Guinea Coast, on the Mediterranean, and around Lake Victoria, where rural densities exceed 3,000 persons per square mile (1,155/km^2). In most areas, annual growth rates are high, averaging 2.4% for the continent as a whole, with the crude birthrates and crude death rates being 38 and 15 per thousand, respectively—down from 46.4 and 19.8 in 1977. Up until the 1990s African crude death rates fell more rapidly than birth rates, at which time their decline slowed. Since that time, with the exception of the countries hard-hit by HIV/AIDS, birth rates have declined faster than death rates. In most countries population growth has not been accompanied by an equal or faster growth in social services, housing, employment opportunities, industry, investment, food supply, and general well-being. As the chapter will indicate, there have been big changes in African demography in recent decades, and we now know that African populations are increasing at a decreasing rate. Cause for guarded optimism can be found in some success stories in people–environment relations, particularly in the conservation of the environment. On the other hand, the AIDS epidemic is profoundly altering demography in some countries.

This chapter describes and interprets several important aspects of Africa's populations: the distribution and density patterns, population pressure, growth rates, migration, and urban concentrations, and the problems these factors pose to individual governments as they search for economic development and political stability. Caution is advised in interpreting the data since they are in most cases estimates compiled from incomplete information, often censuses and surveys whose accuracy is not known. The Nigerian censuses serve as an illustration. The 1952 census showed a population of 30.7 million, and the 1963 census reported 55.7 million. If these figures are correct, then Nigeria's population grew yearly by 6.1% per annum (Caldwell 1968). Most demographers feel the 1963 total is unrealistically high, having been inflated for political purposes. In 2000, U.S. analysts estimated much lower growth rates for Nigeria during the same time periods: 2.3% for 1950–1960; 2.5% for 1960–1970; 3.1% for 1970–1980; and 2.9% for 1990–2000 (U.S. Bureau of the Census 2000). Many demographers feel that the current estimate for Nigeria's population, 132 million people, has been inflated for political purposes, like the 1963 figure.

African census returns are commonly inaccurate for several reasons in addition to political motivations: accurate base maps, necessary for the delineation of enumeration areas, are often lacking; there is a shortage of trained personnel; rural populations are frequently dispersed and far from transport routes and are migratory in any event; many believe that census data will be used as a basis for increasing taxes, conscription, demands that wives of polygamous marriages be relinquished; or suppression of a political nature. Moreover, the costs of a nation- or province-wide enumeration are prohibitive; census boundaries are poorly delimited and subject to change, making comparison difficult: Few countries take censuses at regular intervals. Ethiopia, undertook its first census in 2000. (Table 6-1).

TABLE 6-1 MID-2005 POPULATION ESTIMATES (MILLIONS).

Region	Country	Population	Region	Country	Population
Northern Africa	Algeria	32.8	Eastern Africa	Burundi	7.8
	Egypt	74.0		Comoros	0.7
	Libya	5.8		Djibouti	0.8
	Morocco	30.7		Eritrea	4.7
	Sudan	40.2		Ethiopia	77.4
	Tunisia	10.0		Kenya	34.0
	Western Sahara	0.3		Madagascar	17.3
Western Africa	Benin	8.4		Malawi	12.3
	Burkina Faso	13.9		Mauritius	1.2
	Cape Verde	0.5		Mozambique	19.4
	Côte d'Ivoire	18.2		Réunion (Fr.)	0.8
	Gambia	1.6		Rwanda	8.7
	Ghana	22.0		Seychelles	0.1
	Guinea	9.5		Somalia	8.6
	Guinea-Bissau	1.6		Tanzania	36.5
	Liberia	3.3		Uganda	27.0
	Mali	13.5		Zambia	11.2
	Mauritania	3.0		Zimbabwe	13.0
	Niger	14.0	Southern Africa	Botswana	1.6
	Nigeria	132.0		Lesotho	1.8
	Senegal	11.7		Namibia	2.0
	Sierra Leone	5.5		South Africa	46.9
	Togo	6.1		Swaziland	1.1
Middle Africa	Angola	15.4			
	Cameroon	16.4			
	Central African Republic	4.2	Total		906
	Chad	9.7			
	Congo	4.0			
	Equatorial Guinea	0.5			
	Gabon	1.4			
	São Tomé and Principe	0.2			
	Democratic Republic of the Congo	61.0			

Source: Population Reference Bureau (2005).

DISTRIBUTIONS AND DENSITIES

The uneven distribution and densities across Africa reflect an unequal geographic distribution of resources and opportunities, differences in the ability to adapt to the environment, and various historical events and situations. Figure 6-1 shows that the major concentrations are situated in six areas. These six islands of population constitute about 2% of the land area of Africa but contain about half of the population.

1. Nile Valley and Delta
2. Gulf of Guinea
3. Hausaland
4. Lake Victoria region
5. Ethiopian Highlands
6. Witwatersrand industrial region

The *Nile Valley and Delta* comprise a well-watered and fertile linear oasis that supported about 72 million people in 2005 based on projections from Grove (1989) and the Population Reference Bureau (2005). Almost all of Egypt's population is crowded into less than 5% of the area, and rural densities in the lower Nile exceed 3,700 persons per square mile (1,425/km^2). These are some of the most fertile farmlands in the world, where

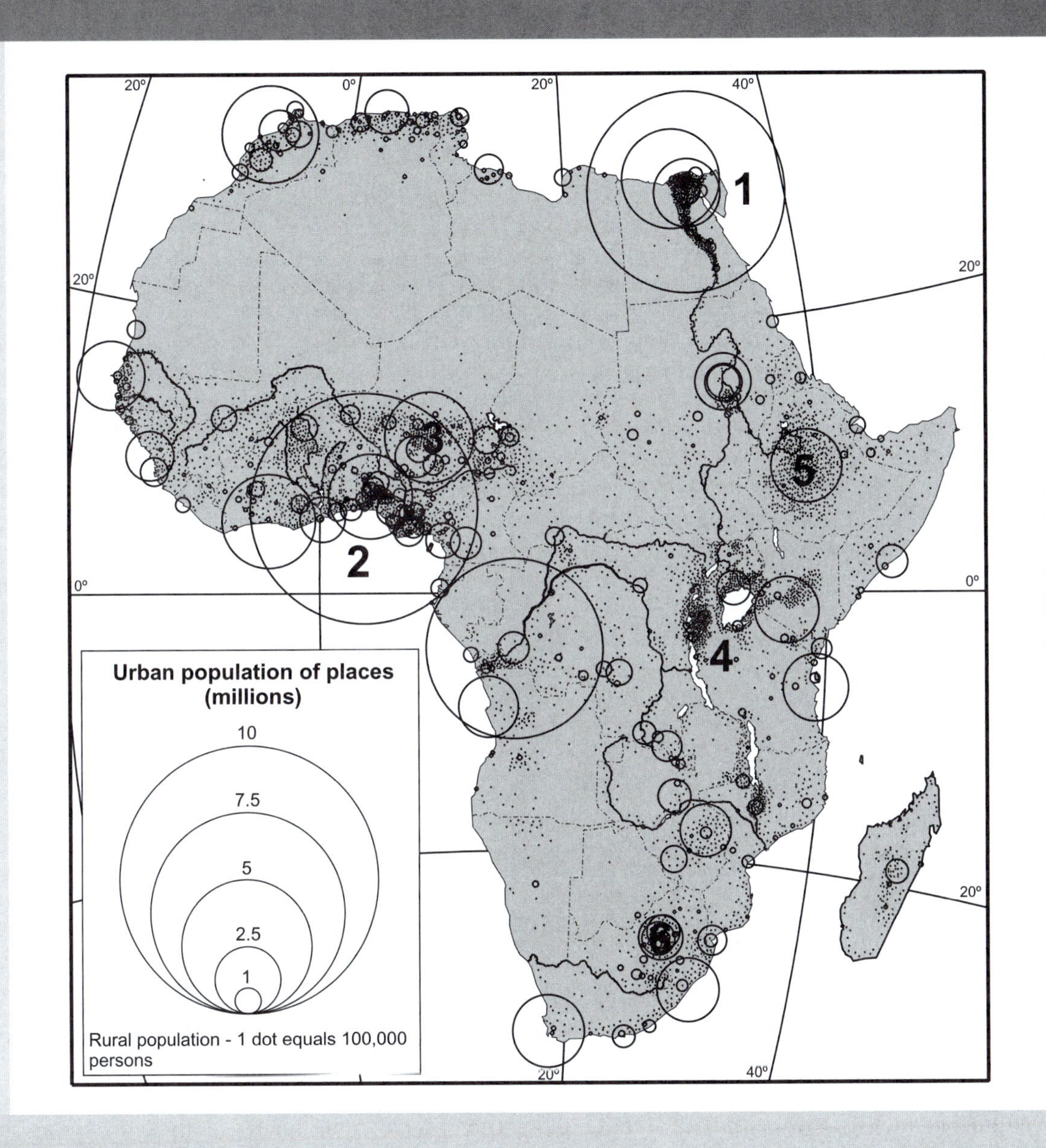

FIGURE 6-1 Urban population and rural density in Africa, 2005.

the exceptionally rich alluvial soils, controlled irrigation, and a long tradition of relatively sophisticated farm practices have contributed to high densities. Nevertheless, the flow of the Nile has been so constrained by dams and barrages that the famously rich silt, as in the past, is no longer deposited and the Nile Delta is disappearing into the Mediterranean Sea. Fertility has to be maintained by means other than the annual flood, such as the application of organic and inorganic manures. Large urban areas include Cairo at the apex of the delta and Alexandria at the northwest angle.

The largest concentration of people in Africa is found along the *Gulf of Guinea* including southern Nigeria, Benin, and Togo. The estimated population of this sea-coastal cluster, which extends from Côte d'Ivoire to Nigeria, is 150 million people. The region receives a significant amount of rainfall, which supports root crops (yams and cassava) and plantains in the south, cereals in the north, and a variety of cash and industrial crops throughout. Lagos is the most important conurbation in the region in terms of population and industry, and in economic and political dominance.

Hausaland is a fascinating area, located in Northern Nigeria, where the principal occupation is farming. Often called the "Kano close-settled zone" (Mortimore and Adams 1999) because of the unusually high density of population, it supports about 40 million people. A combination of permanent cultivation, manuring, exploitation of a variety of ecological niches, and crop rotation have enabled this high population to be supported in a very dry climate.

In the highland *Lake Victoria region* and in the highlands to the east and southwest, about 54 million people live mainly as agriculturalists. The region extends from southern Uganda, Rwanda, Burundi, western Kenya, northwest Tanzania, and the eastern Democratic Republic of the Congo. Over 80% of Kenya's population of 30 million lives on only 10 percent of the land, and two-thirds of the Tanzanian population (36 million) is concentrated on less than a fifth of the land. The Lake Victoria region is located in and around the African Rift Valley: the soils are rich, the climate is temperate, and a variety of food and cash crops are grown.

The *Ethiopian Highlands* support relatively dense populations of agriculturalists and livestock on the well-watered, volcanic soils. The estimated population in 2005 was 40 million people.

The *Witwatersrand* region is the only region in Africa that compares with industrial developments in the United States or Europe. Grove (1989) estimated that this region contributes as much as 25% of the industrial output of Africa as a whole. About 9 million people live in the region, many coming from neighboring countries as migrant labor. Gold, coal, and proximity to markets drew European capital and African labor here at the end of the nineteenth century.

Elsewhere there are smaller zones of high concentration, especially around active mineral resources (Democratic Republic of the Congo and Zimbabwe) and successful agricultural schemes (Malawi, Sudan, Angola, and Tanzania), and along the coast of the Maghreb in northern Africa. In the northern desert states, populations are concentrated around water and tend to occupy only a small proportion of the total land areas. For instance, approximately 95% of 32.8 million Algerians live on only 15 percent of the land, and half the population is crowded into only 3% of the land, especially the coastal lowlands and the fertile valleys of the Atlas mountain range. Environmental factors have clearly influenced the distribution and density of population in Africa: the drier and disease-prone regions in general support fewer people than the more humid and temperate zones, and there is a strong correlation between high density and rich soils and mineral workings.

Despite the dismantling of the apartheid system in the 1990s, in South Africa today there remain sharp contrasts in population density and ethnic homogeneity. The pattern is repeated in certain ways in Zimbabwe. In both countries the areas under customary African tenure are overpopulated, predominantly rural, and lacking in development resources; they are labor reservoirs for the commercial regions, mainly controlled now or formerly by Europeans. The European freehold areas, in contrast, are frequently sparsely populated and commercially oriented. Furthermore, both South Africa and Zimbabwe have large urban concentrations, the largest in South Africa is the Witwatersrand. Malawi has "islands" of high density similar to East Africa, and in Madagascar there are also sharp contrasts, the greatest concentration being on the agriculturally superior massif around Antananarivo, center of the once powerful Merina kingdom.

UNDERSTANDING POPULATION GROWTH

Research on the impacts of population on resources in Africa has taught us much over the last 25 years. In this section we will look at a variety of theories about the relationship of population to resources. Three principal theoretical strands have emerged: the *environmentalist* strand, the *transformationalist* strand, and the *distributionalist* strand. The environmentalist strand is represented by the Malthusian and neo-Malthusian theories, which are based on a 1798 conclusion by Thomas Malthus that population increases geometrically and food resources increase only arithmetically. The environmentalist thinker holds that the land

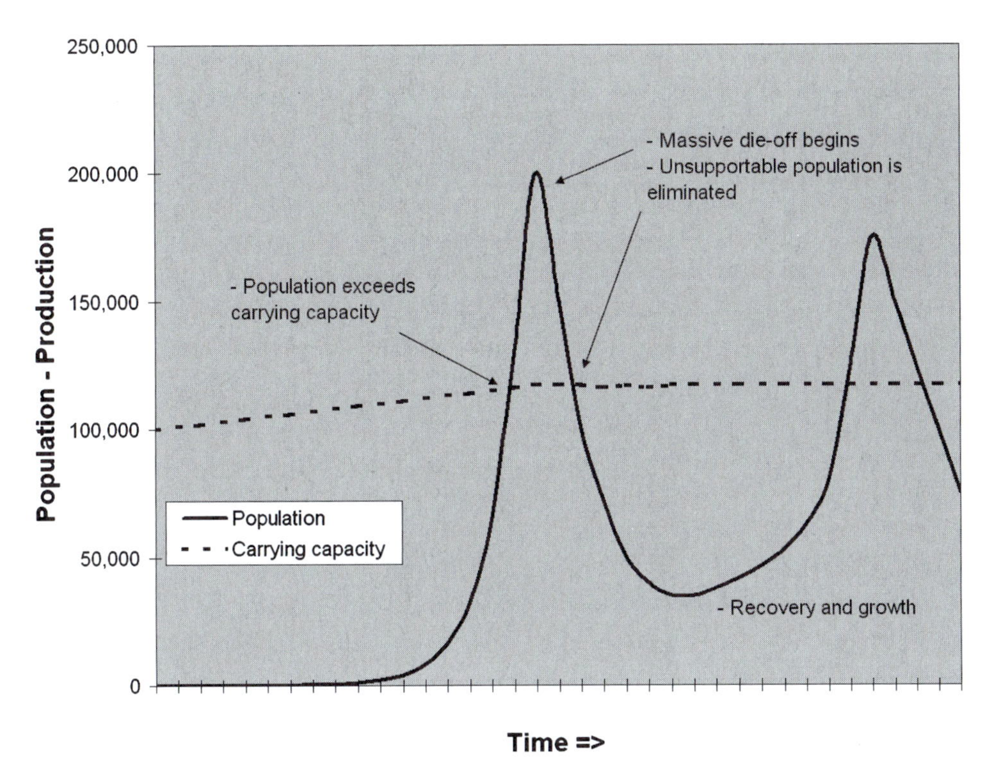

FIGURE 6-2 The Malthusian environmentalist model of carrying capacity versus population growth.

possesses a carrying capacity for people, and beyond that capacity, human reproduction causes increasing impoverishment, land degradation, misery, disease, and ultimately die-off of excess population (Figure 6-2).

Malthus, who died in 1834, believed that population was limited by the resources at hand (carrying capacity) and that these resources in turn were limited. Malthus believed that only by increasing the area under crops could production increase and more people be supported. Once the supply of new land became exhausted, Malthus envisioned the beginning of a downward spiral. Malthusians and neo-Malthusians differ only in that the latter recognize that carrying capacity can be raised by large-scale historical events such as the Industrial Revolution, and by technological change. The neo-Malthusian holds that such changes merely delay the inevitable population crisis.

The neo-Malthusian view is typified by the biologists Paul Ehrlich (1968) and Garrett Hardin (1968). Hardin popularized the phrase "tragedy of the commons," which implies that if individuals' personal goals are not consistent with what is best for society and the environment, society and the environment will suffer. In his famous analogy involving a hypothetical grazing area owned in common, Hardin showed that it was in the interest of each herder to maximize the number of his own livestock on the commons because the added benefit to the individual of a sale of an additional animal outweighed the chance that the pasture would become overgrazed.

In opposition to the environmentalist strand is what critics have perhaps unfairly called the cornucopian (Miller 2001) or technocratic (Clawson and Fisher 1998) theory of population and resources, as exemplified by Julian Simon, an economist. Simon (1998) propounded the view that changes in market supply and demand, substitution of one resource for another (plastics for increasingly scarce metals, for example), and the inventive genius of people have worked together in the past (and will work in the future) to avoid the crises predicted by Ehrlich, Hardin, and other neo-Malthusian environmentalists.

Empirically, but not necessarily philosophically, associated with Simon's view are Ester Boserup and Warren Thompson. Ester Boserup was a Danish economist who found in her examination of agricultural change and population (1965) that population growth could stimulate agricultural change in such a way that more

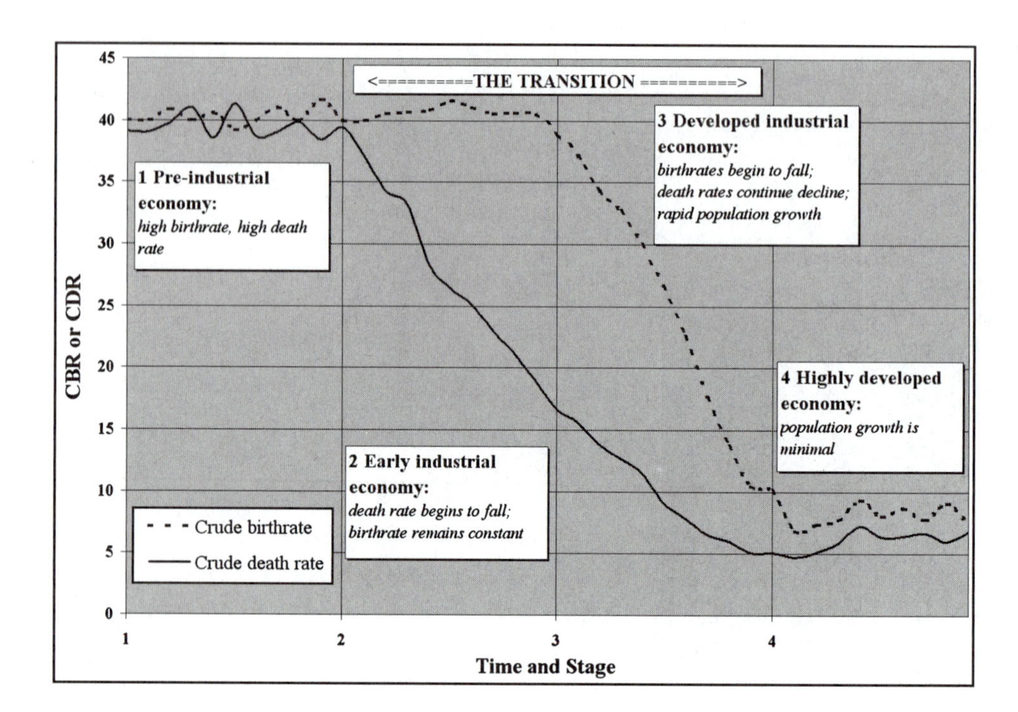

FIGURE 6-3 The demographic transition model.

population could be supported. Her research helped support critics who felt that Malthus his successors are too pessimistic.

Married to the empirical and theoretical findings of Boserup and Simon is the demographic transition theory. Examining demographic data from around the world between 1908 and 1927, Sociologist Warren Thompson (1929) noticed that countries of the world fell into three groups: those that had experienced a rapid decline in birth and death rates (northern and western Europe and the United States), those that had had a moderate decline (Italy, Spain, and central Europe), and those in which there had been no decline at all for the years of the study. The early model for the demographic transition theory explained high fertility as a reaction to high mortality, but 30 years later other analysts explained the changes in demographic behavior (birth and death rates) with reference to modernization (Weeks 1992) or economic development. In countries with low birth and death rates there were also high levels of economic development, and in less developed countries there were high birth and death rates. The model that eventually emerged possesses four stages: a start stage, an end stage, and two periods of transition. The transition apparent in the pattern discovered by Thompson was one from high to low birth and death rates and was caused by variables associated with economic development: improved sanitation, urbanization, the rise of industry, technological change, wealth, and so on (Figure 6-3). Over time, according to Weeks (1992), the demographic transition theory was refined and became less descriptive and more a predictive model of change. The early formulations of the model

> drew on the available data for most countries that had gone through the transition. Death rates declined as the standard of living improved, and birth rates almost always declined a few decades later, eventually dropping to low levels, although rarely as low as the death rate. It was argued that the decline in the birth rate typically lagged behind the decline in the death rate because it takes time for a population to adjust to the fact that mortality really is lower, and because the social and economic institutions that favored high fertility require time to adjust to new norms of lower fertility that are more consistent with the lower levels of mortality. Since most people value the prolongation of life, it is not hard to lower mortality, but the reduction of fertility is contrary to the established norms of societies that have required high birth rates to keep pace with high death rates; such norms are not easily changed, even in the face of poverty. Birth rates eventually declined . . . as the importance of family life was

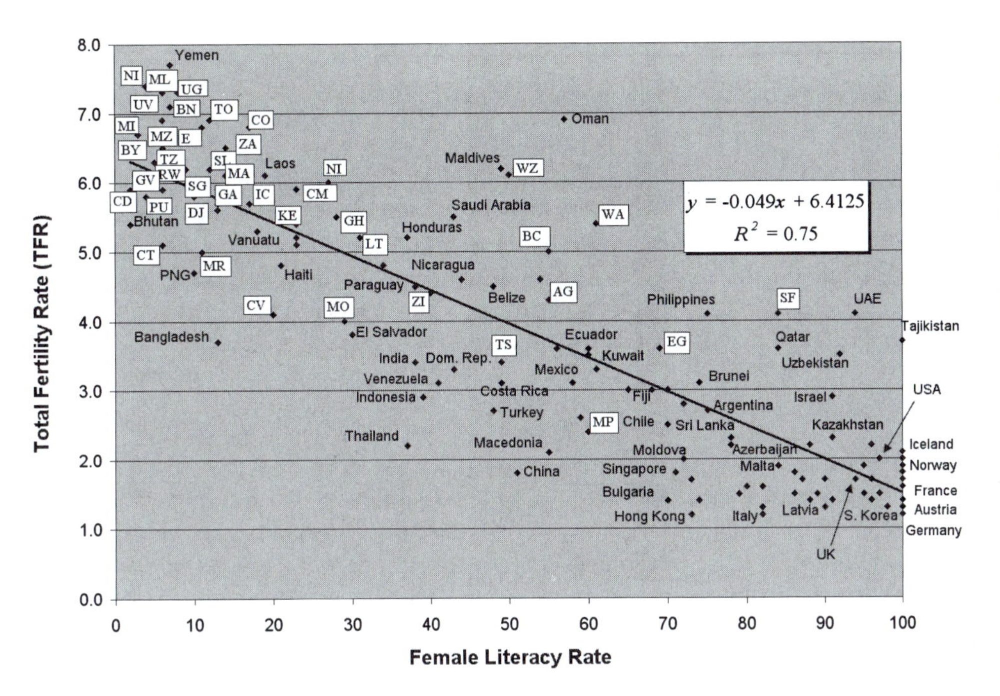

FIGURE 6-4 The relationship between the education of women and the fertility rate by country. African countries are represented by boxes containing their country codes [in the Federal Information Processing Standard (FIPS) system].

> diminished by industrial and urban life, thus weakening the pressure for large families. Large families are presumed to have been desired because they provided parents with a built-in labor pool, and because children provided old-age security for parents. The same economic development that lowered mortality is theorized to transform a society into an urban industrial state in which compulsory education lowers the value of children by removing them from the labor force, and people come to realize that infant mortality means that fewer children need to be born to achieve a certain number of surviving children. (Weeks 1992: 77)

Later it was discovered that a focus on the development of women can be enough to change demographic behavior. Figure 6-4 displays a scatter chart of the relationship between female literacy and total fertility rate (TFR) for the countries of the world in 1996. Female literacy was measured as the number of females in secondary school as a proxy, or stand-in, for literacy. The TFR for a country is the average number of children born to a woman of childbearing years. Clearly there is a relationship between education and the average number of children born per woman. From regression analysis studies, statisticians have found that literacy explains about 75% of the variation in TFR (R^2 = 0.7478 in Figure 6-3). The reader may point out that a rise in literacy seems to be part of the development process, reasoning that as a country becomes wealthier, there will be more investment in its human capital. It has also been observed that the education of women in the absence other changes has a profound impact on the number of children a woman bears. For one thing, a woman who is educated through high school is more likely to be unmarried and to be childless until marriage. Moreover, employment and income possibilities undreamt of by her sisters may further delay her entry into a sexual union and childbearing. In many African countries, especially in rural areas, women are married by the age of 14, perhaps even younger, and there is considerable pressure on them to have children as soon as possible. The cluster in the upper left of Figure 6-4 is so dense because TFR is very high in many African countries.

In addition to being different in female literacy and TFR, the African countries to the right of the major African cluster in Figure 6-4 are geographically distinct: they are located either in North Africa, or southern Africa, and their position on the graph indicates that they are more like the countries around them on the graph than their geography would indicate.

The distributionist strand of population theory is associated with the political economist Karl Marx and his collaborator Friedrich Engels, who believed that there was no population problem as stated by Malthus. Instead, Marx and Engels saw a problem in the distribution of resources under capitalism. They argued that if resources could be redistributed equally (socialism), then whatever had appeared to be an imbalance between population and resources would disappear. Marx associated the population problem with the English class system in which the vast majority of resources and wealth were controlled by a small numbers of aristocrats and capitalists, while the masses lived in poverty.

In a famous example in his book *Capital*, Marx described what had happened in northern England when the landowners stopped the labor-intensive production of wheat and began raising sheep. Drawn by potential profits from the sale of wool to the newly emerging woolen industry, landowners evicted their tenant farmers to be able to raise sheep on the wheatlands. This forced masses of former tenants to live in the forests under difficult conditions, creating for Marx an example of unjust relations between classes rather than representing a population issue. Marx also felt that technology and socialism together would diminish the importance of the population. Marx, like Malthus, was a nineteenth-century thinker whose theories have been much debated, revised, and ultimately rejected over the years.

RATES OF BIRTH, DEATH, AND GROWTH AND AFRICAN POPULATION POLICIES

There are important regional differences in Africa's birthrates, death rates, and growth rates. The lowest birthrates (Figure 6-5) occur in northern and southern Africa and on the islands of Mauritius and Seychelles, while the highest are in West Africa, equatorial Africa, Uganda, Madagascar, and Zambia. Some of the most populous countries have the highest birthrates. Birthrates vary considerably within countries according to race, socioeconomic achievement, and residence (rural or urban). In South Africa, Zimbabwe, and Kenya, for instance, there are extremes in lifestyles, opportunities, value systems, and other factors that affect birthrates. In these countries, people of European ancestry, Asian immigrants, and the indigenous Africans have different birthrates. People of European ancestry are predominantly urban, enjoy a high standard of living, and are part of the industrialized world; most black Africans, on the other hand, being more rural than urban, are part of the "developing" world, and their higher birthrates reflect this reality.

Figure 6-5 shows that death rates are generally lowest in the northern tier of states and on the islands and are highest in southern Africa. Death rates vary considerably within countries according to nutritional standards, rural or urban residency, access to clinics and doctors, population density, and the quality of housing and sanitary facilities present. Prior to the AIDS epidemic in Africa, the spatial pattern of infant mortality rates (Figure 6-5) was very similar to that of crude death rates because much of that mortality was in the early years. Today the pattern diverges, and this is no longer the case. Although it is a complex picture, higher infant mortality seems to be associated today with conflict in Africa—Angola, Liberia, Rwanda, Sierra Leone, and Somalia.

Africa's population growth rate has been increasing over the past several decades; at the turn of this century it was less than 1% per annum; from 1930 to 1950 the annual rate was 1.3%, from 1950 to 1970 about 2.3%, and from 1980 to 2000 it was 2.7%. The accelerating growth is due to falling death rates combined with consistently high, and in some cases rising, birthrates. Overall, for the entire continent, birthrates began to decline in the early 1990s (Figure 6-6). As recently as the 1940s and 1950s, death rates were generally between 30 and 40 per thousand, while birthrates were about 48 per thousand. For example, in Côte d'Ivoire, birth and death rates in 1940 were 55 and 38 per thousand, respectively. In 1975 the respective rates were 45.6 and 20.6, and in 2005 they were 39 and 17 per thousand. In contrast, the 2001 rates for the United States, a highly developed country by many measures, were 14 births and 8 deaths per thousand population. The rate of natural increase was 0.6%. As Figure 6-6 indicates, Africa's demographic variables are projected to change significantly over the next 50 years.

The demographic transition model introduced in the preceding section supposes that over time, birth and death rates change because of changing social, economic, and political forces, and that we can see a number of distinct stages. In less developed economies, where health facilities and schools are generally lacking, the level of technology is low, and people live close to

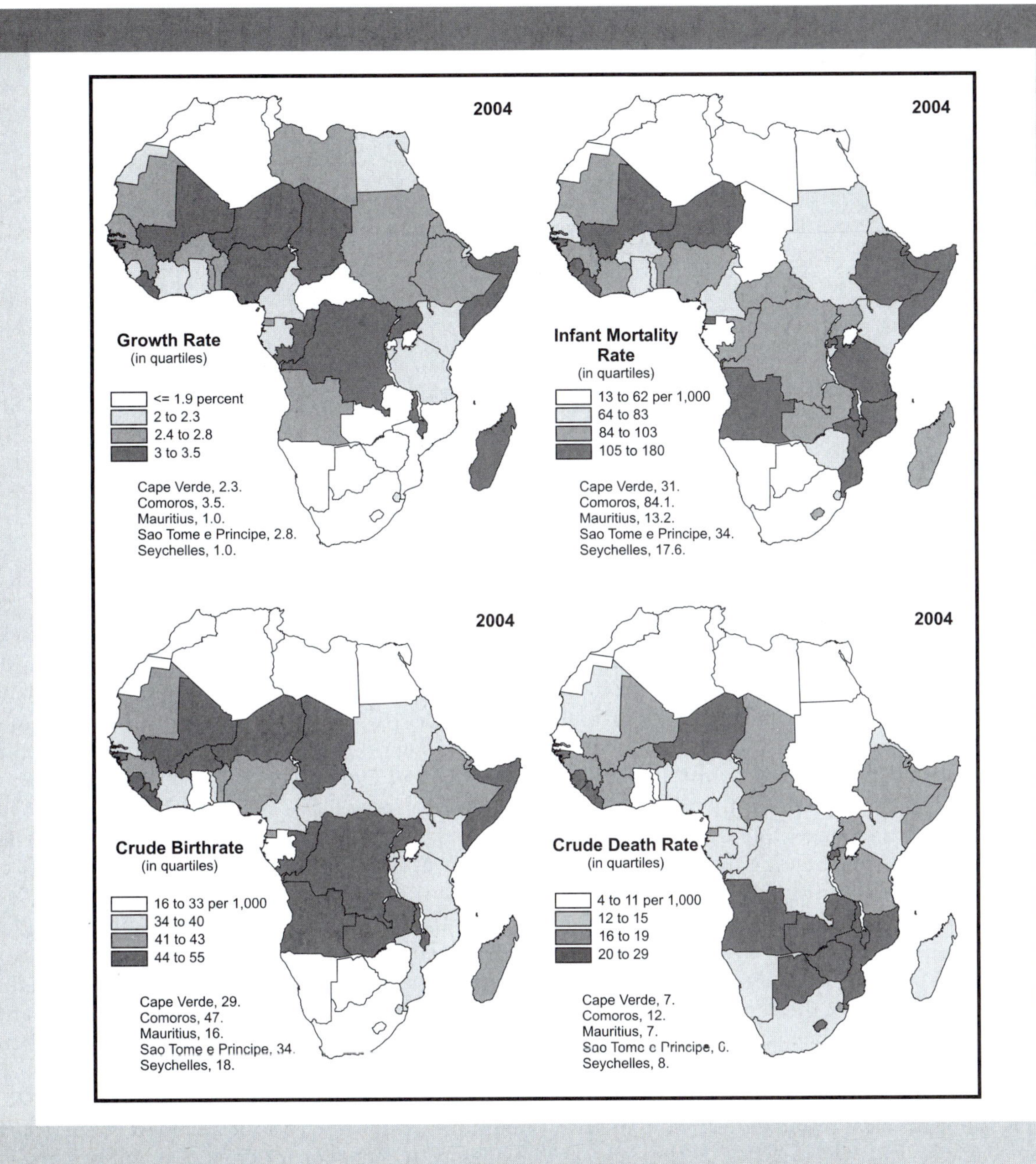

FIGURE 6-5 Rates of growth, birth, death, and infant mortality for Africa by country, 2004.

the environment, crude birthrates and death rates are high and fluctuate widely from year to year, so that on occasion death rates may even exceed birthrates (stage 1 in Figure 6-3). The overall effect is a low growth rate, generally less than 1%. As the economy changes and improved health facilities and technology are introduced, death rates begin to fall (stage 2), but birthrates remain high, and the net result is a high growth rate, generally between 2 and 4%. If development increases in a way that serves to enhance education, raise standards of living, and increase urbanization, and if awareness of population growth

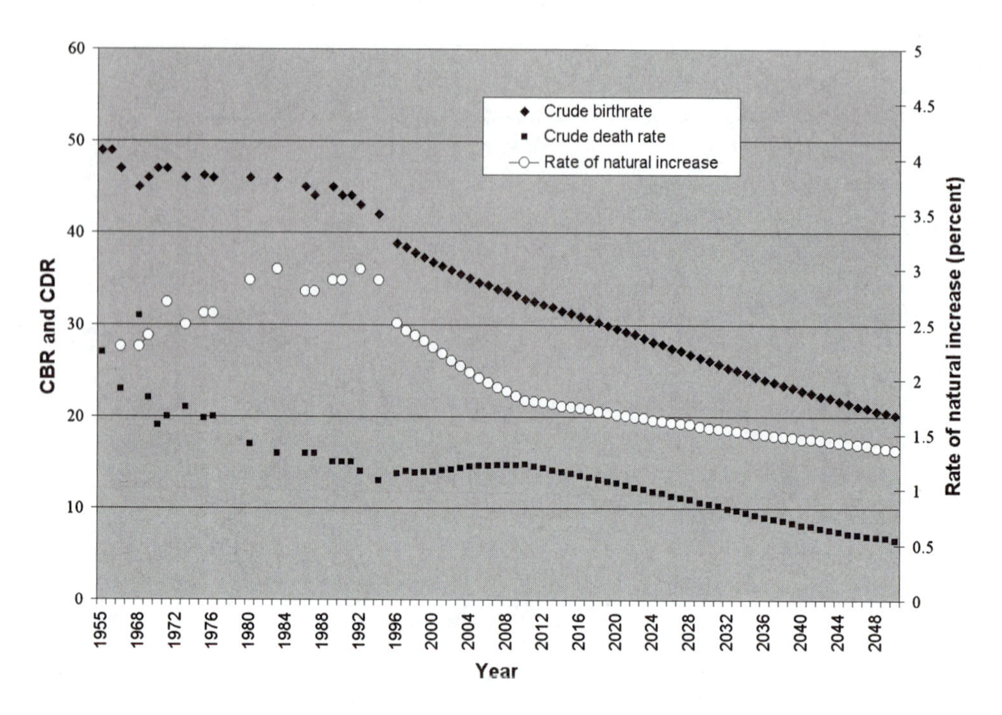

FIGURE 6-6 Crude birthrate, death rate, and rate of natural increase in Africa from 1955 to 2050 (projected data begin in 2004).

increases and population control measures are adopted, birthrates may then decline (stage 3 of the model). In stage 4, the rates stabilize slightly higher than the mortality rates and result in a small but stable population growth. This is the pattern of the industrialized world.

Although the variability from country to country can be great, in general, some countries in Africa are beginning to move into stage 3 of the demographic transition. They left the first during the colonial era as a result of the assault on infectious diseases and the assimilation of technological changes; they left the second in the early 1990s. This progress was accomplished by means of a gradual change in the traditional social and economic lives of the people, and within an extremely short period in comparison to Europe and the United States. Africa still has one of the world's highest crude death rates (14/1,000), and while this number is falling for most countries, birthrates remain high on average. Thus most African countries are growing (Table 6-2), and Africa's fertility rates remain the highest in the world, as suggested by the photos of the Traoré brothers and their families, in Mali (Figures 6-7 and 6-8).

By 2025 there will be three demographic clusters of African countries: those that have made the transition to stage 3 of the demographic transition (Algeria, Cape Verde, Egypt, Libya, Mauritius, Morocco, Seychelles, and Tunisia); those that continue to grow but at a decreasing rate (Ethiopia, Guinea, Madagascar, Mali, etc.); and those, mainly found in southern Africa, hard hit by the AIDS epidemic (Botswana, Kenya, Lesotho, Mozambique, Namibia, South Africa, Swaziland, Zambia, and Zimbabwe). In these latter countries the crude death rates have actually risen between 1975 and the present and will continue to rise well into the foreseeable future, rather than decline as would be the case without the disease.

Of the various death rates, the infant mortality rate has shown the most dramatic decline. In the 1950s, the infant mortality rate was commonly over 200 per thousand, while in the mid-1970s it averaged 156 and ranged from a low of 103 in Egypt to 229 in Gabon. Although there will be variation from country to country, in 2005 there were 88 infant deaths per thousand; by 2025 an improvement to 53 per thousand is expected. The widespread reduction of infant mortality combined with the sustained high birthrates resulted in a large proportion of many African countries (40–50 percent) of the total population being under 15 years of age. For the continent as a whole, the proportion (percentage) of children less than 15 years old was, 42% in 2005, but is predicted to be 37% in 2025. The population pyramids for the entire continent for the years 1950 through 2025 illustrate this (Figures 6-9 and 6-10).

TABLE 6-2 SELECTED POPULATION DATA, 1975, 2005, AND 2025.

	CBR			CDR			Growth rate		
Country	1975	2005	2025	1975	2005	2025	1975	2005	2025
Algeria	46.1	20	15.7	15.6	4	5.2	3.1	1.5	1.1
Angola	45.9	49	36.2	26.1	24	16.9	2.0	2.6	1.9
Benin	50.5	42	31.1	24.8	13	11.1	2.6	2.9	2.0
Botswana	45.6	25	21.5	23.0	28	33.2	2.3	−0.3	−1.2
Burundi	46.8	43	31.2	18.2	15	12.6	2.9	2.8	1.9
Cameroon	46.2	38	25.6	15.8	15	10.6	3.0	2.3	1.5
Central African Republic	43.4	37	25.6	22.5	19	15.7	2.1	1.7	1.0
Chad	49.7	45	37.3	25.9	17	8.9	2.4	2.7	2.8
Congo Republic	45.5	49	29.2	20.8	13	11.8	2.0	3.1	1.7
Côte d'Ivoire	48.5	39	28.2	16.6	17	14.1	3.5	2.2	1.4
Djibouti	nd	32	29.9	nd	13	9.6	nd	1.9	2.0
Democratic Republic of the Congo	45.2	45	35.3	20.5	14	9.8	2.5	3.1	2.6
Egypt	43.2	26	16.8	15.4	6	6.9	2.8	2.0	1.0
Eritrea	nd	39	30.0	nd	13	7.2	nd	2.6	2.3
Ethiopia	49.4	41	35.5	25.8	16	14.1	2.4	2.5	2.1
Gabon	32.8	33	23.7	19.7	12	16.6	1.3	2.1	0.7
Ghana	43.5	33	19.4	13.7	10	10.3	3.0	2.3	0.9
Guinea	48.4	43	29.6	27.2	16	11.5	2.1	2.7	1.8
Guinea-Bissau	47.6	50	28.6	25.1	20	10.0	2.3	3.0	1.9
Kenya	51.0	38	19.3	11.6	15	17.0	3.9	2.2	0.2
Lesotho	41.1	26	23.9	15.2	28	20.7	2.6	−0.1	0.3
Liberia	45.7	50	34.2	17.3	22	10.0	2.8	2.9	2.4
Libya	44.8	27	17.4	10.9	4	3.8	3.4	2.4	1.4
Madagascar	46.2	40	36.9	20.2	12	7.4	2.6	2.7	3.0
Malawi	53.3	50	24.0	26.7	19	20.3	2.7	3.2	0.4
Mali	50.1	50	36.2	25.9	18	10.8	2.4	3.2	2.5
Mauritania	44.9	42	32.1	25.9	15	8.0	1.9	2.7	2.4
Mauritius	25.3	16	12.6	8.1	7	8.1	1.7	0.9	0.4
Morocco	46.2	21	16.4	15.7	6	5.5	2.9	1.6	1.1
Mozambique	43.1	42	25.1	20.1	20	25.0	2.3	2.2	0.0
Namibia	nd	27	28.0	nd	17	24.3	nd	1.1	0.4
Niger	52.2	56	35.9	25.5	22	14.2	2.7	3.4	2.2
Nigeria	49.3	43	29.2	20.5	19	12.4	2.9	2.4	1.7
Rwanda	51.2	41	21.4	19.9	18	20.8	3.1	2.3	0.1
Senegal	52.0	37	25.3	23.9	12	5.2	2.4	2.6	2.0
Seychelles	30.5	16.9	12.7	7.3	8	6.5	2.3	1.0	0.6
Sierra Leone	48.2	47	34.0	30.1	24	12.0	1.8	2.3	2.2
Somalia	44.5	46	37.5	18.2	18	11.4	2.6	2.9	2.6
South Africa	42.9	23	17.0	18.2	16	24.9	2.6	−0.1	−0.8
Sudan	47.8	37	23.2	17.5	10	6.7	3.0	2.7	1.7
Tanzania	49.5	42	28.0	18.5	18	10.9	3.1	2.4	1.7
Tunisia	36.0	17	12.1	9.8	6	6.1	2.6	1.1	0.6
Uganda	49.0	47	36.9	23.4	15	11.0	2.6	3.2	2.6
Zambia	51.5	41	30.0	20.5	23	16.8	3.1	1.9	1.3
Zimbabwe	47.9	31	18.9	14.4	20	28.9	3.4	1.1	−1.0

Source: United Nations (2002), Population Reference Bureau (2005, 1975).

FIGURE 6-7 Musa and Yahya Traoré and their wives, 1978.
Roy Cole.

As social and economic development increases, the population pyramids will narrow at the base and broaden at the apex. In the more developed countries of the world, only 16 percent of the children are under 15 years of age. The population pyramid for more developed countries displays today the pattern projected for African countries after 2025: most of those born live until old age (Figure 6-11).

FIGURE 6-8 Musa Traoré and wives and children, 1986 (Yahya not pictured).
Roy Cole.

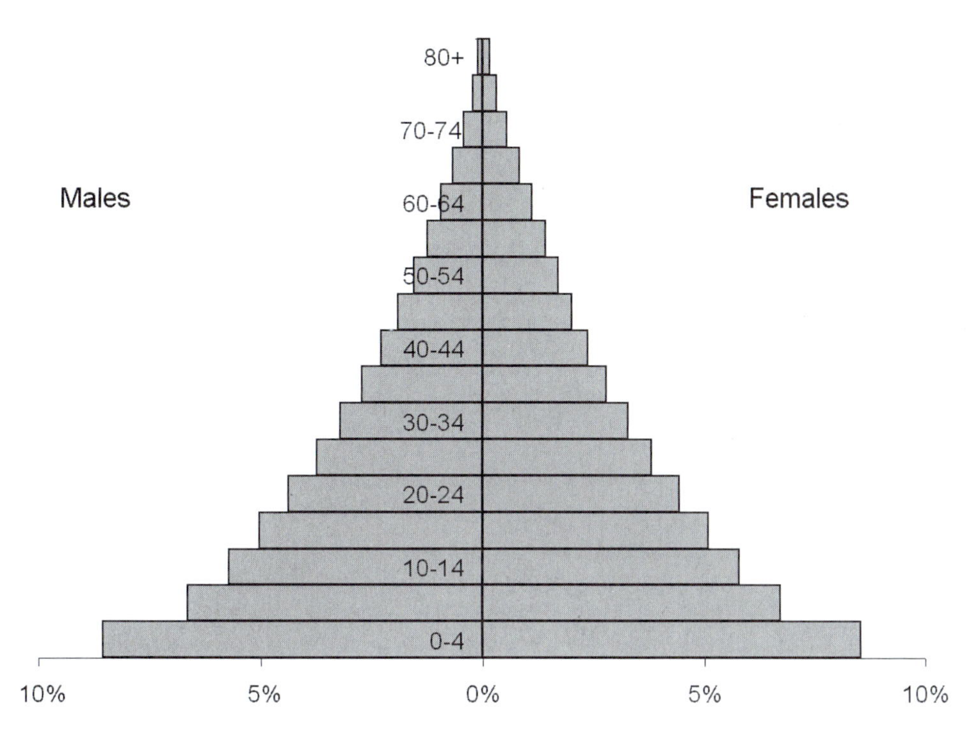

FIGURE 6-9 Population pyramid for Africa in 1950.
Source of data: US census.

Many African countries are feeling and will continue to feel the strains of increasing population. Among other problems, such countries experience low levels of subsistence in rural areas, rising rural and urban unemployment, an increasing drift to urban areas, the spread of slums and shantytowns, and increasing claims on government resources for social services in education and health. Chronically rising

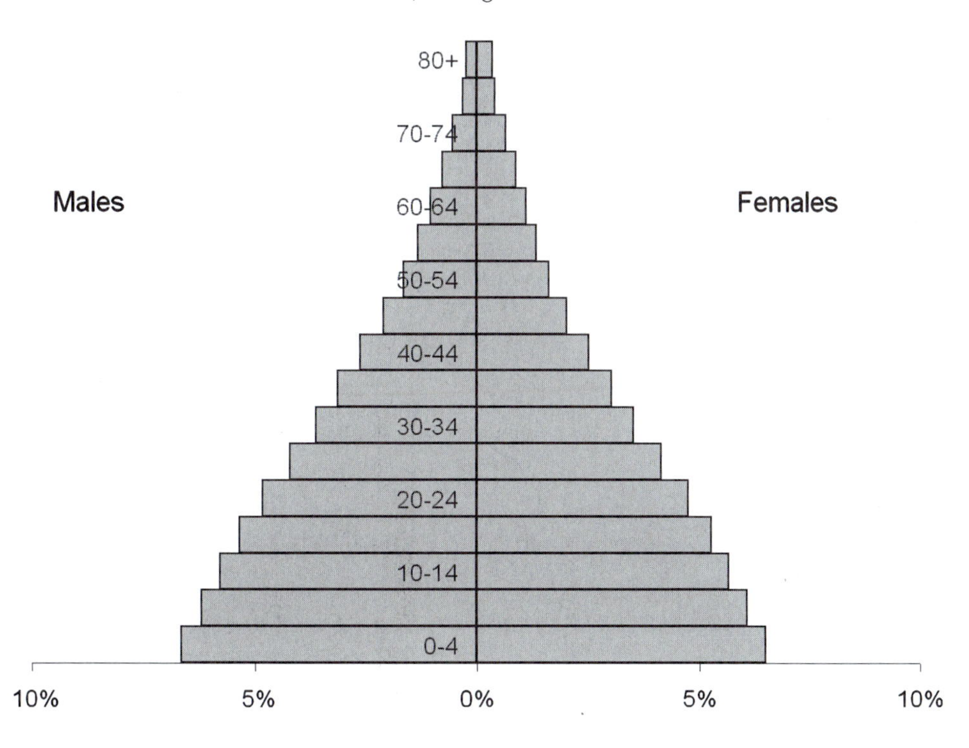

FIGURE 6-10 Population pyramid for Africa in 2025.
Source of data: US census.

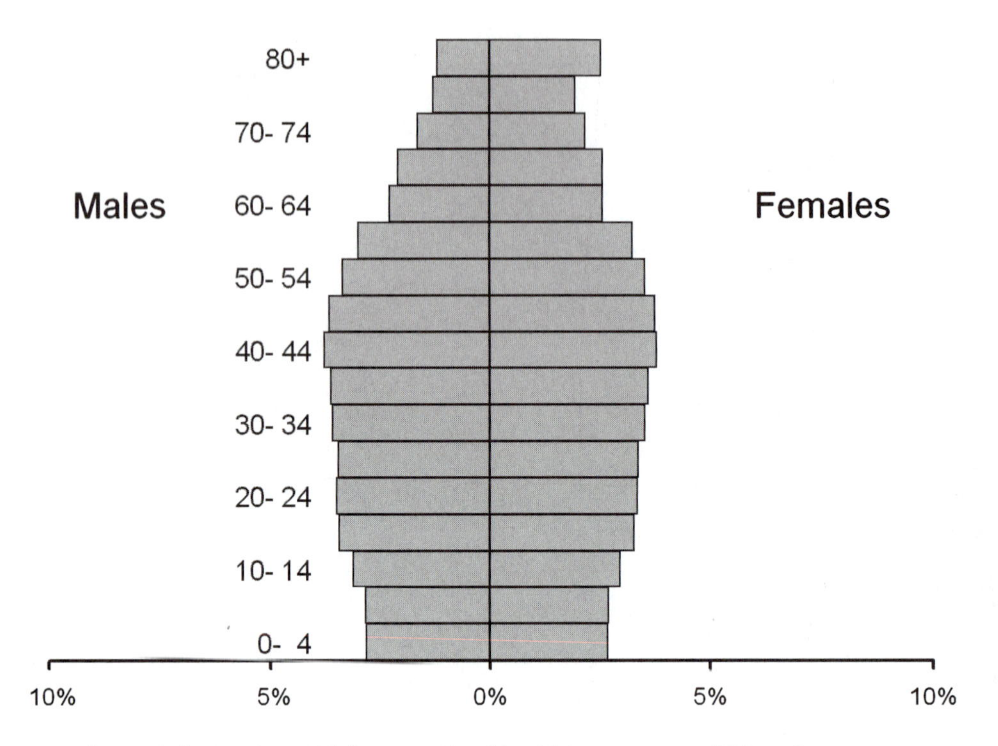

FIGURE 6-11 Population pyramid for developed countries, 2005. *Source of data: US census.*

population is becoming widely recognized in Africa as a contributing factor to underdevelopment.

Today, most African governments view rapid population growth as a problem to be solved through the adoption of official population policies such as fertility control and family planning. The notion that rapid population growth jeopardizes national development objectives has gained wide but by no means universal acceptance. In the past, official emphasis has been placed almost wholly on increasing the pace of national economic growth to achieve higher standards of living, not on taking care of a growing population.

In the 1970s, Kenya, feeling the strains of rising numbers, was one of the first African countries to initiate a program of population control. The project had three broad goals: training field personnel to extend existing family planning and maternal and child health services, strengthening the rural health system, and developing an appropriate institution to support family planning services. The International Development Association (IDA), an affiliate of the World Bank, awarded Kenya a $12 million credit to assist in implementing its Five-Year Family-Planning Program (1975–1979). Kenya was the first country in Africa south of the Sahara to receive World Bank aid for population control (Mafukidze, 1974).

Thirty-five years ago, most African governments did not share Kenya's views and were unenthusiastic about family planning. By 2003, in stark contrast, only two African countries, Angola and Gabon, have pursued a policy of raising their population. Equatorial Guinea has a policy of maintaining its current population. Only 13 countries (Mozambique, Somalia, the Central African Republic, Chad, the Democratic Republic of the Congo, São Tomé and Principe, Libya, Benin, Guinea, Guinea-Bissau, Mali, Sierra Leone, and Togo) have a no-intervention population policy. The rest are endeavoring to lower population growth. Table 6-3 and Figures 6-12 and 6-13 illustrate the shift in population policy of African countries from 1976 to 2003.

Family planning services are becoming more widely available through private and government programs, but the problems of instituting birth control plans are many, and they vary from society to society. Traditional family planning practices are breaking down in the urban areas, so the demand for modern contraceptives is growing with rapid urbanization. In rural areas the use of contraceptives is not widespread. In most African countries there are strong religious and cultural biases against the liberalization of abortion laws, few condone the practice as a birth control measure.

TABLE 6-3 CHANGE IN POPULATION POLICY OF AFRICAN COUNTRIES, 1976-2001.

	Policy									
	Lower population		Maintain population		Raise population		No intervention		Totals	
Year	Freq	%	Freq	%	Freq	%	Freq	%	Freq	%
1976	12	25	0	0	7	15	29	60	48	100*
1986	20	39	3	6	4	8	24	47	51	100
1996	31	58	2	4	2	4	18	34	53	100
2001	37	69.8	1	1.8	2	3.7	13	24.5	53	100

Source: UN Population Division (2003).
**Note: Numbers may not add up to 100 because of rounding.*

In a study that supports the notion that the market and development can act together as to help resolve people and environment problems in Africa, Tiffen, Mortimore, and Gichuki (1994) found in Kenya that a combination of sound government policy, farmer-led technology change, and better market linkages and incentives enabled a semiarid area once categorized as an "environmental disaster" to support five times as many people a half-century later, with higher yields under a more labor-intensive system. Land, which had once been abundant, became a scarce and highly valued resource by virtue of population growth. The government made family planning choices available, and the people in the region later began limiting their numbers in response to a variety of economic factors.

> On the basis of their study, Tiffen and Mortimore conclude that the experience of Kenya between 1930 and 1990 lends no support to the view that population growth, even rapid population growth, leads inexorably to environmental degradation. It is impossible to show that a reduced rate of population growth might have had a more beneficial effect on the environment. On the contrary, it might have made less labour available for conservation technologies, resulted in less market demand and incentives for investment, and reduced the speed at which new land was demarcated and conserved after being cleared. Population growth has made land a scarce and increasingly valued asset. Falling fertility, reflected in a lower rate of population growth, suggests a spontaneous response to changed economic conditions in the period 1979–1989. An extreme shortage of land in some parts of the Machakos District, diminished opportunities for income diversification with the national economy in recession, and the high educational and other costs of raising children appear to be leading to voluntary family limitation. The provision of family planning information and the making accessible of supplies can be justified as adding to peoples' choices and the control which they have over their circumstances. To argue for population limitation on environmental grounds weakens the case for it both theoretically and practically. (Tiffen and Mortimore 1994: 284)

Tiffen and Mortimore's evidence seems to indicate that a partnership between responsible government and local populations can result in genuine development. Nevertheless, according to the United Nations, many African countries face daunting constraints to food security in the coming years: limited arable land, the shrinking size of family farms as plots are subdivided for each growing generation of heirs, and increasing overuse of the soil.

URBAN GROWTH

By whatever definition, Africa is the least urbanized of the continents; yet its urban growth rate is very high, and its "hierarchy of urban places" is developing.

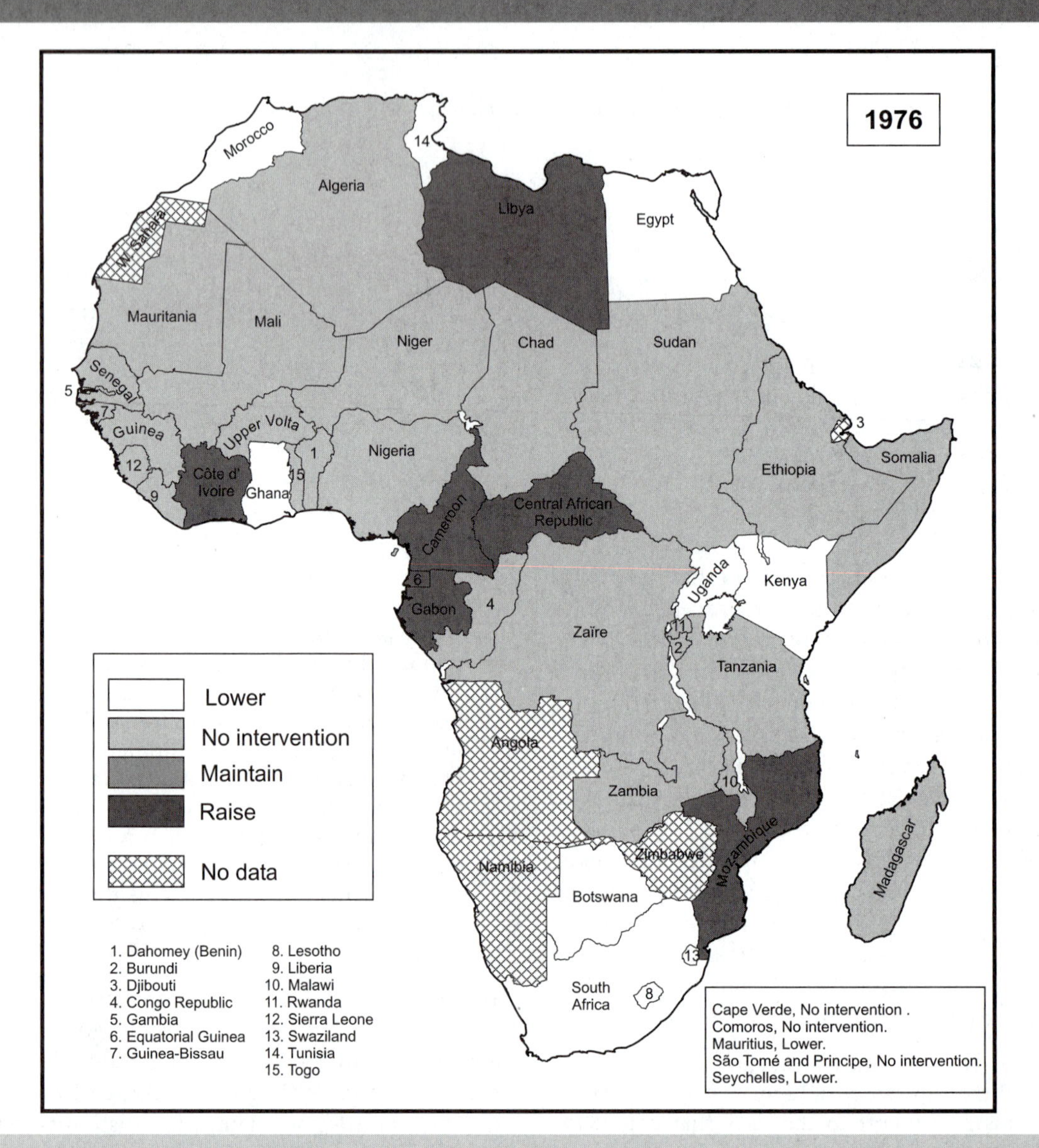

FIGURE 6-12 African government population policy by country, 1976.
Source of data: UN 2003 World Population Policies 2003. United Nations Population Division, New York.

Although the city is a dominant feature in African life, Africa is not a continent of cities or towns but of villages (Figure 6-14). According to UN estimates, approximately 40% of the African population resided in an urban area in 2003. This figure was 25% in 1975, compared with 19% in 1960, and only 15% in 1950. It is expected to pass 50% by 2030, as indicated by Figure 6-15.

South Africa has the most developed urban hierarchy but only 55% of its population live in towns and cities. In North Africa, where each state has its own distinct urban clusters, the urban populations range from 43% in Egypt to 86% in Libya (up from 29% in the 1970s). While there are several large cities in West Africa, and especially in Nigeria, the population is still predominantly rural, with more than a

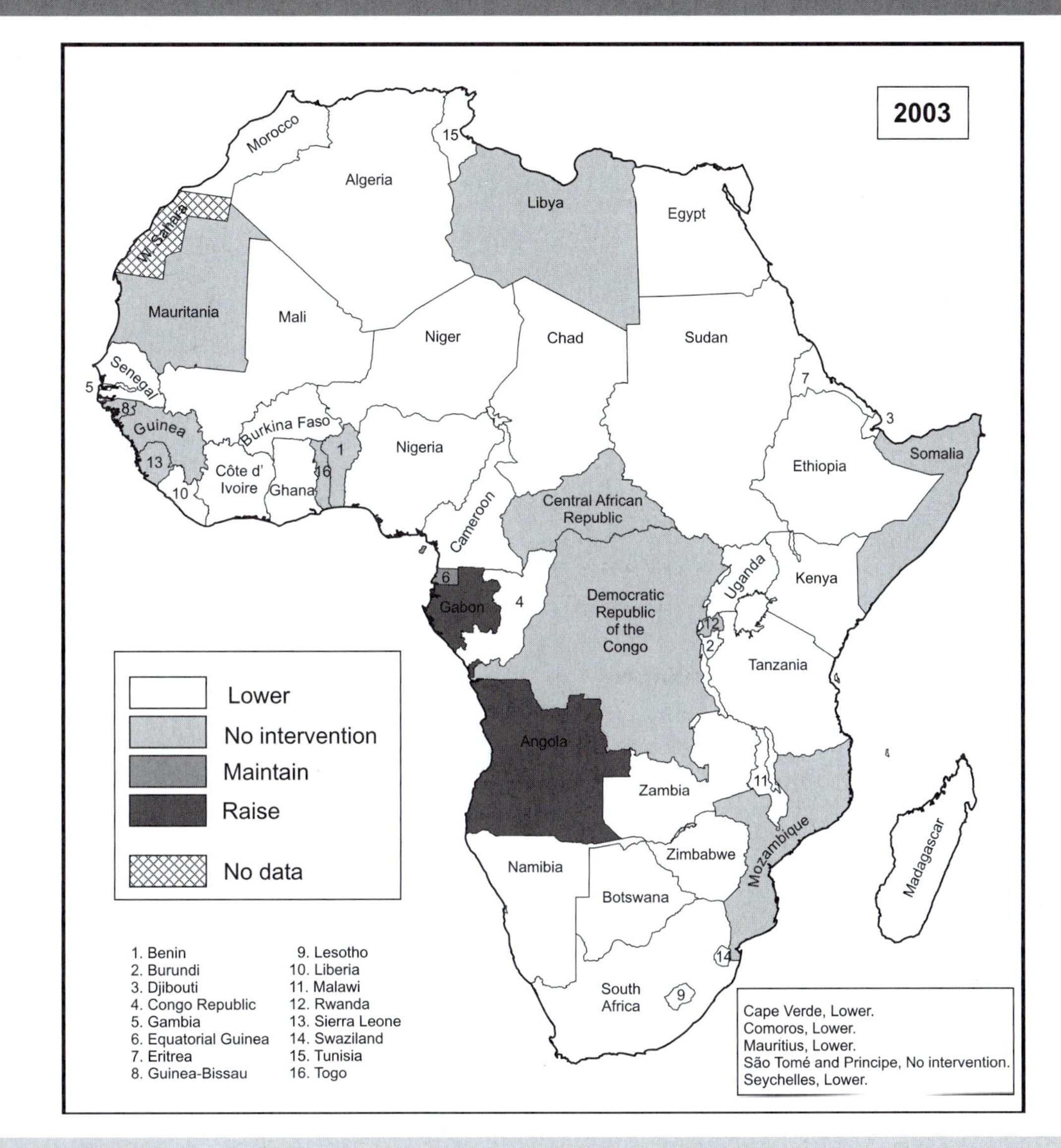

FIGURE 6-13 African government population policy by country, 2003.
Source of data: UN 2003 World Population Policies 2003. United Nations Population Division, New York.

third of the population of many states classified as urban. The highest urban ratios occur in the coastal states (Ghana, 37%; Senegal, 43%; Liberia, 45%) and the lowest occur in the interior (Burkina Faso, 15%; Niger, 17%; Chad, 21%). Other figures include Ethiopia, 15%; Uganda, 15%; and the Democratic Republic of the Congo, 29%. The lowest urban population is in Rwanda (5%) and Burundi (8%). Figures 6-16 and 6-17 illustrate the change in urbanization from 1976 to 2002.

From 1950 to 1995, Africa's urban population grew at about 5% per year, but since that time its growth has slowed. The United Nations (2004) projects that by 2030, Africa's urban populations will be growing at 3% annually. Natural growth rates, that is, births less deaths, are generally higher in urban than rural areas

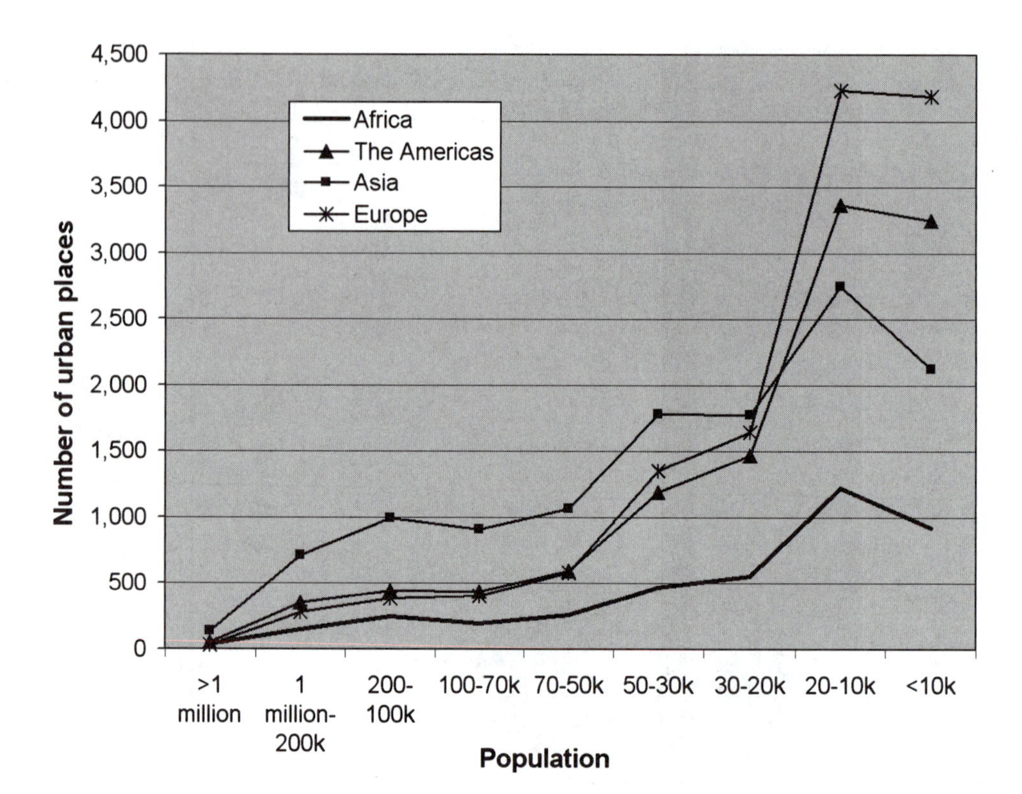

FIGURE 6-14 Number of urban places of different sizes, world regions.

because there is some control over death, especially the death of infants and children, whereas control over births tends to be lacking. Most urban growth is accounted for by in-migration, the migrants themselves contributing to the natural increase.

While there are obvious differences in the size and growth of urban centers, in the wealth and occupations of the urban population, and in the quality of housing and social services available, there are important similarities in the urban patterns from one state to another. First, most growth occurs in the respective primate cities (Figure 6-18). Table 6-4 shows this phenomenal growth in absolute terms over the last seven decades. Some cities, for example Cairo, have slowed in recent years; but even so, Greater Cairo grew to 16 million by 2005. Likewise, although still growing at an incredible rate, Greater Brazzaville–Kinshasa had 9.7 million people in 2005, while Greater Lagos and Greater Johannesburg had over 10 millions and 6 million people, respectively. Almost all of Africa's primates continue to double their populations every 20 years, and migration is a major contributing factor to this growth. Populations from the surrounding rural areas and smaller towns are attracted by the economic opportunities (real or perceived) of their country's major industrial–commercial–political center, leading to unplanned neighborhoods of poorly constructed housing near the larger cities (Figure 6-19).

Much of the economic growth, capital investment, and political and economic foundation is concentrated in the primates, largely a carryover from colonial times. The primates are the meeting ground between the national and international economies, and are situated in strategic and convenient locations. Moreover, in all but a handful of states, the primate cities account for more than half the total urban populations: Addis Ababa has 59%, Conakry 59%, Tripoli 63%, Kampala 68%, and Banjul 100%. The important exceptions are Lagos, 14%; Johannesburg, 14%; Algiers, 29%; and Kinshasa, 34%. A high percentage of the urban population is located in cities of 100,000 and more, and there is a large disparity between the primate cities and the smaller towns. The rank–size rule applies very poorly, which means that the hierarchical arrangement of towns and cities associated with developed countries is generally absent, and the spatial distribution of wealth and development is highly concentrated. Finally, as migration adds to the local population increase, the major urban areas are witnessing acute shortages of housing and rising

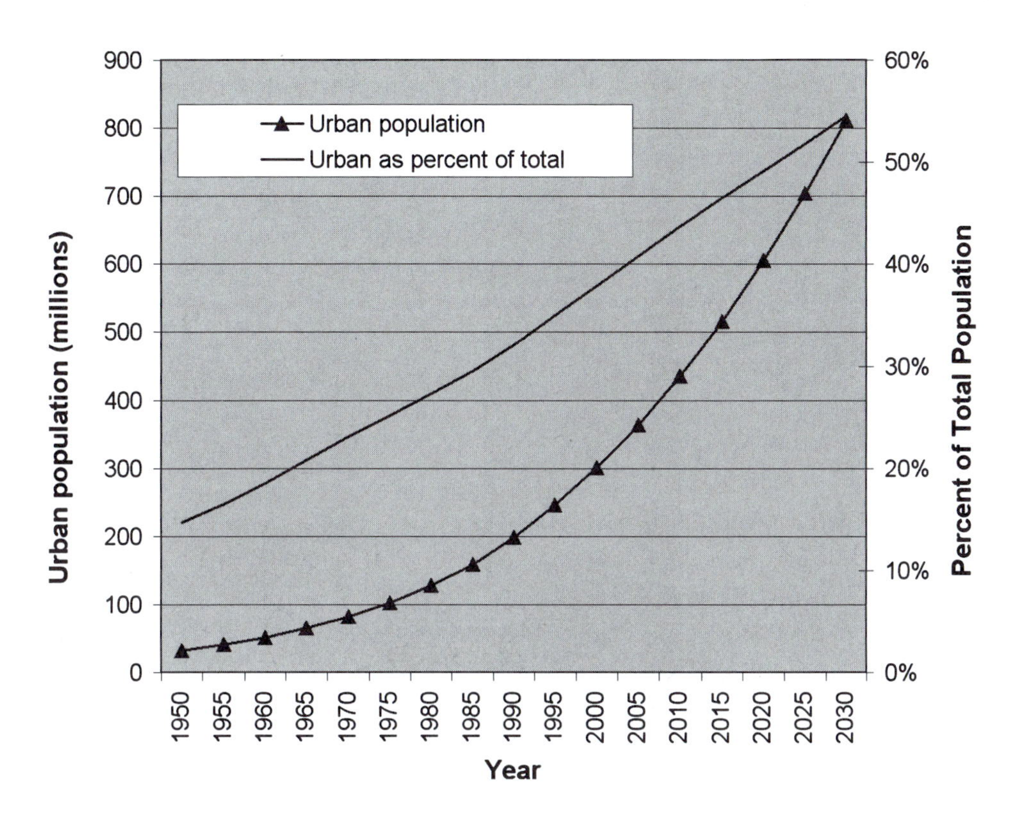

FIGURE 6-15 African urban population alone and as percentage of total population, 1950–2030.

unemployment. For most, employment is to be found only in the informal sector (Figure 6-20).

The problems associated with urban growth and the regional disparities of wealth and development are known to most governments, but few undesirable situations are being rectified. Most governments wish to control the growth of cities by making the rural areas more attractive places in which to live, with jobs and adequate social services. Commonly adopted strategies include decentralizing industry by encouraging high-manpower enterprises to locate at prescribed rurally based "growth centers," introducing large-scale, high-manpower agricultural projects, improving agricultural techniques and revising land tenure, and providing improved medical, educational, and social facilities in rural areas. Specific examples are taken up in the regional chapters that follow.

MIGRATIONS

Migration can be both the cause and consequence of population growth. In Africa such movements of people play an especially important role in urban growth, the circulation of income, and the diffusion of ideas and values. In addition, they have contributed to the concentration of industry, economic growth, and political power in relatively small areas at the expense of the surrounding regions. While important in the development of contemporary Africa, migrations have occurred throughout history and have contributed to the contemporary patterns of distributions, densities, and economies. They can be voluntary or involuntary, permanent or temporary, within or between states, rural–urban, urban–rural, rural–rural, urban–urban, involving individuals or groups. They are the response to changes (either real or perceived) in the natural environment, and social, political, and economic systems in a migrant's place of origin or place of destination, or both.

Among the major precolonial migrations have been the Arab invasions across the Sahara into the Sudanic belt; the eastward pilgrimages to Mecca; Bantu movements from the Cross River region of Nigeria to East, and central southern Africa; the southward migration of Nilotic and Cushitic groups in eastern Africa, and the wave effect of Khoisan peoples into the less hospitable areas of southern Africa; the trans–Indian Ocean movements from Southeast Asia that account for the Asian names, rice culture, and other distinctly Asian influences in Madagascar; the

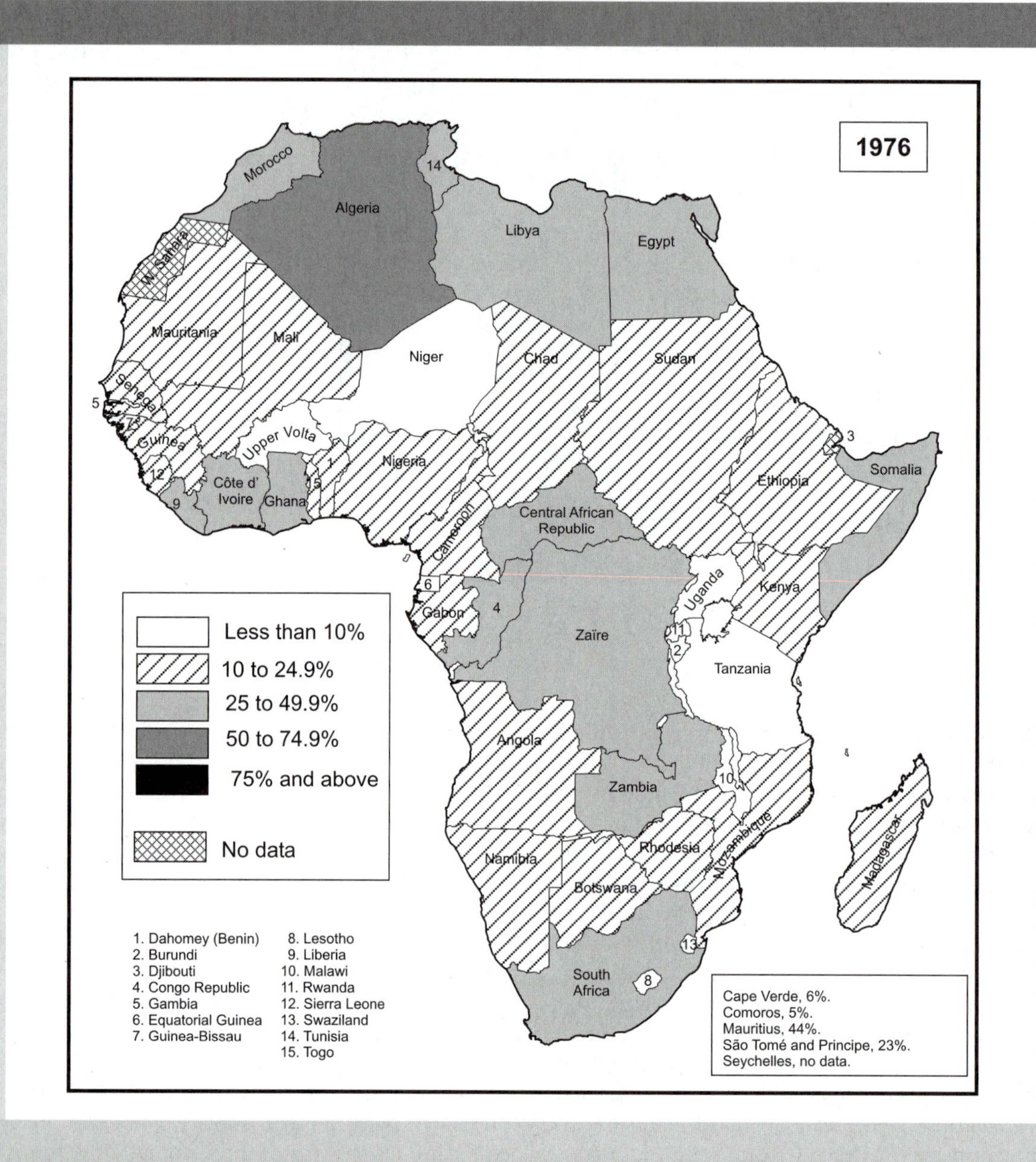

FIGURE 6-16 Percent urban by country, Africa, 1976.

migrations and countermigrations of pastoral nomads throughout the savanna zones and certain highland regions; and the wholesale removal of Africans to the New World, North Africa, and Asia.

Migrations during the colonial era differed markedly in their scale, orientation, incentives, and consequences, although of course many basic patterns of local migrations by pastoralists and traders persisted, as they do in postcolonial Africa. Tens of thousands of colonial administrators, settlers, and entrepreneurs poured into the continent from the metropoles and other colonial possessions. The British, for instance, brought with them indentured laborers from India and Mauritius to clear the land for crops

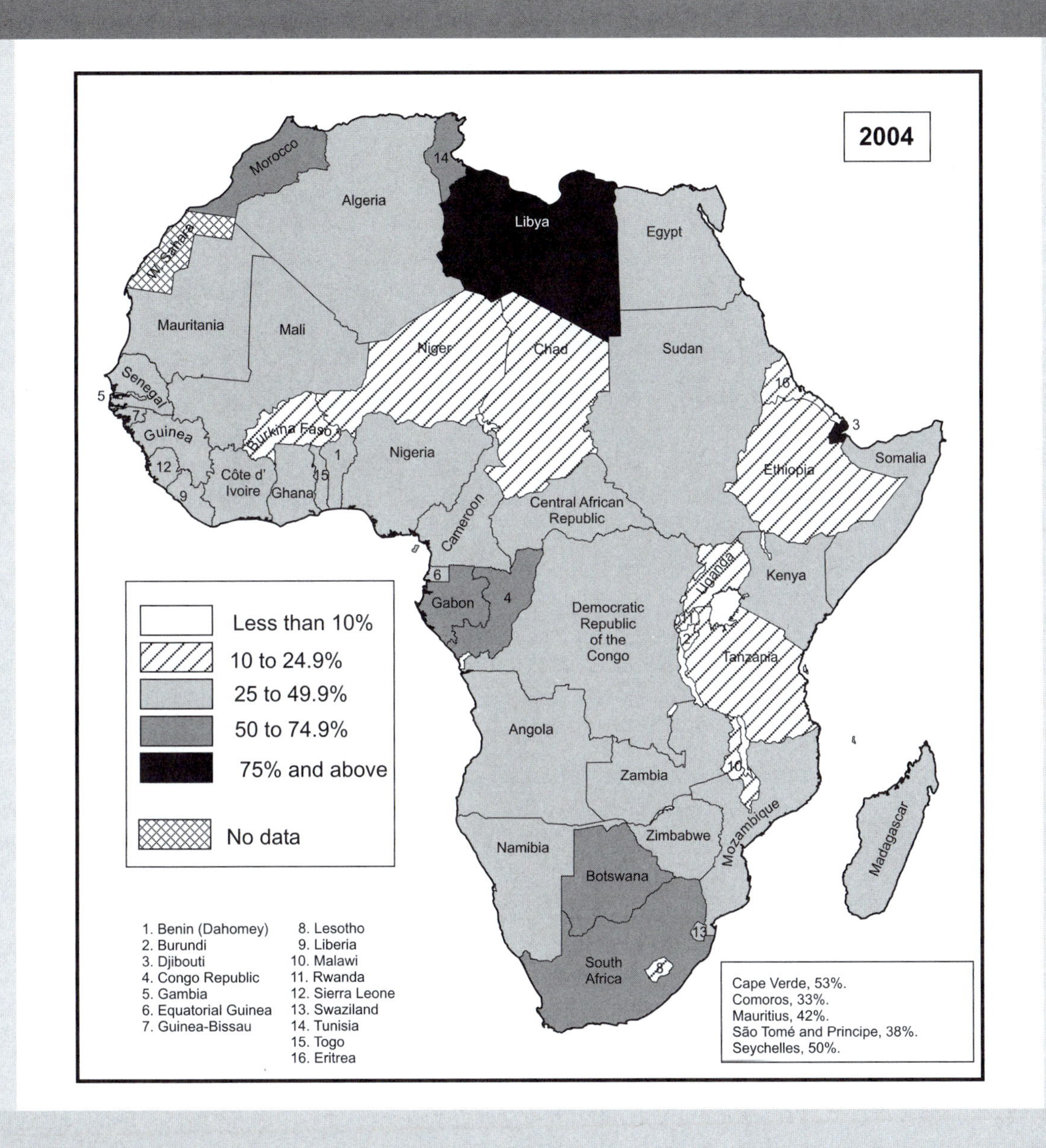

FIGURE 6-17 Percent urban by country, Africa, 2004.

and to build railways and cities in Kenya, Uganda, and South Africa. Syrians, Lebanese, and West Indians migrated to West Africa to become traders, bankers, and builders. French settlers occupied valuable farmlands in Algeria and Morocco, and the Portuguese opened up parts of Mozambique and Angola. This colonial input set in motion countless migrations by individuals in search of economic gain and by groups to meet the requirements of the colonial *sys*tems.

Thousands migrated to the new cities, plantations, and mines, and entire ethnic groups adjusted to the new administrative boundaries. The first labor migrations were frequently involuntary. Africans were forced to build roads, railways, and dams and to work on

FIGURE 6-18 Abidjan, primate city of Côte d'Ivoire.
Christine Drake.

other public projects. Compensation was either lacking or inadequate, and thousands died of malnutrition, disease, and unsafe work conditions. Government-controlled recruiting organizations supplied labor to the mines of Katanga, Rhodesia (Zimbabwe), and South Africa and to the peanut and coffee plantations of Senegal and Côte d'Ivoire, respectively. Although taxes were not new to Africa colonial governments

TABLE 6-4 ESTIMATED GROWTH OF SELECTED PRIMATE CITIES IN THE 1930s, 1960s, AND 1990s, AND IN 2005.

	1930s		1960s		1990s		2005
City	**Year**	**Population**	**Year**	**Population**	**Year**	**Population**	**Population**
Abidjan	1930	10,000	1968	500,000	–	–	3,677,000
Addis Ababa	1938	300,000	1968	620,000	1994	2,085,800	2,758,000
Brazzaville	1936	24,000	1968	200,000	–	–	1,285,000
Cairo	1937	1,312,000	1966	4,220,000	1996	6,790,000	7,735,000
Casablanca	1936	257,000	1966	1,120,000	1994	2,941,000	3,145,000
Conakry	1931	7,000	1967	197,000	–	–	1,871,000
Dakar	1931	54,000	1968	600,000	1997	1,968,000	2,352,000
Dar es Salaam	1931	34,000	1967	273,000	–	–	2,699,000
Harare	1931	32,000	1968	380,000	1992	1,189,000	1,543,000
Johannesburg	1936	517,000	1960	1,153,000	1996	1,481,000	2,026,000
Kinshasa	1930	33,000	1966	508,000	–	–	7,786,000
Lagos	1931	127,000	1963	665,000	1991	5,195,000	8,789,000
Lusaka	–	–	1963	119,000	1990	769,000	1,267,000
Monrovia	1934	10,000	1967	100,000	–	–	940,000
N'Djamena	1936	8,000	1967	150,000	1993	531,000	721,000
Nairobi	1936	50,000	1962	315,000	–	–	2,751,000
Tripoli	–	–	1968	330,000	–	–	1,151,000

Source: Best and de Blij (1977), World Gazetteer (2005).

FIGURE 6-19 Unplanned urban settlement near Bamako, Mali.
Roy Cole.

introduced some form of head tax that had to be paid in cash, forcing the African into new regions and into new lifestyles. As these and other injustices were exposed by private and government inquiries, and even by the International Court of Justice, labor laws were relaxed and migrations became more voluntary and individual.

In West Africa, where the general flow of migratory labor was and is from the interior to the coast, the French instituted a system of forced labor to

FIGURE 6-20 Street hawking, emblematic of the urban informal sector. Abidjan, Côte d'Ivoire.
Christine Drake.

ensure the success of their plantations in Senegal and Côte d'Ivoire, and the construction of the Kayes and Abidjan–Bobo Dioulasso railways. In the middle 1930s French-owned coffee and cocoa plantations in Côte d'Ivoire employed about 20,000 Upper Voltans (Burkina Faso), half of whom were forced laborers. Between 1936 and 1939 forced labor was formally abolished, but during World War II compulsory recruitment was reintroduced. Between 1940 and 1944, some 277,000 migrants entered Côte d'Ivoire from Upper Volta, 171,000 under this program. With the abolition of forced labor in 1946 and the movement of labor becoming unpredictable, the Côte d'Ivoire planters formed a labor-recruiting agency in Upper Volta, which recruited 163,000 workers between 1952 and 1959. The program was carried out in such a ruthless manner that in 1960 the government of Upper Volta prohibited all further operations. While migrations continue—less than half of Côte d'Ivoire's unskilled work force in the private sector are Ivorians—work conditions have improved, and Côte d'Ivoire helps cover the costs of recruitment. An equal number of Malians and almost as many Guineans migrate each year to Côte d'Ivoire. The impact of these migrations into Côte d'Ivoire is at the root of the political impasse it is in today: Half the country in the hands of rebellious northerners and migrants, half in the hands of southern. The problem of integrating migrants is not unique to Côte d'Ivoire we find this cultural issue everywhere. In the case of Côte d'Ivoire the problem is exacerbated by the fact that migrants are Muslim and the Ivoirian establishment is Roman Catholic.

In adjoining Ghana, the cocoa industry has long depended on migrants from northern Ghana, Togo, Burkina Faso, Côte d'Ivoire, and even parts of Nigeria. Forty percent of the work force in southern Ghana comes from these regions. Migrants do most of the menial work such as clearing the forest and harvesting the cocoa. Most migrants leave the north during the dry season, when local labor requirements are low. In South Africa and adjoining countries, where employment opportunities are low and populations are burgeoning, the mining economy is based on migrant labor. Most migrants are contracted to work for periods of nine months or a year. Others seek seasonal work on neighboring farms, while over the last century hundreds of thousands have left permanently for the cities. Thousands of migrants are recruited each year by agencies in Lesotho, Swaziland, Botswana, Mozambique, and Malawi, and elsewhere. In South Africa as elsewhere in the continent, the migrant is usually male, unskilled, and a temporary participant in the modern exchange economy who must support a family in his village. If he succeeds in breaking the circulatory system and becomes a permanent resident in the city, his family will follow.

Independence brought migrations of a different nature: the exodus of colonial settlers, and the return of African exiles. Over a million Europeans left Algeria following the Franco-Algerian War, to be replaced by an even greater number of Algerians who had earlier fled to Morocco and Tunisia. Some 330,000 Portuguese left Angola, and 150,000 others left Mozambique in the months immediately before these territories gained their independence. Thousands of British settlers moved from East Africa (especially Kenya) during and after the Mau Mau uprisings and following independence, and settled in Rhodesia (Zimbabwe), the Republic of South Africa, and Canada. An estimated 75,000 Asians were expelled from Uganda in 1972, and still others were forced out of Kenya and Tanzania. Thousands of Syrian and Lebanese traders lost their licenses (and hence their livelihoods) in Ghana, Liberia, Côte d'Ivoire, and Nigeria as these countries moved toward localizing their economies.

But the most interesting aspect of African migration in coming years is how the demographic momentum of Africa will change Europe. Africa, especially Subsaharan Africa, is the last large area of the earth to enter the demographic transition. Over the coming decades it will possess the youngest, most vigorously growing population on earth. Adjacent Europe, already in population decline, has been importing African workers over the past few decades, and there are easily recognized African sectors in most European cities. Immigrants typically cluster together in the same neighborhoods for mutual support. By 2050, the African share of world population will be over 20% while Europe will have declined from 16% to 7%. For Europe to maintain its economically active population and high standards of living, millions of immigrants much be accepted in Europe. Added to these pull factors is the push factor of the historic development differential between Africa and Europe. If the levels of economic and human development (standard of living, quality employment, political and economic stability) remain lower in Africa than in adjacent regions, then there will be a significant push powering outmigration of Africans, which is likely to dramatically transform the demographic profile of nearby Europe, the likely destination.

Can one speak of the Africanization of Europe? Perhaps not just yet, but time will tell. If this idea seems far-fetched, one need only consider that most of the Caribbean and South America, and the southern states and urban places of the United States of America have been transformed by the African presence. Afro-Caribbean music is popular all over the world, and American popular culture has an African (American) basis. More recently the United States has experienced the immigration of millions of Mexicans and other Latin Americans. This in-migration is profoundly changing American culture and is in large part the cause of the continuing robust growth in U.S. population compared to Europe and Japan. Although we cannot know with precision at the present moment, we do know with confidence that demography is destiny. Population growth in every major region of the world except Africa is past the peak or is peaking. Africa, the last region of the world to enter the later stages of the demographic transition, may possess the demographic momentum to profoundly change the make up of some of the regions that entered it first.

CHAPTER REVIEW QUESTIONS

1. Discussion of the impacts of population growth is often lively. Use evidence from this chapter to compare and contrast different ways of understanding population growth.
2. What is a population pyramid, and how does it help us understand the demography of an African country? How can comparing population pyramids help us understand demographic change over time?
3. What is the demographic transition model suggested by Warren Thompson? Using time-series data from the chapter for a sample of African countries, develop an argument in support of the demographic transition hypothesis.
4. What is a urbanization? What impacts might urbanization have on the development of a country and its demography?
5. What is the role of government in demographic behavior?

GLOSSARY

Crude birthrate (CBR) The number of births per thousand population in a country or region. Although easy to calculate, the CBR does not account for the age and sex structure of the population.

Demographic transition model A transformation from high to low birth and death rates that has been linked to changes in standards of living associated with a country's economic development.

Economic development Improvement occurring unevenly over time and space, in the political, economic, social, and environmental conditions of life for people in a country or region.

FIPS code A two-letter abbreviation for a country's name. Most are self-evident, but for countries that have undergone a name change, the FIPS code bears no relation to the present name (e.g., Burkina Faso is coded UV because when the code was established, the country was called Upper Volta).

Malthusian theory Population growth is limited by the available food; and when population grows, it tends to overshoot carrying capacity (food supply), resulting in a mass die-off of excess population. This theory is based on the work of Thomas Malthus in the eighteenth and early nineteenth centuries.

Neo-Malthusian theory The idea that Malthus was essentially correct, but the Industrial Revolution delayed the mass die-off that, according to the original hypothesis, was inevitable.

Primate city An urban place that is much larger than the next-sized city in a country.

Crude death rate (CDR) The number of deaths per thousand population in a country or region. Although easy to calculate, the CDR does not take into consideration the age and sex structure of the population.

R^2 A correlation coefficient denoting the percent variation of the dependent variable that is "explained" by the independent variable.

Technocratic theory The hypothesis, propounded mainly by the late Julian Simon, that innovations and the market will raise the ability of the earth to support more people. Opposed to the Malthusian thinking.

Total fertility rate (TFR) The number of births a woman of child bearing age would be expected to have on average in given a country.

BIBLIOGRAPHY

Best, A., and H. J. de Blij. 1977. *African Survey*. New York: Wiley

Boserup, E. 1965. *The Conditions of Agricultural Growth: The Economics of Agrarian Change Under Population Pressure*. Chicago: Aldine De Gruyter.

Caldwell, J. C. 1968. The Control of Family Size in Tropical Africa. *Demography*, 5(2): 598–619.

Clawson, D., and Fisher, J. 1998. *Geography and Development: A World Regional Approach*. 6th ed. Upper Saddle River, N. J.: Prentice Hall.

Ehrlich, P. 1968. *The Population Bomb (Is Everybody's Baby)*. Cutchogue, N. Y.: Ballantine.

Grove, A. T. 1989. *The Changing Geography of Africa*. Oxford: Oxford University Press.

Hardin, G. 1968. The Tragedy of the Commons. *Science*, 162: 1243–1248.

Helders, S. 2005. World Gazetteer. Online Leverkusen, Germany.

Mafukidze, T. S. 1974. Africa's need for Planning: Conspiracy or Myth? *Africa Report*, 20(4): 38–41.

Malthus, T. [1798] 1999. *An Essay on Population*. Reprint of 1798 edition. Oxford: Oxford University Press.

Miller, G. T. 2001. *Living in the Environment: Principles, Connections, and Solutions*. Stamford, Conn.: Brooks/Cole.

Mortimore, M. J. and W. M. Adams. 1999. *Working the Sahel: Environment and Society in Northeast Nigeria*. London: Routledge.

Population Reference Bureau. 2005. World Population Data Sheet, book ed. Washington, D.C.: Population Reference Bureau.

Population Reference Bureau. 1975. World Population Data Sheet. Washington D.C.: Population Reference Bureau.

Rostow, W. W. 1960. *The Stages of Economic Growth: A Non-Communist Manifesto*. Cambridge, U.K.: Cambridge University Press.

Simon, J. 1998. *The Ultimate Resource*, Vol. 2. Princeton, N. J.: Princeton University Press.

Tiffen, M., M. Mortimore, and F. Gichuki. 1994. *More People, Less Erosion: Environmental Recovery in Kenya*. New York: John Wiley & Sons.

Thompson, W. 1929. Population. *American Journal of Sociology*, 34(6): 959–975.

United Nations Population Division. 2003. World Population Policies. New York: United Nations.

United Nations. 2004. World Population Prospects: the 2004 Revision. New York. United Nations.

United Nations. 2003. World Populations Polices. Data available online at www.un.org/esa/population/publications/wpp2003/

U.S. Bureau of the Census. 2000. IDB Summary Demographic Data for Nigeria. Washington, D.C.: Census. http://www.census.gov/cgi-bin/ipc/idbsum?cty=NI.

Weeks, J. R. 1992. *Population: An Introduction to Concepts and Issues*, 5th ed. Belmont, Calif.: Wadsworth Publishing Company.

APPENDIX 1 WORLD WIDE WEB LINKS

International Society of Malthus	http://www.igc.org/desip/malthus/
Julian Simon. Books on the Web.	http://www.juliansimon.com
Population Reference Bureau.	http://www.prb.org
U.S. Census International Programs Center (IPC). (Population data and trends for the world, countries and regions.)	http://www.census.gov/ipc/www/idbnew.html
United Nations. The State of World Population, 2001: Population and Environmental Change	http://www.unfpa.org/swp/swpmain.htm.
World Bank Poverty Net (Trends in social indicators over time, world and world regions. Life expectancy, infant, child and maternal mortality.)	http://www.worldbank.org/poverty/data/trends/mort.htm.
World Gazetteer.	http://www.world-gazetteer.com.

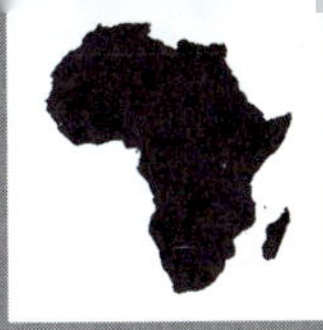

CHAPTER 7

The Geography of Health and Disease

According to the United Nations' normative definition, "health is a state of complete physical, mental, and social well-being and not merely the absence of disease or infirmity" (Meade and Earickson 2000: 2). To be healthy implies an ability to respond to and withstand "insults" from the environment such as those posed by disease and by chemical and other nonorganic agents. Good and bad health also extends to the psychological and social dimensions; just as a person's body may be attacked by an infectious agent, psychological (internal) or social (external) events may have a negative or positive impact on health.

A variety of factors interact to influence and determine health: geographic (where one lives on the surface of the earth), social (wealth and access), and biological (sex, age, and genetic makeup), as well as level of development or standard of living. Figure 7-1 presents a conceptual model of interacting triangles of factors that have an impact on a person's state of health. Three clusters of health-related variables, the environment, behavior, and population characteristics, condition the impact of threats to health. Cluster 1, population, is concerned with people as biological entities who are potential hosts of disease. One's genetic makeup, age, and sex are all interacting variables that have been shown to be important in health. Cluster 2, behavior, is rooted in cultural beliefs, social norms, and the opportunities and constraints posed by socioeconomics and technology. Nutritional status, which consists of the range of food people of a given culture consider acceptable to eat, is a population characteristic affected by behavior and also by the environment. Cluster 3, environment, comprises the physical or natural environment, the social environment, and the artificial or "built" environment that people use as shelter from the natural, and sometimes social, environment.

Living close to the natural environment implies greater risk of insults to one's health. A social environment where a small minority of people control access to resources may mean that certain groups face higher risk of disease than others. People create their environment through transforming it to meet their needs, and these created environments may increase or decrease exposure to health risk. Window screens have a measurable impact on exposure to risk from certain diseases, while central heating, air conditioning, and electric lighting create an artificial environment far removed from the natural environment and the risks posed by many disease vectors, or transmission agents.

Education is considered to be a behavioral variable that conditions one's relationship to the environment. Meade and Earickson (2000) imply that as a person's education and knowledge of health issues increases, potentially hazardous health threats in the environment will be buffered by behavior change. On the other hand, cultural factors can override knowledge, and the environment may provide limited options. For example, Beja pastoralists of Red Sea Province in the Sudan believe that eggs are unclean and therefore do not eat this food of known nutritional value. Moreover, life circumstances put people in harm's way, as in the case of the Egyptian farmer who may know that wading in the paddy puts her at risk of contracting schistosomiasis but nevertheless must transplant rice seedlings from the nursery.

Flores (2001) looks at disease, risk, causation, and outcomes in developing countries as a recursive

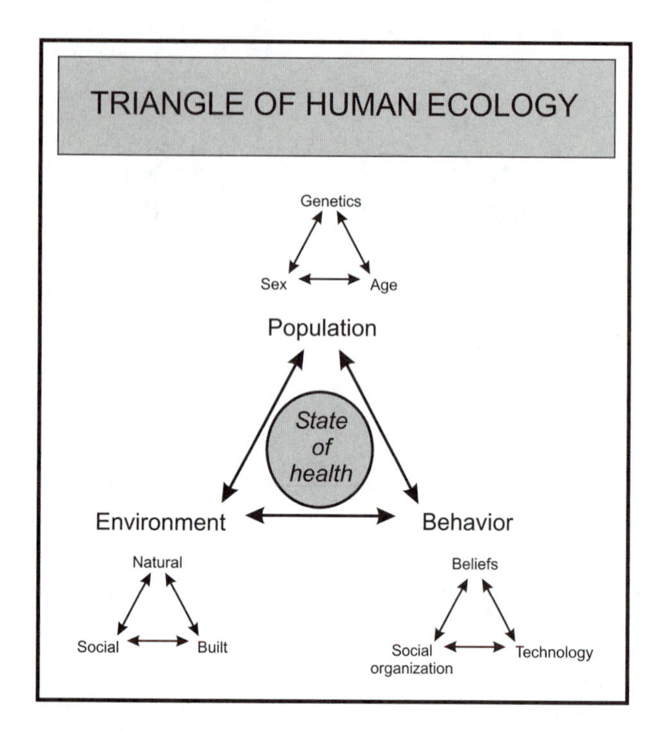

FIGURE 7-1 Triangles of influences on health.

process that may lead to a multithreaded, downward spiral:

> While the interaction of inadequate dietary intake and disease leads to malnutrition, disability, and death, it is also clear that insufficient access to food, inappropriate caring practices, a poor environment, inadequate health services, low women's status, and poverty play a major role in catalyzing the whole process. Each of these factors can be a cause, a risk, and/or an outcome; the pathways toward and away from good health and nutrition go in multiple directions. Take the impact of poverty on HIV/AIDS and TB. Poverty increases the exposure to, as well as the impact of, HIV/AIDS. It diminishes the perceived value of avoiding HIV/AIDS ("we will die soon anyway"), increases the relative costs of preventing and treating the illness, and worsens the impact of weakened immunity because it commingles with a more hostile bacterial and viral environment. Under these circumstances, TB reemerges rather easily, augmenting the negative impact of poverty. Poverty and TB in tandem increase the already deleterious impact of HIV/AIDS on family, friends, community, and state. Looking at the causal process in reverse, HIV/AIDS and tuberculosis increase poverty in the short to medium run by stripping assets—human, social, financial, physical, natural, informational, and political. Asset rundown leaves individuals, families, and communities more exposed to future health and nutrition shocks. In the meantime, public health resources are increasingly diverted away from prevention and rural primary care to the treatment of HIV/AIDS-infected individuals. Flores (2001: 3)

The picture that emerges from examining the medical geography of Africa creates concern. African overall life expectancy is low (Table 7-1 column 3); life lived in good health, free of debilitating diseases, is even lower (Table 7-1 column 4). Compare the health-adjusted life expectancy of the African countries to that in France, Japan, the United Kingdom, or the United States. The health-adjusted life expectancy (HALE) is based on life expectancy at birth but includes an adjustment for time spent in poor health. It is most easily understood as the equivalent number of years in full health that a newborn can expect to live, based on current rates of ill health and mortality.

The data indicate that Africans live much shorter lives than the French, the Japanese, the British, and Americans. The average Briton born today can expect to live almost twice as long as the average Botswanan, Malawian, Ugandan, Zambian, or Zimbabwean, and most Africans live at least 20 years less than the French, Japanese, British, and Americans on average. With regard to living a life without disability, the picture is grimmer. The average HALE for France, Japan, Britain, and the United States is 72.3 years, while for the African countries it is 41.7 years. Furthermore, women in Africa can expect roughly the same number of healthy years as men, unlike women in France, Japan, the United Kingdom, or the United States, who enjoy; 75.1 years, as opposed to 69.6 years for men. What is also remarkable about the data in Table 7-1 is that for some countries, Sierra Leone for example, the average life expectancy is very low, and the average newborn today can expect to live 20 years, almost half his or her life, in poor health, pain, and disability.

THE PREDOMINANCE OF INFECTIOUS DISEASES IN AFRICA

Infectious and parasitic diseases may be spread from human to human, directly from animals to people (zoonosis), or from human to human via an agent of

TABLE 7-1 HEALTH-ADJUSTED LIFE EXPECTANCY FOR 53 AFRICAN AND 4 NON-AFRICAN COUNTRIES THAT PARTICIPATE IN THE WORLD HEALTH ORGANIZATION.

Rank	Country	Life Expectancy (Years)	HALE
1	Japan	81	74.5
3	France	79	73.1
14	United Kingdom	77	71.7
24	United States	77	70.0
78	Mauritius	71	62.7
84	Algeria	69	61.6
90	Tunisia	72	61.0
107	Libya	75	59.3
108	Seychelles	70	59.3
110	Morocco	69	59.1
115	Egypt	66	58.5
118	Cape Verde	68	57.6
132	São Tomé and Principe	65	53.5
143	Gambia	52	48.3
144	Gabon	52	47.8
146	Comoros	56	46.8
149	Ghana	58	45.5
150	Congo	50	45.1
151	Senegal	52	44.6
152	Equatorial Guinea	50	44.1
154	Sudan	56	43.0
155	Côte d'Ivoire	46	42.8
156	Cameroon	55	42.2
157	Benin	50	42.2
158	Mauritania	51	41.4
159	Togo	55	40.7
160	South Africa	53	39.8
161	Chad	50	39.4
162	Kenya	48	39.3

Rank	Country	Life Expectancy (Years)	HALE
163	Nigeria	52	38.3
164	Swaziland	40	38.1
165	Angola	38	38.0
166	Djibouti	46	37.9
167	Guinea	45	37.8
169	Eritrea	55	37.7
170	Guinea-Bissau	45	37.2
171	Lesotho	53	36.9
172	Madagascar	54	36.6
173	Somalia	46	36.4
174	Democratic Republic of the Congo	48	36.3
175	Central African Republic	45	36.0
176	Tanzania	53	36.0
177	Namibia	46	35.6
178	Burkina Faso	47	35.5
179	Burundi	47	34.6
180	Mozambique	72	34.4
181	Liberia	50	34.0
182	Ethiopia	52	33.5
183	Mali	46	33.1
184	Zimbabwe	40	32.9
185	Rwanda	39	32.8
186	Uganda	42	32.7
187	Botswana	41	32.3
188	Zambia	37	30.3
189	Malawi	39	29.4
190	Niger	41	29.1
191	Sierra Leone	45	25.9

Source: WHO (2001a), Population Reference Bureau (2001).

transmission (vector); sometimes there is a combination of zoonosis and vectors involving both people and animal populations (anthropozoonosis) (Figure 7-2). As Table 7-2 indicates, most people in Africa die of infectious diseases. Except for the upper elites, Africans do not live long enough to die of the lifestyle diseases (heart disease, cancer, degenerative diseases such as Alzheimers, etc.) prevalent in countries such as Canada, France, Japan, Singapore, or the United States. An infectious disease is one that is passed from person to person. These diseases are transmitted in a variety of ways: coughing, spitting, and sneezing are very common, and many infectious diseases are transmitted when food or water has been contaminated by fecal matter from infected persons. Ordinarily, flies, cockroaches, or other insects carry the fecal material that contaminates food that is later eaten by people. Likewise, the presence of flies throughout the year poses the threat of contamination. Mosquito-borne diseases take a heavy toll of life in Africa, particularly among the young and the very old. Sexually transmitted diseases (STDs), such as HIV/AIDS and gonorrhea are widespread. Biting flies transmit two major infectious diseases from person to person, river blindness and sleeping sickness.

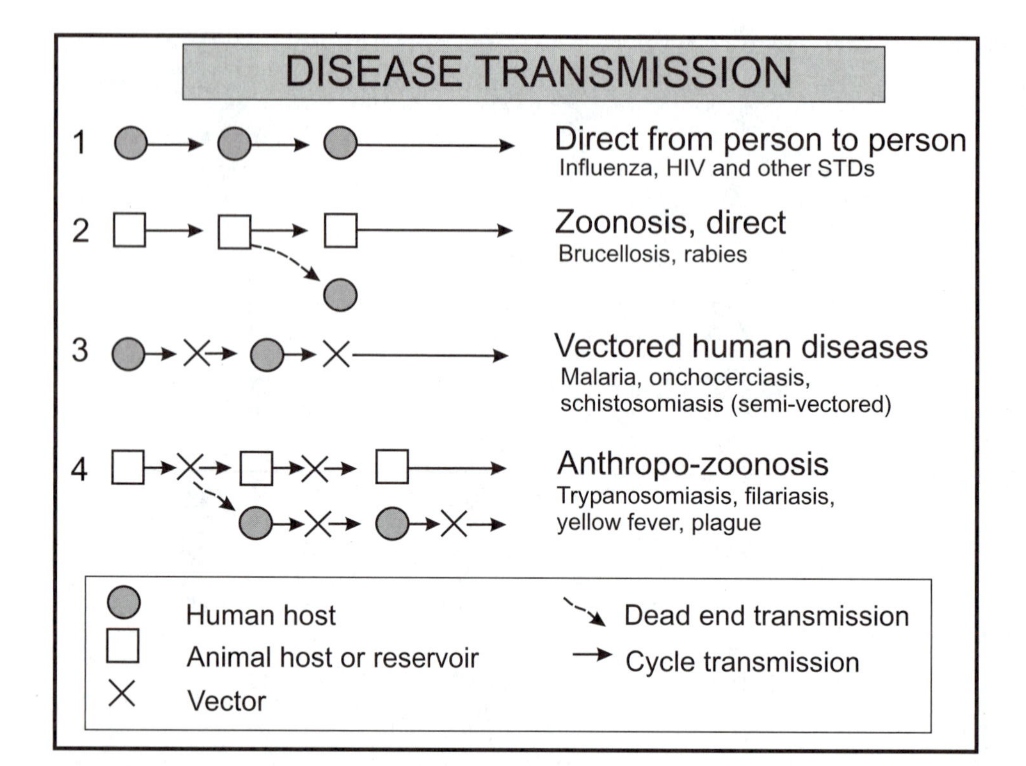

FIGURE 7-2 Chains of disease transmission for infectious and parasitic diseases.

Why are Africans more likely to die of infectious and parasitic diseases than people living elsewhere? Although population, environment, and behavior (Figure 7-1) offer many clues, the level of development (the socioeconomic environment) and climate (the physical environment) may explain the rest, with socioeconomic environment being the most important factor. Areas of higher levels of development have

TABLE 7-2 PERCENT OF ALL DEATHS BY CATEGORY AND CAUSE FOR TWO WORLD REGIONS.

		Deaths in region (%)	
Category	**Cause of Death**	**Africa**	**Americas**
Infectious and parasitic diseases	Lower respiratory	10.11	4.42
	Diarrheal	6.67	1.33
	HIV/AIDS	22.63	1.23
	Malaria	9.14	0.03
	Tuberculosis	3.60	0.96
	Other infectious and parasitic	9.42	1.00
	Total	61.57	8.97
Noncommunicable diseases	Cancer	4.29	18.28
	Cardiovascular	9.22	34.42
	Other noncommunicable	8.52	23.08
	Total	22.03	75.78
All other diseases		16.40%	15.25%
Total		100.00%	100.00%

Source: WHO (2001a).

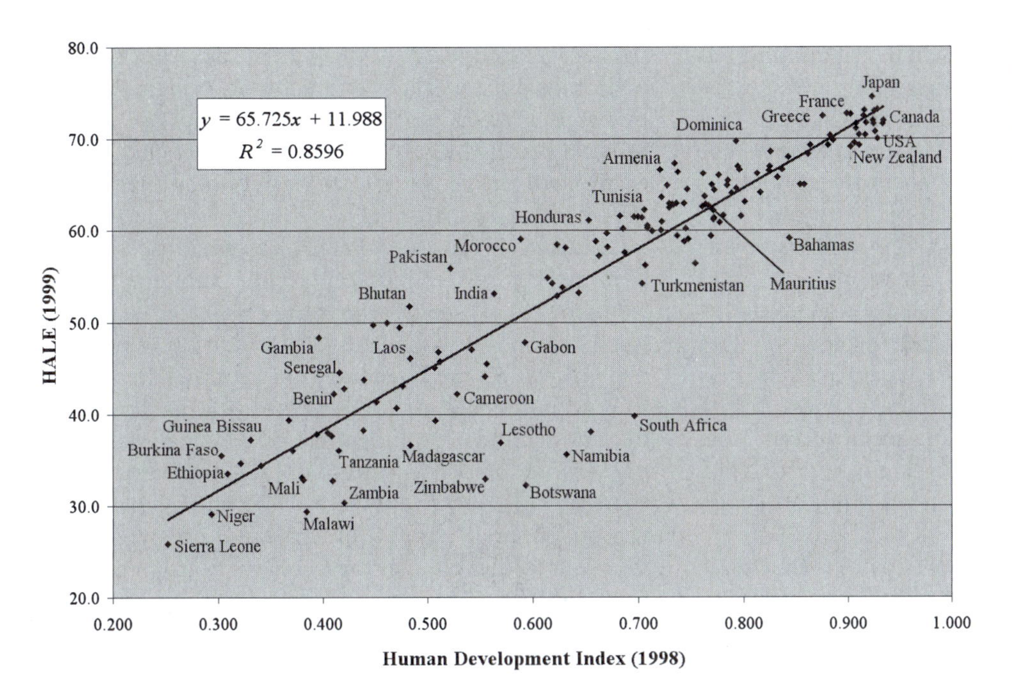

FIGURE 7-3 Health adjusted life expectancy (HALE) and the human development index (HDI) for 156 countries.

made an epidemiologic transition from infectious and parasitic to noncommunicable diseases. This is because as development occurs over time and across space, the political, social, economic, and environmental conditions of a region or country are improved, furthering the well-being of the people who live there. Among the characteristics of the process of development are the following: the relations between people and the environment become more complex and distant; economic activity shifts away from primary activities to secondary to tertiary to quaternary activities, and becomes more specialized; people migrate to urban areas; and enormous wealth is generated.

Development has profound health consequences, most positive but some negative. In less developed regions of the world people live very close to the environment, and direct contact increases the risk of infection from other organisms that have evolved to profit from the availability of human hosts. By the same token, in more developed places in the world people have greater access to education and modern medicine, and more often live in screened and sealed dwellings in cities far from the natural environment. Even so, the city, and the large concentrations of people and animals in it, may pose additional risk to people.

Figure 7-3 shows how development affects health: the graph depicts the relationship between development as measured by the Human Development Index (HDI) of the United Nations and the Health-Adjusted Life Expectancy (HALE) of the World Health Organization. The HDI is a composite measure of three variables, one economic, one social, and one related to health. Figure 7-3 clearly demonstrates that the fit between health and development is close. Almost every African country is found on the lowest quarter of the distribution; the only non-African countries there are Cambodia, Haiti, Laos, and Papua New Guinea, which are on the high end. The six African countries that do not fall within the lowest quartile are, from lowest to highest, Cape Verde, Egypt, Morocco, Tunisia, Algeria, and Mauritius. The geographic distribution of these countries is of interest; four of the countries are in North Africa, and two are island nations far from mainland Africa. The countries having the highest HIV infection rate (South Africa and Botswana, among others) have been overpredicted by the regression line, meaning that their HDI is relatively high but the HALE is unusually low.

Malaria

Malaria is the most common disease in tropical Africa and is the most serious vectored disease in the world (Meade and Earickson 2000). Ninety percent of all cases of malaria occur in Africa. This debilitating disease reduces a person's functioning and lowers resistance to other diseases; it completely incapacitates tens of millions of people and kills close to a million people each year in Africa, mainly children and the elderly. Its impact on the household and national economy alike is great: in 1995, in countries where malaria is severe, incomes were 33% of those in countries without malaria (Gallup and Sachs 1998). Malaria has a particularly pernicious effect on pregnant women. During pregnancy, in the areas of Africa where malaria is most prevalent, women are more than four times as likely to suffer a clinical attack of malaria than during other times (WHO 2001b). Spiking fever, chills, joint pain, headache, vomiting, convulsions, and coma are symptoms of malaria.

Although there are four kinds of parasite that cause malaria, the major African parasite is *Plasmodium falciparum*, the most deadly form. Malaria is spread from infected to uninfected people by the female anopheles mosquito (especially *A. gambia:* Figure 7-4), which must feed on blood to reproduce. When introduced into the bloodstream of an uninfected person, the parasite colonizes the liver and reproduces in enormous numbers in the blood. It consumes red blood cells as part of its reproductive cycle, which causes high fever and chills, and weakens the body. Although *P. falciparum* itself has a mortality rate exceeding 10%, other diseases usually finish off the victim. Where malaria is controlled, fertility rates increase and infant mortality rates decline dramatically.

Malaria can be controlled by spraying insecticides, draining swamps, and pouring oil on standing water. Nevertheless, *A. gambia* is hard to eradicate because it is a survivor: it feeds on animals in addition to people, and it has become resistant to pesticides. The use of mosquito nets decreases the chances for infection but does not eliminate it altogether. Drugs such as chloroquine work to kill the parasite in the human body, but there is widespread resistance (Oloo et al. 1996). On the other hand, many West Africans themselves are resistant to malaria, having been previously exposed to it. In addition, many Africans (roughly 30–40% in some areas) have some protection from malaria because they are carriers of "sickle shaped" blood cells that the parasite cannot attack. The anemia that goes with sickle-shaped blood cells can be fatal, especially to children.

Although the relationship between migration and disease has been known for quite some time

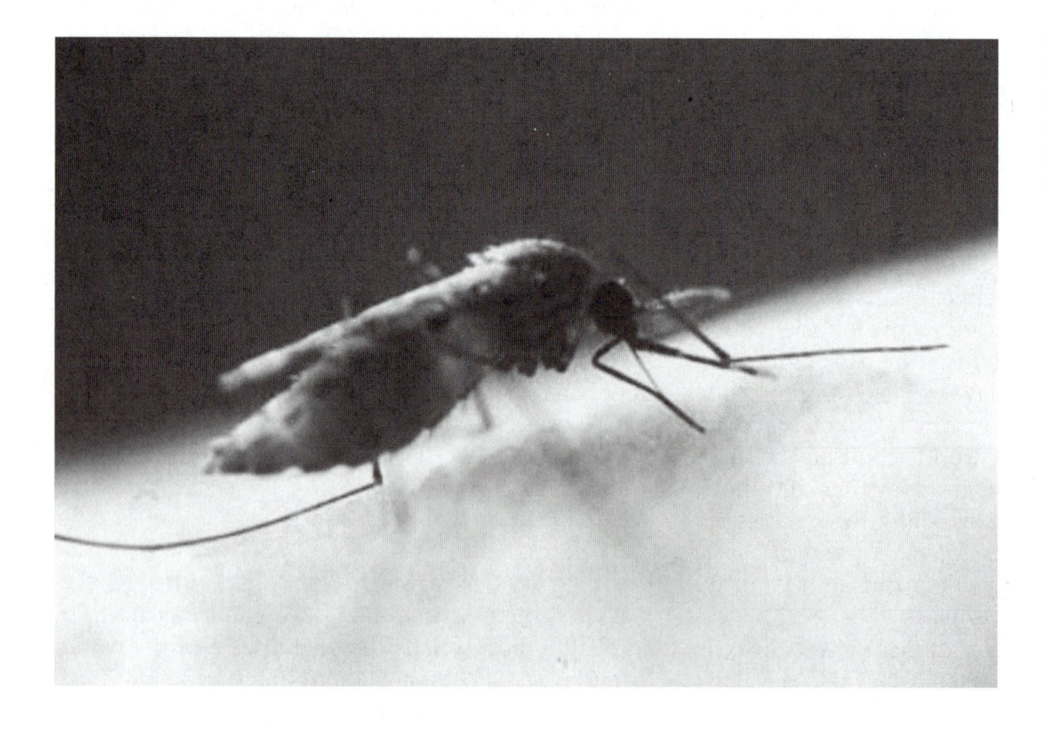

FIGURE 7-4 *Anopheles gambia*, malaria vector. *World Health Organization.*

(Prothero 1965), migration of infected individuals has been found to play an important role in the recent resurgence of malaria in addition to the biological causes. Martens and Hall (2000) have identified migration to urban areas, the colonization of new land for farming, large-scale agricultural development involving irrigation, refugee migrations, and intercontinental travel as factors in the resurgence of malaria today.

Recent research has focused on chemotherapy to disrupt the life cycle of the parasite in the blood (Rosenthal 1998), but the prospect of a world free of malaria has receded over the last 30 years. In fact, the World Health Organization is now speaking of the spread of malaria where it has not been endemic as a result of global climate change. The optimism of the 1960s has given way to pessimism.

Onchocerciasis

Also called river blindness, onchocerciasis is found in 30 countries of Africa, affecting people who live within 12 (19 km) miles of rapidly moving river water. The disease is most prevalent from Senegal to Kenya, where 115 million people are at risk and 18 million are infected. Of those infected, 6.5 million suffer from severe itching or dermatitis and 270,000 (1.5%) are blind (WHO 2000a).

River blindness is hyperendemic along stretches of the upper tributaries of the Volta River of Ghana, where it has been estimated that the incidence of blindness before spraying began in the 1970s was 3,000 per 100,000, versus about 200 cases of blindness per 100,000 in Europe (Hunter 1966). Almost half of the cases of blindness in northern Ghana were the result of onchocercal infection, and where the local incidence was 10% or more (closer to the rivers), 9 out of 10 cases were due to onchocerciasis. Furthermore, the World Health Organization found that up to 20% of those over 30 years of age along some stretches of the Volta tributaries were blind (WHO 1997).

Onchocerciasis is caused by the tiny parasitic worm *Onchocerca*, which is transmitted from infected to uninfected humans by the small, bloodsucking black fly *Simulium damnosum*, which needs a blood meal after mating to lay eggs (Figure 7-5). The fly breeds along river-banks, and when the rivers are not flowing it disappears without trace; but once the summer rains begin, the fly reappears and usually concentrates within 12 miles of the rivers themselves. The incidence of onchocerciasis is highest close to the rivers and decreases with distance from water into the interfluves. Before the spraying campaigns that began in the 1970s, lands adjoining the infected regions were frequently abandoned, and there is ample evidence that settlement had retreated as onchocerciasis encroached in the past (Figure 7-6). Thus valuable arable land used to lie idle while undernourished and malnourished people lived only short distances away, unable to farm and irrigate because of the certainty of contracting the disease.

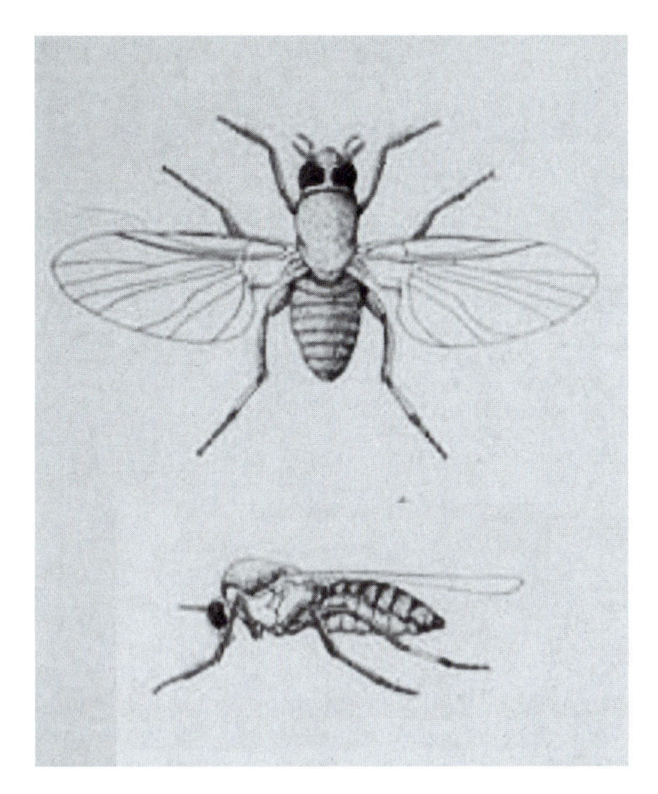

FIGURE 7-5 *Simulium damnosum*, river blindness (onchoceriasis) vector.
World Health Organization.

The female worm, which can be more than half a meter (20 in.) in length, can live for up to 14 years in the human body. Each worm produces millions of microscopic larvae, or microfilariae (Figure 7-7), that migrate throughout the body causing a variety of problems, including rashes, itching, skin depigmentation, lymphadenitis (hanging groin and elephantiasis of

FIGURE 7-6 Villages along the Volta River that were abandoned owing to river blindness.
World Health Organization.

the genitals), visual impairment, and ultimately blindness. The microfilariae are not able to develop further unless they are taken up by the fly *S. damnosum,* which then bites the host. Once a new person has been infected, the worm's life cycle is complete. Larvae of *Simulium* need to be in fast-flowing water to filter out organic particles for food. *Simulium* lays its eggs on rocks or plants located in turbulent water around rapids or falls. The adults emerge after about 10 weeks and live for approximately one month.

FIGURE 7-7 Microfilariae of *Onchocerca.*
World Health Organization.

FIGURE 7-8 Larvicide spraying, part of the Onchocerciasis Control Program.
World Health Organization.

The first major effort to control river blindness, the Onchocerciasis Control Program (OCP), was originally sponsored and developed by the United Nations and the World Bank but has been increasingly taken over by African governments. Based in Ouagadougou, Burkina Faso, OCP began spraying in 1974 in 11 heavily infested West African countries and continued until 2002. The logic of spraying (Figure 7-8) was to prevent new infections for 14 years so that the adult worms would die out and the human reservoir would become disease free, thus breaking the cycle of infection. This goal had been accomplished for seven of the OCP countries by 2000 (WHO 2000c) and the remaining four countries are now onchocerciasis free. A drug to combat the parasite in humans was developed in the 1980s, Ivermectin, which kills the microfilariae in the body and need be taken only once a year. The manufacturer of ivermectin, Merck, provides the drug free of charge to those who need it. Twenty million received the drug in 2000 (WHO 2001d).

In 1995 the African Program for Onchocerciasis Control (APOC) was created to extend the successes of the OCP to other areas in Africa where onchocerciasis was endemic. The purpose is to create a community-based response to river blindness consisting principally of the use of Ivermectin with some targeted spraying to kill *Simulium damnosum*. By 2007, it is envisioned that the program will involve over 60 million people in 17 non-OCP countries (WHO 2001d).

The responsibilities of the OCP have been devolved to the public health services of the 11 countries where onchocerciasis had been endemic. The achievements of the program are many. River blindness has been eliminated for all practical purposes from the OCP countries, and for the first time children are being born who do not have to face the risk of infection. According to the World Health Organization (1997), 30 million people have been protected from the disease, 100,000 have been saved from blindness, and 1.25 million have become free of the onchocercal infection. Millions of acres that were previously too dangerous to farm have been brought into cultivation.

Trypanosomiasis

Trypanosomiasis, or sleeping sickness, is a very dangerous disease transmitted from game animals to humans

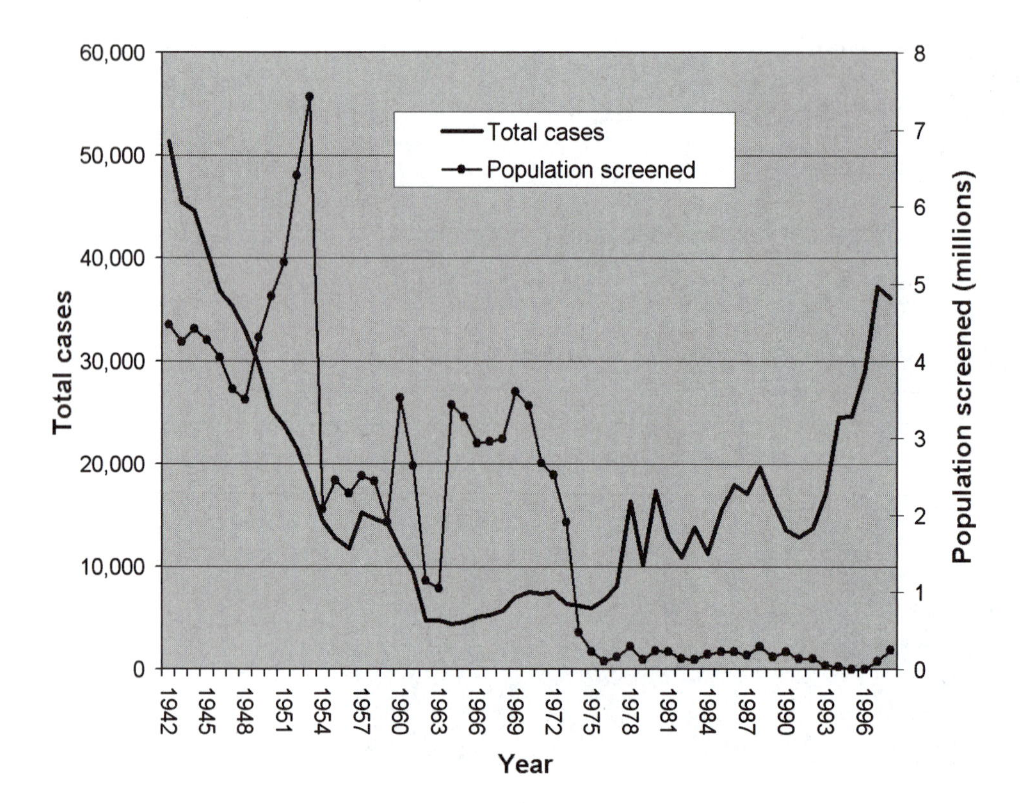

FIGURE 7-9 Total cases of sleeping sickness and screened population, 1942 to 1998.

or from human to human by the tsetse fly (*Glossina* spp.). The ravages of this debilitating and deadly disease have influenced the distribution of human and cattle populations in Africa. In 2000, about 50,000 cases were reported to the World Health Organization (WHO 2001a) although it is estimated that the disease is grossly underreported and that between 300,000 and 500,000 persons are actually infected by the disease today (Figure 7-9). In some areas of high endemicity, with 20 to 50% of the population infected, sleeping sickness is the greatest or second greatest cause of mortality, ahead of AIDS. Despite the efforts of governments and international agencies, there has been a resurgence of the disease over the last 35 years, particularly where political instability, war, and poverty compromise efforts to combat it (WHO 2001c).

The tsetse fly is a large brown fly with gray stripes on the thorax and overlapping wings (Figure 7-10). It is found in much of the tropical rain forest and savanna areas and is concentrated in belts along the margins of forest and bush, near rivers and lakes. It is less common in the more open savanna, where summer temperatures are very high, and in the higher elevations of East Africa (above 4,000 ft; 1,220 m), where it is too cool.

There are about 20 species of tsetse fly that transmit the single-celled organism known as *Trypanosoma brucei gambiense*. When the tsetse fly takes blood from an infected person or animal, the organisms (trypanosomes) pass into the fly, where they develop; then they are passed into the blood of the person or animal on which the fly next feeds. The resulting disease in humans is called trypanosomiasis, while that of livestock is called nagana. Once in the blood, the parasite multiplies in the blood and lymph glands. It later invades the central nervous system and causes major neurological disorders. Sleeping sickness is always fatal without treatment; even after successful treatment, however, the neurological damage is often irreversible.

Sleeping sickness is known to have existed in West Africa during the fourteenth century and was probably confined to that region until the late 1800s, when it spread east and south. Three severe epidemics of sleeping sickness have been recognized since the late nineteenth century: one that ravaged in Central Africa between 1896 and 1906 in East Africa, one in 1920 in several countries in Central Africa, and one that started in 1970. The disease was first recognized in Uganda in 1901, and 4 years later about 200,000 persons died of sleeping

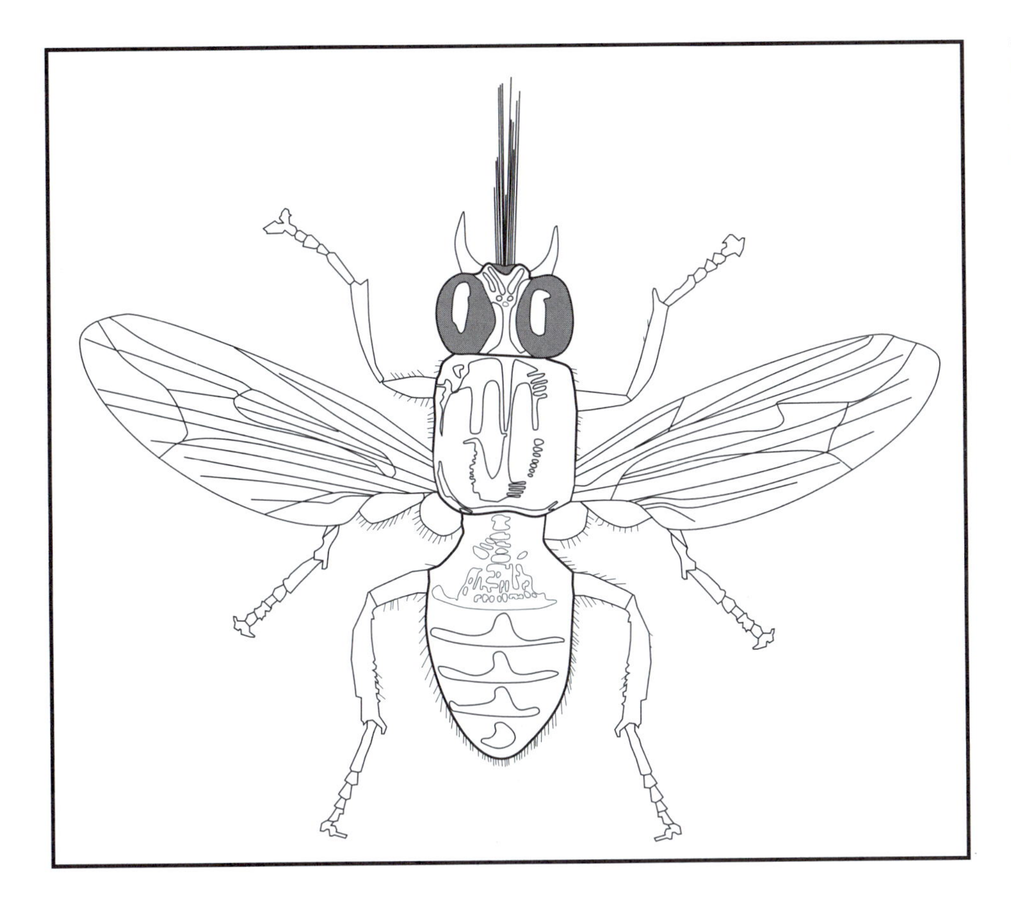

FIGURE 7-10 Tsetse fly (*Glossina* spp.). *Adapted from Livingstone (1857).*

sickness in the Busoga Province of Uganda alone. Since then it has spread as far south as Mozambique and South Africa, leaving in its wake extensive areas of sparse population and limited pastoral activity. The Gambian variety most often has a human host, while Zimbabwean sleeping sickness (carried by *Tripanosoma brucei rhodesiense*) is usually associated with game (Figure 7-11).

The presence of the tsetse fly has profoundly influenced the geographical distribution and density of pastoralism and of mixed crop and livestock farming. Cattle have been restricted to tsetse-free areas of Africa and, in the past, agriculture was commonly concentrated on the arid but tsetse-free margins of the savannas, where food shortages were not infrequent. The fly has been a determinant of seasonal migration patterns in both East and West Africa. For example, the Fulani (Fulbe) of West Africa trek their cattle northward as the rains and tsetse fly spread from the south, returning to southern areas of permanent pasture during the dry season. Centuries ago, cattle were driven south of the Congo Basin, the major route being in the highlands between Lakes Tanganyika and Malawi and onto the southern plateaus (Figure 7-12).

Sleeping sickness may explain why the plow and animal-drawn carts did not diffuse or develop independently in precolonial Africa (Knight 1971). Sleeping sickness also limited the extent of the migration of outsiders into tropical Africa. For example, the movement of Europeans and North Africans was checked in the tsetse-infested areas of precolonial West Africa, where sleeping sickness originated. The disease has been spread inadvertently by shifting cultivation, by the expansion of plantation agriculture into new areas, by the forced relocation of Africans to the New World during the era of the Atlantic slave trade, and by modern instances of migration. During the 1920s, for example, the British government resettled Napore and Nyaneya pastoralists from Uganda's northern border and the Sudan to control communal violence and political instability. As the pastoralists left, both game and the tsetse fly moved in. Within a very short time,

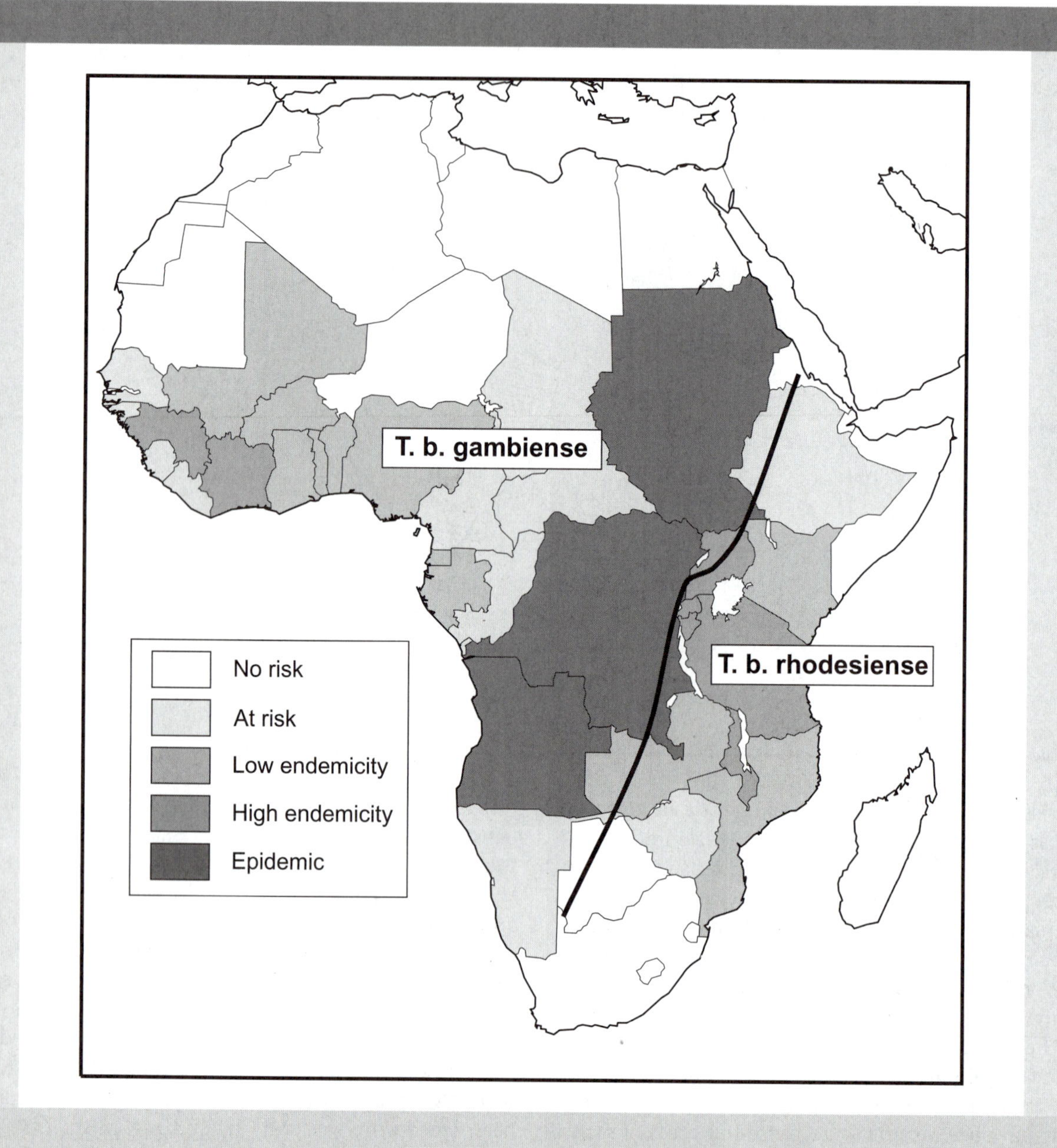

FIGURE 7-11 Distribution of trypanosomiasis in Africa.

about 1,500 square miles (3,885 km^2) became infested, forcing the pastoralists to migrate into poorer country. Following a tsetse fly clearance program two decades later, the land was reclaimed and both tillage and grazing resumed (Deshler 1960).

Control of the fly and disease is both difficult and costly, and different methods of eradication must be adopted according to the species of *Glossina* present. In the past the insecticide dieldrin was completely successful in controlling tsetse fly in parts of northern Nigeria; in Zimbabwe, DDT has been used successfully.

In addition to introducing the problem of insect resistance, the use of insecticides sometimes kills useful insects and may be ineffective against pupae that lie

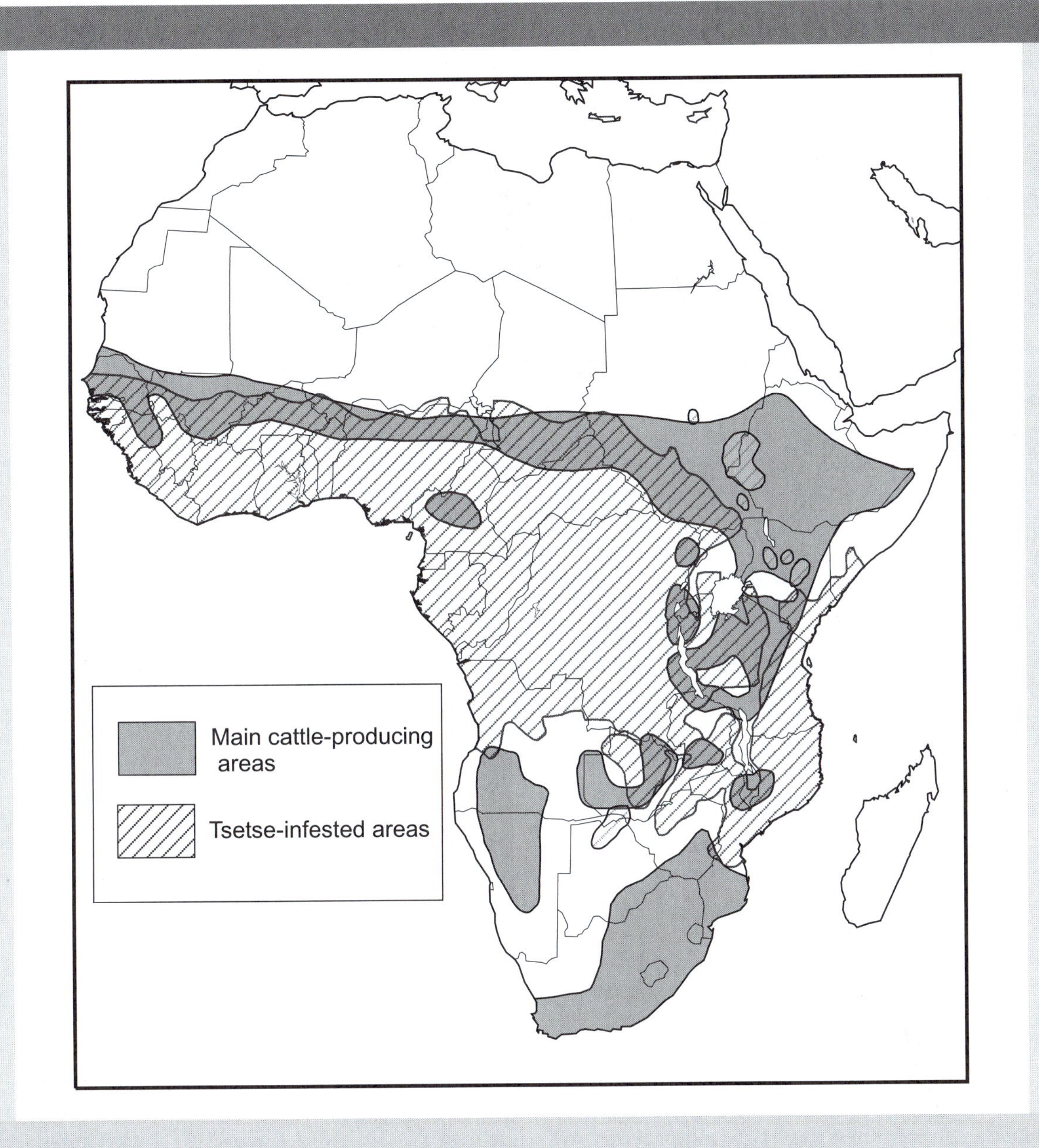

FIGURE 7-12 Areas of tsetse infestation and cattle production areas in Africa, circa 1960.

hidden in the soil. The clearing of woodland is common and involves the destruction of vegetation in a band perhaps 3 miles (5 km) wide and scores of miles in length. This has been done in the past in parts of Uganda, Tanzania, and Zimbabwe, where in conjunction with the clearing, thousands of game animals (especially antelope), carriers of trypanosomes, have been destroyed.

Treatment with drugs to kill the parasite in a patient's body is the immediate but costly solution, and this treatment is difficult to administer in the less developed regions of Africa. Early diagnosis, of paramount importance to avoid neurological damage, is seldom achieved. The ultimate control may come only with the destruction of the tsetse habitat and wildlife

FIGURE 7-13 Tsetse fly trap. *World Health Organization.*

reservoirs through the clearing of thickets and the establishment of denser human and livestock populations. Efforts have also focused on eliminating the disease in cattle, and traps impregnated with chemicals that attract tsetse flies to their death have been developed (Figure 7-13). We know that the disease can be reduced; the epidemic in 1920 was stopped as a result of a massive screening effort of millions of people. In the early 1960s, the number of reported cases of sleeping sickness declined to the lowest levels ever but began to rise in the decades that followed.

Today sleeping sickness has a major impact on rural development by debilitating entire populations and decreasing the productivity and work capacity of the labor force. The third severe epidemic of sleeping sickness, which began in 1970, continues unabated and is a serious obstacle to the development of some regions. In the three worst cases, today Angola, the Democratic Republic of the Congo and the Sudan, the national capacity to respond to the epidemic situation has been overwhelmed, and in certain areas the prevalence continues to be very high. In these worst-case scenarios, the ability of government to respond has been compromised by years of political instability and war.

Schistosomiasis

One of the most debilitating and widespread diseases is schistosomiasis, or bilharzia, which ranks second behind malaria in its socioeconomic and public health importance (WHO 1996). The 170 million people in Africa who have been infected with schistosomiasis represent 85% of the world's schistosomiasis infections. The disease is endemic in almost all countries of Africa except Lesotho, possibly affecting one out of two persons in the continent (Figure 7-14). Meade and Earickson (2000) call it the most rapidly spreading serious infectious disease in the world. The impact of schistosomiasis is profound. It is particularly debilitative in its effects, causing weakness and lethargy; performance of children in school is retarded, and the ability of farmers to work is seriously eroded.

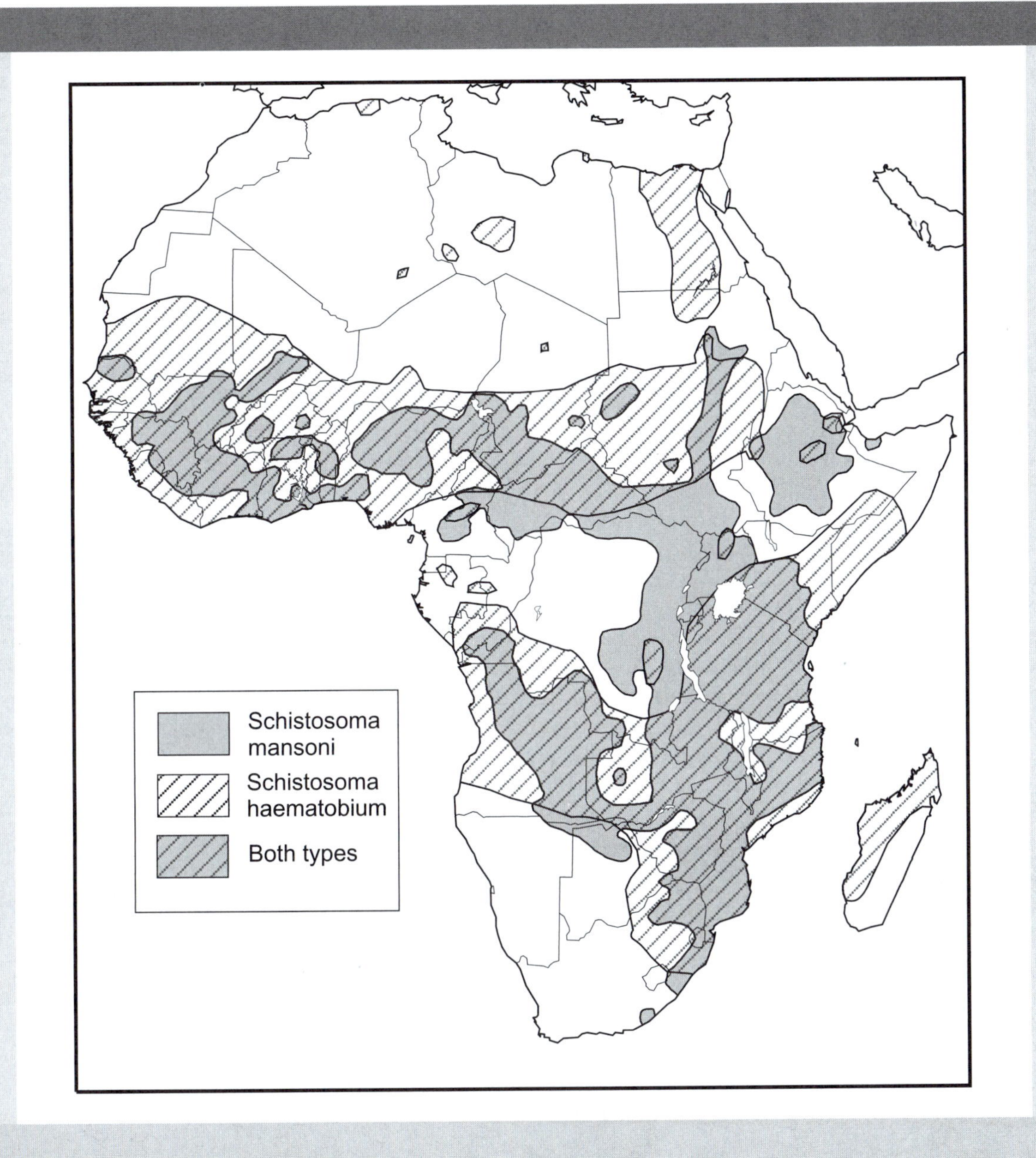

FIGURE 7-14 Distribution of two major types of schistosomiasis (bilharzia) in Africa.

In the 1960s Egypt was one of the most heavily infected states, with more than 15 million of its 32 million population affected, roughly 42%. One study in Mozambique showed infection in 66% of a sample population aged 3 to 24 years, while another showed that in children 3 to 16 years, 80% were infected, and parasitic association was present in half of the cases (Hughes and Hunter 1970). Very high incidences have been confirmed in the West Nile District of Uganda and in Mauritania, Nigeria, Tanzania, Côte d'Ivoire, Natal, South Africa, and parts of the Sudan. In Djibouti and Somalia in the 1990s, schistosomiasis of the intestines was introduced through the forced migration of people due to political instability and war. In Senegal and Mauritania, intestinal schistosomiasis infections associated with the construction of the Diama Dam on the Senegal River 20 years ago and the consequent development and spread of irrigation have reached epidemic proportions and continue unabated. In Ghana, the construction of dams, especially the Akosombo Dam, has caused intestinal schistosomiasis to increase to such a degree that over 90% of the children around Lake Volta have been infected. On the other hand, the combination of chemotherapy and mass education via the mass media in Egypt has caused a significant decrease in the morbidity and prevalence of schistosomiasis. In Morocco, schistosomiasis has been controlled. It has disappeared altogether in Tunisia. Recent research has revealed that unsterile hypodermic needles used in controlling schistosomiasis in Egypt caused the spread of hepatitis C (Whyte 2000).

Schistosomiasis is caused by several strains of a parasitic blood fluke of the genus *Schistosoma,* the most common being *S. haematobium, S. mansoni,* and *S. intercalatum* (Figure 7-15). The former two varieties are found in almost every African country, while the latter has been reported in ten central African countries (WHO 1996). Both *S. haematobium* and *S. intercalatum* infect the intestines, while *S. mansoni* infests the urinary system. The schistosoma are carried from person to person through an intermediate host, a water snail, that lives in rivers, lakes, and irrigation canals. They are readily picked up while bathing or washing clothes, or by drinking infected water.

The worm is slightly more than a half inch (15 mm) in size and lives in the human bladder, intestines, or liver, feeding on blood. Its life cycle involves people and snails as hosts and two periods in water in which larvae swim to a host. Adult worms in an infected person lay eggs that are passed into the environment in course of excretion. Eggs that are passed into water hatch, producing larvae called miracidia, which penetrate *Bulinus* (Figure 7-16) or *Biomphalaria* snails if any are present. These snails prefer slow-moving, weedy, water. If these snails are not present, the larvae die. If the larvae find such snails, they penetrate them, multiply profusely, and release into the water large numbers of a different stage of the larvae (cercariae). These larvae can easily penetrate the skin of a human, whereupon they develop into schistosome worms that eventually reach the liver, bladder, or intestines to multiply and complete the life cycle. The parasite develops into a long worm in a month or two, laying from 200 to 2,000 eggs per day over the average 5-year life span. Symptoms of schistosomiasis are

FIGURE 7-15
Schistosome worm.
World Health Organization.

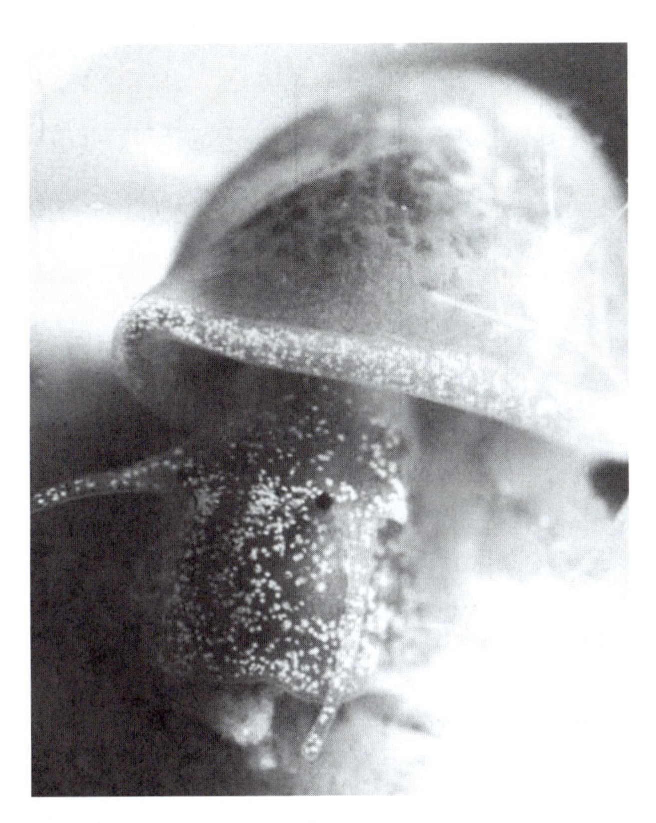

FIGURE 7-16 *Bulinus* snail.
World Health Organization.

blood in the urine or feces. Incidence of bladder cancer associated with schistosomiasis in some parts of Africa is 32 times higher than that of bladder cancer in the United States (WHO 1996).

One of the unfortunate corollaries of development is the spread of schistosomiasis. Irrigation works in particular tend to foster dense populations of both humans and snails, and to attract migrant workers who may come from endemic regions and thus act as agents in the diffusion of the disease. Schistosomiasis probably began to spread in a significant way in Egypt in the early 1800s with the expansion of irrigated cotton cultivation under the modernization program of the ruler, Muhammad Ali. In the 1960s when the Aswan High Dam was built, it was found that in four selected areas in a 3-year period, schistosomiasis infection rates increased as follows: from 10% to 44%, from 7% to 50%, from 11% to 64%, and from 2% to 75%. Similarly, at the Sudanese Jezira Irrigation Scheme near the confluence of the White and Blue Niles, field investigations in 1947 showed a mean incidence of 21% among adults and 45% among children, while prior to construction, neither schistosomiasis nor the host snail had been present. It is believed the disease was introduced to the areas by migrant workers from West Africa (Hughes and Hunter 1970). Other major irrigation schemes reporting schistosomiasis include the Diema Dam project on the Senegal River; the Mbarali on the Rufiji River, Tanzania; the Bacita near Jebba, Nigeria; l'Office du Niger on the dead delta of the Niger River in Mali, and the Volta Project in Ghana.

Controlling the disease in the environment has not been very successful in the past, but chemicals have been used both to kill snails and to destroy their habitat. Drugs to kill the parasite have been the most effective treatment for people, and currently there are three drugs that are effective and inexpensive. Most countries of Africa have mounted education campaigns to persuade people to keep out of the water. The World Health Organization suggests that prevalence can be reduced by 75% within 2 years in endemic areas, but surveillance must continue for 10 to 20 years to prevent reinfection (WHO 1996). Although there have been some successes, where primary health care personnel are few in number the effective treatment of schistosomiasis (and other infectious diseases) will be difficult.

Human Immunodeficiency Virus (HIV)

Human immunodeficiency virus (HIV) and its disease, acquired immune deficiency syndrome (AIDS), pose one of the greatest threats to human well-being in history. Currently 38 million people are infected with HIV/AIDS, of which 25 million live in Subsaharan Africa. Each year about 5 million new cases occur, of which 3 million are in Africa South of the Sahara (UNAIDS 2004) (Figure 7-17). At the start of the new millennium, Peter Piot, the executive director of the Joint United Nations Program on HIV/AIDS, included the following statement in the United Nations Report on the Global HIV/AIDS Epidemic.

> A decade ago, HIV/AIDS was regarded primarily as a serious health crisis. Estimates in 1991 predicted that in Subsaharan Africa, by the end of the decade, 9 million people would be infected and 5 million would die—a threefold underestimation. Today it is clear that AIDS is a development crisis, and in some parts of

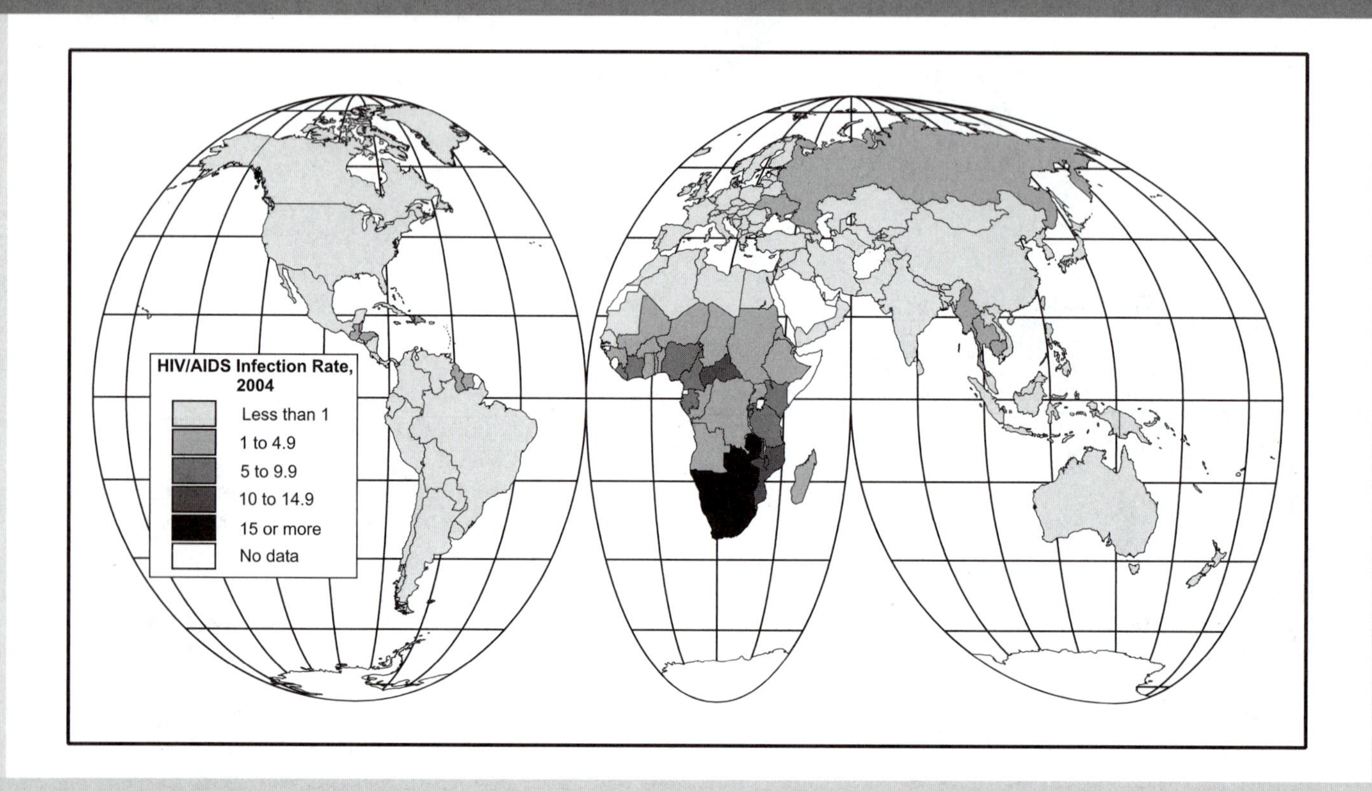

FIGURE 7-17 World HIV/AIDS infection as a percentage of total population by country 2004.

> the world is rapidly becoming a security crisis too. There is now compelling evidence . . . that the trend in HIV infection will have a profound impact on future rates of infant, child and maternal mortality, life expectancy and economic growth. These unprecedented impacts at the macro-level are matched by the intense burden of suffering among individuals and households. AIDS is unique in its devastating impact on the social, economic, and demographic underpinnings of development. (UNAIDS 2000c: 7)

HIV/AIDS originated in central Africa, and in countries where it is most widespread, the population has been altered in a manner unprecedented in human history because it kills young, heterosexual adults who are in their most productive years. No other disease even comes close to the pernicious impacts of AIDS. Over 20 million people have died worldwide (mostly in Africa) as a result of AIDS since 1981 when the first cases were identified. According to the United Nations (2004), deaths from AIDS are expected to double over the next decade. This is because, as noted earlier, more than 5 million new infections occur each year. AIDS is expected to kill at least one-third of the young men and women in many southern African countries and, in the countries where it is most prevalent, up to two-thirds.

Table 7-3 places HIV/AIDS in the context of previous epidemics. It is clear that HIV/AIDS is in a class with the Black Death of the fourteenth century. HIV/AIDS spreads through the transmission of human bodily fluids, principally through sexual relations. HIV is particularly dangerous because it is always fatal, although there have been reported cases of people successfully fighting off HIV infection in the past. Most importantly, its spread can go undetected—ten or more years may pass between initial infection and manifestation of symptoms.

HIV is a virus that attacks white blood cells and compromises the victim's immune system, opening the body up to attack from a variety of opportunistic diseases such as tuberculosis, pneumonia, and certain

TABLE 7-3 PROFILES OF MAJOR EPIDEMICS THROUGHOUT HUMAN HISTORY.

Epidemic and Date	Origin	Method of Transmission	Mode of Introduction and Spread	Description and Its Effects on Population
Black Death in Europe; fourteenth century	Asia	Rat-borne fleas; coughing and sneezing	Spread along trade routes from Middle East	25% of European population died (25 million people)
Smallpox in the Americas; sixteenth century	Europe	Coughing and sneezing; physical contact	European migration	Killed 10 million to 20 million
Influenza, worldwide; 1917–1919	Asia	Coughing and sneezing	Brought to United States by U.S. troops who fought in World War I	25 million to 40 million deaths worldwide
HIV/AIDS, worldwide; 1980 to present	Africa	Bodily fluids, mainly through sexual intercourse	A primate virus that jumped species	Over 20 million deaths; 38 million infected, mainly in Africa as of Dec. 2003

Source: Adapted from Meade and Earickson (2000).

cancers. Two main varieties of HIV have been identified, HIV1 and HIV2. The most virulent of the strains, HIV1, is found principally in southern Africa and is the virus of the world pandemic. HIV2, a West African variety, lower in virulence and transmissibility, may not manifest symptoms for 20 years. It is agreed that HIV is probably a simian immunodeficiency virus (SIV) that jumped species. HIV1 is probably associated with a chimpanzee simian immunodeficiency virus, while HIV2 seems to be closely related to the habitat of the sooty mangabey in West Africa (Weiss 2002). It was transmitted to other parts of Africa by migration, particularly from its rural home to urban areas, where it spread very rapidly.

There is some discussion about how SIV migrated to human beings from monkeys. The most common explanation is that it spread to people through a bite, but there are a variety of other ways such as through butchering monkeys and eating their cooked flesh, or even keeping them as pets. The "cut hunter" hypothesis, another variant, suggests that blood from a monkey killed in the hunt infects a person through a new or existing lesion.

Another possible explanation that has raised volumes of debate is that HIV came to the human population as a result of the oral polio vaccination (OPV) campaigns in central Africa in the late 1950s and early 1960s. Polio vaccines are cultured on monkey kidneys which, the theory goes, could have been contaminated by SIVs (Hooper 2000). Proponents of this theory argue that it is possible that HIV was transmitted to people via polio vaccines because, reasoning analogously, another monkey virus, SIV-40, was passed to humans that way. In addition, they hold that an isolated bite or contamination of a wound is unlikely to result in the widespread and rather sudden appearance of HIV/AIDS in central Africa—although the mass inoculation of young children and infants who have less developed immune systems might have this effect.

Weiss (2002) believes that the evidence does not indicate that vaccines were contaminated by SIV. Rather, HIV became epidemic at least partly as a result of the mass inoculations in the middle of the twentieth century and the widespread use of unsterile needles to inject antibiotics. Such a scenario is analogous to the epidemic of hepatitis C in Egypt which was spread by injection for schistosomiasis during the 1970s (Weiss 2002, Whyte 2000). HIV could also have spread through widespread practices which use unsterilized instruments such as circumcision, clitoridectomy, scarification, and tattooing.

Most authorities on HIV/AIDS believe that HIV jumped species about 70 years ago, assuming a

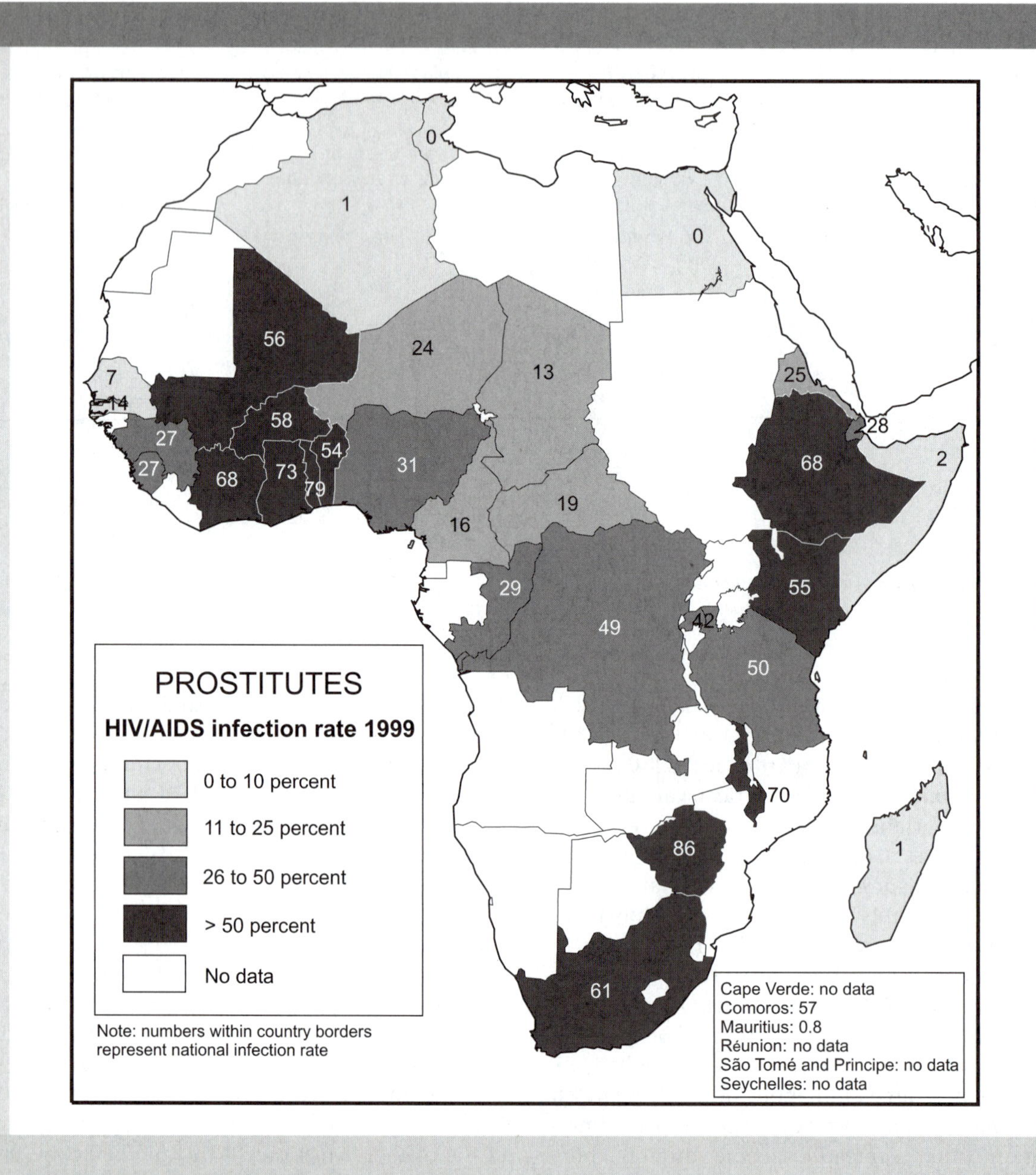

FIGURE 7-18 HIV/AIDS infection among prostitutes in Africa by country, 1999.

constant rate of genetic change in the virus. If this assumption is incorrect, the virus may be much older. In any case, cross-species transfer happens often with viruses. The virus that causes yellow fever came from Asian monkeys, and the malaria parasite *Plasmodium falciparum* is closely related to the parasite of chimpanzees (Weiss 2002).

The HIV/AIDS epidemic began in the late 1970s in Subsaharan Africa and in the late 1980s in North Africa. HIV1 was first identified in 1982 and HIV2 in 1985 (UNAIDS 2000a). Originally thought to have been brought to the city by rural migrants, HIV became established and has risen to incredibly high levels in high-risk populations such as prostitutes (Figure 7-18).

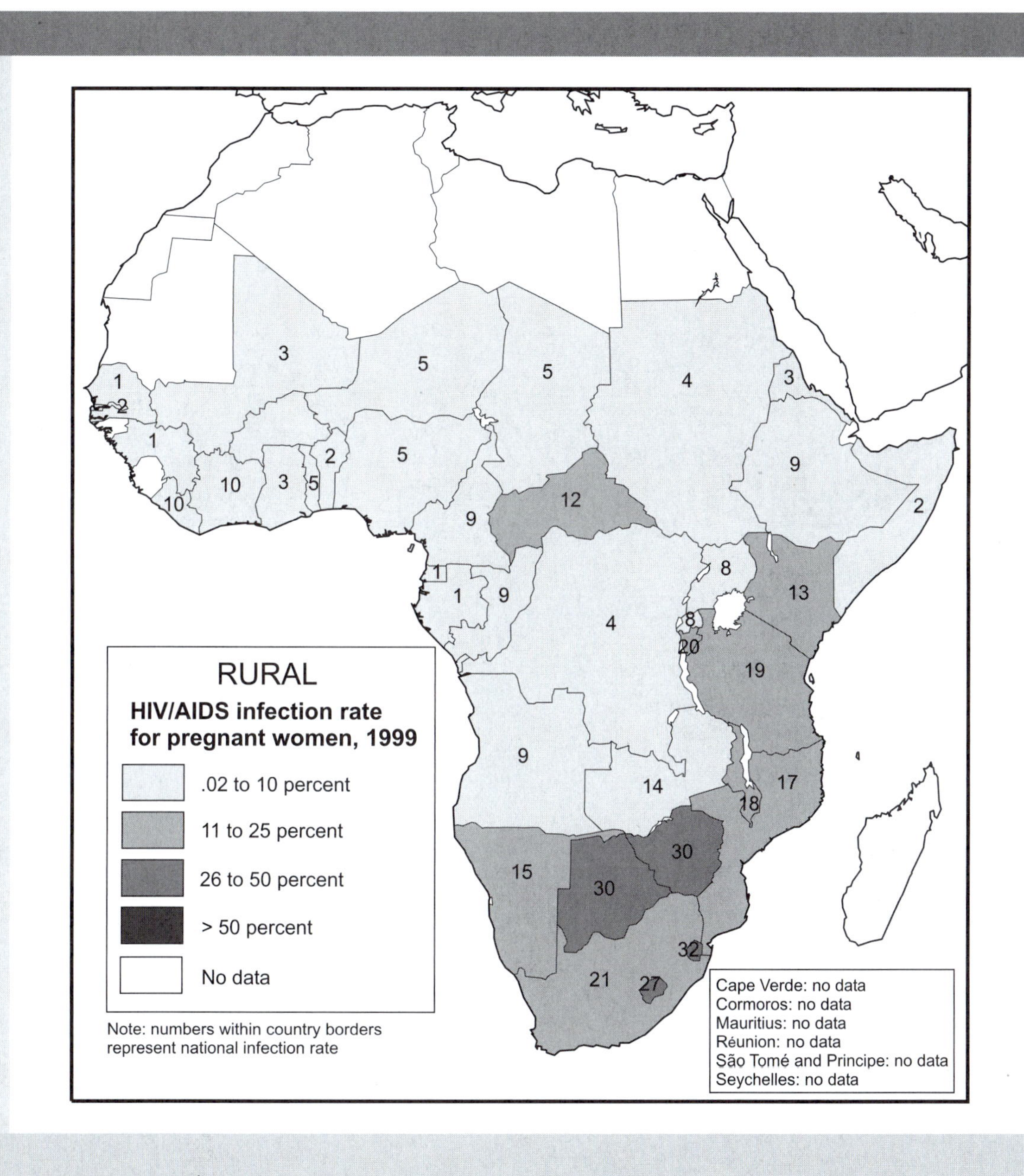

FIGURE 7-19 HIV/AIDS infection among pregnant women in rural Africa by country, 1999.

Presently, the pattern of transmission involves the migration of men from rural areas to seek employment in urban areas, in mines, and on distant plantations. Often the migration is seasonal, with the migrant returning home in the rainy season to cultivate; on the other hand, his migration may be relatively permanent (at least for certain age groups), with several visits to the home area throughout the year. Uninfected men get HIV from prostitutes and later take it back to their villages. In some areas today, the rate of infection among rural pregnant women is higher than the urban rate (Figures 7-19 and 7-20). Because of the late onset of

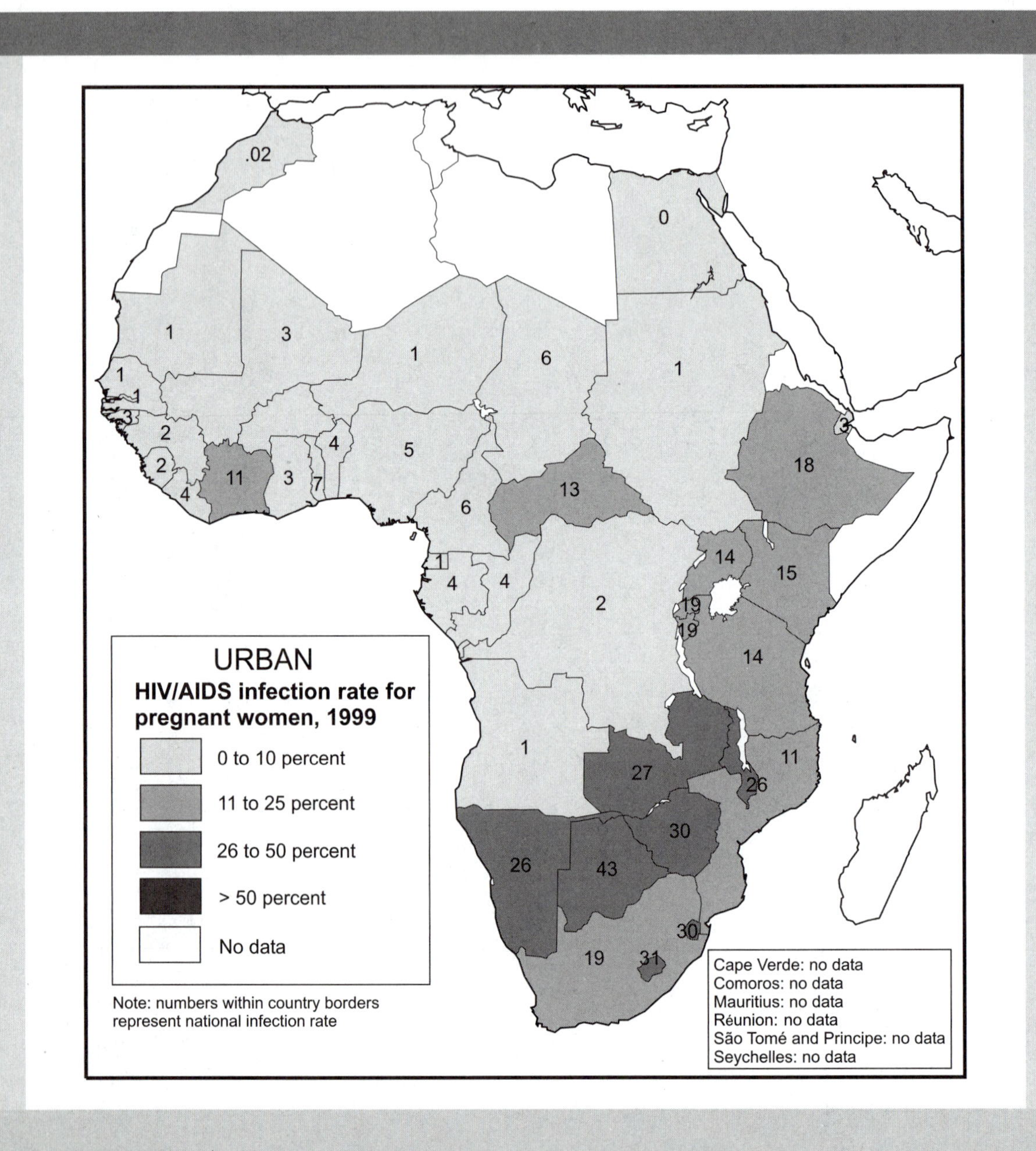

FIGURE 7-20 HIV/AIDS infection among pregnant women in urban Africa by country, 1999.

symptoms, infected people have no idea that they have caught a deadly disease until it is too late, and by that time the infection has spread even further.

At the present stage of the world HIV epidemic, the pattern of HIV infection in Africa appears to differ from other parts of the world in terms of who is most affected. Outside Africa, men are the main victims, either through homosexual sex or as intravenous drug users. In Africa, however, women and men are affected almost equally. Fifty-five percent of the HIV-positive cases in Subsaharan Africa are women compared with only 20% in North America (Table 7-4).

TABLE 7-4 REGIONAL HIV/AIDS STATISTICS BY SEX, DECEMBER, 2000.

Region	Epidemic Started	Number of HIV-Positive Persons	New Infections	Adult Prevalence (%)	Women HIV-Positive Adults Who Are Women (%)
Subsaharan Africa	Late 1970s-early 1980s	25.3 million	3.8 million	8.80	55
North Africa and Middle East	Late 1980s	400,000	80,000	0.20	40
South and Southeast Asia	Late 1980s	5.8 million	780,000	0.56	35
East Asia and Pacific	Late 1980s	640,000	130,000	0.07	13
Latin America	Late 1970s-early 1980s	1.4 million	150,000	0.50	25
Caribbean	Late 1970s-early 1980s	390,000	60,000	2.30	35
Eastern Europe and Central Asia	Early 1990s	700,000	250,000	0.35	25
Western Europe	Late 1970s-early 1980s	540,000	30,000	0.24	25
North America	Late 1970s-early 1980s	920,000	45,000	0.60	20
Australia and New Zealand	Late 1970s-early 1980s	15,000	500	0.13	10
Total		36.1 million	5.3 million		
Average				1.51%	28%

Source: UNAIDS (2001c).

Why would HIV infect women in Subsaharan Africa more than elsewhere? One reason is that females are more easily infected than males during vaginal intercourse with an infected partner. Women and men are infected about equally in Africa, but young women are infected far out of proportion to their numbers. This is due to sexual age-mixing between young women and older men (intergenerational sex), who are more experienced sexually, hence are much more likely than the young girls to be infected with HIV.

Figure 7-21 presents the relationship between the male rate of infection and the female rate for 15- to 24-year-olds. Only in Mauritius are the rates close. Over most of Africa, the male rate in 15- to 24-year-olds is less than half that of the female rate. The lowest rates for male 15- to 24-year-olds, about 30% of the female rate for the same populations, are mainly in West Africa.

The majority of factors influencing the spread of HIV in Africa are social. For example, the widespread practice of having multiple concurrent sexual partners and overlapping sexual relationships (as opposed to serial) causes the virus to spread throughout a sexual network very rapidly. Where there is labor migration, over all of Africa but particularly in southern Africa, there are large sexual networks, and the infection spreads from the distant workplace to the home area quickly. As already noted age mixing between older men and young women and girls, occurring because, among other reasons, older men control more resources, is responsible for the disproportionate infection rate in young women. Condoms, which are expensive and are negatively associated with prostitution, are little used. Poor education about sexuality and the risk of infection perhaps is the most fixable of the social influences, and governments are mounting education campaigns.

Of the biological factors involved in the spread of HIV in Africa, the high rates of sexually transmitted diseases, particularly those that cause genital lesions, increase the chances of transmission of HIV. In addition, the transmission of HIV from mother to child across the placenta is significant. Circumcision of the penis seems to confer on the circumcised a degree of protection from infection, probably having to do with the toughening of the skin of the penis.

There is also a difference between HIV viruses: HIV1, found throughout Africa but mainly in central and southern Africa, is highly virulent, while HIV2, found in West Africa, is much less virulent and harder

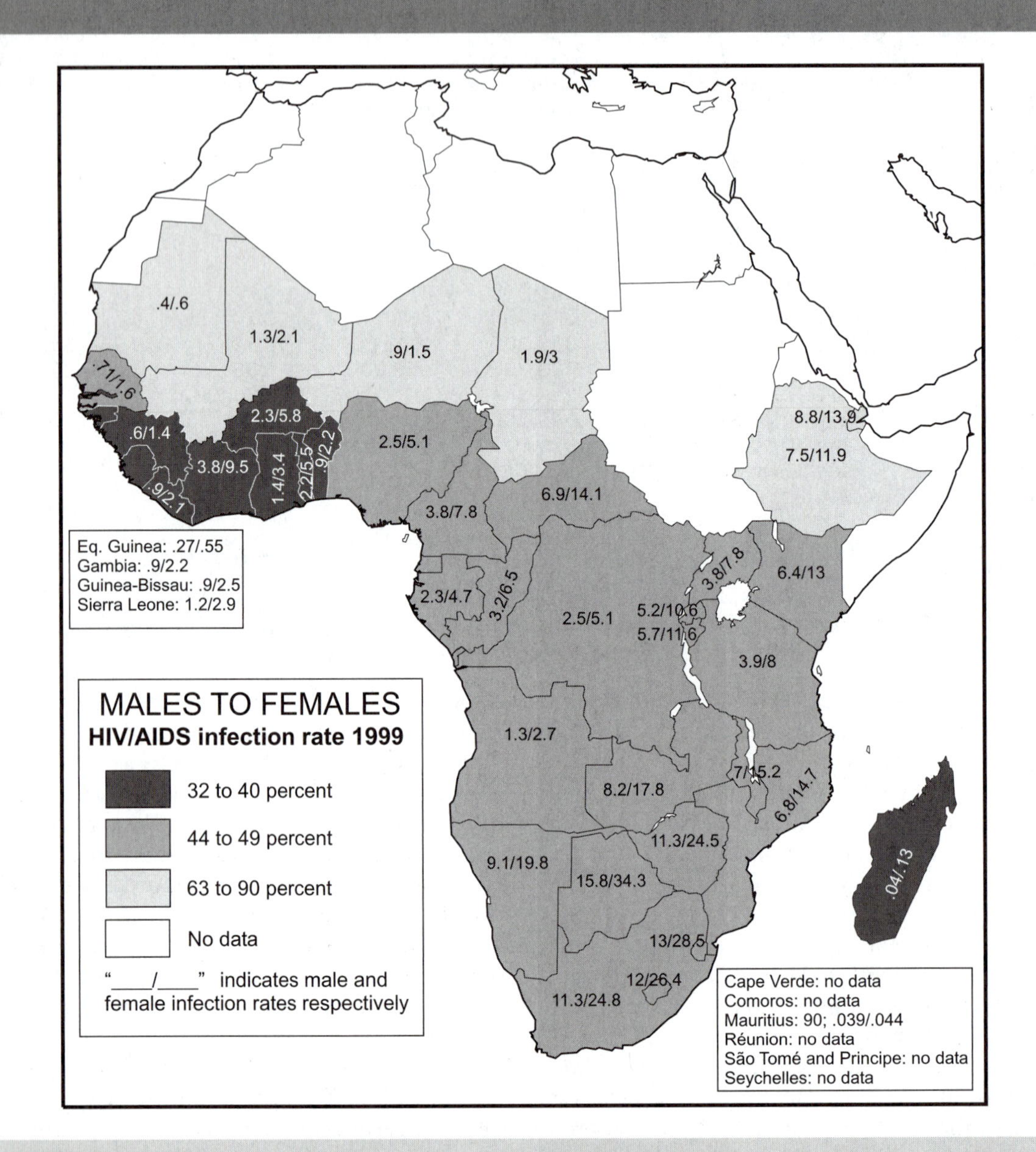

FIGURE 7-21 Ratio of male-to-female HIV infection rates for the 15- to 24-year-old population in Africa, by country.

to contract. An infected person can live up to 20 years with HIV2 while HIV1 kills within 10 years. Poverty or underdevelopment is a consequence of HIV infections as well as a cause. Figure 7-22 illustrates the recursive relationship between poverty and HIV infection. At the heart of the vicious circle is behavior.

Impacts of AIDS

Clearly, HIV/AIDS is a problem for Africa today, particularly Subsaharan Africa, in a way that it is not for most other regions of the world, at least not at this time (Figure 7-17); and profound demographic consequences

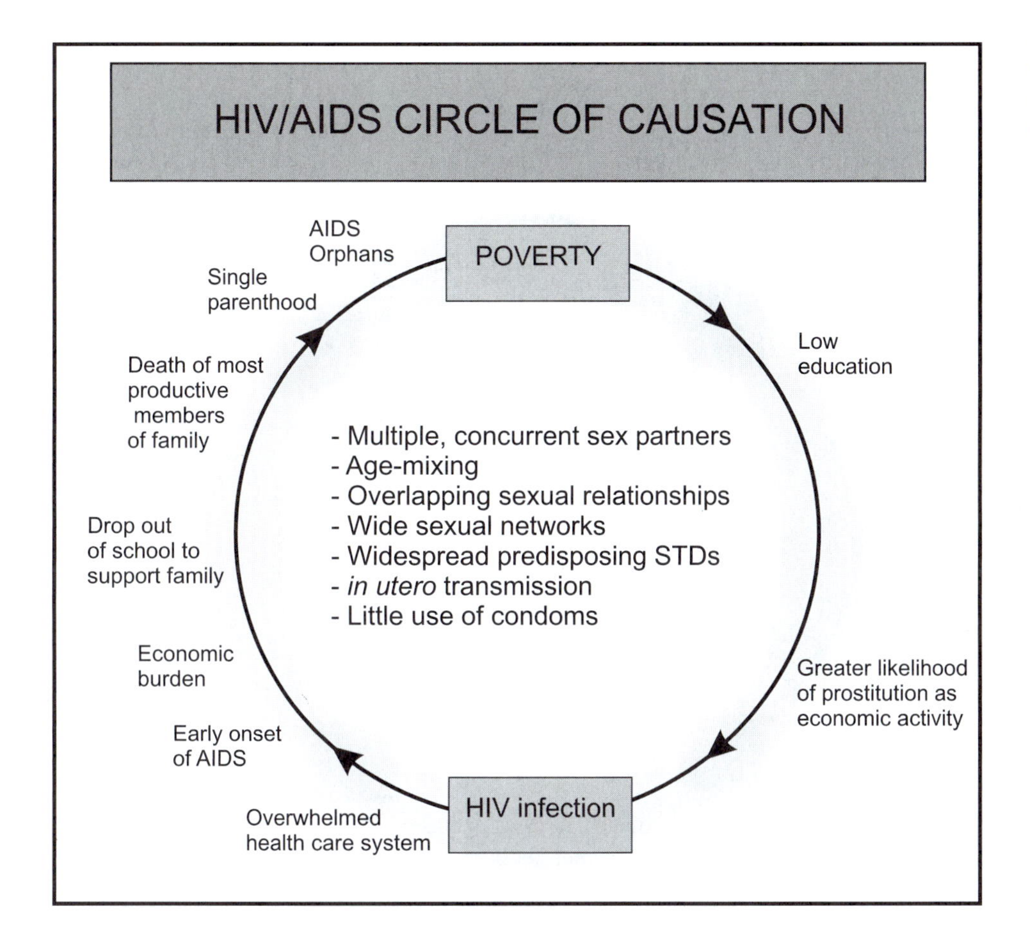

FIGURE 7-22 Vicious circle of poverty, behavior, and HIV infection.

are projected for African countries with high infection rates. Life expectancy, which in some countries was approaching the levels found in Europe and other more developed areas, has plummeted. Indeed, Botswana, South Africa, Zimbabwe, and Zambia will soon have negative population growth rates.

Figure 7-23 depicts such a negative growth rate for South Africa beginning in 2003 and continuing for 40 years, and Box 7-1 describes the impact on Botswana's population pyramid. Projections are linear and reflect the AIDS epidemic in the future based on today's data. It is noted for completeness that behavior may change, or a vaccine may be found, changing the future impact of AIDS on population growth. The prospects for a vaccine in the near future look dim, however; researchers say that a vaccine is not on the horizon and that the virus is developing resistance to the conventional treatments used in rich countries. In the United States, where a "cocktail" of three drugs to control AIDS is commonly prescribed, 78% of AIDS patients have a variety of the viruses resistant to one of the three drugs, and more than 50% have a variety resistant to more than one of the drugs. The symptoms of AIDS appear when the virus evolves to develop resistance to all three drugs.

The impacts of AIDS are not equally distributed throughout Africa. A glance at any of the figures already presented bears this out. The hardest hit countries are located in southern Africa. The impacts of AIDS are obvious on health and mortality (Box 7-1), but there are other impacts, such as increasing poverty, especially for children, as parents waste away and die. In urban Côte d'Ivoire, among households where a family member had AIDS, food consumption declined 41% per person, spending on school fell by half, and spending on health care more than quadrupled. Since the AIDS epidemic began, 13 million children under the age of 15 in Africa have lost their mother or both

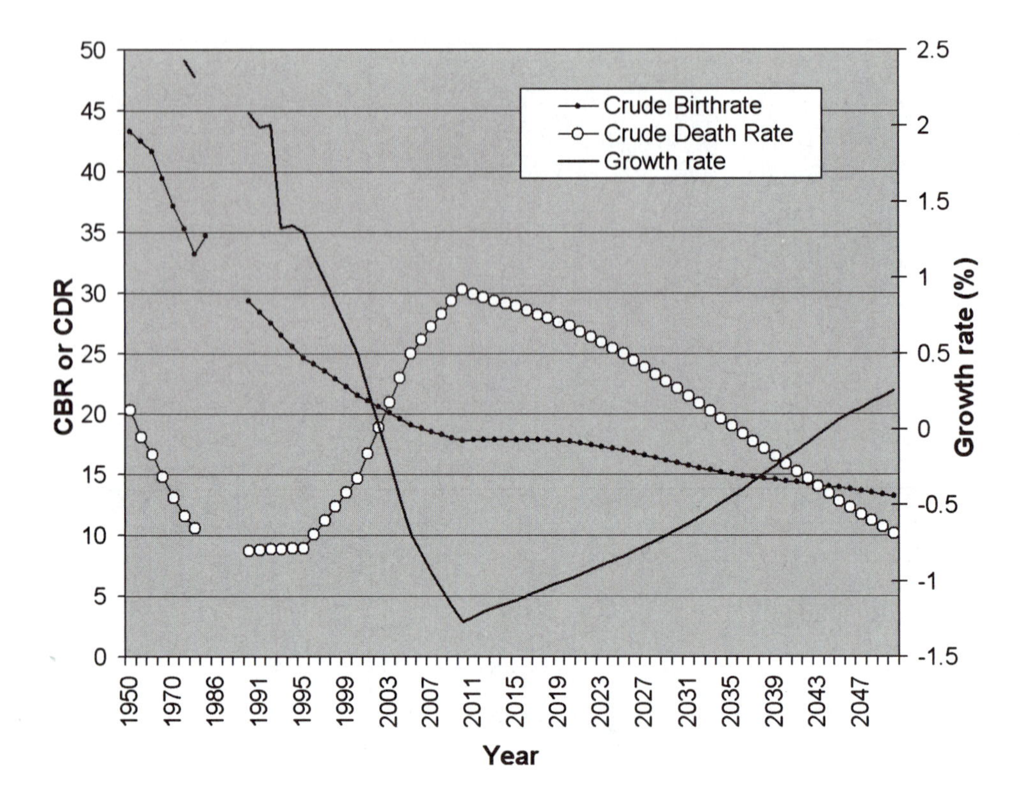

FIGURE 7-23 Population growth and crude birth and death rates, South Africa, 1970 to 2050.

parents to AIDS. The cost of AIDS eats into the family budget by reducing funds available for school fees and even by forcing children to drop out of school and join the workforce to make ends meet. In Zimbabwe, 48% of AIDS orphans of primary-school age dropped out of school, and none completed secondary school. Teachers have been affected by AIDS, resulting in increases in the size of classes already too large. Almost as many teachers died as retired from 1996 to 1998 in the Central African Republic, and of those who died, over 85% were infected with HIV. In 1998 Zambia lost 1,300 teachers to AIDS: the annual equivalent to two-thirds of new teachers in the country (UNAIDS 2000b).

Public health services in countries with high levels of AIDS cases have been simply overwhelmed by the enormity of the epidemic, which is expected to increase in the future. In 7 of 16 African countries where total health spending from public and private sources accounted for between 3 and 5% of GDP, spending on AIDS accounted for over 2% of GDP in 1997 (UNAIDS 2000b). The fight against AIDS is woefully underfunded—national budgets just cannot generate the kind of money that a more developed country like the United States can. In Subsaharan African countries such as Burundi and Kenya, HIV patients occupy half the big-city hospital beds. Furthermore, the presence of HIV-positive patients puts health care personnel at risk. The United Nations reported (2000b) that at one hospital in Zambia, AIDS-related deaths of hospital workers increased 13-fold from 1980 to 1990.

AIDS has had profound economic impacts on business. In addition to replacing retirement as the principal cause of employee attrition, UNAIDS (2000b) found enormous losses.

> Managers at one sugar estate [*in Kenya*] quantified the cost of HIV infection as follows: 8000 days of labour lost to illness between 1995–1997; 50% drop in processed sugar recovered from raw cane between 1993–1997; higher overtime costs; fivefold increase on funerals between 1989–1997; tenfold increase in health costs. The company estimates over three-quarters of all illness is related to HIV infection. (UNAIDS 2000b: 2)

There have been some successes in the fight against HIV/AIDS. Uganda has achieved a decline in HIV infection from 14% in 1991 to 8% in 1999, because of a

BOX 7-1

THE DEMOGRAPHIC IMPACT OF AIDS

LIFETIME RISK OF DYING OF AIDS

The HIV prevalence rate among 15- to 49-year-olds has now reached or exceeded 10% in 16 of the world's countries, all of them located in Subsaharan Africa. These rates greatly understate the demographic impact of AIDS. For example, in southern Africa, AIDS is set to wipe out half a century of development gains as measured by life expectancy at birth. From 44 years in the early 1950s, life expectancy rose to 59 in the early 1990s. Now, a child born between 2005 and 2010 can once again expect to die before age 45. According to conservative analyses of HIV/AIDS risk, in countries where 15% of adults are currently infected, around a third of today's 15-year-olds will die of AIDS. Where adult prevalence rates exceed 15%, the lifetime risk of dying of AIDS is much greater: in countries such as South Africa and Zimbabwe, where a fifth or a quarter of the adult population is infected, AIDS is will kill about half of all 15 year olds and in Botswana, where about one third of the adults are already HIV-infected, two-thirds of today's 15 year old boys will die prematurely of AIDS.

THE POPULATION CHIMNEY

In developing countries, the population structure is generally described as a pyramid. The youngest age groups—babies, children and healthy young people up to 19—form the broad base of the pyramid, which then tapers up gradually through the older age groups which have begun shrinking through illness and death. Now, AIDS has begun to introduce a completely new shape—the "population chimney"—for Botswana's projected population structure in 2020. Compared with the population structure that Botswana would have in the absence of an AIDS epidemic, the base is less broad. Many HIV-infected women die or become infertile long before the end of their childbearing years, so fewer babies are born. And up to a third of the infants born to HIV-positive women become infected themselves before or during birth or through breast milk, so fewer babies survive. The most dramatic change in the pyramid occurs at the ages when young adults infected early in their sex lives begin to die of AIDS. In Subsaharan Africa, young women are typically two or three times more likely to have become infected by age 24 than young men. Thus, starting with women in their mid-20s and men in their mid-30s, the adult population shrinks radically.

Only those adults who escape HIV infection can survive to middle and old age. In 2020, Botswana is expected to have more women in their 60s and 70s than women in their 40s and 50s.

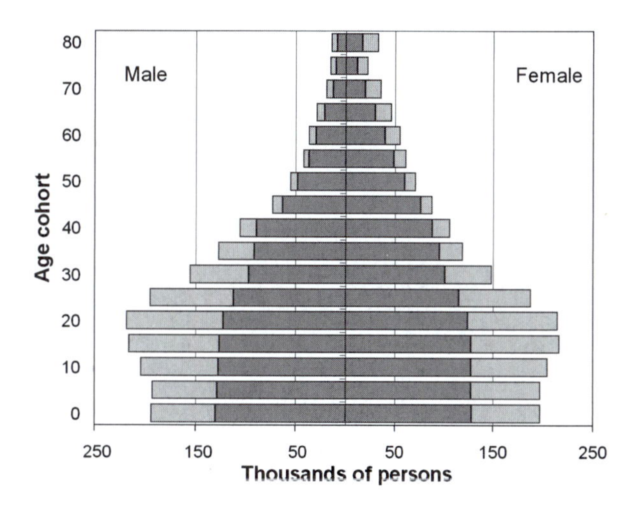

FIGURE 7-24 Projected population of Botswana in 2020 with and without AIDS.
Source: UNAIDS.

Source: UNAIDS (2000b).

rise in the age at first intercourse and increased use of condoms in cities (UNAIDS 2000a). In Zambia, both government agencies and nongovernmental organizations (NGOs) are using education to help people change behavior. The HIV prevalence rate among pregnant 15- to 19-year-olds has declined from about 27% to 15% from 1993 to 1998 (Zambia 1999).

Although it is difficult for scientists to predict the peak of the AIDS epidemic, the long-term prospects are frightening. The secretary general of the United Nations, Kofi Annan, announced that AIDS is Africa's greatest development challenge, and pointed out that AIDS has killed more people than all the wars in Africa. There has been some progress in some areas, but the data look pessimistic. AIDS, like most infectious diseases, is a development issue and Collins and Rau (2000: 60) "suggest rapid and equitable economic growth as an essential element in slowing AIDS." It is difficult not to be pessimistic about the tragic impact of HIV/AIDS in Africa. Although the UN secretary general has called for a Marshall Plan for AIDS, this is unlikely to materialize. Countries can reduce new infections through prevention programs that encourage abstinence, fidelity, and sex with condoms. Of these three, of critical importance is the use of condoms because there are many cultural barriers to changing sexual behavior.

A MEDICAL MARSHALL PLAN FOR AFRICA?

Most Africans rely on traditional medicines and therapies, but in most African countries modern biomedical services can be found provided by a variety of organizations: military, governmental, and nongovernmental organizations, as well as private voluntary organizations (PVO).

Geography, political instability, and poverty work against equality of access to national public health facilities. The vast majority of Africans live in villages far removed from sources of biomedical health care intervention. Added to the impacts of war and drought in some African countries is the sheer magnitude of debilitating diseases almost everywhere on the continent, with the result that the national public health services have been overwhelmed.

The World Health Organization (2001e) estimates that annual spending on health in Africa is less than $50 per capita and recommends more developmental assistance by international donors to help combat infectious disease, especially the two deadliest diseases in Africa: malaria and HIV/AIDS. Current spending for AIDS alone is inadequate: only $165 million spent in Africa in 1999 (excluding South Africa); and WHO believes that at least $2.5 billion should go for prevention alone. When the costs for the care of AIDS patients is added in, the amount is over $5 billion. For malaria, WHO estimates a need for at least $1 billion a year.

HIV/AIDS and malaria can be prevented or treated for less than $10 per person according to the WHO, and HIV infections can be reduced by 80% and malaria death rates halved if the efforts are made (2000e). Against these costs for malaria is a payoff of between $3 billion and $12 billion per year as a result of removing the economic burden of malaria to national GDPs. For HIV/AIDS, the economic burden is 1% of GDP in countries where 20% of the adult population is infected. Across Africa this amounts to some $4.3 billion of the 2000 GDP of $433 billion for countries with a 20% or more HIV infection rate: Botswana (36%), Lesotho (25%), Namibia (20%), South Africa (20%), Swaziland (24%), Zambia (20%), and Zimbabwe (25%).

WHO has had great success in eliminating river blindness in some parts of Africa, and smallpox has been eradicated across the world. The economic benefits of a Marshall Plan for HIV and malaria appear to be tangible. Furthermore, balanced against the costs and benefits of massive international assistance is the issue of national and global security. AIDS is a national security issue; rampant infection erodes society, hobbles agriculture and industry, undermines social, economic and political stability, and contribute to regional insecurity. Given the apparent formidable economic, social, political developmental, environmental, and logistic constraints to good health in Africa, there is reason not to expect a sudden near-term improvement in public health.

CHAPTER REVIEW QUESTIONS

1. African countries have made progress over the last few decades in combating some diseases. Discuss some African successes in the fight against infectious diseases.

2. What are some challenges to the health of Africans, and where is risk highest?
3. Why are Africans more likely to die of infectious diseases than Americans or Europeans?
4. Define "health" and "development." Describe the relationship between development and health, using examples from the text.
5. What explanations may account for the fact that the health-adjusted life expectancy (HALE) for North African countries, the Cape Verde Islands, and the island of Mauritius are so much higher than for the rest of Africa?
6. Africa has seen the resurgence of malaria and sleeping sickness, which were once thought to be under control. What is the outlook for the future for infectious diseases in Africa?
7. Why is it so difficult to change sexual behavior to prevent HIV infection in Africa? How is HIV transmitted in your country, and has behavior change been successful in halting HIV infections?
8. Describe the spread of HIV/AIDS in Africa. Why has it been so rapid? What solutions can you propose to cope with the disease?

GLOSSARY

Agent Organism that causes a disease. The agent may be called a germ, parasite, microbe, or pathogen. It could be a bacterium, protozoan, virus, or worm.

Disease vector Arthropod (insect, tick, or mite) that transfers an agent between hosts. The agent often goes through life-cycle changes while in the vector.

Epidemiologic transition The passage, which comes with socioeconomic development, from a state where most people die of communicable and parasitic diseases at a relatively young age to one in which death results from lifestyle diseases such as cancer and heart disease at older ages.

Host A living organism from which a disease agent obtains nutrition and in which it carries out all or part of its life cycle. An agent can live until maturity in a primary host. An agent that spends the larval stages of life in an intermediate host, must seek a primary host to reach maturity.

Incidence The number of new cases of a disease being reported for a population during a defined time period.

Morbidity The proportion of population in a geographic area with a specific disease.

Prevalence The number of people in a population sick with a disease at a particular time, regardless of when the outbreak began. Compare to **incidence.**

Zoonotic Describing diseases communicable to humans that originate in animal populations.

BIBLIOGRAPHY

Collins, J., and B. Rau. 2000. *AIDS in the Context of Development.* UNRISD Programme on Social Policy and Development, paper number 4, New York: United Nations Research Institute for Social Development.

Deshler, W. 1960. Livestock Trypanosomiasis and Human Settlement in Northeastern Uganda. *Geographical Review,* 50(4): 541–554.

Flores, R. 2001. Health and Nutrition: Emerging and Reemerging Issues is Developing Countries. *Focus,* 5(1): 3–4. Washington, D.C.: International Food Policy Research Institute.

Gallup, J. L., and J. D. Sachs. 1998. *The Economic Burden of Malaria.* Cambridge, Mass.: Harvard Center for International Development.

Hooper, H. 2000. *The River: A Journey Back to the Source of HIV and AIDS.* New York: Penguin Books.

Hughes, C., and J. M. Hunter. 1970. Disease and "Development" in Africa. *Social Science and Medicine,* 3(4): 443–493.

Hunter, J. M. 1966. River Blindness in Nangodi, Northern Ghana: A Hypothesis of Cyclical Advance and Retreat. *Geographical Review,* 56(3): 398–416.

Knight, C. G. 1971. The Ecology of African Sleeping Sickness. *Annals of the American Association of Geographers,* 61(1): 23–44.

Kusinitz, M. 1995. HIV Survivor Could Provide Clues to Assist in AIDS Battle. The Gazette, 24(16). January 9, 1995. Johns Hopkins University, Rutgers.

Livingstone, H. 1857. Missionary Travels and Researches in South Africa. London: Murray.

Martens, P., and L. Hall. 2000. Malaria on the Move: Human Population Movement and Malaria Transmission. *Journal of Emerging Infectious Diseases,* 6(2).

Martin, B. 1998. Political Refutation of a Scientific Theory: The case of polio vaccines and the origin of AIDS. *Health Care Analysis,* 6: 175–179.

Meade, M. S. and R. J. Earickson 2000, Medical Geography. New York: Guilford Press.

Oloo, A. J., J. M. Vulule, and D. K. Koech. 1996. Some Emerging Issues on the Malaria Problem in Kenya. *Journal of East African Medicine*, 73(1): 50–53.

Packard, R. M. 1986. Agricultural Development, Migrant Labor and the Resurgence of Malaria in Swaziland. *Social Science and Medicine*, 22(8): 861–867.

Population Reference Bureau. 2001. World Population Data Sheet. Washington D.C.: Populations Reference Bureau.

Prothero, R. M. 1965. *Migrants and Malaria in Africa*. Pittsburgh: University of Pittsburgh Press.

Rosenthal, P. 1998. Proteases of Malaria Parasites: New Targets for Chemotherapy. *Journal of Emerging Infectious Diseases*, 4(1): 49–57.

UNAIDS. 2000a. *AIDS Epidemic: Global Update*. New York: United Nations.

UNAIDS. 2000b. AIDS and Development. UNAIDS fact sheet. July 2000. New York: United Nations.

UNAIDS. 2004c. *Report on the Global HIV/AIDS Epidemic, July 2004*. New York: United Nations.

Weiss, R. A. 2002. Reflections on the Origin of Human Immunodeficiency Viruses. *AIDS and Hepatitis Digest*, January 2002.

WHO. 1996. Schistosomiasis. Fact Sheet 115, Geneva, Switzerland: World Health Organization.

WHO. 1997. *OCP Facts and History*. Geneva, Switzerland: World Health Organization.

WHO. 2000a. Onchocerciasis. Fact Sheet 95. Geneva, Switzerland: World Health Organization.

WHO. 2001a. *The World Health Report*. Geneva, Switzerland: World Health Organization.

WHO. 2001b. Pregnant women are one of the Groups Most at Risk from Malaria. *Action Against Infection*, 2(4): 3. Geneva, Switzerland: World Health Organization.

WHO. 2001c. Sleeping Sickness (African Trypanosomiasis). *Action Against Infection*, 2(5): 1. Geneva, Switzerland: World Health Organization.

WHO. 2001d. Onchocerciasis. *Trends in Parasitology*. Geneva, Switzerland: World Health Organization.

WHO. 2001e. *HIV, TB and Malaria—Three Major Infectious Disease Threats*. Backgrounder No. 1 for the G8 discussions. New York: United Nations.

Whyte, B. 2000. Spread of Hepatitis C Linked to Unsafe Injection Practices in Egypt. *Bulletin of the World Health Organization*. Geneva, Switzerland: World Health Organization.

Zambia. 1999. Government Report: *HIV Prevalence Rate Among Pregnant 15–19 Year Olds, Lusaka, Zambia, 1993–1998*. In UNAIDS 2000a.

CHAPTER 8

Land Use in Africa

The purpose of this chapter is to investigate the geography of land use in Africa, focusing first on agriculture and then on industry. Land use in Africa is complex and is related to the economic development, national policy, and local customs of a country or a region. Level of development accounts much for the geographic variation that one observes from region to region in Africa, and the way in which land is used has changed over time as well as over space. Thus one thing is constant and certain: land use is best understood as a historically and geographically dynamic phenomenon.

The way that Africans use the land is similar and at the same time different from land use in other parts of the world. One big difference is that far more people are involved directly in food production in Africa than in more developed parts of the world—Europe, Japan, and North America, for example. Less than 2% of Americans but 54% of all Africans—and 62% of Subsaharan Africans—were involved in agriculture in 2003 (FAO 2005). Moreover, there is variation between countries in their involvement with agriculture; for example, from 3% of the population in Réunion, and 6% in Libya to 90% in Burkina Faso, Burundi, and Rwanda. There is also far less industrial and commercial use of the land in Africa than in more developed parts of the world, although this, like agriculture, varies from country to country and from region to region within a country.

On a subsistence basis, Africans may obtain the nutrients necessary for life in a variety of ways. Three ideal types are hunting and gathering, pastoralism (livestock), and farming. In actual fact Africans rarely rely on a single strategy to produce food but instead diversify their activities in order to minimize the risk of loss in any one activity (Figure 8-1). When developing their risk-minimizing "portfolios," Africans adopt diversification to help assure an adequate and predictable food stream to their families throughout the year. That living close to the land is a risky business is evident in a figure reported by the Food and Agriculture Organization of the United Nations (FAO 2001), namely, that 24% of all people in Subsaharan Africa (194 million) are undernourished; most of these are children. The reason for the undernourishment of almost a quarter of Africans south of the Sahara is a complex one rooted in economic, political, social, technological, and environmental dimensions.

Although the environment alone cannot explain the challenge of obtaining sustenance from the land in Africa, it will be the point of departure in this chapter. The environment provides opportunities, constraints, and even challenges to people and their activities. People are inventive and have devised a variety of methods to gain a living, some of which involve living close to the environment. The reader of this book probably lives in an artificial, urban environment, not winning his or her daily sustenance from the morning tide or the annual harvest and living a life very different from that of most Africans.

Africa is a relatively dry region in comparison to other areas of the world, and this is exemplified in how people use the land. For the continent of Africa as a whole, 37% of the land was classified as "agricultural" by the FAO in 2001. The reader should not assume that all of this land is suitable for farming, since "agricultural land" also includes land used for pasture—the raising of livestock is part of agriculture as is the cultivation of the soil. Of the 37% of Africa that is agricultural (4,251,626 mi^2, 11,012,100 km^2), 7% (791,887 mi^2, 2,050,978 km^2) was used for arable and permanent crops and 30% (3,459,738 mi^2, 8,960,680 km^2) represented permanent pasture (Figure 8-2). The 7% of the African landmass that is used for arable and permanent crops is roughly equal to the combined area of Libya and Tunisia in North Africa or

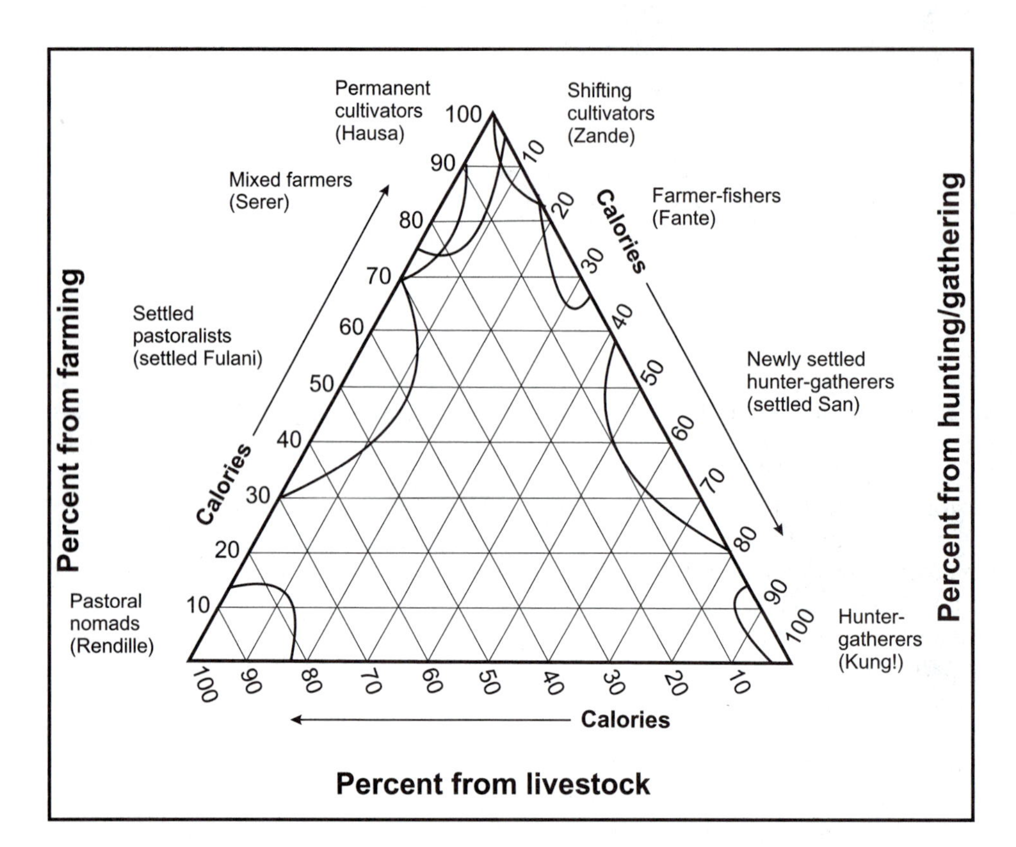

FIGURE 8-1 Percentage of calories obtained from variations on three African subsistence strategies.
Adapted from Stock (1995).

the combined area of Arizona, California, Colorado, Idaho, New Mexico, Oregon, and Utah.

Of the land that is currently agricultural, over 80% is used as permanent pasture (Table 8-1), and the rest is dryland (rain-fed) farming. Not surprisingly, only a small fraction of Africa is irrigated; opportunities for irrigation are rather limited in the dry African climates. Where temperate (mesothermal, or C) climates prevail, a much greater proportion of the land is under cultivation. This is principally in the moist midlatitudes of the earth, where evapotranspiration is relatively low. Europe, with 63% of its agricultural area cultivated, is an excellent example of the midlatitude agricultural advantage. Although the cultivated (arable and permanently cropped) area of Africa accounts for only a small area overall, over the last 40 years, it has expanded by 29%, mainly into forest and woodland.

This chapter will discuss major land uses, beginning with pastoralism and the raising of livestock and ending with farming. Our focus will be on the distribution of activities across geographic space and how the activities have changed over time. Human activity in Africa is constantly changing, and those wishing to understand African affairs must be ready to adopt an analytic rather than a simply ideographic perspective. The bulk of the presentation is on farming, the activity that occupies most Africans.

LIVESTOCK RAISING AND PASTORALISM

Pastoralism is a traditional way of life in which the main source of nutrition comes from animal products. As a subsistence strategy, pastoralism is among the oldest. People in most African cultures practice some form of animal husbandry, and many peoples described as strictly herders actually may also fish, hunt, gather wild foods, or even grow a few crops. Sources of calories may vary throughout the year. Nevertheless, above all else, pastoral peoples' lives revolve around their animals. Pastoralists consume milk and meat principally, but some East African pastoralists also periodically

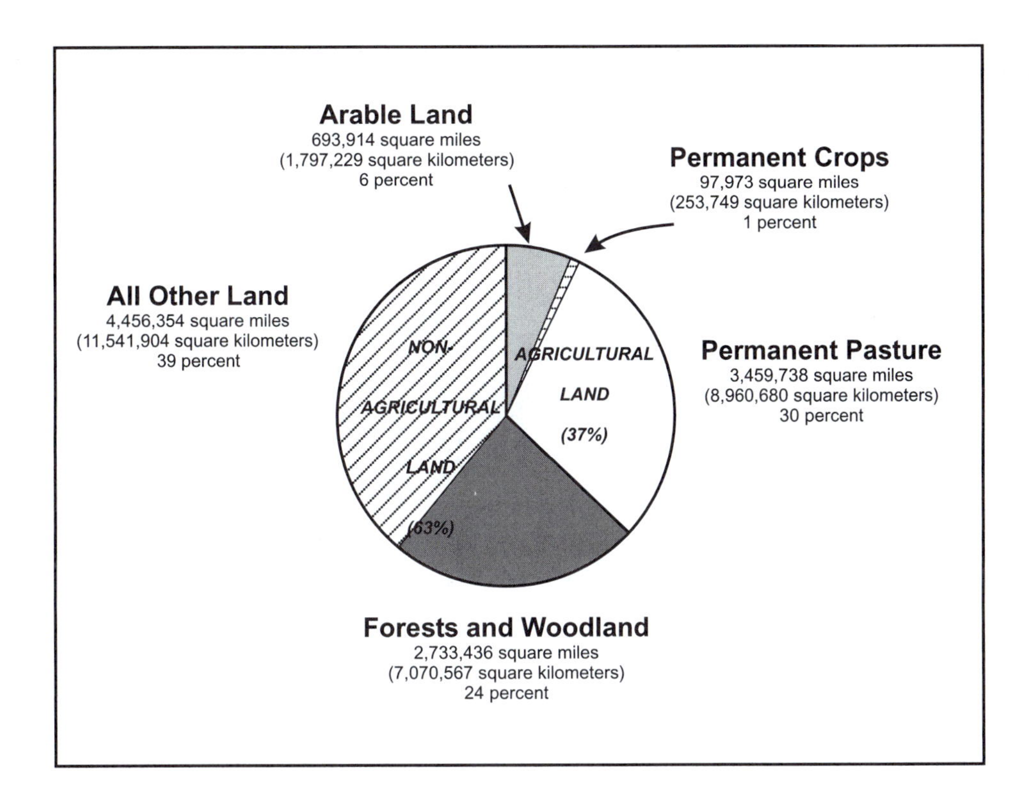

FIGURE 8-2 Land use in Africa: 2001.

bleed their animals and make a tasty, frothy mix of milk and blood.

In an economic sense, nutrition based on animal products is more expensive than that based on plants. As a rule of thumb, it take 10 units of grain or vegetable energy to produce one unit of meat in livestock, and where conditions are more humid, farming will outcompete pastoralism as a way of life. As a consequence, in the wetter areas of Africa people produce grain and/or starchy crops as a cheap source of calories and some protein, which they supplement with an edible fat, generally obtained from plants. In terms of the geographic distribution of animal raising and farming, farming has a competitive advantage in

TABLE 8-1 AGRICULTURAL STATISTICS FOR SELECTED REGIONS OF THE WORLD.

	Land Area (mi^2)			Use of Agricultural Land	
Region or Country	Total	Agricultural	Amount Agricultural (%)	Cultivated (%)	Permanent Pasture (%)
Africa	11,441,415	4,251,626	37	19	81
Africa south of Sahara	8,754,365	3,487,584	40	18	82
Asia	11,962,271	6,393,311	53	33	67
Europe	8,729,708	1,886,252	22	63	37
South America	6,768,162	2,390,436	35	19	81
United States	3,536,294	1,614,872	46	43	57
World	50,431,201	19,205,690	38	30	70

Source: FAO (2002).

TABLE 8-2 CHARACTERISTICS OF THE MAJOR TYPES OF LARGE LIVESTOCK IN AFRICA.

Animal	Food	Water	Reproduction	Resources	Value
Camel	Graze and browse	Best adapted to arid zones Able to go for a week without food or water Able to travel relatively long distances in a day	Once a year	Means of transport, milk, meat, hides; in social exchange; as an investment	High
Goat	Omnivorous herbivore	In the cool season can do without water for up to 2 weeks	Once or twice a year	Milk, meat, hides, investment	Low
Sheep	Grass	Can go for only 2 days without food	Once or twice a year	Wool, meat; as an investment	Low to medium
Cow	Grass	Can go for only 2 days without food	Once a year	Milk, meat, hides; in social exchange; as an investment	High

Source: Adapted from Franke and Chasin (1980).

more well-watered areas and livestock raising has a competitive advantage in the drier lands. Livestock are found in the wetter areas, however; in fact several other constraints limit the distribution of livestock in the wetter areas (trypanosomiasis, for example).

Farmers supplement their diets with a variety of edible fats that take the place of animal fat: four of the most well known sources are peanut (groundnut), sesame, palm, and shea nut oils—only the latter two are indigenous to Africa. In the past, raising livestock has been most competitive on the edges of the sown in the drylands, where crops cannot be grown reliably. Like domesticated plant species and varieties, domesticated livestock have specific breeds for specific biomes. Table 8-2 illustrates the broadest distinctions between livestock species but not breeds within species. Pastoralism is a type of land use that predates cultivation and makes extensive and seasonal use of drylands, places in which settled human life would be otherwise impossible.

Livestock Populations

The geographic distribution of livestock raising and pastoralism as a way of life almost defies cartography in the sense that there are no precise boundaries at which pastoralism ends and farming begins. The limitations imposed by the presence of the tsetse fly upon cattle raising and the pathology of the camel are well known, but the "natural" boundary between "nomadic" and "sedentary" groups is apparent rather than real. Even the range of the tsetse fly is being reduced in some areas by clearing wooded areas. Although maps with neatly bounded regions are produced and included in many textbooks, the boundaries are flexible:

> [Boundaries are] being crossed to varying extents alternately by one or the other according to local conditions that are socio-political rather than physical: Tuareg tents may be seen pitched at the edge of floating *Echinochloa* meadows of Lake Debo at latitude 15° north, where with 400 mm of rain, spiked millet will ripen under rainfall cultivation; and on the other hand, Fulani and Soninke peasants of Mauritania, moving northward, today push the limits of their rainfall farming as far as latitude 17° north. . . . (Monod 1975: 104–105)

The geography of livestock population is complex, and representing the changing underlying reality is a more difficult task than it might at first appear. We know that like the human population, livestock numbers are increasing (Figures 8-3, 8-4, and 8-5) but not everywhere at the same rate. We can understand the geographic distribution of livestock in Africa in a number of ways. The method that is most obvious, and common, but perhaps not the best—namely, to

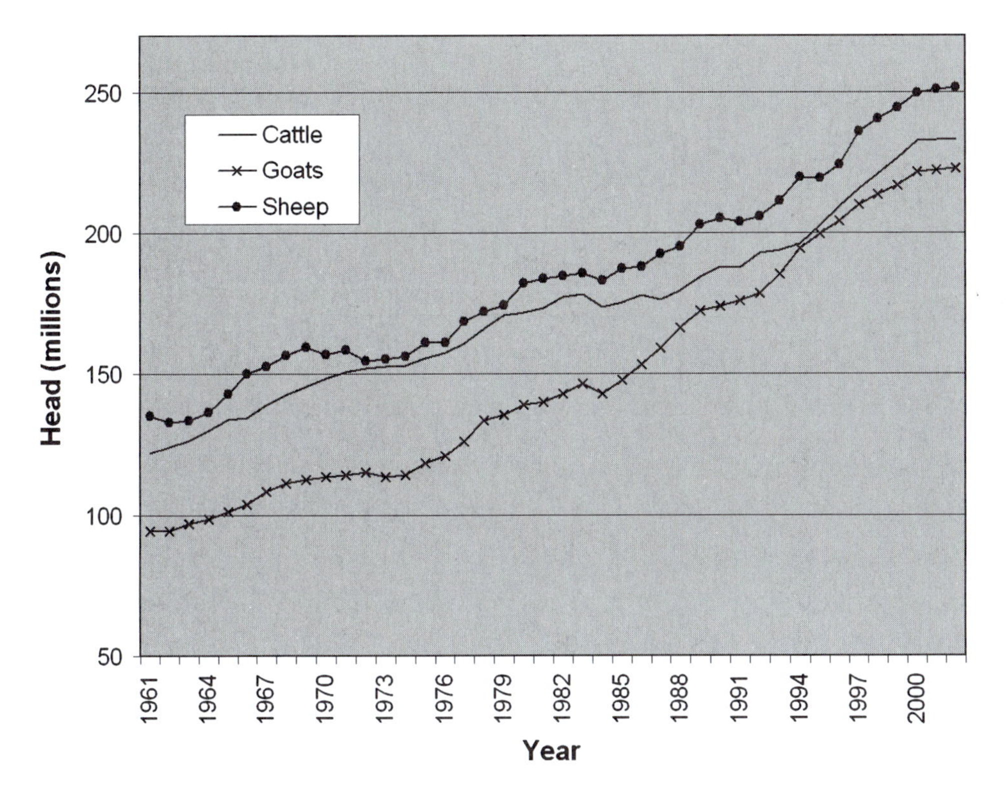

FIGURE 8-3 Cattle, goat, and sheep populations for Africa, 1961 to 2002.

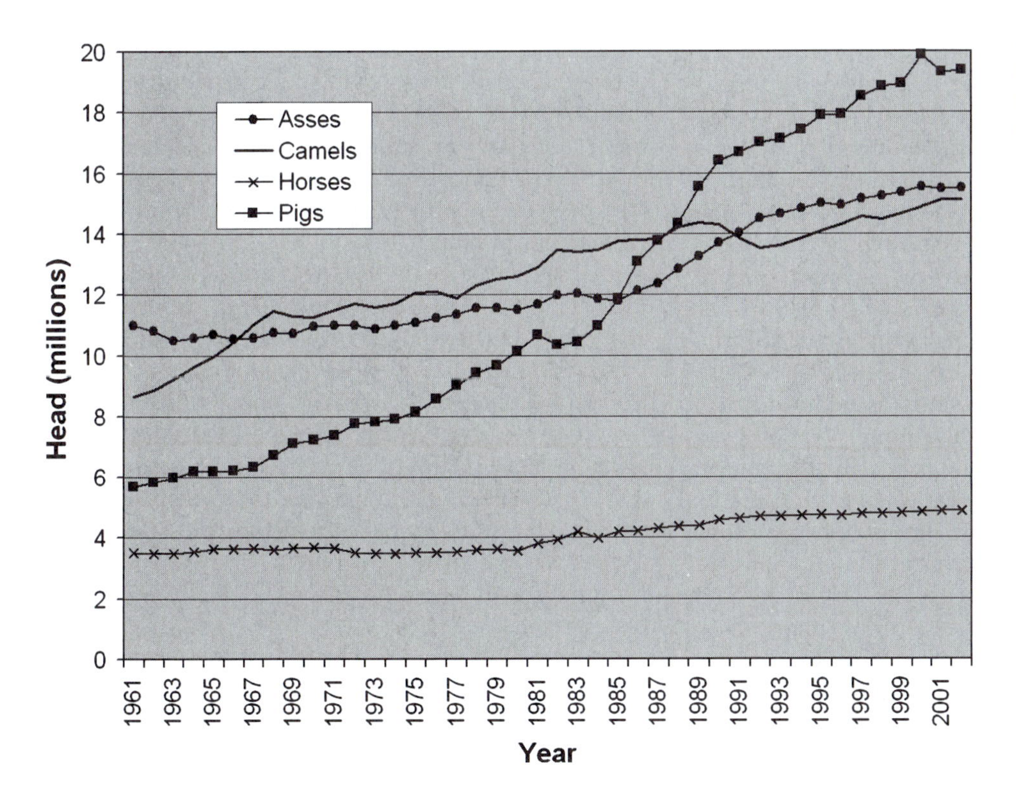

FIGURE 8-4 Ass, camel, horse, and pig populations for Africa, 1961 to 2002.

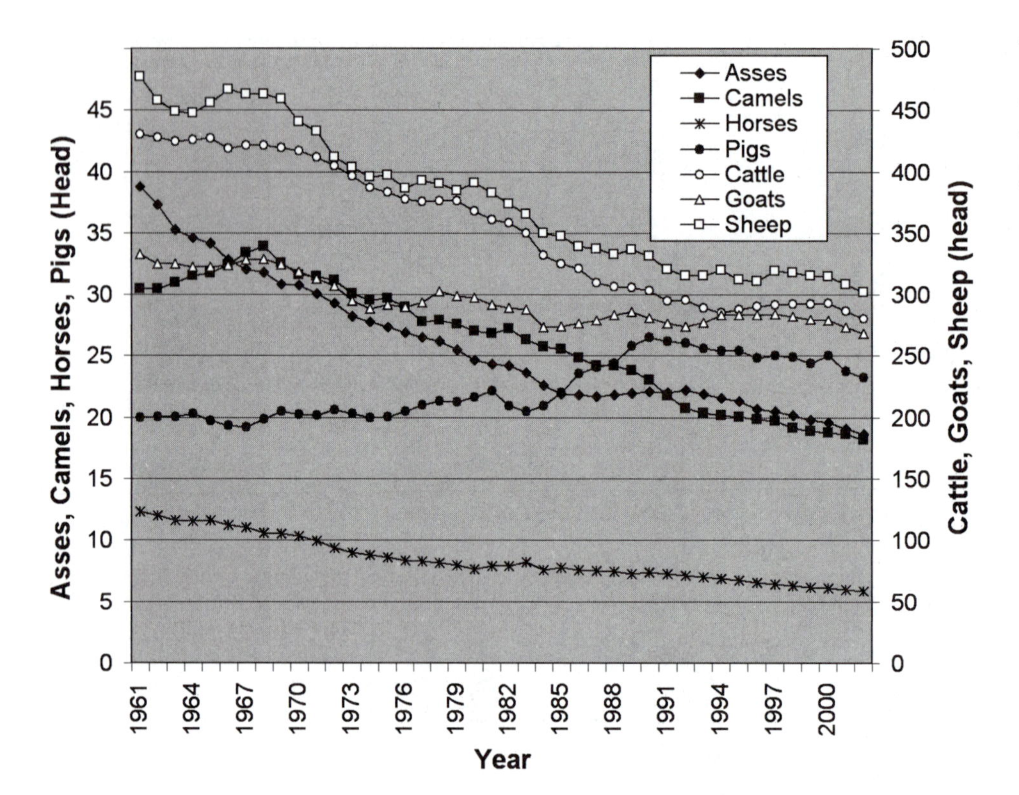

FIGURE 8-5 Ass, camel, cattle, goat, horse, pig, and sheep populations per thousand population for Africa, 1961 to 2002.

enumerate the numbers of livestock for each of the 54 African countries—may be misleading because it does not take into account the relative size of particular countries. For example, larger countries could have higher numbers of livestock simply because there is more territory. To account for the size differential, the density of livestock populations is often calculated per square mile or kilometer for a country. Although on the face of it this measure of density seems to be an improvement over merely reporting numbers per country, this sort of "standardization" also is misleading because it treats the space within a country as homogeneous, and the resulting maps give a false impression of the actual density within a country. A glance at a map of the climates of Africa shows that a variety of environments are found within each country. For example, countries that straddle the Sahara will have both humid and arid regions. To understand the density of livestock, data must be localized so that the resulting maps show where the density is highest, lowest, and in between.

The Food and Agricultural Organization of the United Nations calculates the number of livestock per thousand population in a country. This is a standardization that enables useful comparisons to be made between countries. But what if a country that is highly urbanized is to be compared with one that is deeply rural? One may also standardize livestock populations by the agricultural area of a country. The agricultural area of a country includes arable land (land that is farmed) and pasture.

A useful way to answer the question of where the livestock are in Africa is to look at small units of land and calculate the numbers of livestock for those specific areas. The data for such calculations come largely from aerial surveys of livestock populations and on-the-ground censuses. One measure of the total livestock presence in an area is the tropical livestock unit, or TLU. The TLU is a composite measure of the most populous livestock in Africa: camels, cattle, sheep, and goats. It is calculated by multiplying camels by 1, medium-sized stock like cattle by 0.7, and small stock like sheep and goats by 0.10.

Cattle

Cattle may have been domesticated in Africa; although the evidence is scanty and even questionable,

FIGURE 8-6 Zebu cattle (*Bos indicus*). *Gritty.org.*

it has been suggested that "the North African subspecies of wild cattle or aurochs *Bos primigenius* may have undergone an indigenous African domestication around 10,000 years ago" (Hanotte et al. 2002: 336). It is likely that cattle pastoralism was well-established in North Africa and the Sahara 6,000 years ago: *Bos primigenius* mixed with *Bos taurus* was domesticated in Southwest Asia and introduced through Egypt. Pastoralists migrated south with the drying of the Sahara until they reached the tsetse fly zone. The Ndama breed of cattle, found only in West Africa in the tsetse fly zone, is resistant to trypanosomiasis, sleeping sickness.

Over the ensuing millennia, the diffusion of cattle increased and another breed of cattle, *Bos indicus* (Figure 8-6), or Zebu cattle, was introduced—probably first to the Horn of Africa. Later these animals are believed to have come to East Africa through the Indian Ocean trade on ships and, much later, during the Arab expansion. *Bos indicus* diffused across the Sahel Corridor into West Africa (Figure 8-7), where they constitute the Fulani breed of cattle. Cattle pastoralism was probably established in East Africa over 4,000 years ago and slowly spread south into the Tanzania region. During the first millennium of the common era, cattle (and sheep) diffused to the southern half of Africa with the spread of Bantu-speaking agriculturalists (Hanotte et al. 2002). The colonial episode ushered in more livestock breeding to improve milk and meat production of African animals.

There are today more than 150 breeds of cattle in Africa derived from *Bos taurus*, *Bos indicus*, and mixes (Hanotte et al. 2000). There were about 230 million head of cattle in Africa in 2001 according to the FAO. Cattle populations by country are presented by number in Table 8-3 and by density in Figure 8-8. As would be expected, cattle are most numerous in the larger, drier countries, where raising cattle is a productive use of land that might otherwise stand idle. Forty-three percent of all African cattle were found in East Africa in 2001, with 73% of these in Sudan and Ethiopia alone (38 million and 35 million, respectively).

Sheep and Goats

The most populous stock in Africa is sheep and goats, known collectively as "shoats." There are no known progenitors of sheep (*Ovis aries*) or goat (*Capra hircus*) in Africa; both species were domesticated in western Asia and brought to Africa, where they flourished.

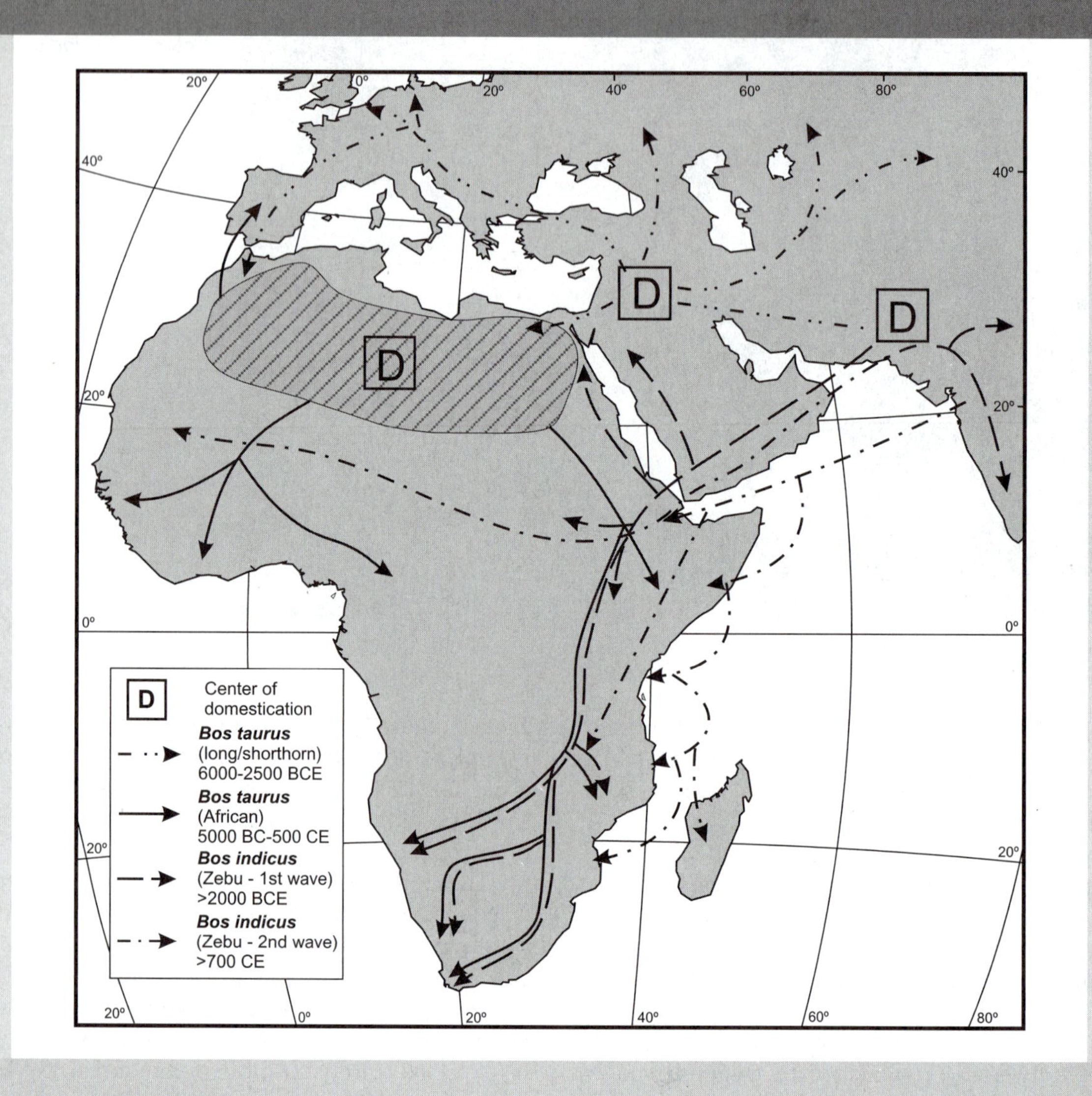

FIGURE 8-7 Centers for the domestication of cattle and diffusion routes.

The earliest evidence for sheep and goats in Africa is in northern Libya, where remains have been dated to 7,000 years ago. Clutton-Brock (1993: 69) writes that "well-defined breeds of domestic sheep were common in ancient Egypt from 3000 BC [and that] during the Early Dynastic period (3100–2613 BC), screw-horned sheep with lop-tails were developed, and fat-tailed sheep began to appear during the Middle Kingdom (1991–1633 BC)." Three-thousand-year-old evidence of domestic sheep/goats and cattle has been found near Lake Turkana. The raising of these animals slowly spread south, probably with the migration of Bantu-speaking farmers. It is thought that sheep reached the western Cape region of South Africa about 2,000 years ago and the eastern Cape about 1,700 years ago.

There are many breeds of African sheep and goats today (Figure 8-9). African goats may be divided into four distinctive groups: dwarfs of the equatorial forest, savanna and Sahelian goats, Nubian goats, and Atlas

TABLE 8-3 AFRICAN LIVESTOCK POPULATION BY COUNTRY IN 2002.

	Population (thousands)						
Country	**Cattle**	**Sheep**	**Goats**	**Pigs**	**Camels**	**Horses**	**Asses**
Africa	230,047	250,147	218,625	18,467	15,124	4,879	15,374
Algeria	1,700	19,300	3,500	6	240	48	180
Angola	4,042	350	2,150	800	0	1	5
Benin	1,500	645	1,183	470	0	6	1
Botswana	2,400	370	2,250	7	0	33	330
Burkina Faso	4,798	6,782	8,647	622	15	26	501
Burundi	315	230	600	70	0	0	0
Cameroon	5,900	3,800	4,400	1,350	0	17	38
Cape Verde	22	8	110	200	0	0	14
Central African Republic	3,100	220	2,600	680	0	0	0
Chad	5,900	2,400	5,250	22	725	205	360
Comoros	52	21	172	0	0	0	5
Congo Republic of the	90	96	280	46	0	0	0
Côte d'Ivoire	1,409	1,451	1,134	336	0	0	0
Djibouti	269	465	513	0	70	0	9
Democratic Republic of the Congo	793	911	4,067	1,000	0	0	0
Egypt	3,636	4,545	3,527	30	120	46	3,050
Equatorial Guinea	5	38	9	6	0	0	0
Eritrea	2,200	1,570	1,700	0	75	0	0
Ethiopia	34,500	22,500	17,000	25	1,070	2,750	5,200
Gabon	36	198	91	213	0	0	0
Gambia	365	106	145	14	0	17	35
Ghana	1,302	2,743	3,077	324	0	3	15
Guinea	2,679	892	1,012	98	0	3	2
Guinea-Bissau	515	285	325	350	0	2	5
Kenya	12,500	6,500	9,000	315	830	2	0
Lesotho	510	730	570	60	0	100	154
Liberia	36	210	220	130	0	0	0
Libya	220	5,100	1,950	0	72	46	30
Madagascar	10,300	790	1,350	850	0	0	0
Malawi	750	110	1,450	250	0	0	2
Mali	6,819	6,400	9,900	66	467	165	680
Mauritania	1,500	7,600	5,100	0	1,230	20	158
Mauritius	28	7	95	21	0	0	0
Morocco	2,663	17,300	5,200	8	36	155	985
Mozambique	1,320	125	392	180	0	0	23
Namibia	2,100	2,200	1,700	18	0	63	69
Niger	2,260	4,500	6,900	39	415	105	580
Nigeria	19,830	20,500	24,300	4,855	18	204	1,000
Réunion	28	260	700	77	0	0	0
Rwanda	800	2	38	180	0	0	0
São Tomé and Principe	4	3	5	2	0	0	0
Senegal	3,227	4,818	3,995	280	4	492	410
Seychelles	2	0	5	18	0	0	0
Sierra Leone	420	365	200	52	0	0	0
Somalia	5,200	13,200	12,500	4	6,200	1	20
South Africa	13,740	28,800	6,550	1,540	0	258	210
Sudan	38,325	47,000	39,000	0	3,200	26	750
Swaziland	615	32	445	34	0	1	15
Tanzania	14,400	4,250	10,000	355	0	0	180
Togo	277	1,000	1,425	289	0	2	3
Tunisia	795	6,600	1,450	6	231	56	230
Uganda	5,900	1,100	6,200	1,550	0	0	18
Zambia	2,400	150	1,270	340	0	0	2
Zimbabwe	5,550	535	2,800	278	0	26	107

Source: FAO (2003).

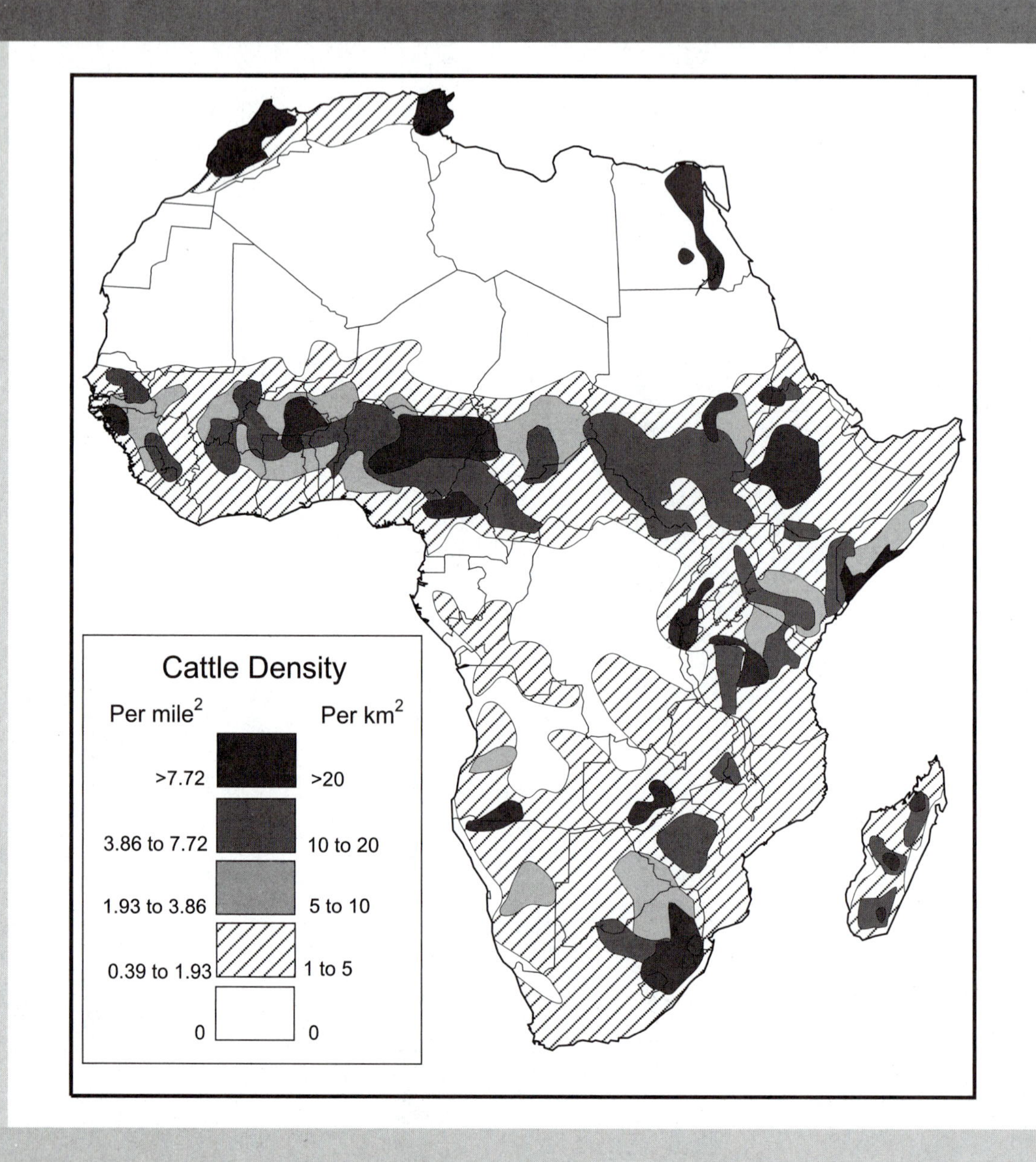

FIGURE 8-8 Cattle density, Africa.

Mountain breeds (Epstein 1971). Sheep can be classified in a number of ways; for example according to the kind and quality of wool they produce (fine, medium, long, and carpet wool breeds). There are sheep, especially in hot and humid climates, that produce hair rather than wool. Sheep can also be classified by where they store their fat; there are fat-tailed and thin-tailed, as well as fat-rumped and thin-rumped, sheep. Breeds of both fat- and thin-tailed sheep are known for their production of wool and hair, but the fat-tailed sheep are so named because each one carries in its tail 6 to 12 pounds of thick oil that is a very important source

FIGURE 8-9 Sahelian goats.
Vera Matlova.

of fat to pastoralists. Naming conventions with regional breeds of sheep are similar to those for goats. Often they take their name from the group of people who bred them. For example, the Fulani (Fulbe) and Maasai sheep breeds are named after the Fulani and Maasai peoples. Dwarf varieties also exist.

There were 250 million sheep and 219 million goats in Africa in 2001 (FAO 2002). Sheep and goats have been bred for all climates in Africa; there are even forest breeds; but most are found in the drier parts of the continent. People in the North African countries of Algeria and Morocco raise a significantly greater number of sheep than goats (47 million sheep to 8.7 million goats). Thirty-three percent of Africa's sheep are raised in Sudan, Ethiopia, and Somalia. Although sheep are found in every country and culture, they have a special ritual importance in the Islamic religion, where they are consumed on certain religious holidays (the prophet Muhammad's birthday; at the end of the fast month of Ramadan, etc.). Thirty-one percent of the goats in Africa are raised in Ethiopia, Somalia, and Sudan. Significant numbers exist in Nigeria and Tanzania. Goats are omnivorous eaters and will even climb trees to get at leaves and green thorns. They have been known to debark and destroy trees.

Camels

Like most domesticated stock in Africa, the camel, *Camelus dromedarius* (Figure 8-10), originated elsewhere. Evidence indicates that the camel was first domesticated in Arabia and probably diffused from the Arabian peninsula to the Horn. The presence of camels in Egypt has been attested during Roman times, but it was probably in the first centuries of the common era that the camel diffused to any extent from its points of introduction (Bulliet 1990). The camel is the most valuable large stock in Africa, and many pastoralists use it as an investment. It provides milk, transportation, leather, and meat, and the wealth of people is often measured in the size of their camel herds.

The camel is well known for its toleration of aridity but less well known is its adaptability to local forage and browse conditions—it can be trained to eat almost anything, even the saltiest of saltbushes. The FAO (2002) reported over 15 million camels for only 18 African countries in 2001. Most of these countries are located in the drylands around the Sahara. Sixty-nine percent of the camels in Africa are found in Ethiopia, Somalia, and Sudan. Somalia has more camels than any other country in Africa, 41% of the African total (Table 8-3).

FIGURE 8-10 Camels on their way to Kassala, Sudan.
Roy Cole.

Other Livestock

Pigs (*Sus domesticus*) are raised in almost all African countries, although there is a taboo against eating pork among Muslims. There is evidence for the existence of domestic pigs in Morocco and along the Guinea Coast in ancient times (Gilman 1976), as well as in early dynastic Egypt. Clutton-Brock (1993) suggests that the prohibition against eating pork, familiar in Jewish and Muslim cultures, may have actually arisen in ancient Egypt. In 2001 there were almost 18.5 million pigs in Africa, absolute numbers being highest in Cameroon, Nigeria, South Africa, and Uganda (FAO 2002).

Smaller stock not discussed in this chapter include chickens and guinea fowl. Chickens were first domesticated in India and Southeast Asia. Found virtually everywhere in Africa, chickens were probably introduced into Egypt from India and to East Africa by Austronesian-speaking migrants from Southeast Asia who settled on the island of Madagascar and probably along the East African coast during the last 2,000 years. Guinea fowl (*Numida meleagris*) are indigenous to Africa and represent an African domestication. The domesticated breeds of guinea fowl are probably derived from West African wild stock, and several distinctive guinea fowl breeds are found in Africa. Although some guinea fowl are white, most are spotted gray. Guinea fowl are meaty compared with chickens, weighing on average about 3.5 pounds (1.6 kg). The meat is white like chicken but has a gamey taste similar to pheasant. Eggs are usually laid in a field and are often placed under a chicken for hatching. The cock of the guinea fowl makes a distinctive shrieking call that is effective in scaring away potential thieves much as a barking dog would.

Livestock in Perspective

To put the data in Table 8-3 a little more into perspective, the reader should look at the mean holdings of livestock in Table 8-4 and then look back at the country-by-country numbers in Table 8-3. Countries that are two or more standard deviations above or below the mean have unusually high or low numbers of livestock. Sudan, for example, is very unusual; it has the highest TLU and has the maximum of all the countries in cattle, sheep, and goats (Table 8-3). By almost any measure, Sudan has extraordinary numbers of cattle, sheep, and goats. Its cattle population is 5 standard deviations above the mean for all countries. It goes without saying that its cattle population is in the upper quartile. But Sudan is a large country and comparing the stock populations of such a large country with those in, for example, Djibouti, could be

TABLE 8-4 LIVESTOCK POPULATIONS PER THOUSAND RURAL POPULATION IN 2002.

Country	Cattle	Sheep	Goats	Pigs	Camels	Horses	Asses	TLU
Africa	450	489	428	36	30	10	30	467
Algeria	135	1,532	278	0	19	4	14	307
Angola	593	51	315	117	0	0	1	464
Benin	394	170	311	124	0	1	0	338
Botswana	3,045	469	2,854	9	0	42	419	2,787
Burkina Faso	480	678	865	62	1	3	50	535
Burundi	56	41	106	12	0	0	0	55
Cameroon	731	471	545	167	0	2	5	634
Cape Verde	140	55	718	1,306	0	3	88	370
Central African Republic	1,474	105	1,236	323	0	0	0	1,198
Chad	889	362	791	3	109	31	54	907
Comoros	131	53	432	0	0	0	13	149
Congo, Republic of	83	88	258	42	0	0	0	97
Côte d'Ivoire	160	165	129	38	0	0	0	145
Djibouti	3,496	6,044	6,668	0	910	0	113	4,708
Democratic Republic of Congo	21	24	109	27	0	0	0	31
Egypt	95	119	93	1	3	1	80	148
Equatorial Guinea	16	123	29	20	0	0	0	29
Eritrea	630	449	486	0	21	0	0	556
Ethiopia	635	414	313	0	20	51	96	640
Gabon	158	872	401	938	0	0	0	332
Gambia	383	111	152	15	0	18	37	334
Ghana	106	224	251	26	0	0	1	126
Guinea	524	174	198	19	0	1	0	406
Guinea-Bissau	513	284	324	349	0	2	5	459
Kenya	607	316	437	15	40	0	0	542
Lesotho	325	466	364	38	0	64	98	428
Liberia	20	118	124	73	0	0	0	46
Libya	339	7,848	3,001	0	111	71	46	1,515
Madagascar	915	70	120	76	0	0	0	667
Malawi	95	14	183	32	0	0	0	89
Mali	885	831	1,285	9	61	21	88	969
Mauritania	1,291	6,540	4,389	0	1,058	17	136	3,161
Mauritius	40	10	136	30	0	0	0	46
Morocco	198	1,286	387	1	3	12	73	368
Mozambique	114	11	34	16	0	0	2	87
Namibia	1,691	1,771	1,369	14	0	51	56	1,573
Niger	275	547	839	5	50	13	71	440
Nigeria	280	289	343	68	0	3	14	278
Réunion	131	1,220	3,284	361	0	2	0	580
Rwanda	117	0	6	26	0	0	0	85
São Tomé and Principe	47	30	55	25	0	3	0	46
Senegal	597	891	738	52	1	91	76	703
Seychelles	52	0	182	638	0	0	0	118
Sierra Leone	122	106	58	15	0	0	0	103
Somalia	958	2,431	2,302	1	1,142	0	4	2,288
South Africa	636	1,332	303	71	0	12	10	631
Sudan	1,662	2,039	1,692	0	139	1	33	1,699
Swaziland	757	39	547	42	0	2	18	606
Tanzania	592	175	411	15	0	0	7	480
Togo	81	291	415	84	0	0	1	136
Tunisia	237	1,971	433	2	69	17	69	536

(Continued)

TABLE 8-4 (Continued)

Country	Cattle	Sheep	Goats	Pigs	Camels	Horses	Asses	TLU
Uganda	287	53	301	75	0	0	1	244
Zambia	407	25	215	58	0	0	0	315
Zimbabwe	755	73	381	38	0	4	15	590
Mean	544	812	781	101	70	10	33	650
Median	332	199	353	26	–	1	4	417
Standard deviation	681	1,592	1,217	234	240	20	65	867
Minimum	16	–	6	–	–	–	–	29
Maximum	3496	7,848	6,668	1,306	1,142	91	419	4,708
Lower quartile	118	59	182	6	–	–	0	139
Upper quartile	636	792	733	71	3	10	53	633

Source: FAO (2003).

misleading because the areas are so different. The high numbers in Sudan depicted in Table 8-3 may be more apparent rather than real if we take other factors such as rural population into consideration (Table 8-4).

Although it's not to be disputed that Sudan has a lot of livestock by almost any measure, a rather different picture of the livestock distribution emerges when the numbers of livestock and the numbers of people in a country is investigated. Table 8-4 breaks down our seven main livestock populations according to their distribution per thousand rural people. Although there are exceptions in Africa, it is the rural population that is occupied with raising livestock, and it makes good sense to standardize the livestock population across the rural rather than the total population of a country. Looking at the data in this way, we see that tiny Djibouti has much higher TLUs per thousand rural population than all other countries in Africa, and the highest numbers of cattle, sheep, and goats per thousand rural people.

Livestock numbers have increased across the continent since 1961 (Table 8-5 and Figures 8-3 and 8-4), the earliest year for which there is reliable data, but have not kept pace with human population (Figure 8-5 and Table 8-6). Should livestock populations keep pace with the growth in human population in a country? The answer to this complex question is not known. In fact it may be that only a small minority of people in a country raise livestock, and since pastoralists do not require abundant farm labor, they have traditionally had smaller families than farmers. Is it desirable for the animal population to surpass the growth of human

TABLE 8-5 CHANGE IN AFRICAN LIVESTOCK POPULATIONS OVER TIME.

	Average		Change	
Stock	1961-65	1997-2001	Number	Percent
Asses	10,717,157	15,327,771	4,610,614	43
Camels	9,257,155	14,853,626	5,596,471	60
Cattle	127,313,943	233,019,751	105,705,808	83
Goats	97,020,824	216,730,580	119,709,757	123
Horses	3,521,145	4,836,783	1,315,638	37
Pigs	5,961,526	18,721,542	12,760,016	214
Sheep	136,223,825	246,865,248	110,641,423	81

Source: FAO (2003).

TABLE 8-6 PERCENTAGE CHANGE IN AFRICAN LIVESTOCK PER THOUSAND PEOPLE, 1961-1965 TO 1998-2002.

	Head per 1,000 People		
Livestock	1961-65	1998-2002	Change (%)
Asses	36	19	−46
Camels	31	19	−40
Cattle	427	289	−32
Goats	326	277	−15
Horses	12	6	−48
Pigs	20	24	21
Sheep	457	312	−32

Source: FAO (2003).

population? Should livestock raising be encouraged among peoples who do not have a history of it?

In a rapidly urbanizing region, decline in per capita livestock numbers may not be a cause for concern. Where the vast majority of people live on the land and depend on livestock for their nourishment, such declines may represent real hardship, but it's difficult to make continental generalizations from macro-level data. Micro-level studies are needed to show the relationship between stock raising, nutrition, and wealth. In any case it is clear that livestock populations are increasing, although some at a greater rate than others.

The results in Table 8-6, which presents in percentages the change in certain stock populations in relation to human population from the early 1960s to the early twenty-first century, are also clear: with the exception of pigs, the per capita numbers of African livestock declined over the time period. Interestingly the per capita decline in the numbers of goats was half that of the other types of stock.

Chickens appear to be the only stock other than pigs that are increasing relative to population. Chickens, not depicted in the tables or figures for quadruped livestock, increased by a billion between 1961 and 2002: from 294 million to 1.3 billion. The absolute percentage increase was 318%, or 61% per thousand population. Are Africans seeking alternative sources of animal products and/or protein? Perhaps. The data are suggestive but too aggregated to tell definitively. Moreover, so much about the African diet remains unmeasured—for example, the contribution to the diet of wild foods, or the use of beans and other sources of vegetable protein to replace animal protein. The FAO keeps no data on many plants consumed by Africans that may be key to understanding African diets.

Characteristics of the Pastoral Economy

The vast majority of Africans who raise livestock are by no means nomadic pastoralists. Rather, they are farmers who raise stock for milk, manure, meat, leather, and transportation, and as animal traction to pull plows, carts, and so on. Africans raise a wide variety of animals, most of which were domesticated elsewhere and diffused to Africa. Small stock like goats, sheep, and pigs and large ruminants like cattle, camels, donkeys, and horses are commonly raised in many parts of Africa. Chickens, guinea fowl, and ducks are found both north and south of the Sahara. The characteristics of the major types of large livestock in Africa that are used for subsistence, for transport and as investments were presented in Table 8-2.

Livestock are important to Africans for several economic and social reasons. First, livestock represent a way to amass and reproduce wealth. Livestock have a high value, particularly cattle and camels; the animals reproduce themselves and grow in number, often representing the best investment opportunity for the farmer as well as for the pastoralist. Livestock are also a visible measure of family or personal wealth, as a stock portfolio (or talk of one) may be in Western industrialized societies. Livestock in Africa are also important in exchanges. For most general purposes, small stock like goats and sheep are used almost like a currency. Cattle and camels play an important role in many parts of Africa as bridewealth or dowry.

Moreover, most pastoral peoples who live on the edge of a town depend on exchanges of animal products for cereals and other farm products with farming peoples. The Sudanese Hadendowa Beja, for example, produce only about one-third of their cereal needs. They depend on trade with their Arab neighbors to supply the rest. The Fulani (Fulbe), scattered across Subsaharan Africa from Senegal to Sudan, have developed complicated relationships with their farmer neighbors. Many Fulani women sell milk and yoghurt in periodic markets, and the cash is used to purchase cereals and other necessities. Fulani herders trade with sedentary farmers, exchanging manure for cereals and grazing for their animals. Often a herder will pasture his animals on a farmer's field for a period of time, grazing on the stubble left over from the harvest. In the past (and in some areas continuing to the present) the Fulani of Macina and elsewhere in Mali have settled farmer-captives in villages and visit these villages annually as part of the seasonal migration, at which time cereals are taken from the "clients" by the herder-masters. Historically, the Kikuyu in Kenya have traded cereals, iron implements, tobacco, and other products to the pastoral Maasai, and the Maasai have given the Kikuyu livestock and livestock products in exchange.

Pastoralism as a land use has several interesting characteristics. With few exceptions, the drylands where pastoralists earn their living are not capable of supporting settled populations of people or livestock. Pastoralists use *movement* across space to enable the sparse and seasonal vegetation to nourish their animals. Movement is a risk-minimizing strategy and the most important characteristic of pastoralism in Africa. Movement ensures that animals get fresh and protein-rich pasture, avoids overgrazing resources, helps to avoid disease and insects that might plague sedentary stock, and gets people and stock away from drought-stricken areas. Movement enables a family to possess a large number of animals in a semidesert environment. Many pastoralists have a home area where they spend part of the season and where a permanent water source is usually found. Pastoralists migrate out of the home area for part of the year with their livestock to exploit seasonal pastures.

The movements of individual members of a pastoral family are rather complex and depend on the person's age, sex, marital status, desires, and abilities. Some young men do not follow the herd and are instead involved with legal trade or smuggling; others migrate to agricultural schemes to work as seasonal labor. Marriageable women may have a different circuit from their brothers to keep them close to immediate kin. In the driest parts of Africa, pastoralists are opportunists who will go where the pasture is to be found; in other areas, however, relatively fixed migration routines exist. Opportunistic herders will migrate from the mountains to the plain or from the plain to the mountains, provided there is grazing for their livestock.

As a simplification, pastoral movement can be characterized as either vertical or horizontal. Some groups move from seasonal grazing in the highlands to seasonal grazing in the lowlands in what is called vertical transhumance. Others practice horizontal transhumance (Figure 8-11), moving from the southern Sahel to the north with the annual summer rains. Pastoralists also herd a variety of animals to exploit several ecological niches, carry baggage, and stagger and extend milk production for consumption or market. A disease outbreak may affect only one type of stock herded (cattle rather than goats for example), reducing a family's risk of total loss.

Herd *diversification* involves temporary or permanent combinations of animals—cattle, camels, sheep, goats, according to what the local mix of animals—to minimize risk. Camels are the most drought-resistant animal and the most valuable. They are milked, used as an investment, and serve as pack animals. Cattle are like camels, valuable; on the other hand, they are the least drought resistant and must be watered every other day. Cattle are used mainly for milk. They function as an investment for farming peoples who keep cattle. Goats are drought resistant and the least valuable of ruminants. They reproduce quickly are easily sold for cash. Sheep have a higher value than goats but are less drought resistant. In Islam, rams have a ritual value that is reflected in their market price. Rams are often purchased for fattening months before the holiday on which they will be slaughtered.

Herd *maximization* is the third fundamental characteristic of pastoralism. The logic of this practice is manifold. On one level, herd maximization is a form of risk avoidance: if drought, for example, strikes, the larger the herd going into the drought, the greater the number of survivors. On another level, herd maximization generates surpluses of animals and animal products that are used in trade, to satisfy social obligations, to maintain human subsistence, to obtain a return on investment, and to achieve economies of scale.

Pastoralists also practice *herd splitting*. The most fundamental division is between lactating females and dry females. African pastoralists keep only a few male animals of whatever species. They raise a vast majority of females in the herd, selling off the males at an early age. Females are more important because they produce milk and young. Lactating females are kept around the herder's residence for milking. The dry herd is usually sent some distance away to exploit green pastures. To

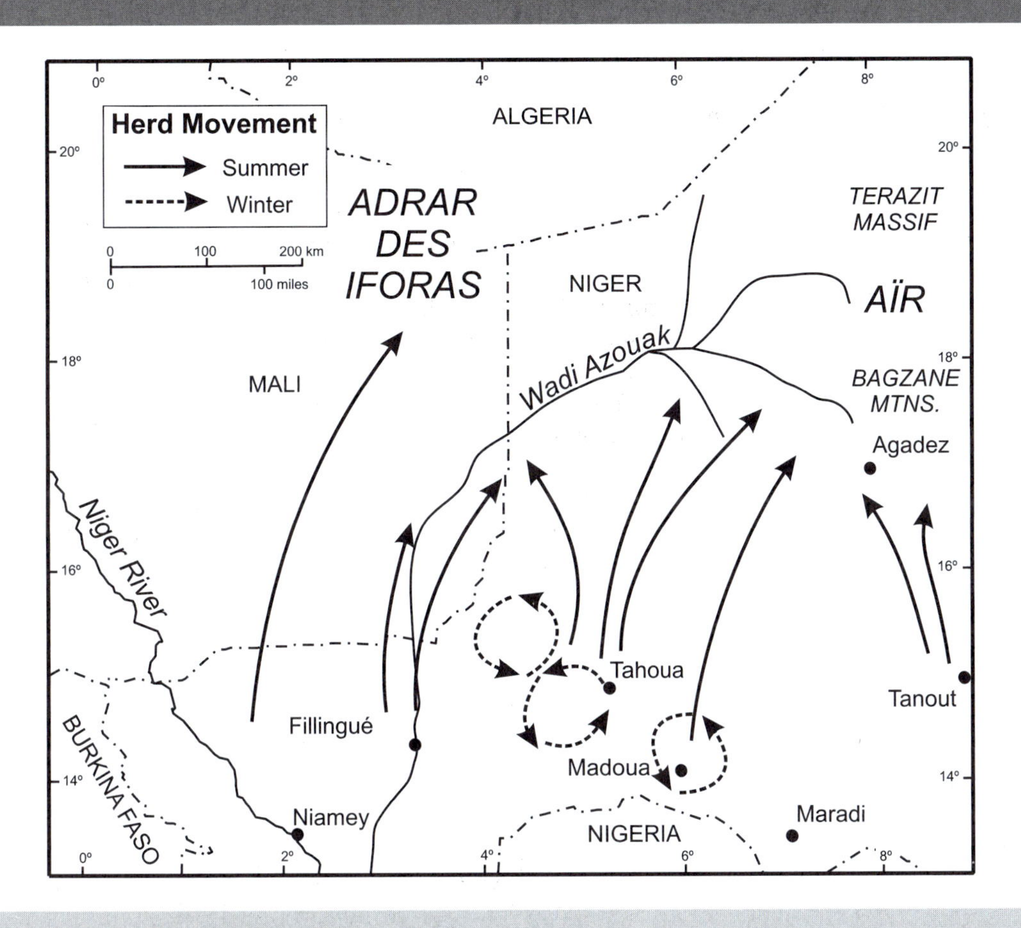

FIGURE 8-11 Transhumance in Niger and Mali.
Adapted from Johnson (1969).

spread risk of loss from disease, drought, and theft, large herds are often split into smaller units if labor is available. Herd splits are sometimes accomplished by using family labor, through stock loans to relatives, friends, or clients, or by employing hired labor.

Pastoralists have systems of *mutual aid and support* that are equivalent to some degree to insurance policies in industrialized economies. Pastoralists support each other through social linkages that provide help during and after any disaster that may strike. Although mutual aid is generally within a clan, it is also known to occur between clans. It most commonly involves stock loans to kin who have lost their herds.

Although all livestock numbers are growing absolutely and some relatively in Africa today, pastoralists are on the wane as a percentage of total population. Pastoralists face incentives to settle posed by loss of livestock, competition with farming for pasture and water resources, sedentarization programs developed by national governments, and even the attractions of the city. Although many pastoralists have settled with some of their livestock and now practice farming, others have continued to migrate with their animals. Some have combined livestock raising with the lure of the town and city and have settled with their animals on the urban periphery, selling milk and other animal products to city dwellers. Examples can be found throughout the drylands, but Port Sudan, Sudan is a notable one. There, cattle and goats are penned on the edge of the city, and boys riding donkeys ferry jugs of milk to their subscribers every morning. It is uncertain what the future of pastoralism will be in Africa. Ranching has been tried as a private property alternative to what Garrett Hardin has called the "tragedy of the commons" (1968) in Kenya and South Africa. Results have been mixed.

FARMING IN AFRICA

The vast majority of Africans are farmers (Figure 8-12). African farmers are vegetation managers who manipulate a variety of domesticated, semidomesticated, and wild plant life to produce a constant stream of food for their families and a marketable surplus. For the most part, the family is the unit of production and consumption in Africa. This means that most of the labor supply is provided by the family and most of the food produced is consumed by the family. The willingness to take risks is a large part of successful farming in Africa, as elsewhere in the world. The African farmer, however, generally lacks the control over the environment that the farmer in more developed parts of the world possesses. First, most of Africa is dry and subject to periodic drought, and opportunities for risk-minimizing strategies such as irrigation are limited. Furthermore, unlike conditions in the United States, for example, where crop insurance is used to hedge risk and protect farmers, in Africa crop insurance is rare. Thus for the African farm family, crop loss and failure can represent real threats to well-being and even physical survival. Furthermore, in the more developed regions of the world, governments have created price supports and subsidies that sometimes pay farmers above the world price for a particular commodity and sometimes pay farmers not to grow certain crops!

African agriculture, although never static, has been changing at an increasing pace since the beginning of the recent colonial experience—new technologies are being adopted, agricultural organization has changed

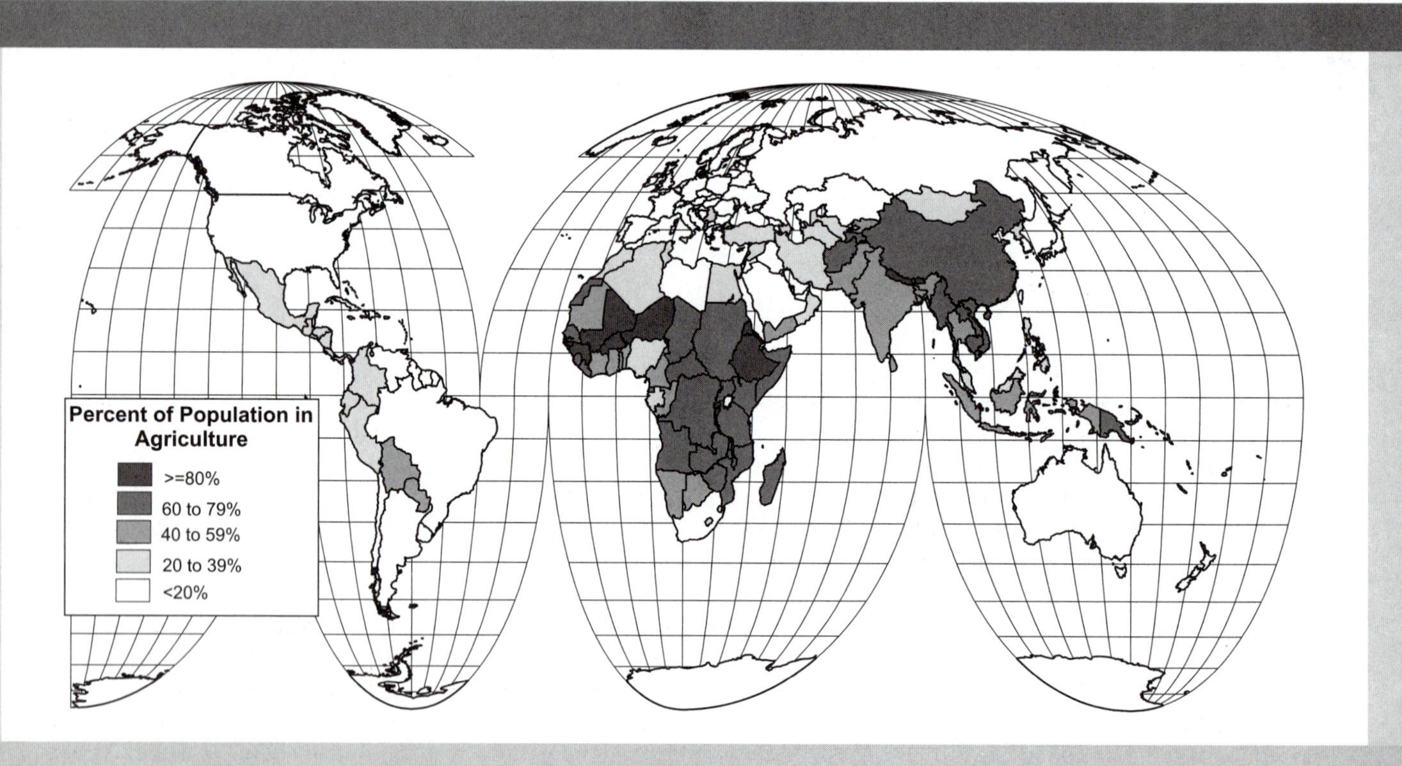

FIGURE 8-12 Percent of population in agriculture by country in 2002.

in many areas, and farming for the market is growing. Africans have marketed agricultural products since ancient times, but their involvement with the market has been increasing at an increasing rate since the twentieth century. Adding to local marketed crops, new food and cash crops, introduced from abroad over the past 500 years, and especially over the last 100 years, have been integrated into the agricultural calendar, sometimes competing against one another for scarce family labor.

There is a market for all cultivated crops and many collected foods in Africa. Perhaps among the only true cash crops are castor, cocoa, cotton, gum arabic, pyrethrum, tobacco, and sisal. Agriculture in Africa has changed dramatically in most places yet has remained much the same in other, usually more environmentally marginal, locations. In Sudan's Red Sea Province, some Beja agropastoralists still plant their durra sorghum with crude dibble sticks and store their harvests in the earth in the manner depicted in ancient Egyptian hieroglyphics. Sudanese entrepreneurs have developed pump irrigation of food crops destined for the regional market on alluvial fans along the Red Sea Hills, not far from the regional center, Port Sudan. In West Africa, it was powerful cocoa planters in the Caribbean and Latin American tradition who led Côte d'Ivoire to independence in 1960. Other Africans farm in much the same way as their grandparents but have adopted some additional mechanical technology.

It is difficult to use the word "traditional" to describe what African farmers do because their world is constantly changing. What has become clear after decades of research is that the African farmer's understanding of the land and ecology is profound and that with hindsight, many early statements of outside observers about the irrationality of African farming now appear to have been made in error. This chapter is meant to be a close look at farming practices around the continent.

The African farmer uses a variety of strategies to minimize risk that are different from and definitely more numerous than the strategies used by farmers in the United States. Most farmers do more than just farm; they supplement their agricultural activities with other activities that may take place in the home area or elsewhere. For example, members of a family may fish, hunt, gather wild foods, produce craft goods, process farm products (e.g., spinning thread, weaving cloth, making mats), and engage in petty trade; many seek work outside the home area on a seasonal basis to earn money. In the off-season, many farmers practice a trade or profession, such as weaving, tailoring, carpentry, smithing, or masonry. Some farmers are tangentially involved in farming and devote most of their time to raising livestock; yet others mix livestock raising and farming. Some African farmers produce for the market and are little different from commercial farmers in the United States or other more developed region of the world. Privately owned, mechanized, large-scale commercial farms and ranches exist in the former settler colonies of Africa such as Zimbabwe, South Africa, Kenya, and Malawi. Multinationals are active in favorable regions as well. African governments and international organizations have been involved in the development of large-scale, mechanized farming schemes such as the Jezira in Sudan or the Office du Niger in Mali.

Most African farmers work independently of large organizations (except perhaps the local extension service), using mainly human and animal power. They are often called "subsistence" or "traditional" farmers, although both terms imply a primitiveness that masks the complexity of actual African agriculture. If traditional farming exists in Africa, it is in highly modified form, and those who produce for their own consumption, marketing nothing, are extremely few. Such subsistence production is found only in areas where market networks have been disrupted by war or outbreaks of disease.

D. G. Grigg, an agricultural geographer, developed a model of farming in which bipolar opposites anchor a continuum of different possibilities. The model is reproduced in Table 8-7 to contrast the traditional and the modern, commercial farmers.

Grigg characterizes the traditional farmer as a risk avoider rather than an innovator, one who is interested in maximizing the total return to land rather than labor. The traditional farmer is more interested in producing food than in profit and will work until the silos are full. Traditional farmers will grow a variety of crops because an adequate diet requires cereals, legumes, and a source of edible fat (plant or animal). Traditional farmers spread the risk of crop loss or failure by planting diverse crops, in a variety of ecological niches, and in different locations. The commercial farmer, on the other hand, is interested in maximizing profit under

TABLE 8-7 FOUR DIMENSIONS OF TRADITIONAL AND MODERN COMMERCIAL FARMING.

Variable	Traditional, Subsistence	Modern, Commercial
1. Proportion of output sold off the farm	Low	High
2. Destination of foods	Family consumption, some processed locally	Market and processing
3. Origin of inputs		
Power	Animal, human	Petroleum, electricity
Plant nutrients	Legumes, ash, manure	Chemical fertilizers
Pest control	Crop rotations, intercropping	Insecticides, fungicides
Weed control	Rotations, hoeing, plow	Herbicides
Implements and tools	Hoe, plow, sickle, scythe	Machinery, self-propelled
Seed	From own harvest	Purchased from seed merchants
Livestock feeds	Grass and fodder crops grown on farm or on common land	Purchased from feed merchants
4. Economic aims	Risk avoidance Provide family food Maximize gross output per acre	Profit maximization Specialized production Maximize output per head Minimize production costs

Source: Grigg (1995).

competitive circumstances and will innovate if it can be shown to reduce costs and/or increase production. Commercial farmers generally specialize in one or two crops and produce them mainly with machines.

Compared with the rest of the world, the use of machine technology on the farm is the least developed in Africa, especially Africa south of the Sahara (Table 8-8). In the labor-scarce economies of Africa, the mechanization that has taken place, mainly since the Second World War, has been the adoption of the animal-pulled plow. This technological package, called "animal traction," has vastly increased the productivity of labor.

Inorganic fertilizer use in Subsaharan Africa is the world's lowest, and there even is not enough organic

TABLE 8-8 MECHANIZATION IN AFRICA AND ELSEWHERE IN 2000.

World Region	Arable Land Hectares (1,000 ha)	Tractors in Use	Tractors per Hectare of Arable Land (ha)	Harvesters and Threshers in Use	Harvesters and Threshers per Hectare of Arable Land (ha)
Africa	179,723,000	525,998	0.29	37,859	0.02
Africa south of Sahara	140,979,000	162,372	0.12	5,066	0.00
Asia	486,254,000	7,279,927	1.50	2,064,476	0.42
Europe	289,795,000	11,088,726	3.83	1,014,031	0.35
Latin America and Caribbean	133,494,000	1,602,038	1.20	163,548	0.12
North and Central America	259,216,000	5,820,467	2.25	830,066	0.32
Oceania	52,459,000	401,512	0.77	60,085	0.11
South America	96,791,000	1,293,036	1.34	127,935	0.13

Source: FAO (2002).

fertilizer to go around. Most African farmers use organic fertilizers in the form of animal manure, composted household wastes, agricultural wastes, night soil, and ashes; but fertilizers, like labor, are scarce in Africa. Because of declining soil fertility under conditions of permanent cultivation, many African farmers leave some of their land fallow for several years to regain fertility, while others practice shifting cultivation. Shifting cultivation involves a group of people, usually related by blood, clearing plots of land and farming for a few years until the land is no longer able to produce a crop. The family then moves to another, sometimes distant, site. Except for the ash of the trees that are burnt during the clearing of the land, the land is not fertilized. The second area is cleared and cultivated for a time and then abandoned for another site. Groups of shifting cultivators may return to the original plots of land after 15 or 20 years—or even longer.

Many African farmers leave useful tree species on the fields, for example, the nitrogen-fixing *Acacia albida*. African farmers also intercrop leguminous plants such as beans with their staple food crops to increase the fertility of the soil. Table 8-9 shows the overall use in metric tons of NPK (nitrogen, phosphate, and potash) fertilizers, as well as the use per arable area in hectares.

TABLE 8-9 USE OF INORGANIC FERTILIZERS PER REGION AND PER ARABLE AREA.

Region	NPK Fertilizers Used (metric tons)	Metric Tons of NPK Fertilizers Used Per Hectare of Arable Land (ha)
Africa	3,880,516	2.16
Africa south of Sahara	1,230,296	0.87
Asia	72,854,881	14.98
Europe	22,382,138	7.72
Latin America and Caribbean	13,167,843	9.86
North and Central America	23,867,956	9.21
Oceania	3,165,607	6.03
South America	10,284,036	10.62

Source: FAO (2002).

Although Grigg's typology is useful for heuristic purposes, reality is much more complex than his two ideal types would suggest. Traditional farmers do innovate; they experiment with new crops, crop combinations, and varieties to try to improve yields. Every African country has an agricultural extension agency to get new technologies to farmers. Traditional African agriculture is probably adequate under conditions of stable population growth; since African populations began vigorously expanding in the 1950s, however, there have been stresses on the land.

Today, researchers are not quite sure what to make of the relationship between population and land use in Africa. On one hand, some believe that the available evidence indicates that population growth entails land degradation and a decrease in the ability of farmers to feed themselves. Others believe that such pressure will foster innovation that will ultimately stimulate the agricultural economy toward land-use intensification. There are other serious issues that have an impact on success or failure in African agriculture. These will be taken up later in this chapter, after we have discussed the distribution and characteristics of most major and several minor but interesting crops.

Distribution of Major Crops

For many of the crops of Africa, a discernible geographic pattern exists. The pattern may be caused by climate, chance, or culture—or all three. Wheat, for example, is a dryland crop, domesticated in Mesopotamia thousands of years ago, that spread into similar Mediterranean climates in North Africa and the Horn. It is consumed principally in the form of bread, and its later spread has been associated with the influence of bread-eating peoples from southern Europe such as the Romans and the later migration, over the last 1,400 years, of bread-eating people from Southwest Asia (principally Arabs) to Africa. Although wheat has spread to the dry regions in Africa just south of the Sahara with the diffusion of Islam—one can buy flatbread in towns along the Sahel Corridor—it actually is little grown because customarily bread is not eaten in Subsaharan Africa, and there is little room in the agricultural calendar for a dryland cereal in addition to millet and sorghum, local domesticates. During the colonial period bread eating became a habit in African towns and cities among all social strata. Europeans

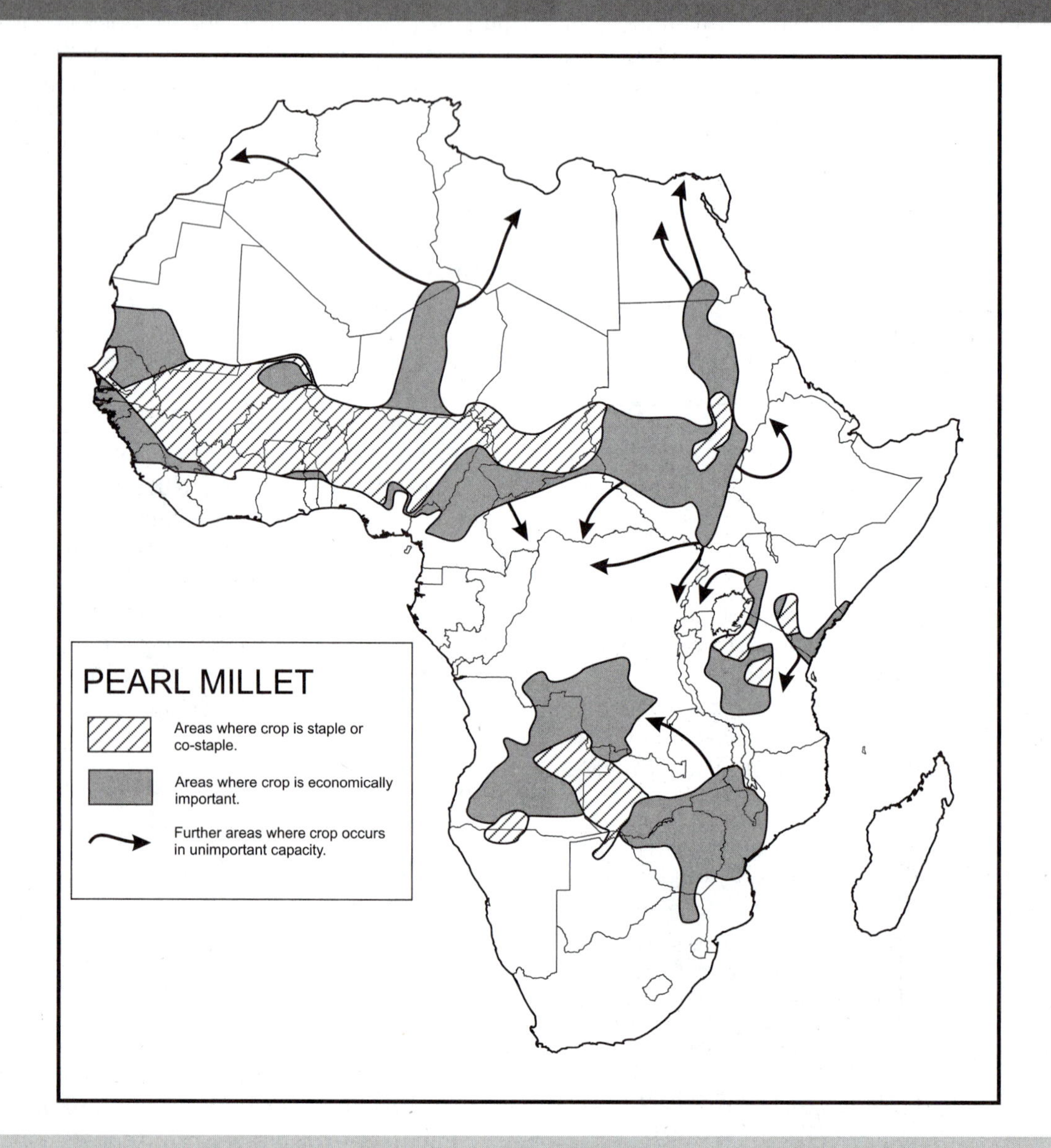

FIGURE 8-13 Distribution and importance of millet in Africa.
Adapted from Murdock (1960).

built bakeries, African urban elites acquired the taste for wheat bread, and the habit diffused to the general urban populace. Eating wheat bread has an air of modernity about it, and the food itself is cheap (sometimes subsidized by African governments after independence), as well as very portable. Where there have been significant settler-colonies of wheat-eating Europeans (South Africa and Zimbabwe) and an appropriate climate, wheat is grown in significant quantities today.

In the following discussion of the major crops of Africa, the geographic distribution of each crop will be presented in maps based on those made by George Murdock in the 1950s, using data compiled from over 2,000 ethnographic sources (Murdock 1960). Data from the Food and Agriculture Organization of the United Nations is presented by country to list the major producers today. Although not free of error, Murdock's maps are useful in that they show us where crops are grown; furthermore, although they depict areas rather than volumes, they provide us with a baseline with which to compare the present cultivated areas with the past.

Pearl Millet

Wild forms of pearl millet (*Pennisetum glaucum*) are found today deep in the Sahara, and pearl millet was probably domesticated somewhere in the Sahara—probably on the western side—during a wet phase (Harlan 1993). As the Sahara dried out, cultivation of pearl millet survived in the Sahel, where it is commonly cultivated today. It is grown throughout the drylands of southern Africa and is important in the Nile Valley (Figure 8-13). The leaves of pearl millet resemble those of maize but are narrower, while the plant itself is much taller than maize, growing up to 13 feet (4 m) tall. The millet ear is long (up to 6.5 ft; 2 m) and forms at the top of the stalk (Figure 8-14). Africans have bred a great many varieties of millet, each with a certain ecological niche in mind; sometimes several varieties are grown together in the same field (Figure 8-15). Some varieties mature in 4 months, others in 60 days. In many parts of Africa the labor-intensive segments of millet (and other crops) processing are done by gang labor. Figure 8-16 depicts the young men's association threshing millet in a village in central Mali. African women generally mill millet using a mortar and pestle (although grain grinders powered by fossil fuels are spreading) and then cook it to make a thick paste, which is eaten with fish, meat,

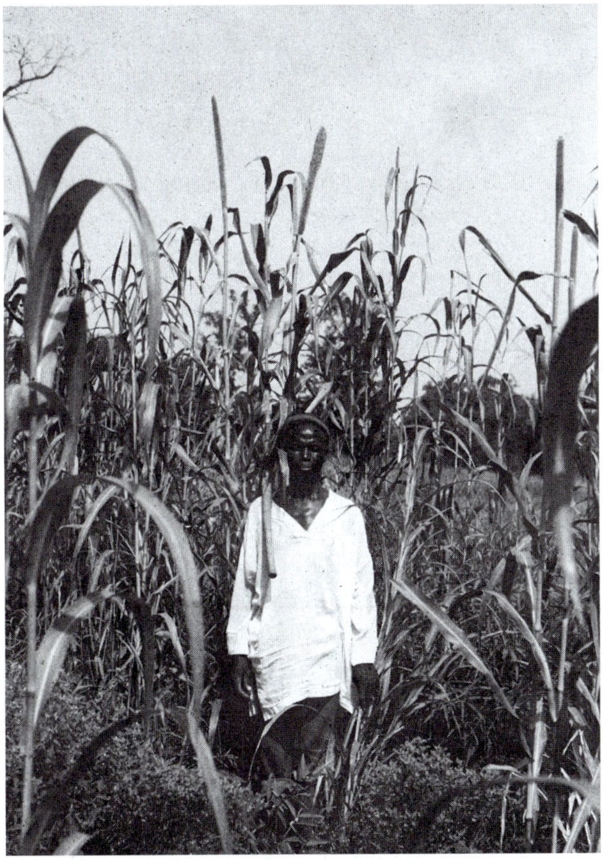

FIGURE 8-14 Pearl millet.
Roy Cole.

FIGURE 8-15 Two varieties of pearl millet grown together.
Roy Cole.

FIGURE 8-16 Threshing a family's harvest: young men's gang labor brigade.
Roy Cole.

or vegetable sauce. Millet is sometimes used in the preparation of a delicious beer. Outside the gleaming modern breweries located in the capital cities of Africa, where imported ingredients are used, in the small towns and rural areas, beer brewing is principally a women's occupation.

There are many varieties of pearl millet, each developed to do well in a particular environment. Finger millet (*Eleusine coracana*), although technically not a millet (discussed here because of the common vernacular name), used to be widely grown in eastern Africa. Murdock argues that it probably originated in Ethiopia and spread far to the south, particularly around Lakes Victoria, Tanganyika, and Malawi, as a staple crop. In Murdock's day it was also grown in important quantities in Zambia, Zimbabwe, northeast Angola, and central Tanzania. Finger millet is a short plant growing only waist high. The ear looks like fingers loosely bunched together in a fist. There is some debate on the palatability of finger millet. Murdock (1960) and Harlan (1993) state that it is not palatable but is used in beer brewing, while the National Research Council (1996) indicates that it is highly palatable and nutritious, and stores well. Other crops have replaced finger millet in its range that require less work, for example, maize, pearl millet, sorghum, and cassava (manioc).

According to the Food and Agricultural Organization of the United Nations (2002), Africa produced almost half of the world millet crop in 2001: 14 million metric tons. The average yield for Africa was 680 kg/ha, although yields for the top 10 producers in Africa ranged from 220 (Sudan) to 1,500 (Uganda) kg/ha (Table 8-10). Differences in yield reflect the vastly different conditions under which millet is grown in Africa from the edge of the deserts to much wetter conditions closer to the equator or in highlands (Figure 8-17). The major producer of millet in Africa in 2001 was Nigeria, which harvested 44% of all the millet produced in Africa. Nigeria and Niger, located just to the north, accounted for almost 62% of total African production. The world average yield for millet was 860 kg/ha in 2001. The world's highest yield in 2001 was 3,600 kg/ha, produced under humid climate conditions in Europe (Croatia), while the lowest was 120 kg/ha in southern Africa (Botswana).

Murdock's map of the distribution of millet in Africa (Figure 8-13) shows a wide distribution—recall that Murdock was measuring the role this widely traded

TABLE 8-10 AREA HARVESTED, YIELD, AND PRODUCTION FOR AFRICA AND THE TOP 10 MILLET-PRODUCING COUNTRIES IN AFRICA IN 2001.

Country	Area Harvested (ha)	Yield (kg/ha)	Production (metric tons)	Percentage of Total
Africa	20,349,540	679.9	13,835,673	100.0
Nigeria	5,914,000	1,032.3	6,105,000	44.1
Niger	5,231,937	461.5	2,414,394	17.5
Burkina Faso	1,520,000	629.9	957,457	6.9
Mali	1,032,436	835.6	862,715	6.2
Senegal	842,124	712.7	600,221	4.3
Uganda	389,000	1,501.3	584,000	4.2
Sudan	2,200,000	219.5	483,000	3.5
Chad	1,026,785	312.3	320,701	2.3
Ethiopia	346,780	916.0	317,650	2.3
Tanzania	170,150	1,001.5	170,400	1.2

Source: FAO (2002).

crop played in the local food budgets and economy, and the geographic center or pivot of millet production in Africa seems to be in the area of Nigeria. This centrality has remained constant over time, although quantities produced have risen dramatically; in 1961 Nigeria produced 40% of Africa's millet while Niger produced 12%. In 2001 the figures were 44.1 and 17.5%, respectively (Table 8-10).

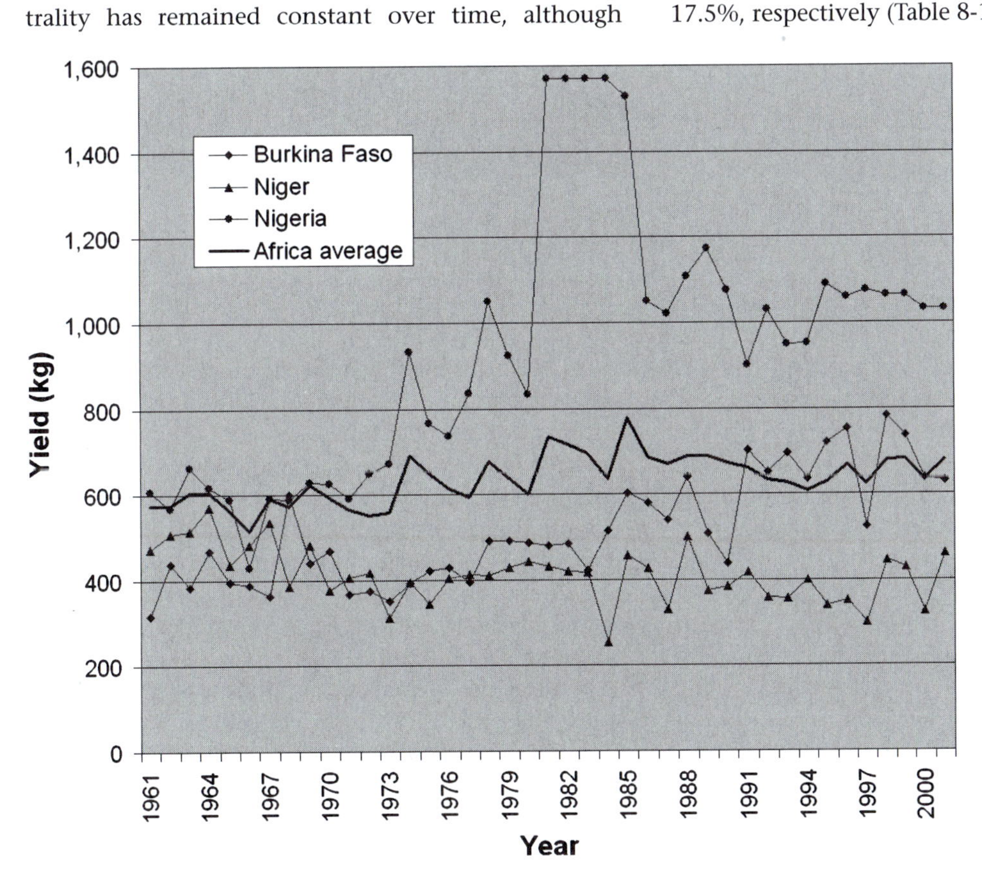

FIGURE 8-17 Variability in millet yield from 1961 to 2001 for Africa's three top-producing countries.

Yields for a rain-fed crop like millet can vary dramatically from year to year, but there seem to be some long-term trends. For Africa in general, yields have gone up since 1961: from an average of 576 kg/ha in 1961 to 680 kg/ha in 2001. Country to country (and region to region) results for the two time periods are more mixed. Nigeria increased the yield per hectare from 606 kg in 1961 to 1032 kg in 2001, while Burkina Faso added roughly 200 kg to its yield per hectare, on average. Niger's yield remained roughly the same over the two time periods: 473 and 462 kg/ha, respectively. These figure should be interpreted with caution, but the long-term data depicted in Figure 8-17 do show a trend.

Sorghum

Sorghum (*Sorghum* spp.) (Figure 8-18) is the most important cereal in Africa. Production and area harvested are greater for sorghum than for any other cereal, including millet. Together, more area is devoted to sorghum and millet in Africa than to all other food crops combined (National Research Council 1996). Sorghum was domesticated in the Sudan/Chad Basin, where the wild forms are most abundant today, and spread to most of the drylands of Africa. It is widely cultivated outside of Africa as a food and feed crop. Evidence for the date of its domestication is sketchy, but it probably occurred well over 4,000 years ago.

FIGURE 8-18 Durra sorghum.
Roy Cole.

There are five major types of sorghum and, with one exception, each has been associated with a different ethnolinguistic group of Africa. The types of sorghum spread with the growth and movement of the people who domesticated them. *Sorghum bicolor* is the most ubiquitous type and, according to Harlan (1993), the closest to the wild ancestor. This type has not been linked to the movements of any particular groups of people but is instead thought to be the common ancestor of all other subsequent types. The *guinea* type of sorghum, is mainly found in West Africa and has been modified to grow well under higher precipitation. It is associated with people who speak languages of the Niger–Congo family. *Sorghum caudatum* is found from Lake Chad to eastern Sudan and is associated with people who speak languages from the Chari–Nile language family. Kafir, the fourth type of sorghum, sometimes called kafir corn, is grown in southern Africa. It is associated with Bantu speakers of the Niger–Congo language family, who migrated into the southern half of Africa a few thousand years ago. The last type of sorghum is called *durra* and is grown mainly in Sudan and along the edges of the Sahara (Figure 8-19).

Sorghum is principally a rain-fed crop but in favorable areas it's grown in a flood-retreat system. Flood-retreat farming is done along the floodplains of a river or on a lake bed and involves the sowing of sorghum as the annual flood diminishes.

African production of sorghum was 19.5 million metric tons in 2001, 34% of the world total. The top 10 African producers of sorghum in Africa are presented in Table 8-11. It should be no surprise that almost all the countries that were important millet producers are also important sorghum producers. Both crops were

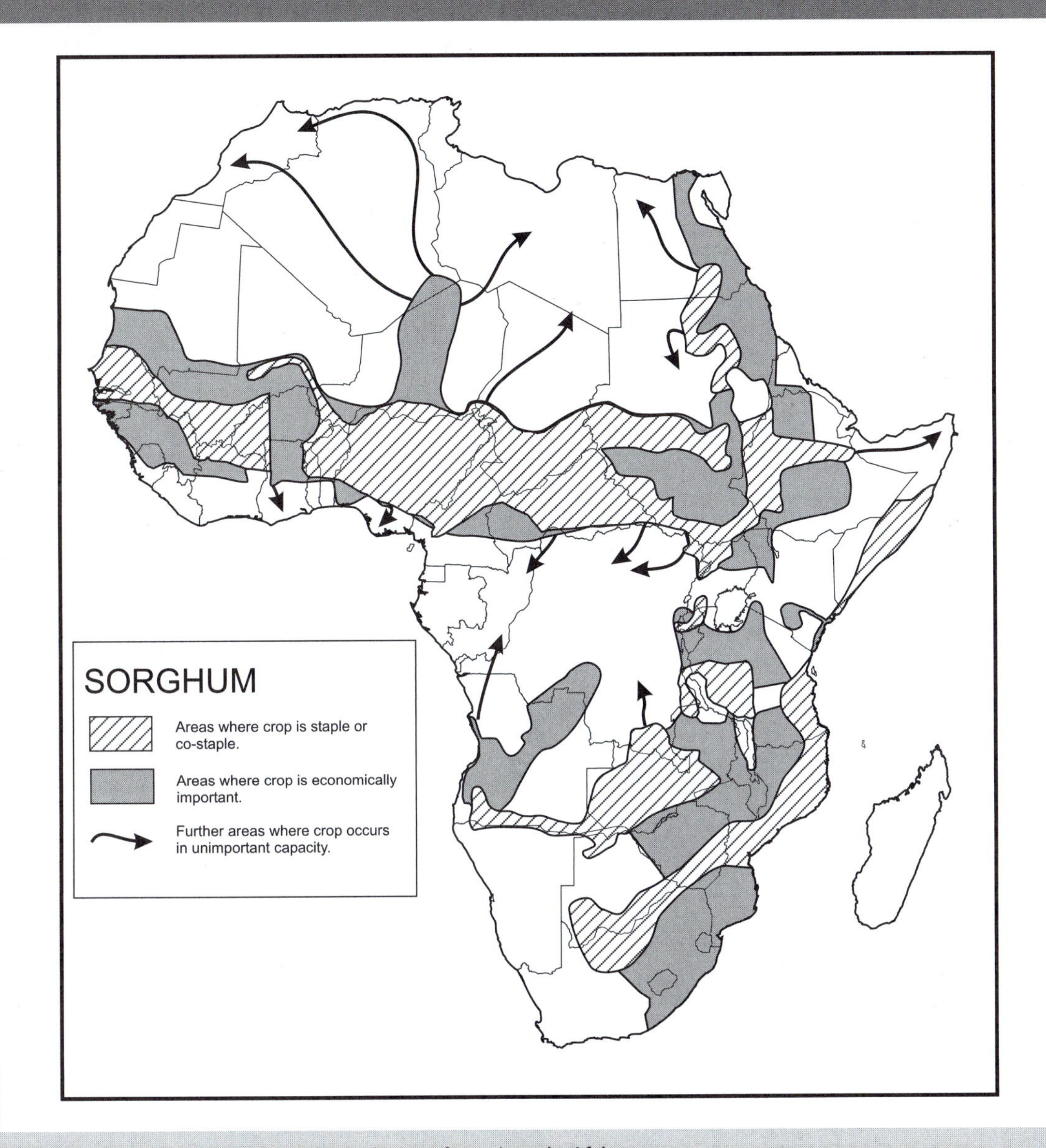

FIGURE 8-19 Distribution and importance of sorghum in Africa.
Adapted from Murdock (1960).

domesticated in the Saharan area and, although it may not be entirely evident from Murdock's maps of millet and sorghum (Figures 8-13 and 8-19), we can speak of a "cereal belt" that extends across the drylands of Africa just south of the Sahara from Senegal to Ethiopia.

Sorghum is eaten in the form of a very thick porridge called *to* in Manding languages of West Africa and *ugari* in East Africa. It is also made into dumplings, couscous, and a delicious beer (Box 8-1).

African sorghum yields can vary enormously depending on the local conditions. For example in 2001

BOX 8-1

SORGHUM BEER

In Africa, as in many parts of the world, brewing uses vast amounts of grain. However, in Africa the raw materials are sorghum, maize, pearl millet, and finger millet, not barley, rice, or wheat. Also, the basic process is unique. African brewing includes a lactic acid fermentation, known as souring. The resulting beverage is something like a fermenting gruel and has the consistency of malted milk.

Normally called "sorghum beer" or "opaque beer," this drink already constitutes a considerable part of the diet in many areas, and it will likely become an ever bigger commodity. With so many people moving into the cities, brewing is even now shifting from an exclusively family enterprise to an industrialized one. In South Africa, for instance, sorghum beer brewing is already a highly specialized industry. Annual production is about one billion liters.

Brewing raises the nutritional value of sorghum. It adds vitamins, neutralizes most of the tannins, hydrolyzes the starch to more digestible forms, and increases the availability of minerals and vitamins. South African studies indicate that iron is 12 times more available in sorghum beer than in a boiled sorghum gruel.

The beer is more than a mere drink. The whole social system of the people is inextricably linked up with this popular beverage: the first essential in all festivities, the one incentive to labor, the first thought in dispensing hospitality, the favorite tribute of subjects to their chief, and almost the only votive offering dedicated to the spirits. Beer is a common means of exchange or payment for services rendered, and in time of plenty it is not only freely consumed, but often is the principal or sole food of many men for days on end. It is evident in all ritual and ceremonial occasions binding together different groups or individuals and affecting a reconciliation when things go wrong.

Source: The Lost Food Crops of Africa: Grains. *Washington, D.C.: National Academic Press. National Research Council (1995).*

TABLE 8-11 AREA HARVESTED, YIELD, AND PRODUCTION FOR AFRICA AND THE TOP 10 SORGHUM-PRODUCING COUNTRIES IN AFRICA IN 2001.

Country	Area Harvested (ha)	Yield (kg/ha)	Production (metric tons)	Percentage of Total
Africa	22,478,707	870.6	19,569,028	100.0
Nigeria	6,885,000	1120	7,711,000	39.4
Sudan	4,195,000	593.1	2,488,000	12.7
Ethiopia	1,200,000	1,166.7	1,400,000	7.2
Burkina Faso	1,350,000	886.7	1,197,060	6.1
Egypt	160,000	5,875.0	940,000	4.8
Tanzania	618,200	1,190.1	735,700	3.8
Mali	747,840	929.2	694,862	3.6
Niger	2,604,308	251.8	655,729	3.4
Chad	986,162	574.5	566,577	2.9
Cameroon	360,000	1,250.0	450,000	2.3

Source: FAO (2002).

yields varied from 252 to 5,875 kg/ha with an average for Africa of 871 kg/ha (Table 8-11). The world average is 1,363 kg/ha; the highest yield in the world was 12,792 kg/ha in Israel.

Rice

African rice (*Oryza glaberrima*), probably domesticated in West Africa, is descended from an annual grass that grows in depressions that fill up during the rainy season (Harlan 1993). Although it was very important in the past, it has been largely replaced by Asian varieties of rice (Figure 8-20). Murdock (1960) identified two areas of Africa where rice is a staple food: West Africa and Madagascar (Figure 8-21). West Africa has already been mentioned as the most likely hearth for African rice. As for Madagascar, it is believed that the people who migrated to the island from Indonesia over a thousand years ago brought with them early forms of Asian rice, their staple crop. Rice is a preferred food in many parts of Africa, particularly in West Africa, and throughout Africa it is cultivated in a variety of ways: in paddies, terraces, and depressions. The scale of production ranges from the small farm using traditional technology to highly mechanized rice production schemes run by the state or state corporations.

FIGURE 8-20 Panicle of rice.
Roy Cole.

During the colonial period and in the decades following independence, rice was of great interest to governments because, in theory, it permitted the farmer greater control over one factor of production, water, than was possible with rain-fed millet and sorghum. Rain-fed crops depended on chance, while irrigated crops depended on the technology to get the water where the farmer needed it—or so it seemed at the time, according to the developmentalist logic. Additionally, rice was a crop that enabled the newly independent states of Africa to exercise an important leadership role in agricultural development that they did not have with traditional rain-fed crops such as millet and sorghum, which seemed bound up in traditional practices and didn't respond well to applications of inorganic fertilizer. Donor countries had millions of dollars to spend on agricultural development as well. Colonial powers and later the independent West African governments and their international assistants built dikes, dams, and innumerable engineering works to gain control of water during the wetter-than-normal 1950s and early 1960s. When more normal conditions returned at the end of the 1960s, there was a period of "retrenchment" and "triage" as canals had to be dug deeper to accommodate lower seasonal floods and some fields had to be retired from rice production for lack of water.

Africa is a dry continent, and suitable areas for irrigated agriculture are more limited than in other large regions of the world. Africa is mostly dry because of its geographic location. Almost all of Africa is located between the Tropics of Cancer and Capricorn, and the continent sits squarely astride the equator. Although the extreme northern and southern tips of Africa have mesothermal climates (C climates—see Chapter Two), the northern and southern thirds of the continent (outside the northern and southern extremities) are dry because of their position relative to the subtropical high-pressure belts. Continentality exacerbates the effects of the high-pressure belts. In addition, evapotranspiration in Africa is relatively high compared with

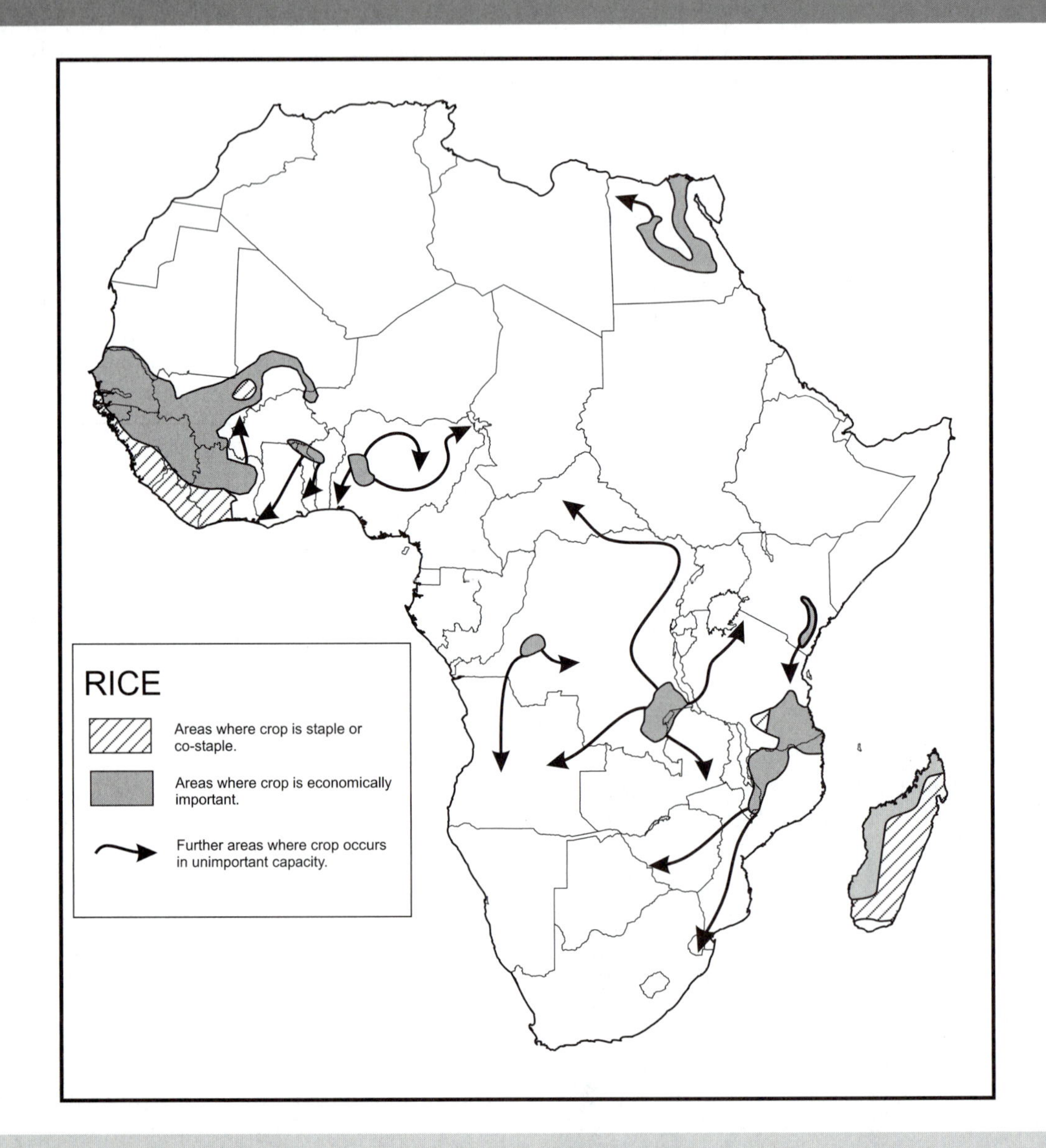

FIGURE 8-21 Distribution and importance of rice in Africa.
Adapted from Murdock (1960).

northerly regions located outside the tropics. As a consequence, less can be done agriculturally with any precipitation received. After the 1960s, African governments learned that rice is not a drought-proof crop but is subject to the vagaries of the weather just like millet and sorghum.

The major producers of rice in Africa are listed in Table 8-12. Note that only one top rice-producing country does not possess a major river. Madagascar's paddy culture was brought from Indonesia centuries ago, and the rain- and stream-fed terraces resemble those in Southeast Asia. Egypt produced a third of

TABLE 8-12 AREA HARVESTED, YIELD, AND PRODUCTION FOR AFRICA AND THE TOP TEN RICE-PRODUCING COUNTRIES IN AFRICA IN 2001.

Country	Area Harvested (ha)	Yield (kg/ha)	Production (metric tons)	Percentage of Total
Africa	7,672,751	2,212	16,974,457	100.0
Egypt	650,000	8,769	5,700,000	33.6
Nigeria	2,199,000	1,500	3,298,000	19.4
Madagascar	1,206,940	1,906	2,300,000	13.5
Côte d'Ivoire	600,000	1,667	1,000,000	5.9
Guinea	580,000	1,500	870,000	5.1
Mali	365,646	2,297	840,051	4.9
Tanzania	401,070	1,282	514,000	3.0
Democratic Republic of the Congo	431,821	755	326,025	1.9
Ghana	115,200	2,159	248,700	1.5
Senegal	86,252	2,345	202,293	1.2

Source: FAO (2002).

Africa's total rice production in 2001 and 26% of Africa's rice in 1961.

Africa produced a mere 2% of the world's rice in 1961 and almost 3% in 2001, a surprising difference given the dry climate in Africa and the fact that most countries of the world able to produce rice were scrambling to increase their production over the same time period as well. Rice production increased in most African countries that produced it in 1961, but most started from a very low base. For example, rice production increased from 500 metric tons to 80,000 metric tons in Mauritania from 1961 to 2001, a 160-fold increase, and the most extreme increase in Africa. Mauritania is situated in the Sahara, but its southern border is anchored by the Senegal River, the rice production area. For Africa as a whole, the average increase was 393% over the 40-year period. The phenomenal rise in the volume of rice production in Nigeria and Egypt (Figure 8-22) makes the greatest contribution to the Africa's growth in rice production. Nigeria's output increased 2,500% from 133,000 metric tons to 3,298,000 metric tons from 1961 to 2001. Over the same 40-year period, Egypt, already a big producer, jumped over 100%, from 2,442,000 metric tons to 5,700,000 metric tons. Egypt brought thousands of hectares under irrigated cultivation as a result of the building of the Aswan High Dam in southern Egypt and the double and triple cropping it permitted. Yields have increased as well for Egypt; other countries have remained constant or declined slightly in rice yield. Nigeria had an oil boom in the 1980s and, although the Nigerian government has had a reputation for being little interested in investment in developing its dryland agriculture, the evidence seems to indicate that some investment was made in rice irrigation projects.

Fonio and Tef

We shall discuss together fonio and tef, two indigenous African small grains, because of their similarity and limited geographic distribution (Figure 8-23). Sometimes, and erroneously, called "hungry rice," fonio (*Digitaria exilis*) is a highly palatable small grain, often consumed as couscous and probably domesticated in West Africa. The National Research Council (1996) hypothesizes that it was probably the first cereal to be domesticated in Africa. Fonio is one of the smallest grains grown in Africa. Figure 8-24 shows fonio seed ready for planting via broadcasting. Note the size of the rice grain in the middle of the photograph in comparison to the darker fonio seeds around it.

Fonio is called *fini* by the Bambara and other Mande-speaking peoples, and *acha* in Nigeria. Guinean varieties mature in 120 days, while varieties developed in drier climates mature in 6 weeks to 90 days. Fonio contributes to the household food budget at a critical

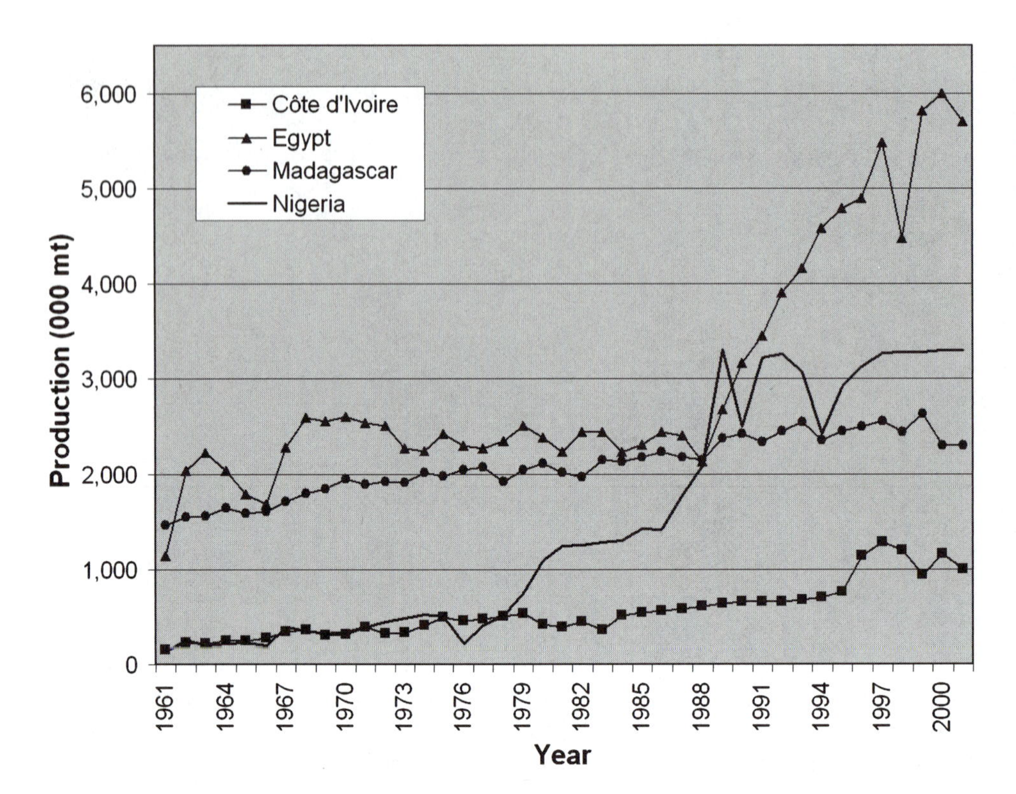

FIGURE 8-22 Rice production for Africa's top-producing countries, 1961 to 2001 (thousands of metric tons).

time in the year, the hungry season, the time of year just before the new harvest, when stores of food are low. Because fonio matures quickly, it is often the first crop of the season to be eaten.

Fonio grows on almost any soil and is very drought resistant. Although it does respond to organic fertilizer, it needs no fertilizer to grow and it is often grown on poorer soils. Despite its humble locale, fonio is more expensive than millet because it is very tasty, scarce, and the ear is difficult to husk.

Fonio is grown in only eight African countries and none outside of Africa (Table 8-13). With the exception of the three countries presented in Figure 8-25, Burkina Faso, Guinea, and Nigeria, fonio growing has been

TABLE 8-13 AREA HARVESTED, YIELD, AND PRODUCTION FOR AFRICA AND EIGHT FONIO-PRODUCING COUNTRIES IN AFRICA IN 2001.

Country	Area Harvested (ha)	Yield (kg/ha)	Production (metric tons)	Percentage of Total
Africa	361,987	734	265,700	100.0
Guinea	150,000	820	123,000	46.3
Nigeria	133,000	571.4	76,000	28.6
Burkina Faso	22,000	1,085.5	23,882	9.0
Mali	27,287	784.9	21,418	8.1
Côte d'Ivoire	22,000	681.8	15,000	5.6
Guinea-Bissau	3,500	1,114.3	3,900	1.5
Benin	2,600	576.9	1,500	0.6
Niger	1,600	625	1,000	0.4

Source: FAO (2002).

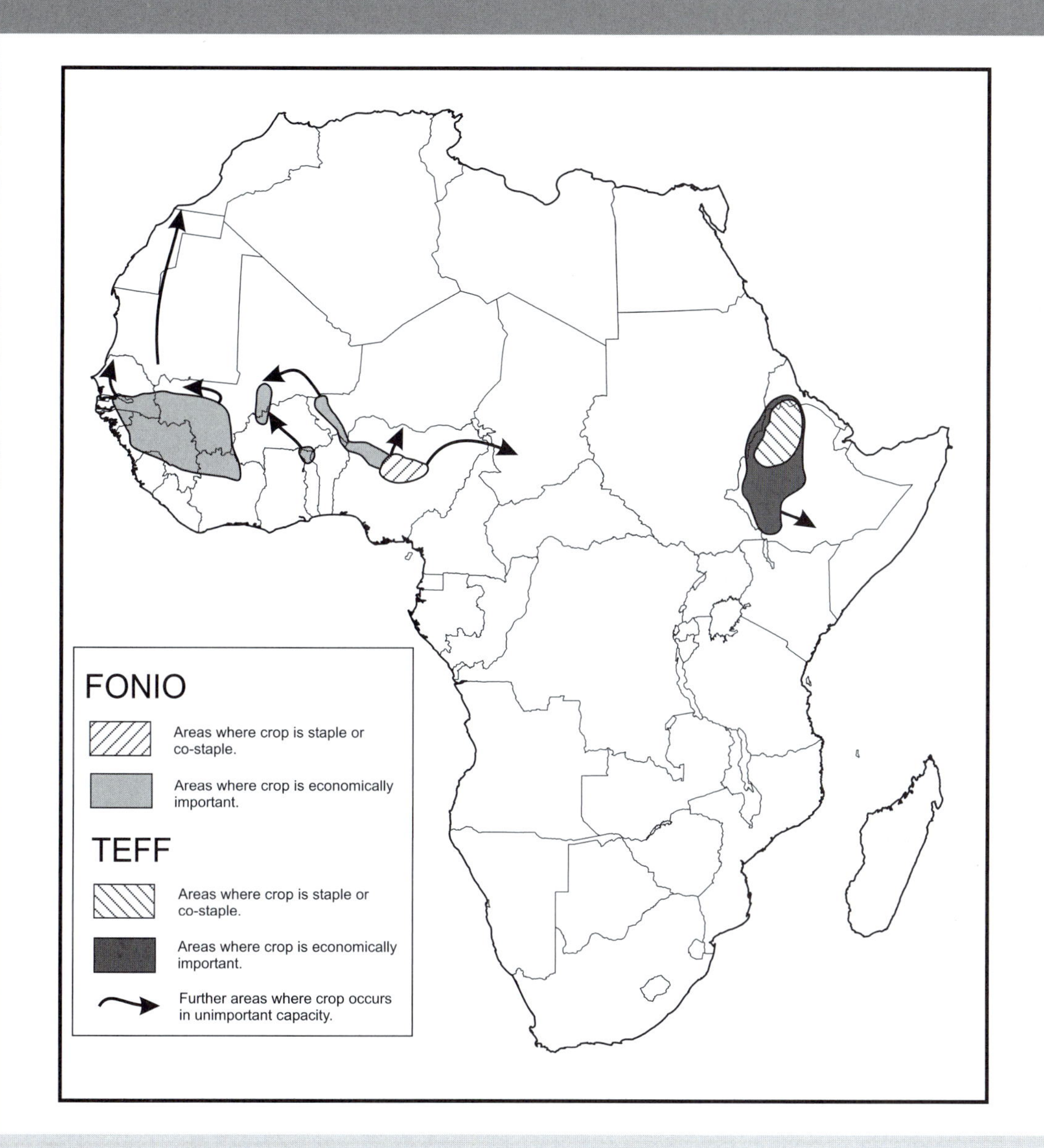

FIGURE 8-23 Distribution and importance of fonio and tef in Africa.
Adapted from Murdock (1960).

declining since the 1960s. Recurrent drought in West Africa since the late 1960s, however, has prompted many governments to recommend that farmers plant more fonio.

Fonio is grown with a minimum of machinery. It is broadcast and cultivated with a hoe-like implement called a daba, as indicated in Figures 8-26 and 8-27. The introduction of the plow and animal traction in central Mali after the Second World War transformed the ability of families to feed themselves, but the plow is not used to prepare the soil for fonio. The continuance of the backbreaking labor of hoe cultivation may be due to the small areas put to fonio in this area, or it may represent adherence to tradition.

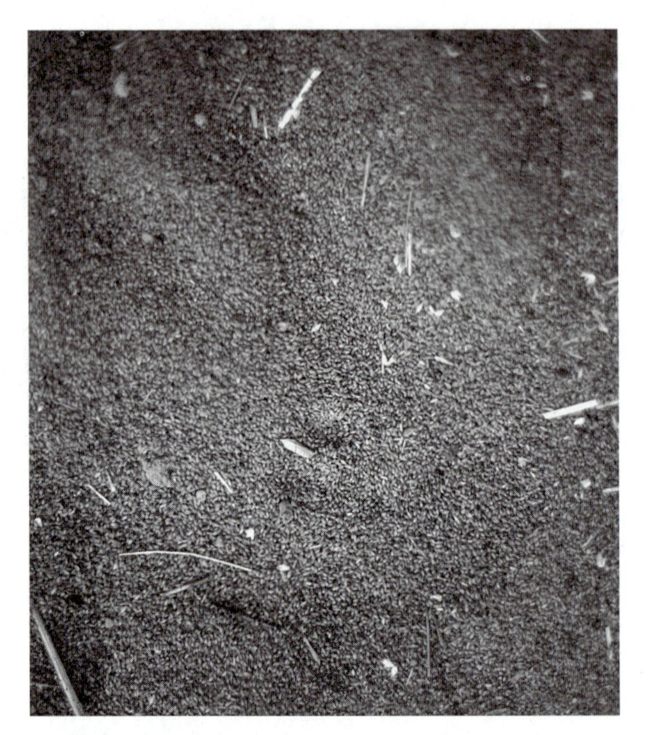

FIGURE 8-24 Fonio seeds. *Roy Cole.*

Africa's rural economy has always been labor scarce. This means that there were never enough hands to do the work that needed to be done for a family. The infant mortality rate was high, and life expectancy was low. Moreover, debilitating diseases such as malaria dramatically cut the pool of available family labor just as the rainy season began. Chronic labor shortages resulted in efforts by individuals, families, groups of families, and states to increase the pool of available workers by other means. The most pernicious manifestation of the drive to control labor for agriculture was African slavery. Typically, members of the owning family would work side by side in the fields with the captives. In addition, families would endeavor to capture or purchase young people, particularly women, to increase the size of the family and the hands available for farm work. Slave labor is rare in Africa today, and although it still exists it is against the law in every African country. In the farm economy, gangs of slave laborers and the hoe have been replaced by free labor and the plow.

Teff (*Eragrostis tef*) historically has been grown in only two countries in the world, Ethiopia and Eritrea

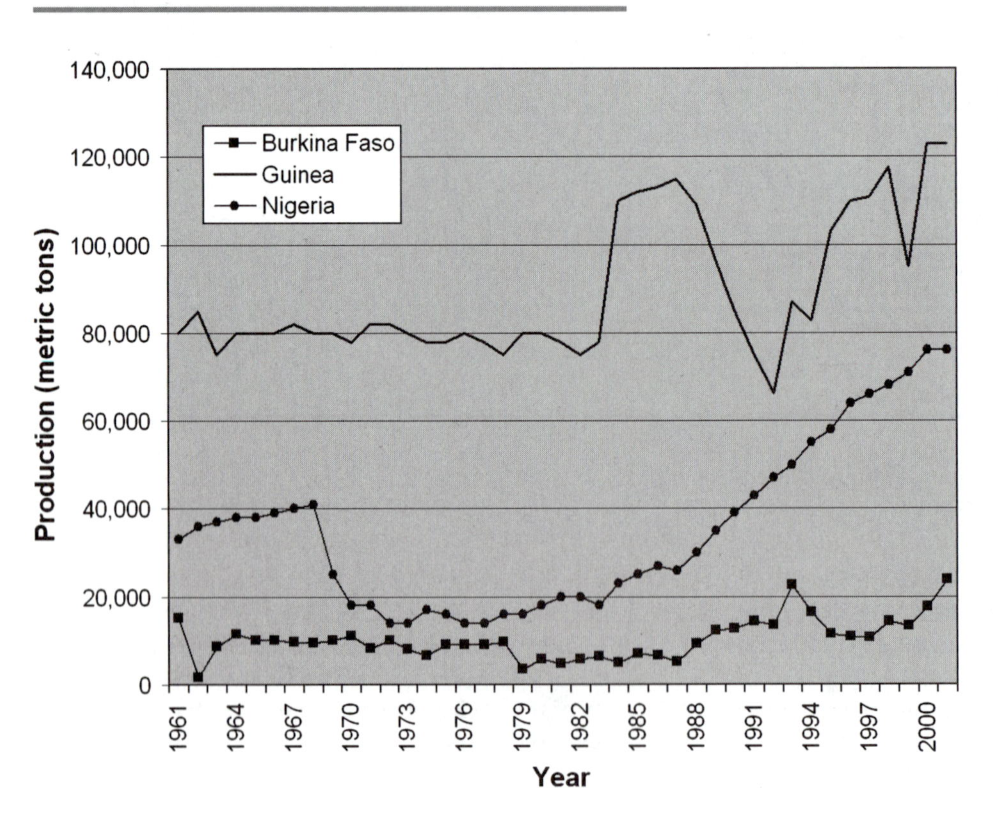

FIGURE 8-25 Fonio production for top producing countries, 1961 to 2001.

FIGURE 8-26 Broadcasting fonio seed in central Mali.
Roy Cole.

FIGURE 8-27 Working fonio seed into soil with a daba.
Roy Cole.

(Figure 8-23), but along with the Ethiopian diaspora that occurred as a result of drought and brutal dictatorships, the taste for teff spread to other parts of the world. Ethiopia produces by far the lion's share of teff. In Ethiopia it is grown more widely than any other crop, particularly in the western part of the country. Like fonio, teff is very drought resistant; unlike fonio, however, teff plays a major role in the daily diet of Ethiopians. Although many Eritreans eat sorghum in place of teff, most Ethiopians eat it at least once a day, usually in the form of a soft, spongy, round bread, *injera,* which can have the diameter of 2 feet (National Research Council 1996). During a meal, *njera* is dipped into meat sauce and functions as a sop to get the sauce to the mouth while keeping the hands clean. It is also used to pick out small chunks of meat from the sauce.

FIGURE 8-28 Maize drying in baobab tree after harvest, Côte d'Ivoire.
Christine Drake.

Maize

Maize (*Zea mays*), an import from the Americas, was produced by no less than 50 African countries in 2001, which demonstrates that the crop fills an ecological as well as a cultural niche (Figure 8-28) in the agricultural economies of Africans, especially south of the Sahara, where it is cultivated the most (Figure 8-29). If wheat and barley are associated with Middle Eastern influences, maize is associated with Europeans and the Americas. Europeans, who learned how to cultivate maize in the Americas, brought it to Africa during the colonial period, and it diffused across the continent.

Maize is more demanding of soil and water than millet and sorghum and can best be grown in the wetter parts of Africa, although agronomists have bred short-cycle, drought-resistant varieties that are suitable for drier areas. In most of West Africa, maize is grown in small garden plots, while in southern Africa it is grown as a field crop (Harlan 1993). In southern Africa it is made into cornmeal and eaten in a variety of ways, while in the drier fringes it is roasted on the cob. Maize has a lower protein content than either of the indigenous cereals, millet and sorghum, but generally has a higher yield (Table 8-14). Sorghum's protein content is highly variable (7–15%), and protein and starch availability is compromised by the presence of tannins, which are "locked up" unless the grain is milled, made into beer, or fermented in wood ash. Plant geneticists believe that African farmers selected varieties of sorghum with high tannin content because birds avoid these varieties. Perhaps these drawbacks made maize preferable to sorghum in suitable climates, but surely the influence of European settlers in Kenya, Zambia, Zimbabwe, and South Africa was important as well.

African maize production spread across the continent, with the top-producing South Africa, Egypt, and Nigeria anchoring the southern, northern, and western regions of the continent, respectively. Although Egypt devotes the smallest area to maize, it has the second highest production because of high yields (Table 8-14). African production was 6.93% of world

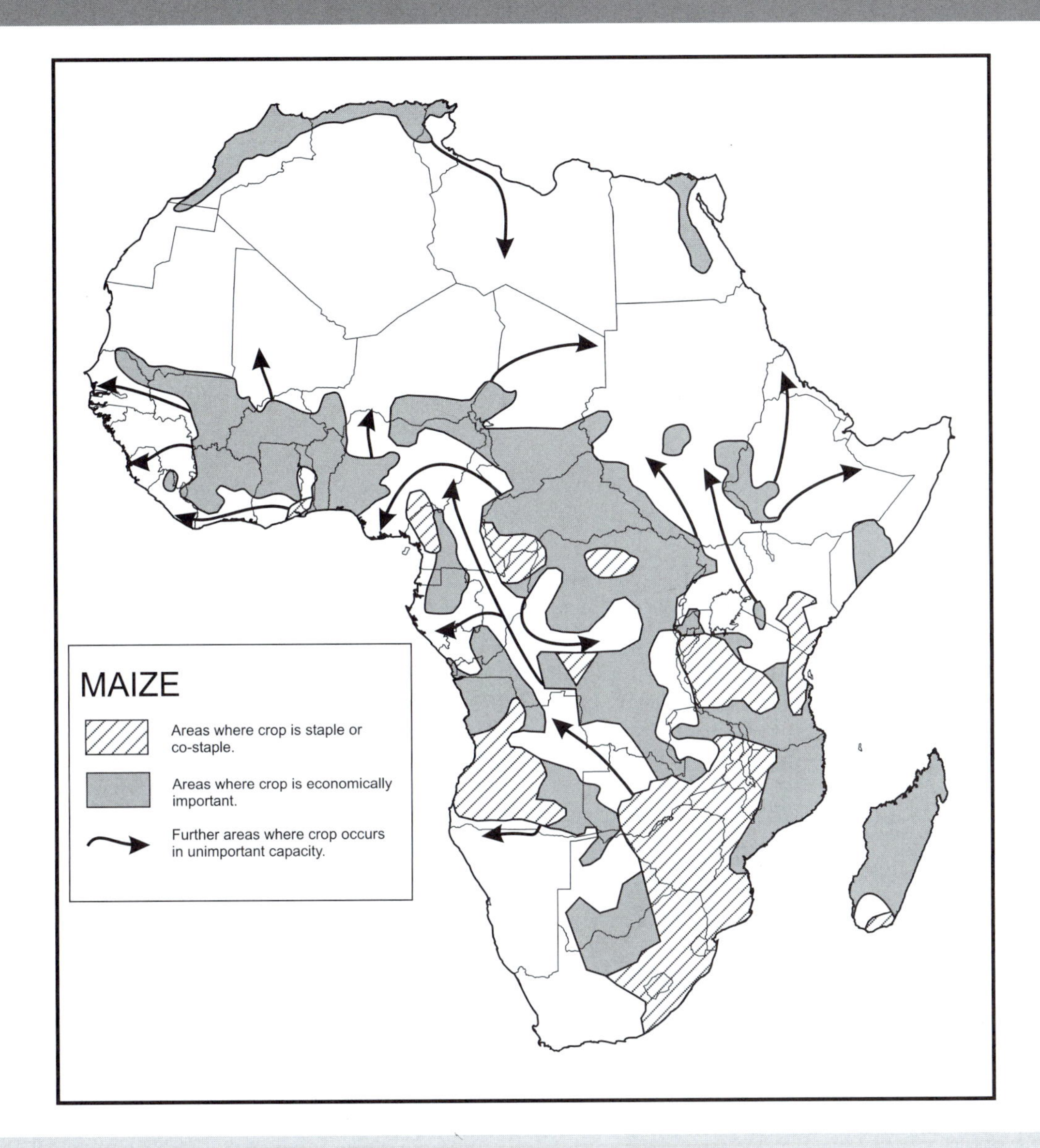

FIGURE 8-29 Distribution and importance of maize in Africa.
Adapted from Murdock (1960).

production in 2001, declining almost 12% from 1961 when it was 7.87%.

Yields from 1961 to 2001 remained relatively constant for all top-producing countries except Egypt (Figure 8-30). Note the fluctuation from year to year in South Africa, where maize is grown mainly without irrigation. Egypt is truly an oasis, with a control over water that other African countries envy.

Wheat and Barley

Wheat (*Triticum vulgare*) and barley (*Hordeum vulgare* and *H. distichium*) were domesticated in Southwest Asia and spread into Egypt, North Africa, and Ethiopia during the Neolithic. The ancient emmer wheat (*Triticum dicoccum*) is still cultivated in parts of Ethiopia today. Wheat cultivation is little established elsewhere except where bread-eaters settled. For example, South Africa,

TABLE 8-14 AREA HARVESTED, YIELD, AND PRODUCTION FOR AFRICA AND THE TOP 10 MAIZE-PRODUCING COUNTRIES IN AFRICA IN 2001.

Country	Area Harvested (ha)	Yield (kg/ha)	Production (metric tons)	Percentage of Total
Africa	25,372,233	1,663.2	42,200,324	100.0
South Africa	3,500,000	2,028.6	7,100,000	16.8
Egypt	840,000	7,678.6	6,450,000	15.3
Nigeria	3,999,000	1,399.8	5,598,000	13.3
Kenya	1,500,000	1,800.0	2,700,000	6.4
Tanzania	1,457,100	1,795.3	2,616,000	6.2
Ethiopia	1,450,000	1,724.1	2,500,000	5.9
Malawi	1,500,000	1,666.7	2,500,000	5.9
Zimbabwe	1,300,000	1,247.7	1,622,000	3.8
Uganda	652,000	1,800.6	1,174,000	2.8
Democratic Republic of the Congo	1,463,314	799.0	1,169,188	2.8

Source: FAO (2002).

which began to be colonized by Dutch and English settlers over 400 years ago, is the second largest producer of wheat after Morocco. The consumption of wheat bread is increasing in Africa south of the Sahara, especially in urban areas, where colonial Europeans built bakeries to make bread from imported flour. This tradition continued after independence.

Although Murdock (1960) mapped wheat as being grown in only 10 countries of Africa during the 1950s, it was being produced by 33 African countries in 2001. Only six countries produced more than a million metric tons, however, as can be inferred from Figure 8-31. In North Africa, Egypt, Morocco, Algeria, and Tunisia are the major producers; elsewhere,

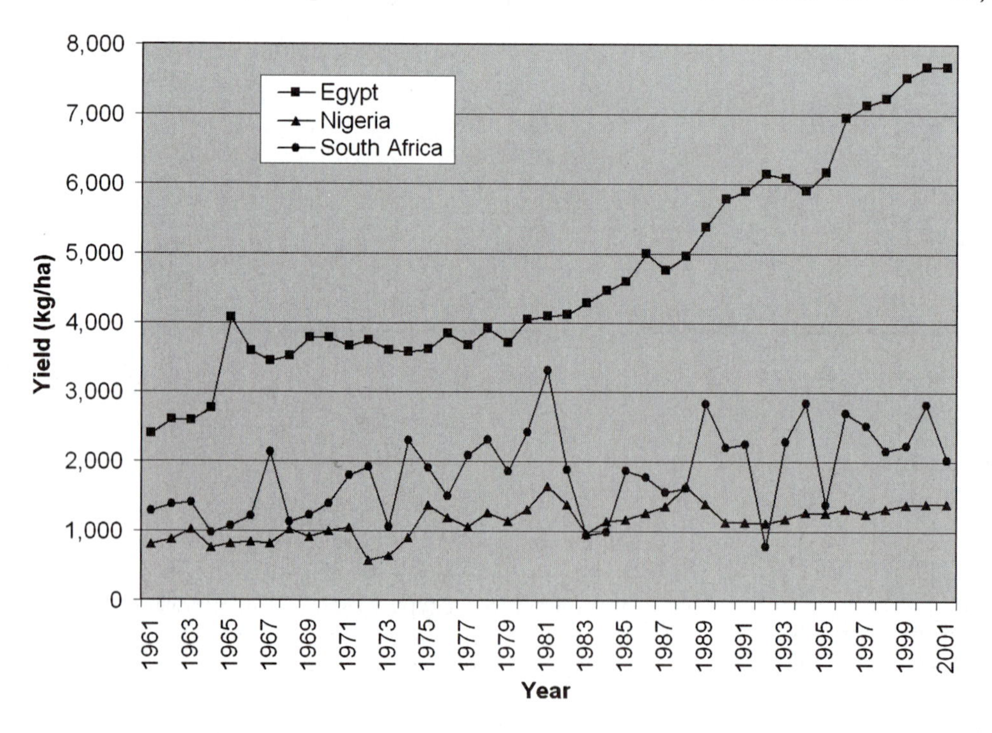

FIGURE 8-30 Maize yield for top African producers, 1961 to 2001.

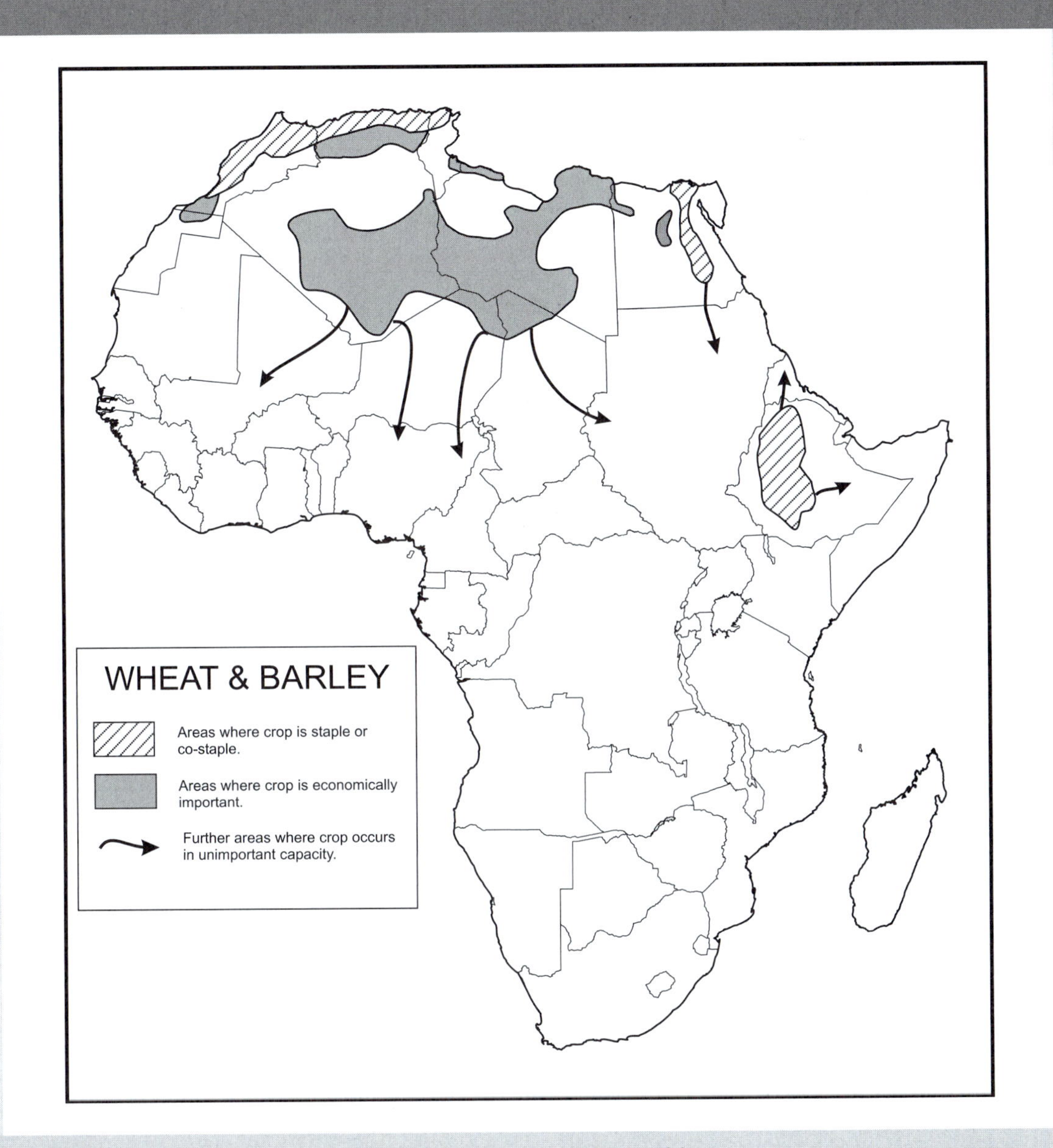

FIGURE 8-31 Distribution and importance of wheat and barley in Africa. *Adapted from Murdock (1960).*

South Africa and Ethiopia were significant (Table 8-15). Egypt produced 35% of the African production while the closest-ranking competitor in Africa, Morocco, produced just over half as much. African wheat production in 2001 was 3% of world production while in 1961 it was 2.3%.

African yields remained relatively constant over the 40-year period, at least as a percentage of the world average: 64 and 66% of the world average in 1961 and 2001, respectively. In actual fact, yields increased—but not for all countries. Egypt and Zimbabwe saw the greatest increase in yields from 1961 to 2001

TABLE 8-15 AREA HARVESTED, YIELD, AND PRODUCTION FOR AFRICA AND THE TOP 10 WHEAT-PRODUCING COUNTRIES IN AFRICA IN 2001.

Country	Area Harvested (ha)	Yield (kg/ha)	Production (metric tons)	Percentage of Total
Africa	9,949,963	1,804.6	17,956,095	100.0
Egypt	983,947	6,356.6	6,254,580	34.8
Morocco	2,700,600	1,228.0	3,316,380	18.5
South Africa	860,000	2,478.9	2,131,870	11.9
Algeria	2,400,000	825	1,980,000	11.0
Ethiopia	1,300,000	1,230.8	1,600,000	8.9
Tunisia	990,000	1,354.5	1,341,000	7.5
Sudan	136,920	2,439.4	334,000	1.9
Zimbabwe	36,000	6,944.4	250,000	1.4
Kenya	132,000	1,363.6	180,000	1.0
Libyan Arab Jamahiriya	165,000	787.9	130,000	0.7

Source: FAO (2002).

(Figure 8-32). Fluctuating yields in the late 1990s in Zimbabwe, where wheat is grown as a rain-fed crop, were due to drought.

Although only 16 African countries produced it in 2001 and production is significantly less than wheat, barley, another cereal of Middle Eastern origin, follows essentially the same geographic distribution as wheat. Morocco and Ethiopia are the major African producers, and Algeria and Tunisia produce considerably less (see Figure 8-31).

Yams

The yam (*Dioscorea alata, D. bulbifera, D. esculenta*) is a climbing perennial vine that produces a tuber. The tuber

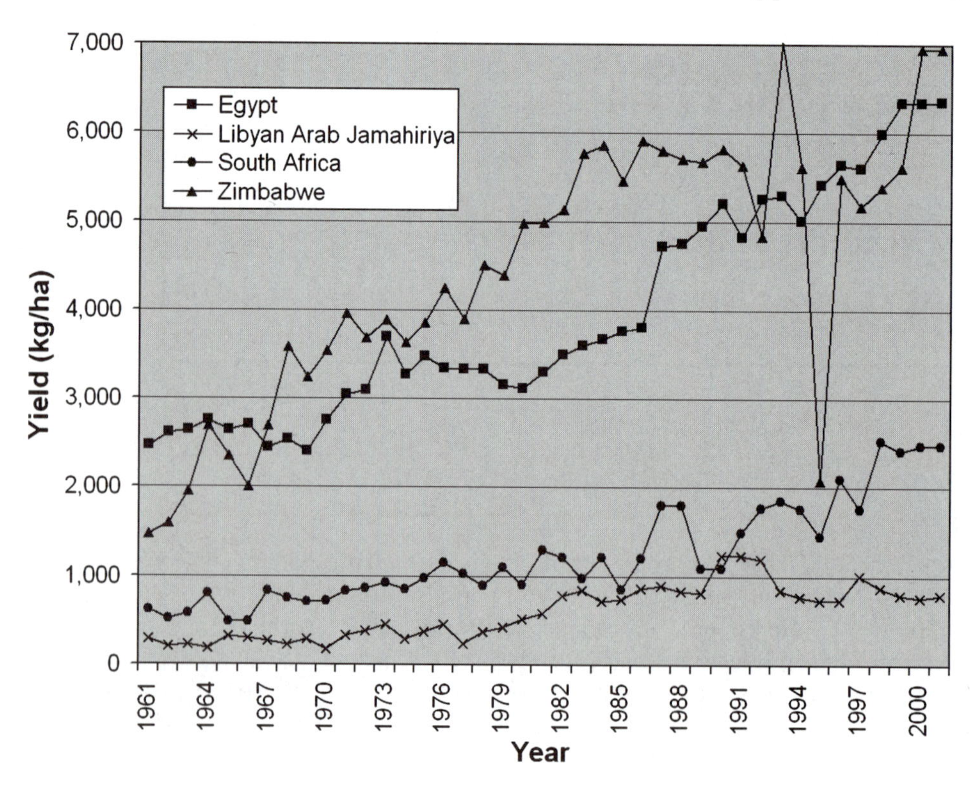

FIGURE 8-32 Wheat yields for major African producers, 1961 to 2001.

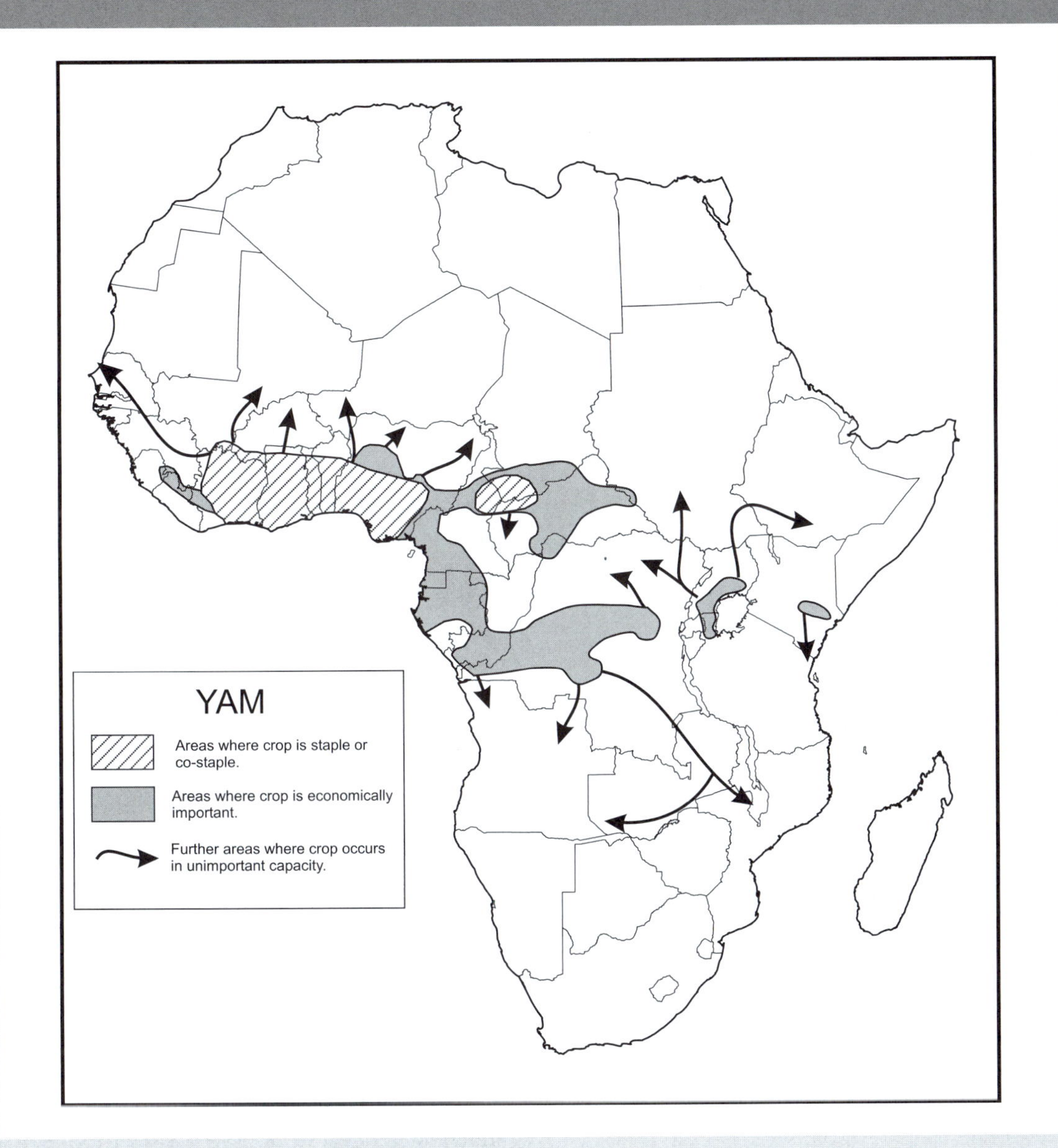

FIGURE 8-33 Distribution and importance of yams in Africa.
Adapted from Murdock (1960).

grows principally underground but sometimes appears on the vines. The African yam (*Dioscorea rotundata*) was domesticated in the forest–savanna zone of West Africa between 4,000 and 5,000 years ago (Okigbo 1980) but has been largely replaced by the three Asian varieties just mentioned, which were brought by the Portuguese in the early years of their involvement in Africa (Newman 1995). Leaves are distinctive and heart shaped. The underground tubers range in size from 3 to 10 pounds, but tubers 10 times as big are not uncommon (Stevens 1994). Ninety percent of the world's production of yams is cultivated in a belt that extends from Côte d'Ivoire in the west to central Africa in the East. The yam is also important in the equatorial area (Figure 8-33).

FIGURE 8-34 Yam mounds, Côte d'Ivoire. *Christine Drake.*

Yams are often intercropped with maize and vegetables in the yam belt. There are many constraints to the successful cultivation of yams. Yields of yams may be drastically reduced by weeds, and the seed yams are expensive, accounting for up to 50% of the cost of production (IITA no date). Growing yams is very labor intensive; indeed, labor can account for up to 40% of the cost of production. The yam mounds shown in Figure 8-34 suggest the amount of work required in cultivation and harvest. The yam vines must be staked for a good crop to be produced. Furthermore, the yam is difficult to store because, like the potato, it sprouts after a dormant period in storage; cereals may be stored for up to 5 years but yams only for a season. In addition, the yam (as well as other tubers) has a high moisture content, hence is susceptible to a variety of pests while stored.

Poisonous types of wild *Dioscorea*, sometimes eaten as a famine food, must be "washed" to remove the toxins. The preparation of domestic yams as a food involves the "boiling, peeling, slicing and pounding, and sometimes also steeping them in running or preferably salt water, a detoxification process which generally requires about three days" (Andah 1993: 247).

The data in Table 8-16 demonstrate that there truly is a yam belt in Africa extending along the Gulf of Guinea from Nigeria to Benin, where 93% of Africa's production was concentrated in 2001. Nigeria alone accounts for over 70% of Africa's yam production.

The pattern in 1961, the earliest year for which there are data, was similar to 2001 at least in terms of the top 10 countries. The volume of production did change dramatically over the 40-year period, particularly for Nigeria, which increased production over sevenfold. The Democratic Republic of the Congo more than quadrupled its production, although its output was rather modest compared with the top-producing countries. Ghana and Côte d'Ivoire tripled their production. An explanation for the prodigious increase in some countries' production of yams can be had from comparing Tables 8-16 and 8-17. In Ghana, area harvested increased by over 100,000 ha and yield per hectare doubled, while in Nigeria the area in which yams were grown was expanded so massively that its share of Africa's total output increased from 46% in 1961 to 71% in 2001. In a few countries yields actually decreased, but area was expanded during the time period.

Cassava and Ensete

Cassava, or manioc (*Manihot esculenta*), is a starchy South American tuber that was brought to Africa

TABLE 8-16 AREA HARVESTED, YIELD, AND PRODUCTION FOR AFRICA AND THE TOP 10 YAM-PRODUCING COUNTRIES IN AFRICA IN 2001.

Country	Area Harvested (ha)	Yield (kg/ha)	Production (metric tons)	Percentage of Total
Africa	3,908,417	9,499	37,123,928	100.0
Nigeria	2,742,000	9,555	26,201,000	70.6
Ghana	261,000	14,061	3,670,000	9.9
Côte d'Ivoire	350,000	8,571	3,000,000	8.1
Benin	155,000	11,441	1,773,363	4.8
Togo	51,220	10,997	563,286	1.5
Central African Republic	53,000	6,793	360,000	1.0
Ethiopia	68,000	3,971	270,000	0.7
Cameroon	58,000	4,483	260,000	0.7
Democratic Republic of the Congo	39,000	6,539	255,000	0.7
Chad	24,000	9,583	230,000	0.6

Source: FAO (2002).

TABLE 8-17 YAM AREA HARVESTED, YIELD, AND PRODUCTION FOR AFRICA AND THE TOP 10 PRODUCERS IN 1961.

Country	Area Harvested (ha)	Yield (kg/ha)	Production (metric tons)	Percentage of Total
Africa	1,058,903	7,166	7,587,702	100.0
Nigeria	450,000	7,778	3,500,000	46.1
Côte d'Ivoire	150,000	7,667	1,150,000	15.2
Ghana	150,000	7,333	1,100,000	14.5
Benin	69,138	8,883	614,172	8.1
Togo	30,000	10,000	300,000	4.0
Ethiopia (includes Eritrea)	54,000	3,648	197,000	2.6
Cameroon	36,000	4,167	150,000	2.0
Chad	14,500	7,586	110,000	1.5
Central African Republic	20,000	5,000	100,000	1.3
Democratic Republic of the Congo	8,000	7,500	60,000	0.8

Source: FAO (2002).

by the Portuguese centuries ago. It gradually diffused and is firmly established in equatorial Africa and Madagascar, and is important throughout West and East Africa. Cassava will grow in a variety of soils, including those in which cereals and other crops do poorly. Compared with cereals, cassava yields are very high, although protein content is low. In regions of Africa where tubers are the principal food consumed, children sometimes suffer protein-calorie malnutrition (kwashiorkor); they receive starchy calories (carbohydrates) but little protein (Meade and Earickson 2000). An advantage of cassava over other tubers is that it can be left in the ground for 2 years or more rather than being immediately harvested when ripe, allowing people to balance food production and consumption throughout the year. Cassava is cultivated widely in tropical Africa, from the equatorial region to the drylands (Figure 8-35).

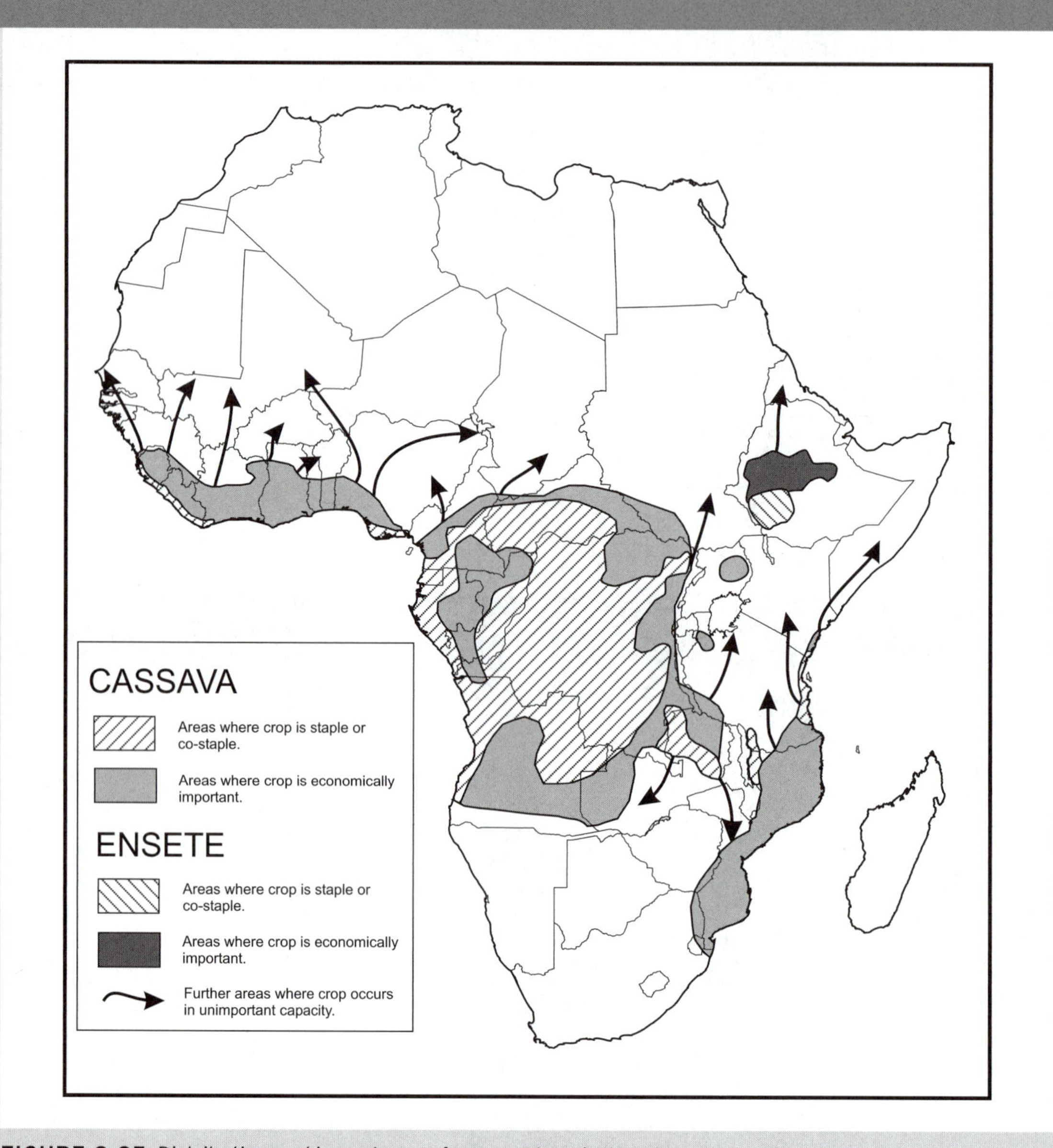

FIGURE 8-35 Distribution and importance of cassava (manioc) and ensete in Africa. *Adapted from Murdock (1960).*

Africa today produces about 53% of the world's cassava, mainly in Nigeria, the Democratic Republic of the Congo, and Ghana (Table 8-18). In 1961 Africa produced 44% of the world's cassava.

Ensete (*Ensete edulus*), sometimes called the Ethiopian banana or false banana, is not a banana at all. Although the vegetative part of the plant looks like a banana plant, it is the tuber which is harvested and eaten. Ensete is a staple crop in southern Ethiopia only. The United Nations Food and Agricultural Organization does not collect data on this geographically limited crop, and there are few figures on its production.

Date Palm and Banana

The data palm (*Phoenix dactylifera*) came to Africa from southwestern Asia and is a primary crop in North Africa.

TABLE 8-18 AREA HARVESTED, YIELD, AND PRODUCTION FOR AFRICA AND THE TOP 10 CASSAVA-PRODUCING COUNTRIES IN AFRICA IN 2001.

Country	Area Harvested (ha)	Yield (kg/ha)	Production (metric tons)	Percentage of Total
Africa	10,894,383	8,742	95,238,579	100.0
Nigeria	3,135,000	10,799	33,854,000	35.5
Democratic Republic of the Congo	1,902,359	8,114	15,435,700	16.2
Ghana	600,000	14,187	8,512,000	8.9
Tanzania	761,100	7,424	5,650,000	5.9
Mozambique	925,902	5,791	5,361,974	5.6
Uganda	390,000	13,500	5,265,000	5.5
Angola	530,000	6,226	3,300,000	3.5
Benin	260,000	10,769	2,800,000	2.9
Madagascar	349,750	6,370	2,228,000	2.3
Côte d'Ivoire	330,000	5,758	1,900,000	2.0

Source: FAO (2002).

Although Figure 8-36 seems to indicate that it is widely present in the central Sahara, it is only grown in oases. The date was one of the earliest fruits in trans-Saharan trade, where millet, captives, kola, and other forest products from the south were traded for salt and dates from the north (Curtin 1984). The date is still commonly traded across the Sahara to the Subsaharan countries, but it is transported more by truck today than by camel. The date palm needs temperatures around 104° F (40° C) and plenty of water for the fruit to mature properly. Irrigation rather than rainfall is what the date requires; rainfall can actually damage the fruit. Curtin (1984) called the date the perfect commodity for long-distance trade because, although it is prized for its high sugar content, one cannot subsist on dates alone. It is a natural candidate for long-distance exchange for cereals but could not traded without transportation.

The camel, introduced into Egypt around 1,800 years ago and subsequently spreading across the Saharan region, provided a most efficient pack animal. The rise of the trans-Saharan trade, West Africa's "first and most important outside contact" (Curtin 1984: 21), was made possible by the camel and by the existence of tradable commodities (dates and cereals). A commodity is tradable when it is abundant in one place but scarce (and desired) in another. When that second region produces a product that is needed in the first region, complementarity is said to exist for those products. The trade that existed (and still exists, although the camel has mostly been replaced by the truck) between the date-producing oases of the Sahara and the cereal-producing regions in North Africa or Subsaharan Africa is a classic textbook case of regional complementarity. Africa produced 37% of the world date crop in 2001, down from 45% in 1961. In 2001, Egypt and Algeria accounted for almost 75% of African production (Table 8-19). Egypt produced about three times as many dates as Algeria on one-third the land.

Bananas and Plantains

The banana and plantain (*Musa* spp.) were introduced to Africa from Asia over a thousand years ago. It is still unclear where the banana was first domesticated, India or Southeast Asia. Nevertheless, the first hard evidence for the existence of the banana in Africa is a text and sketch dating to 525 CE of the plant and fruit, found at Adulis, a former trading town associated with Cush, located on the Red Sea coast of present-day Eritrea (Vansina 1990). We do not know how the banana entered Africa exactly but there are two scenarios. The first scenario has the banana being brought by Indonesians, who colonized the island of Madagascar and probably parts of the East African coast. It is thought that they brought to Africa a variety of other crops: for example, cocoyams (taro), mangoes, perhaps sweet potatoes, and even the chicken. The second and more recently developed hypothesis suggests that the banana diffused

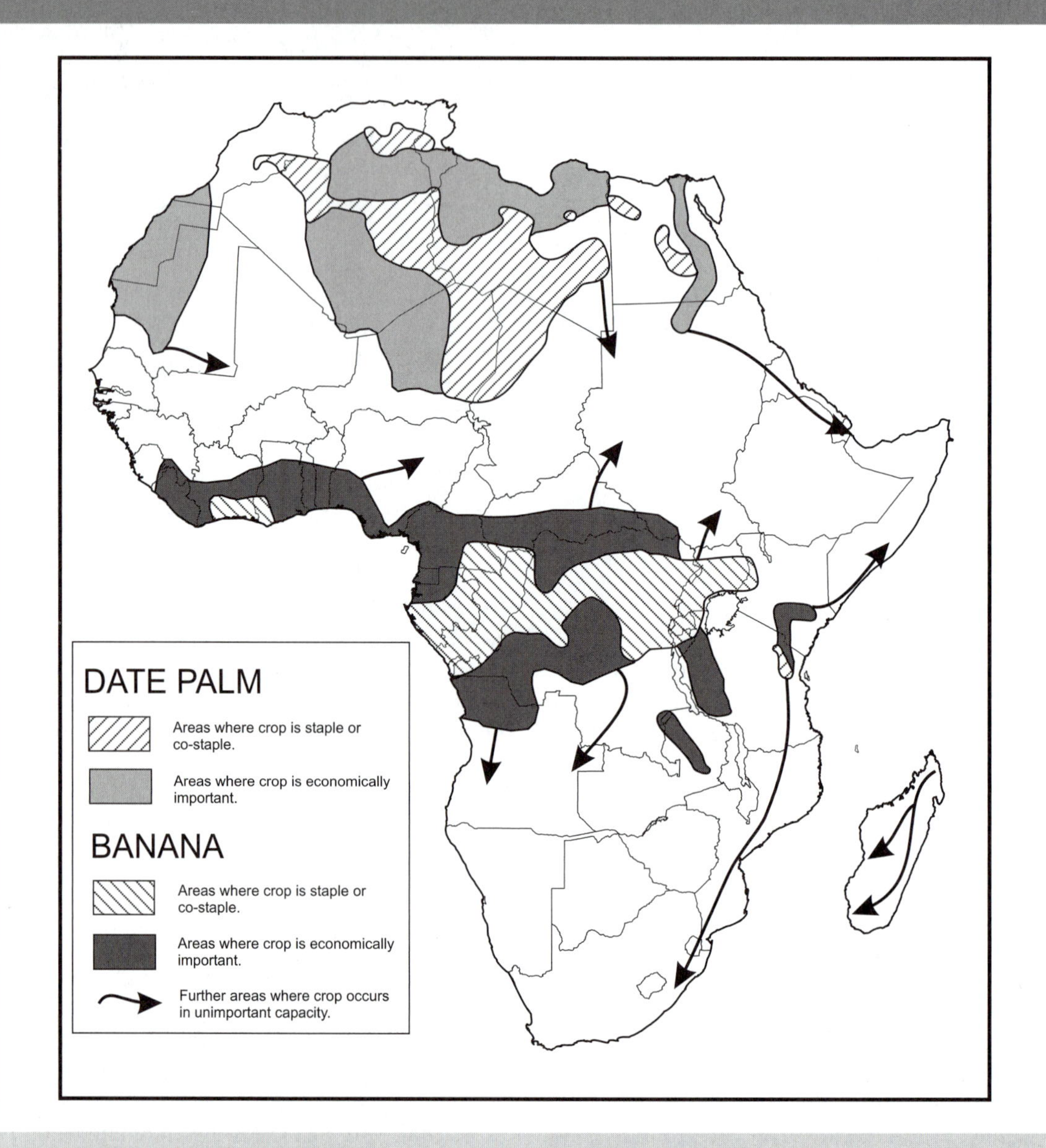

FIGURE 8-36 Distribution and importance of the date palm and banana in Africa.
Adapted from Murdock (1960).

from the Upper Nile Valley as well, probably very slowly from patch to patch (Vansina 1990).

In relation to the yam, an early stable of Bantu farmers, the banana (banana and plantain) increased agricultural production where it was adopted by a factor of 10 and replaced the yam in many areas where both crops could have been planted. In addition, the banana can be grown with far less labor than is needed for the yam. Vansina calls it the "ideal staple crop for the rain forests," which allowed farmers to extend their colonization of the equatorial rain forest to its very edges. The banana is sweet and usually eaten raw, while the plantain has a low sugar content, tastes starchy, and is cooked. Almost three times as many plantains are produced in African than bananas, and the geographic distribution differs for each crop.

TABLE 8-19 AREA HARVESTED, YIELD, AND PRODUCTION FOR AFRICA AND THE TOP 10 DATE-PRODUCING COUNTRIES IN AFRICA IN 2001.

Country	Area Harvested (ha)	Yield (kg/ha)	Production (metric tons)	Percentage of Total
Africa	275,635	7,188	1,981,260	100.0
Egypt	32,000	34,448	1,102,350	55.6
Algeria	100,000	3,700	370,000	18.7
Sudan	19,500	9,077	177,000	8.9
Libyan Arab Jamahiriya	26,500	5,000	132,500	6.7
Tunisia	30,000	3,567	107,000	5.4
Morocco	46,650	695	32,400	1.6
Mauritania	8,000	2,750	22,000	1.1
Chad	7,600	2,368	18,000	0.9
Somalia	2,400	4,167	10,000	0.5
Niger	2,200	3,455	7,600	0.4

Source: FAO (2002).

Plantains are strictly a Subsaharan crop produced in only 18 Subsaharan countries. Almost 60% of the 2001 plantain output was from East Africa and about 30% from West Africa (Table 8-20). Bananas, on the other hand, were produced in 37 African counties: 45% of the production comes from East Africa, 11% from North Africa, principally Egypt (Table 8-21). In East Africa, the cultivation of plantain is most intense in the Lake Victoria area of Uganda and Tanzania. Here, the introduction of the plantain and the banana enabled a high population density to be supported. The general pattern may be observed in Figure 8-36. Bananas and plantains grow in shoots, each shoot producing only one hand of fruit. The entire plant, however, produces about 8 hands of 15 bananas each. Average yield is 44 pounds (20 kg) per plant. Domesticated bananas and plantains have no seeds; propagation is by cuttings, which are planted.

The banana is the world's second most traded fruit after citrus. World production in 2001 was 69 million

TABLE 8-20 AREA HARVESTED, YIELD, AND PRODUCTION FOR AFRICA AND THE TOP 10 PLANTAIN-PRODUCING COUNTRIES IN AFRICA IN 2001.

Country	Area Harvested (ha)	Yield (kg/ha)	Production (metric tons)	Percentage of Total
Africa	3,690,135	5,577	20,578,578	100.0
Uganda	1,598,000	5,966	9,533,000	46.3
Ghana	244,400	7,905	1,932,000	9.4
Nigeria	281,400	6,759	1,902,000	9.2
Rwanda	357,080	4,404	1,572,661	7.6
Côte d'Ivoire	400,000	3,750	1,500,000	7.3
Cameroon	255,000	5,490	1,400,000	6.8
Tanzania	85,000	7,675	652,378	3.2
Democratic Republic of the Congo	123,155	4,277	526,735	2.6
Guinea	83,000	5,180	430,000	2.1
Kenya	85,000	4,353	370,000	1.8

Source: FAO (2002).

TABLE 8-21 AREA HARVESTED, YIELD, AND PRODUCTION FOR AFRICA AND THE TOP 10 BANANA-PRODUCING COUNTRIES IN AFRICA IN 2001.

Country	Area Harvested (ha)	Yield (kg/ha)	Production (metric tons)	Percentage of Total
Africa	1,028,290	7,485	7,697,121	100.0
Burundi	300,000	5,163	1,548,897	20.1
Uganda	162,200	6,000	973,200	12.6
Cameroon	57,000	14,912	850,000	11.0
Tanzania	85,000	9,415	800,300	10.4
Egypt	22,734	32,374	735,999	9.6
Democratic Republic of the Congo	83,674	3,737	312,690	4.1
Angola	31,000	9,677	300,000	3.9
South Africa	16,000	16,752	268,026	3.5
Madagascar	45,000	5,778	260,000	3.4
Côte d'Ivoire	16,500	13,152	217,000	2.8

Source: FAO (2002).

metric tons, of which Africa produced almost 8 million, or 11%. On the other hand, Africans produce the vast majority of the world's plantains; in 2001 Africa produced 71% of the 29,120,888 metric tons of world output. What is surprising about the data is the small role that Madagascar, the probable point of origin, plays in both banana and plantain production.

Stimulants (Coffee, Kola, Qat, Tea, Tobacco)

Coffee (*Coffea arabica, C. robusta*) is grown in 28 African countries, but 8 countries produce almost 85% of the crop. The top producers, the Côte d'Ivoire, Ethiopia, and Uganda, grew 60% of the crop in 2001 (Table 8-22). Coffee was first domesticated in southwest Ethiopia and was brought to the Middle East by the Turks in the fifteenth century. Until modern times coffee was consumed in the same way as qat, by chewing (Purseglove 1976). In the sixteenth century it began to diffuse to other parts of the world and eventually became a global product. As a percentage of world production, Africa's share is small: only 16% in 2001, down from almost 25% in 1961. Over the 40-year time period Africa's

TABLE 8-22 AREA HARVESTED, YIELD, AND PRODUCTION FOR AFRICA AND THE TOP 10 COFFEE-PRODUCING COUNTRIES IN AFRICA IN 2001.

Country	Area Harvested (ha)	Yield (kg/ha)	Production (metric tons)	Percentage of Total
Africa	2,786,828	416.3	1,160,114	100.0
Côte d'Ivoire	1,000,000	280.0	280,000	24.1
Ethiopia	250,000	912.0	228,000	19.7
Uganda	264,000	747.8	197,410	17.0
Cameroon	300,000	256.7	77,000	6.6
Kenya	165,000	454.5	75,000	6.5
Madagascar	193,125	332.4	64,200	5.5
Tanzania	118,000	492.4	58,100	5.0
Democratic Republic of the Congo	145,000	254.3	36,870	3.2
Burundi	23,000	1,087.0	25,000	2.2
Guinea	50,000	410.0	20,500	1.8

Source: FAO (2002).

TABLE 8-23 AREA HARVESTED, YIELD, AND PRODUCTION FOR AFRICA AND THE TOP KOLA NUT PRODUCING COUNTRIES IN AFRICA IN 2001.

Country	Area Harvested (ha)	Yield (kg/ha)	Production (metric tons)	Percentage of Total
Africa	382,800	568.2	217,500	100.0
Nigeria	91,000	901.1	82,000	37.7
Côte d'Ivoire	110,000	681.8	75,000	34.5
Cameroon	95,000	378.9	36,000	16.6
Ghana	70,000	285.7	20,000	9.2
Sierra Leone	15,000	266.7	4,000	1.8
Benin	1,800	277.8	500	0.2

Source: FAO (2002).

production has remained relatively constant, while other countries have increased their output.

Arabica coffee needs at least 31.5 inches (800 mm) of rainfall and grows best between 1,500 and 2,300 m (4,921–7,546 ft) at temperatures between 66 and 73° F (19–23° C). Arabica is the variety of coffee grown most often commercially, while robusta is more often seen on small holdings. Robusta coffee can tolerate higher temperatures than arabica (20–26° C), grows at a lower elevation (1,000–1,500 m; 3,281–4,922 ft) but needs more rainfall (1,000 mm minimum). Commercial coffee plantations generally wet-process coffee: the cherry is fermented to remove pulp, the bean is sun-dried, and then roasted. After roasting, the beans are graded and packaged. Arabica is dry-processed, and pulp is removed by machine; then the beans are roasted, graded, and packaged for shipment.

Kola nuts (*Cola acuminata* and *C. nitida*) are produced in only a handful of countries, all of which are in West Africa (Table 8-23). The kola nut, actually a seed, contains caffeine and is used as a stimulant, is widely traded in West Africa, and plays an important social role as a token of respect or esteem from one person to another. The nuts are often used as the first step in negotiations of any kind: for example, the first gift that opens marriage talks between a suitor and the parents of his intended wife. Kola nuts are consumed socially at almost any time, but because chewers have conspicuous and unsightly reddened teeth and gums and must spit regularly, these seeds have a greater role to play in traditional and rural culture than in modern, professional life. Only six countries produced kola nuts in 2001 according to the Food and Agriculture Organization of the United Nations (2002). Nigeria and Côte d'Ivoire account for over 72% of the production alone (Table 8-23). Historically, kola was part of the trade in gold and captives between the forest lands of West Africa and the north, where they were traded for dry land and desert-side products such as salt, cereals, and dates. Kola has been used in West Africa since ancient times; it is significant that in Bambara, one of the Manding languages of central Mali, the word for the cardinal direction "south" is *worodugu*, "land of the kola nut."

Qat (*Catha edulis*) is a stimulant that contains the natural amphetamine cathinone. It was domesticated in Ethiopia, and cultivation spread to neighboring countries. Today qat chewers may be found in Kenya, Malawi, Uganda, Tanzania, Saudi Arabia, the Congo, Zimbabwe, Zambia, Madagascar, South Africa, and Yemen. It is the national drug of Yemen, where thousands of Yemenis take an afternoon "chewing" break. The drug has diffused to Europe with the migration of Yemenis. Detractors criticize the the qat habit as a waste of money and a direct cause of low productivity. It is estimated that qat chewers spend up to one-third of their annual income on the purchase of qat leaves.

> Qat sessions are a major part of Yemeni life: participants regard the time spent chewing qat as productive time, when business deals are arranged, communication is strengthened and verbal skills are improved. Information and ideas are easily exchanged, and culturally "desirable behavior" is reinforced. In general, women and men hold qat sessions separately. Women's qat sessions, with dancing and music, are often more lively than men's. (Maran Brooks 1998: 1)

TABLE 8-24 AREA HARVESTED, YIELD, AND PRODUCTION FOR AFRICA AND THE TOP 10 TEA-PRODUCING COUNTRIES IN AFRICA IN 2001.

Country	Area Harvested (ha)	Yield (kg/ha)	Production (metric tons)	Percentage of Total
Africa	215,102	1,958.3	421,225	100.0
Kenya	113,000	2,123.9	240,000	57.0
Malawi	19,000	2,368.4	45,000	10.7
Uganda	15,761	2,084.7	32,857	7.8
Tanzania	19,000	1,342.1	25,500	6.1
Zimbabwe	6,000	3,666.7	22,000	5.2
Rwanda	12,000	1,250.0	15,000	3.6
South Africa	6,700	1,783.7	11,951	2.8
Mozambique	5,631	1,858.6	10,466	2.5
Burundi	8,900	993.3	8,840	2.1
Cameroon	1,550	2,903.2	4,500	1.1

Source: FAO (2002).

Supporters of qat point to the contribution it makes to the economy of the exporting countries. It is Ethiopia's fastest growing export and is grown on close to 100,000 hectares (247,100 acres). The drug is illegal in the United States.

Tea (*Thea sinensis*) was domesticated in China and was introduced to other parts of the world by Arab traders and most recently by Europeans. It is the national drink in many countries around the world. The United Kingdom, former colonial power in Africa, encouraged cultivation in Africa and is one of the largest markets for African tea. Tea is grown in 18 African countries today, but its production is insignificant except in a handful. Kenya, a former British settler colony, produced almost 60% of the total Africa output in 2001 (Table 8-24).

Africa's share of world tea production more than doubled from the 1960s to the present. Average annual production in Africa represented only 6% of world production from 1961 to 1965 but 13% for the 1997–2001 period. Production of tea in Africa has changed dramatically over the last 40 years, particularly among the more market-oriented economies such as Kenya. Figure 8-37 shows the remarkable growth of this crop from 1961 to 2001 for the African countries (all former British colonies) that produced more than 5% of African output in 2001.

Tea grows best at altitudes between 1,000 and 2,200 m (3,281–7,218 ft), preferring rainfall between 1,600 and 2,300 mm (5,250–7,550 ft) and deep, well-drained acid soils. Leaves are plucked when the bush is between 3 and 4 years old, and harvesting of leaves is done throughout the year. In East Africa, tea bushes are pruned to waist level to form a "plucking table" (Hickman 1990) and are plucked for up to 50 years.

Tobacco (*Nicotiana tabacum*) is a widely consumed, addictive leaf that originated in the Americas and was brought across the Atlantic by Columbus and other Europeans. It was grown by 38 African countries in 2001, but Zimbabwe and Malawi accounted for most of the output and exports (Table 8-25). African tobacco production accounted for 5.4% of world production on average for 1961 to 1965 and increased 226% from 1961–1965 to 1997–2001. For the same time period, its share of world tobacco production increased by 7%. Although most producers doubled their output over the 40-year period, Kenya and Malawi had the greatest relative change (Figure 8-38).

Tobacco is a soil-exhausting crop and is generally grown in only one year in five on a particular plot of land. According to Hickman (1990), a tobacco plantation needs about 200 hectares (494 acres) of reserve land for every 50 hectares (125 acres) or so in tobacco. Tobacco is a capital-intensive crop requiring expensive machinery for heavy agricultural operations that, at the same time, demand a lot of labor. Labor costs are high because leaves must be picked by hand, and the

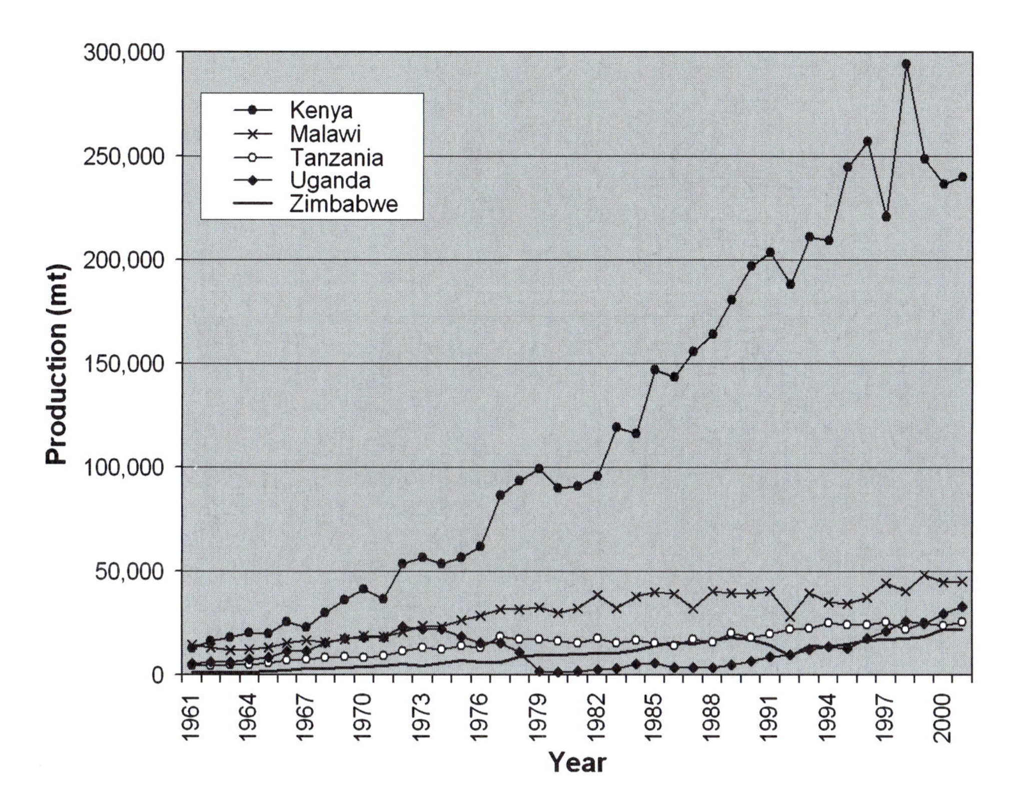

FIGURE 8-37 Growth in tea production for major African producers, 1961 to 2001.

overall cost of producing tobacco is high because hybrid seeds, fertilizer, insecticides, and herbicides must be purchased to ensure high-quality production. Tobacco must be cured with heat or air-dried, processed, stored, packed, and shipped to consumers. The income from tobacco, although variable from year to year, is five times higher than maize and twice to three times that of cotton or peanuts (Hickman 1990). Tobacco grows best at temperatures between 13 and 27° C (55–81° F) and prefers an altitude between 900

TABLE 8-25 AREA HARVESTED, YIELD, AND PRODUCTION FOR AFRICA AND THE TOP 10 TOBACCO-PRODUCING COUNTRIES IN AFRICA IN 2001.

Country	Area Harvested (ha)	Yield (kg/ha)	Production (metric tons)	Percentage of Total
Africa	383,257	1,189.4	455,853	100.0
Zimbabwe	81,310	2,156.4	175,335	38.5
Malawi	120,000	833.3	100,000	21.9
South Africa	14,100	2,106.4	29,700	6.5
Tanzania	35,500	683.7	24,270	5.3
Uganda	9,500	2,376.0	22,572	5.0
Kenya	14,500	1,241.4	18,000	3.9
Côte d'Ivoire	20,000	600.0	12,000	2.6
Mozambique	7,000	1,352.9	9,470	2.1
Nigeria	22,000	418.2	9,200	2.0
Algeria	5,700	1,263.2	7,200	1.6

Source: FAO (2002).

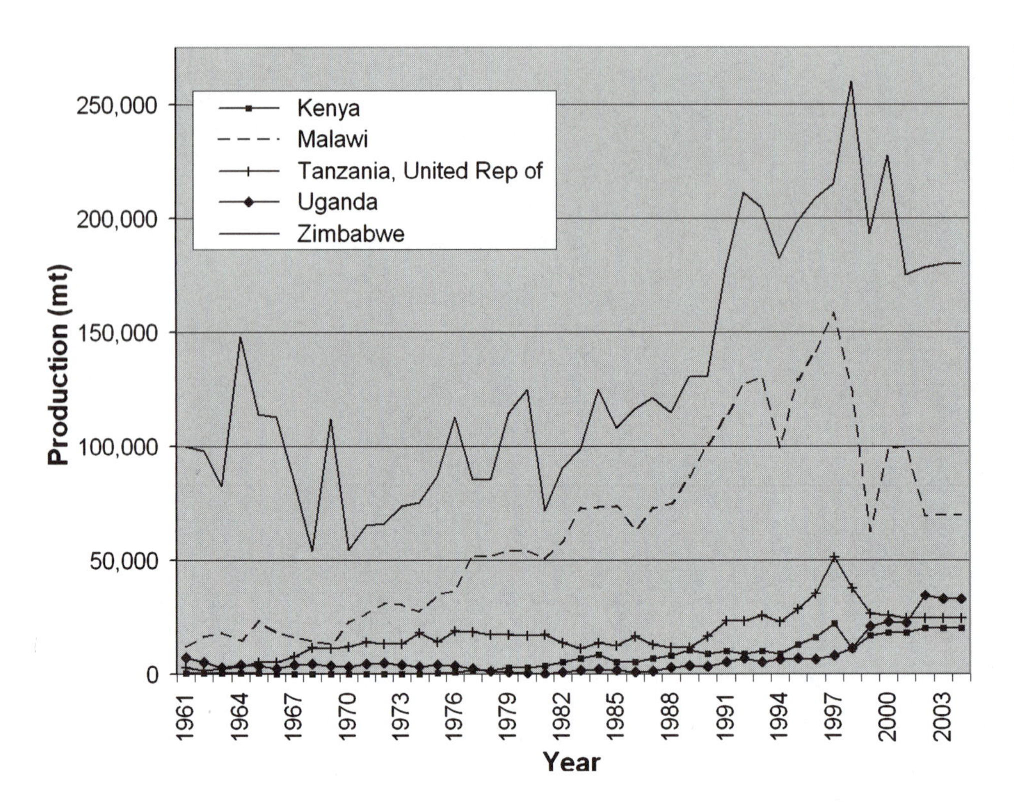

FIGURE 8-38 Tobacco production for major African producers, 1961 to 2004.

and 1,500 m (2,953–4,921 ft). Rainfall should be at least 650 mm (26 in.).

Tobacco is grown on estates in Zimbabwe and Malawi but in small farms elsewhere in East Africa. As part of its land reform program, the government of Zimbabwe has encouraged small farmers to cultivate tobacco by buying large white-owned estates and resettling black Zimbabweans on small farms carved out of the estates. A cooperative system is used to develop economies of scale in production and marketing. A cooperative is a group of farmers who share the costs and use of expensive farm equipment that single individuals would not be able to afford. A cooperative may market its produce as a group to get higher prices.

Cotton

Cotton (*Gossypium* spp.) is a crop that has long been grown in Africa. Arnon (1972) believes that cotton originated in equatorial Africa when its climate was much different from today. Cotton is a perennial shrub that does not tolerate cold at all.

Although cotton is indigenous to Africa, it has been a crop that colonial and local powers encouraged, even forced, Africans to grow. The importance of securing a reliable source of cotton was a point made clear to Britain and France when cotton exports from the southern states of the United States were disrupted from 1861 to 1865 during the American Civil War (Roberts 1996). The British, French, and Turko-Egyptians set up cotton schemes at the confluence of the Blue and White Nile rivers in the Sudan, the Niger Inland Delta in Mali, and in eastern Sudan (Gash and Tokar deltas), respectively. Muhammad Ali, pasha of Egypt in the early nineteenth century, embarked on an agricultural modernization program that involved the expansion of Egyptian long-staple cotton, the finest cotton in the world. In 1825 cotton cultivation was extended to the Sudan's Tokar Delta by Egypt's Turko-Egyptian rulers. In Banamba and Sokoto, West Africa, cotton was grown on plantations that were remarkably similar to cotton plantations in the southern United States in the early 1800s, especially in their use of captive workers, owned by aristocratic elites, as the principal source of labor. Encouraged by the windfall profits made by exporting cotton during the U.S. Civil War, Muhammad Ali's successors in Egypt invested heavily in the expansion of cotton.

Cotton cultivation has expanded greatly in Africa since independence. Newly independent developing

FIGURE 8-39 Smallholder cotton drying in the field prior to harvest.
Roy Cole.

countries of Africa saw cotton as a crop that could return the high investment costs of irrigation. The textile industry was also viewed as a necessary first step in the industrialization of developing countries (Arnon 1972). In many countries cotton is grown by smallholders (Figures 8-39 and 8-40). The top 10 cotton-producing countries of Africa account for 78% of African output. Egypt is by far the most efficient producer. Its yield per hectare is twice that of the other cotton producers in Africa (Table 8-26). Cotton is both

FIGURE 8-40 Smallholder cotton collection point for the Malian cotton parastatal.
Roy Cole.

TABLE 8-26 AREA HARVESTED, YIELD, AND PRODUCTION FOR AFRICA AND THE TOP 10 COTTON-PRODUCING COUNTRIES IN AFRICA IN 2001.

Country	Area Harvested (ha)	Yield (kg/ha)	Production (metric tons)	Percentage of Total
Africa	4,142,902	952.5	3,946,235	100.0
Egypt	217,781	2612.0	568,838	14.4
Côte d'Ivoire	330,000	1,209.5	399,138	10.1
Nigeria	538,000	741.6	399,000	10.1
Benin	372,427	974.3	362,841	9.2
Zimbabwe	369,935	883.9	327,000	8.3
Mali	227,805	1,065.7	242,772	6.2
Sudan	169,680	1,367.3	232,000	5.9
Burkina Faso	209,113	1,016.4	212,545	5.4
Cameroon	198,558	1,027.4	204,000	5.2
Chad	240,000	595.8	143,000	3.6

Source: FAO (2002).

irrigated and rain-fed in Africa, probably accounting for the wide disparity in yield. African production of cotton has remained constant over the last 40 years at 8% of world production, although African output has increased 176%.

Cotton needs 6 months to reach maturity, about 750 mm (30 in.) of rainfall or irrigation to grow well and a dry season for drying and picking. It does best below 1,400 meters (4,593 ft) in elevation on a well-drained soil (Hickman 1990). Cotton lint is separated from the seed by hand or at local cotton gins, and the lint is then baled and sent to textile mills. Oil from the seed may be extracted locally or by presses elsewhere. The residue from oil extraction, cotton seed cake, is a superb livestock feed.

Oil Palm

Oil palm (*Elaesis guineensis*), a very useful tree providing fruit, oil, wine, building materials, and rope, probably originated along the forest fringe of Guinea. According to Andah (1993), its present distribution is probably due to land clearance for shifting cultivation. The red oil, derived from the palm kernel, is widely used in cooking and is exported. Oil palm production is clearly a West African enterprise—Nigeria and Côte d'Ivoire accounted for over 60% of oil palm production in Africa, while West Africa as a whole produced 83.2% of all African oil palm output in 2001 (Table 8-27).

Oil palms are usually grown on small farms, although there are plantations from Ghana to the Democratic Republic of the Congo (Figure 8-41). Since independence, the production of palm oil has become globalized, reflecting the increasing use of this commodity as an ingredient in processed food. As a result, Africa's share of world production declined dramatically: 83% of world oil palm output on average for the years 1961–1965 but only 13% for the years 1997–2001.

Cocoa

Cocoa originated in Central and South America and was brought to the rest of the world by the Spanish. In Africa, it is grown mainly in West Africa as a cash crop, where it was introduced by the British. It is grown principally on small farms in 19 African countries, although in 2001 over 92% of cocoa production came from Côte d'Ivoire, Ghana, and Nigeria (Table 8-28). Africa's share of world cocoa production declined from an average of 73% in 1961–1965 to 67% in 1997–2001, and within Africa, there has been competition between countries in the production of cocoa. World prices for this export crop have been declining; nevertheless, cocoa production has increased by 249% since the early 1960s.

There are two main types of cocoa, Criollo and Forastero. The former, grown in Central America and

TABLE 8-27 AREA HARVESTED, YIELD, AND PRODUCTION FOR AFRICA AND THE TOP 10 OIL PALM-PRODUCING COUNTRIES IN AFRICA IN 2001.

Country	Area Harvested (ha)	Yield (kg/ha)	Production (metric tons)	Percentage of Total
Africa	3,983,200	3,785.9	15,080,000	100.0
Nigeria	3,000,000	2,666.7	8,000,000	53.1
Côte d'Ivoire	141,000	12,560.3	1,771,000	11.7
Cameroon	52,000	20,192.3	1,050,000	7.0
Ghana	115,000	9,130.4	1,050,000	7.0
Democratic Republic of the Congo	220,000	4,090.9	900,000	6.0
Guinea	310,000	2,677.4	830,000	5.5
Angola	23,000	12,173.9	280,000	1.9
Benin	21,000	10,476.2	220,000	1.5
Sierra Leone	21,800	8,027.5	175,000	1.2
Liberia	16,200	10,740.7	174,000	1.2

Source: FAO (2002).

in the Caribbean, accounts for only 15% of world production. The latter is what is grown in Africa. It is higher yielding than Criollo but not as flavorful. Each cocoa tree produces 20 to 40 pods, which contain cocoa beans. When the pods become yellow they are ripe and are harvested. Cocoa beans, like all other nut crops, must be fermented to remove pulp from around the bean. Smallholders cover them and ferment them on the ground. When dried, the beans are ready for shipping.

Although some have observed that the benefits of cash cropping have accrued disproportionately to men in Africa (Stock 1995), women are as involved in cocoa production as men in Ghana.

FIGURE 8-41 Oil palm plantation in Côte d'Ivoire. *Christine Drake.*

TABLE 8-28 AREA HARVESTED, YIELD, AND PRODUCTION FOR AFRICA AND THE TOP 10 COCOA-PRODUCING COUNTRIES IN AFRICA IN 2001.

Country	Area Harvested (ha)	Yield (kg/ha)	Production (metric tons)	Percentage of Total
Africa	5,114,786	4,135.0	2,114,960	100.0
Côte d'Ivoire	2,200,000	5,455.0	1,200,000	56.7
Ghana	1,350,000	3,037.0	410,000	19.4
Nigeria	966,000	3,499.0	338,000	16.0
Cameroon	370,000	3,108.0	115,000	5.4
Sierra Leone	30,000	3,640.0	10,920	0.5
Togo	21,400	3,738.0	8,000	0.4
Democratic Republic of the Congo	23,000	2,617.0	6,018	0.3
Equatorial Guinea	60,000	750.0	4,500	0.2
Madagascar	4,650	9,677.0	4,500	0.2
Uganda	14,200	2,782.0	3,950	0.2

Source: FAO (2002).

> There are many women landowners . . . ; indeed in the western area there are as many women as men farmers. [Women] were originally food farmers who planted cocoa and entered commerce. Some inherited cocoa farms from their mothers so the tradition of women growing cocoa is well established. (Hickman 1990; 69)

Up until the 1970s, Ghana was the world's largest cocoa producer, but political and economic instability in Ghana and an aggressive expansion program by the Côte d'Ivoire changed all that. Ghana's average production was 49% of Africa's total production from 1961 to 1965, while Côte d'Ivoire's was only 12%. Their positions were almost exactly reversed 40 years, later with 18% and 58% for Ghana and Côte d'Ivoire, respectively (Figure 8-42). Nigeria's share of Africa's cocoa production declined almost 10% over the same time period.

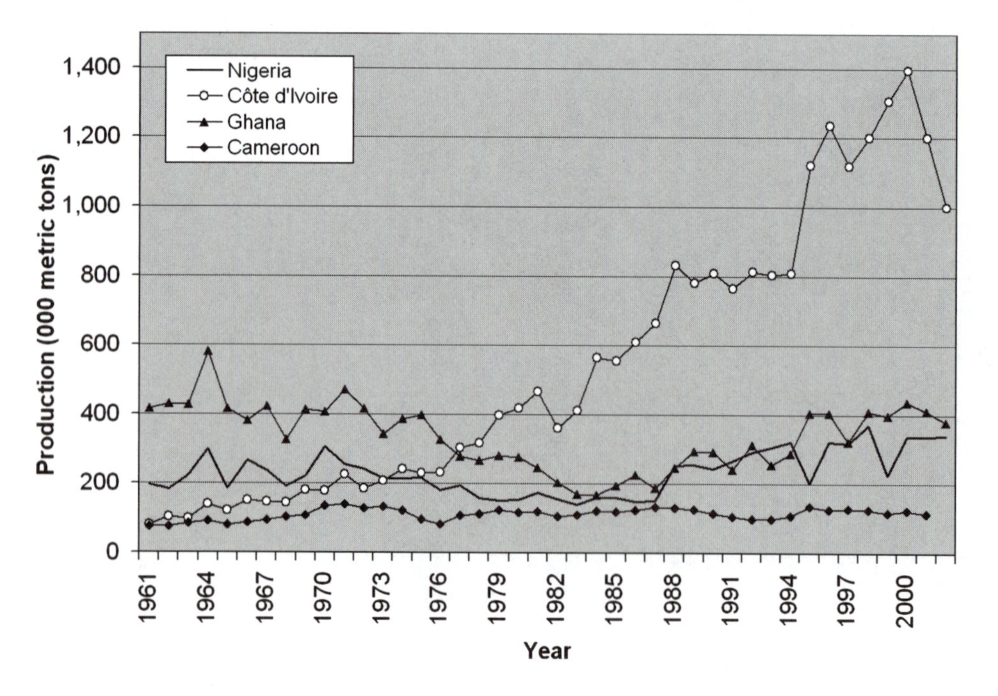

FIGURE 8-42 Cocoa production for major African producers, 1961 to 2002.

Peanuts

The peanut (*Arachis hypogaea*), or groundnut, is an oil-producing plant that originated in the Americas, most likely the Gran Chaco area of Argentina and Paraguay, and then spread to the tropical and subtropical regions of both North and South America (Arnon 1972). Today the peanut is grown throughout the tropics and subtropics, mainly in the more humid parts. The peanut was brought to Africa as a cash crop by various nations of Europe that were involved in the slave trade and colonialism.

The peanut as a food is high in protein and fat and when pressed produces an oil that is edible and used in cooking. Peanut oil production is done on a small scale, using manual presses on the farm in many areas of Africa, and has become a large, automated industry mainly in the cities. Forty-seven African countries produced peanuts in 2001, but the vast majority produced only small quantities. Nigeria, Senegal, and Sudan produced over 50% of all African peanuts in 2001 (Table 8-29).

Africa's average annual production of peanuts grew 49% over the past 40 years, from 5.4 million to 8.1 million metric tons, while world production grew 115%. Africa's average share of world production was 35% from 1961 to 1965 and 24% from 1997 to 2001, a decline of almost 31%. Most of the world's increase was associated with increases in yield (intensification) rather than increases in the areal extent of peanut cultivation. The world average annual yield increased almost 64% from 1961–1965 to 1997–2001, while Africa's average annual yield for the same period increased only 4%. Change in yield varies from country to country, and it's difficult to make any continent-wide generalizations about stagnant or declining yields of peanuts. Figure 8-43 shows the dramatic increases and decreases in yields over the 40-year period for all African producing countries. The countries in the middle show relatively little change in yields, while those on the left show a decrease and those on the right an increase. Of the top 10 producers, only Sudan has experienced a decline in yield from 1961–1965 to 1997–2001. The increase in the Egyptian yield of peanuts has been dramatic, probably associated with the increase in irrigated area caused by the Aswan High Dam. For good yields, peanuts need rainfall of at least 550 mm (22 in.), which must be evenly distributed over the growing season (Arnon 1972), since any moisture stress has a negative impact on yield. The growing season varies from 3 to 5 months, but some varieties have been reported to mature in as little as 60 days. The peanut has a reputation as a soil-exhausting crop, and farmers may apply organic and inorganic fertilizer to maintain or increase yields. Many farmers intercrop this demanding crop with nitrogen-fixing plants like beans (Figure 8-44).

TABLE 8-29 AREA HARVESTED, YIELD, AND PRODUCTION FOR AFRICA AND THE TOP 10 PEANUT-PRODUCING COUNTRIES IN AFRICA IN 2001.

Country	Area Harvested (ha)	Yield (kg/ha)	Production (metric tons)	Percentage of Total
Africa	9,665,798	891.2	8,614,577	100.0
Nigeria	2,668,000	1,087.3	2,901,000	33.7
Senegal	1,095,390	969.1	1,061,540	12.3
Sudan	1,460,000	684.9	1,000,000	11.6
Democratic Republic of the Congo	473,644	777.9	368,445	4.3
Chad	350,000	951.4	333,000	3.9
South Africa	162,250	1,644.2	266,776	3.1
Guinea	210,000	1000	210,000	2.4
Ghana	195,000	1,051.3	205,000	2.4
Mali	270,000	725.9	196,000	2.3
Egypt	60,329	3,102.5	187,169	2.2

Source: FAO (2002).

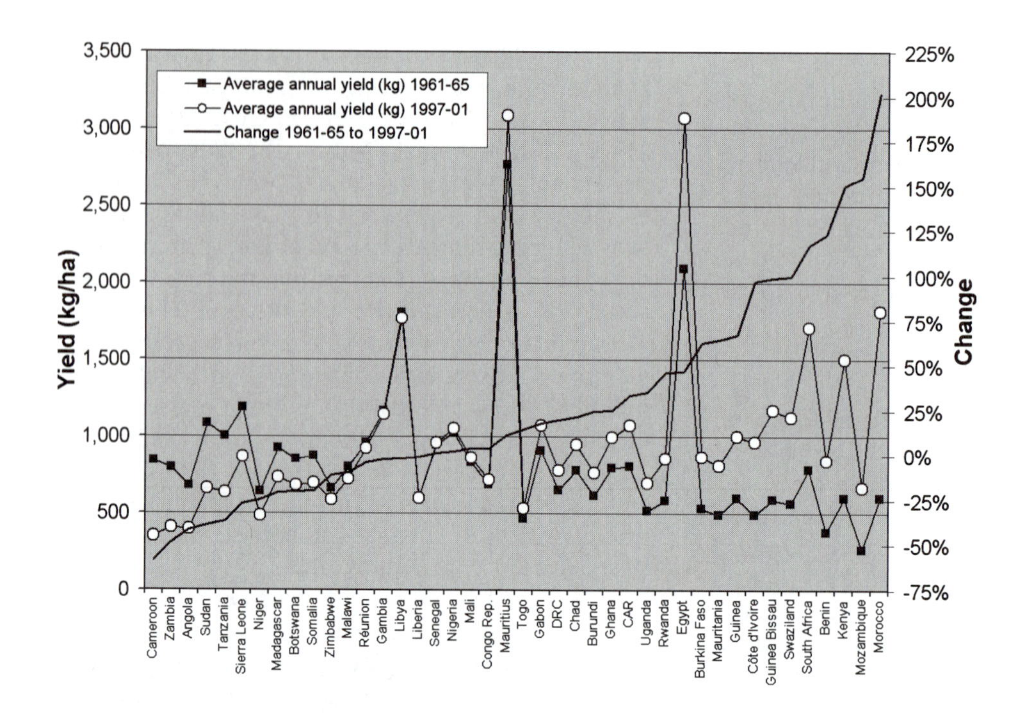

FIGURE 8-43 Average annual yield of peanuts by country, 1961–1965 and 1997–2001, and percent change.

FIGURE 8-44 Peanut plant intercrop with beans.
Roy Cole.

Sesame

Sesame (*Sesamum indicum*) is, like the peanut, an oil crop, and perhaps one of the earliest domesticated plants on earth. Using the number of wild forms as an indicator suggests a probable African origin. This dryland crop is usually planted with sorghum or maize in Africa (Arnon 1972). Sudan's production of sesame in 2001 dwarfed that of Uganda, the next largest producer. These countries accounted for 54% of Africa's production of sesame in 2001 (Table 8-30). In relation

TABLE 8-30 AREA HARVESTED, YIELD, AND PRODUCTION FOR AFRICA AND THE TOP 10 SESAME-PRODUCING COUNTRIES IN AFRICA IN 2001.

Country	Area Harvested (ha)	Yield (kg/ha)	Production (metric tons)	Percentage of Total
Africa	2,792,828	264.5	738,632	100.0
Sudan	1,900,000	157.9	300,000	40.6
Uganda	203,000	502.5	102,000	13.8
Nigeria	151,000	457.0	69,000	9.3
Tanzania	100,000	390.0	39,000	5.3
Egypt	30,401	1,210.8	36,811	5.0
Central African Republic	44,000	736.4	32,400	4.4
Somalia	70,000	328.6	23,000	3.1
Ethiopia	45,000	488.9	22,000	3.0
Niger	41,552	523.2	21,740	2.9
Chad	52,000	326.9	17,000	2.3

Source: FAO (2002).

to the rest of the world's production of sesame, Africa's output has remained relatively stable at around 25% (23.8% on average for 1961–1965 and 25.5% for 1997–2001). World sesame production is becoming more intensive, while in Africa it has become more extensive over the last 40 years. The harvested area of sesame in Africa increased 173%, while yield has declined 30%. The world, on the other hand, increased yields by 38% while increasing area only 28%. In other words, Africa's output between 1961 and 1965 represented 130% of the world's output while between 1997 and 2001 it represented only 66%.

Sesame seeds are pressed in the same way as peanuts to extract the oil, which is used in cooking. The sesame cake that remains after the oil has been pressed is used for animal feed.

Shea Nut Oil

The shea, shé nut, or karité tree (*Vitellaria paradoxa*) produces a fruit (Figure 8-45) with an edible outer pulp and a nut that contains an edible oil. The nut is processed to extract the distinctive tasting oil, which is used in cooking. The shea nut tree is ubiquitous throughout West Africa and forms part of managed farm vegetation. Africa produces 100% of the shea nuts in the world (Table 8-31), and the oil is marketed internationally for use in cosmetics and as a substitute for cocoa in chocolate. Locally, shea oil is used in cooking, soap making, as

TABLE 8-31 AREA HARVESTED, YIELD, AND PRODUCTION FOR AFRICA AND THE TOP SHEA NUT PRODUCING COUNTRIES IN AFRICA IN 2001.

Country	Area Harvested (ha)	Yield (kg/ha)	Production (metric tons)	Percentage of Total
Africa	340,300	1,993.3	678,333	100.0
Nigeria	232,000	1,724.1	400,000	59.0
Mali	33,000	2,575.8	85,000	12.5
Burkina Faso	28,000	2,500.0	70,000	10.3
Ghana	22,000	2,954.5	65,000	9.6
Côte d'Ivoire	16,500	2,171.7	35,833	5.3
Benin	6,000	2,500.0	15,000	2.2
Togo	2,800	2,678.6	7,500	1.1

Source: FAO (2002).

FIGURE 8-45 Shea nuts just prior to boiling.
Roy Cole.

a fuel for lighting, and as a cosmetic. Higher-quality oil is used in cooking and cosmetics, while that of lower quality is used to make soap; rancid shea oil is used in lamps. In most parts of West Africa, kerosene has replaced shea oil as the fuel of preference in lamps.

Shea nut production is a women's industry in most parts of West Africa. Village women and their children collect the nuts, ferment the pulp, and soften the shell in underground silos about 2 m (6 ft) deep and 1 m in diameter (Figure 8-46). Once the shells have been softened, the nuts are removed and the long process of boiling begins. When the oil has been removed from the nuts, it is left to cool. Shea oil is solid at room temperature, and when it has cooled it is wrapped in fig or other leaves and stored for market. When the nuts are processed in a factory, the oil is removed chemically or with a mechanical press.

Sugarcane

Although first domesticated in the South Pacific, sugarcane (*Saccharum officinarum*) was introduced to the Mediterranean region from India, where the sugar was first extracted from the cane (Arnon 1972). The plant, a member of the grass family, is made up of a number of stems and grows to a height of 2 to 6 m (6–20 ft). It does best at temperatures above 68° F (20° C). Roots can reach to a depth that is almost as deep as the plant is tall. Sugarcane is a perennial crop that requires 10 to 24 months to produce a crop. It is propagated vegetatively through the planting of cuttings (segments of stalk) rather than seeds.

Over the last 40 years, African sugarcane production increased 276%, from 31 million to 85 million metric tons, while world production increased from 468 million to 1.26 billion metric tons, roughly the

FIGURE 8-46 Shea nut fermentation silos. *Roy Cole.*

same percentage increase. Since global sugar production has grown in tandem with Africa's, Africa's share of world production has not changed in 40 years, remaining at 6%. Thirty-nine African countries grew sugarcane in 2001, but 45% of it was produced by only two countries, South Africa and Egypt (Table 8-32). Most sugarcane-growing countries in Africa dramatically increased their production over the last 40 years, but war-torn Mozambique was an exception. Sugar, like cotton, is another high-value crop important to nascent agricultural processing and industry. Economic development in the early years of independence focused on substituting locally produced products for imported ones. Farm output of cotton and sugar was linked to the development of import-substituting textile and sweetener industries.

TABLE 8-32 AREA HARVESTED, YIELD, AND PRODUCTION FOR AFRICA AND THE TOP 10 SUGARCANE-PRODUCING COUNTRIES IN AFRICA IN 2001.

Country	Area Harvested (ha)	Yield (kg/ha)	Production (metric tons)	Percentage of Total
Africa	1,393,571	62,791	87,503,715	100.0
South Africa	321,913	74,232	23,896,100	27.3
Egypt	133,990	116,579	15,620,400	17.9
Mauritius	78,000	70,513	5,500,000	6.3
Kenya	60,000	85,833	5,150,000	5.9
Sudan	65,000	76,923	5,000,000	5.7
Zimbabwe	42,000	97,619	4,100,000	4.7
Swaziland	36,500	106,427	3,884,600	4.4
Ethiopia	20,000	120,000	2,400,000	2.7
Madagascar	67,200	32,738	2,200,000	2.5
Malawi	18,000	105,556	1,900,000	2.2

Source: FAO (2002).

FIGURE 8-47 Rubber plantation, Côte d'Ivoire.
Christine Drake.

FIGURE 8-48 Tapping a rubber tree.
Christine Drake.

Raw sugar is produced from cane by extracting the juice in a press, boiling the juice until it thickens and crystallizes, and spinning the crystallized sugar in a centrifuge to remove the syrup. The raw sugar is then sent to a refinery, where it is washed and filtered, crystallized, dried, and packaged.

Rubber

The rubber tree (*Hevea brasiliensis*) is indigenous to the Amazon rain forest and was brought to Europe by Christopher Columbus. No systematic use was made of it until several centuries later, when in 1839 Charles Goodyear discovered a way to make it stable in heat and cold (the vulcanization process). At the end of the nineteenth century 70,000 plants were smuggled out of Brazil by an English adventurer and later used to develop plantations in Indonesia and in today's Sri Lanka and Malaysia (Watson no date). About 90% of the world's natural rubber production comes from Southeast Asia today. Synthetic rubber, derived from petroleum, provides 60% of all the rubber used today.

Natural rubber is a low-altitude crop that thrives in tropical rain forest conditions (Figure 8-47). Most of it is grown north and south of the equator to about 15° latitude. It grows best in high temperatures, where the annual average is at least 70° F (20° C) and the climate is rainy throughout the year. If rainfall is deficient, sap production will be low (Hickman 1990). After planting, the tree requires 6 to 7 years of growth before it can be tapped. The bark is cut at an angle, and a container is affixed to the lower end of the cut into which sap (latex) runs (Figure 8-48). Every few days a cut is made a few inches below the previous cut, and

TABLE 8-33 AREA HARVESTED, YIELD, AND PRODUCTION FOR AFRICA AND THE TOP 10 RUBBER-PRODUCING COUNTRIES IN AFRICA IN 2001.

Country	Area Harvested (ha)	Yield (kg/ha)	Production (metric tons)	Percentage of Total
Africa	638,200	707.3	451,400	100.0
Liberia	140,000	964.3	135,000	29.9
Nigeria	330,000	348.5	115,000	25.5
Côte d'Ivoire	67,000	1,611.9	108,000	23.9
Cameroon	40,000	1,500.0	60,000	13.3
Ghana	17,500	754.3	13,200	2.9
Gabon	11,000	1,000.0	11,000	2.4
Democratic Republic of the Congo	30,000	233.3	7,000	1.6
Republic of the Congo	1,500	800.0	1,200	0.3
Central African Republic	1,200	833.3	1,000	0.2

Source: FAO (2002).

the process continues until the ground has been reached, after which time the tree is taken out of production for a year. Each tree produces about 4 kg (8.8 lb) of latex a year (Watson no date). Most rubber in Africa is produced on large estates such as the Firestone Plantations in Liberia, and tappers are given a fixed number of trees to tap. Some rubber is produced on small farms, which then sell it to the large estates just as maple syrup or milk is often sold in the United States and Canada. Rubber is a crop that lends itself to plantation agriculture because successful production for the world market requires a large investment of capital. Africa contributed only 7.2% to the world production of rubber in 2001. Its share of world rubber production declined 14% over the last 40 years from 7.2% to 6.2%. Within the continent, three West African countries produced over 79% of output in 2001: Liberia, Nigeria, and the Côte d'Ivoire (Table 8-33).

Figure 8-49 illustrates the change in rubber production from 1961 to 2001 for the top producers. What is fascinating about Figure 8-49 is how quickly Liberia's production crumbled after 1990 and how rapidly Nigeria took the lead. The dramatic decline in Liberia's output had nothing to do with world prices and everything to do with political instability and civil war. A rebel group, the National Patriotic Front of Liberia (NPFL), led by Charles Taylor, took over a Firestone Plantation in 1990 and used its facilities as headquarters. Freelance tappers were permitted to take over the plantations. Monrovia, the capital of Liberia, was under siege by July 1990, and the president of Liberia, Samuel Doe, was captured on September 9 and brutally killed the next day. During the 1990s, "[a]ll of Liberia's natural resources were open to exploitation by international traders in league with the warring factions and Ecomog" (Ellis 1999: 166). ECOMOG, the Economic Community of West African States Ceasefire Monitoring Group, was a military force led by Nigeria that was deployed in Liberia by ECOWAS, the Economic Community of West African States, to keep the peace. Firestone Corporation protested the illegal use of its rubber, the "bouncing diamond" as locals called it, and a series of ineffectual export bans was put into place.

> In response to complaints from the Firestone company, the Liberian government imposed a ban on rubber exports, but this had little effect because Ecomog officers and Liberian government ministers, the latter being nominees of the warring factions, could be bribed to overlook such restrictions. The ban was effective, though, in clearing out the private tappers and allowing the armed factions themselves to work the Firestone rubber plantation. The factions then trucked the rubber to Monrovia where it was sold generally to Malaysian, Korean and Lebanese traders. . . . (Ellis 1999: 166)

Minerals and other natural resources have been accorded the same treatment in other African countries

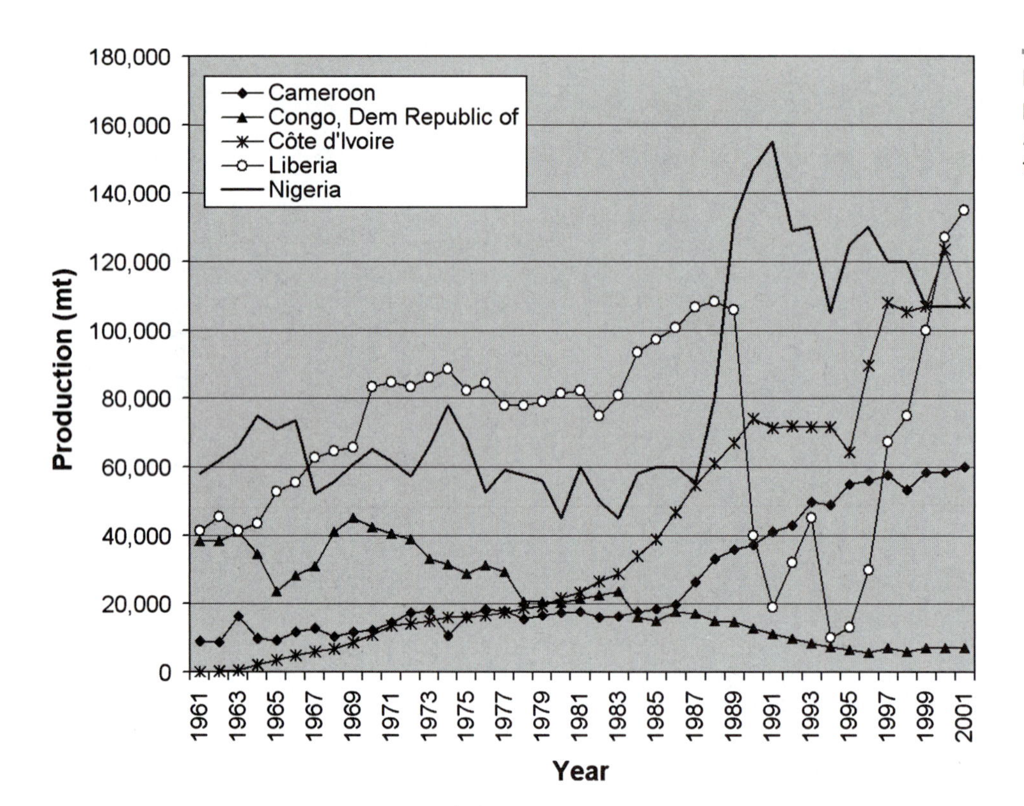

FIGURE 8-49 Rubber production for major African producers, 1961 to 2001.

during civil war. The most well-known cases involve the "conflict diamonds" from several countries engaged in civil war in Africa during the 1990s.

Sisal

Sisal (*Agave sisalana*) fibers are used in the manufacture of rope and twine. Sisal, important today principally in former British East African colonies, is a crop of declining importance in both Africa and the world because it is being replaced by artificial fibers. The principal producing countries in 2001 were Kenya, Tanzania, and Madagascar (Table 8-34).

Decline in the production of sisal is evident from a glance at Figure 8-50. For Africa as a whole, there was an 83% decline in production from 1961–1965 to 1997–2001, while for the world in the same time frame the decline was almost 60%. Africa's share of world production fell from 50% to 20%, and the average yield fell 21% as well. In Angola, sisal was mainly the domain of Portuguese colonists, and its cultivation was abandoned in the mid-1970s, owing to decolonization, civil war, and a growing subsistence focus.

The Spatial Organization of the African Farm

The geography of the African farm is variable from place to place and yet based on similar principles. The principal influences on the organization of space around a village or farm are the transportation costs in time and labor of critical factors of production. Factors of production include the necessities of farming: for example, labor, equipment, seeds, and animal manure or inorganic fertilizer. Most factors of production are mobile—they can be brought to the farm field; but one is immobile: land. African farmers overcome the immobility of this factor of production by cultivating scattered plots, to spread risk over space.

We shall examine the geography of the farm by means of three types prevalent in Africa: farms associated with nucleated farm villages, independent farmsteads, and the modern, commercial farm. The nucleated farming village is the spatial organization most frequently seen and is found throughout Africa. Independent farmsteads can be found in any country but are the dominant type in East Africa, while the commercial farms can be found in any country.

TABLE 8-34 AREA HARVESTED, YIELD, AND PRODUCTION FOR AFRICA AND THE TOP 10 SISAL-PRODUCING COUNTRIES IN AFRICA IN 2001.

Country	Area Harvested (ha)	Yield (kg/ha)	Production (metric tons)	Percentage of Total
Africa	93,130	726.4	67,645	100.0
Kenya	20,000	1,250.0	25,000	37.0
Tanzania	46,120	444.5	20,500	30.3
Madagascar	14,200	1,056.3	15,000	22.2
South Africa	3,500	742.9	2,600	3.8
Morocco	2,500	880.0	2,200	3.3
Ethiopia	1,000	700.0	700	1.0
Mozambique	3,000	200.0	600	0.9
Angola	450	1,111.1	500	0.7
Central African Republic	810	401.2	325	0.5
Guinea	1,100	118.2	130	0.2
Malawi	450	200.0	90	0.1

Source: FAO (2002).

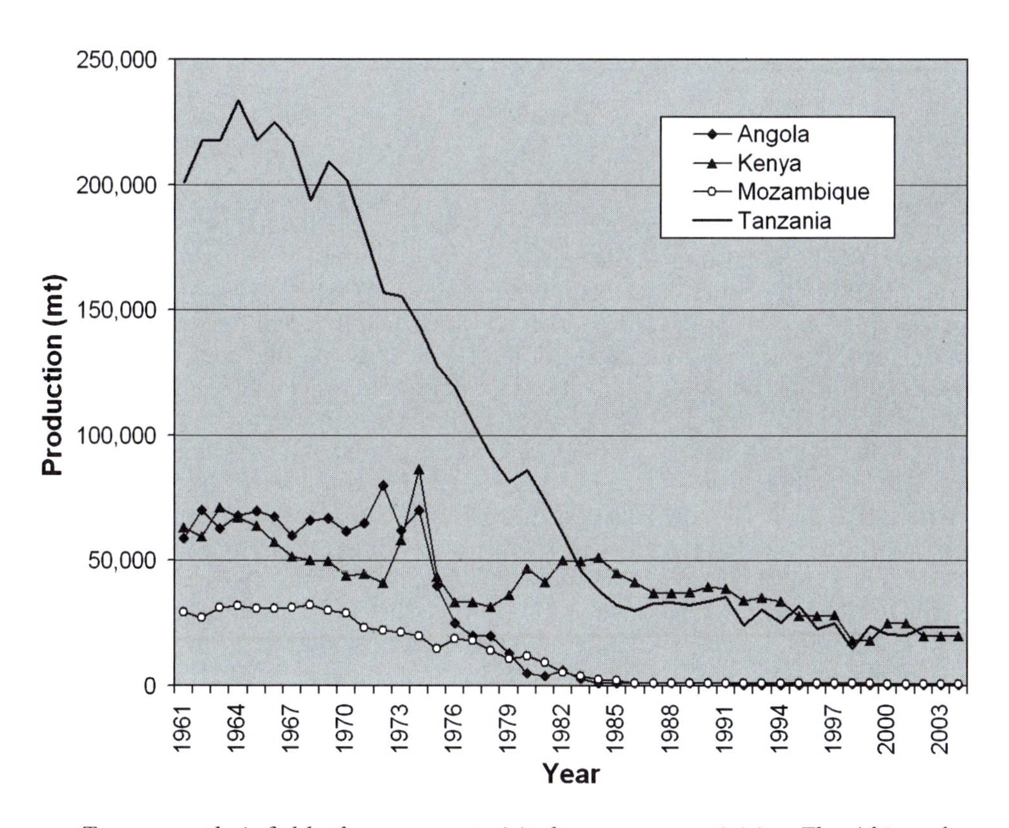

FIGURE 8-50 Sisal output for Africa's top producers, 1961 to 2001.

To prepare their fields, farmers must visit the areas many times in a season, turning over the soil, planting and replanting, weeding, and finally harvesting the crop. If a field is distant, farmers will often erect temporary housing near the site of the most labor-intensive activities. The African farmer's job is logistically difficult because fields in a variety of locations are cultivated by one family, and some may be rather distant from the owner's farmstead. Rainfall is not distributed uniformly across space, and the farmer wants to increase

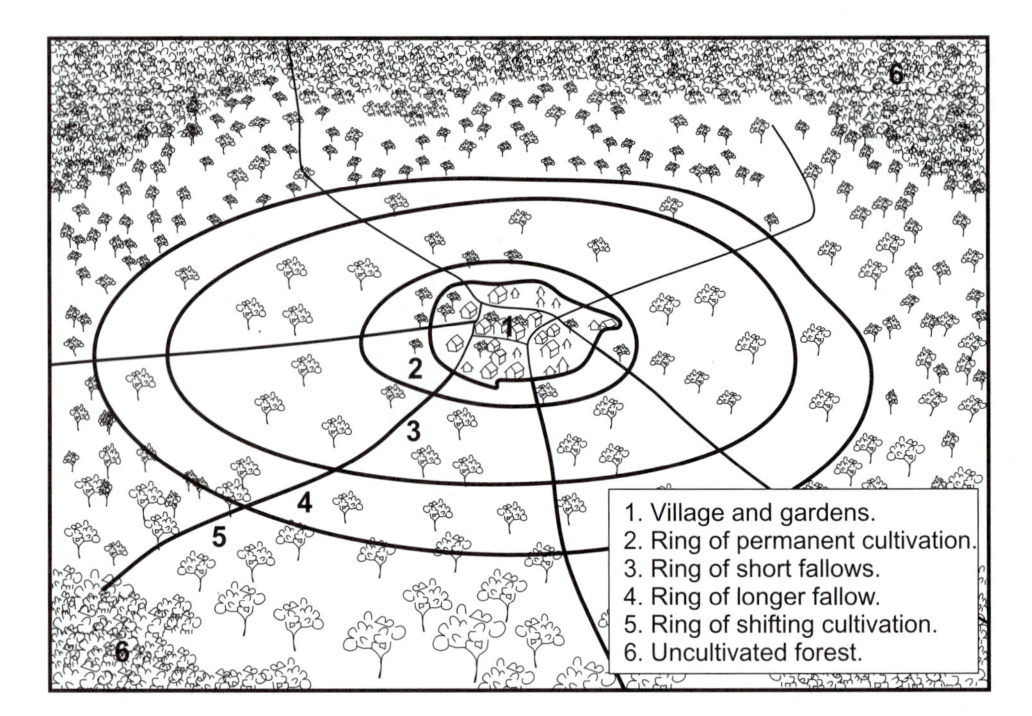

FIGURE 8-51 Model of traditional Serer land use.
Adapted from Pelissier (1964).

the chance of getting rainfall and to minimize the likelihood of a total loss if there is an infestation of insects or an outbreak of disease. Farmers may also plant in more than one ecological niche. For example, millet will be planted on hills and in low spots so that in a well-watered year the millet on the higher ground will do well, although the same crop planted in depressions will be flooded. In the event of a year of mediocre or poor rainfall, the crop planted in the low spots (where there is additional moisture from runoff) will produce, while plants in higher areas will not survive.

Several models of the African farm geography have been developed from micro-level studies. All models have at their core the work of von Thünen, who was one of the first theorists of space. His empirical research indicates that space is organized around a principle of centrality and that there are costs, the "friction of distance," to moving across space. For von Thünen's farm model, the distance to the market and the costs of transportation of products from the farm were the most important factors in organizing space. One such model was developed by the geographer Paul Pelissier in the 1960s. In Pelissier's model, based on his research in Senegal in the 1950s, labor and manure are centrally located in a village. Manure, essential for maintaining yields, is a scarce factor of production in Africa; there is not enough of it to go around.

The Pelissier model (Figure 8-51) consists of five rings, each of which involves a different agricultural activity or level of activity; beyond the rings lies uncultivated forest. Ruthenberg (1976) notes that depending on circumstances, nonannular patterns may develop, for example, cultivation of plots in valley bottoms or depressions. Crop location, crop type, and cultivation intensity therefore may vary from Pelissier's model. The two inner rings are often called the home field or the infield. The outer circles are called the bush field or the outfield. The key to the different activities/land uses in the figure is as follows:

1. **The village and gardens:** the use of manure allows the permanent cultivation of gardens, fruit trees, and millet. High fertility.
2. **The main fields:** mostly permanent cultivation of, millet, and sorghum. Manured. Lower fertility.
3. **Intensive fallow system of millet or sorghum:** no manure, short fallows of 1–2 years, principally as pasture. Lower fertility.

4. **Intensive shifting cultivation of millet, or sorghum:** no manure, longer fallows than in ring 3. Lower fertility.
5. **Bush and extensive shifting cultivation:** long fallows, forest. High fertility.

A major undertaking for any farm is the maintenance of soil fertility. In the traditional farming system of the Serer-speaking people on which Pelissier based his research, the *Acacia albida* tree is used to help maintain soil fertility around the main fields. In addition to planting *A. albida*, farmers transport household wastes and scarce manure to the main fields. *Acacia albida* has several beneficial effects for agriculture that most other trees do not possess: it fixes nitrogen, produces pods edible by livestock, and sheds its leaves in the rainy season (Felker 1981, 1978; Watt and Beryer-Brandwijk 1962). The wooded parkland formed by *Acacia albida* minimizes evaporation losses and creates a humid microclimate for farming. The permanent vegetation in the fields acts as baffles to the wind, minimizes diurnal temperature and moisture variations, protects crops from burning, recycles nutrients from the soil, and produces useful and marketable products (Pullen 1974, Wilkin 1972). The tree is so central to the traditional agriculture that, as one Serer farmer put it, "five acacias fill a store with millet" (Lericollais 1972: 29). Monteil, in his ethnography of the Bambara in Mali (1977), reports that the tree is central to Bambara identity itself.

The cost of transporting manure and labor (both to and from the fields over the agricultural year) also has an impact on the traditional farm geography. Manure and compost are collected in or near the village and are transported only to the main fields at the start of the growing season. The customary method of transporting these scarce materials is by head porterage. Furthermore, there are a variety of farm operations that must be accomplished throughout the agricultural year involving many people and tools. Given locomotion by foot, it is axiomatic that as distance increases from the settlement, the cost of fielding people and inputs will increase and, as Pelissier indicates, the intensity of cultivation will decrease. Land-use intensity decreases from the center to the periphery, from the gardens to the bush, in a distance–decay relationship. Soil fertility is highest closest to the houses because of the presence of household wastes, livestock feces, and litter from trees; it declines with distance from the houses. The distance decay of soil fertility corresponds to that of land-use intensity except at the periphery, where soil fertility is higher than the adjacent rings because of long fallows.

Hickman (1990) has sketched farm models from all around Africa. In the more traditional models, Pelissier's fertility–friction-of-distance model applies, although the specific circumstances and scales may differ considerably. Northern Ghana is a transition zone between the drier area to the north, where mainly cereals are grown, and the yam belt to the south. Farms are small, averaging 2.5 to 5 acres (1–2 ha). The farm depicted in Figure 8-52, worked by one nuclear branch of a larger, extended family, is located in a tsetse-free area and, as a consequence, cattle are kept. The manure has enabled an intensive agriculture to develop, and the house fields produce twice as much as the bush fields (not depicted). Families live in individual farmsteads on the infields. A variety of crops are grown on

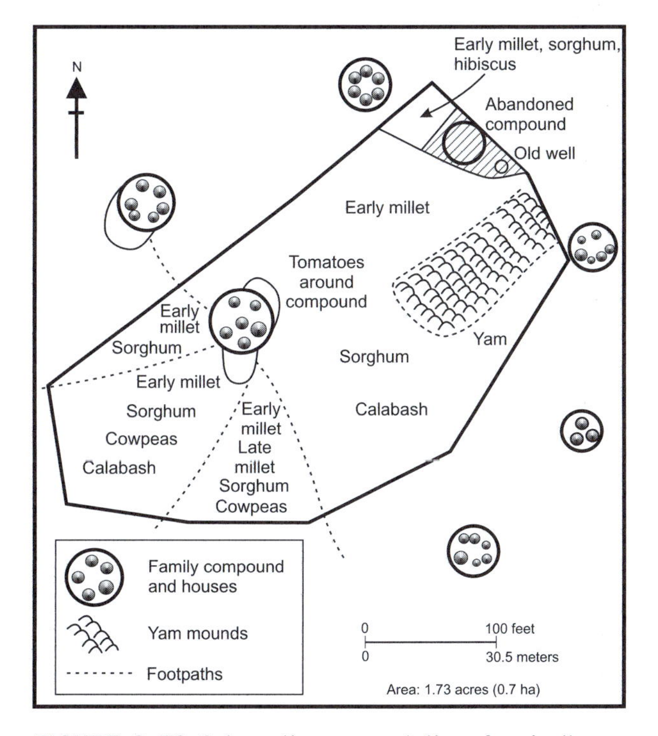

FIGURE 8-52 Schematic representation of agriculture around a small family compound in northern Ghana. *Adapted from Hickman (1990).*

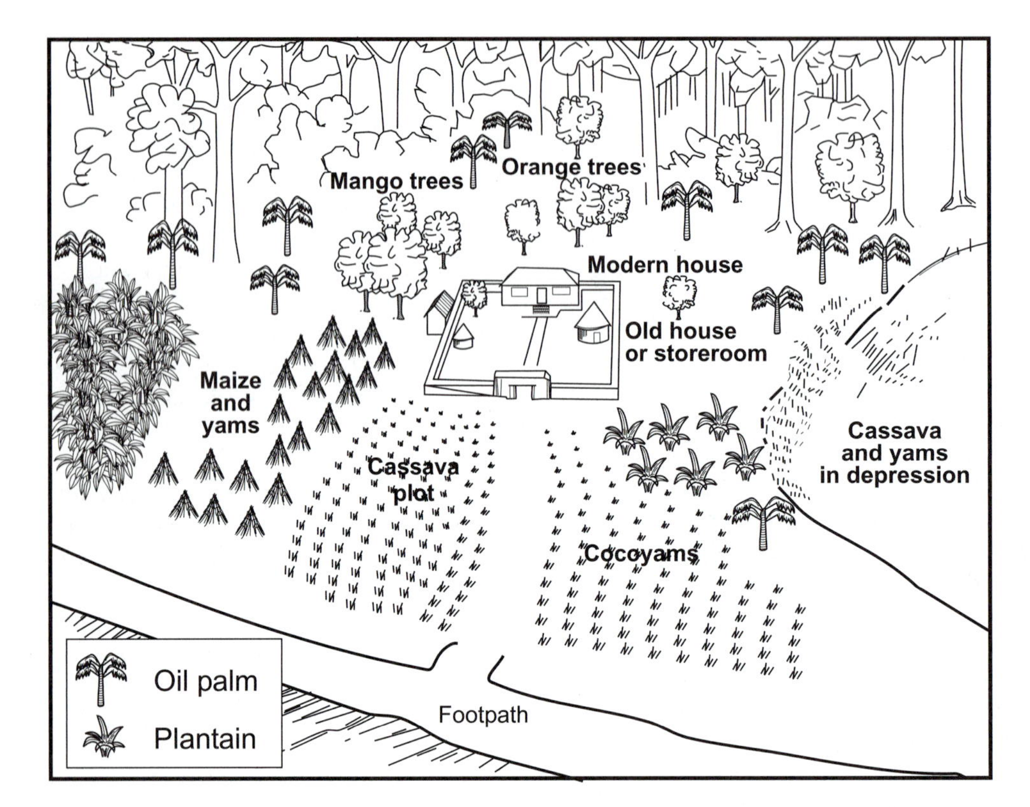

FIGURE 8-53
Schematic representation of an Ibo mixed farm in the Niger Delta, Nigeria.
Adapted from Hickman (1990).

this farm: cereals, root crops, and vegetables. Note that the tomatoes, which need almost daily maintenance, are planted very close to the compound. A variety of crops are grown to stagger food availability from early summer to the dry season. Early millet is primarily a hungry-season crop that is eaten after last year's stocks have been exhausted or are near exhaustion. Hibiscus flowers are used in making sauces to go with meals.

In the Niger Delta of Nigeria, the heart of the yam belt, Ibo farmers mix oil palms along with their food and cash crops (Figure 8-53). In this densely populated region, manure is applied to the fields around the house. On average families own 80 to 90 oil palms in this region, and the oil is used in cooking and is sold in the market. The outfields are not manured, and shifting cultivation is practiced. The length of fallowing has been shortening as population pressure has increased.

Modern, commercial farms are larger in scale and more mechanized; from a spatial perspective, they appear to be more organized and compartmentalized than traditional farming in Africa. Figure 8-54 depicts a modern, South African commercial fruit farm that is privately owned. Commercial farms may seem to be more spatially organized than more traditional African agriculture because of the practice of monoculture (planting a single crop in a field), the cultivation of vast, largely rectangular, neat-looking rectangular fields, bordered by roads, and the use of fencing.

Traditional African farmers practice polyculture (growing a variety of crops) and intercropping (growing two or more crops in a field together) to enhance yields. Intercropping can reduce the risk of total loss due to pest infestation of one crop. The relatively dense layers of plants can shade out and reduce weed growth. In one field may be one or more varieties of beans (to fix nitrogen) and several varieties of sorghum, watermelon, and gourd. Commercial farming is heavily dependent on fossil fuels and imported machinery. In states that tried the socialist organization of production, large state farms were set up to capture economies of scale similar to the large-scale, privately owned, commercial farm. The principal difference was that the state owned the means of production and the farmers worked for a wage and/or part of the crop. This model was largely unsuccessful where it

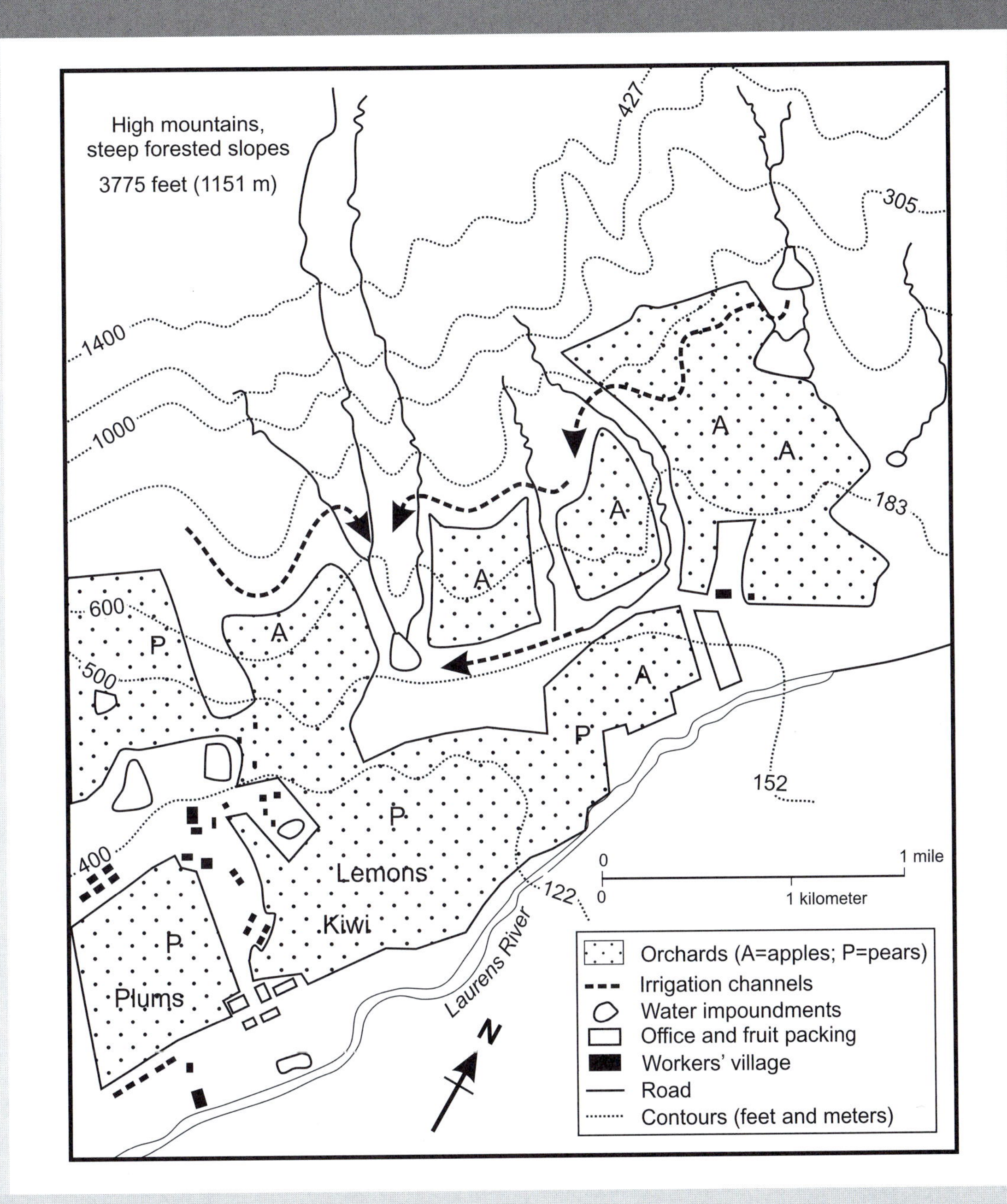

FIGURE 8-54 Lourenford estate, Cape Province, South Africa.
Adapted from Hickman (1990).

was tried in Africa, principally Ghana, Ethiopia, Mozambique, and Tanzania, as well as many other countries. Machinery broke down and the foreign exchange needed to purchase replacements was not available; fuel ran out; bureaucrat-managers were corrupt. Ultimately, the inefficiencies inherent in state control over large-scale productive enterprises in poor, African countries became clear.

On the other tip of the continent, in a similar but drier climate, North African fruit farmers produce

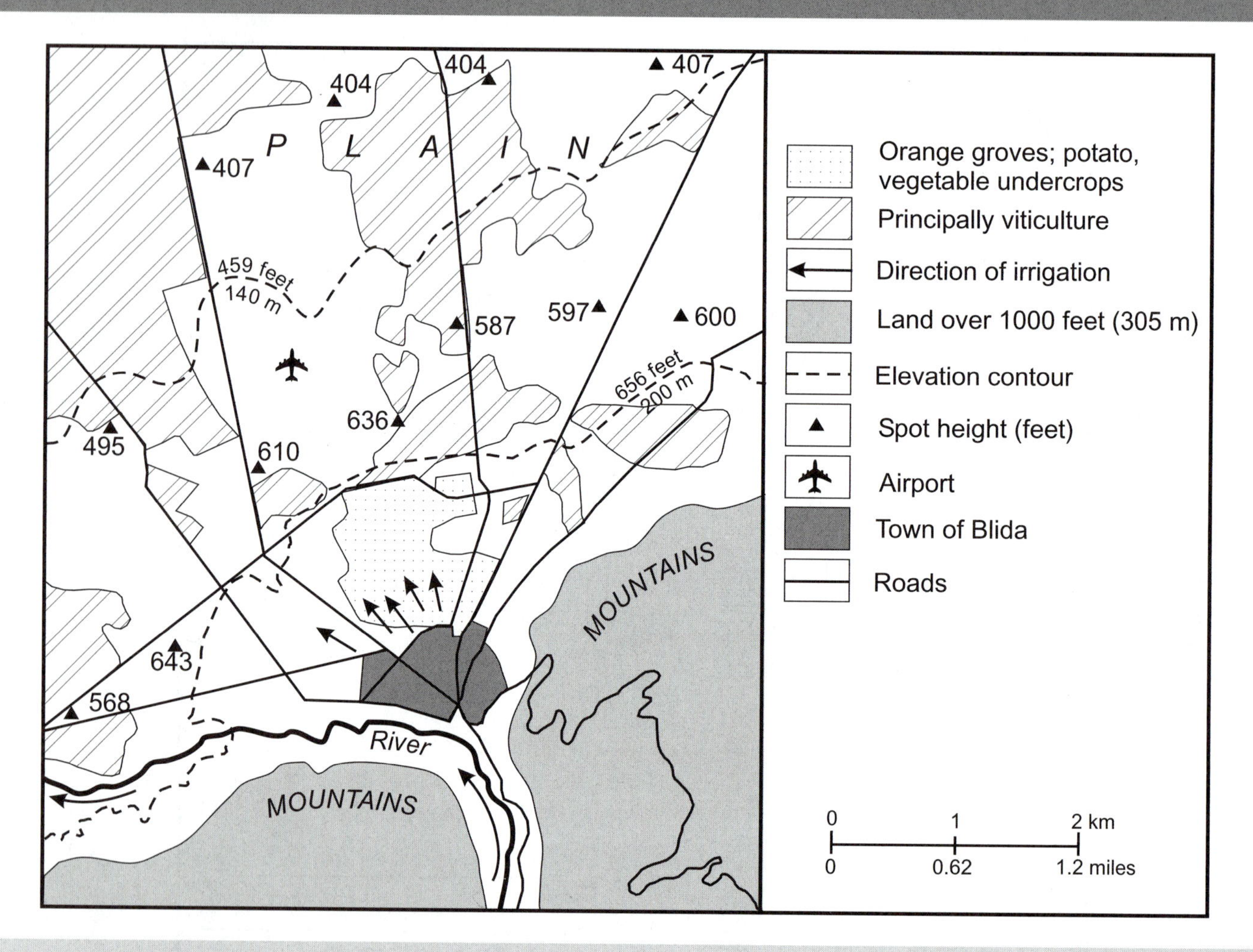

FIGURE 8-55 Agriculture around the village of Blida, Algeria.
Adapted from Hickman (1990).

much the same crops as at the southern tip. Blida, Algeria, is located about 60 miles (100 km) south of the capital, Algiers. It is located in the foothills of the Atlas Mountains on an alluvial fan, containing deep and permeable layers of soil deposited where a river leaves its mountain bed. Once out of a constraining mountain valley, a river meanders across the plain, building up fertile deposits in the shape of a fan. Because of the constant water supply from the river, water is never far from the surface in an alluvial fan, which in many respects is similar to a river delta.

Cash crops in the region focus on citrus fruits and grapes as in other parts of Mediterranean Africa and in like climates of South Africa. The village of Blida is located at a critical junction between mountain and plain where water is available for irrigation and the water table is relatively high (Figure 8-55). The orange groves are located higher on the alluvial fan, where there is less likelihood of frost because the cold air drains to the plain below. Grape vines, which are grown on the plain, are more tolerant of winter frost than oranges.

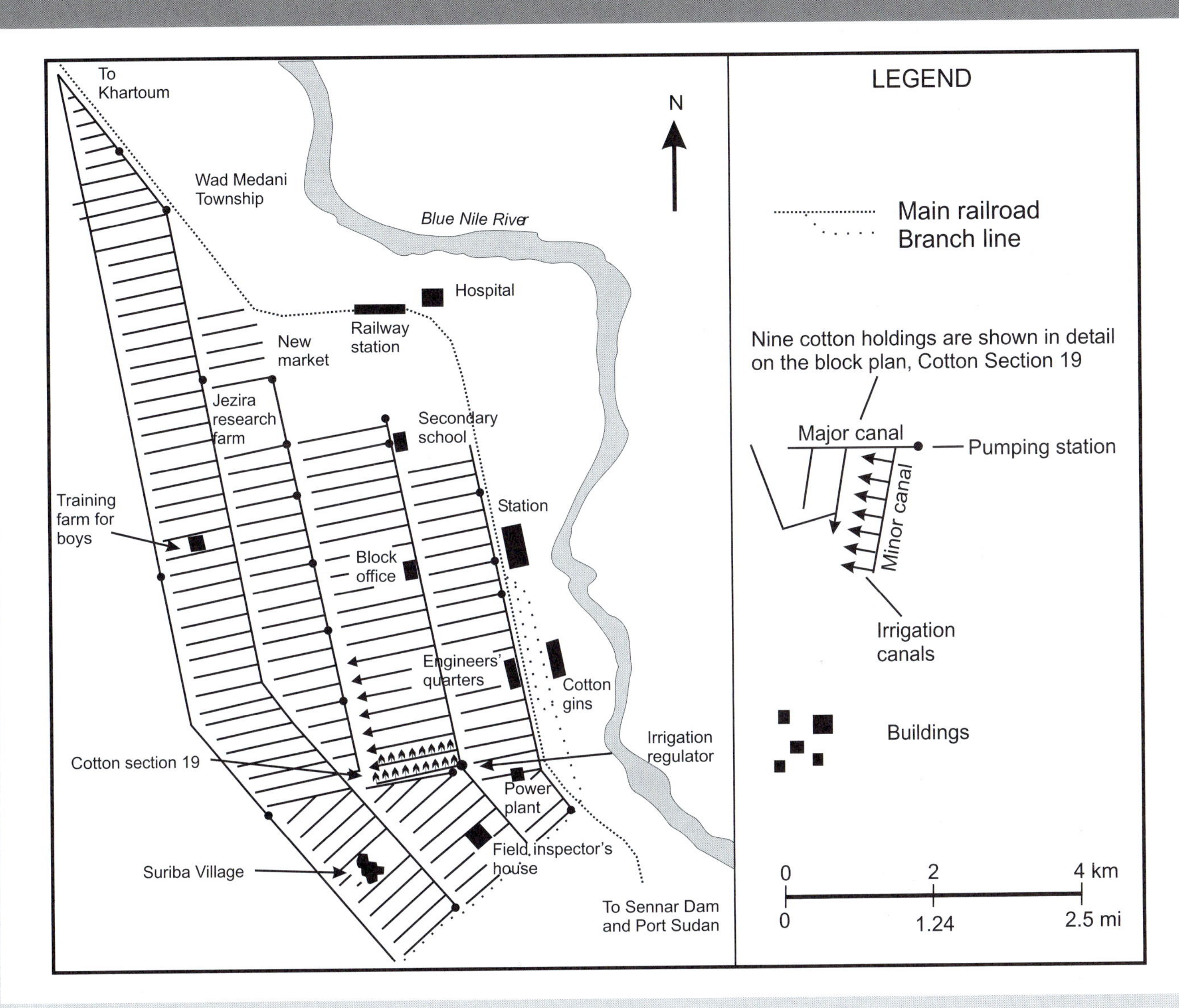

FIGURE 8-56 The Jezira scheme for cotton production: Block 10, Wad Medani.
Adapted from Hickman (1990).

Several state-run agricultural projects or "schemes" in several African countries have had some success, although the project-oriented approach to the development of agriculture has been problematic in Africa. The Jezira scheme in Sudan is an example of project-led development that has had some success (Figure 8-56). The Jezira has a colonial past. It was begun in 1911 by the British to produce cotton for British textile mills. Although heavily subsidized in the beginning, the Jezira scheme is self-sufficient today. Cotton production has continued, but production has diversified to include wheat, sorghum, sunflower, peanuts, and sesame. Food-processing plants have been set up, as well as a cigarette factory. Hickman (1990) calls the Jezira "a remarkable partnership combining state ownership and individual enterprise" (1990: 117). Tenant farmers, living in nucleated villages, work the land for a share of the profits from the sale of cotton and other cash crops. Food crops (sorghum and vegetables) are grown by the tenants on the scheme.

African agriculture is practiced in a variety of ways and circumstances: on isolated farmsteads, in nucleated villages, and on large-scale estates run by the state or by multinational agribusinesses. African farmers produce for the market and for home processing and consumption. The old cliché that the only constant is change is clearly true about African farming. In Chapters Ten through Thirty, which focus on the regional geography of Africa, we will look at the spatial characteristics of agriculture within each African country.

THE FUTURE OF AFRICAN AGRICULTURE

African agriculture is dynamic and changing. Nevertheless, African agriculture as an economic activity and as a way of life faces many challenges. A major challenge is keeping output ahead of a rapidly growing population. Keeping output up is difficult in traditional agriculture, where human muscle is the principal source of power, capital to invest in technology is scarce, inputs to enrich the soil are few, and there is population pressure to shorten fallowing. The weakening of many farmers who have been infected with HIV/AIDS adds a further burden to the farm economy. A highly variable environment adds uncertainty to farm production from one year to the next.

Problems in African agricultural policy have been publicly stated for over a decade:

> African farmers have faced the world's heaviest rates of agricultural taxation . . . explicitly through producer price fixing, export taxes, and taxes on agricultural inputs. They were also taxed implicitly through overvalued exchange rates and through high levels of industrial protection. . . . [These policies] contributed to Subsaharan Africa's alarming decline in . . . agricultural growth. (World Bank 1994: 76)

In addition, there is evidence that protectionism in more developed countries (mainly the Americas and Europe) keeps out cheaper African farm exports (United Nations 2002). Protecting farmers through import tariffs and subsidies is big business in many European countries and the United States. This is a view that is seconded by many African leaders. Nevertheless, farm exports from Africa to Europe have been and will continue to be largely focused on commercially grown fruit, flowers, and vegetables from the big producers such as Egypt and South Africa, as opposed to the crops grown by the average farmer in the yam or cereal belt.

It does not seem likely that Africa will become a major maize exporter because such a role cannot be played unless maize production in large areas of Africa becomes a large-scale, mechanized, commercialized operation. If African countries follow the European and American model, mechanizing maize production would involve a massive decline in agricultural employment, followed by rural poverty and rural depopulation. African governments and international agencies have worked with small-scale African farmers in an attempt to develop meso-scale commercial farmers with limited success. Perhaps the free market and price incentives will succeed where policy has had mixed results.

The authors of International Food Policy Research Institute report on global food projections suggest that structural and institutional changes along these lines in African agriculture are needed to foster agricultural intensification (higher yields) because the limits to expanding the cultivated area (extension) are being reached.

> [A]rea expansion cannot be the solution to Africa's production problems. Yields must grow at relatively high rates, and smallholders must move more rapidly from subsistence to commercial production. The last three decades have left a legacy of increasing degradation and desertification, with insufficient investment in soil improvement causing severe nutrient depletion. Poor policies and institutional failures have removed the incentive for farmers to care for the land and to invest long-term capital in yield-boosting technologies. . . . [T]he loss of the natural resource base in Subsaharan Africa is of [great] concern because of the seeming inability of governments in the region to address the growing problems they face. (Rosegrant et al. 2001: 44)

Having said all that, though, is it time to write off African agriculture? The evidence for African agriculture seems to be mixed at least in regard to country-level data, as Table 8-35 indicates. Although the authors concur with the findings of the International Food Policy Research Institute (Rosegrant et al. 2001), an

TABLE 8-35 PERCENT DIFFERENCE IN THE PER CAPITA PRODUCTION OF ALL CEREALS AND ALL TUBERS, 1961-1965 AND 1997-2001.

Country	All Cereals	All Tubers	Country	All Cereals	All Tubers
Algeria	40.6%	186.3%	Madagascar	58.6%	82.5%
Angola	40.3%	88.4%	Malawi	85.8%	414.7%
Benin	118.1%	166.4%	Mali	102.2%	54.2%
Botswana	29.7%	95.5%	Mauritania	74.4%	39.0%
Burkina Faso	114.5%	35.0%	Mauritius	139.5%	191.6%
Burundi	96.9%	80.2%	Morocco	58.3%	233.0%
Cameroon	75.2%	82.5%	Mozambique	111.0%	90.5%
Cape Verde	64.8%	36.6%	Namibia	108.9%	62.8%
Central African Republic	90.2%	55.3%	Niger	69.1%	36.1%
Chad	64.6%	127.8%	Nigeria	99.5%	166.9%
Comoros	54.9%	46.2%	Réunion	58.7%	47.7%
Democratic Republic of the Congo	120.2%	59.9%	Rwanda	56.8%	127.5%
Republic of the Congo	42.2%	57.3%	São Tomé and Principe	213.3%	157.3%
Côte d'Ivoire	119.4%	72.9%	Senegal	56.7%	22.9%
Djibouti	NA	NA	Seychelles	NA	19.0%
Egypt	140.9%	191.7%	Sierra Leone	46.8%	177.3%
Equatorial Guinea	NA	94.0%	Somalia	39.5%	128.9%
Eritrea	NA	NA	South Africa	75.9%	176.5%
Ethiopia	NA	NA	Sudan	97.6%	18.3%
Gabon	116.0%	95.3%	Swaziland	89.3%	35.9%
Gambia	52.0%	30.5%	Tanzania	110.1%	69.7%
Ghana	164.0%	159.6%	Togo	125.1%	69.8%
Guinea	81.9%	79.2%	Tunisia	77.4%	298.1%
Guinea-Bissau	91.9%	114.7%	Uganda	72.6%	124.4%
Kenya	53.1%	76.2%	Zambia	45.2%	177.1%
Lesotho	56.8%	1324.5%	Zimbabwe	64.3%	103.1%
Liberia	63.5%	61.8%	Africa	85.0%	119.3%
Libya	49.2%	481.3%	World	117.3%	76.6%

Source: FAO (2002).

examination of the data at the country level indicates that some countries are making progress in increasing output in food crops above population growth despite many challenges. Thirty-nine percent of the 54 countries listed in Table 8-35 increased their per capita production of roots and tubers between 1960–1965 and 1997–2001. Twenty-six percent increased their per capita production of cereals over the 40-year period.

Food insecurity and famine in Africa are regional rather than continental phenomena that are linked to environmentally marginal environments, politics, and conflict. As de Walle (1996) pointed out, food shortages and famines occur, but the famines that kill are those that are exacerbated by other factors, for example, war and political and economic instability. Drought exacerbated by war and Stalinist policy in Ethiopia in the 1980s caused countless deaths. The same may be said for Angola from the early 1970s until recently (Figure 8-57). In both countries the negative impact of the radical reorganization of agriculture was exacerbated by war. More recently, political upheaval in Zimbabwe has caused people in this once-stable state to teeter on the brink of famine. Almost everywhere, HIV/AIDS will continue to weaken and kill agriculturalists, with serious consequences for production.

International trade barriers that keep more competitive African products off the world market should be revised. It surely is important for African governments to "get prices right" on farm products to provide incentives to farmers to increase production, but economic activity of any sort needs economic and political stability to grow. There are many studies of

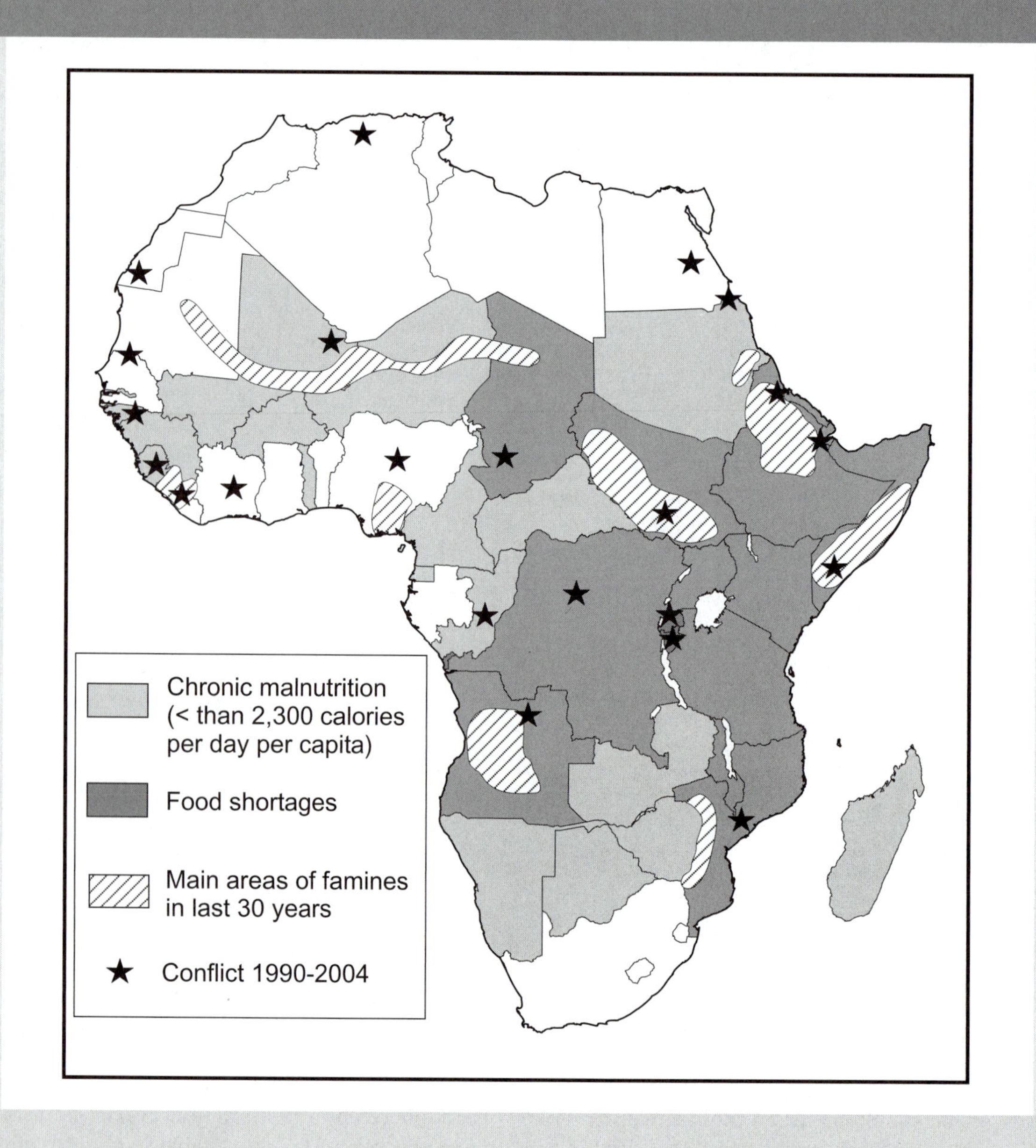

FIGURE 8-57 Regions of chronic malnutrition, food shortages, famine, and conflicts.

African famine but comparatively few of African agriculture.

Geographers have used drought and famine as instances in which the workings of people and society under stress could be studied. This kind of research is important; but it is equally, if not more, important to get a micro-level view of African agriculture outside times of crisis. Micro-level studies will help sweep aside the continent-wide generalizations that have been the stuff of stereotype and give us an empirical picture of the reality on the ground. An excellent example of a micro-level analyses is provided by Tiffen and Mortimore (1994), who found in the Machakos District of Kenya that a combination of sound government policy, farmer-led technology change, and better market linkages and incentives enabled a semiarid area that had been categorized as an "environmental disaster" in the 1930s to support five times as many people in the 1990s, with higher yields under a more labor intensive system. [See also Mortimore (1998) for an evaluation of agriculture in the drylands of Africa.] In the longer term, we can also expect the populations of Africa, encouraged by their governments, to make their demographic transition from high to low birth and death rates.

CHAPTER REVIEW QUESTIONS

1. What cultural and historical factors might explain why the countries located along the Gulf of Guinea produced and continue to produce prodigious quantities of yams but relatively little cassava (manioc)?
2. Nigeria is clearly the agricultural powerhouse of Subsaharan Africa. Discuss the relationship between population and agriculture and the future of Nigerian farming.
3. What crops were domesticated in Africa, and how were other crops later introduced? Discuss the impacts such agricultural diversity might have had on Africans.
4. What hypotheses might explain why Africa's contribution to the world banana production is so small, yet its plantain production is highest in the world?
5. How does standardizing livestock data, versus reporting absolute numbers of stock per country, influence the interpretability of country-by-country comparisons? To extend your answer into another direction go to the FAO Web site (FAOSTAT) at http://faostat.fao.org and download data for agricultural area for all countries of Africa. Note that the land-use data are reported in units of 1,000 hectares. Use the spreadsheet to convert 1,000 hectares to square miles. Put the data in a spreadsheet form, and calculate the density of cattle, sheep, and goats per square mile of agricultural land. How do these results differ from that standardized by rural population? Which is the better way to represent livestock populations?
6. Discuss the similarities and differences in objectives and strategies of raising livestock and investing in stocks listed on the Dow Jones Industrial Average.
7. Where were cattle first domesticated, and how did they become so widespread in Africa? What limiting factors exist to the distribution of cattle in Africa.
8. Describe the characteristics of some of the most common stock in Africa: cattle, camels, goats, and sheep.
9. What strategies do pastoralists use to make a living in the drylands of Africa?
10. Write a paragraph describing how you would imagine the sexual division of labor to be organized in a pastoral society. Use that paragraph as the basis for a class discussion about gender and land use in Africa.

GLOSSARY

Alluvial fan Usually a delta-shaped landform found in locations where a mountain valley meets a plain and the water coursing from the valley slows down and spreads out, dropping much of its suspended load of sediment. The alluvial fan consists of accumulated sediment.

Capital-intensive Describing a system that uses machinery (capital goods) to replace labor. In agriculture a capital-intensive farm would be one that used tractors in the place of human labor in most farm operations.

Cooperative system A way to organize agricultural production such that different families or villages form a larger unit to gain economies of scale. Often the "cooperation" relates to the purchase and use of farm machinery that would be prohibitively expensive for a single farm family to purchase. Cooperatives are also used in the bulk purchase of farm inputs such as seeds and fertilizer and in the marketing of produce.

Economies of scale An economy of scale occurs when it is cheaper per unit of output to produce in a large farm or factory than in a small farm or factory.

Factors of production The social, economic, and technical elements that are assembled to produce a product. In agriculture, the principal factors of production are land, labor, and capital.

Fallow Periodic rest for a piece of land to allow it to regain fertility.

Friction of distance The cost of movement over space. There is more "friction" over land than water. The cost of movement varies with the mode of transport and the condition of the transportation infrastructure.

Ideographic Description, describing. An ideographic way of knowing is focused on describing (painting a picture of) the phenomenon being examined. Compare with nomothetic.

Infield/outfield farming The concentration of labor and other inputs in the fields close to the village to reduce the costs of movement.

Leguminous crops Those that fix nitrogen in the soil, for example, beans.

Mean The arithmetic average of a series of numbers. The mean is one of the principal measures of central tendency used in statistics to describe or summarize a data series.

Median The value in a series of numbers located such that 50% of the values lie above it and 50% of the values lie below it. A measure of central tendency that is useful in showing whether the mean is a good descriptor of a data series.

Nomothetic Science or inquiry that focuses on understanding the laws that govern or influence a particular phenomenon.

Parastatal A state corporation involved in some aspect of the economy, whether for the promotion and management of a certain commodity or for development. British Gas was a parastatal owned by the British government whose responsibility was providing natural gas to Britons. Opération Riz is a Malian parastatal that is responsible for the promotion and development of rice in Mali.

Periodic market A periodic market is held intermittently over a given time period, usually a week. In West Africa many periodic markets meet once a week. In other areas the market cycle is every 4 days. A periodic market occurs where demand is not sufficient to support permanent retailing.

Quartile A measure of the location of a given point within a data series. Quartiles are generally used to determine what is unusual (above the 75th or below the 25th quartile) from what is common (between the 25th and 75th quartiles).

Regional complementarity One of the bases of spatial interaction: when one region or country produces a commodity of which another region or country has a deficit, the potential exists for exchange between the two areas.

Standard deviation A statistic that measures the distribution of values around their mean. When the standard deviation is disproportionately large, the mean may not be a good summary of the data series.

Statist, statism A political condition in which the national government has an inordinate amount of power and responsibility in the national and local economy, the process of socioeconomic development, culture, and people's day-to-day lives, leaving little or no room for individual initiative or enterprise.

Transhumance Seasonal movement with livestock, which can be vertical (from lowland to highland), horizontal (from river valley to the edge of the seasonally rain-fed desert, or some combination.

Tropical Livestock Unit (TLU) A composite measure that standardizes and summarizes livestock numbers in a region or country. It is most often calculated by multiplying camels by 1, cattle by 0.7, and sheep and goats by 0.10.

BIBLIOGRAPHY

Andah, B. 1993. Identifying early farming traditions in West Africa. In *The Archaeology of Africa: Food, Metals, and Towns*, T. Shaw, P. Sinclair, B. Andah, and A. Okpoko, eds. London: Routledge, pp. 240–254.

Arnon, I. 1972. *Crop Production in Dry Regions: Vol. 2, Systematic Treatment of the Principal Crops*. London: Leonard Hill Books.

Brooks, M. 1998. *Qat, Environmental Impacts, Social Effects, and Trade*. Trade and Environment Case Studies, 8(1), January, http://www.american.edu/ted/index.htm.

Bulliet, R. W. 1990. *The Camel and the Wheel*. New York: Columbia University Press.

Clutton-Brock, J. 1993. The spread of domestic animals in Africa. In *The Archaeology of Africa: Food, Metals, and Towns*, T. Shaw, P. Sinclair, B. Andah, and A. Okpoko, eds. London: Routledge, pp. 61–70.

Cole, R. 1982. The sedentarization of pastoral nomads in Africa: An examination of twenty-three settlement schemes. Master's thesis, Michigan State University.

Curtin, P. 1984. *Cross-Cultural Trade in World History*. Cambridge, U.K.: Cambridge University Press.

de Walle, A. 1996. *Famine That Kills*. London: Clarendon Press.

Ellis, S. 1999. *The Mask of Anarchy: The Destruction of Liberia and The Religious Dimension of an African Civil War*. New York: New York University Press.

Epstein, H. 1971. *The Origin of the Domestic Animals of Africa*. New York: Africana Publishing.
FAO 2001. *The State of Food Insecurity in the World*. Food and Agriculture Organization of the United Nations. Rome: United Nations.
FAO 2002. *The State of Food Insecurity in the World*. Food and Agriculture Organization of the United Nations. Rome: United Nations.
FAO 2003. FAOSTAT. Rome: United Nations. Available online at http://www.FAO.org
FAO 2005. FAOSTAT. Rome: United Nations. Available online at http://www.FAO.org
Felker, P. 1978. State of the art: *Acacia albida* as a complementary intercrop with annual crops. Washington, D.C.: USAID Information Services.
Felker, P. 1981. Uses of tree legumes in semiarid tropics. *Economic Botany*, 35(2): 174–186.
Franke, R., and B. Chasin. 1980. *Seeds of Famine*. New York: Rowman & Littlefield.
Gilman, A. 1976. *A Later Prehistory of Tangier, Morocco*. Cambridge Mass.: Harvard University Press.
Grigg, D. G. 1995. *An introduction to agricultural geography*, 2nd ed. London: Routledge.
Hanotte, O., C. L. Tawah, D. G. Bradley, M. Okomo, Y. Verjee, J. Ochieng, and J. E. O. Rege. 2000. Geographic distribution and frequency of a taurine *Bos taurus* and an indicine *Bos indicus* Y-specific allele amongst Subsaharan African cattle breeds. *Molecular Ecology*, 9: 387–396.
Hanotte, O., D. G. Bradley, J. W. Ochieng, Y. Verjee, E. W. Hill, and J. E. O. Rege. 2002. African pastoralism: Genetic imprints of origins and migrations. *Science*, 296: 336–339.
Hardin, G. 1968. The tragedy of the commons. *Science*, 162: 1243–1248.
Harlan, J. R. 1993. The tropical African cereals. In *The Archaeology of Africa: Food, Metals, and Towns*, T. Shaw, P. Sinclair, B. Andah, and A. Okpoko, eds. London: Routledge, pp. 53–60.
Hickman, G. 1990. *The New Africa*. London: Hodder & Stoughton.
International Institute of Tropical Agriculture. no date. Yam. http://www.iita.org/crop/yam.htm.
Johnson, D. L. 1969. *The Nature of Nomadism: A Comparative Study of Pastoral Migrations in Southwestern Asia and Northern Africa*. Department of Geography Research Paper no. 118. Chicago: University of Chicago.
Lericollais, A. 1972. *SOB. Étude géographique d'un terroir Sérèr*. Paris: ORSTOM.
Meade, M. S., and R. J. Earickson. 2000. *Medical Geography*. New York: Guilford Press.
Monod, T. 1975. *Pastoralism in Tropical Africa*. Oxford: Oxford University Press.
Monteil, C. [1924] 1977. *Les Bambara du Ségou et du Kaarta: Étude Historique, Ethnographique et Littéraire d'une Peuplade du Soudan Français*. Paris: Maisonneuve and Larose. Reprint of original 1924 edition.
Mortimore, M. 1998. *Roots in the African Dust: Sustaining the Drylands*. Cambridge, U.K.: Cambridge University Press.
Murdock, G. P. 1960. Staple crops of Africa. *Geographical Review*, 50: 523–540.
National Research Council. 1996. *Lost Crops of Africa: Vol. 1, Grains*. Board on Science and Technology for International Development. Washington, D.C.: National Academies Press.
Newman, J. 1995. *The Peopling of Africa: A Geographic Interpretation*. New Haven, Conn.: Yale University Press.
Okigbo, B. N. 1980. Plants and food in Igbo culture. Ahiajoku Lecture. State Ministry of Information, Culture Division, Owerri State, Nigeria. Cited by Andah, B. W. 1993. Identifying early farming traditions of West Africa. In *The Archaeology of Africa: Food, Metals, and Towns*, T. Shaw, P. Sinclair, B. Andah, and A. Okpoko, eds. London: Routledge, pp. 240–254.
Pullen, R. 1974. Farmed parkland in West Africa. *Savanna*, 3(2): 119–151.
Purseglove, J. W. 1976. The origins and migrations of crops in tropical Africa. In *Origins of African Plant Domestications*, J. R. Harlan, J. M. de Wet, and A. B. L. Stemler, eds. The Hague: Mouton, pp. 291–309.
Roberts, R. L. 1996. *Two Worlds of Cotton: Colonialism and the Regional Economy in the French Soudan, 1800–1946*. Stanford, Calif.: Stanford University Press.
Rosegrant, M. W., M. S. Paisner, S. Meijer, and J. Witcover. 2001. *Global Food Projections to 2020: Emerging Trends and Alternative Futures*. Washington, D.C.: International Food Policy Research Institute.
Ruthenberg, H. 1976. *Farming Systems in the Tropics*. Oxford: Clarendon Press.
Stephens, J. M. 1994. Yams. Fact Sheet HS-686. Horticultural Sciences Department, Florida Cooperative Extension Service, Institute of Food and Agricultural Sciences. Gainesville: University of Florida.
Stock, R. 1995. *Africa South of the Sahara: A Geographical Interpretation*. New Haven, Conn.: Yale University Press.
Tiffin, M., M. Mortimore, and F. Gichuki. 1994. *More People, Less Erosion: Environmental Recovery in Kenya*. New York: Wiley.
United Nations. 2002. *Economic Development in Africa, from Adjustment to Poverty Reduction: What Is New?* UN Conference on Trade and Development. New York: United Nations.
Vansina, J. 1990. *Paths in the Rainforest: Towards a History of Political Tradition in Equatorial Africa*. Madison: University of Wisconsin Press.
Watson, P. J. no date. *History of Rubber*. Heron House, Wembley, U.K.: International Rubber Study Group.
Watt, J. M., and M. G. Breyer-Brandwijk. 1962. *The Medicinal and Poisonous Plants of Southern and Eastern Africa*, 2nd ed., London: Livingstone.
Wilkin, G. 1972. Microclimate management by traditional farmers. *Geographical Review*, 62: 544–560.
World Bank. 1994. Quoted in United Nations (2002). *Economic Development in Africa, from Adjustment to Poverty Reduction: What Is New?* UN Conference on Trade and Development, New York: United Nations.

CHAPTER 9

African Development, Manufacturing, and Industry

The topic of development has been woven through several of the preceding chapters. This is no accident, for African regions, states, and localities are not static entities but are best understood as being located on a continuum (or continua) of change. The location of any area on a continuum of development is dependent on global economics, national policies, resource endowments, culture, geography, and history.

This chapter explores the conceptual and empirical dimensions of development and the possibilities and problems for African nations more closely than we have done thus far. During the heady days at the dawning of independence, Africans believed that ridding themselves of colonial powers would usher in a period of development, prosperity, national growth, and well-being. Colin Leys (1996) calls this a phase of optimism in which newly independent Africans commonly and, with hindsight, naïvely assumed that their economies would grow by at least 5 or 6% each year.

The phase of optimism was dashed by postindependence realities that led to the articulation and elaboration of another theory of Africa's failure to develop, neocolonialism. This formerly popular theory postulated that what African countries received at independence was not total independence from their respective European colonizers but merely political independence. Economic control was still tightly held by the former colonial powers. Many thought at the time that the putative economic control had to be broken and that African development would be along lines different from Europe's. Current thought, quite a turnaround, holds that the economic relations with the former colonial power need to be deepened and at the same time broadened to include other industrialized countries and that African countries will follow essentially the same developmental path as Europe and other presently more developed regions of the world.

FACETS OF DEVELOPMENT

Development is a process that is influenced by geography and culture. Even if we define geography in a limited way, referring only to location, geography clearly matters in development, and a glance at the countries of Singapore and Burkina Faso will help explain why. Singapore is an island city-state located in one of the most strategic straits in the world; through the Strait of Malacca passes much of the world's oil supply from the oil-producing economies in the Middle East to the oil-consuming economies in East Asia, Europe, and North America. Singapore is conveniently located between the growing economies of South and East Asia and, in addition to petroleum, other raw materials and manufactured goods worth billions of dollars pass through the Strait of Malacca each year. Geographically, Singapore is well placed as an economic pivot between East and West, serving both markets in manufactured goods, particularly electronics. Burkina Faso, on the other hand, has a poor location relative to the commodity flows of the world economy; it is a totally landlocked state with absolutely no access to the sea. Furthermore, it is not located in any beneficial proximity to a large or growing economy, as Mexico is to the United States, for example.

The role of geography becomes ambiguous if we extend our definition to include climate, the environment, and natural resources. Although the climate of Singapore is tropical moist, 100% of the population lives in urban areas in an artificial environment very removed from the daily challenges of climate. In the not very distant past, however, people earned their living from the tide and agriculture, and one aspect of Singaporean development has been urbanization. Surprisingly for a country as rich as it is, Singapore has no natural resources to sell to the world. In contrast to the climate of Singapore, Burkina Faso's climate has no moist maritime influences. It straddles the Sahara, and the vast majority of Burkinabe are dryland farmers or pastoralists living in small villages or hamlets close to the natural environment. People in Singapore live long lives, are well educated, and have a high standard of living, in vivid contrast to the people in Burkina Faso who are poorly educated, have a low standard of living, and die young. Judging from this comparison of two countries, some elements of geography seem to matter in development and others do not.

TABLE 9-1 PERCENTAGE OF PEOPLE LIVING ON LESS THAN $1.00 PER DAY FOR FIVE WORLD REGIONS.

	Year		
Region	**1987**	**1990**	**1998**
East Asia and the Pacific	26.6	27.6	14.7
(excluding China)	23.9	18.5	9.4
Europe and Central Asia	0.2	1.6	3.7
Latin America and the Caribbean	15.3	16.8	12.1
Middle East and North Africa	4.3	2.4	2.1
South Asia	44.9	44	40
Subsaharan Africa	46.6	47.7	48.1
Total	28.3	29	23.4
(excluding China)	28.5	28.1	25.6

Source: World Bank (2001).

Culture is important in development because people who have been raised in different cultures may have very different ideas about what in important in their lives and, concomitantly, what should and should not be part of the changes that development represents. Africans tend to be more conservative than Americans when it comes to embracing change, seeing much in their own behavior, values, and institutions that is valuable and worth preserving—and rightly so! Many Africans want the benefits of Western technologies, such as television and the Internet, but not what they view as the associated Western social pathologies, for example, hyperindividualism, the breakdown of the family, social fragmentation and alienation, pornography, and Western media models of behavior that seem to run counter to African values concerning the family, sex, and children's and women's behavior.

It is no secret that development is about reducing poverty, a constant companion to most Africans (Table 9-1). But development is much more than that—it is a complex phenomenon that almost defies definition; surely there is no definition that pleases everyone, and in some sense development is in the eye of the beholder.

A general, conceptual definition of development might be "a process by which the political, social, economic, and environmental conditions and well-being of the people living in a region or a country are improved." What the improvements might be is not specified in this purely conceptual definition. The Human Development Institute (HDI) created an empirical (measurable) definition of human development that can be used as a yardstick with which to measure and compare countries and regions within countries. The HDI defines development as a process of increasing people's choices; the three most basic and essential ones are those that allow people to lead a long healthy life, to acquire knowledge, and to have access to the resources needed for a decent standard of living. These three choices are measured as individual variables for any country by life expectancy in years, the literacy rate, and per capita gross national product, respectively, and then combined in the index (0 being low and 100 high).

The Human Development Institute holds that if these essential choices are not available, many other opportunities are out of reach, for example, "political,

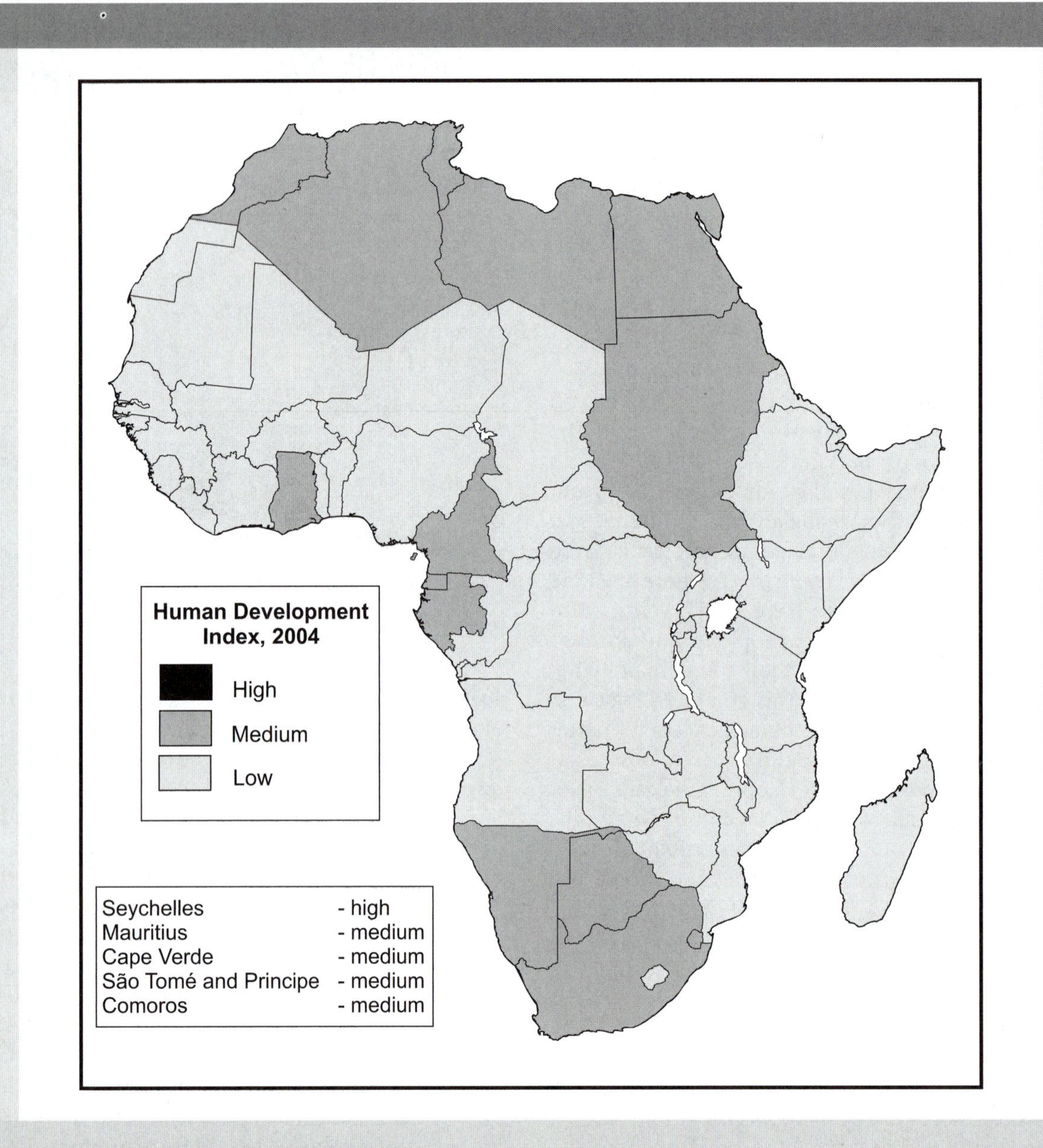

FIGURE 9-1 Level of human development in Africa in 2004.

economic, and social freedoms, opportunities for being creative and productive, and enjoying personal self-respect and guaranteed human rights" (UNDP 1990). While only one African country (Seychelles) is in the high human development category, some are in the medium category, and most are in the low development category (Figure 9-1), human development in Africa as a whole is improving—albeit at a growth rate slower than that in any other world region.

Some African countries have lost ground over the short and long term. For example, Zambia had a lower HDI score in 2000 than it did in 1975, reflecting the downturn in the price of copper, the principal export, as well as lowered copper exports, decreased life expectancy due to the HIV/AIDs pandemic, and the declining effects of education as children are forced to leave school to become breadwinners for their sick parents. Zimbabwe's HDI was lower in 2000 than at any time since independence in 1980 as a result of HIV/AIDS, political instability, and drought. Botswana, Burundi, the Democratic Republic of the Congo, and Lesotho had lower scores in 2000 than in 1985; Cameroon, Kenya, South Africa, and Swaziland had lower scores in 2000 than in 1990; and Malawi and Namibia had lower scores in 2000 than in 1995.

There are other ways in which to measure development in addition to the Human Development Index. Some common indicators of development are the per capita gross national product (GNP), per capita gross domestic product (GDP), life expectancy, crude birthrate (CBR), crude death rate (CDR), infant mortality rate, total fertility rate (TFR), literacy rate, and number of physicians per million population. More recently, researchers have begun to look at the relationship between gender and development as measured by the female literacy rate, the number of women in governance, and the amount of female participation in other spheres of life.

Political freedoms and democracy are also part of development. Democratic systems seem to be associated with economic growth and richer countries tend to be more democratic. Although it is true that the fastest-growing economies in the world are dictatorships, the worst performing economies are also dictatorships (UNDP 2002). Some analysts feel that democratic countries encourage growth by being good guarantors of property rights, and where property rights of individuals are poorly backed up by the state, entrepreneurs invest elsewhere (UNDP 2002). It is said that unlike dictatorships, democracies engage in far fewer unwise military adventures and do not wage war on one another. A wide variety of measures of political freedom and democracy have been developed and used to rank countries of the world. Freedom House has developed three indices of freedom: civil liberties, political rights, and freedom of the press. The indices class countries as "free," "partly free," or "not free" and may be combined to give an overall score that can be used to compare one country with another. Transparency International has developed a Corruption Perception Index, which measures the degree of official corruption perceived by academics, analysts, and businesspeople in a country.

Other measures of development have to do with geographic and social connectivity and the state of infrastructure in a country or region. Examples include miles of paved road per capita, telephone lines per capita, and televisions and radios per capita. In all the measures of development, listed in Table 9-2, Africa lags behind most other parts of the world.

STRATEGIES FOR DEVELOPMENT

There is much about defining development on which geographers and other social scientists, theorists, and practitioners disagree. All observers, however, agree that development is about generating wealth (or capital) at the national level to invest in agricultural, industrial, and social transformations and to pay for the improvement in the standard of living of the citizenry. The United Nations (UNCTAD 2000a, 2000b) recognizes that poverty reduction is a very important development issue.

Although there have been many disagreements on the best way to implement a vision of development, the major area of disagreement has been growth versus equity. When one speaks of development as growth, economic growth is what is meant: higher output, incomes, savings, and investment. Growth-oriented development is focused on competitive free markets, privatization of state-owned industry and enterprise, opening up a nation's domestic economy by encouraging free trade (through the abolition of tariffs and quotas), and foreign direct investment (FDI) rather than aid. Growthists believe that social inequality is a necessary part of the market economy, but a rising tide lifts all ships—that economic growth will benefit the poor as well as the rich. There is evidence to support this contention. Dollar and Kraay (2000), for example, found in their research on 80 countries over a 40-year period that contrary to the commonly held view that economic growth benefits only the rich and may even exacerbate poverty, the incomes of the poor rise one-for-one with economic growth in a country.

For a growthist, development might be defined very simply as the application of capital and labor to create value or generate a profit. This view is overly simplistic, however, since capital and labor are applied

TABLE 9-2 SELECTED WORLD DEVELOPMENT INDICATORS.

	Region					
Development Indicators	**Subsaharan Africa**	**Latin America and Caribbean**	**East Asia**	**Europe**	**United States**	**World**
Life expectancy at birth (years)						
Females	49	75	75	79	80	69
Males	47	69	71	71	75	65
GNI PPP per capita[1], 2004 (US$)	$1,830	$7,530	$7,990	$19,980	$39,710	$8,540
Infant mortality rate[2]	93.8	27.5	24.8	7.1	6.6	54
Literacy rate[3]						
Female	69	96	99	–	97	82
Male	80	95	99	–	97	89
Population with access to improved drinking water source, 2002 (%)						
Rural	50	69	69	–	100	71
Urban	85	96	94	100	100	94
HIV/AIDS among adult population, ages 15–19, 2003–04 (%)	6.8	0.7	0.1	0.5	0.6	1.2

1. Gross national income at purchasing power parity per capita. Purchasing power parity is a standardized measure that refers to a common basket of goods referenced from country to country.
2. Mortality per thousand live births among children under the age of 5.
3. Percentage literate for 15–24 age group, 2000–2004.

Source: Population Reference Bureau (2005).

not in a vacuum but in a social, cultural, environmental, and political context. Although early growth-oriented developmental economists envisioned a central role for the state in creating a favorable economic and social climate for growth, managing markets and people, and guiding development (Hameso 2001), mainstream professional opinion changed. Economic growthists are advocates of free markets, free enterprise, free trade, and small, weak but honest governments that do not interfere in the marketplace. Deepak Lal, arguing against centrally planned national economies such as those in place at the time in the Soviet Union, which had provided models for many African countries, stated; "The strongest argument against planning . . . is that, whilst omniscient planners might forecast the future more accurately than myopic private agents, there is no reason to believe that flesh-and-blood bureaucrats can do any better—and some reason to believe that they may do much worse" (1983: 75). Neoliberal (free-market) economists believe that the market is the answer and that growth problems in developing countries are the result of internal rather than external causes, such as large and overstretched public sectors and rigid economic controls that reduce incentives to private initiative.

Opposing the growthists along a theoretical divide on development are those who feel that economic growth can be very dangerous and must be controlled by the state or party lest it create or exacerbate social cleavages. These economists raise issues of equity: social equity between rich and poor, dominant and dominated, center and periphery. Development theorists who believe that social equity is of critical importance say that to minimize the gulf between rich and poor and to ensure that the benefits of development are shared by all, markets, enterprises, and inequality must be controlled and managed, generally by a large and powerful state. The objective of both the growth and the equity development theory is to improve the well-being of people, but each has its own recipe for success. The free marketeers believe that a free market generates the greatest benefit for the greatest number of people; statists and socialists want government to take a firm and active hand in the planning of development to reap the most benefits.

It is indisputable that people living in more developed countries (MDCs) that are organized along free-market principles live longer, lead healthier lives, have more economic security, exercise more freedoms, and are more educated than their counterparts in less developed countries (LDCs). A country that is more developed is better able to cope with, for example, economic downturns. A less developed country often appears to be at the mercy of the monsoon, drought, disease, or other outside forces. The more developed countries of the world appear to have stable governance and democracy, while many less developed countries, particularly in Africa, can be characterized as failed states with dictatorial and patrimonial rule and widespread corruption.

The gulf between rich and poor countries is so great that a tripartite taxonomy comprising First, Second, and Third Worlds, was developed after decolonization over 40 years ago. The First World referred to the market-oriented, industrialized economies such as those of the United States, Japan, Australia, Canada, and Western Europe. The Second World referred to the planned (command) economies of the socialist world as represented by the Soviet Union, China, East Germany, and Eastern Europe. The Third World consisted of the rest of the countries of the world, that is, the LDCs. The poorest of the poor in Africa, mainly the landlocked Sahelian countries, sometimes have been referred to as the Fourth World.

With the collapse of the Soviet Union in the 1990s and the shift away from communism and the planned economy in other countries, the Second World has largely disappeared; what is left is a similar classification having three parts. The First and Third Worlds remain the same, but the second category has been replaced by another category, the "newly industrializing countries" (NICs), "emerging economies" generally with an export orientation. Some NICs (or newly industrializing economies, NIEs) are South Korea, Taiwan, Brazil, Argentina, Vietnam, and China (especially coastal China). Significantly, there are no African countries among the NICs.

A variety of concepts have been put forward by economists over the past 50 years such that one can almost speak of prescriptions or recipes for development. Undoubtedly one of the most persistent has been the stages-of-growth model developed by Rostow (1959) during the Cold War. In this modernization paradigm, Rostow described the process of development from a traditional agrarian, inward-looking, poor society to a modern, industrial, urban, democratic, affluent consumer economy. Rostow's historical model was Great Britain which, he felt, other countries would inevitably emulate. The Industrial Revolution occurred first in Great Britain, and as a consequence Britain was the first country to develop along modern lines. Where Britain's development had been endogenous (internally generated), other countries' experience would be exogenous, stimulated from outside by other, more advanced, countries like Great Britain.

Early developmental economists thought that a developing economy needed a large infusion of capital, a "big push," to jump-start self-sustaining industrial development. The "big push" model for development originated in the 1950s and involved the massive investment of capital by the state in order to industrialize a country. With industrialization, manufactured goods produced for domestic consumption begin to replace more expensive imports. Since African countries did not have the capital to invest, funds were provided by donor countries, often the former colonial powers or the World Bank. In his theory of economic development with unlimited supplies of labor, W. A. Lewis (1954) postulated the existence of two sectors in a developing economy, modern and traditional. The modern sector was characterized by up-to-date technology, western corporate organization, and a market orientation. The capital-intensive production of monocrops or plantation crops exemplified the modern agricultural sector in Lewis's analysis, as did capital-intensive urban factories. In stark contrast was the family-centered, labor-intensive, and capital-scarce traditional sector, where industrial production took place at home or in small shops or cottages and could better be termed "crafts" or "artisanal production." Agriculture in the traditional sector was based on human or animal motive power. Lewis felt that human labor was not being used as efficiently as possible in the traditional sector and that productivity could be increased by tapping into this "hidden" labor pool. He believed that labor could be released from the traditional sector with the right wage incentives to draw workers into the urban areas for employment, and that this relocated population could power urban industrial development and constitute a urban market.

In contrast to market-centered approaches to organizing development, the Soviet Union and China put forward revolutionary socialist models. The Soviet state nationalized all private property and took control of all means of production including banking and finance. The one-party Soviet model of development involved taxing the agricultural sector to pay for urban industrialization. The state taxed farming by fixing prices for agricultural products at low levels and was supposed to invest the surplus in building large-scale factories, feeding workers, and furthering urban development. Agricultural organization was radically changed by coercively organizing farmers into cooperatives and collectives. It was thought that large-scale organization would enable economies of scale and increased production. Extensive 5-year development plans were elaborated in Moscow to manage the production of goods and provision of services across the entire country. Although the Chinese model also involved the seizure of the means of production, nationalization, and the creation of collective units of production, it differed from the Soviet model in that the rural areas were the main focus of industrial development, with the goal of agricultural transformation through collectivization.

Like their European and Asian counterparts, African manifestations of revolutionary socialism claimed to represent the people (African socialism), thereby constituting bottom-up development, or "development from below." In reality such claims represented mere ideology; both were top-down forms of development, led for the most part by European-educated, urban-bound bureaucrats and ideologues who imposed their vision on an unconvinced and often uncooperative populace. In Africa, many countries followed the Soviet and Chinese models only partially, since African states lacked the coercive power and the money to enforce revolutionary socialist policies on the ground. In addition, and this was true of the more market-oriented states as well, the cultural diversity and wobbly infrastructure of the newly independent states led to the differential application of policy. It was said of many Subsaharan African countries at independence that the paved roads ended at the edge of the capital city. It is also certainly true that policy formulated in many capital cities was more strongly applied in and around the capital: with distance from that city and the major roads leading to it, the effects of developmental policy were less noticeable. The weak application of the villagization and collectivization of agriculture policy in Tanzania, for example, lessened the negative impact of those concepts on Tanzanian society. Had the Tanzanian government at the time been strong enough to effectively enforce the policy, however, there would have been much more social dislocation and human suffering than there was (Scott 1998). In a similar vein, Falola describes the decline of the Ghanaian economy from independence to 1966, when the revolutionary socialist government was overthrown in a military coup:

> In the Ghanaian economy, public enterprises were allowed to collaborate with foreign ones. By using state agencies to buy and market cocoa, the leading export commodity, the government was able to make a profit by underpaying the farmers. Angry and disappointed by the excessive government control, many farmers cut down on cocoa production. In the early 1960s, cocoa prices declined on the world market, the economy of Ghana began to suffer a decline, and the country resorted to external borrowing. By 1966 when Nkrumah was overthrown, the price of cocoa was low and the economy was in serious trouble. (Falola 2002: 258)

In general, state-led, state-controlled development was the norm in the early development decades in Africa whether the ideology was communist, socialist, or capitalist. The size and role of the state in the African versions of the socialist models was larger than the market-oriented approaches to development in Africa; nevertheless, the state was the principal actor in development in all models. Development was largely focused on projects or "schemes" funded by outside donors.

By the early 1970s, the Second Development Decade as declared by the United Nations, it was becoming increasingly clear that the development plans in Africa, especially Subsaharan Africa, were not achieving the intended results (Quinn 2002). Political instability became chronic, such that 21 countries were under military rule after coups d'état by 1975 (Falola 2002). Drought and famine in the Sahel, conflict and civil war in several African countries, economic stagnation, and well-supported critiques of the modernization paradigm led development planners to formulate the "basic needs" strategy to address the plight of the poor.

The basic idea behind the basic needs strategy was to raise the productivity of the poor so that they could become part of the economic system rather than exist on the margins. The central concept of basic needs seemed

> so simple that one wonders why it took so long to see the light of day. As 40 per cent of people in the South [third world] live in "absolute poverty," does it not make sense to attend to the most urgent things first by enabling them to gain (or regain) decent conditions of existence? In order to live, everyone obviously needs to eat, to have a roof over their head, to clothe themselves, to have surroundings conducive to health, and to receive some education so that they can "earn their living." Should "development" priorities reflect these invariable basic needs of "human nature"? (Rist 1997: 163).

The theoretical foundations of the basic needs strategy were weak, and there were few practical applications (Rist 1997). However the ideology of addressing the basic needs of the poor was fashionable into the 1980s, when structural adjustment programs (SAPs) were formulated and implemented in Africa. Structural adjustment involved the recognition of internal, structural impediments to development in African countries. African governments, in collaboration with the World Bank and the International Monetary Fund (IMF), authored SAP documents to reform macroeconomic government policy that was thought to work against economic growth, and plans were outlined to reduce the size of the state sector and privatize state corporations. Adjustment involved devaluation of currency (kept artificially high by African governments to make foreign imports cheaper, with the unintended effect of making African exports expensive), reduced spending on social services such as health and education, and the encouragement of entrepreneurialism. Most African countries underwent structural adjustment, but 20 years later, the evidence is mixed, especially with respect to the impact of such policies on the poor (Saasa 2002). There has been growth, but structural adjustment did not work as well as it should have because it was imperfectly implemented by African governments.

A new poverty reduction strategy has been developed to complement structural adjustment in order to soften the negative impact of structural adjustment on the poor. This approach, which is expressed in strategy papers, authored by national governments and national stakeholders, is concerned with macroeconomic policy as well as good government. It is focused on "rapid and sustained growth and job creation" as well as "public provision of education and health services" (UNCTAD 2002a: 4).

By 2001, the lessons learned from 40 years of development efforts resulted in an initiative from African leaders that articulated a new policy for African development (Melbar et al. 2002). The New Economic Partnership for Africa's Development (NEPAD) is at one and the same time an ideology, a strategy, and a policy framework for change in Africa that appears to be informed by the policy and implementation problems in the past and built upon the concepts and recommendations of structural adjustment and the poverty reduction strategies of the World Bank.

> The New Partnership for Africa's Development is a pledge by African leaders, based on a common vision and a firm and shared conviction, that they have a pressing duty to eradicate poverty and to place their countries, both individually and collectively, on a path of sustainable growth and development and, at the same time, to participate actively in the world economy and body politic . . . (NEPAD 2001: 1).

Fundamental to the strategy are good governance, democracy, regional development, and a central role for private investment capital. According to Abdoulaye Wade (2002), president of Senegal, there are eight focus areas for investment and development: infrastructure, education, health, agriculture, the environment, information technology, energy, and access to the markets of MDCs. It is not an exaggeration to say that the NEPAD appears on the surface to be a departure from the traditional, statist model of African development. It is significant that instituting government and judicial reform, combating corruption, and insisting on accountability have central importance to the strategy. In addition, NEPAD calls for the building of civil society in Africa, a goal that includes the replacement of the one-party states and presidents-for-life by political pluralism, free elections, civilian control of the military, and parliamentary democracy.

Gathaka and Wanjala (2002) are skeptical of the New Economic Partnership for Africa's Development and dismiss it as merely a statement of principles by African leaders, the very same group of people who, they hold, are responsible for Africa's problems. It also remains to be seen how the plan will be implemented by the often deeply divided African leaders, and whether foreign investment capital will be attracted to Africa. Cornwall (2002) opines that the NEPAD may be Africa's last chance for development. Although that may be an exaggeration, his concern that NEPAD is already too top down and statist is a serious criticism that bodes ill for the plan's chances for success.

With or without NEPAD, foreign investment will initially go where it has always gone in Africa: to petroleum, strategic metals, precious stones. Although Zambia has privatized its copper industry, it still is a one-product exporter with only a dream of economic diversification.

A first step in reducing the importance of the state in the economy is privatization: shedding government ownership of breweries, cigarette factories, plantations, vehicle assembly plants, oil fields, flour mills, grain storage, state corporations and commodity parastatals, port facilities, mines, hotels, shops, factories, and the like. Also essential to the vision of NEPAD is the opening up of industrialized economies to African exports. NEPAD says that European Union protectionist tariffs are an impediment to African agricultural development because they make cheaper African food prohibitively expensive in foreign markets. It is unclear, however, that Europe has effective barriers to African farm products. Bananas and citrus and other tropical fruit grown in Africa are routinely available in Europe. France buys Senegalese peanuts and peanut oil at above world market prices and has done so for many years. On the other hand, yams, cassava, millet, and sorghum, some of Africa's major agricultural products, are not consumed to any degree in Europe. Europeans prefer the potato to the yam and would rather eat wheat than African cereals, even though the latter are quite tasty. There has been a lively market for many years in cut flowers, which are exported to Europe from several African countries. African vegetables also are sold in Europe. Thus much opportunity already seems to exist for trade in oil crops, vegetables, and tropical fruits.

In any event, private foreign investment is much different from government-to-government foreign aid in a couple of respects. First, private foreign investment is about making money, and government-to-government foreign aid is about politics. This was especially true during the Cold War, when superpowers were interested in winning the hearts and minds of Africans to particular economic and social ideologies. Second, although there is corruption in the business world—Enron is but one example—businesspeople in general are skeptics who demand accountability and will not throw shareholder value away, as superpower governments did on impractical and unprofitable activities that could ultimately undermine the sponsoring organization.

As the preceding discussion has indicated, there have been a variety of development strategies employed in Africa. In the pages that follow, we will examine the performance of Africa's economies, look at some key industries, and then consider two failed theories of development that were influential at one time in the past.

AFRICA'S ECONOMIC PERFORMANCE

Africa's economic growth in 2001 was faster than that of any other developing area of the world, a respectable 4%, "reflecting better macroeconomic management, strong agricultural production, and the cessation of conflicts in several countries" (United Nations Economic Commission for Africa 2002: 1). There were large disparities in the distribution of growth within Africa (Table 9-3). Africa's exports to the United States increased from $1.9 billion per month in 1999 to $2.3 billion a month in 2001, principally as a result of the African Growth and Opportunity Act, approved by the U.S. Congress in January 2001. Private capital flows to Africa have increased modestly; but foreign direct investment (FDI), the most important source of external capital for Africa, declined dramatically from 25% of all FDI to developing countries in 1975 to only 5% in 2000. Foreign aid to Africa, principally in the form of government-to-government grants and low-interest loans, is increasingly unattractive to donor countries, having fallen from $32 billion in 1991 to $17 billion in 2000.

TABLE 9-3 PER CAPITA GDP GROWTH IN AFRICA BY ECONOMIC GROUP, 2001-2002.

Group	Share of Africa's GDP in 2000 (%)	Share of Africa's Population in 2000 (%)	Mean Growth Rate	
			2001	2002
Top 5 African economies[1]	58.8	38.5	3.9	2.8
African oil exporters	49.1	35.0	4.1	2.8
Non-oil exporters	50.9	65.0	3.7	3.9
Least developed countries	16.5	48.1	4.8	6.2
Island economies	1.6	2.4	5.3	5.3
Landlocked countries	9.8	23.0	3.5	4.0
Subsaharan Africa	59.5	77.7	3.2	3.7

1. South Africa, Egypt, Algeria, Nigeria, and Morocco.

Source: United Nations Economic Commission for Africa (2002).

In 1975 per capita GDP growth in Subsaharan Africa averaged 1%, much lower than 2001–2002 share reported in Table 9-3. Up to 1999 the per capita GDP growth rates of 18 Subsaharan African countries were lower than in 1975, undermining the optimism generated by the figures in Table 9-3 after 20 years of economic reform in Africa.

ENERGY IN AFRICA

Development is predicated on the use of energy, principally electricity. The use of energy releases people from backbreaking drudgery, extends the length of the productive workday by lighting the night, powers modern industrial enterprises, and enables rapid communication among people. There are obvious qualitative and quantitative differences between working by light from a cotton rag soaked in shea nut oil and burning in a clay lamp, light from a kerosene lamp, and electric light.

The major source of energy in Africa has been biomass—wood and charcoal—and Africa is the world's largest consumer of biomass energy as a percentage of total energy consumption (UNEP 2002). Biomass as a source of energy has just a few uses in Africa: cooking and heating. Wood is the fuel of choice in the countryside for cooking and heating because it is readily available, while charcoal is used throughout urban areas. Fossil fuels are increasingly being substituted, particularly natural gas, among the more affluent. But for many families in the countryside and city alike, the quest for fuel becomes more demanding as population grows and forest is cut for farmland. The geographic pattern of fuel wood use in Africa is varied. In North Africa fuel wood use accounts for only 5% of the total energy consumption, while south of the Sahara (excluding South Africa) it accounts for 86% (EIA 1999). In South Africa, 15% of the energy consumed comes from biomass. The FAO (1995) has estimated that 130 million people in Subsaharan Africa live in areas where the consumption of fuel wood is greater than forest regeneration. Biomass consumption is increasing with population growth; but rather than fueling development, it is instead an indicator of low development.

The principal options for the generation of electricity in Africa are coal, oil and gas, hydroelectric power, and nuclear power. Only 2% of Africa's electricity was generated by nuclear power plants in 2001. Alternative methods of producing power such as solar and wind are found locally at the micro scale in Africa, but national development requires output at a larger scale than these technologies can provide at the present time. There is potential for the generation of electricity from geothermal power in the East Africa Rift Valley (Coakley et al. 2003).

Oil and gas are found in a handful of African countries and account for only 6% of world oil reserves (EIA 2003), but it seems as though more reserves are found every year. Africa obtained 79% of its electrical energy from thermal generation (heating water to make steam to turn turbines) in mainly oil

FIGURE 9-2 Kariba Dam, Zimbabwe. *Christine Drake.*

and gas powered plants (EIA 2003). Nevertheless, these resources could be exported to earn tremendous amounts of foreign exchange, which in turn could be invested in development.

Hydroelectric power is used to a greater extent in West and East Africa than elsewhere in Africa, but great potential exists in central Africa in the Democratic Republic of the Congo. The Congo River is a great, relatively untapped source of hydroelectric power. Countries that rely on hydroelectric power are normally small and nonindustrial, with little demand for electricity. Only 18% of Africa's electricity was generated by water in 2001 (EIA 2003). Every year hundreds of small dams are built to impound water in semiarid areas. Large hydroelectric projects like Zimbabwe's Kariba Dam (Figure 9-2) are being discussed in many countries. A few large hydroelectric dam projects were in development in Africa as the twenty-first century opened. In 2002 the Ethiopian government gave China a contract to build a dam on the Tekeze River, a tributary of the Blue Nile (BBC News 2002). Located in the drought-prone north of the country, the dam, in addition to producing electricity, is expected to irrigate farmland and help increase agricultural output. The dam will be the largest in Africa and, at 185 meters (607 ft) in height, will be 10 meters (32.81 ft) higher than China's Three Gorges project. It will be equipped with generators capable of producing 300 megawatts (MW) of electricity. It represents the largest hydro project in Africa and the largest cooperative project between China and an African country (People's Daily 2003). It is expected to be completed by 2007.

Ethiopia's neighbor, Sudan, is planning to build a dam on the White Nile River that will produce 1,250 megawatts a day. The dam, located near the Iron Age ruins of Meroë, will also provide water to irrigate the parched north of the country (*Arab-American Business* 2003). The dam will have profound human and cultural costs, as thousands of people will have to be relocated and the famous ruins at Meroë will likely be inundated.

Uganda is planning a dam on the White Nile near the Bujagali Falls, to be financed in part by the World Bank and built by an American company. The dam will be located about 50 miles east of the capital, Kampala, and is projected to generate about 250 megawatts a day. The project has been delayed for over 7 years by opposition. Although it will displace fewer than a hundred families, it will submerge the picturesque Bujagali Falls, a national treasure and popular tourist destination. Questions have been raised by Ugandan environmentalists regarding the inundation of productive agricultural land, as well as corruption ("rent-seeking" by those in positions of power) and the ultimate costs that will be borne by Ugandan consumers (IRN 2002). Given the limited possibilities for other energy sources, alternative (solar, wind) or not (coal, petroleum, gas), it is reasonable to assume that more hydroelectric plants will be built in Africa.

INDUSTRY

As a sector, industry includes mining and other extractive enterprises, manufacturing, construction, and the provision of electricity, gas, and water. Industry in 2000 accounted for just over a quarter (26.3%) of Africa's GDP, and manufacturing accounted for over half of the total (Table 9-4). The contribution to GDP of industry declined from 1980 to 2000 by almost 13%.

Manufacturing Industry

Manufacturing, regarded as key in development, is low throughout the continent but growing impressively in southern Africa (Table 9-4). The contribution to GDP by industry declined in North Africa, particularly in Algeria and Egypt, countries that were affected by the decline in petroleum prices during the 1980s and 1990s and plagued by Islamic extremism and civil war over the last 20 years. Industry is not as prevalent in Africa as in more developed regions. In 2003 only 13% of economically active Africans were employed in industry (of which 7% was manufacturing); 58% were employed in agriculture, and 29% were employed in the service sector but there are wide regional differences.

Whereas agriculture includes farming and pastoralism, industry concerns the direct extraction of raw materials and includes mining, forestry, fishing, and manufacturing (where raw materials are processed to produce goods). "Services" are activities that do not involve the direct processing of physical materials, such as retail and wholesale sales, marketing, banking, government services, and tourism.

Most modern African industries were found in or near the capital city at independence, but that pattern has been changing. There has been some regionalization of, for example, textile mills, which are located in cotton-producing regions. Extractive industries such as mining are, of course, located on the resource itself. In terms of gross domestic product in 2001 (World Bank 2002), Egypt, Morocco, and Tunisia in North Africa account for 39% of all African GDP in industry. South Africa's GDP accounts for 22% of the entire continent and 36% of Subsaharan Africa. Only in South Africa does one find the type of integrated manufacturing and processing operations that exist in the industrialized world. For the rest of Africa, industrial development since independence has been rather disappointing.

Zimbabwe, once one of the more industrialized African countries, has been experiencing political instability and rampant inflation associated with the government of President Robert Mugabe and faces deindustrialization because of the political climate and dangerous government policies to tax industry.

TABLE 9-4 DISTRIBUTION OF GDP BY SECTOR IN AFRICA AND ITS MAJOR REGIONS IN 1980 AND 2000.

	Agriculture		Industry		Manufacturing[1]		Services	
Region	1980	2000	1980	2000	1980	2000	1980	2000
Africa	22.3	20	39	26.3	8.7	13.2	38.7	53.7
North Africa[2]	13.5	16.6	48.5	37.6	8.8	11.3	38	45.8
West Africa	33.7	36.6	18.6	28.6	5.9	7.7	47.7	35.1
Central Africa	28.9	20.9	32.7	38.2	6.8	10.1	38.4	40.9
East Africa	32.6	38.3	16.6	18.2	8.3	7.5	50.8	43.5
Southern Africa	22.9	11	28.3	37.4	10.8	20.5	48.8	51.6

[1] Manufacturing is a subset of industry.
[2] Includes Mauritania and Sudan.

Source: United Nations Economic Commission for Africa (2002).

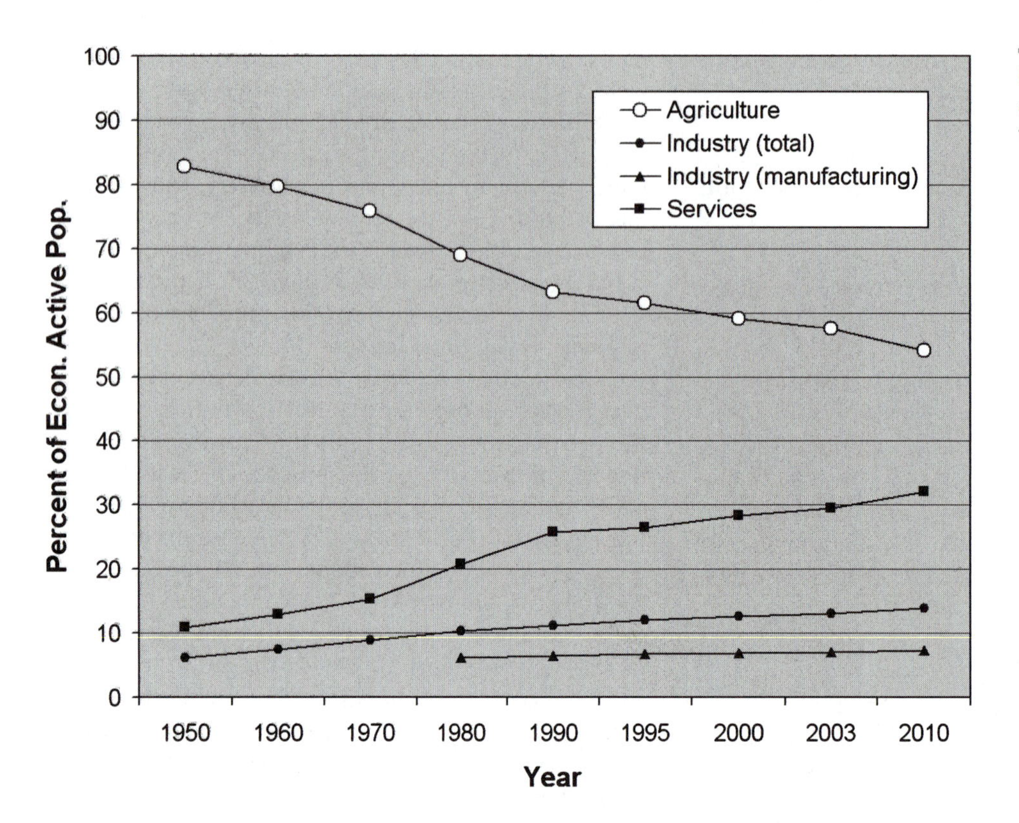

FIGURE 9-3 African employment by sector, 1950 to 2010.

Agricultural development in Zimbabwe has stalled in some areas and declined in others because of government ineptness in handling the white minority farm issue. Nigeria accounted for 13% of Subsaharan Africa's GDP in 2001.

African participation in industry has been increasing, almost doubling from 1950 to 2003 as a percentage of the economically active population, with continued increases projected (Figure 9-3). More developed regions of the world, in contrast to Africa, have most of their economically active population working in industry rather than in agriculture. Data for more developed regions of the world (North America, Europe, East Asia) from the International Labor Office of the United Nations for 2003 indicate that 36% of the active population in MDCs was in industry (19% of that in manufacturing), 63% in services, and less than 1% in agriculture. Females were disproportionately represented in services in these economies. For example 76% of active females worked in the service sector, while only 53% of the active male population did. Forty-four percent of active men worked in industry, while only 27% of women did. In manufacturing the difference was less marked than for industry as a whole but still appears significant: 23% of active men worked in manufacturing, as opposed to only 16% of women.

There is a gender difference in employment in Africa as well. Seventy percent of economically active women but only 49% of the men were engaged in agriculture in 2003. Eighteen percent of men were involved in industry and only 6% of women. For manufacturing the differences were small in contrast to the more developed regions of the world: 5% for women and 8% for men. Women may possess an advantage in winning light manufacturing jobs over the next few decades in Mediterranean Africa if free-trade and tariff-free zones develop similar to northern Mexico, Dubai, and the Mumbai region. Women are preferred for light manufacturing jobs because they are thought to have better fine motor skills than men (nimble fingers), are thought to be more steady on the job, to be more passive than men (fewer strikes), and to require less pay. Skepticism regarding the merits of women in the workplace in these Islamic countries may dampen this possibility. It is nevertheless unlikely that such zones will develop in East, southern, and West Africa because these areas are out of major world shipping lanes, and countries with easy access to Suez and Indian Ocean shipping lanes (Dubai, Oman, Yemen, and India) have taken the lead and will make stiff competition. Africa may be more attractive however because of lower labor costs. If such development is a result of policy generated in investor countries, then there is a credible chance of it happening. Such investment may also take place through a two-step process: risk-averse MDCs invest in the oil economies of Africa (Nigeria, for example) and the industrial powerhouses (South Africa), and these countries then invest in manufacturing their respective neighborhoods.

Many African countries have experienced industrial growth in recent years. For example, in 2000 Uganda's industrial output grew by 15.2%, Lesotho's by 11.8%, Angola's by 7.9%, Mozambique's by 7.8%, Ethiopia's by 7.5%, Burkina Faso's by 7%, Benin's by 6.8%, and Algeria's by 5.9%; outputs of Tunisia, Tanzania, and Botswana grew by 5.7% apiece, and Senegal's by 3.5% (United Nations Economic Commission on Africa 2002).

Mining constitutes a large segment of industrial growth and accounts for substantial industrial prosperity in many African countries. Other extractive industries that contribute to Africa's economy are fishing, forestry, and petroleum and natural gas. We turn next to a brief consideration of the continent's marine environment and the fishing industry.

Extractive Industries

Fishing

Africans obtain the majority of protein from vegetable sources. Protein from fish contributes less than 5% to the total protein content of the African diet but represents about 18% of the animal proteins consumed by Africans (UNEP 2002). In regions where there is a marine or lacustrine coast, fish may constitute the major input of animal protein in the people's diet. In general, coastal ecosystems are among the most biologically productive of earth, and many coastal ecosystems virtually teem with life. The density of life in coastal waters is much higher than on land because except for birds and insects, on land life is two-dimensional—restricted to the earth's surface. In the aquatic coastal environment, however, the medium of water makes three-dimensional life and navigation in search of food possible for all but the rooted plants and bottom feeders. The upper layers of the ocean teem with phytoplankton and zooplankton, which nourish countless larger organisms, which in turn nourish higher carnivores. The opportunities for life—and to make a living—are more numerous in coastal waters than on land. In addition, although most fish do not spend their entire lifetimes close to the shore or in estuaries, they must visit these areas for reproduction.

On the surface, the size of the African continent, the second largest in the world, and its long coastline, would seem to indicate that fish as a resource would have great potential for development. On closer inspection, the potential to be realized is seen to be restricted politically, economically and, more importantly, geographically. Access to coastal waters is an issue because 14 African countries are landlocked. Although these countries do have freshwater industries, they have no marine fisheries, and the potential of freshwater fisheries is limited except in certain lacustrine regions. African coastal countries claim an exclusive economic zone (EEZ) of 200 nautical miles (371 km) from the edge of land out to sea. In practice, this makes conflicting claims inevitable, but no wars have been fought over fishing: few African countries have invested in their marine fisheries, and most fishing remains restricted to the near-coast zone.

The physical geography and climate of the continent itself present another constraint to the development of fishing in Africa. The continental shelf is narrow and drops off rapidly to the abyssal plain, a consequence of the deep rifting and movement of continental crust away from Africa and Africa's relative stability since the breakup of Pangaea 225 million years ago (Chapter One). The continental shelf, the submerged, shallow area on the margins of a continent, which slopes gently toward the abyss or deeps, is the most productive part of an ocean; nutrients flow in from rivers and support diverse plant and animal populations. Well-known examples of rich shelf fisheries are the Grand Banks of the northeast coast of North America and the North Sea in northern Europe. In addition, because the African continent extends from the lower midlatitudes of the Northern Hemisphere to the lower midlatitudes of the Southern Hemisphere, about a third of the continent comes under the influence of the dry, descending air of the subtropical high-pressure belts located about 30° north and south of the equator. Aridity in northern Africa is exacerbated by continentality as well (Chapter Two).

Despite the dryness, there are major rivers in Africa, many lakes impounded behind dams, and two lake districts (in Mali and along the Rift Valley). Africa's total fish catch (marine and freshwater) in 2001, 7.3 million metric tons, accounted for a modest 5.6% of the world's total catch, down from 6.2% in 1961. Sixty-one percent of Africa's 2001 catch came from Subsaharan Africa, up from 45.9% in 1961. The African marine fishery contributed 6.6% to world production in 2001 and of that, Subsaharan Africa's share was 3.6%. Freshwater fish production was 4.4 and 3.2%, respectively. Exports of fish contributed $450 million dollars to the economies of African countries in 2001 (FAO 2005), 4.4% of the world total export value, up from 1.7% in 1961.

Contrast Africa's figures for 2001 with the large producers in the world: the value of European Union fish exports was $3.5 billion, Canada and the United States was $1.3 billion, and East and Southeast Asia was $1.3 billion. European Union fisheries contributed 35% of the total world fish export value in 2001, Canada and the United States 13.3%, and East and Southeast Asia 13.1% (FAO 2005).

African total fish production has doubled since the 1960s according to data made available by individual countries. It must always be borne in mind that with African data significant production may remain unmeasured because of the remoteness of some localities, the often light presence of government on the ground, and the willingness of producers to conceal their activities from the government to avoid taxation. The 5-year average fish production for the periods 1961 to 1995 and 1997 to 2001 was 2.8 and 6.7 million metric tons, respectively. Most of that increase has been in the freshwater fishery, which increased from 733,000 metric tons average from 1961 to 1965 to an annual average of 2.5 million metric tons for 1997 to 2001. Marine fish production increased by 159% over the same time period. Part of the increase may be due to the availability of more fish as a result of the increased number of lakes: many dams and water impoundments have been built since independence, including some of the largest and most famous, Lakes Nasser, Volta, and Kariba. Some of the increase may be explained by a more scientific approach to fish and fish populations, exemplified by production through the planting of fish in rivers and lakes, for example, and by fish farming.

As in agriculture, certain African waters are more productive than others, and the fishing industry has a few African giants who account for most of the production and exports. The greatest single country producer was Egypt, which produced over 520,000 metric tons of freshwater fish in 2001. Regionally, 39% of the 2.3 million metric tons produced by the top 25% of African countries in 2001 was came from three countries: Tanzania, Uganda, and Kenya, of the Great Lakes Region of the African Rift Valley, and the Democratic Republic of the Congo in the Congo Basin. Nigeria was the fourth largest producer of fresh fish in 2001.

The marine industry was also dominated by a few big producers. The top 10 marine fishing nations produced 73% of all marine fish caught in 2001 (FAO 2005). Morocco and Egypt were the largest producers of marine fish, with 1 million and 772,000 metric tons each. Marine production was similar for North and West Africa (1.4 and 1.3 million metric tons respectively) while that of Central, Eastern and Southern Africa was much lower (350,000, 265,000 and 542,000 metric tons respectively). Production from eastern Africa was just over a quarter of a million metric tons.

Unfortunately, the world's fisheries are being depleted from overuse and damming, and Africa's are no exception. The FAO estimated (2000) that 38% of Africa's coastal ecosystems were under severe threat from development. Damming rivers reduces the nutrients that are available for marine life because there is no more flow to the river delta and sea. The Egyptian Mediterranean sardine catch was halved in the first 12 years after the construction of the Aswan High Dam, which reduced the nutrient load of the Nile River into the Mediterranean Sea (UNEP 2002).

Forestry

In some way, forests are an important resource for almost African every household. They provide fuelwood, material for construction, food for livestock, gathering and hunting opportunities, and medicine. Environmentally, they help reduce runoff and erosion and are the environmental niche for 1.5 million species of plant (Groomsbridge and Jenkins 2000). The forested area in Africa has declined by 10.5% since 1980 according to United Nations estimates (UNEP 2002),

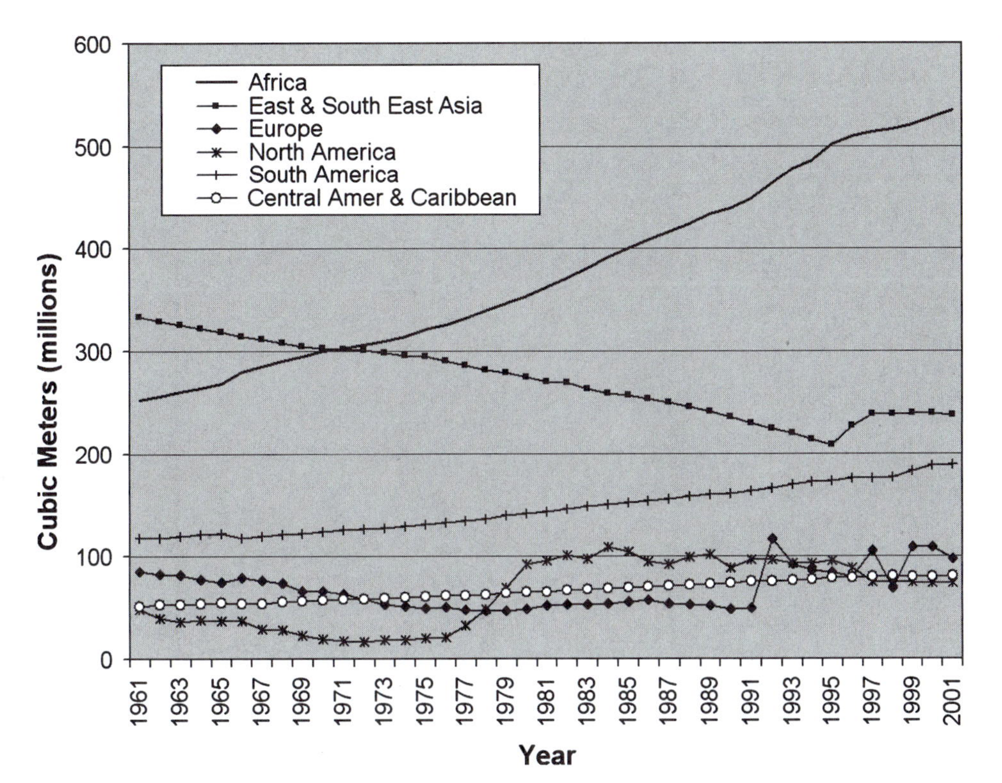

FIGURE 9-4 Fuelwood production for major world regions, 1961 to 2001.

mainly as a result of commercial logging, fuelwood consumption, and the clearing of land for agriculture. The production of fuelwood as either wood or charcoal is a complex industry in itself that deserves more study. With the exception of South Africa, most fuelwood production and consumption in Africa takes place in Subsaharan Africa. While North African countries have alternatives in natural gas, the demand for wood and charcoal in Africa has been increasing more than in any other world region, reflecting the choices dictated by the proximity of the vast majority of the population to the land (Figure 9-4). Ethiopia, the Democratic Republic of the Congo, and Nigeria produced over 40% of the fuelwood in Africa in 2001 (FAO 2002). Fuelwood demand in Africa is projected to increase by close to 50% over the next 30 years (UNEP 2002) as population increases.

Africa's logging industries produce sawn timber in the form of roundwood. The continent produced 17% of the world's roundwood in 2001, worth almost a billion dollars (FAO 2002). Ethiopia, the Democratic Republic of the Congo, and Nigeria accounted for 38% of the total African production. Africa is principally an exporter of timber and other primary forest products rather than manufactured or finished goods. Roundwood is used mainly in construction, but the better grades are used in making veneer, the thin layer of high-quality wood that is glued to wood of lower quality to give a piece of furniture a richer look. Africa's exports of roundwood have declined over the last 40 years from 12% in the early 1960s to 4% in 2001. The difference illustrates the relative decline of African exports in relation to the increase in the exports of other countries. Absolutely, Africa's exports of roundwood increased by over half a million cubic meters (21 million ft^3) over the last 40 years, from 4.8 million to 5.4 million m^3 (164.5 million to 191 million ft^3). European roundwood exports, in contrast, increased from about 30% to over 60% of world exports.

Although the manufacture of furniture to satisfy local demand is evident in the small workshops visible in almost every African town or city, exports of finished products remain limited. African production of paper was less than 1% of world production in 2001 (FAO 2002) and was concentrated in two regions, southern Africa and North Africa. South Africa is the major paper producer in Africa; its production—70% of all paper produced on the continent—simply dwarfs

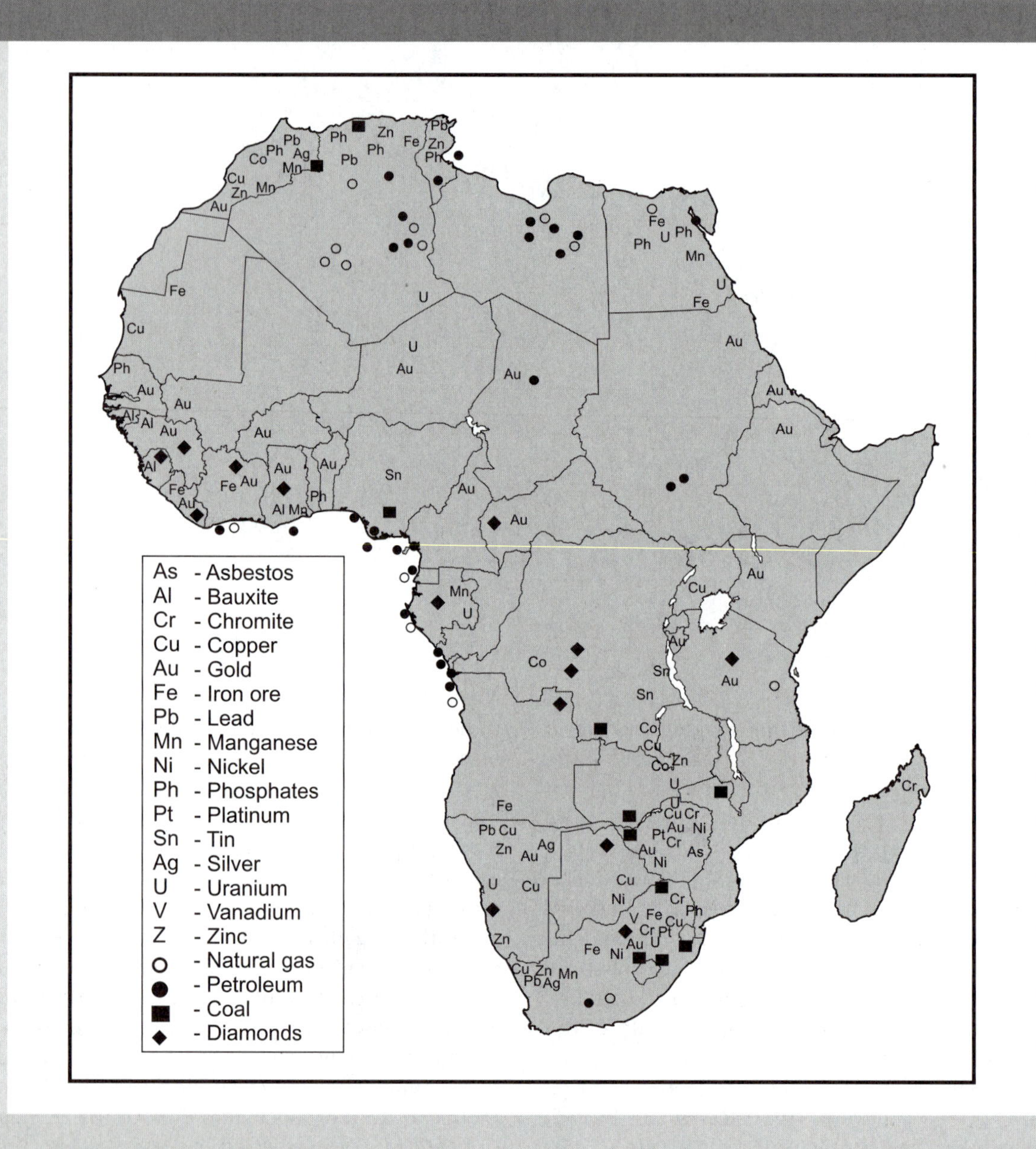

FIGURE 9-5 The mineral resources of Africa.

that of all other producers combined. In North Africa, Egypt, Morocco, and Tunisia account for most of the rest of African production.

Minerals and Mining

Africa possesses many mineral resources (Figure 9-5) that have been exported to earn foreign exchange. Some African minerals are of world strategic importance, particularly iron and ferrous ores, and other metals such as chrome, cobalt, manganese, molybdenum, nickel, and tungsten. The nonferrous minerals are combined with iron to make tough specialty steels. Manganese can be combined with aluminum to make light and strong alloys. Africa ranks first or second in the world in its reserves of bauxite, chromite, cobalt, diamonds, gold, manganese, phosphate rock, platinum and related metals, titanium, vanadium, vermiculite, and zirconium (Coakley 2003). Currently world interest in African minerals has been encouraged by the diminution of political instability and conflict and the increase in responsible government on the continent, as well as the new discoveries that are made every year.

Africa's mining history can be divided into several periods: precolonial mining, the colonial boom, the post–World War II boom, followed by a decline from the 1980s, and revival near the end of the Cold War (Ericsson and Tegen 1999). There is plenty of evidence for mining in Africa deep into antiquity, principally for gold, copper, and iron from scattered locations. Egyptian gold mining is well known from evidence in the Red Sea Hills of the Sudan (Cole 1990). The first European mines in Subsaharan Africa were probably the Portuguese coal mines at Moatize, in present-day Mozambique (Gleave 1992).

Exploration and mining were an imperative if not imperial monopoly during the European colonial period. Gold and diamonds had begun to be mined in South Africa during the latter quarter of the nineteenth century, and as the twentieth century dawned, colonialists began exploiting copper in central and southern Africa. Nigeria produced tin from mines near Jos, and iron was exported from Algeria. In German South-West Africa, today's Namibia, copper, lead, zinc, and diamonds were discovered during the first decade of the twentieth century. Colonial mining interests had railroads built to haul the resource to the coast for processing or, more often, shipping to the metropole, but seldom did a rail transportation network suitable for national development emerge. Rather, Europeans without any other connection to the colony other than mining built multiple rail lines, even of different gauges within the same territory, from the coast to the mineral deposits. Although many of these railways are stranded today, just as they were during the colonial period, the early rail system provided the first modern transportation infrastructure in Africa.

In addition to providing transportation infrastructure, the early European mining ventures, particularly in southern Africa, redirected African labor from traditional activities into the mines, most often through the poll tax or through physical force. These early industries

> taught African workers a basic range of industrial skills within limits of the job colour-bar. They introduced for the first time the discipline of industrial working. They created urban markets where food produce from the rural sector could be sold for cash. Mineral production enabled some colonial administrations to show a financial surplus and so to allow some money to be spent on long-term development. (Gleave 1992: 233)

The large mining expansion of the post–World War II period was almost entirely controlled from outside Africa or from settler-colonies within Africa. British and South African interests were strongest in southern Africa, while companies from other European colonial powers had control in their respective spheres. Transnational corporations from Canada and the United States were involved in iron mining in Liberia and in ferrous alloy mining in Namibia and South Africa, bauxite mining in Guinea, and coal mining in South Africa (Ericsson and Tegen 1999). Uranium mining boomed after the Second World War in the Congo, Niger, Gabon, and later Namibia, as interest in uranium and nuclear power grew in the industrialized world. The international (mainly European) demand for iron and metals to use as alloys in making stainless steel also grew after the war, and African sources of these ferroalloys were central to world production. The principal mines for ferroalloys were located in southern Africa in South Africa (vanadium and manganese), Zimbabwe, or Southern Rhodesia (chromite), and cobalt from along the Copper Belt between the Democratic Republic of the Congo and Zambia, the then Northern Rhodesia. Iron was exported to Europe from mines in Angola, Liberia, and Mauritania. Phosphates for use as fertilizer also began to be exported from Morocco and Senegal.

During the late 1960s independent African countries began to nationalize mining enterprises with the intention of financing their economic and social development, and thus began an almost 25-year decline.

> The decline of African mining started in the 1980s after the first enthusiasm of domestic control over the mining production had waned. Due to lack of trained staff, poor access to final markets, lack of technology and several other reasons, the mines, albeit owned by the host governments, were still to a large extent run by the mining transnationals and expatriate staff. Excessive draining of profits from mining into the general public domain, and too little reinvestment in plant, exploration, training and research, gradually

led to the mines coming to a standstill. In some cases, such as Zaïre, corruption added to the problem, and Zairean copper production dropped from 500,000 tons in 1985 to virtually zero in 1995. The lack of domestic exploration efforts by the state owned companies and total absence of foreign capital led exploration in Africa outside of South Africa to decline to very low levels in the 1980s. (Ericsson and Tegen 1999: 2)

Africa became a subject of renewed interest by the international mining community and investors in the 1990s, when the perils of nationalization and state-run industries had become well apparent and moves had been made to privatize the mining industry in many countries. Africa has rich deposits of many minerals and hydrocarbons, and as governments have become more receptive to market-based alternatives to government parastatal corporations, investors are returning. This trend is expected to continue into the future. For the strategic minerals that Africa possesses in great abundance in comparison to other regions of the world, Africa's advantage will remain unchallenged except where substitutes can be found. With regard to other minerals that African countries do not monopolize, copper being a good example, increasing production from China will dampen prices.

Table 9-5 presents available data on the African production of minerals and metallic resources and the percentage of world production that each represents. The African share of production for most resources has declined for reasons cited by Ericsson and Tegen (1999). For example, African gold production declined from 63% of world production in 1975 to 27% in 1996. Copper declined from 20% to 6% over the same time period. Only bauxite and iron ore remained relatively constant: 12 to 13% and 7 to 5%, respectively. With regard to smelted copper, aluminum, and pig iron, copper smelting has followed the same dramatic decline as copper mining, aluminum production, on the other hand, has increased from 2% to 5% of world production, while pig iron production has remained the same, at 1% of world production. Titanium production increased from to 11.5% to 18% of world production from 1986 to 1990. Africa's share of world chromite production increased from 37% to 41% over the same time period. The picture for iron ore is mixed. South Africa, for example, increased its iron ore production by 24% from 1986 to 1990, but production fell by 74% in politically unstable Liberia over the same time period. The production of some other minerals, for example, manganese, remained constant.

TABLE 9-5 AFRICAN PRODUCTION OF MINERALS AND OTHER RESOURCES AS A SHARE OF WORLD PRODUCTION.

Mineral	African Share of World Production (%)	Share of World's Reserves (%)
Bauxite	13.7	33
Chromite	44	95
Cobalt	37.2	55.8
Copper	6	22.2
Gold	29	40
Iron ore	3.1	8
Lead	5.5	NA
Natural gas	8.3	7.7
Manganese	33	85
Nickel	5.2	NA
Petroleum	11.1	7.1
Platinum	67.1	90
Phosphate rock	25.2	64.4
Silver	2.9	NA
Tin	3	NA
Uranium	23	21.4
Vanadium	36.5	NA
Zinc	6.3	NA

Sources: Goode's World Atlas (2000), Mbendi Information for Africa (2003), U.S. Geological Survey (2003).

Petroleum and Natural Gas

The North African countries of Algeria and Libya are well endowed with oil and gas. Egypt, Morocco, and Tunisia have modest deposits. Africa's major Subsaharan producers of oil are Nigeria and Angola, with reserves of 17 billion and 1.2 billion barrels, respectively. Other oil producers are Sudan, Gabon, the Congo Republic, Equatorial Guinea, São Tomé and Principe, Chad, the Democratic Republic of the Congo, Cameroon, and Ghana. Although Côte d'Ivoire has petroleum deposits, its reserves are not proven. A number of Subsaharan countries have natural gas deposits but little oil—at least until oil is found. These countries are Tanzania, Mozambique, Rwanda, Guinea, Somalia, and Madagascar. There have been concerns regarding oil and corruption; Nigeria is a case in point, as well as

Angola, where it has been estimated that rulers pocketed a billion dollars a year in oil revenues (*Economist* 2002).

African oil reserves were estimated to have increased 34%, from 57 billion to 76 billion barrels between 1980–1984 and 1999–2003. Most of the increase was in Subsaharan Africa (65%). Nevertheless, Africa's relative share of world oil reserves has been declining over the last 20 years, from 11% to 7% for the continent as a whole and from 3.8% to 2.9% for Subsaharan Africa. Nevertheless, there are some indications that discoveries in the large sedimentary basins of Subsaharan Africa may increase in the near future. Production of oil in Africa has increased slightly over the last 20 years, from 9% to 11% of total world output for the entire continent and from 4% to 6% for Subsaharan Africa. U.S. oil companies are increasingly active in the Subsaharan region, and, according to the *Economist* (2002), the U.S. consumption of Subsaharan Africa oil is set to increase because of the attractiveness of the product, owing to its good quality, as well as to the region's relative proximity to North America and, with the exception of Nigeria, its independence from the Organization of Petroleum Exporting States (OPEC). Although Subsaharan African imports amounted to only 15% of U.S. oil imports in 2003, they are likely to rise to 25% by 2015.

The continent of Africa's natural gas production in 2000 represented only 7.7% of the world's output; Subsaharan output was 2.9% of that. Africa marketed only 54% of its gross output in 2000, and Subsaharan Africa marketed even less (33%) of its production. The reason is that over 50% of the natural gas produced in Subsaharan Africa is flared (burnt off), whereas in most other regions of the world this resource is captured and marketed.

Wealth in natural resources can be an enormous boon in development, but sometimes it is a two-edged sword leading to political destabilization as groups compete to control the resource (World Bank 2001). Although the term "conflict diamonds" is part of the world's vocabulary today, almost every natural resource in Africa has been sold by warlords and rebels to finance their rebellion, even rubber (the "bouncing diamond") and timber in Liberia. One of the first resource crises of the twenty-first century was the looting of coltan in the Democratic Republic of the Congo by its neighbors and armed rebels. Coltan (columbite–tantalite) is a metallic ore found in the eastern region of the DRC. When refined, coltan becomes highly conductive of electricity and heat resistant. These properties make coltan attractive in small electronic devices such as cellular telephones and laptop computers. Presidents-for-life and other dictators have been known to exploit national resources for personal gain and political patronage and to use disorder as a governing tool (Chabal and Daloz 1999). On the other hand, when a country's government is accountable to the citizenry through the democratic process, corruption is generally low—but generally not absent altogether.

DEVELOPMENT, GROWTH, AND EQUITY REVISITED: EXPLANATIONS OF DEVELOPMENT

In the late 1970s, there was much debate among academics and practitioners about the lagging performance of African economies after independence; controversy raged around the world about the causes and consequences of African development. Several extreme positions were articulated at the time that are worthy of note in this book. One explanation held that the potential for development in Africa was so constrained by the harsh and meager environment (climate and resources) that the probability that many African countries could transcend it was low. Perhaps this view was best encapsulated in Bryson and Murray's *Climates of Hunger: Mankind and the World's Changing Weather*, published in 1977.

> The crucial question for the Sahel . . . then, is where the border between moist and dry air lies. If the whole system moves north in the summer, the moist monsoon air goes north and people in those regions prosper. If this movement comes every summer, people of the monsoon lands prosper. If the system stays farther south for most or all of a couple of summers, green fields become desert and thousands die. If the monsoons stay south for years, how many millions will die? (Bryson and Murray 1977: 106)

Bryson and Murray simultaneously pose a question and offer an answer about people and environmental variability. Although the question posed is an important one, the offered answer, if not entirely wrong, is not as clear as these geographers seem to indicate. People are not a constant; they are always altering their relations with the environment to survive and prosper. We can surely expect that if the climate changes, people will adapt in some way, even if it means relocation. To imagine otherwise is to give climate and the environment too much credit in the determination of human behavior. For the long duration of human history, all climates have been climates of hunger—no region has been immune from famine, disease, disruption, and dislocation. Only in relatively recent times have human prospects been improving.

At a more macro level, opposed to environmental and behavioral reasoning, are structural explanations of development suggesting that it is Africa's relationship between Europe and other more developed regions of the world that has undermined, subverted, and "under"-developed the continent. In the dependency theory of Andre Gunder Frank (1972) and the world systems theory of Immanuel Wallerstein (1979), these concepts are most elaborated and developed. Both Gunder Frank and Wallerstein hold that 500 years ago, most people around the world lived rather similar lives: the vast majority of people lived in small villages, the day light hours were occupied by agricultural labor, and life was centered around family and religion.

The Industrial Revolution occurred first in Britain and then spread to Europe. Gunder Frank and Wallerstein believe that Europe was transformed by the Industrial Revolution and ultimately soared in its development to an extent that permitted enormous amounts of wealth to be created and standards of living to be vastly improved in a relatively short time. But at the same time that Europe was developing, it was also influencing the development of peripheral regions, causing them to "develop" subordinate economic and political roles to serve Europe's economic and political interests.

For Gunder Frank and other dependency school theorists, a country could not "develop" beyond the role assigned it by the workings of international capitalism centered in the MDCs. Thus these theoreticians counseled the developing states of Africa and elsewhere to cut their economic ties to Europe and other industrialized states in order to pursue "genuine" development. The dependency school believed that the developmental path of the "underdeveloped" world would be separate and different from Europe's path. Neither Gunder Frank nor Wallerstein specified empirically how underdevelopment occurs; equally opaque is what was meant by "genuine" development. Moreover, where the prescriptions of the dependency and world systems theorists have been applied, success has been problematic. Critics of these and similar theories point out that the excessive externalization of responsibility for the poverty and chaos in Africa effectively absolves the national political elite in any African country of blame (Hameso 2001). Dependency and world systems explanations are political statements rather than any basis for concrete economic policy.

Most observers today have moved away from macro-level explanations like environmentalism and the rigid economic structuralism of neo-Marxist world systems/dependency theory in favor of an understanding less wedded to grand, totalizing, historical narratives and closer to the political, social, economic, and environmental conditions in Africa. This is not to say that the environment or relations with rich countries are not important—far from it—but the overarching emphasis placed on the environment by one group of proponents or on economic structure by the other is inadequate today because these are not ultimate causes of African inadequate development.

The causes for the differing developmental paths of states and regions are sought closer to the ground today in variables such as location, political system, political organization, degree of political freedom, conflict, and culture. The new consensus is that there is no such thing as separate or different development for Africa. We know now that culture is one of the most important factors in development and that people in other parts of the world have overcome meager resource endowments and unforgiving climates, even the "international division of labor" much touted by world systems theory, to achieve high levels of human development.

Japan is a country that initially overcame its resource poverty through trade. After the terrible mistakes that led to World War II, Japan again pursued trade as an engine of development. Singapore and South Korea are two more recent examples. In 1995 Singapore was declared by the United Nations to be a more developed country, on a par with France and the United States. A malarial backwater just a few short decades before, Singapore had been a colony of both the British and the Japanese; and yet it managed to become highly developed as a result of fostering an

TABLE 9-6 INTERNAL AND EXTERNAL IMPEDIMENTS TO DEVELOPMENT IN AFRICA.

Internal Impediments	External Impediments
Failure to industrialize High rate of population increase Political instability caused by wars of succession, interethnic discrimination, tension, and violence, and confessional (religious) violence ultimately resulting in ungovernable political entities Lack of a national identity Overvalued currencies Environmental instability Decline in agricultural production Debilitating diseases Inadequate health services Corruption at all levels, but especially in the elites Governments that are too statist and also antientrepreneurial	Decline in the value of mineral exports Decline in the value of tropical agricultural products Oil shocks since the 1970s The end of the Cold War and declining interest by the remaining world powers

export orientation and trade. Less developed countries around the world have been looking very carefully at the Singapore model. But why not Africa? The potential for development in Africa is high, but at the same time the impediments seem to be manifold and mainly internal (Table 9-6).

Some of the obstacles mentioned in Table 9–6 are environmental, others are related to the world political economy, and others are related to the local culture. Culture means many things: religion, language and dialect, historical experience, and race and ethnicity are all elements of culture. But one can also speak of political and economic culture, or "traditions"; and where traditions are inimical to development, it will not occur. Sen (2003) has pointed to political and economic freedom as the most significant variables in development. De Soto (2003) found that where people do not have secure title to property, it is difficult to build the capital necessary for economic growth and development. The Human Development Institute (Wade 2002) found that there is no automatic link between democracy and equitable development; rather, the rich usually benefit more than the poor from development—at least at first. Moreover a democracy can be organized in a variety of ways, some offering citizens a greater voice in government than others, without forfeiting the label "democratic."

Furthermore, from the 1980s to the present, the role of the state in African development has increasingly been found wanting. Van de Walle (2001) has argued that the principal obstacle to development in Africa has been the African state itself not international capital. Reno (1998) described the weak African state and the rise of warlordism in the struggle to control national natural resources in West and central Africa. Bayart, Ellis, and Hibou (1999) have indicated that the rise of the "kleptocratic" and criminal state in Zaïre was responsible for the enrichment of the President-for-Life Mobutu and his associates and the impoverishment of the people of what is now the Democratic Republic of the Congo. Chabal and Daloz (1999) suggest that governments of African weak and failed states use disorder to maintain control and enrich themselves. Van de Walle (2001) has found that structural adjustment programs by the IMF and aid from other international agencies have only served to strengthen the hand of those in power in African states.

CHAPTER REVIEW QUESTIONS

1. In this chapter it was mentioned that FDI (foreign direct investment) and government-to-government foreign aid from donor countries had declined over the last decade in Africa. Discuss the reasons for the decline in both activities. What shift in development policy has been associated with the decline in foreign aid?
2. Dependency theory placed the blame for Africa's disappointing development with the capitalist world system. Neoliberal theorists place the blame on national leaders and policy. Discuss how it might be that both theories are partly right.
3. Which regions of Africa are best poised for rapid development?
4. Discuss the proposition that the African state should maintain an active role in development.

GLOSSARY

Development A process by which the political, economic, social, and environmental conditions in a country or region improve. One may distinguish economic from human development in that economic development may involve aggregate economic growth and output in a country. Human development involves raising the standard of living, quality of life, and life chances for people in addition to economic growth.

Indicators of development An indicator is an empirical measure of a concept. Complex concepts like development, which cannot be directly measured, are "indicated," usually by several measures that capture the essence of the concept. For example, generally recognized characteristics of a highly developed country are a high national wealth, a long life expectancy, and a high level of education. The Human Development Index (HDI) correspondingly takes various economic, health, and social indicators and combines them into a single measure to summarize the development of a country. The HDI is useful in comparing the relative scores of countries. The three indicators used in the HDI are GNP per capita income, life expectancy, and literacy rate.

Modernization A theory of development envisioning that the state in developing countries would transform national infrastructure and institutions to build a modern-looking country on the European model. Key to the modernization paradigm were large foreign investments in key sectors of the economy to jump-start development.

Neoconservatism A political philosophy that holds the basic premises of neoliberalism with respect to the central role of free market in development. In addition neocons believe that democracy and civil society are key to development. Neocons differ from neoliberals in that they feel that the United States must be the instrument for the achievement of democracy and liberal economics around the world, even to the extent of forcibly intervening in the internal affairs of other countries.

Neoliberalism The political and economic philosophy that holds that small government and a free market are the best medicine for development. (see Neoconservatism)

Veneer Thin layers of high-quality wood used to cover the visible areas of furniture made of lower-quality wood to give it a more expensive look.

BIBLIOGRAPHY

Arab-American Business. 2003. Kuwaiti fund provides $100 million loan for dam in Sudan. *Arab-American Business*, April.

Bayart, J.-F., S. Ellis and B. Hibou. 1999. *The Criminalization of the State in Africa*. Oxford: James Curry.

BBC News. 2002. Work starts on giant Ethiopian dam. BBC, Monday, August 12. http://news.bbc.co.uk/1/hi/business/2188785.stm.

Bryson, R., and T. Murray. 1977. *Climates of Hunger: Mankind and the World's Changing Weather*. Madison: University of Wisconsin Press.

Chabal, P., and J.-P. Daloz. 1999. *Africa Works: Disorder as Political Instrument*. Oxford: James Curry.

Coakley, G. 2003. *The Mineral Industries of Africa*. U.S. Geological Survey Mineral Report. Va.: USGS. http://minerals.usgs.gov/minerals/pubs/country/.

Cole, R. ed. 1990. *Measuring Drought and Drought Impacts in Red Sea Province*. Oxfam Research Paper Number 2. Oxford: Oxfam House.

Cornwall, R. 2002. The new partnership for Africa's development: Last chance for Africa? In *The New Partnership for Africa's Development—African Perspectives*, H. Melbar, R. Cornwall, J. Gathaka, and S. Wanjala, eds. Nordic African Institute Discussion Paper Number 16, Uppsala, Sweden, pp. 23–32.

Desoto, H. 2003. The Mystery of Capitalism: Why Capitalism Triumphs in the West and Fails Everywhere Else. New York: Basic Books.

Dollar, D., and A. Kraay. 2000. *Growth Is Good for the Poor*. World Bank Development Research Group. New York: World Bank.

Economist. 2002. Black gold: Subsaharan African oil. *Economist*, October 26.

EIA. 1999. *Energy in Africa*. Washington, D.C.: U.S. Energy Information Administration.

EIA. 2003. *Annual Energy Review*. Washington, D.C.: U.S. Energy Information Administration. http://www.eia.doe.gov/emen/aer/contents.html.

Ericsson, M., and A. Tegen. 1999. *African Mining in the Late 1990s—A silver Lining?* CDR Working Paper Subseries, 99(2). Copenhagen: Institute for International Studies, Department for Development Research.

Falola, T. 2002. *Key Events in African History: A Reference Guide*. Westport, Conn.: Greenwood Press.

FAO. 1995. *Future Energy Requirements for Africa's Agriculture.* Food and Agriculture Organization of the United Nations. New York: United Nations.

FAO. 2000. *State of the World's Fisheries and Aquaculture.* Rome: United Nations Food and Agriculture Organization.

FAO. 2005. FAOSTAT. Rome: U.N. Food and Agriculture Organization.

Gathaka, J., and S. Wanjala, 2002. Kenya and NEPAD. In *The New Partnership for Africa's Development—African Perspectives*, H. Melbar, R. Cornwall, J. Gathaka, and S. Wanjala, eds. Nordic African Institute Discussion Paper Number 16. Uppsala, Sweden: Nordic African Institute, pp. 14–22.

Gleave, M. T. 1992. *Tropical African Development: Geographical Perspectives*. Essex, U.K.: Longman.

Goode, J. P. 2003. *Rand McNally Goode's World* Atlas. Skokie, Ill.: Rand McNally.

Groomsbridge, G., and M. D. Jenkins. 2000. *Global Biodiversity: Earth's Living Resources in the 21st Century.* Cambridge, U.K.: World Conservation Press.

Gunder Frank, A. 1972. The development of underdevelopment. In *Dependence and Underdevelopment*, J. D. Cockcroft, A. Gunder Frank, and D. Johnson, eds. Garden City, N.Y.: Anchor Books.

Hameso, S. Y. 2001. *Development, State and Society: Theories and Practice in Contemporary Africa*. Lincoln, Neb.: Authors Choice Press.

International Labor Office. 2003. Statistics on the Web. http://laborsta.ilo.org/.

IRN. 2002. *World Bank Dam in Uganda Overpriced by $280 Million*. International Rivers Network, November 20. www.irn.org/programs/bujagali.

Lal, D. 1983. *The Poverty of Development Economics*. London: Institute of Economic Affairs.

Lewis, W. A. 1954. Economic Growth with Unlimited Supplies of Labour. Manchester School of Economics, 22(2): 139–191.

Leys, C. 1996. *The Rise and Fall of Development Theory.* Oxford: James Curry.

Mbendi Information for Africa. 2003. Overview on African bauxite mining. http://www.mbendi.co.za/indy/ming/baux/af/p0005.htm. Accessed May 20, 2003.

Melbar, H., R. Cornwall, J. Gathaka, and S. Wanjala, eds. 2002. *The New Partnership for Africa's Development—African Perspectives*. Nordic African Institute Discussion Paper Number 16, Uppsala, Sweden: Nordic African Institute.

NEPAD. 2001. *New Partnership for Africa's Development.* www.nepad.org.

People's Daily. 2003. China wins contract to build another "Three Gorges" dam in Africa. *China People's Daily*, June 1, 2002.

Population Reference Bureau. 2005. *World Population Data Sheet*. Washington, D.C.: PRB

Quinn, J. J. 2002. *The Road Oft Traveled: Development Policies and Majority State Ownership of Industry in Africa.* Westport, Conn.: Praeger.

Reno, W. 1998. *Warlord Politics and African States*. Boulder, Colo.: Lynne Rienner.

Rist, G. 1997. *The History of Development: From Western Origins to Global Faith*. London: Zed Press.

Rostow, W. W. 1959. *The Stages of Economic Growth: A Non-Communist Manifesto*. Cambridge, Mass.: Cambridge University Press.

Saasa, O. 2002. *Aid and Poverty Reduction in Zambia: Mission Unaccomplished*. Uppsala, Sweden: Nordic Africa Institute.

Scott, J. C. 1998. *Seeing Like a State: How Certain Schemes to Improve the Human Condition Have Failed*. New Haven, Conn.: Yale University Press.

Sen, A. 2003. *Development as Freedom*. New York: Anchor Books.

UNCTAD. 2002a. *Economic Development in Africa. From Adjustment to Poverty Reduction: What Is New?* United Nations Conference on Trade and Development. New York: United Nations.

UNCTAD. 2002b. *Trade and Development Report, 2002: Developing Countries in World Trade*. United Nations Conference on Trade and Development. New York: United Nations.

UNEP. 2002. *Africa Environment Outlook: Past, Present, and Future*. United Nations Environment Program. Nairobi, Kenya: United Nations.

United Nations Development Program. 1990. *Human Development Report, 1990*. Oxford: Oxford University Press.

United Nations Development Program. 1990. *Human Development Report 1990: Concept and Measurement of Human Development*. New York: UNDP.

United Nations Development Program. 2002. *Human Development Report, 2002*. Oxford: Oxford University Press.

United Nations Economic Commission for Africa. 2002. *Economic Report on Africa, 2002: Tracking Performance and Progress*. Addis Ababa, Ethiopia: United Nations.

U.S. Geological Survey. 2003. *Mineral Reports*. Reston, Va.: USGS. http://minerals.usgs.gov/minerals/pubs/country/.

van de Walle, N. 2001. *African Economies and the Politics of Permanent Crisis, 1979–1999*. Cambridge, U.K.: Cambridge University Press.

Wade, A. 2002. *Special Report to the Human Development Institute, 2002*. Human Development Institute, United Nations Development Program. Oxford: Oxford University Press.

Wallerstein, I. 1979. *The Capitalist World Economy*. Cambridge, U.K.: Cambridge University Press.

World Bank 2001. *Global Economic Prospects and the Developing Countries*. New York: World Bank.

WEST AFRICA

West Africa includes all the countries located in the Subsaharan part of the West African bulge of the continent from the Atlantic along the west and southern coast of the bulge to Chad and Nigeria in the east (Table 1). There are two tiers of countries, one comprising mainly landlocked countries located in the drier north, facing the Sahara, and one facing the South Atlantic in the more humid south (Figure 1a and Table 1).

West Africa is a mosaic of peoples and is arguably the most culturally diverse region in Africa. On linguistic criteria alone, it can be considered a separate province of Africa. It is home to the oldest and most diverse languages of the Niger–Congo language phylum, the most numerous and widespread family of languages spoken in Africa (Figure 1b). Speakers of this constellation of related languages are more numerous than for any other African language phylum. Linguistically, West Africa can be roughly bounded in the north and northeast by the Afroasiatic and Nilo–Saharan language phyla. These two language phyla are very different from Niger–Congo and, although they must be related in the deep and distant past, no relationship has yet been shown to exist between these languages and Niger–Congo. To the east, two groups of Niger–Congo languages, Adamawa–Ubangi and Bantu, spill out of West Africa into central, southern, and eastern Africa. Adamawa–Ubangi comprises a small group of languages whose speakers migrated out of West Africa into present-day Chad, the Central African Republic, and Sudan a few thousand years ago. The numerous and widespread Bantu languages, of which there are over 500 today, make up the dominant languages of the southern half of Africa. They appear so similar that they almost seem to be simply dialects of the same language. Speakers of the early Bantu languages were iron-using farmers who spilled out of the Cross Rivers area of Nigeria near the Cameroon border at about the same time as speakers of Adamawa–Ubangi migrated east, to colonize almost the entire southern half of the continent. Because the Bantu languages are so similar to the other Niger–Congo languages, it is convenient to group them together as part of other regions, where the impact of more recent phenomena, for example colonialism, have left an impress on the cultural landscape sufficient to allow reasonably coherent regions to be built.

TABLE 1 CAPITALS, AREAS AND RELATIVE SIZES OF THE COUNTRIES OF WEST AFRICA.

		Area			
Country	Capital City	Square Miles	Square Kilometers	Percent of Region	Percent of Continent
Benin	Porto Novo	43,483	112,620	1.8	0.4
Burkina Faso	Ouagadougou	105,792	274,000	4.5	0.9
Cape Verde Islands	Praia	1,556	4,030	0.1	0.0
Chad	N'Djamena	495,753	1,283,994	20.9	4.2
Côte d'Ivoire	Abidjan	124,502	322,459	5.3	1.1
Gambia	Banjul	4,363	11,300	0.2	0.0
Ghana	Accra	92,100	238,538	3.9	0.8
Guinea	Conakry	94,927	245,860	4.0	0.8
Guinea-Bissau	Bissau	13,946	36,120	0.6	0.1
Liberia	Monrovia	43,000	111,369	1.8	0.4
Mali	Bamako	478,838	1,240,185	20.2	4.1
Mauritania	Nouakchott	395,954	1,025,516	16.7	3.4
Niger	Niamey	489,189	1,266,994	20.6	4.2
Nigeria	Abuja	356,668	923,766	15.1	3.1
Senegal	Dakar	75,954	196,720	3.2	0.7
Sierra Leone	Freetown	27,699	71,740	1.2	0.2
Togo	Lomé	21,927	56,791	0.9	0.2
Total		2,369,898	6,138,008	100.0	20.3

The superimposition of religion on the distribution of language phyla adds another variable to the West African mosaic. The northern half of West Africa is Islamic, representing the fastest growing religion in Africa, while the southern half is Christian and animist. Religious lines of tension have appeared from time to time, exacerbated by ethnic national identity, and vice versa. The conflict in Côte d'Ivoire, Africa's first conflict of the twenty-first century, concerned the unwillingness of the politically and economically dominant Roman Catholic and Baoulé south to admit into the political process candidates from the poorer, less developed, and predominantly Muslim groups in the north.

On physical and environmental criteria, the boundaries of West Africa are more and less well defined. In the west and south, the Atlantic Ocean makes a natural boundary. In the north, the Sahara forms another boundary—at least at first glance. The Sahara has indisputably divided peoples of North Africa from West Africa. The cultural and racial divide along the Senegal River between Mauritania and Senegal provides an extreme example. No natural corridor like the Nile River transects the Sahara, facilitating the movement of people and ideas. Nevertheless, the Sahara is a bridge as well as a barrier. Since the development of the trans-Saharan trade around 1,800 years ago, the desert has been the backdrop to cultural interaction and change and, despite cultural cleavages in several dimensions today,

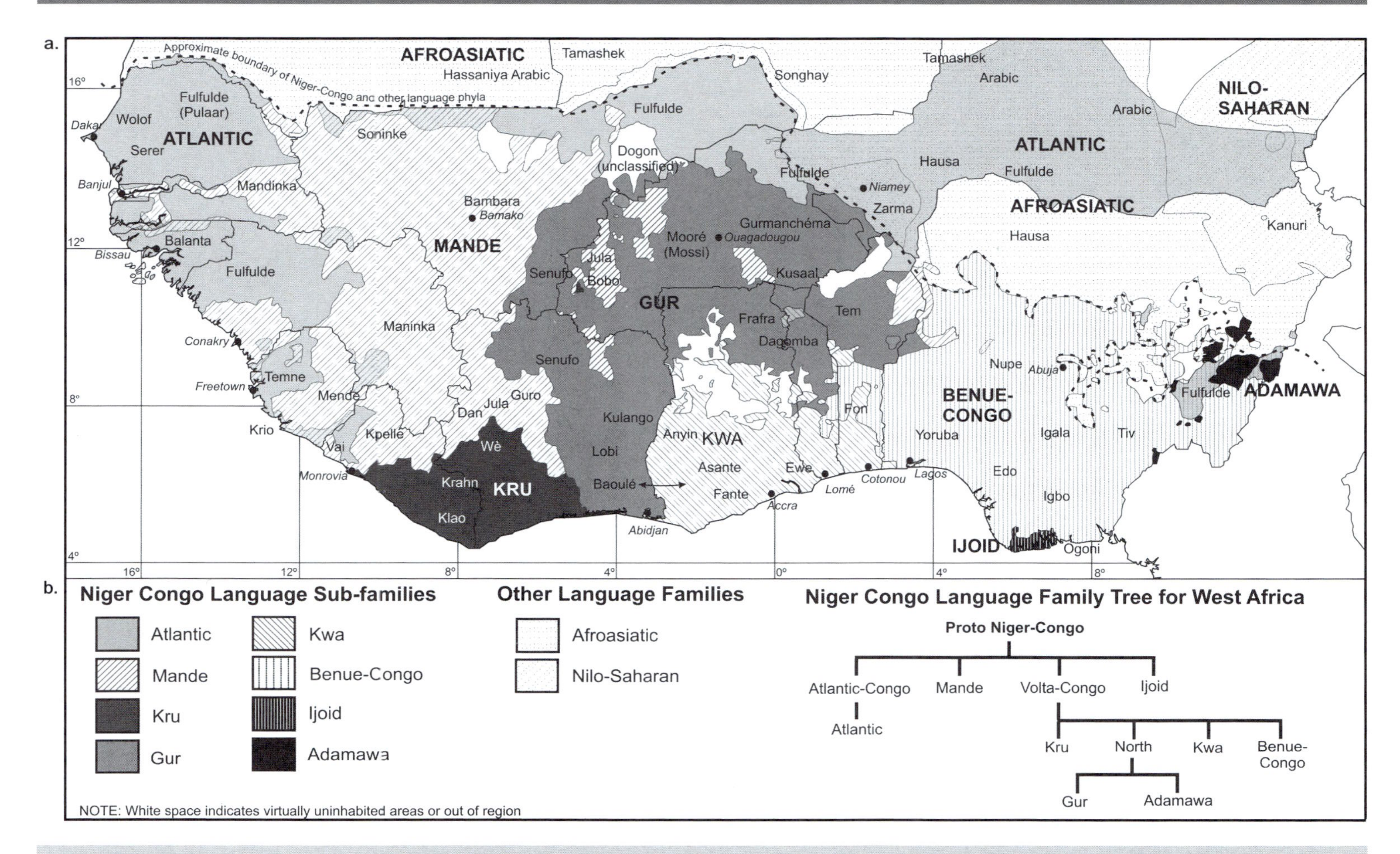

FIGURE 1 Languages in West Africa. (a) The dominant Niger-Congo language phylum. (b) Subfamilies and selected spoken languages.

the peoples of North Africa, and especially those along the northern tier of West Africa, are becoming culturally and genetically more alike. Grouping Chad into West Africa is problematic. Its South is more culturally akin to Central Africa while the North has been dominated by Arabs. The country of Chad illustrates the problems of regionalizing the diversity of Africa.

In Chapters Ten through Fourteen we have organized the countries of West Africa to facilitate their discussion. Small countries with similar characteristics have been grouped together to maintain similar levels of detail in the figures, tables, and text.

CHAPTER 10

Ghana

An African Epitome

When the Gold Coast achieved its independence under Kwame Nkrumah on March 6, 1957, and took the name Ghana, a decisive new era of African history began. True, Egypt, Libya, Morocco, Tunisia, and Sudan had gained their independence earlier, but Ghana was the first of many colonies in Africa south of the Sahara to be liberated from colonial rule. As a Subsaharan African state, its socioeconomic policies and performances, and its international relations and domestic politics, would be critically monitored and appraised both in Africa and overseas. Would Ghana become the model to be emulated or avoided by other territories scheduled for independence? Would political independence bring economic independence or at least greater economic self-sufficiency? Would economic ties with Britain be retained, or would Ghana seek new partners in the West, in Africa, or possibly the East? Would Nkrumah's new welfare state, which aimed at transforming an essentially agricultural, traditional society into a modern semi-industrial state, be guided by democratic principles? Would it be possible to reject tribalism and sectionalism, and cultivate in their stead Pan-Africanism? Nkrumah had announced his intent to establish "fraternal relations with, and offer guidance and support to all nationalist, democratic, and socialist movements in Africa and elsewhere which were fighting for national independence and self-determination on the one hand and whose programmes were opposed to imperialism, colonialism, racialism, tribalism, and religious chauvinism and oppression, on the other" (Nkrumah, 1961). Would he achieve it?

These and similar questions were raised by numerous governments and individuals alike, and many have now been answered. Certainly Ghana's economy has diversified and reoriented itself somewhat; there has been substantial progress in education and social welfare; and there have been several changes in government both by popular election and military coup. Ghana's recent peaceful transition to democracy and free-market principles provides a model for others to follow. Indeed, in its political and economic spheres, Ghana epitomizes the African experience and is very much a "typical" African state in many respects, and yet atypical in others.

THE POLITICAL KINGDOM

Centuries before Portuguese merchants established a fort and trading post at Elmina on the Guinea Coast in 1482, gold, ivory, slaves, and other commodities had been shipped northward across the Sahara from parts of what is today central Ghana. Within a century the so-called Gold Coast became a hub of European activity, where almost 70 forts and innumerable trading posts had been built by the English, Dutch, Danes, Swedes, and Brandenburgers, besides the Portuguese, and keen international competition for the trade of the interior had developed. The success of the Portuguese in diverting southward this trans-Saharan trade was due largely to the proximity of their Elmina trading station to this important supply area. Eventually that area became consolidated politically as the Ashanti Union of Akan States. The people on the coast, the Fante, brought goods from the interior to the coastal trading posts, and thus a north–south flow of exports was initiated which has become a permanent feature of the coastal states of West Africa. The Fante middlemen jealously guarded their role in this trade pattern and resisted efforts by the Europeans to penetrate the interior themselves. Meanwhile, they bartered with the interior producers at one end of their route and with the Europeans at the other.

Consolidation in the interior began when a number of the Akan states in the northern margins of the forest formed the Ashanti Union. The first steps were taken in the 1600s, but during the eighteenth century the union became the most powerful political and economic unit in the region. Ashanti began to encroach upon the middlemen of the south. Eventually, this led to contact with the Europeans, who failed to protect the peoples of the coastal states and were unsuccessful in attempts to come to terms with Ashanti. The abolition of the slave trade and British efforts to eradicate its remnants brought economic ruin and strife to the region. Legitimate trade was brought to a virtual standstill by intermittent hostilities. The Fante peoples tried to resist Ashanti encroachment by forming a Fante Confederation, but the British did not retaliate until 1874, when the Ashanti army again crossed the Pra River—the traditional boundary. The Ashanti capital, Kumasi, was burned, and a treaty was forced upon the empire.

The events of 1874 did not end Ashanti imperialism in the area. Resentful of their defeat and the consequent liberation of the southern Fante states, the Ashanti rulers once again prepared for war, but the British in 1896 expelled the king and forestalled the outbreak of war by a show of force. At that time, Ashanti was made a protectorate, but in 1901 a new revolt erupted. When this outbreak was quelled, the British established a colony over the Ashanti Empire and included as a protectorate the areas beyond the Volta River, the Northern Territories.

Thus, the heart of African organization and power long was in Ashanti, and Ashanti expansionism was contained by the British, just as friction in Nigeria between the northern Muslims and southern peoples who practiced traditional religions and Christianity was terminated by British intervention during the colonial period. But in Ashanti, indirect rule did not work, and thus a colony was established there, while the Nigerian north became a protectorate. The definition and delimitation of the Gold Coast's boundaries (with the French in the west and north, and with the Germans in the east) took place mainly between 1896 and 1901; as elsewhere in Africa, there was little regard for the ethnic groups. In the southeast, the Ewe people were divided; in the west, the Nzima, among other peoples, were fragmented by the new political order. The Dagbani (Dagomba) were split between Ghana and Togo. These divisions were to create future problems, but the incorporated area itself included many small, diverse African political units that suddenly found themselves within a political boundary and part of a larger political entity. The Gold Coast thus faced the problems common to most modern political entities in Africa: there were perhaps over a hundred distinct African states and chieftaincies, each with its own traditions. The British faced a formidable task in helping the inhabitants of the country forge a nation; it should not be surprising that Ghanaian leaders are still struggling with the task of nation building almost 50 years later.

In this effort during the early years of British administration, the latent enmity between the Ashanti, whose nationalistic expansionism had been halted, and the peoples of the south formed an ever-present obstacle. The south increasingly became the focus of activity in the Gold Coast, both administratively and economically. Just as Lagos was the capital of Nigeria in earlier days, Ghana has a coastal capital, Accra. As in Nigeria, much of the development that took place as a result of colonial control occurred in the south, where education made most progress, communications were most rapidly improved, and the introduction of new crops proved most successful. Eventually, the country achieved sovereignty under mainly southern leadership and as a result of pressures exerted by southern politicians and citizenry.

In many ways, Ghana repeats on a smaller scale the regional differences of Nigeria. Like its larger neighbor, Ghana has a Muslim north and a dominantly non-Muslim south. As in Nigeria, where Fulani (Fulbe) expansionism was halted by British intervention, Ashanti imperialism was contained by British control. Both states have a highly productive southern belt, which felt most strongly the impact of colonial administration. Political sophistication came to the south at a comparatively early stage, and pressures for independence rose there, while the far north remained little changed and under indirect, protectorate administration.

Northern suspicions of southern designs were reflected by a northern desire for prolonged British involvement in local administration. But there were important differences: unlike Nigeria at the time, Ghana's northern regions did not have a majority population, and northern Ghana did not have a core region that was as significant economically as that of northern Nigeria. Southern Ghana still leads economically and politically, which is reflected in the high degree of concentration of all types of communications and industry in the southern third of the country. It almost defies the imagination to think that in the twenty-first century, 50 years after independence, there still is no railroad link anywhere north of Kumasi.

These differences are reflected also in Ghana's political organization and geography. Unlike Nigeria, which in response to its internal diversity adopted a federal constitution, Ghana became a unitary state with a highly centralized government. Ethnic nationalism was attacked, the power of the chiefs reduced, education promoted vigorously, and tight control rigorously maintained. Efforts have been focused on the eradication of divisive politicogeographical characteristics within the country rather than adjustment to them. From a geographical point of view, Ghana clearly lends itself better to this attempt than does Nigeria. The country and its population are much smaller: for all practical purposes, the main productive capacity, the major administrative center, and the best amenities for education and communication are all located in just three southern regions. But the modern country of Ghana is centered on the former Ashanti state and its capital, Kumasi. This has given Ghana a political integrity that, with few exceptions (Uganda, Botswana, Lesotho) is unmatched in Africa. The language of the Ashanti state, Twi, has effectively become the national language of Ghana. In other words, Ghana has one core area of activity rather than several. This is not to underestimate the country's internal variety, which is almost as great as that of Nigeria. But all of Ghana focuses on the productive south, the source of many of the political ideas being applied throughout the country, its main outlet, and area of contact with the world.

While Ghana adopted a unitary form of government, most Ashanti leaders preferred a federal system, which they believed would protect Ashanti custom, tradition, and the chieftainship system, and provide balance between themselves and the coastal colony. Nkrumah and his party drew their support mainly from the colony, which contained the greater part of the population, possessed a higher standard of education, and was strategically positioned to control the interior. The federal–unitary debate in the immediate preindependence period was further aggravated by the inclusion in Ghana of British Togoland. A plebiscite was held under United Nations supervision in that British-administered former German territory in 1956, to determine the people's wishes concerning their political future. The choice was between union with an independent Ghana and continuation as a British trusteeship pending an alternative arrangement. A majority (58%) of the registered voters favored union with Ghana (Coleman 1956).

The inclusion of British Togoland in Ghana profoundly affected the Ewe people, who live in the south. Colonial penetration and subsequent division of the land in this area left the majority of the Ewe in the Gold Coast, but there were significant numbers in both British Togoland and French Togoland. In the vote concerning British Togoland's merger with Ghana, the Ewes voted two to one in favor of continued trusteeship by Britain, implying a desire for a future merger with French Togoland. But the overall majority of British Togoland's voters favored union with Ghana, so that the Ewes were reluctantly joined with that state. Irredentist problems have arisen since the implementation of the merger, and relations between Ghana and Togo have not always been good.

When Ghana attained its independence in 1957, it was initially divided into five administrative regions: Western, Eastern, Ashanti, Northern, and Trans-Volta Togo. Several changes have since been made in the interests of more efficient government and with a view of historical and political realities. Later the administrative regions were increased to nine, and later still another was added. Except for Ashanti, however, these regions do not coincide with ethnolinguistic areas (Figure 10-1). The four northern regions (Upper West, Upper East, Northern, and Brong-Ahafo: Figure 10-2) have the greatest ethnic and linguistic diversity, with the Gonja, Mampruli, and Dagbani speakers being the largest groups speaking languages in the Gur subfamily of Niger–Congo. The Volta Region, formerly part of British Togoland, is similarly ethnically heterogeneous, with the south being densely settled by Ewes. Various Akan groups are dispersed through the Western, Central, Ashanti, and Eastern regions; and the Greater Accra Region, once the homeland of the Ga-Adangbe

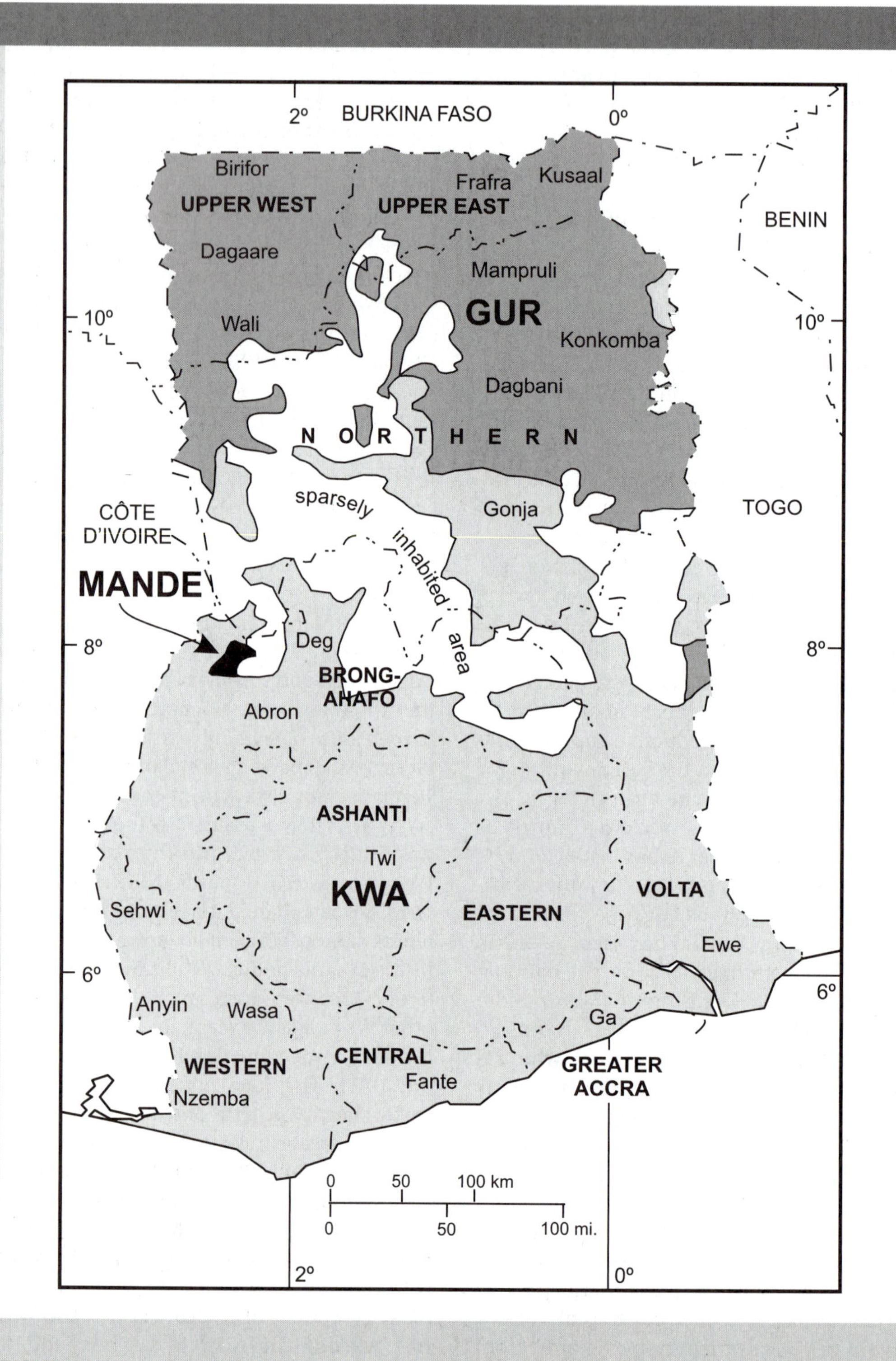

FIGURE 10-1 Language subfamilies and selected major languages of Ghana.

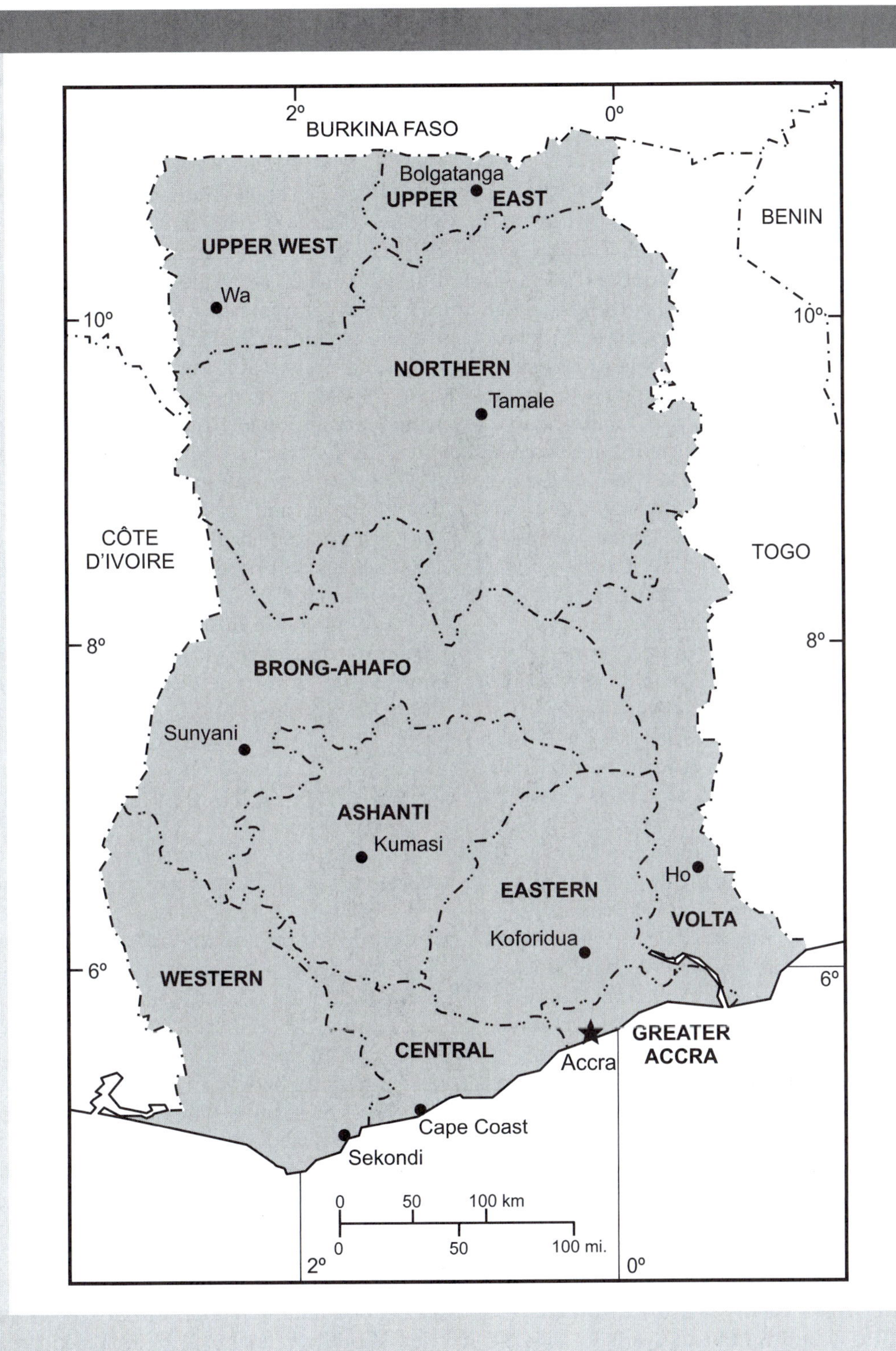

FIGURE 10-2 Administrative regions of Ghana.

is, because of considerable migration to the capital and to Tema, its industrial satellite town, one of the most ethnically diverse regions today.

ENVIRONMENTAL CONTRASTS

Ghana, like its neighbors in coastal West Africa, lies across several of the region's east–west climatic and vegetation belts. The country does not extend as far into the interior as does Nigeria, so that the total range of its environmental variety is somewhat less; indeed, the northern productive regions of Nigeria largely lie beyond 11° north latitude, which happens to be the line marking Ghana's northern border. Ghana receives most rainfall in the southwest, where its forest zone is best developed; in the southeast lies the anomalous dry zone (Figure 10-3). In the southwestern lowlands the rainfall exceeds 80 inches (2032 mm) annually, and most of the rest of the southwest receives over 50 inches (1270 mm). The forest that has developed lies in a triangular area that is broadest in the west and narrows eastward; its northern boundary corresponds generally to the Mampong Escarpment. This escarpment is the south-facing edge of a narrow plateau that separates the productive and densely populated south from the more empty, savanna-covered north. Along the coast is a stretch of scrub and, at the water's edge, a narrow zone of mangroves.

In the extensive savanna lands, the rainfall, although mostly exceeding 40 inches (1016 mm), is rather variable and often comes in severe storms, leading to excessive runoff and problems of erosion. As elsewhere in West Africa, a relatively unproductive "middle belt" lies to the north of the forest zone, its character reflected by the nature of its transport routes. In the south, a dense network of such routes has developed, but in the "middle belt" there are mainly north-south linking roads with very few feeders. Beyond, there is a cluster of dense population around Tamale and Bolgatanga (see Figure 10-2); the northern half of the country is suited to cattle raising as well as the cultivation of peanuts and cereal crops. No extensive northern areas approach the productivity of the south.

The pivotal physiographic feature of Ghana is the Black Volta River (Figure 10-4), which traverses the length of the country from northwest to southeast. It forms the boundary with Burkina Faso and Côte d'Ivoire before turning east and being joined by its major tributary, the White Volta. The entire northeastern region consists of the basin of this river system, about 40,000 square miles (103,600 km^2) in area, and is underlain by near-horizontal sandstones, which have produced generally infertile lateritic soils. Some 60 miles (97 km) from the coast, the Volta breaks through the Akwapim–Togo fold mountains (1,000–1,500 ft; 305–457 m) and then flows across the gently undulating Accra Plain before fanning out into a small delta. In the 1960s the Volta was dammed at Akosombo to provide hydroelectric power and water for the Accra–Tema core region, and to irrigate the adjoining Accra Plain. Impounded behind the dam is Lake Volta, one of the world's largest artificial lakes, which extends some 250 miles (400 km) inland. Flowing out of the Kwahu and Ashanti uplands (average elevation 2,000 ft; 610 m) to the west, and across the densely settled Akan lowlands and coastal plains, are several small but important rivers including the Bia, Tano, Ankobra, and Pra. The region of northern plateaus is part of the same plateau structure found in Côte d'Ivoire. The average elevation ranges from 600 to 1,200 feet (183–366 m). The soils, developed on granitic parent material, are more fertile than those developed on the sandstones of the Volta Basin.

TRANSPORT AND DEVELOPMENT

The outstanding geographical aspect of Ghana is the concentration of its productive capacity in the south and southwest, corresponding largely to the area under forest cover. Most of the agricultural exports come from this zone, with crops to the north being consumed locally, for the most part (Figure 10-5). The majority of the important mineral deposits also lie in the south and southwest. In fact, the area around Dunkwa (about halfway between Takoradi and Kumasi) is one of Africa's most important mining zones. Several factors have favored Ghana's relatively rapid economic development, among which the juxtaposition of its mineral and agricultural resources is a major one; transport routes served both industries at once. The productive area is located near the coast (Figure 10-6), and although Ghana has not enjoyed the benefits of a really good natural harbor, the volume of exports has risen steadily. Palm products, gold, and rubber were important substitutes for slaves during the latter half

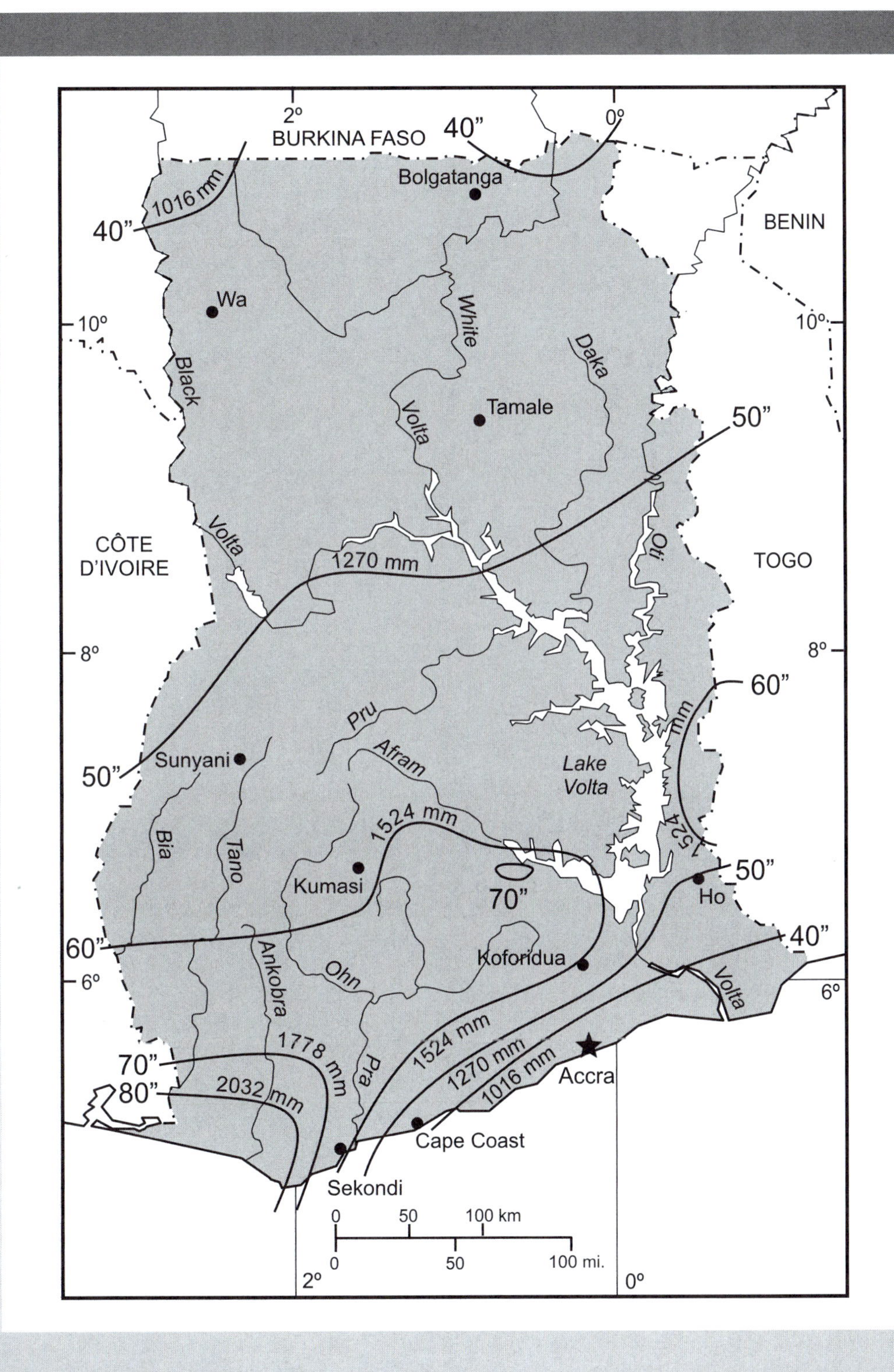

FIGURE 10-3 Precipitation patterns in Ghana.

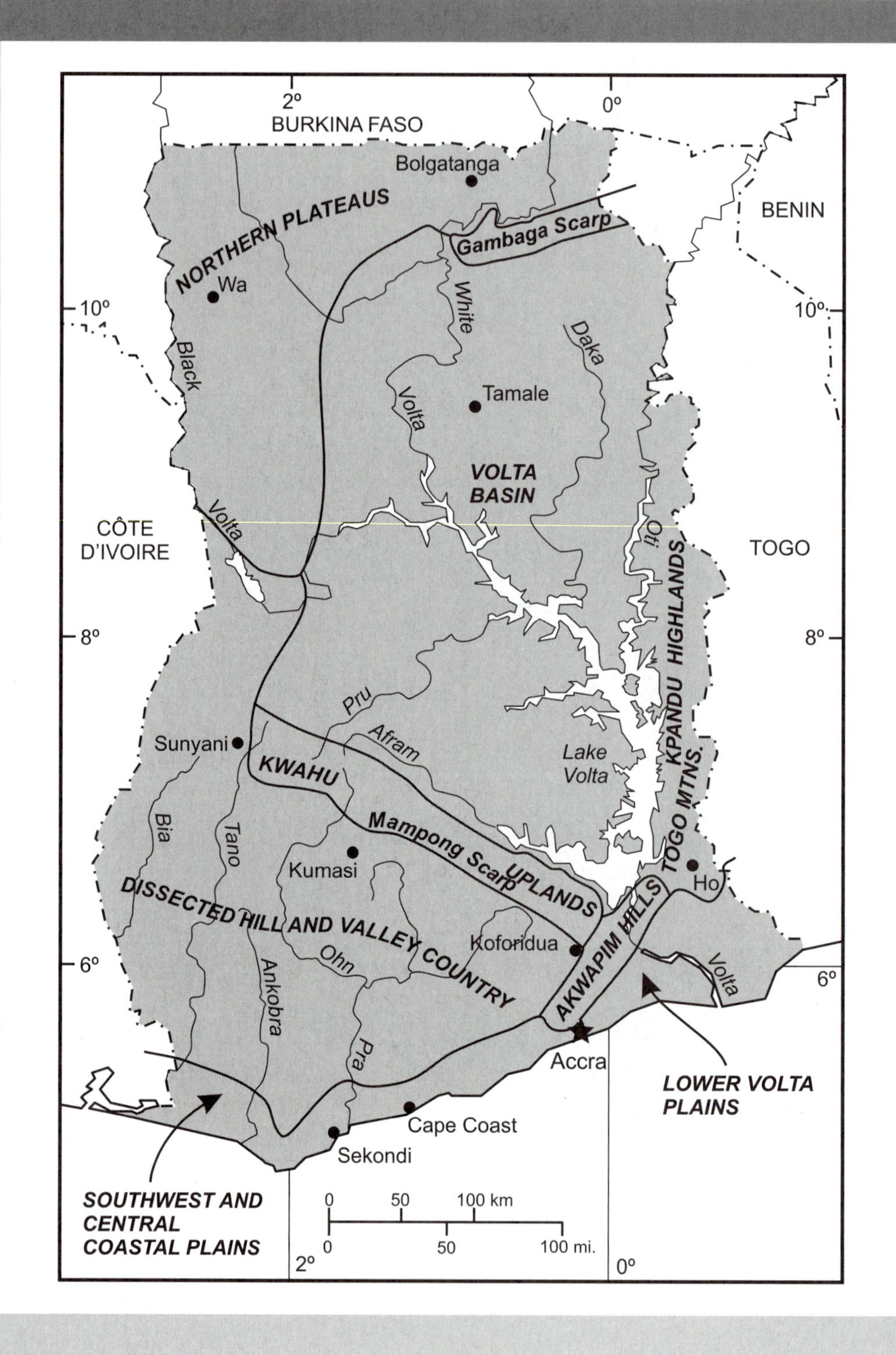

FIGURE 10-4 Physiographic regions of Ghana.

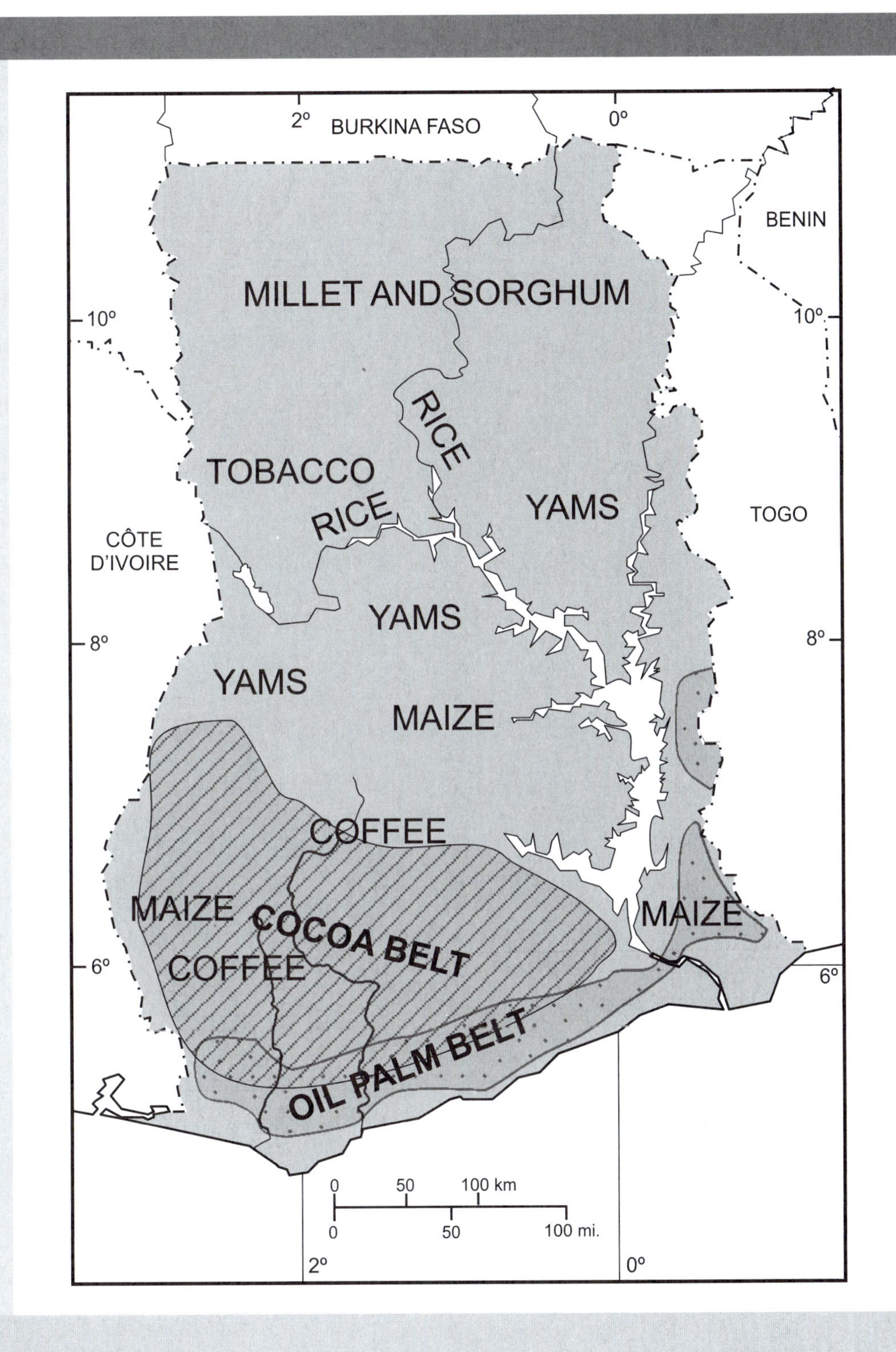

FIGURE 10-5 Agriculture in Ghana.

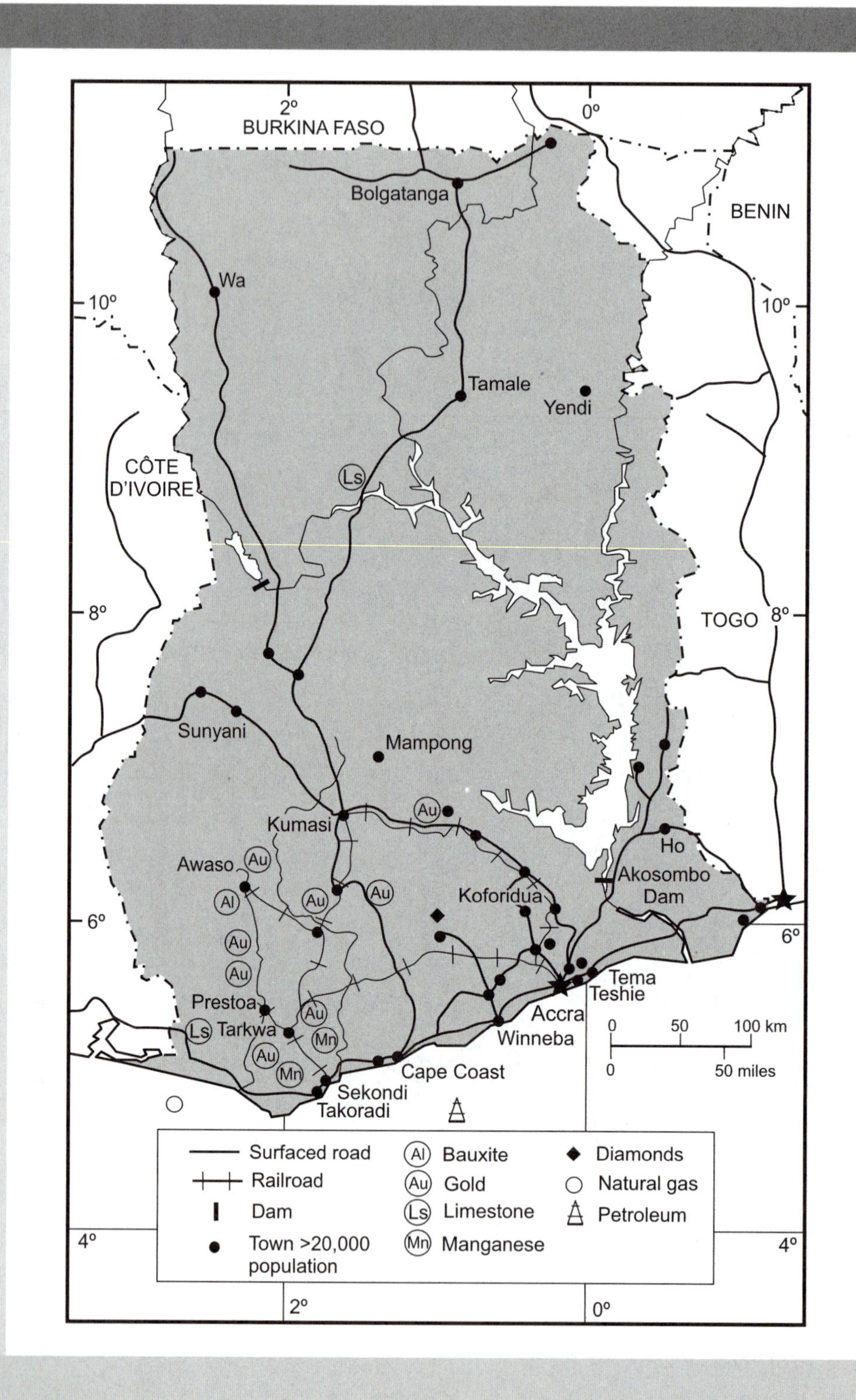

FIGURE 10-6 Infrastructure, mining, and towns in Ghana.

of the nineteenth century, but the strongest impact was made by cocoa, introduced in 1879. By 1891, this crop had begun to figure among the exports, and a half-century later, Ghana was poised to reap the spectacular profits of the postwar cocoa boom. Although in the years right after independence 50% of all the cocoa produced in the world came from Ghana, and the Ghanaian economy rode the end of the first cocoa boom to the 1970s, by 2001 Ghana's contribution to world production was only 18%. The cocoa market was taken over by a more competitive neighbor, Côte d'Ivoire, before the civil war set its own industry back, but Ghana is still the second most important producer of cocoa in the world.

Most of the cocoa is exported as raw, unprocessed beans to the North American and European markets, where it is made into chocolates. Ghana and Côte d'Ivoire are alike in that in both countries cocoa production is, and has always been, almost exclusively in African hands, both men and women owning cocoa farms. Ownership patterns will change with increasing privatization, market liberalization, and globalization. One may expect increasing outside ownership when world prices increase.

The modern agriculture (and mining) economy of Ghana rests squarely on what was built during colonial rule. The rapid economic growth after independence could not have taken place without the establishment of a transport network. Having been contained at the coast for a long time, and succeeding in finally subduing all the peoples of Ghana only after the turn of the century, the British could not address the problem of communications until a late stage, by which time the absence of good roads and railroads was impeding the development of mining and was affecting the export quantities of rubber, timber, and palm oil. Between 1890 and 1900, mining companies exerted pressure upon the government to provide bulk transport facilities, and the first surveys for projected lines were made during that decade. In 1898 the Sekondi–Tarkwa line was begun, and it was completed in 1901, connecting the goldfields at Tarkwa to the coast. Immediately, heavy use was made of the line, and its extension toward Kumasi, hastened by the need for rapid transport to the rebellious Ashanti capital, was finished by late 1903.

In the east, agricultural and forest production benefited greatly from the railroad begun in 1907 at Accra. By 1915 it had reached Koforidua, and carried no less than 40,000 tons of cocoa. The process of railway construction was interrupted by World War I, but soon after 1918, it began again, and Kumasi and Accra were linked in 1923. Another line was started eastward from the Sekondi–Kumasi railroad, through the Central Region, reaching its terminal, Kade, in the Eastern Region, in 1927. A later phase of construction (1953–1956), completed the triangle from Sekondi to Kumasi to Accra during which period the new, important port of Tema was linked to the system at Achimota.

Throughout this period of railroad construction and operation, and except only for wartime periods of gasoline rationing, railroad transportation faced the competition of roads. Furthermore, both road and rail had to compete with riverboats, especially on the Volta, and even with headload transport.

Surf boats operated between ship and shore, at Sekondi and Accra, and for some time these ports vied for the lead in handling the country's external trade. Originally, owing to its position at the coastal terminal of the country's only railroad, Sekondi was the first port; the Accra–Kumasi connection was completed in the 1920s. Soon afterward the more modern port of Takoradi, near Sekondi, became operative, and the western terminal began to forge ahead of its competitors (Dickson 1969). Takoradi and Sekondi have since become virtually a single metropolitan area. With a population of 170,000, they support several large industrial establishments, including an automobile assembly plant and plywood and paper mills. About 50% of the total export tonnage, mainly minerals and timber, is shipped from Takoradi.

The effect of the transportation network in stimulating Ghana's economic development has been spectacular. In addition to gold, which until 1910 was the only significant mineral export, Ghana possesses significant deposits of bauxite, manganese ore, and diamonds (Figure 10-6). Of these, all but the diamonds are found near the Takoradi–Kumasi railway; manganese near Takoradi and Tarkwa, bauxite near the terminal of the Awaso extension and in larger quantities not far from Kumasi itself, and gold near Tarkwa. Only railroads could have provided the means to carry the heavy equipment required by the mining operations, and without rail transport, the ores could not have been exported easily.

Equally impressive has been the effect of improved communications on the agricultural industries, especially cocoa cultivation. Cocoa dominated the agricultural export list to the exclusion of practically all else. Before the advent of economic diversification, cocoa contributed between one-half to two-thirds of the annual export revenue, but in recent years its contribution has fallen to 20 percent. Although sometimes cocoa cultivation existed before adequate roads were pushed into an area, the actual road-building process was immediately accompanied by the staking out of farms, while elsewhere the knowledge of plans for future road construction led to the preparation of such farms in advance.

In Ghana, the forest belt is the most densely populated area, economically the most important, and in terms of communications, by far the best endowed. Lying between the narrow coastal region of scrub in the south and the savanna region of the north, it covers most of the Western and Central regions, virtually all of Ashanti (although the forest thins out to the north), and extends into the northern part of the Eastern Region. As a result of the success of cocoa in this belt and the concentration of other economic activities there, population has moved into Ashanti from the north and even from adjoining countries. With the rise in demand for land and forest products has come the destruction of large parts of the forest, soil erosion in serious measure, and the loss of the vegetative protection against the drying effect of the harmattan.

Toward the northern fringe of the forest belt is Kumasi, Ghana's largest city of the interior and the northernmost point of the national core region; its population in 2003 was 1.3 million. Founded some 300 years ago, the city became the political, economic, and cultural center and capital of the Ashanti Kingdom. All major north–south trading routes across Ashanti converge on the capital, and despite British efforts to divert trade from the town following the wars of 1874, Kumasi quickly regained its commercial importance and became not only the heart of Ghana's cocoa belt, but also a light manufacturing center. From the surrounding forest came the country's initial exports: palm oil, rubber, and timber. But with the introduction of cocoa, much of the land was eventually taken up by this cash crop, and so lucrative was the trade that food staples had to be imported.

Ghana's capital, Accra (1.7 million within the city limits but 2.8 million for the urban agglomeration), has more than doubled its population in the last 25 years. Yet Accra has not grown as rapidly as other African capitals or become the country's leading industrial city, largely because of its unfavorable transport system and limited water supplies. It lacks a natural deepwater harbor; in fact, it has no harbor facilities to speak of. Cargo vessels had to anchor offshore and load their goods onto small surf boats, which was costly, time-consuming, and risky. Nevertheless, Accra is an important transport node and, until the growth of Tema, it was Ghana's largest industrial city, specializing in light industries such as textiles, food processing, leather goods, and printing.

Twenty miles (32 km) east of Accra is the port of Tema (250,000), opened in 1962 and originally built as a requirement of the Volta River Project (Box 10-1). It serves the whole country and handles about 83% of all imports, and about 40% of Ghana's cocoa exports. It has become Ghana's principal manufacturing center, utilizing both imported and natural resources. Its products are geared not only to local markets but those of Burkina Faso, Mali, and Togo. Indeed, Tema is now the largest industrial town on the Guinea Coast west of Lagos. The aluminum smelter at VALCO (Volta Aluminium Company), which processes imported alumina, is the single largest processing industry in Ghana and the third largest producer of aluminum metal in Africa. A steel mill, using scrap iron from Ghana's urban centers and mines, and chemicals and ferroalloys imported from Britain and Sweden, was built in 1962 in response to the expanding demands for iron and steel products. Like VALCO, this facility depends on relatively cheap hydroelectric power from the Akosombo Dam. Tema's third capital-intensive industry is the oil refinery, built and maintained by Ghanaian and Italian concerns. It produces fuel oil, kerosene, and gasoline from crude oil imported mainly from Nigeria. Also manufactured in the city are chemicals, paints, textiles, aluminum products, beverages, furniture, cigarettes, cement, clothing, and pesticides. The combination of Tema's close proximity to markets and to Accra, the city's special tax structures, the modern harbor facilities, good rail connections with the interior, and the availability of cheap power have made Tema Ghana's most diversified and largest industrial city.

Industry contributes 25% to the Ghanaian gross domestic product, while agriculture and services account for 36 and 39%, respectively. Although the

BOX 10-1

THE VOLTA RIVER PROJECT

Ghana's single most important development project undertaken since independence is the multipurpose Volta River Project, financed by long-term loans from the World Bank, the U.S. Agency for International Development (USAID), the U.S. Export-Import Bank, and British banking institutions. Conceived as early as 1925 but abandoned for economic reasons, the idea was revived in the late 1950s, and following 4 years of political maneuvering, construction began in 1962. The project has four major components: the Akosombo Dam and hydroelectric plant, whose generating capacity is 900 megawatts; a 225,000 metric ton capacity aluminum smelter at Tema, which places Ghana among the world's top 10 producers of aluminum; an extensive transmission system to relay electricity to Accra, Tema, Kumasi, Sekondi-Takoradi, the mining areas, and even Burkina Faso, Togo, and Benin; and a modern deepwater port at Tema and related road and railway improvements. Additional but secondary benefits of the project include increased water supplies for the Accra-Tema area, development of an inland fishing industry, low-cost water transport between northern and southern Ghana, and irrigation of the Accra Plain.

Generation of electricity began in 1966, and the project accounted for 70% of Ghana's total electricity production in 2001; VALCO's aluminum smelter is the single largest consumer (70% of the total). Additional dams and electrical generating facilities may be built if markets can be found. Some 80,000 persons were removed from the land now flooded by Lake Volta and resettled in 52 villages in the same general area. Each family that opted for resettlement into agricultural villages was promised 6 acres (2.4 ha) of cleared farm land; some 10,000 individuals chose a cash payment instead (Lumsden, 1973).

Although the area was sparsely populated prior to the creation of Lake Volta, people settled along the lake to fish, the land being rather infertile. Some of the migrants brought the debilitating disease, schistosomiasis (bilharzia), which became established in the region along the lake. Malaria, a mosquito-borne disease, also has increased in the populations along the lakes. Ghana added a 160 MW hydroelectric plant at Kpong, downstream of Akosombo and is negotiating the construction of a hydroelectric plant on the Black Volta at Bui. Thermal generation plants are located at Tema and Takoradi. The Ghanaian government would like to reduce its dependence on hydroelectric power as a result of fluctuations in output caused by drought.

percentage is lower than in many other African countries, a still significant 60% of the population is engaged in agriculture. But industry has been growing faster than agriculture. Ghana is the second-largest producer of gold in Africa, next to South Africa. Gold production increased from 16.8 metric tons in 1990 to 72 metric tons in 2000, and is expected to remain at the higher level for the near future (Coakley 2001).

Petroleum, of which Ghana had no production until 1996 was 2.2 million barrels in 2000. Silver has increased from 1,850 pounds (840 kg) in 1990 to 8,000 pounds (3,630 kg) in 2000. Ghana's production of steel has grown from 26,000 metric tons in 1990 to 75,000 metric tons in 2000. Ghana is also one of the world's most important producers of arsenic.

The resurgence of mining activity has not been without costs: increased water pollution, deforestation, and erosion. There has been sulfur and arsenic oxide pollution from gold processing, as well as mercury pollution of rivers and streams. Mining companies have added arsenic recovery systems to processing plants. It has been difficult to control mercury pollution because this element is widely used by artisanal miners to amalgamate gold. The Environmental Protection Agency, established by the government in 1994, has developed environmental regulations for mining in Ghana. Environmental impact studies and pollution-minimization plans are required of all new mining operations; existing mines are being rehabilitated. In addition, a percentage of mining royalties is put into a fund that is used to clean up polluted sites (van Oss 1994).

DEVELOPMENT CONSTRAINTS, STRATEGIES, AND OPPORTUNITIES

When Ghana gained its independence in 1957, it was the envy of many developing countries. It had a strong agricultural and light industrial base, the highest per capita income of any tropical African country, better than average education and medical services, an efficient transport system, substantial foreign currency reserves from cocoa earnings, a dynamic political leadership, and other assets normally taken as requisite ingredients of national viability. Yet these assets were no guarantee of economic prosperity and political stability. Indeed, Ghana has experienced considerable difficulty in meeting its national objectives.

Between 1957 and 1966, President Kwame Nkrumah and his Convention People's Party (CPP) adopted a policy of "African socialism" at home and nonalignment and Pan-Africanism abroad. Nkrumah tightened his control over the party and country, and by 1964 Ghana was a legalized single-party state that was characterized by graft and corruption within the party and civil service, economic mismanagement, a rapidly deteriorating balance of payments position, huge losses of external reserves, and the flight of foreign investment. Considerable money was spent on prestige projects such as government buildings, the national airways, superhighways, and national edifices, bringing the country to the verge of economic collapse (Guyer 1970). A 7-year development plan announced in 1964 was designed to revolutionize and collectivize agriculture, industry, and education. The aim was to build a socialist state in which there would be state industry, state farms, and a state-sponsored workforce, while investment by foreigners would be encouraged to obtain foreign exchange and to increase productive efficiency. Indeed, the regime emphasized total dependence on foreign capital to industrialize the country. The Volta River Project was the largest and most successful of these commitments, undertaken at a time when world cocoa prices were falling, severely curtailing government revenues.

On February 24, 1966, while Nkrumah was en route to Beijing, the Ghanaian army and police staged a successful coup d'état. The constitution was suspended, the CPP and all other political parties were outlawed, the national legislature was dissolved, and the army established a National Liberation Council (NLC) of four army and four police officers under the chairmanship of General Ankrah. The NLC inherited a national debt of more than £400 million compared with £20 million when independence was granted, and by 1968, this had risen to £652 million. During the last 3 years of Nkrumah's administration, there was a 66% rise in the cost of living. Ankrah reduced government spending and cut imports of luxury goods, transferred several state corporations to private enterprise, and moved toward "constructive partnership" between the public and private sectors. Former Commonwealth ties were reaffirmed, and a more pro-West foreign policy was adopted, while the NLC maintained its support of Pan-African aims. Many of the old regime's links with Eastern Europe and the People's Republic of China were severed, and Ghana's relations with its immediate neighbors improved.

In 1969 Ghana was returned to civilian rule under the leadership of President Busia. But the country still suffered from foreign indebtedness, low cocoa prices, and rising imports. The government decided to deport aliens without residency permits in an effort to reduce the outflow of remittances, and to provide jobs for Ghanaians. At the time there were 2 million aliens, including 700,000 Nigerians, 500,000 Burkinabe, and others from Benin, Niger, Togo, and Liberia, of whom 800,000 had no residency rights. Most were engaged in petty trading, or held seasonal jobs in the cocoa industry. In 1971 Busia ordered a 44% devaluation of the national currency, the cedi, which brought huge rises in the cost of living, general public discontent and, eventually, the president's downfall. In January 1972, Colonel Acheampong, the leader of a second military coup, established an ethnically balanced National Redemption Council of army, navy, air force, and police officers. The cedi was revalued by 42%, and an austerity program was introduced. The NRC's main economic policy was ostensibly one of "self-reliance," which aimed at increasing domestic food production and use of local raw materials for industry. But the government failed to provide new roads and other services needed for the realization of these goals, the cocoa industry continued to suffer despite enormous rises in world cocoa prices, and economic hardship increased for ordinary citizens of Ghana. The 7 years of NRC rule were characterized by the personal accumulation of wealth by officials.

The military regime revived trade with East European and other socialist countries and secured their assistance in the reactivation of numerous state manufacturing and processing projects that had been abandoned in 1966. Trade and aid agreements were signed with the Soviet Union, and the People's Republic of China began to assist in large-scale socialist agricultural projects. The government also assumed a 55% participation in the country's mining enterprises, and 55% interest in foreign timber, oil, and fertilizer companies. The military appeared to be well entrenched in the machinery of administration at all levels, and in the mid-1970s a return to civilian rule seemed unlikely. In the face of deteriorating economic and social conditions and rising popular opposition to the regime, the government was removed from power in 1978 in a palace coup and replaced by Acheampong's second in command, General Fred Akuffo, who was interested in moving the country to civilian rather than military rule.

Akuffo legalized political parties and organized an assembly to formulate a new constitution. Amid rumors that Acheampong and Akuffo were manipulating the political process to secure for themselves immunity from prosecution after power was handed over to a civilian government, Flight Lieutenant Jerry Rawlings led an army mutiny in May 1979. Although the small-scale mutiny was unsuccessful and Rawlings was imprisoned, two weeks later there was a general military uprising against the regime. Then Rawlings was appointed the head of state and Acheampong and Akuffo were executed (Haynes 1995). In September 1979 Rawlings handed power over to an elected government led by Hilla Limann. The corruption and incompetence of this government led to its overthrow in 1981 and the return of Rawlings.

As the uncontested leader of the Provisional National Defense Council (PNDC) from 1981 to 1992, and after some false starts, Rawlings led Ghana to sustained and significant economic growth. It appeared when he took power for the second time in 1981 that he would revive the populist socialism of his predecessor Kwame Nkrumah. There were some executions and mass rallies; there was also revolutionary talk, purging of the opposition, violence, and nationalization. But in 1984, confronted by inflation and economic instability, President Rawlings made an about-face and began to liberalize the market and encourage privatization of public enterprises. He also embarked on a structural adjustment program of the International Monetary Fund (IMF). Although later there were economic problems caused by a downturn in the prices for cocoa and gold, Ghana became one of the success stories for the IMF. Rawlings's strength lay in his courage to confront Ghanaian statism and patrimonialism head-on and lead the country in a market-oriented direction (Hutchful 2002). He began a process of "guided democratization" or "democratization from above" that led to multiparty elections in 1992 (Box 10-2).

Under the Rawlings government, the country slowly liberalized its institutions, wrote a new constitution, legalized political parties and a free press, and created independent national human rights organizations (Schaefer 2000). In an election marred by some fraud, Rawlings won the presidency and his political party swept into power across the country. In 1996 he won again in a free election. He was barred from running for election for a third term by a constitutional provision. In 2000 John Kufuor was elected president and he was reelected in 2004.

Prior to the changes instituted by Jerry Rawlings, Ghana's economy operated under a set of sociopolitical constraints not very conducive to growth and diversification. Investors, both local and foreign, lacked confidence in Ghana's political leadership and economic climate. Venture capitalists were not attracted to invest in Ghana because of the monopolistic control of production and marketing by the government, and high taxes on exports of raw materials were not conducive to investor confidence.

Statism and patrimonial rule are no longer the practice in Ghana, but the economy has several structural and spatial weaknesses that impede development, the principal being the high emphasis on agricultural and mineral exports, the balance of payments problem due to heavy reliance on imported consumer goods and equipment, the lack of basic capital goods industries, the high degree of industrial concentration in the metropolitan centers of Accra–Tema,

BOX 10-2 PORTRAIT OF A PRESIDENT: JERRY RAWLINGS

The man who has been so instrumental in the recent transition of Ghana to democracy and private enterprise, Flight Lieutenant Jerry Rawlings, was born in 1947 to an Ewe mother and a Scottish father. He was educated at Achimota College, in the Ghanaian air force cadet school, and in the Military Academy at Teshie, where he trained as a combat pilot and from which he graduated with the rank of Flight Lieutenant in 1969. Uncharacteristically for African military leaders, Mr. Rawlings never had himself promoted to the rank of general but remained a flight lieutenant throughout his tenure in office. After retirement from office in 2000 he became a leader of humanitarian causes and development for a brief period and is now again involved in party politics. Although he violently overthrew a constitutionally-elected government and was an avowed critic of constitutional rule and private enterprise, he fostered the transition to democracy and constitutional law after 11 years of rule and was himself twice winner of the presidency.

Sekondi–Takoradi, and Kumasi, and the weak interregional and intersectoral linkages. Some of these problems have existed since early independence; some date back to colonial days.

Despite the creation of a climate conducive for growth since 1981, Ghana must continue to diversify its economy and encourage the spread of industry across the country. But industrial decentralization, while desirable, is rarely easy to promote under any macroeconomic framework. Vast areas of northern Ghana, where cereals and cattle have considerable potential, need to be integrated with the more populous and prosperous south. But the northern regions are handicapped by inefficient techniques of food production, conservative cultural values and preferences, and a sparse transport network, all of which have defeated various efforts since independence to promote the region's economy and to integrate it more with the south (Dickson 1969).

Poverty and underemployment prevail in the Volta Region, where discontent comes and goes in various guises and secessionist aspirations are probably not yet dead. But disparities between rich and poor are not confined to these broad regions; they also exist at the local level and within metropolitan areas. Recent research, however, appears to indicate that Ghana's Structural Adjustment Program (SAP), begun in the mid-1980s, is now paying off in providing a liberal economic framework to encourage the diversification of production and exports other than the traditional triad of gold, cocoa, and timber. In 1999 nontraditional exports accounted for over 12% of exports, and they are projected to increase to 30% by 2030 (Konadu-Agyemang and Adanu 2003). Furthermore, the diversification of the economy stimulated by the SAP toward nontraditional exports has had positive impacts on the poorer northern regions of Ghana, where the timber industry is small, cocoa is not grown, and the deposits of gold are meager.

In just a few years after independence Ghana, once the model of development in Africa during the postwar cocoa boom, came to represent what had gone wrong with African states. Beset with internal difficulties that seemed almost impossible to control, Ghana came to epitomize the problems of development so common in Africa after independence: political instability, bad government, mismanagement, and corruption. Undeniably Ghana has social and economic problems today, yet it has accomplished much: it has made the transition to a representative government that is reasonably transparent; it has a growing and diversifying market economy, and a free and active press; it has developed a judicial system that is a pragmatic and clever combination of English common law and Ghanaian customary law. Although it is too early to predict the future, Ghana has become a model of development that other states in Africa could emulate.

BIBLIOGRAPHY

Coakley, G. 2001. *The Mineral Industry of Ghana*. U.S. Geological Survey Mineral Report. Reston, Va.: USGS. http://minerals.usgs.gov/minerals/pubs/country/.

Coleman, J. S. 1956. Togoland. *International Conciliation*, 509: 3–91.

Dickson, K. B. 1969. *A Historical Geography of Ghana*. Cambridge, U.K.: Cambridge University Press.

Guyer, D. 1970. *Ghana and the Ivory Coast*. Jericho, New York: Exposition Press.

Hutchful, E. 2002. *Ghana's Adjustment Experience: The Paradox of Reform*. Geneva: United Nations Research Institute for Social Development.

Konadu-Agyemang, K., and S. Adanu. 2003. The changing geography of export trade in Ghana under structural adjustment programs: Some socioeconomic and spatial implications. *Professional Geographer*, 55(4): 513–527.

Lumsden, D. P. 1973. The Volta River Project: Village resettlement and attempted rural animation. *Canadian Journal of African Studies*, 7(1): 115–132.

Nkrumah, K. 1961. *I Speak of Freedom*. New York: Praeger.

Schaefer, P. J. 2000. *African Politics and Society: A Mosaic in Transformation*. Boston: St. Martin's Press.

Van Oss, H. 1994. *The Mineral Industry of Ghana*. U.S. Geological Survey Mineral Report. Reston, Va.: USGS. http://minerals.usgs.gov/minerals/pubs/country/.

CHAPTER 11

Experiment in Federation in Nigeria

The Federal Republic of Nigeria is Africa's most populous state (134 million in 2003), and its thirteenth largest in area (about the combined area of California, Oregon, and Nevada). It is a country of great cultural and environmental diversity, and one of enormous economic potential. There are some 250 ethnic groups, although only four (Hausa, Fulani, Yoruba, and lbo) account for 68% of the total population. A federal constitution was adopted in partial recognition of the country's human and environmental diversities and, in this respect, Nigeria differs from most African states.

Despite its putative advantages, federation has not been an unqualified success and periodically, Nigeria has been beset by regional crises, the most devastating and most publicized being the Biafran war of 1967–1970. The challenge to Nigerian unity of the first decade of the twenty-first century, however, may very well be the recent institution of the Islamic legal system, or Sharia, in 12 northern states. In any event, a major politicogeographic issue to be resolved is the establishment of meaningful administrative units or states. Since independence, Nigeria's internal political boundaries and administrative organization have undergone considerable adjustment in response to both regional and national interests. This chapter focuses on these specific issues and on the general questions of regionalism and the federal experiment.

DIVERSITY

Nigeria is a country of immense geographic diversity. Its climates are more varied than any other West African state, its vast area encompassing many distinct and varied landscapes and numerous different ethnic groups. The pivotal physiographic feature is the Niger River, which, with its major tributary, the Benue, forms a Y-shaped system that divides the country into three parts. Below Onitsha, the Niger is navigable year round by shallow craft, and during high-water periods the Niger and Benue are navigable as far as Jebba and Yola, respectively. Both rivers have figured prominently in the historicogeographical development of Nigeria. When the jihad waged by the Muslim Fulani of the north brought Fulani hegemony to the Niger's banks, the river became the natural boundary between the Yoruba and Fulani empires, but it did not stop the spread of Islam, which was adopted by many Yoruba to the south.

Nigeria possesses a number of distinct physiographic regions (Figure 11-1). The highest elevations (6,700 ft; 2,040 m), occur in the deeply incised volcanic Adamawa Highlands along the Cameroon boundary. The most extensive elevated region is the Jos Plateau and its surroundings, and the high plains of Hausaland, which lie north of the Niger–Benue confluence. Together, these old crystalline plateaus make up the core of northern Nigeria from which several important rivers flow: the Komadugu and Hadejia northeast into Lake Chad, the Gongola and Mada south into the Benue, and the Sokoto and Kaduna southwest into the Niger. Toward the northeast, the land drops gradually into the Chad Basin, and in the northwest lies the Sokoto, or Sokoto–Basin, which extends into central Niger. The Niger–Benue Lowland is separated from the coastal areas, including the extensive Niger Delta, by the Oyo–Yoruba Plateau in the west and the Udi Plateau in the east, both regions lying mainly between

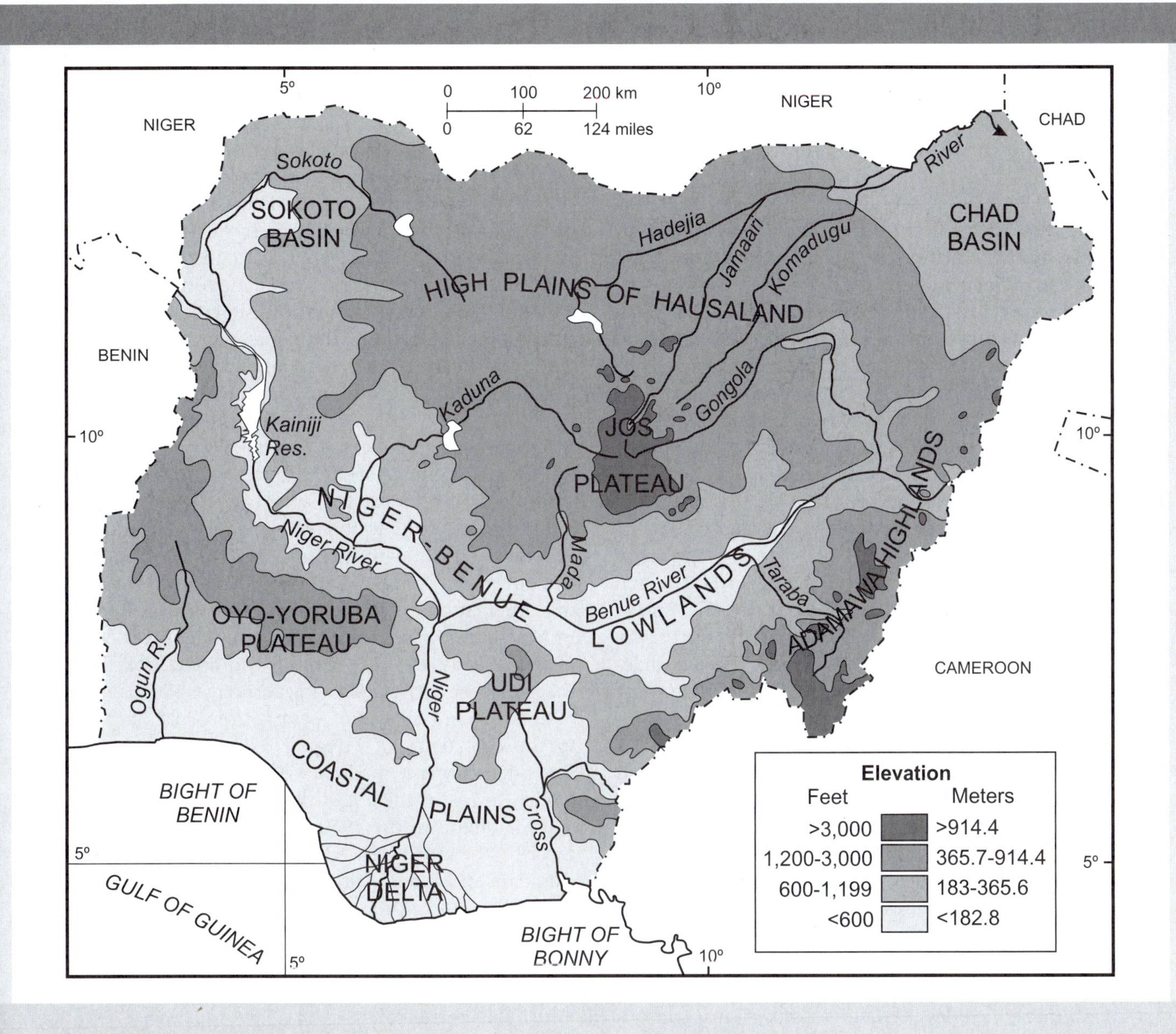

FIGURE 11-1 Physiographic regions of Nigeria.

1,000 and 2,000 feet (300–600 m). In the south, the coastal plains decline gently toward a shoreline of spits and bars and lagoons and luxuriant mangroves, where the Niger has built an extensive delta.

Nigeria possesses a wider range of climates than any other West African state, but it is similar in its general pattern of climate to other West African coastal states. Isohyets, or lines on a map connecting areas of equal rainfall, trend generally east–west across the country, and the extreme south receives over 140 inches (3,550 mm) annually, while the northern margins record less than 20 inches (508 mm). In central Nigeria, including the southern portion of the high plains of Hausaland and the northern sections of the Oyo–Yoruba and Udi plateaus, rainfall is between 45 and 60 inches (1,140–1,525 mm). The northern periphery is steppe and thorn forest and, toward the north, rainfall variability increases as the length of the rainy season decreases.

Thus a series of vegetation zones lies parallel to those of rainfall, reflecting the increasing dryness from south to north. Inland from the coastal mangrove and freshwater swamps lie areas of rain forest (better developed in the central and eastern parts than in the west), which were formerly more extensive than they are today. As a result of human activity, this inner margin of rain forest has become a zone of forest and savanna, sometimes referred to as "derived savanna." North of this belt, the country becomes more open, the bulk of it being covered by a savanna that is characterized by increasing dryness (Figures 11-2 to 11-4). Nigeria's physiographic variety is paralleled by the diversity of its population groups, and their religions, languages, and modes of living. South of the Niger and Benue rivers, two peoples dominate numerically: the Yoruba in the west, and the Ibo in the east. To the north, the majority peoples are the Hausa and the Fulfulde-speaking Fulani (see Figure 1a, p. 281). Unlike most African countries, in Nigeria there are three clusters of high density population, each associated with a historically dominant group: the Hausa–Fulani core located north of the Niger and Benue rivers, the Yoruba core in the southwest, and a southeastern core dominated by the Ibo (Figure 11-5).

Depending on the bases employed for dividing Nigeria's population into ethnic groups, perhaps 250 distinct ethnolinguistic groups might be recognized, numbering from a few hundred individuals to several million. The region's history, like that of other parts of West Africa, includes alternate periods of local consolidation and ethnic fragmentation; both the plains of the north and the forests of the south were scenes of efforts to create lasting states and empires. Some of these, like Benin, became powerful entities, surviving several turbulent centuries, leaving their mark on the country to this day. Others failed in the course of feudal rivalries and were absorbed by more successful contemporaries. From the north came the impact of Islam, as the Fulani, living in the land of the numerous Hausa farmers and traders, rose to power and established strong emirates. From the south came the impact of Europe, and the Atlantic slave trade (and the trans-Saharan trade as well) gave way to legitimate trade and to territorial control.

Nigeria, then, incorporates a wider range of physical and cultural conditions than any other West African country. Some features are shared by several parts of the country; others are confined to one area. Sections of the south to the east of the Niger River, for instance, possess soils and vegetation, as well as climatic conditions, that are similar to those found to the west of the river. But in terms of their culture and development, these two areas are quite different. And both differ greatly from the area lying to the north of the Niger and Benue.

In many countries of the world transportation infrastructure has often acted to help overcome regional differences and to bridge the gap between peoples and culture who live under the same flag. The colonialists built from the mine to the port, and the infrastructure that most African countries inherited at independence was outward looking. A glance at the map will confirm this observation for most African countries even today. In Nigeria, the railroad from Kano to Lagos was completed as early as 1912, with an additional link from that northern city to the navigable Niger River. Other railroads were built subsequently, from Port Harcourt to the coal fields of Enugu (1916), and on to the Lagos–Kano line at Kaduna (1932). The line was extended to Jos in the 1920s, Bauchi in the late 1950s, and Maiduguri in the 1960s (Odeleye 2000). The railways in the north served to foster urbanization, to stimulate enormously the production of peanuts, cotton, hides, and skins, and to promote interregional trade and contact. The Nigerian rail system is managed by the National Rail Corporation (NRC), a parastatal established by parliament in 1955 and modeled on the British Rail parastatal.

At independence, transport infrastructure was more developed in Nigeria than in most African countries, and the government has made some efforts to extend it. The road and railroad networks are best developed within the three major producing regions (the south, the areas east and west of the Niger River, and the north), however, and connecting links have long been tenuous. Kaduna is 561 and 569 miles (903 and 916 km), from the two seaports by the western and

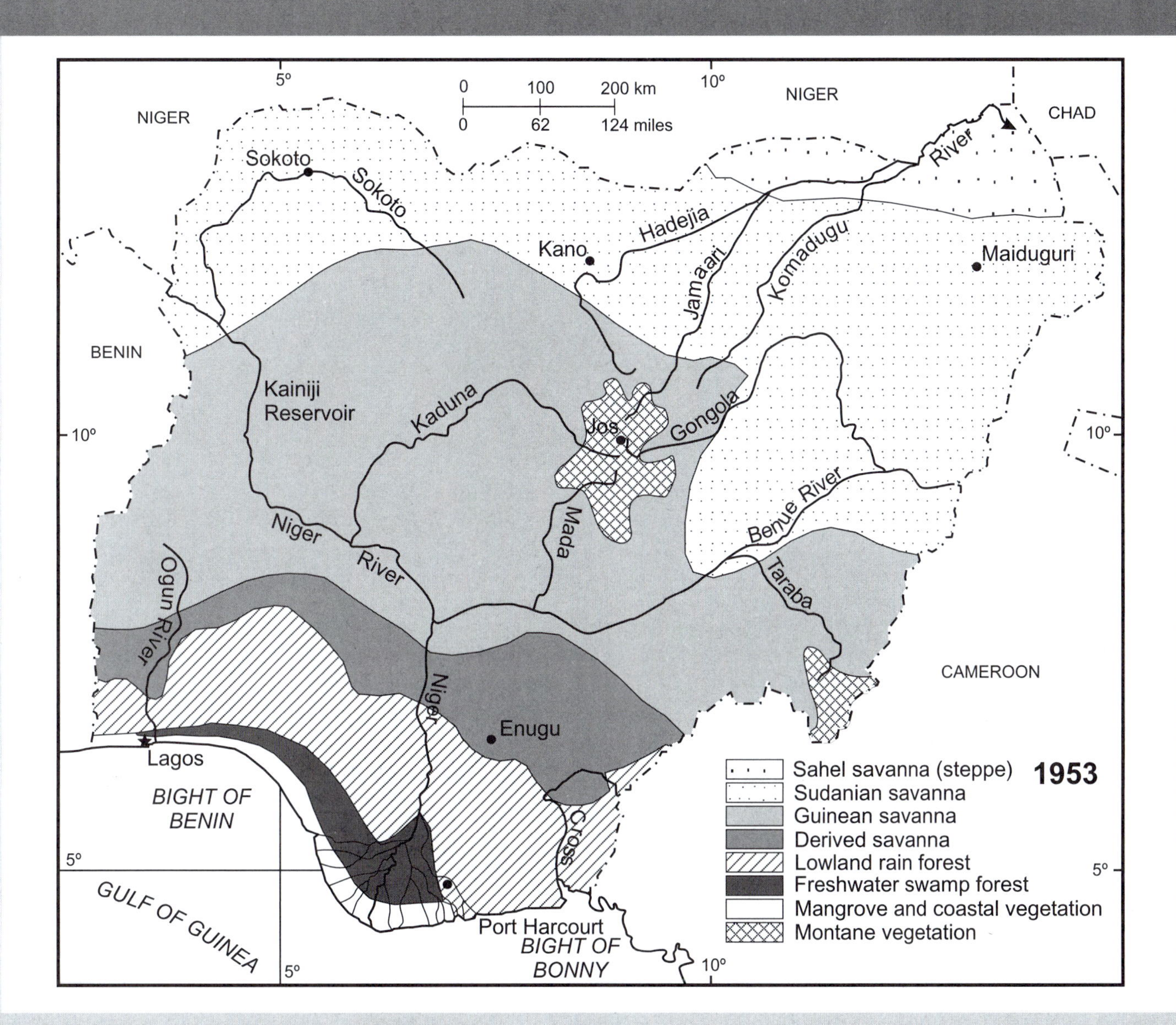

FIGURE 11-2 Vegetation regions of Nigeria in 1953.
Adapted from Geomatics (1998).

eastern lines, respectively, and the lines cross the less productive Middle Belt. In over 40-some years since independence, there is still no railway connecting the eastern and western parts of the south across the Niger River, although a line from Lagos to Calabar through Benin City has been discussed for years. Nigeria added 32 miles (51 km) to the rail network in the 1990s to link the iron ore at Itape with the Ajaokuta steel mill, and there have been discussions of linking Itape with the steel production center near Warri. But after years of

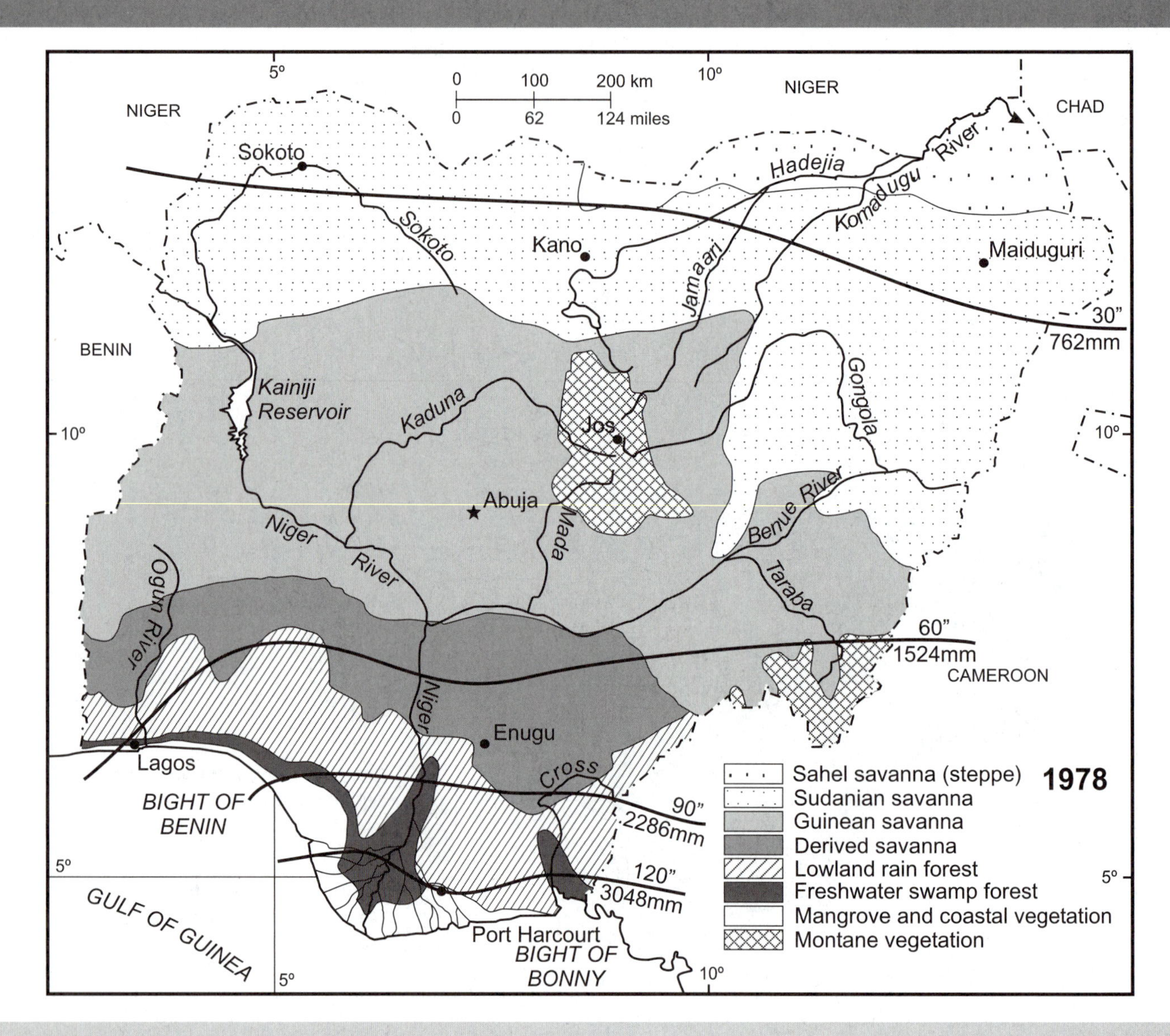

FIGURE 11-3 Vegetation regions of Nigeria in 1978.
Adapted from Geomatics (1998) and Best and de Blij (1977).

underinvestment, the existing rail network and rolling stock have deteriorated, and the Nigerian government is seeking outside investors to revitalize and expand the railroads (Odeleye 2000).

Although Nigeria has not expanded the rail network much since independence, since the oil boom of the 1970s it has developed the road network significantly. There is an emerging national network of surfaced roads (see later: Figure 11-16), superhighways connecting Lagos to Ibadan and Benin City in the southwest, Port Harcourt and Enugu in the southeast, and Kano and Kaduna in the north, and

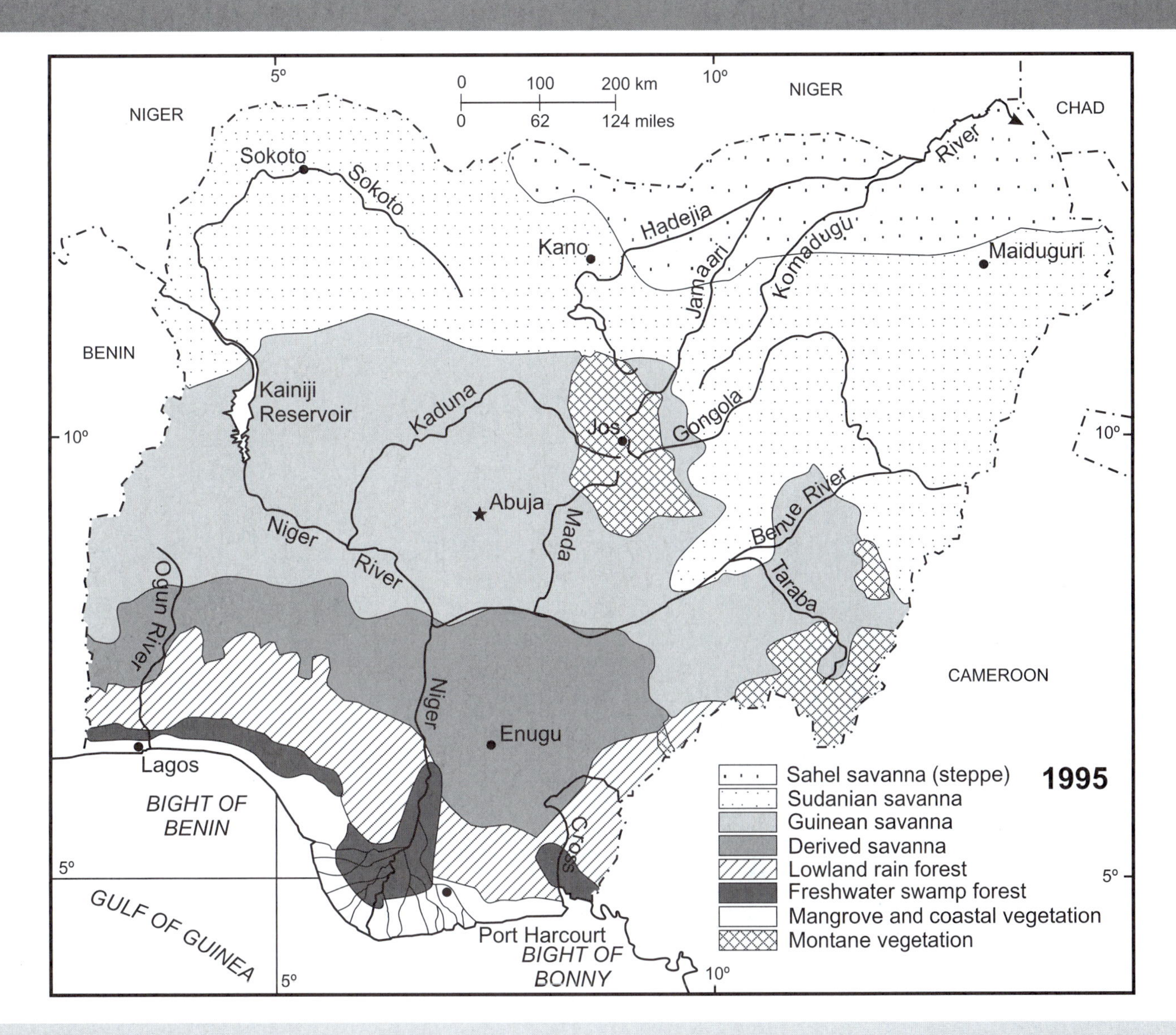

FIGURE 11-4 Vegetation regions of Nigeria in 1995.
Adapted from Geomatics (1998).

four road bridges across the Niger and two over the Benue. The government seems to be pursuing a balanced policy toward building its superhighways within the three cultural core regions of the country and, at the national level, east-to-west connectivity is developing. The government has not been maintaining the national road network, however, and much of what the government calls surfaced road has deteriorated into gravel. This is in part a result of the deterioration of the rail network as the heavy shipping load on the railway has shifted to motorized vehicles on roads.

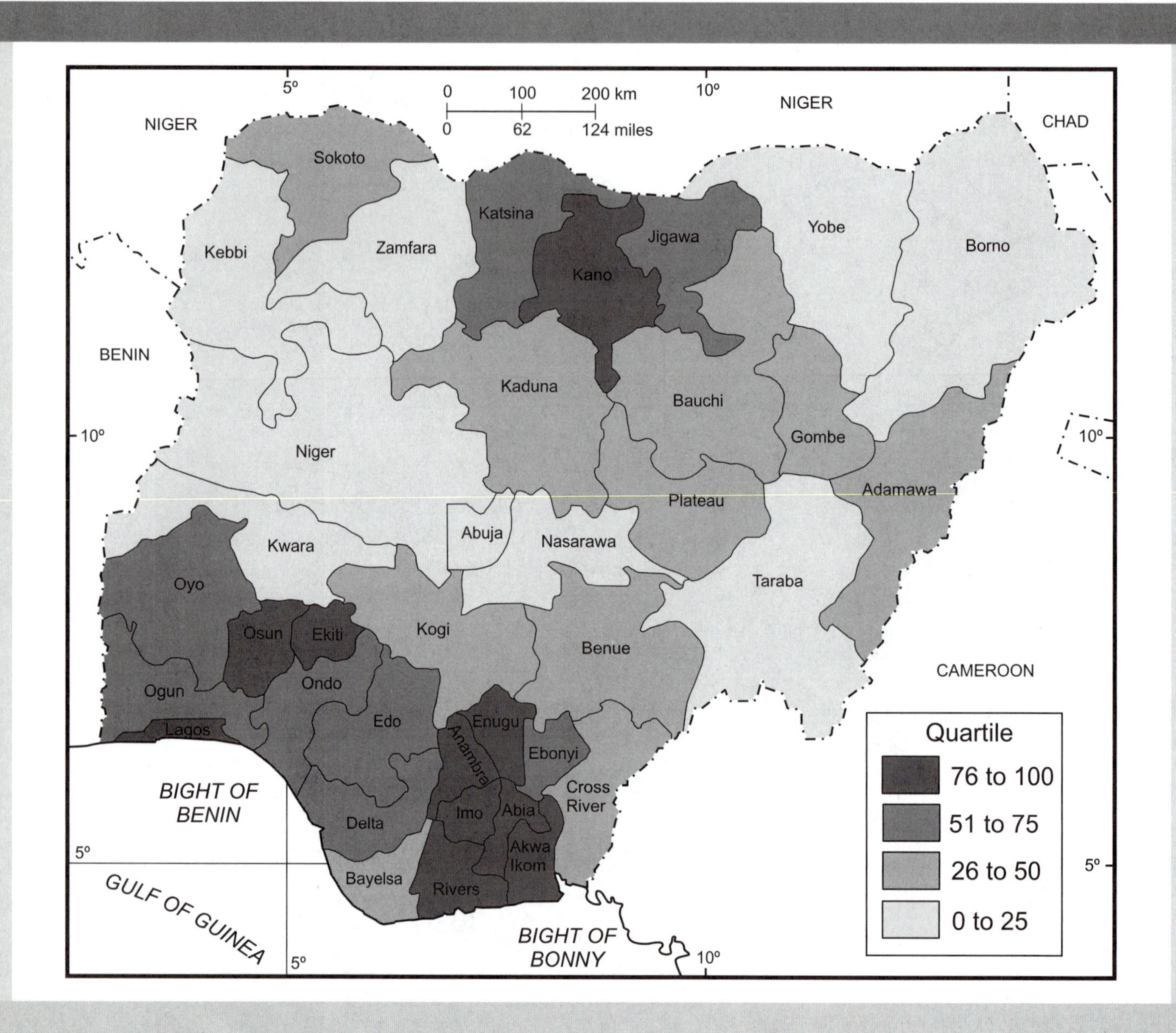

FIGURE 11-5 Population density of Nigeria by state.

URBANIZATION

Despite a long history of urban life in the area that we now call Nigeria, the population is overwhelmingly rural (64% according to the 1991 census)—but less so than in other parts of Subsaharan Africa and especially West Africa, as suggested by Figure 11-6. The process of urbanization increased after the oil boom of the 1970s, and Nigeria has experienced some of the fastest urbanization rates in the world. But within Nigeria itself the urbanization process has been uneven: some places

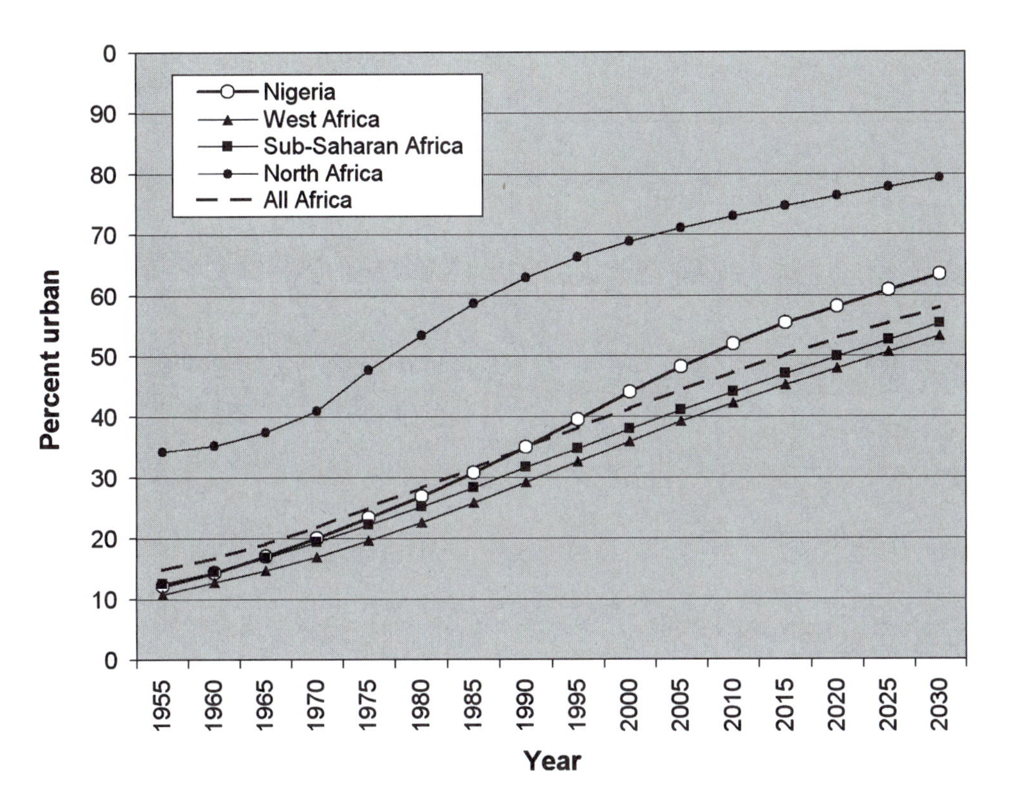

FIGURE 11-6 Urban dwellers as a percentage of total population for Nigeria, regions, and Africa, 1955 to 2030.

seem overurbanized, yet others are disconnected and profoundly rural. Nigeria is large, populous, and possesses a hierarchy of urban places (Figure 11-7), but it has only one distinctly urban region, the Lagos–Ibadan corridor in the southwest. The five southwest states of Lagos, Ogun, Ondo, Osun, and Oyo represent 7.8% of the national area, but the cities in these states accounted for 41% of the entire Nigerian urban population in 1991 (12 million people) and, according to estimates (World Gazetteer 2003), 43% of the 19 million urban Nigerians in 2003. Stretching northeast from Lagos, Nigeria's primate city, along the line of rail, is the largest concentration of towns and cities in tropical Africa: Ibadan, Abeokuta, Oyo, Ife, Ede, Ogbomosho, Iwo, Ilorin, and others—all Yoruba towns dating back in some instances many centuries. Historically, many of these Yoruba towns performed the special function of religious center, and all of them afforded protection in times of war. Most were walled and contained within their boundaries land devoted to food crops, although today the principal cultivated areas are some distance from the towns themselves.

Many Nigerian towns today have two distinct parts: an old and highly congested section made up of traditional compounds and wards, and a newer quarter with some vertical development, modern buildings, and wider streets. In the older sections, the population is engaged primarily in agriculture or farm-related activities such as marketing, food processing, and the crafts. With the exception of the north, marketing is dominated by women (Box 11-1). The newer quarters, begun during colonial times and generally situated on the outskirts of the traditional city, contain larger businesses and industries and house the more affluent and more educated Nigerians; today, with the possible exception of the communities of the elite, the suburbs are likely to be just as clogged with vehicles, pedestrians, and livestock as the old part of town.

The largest Yoruba town is Ibadan, founded as a military camp during the early nineteenth century by refugees from Old Oyo and other parts of Yorubaland laid waste by Fulani peoples from the north; it ultimately became the most powerful Yoruba city-state, halting the advance of the Fulani jihad into the

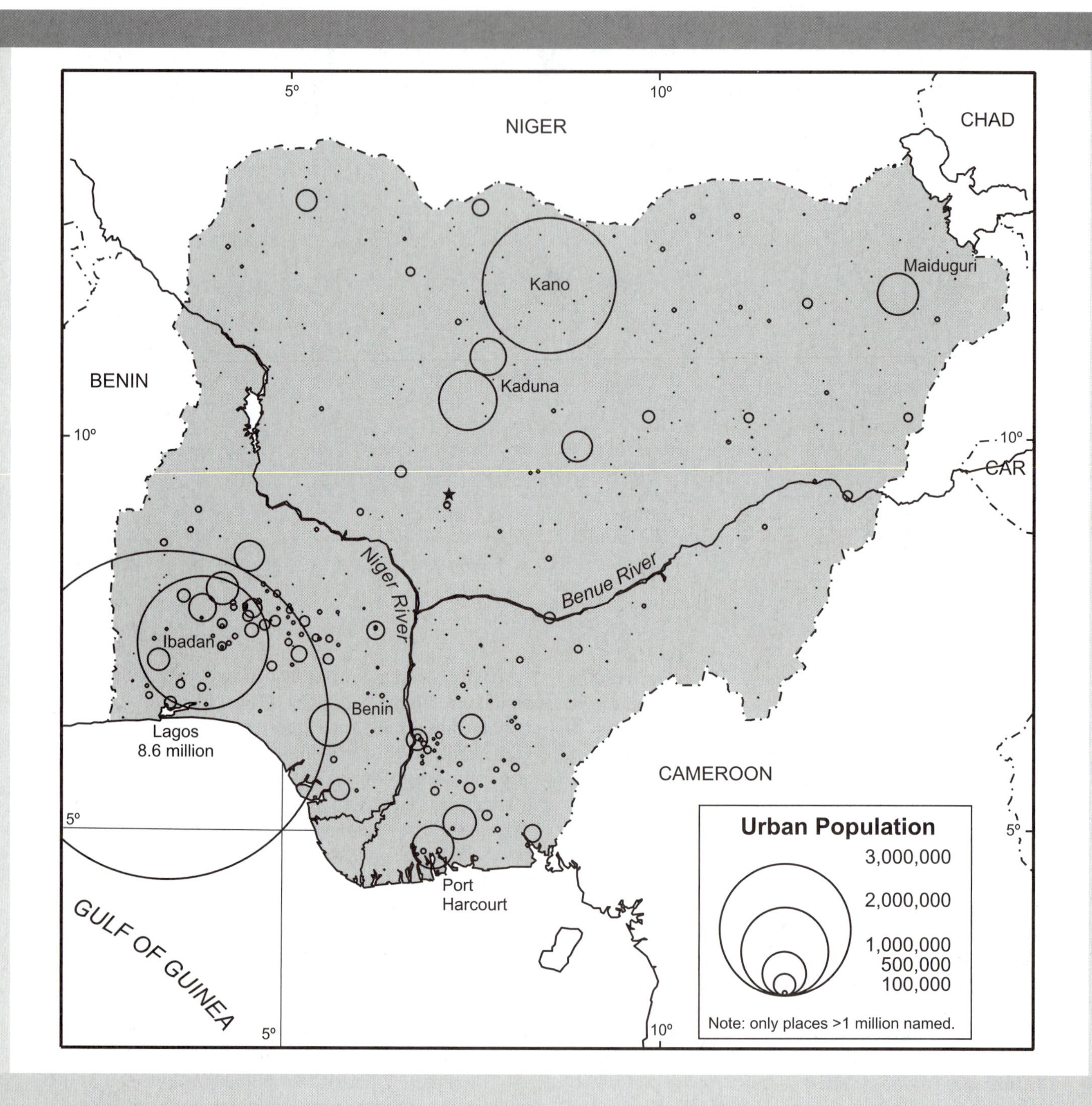

FIGURE 11-7 Distribution of urban populations in Nigeria.

BOX 11-1

MARKET WOMEN IN NIGERIA

There is a great deal of information on the economic activities of Yoruba women in the precolonial era. Apart from oral information, there are also various reports by the nineteenth-century European travelers across Yorubaland who described the economic activities of Yoruba women. Clarke and the Lander brothers, who followed some of the caravans during their journey in Yorubaland, reported the predominance of women entrepreneurs, whom they described as shrewd businesswomen.

Yoruba women were engaged in commercial activities in the precolonial era for many reasons, including providing supplementary income for their families. Their involvement brought enormous wealth to some of them. We must stress that most of the popular Yoruba women in this period became popular because of the wealth derived from their commercial activities. Their wealth, in turn, gave them political positions in their respective communities. Two prominent women were Efunsetan Aniwura of Ibadan and Madam Tinubu of Lagos and later Abeokuta.

In the colonial period, in spite of the negative impact of colonial rule on the status of Nigerian women, women continued to dominate small-scale businesses. One should also mention that the cash crops introduced by the colonial government offered Yoruba women entrepreneurs more opportunities to diversify their economic activities. However, in comparison to their male counterparts the period did not fetch them much wealth.

In the postcolonial period, there was a decrease in the percentage of women in the commercial sector, especially in the informal sector. This was as a result of government policy and the oil boom. However, from the 1980s, there was a reversal following the global oil glut and the downward trend in developing economies, including Nigeria. The various economic policies introduced by the military governments of Buhari/Idiagbon and Babangida led to the retrenchment of many women, including their husbands, from the civil service. The need for survival led many women to take to commercial activities to support themselves or their families.

Source: Dr. Olayemi Akinwumi, 2000. Women entrepreneurs in Nigeria. Africa update, 7(3): 2–3.

heart of Yorubaland. When Ibadan came under British influence in 1893, its population was already over 100,000; and in the census of 1991 it was 3.1 million. Today Ibadan is an administrative capital, major commercial center, and an important light manufacturing and educational center (University of Ibadan). Metal products, furniture, and soap are manufactured in Ibadan, and it is an important market for cocoa and cotton.

Benin City, the center of the rubber industry, also processes oil palm products and timber, and manufactures furniture and carpeting. The Kingdom of Benin, of which Benin City was the capital, was at its height from the fourteenth to seventeenth centuries. The kingdom is best known for figures made in iron, ivory, and brass. Ife, relatively small compared with the other Yoruba cities, is considered the oldest Yoruba city. Yoruba chiefs trace their lineage from Oni (King) Oduduwa, the mythical king of Ife. Although it is difficult to perceive today, Ife was the most powerful Yoruba kingdom until it was eclipsed by Oyo (Old Oyo) in the seventeenth century. Ife is famous for its twelfth-century and later terra cotta and bronze sculptures. Ife today is a marketing center for cocoa products and is the site the University of Ife.

Lagos occupies a strategic position and plays an important role in the region and federation. Socially, it represents an exciting (and deracinating) force for Nigerians unaccustomed to large urban areas, and in Lagos and in some of the larger cities, a distinctly urban culture has developed that transcends regional cultural identities. Founded more than 300 years ago by a Yoruba subgroup, Lagos became one of the leading slave ports of West Africa. It depended on the adjoining Yoruba towns for its captives, and following the termination of the slave trade, and with the British occupation in 1861, it became a refuge for freed slaves from Brazil and the immediate interior. Under the British, Lagos became an administrative center and port, although its port functions were initially handicapped by offshore bars and a shallow harbor approach. The turning point in its development came when the British selected it in preference to Warri or Sapele as the coastal terminal of the railway to the northern regions (Udo 1970a). Modern port facilities were completed in the 1920s and have since been improved to handle the ever-increasing exports, including cocoa, cotton, groundnuts, palm kernels, animal products, tin, and columbite, many of which originate in distant areas.

Lagos is Nigeria's major port and industrial city. Its industries include automobile assembly, flour milling, metal fabricating, cement, textiles, and chemicals. The selection of Lagos as the federal capital in 1954 did not meet with universal approval, for the relative strengths of the regional parties represented in the federal parliament were related to the numerical strength of the population in each state. By separating densely populated territory from the former Western Region, the region's leading party lost a sizable number of voters. Several grounds were put forward to justify the action: first, Nigeria's administration had long been headquartered in Lagos, so that it was by far the best equipped city for the continuation of government. Second, the inclusion of the port facilities in the federal territory constituted an assurance for the former Northern Region, which thus came to depend for its natural outlet on a federal rather than a Western Region harbor. Third, the selection of another western city would have had political consequences similar to those involved in the choice of Lagos, and the selection of any other city would have required the construction of new facilities there. Finally, Nigeria could ill afford the building of a new capital. In 1976, however, the decision was made to build a new federal capital at Abuja, the geographic center of Nigeria.

The impact of oil on the development of Lagos was profound. Skyrocketing oil prices of the early 1970s caused the Lagos metropolitan area to boom to such an extent that it accounted for 40% of Nigeria's external trade by 1978 and possessed 40% of the skilled population of the nation (United Nations no date). The oil glut of the early 1980s pushed Lagos into debt and the economy into inflation; and investment in infrastructure and social services in the city stopped.

Today Lagos is still a chaotic, polluted and congested place to live—a challenge to Nigerians. The center of Lagos is still Lagos Island (Figure 11-8), where the central business district is located, but the city has sprawled north and west. The development of Lagos has been haphazard. Although 90% of residents have access to electricity, there are water shortages, most of the sewerage is open and discharges directly into the lagoon and harbors, and traffic congestion is a daily headache; a 5- to 10-mile (8–16 km) trip across the city can take 2 to 3 hours. Although Lagos is chaotic today, the important question is what will it be like in the future. Is Lagos a nascent London—a London of the seventeenth or eighteenth century—that will become a world-class city with time, or will it remain chaotic, corrupt, dysfunctional, unlivable, changing ultimately into something completely different? Such comparisons are fraught with peril, but let's look at some similarities and differences.

Lagos and London are roughly the same in two important respects: area and population. Lagos is about 116 square miles (300 km^2), and London is 93 square miles (242 km^2) in area; the populations are similar as well (8.3 million and 7.1 million in 2003, respectively). The metropolitan areas of Lagos and London are also similar in population. Moreover, the cities were similar in the geography and history of the early settlements. Both had native settlements; both were slave ports and imperial outposts. Both were chosen by outsiders to be ports because of their strategic locations. Only London was the seat of empire.

London was founded by Romans almost 2,000 years ago as a ford across the Thames River. The location commands the land, river, and sea routes of southeastern Britain and is the lowest bridgeable point on the Thames before it widens into the Thames Estuary. Lagos possesses a very good geographical location for a port, although it is swampy. The lagoon is wide and calm; there are several islands that are easily defended; and the permanent outlet to the sea fed by the Ogun River provides an avenue for seaborne trade. The island of Lagos was populated by farming and fishing villages

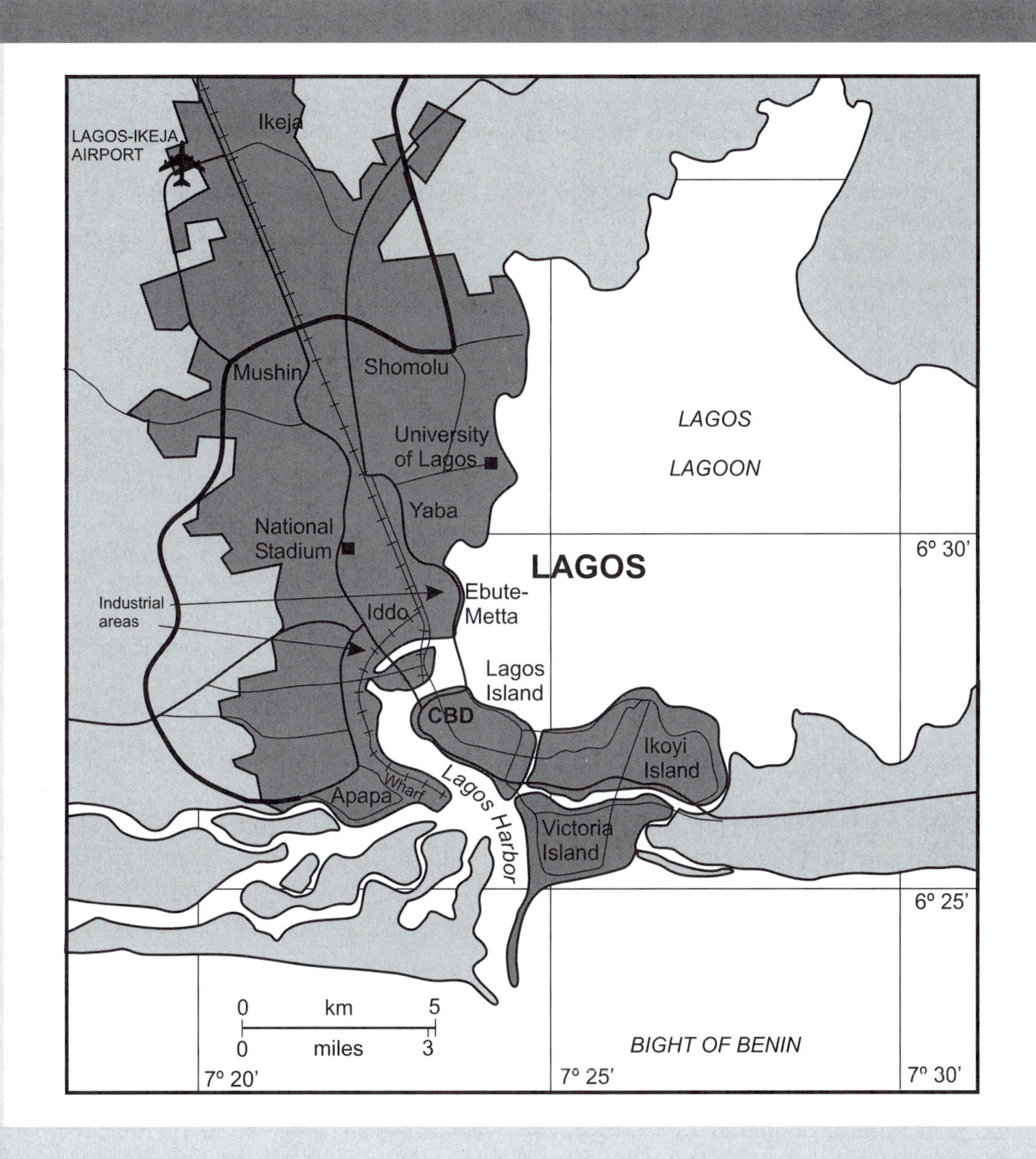

FIGURE 11-8 City of Lagos, Nigeria, showing the central business district (CBD) on Lagos Island.

by groups of Yoruba who first established themselves at Isheri and Ebute Metta on the mainland and later moved to the islands. There were probably seasonal settlements of fishers on the islands and along the coast prior to the establishment of the settlement that was to grow into Lagos. The Portuguese entered the picture in the fifteenth century and called the lagoon Lake Curamo (Lago de Curamo), later shortened to Lagos. "It was an excellent lair for slave traders, near the sea, yet easily protected from sea attack. To it slaves could be sent by inland waterways from the Dahomey, Yoruba and Benin kingdoms which were willing to sell their military and political captives" (Harrison-Church 1980: 426).

Historically London was compact, walled (after about 200 CE), and socially exclusive. Like London, Lagos was compact and stratified residentially; because the island location was easily defendable, however, the African city was never walled. In the open savanna of northern Nigeria, in contrast, cities were traditionally walled for defense. Although many residential areas in Lagos were and are segregated by ethnic and even village origin, wealth seems to be a major factor in determining residency. Rich people live in large houses in more exclusive neighborhoods, and within poor areas people of similar ethnic and village backgrounds appear to cluster together. Ethnic neighborhoods did not develop to exclude others; rather, they arose naturally as part of the process of integrating outsiders into city life. The same may be said of London in the past and London today, although London is much more cosmopolitan than Lagos.

Although Lagos and London have roughly equal populations today, London grew more slowly than Lagos, and much of the chaos of Lagos is due to sharp social cleavages between rich and poor and the inability of city planners to catch up to rapid growth. For much of its history, London was chaotic, polluted, and unhealthy, with the masses of Londoners living in poverty. Only in recent times has London become a healthy and clean place to live for ordinary people.

A location in the densely populated part of West Africa along the Gulf of Guinea is important for Lagos in terms of potential for transoceanic trade and, like London, Lagos is well positioned geographically to trade with the Americas. Indeed, it has been trading with the Americas for centuries. As development increases in Africa and the less developed countries of the Americas, can we expect that Lagos will emerge as a South Atlantic manufacturing hub? It is something that Nigerian planners may have been thinking about with the establishment of the Free Trade Zone (FTZ) near Calabar. A Free Trade Port for oil and gas was recently created on the Bonny River, downriver from Port Harcourt and 20 miles (31 km) from the open sea. The Federal Ocean Terminal (FOT) was completed in 2004, both created to reduce impediments to production and trade.

The United Nations projects that the population of Lagos will be close to 25 million by 2050, almost three times its size at the turn of the twenty-first century. Thus the Nigerian government can little afford to ignore the negative impact of haphazard urbanization, and in this urban planning effort Nigeria's large petroleum resources will give it an asset that its neighbors do not possess. If the Nigerian government makes wise decisions and investments, Lagos will assume a much greater importance in international trade, generate millions in additional revenues, and diversify the Nigerian economy, giving a much needed boost to industrialization. If the Nigerian government makes poor decisions, then the sobriquet "Calcutta of Africa" will be aptly descriptive.

Southeast Nigeria is the only major region in the country not to have experienced a lengthy period of indigenous urbanization; there is no Ibadan, Abeokuta, Kano, or Zaria (Zazzau). Although urbanization is taking place and urban structure is developing, even today the major towns are fewer in number and smaller in size than those in southwest and north central Nigeria; there are a few larger cities (Aba and Enugu, for example, and the port cities of the Niger Delta). Enugu (408,000) was founded as recently as 1915 following the discovery of coal in the Udi Plateau, and it owes much of its growth to the coal industry and to its administrative functions. Port Harcourt and Calabar had sizable populations in 1991, at 703,400 and 310,800 persons, respectively. Port Harcourt was established by the British in 1915 and Calabar, an old slave port, became important in the nineteenth century as a center for the palm oil trade. Petroleum, palm products, rubber, cacao, and timber are exported from Calabar.

A Nigerian Export Processing Zone (NEPZ) was set up in Calabar in 1992. Port Harcourt, a deepwater port, and third largest in the country, located on the Bonny River in the Niger Delta, is a commercial and industrial center and headquarters of the Nigerian petroleum industry. Port Harcourt rose in importance when it was selected by the British to be the rail terminus for a line of rail serving southeast and central Nigeria. Bicycles, cars, and trucks are assembled there, and manufactured goods made of steel, aluminum, concrete, and glass are produced. Automobile and truck tires are manufactured in Port Harcourt as well as paint, shoes, furniture, cigarettes, and beer. Oil is refined and sent to Bonny for export. Palm products, cacao, tin, coal, and peanuts are its principal exports.

Urban centers in northern Nigeria have a long history and some are over a thousand years old. The seven Hausa city-states of the north (Kano, Zaria, Gobir, Katsina, Rano, Daura, and Biram) emerged as power centers from the thirteenth to fifteenth centuries and were involved in the trans-Saharan trade. The cities were walled, and military force was used to maintain their power, but eventually the city-states began to fight among themselves. Gobir, Zaria, Katsina, and Kano were the greatest of the Hausa city-states. According to Davidson (1995), the Hausa city-states never came together to form a larger polity but instead formed a loose association of partners and sometimes rivals until the arrival in the early nineteenth century of the Fulani (Fulbe) under the command of the emir, Uthman dan Fodio, and his son Muhammed Bello. In 1809 the Hausa were conquered by the Fulani, who established an Islamic state, the Sokoto Caliphate. The Fulani, originally from the Senegal River valley, went on to found many new settlements across Nigeria, Cameroon, and beyond. It has been estimated that over half the population of Hausaland was urbanized in the nineteenth century (Anderson and Rathbone 2000), although the distinction between town and country was blurred in these traditional cities, where the majority of city dwellers were in fact farmers. The population of the northern cities is significant and growing rapidly (Table 11-1).

Kano's commercial importance reached great heights during the fifteenth century, when it was one of the Subsaharan "ports" in the trans-Saharan trade. Today it is the principal commercial center of northern Nigeria and an industrial town producing textiles, leather goods, food products, and other items based on local resources. In its immediate surroundings, population densities exceed 1,000 per square mile, and up to 40 miles (64 km) away, densities frequently exceed 500 persons per square mile. Thus, well over a million people are concentrated in Kano's immediate environs, cultivating almost every available patch of land, which is constantly fertilized with night soil brought from the city. Still, Kano is a food-deficit area.

TABLE 11-1 POPULATION OF SELECTED NORTHERN CITIES IN 1991 AND 2003.

	Population (000)		
City	1991	2003	Change (%)
Kaduna	993.6	1,510.3	52.0
Kano	2,166.6	3,329.9	53.7
Katsina	259.3	396.8	53.0
Maiduguri	618.3	993.9	60.7
Sokoto	329.6	512.8	55.6
Zaria	612.3	930.6	52.0

Source: World Gazetteer (2003).

Although densities are less and the extent of settlement is smaller, the basic pattern is repeated at Katsina, Zaria, and Sokoto. These and other towns serve the surrounding areas, and in the Kaura–Namoda–Zaria–Kano–Nguru region, they are part of an expanding core area. But the north is vast and has several state capitals, each of which is a locally important growth center.

Zaria produces a wide variety of products and is a university town (Ahmadu Bello University). It is a center for ginning cotton, and processing shea nut oil (karité), and a transshipment point for northern agricultural products. Kaduna, located just south of Zaria, was founded by the British in 1913 and was the capital of the then Northern Region until 1967. It is an industrial and commercial center possessing petrochemical, textile, furniture, and beverage factories and an oil refinery. The oil refinery is linked by pipeline to the Niger Delta oil fields. Kaduna is an educational center; two universities, a technical institute, and three professional training schools are located there.

Sokoto was founded by the Fulani leader Uthman dan Fodio in 1809, became the capital of his empire, was developed by Muhammed Bello, Uthman's son, and was conquered by the British in 1903. Today it is a commercial center and bulking point for agricultural products such as animal skins and peanuts. Local industry revolves around the production of cement and tanning and dyeing. Maiduguri, founded as a British garrison town in 1907, is an industrial center processing food and producing aluminum, steel, and cement. There is a tanning and dyeing industry as well. As the rail terminus on the Port Harcourt line, Maiduguri collects and exports peanuts, cotton, and skins. It is also a transshipment center for the rest of northeast Nigeria and for landlocked Chad.

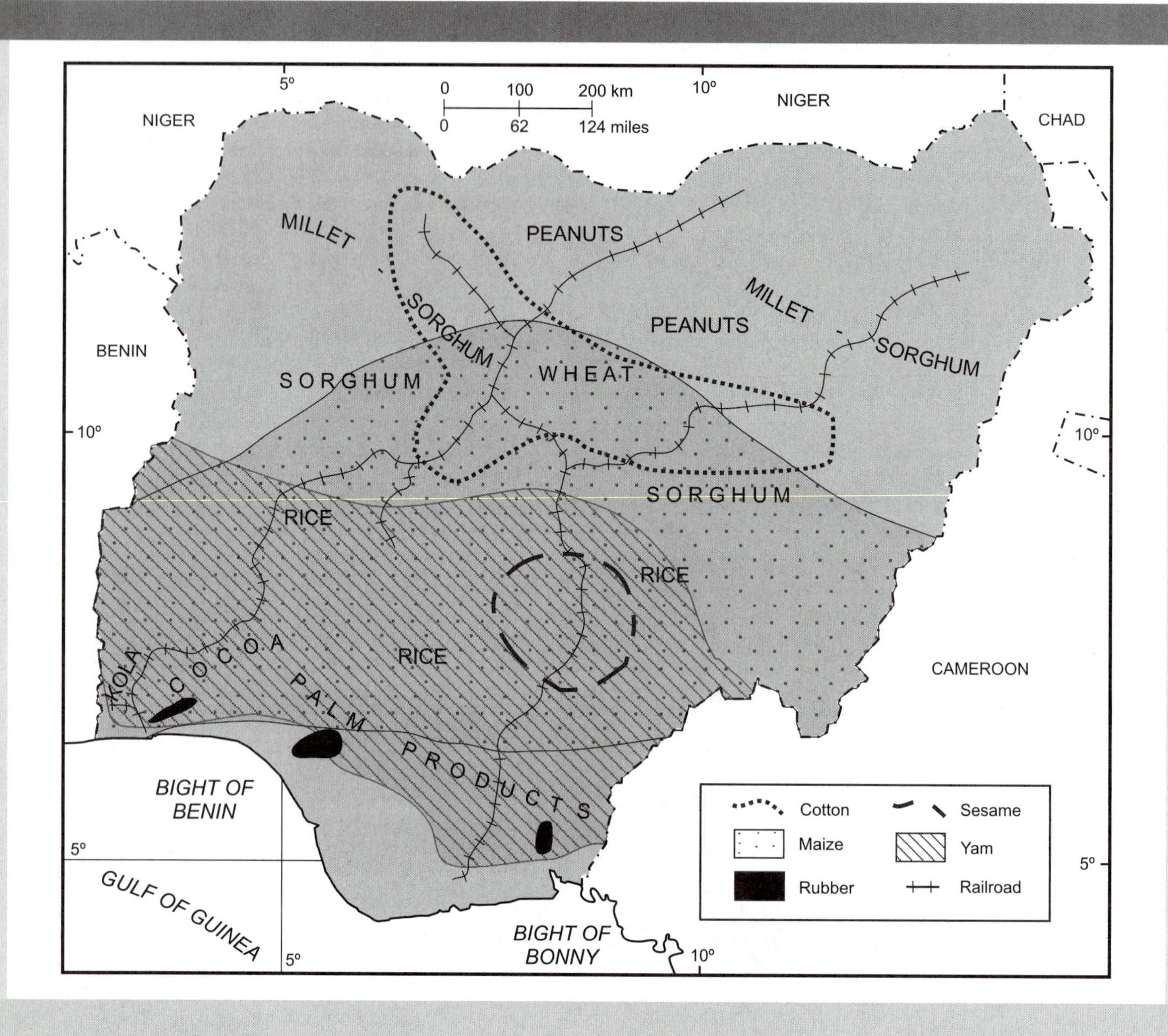

FIGURE 11-9 Agricultural regions of Nigeria.

AGRICULTURE

Nigeria, along with Egypt and South Africa, is one of the agricultural powerhouses of Africa. The main agriculturally productive areas lie in two rather distinct belts (Figure 11-9): one to the south of the Niger–Benue Lowland (Figure 11-1), the other to the north. The southern belt, corresponding approximately to the forest zone, extends from the cocoa-growing west to the palm-supported east. North of the lowland, the most productive zone lies east–west in the Kano–Zaria latitude, extending to the margins of the Sokoto and Chad basins. The two zones are divided

BOX 11-2

WOMEN AND AGRICULTURE IN NIGERIA

The roles of women farmers in Nigeria vary considerably according to the three major ethnic groups. The Hausa/Fulani women do little work in the fields because of the plough/grain culture and the restrictions on women of the Islamic religion. Nevertheless, poor Muslim women are heavily involved in farming, food processing, and marketing of farm produce. The Yoruba women are becoming more and more involved in agricultural work due to increase in value of cash crops coupled with the expansion of food production and raw materials for industries. Among the Ibo, there is traditional gender division of labour; men are responsible for land preparation while women engage in planting, weeding, harvesting, animal husbandry and food processing. With the involvement of men in cash crops, women are increasingly performing all the tasks involved in food production, processing and marketing. In Ibadan, 50% of the women sellers and hawkers interviewed during our market survey indicated that they own farms while 66% of these women say that their husbands also practice farming.

The empowerment of women is a vital key to agricultural development and food security in Nigeria. It aims to give women equal access to and control of land and other productive resources, increase their participation in decision and policy-making, reduce their workloads and improve their socio-economic status.

Source: Ajani, O.O. and O.O. Aina. 2002. Paper presented at the 8th International Congress on Women, Kampala, Uganda. Department of Women and Gender Studies, Makerere University (Kampala), July 21–26, 2002.

by the comparatively unproductive "Middle Belt," characterized previously as having been a buffer between the north and the south. The Middle Belt is an area infested by the tsetse fly, precluding the raising of livestock (except on the high Jos Plateau and in the Adamawa Highlands). Its transitional location permits the cultivation of both the grain crops of the north and the root crops of the south, but no important cash crops have been added to these staples. Labor in agriculture varies geographically by sex and social class, with fewer Muslim women in the north participating except among the poorer households (Box 11-2).

Cocoa, which was brought to Nigeria from Bioko Island (Fernando Póo) in 1874, came to occupy first place among the cash crops of the region and country. Before the development of the petroleum industry, which made great strides following the conclusion of the Biafran war, cocoa frequently produced a quarter of all Nigerian export revenues. Grown by small farmers whose plots average only 3 or 4 acres (1.2–1.6 ha), cocoa has long been the basis of the comparative wealth of the west, especially in the years after World War II.

Although Nigeria began exporting cocoa earlier (1895) than Ghana, its total production lagged behind Ghana until the 1980s, when Nigeria began to produce about the same as Ghana for the first time. Nevertheless, Nigerian cocoa fields have almost always been more productive than Ghana's, yielding 53 pounds per acre (60 kg/ha) more on average from 1961 to 1965 and 63 pounds per acre (70 kg/ha) from 1998 to 2002. Production by any measure, volume, area, or yields, of both Nigeria and Ghana, was eclipsed by Côte d'Ivoire in the late 1970s, with no turning back for the Ivorians. Ivorian cocoa yields were 52 pounds per acre (59 kg/ha) more than Nigeria's on average from 1961 to 1965 and 159 pounds per acre (178 kg/ha) more from 1998 to 2002. The political and economic stability in Côte d'Ivoire and an entrepreneurial outlook enabled the Ivorians to enter the world cocoa market in an aggressive way. Recent civil war is beginning to reverse the fortunes of Côte d'Ivoire.

Most of the Nigerian cocoa crop comes from a belt that centers on Ife (Figure 11-9). The introduction of cocoa into the region involved not just a few plantation owners but a large cross section of the farming

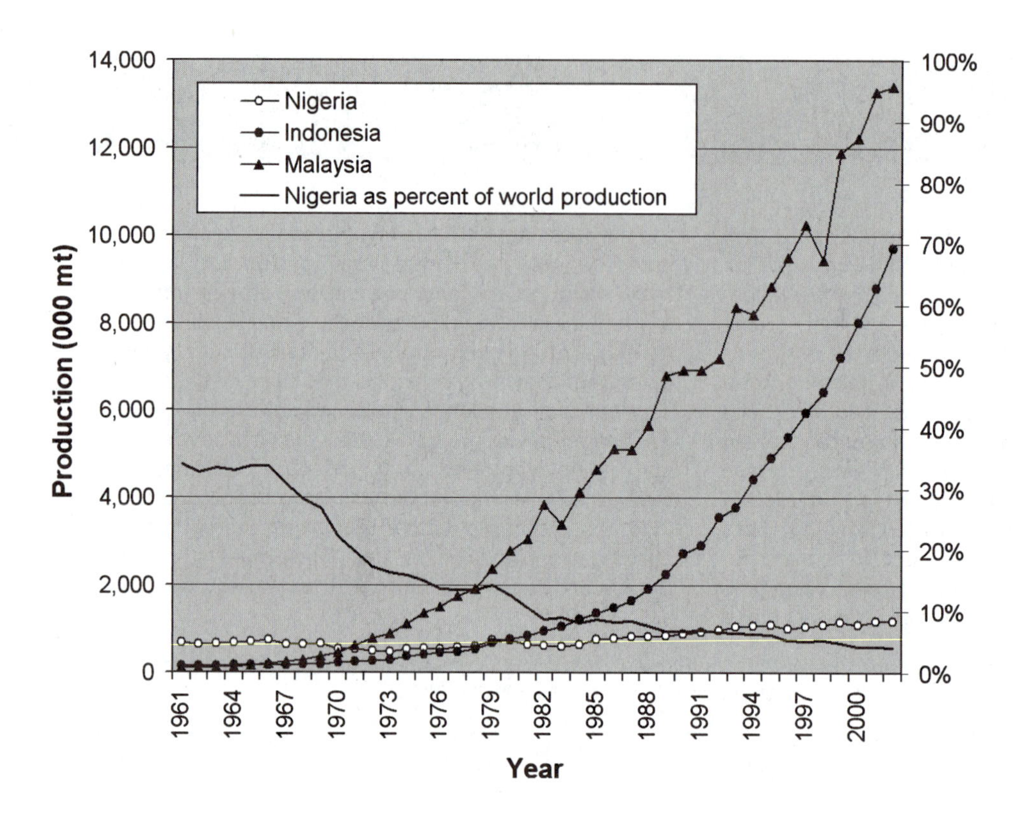

FIGURE 11-10 Palm oil production in Nigeria, Indonesia, and Malaysia, 1961 to 2002.

population. After its introduction, standards of living rose, communication lines were extended, long-isolated areas were drawn into the exchange economy, cooperative and other organizations were formed, and internal trade was stimulated. Furthermore, the cocoa industry provided both seasonal and permanent employment for thousands of workers from the northern states.

The oil palm is indigenous to West Africa; it occurs naturally in the forest, in a semidomesticated state, and in plantations. The vast majority of the harvest comes from smallholders who plant food crops such as cassava, yam, and maize under the trees, rather than from plantations. The oil palm is unique among oil-producing plants in that produces more vegetable oil per acre than any other plant. In addition to oil, the oil palm produces a variety of useful products. The unrefined oil is used in cooking and can be processed in factories to make margarine, soap, cosmetics, glycerin, and candles. Refined oil is marketed internationally. The glycerin derived from the oil is used as a solvent, skin lotion, and food preservative, and in the manufacture of explosives. The trees are used for building material; leaves are woven into mats and are also used in roofing; and an excellent wine is made from the sap.

Palm oil was once important in Nigeria as an earner of foreign exchange, but exports dropped dramatically in the late 1960s and continued to decline for three reasons: the war in Biafra, the rise in local consumption, and foreign competition. The Biafran war took place in and around the major palm oil–producing region of Nigeria and severely disrupted the oil palm industry. With population growth, the development of infrastructure and trade, and urbanization, the internal market for palm oil (and other palm products) has grown significantly, capturing most of the increasing domestic production. But the major factor has been the loss of the export market to Southeast Asia after Malaysia and Indonesia massively invested in the development of oil palm plantations (Figures 11-10 to 11-12). During the 1960s Nigeria produced about a third of the world's palm oil (and about 20% of world exports). In conjunction with government development programs, large multinational companies established efficient, modern plantations in Malaysia and Indonesia, production

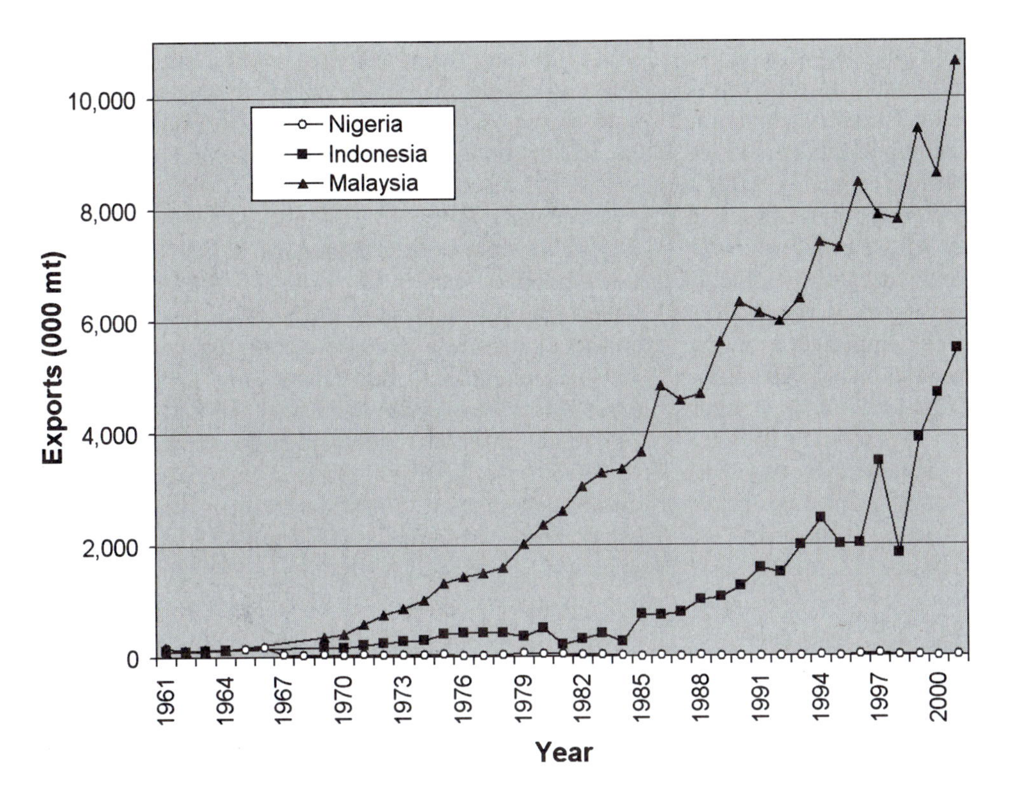

FIGURE 11-11 Palm oil exports from Nigeria, Indonesia, and Malaysia, 1961 to 2001.

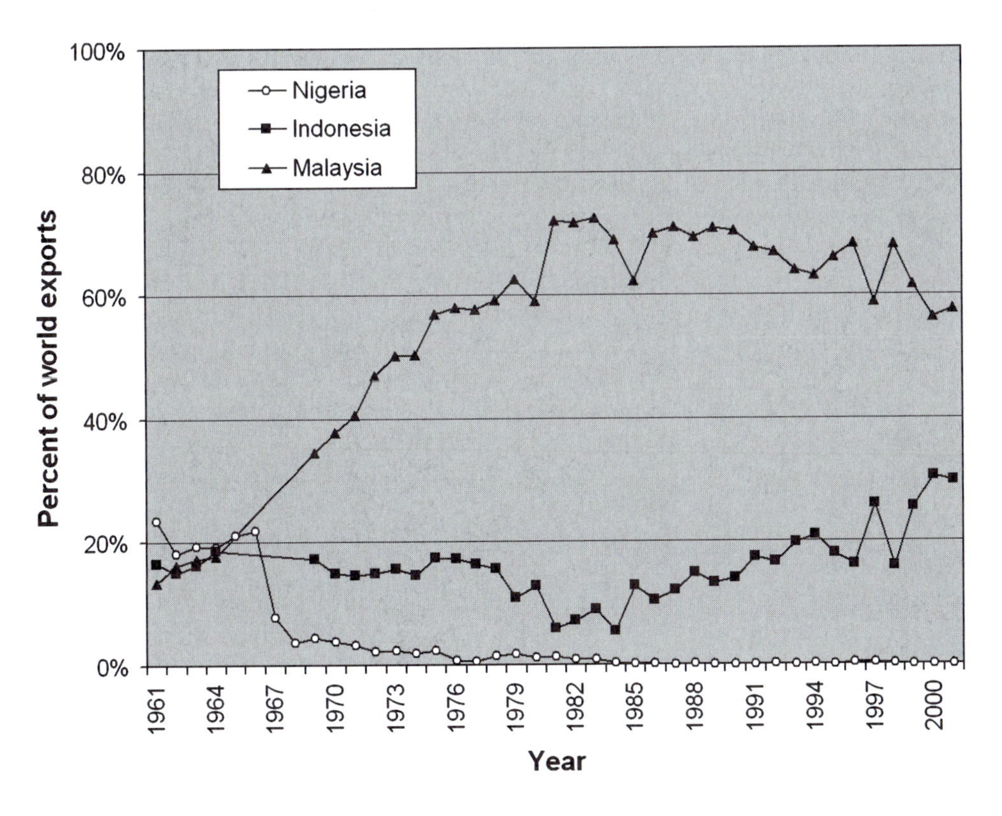

FIGURE 11-12 Palm oil exports as a percentage of world exports for Nigeria, Indonesia, and Malaysia, 1961 to 2001.

skyrocketed, prices declined, and Nigeria lost its comparative advantage. Other Southeast Asian and South Asian countries are currently establishing large plantations, and it is expected that competition will grow. Prices may not fall, however, because the uses to which palm oil are being put are growing; for example, in Malaysia it is being combined with low-grade diesel fuel and burned to generate electricity. Rising demand for palm oil in Nigeria has led to palm oil imports over the last few years, but the amount imported is only a tiny proportion of Nigerian production. Nevertheless, there may be some danger of loss of domestic market share to cheap imports.

Globally there has been a movement by environmentalists to fight the spread of oil palm plantations because tropical rain forest is destroyed to make vast oil palm monocultures. This is not a large issue in Nigeria today because most production has been outside plantations, although it may be an issue in the future as Nigeria rationalizes its oil palm production in an attempt to remain competitive. The stimulus for change may come from outside: Malaysia is developing plans to invest in a large plantation in southeast Nigeria and set up oil processing facilities in the Free Trade Zone located near Calabar.

Rubber and kola are grown in southern Nigeria as cash crops, the rubber being exported either through the port of Lagos or Port Harcourt, while the kola is shipped north and northwest. Rubber has long been an important Nigerian export and, with the rapid pace of local industrialization and rising per-capita income, domestic demands are increasing. Thus the industry, although subject to price fluctuations, appears to have a secure future. The Nigerian government has taken an active role in encouraging the growth of Nigerian rubber processing by prohibiting the export of unprocessed rubber. Unlike production in Liberia, where large-scale plantations predominate, approximately 90% of the rubber comes from small farms of between 1 and 25 acres (0.4–10 ha) run by family farmers. The large plantations are owned by foreign companies, Nigerian companies, and Nigerian individuals. Nigeria produces some very high-grade rubber but in general Nigerian rubber is of lower quality and is used in the local production of tires. Over the last 40 years the area harvested has increased more than 2.5 times, and Nigeria's output of rubber has tripled reaching a peak in the 1990s. Nigeria produces about 25% of all African rubber, but year-to-year production varies on prices. As with other cash crops that were once exported to more developed countries for processing, exports of rubber have been declining as local industry has increased.

Sugarcane production is not very significant in Nigeria in absolute terms, but it has increased 330% since independence and in 2002 almost 100,000 acres (40,000 ha) was harvested. In the 1960s less than one-quarter of that was harvested, but this increased to 50,000 acres (20,234 ha) in the early 1970s and remained between 50,000 and 60,000 acres (20,000 and 24,000 ha) until 2001 when a new sugar project added 42,000 acres (17,000 ha). Nigeria processes and consumes all of its domestic production and imports to satisfy domestic demand: over 1.2 million metric tons of sugar was imported in 2001. On a per capita basis, sugar imports accounted for less than 3.5 pounds (1.6 kg) per person from independence until the oil boom of the 1970s and increased steadily thereafter to almost 24 pounds (10.8 kg) per person in 2001.

The possibilities for agricultural development in the north are considerable, but rainfall is less than in the south, and it is more seasonal and quite variable along the northern border. In the Sokoto and Chad basins the lack of water has restricted settlement. North of the tsetse area that covers much of the Middle Belt, however, the country affords opportunities for cattle raising. The absence of fodder in southern Nigeria and the presence of diseases in the Middle Belt make the northern savanna zone even more important from a cattle-raising perspective. The agricultural opportunities of the north are not restricted to cattle raising, though. Peanuts, cotton, tobacco, and sesame are important cash crops, while sorghum, millet, fonio, and cassava are common staples. Sugarcane and rice are grown along the rivers and on fadamas (wide valleys that are seasonally flooded).

Cotton has been grown for several centuries in northern Nigeria and today about 90% of Nigeria's cotton is produced there. Mostly imported cultivars are grown, especially the North American Allen variety. The principal cotton-growing region is in the drier north central area, although varieties introduced by the Portuguese are grown in some southern localities. Cotton, the quintessential colonial cash crop, grown to feed the cotton mills in Britain and excoriated for taking the place of food crops in the field, now supplies

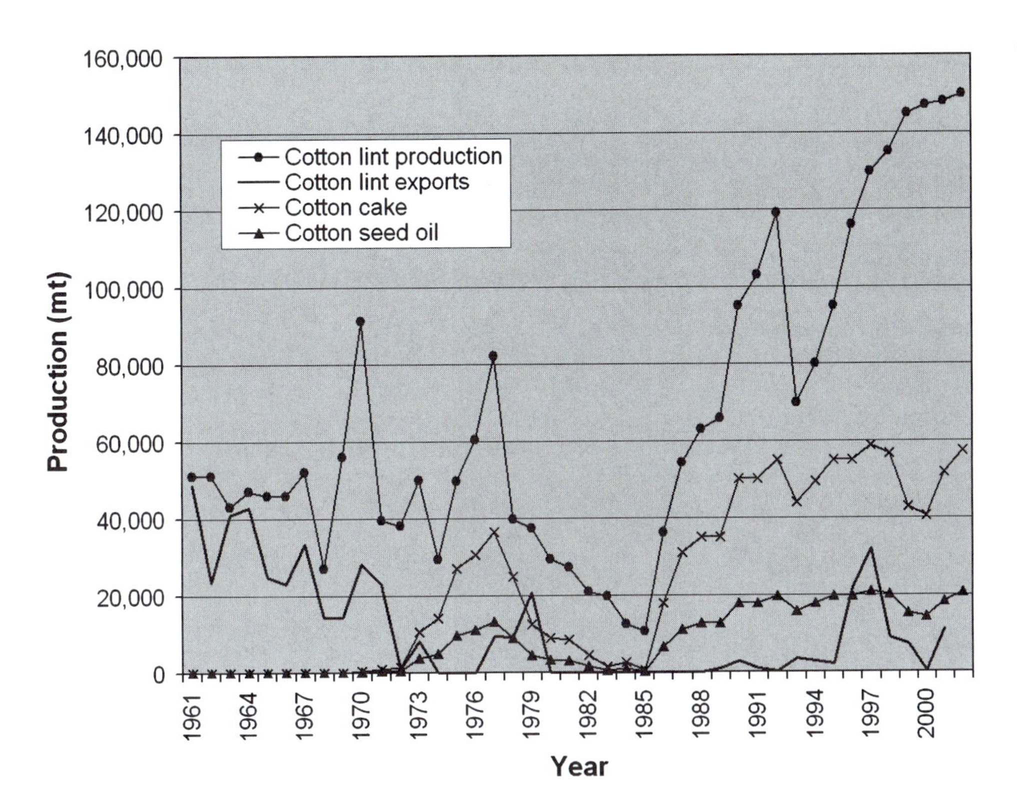

FIGURE 11-13 Nigerian cotton and cotton product production, 1961 to 2002.

Nigerian textile mills producing Nigerian cloth for local sale and export (Figure 11-13).

Textile mills located in most northern and several southern cities produce a wide variety of cloth. Although local textile manufacturing is an important step in the manufacturing development in a country, Nigerian cotton production and exports might even be greater if the United States did not heavily subsidize its own cotton crop. Some areas of northern Nigeria are economically dependent on cotton sales for their well-being and would benefit from the loosening of North American trade barriers. That said, the U.S. subsidies may have been an incentive for local manufacturing in Nigeria rather than export of the primary product. Nevertheless the United States and the European Union agreed in 2003 that export subsidies would be phased out (*Economist* 2003b), although no timetables were set.

Nigerian peanut, or groundnut, exports accounted for half of all African peanut exports and 41% of world exports in 1969. When exports peaked in 1969, peanuts were Nigeria's leading export crop and Nigeria was the world's largest exporter, generally accounting for about a third of the world's commercial production. This all changed in the early 1970s when exports fell to record low levels partly because of drought conditions and a peanut virus; in addition, low prices paid for the crop induced farmers to smuggle their harvest into neighboring states such as Niger, where higher prices were obtainable. In 1976 peanut exports were prohibited by the Nigerian government because of drought-induced food insecurity. Since that time exports of peanuts have not resumed, but the data indicate that peanut production has almost doubled from the average production in the 1960s and that there has been a significant shift from exporting to local processing of peanuts into oil and cake (Figure 11-14).

Peanut oil is used in cooking and in the manufacture of margarine, while peanut cake, produced in the process of pressing peanuts for oil, is a high-protein animal feed. Locally manufactured cooking oil and margarine are cheaper than expensive imports but the imported oil may have a prestige value, despite the irony that it may have been produced from exported African peanuts in the first place. Locally produced peanut cake is an otherwise unavailable and greatly

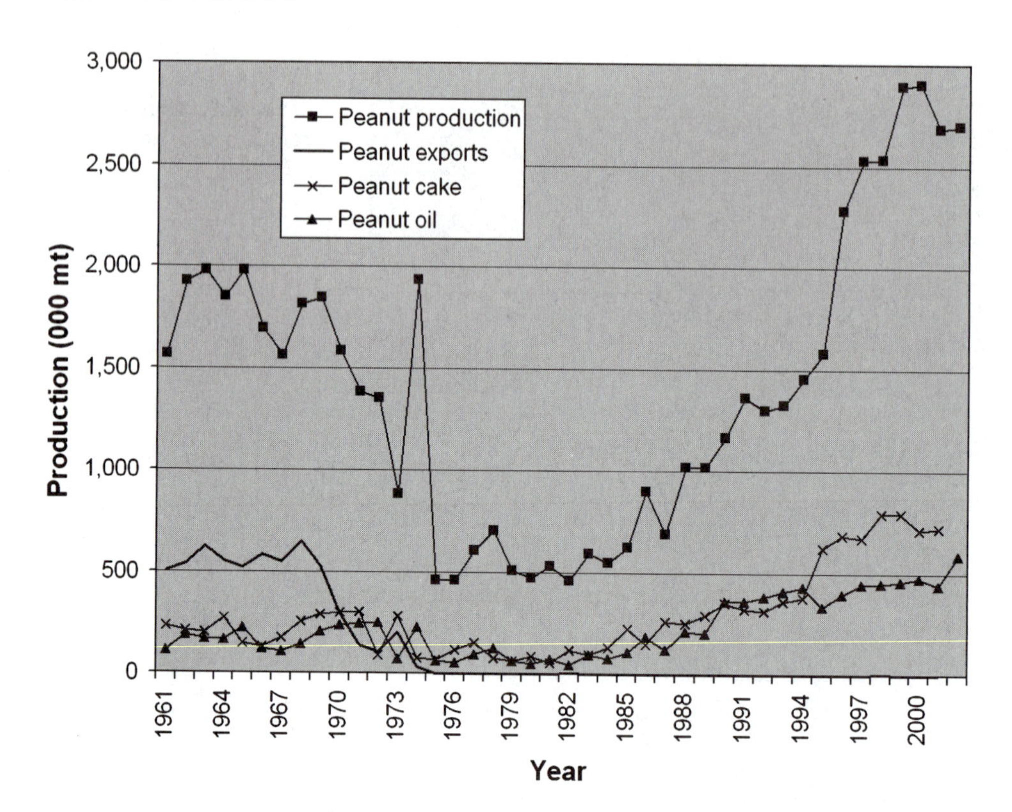

FIGURE 11-14 Nigerian peanut and peanut product production, 1961 to 2002.

needed source of nutrition for livestock and augments the diet of the mainly range-fed animals.

Nigeria is one of the most important producers of livestock in Africa and, although the export of raw hides has been prohibited by the Nigerian government to spur local manufacturing (U.S. Department of Commerce 2002), Nigeria has exported hides for many centuries across the Sahara to be processed in North Africa and Europe. Goat skins from northern Nigeria were and continue to be the raw material of the famous "morocco," a soft-dyed leather made in the country of the same name. In Nigeria most livestock are raised in the drier north, where the tsetse fly is not a problem. The animals are most important for the internal market in milk and meat, and in the north they provide manure essential for the intensive farming that exists around Kano. Meat is consumed principally by urbanites, and livestock make an annual trek from the producing north to the consuming south every year. Although road and rail links exist, most livestock walk to market, incurring substantial weight losses and attritition due to predation and disease. Cattle, sheep, goats, and pigs have always been more populous than other large domestic livestock in Nigeria according to the available data. Populations of horses and asses are about 72% of what they were in the 1960s; the numbers of camels, never more than 20,000 head, have increased by about 30%; but numbers of all other stock, particularly goats, have increased substantially (Figure 11-15).

MINING

Tin and columbite from the Jos Plateau have long been the region's most important minerals, and prior to the extraction of petroleum, tin was Nigeria's leading mineral export. Almost all of it is smelted locally at Jos and then railed to Port Harcourt or Lagos. The ore smelted in Nigeria has declined in recent years because tin miners have preferred to sell to smugglers for hard currency (Izon 1994). Iron and phosphates are mined in several locations around the country (Figure 11-16). In the mid-1970s the Nigerian government was assisted by the Soviet Union in the development of an iron and steel industry. The Soviets helped in the

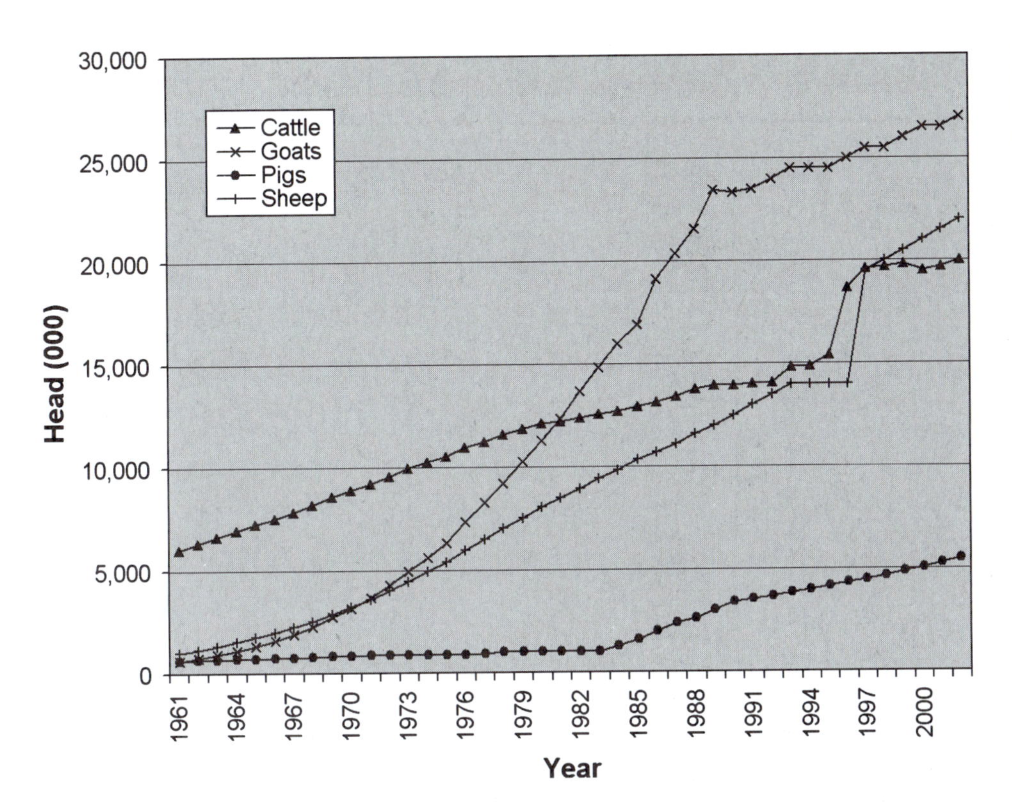

FIGURE 11-15 Selected large domestic livestock populations in Nigeria, 1961 to 2002.

construction of Nigeria's first iron and steel mill at Ajaokuta along the Niger River, north of Onitsha. By 1998 the mill was 98% complete, with plans for expansion. It uses locally mined, low-grade iron ore, coal from Enugu, and local limestone in the production of steel. Ajaokuta is centrally situated with respect to the nation's major population centers, but at present has inadequate transport facilities other than the Niger itself. In 1990 a 32-mile (51 km) rail line was completed between Ajaokuta and the Itakpe iron mines to the northwest. Three government-controlled rolling mills were built, at Jos, at Katsina, and at Oshogbo, but have been in and out of operation. A second steel mill was later built near Warri in the Niger Delta. Chinese companies helped build a cement reinforcing rod plant near Abuja and were negotiating with the Nigerian government in 2003 about building a large steel mill in the Calabar Free Trade Zone. As of 2001 most of the government-owned steel facilities were not producing (Table 11-2). Nigerian coal and iron ore reserves were estimated at 1.5 billion and 2.5 billion metric tons, respectively (Mobbs 2001).

Nigeria's industrialization program included the development of a large aluminum smelter at Ikot Abasi in Cross River State. The plant was completed in 1996 but suspended operation in 1999. The smelter was costly to build because basic infrastructure was lacking, a common problem in the less developed world. Aluminum smelting is very energy intensive, and aluminum is expensive to produce. Most smelters around the world are associated with relatively cheap hydroelectric power, but the Ikot Abasi plant is powered by costly hydrocarbons rather than hydroelectricity. A natural gas–fired power plant, planned and built for the smelter when gas prices were low in the late 1980s and early 1990s, has since closed. Production is expected to rise in the first decade of the twenty-first century. Although official sources give no reasons for the closing of this plant in an era of rising gas prices, the high costs of production may well have been determining. Expenses are also high because all raw materials must be imported.

Petroleum dominates the Nigerian economy and is by far the greatest source of Nigeria's foreign revenues.

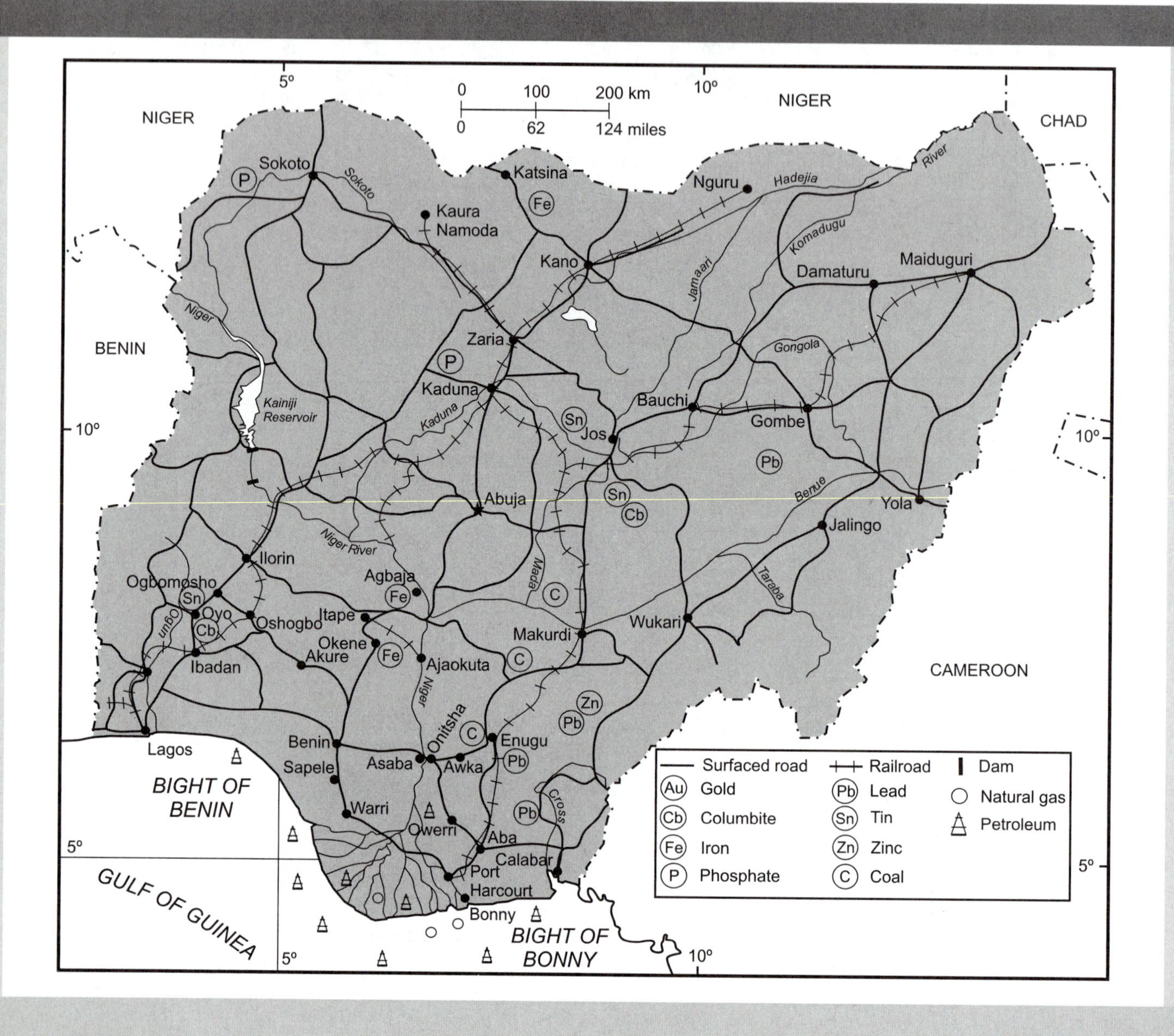

FIGURE 11-16 Infrastructure and minerals in Nigeria.

In 1958 petroleum formed less than 1% of total export earnings; by 1965 it accounted for almost 26%; and by 1974 it represented 92% of all export earnings (Nigeria 1975). Over the last decade, about 95% of Nigeria's foreign exchange earnings have been consistently generated by the export of oil. Today, petroleum accounts for 20 to 40% of GDP and at least 65% of national budget revenue (CIA 2003). The contribution to the national budget, it should be noted, has been reported to be as high as 80% (Mobbs 1997). Nigeria possesses an estimated 25 billion barrels of reserves, and it is expected that exploration, stimulated by high oil prices

TABLE 11-2 FOSSIL FUEL AND SELECTED METALS PRODUCTION IN NIGERIA, 1990 TO 2001.

	Fossil Fuels			Metals (metric tons)		
Year	Crude Oil (thousands of 42-gallon barrels)	Natural Gas (millions of cubic meters)	Coal (metric tons)	Aluminum	Crude Steel	Tin, Smelted
1990	660,000	27,600	78,000		220,000	130
1991	689,000	31,300	138,000		250,000	230
1992	715,000	32,000	140,000		200,000	370
1993	748,000	31,300	50,000		150,000	175
1994	743,500	31,300	130,000		105,000	185
1995	740,000	35,000	29,000		36,000	250
1996	798,620	37,000	7,116		0	100
1997	845,000	30,000	7,000	2,500	0	100
1998	788,000	30,000	30,000	20,000	2,000	150
1999	777,000	30,000	30,000	16,000	0	50
2000	783,000	30,000	35,000	0	0	25
2001	823,000	35,000	35,000	0	0	25

Sources: Izon (1994), Mobbs (1997, 1999, 2001).

since 2000, will reveal about 15 billion more. Natural gas reserves have been estimated at over 28 trillion cubic meters (U.S. Department of Commerce 2002).

The search for oil in Nigeria began as long ago as 1937 but was interrupted during World War II. In 1947 the exploratory efforts were resumed, initially through shallow drilling, until the geologic mapping of the Niger Delta area was completed in 1953. Then deep wells were sunk, and in 1956 wells near Oloibiri and Afam, within the delta proper, showed commercially viable oil accumulations. Several new deposits were subsequently located, and activity increased greatly in the whole region. Petroleum production in 1959 amounted to 4 million barrels; in 1962 it exceeded 25 million; and early in 1970 production reached one million barrels per day. During 1973 it exceeded the 2-million-barrels-per-day mark. In 1974 and 1975, following the world oil crisis, daily output was deliberately kept at 2-million barrels a day, but since 1993, crude oil production has consistently exceeded this level (Table 11-2). In northern Nigeria, the federal government has been looking for oil in the Lake Chad Basin without success since the mid-1970s. The recent discoveries in Chad have encouraged further exploration in Nigeria.

Until 1964, all of Nigeria's crude petroleum was exported, but with the completion in 1965 of an oil refinery at Port Harcourt, a small amount started to be refined for domestic use. Three additional refineries were later built: one at Warri, one at Kaduna, and a third at Port Harcourt. This could give Nigeria the overall capacity to be an exporter of refined petroleum but almost all of the refined products are consumed locally. As part of the current emphasis on market forces, the Nigerian government is looking for private investors to develop additional crude oil refineries (Mobbs 2001).

Production costs for Nigerian petroleum are from three to seven times higher than in Middle Eastern countries, but the oil has an exceptionally low sulfur content. From year to year, Nigeria has been Africa's largest producer of oil; its rank as a world producer has varied from eighth to eleventh over the last few year, representing about 3 to 3.5% of the world total. Major purchasers of Nigerian oil are the United States and Western Europe. In 1972 Nigeria became a member of the Organization of Petroleum Exporting Countries (OPEC), but it has been less militant than many OPEC states in its relations with consumer countries over the years. Nevertheless, it followed the general trend set by OPEC toward higher royalties.

In 1994 Nigeria was flaring 79% of its natural gas, a waste of a nonrenewable resource and a serious environmental problem in the Niger Delta. In the 1990s the Nigerian government resolved to phase out flaring by 2008 and to direct natural gas from the well head pipelines to a liquefied natural gas (LNG) facility located along the Niger Delta coast near Bonny and from there directly to Togo and Benin for use in these countries. Algeria has proposed a trans-Saharan pipeline to its northern coast in order to ship Nigerian natural gas to Europe. Two additional LNG plants are being discussed, one near Brass to be opened in 2007 and the other in the West Niger Delta to open in 2008.

Although petroleum has brought vast wealth to Nigeria, it has caused many problems as well. It was a factor in the Biafran secession and civil war. The federal government controls oil revenues, and years of neglect of the people in the oil-producing regions led to much political protest by the have-nots in the oil-producing states of Nigeria. In 1999 President Olusegun Obasanjo announced a 13% oil revenue allocation to the oil-producing regions from the federal budget and in 2000 the government passed the Niger Delta Development Bill, intended to increase investment and development in the Niger Delta. People in the oil-producing regions remain skeptical (U.S. Department of Commerce 2001). Local people continue their violent protests and oil pilfering. Retaliation by the state has been violent—even including executions. The dynamic tension between state and federal governments is rapidly evident when the discussion turns to the control of natural resources. The federal government has always maintained that it, rather than the states, controls all Nigerian mineral rights. But individual states and even local political units have been pressuring the federal government to give all control of minerals to them. In 1999 the federal government stated once again its intention to control mining and minerals through the enactment of Mining and Mineral Decree Number 34, which states that all mineral rights belong to the federal government.

FORESTRY

Although use of trees is ubiquitous, the commercial forests of Nigeria are located in the south of the country and the forest industry is centered on Sapele located in the western Niger Delta (Figure 11-16). Nigeria has long had a timber industry, which has grown steadily in importance over the years. Roundwood production increased from 39 to 70 million cubic meters from 1961 to 2002, representing roughly 10 to 15% of African output. In an effort to stimulate local industry, the federal government prohibits the export of timber and unfinished wood products with some exceptions: railway sleepers (ties), pallets, wood flooring and ceiling tiles, and doors and window frames. As a result of the oil boom, a large internal market for wood products has developed, and almost all wood harvested in Nigeria is processed locally for local consumption. For example, Nigeria exported about 25% of its sawn wood during the early 1960s before the oil boom; it exports less than 2% of its sawn wood products today despite a fivefold increase in output. Before the oil boom, 75% of Nigeria's plywood production was exported, but exports thereafter declined dramatically and are less than 1% today. Although Nigeria has had a reputation for replanting cleared areas, recent research has shown than significant deforestation is taking place not only in the commercial forest areas but throughout Nigeria. Most deforestation is a result of the expansion of agriculture rather than the forest industry. There are almost 23 million acres (93,000 km^2) of forest reserves throughout Nigeria.

NATIONAL INTEGRATION AND REGIONAL INDIVIDUALISM

Although political entities existed in the area today occupied by Nigeria prior to the spread of effective European control, there never was a unified state covering all of the territory. In a political sense, Nigeria is a European creation, a piece of West Africa's physiographic and ethnic diversity around which boundaries were drawn and within which the course of change has been one of constant adjustment. The basic elements of Nigeria's internal divisions were already there before European conquest and consolidation took place. But such divisions are necessarily fluid over time and space. There has been expansion of powerful regional ethnic identities. In northern Nigeria, the Hausa language and identity has been expanding since well before the appearance of the British and is being assumed by speakers of minority Afroasiatic languages, many of which are on the verge of disappearing (Grimes 2000). In some cases the influence of Europeans very early on facilitated the consolidation of ethnic identity. For example, in the nineteenth century the Yoruba identity began to coalesce around the concepts of Christianity and Western education, and fragmented, tangentially related groups in southwestern Nigeria began to identify themselves as Yoruba (Mustapha 2002) as they adopted cultural characteristics from the British.

Internal divisions were not unique to the Nigerian part of West Africa; they relate to the cultural, environmental, and locational aspects of the region. The peoples of the northern savanna plains traded with the far north and felt the spread of unifying ideology from that direction. Their semiarid terrain required (and permitted) movement to a far greater extent than in the densely forested south, and their states and empires grew larger. They adopted Islam and propagated it vigorously, and they penetrated the lands of the southern peoples in search of human and natural resources. It was the Fulani upheaval, as well as the desire to terminate the slave trade, and competition with other European powers that contributed much to the British decision to intervene in the Nigerian interior late in the nineteenth century.

The involvement of Europe in Nigeria, as in other parts of West Africa, changed the whole economic and social orientation of the area. Both the north and south had their own political organization prior to the British, and both had engaged in the slave trade; but the north had long experienced more of the effects of circulation and contact with an outside world. In a geographic sense, the south was a hinterland, separated from the north by a middle belt comprising many ethnic groups that shared only a fear of domination by their more populous and powerful neighbors. British posts of trade and administration along the southern coast (which in Nigeria presented major obstacles to permanent settlement) began to develop in earnest toward the middle of the nineteenth century. In 1861 the site of Lagos was ceded to the Crown by King Dosunmu. But in the early years, Lagos Colony remained only a minor part of the British West African Settlements, administered from Sierra Leone. In 1874 it became a part of the Gold Coast Colony, and only in 1886 did it cease to be administered from another British post.

The decade from 1880 to 1890 saw important changes initiated. The British, desiring to end the hostilities prevailing in the interior, now intervened, and in the south, a rudimentary protectorate administration was established. Also founded were several chartered companies, of which the most important was the Royal Niger Company which did much to extend British influence northward. The process of penetration was completed by Frederick Lugard, who repelled French encroachment while subjugating northern Nigerian powers with the aid of locally drafted forces. The result was a British-controlled territory consisting of three parts: Lagos Colony, Southern Protectorate, and Northern Protectorate. Actually, the two major parts remained practically separate, each under its own administration; Lagos and the Southern Protectorate were merged in 1906. Soon after his conquest, Lugard advised the British government to unite the country, but only in 1914 did this step take place, at which time the Nigeria Colony and Protectorate was officially established. Even then, the south and north remained under different forms of administration. Southern Nigeria was governed with the aid of a partially elected legislative council, but the protectorate to the north remained under the jurisdiction of a governor. Thus, "the North and the South . . . remained almost complete strangers to each other. The . . . South was looking toward England and Western Europe. The Islamic North fixed its gaze on distant Mecca" (Davies 1961: 92).

The political entity now consolidated, internal differences began to become increasingly well defined. The most obvious were those between the north and south; the Southern Provinces were not divided into an eastern and a western section until 1939. The exposure of the south to European influence brought change there, from which the north remained shielded. Most significant perhaps was the contrast in how the two regions were administered. In the north, the Fulani emirates were left intact, and indirect rule prevailed. Traditional authorities and ways of life were little disturbed. Slavery was prohibited, but Western education did not become established in the north because the settling of Christian missionaries, who commonly built schools at their missions, could take place only with the consent of the emirs, who were generally anti-Christian. In the south, on the other hand, colonial administration quickly led to the development of an English-educated elite and African participation in government: as early as 1923, the legislative council included African members. Nigeria was the first British African territory to include African representatives in its governing body. The important point, however, is that this step was taken in southern Nigeria and little had changed in the north. The difference was to become even greater, for political activity rapidly grew more intense in the south, which began to produce modern political leaders. These leaders came to demand changes in Nigeria's political situation, with which the northern traditionalists did not always associate themselves.

Southern political sophistication and northern traditionalism inevitably emerged early on as major centrifugal forces in Nigeria's political geography. Now, the north was the distant hinterland, and the south had contact with the outside world; surrounded by other colonial territories, the north to a large extent lost its access to the sources of its cultural and religious heritage. Having once propagated Islam to the south of the Niger and Benue, the Muslims saw Christianity gain in the lands of the coast, and with Christianity came a spread of education to which the north had no access. The difference, as reflected by literacy rates, became ever larger; as late as 1931, less than 3% of the population in the northern part of the country could read.

The developing transport network made the consolidation of the area possible, and it is a unifying element in the divided country. Prior to the arrival of the British, there had been trade across the Niger River between the Fulani emirates and the Yoruba kingdoms. As soon as the area had been brought under British rule, however, the importance of economic links in forging the whole was recognized, as was the usefulness of good communications in effective control and administration. But improving communications links did not eliminate the regional differences in the country's economy. The separation of the major areas of productive capacity, resulting from a combination of topographic and climatic factors and coinciding with a historicogeographical separation of the region's major peoples, has promoted regionalism. Thus the Niger and Benue rivers formed major obstacles to effective interregional contact. In the south, the Niger became an internal political boundary when, in 1939, the Southern Provinces were divided into an Eastern and a Western province. Although the boundary between northern and southern Nigeria never ran along the Niger and Benue rivers (above their confluence), but somewhat to the south, the divisive effect of these valleys remained. The core region of the north lies in the high plains of Hausaland, and it has always been separated from the southern regions by the lagging Middle Belt, including the river lowlands.

As Nigeria developed economically and politically, the centrifugal forces of regionalism asserted themselves: they were strong enough to play a major role during the time that produced unity in most colonial territories, the preindependence period. It is a measure of the intensity of regional differences, notably between north and south, that there were northern requests for delays in the attainment of sovereignty, which in the south was viewed as a national aim. In 1946 and 1947 the first real steps were taken in the creation of the modern political framework. A legislative council was established to deal with the whole of Nigeria rather than with the south alone, and regional legislatures were formed for the Northern, Western, and Eastern regions, as the former provinces were then called. The internal diversity of the country clearly called for a politicogeographical arrangement permitting a considerable amount of local autonomy; without it, the cooperation of most of the people of the north, and minority groups elsewhere, would have been lost. In 1954 a federal structure was put into effect, and after repeated adjustments, the country became independent in 1960.

FEDERAL READJUSTMENTS

When independence was granted, Nigeria was composed of three political regions: the Northern, Eastern, and Western (Figure 11-17). Each had the ingredients of a viable political entity: a sizable population, an economic base, a strong sense of nationhood, and political parties with strong leadership. Despite differences over some issues, all the parties stood for the preservation of self-identity against political domination by neighboring states and for rapid modernization in the social and economic spheres. The Northern Region represented 79% of the area and contained 53.6% of the total population, while the Eastern and Western Regions had 22.3 and 24.1% of the population, respectively. Because representation in parliament was based on population,

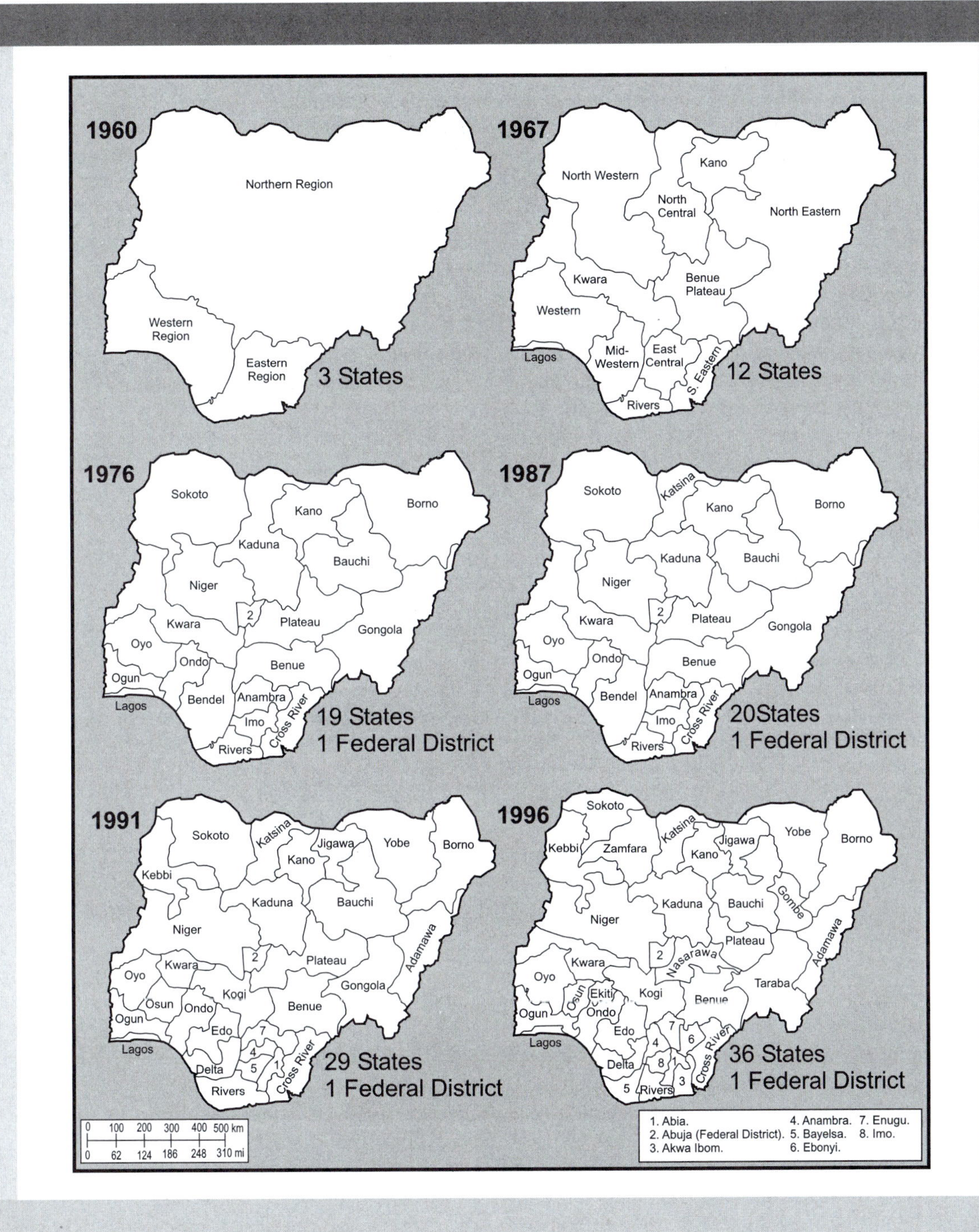

FIGURE 11-17 The emergence of Nigerian states 1963 to 1996.

and because political parties were organized on ethnic and state lines, the Northern Region was in a position to dominate the country politically.

The major peoples of the north are the Hausa and Fulani, who have historically been allies. The urban-based Hausa are numerically superior, and the more rural and traditionally cattle-raising Fulani supply the ruling element. The Hausa, in contrast, are primarily settled cultivators and traders, who, before the Fulani conquest of the early nineteenth century, were organized into large states, the most prominent being Zaria, Kano, and Gobir. Today, both peoples are widely distributed but are especially numerous in and around Kano, Katsina, Sokoto, and Kaduna. Other peoples joined together in the three-state arrangement after independence were the Kanuris, Jukun, Tiv, Igbira, Nupe, Gwari, and Bariba. Northern Nigeria has been described as a world apart from the rest of Nigeria, and with much justification: this is Muslim Nigeria, where the legacy of a feudal social system, conservative traditionalism, and resistance to change hang heavily. Literacy is lower than elsewhere, and per capita incomes are generally little more than half the national average. Proportionally fewer people have been drawn into the modern sector of the economy than elsewhere in the country.

The Eastern Region that existed at the time of the three-region setup had a population of 12.6 million, of whom Ibos constituted 64%, Efik and Ibiblo 17%, Annang 5.5%, Ijaw and Ogoni 7.5%, and Ekoi, Yalia, and others the balance. The Ibo dominated the economy in most parts of the region, including the petroleum industry centered on Port Harcourt. Ibos provided most of Port Harcourt's labor force and owned 95% of the city's private property and business establishments. Throughout the delta and to the east, Ibos were traders, teachers, and civil servants in almost every town and village. They also dominated government-owned plantations. Ibo domination beyond Iboland came about gradually as more and more Ibos left the land in search of new opportunities. They were prepared to do work others would not, and because of their generally higher education, they acquired both skilled and responsible jobs. From independence to 1967 during the time of the three regions, northern dominance meant keen interregional competition for increased populations, which led to charges and countercharges of falsification of population data. The 1962 census, which showed the Northern Region to be the most populous, was rejected by southern politicians. A recount in 1963 confirmed the 1962 results and similarly came under much criticism from those seeking increased representation in parliament and a greater share of the revenues. The census controversy contributed much to the worsening relationships between the north and south, and especially between the Hausa–Fulani group and the Ibo. The 1964 elections were won by the government party in the Northern Region.

The Ibos in the Eastern Region and the Yorubas in the Western Region, fearing the power of the Hausa–Fulani who dominated the Northern Region, wanted Nigeria reorganized into smaller states. Minority groups throughout the country, which together were numerically superior to the three majority groups combined, supported this notion. In 1963, the Edo (Bini) Ika, Western Ijaw, and Ibo, who comprised the opposition minorities in the Western Region, took advantage of a political crisis there and, following a referendum, created for themselves a new state, the Mid-West State. But the creation of this state, centered on Benin City, capital of the once-powerful Benin kingdom, did not receive universal support in that area. Ibos living in the region were divided on the matter: the educated people favored the proposition because they saw themselves taking a leading role in government and administration, while the uneducated Ibos preferred union with the Eastern Region, which was still in existence. The majority Edo (Bini) and the smaller Ijaw and Itsekiri minorities supported the new state. The state's ethnic diversity had been considered an important ingredient in the Mid-West's political stability, since no single group was large enough to dominate the rest. Its administrative divisions were increased from 8 to 14 after 1963 to respect the ethnic diversity, each division receiving the same amount of government revenues for roads and social services.

Ibos saw in petroleum a solution to their otherwise precarious and stagnating rural economy. Under the constitutional arrangements, control of petroleum rested with the federal government rather than with the regions, and the major share of the public revenue derived from this source went to the nation as a whole rather than to the region of origin. Those immediately responsible for the production of oil, however, wanted greater if not absolute control of the revenues. Since the petroleum fields, the refinery, and the port facilities all lay south of Iboland, and the two subareas were economically and politically interdependent, it was essential that the region remain one. Indeed, so important was Port Harcourt to the changing Ibo economy that, at a time when secessionist leaders still had the option for a negotiated peace settlement, the question of making the port a part of the East Central State came up repeatedly (Udo 1970b).

Following the October 1965 elections in the Western Region, which were manipulated in favor of the government party, the country was reduced to a state of anarchy and widespread disorder that resulted in a military coup in January 1966. The Yoruba and Fulani premiers of the Western and Northern regions, respectively, were killed, as was the federal prime minister, who was Hausa. The two Ibo premiers of the Eastern and Mid-Western regions were spared. The leaders of the coup were mainly Ibos, who established a military regime under Major General Ironsi, also an Ibo. Most Nigerians saw the coup as a bid for Ibo domination, especially since Ironsi suspended the federal constitution and the elected governments of all four regions that existed at the time (Northern, Western, Mid-Western, and Eastern), and attempted to establish a unitary rather than a federal form of government.

In July 1966, a second military coup took place, dominated this time by northerners, who killed Ironsi and many of his officers and established General Gowon (a northerner although not a Muslim) as head of state. The atrocities against Ibos, which began soon after the Ironsi coup, intensified. An estimated 30,000 Ibos, who had long dominated the civil service and the commercial and industrial life in the Northern Region, were slaughtered; over 600,000 others fled to their homeland. Not only were the immigrant Ibos the wealthiest community and political elites in the north, but they were arrogantly self-conscious of their superiority, and were leaders of progressive politics aimed at destroying the Northern Region's feudal structure. They looked down on their Hausa hosts as "unenterprising, lazy, backward, and feudal," while northern peasants had long complained of Ibo exploitation, and educated northerners spoke of the Ibos as "vermin, criminals, money grabbers, and subhumans without genuine culture" (Legum 1966).

As reprisals against Ibo civilians in the Northern Region continued, and as Ibo military officers serving outside of the Eastern Region were slaughtered, the numbers of Ibo refugees from the Western and Mid-West regions increased, and the Ibo governor ordered all Northerners and Westerners out of his region. Reprisal brought more reprisal, with thousands of killings and widespread destruction of property in both the Eastern and Northern regions.

By late 1966, the Ibo-controlled government of the Eastern Region was contemplating secession from the federation still led by Gowon, and there followed 6 months of open defiance of the federal government. This instability spurred minority groups throughout Nigeria to become more vocal in their demands for separate states. In May 1967, to stave off secession, the federal government divided the country into 12 states in place of the four regions (Figure 11-17). The Ibo representation rejected this plan and proceeded to proclaim Eastern Nigeria as the independent Republic of Biafra, which plunged the Nigerians into 30 months of civil war.

The Biafran war brought widespread devastation to crops and property, and completely disrupted the petroleum and related industries. In January 1970 the rebel state of Biafra capitulated to the federal government, and Nigeria embarked on a program of reconstruction and redevelopment within a three-state system that the federal government had instituted in 1967, converting the Eastern Region into the South Eastern, Rivers, and East Central states, the latter being the core of Iboland. For the Rivers people, the creation of their own state and control over their new capital, Port Harcourt, were overwhelmingly important; but Port Harcourt was to all the Ibos an Ibo city and not easily surrendered. While many properties were abandoned, others remained in Ibo hands, and Ibos had great difficulty collecting rent and raising capital that were desperately needed to rehabilitate the economy in what was now the East Central State. In the South Eastern State, Ibos experienced similar difficulties and resentment from Ibiblos and Ekoi people. In both the Rivers and South Eastern States, there were fewer Ibos in civil service and government corporations than before. Indeed, postwar anti-Ibo feeling has periodically been greater in the Rivers and South Eastern states than elsewhere in Nigeria.

The 1967 redistricting plan that divided the Eastern Region into three states also created other smaller governmental units. The Northern Region was divided into

BOX 11-3

MILITARY GOVERNMENT IN AFRICA

There have been over 70 military coups d'état in Africa since independence, and 60% of African countries have experienced military rule at one time or another. The number of coups has been declining, however: there were only 15 in the 1990s, versus 29 in the 1960s. The usual pattern in a military coup is that junior-ranking officers occupy the presidential palace and capture or kill the president and perhaps other key officials. The rebels hold the airport, and once they have captured the national radio station, they announce the coup and pledge to "clean up" government and then return the country to civilian rule as soon as possible (Thomson 2000).

Nigeria has experienced six successful coups, tied with Benin and Burkina Faso for the greatest number in Africa. There were two coups in Nigeria in 1966, and one each in 1975, 1983, 1985, and 1993.

six states: North Eastern, Kano, North Central, North Western, Benue Plateau, and Kwara (West-Central). The Mid-Western Region retained the boundaries established in 1963, and became the Mid-Western State, while the Western Region, which contained the federal capital at the time (Lagos), was divided into the Western and Lagos states. These states varied considerably in area and population. Changes in state boundaries were accompanied by changes in the constitution.

The power of the federal government was increased, while that of the state governments was decreased. A new principal organ of government, the Federal Executive Council, was established. Its composition of 11 military officers and 9 civilians reflected a power balance within and among the military establishment, the regional ethnic alliances, and the elite in the modern sector of the economy. Political parties were banned. The government's decision to revert to civilian rule in 1976 was revoked as a safeguard against the reemergence of social strife resulting from a premature withdrawal of military involvement in the political process.

In 1976 a new federal redistricting plan was announced and subsequently implemented that divided the country into 19 states and a federal district and authorized the building of a new federal capital at Abuja. Lying approximately 100 miles (160 km) north of the Niger–Benue confluence, Abuja has the advantage over Lagos, the former capital, of being centrally located and within a region free from control by any major ethnic group (see Figure 1a, p. 281). It lies in a sparsely populated section of the Middle Belt, where the federal government envisioned major agricultural and industrial development.

The military regime that had begun in 1976 was ended in 1979 by Lieutenant General Olusegun Obasanjo. The future president of Nigeria returned the country to civilian rule, under Shehu Shagari, who gave the term "Second Republic" to this second effort at civilian government. The Second Republic ended in 1983 following a military coup d'état led by Major General Muhammed Buhari, which aimed to stem corruption and mismanagement of the economy (Box 11-3). Another coup, in 1985, was prompted by economic problems. The military government of Major General Ibrahim Babangida drew up a new federal constitution and promised civilian rule. The Babangida government attempted to quell ethnic fragmentation by allowing political parties to exist, provided they had a national, rather than a regional (or local) following. The election in 1993 of Moshood Abiola was annulled by the military, and Chief Ernest Adegule Shonekan took power, only to be replaced by General Sani Abacha. The repressive rule of General Abacha brought unwanted international attention to Nigeria. When Abacha died unexpectedly in 1998, he was replaced by General Abdulsalami Abubakar, who led the country to elections and civilian rule in 1999.

Olusegun Obasanjo, a Christian and former military ruler (1976–1979), was elected president of the Third Republic. Elections are held every 4 years, and Obasanjo was elected for a second, and last, term in April, 2003. His major opponent was a northern Muslim, former coup leader and head of state Muhammed Buhari, who made threats but ultimately took no action to contest what appeared to be an election that was flawed but free of systematic cheating (*Economist* 2003a).

In the years since 1976 when the 19-state arrangement was instituted, the role of ethnic patronage has became central to the system, and asserting an ethnic "identity" a necessary part of acquiring resources from the federal government. By 2002, an additional 16 states had been added to the 19 that existed in 1976, and within almost every state, new local districts were created to address the aspirations for political power by smaller and smaller groups. Thomson (2000) states that the federal Nigerian state, unable to assert its control over many ethnic groups, bought the compliance of uncooperative ethnic groups by the distribution of resources and wealth in a process of "hegemonial exchange."

> Politics had become centered on the short-term winning of state resources, and gaining access to the levers of power. Little long-term strategic political or economic planning could survive in this institutionalized system of political exchange. Resource capture and distribution had become more important to politicians and bureaucrats than the actual development of the economy that produced these resources. Nigeria had hit, head-on, the problems of inefficiency and legitimacy associated with the hegemonial exchange model. (Thomson 2000: 70)

Hegemonial exchange encourages both centrifugal and centripetal processes. Rather than build national unity, the implicit corruption that is a key part of hegemonial exchange encourages fragmentation as smaller and smaller groups come forward to get their piece of the pie. Not surprisingly, the political map of Nigeria has continued to fragment, and presently there are 36 Nigerian states plus the federal district. Within states themselves, local political units appear to be drawn on ethnolinguistic lines, and there is likely to be further division along political boundaries as Nigerians migrate to where their people are in the majority from places where they are not.

Moreover, at the national level, precolonial cleavages have continued to widen; divergence between the north and the south has continued along sectarian lines. From the late 1990s to 2002, twelve northern states opted for Islamic law (Sharia) in place of the English common law and customary law that once prevailed everywhere, and the outcome looks remarkably similar to the old Northern Region that existed before 1967 (Figure 11-18). The implementation of Islamic law has brought sectarian strife, many deaths of both Christians and Muslims, and the migration of many southerners, who had earlier migrated to the north, back to the south. A dual judicial system has also created confusion about which legal system is to be used in cases that involve both non-Muslims and Muslims. The implementation of Islamic religious law (Sharia) in the north has brought international condemnation of Nigeria for judicial decisions that involve such cruel and unusual punishments as death by stoning. The federal government has stated that it will not allow such controversial Sharia rulings to be carried out, and in September 2003 the sentences of execution by stoning, pronounced on a woman who had been charged with adultery, were rejected by the Sharia Court of Katsina (Amnesty International 2003). Thus in one sense, federalism permits a diversity of "systems" to coexist within the federal structure. On the other hand, where "states' rights" diverge markedly from the national consensus, centrifugal processes begin to dominate the national dialogue and have to be addressed.

POLITICOGEOGRAPHIC FORCES

The basic geographic premise of federalism is the existence of regionally grouped diversities; the units, while desiring to preserve their autonomy, prosper by a union. A diversity of resources and peoples does not of itself constitute an asset, nor does it guarantee the possibility of federation, but it may be useful in its attainment. What is necessary is for people to have a common raison d'être that can be supported by the resources. A viable federation exists where there is a strong sense of national purpose and local interests are secondary to national ones. Ideally, the federation should have a high degree of symmetry at both the

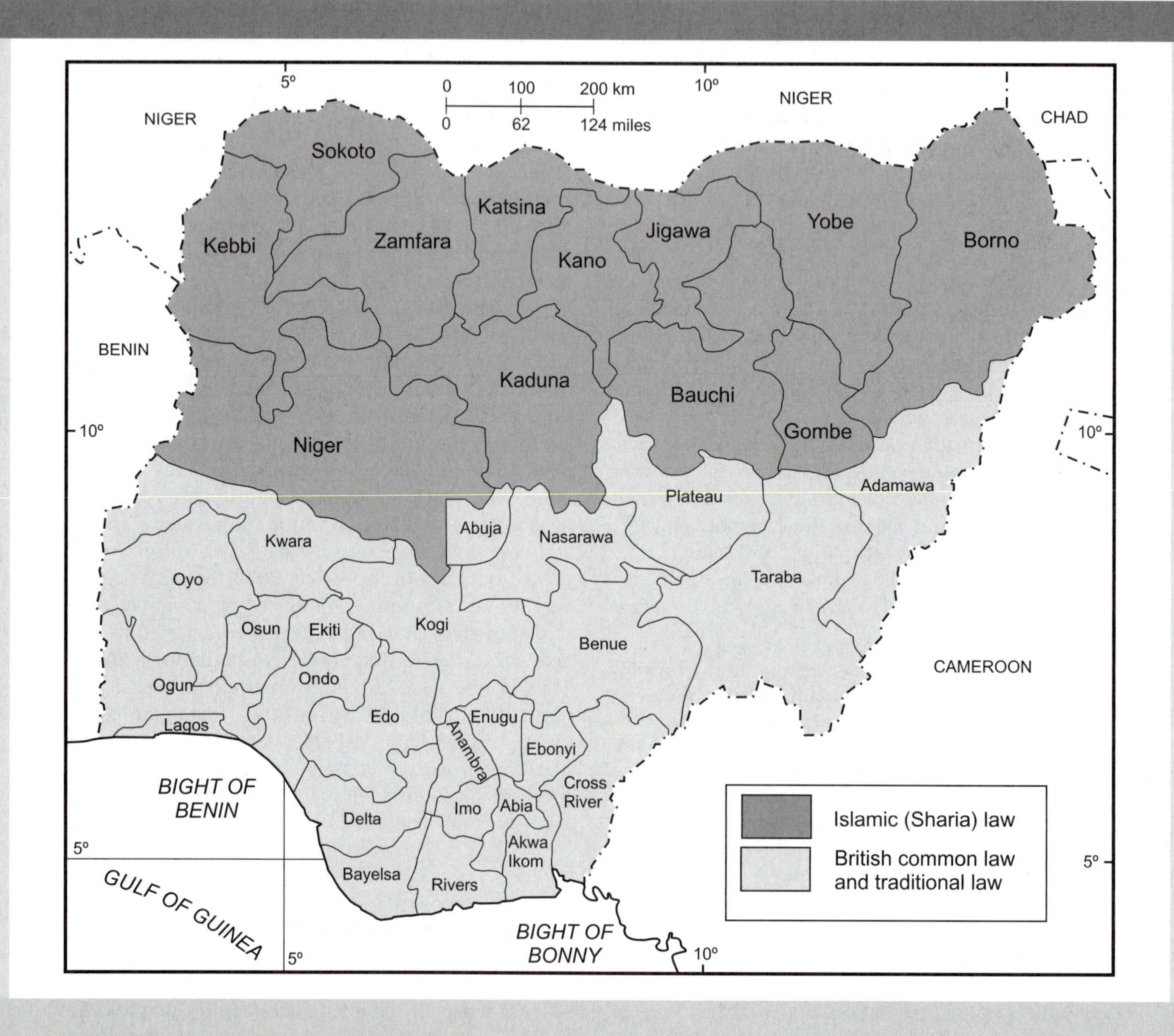

FIGURE 11-18 Sharia states in Nigeria in 2005.

national and regional levels, measured (*inter alia*) in terms of language, cultural heritage, economic well-being, and political awareness. If the system is highly asymmetrical in its component parts, and the common national ideals and interests are weak, a durable and harmonious federation is unlikely.

The politicogeographical forces that played a role in the evolution of the federal state in Nigeria continue to influence the state today, although some now exert more influence than they did previously and others have declined in importance. The centrifugal forces that were strong enough to delay the coming of independence remain, and they include the separatist feelings of those living in the peripheral zones around the major regional cores. The fear of domination by the majority is strong among many of the smaller groups in all regions. The creation of new states is an attempt to reduce this domination, but ethnically based states rekindle ethnic nationalism and run counter to Nigerian nationalism.

There is also the problem of revenue allocation. The distribution of federal revenues to the various states is a compromise among many demands. Before the 1976 redistricting, half the available funds were disbursed in 12 equal parts, and half in proportion to population. The Rivers State, then the smallest according to population, generated a disproportionately large amount of the federal funds (from petroleum), while the most populous state (North Eastern), with few economic resources, was heavily dependent on its richer partners for much-needed development monies. This problem continues today.

In recent years, especially in the post-Biafran era, the centripetal forces that bind together Nigeria's diversities have become stronger. Nationalism is emerging, as more and more Nigerians break with tradition and identify increasingly with the values and ideals of the modern state. As the circulation of people, goods, and ideas increases, regional political and economic disparities should lessen. The states have become more interdependent in many respects, and interregional economic transactions are growing. Indeed, these interregional bonds reduce the likelihood of politicogeographic fragmentation for Nigeria and strengthen the federal ideal. On the other hand, although it is unlikely that a state or states will attempt to secede today, the religious lines of cleavage have become deeper in Nigeria over the past 25 years and are making for a lively dialogue. It is difficult to predict whether the outcome of that dialogue will involve the decentralization of some federal powers and the increase of regional autonomy, the centralization of power by the federal government, or a little of both in an effort to please everyone. But it does seem clear that the continued existence of Nigeria as a political entity owes much to its brand of federation. Petroleum has been the glue that has held Nigeria together, and as long as the federal government maintains a reasonably equitable distributive role with respect to the various regional stakeholders, fragmentation will probably be avoided.

BIBLIOGRAPHY

Ajani, O. O., and O. O. Aina. 2002. *Agricultural development and food security in Nigeria: The constraints that women face.* Paper presented at the Eighth International Interdisciplinary Congress on Women, Department of Women and Gender Studies, Makerere University, Kampala, Uganda, July 21–26, 2002. www.makerere.ac.ug/womenstudies.

Akinwumi, O. 2000. Women entrepreneurs in Nigeria. *Africa Update,* 7(3): 2–3.

Amnesty International. 2003. Nigeria: Amina Lawal's death sentence quashed at last but questions remain about discriminatory legislation. Amnesty International Press Release AFR 44/032/2003, 25 Sept. 2003.

Anderson, D. M., and R. Rathbone, eds. 2000. *Africa's Urban Past.* Oxford: James Curry.

Best, A., and H. J. de Blij. 1977. *African Survey.* New York: Wiley.

Central Intelligence Agency. 2003. *CIA World Factbook.* www.odci.gov/cia/publications/factbook/.

Davidson, B. 1995. *Africa in History.* New York: Simon and Schuster.

Davies, H. 1961. *Nigeria: The Prospects for Democracy.* London: Weidenfeld & Nicholson.

Economist. 2003a. Elections in Nigeria: The people disagree. *Economist*, April 24.

Economist. 2003b. A small step on the road to Cancún. *Economist*, August 14.

Geomatics. 1998. The assessment of vegetation and land use changes in Nigeria between 1976/78 and 1993/95. Report submitted to the Forest Management, Evaluation and Co-ordinating Unit (FORMECU) of the Nigeria Federal Department of Forestry as part of the World Bank Funded Environmental Management Project (EMP) for Nigeria: Abuja.

Grimes, B. 2000. *Ethnologue. Languages of the World*, 14th ed. Dallas: SIL International.

Harrison-Church, R. J. 1980. *West Africa: A Study of the Environment and Man's Use of It*, 8th ed. London: Longman.

Izon, D. 1994. *The Mineral Industry of Nigeria.* U.S. Geological Survey Mineral Report. Reston, Va.: USGS. http://minerals.usgs.gov/mincrals/pubs/country/.

Legum, C. 1966. The tragedy in Nigeria. *Africa Report*, 11(8): 23–24.

Mobbs, P. 1997. *The Mineral Industry of Nigeria.* U.S. Geological Survey Mineral Report. Reston, Va.: USGS. http://minerals.usgs.gov/minerals/pubs/country/.

Mobbs, P. 1999. *The Mineral Industry of Nigeria.* U.S. Geological Survey Mineral Report. Reston, Va.: USGS. http://minerals.usgs.gov/minerals/pubs/country/.

Mobbs, P. 2001. *The Mineral Industry of Nigeria.* U.S. Geological Survey Mineral Report. Reston, Va.: USGS. http://minerals.usgs.gov/minerals/pubs/country/.

Mustapha, A. R. 2002. Coping with diversity: The Nigerian state in historical perspective. In *The African State: Reconsiderations*, A. I. Samatar and A. I. Samatar, eds. Portsmouth, NH: Heinemann, pp. 149–176.

Nigeria 1975. *Guidelines for the Third National Development Plan, 1975–80*. Lagos: Government Printer.

Nixon, C. R. 1972. Self determination: The Nigeria/Biafra Case. World Politics, 24(4): 473–497.

Odeyele, A. J. 2000. Public–private participation to rescue railway development in Nigeria. *Japan Railway and Transport Review*, 23: 42–49.

Thomson, A. 2000. *An Introduction to African Politics*. London: Routledge.

Udo, R. K. 1970a. *Geographical Regions of Nigeria*. Berkeley: University of California Press.

Udo, R. K. 1970b. Reconstruction in the war-affected areas of Nigeria. *Area*, 3: 9–12.

U.N. No date. United Nations Cyber Schoolbus. http://www.un.org/cyberschoolbus/habitat/profiles/lagos.asp

U.S. Department of Commerce 2002. *Country Commercial Guide: Nigeria, Fiscal Year 2002*. International Trade Administration. Washington, D.C.: U.S. Department of Commerce.

World Gazetteer. 2003. Population of Nigerian cities. http://www.world-gazetteer.com/fr/fr_ng.htm.

CHAPTER 12

Liberia and Sierra Leone

The End of Anarchy?

Liberia and Sierra Leone are two West African coastal countries that, in addition to their common geographies, possess very similar histories. Coastal parts of both Liberia and Sierra Leone served as havens for freed slaves and captives. In the case of Liberia, the vast majority of settlers were returning to Africa after having been freed in North America, while most of those who were set free in Sierra Leone came from slave ships that had been stopped by British warships intent on terminating the Atlantic slave trade. Sierra Leone achieved independence from Britain in 1961, while Liberia has been independent since 1847. If independence were the only significant variable influencing social and economic development, we should observe some remarkable differences between these two countries, but that has not been the case. The societies and economies of Sierra Leone and Liberia were bifurcated from the beginning: only short distances from Freetown, Monrovia, and other coastal settlements, there was little evidence of a modern society. Rather, a subsistence economy prevailed, hospitals and schools were few and far between, roads were almost nonexistent, and the indigenous peoples were poorly represented in government.

After World War II, Liberia's economy diversified and assumed many of the structural attributes of colonial Africa, and the interior regions became integrated more efficiently with the rest of the country. It seemed at the time that Liberia was on the verge of considerable economic growth and development, financed substantially by external sources of capital. Sierra Leone seemed to be modernizing; infrastructure was bringing innovation and new ways of thinking far from the capital city.

Yet, despite the natural wealth of both countries and hardworking peoples, both Liberia and Sierra Leone remained on the verge of development throughout the 1960s and 1970s and then descended into many tumultuous years of military and civilian violence characterized by unimaginable brutality, civil war, and anarchy. Writer Robert Kaplan (2001) has called the recent past in countries like Liberia and Sierra Leone a "coming anarchy": a dark, violent, overpopulated, resource-scarce, and uncertain future for many weak states in Africa and elsewhere in the world. We will return to this theme at the end of this chapter, but let us first take a look at the Liberian and Sierra Leonian experiences.

LIBERIA

Beginnings

Liberia owes its origins to the movements aimed at the abolition of slavery and slave trade prevailing in the United States around the turn of the nineteenth century. In 1818 representatives of the American Colonization Society crossed the Atlantic in search of African land to which freed slaves could be repatriated. Gradually, by purchase and conquest, settlement expanded under the guidance of white governors and with aid from American organizations and individuals. The burden of administration and financial support became too much for the American Colonization Society, which in the 1840s indicated a desire to withdraw its administrative assistance. In response to these developments and to pressure from within what essentially was the colony of a private society, the Liberian legislature declared the colony independent and Joseph Jenkins Roberts, the first black governor of Liberia, was elected Liberia's first president in 1847. Thus began the fight for survival of the tiny state in turbulent Africa.

Liberia did not solve many problems by attaining sovereignty. Its status as an independent country was no guarantee of economic prosperity or stability. Up until World War II, Liberia's economy lagged behind its neighbors, and most of the development infrastructures provided for by colonial administrations elsewhere in Africa were absent in Liberia. Indeed, from several points of view the situation deteriorated. The support from the American Colonization Society was terminated, and efforts to collect duties on goods exported had little success. In addition, European interest in the African west coast was on the increase, and frontier disputes with the French in Côte d'Ivoire and the British in Sierra Leone were frequent. The "Independent African State of Maryland" in the east was absorbed by Liberia in 1857, but the problem of effective national control remained. The African peoples of the interior retained their traditional ways of life and religion and did not pay allegiance to the new rulers on the coast. This induced Britain and France to bring pressure to bear upon the Liberian government to cede certain areas, and boundary treaties were signed with Britain in 1885 and with France in 1892. Liberia lost still more territory as recently as 1910, when the French, claiming Liberian failure to exercise control over certain peoples, took 2,000 square miles (3,220 km^2) of the nominally Liberian interior.

The cultural, linguistic, religious, and economic differences between the Americo-Liberians (who by 1867 probably numbered about 19,000) and the indigenous peoples led to the establishment of two individual politicogeographical regions in the state of Liberia. The Americo-Liberian element became an important force in the politicoeconomic growth of Liberia. Having long lived in North America and possessing principally English surnames, the Americo-Liberians were often as foreign to Africa as were the European colonists elsewhere. They may have adapted more quickly, but they separated themselves from the local inhabitants, who were slow to cooperate in the development of the Liberian state.

The division between the settlers and the local people came to be expressed politicogeographically too, for the Liberian government ruled a coastal strip to about 40 miles (64 km) inland but was ignored by the people of the interior. The coastal belt was divided, for administrative purposes, into five counties (one of which was Maryland County, the former "independent state"). The interior was divided into three provinces and nine districts, but for a century after the arrival of the first colonists, there was little local participation other than by those who had come under the immediate influence of the new settlement within which they resided. Beginning at the end of the nineteenth century and spurred by its colonial rivals, the Liberian government extended its authority over the interior peoples by sometimes brutal force and by establishing an indirect rule very similar to that of the British: the central government appointed paramount chiefs and commissioners who were required to support the policies and programs of Monrovia. In 1963 indirect rule was abolished and the interior region was divided into four new counties (Lofa, Bong, Nimba, and Grand Gedeh), and the country was ruled with a uniform administrative structure for the first time.

From the beginning, some of the repatriates gave expression to their feeling of superiority over the native people (who are called "aborigines" in the constitution), a practice that did not enhance the prospects for unity. The Americo-Liberians, for example, went to churches and social clubs from which the African populations were excluded.

> As late as the 1960s, Liberian politicians dressed for formal occasions in top hats and tail-coats. Elite Liberians did not speak African languages. Office workers sweltered in three-piece suits, collars and ties. Young Liberian sophisticates danced to Tamla Motown at discos and parties and despised African music. The sons and daughters of the wealthy, educated in the United States and Britain, referred to other West Africans as "coasters." (Ellis 1999: 43)

It was not until the 1920s that the Liberian government attempted to unify the two regions and to try to change the attitude of indifference of the colonists toward the indigenous populations. For the first 80 years of Liberia's existence as an independent state, the Liberian economy was almost totally subsistent. Only limited amounts of coffee, palm oil, and camwood were exported to the United States and Europe. A small industry developed around the piassava, a fiber obtained from the raphia palm, which was used to make brushes. But there was competition because of easily accessible supplies in Sierra Leone (notably Sherbro Island). Similarly, coffee exporters could not compete with the rapidly expanding industry of Brazil. By the end of the nineteenth century, Liberia was debt ridden and had few prospects of emerging from this condition.

In the years prior to the First World War, the country appeared on the verge of losing its sovereignty. American loans were provided, but under conditions that Americans should be in charge of customs duty collections, Americans should train a frontier police force, and an American should "advise" the Liberian government on all monetary matters. Meanwhile, the government was exploring the possibility of granting concessions of land to foreign companies, as was happening elsewhere in Africa. Thus, from 1904, foreign companies inspected Liberian possibilities, and in 1906 a British company established a sizable rubber plantation near Monrovia. Britain, the first nation to recognize the Republic of Liberia in 1848, was then the leading supplier of Liberia's imports, and soon after World War I, Germany and the Netherlands became the leading trade partners.

The major breakthrough did not come until 1926, when one million acres was leased to the Firestone Company for a 99-year period. Previous investigations had proved Liberia to be very suitable for rubber cultivation, and the American company laid out extensive plantations, began to employ thousands of Liberian workers, and spent much money within the country. By the mid-1930s, rubber occupied a place of importance in the list of Liberian exports, and from 1940 to 1961, rubber was the country's leading export. Firestone in effect put Liberia under American financial supervision and provided a regular income to both Liberians and the government. Its system of labor recruitment, while approved by the Liberian government, was frequently criticized overseas.

American involvement in Liberia's development increased during World War II. Pan American Airways opened up Roberts Field east of Monrovia, and the U.S. Army Air Corps began to use the civilian facility to stage American assistance to the Middle Eastern and North African theaters. The U.S. Navy built the country's first modern harbor at Monrovia, which enabled iron ore development to follow. In 1944 the Liberian dollar was raised to parity with the U.S. dollar, and American models were introduced in government and business. Natural resource surveys were undertaken for the first time, and the value of iron ore deposits was recognized.

The Era of Change

Liberia's first century of independence was a "century of survival" and near isolation, not one of spectacular economic accomplishment or social and political enlightenment. Indeed, Liberia was in many ways less prepared to deal with a changing world than most African countries of similar size at the close of World War II. In 1939 Monrovia, Liberia's capital and largest urban center, had no telephone system, piped water, or sewage disposal. In 1950 the country had only 8 hospitals (426 beds) and 15 physicians, of whom only 2 were Liberians. For a population of 750,000 there were only 253 schools, with a total enrollment less than 21,000, of whom 96% were in elementary school (Von Gnielinski, 1972). There was no public railway and only 220 miles (354 km) of public roads, mainly located around Monrovia and the Firestone plantations. The interior was isolated politically, economically, and geographically from the coastal areas, and there was a need to unify and assimilate the various population groups. Other than Firestone, there was no large-scale foreign investment in the country.

To launch Liberia into an era of change, President William V. S. Tubman initiated two important policies in 1944 (Clower et al. 1966). First, he aimed at the assimilation and unification of the Liberian people. This policy of "unification" required constitutional changes that gave the right of suffrage to all Liberians and revised the regulations for governing the interior. And then came an "open-door policy," which encouraged the investment of foreign capital in the development of Liberia. In 1949 the first ships bearing the Liberian "flag of convenience" set sail, and since that time Liberia has grown remarkably as a nation of registry for foreign-owned oceangoing vessels. Low rates and lenient laws made the Liberian flag attractive and, up until the civil war years, Liberia had the largest registered tonnage in the world. During the war, revenues earned through the registry program accounted for 70% of government income, generating roughly $25 million annually (Pike 2003).

As a result of the open-door policy, the pace of development picked up markedly, especially in the mining of iron ore, the construction of railways, the expansion of forestry resources, and the improvement of agriculture in the interior. As part of the Tubman policy, the focus of economic activity shifted from the coast toward the interior. A network of roads and railways was built to link these activities with the coast and especially Monrovia, which had grown from an insignificant town of 27,000 people in 1950 to an important city that had a population of 180,000 in 1974 and over 900,000 in 2005. Along these arteries commercial crops, especially coffee, rice, oil palm, and rubber, were encouraged, as well as commercial forestry. But Liberia needed transport routes connecting more parts of the country with the emerging core region. For example, as recently as 1968, there was no road connection between Monrovia and Robertsport, Greenville, and Harper, three of the five county headquarters along the coast. Furthermore, thousands of square miles of the interior, especially in the east, do not enjoy the benefits of even a passable road, and Liberia has limited connections with the neighboring states. Liberia's rivers do not serve as countrywide communications, for they are obstructed by rapids: the St. Paul River is navigable from its mouth at Monrovia to White Plains about 15 miles (24 km) inland; in the east, the Cavalla River is navigable for 50 miles (80 km) from its mouth. In any event, the rivers run directly from the interior to the coast and thus parallel any existing land transport routes. Stanley's description of the transport network is as true today as it was over 30 years ago: the present transport network is sparse and is oriented to the coast, with only rudimentary internal connections (Stanley 1970a). As rudimentary and recent as much of the road network was over 30 years ago, things have gotten worse: in 2002 none of the railroads was working and roads were in a poor state because of heavy rain and little maintenance. When transportation and communications infrastructure is built and improved, places can seem to be closer together in a relative sense. This process of space–time convergence may also be reversed—places can become unconnected or diverge from one another. Today, for example, it takes longer than formerly to travel from one point to another in Liberia because of the increasingly poor state of the roads and rails; relatively speaking, therefore, places are becoming further and further apart. Liberia seemed to hold so much promise in the 1960s. The country's failure to live up to that promise is discussed next.

Rising Political Instability and the Roots of Civil War

Prior to the Tubman administration (1944–1971), the Liberian government used the Liberian Frontier Force (later changed to "Armed Forces of Liberia") to quell dissent in the interior. Government revenues were enriched by hut taxes used to force the indigenous Africans to work on the growing numbers of plantations. President Tubman used the revenues generated from his open-door enterprises in an attempt to unify the Liberian state through a hierarchy of patronage linked ultimately to himself, rather than by brute force. Connections became paramount in securing appointments. Indigenous Africans discovered that joining one of the Americo-Liberian churches or even a Masonic lodge would help them get a job and achieve social mobility. Many indigenous Liberians were educated (many studied in the United States and Britain) and worked their way into the lower levels of the bureaucracy. But inequality between the Americo-Liberians and the indigenous Africans did not magically go away with prosperity, and during the administration of the next president, William R. Tolbert (1971–1980), many indigenous as well as settler Liberians were calling for political reform of a political system that increasingly resembled a "corrupt and ramshackle neo-colony managed on behalf of the US government and the Firestone rubber company" (Ellis 1999: 50). Indeed, the benefits of prosperity still accrued disproportionately to the small minority of Americo-Liberians and Africans who had been assimilated into the system.

Things came to a head in 1980 when the Americo-Liberian regime was violently removed from power by low-ranking and politically inexperienced soldiers mainly of African origin—an event from which the country has yet to recover. President Tolbert was assassinated, and coup plotter Master Sergeant Samuel Doe emerged early on as leader of the People's Redemption Council (PRC), the ruling party of Liberia after the coup. Although Doe was almost illiterate, he quickly mastered the nuances of corrupt Liberian patronage politics, a system that he helped perpetuate rather than reform. He regained the political and financial support of the United States that Tolbert had lost, rigged a presidential election in 1985 which he won, brutally eliminated his enemies, lost the support of key ethnic constituencies, and was himself deposed and murdered in 1990.

The seeds of Doe's destruction were sown in his own bloody coup against Tolbert. Doe had promised that in ending Americo-Liberian control he was ushering in a new era in which the focus of development would be on the indigenous peoples of Liberia. Yet Doe's regime was more corrupt than previous governments, and he surrounded himself with people of his own Krahn ethnic group, alienating vast segments of the indigenous populations, particularly the Gio and the Mano. Also, although Doe had been successful in obtaining the Cold War support of the United States, he quickly alienated his neighbor, Côte d'Ivoire. Despite Doe's promise to spare the life of Adolphus Tolbert, the murdered president's son and a relative through marriage to Félix Houphouët-Boigny, president of Côte d'Ivoire, the young man was abducted from the French Embassy and killed.

It was no accident that Charles Taylor, an American-educated Americo-Liberian with broader Liberian appeal than Doe, particularly among the disaffected Gio and Mano peoples, led his forces into Liberia in 1989 from Côte d'Ivoire.

Taylor's path to power exemplified African regional intrigue and power politics. Taylor was part of a small Liberian exile community in Burkina Faso that helped Blaise Campaoré overthrow Thomas Sankara, the then president of Burkina Faso. In return for his support for his coup, Campaoré promised to support Taylor's bid against Doe. Through Campaoré, Taylor gained Libyan president Muammar Gadhafi's support and a legitimacy other putative Liberian exiles with aspirations did not possess. Taylor and many of his men trained in Libya, where he met other would-be revolutionaries under Libyan tutelage, including the Sierra Leonian Foday Sankoh of the Revolutionary United Front (RUF). Taylor called his group the National Patriotic Front of Liberia (NPFL).

On Christmas Eve, 1989, a small force representing the NPFL marched on iron-rich Nimba County, and by July 1990 they were in Monrovia. In September President Doe was brutally murdered by one of Taylor's allies (later a rival), Prince Johnson. It was at this time that Taylor sent some of his most battle-hardened units to Sierra Leone to help his friend Foday Sankoh in his bid for power; thus widening the theater of war and setting the stage for wrenching population displacements of a magnitude probably never before seen in so short a time in West Africa (Figure 12-1). Although the situation has improved somewhat for Sierra Leone since 1999, it was not so for Liberia: at the end of 2002 over 390,000 people were "of concern" to the United Nations High Commission for Refugees, mainly internally displaced persons (Figure 12-2), returnees or refugees from neighboring countries, and 271,000 Liberians refugees in exile, of which there were 255,000 in neighboring West African countries. In the same year there were 140,000 internally displaced people in Sierra Leone and 122,000 Sierra Leonian refugees in neighboring countries (UNHCR 2003).

The extent of the human displacement is enormous: refugees from Liberia, the internally displaced within Liberia, and refugees from other countries in Liberia represented just over 20% of the Liberian population in 2002. If an equivalent calamity were to befall the United States, relative to its population, 58 million people would have to be either in exile or internally displaced. It is difficult enough for poor governments to respond to the occasional natural disaster, but such man-made disasters are overwhelming. In addition to the enormous burden refugees pose to a country in West Africa, a large refugee population is an inherently destabilizing factor in several respects. Crime and disease are most notable, but one must not discount the political dimension, as the Ivorian government learned when Liberian refugees began their march on Abidjan in 2002–2003.

In 1992 the United Nations imposed sanctions on Liberia to prevent the sale of its natural resources by rebels for the purchase of weapons. Taylor's ambitions in Liberia were temporarily thwarted by a peacekeeping force led by Nigeria on behalf of the Economic Community of West African States (ECOWAS). The political and military situation became even more complex than it had been because some Nigerian peacekeepers sided with local armed factions against the NPFL and engaged in arms selling, diamond dealing, and corruption. In 1993 and 1994 the warlords signed cease-fires in Benin and Ghana, respectively, but neither order held for long, and thousands of people were displaced.

In 1995, after a regime change in Nigeria brought to power General Sani Abacha, a face friendlier to

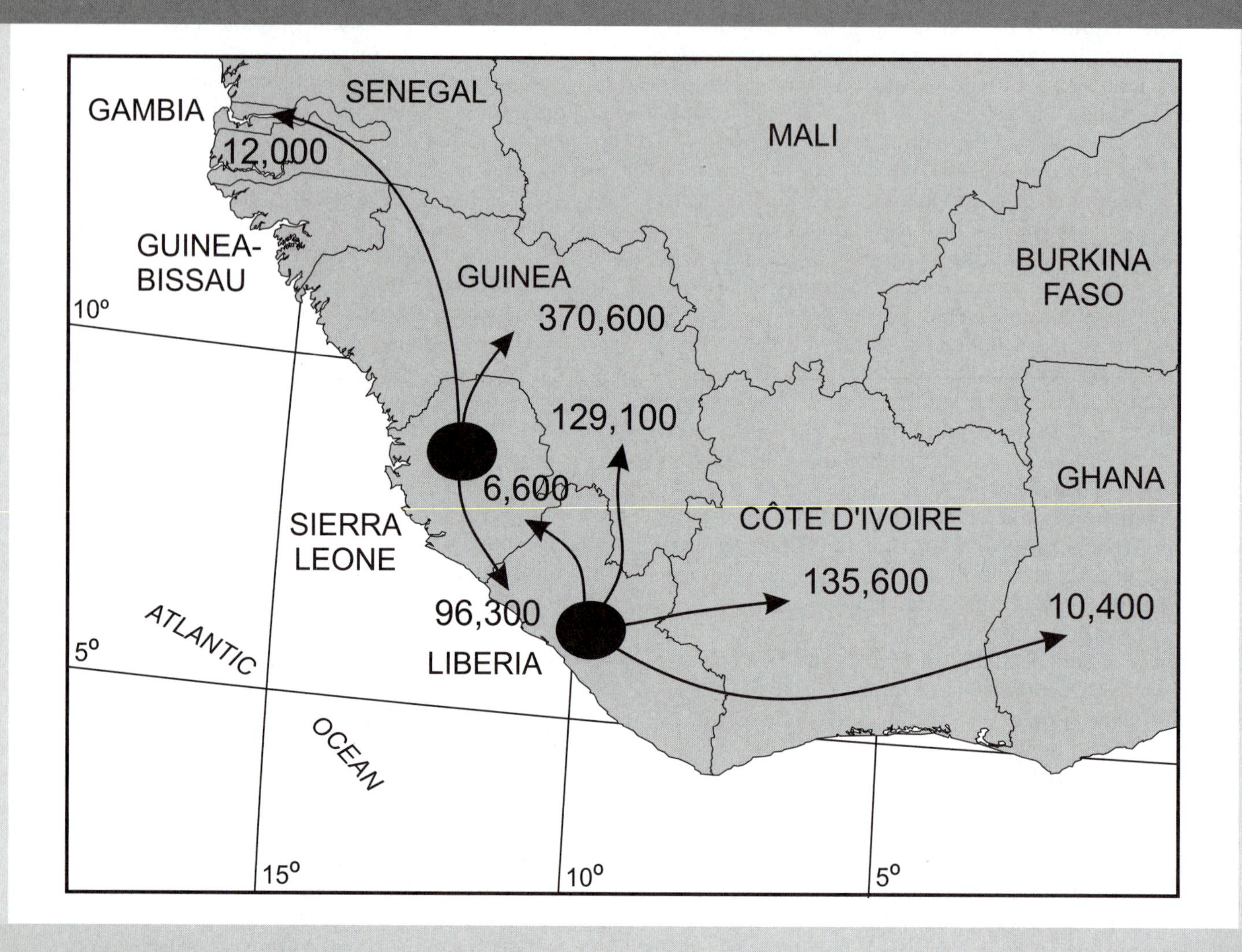

FIGURE 12-1 Refugee dispersals from Liberia and Sierra Leone in late 1999.

Charles Taylor, the Abuja agreement was signed and the way paved for presidential elections in 1997. Charles Taylor won 75% of the vote in a free and fair election against Ellen Johnson Sirleaf, an experienced administrator with international support who had been living in exile. Ellis (1999) contends that one of the factors that may have prompted people to vote for Taylor was his determination to be president and the threat of more civil war if he did not win. In 2001 the UN imposed sanctions on the Taylor regime to prevent it from selling Sierra Leone's diamonds and timber to buy weapons to support the civil war in Sierra Leone. In March 2003, President Taylor was indicted by the United Nations for war crimes as a result of his support for the RUF insurgency in Sierra Leone.

Peace was elusive during Taylor's tenure as president, as government forces clashed on many occasions with two rebel groups backed by Guinea and Côte d'Ivoire, respectively: Liberians United for Reconciliation and Democracy (LURD) and Movement for Democracy in Liberia (MODEL). These anti-Taylor rebels terrorized the countryside and marched on

FIGURE 12-2
Liberians fleeing anti-government rebels in 2003.
UNHCR A. van Genderen Stort.

the major cities. Although the rebels twice came very close to capturing Monrovia in 2003, they were repulsed by government forces. Under international pressure, Charles Taylor stepped down from the presidency in August 2003 for exile in Nigeria, and control of the country passed to his second in command, Moses Blah. Nigeria sent in over 1,000 peacekeepers, South Africa promised to send peacekeepers, and the United States positioned over 2,000 Marines offshore in a warship. The stabilized situation permitted an election in 2005 won by Africa's first female president, Ellen Sirleaf Johnson.

Economic Activities

The mineral resources of Liberia consist of gold in the east and diamonds, gold, iron ore, nickel, manganese, palladium, platinum, and uranium in the rocks of western Liberia. The interior hills contain substantial iron ore deposits (Figure 12-3). Before the recent civil wars, Liberia was the leading producer and exporter of iron ore in Africa, and the fifth largest producer in the world. Since 1951, four of Liberia's ore bodies have been tapped by foreign companies, and from 1961 until the late 1980s iron ore was Liberia's leading export. The Bomi Hills deposit 50 miles (80 km) northwest of Monrovia was the first to be developed. In the early years, reserves exceeded 20 million tons of magnetite and hematite containing 66% pure iron. These high-grade ore bodies were quickly mined out and attention shifted to less attractive ores, which were processed at the mine. In 1975 the Bomi mine closed. In 1961 the Mano River deposit, located 20 miles (32 km) further inland from the Bomi Hills and close to the Sierra Leone border, was opened with the assistance of the Liberia Mining Company (LMC), in which Republic Steel Corporation held a 60% interest.

Both the Bomi Hills and Mano mines required railways to be built and special loading facilities at the deepwater port of Monrovia. Most of the Mano ores were shipped to the United States, the Netherlands, Britain, and Germany. The largest deposits, probably over 300 million tons and including at least 235 million tons of high-grade ore of 65 to 70% iron content, are located in the Nimba Mountains close to the Guinea border. The mining of this ore deposit began in 1963. The Nimba ores have almost no overburden and

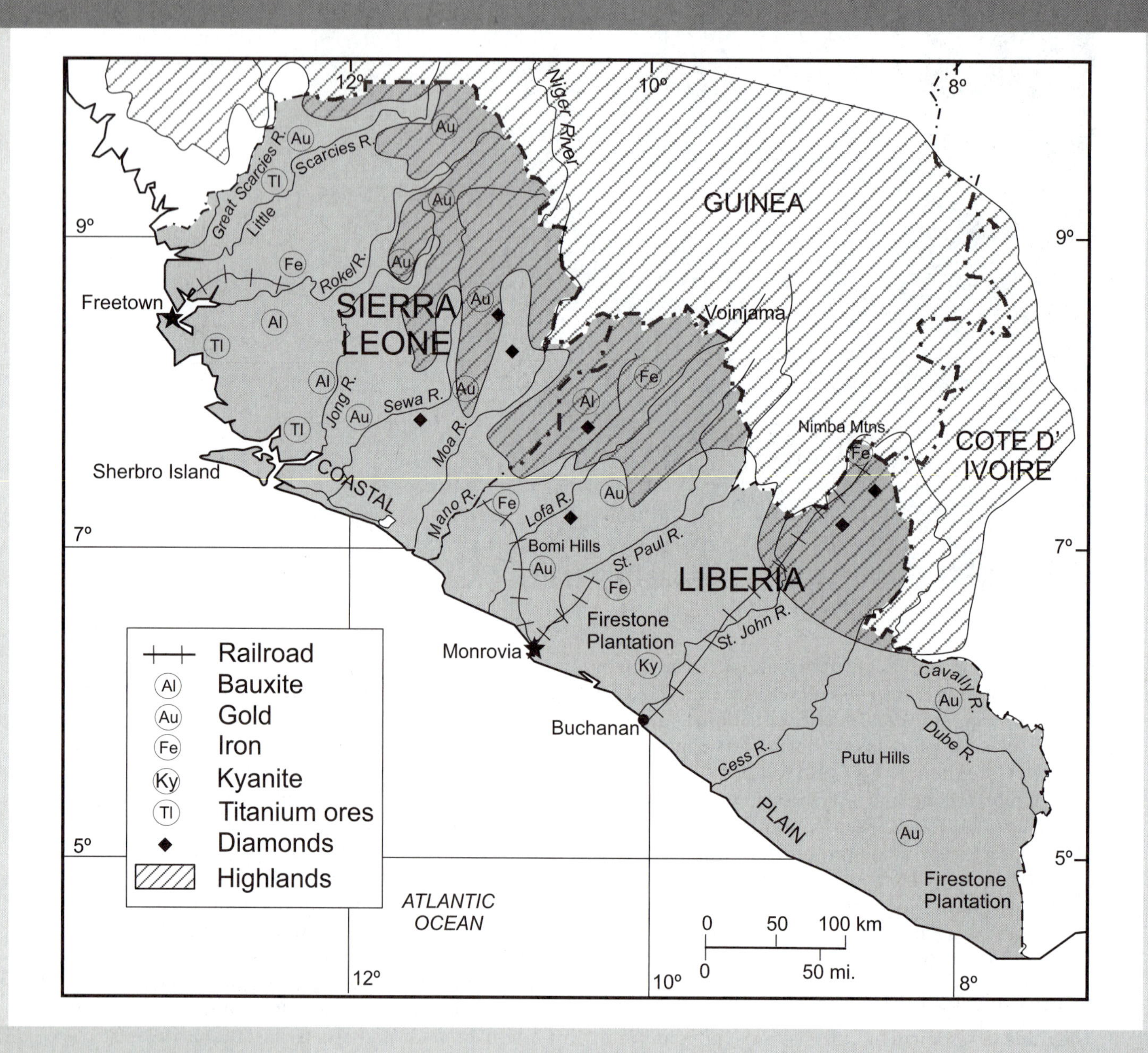

FIGURE 12-3 Environments and natural resources of Liberia and Sierra Leone.

are low in impurities. The Nimba project, which included not only the mining operations, but a pelletizing plant at the port of Buchanan, and a 170-mile (270 km) standard gauge railway from the mine to the coast, was at its inception the largest single private enterprise in tropical Africa. Its main shareholders were Swedish companies and the Bethlehem Steel Corporation, like Republic Steel, a U.S.-owned enterprise. Several additional iron ore deposits (Mount Beeton and Mount Tokadeh) located near the Nimba Mountains were brought into production in the early 1970s, while untapped reserves remain in the Wologisi Range in the far northwest and the Putu Range in the east. Japan has shown interest in the Wologisi ores, which may lead to its production in the future, with associated construction of new railways, expansion of harbor facilities at Robertsport, and provision of new employment. The fourth open-cast mine was operated by the Bong Mining Company (DELIMCO) in the Bong Ranges, a short distance inland from Monrovia. The ores, while extensive (250 million tons), are poorer, having an average iron content of only 38%. The mine was connected with Monrovia by a 50-mile (80 km) standard gauge railway and was operated by German and Italian interests. In the 1980s, a pelletizing plant was built at the Bong fields. All iron ore handling and processing facilities have been destroyed by the war (Coakley 2004).

Liberia is a country rich in natural resources, but "in the past 23 years it has regressed almost beyond measure" (*Economist* 2003: 40). Per capita GDP (1995 dollars) was about $900 during the Tolbert years (1971–1980), declining 76% to less than $600 during the Doe administration. During the first civil war (1989–1997) it declined to less than $100. During the presidency of Charles Taylor (1997–2002), it rose to just under $200 (*Economist* 2003). The insurgency probably caused a further decline, although no figures are available. It is clear, though, that formal economic activity in Liberia ceased at least a decade ago as gangs and warlords took control of the natural resources for their own gain and most businesspeople fled the country. The control of natural resources became an objective in the conflict, both to purchase weapons and to wield the stuff of patronage. Diamonds, iron ore, rubber, and timber have been used in support of insurgency by all sides. Since the election of 1997, formal mining has been slow to recover, and what mining activity took place consisted of the artisanal production of diamonds and gold. Exploration increased after 2000 because investors are interested in Liberia because of its potential mineral richness; Liberia has been said to possess the world's last "virtually unexplored" rocks (Szczesniak 2001).

Manufacturing has remained very undeveloped in Liberia. Up until the Doe coup in 1980, there was virtually no manufacturing other than the production of chemicals and explosives that were required by the mines, a petroleum refinery that used imported fuel, a cement factory, and a handful of smaller and privately owned concerns that produced shoes, beverages, plastics, confectioneries, furniture, and building materials. Some 95% of the raw materials required by these industries had to be imported. Most of these enterprises have closed or drastically reduced operations.

Rubber was an important industry that ranked after iron ore and diamonds in the export market; as with the latter industries, however, the production of rubber has been disrupted by economic and political instability and civil war since 1980. Firestone was the country's main producer (52% of the total), and its estates at Harbel and Cavalla (near the Côte d'Ivoire border) total some 80,000 acres (32,400 ha), or one-third of the country's total. Of the over 14,000 Liberians employed on the estates during their heyday, most worked as tappers. Small producers of rubber have found it hard to compete with the multinationals because of the shortage of efficient tappers, as well as low yields and and high transport costs (Stanley 1968). With its relatively low costs per unit output and modern management, the plantation has an economy of scale that the small producer can rarely match. It is expected that if peace can be maintained, formal rubber production will resume in the near future.

The Liberian economy continues to be dominated by subsistence agriculture as it was throughout the relative calm of the Tubman and early Tolbert years (Beleky 1973). Agriculture as a share of GDP increased significantly after 1980 as civil unrest rose and mining and commercial activity collapsed. Per capita income was less than $200 in 2002, and 80% of the population lived below the poverty line.

Colonialisms and Inequality

Whatever the political indignities of colonial rule, it cannot be denied that colonialism brought certain material and human assets to many African territories that are lacking, or were absent for a long time, in Liberia. What material (and human) assets were developed were squandered in over 20 years of misrule, mismanagement, and civil strife that had been rooted in inequality between the African-American settlers and the local groups of Africans. The African-American settlers and their descendants have in many respects remained as divided and separated from their indigenous countrymen as were Europeans elsewhere, but without bringing to Liberia the techniques, capital, and experience of the Europeans. The pattern of the country's political and economic development, in many respects resembles that of, say, Angola during the time of the Portuguese. Economic activity has been aimed at the exploitation of raw materials. Involvement of the majority of the local people in government has been minimal, and the internal economy has not had much of a chance. Many other amenities colonial rule helped bring, such as agricultural research stations, schools, hospitals, and communications, lacking until fairly recently, have deteriorated. An entire generation of Liberian children, forcibly conscripted into rebel armies and taught to terrorize and kill, has been traumatized and dehumanized by the experience and will be difficult to integrate into the national fabric—if it ever emerges.

When development came to Liberia, it came on terms more or less dictated by those desiring to exploit the country's raw materials; the Liberians had little choice but to welcome whatever benefits accrued. Unlike much of the remainder of Subsaharan Africa, Liberia did not experience the salutary efforts of a colonial power attempting to "prepare" the country for independence while instituting development plans and building dams, roads, and airfields. There was not the centripetal effect of a rising anticolonial nationalism releasing the energies that, with independence, have been put to excellent use elsewhere. In many ways, Liberia suffered many of the negative, divisive effects of colonial rule, although not by Europeans, without receiving a share of its benefits. Such, for the majority of the country's inhabitants, has been the legacy of Liberia's particular brand of independence. But for Africa, an ethnically diverse continent divided into artificial political entities, the Liberian experience raises a broader question that every African government has been struggling with, or ignoring at its peril, since independence. This question has to do with the unequal sharing of power, resources, and opportunities among people of different ethnic nationalities and subethnic identities (castes, for example) within the African countries themselves. In many African countries indigenous colonialisms exist today. Thus, the Liberian question is related to the African question of how to effectively mobilize the aspirations of all citizens—including the historically unrepresented and often powerless and exploited groups—in the nation-building enterprise.

SIERRA LEONE

Sierra Leone, once the center of administration for British West African possessions, became a sovereign state in 1961, later than either Ghana or Nigeria. Although it possessed the oldest British-founded municipality in Africa (Freetown), the first modern institution of higher learning in tropical Africa (Fourah Bay College), and a long history of political activity on the part of local people, Sierra Leone was less prepared for independence than either Ghana or Nigeria. Illiteracy was in the vicinity of 95%, per capita income was about one-third that of Ghana, and political difficulties involving the imposition of proportional representation on a long-privileged elite were unsolved. And, like most colonies, Sierra Leone possessed a dual economy that was regionally imbalanced. Since independence, the dual economy has persisted, and growth has not been accompanied by development.

Sierra Leone, named by Portuguese navigators exploring the monsoon coast of West Africa in the sixteenth century, occupies the zone between the southern divide of the Fouta Djallon (Futa Jallon) Mountains and the coast. The descent, from over 6,000 feet (1,830 m) to the embayed coastline, is accomplished by a series of steps representing, in the upper regions, cyclic erosion surfaces and, in the lower areas, raised marine terraces. Between the southern boundary of Senegal and the western border of Côte d'Ivoire, the West African coastline trends northwest to southeast, thus lying directly in the path of the moisture-bearing air masses. The resultant rainfall is high but is concentrated during the pronounced wet season; Freetown receives an annual average of 157 inches (3,988 mm). During the dry season the dusty harmattan is very much in evidence. This, then, is the prime example of the monsoon coast of West Africa.

Considering its small size (27,700 square miles; 71,740 km^2) and compact shape, Sierra Leone boasts remarkable environmental diversification and a great range of raw materials. In some cases, exploitation has yet to begin. The fishing waters off the country's shores may be the best of West Africa. The coastal regions, with their swampy lowlands, support rice and could produce much more of this commodity than they do. Cocoa can be grown in the southeast, ginger in the south–central region, and peanuts in the north. While the small remnant of hot, wet forest in the southwest is suitable mainly for extractive activities, the north can sustain a cattle industry. Cassava and kola nuts are grown, and coffee (robusta) has joined the list of exports. There are large reserves of medium-grade iron ore at Marampa and in the Sula Mountains, and many river gravels of the east are richly laden with diamonds. Extensive bauxite deposits exist in the Mokanji Hills and at Port Loko, while the titanium ore rutile is found in the swamplands and estuaries along the south coast (Figure 12-3).

Yet Sierra Leone is a poor country, consistently the lowest ranking on the Human Development Index (HDI) since its inception in 1990. The per capita income in purchasing power parity was $490 in 2000, and there is gross inequality in the distribution of income. In 2000 the richest 10% of the population possessed 43.6% of the income, while the richest 20% controlled 63.4% (HDI 2002). The figures for the countries with the highest HDI scores in the world were 23.9 and 39.1%, respectively. Sixty-eight percent of the population lived below the poverty line in 2000, only 28% had access to safe water, and 41% were undernourished. There have been some modest improvements. The infant mortality rate had declined from 206 per thousand to 180 per thousand by 2000. Almost a third of all children die before they reach the age of 5 years, although the mortality rate for this age group has declined since 1970 (HDI 2002). In the section that follows, we will take a look at the history and geography of Sierra Leone to try to piece together an explanation for the present level of underdevelopment.

The People: Kings and Creoles

The oldest numerically important residents of Sierra Leone appear to be the Temne people (see Figure 1a, p. 281), who had settled the coastal regions of the country by the fifteenth century and with whom Europeans made contact soon afterward. Seeking fresh water and other necessities, the Portuguese and other Europeans put in at what is now Freetown, West Africa's best natural harbor, which soon became a trading center for slaves and ivory as well as a supply station for ships. No continuous European rule was established, and the resident traders came under the local African rulers. Meanwhile, the Mende people were settling in the north, and Sierra Leone did not escape the effects of the Fulani and Mandingo holy war, which spread Islam from the north, beginning in the early eighteenth century.

At various times there have been between 100 and 150 chiefdoms in the area of Sierra Leone. When Granville Sharp, the British opponent of slavery, succeeded in establishing in Sierra Leone a small settlement for freed slaves in 1787, one of these rulers gave his permission for the use of a section of his land for this purpose. "Province of Freedom," as the settlement was called, did not survive long, for the donor's royal successor destroyed it. But in 1791 it was revived under the auspices of the Sierra Leone Company, sponsored by British abolitionists, and this time the effort had permanent success. The settlement was rebuilt and named Freetown, and freed slaves were brought there, not only from captured slave vessels but also from the Americas. Some Africans had gained their liberty by joining the British forces during the American Revolution, and others had been emancipated in the West Indies.

Britain outlawed the slave trade in 1807, and in the following year the government took charge of the settlement as a colony, continuing the policy of making it a homeland for emancipated slaves. Freetown became an important base from which operations against the slave trade were carried out. During the half-century after 1808, an estimated 50,000 freed slaves were thus brought to Sierra Leone.

The first groups of former slaves, who had come from the Americas, had in common some knowledge of the English language, and many were Christians. Those who arrived after 1807, never having made the Atlantic crossing, not having been exposed to Anglo-America, and coming from various parts of Africa, had little or nothing in common with the first groups. As a result, an extremely heterogeneous community developed in Freetown and its immediate vicinity. The government then embarked upon a policy designed to provide the residents with a common language, religion, and culture. Schools were built, missionaries sent to the colony, and much was achieved in a remarkably short time. It was during this period, in 1827, that Fourah Bay College was established. The new immigrants responded well, were active and very successful as traders, and began to settle in other parts of the general area of Sierra Leone.

This first generation of settlers in Sierra Leone came to be known as "Creoles," and they soon attained a privileged position in the embryo country, carrying on most of the trade, enjoying educational opportunities not available to the vast majority of the people in the hinterland of Freetown, sharing a common language that has come to be known as "Krio," and enjoying a position of influence in the administration of their part of the region. Indeed, they were responsible for the first efforts to expand the colony's influence into the interior. They failed initially to interest the British in accepting the responsibilities inherent in this move, but success was achieved later.

Colony and Protectorate

What led to the extension of British power into the hinterland was not the Creoles' desires, but French activities in West Africa and the realization in London that Freetown's strategic qualities would be endangered if French encroachment went unchecked. Hence, between 1890 and 1896, treaties were signed with chiefs in the interior, and in the latter year, boundaries were defined as agreed upon by the French and Liberian governments. While the coastal region remained a colony, the interior became a protectorate, and the administrative division of Sierra Leone had become a fact. In Sierra Leone as elsewhere, the protectorate status of the bulk of the country required the introduction of the principle of indirect rule, and so the chiefs remained in power, traditional ways were encouraged, and few modern amenities were introduced. But the Creoles, who had been involved in government in Sierra Leone ever since the colony obtained a legislative council in 1863, found their influence waning rather than increasing after the establishment of the protectorate. British district commissioners and administrators governed both the protectorate and the colony, and Creoles were gradually removed from those offices they held.

In 1924, when a new constitution was drawn up, the legislative council consisted of a few elected Creoles and, for the first time, protectorate representation in the persons of nominated chiefs. This arrangement foreshadowed independence and eventual proportional representation for all citizens of Sierra Leone, which inevitably meant the loss of privileges and the end of cultural isolation for conservative Creoles. A unitary constitution introduced in 1951 gave political power to candidates who won a majority of votes. A decade later independence was granted, and a predominantly protectorate party won the elections and formed the government of the new state. In 1975 Sierra Leone became a one-party state.

Throughout the colonial era, British policies toward the development of Sierra Leone were unimaginative and largely uncoordinated. Little money was spent on basic education, improving staple food production, or large-scale development projects and communications. Moreover, the Western Area (the former colony) usually received a disproportionate share of the limited development funds. While cash crops such as palm kernels, cocoa, piassava, and coffee became important exports, little effort was made to ensure efficient production, marketing, and distribution. A narrow-gauge railway was built from Freetown east through Bo (the protectorate's former capital), and Kenema to Pendembu by 1908, and a branch line reached the northern trading center of Makeni by 1916 (Figure 12-4). These cut across traditional sea-bound porterage routes and diverted much of the trade, especially palm produce, to Freetown. Roads complemented the railway until the late 1920s, when feeder roads were linked. But the line was poorly maintained, costly to operate, and unsuitable for transporting modern consumer goods, and so was phased out between 1971 and 1974. Its closure is of limited significance, since the railway served few areas for which adequate road transport alternatives are not now available (Williams and Hayward 1973). In 1933, a special-purpose private railway was built from Pepel to the iron ore mine at Marampa.

The lack of commitment that characterized Britain's administration of Sierra Leone is well illustrated by the diamond industry. In 1935 a British company was granted a 99-year concession over the entire country for diamond mining, mainly alluvial operations. A wave of illicit diamond digging and smuggling took place, especially during and immediately after the lean years of the Second World War. Sierra Leone's diamonds are over 50% gem stones, and the country lost perhaps one-half of the returns for its most valuable product. Eventually, in 1955, the government stepped in, limiting the company to 450 square miles (1165 km^2) in the Yengema area, and established a monopoly on the purchase of

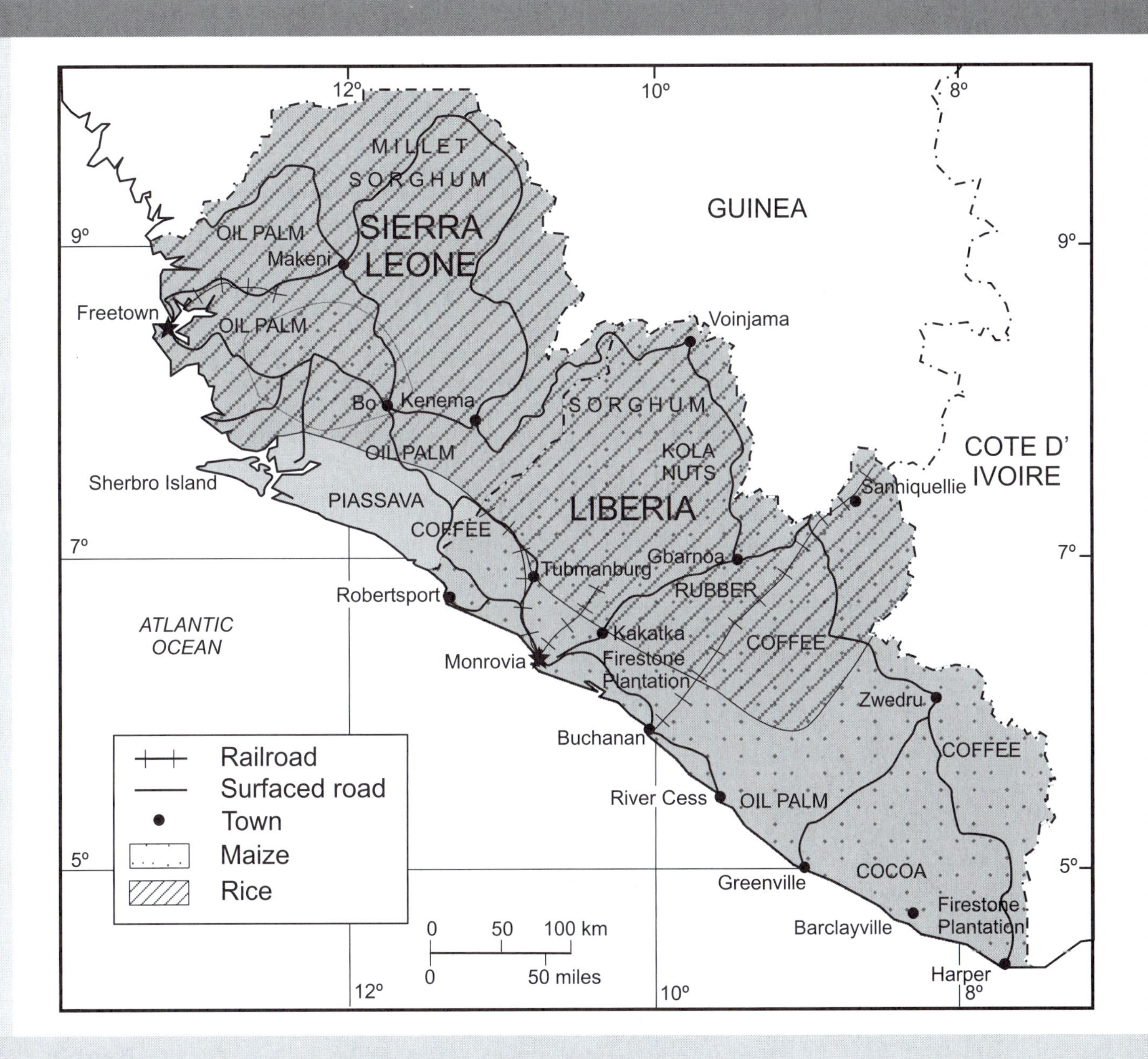

FIGURE 12-4 Agriculture and infrastructure in Liberia and Sierra Leone.

gems. But incalculable losses had already been sustained. Not only had revenues been lost, but much of the land's agricultural potential had been destroyed, and farm outputs declined as labor was diverted to mining areas (Forde 1974).

Development After Independence

When Sierra Leone achieved independence, it inherited a markedly dualistic economy in which diamonds dominated the export sector and subsistence agriculture the domestic. The economy was largely in the hands of Lebanese immigrants, the Creole minority, and a few British expatriates. The Lebanese controlled a disproportionate amount of the country's wholesale trade and most of the retail trade, being closely associated with much of the country's agricultural exports and diamond marketing (Stanley 1970b). Furthermore, there existed a strong regional imbalance in the economy: primacy prevailed in Freetown, while poverty and inequality persisted in the provinces, especially the Northern Province (Riddell 1970). Other disadvantages included a politically inexperienced people and a paternal system of government that supported the authority of the chiefs. These difficulties contributed to a decade of political instability that witnessed military coups, countercoups, the presence of Guinean troops to support the government, and the introduction of a one-party legislature in 1973.

The staple food crops of Sierra Leone are rice and cassava (manioc), but a wide variety of other crops are grown—for example, millet and sorghum in the drier north and maize in the center. The leading agricultural exports have been coffee, cocoa beans, palm kernels, kola nuts, and piassava (Figure 12-4). Most agricultural output is in subsistence production. Agriculture, disrupted by civil war in the 1990s, may be recovering. The FAO estimates that agricultural output dropped by 50% in the mid 1990s as a result of the RUF insurgency (FAO 2003) and fell further since then but has leveled off over the last couple of years. Rice production remains below the 1960s level despite a trebling of population (Figure 12-5).

Before the Marampa iron ore mine closed in late 1975, mining accounted for 85% by value of the country's total exports, yet for only 16% of the GDP, and employed an even smaller percentage of the workforce. Yet Sierra Leone is rich in minerals. It possesses titanium ores, bauxite, gold, iron ore and, of course, diamonds. During the 11-year civil war, formal mining came to a standstill and the economy depended heavily on foreign trade. Diamonds were the main commodity, dictating not only the level of government

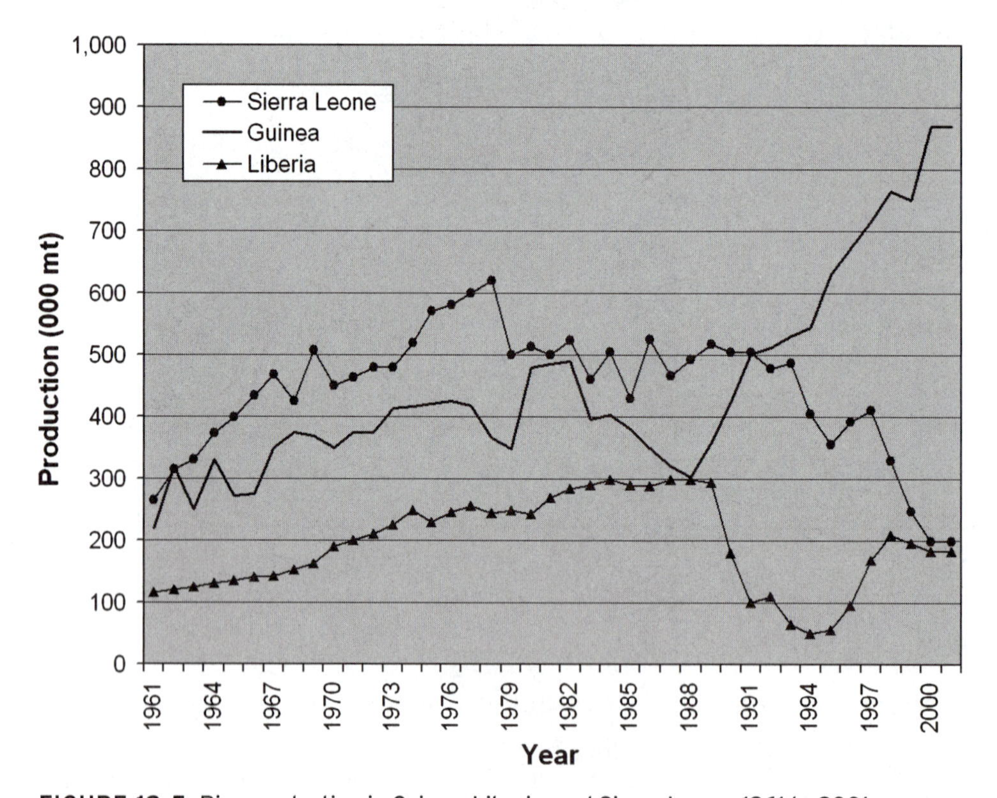

FIGURE 12-5 Rice production in Guinea, Liberia, and Sierra Leone, 1961 to 2001.

revenue but the availability of foreign exchange for importing capital goods for development. Diamond-producing areas that fell under the control of rebels were soon illegally exporting the precious stones for the purchase of weapons. As in Liberia, manufacturing in Sierra Leone is poorly developed and emphasizes the processing of local agricultural and forestry resources.

There are, and have long been, marked disparities in the levels of economic, social, and political–administrative development between the coastal region, especially Freetown, and the interior. Southern Sierra Leone is somewhat more developed than the latosol-covered, repeatedly drought-stricken interior north. This is partly due to the pedological and climatic factors, but the south has also benefited from the magnificent Freetown harbor and the city's contact with the world. These differences had long found expression in the respective political situations of the small colony of Sierra Leone, on the coast, and the large protectorate, the almost untouched, traditional interior. As in Liberia, this separation has its roots in historicogeographical developments that go far to explain the socioeconomic condition of the present-day state.

Beginning just 2 years after Charles Taylor's march on Liberia, the civil war between the Sierra Leonian government and the Revolutionary United Front lasted 3 years longer than the Liberian civil war. As was the case in Liberia, several cease-fires and peace agreements were inked but abrogated by continued disagreement and fighting. The United Nations was humiliated in 2000 when RUF fighters took 500 UN peacekeepers hostage. It was the intervention of British commandos alongside the UN forces that year, that ultimately drove out RUF. The war was officially declared over in January 2002, with the disarming of remaining RUF forces. In May 2002 elections were held, and a new government took power but threats to peace continued. Foday Sankoh, the leader of RUF, was captured in May 2000 and died in July 2003 in captivity.

The recovery of Sierra Leone depends on peace, stability, and foreign assistance. Sierra Leone needs processing industries in the smaller towns and rural areas. These include the processing of sugar, lumber, fruit, and cotton, the objectives being to promote local growth, contain the primacy of Freetown, and reduce the migration of labor to Freetown. But Sierra Leone does not offer a promising field for major modern industries because its internal market is small and the population's purchasing power extremely limited (Box 12-1). The emphasis in development should be placed on industries based on local raw materials for which adequate labor is available. Foreign investment is encouraged by the government, which also will assist private initiative for the establishment of industries in the form of cheap credit, export incentives, arrangements for repatriating capital and profits, and measures for encouraging reinvestment of profits. The government is also committed to creating a wider regional market for industry through the Mano River Union with Liberia and ECOWAS. While these plans are ambitious and take cognizance of existing spatial and structural attributes of the economy, they do not materially alter the fundamental patterns and problems. Sierra Leone is recovering from 11 years of anarchy. It retains the outline of its prewar dual economy and regional imbalances. If it is to regain lost ground, it will have to adopt an export-led path to development, but progress in the social and economic spheres will be slow and costly.

THE COMING ANARCHY?

Today, it seems that both Sierra Leone and Liberia are emerging from a chaotic period in their histories. Yet some see another grim future for Sierra Leone and Liberia that is similar to their recent past of violence, terror, instability, and political fragmentation and to their older past of exclusion, elitism, and inequality. In a widely read and controversial essay published in 1994 in the *Atlantic Monthly*, and expanded in book form in 2001, Robert Kaplan sketched out a chilling vision of the future that he called "the coming anarchy." He referred specifically to West Africa and in particular to Sierra Leone and Liberia to exemplify his vision.

> West Africa is becoming the symbol of worldwide demographic, environmental, and societal stress, in which criminal anarchy emerges as the real "strategic" danger. Disease, overpopulation, unprovoked crime, scarcity of resources, refugee migrations, the increasing erosion of nation-states and international borders, and the empowerment of private armies, security firms, and international drug cartels are now most tellingly demonstrated through a West African prism. West Africa provides an appropriate introduction to the issues, often extremely unpleasant to discuss, that

BOX 12-1

PERIODIC MARKETS

As noted earlier (see pages 191 and 252) a common phenomenon to much of tropical Africa is the rural periodic market. As the term implies, this is a market that operates at intervals, usually in or between villages. The intervals vary from place to place according to the amount of goods produced for sale, the size of the consumer population, tradition (perhaps influenced by religious principles), the density and nature of transport facilities, distance from other markets (especially urban centers), and other factors. In much of northern Nigeria, southern Ghana, Côte d'Ivoire, Liberia, Guinea, and the interior states of West Africa, these markets are held every seventh day. Elsewhere they may meet every second, third, fourth, or eighth day or at another interval. In Sierra Leone, periodic markets are not as common as elsewhere in West Africa, and they are relatively recent phenomena. In southern Nigeria a four-day cycle is common among the Yoruba, but there is no clear spatial pattern of market frequency for Africa as a whole.

Periodic markets form a sort of interlocking network of exchange places that serves even areas that lack roads; as each market in the network or "ring" gets its turn, it will be near enough to one section of the area so that people who live in the vicinity can walk to it carrying whatever they have to sell or trade. In this way, small amounts of produce filter through the market chain to a larger regional market, where shipments are collected for interregional or perhaps even international trade. Itinerant traders travel from one market to another in a market "ring," taking goods and customers to different markets each day. What is traded depends on where the market is located. In the savanna zone, sorghum, millet, groundnuts, and shea butter are common food items; in the forest zone, yams, cassava, and palm oil are commonly sold.

In general, the quantities traded are small and values are low: a bowl of sorghum, a bundle of firewood, a gallon drum of kerosene, a few canned goods, a packet of cigarettes, cooking utensils, and clothing. The marketplaces are arenas not only of trade, but also for social and cultural exchange. Here the latest news can be heard, professional letter writers attend to important business, clothing is fashioned, marriage contracts can be made, and the people barter, discuss, argue, gossip, eat, and drink. At the end of the day, the marketplace is deserted; three, four, seven days later, the market reopens.

will soon confront our civilization. Sierra Leone is a microcosm of what is occurring, albeit in a more tempered and gradual manner, throughout West Africa and much of the underdeveloped world: the withering away of central governments, the rise of tribal and regional domains, the unchecked spread of disease, and the growing pervasiveness of war. West Africa is reverting to the Africa of the Victorian atlas. It consists now of a series of coastal trading posts, such as Freetown and Conakry, and an interior that, owing to violence, volatility, and disease, is again becoming, as Graham Greene once observed, "blank" and "unexplored." (Kaplan 1994: 7–9)

Peace in Sierra Leone and Liberia are encouraging signs that may promise a better future for the peoples of these countries and the region. However, unless outsiders take an interest in ensuring that these states do not descend into chaos once again, the fundamental truth of Kaplan's nightmare scenario may be inescapable. Of primary importance in intervention are humanitarian issues, but there are other, more global ramifications of failed states that must be addressed. Failed states are breeding grounds for international crime and terrorism, as the United States became painfully aware in 2001. Since the attacks on the World Trade Center in New York and the Pentagon in Washington, the United States and its allies have gone on the offensive against international criminals and terrorists who have sought shelter in failed and rogue states. Weak states in West Africa already provide key transit and money-laundering points for the international drug trade to Europe and the Americas. It is imperative that the international community broaden, by force if necessary, its already active role in strengthening peace, democracy, and development throughout the region.

BIBLIOGRAPHY

Beleky, L. P. 1973. The development of Liberia. *Journal of Modern African Studies*, 11(1): 43–60.

Clower, R. W., et al. 1966. *Growth Without Development: An Economic Survey of Liberia*. Evanston, Ill.: Northwestern University Press.

Coakley, G. J. 2004. *The Mineral Industry of Liberia*. U.S. Geological Survey Mineral Report, Reston, Va.: USGS.

Economist. 2003. Liberia, goodbye to all that? *Economist*, August 16, 2003: 39–40.

Ellis, S. 1999. *The Mask of Anarchy: The Destruction of Liberia and the Religious Dimension of an African Civil War*. New York: New York University Press.

Kaplan, R. 1994. The coming anarchy: Shattering the dreams of the post-cold war. *Atlantic Monthly*, 273(2): 44–76.

———. 2001. *The Coming Anarchy: Shattering the Dreams of the Post-Cold War*. New York: Vintage.

Pike, J. 2003. Liberian Ship Registry. www.GlobalSecurity.org.

Riddell, J. B. 1970. *The Spatial Dynamics of Modernization in Sierra Leone*. Evanston, Ill.: Northwestern University Press.

Stanley, W. R. 1968. The cost of road transport in Liberia: A case study of the independent rubber farmers. *Journal of Developing Areas*, 2(4): 495–510.

Stanley, W. R. 1970a. Transport expansion in Liberia. *Geographical Review*, 60(4): 529–547.

Stanley, W. R. 1970b. The Lebanese in Sierra Leone: Entrepreneurs extraordinary. *African Urban Notes*, 5(2): 159–174.

Szczesniak, P. A. 2001. *The Mineral Industries of Côte d'Ivoire, Guinea, Liberia, and Sierra Leone—2000*. U.S. Geological Survey Mineral Report. Reston, Va.: USGS. http://minerals.usgs.gov/minerals/pubs/country/

UNDP. 2002. Human Development Report 2002: Deepening Democracy in a fragmented world. United Nations Development Program. Oxford: Oxford University Press.

UNHCR. 2003. *Refugees and Others of Concern to UNHCR—2002 Statistical Overview*. Geneva, Switzerland: United Nations High Commission for Refugees. www.unhcr.ch.

Von Gnielinski, S., ed. 1972. *Liberia in Maps*. New York: Africana Publishing Corporation.

Williams, G. J. and D. F. Hayward. 1973. The changing transportation patterns of Sierra Leone. *Scottish Geographical Magazine*, 89: 107–118.

CHAPTER 13

A French Legacy

Côte d'Ivoire, Guinea, Togo, and Benin

The geography of former French West Africa may be treated in several ways. For our purposes, this vast francophone area can be divided into two separate units: an interior tier of contiguous, sparsely populated, and mainly landlocked states that border the Sahara, and four smaller states (Côte d'Ivoire, Guinea, Togo, and Benin) that are part of the humid Guinea Coast, where French interests first focused and strong economic dualism still prevails. The two regions are linked by history (both colonial and indigenous), and their economies are, to some extent, interdependent. Literacy in French has increased over the last 25 years from around 10% of the population to 44%. Available figures for women from the U.S. Census International Database indicate that a third of the female population is literate in Côte d'Ivoire, Guinea, and Togo. There are no data available for Benin. Few would deny that the intensity of the cultural and political impact made by France justifies the continued application of the term "francophone," even after independence. The governing elite is French speaking, the political machinery is based on French example, and virtually every modern institution in each country is based on French models. Unlike the Islamicizing northern tier of West African countries, which moved the weekly day of rest to Friday to be in conformity with the Islamic week, the countries of the southern tier are predominantly Roman Catholic and still celebrate Sunday as the day of rest. French feeling prevails because the territories participate with France in several supranational organizations and French technical assistance is important in the region. France's role in West African development has differed markedly from place to place, and its role in contemporary francophone West Africa is changing. This chapter examines the French legacy of Côte d'Ivoire, Guinea, Togo, and Benin (formerly Dahomey).

THE FEDERATION ESTABLISHED

French lines of penetration into the interior of West Africa were to some extent controlled by the consolidation of British, German, and other interests along the coast. Eventually, the French obtained five corridors between the coast and the inland areas: Dahomey (Benin) between German Togoland and British Nigeria, Côte d'Ivoire between British Gold Coast (Ghana) and independent Liberia, Guinea between Liberia and Portuguese Guinea (Guinea-Bissau), and the stretches on either side of British-held Gambia, now combined in the Republic of Senegal.

The first decree with the aim of establishing a central government for French West African possessions was issued in 1895. At this time, Senegal, Guinea, Soudan (Mali), and Côte d'Ivoire were placed under the jurisdiction of the governor of Senegal, where France's oldest West African colonial settlement was located. Dahomey was added in 1899, but there was much competition between the administrations of the various individual colonies, preventing the introduction of an all-encompassing budget, a step that Paris regarded as an essential element in French West African unity. The system of administration that was to survive until the late 1950s was established by a decree issued in 1904, by which time central control had become more effective. Dakar became the seat of the governor general of the federation, and as such the capital of the vast French West African region.

The 1904 decree by no means ended the French campaign for the consolidation of its West African empire. Effective occupation was not complete in Niger until 1906, in Mauritania until 1910, and in Côte d'Ivoire until 1914. Even after World War I, several changes took place. Upper Volta (Burkina Faso) was not created until 1920, when it was excised from parts of Niger, Côte d'Ivoire, and Soudan; it was dismembered in 1932 and recreated in 1947. And in 1922, although not actually a part of the federation, Togo became a League of Nations mandate, the eastern part of which was under French administration.

France's main interest in West Africa was the exploitation of the federation's resources, not the welfare of its people. French interest first focused on the coast, but later extended into the drier interior as roads and railways were built and new opportunities were foreseen. While imports and exports were in the hands of a few large French corporate monopolies like SCOA (Société Commerciale de l'Ouest Africain) and CFAO (Compagnie Française de l'Afrique Occidentale), the administration stimulated economic activity for its own profit in various ways. Taxation was undoubtedly the most important, for it encouraged the production of cash crops for export and provided some revenues for extending the infrastructure (Hopkins 1973).

Coffee, cocoa, and palm oil became the main products of the forest belt, while cotton and groundnuts predominated in the interior. A program of compulsory crop cultivation was introduced, and where cash crops could not be grown, the local farmers were forced to seasonally migrate to areas where such production was possible. Thus, thousands of Mossi and Bobo from Burkina Faso migrated south to Côte d'Ivoire's coffee and cocoa plantations, while the Bambara and others went west to the groundnut plantations of Senegal. Forced labor was also introduced, whereby Africans worked on public works projects like roads and railways, and, if required, on European plantations.

Although marketing was not restricted to crops introduced by Europeans and virtually every African crop was marketed, African ownership in the money economy was minimal, and almost no African entrepreneurs existed before World War II. In precolonial days, Africans controlled the trade between European exporter and African peasant producer. During the colonial era, however, Lebanese merchandisers, content with lower profit margins, took control. It was not until 1946 that France made much effort in developing the basic infrastructure for the federation. Then, considerable expenditures were made on roads, railways, ports, agricultural schemes, water supplies, schools, and hospitals. Even so, it was France, not the federation, that benefited most. Roads and railways linked the ports with their export-oriented hinterlands such as the Sansanding Project (l'Office du Niger) of central Mali, and the Richard Toll Scheme on the lower Senegal River. And the majority of schools and hospitals were situated either in or close to these islands of French exploitation.

Although never able to master the internal economy (Roberts 1996), France controlled the federation's external economy. It provided its territories with manufactured goods, technical assistance, and financial aid in exchange for raw materials and semiprocessed goods. Very little intraterritorial trade was conducted other than the distribution of consumer goods from Dakar and, to a lesser extent, from Abidjan. Trade between the federation and other African territories was almost nonexistent.

Economic policy dovetailed with various social and political policies, those of "assimilation" first, and later "association." Association in turn led in 1958 to the establishment of autonomous states within the French Communauté: European France and its overseas territories. Complete political independence followed in 1960. Only Guinea, whose anti-French sentiment was strongest, chose independence outside the Community in 1958. The other territories chose autonomy buttressed by continuing French economic aid. They gained their independence in 1960.

In contemporary West Africa, seven francophone countries and one lusophone country are linked economically in the Communauté Financière Africaïne (CFA), also known as the Franc Zone, which embraces all countries that meet two requirements: their currencies were linked with the former French franc (and now euro) at a fixed rate of exchange, and they agree to hold their currency reserves mainly in the form of euros and to effect their exchange on the Paris market. Each country has its own central issuing bank, and its currency is freely convertible into euros. All the former French West African territories except Guinea (Guinean franc) and Mauritania (ouguiya) are members. Guinea-Bissau, formerly a Portuguese possession, later joined the CFA. The West African CFA countries, Benin, Burkina Faso, Côte d'Ivoire, Guinea-Bissau, Mali, Niger, Senegal, and Togo, also participate in the West African Economic and Monetary Union, known by its French acronym, UEMOA, which aims to promote the harmonization of tariffs, to encourage trade and rational development strategies, to promote economic integration and the free movement of people, and to coordinate international assistance and cooperation. France's ties with CFA countries involve not only monetary arrangements but also comprehensive French assistance in the form of budget support, foreign aid, technical assistance, and subsidies on commodity exports. The 14 CFA countries are grouped into West African and central African regions and use a slightly different currency that is freely exchangeable.

CÔTE D'IVOIRE: TROUBLE IN PARADISE

Côte d'Ivoire is by far the richest and economically the most diversified territory in former French West Africa. Its production, mainly agricultural, represents 40% of the GDP of the eight members of UEMOA. The geography of agriculture in Côte d'Ivoire typifies the lower tier of West Africa. In the more humid south, oil palm, cocoa, coffee, bananas, and pineapple are grown along with yams, cassava, plantain, rice, corn (maize), and most recently soybeans. In the drier north, millet and sorghum are grown, mainly for food, and cotton and tobacco are grown for cash. Average farm size is larger in the more commercially oriented south–15 acres (6 ha) on average, compared with 8.9 acres (3.6 ha) in the north. The most important export crops are coffee, cocoa, palm oil, lumber, and rubber (Figure 13-1). Côte d'Ivoire is the leading coffee producer in Africa (distantly followed by Kenya), and is the world's third largest producer after Brazil and Colombia. First grown in 1891, coffee now covers almost 3 million acres (1.17 million ha) of the southern forest zone, and almost all the acreage is African owned. Most of the coffee (robusta) is marketed in the United States and Europe as instant coffee. The area devoted to coffee increased steadily since independence but has declined since the early 1990s, especially since 2001 when strife between northerners and southerners began.

The second most important food export is cocoa. Almost 5.5 million acres (2.25 million ha) is cultivated in the south and southeast, where the plant was first introduced from the Gold Coast (Ghana) in 1895. As with coffee, cocoa cultivation has been growing since independence. It is instructive to note that in the early years after independence it was Ghana, not Côte d'Ivoire, that was the leading cocoa producer in Africa. But Ghanaian production for the world market faltered owing to economic and political instability, mismanagement, and the socialist agenda of Ghana's first president, Kwame Nkrumah. World supply shifted across the border to stable Côte d'Ivoire. The dramatic change in cocoa fortunes is best exemplified in figures. In 1961 Ghana produced 415,000 metric tons of cocoa, while Côte d'Ivoire produced only 80,000. Forty years later, the two countries' positions had almost reversed, Côte d'Ivoire having aggressively expanded cocoa cultivation to meet world demand—and build the Ivorian economy. Ghana's misfortunes hit a low point in 1984, with only 166,000 metric tons of cocoa produced. In 2002 Côte d'Ivoire produced over a million metric tons, while Ghana produced only a third as much.

Coffee and cocoa constitute the majority of the exports of Côte d'Ivoire. In the early 1970s both coffee and cocoa had declined in importance, and timber and timber products headed the export list. The country depended on a system of price supports for coffee instituted by France and Ivorian farmers frequently received twice the world market price. This was the situation that led the leaders of the Côte d'Ivoire to desire continued close association with France when the question of independence was under discussion. This system of price supports no longer exists, and from 1997 to 2001 Côte d'Ivoire exported $2.5 billion worth of coffee, $1.3 billion of cocoa, and only $220 million worth of timber, aided by the devaluation of the CFA franc in 1994 and higher worldwide demand for coffee and cocoa products. Agriculture and primary products were almost the exclusive exports in the early years after independence; today, however, manufactured goods account for 15% of exports, something which will give the economy some foreign exchange income stability and should be encouraged (HDI 2002).

In recognition of the vulnerability of the country's economy, depending as it does on crops whose world price varies greatly, agricultural diversification has been encouraged. Pineapples, bananas, and cotton

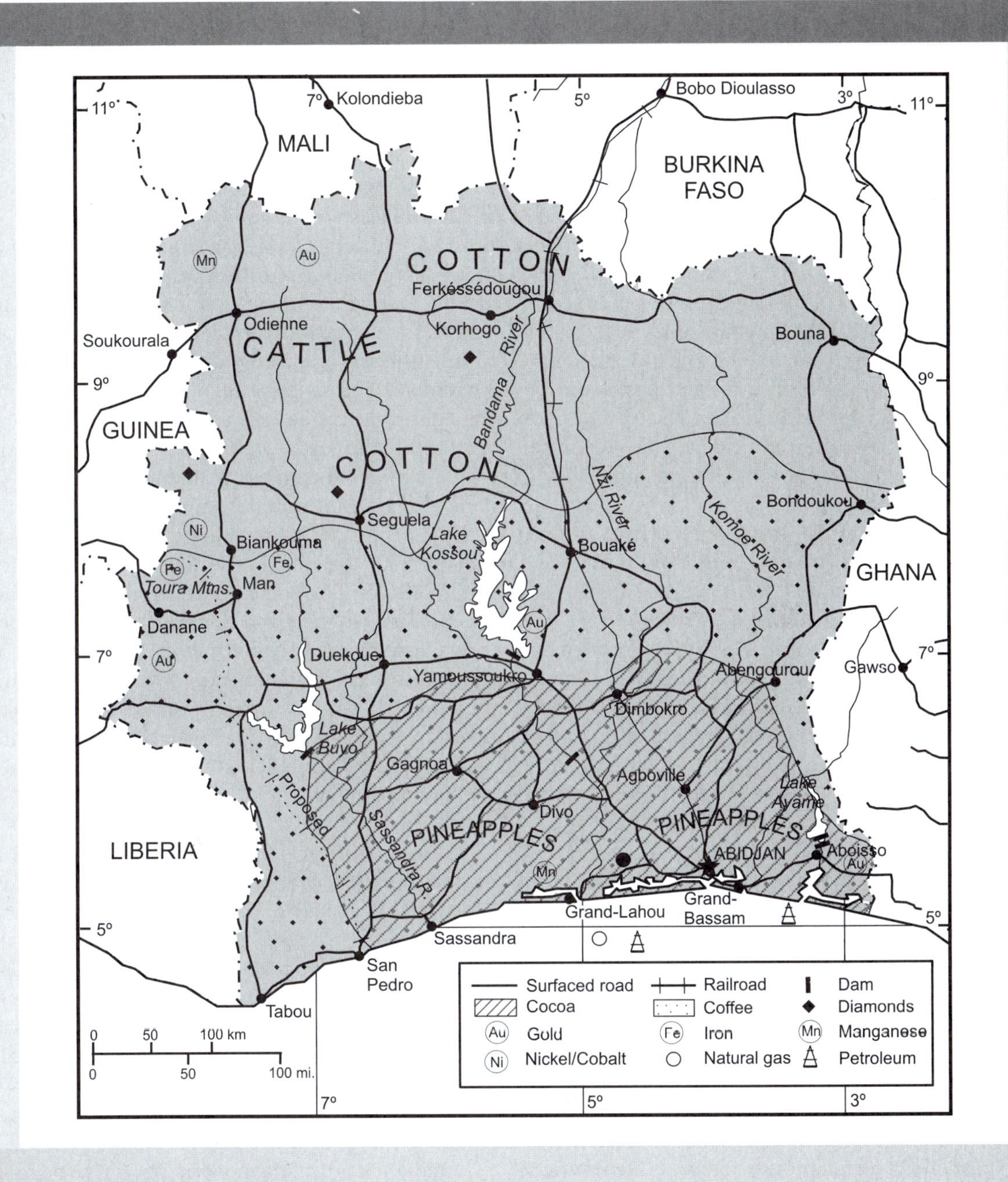

FIGURE 13-1 Economic map of Côte d'Ivoire.

have shown considerable promise and are now the principal exports, after coffee and cocoa. Sugar production is being expanded in the northern regions around Ferkéssédougou to meet domestic requirements. Livestock production is foremost among the agricultural priorities, with the aim of reducing the country's dependence on the highly irregular flow of cattle from Burkina Faso, Mali, and Niger. New areas are being opened up in the north as trypanosomiasis and other diseases are brought under control.

Offshore oil was discovered in Ivorian territory during the 1970s, and great hopes were placed in petroleum as the engine of modernization and development. Today Côte d'Ivoire is a petroleum and natural gas producer, but output has been below initial expectations. There are eight petroleum companies involved in the exploration, production, distribution, and marketing of petroleum products and natural gas. About one billion cubic meters of natural gases was produced a year from 1997 to 2001 (Yager et al. 2002) and about 1.5 billion cubic meters from 2002 to 2004 (Bermúdez-Lugo 2004). About 5.5 million barrels of petroleum a year were produced from 1995 to 2004. The oil refinery at Abidjan is the leading enterprise in Côte d'Ivoire, and the third largest enterprise in francophone West Africa. Its own oil production would hardly keep it at capacity, but it refines oil from Nigeria and other African oil producers in addition to its own. It may be kept even busier with output from offshore fields recently discovered near São Tomé and Principe.

Although most of the electricity in the country is of hydroelectric origin, several gas-fired power plants have been built and contribute significantly to the growing national electric grid. The government is considering the feasibility of constructing an oil pipeline to Ghana. If the project is completed, it will be an important step in developing closer relations with nonfrancophone neighbors.

Côte d'Ivoire, although no South Africa or Zimbabwe, does have other valuable natural resources. There are significant deposits of mainly alluvial gold. Deposits of nickel and cobalt are mined at Touba, near Biankouma. Large iron deposits exist in the Touba Mountains. These are linked to the large iron deposits in the Nimba Mountains along the Liberia–Guinea border. The iron ore, 32.7% iron, is pelletized to raise the iron content to 70% and shipped overseas. There are economically recoverable deposits of bauxite, manganese, tantalite, and diamonds, as well (Bermúdez 1999).

Côte d'Ivoire has three broad climate and vegetation regions: lagoon, forest, and savanna. The lagoon region is a low, narrow strip of land and water that extends from the mouth of the Sassandra River in the west to the Ghana border. It comprises a series of low, frequently flooded, sandy islands and sandbars, often wooded with mangrove. The lagoons develop into wide productive estuaries along the Ivorian rivers that flow to the ocean. Most of the tropical rain forest that once grew along the littoral has been cut to make way for agriculture. The forest region covers about one-third of the country, from the southern lagoons north to about 8 degrees latitude, approximately tracing an east–west line from Man to Bondoukou. Elevations rise in the western extremes of the forest region along the Liberian and Guinean borders to a maximum of 4,265 feet (1,300 m) in the Toura Mountains. Average annual rainfall in the forest and lagoon regions is 100 inches (2,500 mm) and falls in two seasons: March to July, the season of heaviest rains, and September to November. The savanna region, the northern half of the country, may be characterized as a broad plateau with rolling hills, low vegetation, and a low density of trees. The region can be considered to be a transition zone along a continuum from the forest region in the south to the drier climate and vegetation complexes that exist to the north. The density of vegetation as well as its composition declines from south to north, except along streams and rivers, where more moisture is available. Trees become sparser with distance from the forest zone, and moisture-loving species are replaced by those that thrive in drier Soudanian conditions. In the far north, the average annual rainfall is 35 inches (900 mm). The months of heaviest rainfall are July and August, and there is a distinct but short dry season.

Côte d'Ivoire has benefited from its coastal location, although access was initially difficult because of the straight coastline, many offshore sandbars, and constant silting of river mouths and artificial harbors by longshore currents. For many years the largest port was Port Bouet, located on a sandbar across the Ebrie Lagoon from Abidjan. In 1950 the Vidri Canal was cut through this bar, and the large sheltered port of Abidjan was developed. The canal was widened in 1975, and new deepwater quays were opened to make Abidjan the second largest port in West Africa after Dakar.

Abidjan is a thriving modern city whose population has grown from a mere 46,000 in 1945 and a modest 185,000 in 1959 to over 700,000 in 1976, 1.8 million in 1988, and 3.3 million in 2002. In addition, Abidjan and the surrounding urbanized area had close to 4 million people in 2002. It has replaced Dakar as the largest manufacturing city in francophone West Africa, and its industries include textiles, chemicals, oil refining, auto assembly, engineering, and food processing. Its markets are expanding both locally and nationally, and its economic hinterland includes the interior states of Burkina Faso and Mali. Abidjan is the coastal terminus of the nation's only railway, a line that runs north through Bouaké and Ferkéssédougou to Ouagadougou, capital of Burkina Faso. While crossing several agroeconomic regions and passing through some densely settled areas, the line operates well below capacity and suffers from increasing competition from truck traffic. It lacks good feeder roads and branch lines, and other than Bouaké it serves no industrial town.

Côte d'Ivoire needs to overcome regional differences in development and attitudes. The cash crops have always come mainly from the south. In most aspects of development—education, social services, political consciousness, and agriculture—the northern savanna has lagged behind the forest south in the same way that northern Ghana, in Anglophone Africa, has lagged behind southern Ghana. Bouaké, population 550,000 in 2002, a market center benefiting from its location on the railroad, on a main motor road to Abidjan, and near the forest– savanna transition zone, is the most important urban agglomeration in the interior. Yet the city's small size, in spite of its advantageous position, reflects the limited capacities and needs of its region. There are mills to make rope and twine and a cotton mill.

At the time of independence, Côte d'Ivoire chose to rely on free enterprise and foreign investment to accelerate economic growth, and in common with other francophone territories it adopted a monetary system that ruled out inflationary policies. Since independence, the result has been a phenomenal 25 years of economic growth and development, which has made Côte d'Ivoire a regional economic power. Up until 1985, GDP has grown, in real terms, at an annual rate of 8 to 10%. Growth has been recorded in all sectors of the economy, but especially manufacturing, which has emphasized import substitution. Today, 15% GDP is accounted for by manufacturing, the highest in the former French West African region. In the 1960s, light manufacturing and agricultural processing grew by more than 20% per year, and foreign private investors, mainly French, provided most of the capital and technological know-how, with small-scale, rather than large-scale, prestigious projects receiving priority.

Up until the mid-1980s, when the demand for its agricultural exports began to soften, export earnings since independence increased on average about 12% per year. Exports have diversified, mainly in the agricultural sector, and the country enjoyed a favorable balance of trade, with the main markets being France, the United States, and the EU countries. Trade with adjoining francophone states is small but expanding. The French presence in Côte d'Ivoire is obvious but has declined from a peak of 60,000 at independence to about 30,000 today. Nevertheless, Europeans still hold key positions in industry and provide technical, administrative, and economic expertise. Former president Félix Houphouët-Boigny saw merit in greater "Ivorianization" but was reluctant to force the issue. Abidjan has been called an "island of prosperity," the "Paris of Africa," created for the comfort of Europeans and critics claimed that the country was trading its future prosperity and independence for rapid growth in the short term (O'Connor 1972). Time has shown the wisdom of the early Ivorian export orientation. Rapid industrialization has also brought thousands of Ivorians from rural areas in search of employment in Abidjan, Bouaké, Gagnoa, Abengourou, and other urban centers, but nonagricultural jobs are not increasing as fast as urban demands. Furthermore, because Côte d'Ivoire has had such a relatively robust economy, many Malians, Burkinabe, and others from the north have migrated for employment. Today, these predominantly Muslim groups constitute over 40% of the Ivorian population, contribute significantly to the economy, and comprise the groups most dissatisfied with the dominance of the southern Roman Catholic elite.

Unlike Guinea, which severed its ties with France in 1958, Côte d'Ivoire has long cultivated its French connections and has succeeded in attracting much new investment to the country. The country's political stability was, to a large extent, due to the economic growth that has resulted from this strategy. During his 33-year tenure at the helm of the one-party Ivorian state, Houphouët-Boigny advocated strong links with France and continued support of the Five-Nation Entente, consisting of Senegal, Côte d'Ivoire, Gabon, Chad, and Djibouti, and of OCAM and CEAO, which are called in English the African and Mauritian Common Organization and the Economic Community of West African States, respectively. He also proposed dialogue between black African states and apartheid South Africa. He was vehemently anticommunist, abhorred radical governments such as that of neighboring Guinea under President Sékou Touré, and kept Côte d'Ivoire a passive member of the OAU. President-for-Life Houphouët-Boigny died in 1993 and was succeeded by Henri Konan Bédie, who began to pursue an "Ivorization" focused not specifically on French workers and businesspeople but on African immigrants from neighboring countries. His pursuit of an ethnic agenda and his attempt to exclude immigrants from political power, culminated in civil war in 2002. Côte d'Ivoire began to make the transition to democracy after the death of Houphouët-Boigny, with the first elections held in 1995, the second in 2000, and third in 2005. Bédie was elected to the presidency in 1995, but the election was clouded by the refusal of the government to recognize the candidacy of the popular Alassane Ouattara, a Muslim northerner who had been prime minister under Houphouët-Boigny.

Given its early encouraging economic performance and the strong economic ties with France and the West in general, Côte d'Ivoire appeared to have a brighter future than most West African states. The nationality question inflamed by former President Bédie and his Catholic Baoulé political apparatus showed this not to be the case. In 1999 the long-standing political stability of the country was shattered by a military coup d'état and, despite the return of civilian government under President Laurent Gbagbo in 2000, political uncertainty remained. The country then experienced extended periods of instability, culminating in a rebellion in 2002 that continued until French peacekeepers intervened. Rebel groups controlled the northern half of the country above Bouaké through 2005, although they had officially made peace with the Gbagbo government in July 2003. Spillover rebels from Liberia took over Man and Danané and a strip of land just east of the Liberian border. The country has been under a curfew since the 2002 rebellion, and the free movement of people has been restricted and controlled by numerous military checkpoints. Of five international borders, only the Ghanaian one remained open in 2003. The Pretoria Agreement signed in South Africa in April 2005 laid the groundwork for disarmament and elections. In September 2005, just one month before the elections were to occur, the government cancelled them. If peace can be maintained and opportunity for the have-nots in Côte d'Ivoire increased, prosperity may yet be regained. There is much at stake and so many uncertainties.

At the bottom of the crisis in Côte d'Ivoire is the Pandora's box first opened by President Bédie during the transition to democracy. The recent time of troubles in Côte d'Ivoire has shown the world, and more importantly the Ivorians themselves, that they are not immune from the religious and ethnic nationalist crises that have destroyed many states in Africa; one need only look to Côte d'Ivoire's western neighbors. With hindsight, it now seems that such conflict between the haves and the have-nots was inevitable after the charismatic, nonconfrontational, patrimonial rule of Houphouët-Boigny ended and people were given the vote. Although he was of Baoulé Catholic background himself, throughout his long tenure as president, Houphouët-Boigny minimized the role of ethnicity in Ivorian politics. He maintained a balanced ethnic representation in his political appointments but avoided bringing traditional leaders into the top levels of government. He encouraged the formation of multiethnic interest organizations in urban areas. He developed a patronage system in which he provided education and career opportunities to the best and the brightest young men (and some women) from around the country. The national government developed in such a way that "social relations were ordered more by access to status, prestige, and wealth than by ethnic differences, and for most people the locus of this access was the government . . . [of Félix Houphouët-Boigny]. Wealth and government service became so closely linked that one was taken as a symbol of the other" (Handloff 1988: 3).

Democracy holds danger as well as promise, and Côte d'Ivoire has recently stumbled. To avoid the anarchy of its neighbors, the government of Côte d'Ivoire must work to reestablish the unity of the nation, to expand the franchise to include immigrants, to genuinely develop trust between Muslims of the north and Christians and animists of the south, and to revive the economy, which has been shrinking since 2000.

GUINEA: THE DISTANT RELATIVE

Having been a part of the French West African Federation for over a half-century, Guinea, in 1958, was the only prospective French Community state to refuse further participation in the French framework. This was the result of a variety of political and economic, as well as historical, factors and caused the immediate cessation of French aid (then amounting to some $17 million annually), the removal of many essential facilities, and the immediate departure of over half the European population, numbering more than 7,000 at the time.

Guinea withdrew from the Franc zone and declined membership in all major monetary, trade, and cultural agreements with the francophone group. The country turned first to the Soviet Union, and later to the People's Republic of China, for technical assistance and models of social and economic development. Up until recent times, it has remained ideologically and economically distant from its francophone neighbors and adopted a socialist economy with direct state control of production and consumption in every sector except mining. Mining, the most prosperous sector, was developed by multinational companies in conjunction with the Guinean government.

Guinea is a land of variety, physiographically, ethnically, and economically, and it is a country of considerable potential. There are four agro-ecological zones: Coast, Middle, Upper, and Forest. The physiography of the Middle agro-ecological zone is dominated by the Fouta Djallon Highlands, rising several thousand feet and consisting of a dissected plateau with prominent peaks and deep valleys. The Fouta Djallon forms the central backbone of the country, beyond which the land begins its gentle decline north into the Niger Basin. In the north, elevations of over 5,000 feet (1,525 m) are sustained, and southward, the mountains (here referred to as the Guinea Highlands and mainly crystalline), are shared by Sierra Leone and Liberia, and so are the iron ores they contain as, for instance, in Mount Nimba on the Liberian border (Figure 13-2).

The highlands support cattle rearing and plantation agriculture of coffee and bananas. Maize is grown throughout the Fouta Djallon Highlands. The highlands in general draw a large amount of precipitation from the air rising along their slopes, concentrated especially during July and August but lasting from March to December; soil erosion is a major problem. Conakry receives some 170 inches (4,300 mm) annually, but in the lee of the mountains the totals drop rapidly, and savanna conditions prevail. At Siguiri in the Upper zone, the average annual precipitation is 51 inches (1,300) mm. Cereals are the dominant crop here: millet, fonio, sorghum, corn. Rice is grown along the Niger and in depressions. Sorghum is grown in virtually every agro-ecological zone. The coastal plain, less than 50 miles (80 km) in width, is hot and humid, being low in elevation and wedged between the swampy coast and the sudden slopes of the Fouta Djallon. In spite of its character, the plain is densely populated and supports the cultivation of a variety of crops, including rice, corn, cassava, oil palm, and kola nuts. In the vicinity of Conakry, the coastal plain narrows and some banana cultivation is carried on.

Prior to independence, bananas were the leading export, but the industry received a serious blow when France refused to purchase the harvest after the negative vote of 1958 and many European planters withdrew. Pineapples, coffee, citrus, and palm kernels are all grown commercially. Under socialism, however, Guinea experienced difficulty finding markets and controlling distribution. Some Guinean farmers found it more profitable to smuggle their produce over the borders into Sierra Leone and Liberia than to market it through state channels and at state prices. Rice is the principal food in Guinea, and mangrove swamps along the coast and parts of the Niger River floodplain were converted to rice production in the twentieth century. Except for rice, agriculture has declined in its contribution to the GDP.

While agriculture remains the principal occupation for the vast majority of Guineans, mining has become the country's most important foreign currency earner and the principal contributor to the GDP (Table 13-1). Bauxite and iron ore, the principal mineral resources, could have provided the revenues needed to transform the national economy after independence, but this has not been the case. It is most unlikely that the mining centers themselves will become industrial growth points; that has been a rarity in Africa. Instead, they will remain spatially small, specialized units of production, linked by rail to their coastal ports and their overseas markets. Although the potential is there, few benefits are likely to be diffused from the mines into the local regions.

Gold has long been mined from the Siguiri region to the northeast, and a new mine that opened there in 1998 is now the largest gold mine in Guinea (Figure 13-2). Diamonds, mainly of industrial quality, are mined in the valleys of the southeast; but considerable revenues are lost because of illicit traffic into Sierra Leone and Liberia. Over 13,000 kilograms

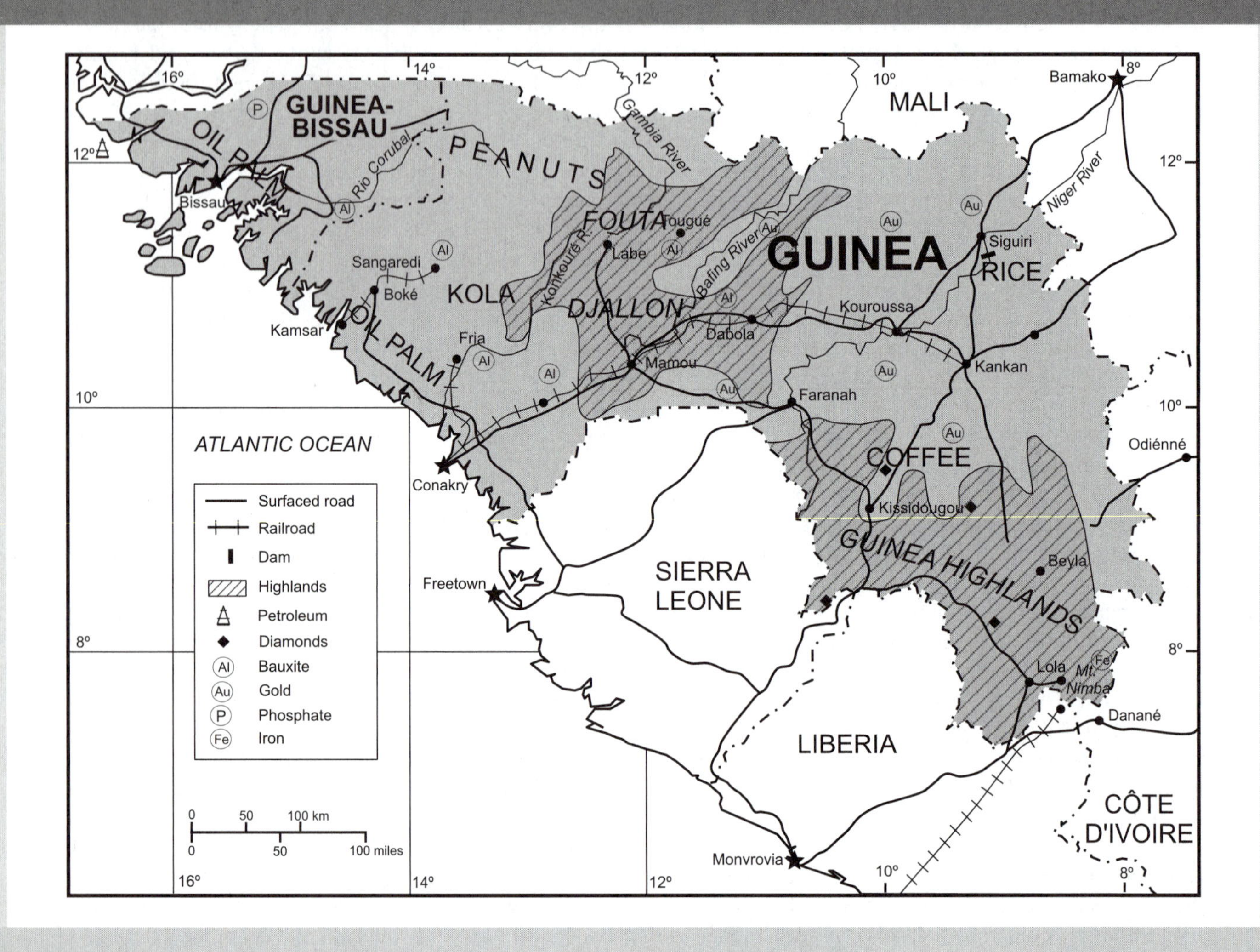

FIGURE 13-2 Economic map of Guinea.

(459,000 ounces) of gold was mined in 1999 (up from 7,860 kg in 1995) as well as 400,000 carets of diamonds, an increase of 55,000 carats over 1995 (Mobbs and Morse 2001). In the southeastern interior, high-grade iron ores have been exploited since 1974. The Nimba deposit, located near the border with Liberia and Côte d'Ivoire, is estimated to contain 6 giga metric tons of 60–68% ore. Until railways are laid to connect with the Conakry–Kankan line, the Nimba ores will continue to be exported through Buchanan, Liberia, the closest outlet to the mines. The ore is contained in four main bodies located in the 15.5 mile (25 km) range.

Guinea possesses over 30% of the world's known bauxite reserves (20 billion tons) and is the world's second largest producer; since 1952 it has been exporting bauxite to France and the United States. It supplies over 40% of the bauxite imported by the United States (Mobbs and Morse 2001). Production approached 3 million tons in 1973 and surpassed 15 million metric tons in 2004 (Bermúdez-Lugo 2004). The Fria (Friguia) deposit, 90 miles (145 km) north of Conakry, was the

TABLE 13-1 COMPARATIVE DATA FOR BENIN, CÔTE D'IVOIRE, GUINEA, AND TOGO 1975 AND 2002 (OR LATEST ESTIMATES).

	Country							
	Benin		Côte d'Ivoire		Guinea		Togo	
Variable	**1975**	**2002**	**1975**	**2002**	**1975**	**2002**	**1975**	**2002**
Population (millions)	3.1	6.6	4.9	16.8	4.4	8.4	2.2	5.3
Percent urban	18	39	20.4	46	19.5	26	13.5	31
Percent labor force in agriculture	84	70	86	68	86	80	78	65
Total literacy rate (%)	24	37.4	22	48.6	37.6	35.9	38.4	57.1
Female literacy rate (%)	NA	23.6	NA	37.2	NA	21.9	NA	42.5
Male literacy rate (%)	NA	52.1	NA	59.5	NA	49.9	NA	72.3
GNP per capita ($U.S.)[1]	110	361	340	637	90	375	160	273
Crude birthrate (births per 000 pop)	49.9	41	45.6	36	46.6	45	50.6	40
Crude death rate (deaths per 000 pop)	23	12	20.6	13	22.9	18	23.5	11
Infant mortality rate (per 1000 pop)	185	85	164	95	216	119	179	80
Human Development Index (2000)	28.8	42	36.9	42.8	NA	41.4	39.4	49.3
Percent of population living in poverty (1987-2000)	NA	33	NA	36.8	NA	40	NA	32.3
Major exports	Oil palm, cotton	Cotton, corn, cassava	Timber, coffee, cocoa	Cocoa, coffee, timber	Bauxite, aluminum, coffee	Bauxite, alumina, gold	Phosphate, cocoa	Phosphate, cotton, coffee
Area (sq. mi.)	43,483		124,502		94,927		21,927	

1. GDP per capita for most recent data.

Sources: Population Reference Bureau (1975, 2002), Human Development Institute (2002), UN Statistics Division, CIA, USAID, and government reports from individual countries.

first to be mined on a large scale, and its aluminum processing plant was producing almost a million tons each year by the mid-1970s. New mines came into production in the Boké District in the 1970s and 1980s. Several Arab states have financed an aluminum smelter at Boké, which handles half the mine's output, the rest begin exported in its raw state. A smaller, but nevertheless substantial deposit (44 million tons) is located at Debélé (Kinia), serviced by a short railway. Other deposits are at Dabola and Togue, located high in the Fouta Djallon north of the Conakry–Kankan railway.

This mineral extraction activity required Guinea to improve its grossly inadequate and poorly maintained transport system, but much work remains to be done. The 375 miles (603 km) of narrow-gauge railway from Conakry to Kankan, whose principal freight was once bananas, was upgraded after independence to handle the bauxite from the Fouta Djallon Highlands but is deteriorating today. There has been talk of extending the line of rail across the border to Bamako, capital of Mali, and become that landlocked state's principal outlet to the sea; no new rail has been laid, however. Iron mining from the Mount Nimba area has relied on the Liberian rail network for the export of iron ore, and it is unlikely that the Guinean government will secure funding for the extension of its rail network to Nimba. Standard gauge rail and a road were built from the port of Kamsar to the Boké mine north of the capital and later extended to extract bauxite at Sangaredi. The country is mountainous, transport is difficult, and both maintenance and construction costs are high, and thus Guinea remains poorly connected. On the other hand, the combination of broken terrain and heavy rainfall has given Guinea tremendous hydroelectric potential. Over 50% of the electricity produced in Guinea is from hydroelectric sources.

The massive investment in minerals, which should have made Guinea one of the richest West African states, was not accomplished without difficulty. Having withdrawn from the French fold, Guinea had the task of seeking foreign support without prior experience. Aid came from several sources, including the former communist bloc, North America, oil-rich Arab states, and most recently from newly industrializing countries of Asia. Most investment has been in mineral-related activities, although Malaysian companies are interested in developing telecommunications. Guinea has attracted some foreign investment to its manufacturing sector, which is presently limited to processing local raw materials, agricultural product processing, and import substitution. Only 10% of its bauxite is processed locally into alumina.

Up until the death of President Ahmed Sékou Touré in 1984, Guinea enjoyed neither economic nor ideological support from its neighbors. Indeed, following the break with France and the francophone group, Guinea followed an almost isolationist course. Internal repression of dissent was routine, and hundreds of thousands of Guineans fled to neighboring countries. In 1970 a force of Guinean dissidents and Portuguese troops led a commando raid from Guinea-Bissau, which at the time was fighting its own liberation war against Portugal (Box 13-1). Most neighboring states were accused by President Touré of aiding and abetting anti-Guinean sentiment, which resulted in Guinea's withdrawal from the Organization of Senegal River States—an organization composed of Senegal, Mali, Mauritania, and Guinea for the development of the Senegal Basin—and further retreat into isolation.

After independence, Guinea did demonstrate to the world that it is possible to sever the umbilical cord with a colonial power, yet survive in a competitive world—for a time. Under the socialist Touré government, Guinea broadened its technical assistance base, diversified its economy, improved its infrastructure, and achieved a favorable balance of trade. On the other hand, the socialist promise did not bear fruit in Guinea: during the height of the Touré regime, the vast majority of its people still lived at the subsistence level, per capita food production was declining, the infant mortality rate was the highest in West Africa, the per capita income was only $125, and the literacy rate was among the lowest in Africa. Touré's strident criticism of neighboring countries, moreover, brought him few friends.

Upon the death of President Touré in 1984, the military took control of the government in Guinea. Then, 5 years before the Cold War had even ended, Guineans began dismantling the socialist economy. In 1993 General Lansana Conté, who had headed the military government, was elected president of the new civilian government, and in 1998 he was reelected. Considerable progress has been made in human development and democracy from 1975 to the present (Table 13-1), but Guineans face many challenges. Forty percent of the population lived below the poverty line in 2000, and material benefits from the massive investment in mining have yet to reach most Guineans. The economy is overly dependent on the mining sector (15% of GDP), which makes it vulnerable to fluctuations in global trade. With the expansion of gold and diamond mining, mining is expected to increase its dominance of the economy (U.S. Embassy Conakry 2001).

Nevertheless, growth in the economy, stimulated by outside investment and loans, has been impressive. Growth in public works such as transport and communications infrastructure grew by over 8% in 1999, manufacturing increased by almost 7%, agriculture grew by 5.5%, services and trade grew by nearly 2%. Even so, the conflicts in neighboring Sierra Leone, Liberia, and Côte d'Ivoire have threatened the investment climate in Guinea. The government of Lansana Conté has begun privatizing state-run enterprises such as mining, banking, communications, and energy, but the country remains one of the International Monetary Fund's list of heavily indebted poor countries (HIPC). With governmental reform there has been some debt forgiveness and rescheduling, but much remains to be done in building the economy, creating employment, reducing poverty, fighting corruption, and building a civil society.

Not only has poverty remained high in Guinea relative to its francophone neighbors discussed in this

BOX 13-1

GUINEA-BISSAU

After 500 years of Portuguese rule, and 10 years of guerrilla warfare, Guinea-Bissau (13,946 square miles; 36,120 km^2) became independent in September 1974. Under the leadership of Amilcar Cabral, the PAIGC (African Party for the Independence of Guinea and Cape Verde) instilled a sense of nationalism among Bissauans and led a successful but costly war of liberation against oppressive Portuguese colonialism. The low-lying terrain, forests, swamps, meandering rivers, and wide estuaries that prevail in all but the eastern interior were assets for the liberation forces (as they later were during the civil war that devastated the country in 1998). In 2000 there were fresh elections, and the country appeared to be making the transition of democracy.

Under Portuguese rule, few advances were made in the social, economic, and political spheres. Education and manufacturing were not encouraged, and few roads and no railways were built. A settler population did not exist, and almost no Portuguese remain. The potential for investment and development after independence was low, given the socialist ideology of the government and the modest resource endowment. Having had its first multiparty elections in 1994, Guinea-Bissau is making the transition away from socialism. Nevertheless, today Guinea-Bissau remains one of the most underdeveloped countries of Africa. Since independence, when the population was estimated to be about 650,000, the population has doubled. The main ethnic groups are the Balante (30%), Fulani (20%), Mandingo (13%), and Mandayiko (14%). Nonindigenous groups include Syrian and Lebanese traders, and Cape Verdian *mesticos*.

Eighty-five percent of the population is engaged in agriculture. Rice, maize, cassava, beans, and sweet potatoes are the staple crops, and small quantities of cashews, fish, groundnuts, and palm oil are exported. Ninety percent of exports are agricultural, but only 11% of the land is arable. There is no mineral production (large bauxite reserves exist in the Boe area near the Guinea border), and industry is negligible. Some offshore petroleum deposits have been located, but recovery costs will be high, and finding companies willing to assume the considerable risk will be difficult. Bissau (288,300), the capital, is the only town of any size. Development is constrained by economic, human, and institutional issues that concern security of land tenure, weak commercial infrastructure and development, and a lack of trained managers and entrepreneurs. Without supranational organization or the incorporation into a larger neighbor, for such tiny, resource-poor, colonial anomalies as Guinea-Bissau, Djibouti, and the Gambia, the potential for development seems low.

chapter, but opportunities to escape from poverty appear to be unequally distributed between the sexes. Although the differential between the sexes in human development is not a matter that concerns Guinea alone, it is a common theme throughout Africa. The available data from Guinea provide an illustrative example of the differential distribution of opportunity between the sexes and across the nation space. The most fundamental opportunity that can be given to any child, which opens the door to almost all other opportunities, is education. In Guinea today 61% of school-age children are in primary school, but only 50% of girls are enrolled. Although this differential is not great relative to some other countries in Africa, there is geographic unevenness across the country for both sexes, but especially for girls (Figures 13-3 and 13-4).

Women make a significant contribution to the economy, but girls have been undervalued in most African cultures and underinvested in by their families, possibly because the return on investment in a girl is lower than that for a boy under the dominant African patrilineal descent system and the rules of residence after marriage. The stark reality is that poor families must make painful decisions given the resources at their disposal and the existing sociocultural framework. In the vast majority of African cultures,

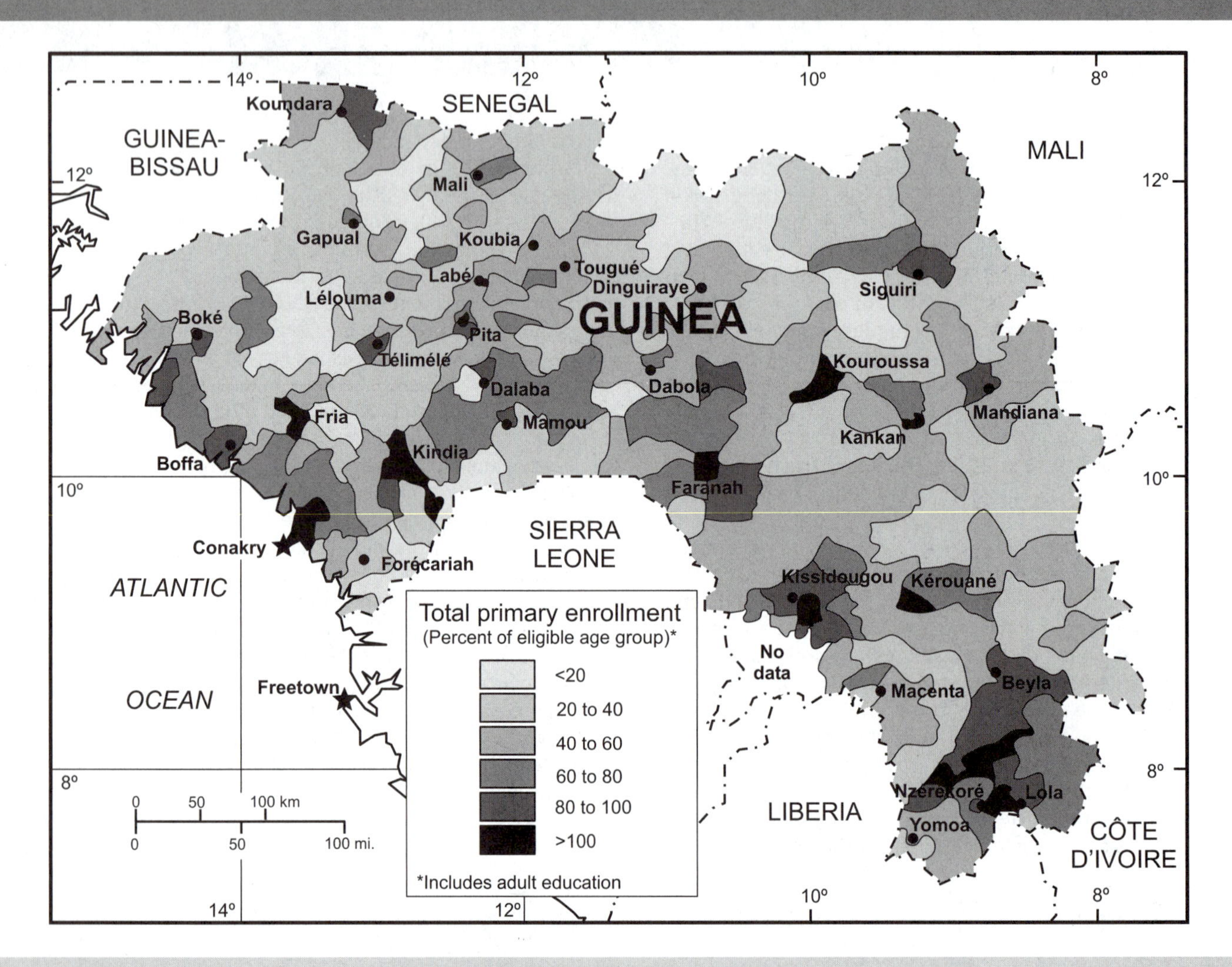

FIGURE 13-3 Total primary school enrollment in Guinea for the 2000-2001 school year. ***Adapted from data provided by USAID Guinea.***

the rules of residence dictate that when a girl is married she moves to her husband's household or village (patrilocality) and becomes a contributing member to that economic unit. She may see her family of origin only on market days and holidays. On the other hand, when a boy is married he brings an adult worker and future farm hands to the family enterprise. Why invest in girls, the logic runs, when they are destined to go elsewhere?

At another level, underinvestment in females makes little sense. A child's first teacher is the mother. An educated mother is a better mother, and the benefits of an educated mother accrue to the husband's family and to society in general. Although sexual inequality is a fact of life throughout Africa, national leaders seem to be getting the message that it is a development issue. The United Nations, the U.S. Agency for International Development (USAID), and other international agencies and nongovernmental organizations are strongly committed to women's development. The figures tell us that opportunities for education are unevenly distributed between urban and

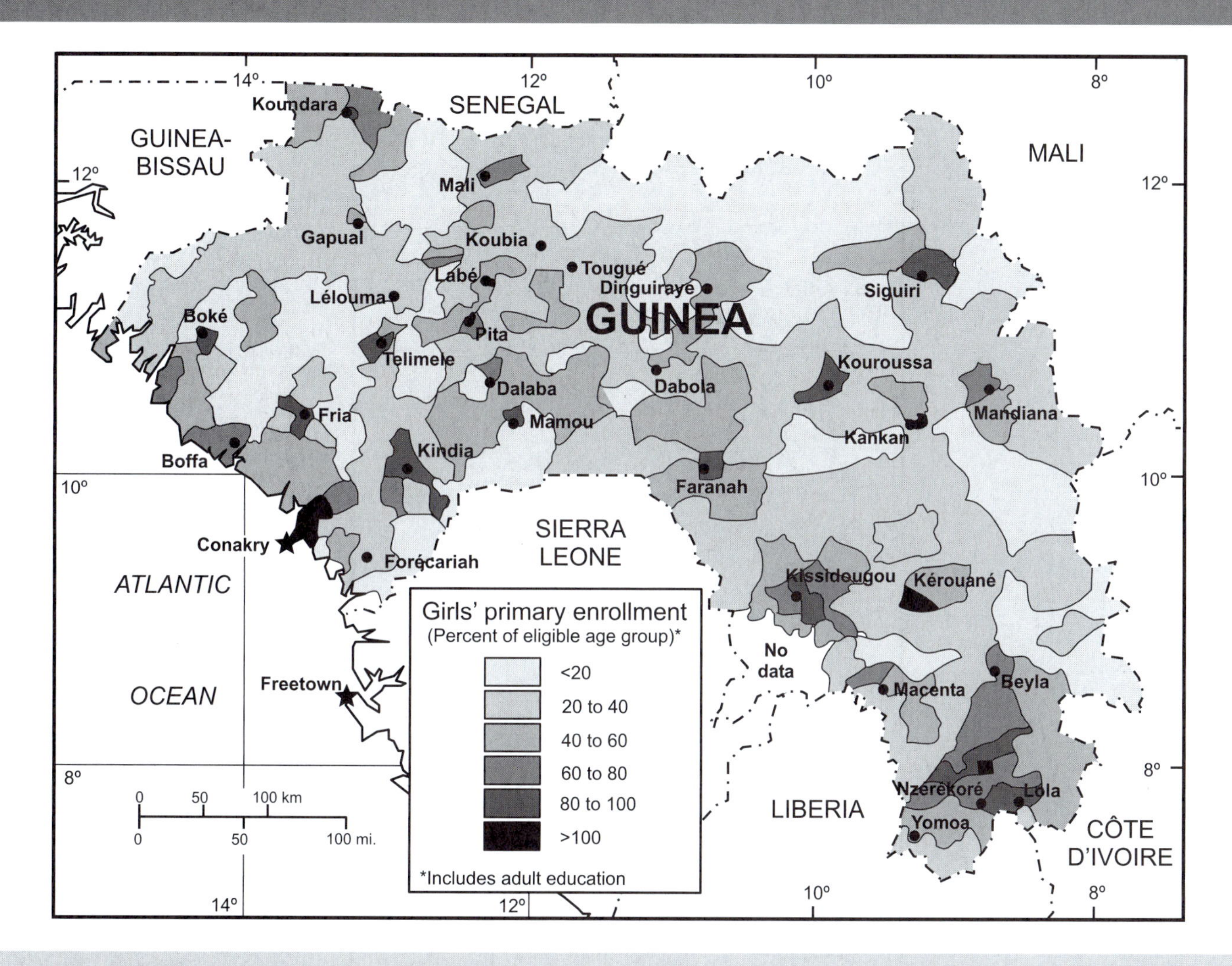

FIGURE 13-4 Total girls' primary school enrollment for Guinea in the 2000–2001 school year. ***Adapted from data provided by USAID Guinea.***

rural areas for both boys and girls in Guinea, but there is much less opportunity for girls.

TOGO AND BENIN: BETWEEN BIG NEIGHBORS

The two small elongated states of Togo (21,927 square miles; 56,791 km^2) and Benin (43,483 square miles; 112,620 km^2), wedged between Ghana and Nigeria, have essentially similar physical environments. Sandbars and coastal lagoons give way to low-lying plains and plateaus, and eventually to the southwest–northeast trending Togo–Atakora ranges, which reach their highest peaks (3,250 ft; 991 m) in western Togo (Figure 13-5). Although deeply dissected and well forested in the south, they are densely populated and produce almost all of Togo's coffee and cocoa. Climatically, the coastal belt falls into the anomalous dry zone that extends west to Accra, yet is sufficiently moist to support the all-important oil palm. Most of the region

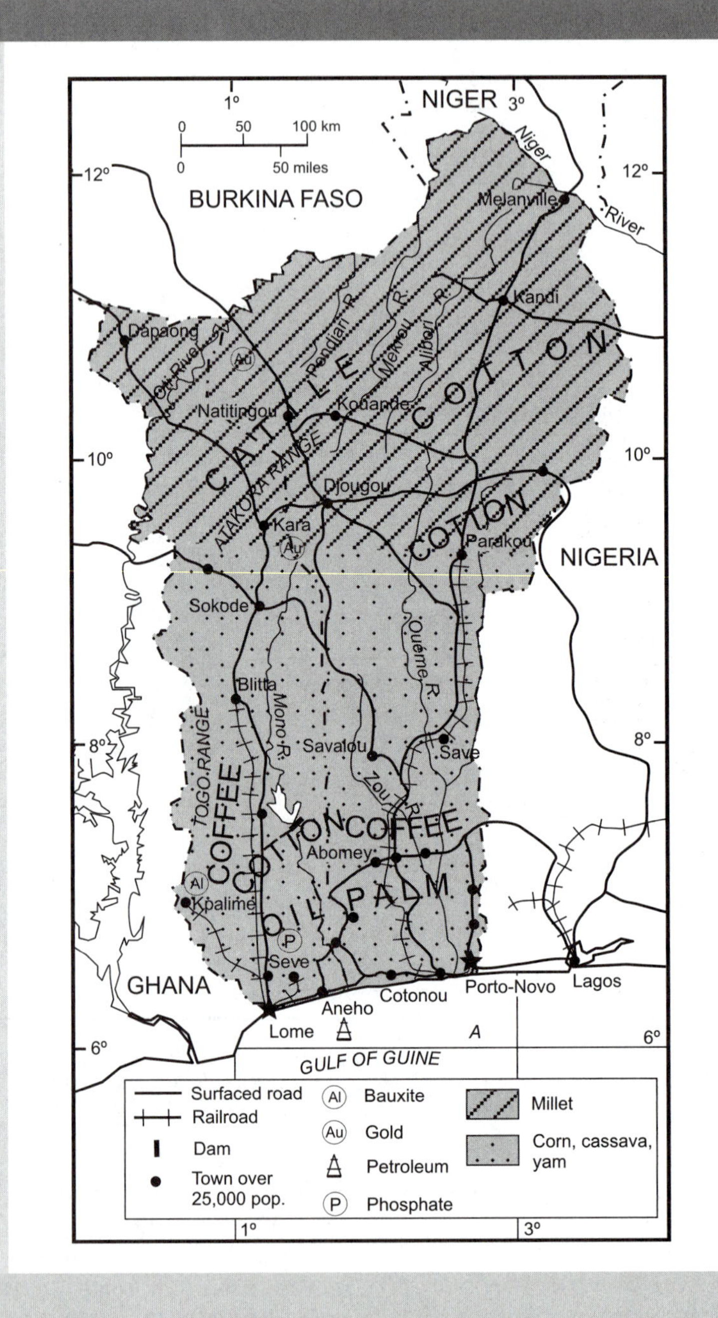

FIGURE 13-5 Economic map of Benin and Togo.

has a savanna climate, but north of the Atakora Range, rainfall diminishes sharply and becomes highly variable in the vicinity of the Niger River.

Togo and Benin were carved out of West Africa as a result of French, German, and British competition. Ouidah had been one of the most important slave exporters of the Gulf of Guinea, and exports from the town continued into the 1860s despite the attempts by the British to end them. The slave economy is evident in the cultural landscape today: the oil palm

BOX 13-2

BENINESE AMBASSADOR APOLOGIZES FOR SLAVERY

The ambassador of Benin to the United States, Cyrille Oguin, toured the country in 2003 apologizing to Americans for the role people in his country played in the Atlantic slave trade.

The state of Dahomey, located on what was formerly known as the Slave Coast, was heavily involved in the regional slave trade, from the capture of free people to marketing. It has been estimated that millions of people were exported from the Slave Coast to the Americas (principally Brazil), through the port of Ouidah (Whydah) on the Beninese coast (Lovejoy 2000).

"The president of Benin, the people of Benin have asked me to come here and apologize for the government, for the Benin people and for Africa for what we all know happened," went the ambassador's speech. "Today, no one wants to take responsibility. It's so easy to say the white man did it to us, but we share in the responsibility" (Wold 2003).

The stigma of slavery still exists in Africa, and the reconciliation organization set up by the Beninese government in 1999 between nations involved in the slave trade is a positive first step in addressing the terrible legacy of slavery.

plantations around the former capital at Abomey were created in the nineteenth century by the aristocracy to profit from captives that would have been exported into slavery in the Americas (Box 13-2), had the institution not been outlawed by the British (Harrison-Church 1980). In 1892 the last king of Dahomey, Behanzin, was conquered by the French, who established the colony of Dahomey, while Germany occupied Togoland, a smaller sliver of land of diverse peoples to its west. Following World War I, Togoland became a League of Nations mandate, divided for administrative purposes between France and Britain. British Togo was administered as though it were a part of the Gold Coast, while French Togo, except for a brief period when it was incorporated into Dahomey between the two world wars, remained a separate political entity. Well after the end of World War II, Dahomey was renamed Benin.

In 1946 the mandate system was superseded by the trusteeship system of the United Nations, and France and Britain continued their administration. In 1956 the people of French Togo voted in a plebiscite to become an autonomous republic within the French Union, and France attempted to tie the trust territory more closely to Paris than the trusteeship agreement appeared to permit. Despite that, the United Nations refused to end the trusteeship status. At the same time, the inhabitants of British Togo voted in favor of union with the Gold Coast. Both Togo and Dahomey received independence from France in 1960.

It is important to recognize the individual contributions made by both Germany and France in the development of the present-day Republic of Togo. During its brief period of rule, Germany established a good railway and road system, focusing on Lomé and its harbor. The Germans introduced crops and developed the economy, founded an educational, judicial, and administrative system, and made considerable progress in the unification of the country. Togo became Germany's model colony (Darkoh 1968). The French, for their part, did not continue the rapid pace of development initiated by Germany, and for several decades the country floated on what the Germans had founded. It was not until the immediate preindependence period that Togo's infrastructure was expanded and its exports, especially cocoa, were increased (see Table 13-1 for basic data).

In Dahomey, France's performance was equally poor, except in the field of education. Large numbers of Beninese received a secondary education, and later they held government positions in several parts of the federation. Following independence, this educated elite was replaced by local civil servants and was forced to return to Dahomey, where jobs were not to be found. As in Côte d'Ivoire, France favored the southern region, but less was achieved in commercial agriculture, forestry, and communications. Both countries have been troubled by political instability, economic stagnation, and strong regional economic imbalances.

Benin has suffered from extreme political instability since independence: there have been six military coups since 1960. Benin's ethnic diversity and unhealthy economy, reflected in the chronic inability to balance the budget, largely explains this instability. The Fon, Adja, and Yoruba form the largest ethnic groups in the south and dominate most branches of government and the cash economy. The less numerous Baribas, Peuls (Fulani), and Sombas of the north believe that they have been inadequately represented in national government and economy.

Sixty-five percent of the Togolese labor force is engaged in agriculture. The main food crops are yams, cassava, corn, and rice in the south and central Togo, while in the north millet and sorghum predominate. Most southerners are small-scale farmers, and a small minority participate in the export economy, producing cotton, oil palm products, cocoa, and coffee. The main cash crops are oil palm in the south and cotton in the north. Livestock are raised principally in the north. In the south, as in most of the West African coast, agricultural activities benefit from two rainy seasons. The main rainy season is from February/March to July, while light rains fall from October to November. The northern region receives its rainfall in only one season, and the rainiest months are July and August.

Manufacturing in Benin is mainly small-scale processing of primary products for export and is concentrated in the capital Porto Novo (227,000), and the major port, Cotonou (721,000). There is also import substitution of consumer goods. Industry, dominated by the state, includes refining, agricultural processing, chemicals, and cement making. Mining plays a minor role in the economy, but there is artisanal mining of gold in the northwest. Benin produced over half a million barrels of crude oil in 1995 and 1996, but production tapered off and the field was closed in 1999. Natural gas reserves were estimated at 80 billion cubic feet (2.27 billion cubic meters) (Mobbs 2001). Petroleum storage was built at the port of Cotonou in 1999 to serve Benin, Burkina Faso, and Mali.

The urban population of Benin has been growing since independence. There are eight towns around the country with a population greater than 100,000. Cotonou has a modern deepwater harbor and is connected by the main line rail to Parakou (195,000), the largest northern town, and by two shorter lines (not shown in Figure 13-5) to Pobe (32,300) and Segboroue. The rail line handles a fourth of Niger's foreign trade, and if it is extended from Parakou to Dosso, Niger, and on to the Arlit uranium deposits, Cotonou may well experience further, but not substantial, growth. The northern region—where cattle, cotton, and groundnuts are the major sources of income—is unlikely to benefit from this extension unless feeder roads are built and marketing systems improved.

The spatial pattern of Togo's development is very similar to that in Benin. Most of the exports, capital accumulation, wealth, urban populations, and political decision makers are concentrated in the south and form the core, while the majority of the country comprises the periphery. The country's three short railways converge on Lomé (658,000), the capital and primate city (Figure 13-5). They draw the resources and peoples to the south but play little role in dispersing goods, ideas, and technology from the core to the periphery. With the exception of Lomé, there are no towns in Togo with a population above 60,000. Togo remains a country of villages and small towns.

Although the contribution of agriculture to the national economy has declined since independence in relation to industry, consisting principally of phosphate mining, about 70% of the population is engaged in agriculture. The agricultural geography of Togo is similar to that of other West African coastal countries. Corn and cassava are grown as food crops along the coast, while oil palm and cocoa constitute the main cash crops. The central region of Togo is really the granary of the country, especially in the higher elevations in the west. Here the cereal and yam belts overlap and with the exception of millet, all major cereals and all tubers can be grown here successfully, as well as coffee and cotton. From the vicinity of Kara to the northern border of Togo there are often food deficits in this drier climate with one rainy season. More drought-resistant varieties of sorghum are grown in this region, and rice is grown along rivers. Millet is the dominant crop, however: 90% of national production takes place here. Drier, more Sahelian climate conditions prevail: there is only a single rainy season, and maximum rainfall is in July and August.

Lomé's manufacturing, like that of Porto Novo and Cotonou in Benin, is dominated by import substitution and agricultural processing. A large cement plant was completed in 1975 to serve the five entente countries and Ghana. The oil refinery, opened in 1976, is supplied by Nigeria and Gabon. Both Togo and Benin have been involved in discussions with Nigeria and Ghana on a Nigeria-to-Ghana oil pipeline. Such a venture would help francophone–anglophone integration and formalize some of the informal economy, represented by the small-scale enterprise shown in Figure 13-6. Cement production, around half a million metric tons throughout the 1990s, has been sufficient to meet Benin's needs. A few miles northeast of Lomé, phosphate is mined near Sévé and railed to the coast for export. Togo is the fourth largest phosphate producer in the world, and phosphate provides more than half of Togo's foreign revenues.

A critical problem in Togo's development has been a shortage of electricity. Togo produces some electricity itself by burning oil, but 83% of its electricity comes from the Akosombo hydroelectric plant in Ghana. Although Togo and Benin cooperate to purchase and distribute electricity from Ghana and Côte d'Ivoire, there have been deficits in both countries. Energy production was taken over by an international company in 2000, and improvements in the provision of electricity are expected.

In 2001, the World Bank announced plans to build a large container facility at the port of Lomé to ease congestion caused by dockers' strikes, crime, and political instability in the ports of Abidjan in the Côte d'Ivoire and Lagos in Nigeria. The $100 million project will make Lomé a regional hub for the transshipment of cargo and stimulate the Togolese economy. Lomé is attractive for development because it's the deepest port along the Gulf of Guinea, and the port is large enough to handle the newest, biggest container ships. It will also be able to load and unload containers much more rapidly than neighboring ports.

Togo lacks industrial towns and cities away from the coast. A few small commercial centers and agricultural collection depots are located along the railways; none is a real growth point, however, and urban population is very low in Togo. Kpalimé (72,000) is a railhead for the cocoa farmers of the southwest, and Blitta serves cotton interests north of Lomé. The northern half of Togo, which lacks railways, good roads, and much profitable commercial farming other than groundnut production, is poorly integrated into the national economy. The mineral industry of Togo is not well developed. Currently phosphates, marble, and limestone are being mined, and there have been indications of small deposits of iron ore near the Ghanaian

FIGURE 13-6 Stock in trade of a small gasoline and oil reseller, Benin. *Steven Montgomery.*

border west of Bassari, as well as manganese and platinum group metals. Gold and diamonds are mined artisanally. As in Benin, the population is predominantly rural and farms are generally small, intensively cultivated, and devoted primarily to subsistence crops.

Togo and Benin have shown an increasing tendency to loosen their ties with the institutions that have been a legacy of French rule. Twenty-five years ago Benin's trade, like Togo's, was still primarily with France. That dependence has lessened since the Lomé Convention presented new possibilities, and with the later widespread acceptance of the concept of globalization. The Lomé Convention, named after the city in which the first agreement was signed in 1975, constitutes a framework that defines the cooperation between the European Union and many African countries. The convention was renewed in 1979, 1985, 1989, and 2000 (ECDPM 2000). Benin is seeking to increase its economic links with both Nigeria and Togo after several years of almost no interchange. The problems of development in Togo and Benin are shared by many other small African states seeking higher standards of living for their citizens. Togo's major export partners are Benin, Nigeria, Belgium, and Ghana, while imports come mainly from Ghana, France, China, and Côte d'Ivoire. Benin's main export partners are Brazil, France, Indonesia, Thailand, Morocco, Portugal, and Côte d'Ivoire; for imports, while France remains its major partner, relationships have also developed with the United States, China, Côte d'Ivoire, Netherlands, and Japan (CIA 2002).

In 1975 Benin adopted Marxism–Leninism as its guiding ideology for development, but in 1989 the country began the long process of transformation to a market economy. A legacy of state involvement in and control of the market persists, although privatization has occurred in some sectors. Although suffering from similar statist problems but without the socialist ideology, Togo, in contrast to Benin, opened a duty-free export processing zone in Lomé in 1989 and has attracted investment from Europe, South Asia, and East Asia, providing needed jobs and opportunities for Togolese.

Benin's transition to democracy was aided by a national conference in 1990, after which power was transferred from the Marxist dictator, Mathieu Kérékou, to a popularly elected government. Without the cooperation of President Kérékou, multiparty democracy would not have been possible in Benin. The national conference process developed in response to a growing crisis in governance. In this process, repeated in many African countries transitioning to democracy, a broad coalition of local and national leaders viewed as independent calls a national conference to discuss the crisis and the way to resolve it. During that conference a transitional government is appointed that works with the existing regime and gradually assumes its powers. The president takes the role of figurehead. The national conference transforms itself into a transitional legislative body that elects a prime minister who becomes the manager of the transition process. The transitional government dissolves itself after adopting a constitution, holding legislative and presidential elections, and inaugurating the newly elected democratic government (Schraeder 2000).

Although Benin moved away from the single-party state and patrimonial rule in 1991, Togo's head of state, President Gnassingbé Eyadéma, was the longest-serving leader of any country in Africa, reelected in 2003 with 56% of the vote for another 5-year term. Eyadéma, a general, seized power in a military coup d'état in 1967 and was maintained in power by the military. There was little international support for Eyadéma's rule, and there has been much political unrest in the country. He died in 2005 and his son was elected president in a flawed election.

Togo has long been known as a point in the international heroin trade route to consumer countries in Europe and has been suspected of laundering conflict diamonds from Angola. With the death in 2002 of Jonas Savimbi, the Angolan rebel leader, and the disbanding of his army, this source of foreign exchange has ceased. International crime and continued questions regarding the legitimacy of the president may make Togo unattractive to international donors and investors.

BIBLIOGRAPHY

Bermúdez, O. 1999. *The Mineral Industry of Ivory Coast.* U.S. Geological Survey Mineral Report. Reston, Va.: USGS. http://minerals.usgs.gov/minerals/pubs/country/.

Bermúdez-Lugo, O. 2004. *The Mineral Industries of Central African Republic, Côte d'Ivoire, and Togo.* U.S. Geological Survey Mineral Report. Reston, Va.: USGS.

Bermúdez-Lugo, O. 2004. *The Mineral Industry of Guinea.* U.S. Geological Survey Mineral Report. Reston, Va.: USGS.

CIA. 2002. *World Factbook*. Langley, Va.: U.S. Central Intelligence Agency. http://www.odci.gov/cia/publications/factbook/index.html.

Darkoh, M. 1968. Togoland under the Germans. *Nigerian Geographical Journal*, 11: 153–168.

Handloff, R., ed. 1988. *Ivory Coast: A Country Study*. Washington, D.C.: Library of Congress.

Harrison-Church, R. J. 1980. *West Africa: A Study of the Environment and of Man's Use of It*. London: Longman.

Hopkins, A. G. 1973. *An Economic History of West Africa*. New York: Columbia University Press.

Human Development Institute. 2002. *Human Development Report*. United Nations Development Program. Oxford: Oxford University Press.

ECDPM. 2000. Implementing the New ACP–EU Partnership Agreement (Lomé Negotiating Brief Number 8). Maastricht: ECDPM. http://www.ecdpm.org.

Lovejoy, P. E. 2000. *Transformations in Slavery: A History of Slavery in Africa*, 2nd ed. Cambridge, U.K.: Cambridge University Press.

Mobbs, P. M. 2001. *The Mineral Industry of Benin*. U.S. Geological Survey Mineral Report. Reston, Va.: USGS. http://minerals.usgs.gov/minerals/pubs/country/.

Mobbs, P. M., and Morse, D. E. 2001. *The Mineral Industry of Guinea*. U.S. Geological Survey Mineral Report. Reston, Va.: USGS. http://minerals.usgs.gov/minerals/pubs/country/.

O'Connor, M. 1972. Guinea and the Ivory Coast: Contrasts in economic development. *Journal of Modern African Studies*, 10(3): 409–426.

Population Reference Bureau. 2002. *World Population Data Sheet*. Washington, D.C.: PRB.

Population Reference Bureau. 1975. *World Population Data sheet*. Washington, D.C.: PRB.

Roberts, R. 1996. *Two Worlds of Cotton: Colonialism and the Regional Economy in the French Soudan, 1800–1946*. Stanford, Calif.: Stanford University Press.

Schraeder, P. J. 2000. *African Politics and Society: A Mosaic in Transformation*. Boston: St. Martin's Press.

U.S. Embassy, Conakry. 2001. *Country Commercial Guide: Guinea*. Washington, D.C.: U.S. Bureau of Economy and Business.

Wold, A. 2003. Benin asks forgiveness for slave trade. *The Advocate*, June 28, 2003. Baton Rouge, La.

Yager, T. R., G. R. Coakley, and P. M. Mobbs. 2002. *The Mineral Industries of Benin, Cape Verde, the Central African Republic, Côte d'Ivoire, and Togo*. U.S. Geological Survey Mineral Report. Reston, Va.: USGS.

CHAPTER 14

The Desert Tier

Problems in the Interior, Climate Change, Population Growth, and Desertification

From Senegal to Chad lies France's legacy across interior West Africa, including Mauritania, Mali, Niger, and Burkina Faso. Once territorial divisions of France's West and Equatorial African empire, these states form a continuous block stretching 2,800 miles (4,500 km) from Dakar in the west to the Darfur Plateau in eastern Chad (Figure 14-1). They extend over 20 degrees of latitude from southern Chad to northern Mauritania, mostly across the Sahara and West Africa's *Sahel*. Their combined area of over 2 million square miles (5.2 million km^2) is equal to two-thirds the continental United States, Chad alone being almost twice the size of Texas. Questions were raised in the middle 1970s, when the combined population of these countries was only 24 million, about the ability of the land to support any more people in view of the low and variable rainfall, the limited agricultural and pasture land, the enormous livestock population (20 million population equivalents), and the sparsity of development resources and industry. Even though population in the region has been increasing rapidly (it currently supports over 60 million people and 35 million livestock equivalents), concern about people and the environment has become more muted over the last 25 years. Nevertheless, serious reservations exist about the long-term sustainability of agriculture and ultimately human development, given current population growth, land-use practices, land tenure, and economic and political organization.

The six states form a distinct region with similarities in ecological characteristics, development problems and prospects, and colonial histories. From 1968 to 1973 the nations of the Desert Tier shared in the worst drought in living memory, a drought that devastated the local economies and brought widespread hunger, disease, and death to people and livestock alike. Drought struck again in the early 1980s and has occurred periodically ever since. Recurrent drought is a characteristic of the region, and the drought of the early 1970s was unusual only in its severity and the attention it received beyond Africa. Since drought figures so prominently in the behavior of millions of pastoralists and sedentary farmers, and plays a central role in the economic policies and development prospects of all six states, it receives special attention in this chapter.

Physiographically, the Desert Tier has considerable diversity, but given the vastness of the area this is not readily apparent. Much of the region is a rocky and sandy plateau, lower in the west and south, and higher in the north and east. In west-central Chad the land rises from 800 feet (250 m) in the Lake Chad Basin to over 11,200 feet (3,400 m) in the volcanic Tibesti Mountains in the north and to over 4,000 feet (1,220 m) in the Kapka Massif along the eastern border. In northern Niger, the Aïr Massif rises abruptly from the surrounding desert plains to heights of 6,600 feet (2,000 m), while in adjacent Mali, the Adrar des Iforghas reach some 2,800 feet (900 m) above sea level. Elsewhere, the land is gently undulating to flat, broken only occasionally by isolated hills, dry river courses, and a few steep escarpments. The Niger River system dominates the region from western Mali to western Niger, while the Senegal flows west from the Fouta Djallon to the Atlantic, and the Volta drains south across Burkina Faso to the Gulf of Guinea. Much of Chad, Niger, Mali, and Mauritania has only interior drainage.

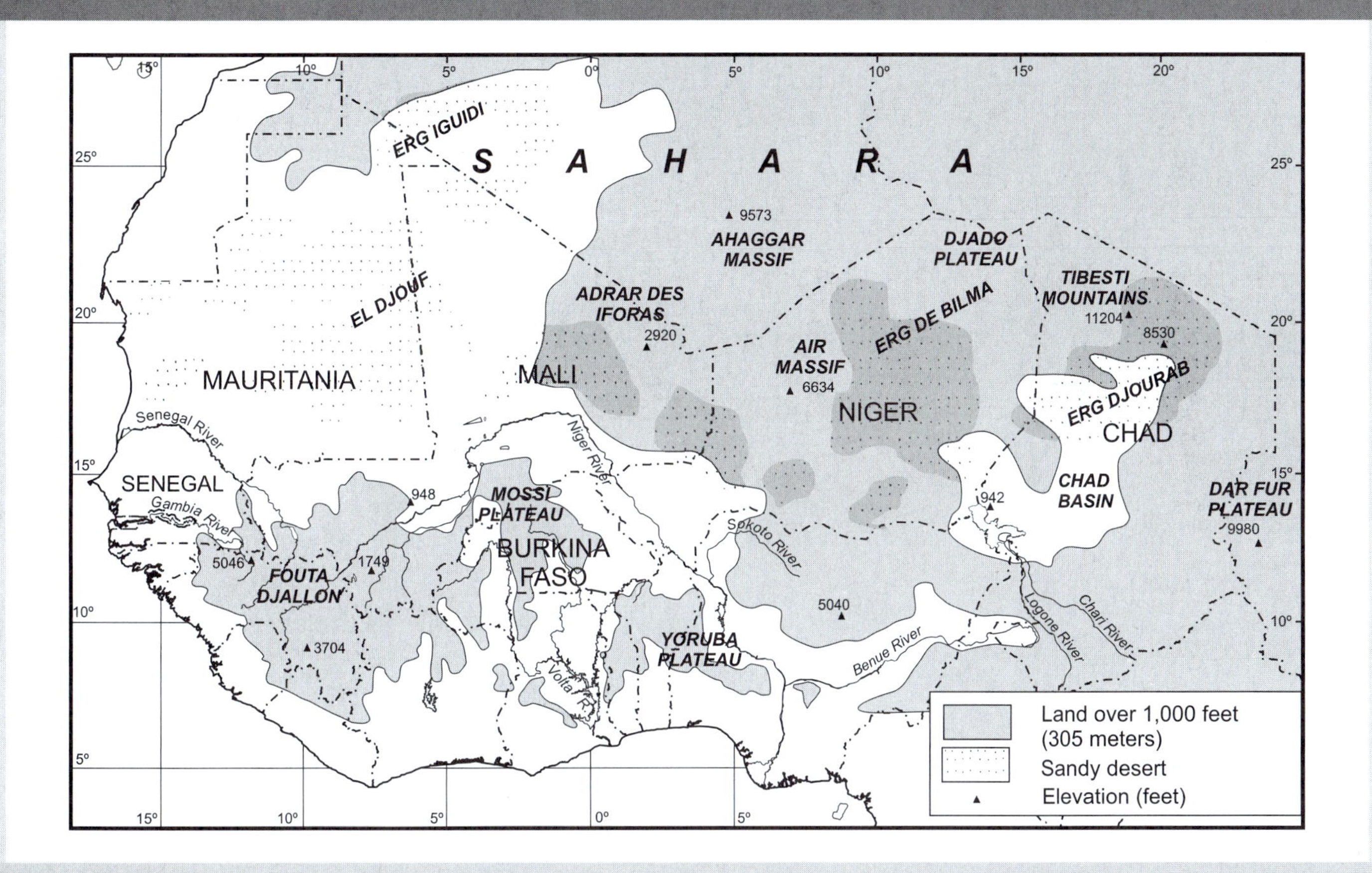

FIGURE 14-1 Physical features of West Africa.

Climatically, this West African subregion falls into three distinct latitudinal belts (see later: Figures 14-8 and 14-16). North of the 4-inch (100 mm) isohyet, which runs from central Mauritania eastward to central Chad, lies desert; here temperatures and evapotranspiration rates are high, rainfall is low and irregular, and the vegetation cover too meager to support permanent grazing. To its south lies the Sahel, or "border" region, where the average annual rainfall increases southward to the 24-inch (600 mm) isohyet; rainfall is highly variable, yet generally sufficient to support short grasses, acacias, and thorn scrub (Figures 14-2 and 14-3). Along the southern border of the six-state region lies the savanna. While three separate climatic regions exist, each with its distinct agropastoral potentials and population densities, the basis of survival in the Sahel has been the interdependence between desert and savanna, pastoral and arable economies, and herder and farmer. It is this interdependence of peoples and economies, so critical for the survival of the region's ecology and people, that has been put to the test since the early 1970s.

DISASTER IN THE SAHEL

The Sahel drought of the early 1970s, a defining moment in the lives of the people who lived through it, was also a critical moment in the development of our understanding of human adaptation. The Sahel is the fringe of the habitable area of the world; literally, the word "sahel" in Arabic refers to the region's location

FIGURE 14-2 Dry season peanut fields in central Senegal in the Sahel.
Roy Cole.

FIGURE 14-3 Sahel in the dry season southwest of Timbuktu near Gossi, Mali.
Roy Cole.

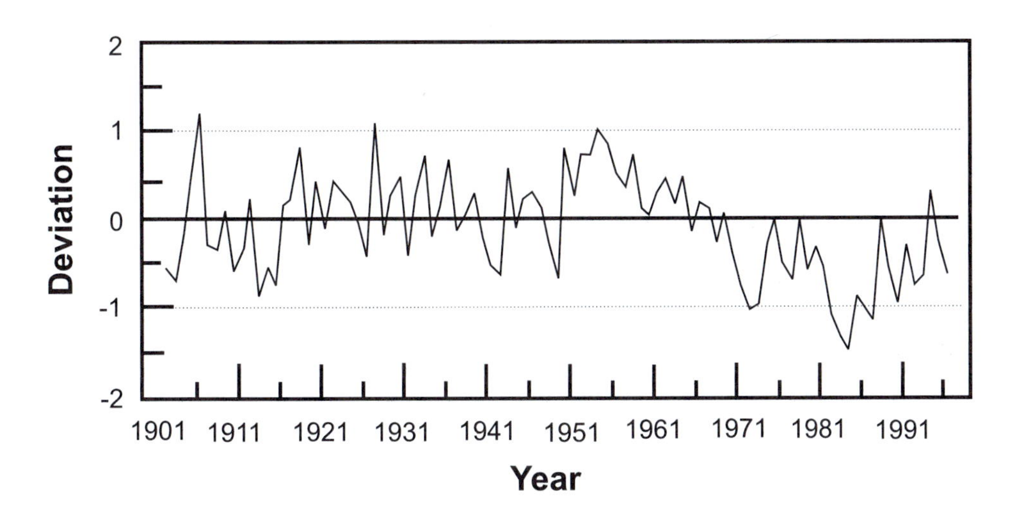

FIGURE 14-4 Precipitation deviation from long-term average for the Sahelian states between 1901 and 1991. *Adapted from McCarthy et al. (2001).*

on the edge of the desert. The Sahel became almost a household word around the world in 1974 following reports of 5 years of unusually light rain, widespread crop failure, the starvation of millions of cattle, sheep, goats, and camels, and the deaths of an unknown number of people. The media portrayed emaciated children suffering from malnutrition standing pitifully before their desert shelters or refugee camps, surrounded by dying cattle and parched landscapes. Fields and pastures lay bare except for helpless herds of emaciated cattle and carcasses, and in the dry river beds or wadis, women and children dug deep into the sands to retrieve precious buckets of water. At the refugee centers, some with 60,000 and more displaced persons, trucks and vans belonging to the International Red Cross and Red Crescent, the FAO, USAID, Oxfam, and other relief agencies, unloaded bags of powdered milk, sorghum, millet, rice, and medicines. Such were the scenes, but how common were they? While the tragedy will never be fully documented, there is ample evidence to suggest that the drought extended across Sahelian West Africa and into Sudan, Ethiopia, Somalia, and Kenya, and that perhaps in all, more than 25 million persons were directly affected. In the six West African Sahelian states, possibly half the population was afflicted.

Drought conditions varied from place to place and generally were at their worst in 1972 and 1973 (Dalby and Harrison-Church 1973). After unusually wet years in the 1950s and early 1960s, when rainfall often averaged 20 to 30% above the mean, a mild drought occurred in 1968, followed by 5 years of abnormally low precipitation (Figure 14-4). The following rainfall data from Niger illustrate some local conditions. Tillabery, a river town north of Niamey, reported a 30-year (1940–1970) mean annual rainfall of 19.2 inches (488 mm). Rainfall in 1971 and 1972 was 10.2 inches (259 mm) and 14.3 inches (364 mm), respectively. Maradi, Niger's third largest town just north of the Nigerian border, had a mean annual rainfall of 24.8 inches (630 mm) prior to the drought; but its rainfall in 1971 and 1972 was 12.5 inches (318 mm) and 11.3 inches (288 mm), respectively. For Agadez, an oasis town toward the northern edge of the Sahel, the corresponding figures are 6.3 inches (160 mm), 3.7 inches (93 mm), and 3.0 inches (72 mm). Four or five years of below-average rainfall and frequent high winds, combined with human impacts associated with farming and overgrazing, resulted in reduced vegetation cover, destruction of the root system, and finally widespread soil erosion. The severity of the Sahel drought can be seen from the following stock losses: Chad 70%, Mali 55%, Mauritania 70%, and Niger 80%. The cattle population in Chad totaled 4.7 million in 1972, and under 3 million a year later. In northern Mali and Chad, up to 90% of the livestock were lost. The loss of livestock is not only a personal catastrophe to the pastoralist, whose whole life is geared to the raising of animals, but a national disaster to states whose economy is dependent on the sale of animals and animal products. Such products figure prominently in each of the Sahelian states.

Agricultural output also showed drastic declines after the drought. In Senegal, where peanuts normally

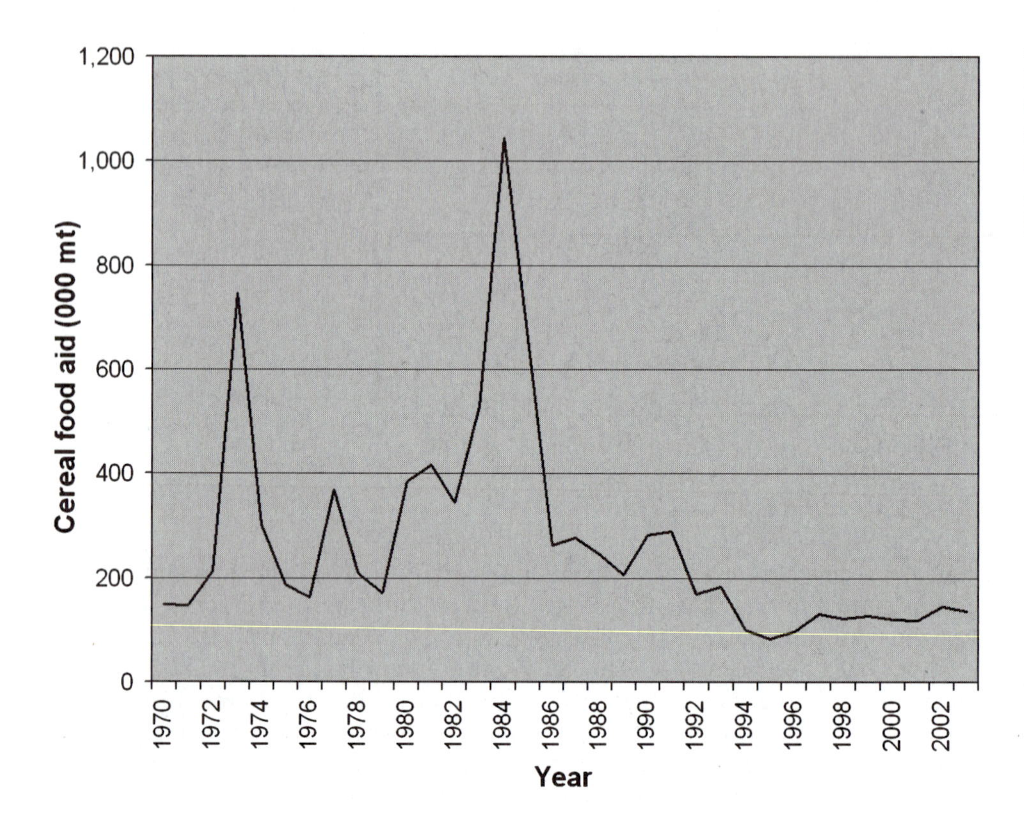

FIGURE 14-5 Cereal food aid from all donors to Burkina Faso, Chad, Mali, Mauritania, Niger, and Senegal, 1970 to 2003.

supply 80% of the foreign exchange, the 1972 output totaled only 587,000 tons, compared with 997,000 tons in the previous year. The drop was even more dramatic in Niger, where the drought reportedly reduced 1974's output to 44,000 tons, while in 1972 some 260,000 tons had been produced, and in 1967, the peak year, more than 312,000 tons had been harvested. Mali's grain production dropped by 50% in 1972, and the millet harvest in the St. Louis region of Senegal plummeted from 63,000 tons in 1968 to 1000 tons 4 years later.

Similar declines were recorded for cotton, cow peas, and other crops across the Sahel. Crop failures, livestock losses, and dried-up wells and river courses forced thousands of refugees to towns and relief centers. Nouakchott, capital of Mauritania, whose population in 1971 was 45,000, had over 120,000 newcomers within the next year. At Mopti, in central Mali, the population swelled from 55,000 to 110,000, and neither town had adequate housing or water. In Niger, at the height of the drought, some 300,000 refugees out of a total population of 4.3 million were registered at official relief centers, and probably twice that number sought refuge elsewhere. Death rates, especially infant mortality and child mortality rates, already among the highest in the world, rose as a result of the aggravated conditions. Reliable data are not available, but during "normal" years, infant mortality rates ranged during the 1970s from 159 per thousand in Senegal to 204 per thousand in Burkina Faso and, on the average, half the children die before their tenth year. Disease and malnutrition accompanied the drought. (Additional data on infant mortality is presented later: see Table 14-3.)

The extent and seriousness of the drought were not formally acknowledged by the six governments concerned until the middle of 1973, partly because of insufficient verifiable evidence, but primarily because the governments were reluctant to admit their inability to handle the crisis. Once the disaster had been officially proclaimed, overseas relief efforts intensified. The U.S. contribution for the fiscal years 1973, 1974, and 1975 totaled $214.7 million: $120.4 million for grains, and the balance for technical assistance, equipment, airlifts, and medical supplies. Total grain aid in 1973 was 750,000 metric tons (Figure 14-5) and the major grain contributors included the United States (256,000 tons), the European Community (110,000), France (70,000),

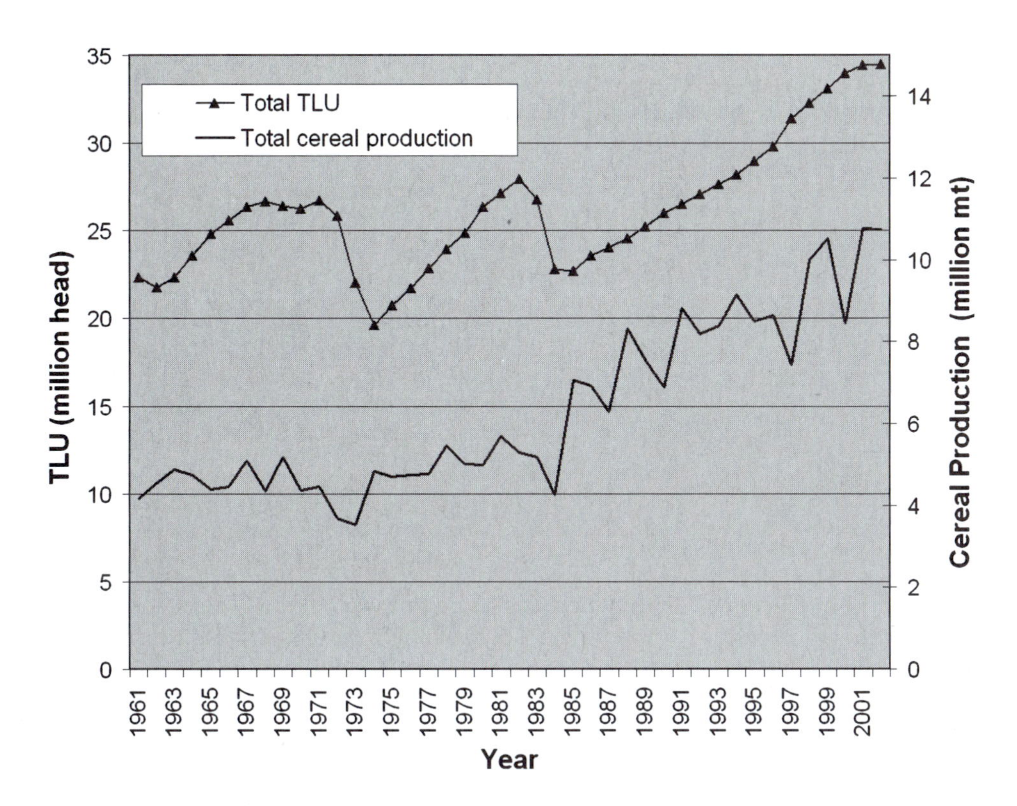

FIGURE 14-6 Total tropical livestock units (TLU) and total cereal production (rice-milled equivalent), 1961 to 2002, for Burkina Faso, Chad, Mali, Mauritania, Niger, and Senegal.

China (50,000), the Federal Republic of Germany (32,000), Canada (26,000), and the Soviet Union (10,000). Many countries, including Nigeria and Zaïre, made cash contributions (Sheets and Morris 1974: 129–130). Relief efforts were hampered by poor communications, inadequate local distribution systems, corruption, and the high cost of transport. More money was spent on airlifting fuel and equipment required by the relief operations than on airlifting grains and medicines. Grain rotted on the docks in Dakar, Abidjan, and Lagos, and at transshipment centers such as Maiduguri, Ouagadougou, Rosso, Niamey, and Koulikoro (Rosenthal 1974). Grains unloaded in interior railheads could not be transported by truck over district roads made impassable by summer rains. Rains, badly needed by the farmer, were a curse to relief agencies.

Persistent Drought and Human Response

After the initial shock of the Sahel drought in the early 1970s, there were a couple of years of average rainfall in the middle and late 1970s; but the worst drought was to come in the early 1980s (Figure 14-4). With few exceptions, below-average rainfall has continued to the present. Relief food donations to the Sahel were high in 1973 and highest in 1985, exceeding one million metric tons. Remarkably, although predrought conditions have not returned, food aid has dropped to the predrought level (Figure 14-5). In an apparent paradox, the human population has increased by millions, livestock numbers have increased by tens of millions of head, and agricultural production has increased by more that 150% under almost continuous drought conditions (Figure 14-6).

Two major processes help explain what is happening in the Sahel: people have adapted to the new rainfall regime, and food insecurity is being addressed in a more targeted way than in the past at the local level by national governments, international agencies, and nongovernmental organizations (NGOs). Overall, the data that are available suggest that except in very localized cases, people have changed their behaviors in a variety of ways and are better able to cope with drought today than they were in 1970 (Cole 1991). This does not mean that all is well along the edge of the Sahara; rather, it indicates that big changes have taken place. Many villages have been abandoned in northern Mali, for example, as farmers move south

TABLE 14-1 PERCENTAGE CHANGE IN SHARE OF NATIONAL PRODUCTION FOR FOUR STAPLE CEREALS AND FOUR PRINCIPAL DOMESTICATED LIVESTOCK, 1961-1965 TO 1998-2002.

	Staple Cereals (%)				Livestock (%)			
Country	Maize	Millet	Rice	Sorghum	Camels	Cattle	Goats	Sheep
Burkina Faso	7	1	0	−8	−0.2	−15.1	4.5	10.8
Chad	9	−8	7	−8	1.2	−8.3	13.6	−6.5
Mali	5	−11	13	−7	0.7	−1.1	5.1	−4.7
Mauritania	1	−4	46	−43	1.7	−14.7	3.9	9.1
Niger	0	4	2	−6	−0.1	−16.0	2.4	13.7
Senegal	3	−9	5	0	−0.6	−18.2	8.9	9.9
Average	4	−5	12	−12	0	−12	6	5

Source: FAO (2002). FAOSTAT.

where rainfall is more abundant and predictable. Urban populations are burgeoning, as shown later (see Figure 14-12). Since 1970, there has been wholesale replacement of long-maturation varieties of such staples as millet and sorghum with varieties that mature in 60 to 70 days. Added to this are new varieties of maize, groundnuts, and beans that all mature months sooner than the older varieties that worked in the more humid 1950s.

Since the 1970s, the production of cereals and the key staple crops of millet and sorghum has increased in the region (see later: Figure 14-14) and, incredible as this may seem, has kept up with rural population growth for the most part. Output varies from year to year and from place to place, and variability increases as one gets closer to the desert. The two most drought-resistant crops, millet and sorghum, have declined as a share of total cereal production (Table 14-1). This relative decrease is linked to the low yields of the crops relative to rice and maize; in addition, varieties of both crops are grown under highly variable rainfall conditions right up to the Sahara and have borne the brunt of recent drought.

The production of rice has significantly increased in both absolute and relative senses in Mali and in Mauritania, doubling in Mali to 176 pounds per capita (80 kg) and to 55 pounds per capita (25 kg) in Mauritania as a result of dam construction and irrigation. It is likely that further gains through irrigation will be difficult to obtain because flooding of the Niger River has decreased (Figure 14-7), and in some cases big dam irrigation has turned out to be a two-edged sword. In the Senegal River valley, for example, traditional flood retreat cultivation of sorghum may be threatened by the impoundment of the annual flood by large dams for rice cultivation. In Chad, Lake Chad was once the sixth largest lake in the world; as shown later, however, it has shrunk to one-twentieth of its recorded size in the 1960s (Figure 14-18). Its rainy season area in 1960 was about 10,000 square miles (25,900 km^2), but in 2001 it was only 839 square miles (1,350 km^2), and discharge from the Chari–Logone rivers into Lake Chad has decreased by 75% over the last 40 years because of irrigation (Coe and Foley 2001). Flood retreat agriculture that was once practiced along the shores of Lake Chad is no longer possible because Lake Chad has changed so dramatically owing to decreased precipitation, but especially to the diversion of its tributaries for irrigation.

There were significant changes in animal husbandry as well. Of the four principal domesticated animals in the region, camels, cattle, goats, and sheep, there was a major shift from cattle to more drought-resistant small stock (mainly goats) throughout the region. In every country of the Desert Tier, cattle as a share of the national herd declined from the early 1960s to the present: in Senegal the decline was 18%, in Niger 16%, in Burkina Faso and Mauritania 15%, in Chad 8%, and in Mali 1%. Only in Mali, where vast grasslands are watered by seasonal inundation of the Niger River, was the change relatively small over the two time periods (Table 14-1).

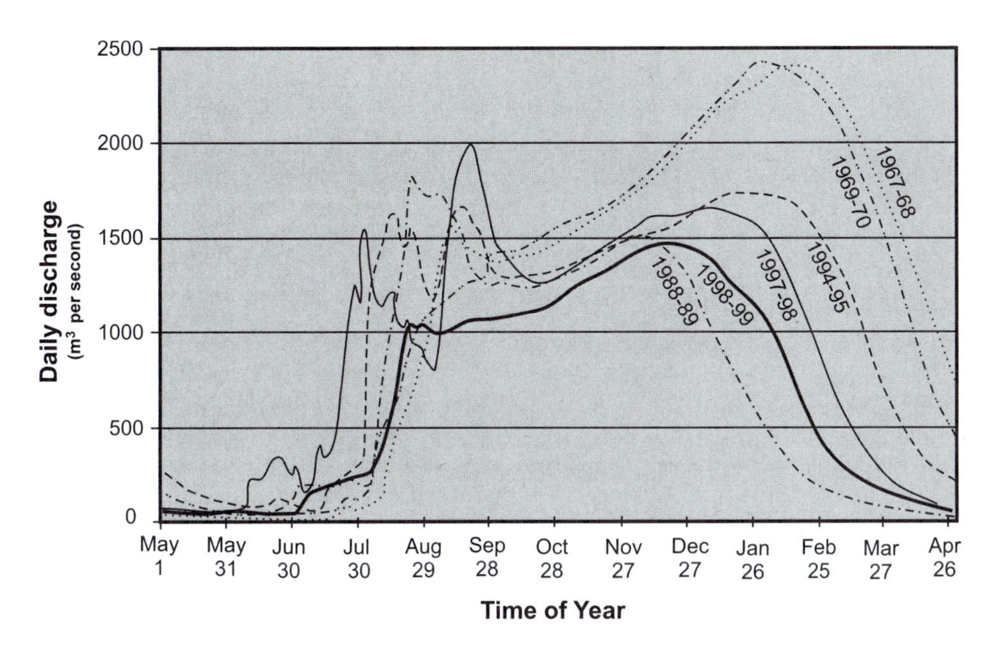

FIGURE 14-7 Daily discharge of the Niger River at Niamey, Niger, for 1967–1968 to 1998–1999. *Adapted from McCarthy et al. (2001).*

In addition to agricultural change, people diversified their economic activities to spread risk. More farmers began to fish, more herders began to farm. People took up trades. More and more people increased their incomes through seasonal labor migration. Recent research has confirmed what previous research had indicated: that a diverse range of socioeconomic activities involving labor migration, off-farm work, social networks, and the raising of livestock are critical in adapting to drought and, furthermore, that investment in women is central to continued response (Roncoli et al. 2001). The opportunity for gainful, nonfarm employment has increased since the 1970s, and especially since the 1980s, when governments began experimenting with market-oriented solutions to the provision of goods and services as an alternative to statism. The demise of statism and government controls over land and marketing has encouraged a freer market for cereals as well and has provided more incentive for farmers.

Donor and receiving countries and organizations have learned many lessons in the Sahel. One of the most important has been that drought is a recurrent phenomenon in the region and that year-to-year variability in rainfall is high. There is ample evidence in written records and archaeological sites that rainfall amounts have varied dramatically in the past (Baier and Lovejoy 1976, Webb 1995). During the fifteenth and sixteenth centuries, when the Songhay Empire was at its height, more humid conditions prevailed that permitted higher densities of people and cattle in the Niger valley. An 11-year drought occurred in the middle of the sixteenth century, and another at the turn of the seventeenth. A 7-year drought occurred at some time between 1690 and 1720, and there were severe famines in the 1740s or 1750s, and again in the 1790s.

According to local chronicles and travelers' accounts, much of the Sudano–Sahel region recorded favorable rains throughout the nineteenth century. Lake Chad was at its highest recorded level in 1880, but 26 years later it was at its lowest up to that point. In 1906 the mean depth was only a few feet (1–2 m), the greatest depth being only 13 feet (4 m), and people on the eastern side of the lake migrated west to find fresh grazing and drinkable water (Grove 1974). Heavy rains in 1916 restored Lake Chad to its mean level for the first half of the twentieth century, and the water rose throughout the 1950s and most of the 1960s. During the late 1960s, however, the lake level plummeted lower than at any time on record: the fishing village of Bol, which once stood on the lake-shore, was 18 miles (29 km) from water by 1973 and 25 miles (40 km) away by 1997 (USGS 2001). The Sahel suffered a serious drought from 1912 to 1914, when in northern Nigeria the isohyets were from 125 to 220 miles (200–350 km) south of their mean position (Grove 1974). Similar conditions prevailed in 1941 and 1942. Unusually heavy rains fell in the 1950s and early 1960s, when most regions also recorded dramatic increases in their human and livestock populations. As a result, when drought conditions returned there was insufficient pasturage, and consequently exceptionally high stock losses.

We have also learned that with the well-known exceptions of the Sahel drought of the early 1970s and 1980s, food insecurity is generally very local and exists in good years as well as bad. Local food insecurity and the day-to-day crisis of the poor is masked by the aggregated data depicted in Figure 14-6, and the use of such data is inadequate in combating hunger. Sen (1982) demonstrated that in many cases famine is caused not by lack of food in the marketplace but the inability to buy it. Lessons learned in the Sahel also made it clear that although food aid may be useful in acute episodes of food insecurity and famine, the long-term mass distribution of cereals is no solution because such artificial abundance in the marketplace ultimately drives prices down and erodes the incentives for local farmers to produce food. Long-term food distribution creates dependency on food aid and encourages rent-seeking (corruption) by officials. Food-for-work programs targeted toward specific families and individuals in specific localities offer an alternative to free distribution.

One of the major problems that emerged in the early 1970s was insufficient information to make appropriate decisions about the impacts of drought and to identify those in need. Since the first Sahel drought, information-gathering networks have been set up throughout the Sahel to collect data on daily rainfall, cereal and livestock prices, unusual migrations of people, infestations of pests such as locusts, begging, and other measures of food insecurity. Most often primary school teachers collect such information on the villages and environs in which they are located, whereupon it is fed into regional, national, and international famine early warning systems (FEWS). Satellite data are also used to estimate the extent of the rains and the vigor of vegetation. The U.S. Agency for International Development (USAID) has been a pioneer in this regard in the Sahel and across much of Africa. Geographic information systems (GIS) are used to combine the various indicators and produce maps that show where food insecurity may be high. The FEWS data collection process has developed over time and is continually getting closer to the ground to those in need (Box 14-1).

Climatic Change or Periodic Drought?

"Desert encroachment" has been defined as the drying up of lakes and rivers, the reduction of vegetation cover, and increased soil exposure and soil erosion along the desert edge. At the time of big drought of the 1970s, however, it was unclear whether the "desertification" of the Sahel was due to secular climate change or a deterioration of the environment *in situ* under the impact of human and animal activity. Hard on the heels of the Sahel drought of the early 1970s many viewed the human consequences of the drought as reflecting an imbalance between population and resources. These observers reasoned that since the carrying capacity of the land was limited, once populations had exceeded the ability of the land to support them, the environment would degrade and there would be a die-off of people and livestock until an equilibrium was reached. Many regarded the increasing imbalance between people and resources as having been set in motion years before by modern medicine and technology brought by European colonialists. Colonialism was also said to have weakened indigenous drought-coping strategies by interrupting traditional migration patterns, disrupting familial control over agricultural surpluses, monetizing the exchange system, and introducing taxes (Copans 1975, Raynault 1975). Some academic analysts of the situation felt that the evidence indicated that ultimate responsibility for the human disaster rested with Europeans through the introduction of cash crops, the modern market economy, and capitalism which, according to this view, "marginalized" the population and increased vulnerability to what otherwise would be considered normal variation in precipitation (Hewitt 1983, Meillassoux 1974, Watts 1987).

Today we know that living near the edge of the Sahara has always been a risky prospect. Although we do not know how many people and livestock can be supported along the desert's edge, it is clear that these numbers will be determined by the level of human and economic development that is established in the region; there is no limit fixed by some innate property of the land or climate. Many highly productive countries and regions of the world that are predominantly semiarid or arid possess a high standard of living. The people of the Sahel need more market mechanisms,

BOX 14-1

FEWS NET SAHEL WEST MONTHLY REPORT

Discussions of food security issues in the Sahelian countries have invariably focused on the question of grain availability, as these food products account for a major share of the typical household diet. For a long time, food security and food self-sufficiency policies have given top priority to boosting grain production.

However, it has now become clear that, in many regions, improvements in grain availability have not necessarily helped the food situation of local households, particularly that of the most food-insecure households.

In rural areas, food insecurity is a complex problem with much closer ties to crop production and to economic, social, and cultural factors. The causes of food insecurity range from shortfalls in crop production to limited food availability, weak purchasing power, food access problems, and the mismanagement of available food supplies.

In urban areas, a large segment of the population has a secure cash income, and for these people, food security problems are more a question of the ability of a given household to buy food on local markets. Moreover, higher-income households have a more diversified diet (including a larger share of meat and imported foodstuffs), while lower-income families, trapped in a vicious cycle of mounting urban poverty, must be content with a diet of so-called famine foods, which are similar to the diet of many in the rural areas.

Ongoing discussions of the feasibility of replacing the grain balance sheet currently used by Sahelian countries with a more comprehensive and inclusive food balance sheet were prompted by repeated entreaties by governments, donors and other users to furnish CILSS [the permanent interstate committee for drought control in the Sahel] member countries with tools capable of assessing the food situation of a given region or population group.

Source: Strengths, weaknesses of cereal balances. Sahel West Monthly Report, Joint Monthly Food Security Report for the Sahel, FEWS Net. April 30, 2003. http://www.fews.net/.

modernity, and articulation with the global economy to widen the economic choices available. To cope with drought, change has to continue.

The scientific consensus is that the climate is changing in West Africa—indeed around the world—and the recent drying trend has been associated with the warming of the Indian Ocean (Giannini et al. 2003). Yet there is uncertainty over the regional manifestations of global climate change. According to the most recent models, some parts of the Sahel will be drier than the long-term mean, and other parts will be moister. The Niger River, according to some models, will increase its flow by 10%, but evaporation will increase by 10% because of an increase in temperature (Hulme et al. 2001).

On the whole, the impact of climate change will probably be gradual and people will adapt as they always have. Without systemic change, however, the environmental implications of such adaptations may be negative and could undermine long-term adaptation. On the plus side, the monitoring of food insecurity has improved markedly since the 1970s and, although not perfect, it is very close to the ground. Prices of cereals and livestock, nutritional surveillance, satellite imagery of vegetation vigor, and firsthand observation by a network of observers are used to map and communicate the food situation to decision makers. Freer trade between countries in cereals and other food crops and internal price structures that reflect market incentives for the production and marketing of food, rather than government policy, will help make response to local and regional crises in West Africa a West African problem rather than a world problem. It is likely that other issues will have to be confronted to maintain sustainable development along the Sahel. One controversial but vitally important issue is the creation of a free land market and increased private land tenure.

TABLE 14-2 AVERAGE ANNUAL FOREIGN DIRECT INVESTMENT (MILLIONS OF $US) IN BURKINA FASO, CHAD, MALI, MAURITANIA, NIGER, AND SENEGAL, 1970-1974, 1980-1984, 1990-1994, 2000-2004.

	Year			
	1970-1974	1980-1984	1990-1994	2000-2004
Economy				
Burkina Faso	1.6	1.6	5.1	21.7
Chad	4.1	1.7	11.6	537.9
Mali	0.8	4.2	1.3	152.0
Mauritania	4.1	12.9	6.9	152.8
Niger	1.2	14.8	20.3	13.0
Senegal	9.0	14.3	27.4	59.1
Total	20.9	49.4	72.7	936.6

Source: UNCTAD (2005).

At the present time it is impossible to identify the factors that will be the most important in adaptation to climate change in West Africa—it is likely that a mix of strategies will be important. Just as with vulnerability to drought and other risks, vulnerability to climate change will be increased because of low levels of development, widespread poverty, dependence on rain-fed farming, low farm labor productivity, high rates of population growth, polygyny and high numbers of female-headed households, high disease burden, agricultural disincentives, and external debt burdens. Tackling these development problems would appear to be a wise long-term approach to overcoming the inherent disadvantages of life in the Desert Tier.

Cheap power is viewed as fundamental to development and new agricultural projects, and thus the impounding of water for the production of electricity and for irrigation is high on the agenda of companies seeking profitable investments, as well as international governments and NGOs. Such projects give governments and agencies an opportunity to transform traditional agriculture. International investment has been encouraged by increasing economic liberalization, economic and political stability, and more transparency in government. Foreign direct investment in the Sahel has increased from an annual average of US $21 million from 1970 to 1974 to US $936 million on average from 2000 to 2004 (Table 14-2). The major recipient of FDI from 2000–2004 has been Chad. The investment is focused on the development of the petroleum deposits located in southern Chad. Major investments in Mali and Mauritania are mineral-related as well.

It should be noted, however, that most big projects involve difficult choices and unintended consequences; moreover they are controversial with environmentalists and unpopular with local people who may suffer. International agencies and African governments have framed the major environmental and developmental issues in a regional rather than national context, although national and local interests sometimes take precedence, particularly with NGOs. To deal specifically with problems directly caused by drought and related ecological problems, the six Sahelian states established the Interstate Committee for Drought Control in the Sahel, headquartered in Ouagadougou, the capital of Burkina Faso. Among the objectives of this organization, usually designated by its French abbreviation, CILSS, are the creation of greenbelts around villages, improved marketing storage and distribution facilities, and FEWs monitoring. CILSS works in conjunction with international agencies to combat the effects of drought.

The Lake Chad Basin Commission, composed of the four states sharing Lake Chad, has expanded the irrigated area along the rivers leading to the lake, but results have been disappointing with continued drought. Mali, Mauritania, and Senegal are members of the Organization for the Development (*mise on valeur*) of the Senegal River (OMVS), the successor to the Organization of River States of Senegal, of which

BOX 14-2 DIAMA AND MANATALI DAMS—A WORLD BANK VIEW

As part of the development of the Senegal River Basin, the ... OMVS constructed two dams. At the river's mouth, the Diama dam inhibits saltwater intrusion into the river to allow its use for irrigation and to regulate water levels for navigation. In the headwaters, the Manantali dam was built to generate hydroelectric power and to provide flows for navigation and irrigation in the middle valley of the Senegal River. Prior to dam construction, which began in 1986, natural inundation covered up to 250,000 hectares of the floodplain of the middle Senegal Valley, and supported up to 125,000 hectares of flood recession agriculture (including maize, beans, watermelon, potatoes, and millet), grazing, forests (which provide fuelwood and construction timber), fisheries, and wildlife habitat. The estimated economic value of the floodplain was: recession agriculture, $56–136 per hectare; fishing, $140 per hectare; and grazing, $70 per hectare. It was feared that cessation of floods through operation of the Manantali dam would have a devastating effect on mid-valley livelihoods. By 1991, the reservoir had filled and the spillway was tested, but the turbines had not been installed. In subsequent years, OMVS conducted managed flood releases to inundate 100,000 hectares of land that provided a cultivation area of 50,000 hectares. This required around 7.5 billion m^3 of water, yielding a value of about $2 per 1,000 m^3 for agriculture, fishing, and grazing. The World Bank has agreed to help finance the installation of turbines at Manatali in 2001, provided that the principle of managed floods is recognized by OMVS as a possible long-term option. The Governments of Mali, Senegal, and Mauritania have now signed a Water Charter that includes the release of the annual flood when sufficient water is available (normally 9 years out of 10).

Source: Davis and Hirji (eds.) (2002).

Guinea was once a member. In 1974 OMVS negotiated a multiproject scheme valued at 800,000 million CFA francs that included a hydroelectric dam and irrigation scheme at Manantali, Mali and an irrigation dam on the Senegal Delta at Diama. The dam designed to produce electricity and irrigate was completed at Sélingué on the Niger River in 1980. The Diama and Manatali dams were completed in 1986 and 1987, respectively; Manatali began generating power in 2001, delayed both by the rift between Senegal and Mauritania and by large cost overruns. Environmental activist groups reported negative effects on traditional flood–retreat agriculture, a decline in the recharge of aquifers, and invasion by nuisance vegetation (NGO Working Group on the Export Development Corporation 2001). But other researchers have found many benefits to such dams (Box 14-2).

A dam along the Niger River in Niger is in the works at Kandadji about 75 miles (120 km) upstream from Niamey. The government of Niger anticipates completion of the dam in 2012 and expects it to generate 165 megawatts of electricity. In 2003 Niger and Nigeria began connecting parts of their national electric grids as part of the West Africa Power Pool (WAPP) project. The power imported into Niger will be significantly cheaper than local oil-generated electricity (EIA 2003). Burkina Faso is linking its national grid to Ghana and Côte d'Ivoire: a 225-kilowatt power line to connect Ouagadougou with Ferkéssédougou in northern Côte d'Ivoire was completed in 2005. Other regional organizations operating in West Africa dealing in one way or another with development issues are the Niger River Commission, the Economic Community of West Africa (CEAO), the African Development Bank, ECOWAS, and the Organization for the Development of the River Gambia (OMVG). The African Development Bank is working with Gambia in the development of hydroelectric power along the Gambia River and in connecting the national grids of Gambia, Guinea, Guinea-Bissau, and Senegal, all members of the OMVG.

TABLE 14-3 SELECTED DEVELOPMENT DATA FOR THE SAHELIAN STATES.

	Burkina Faso		Chad		Mali		Mauritania		Niger		Senegal	
Item	1977	2003	1977	2003	1977	2003	1977	2003	1977	2003	1977	2003
Area (mi^2)		105,792		495,753		478,838		395,954		489,189		75,954
Arable land (mi^2)	10,174	15,444	11,969	14,016	7,951	18,147	772	1,931	12,162	17,375	9,073	9,653
Permanent pasture (mi^2)	23,166	23,166	173,746	173,746	115,831	115,831	151,545	151,545	37,182	46,332	22,008	21,815
Population (millions)	6.4	13.2	4.3	9.3	5.9	11.6	1.8	2.9	4.8	12.1	4.7	10.6
Population density	60.5	124.8	8.7	18.8	12.3	24.2	4.5	7.3	9.8	24.7	61.9	139.6
Agricultural density	629	855	359	664	742	639	2,332	1,502	395	696	518	1,098
Population growth rate (%)	2.2	2.8	2.4	3.2	2.4	3	2.3	2.9	2.7	3.5	2.4	2.7
Infant mortality rate	204	105	175	103	188	126	169	101	200	123	159	68
Life expectancy (years)	36	45	40	49	38	45	46	54	38	45	44	53
Urban population (%)	6.3	16.5	15.6	23.8	16.2	30.2	20.3	57.7	10.6	20.6	34.2	39.6
Population in agriculture (%)	88	84	85-90	72	83-87	79	85-90	63	85-90	86	70	70
Physicians per 100,000 population	1.1	3.0	1.5	3.0	2.6	5.0	3.8	14.0	1.8	4.0	6.9	8.0
GNP per capita ($US)	$70	$250	$85	$230	$70	$260	$175	$440	$110	$200	$285	$540
Literacy rate (%)	6	27	37.75	48	5	46	37.62	42	5	18	37.75	40

Sources: Latest available data from the Population Reference Bureau, USAID, United Nations Human Development Reports.

Although big dams are sometimes controversial, they are few in number. Far more numerous are the hundreds of small impoundments, or microdams, that have been built throughout the drylands. Although the current focus is usually on new microdams, there are ancient impoundments, canals, and dikes in the region, for example, in the Niger Inland Delta of Mali. Modern microprojects are often identified by NGOs and funded by international agencies.

International research organizations such as the International Crops Research Institute for the Semi-Arid Tropics (ICRISAT) have offices in the Sahelian countries and finance research to make local crops more productive. The U.S. Agency for International Development (USAID) and the Organization for Economic Cooperation and Development (OECD) are active in promoting development in the region.

Contrasting Problems and Prospects

In 1999, despite some local shortages, the Sahel had excellent, even record, harvests; 2003 was another good year. Although all evidence points to continuing challenges for people of the Desert Tier, human development has been improving over the past quarter-century (Table 14-3). Nevertheless, life expectancy and other vital statistics are still unacceptably low by the standards of more developed countries. The six Sahelian states share many environmental and institutional characteristics and problems, the most pressing being poverty in a harsh environment; but each state has its individual development priorities, problems, and prospects. Each desires a more balanced economy, greater control over its resources and environment, and additional industry—not merely import substitution manufacturing, but heavy industry that is believed to bring prestige, wealth, and greater economic viability. Several countries desire steel mills, petroleum refineries, aluminum smelters, and automobile factories. But one large integrated iron and steel mill, and only two or three petroleum refineries, for instance, could meet the combined needs of all 16 West African states.

Ideally, a number of integrated economic organizations would operate on a regional basis (Adedeji 1970). A few exist in the form of CEAO, OCAM, OMVS, and ECOWAS, but these agencies do not concern themselves with the real issues of manufacturing specialization, marketing, and distribution on an effective regional basis. Individually, only Senegal shows much developmental potential in the short to near term. Chad, which recently began piping oil into the world market, may have the natural resources to improve the living standards of the majority of Chadians, but it is too early to tell to what use the oil bonanza will be put.

SENEGAL

Of the six Sahelian states, Senegal has the most diversified economy, mainly because of its privileged status during the colonial era, its resource endowment, coastal location, and political stability. Economic diversity helps explain why the percentage of the Senegalese population that is in poverty is roughly half that of Chad (64%), Niger (63%), and Mauritania (57%). Nonetheless, there is cause for concern, particularly with respect to the high unemployment rate of 48%. The city of Dakar became the primate city and federal capital of French West Africa, and thus the most influential, the most prestigious, and the most cosmopolitan. Its economies were geared to and dependent upon a vast hinterland that extended east to Burkina Faso and Niger. When this interior was divided into separate independent states, however, Dakar lost this preeminence. This was underscored in 1960 when Mali broke diplomatic relations with Senegal and closed down its rail links from Bamako to Dakar until 1963.

After independence, Senegal's first president, Léopold Sédar Senghor, pursued a policy of "African socialism" and increased the Senegalization of the economy; but in contrast to some of his neighbors, at the same time maintained close economic relations with France and encouraged other Western capitalist investment with such enticements as the free-trade zone established in Dakar in 1974. Manufacturing, which accounts for a third of the GDP, emphasizes the processing of locally produced groundnuts, cotton, rice, and phosphate ores. Senegal also has a diversified chemicals industry that utilizes imported raw materials, as well as several metal fabricating and truck assembly plants, and the largest textile industry in French-speaking Africa.

Twenty years after the Sahel drought of the early 1970s, Senegal began to dismantle Senghor's socialism and to liberalize its economy. The national currency

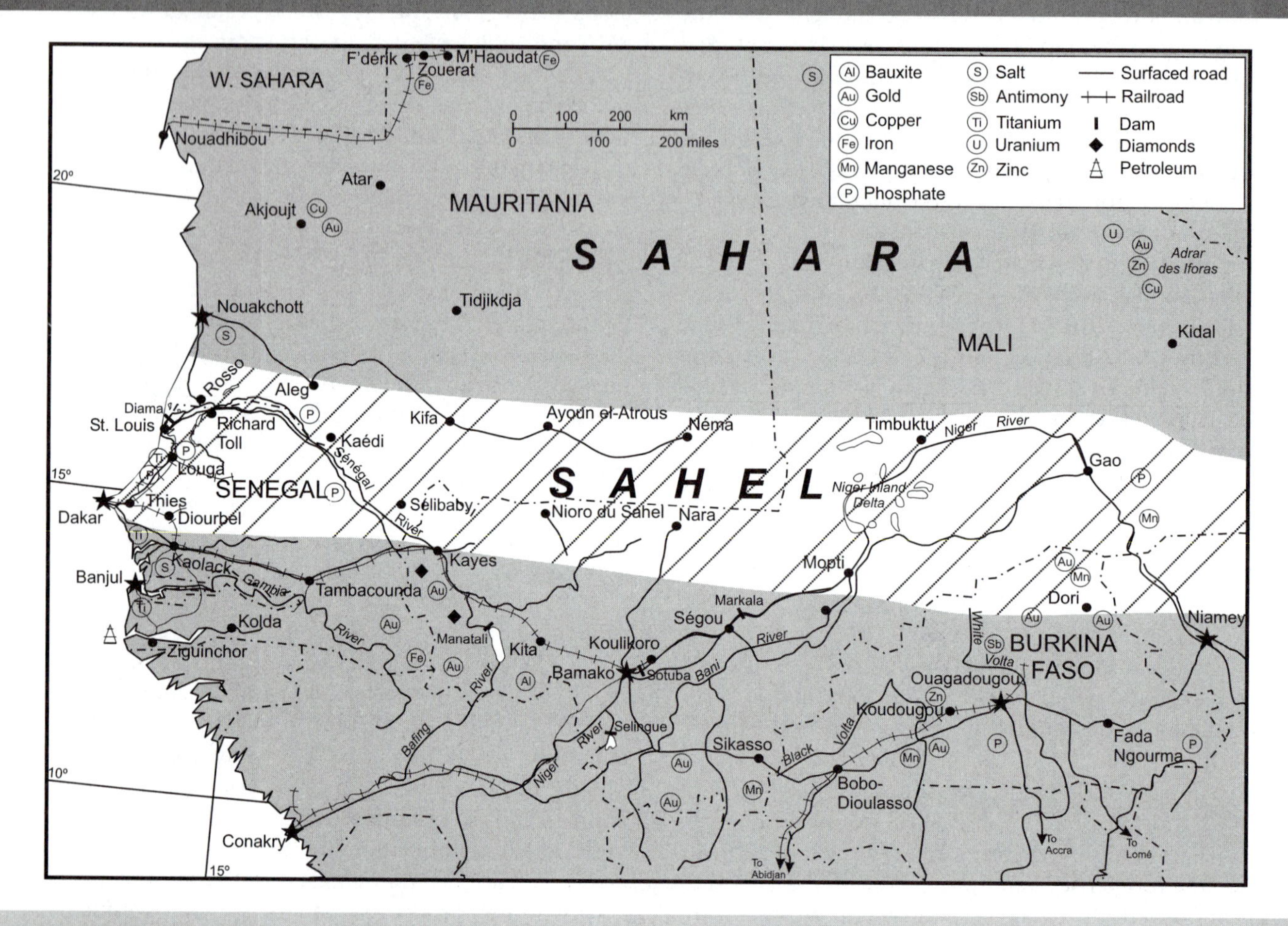

FIGURE 14-8 Natural resources and infrastructure for the Sahelian states of Burkina Faso, Mali, Mauritania, and Senegal.

was devalued, and steps were taken to do away with price controls, government corporations (parastatals), and subsidies. From 1995 to 2002, Senegal's economy grew on an annual basis by 5% on average in real terms (over the inflation rate). Private enterprise and foreign direct invest rose in conjunction with liberalization, and the country has been experiencing growth in information technology services. From 1990 to 2000, manufactured exports increased from 23% to 30% of all merchandise exports, and high technology exports accounted for 13% of manufactured exports in 2000. Most industries are located in metropolitan Dakar (2.4 million), Thies (250,000), Kaolack (177,000), and towns along the railway to St. Louis (160,000), the former capital, and along the main line into the interior (Figure 14-8).

With regard to the three major sectors of the economy, agriculture, manufacturing, and services, Senegal's economy is more balanced than others in the region, and the focus of the economy has been shifting away from agriculture since the 1970s. In the mid-1970s, agriculture accounted for a third of GDP and provided employment for 70% of the economically active population. In 2003, agriculture accounted for only 18% of GDP (industry accounted for 27% and services the rest) but the percentage of the population that was engaged in agriculture remained at 70%. Senegalese agriculture is essentially monocultural, the

FIGURE 14-9 Annual harvest of the peanut cooperative formerly run by the state in Louga Region, Senegal. *Roy Cole.*

overwhelmingly important cash crop being groundnuts (peanuts: Figure14-9). Since independence, the production of groundnuts has trended downward, but output has fluctuated wildly, mainly as a consequence of precipitation variation, reaching a maximum of 1.4 million metric tons in 1975 to a minimum of 501,000 metric tons in 2001 (Figure 14-10). The government has attempted to diversify its rural economy by expanding the production of cotton, rice, sugar, and market garden produce especially along the Senegal

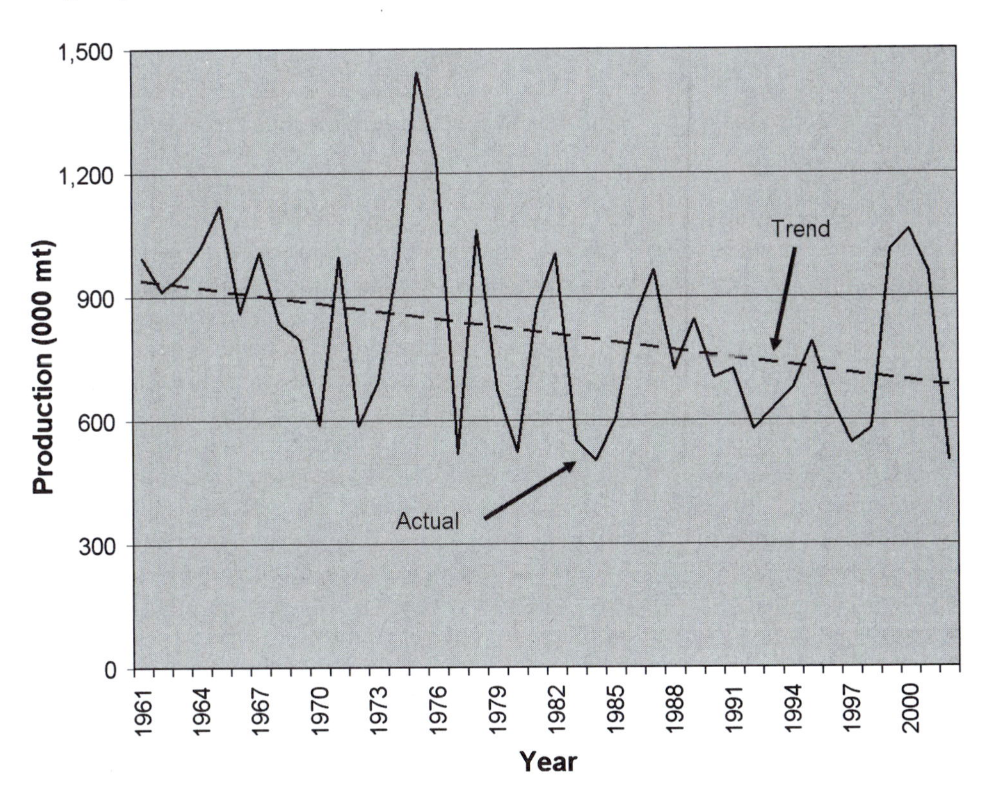

FIGURE 14-10 Trends in groundnut production in Senegal, 1961 to 2002.

BOX 14-3

GAMBIA

The African microstate of Gambia (4,261 square miles; 11,036 km^2) is virtually an enclave within Senegal and is southern Senegal's natural outlet to the Atlantic Ocean. Gambia, which comprises a narrow strip of land on either side of the River Gambia and whose boundaries correspond to no natural or demographic boundaries, was the object of French interests in the eighteenth century until it was reserved for Great Britain by the 1783 Treaty of Versailles. In 1807 the British established a military post at Bathurst (now Banjul, the capital), to control slavery, and acquired additional territory upriver by concluding treaties with local chiefs that established British protection over their lands. For administrative purposes, the territory was divided into a coastal colony and an interior protectorate, and from 1821 to 1843 Gambia was administered from Sierra Leone. Between the 1860s and the 1870s Gambia was a pawn in negotiations with France, with the aim of exchanging it for more desirable land elsewhere. But British merchants convinced London of Gambia's commercial and strategic worth, and final boundaries were drawn in 1889.

The artificiality of Gambia has been detrimental to both Gambia and Senegal in that the boundaries imposed cut ethnic groups and restrict the natural flow of goods and ideas. Since 1973, Gambia and Senegal have been discussing unification. In 1982 Gambia and Senegal joined together in a federation called Senegambia, but it was dissolved in 1989.

Gambia is a smuggling base for manufactured imports into Senegal, and an illegal recipient of Senegalese cattle and groundnuts. Groundnuts and groundnut products normally contribute about 95% of total exports, with the EU being the principal market. Size, shape, and location, combined with a small population (1.5 million), sparsity of resources, and rural economy that is subject to the vagaries of unreliable rainfall and prolonged droughts, make Gambia one of Africa's poorest countries. Development opportunities are extremely limited.

River (Richard Toll Scheme), in depressions (niayes) located along the Atlantic Ocean, and along the Casamance River in the south.

Development in southwestern Senegal is hindered because the Republic of Gambia separates this region from the national core (Box 14-3).

MAURITANIA

The Islamic Republic of Mauritania, a vast compact territory of scanty vegetation, barren surfaces, poor communications, negligible internal commerce, and low population, has been grappling with its past and its environment while at the same time attempting to build modern institutions on which human development can be based. Recent development has focused on minerals, fishing, and irrigated agriculture.

Ecologically less diverse than its more southerly West African neighbors, Mauritania possesses four ecological zones, the Sahara, the Sahel, the Senegal River valley, and the Atlantic coast. Of the four, the Senegal River valley is the anchor of agricultural production and population. The rest of the country is very dry and, except in the towns, supports only a low density of people and livestock. The iron and copper resources of Mauritania are found in the Sahara. Pastoralists seasonally lead their livestock into the Sahel to graze the annual grasses, and traders trek by camel to exchange salt for cereals (Figure 14-11). The Atlantic coast supports a very productive fishery that has been overexploited, but there is a promise of petroleum just offshore.

The country of Mauritania is more racially diverse but less multicultural than most Subsaharan African countries. The people of Mauritania are mixes of Arabs, Berbers, and black Africans. The elite group is made up predominantly of Arabized Berbers who dominate government, industry, commerce, and education. The Arabs are relative newcomers to the region in comparison to the Berbers and black Africans (Soninke,

FIGURE 14-11 Mauritanian camel caravan in Senegal to trade salt for millet and sorghum.
Roy Cole.

Tukulor/Fulani, and Wolof mainly) with whom they have intermarried to a large degree. The period of their conquest and dominance of what is today called Mauritania and the Senegal River valley constitutes the tail end of the migrations of pastoral Arabians into northern Africa beginning in the seventh century and continuing to the middle of the fourteenth. During that time, groups of Arab Bedouin migrated across North Africa to Algeria, Morocco, and beyond, Arabizing the indigenous Berbers (Moors). The official languages are Hassaniya Arabic and French, but Fulfulde, Wolof, and Soninke are also spoken.

Shortly after independence, Mauritania sought to reduce its economic, military, and technical ties with France, and to loosen its ties with other former French West African territories. In 1973 it left the Franc Zone and the West African Monetary Union to join the Maghreb Economic Union. Its bank is supported by several Arab states including Kuwait and Saudi Arabia. This increased identification with the Arab world is not unexpected, since about 70% of the population of Mauritania is Arab or Arab–Berber. But transport links with neighboring Arab states are limited, and the population that resides in the extreme southwest is separated from the major population centers of Algeria and Morocco by thousands of square miles of desert. The non-Arab population, constituting about 30% of the population, comprises mainly sedentary Fulani (Peul), Soninke, Wolof, Tukulor, and Bambara. It has strongly but unsuccessfully opposed Arab domination, yet remains an important centrifugal force in the south. Northern Arabs favor closer union with Morocco.

Mining has underpinned Mauritanian development since independence. Although copper has been mined at Akoujt since ancient times, Mauritanian industry in recent years has been focused on iron mining. In 1952 the Miferma company (Mines de Fer de Mauritanie) was formed to exploit iron deposits in the country. The Miferma iron ore mine at F'dérik (Figure 14-8) came into production in 1963 and, when the world economy is not in recession, can produce over 10 million tons of high-grade ore each year. Three years after the mine opened, the contribution of iron mining to the GDP of Mauritania reached 28% and accounted for a staggering 92% of exports. The economy has since diversified: iron mining accounted for 7% of GDP in 2003, but it still makes up about 44% of export revenues (Bermúdez-Lugo 2003). The iron ore deposits around F'dérik are of exceptionally high quality (65% iron content), and reserves are estimated to be 190 million metric tons. About 700 million metric tons of lower-grade (36–40% iron) magnetite ore is found in the same region. Today over 60% of the high-grade iron comes from mines near M'Haoudat.

Until December 1974, when all holdings were nationalized, iron mining was controlled by French, British, Italian, German, and local interests. With the financial assistance of the oil-rich OPEC states, Mauritania bought out the foreign interests and created a new company, the Société Nationale Industrielle et Minière (SNIM). SNIM was a state corporation, 100% owned by the Mauritanian government. Later, government ownership was reduced to 78%, and international venture capitalists bought the rest.

The contribution of iron mining to Mauritanian development gave the economy a jump-start—the wealth generated by the export of iron ore so impressed the United Nations and the World Bank that Mauritania was reclassified from being a "least developed" to being a "moderate-income" country (U.S. Library of Congress no date). Iron income enabled the rapid urbanization of Mauritania, creating an African anomaly: a population that is almost 60% urban (Table 14-3). At independence in 1961 both Nouadhibou and Zouérat had less than 5,000 population each. By the middle 1970s each had grown to around 25,000, and by 2003 Nouadhibou had grown to over 80,000 inhabitants and Zouérat had risen to over 37,000 people. Nouakchott, the nation's capital, grew from about 5,000 people at independence to 125,000 in the mid-1970s to over 660,000 in 2003.

In 1974 the world demand for iron ore declined with the world recession that began with the sharp oil price increases due to the OPEC oil embargo. Conflict in Western Sahara exacerbated the economic and social trauma of world recession. In 1976 Mauritania and Morocco partitioned Western Sahara (formerly Spanish Sahara), and Mauritania fought for control of its third with the Polisario, an armed, nationalist group that sought to expel both invaders. The Polisario attacked the trains and rail lines that transported iron ore from mine to port and adversely affected the Mauritanian economy, pushing it further into recession. Bloodied, Mauritania withdrew its claims to Western Sahara in 1978.

Long-term drought beginning in the early 1970s reduced the agricultural output and sent thousands of environmental refugees to the cities and towns, creating another economic and political crisis. In 1978 the civilian president, Ould Daddah, was overthrown in a military coup and the country has been ruled by the military (or military in civilian clothing) ever since. In 1989 Mauritania's rigid social cleavages (Box 14-4) separating white Moor from black Moor and both from black African, and the legacy of slavery, came under assault in a rebellion of descendants of slaves, former slaves, and slaves along the Senegal River. The crisis, on the boil since the mid-1980s, almost led to war with Senegal, and thousands of "Senegalese" Mauritanians were expelled across the river to Senegal from 1989 to 1990. In 1991 the military government of a former colonel, Maaouye Sidi Ahmed ould Taya (who had taken over in 1984), permitted the formation of opposition parties, and in 2001 elections were held for the legislature and municipal governments. In 2005 President Maaouye Sidi Ahmed Ould Taya was deposed in a coup. The coup plotters claim that they will hold elections and hand over power in 2 years.

In the 1990s Mauritanian ore production regained the heights of the middle 1970s, and production has continued close to that level to the present; in terms of world production, however, output is minuscule (Table 14-4). The SNIM would like to maintain an iron ore output level of 12 million metric tons a year, close to capacity. Mauritania has few other options. The iron-mining infrastructure was upgraded beginning in 1997 to help meet the 12 million ton target and also to make mining more efficient and competitive. In an attempt to diversify the mining base in the event that iron demand decreases, multinational mining corporations were invited to begin exploration for palladium, titanium, platinum, gold, and diamonds in addition to iron ore. Nevertheless, in 2000, after years of economic and social difficulties, mismanagement, environmental disasters, and the overborrowing of money from world lenders, Mauritania was classified as a heavily indebted country by the World Bank and International Monetary Fund and qualified for debt relief under the World Bank's HIPC initiative—quite a letdown from "moderate income" country of the early 1970s before the era of crises.

BOX 14-4 RACE, ETHNICITY, AND HIERARCHY IN MAURITANIA

The dominant segment in Mauritania's population, the Maures trace their ancestry to Arab-Berber origins, although many have intermarried among African populations over the centuries. Maures occupy scattered areas across West Africa from southern Morocco to Gambia and from the Atlantic Ocean to Mali. The greatest concentration of this group is in Mauritania, which took its name from this dominant segment of its population. Maure society's complex social relationships are based on rigid hierarchical social and ethnic divisions. Social distinctions reflect the interplay of heritage, occupation, and race. Broadly speaking, Maures distinguish between free and servile status on the one hand and between nobles, tributaries, artisans, and slaves on the other hand. Non-Maure populations, termed "black Africans" in this context, are not included in this ranking system.

Two strata, the warriors (hassani) and the religious leaders (zawaya), dominate Maure society. The latter are also known as marabouts, a term applied by the French. These two groups constitute the Maure nobility. They are more Arab than Berber and have intermarried little with black African populations. Tributary vassals (zenaga) are below the hassani and zawaya in status but nevertheless are considered among the elite. They are descendants of Berbers conquered by Arabs, and their Hassaniya Arabic dialect shows a greater Berber influence. Although these three social strata are termed "white" Maures (bidan), the zenaga have intermarried with other groups to a greater degree than have the hassani and zawaya.

Craftsmen and artisans in Maure society are described as members of "castes" because they form closed groups whose members tend to intermarry and socialize only among themselves. Bards or entertainers, called ighyuwa in Mauritania and griots elsewhere in West Africa, are also considered to be members of a caste. At the bottom of the social order are the so-called black Maures, previously the servile stratum within Maure society.

Myths of origin are used to reinforce perceptions of social status and justify elements of this elaborate system of stratification. Craftsmen and musicians in Maure society are said to be of Semitic (Arab) rather than Berber or African ancestry. Imraguen fishermen, a caste group living in the vicinity of Nouadhibou, are thought to be descended from the Bafour, the aboriginal black population who migrated south ahead of the expanding desert. Small hunting groups are considered to be the remnants of an earlier Saharan people and may be of Berber origin.

Source: Mauritania: A Country Study. Washington, D.C.: U.S. Library of Congress.

Mauritania has developed its fishery and fish processing industry, which has made a significant contribution to the economy. Exploration for oil has been carried on since the late 1980s, and there is anticipation of production in the next few years. Mauritania is one of a handful of West African countries with an oil refinery (the others are Côte d'Ivoire, Ghana, Liberia, Nigeria, Senegal, and Sierra Leone). The refinery is operated by Algeria using imported Algerian oil.

Irrigated agriculture, principally rice farming, has grown markedly since the construction of the Manatali Dam along a tributary of the Senegal River. After many years of delays, the dam has begun to produce electricity, and about half of the country's electricity needs are met by hydroelectric power; the rest by thermal plants. Five turbines have been installed: one each for Mauritania, Mali, and Senegal, and two for backup. The Mauritanian economy has recently been growing at 4 to 5% per year, but agricultural potential is so low and the fishery so depleted that prospects for continued human development in Mauritania appear to continue to hinge on the mineral sector. Like Senegal, Mauritania has a higher urban population than its neighbors, and urban population is likely to grow as high as many developed countries in the next twenty-five years (Figure 14-12). The city has been a catalyst

TABLE 14-4 PRINCIPAL PRODUCERS OF IRON ORE AND THEIR PRODUCTION FOR 2002.

Country	Production (millions of metric tons)	Percent of World Total
Australia	190	17.03
Brazil	220	19.71
Canada	35	3.14
China	230	20.61
India	80	7.17
Kazakhstan	16	1.43
Mauritania	10	0.90
Russia	88	7.89
South Africa	37	3.32
Sweden	20	1.79
Ukraine	60	5.38
United States	50	4.48
Other	80	7.17
Total	1,116	100.00

Source: Adapted from Kirk (2003).

for change, innovation, and social transformation in more developed parts of the world and it is likely to be so for Mauritania as well.

MALI

Mali, like Mauritania and Guinea after independence, attempted to become more self-sufficient by reducing its dependence on French financial aid, technical assistance, and trade. But unlike its western neighbors, Mali possesses few known mineral resources, has a less favorable geographic location as a landlocked state, and has suffered a series of economic and political crises that were aggravated by the prolonged drought. Under President Modibo Keita, the government developed a socialist economic policy that saw the nationalization of most enterprises, the creation of a network of rural cooperatives, and close economic ties with both the Soviet Union and China. But increasing economic problems, especially with the state corporations, led to a rapprochement with France, culminating in Mali rejoining the Franc Zone. The economic situation did not

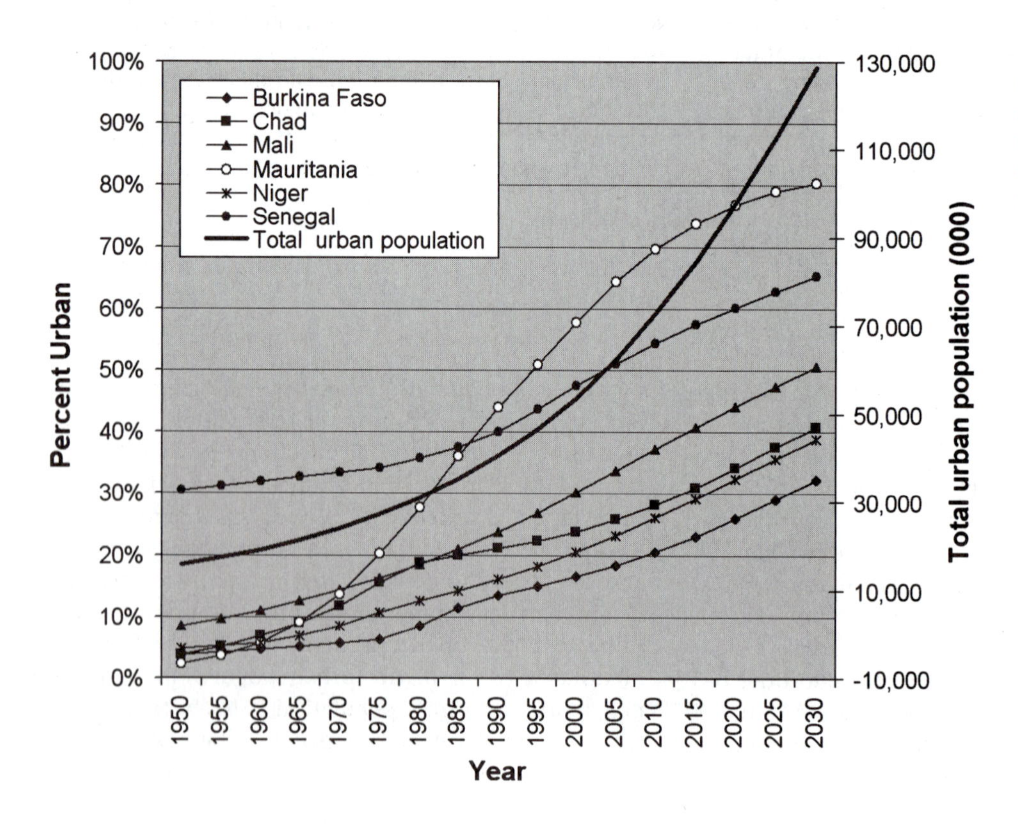

FIGURE 14-12 Percent urban population of Burkina Faso, Chad, Mali, Mauritania, Niger, and Senegal, 1955 to 2030.

improve, and political antagonisms developed between those preferring closer ties with China and those who favored greater rapprochement with the former colonial power. A military coup followed in 1968 and, although civilian government was provided for in the new constitution, which was approved by referendum in June 1974, the military, under Moussa Traoré, stayed in power until 1991. At that time the country began the transition to democracy, and in 1992 a civilian government was elected to power. Civilians have governed ever since.

After cutbacks in infrastructure and social investment, the dismantling of cooperatives, economic liberalization, macroeconomic stabilization, and the closing or privatization of state industries, Mali's economy is beginning to prosper. Mali's GDP growth was estimated to have grown 5% every year since 1995 as a result. Nevertheless, the vast majority of Malians are still engaged in subsistence agriculture, the per capita income is still less than $300, there is a serious balance of payments problem, and the country is one of the poorest in Africa (Table 14-3). Cropland is limited to the southern half of the country, where rainfall is between 25 and 55 inches (635–1,400 mm), irrigation is possible, and where small-scale traditional farming accounts for 90% of the land under cultivation (Figure 14-13). The government is encouraging the planting of cash crops, especially rice along the Niger, Senegal, and Bani rivers, and cotton in the Ségou and Sikasso regions. On the Bafing River, a tributary of the Senegal, near Kayes, Mali built the Manantali Dam in 1987 which currently generates over 200 million kilowatt-hours of electricity and has the potential to irrigate almost a million acres (405,000 ha) of land. In 1980 the Sélingué Dam was built. Although the irrigated area in Mali has increased significantly in comparison to the other Sahelian states, persistent drought has greatly reduced the annual flood (see, for example, that of the Niger River in Figure 14-7). Malian cereal yields have been slowly increasing, and rural per capita production is about the same as it was 40 years ago (Figure 14-14). Progress is needed in diversifying production, increasing yields, and slowing population growth.

Mali did not achieve the economic independence it sought in 1960, and perhaps such a state of self-reliance is mere idealism. Politically it has charted its own course since the end of the Cold War, being one of the first African countries to make the transition to democracy. In 1992 Alpha Omar Konaré was elected president, and he held office for the limit of two terms. President Konaré liberalized the economy, particularly

FIGURE 14-13 USAID field researcher Ambadembé Kassogué inspecting millet planted in Mali after the retreat of the annual Niger Inland Delta flood.
Roy Cole.

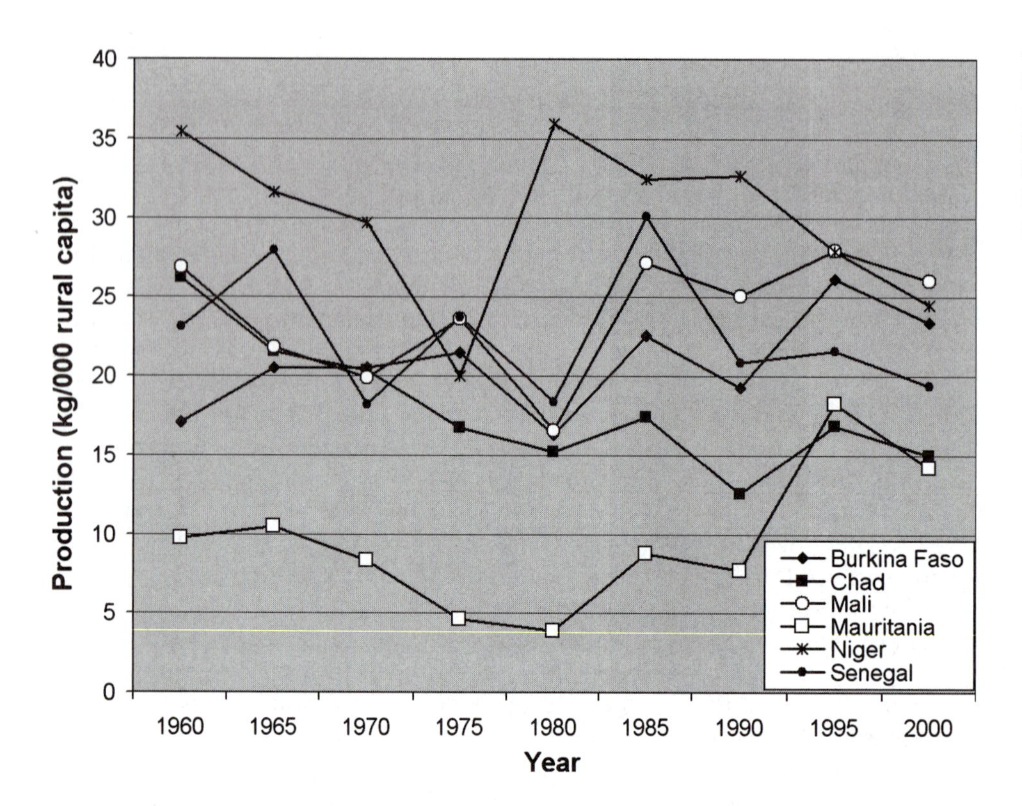

FIGURE 14-14 Cereal production per thousand rural capita in Burkina Faso, Chad, Mali, Mauritania, Niger, and Senegal, 1960 to 2000.

the mining codes, and moved it further from the Marxism–Leninism and statism of the Keita and Traoré years. In the summer of 2002 when the author (Cole) visited Bamako, the contrasts with the early Traoré years were obvious. The level of private economic activity was greatly increased; there were private businesses everywhere, and even more hustle and bustle. New neighborhoods of wealthy home owners had been build on the edges of Niger River valley and in the surrounding hills, but poverty and underinvestment was all too apparent in the center of the city and in unplanned settlements on the periphery.

In 2002 President Touré was elected in a free and fair election. Economic growth of 5% has been relatively consistent for the past decade after the devaluation of the CFA franc and the implementation of an IMF-sponsored adjustment program. For years Mali's principal export has been cotton, while livestock exports contributed significantly to the economy. In 2001 Mali became the third largest gold exporter in Africa after South Africa and Ghana (Szczesniak 2002), and gold is now the principal export accounting for 95% of Mali's mineral exports. Mali also exports quantities of phosphates, salt, and uranium, and there may be potential for bauxite, iron ore, and other minerals. Diamonds are mined artisanally in the south, but no figures are available on their production.

The country's landlocked status and the poor condition of the transportation infrastructure make mining in Mali more costly than along the coastal tier of countries. In 2000 the African Development Fund approved the disbursement of $32 million dollars to improve Mali's road transportation network. The investment is to improve the roadways from Bamako to Kankan, Guinea. The purpose of the 214-mile (344 km) road is to open up Guinea's northeast and Mali's southwest, and benefits include the promotion of economic activities and opportunities for greater investment, the easing of the transit of people, and increased political and economic links between the two countries.

The Malian experience during early independence demonstrates that a single country cannot make it alone. Development in this vast area depends on regional complementarity, increasing intraregional links between countries, and solid economic links with and investment from the global economy. Mali has taken great strides in creating a stable investment climate for economic growth. It remains to be seen whether the economic growth thus far will be accompanied by increased human development.

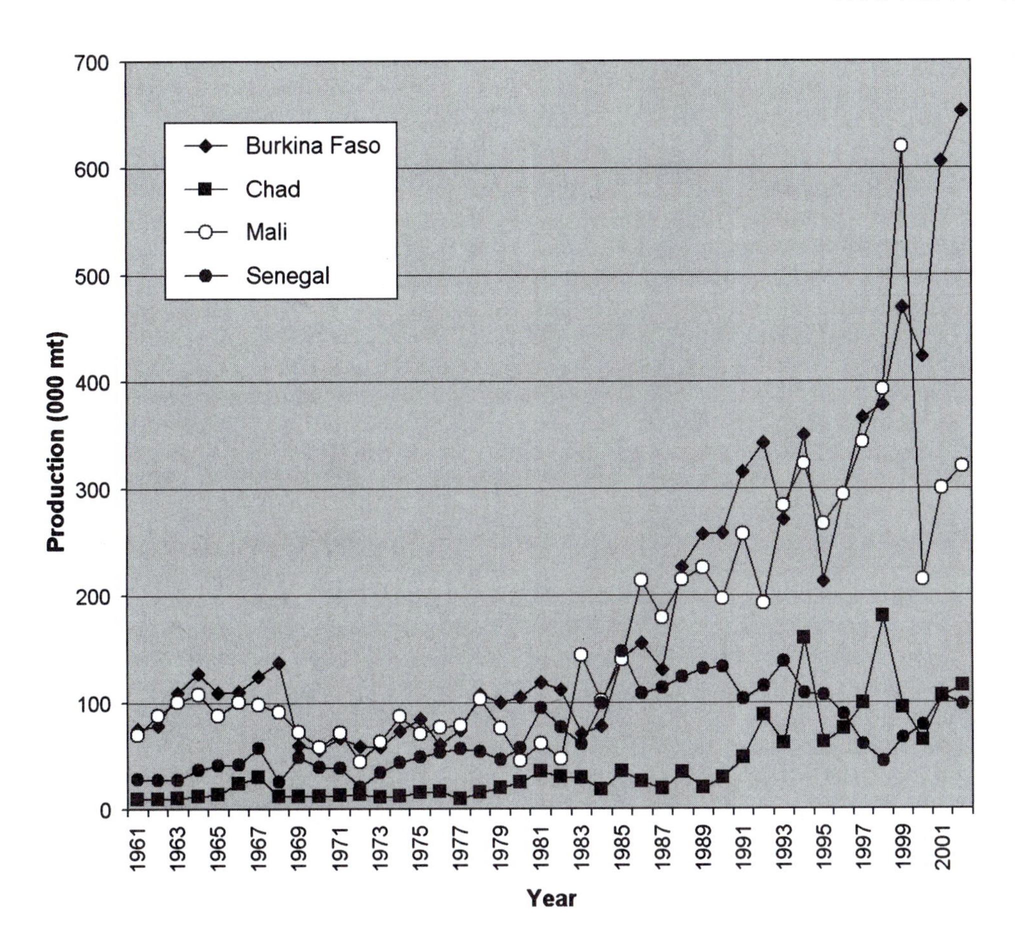

FIGURE 14-15 Maize production in Burkina Faso, Chad, Mali and Senegal, 1961 to 2002.

BURKINA FASO

Many of the same development problems that beleaguer Mali exist in Burkina Faso (formerly Upper Volta), also a primarily agricultural country and one of the poorest in Africa. Since most of Burkina Faso lies in the savanna Sudan belt south of the Sahel proper (Figure 14-8), it receives higher and more dependable rainfall than other Sahelian states. Much of the land has thin and infertile soils, however. Over the past 25 years great progress has been made in eliminating the diseases of river blindness and sleeping sickness, and large fertile areas along rivers that were once uninhabitable are no longer so (McMillan 1995). An international effort spearheaded by the United Nations was able to assist Burkina Faso in eliminating the simulium fly, the vector of river blindness, freeing some of the best land in the country for settlement. With regular spraying of breeding areas, clearance of woodland, and hunting out of game, the habitat for the vector of sleeping sickness, the tsetse fly, was removed. The new lands offered opportunity for the cultivation of maize along the well-watered, fertile river valleys. Production of maize skyrocketed between 1980 and 2002 (Figure 14-15).

The population of Burkina Faso today is more evenly distributed than it was as recently as 1975 when it was concentrated on the Mossi Plateau around the capital city, Ouagadougou (962,100), and in the higher ground around the country's chief market center, Bobo-Dioulasso (319,500), where pressure on the land was great. Nevertheless, the general lack of employment opportunities in mining and manufacturing, and the proximity of jobs in the plantations and mines in Côte d'Ivoire and Ghana, results in an annual out-migration of hundreds of thousands of Burkinabe. The general practice is to travel to Côte d'Ivoire or Ghana after the harvest, look for work, and remain there until field preparation and planting season back home. Quite a few young men remain for 1 to 5 years, usually in an attempt to save enough money for marriage. A minority never return home. Seasonal migration is not

TABLE 14-5 MINERAL PRODUCTION FOR BURKINA FASO, CHAD, MALI, MAURITANIA, NIGER, AND SENEGAL BETWEEN 1996 AND 2000.

	Year				
Country/Commodity	**1996**	**1997**	**1998**	**1999**	**2000**
Burkina Faso					
Gold (kg)	1,063	1,089	1,091	886	1,000
Salt (kg)	7,000	5,000	5,000	5,000	5,000
Mali					
Gold (kg)	4,329	16,323	20,562	23,688	25,000
Salt (kg)	6,000	5,000	6,000	6,000	6,000
Silver (kg)	270	800	900	1,000	1,000
Mauritania					
Gold (kg)	189				
Iron ore (000 mt)	11,363	11,703	11,373	10,401	10,400
Petroleum refinery products (000 42-gallon barrels)	7,060	7,060	7,060	7,060	7,060
Salt (kg)	5,500	5,500	5,500	5,500	5,500
Niger					
Coal, bituminous	140,000	150,000	145,000	168,000	158,000
Gold (kg)	1,000	1,000	1,000	1,000	1,000
Molybdenum concentrate (kg)	10	10	10	10	10
Salt (kg)	3,000	3,000	2,000	2,000	2,000
Tin (kg Sn content)	10	10	10	32	22
Uranium (kg U content)	3,320	3,497	3,731	2,916	2,898

Source: U.S. Geological Survey Mineral Reports (various years).

costless. Although most migrants find work, some do not. Furthermore, in many parts of West Africa the dry season is the time of year when homes are repaired and work around the house and village is carried out. Heads of household generally permit only the young men the family can spare to seasonally migrate. Migrants may face challenges in the host country or region. As was seen in the preceding chapter with respect to Côte d'Ivoire, competition for jobs can result in friction between locals residents and seasonal workers.

Burkina Faso is not rich in minerals, but exploration is continuing and more deposits are likely to be found. As in Mali, gold exports are second in value to cotton exports (Table 14-5). The mining code has been liberalized, and over $100 million dollars has been put into mineral exploration in the last few years. France has begun to invest $40 million over four years as well (Mbendi 2003). Rich manganese reserves at Tambao supply the country with much-needed revenues, but the manganese could be better exploited with a rail link and general improvements to the country's physical infrastructure.

NIGER

The International Monetary Fund classifies the large landlocked state of Niger as a heavily indebted poor country with few prospects for development. Environmentally, Niger is extremely marginal for agriculture and, with the exception of uranium, its mineral endowment is modest. Niger was hardest hit by the drought of the early 1970s, suffering very high livestock losses; perhaps as much as one-quarter of its population either perished or more likely, emigrated to neighboring states. President Hamani Diori, who had been in power since independence in 1960, was blamed for mishandling the economy and emergency relief measures, and for failing to attract foreign capital. He was deposed by a military coup in April 1974. After almost 20 years of military rule, Niger abandoned the one-party state for representative government, holding its first free elections in 1993, not long after its neighbor Mali. Since that time there has been considerable political instability in the north (as there was also in

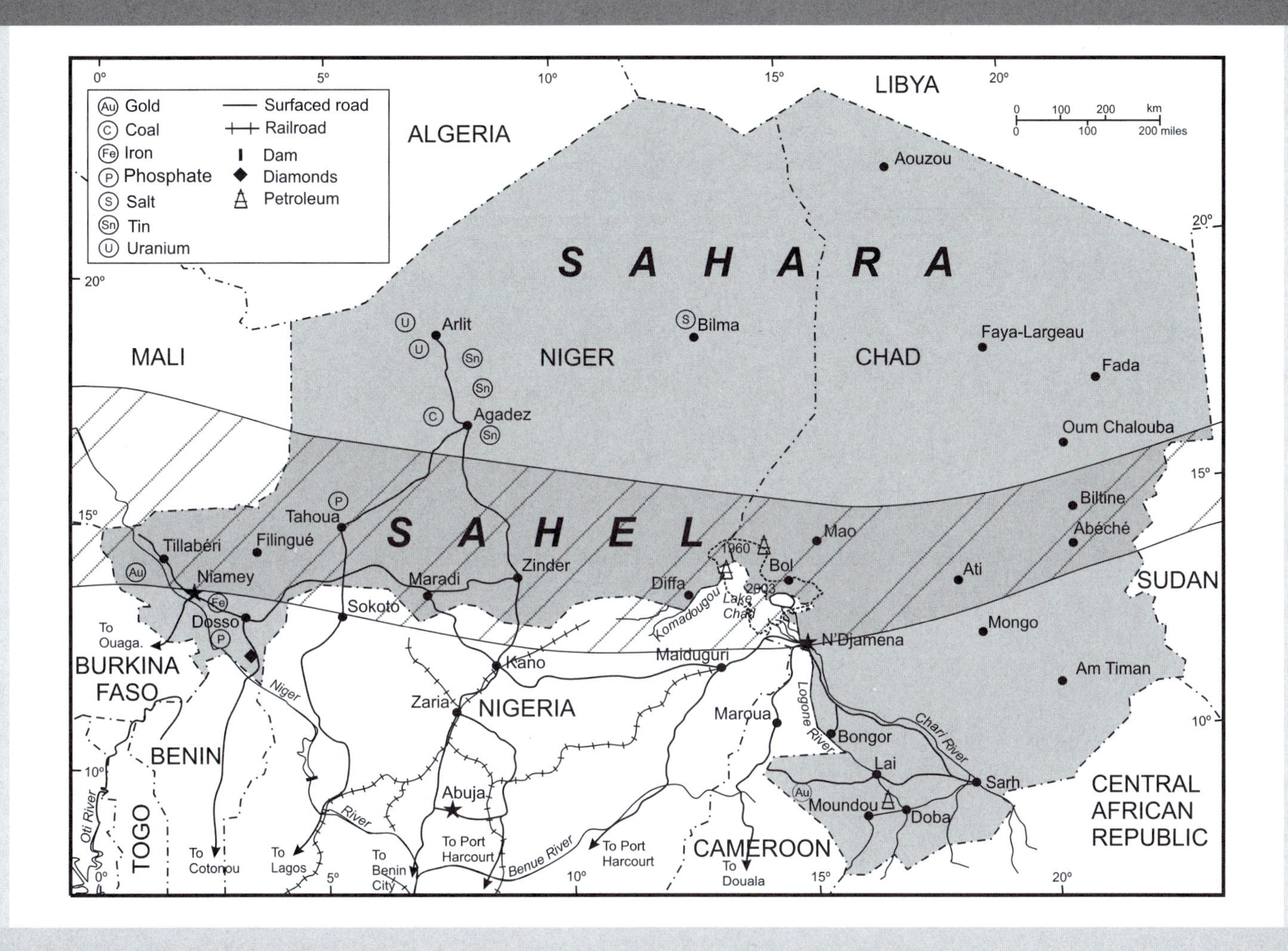

FIGURE 14-16 Natural resources and infrastructure of the Sahelian states of Niger and Chad.

Mali), as Tuaregs sought more autonomy from the central government. In 1996 and 1999 there were coups d'état. A national reconciliation council was organized that paved the way for civilian government in 1999. Niger is poorly connected over the Sahara but is relatively well connected to Benin and Nigeria by road (Figure 14-16), and rail links are not far off in northern Nigeria.

The northern two-thirds of Niger is desert and uninhabited except for scattered oases and mining camps. Agadez, the largest settlement in the Aïr Massif (population 122,200), receives only 6.3 inches (160 mm) of rain on the average; yet Tuaregs and other nomads keep considerable livestock in the vicinity. Only in the extreme south, along the Niger River and in the Maradi District, is there permanent settlement. About 80% of the population is rural (Table 14-3), but only 3% of the land is cultivated. Niamey, the capital, is the largest town (748,600), and only three others—Zinder (202,300), Maradi (189,000), and Tahoua (95,000)—all in the south, have significant populations.

Niger's external economic orientation is toward France and the coastal states, especially Nigeria and Benin. Nigeria takes almost all of Niger's animal exports, is now the principal transit route for Niger's

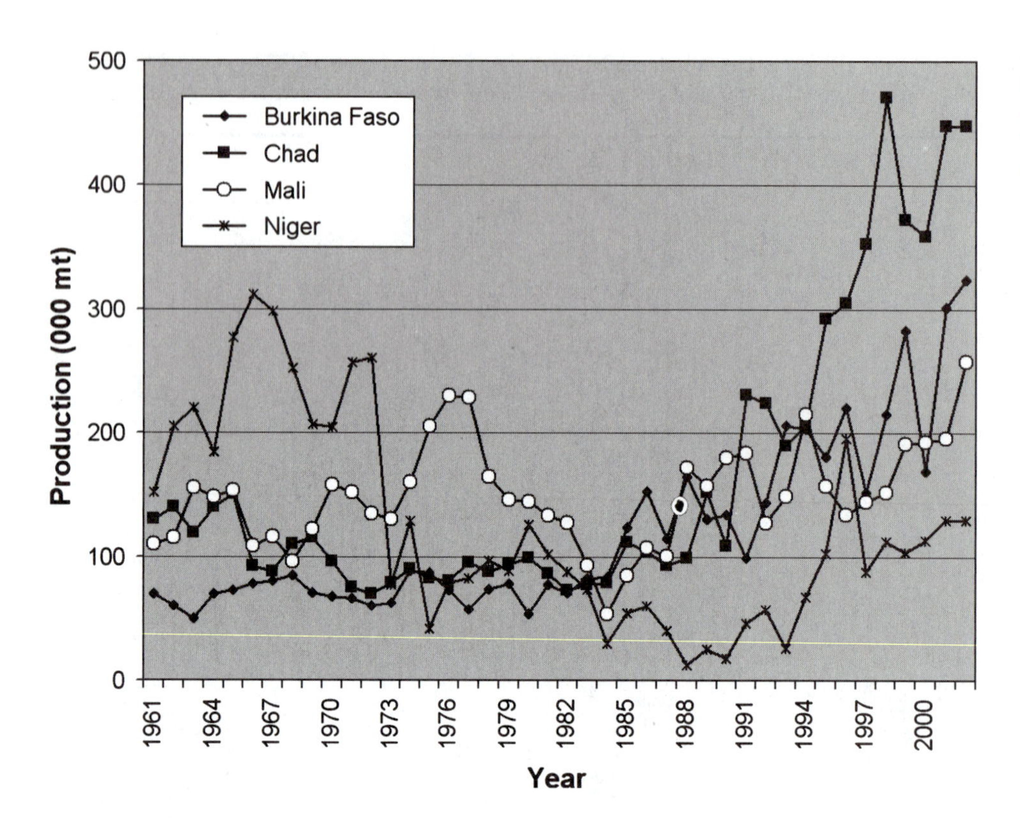

FIGURE 14-17 Peanut production in Burkina Faso, Chad, Mali, and Niger, 1961 to 2002.

foreign trade, and is an important recipient of seasonal labor. Tuareg migrants depend on Nigeria's pastures, and during the last prolonged drought, thousands made Nigeria their permanent home. France has long been Niger's principal source of financial, technical, and military aid, and in the late 1960s France was responsible for developing the uranium ores at Arlit (population 84,100), the largest known deposit in Africa. Production started in 1971, and by 1974 uranium had become Niger's principal export; it accounted for over 32% of Niger's exports in 2003 (Bermúdez-Lugo 2003). While this prominence was due in part to the poor showing of groundnuts and cattle, uranium is expected to remain the country's major source of foreign revenues for the foreseeable future. In the three uranium-producing areas, reserves are estimated at 140,000 metric tons. Niger's uranium production is the world's third highest after Canada and Australia; principal customers are France and Japan, where nuclear energy forms the backbone of national electricity production.

Unlike most African countries, Niger produces coal, although its quality is not high. The coal, estimated at 14 million metric tons of reserves, is found in the Aïr Massif and is used to generate power for the uranium mines located nearby. In the twenty-first century, gold mining appears set to contribute significantly to Niger's economy. Niger's first gold mine opened in 2001, and it is estimated to contain over 19 metric tons (618,000 ounces) of refined gold. At $350 per ounce, that would bring over $216 million into the economy. The gold is located west of Niamey and in the Aïr Massif. The government of Niger provides many incentives to attract foreign mining investment, such as tax holidays, tax exemptions, the right to freely remit dividends, and guarantees against nationalization or expropriation (IDCSA 2003), but the tax on corporate profits is 45%.

Production of Niger's major cash crop and principal export, groundnuts, never really recovered from drought of the 1970s and 1980s. Production dropped from a peak of 312,000 tons in 1966 to only 41,000 tons in 1975 to 13,000 tons in 1988 (Figure 14-17). Since 1988 production has increased, but it has been extremely variable. The production of millet and sorghum, the staple crops, has also been highly variable.

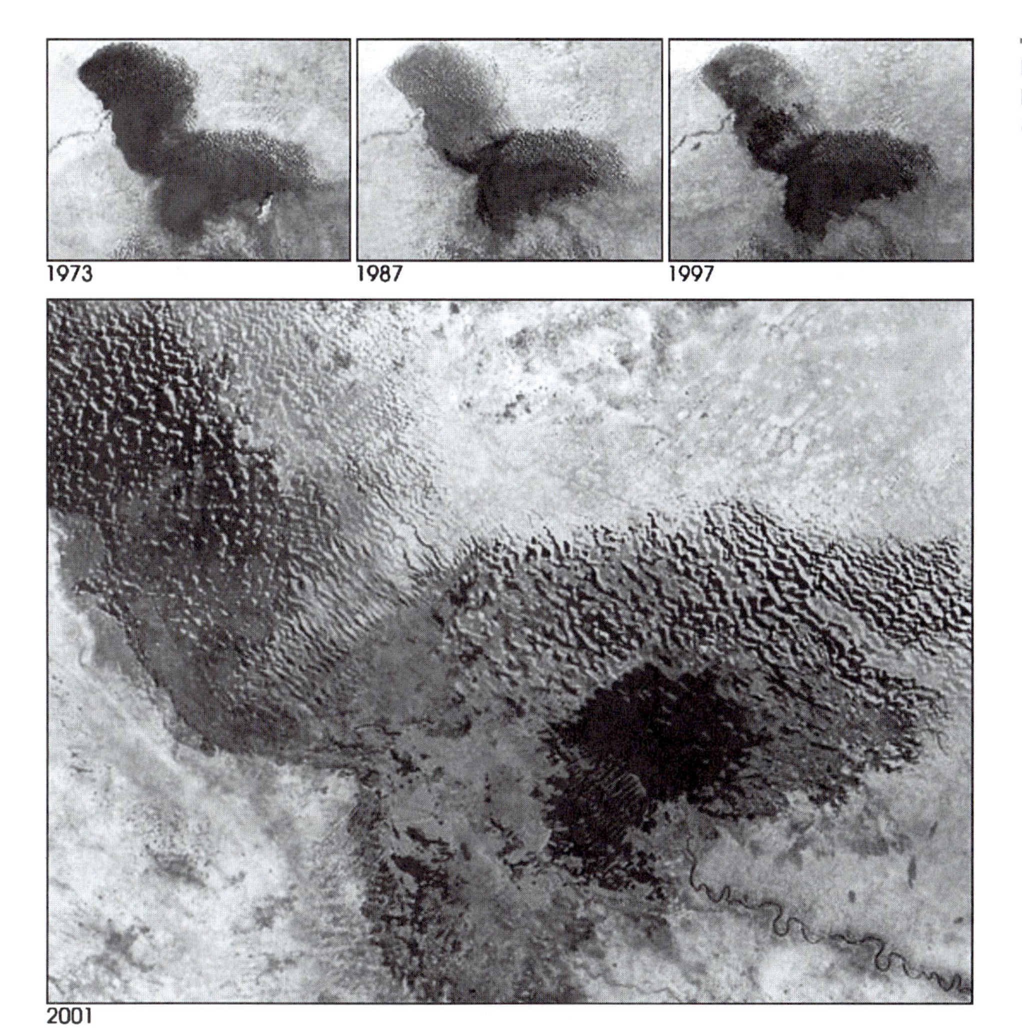

FIGURE 14-18 Shrinking Lake Chad, 1973 to 2001. *USGS.*

Unlike most of the other Desert Tier countries, moreover, production has not kept up with rural population (Figure 14-14).

It is difficult to envision Niger as more than the hinterland of Nigeria, so linked are the two culturally and economically; but union with Nigeria, although some in northern Nigeria would welcome it, is a very unlikely prospect. The Hausa, Niger's dominant culture group, was divided in two by the English and French imperialists between the countries of Niger and Nigeria. The prospect of a "Greater Hausaland" is an intriguing concept: the Hausa cultural identity and Islam would be the glue holding such a polity together, were borders to be redefined. William Miles (1994) has examined this issue and finds that four generations after colonization, many differences persist on either side of divided Hausaland. Nevertheless, he believes that true decolonization will ultimately involve the reemergence of precolonial political entities like Hausaland in a process similar to the breakup of the Soviet Union and the reemergence of formerly independent states. If oil is discovered in northern Nigeria, and if the Sharia states of northern Nigeria give evidence of being able to exist together as a viable entity, then there may be some pressure for unification from both sides of the border.

CHAD

The largest and least accessible of the West African Sahelian states is Chad (Figure 14-16). Chad has no railway, poor roads, a per-capita income less than $250, and a divided people (Table 14-3). While its physical environment has not been conducive to general economic prosperity, especially prosperity based on agriculture, its institutional resources, size, and location have been even less so. But all this may be changing as a result of oil; reserves are estimated to be one billion barrels, and oil exports hold the promise of bringing in billions of dollars of revenues. Exploration for oil begun during the 1970s led to several discoveries (Lake Chad Basin and Doba Basin), but civil war in 1979 forced companies to withdraw. In the mid-1990s further efforts were made to develop the Chadian oil fields, and a pipeline from the Doba fields in southern Chad extending to the Cameroonian coast was constructed; the first oil began flowing in 2003. Chad will probably make about $2 billion from the Doba field if it pumps 250,000 barrels a day for the expected life of the field, which is expected to be 30 years. Cameroon's share of the profits will be half a billion dollars.

It is expected that Chad will soon have an oil refinery built in or near the capital, N'Djamena, to serve domestic needs on the one hand and act as a growth pole for other industries that process and use oil and gas. The government intends to substitute cheaper, locally distilled, fuel oil for costly imports from Cameroon and Nigeria. The oil export pipeline is controversial for some observers, and some international environmental groups have called for a moratorium on the project, asking that the World Bank first invest in the infrastructure needed for sustainable development, good government, rule of law and independent judiciary, democracy, respect for human rights, and protection of the environment (CIEL 2000). Although on the face of it, these goals seem worthwhile, it is difficult to imagine the World Bank being able to effectuate such sweeping change, particularly in a country as impoverished as Chad. It is obvious that no government of an impoverished country such as Chad would refuse an oil windfall, but such rapid change in status may bring some unexpected consequences. In any case, oil and development in Chad should be watched carefully over the next decade. It is quite likely that infrastructural improvements made with the oil windfall will pave the way for cheaper costs for other, as yet undiscovered, minerals.

In any event, greater connectivity will promote regional cooperation. Gold, bauxite, columbite, diamond, uranium, silver, tin, tungsten, and alluvial diamonds have been found in Chad. There are indications of copper, iron, lead, titanium, and zinc as well. Almost all Chadians are anticipating great things, but whether all Chadians will benefit from such development remains to be seen. The presence of land mines dating from the Libyan accupation (1973–1994) poses significant barriers to exploration and mining in the northern region of Ennendi–Borkou–Tebesti and in the two southeastern regions of Moyen Chari and Salamat.

For decades Chad has suffered from regionally based political and economic disorders. The arid north is Muslim, nomadic, sparsely populated, and is isolated from the rest of the country. It has traditionally opposed political domination from the south, and until 1965 (five years after independence), it remained under French military administration. It has periodically received weapons and technical assistance from Libya in its struggles against the south, and during the 1970s, Libya occupied the Aouzou Strip, located in the northernmost part of Chad along the Libyan border. Excluded from government in the early years of independence, northerners found their champion in Muammar al-Gadhafi, who helped them to dominate the country during the civil wars of the 1970s and ultimately to become the ruling group. Libya probably has interfered in Chadian affairs as much as France has since independence, but recently appears to have lost interest in foreign adventures.

Southern Chad is wealthier than the north, more populous, better educated, sedentary in its way of life, and largely animist or Christian. Here lies the capital, N'Djamena, and the densely populated and relatively fertile Longone and Chari valleys, where cotton is by far the dominant cash crop and some areas are subject to seasonal flooding. In 2000 cotton exports comprised 65% of Chad's export earnings (Mobbs 2001), but in 2004 oil exports made up 90.5% of export earnings (Mobbs 2004). Maize and tubers have become more important in the southern areas recently. The region is normally self-supporting in foodstuffs. Indeed, it exports large quantities of meat to Nigeria, Cameroon, and the Democratic Republic of the Congo. The major livestock belt coincides with the Sahel, extending east from Lake Chad to the Darfur Plateau. Economically, southern Chad is geared more to neighboring states than to northern Chad, and this trend will become stronger as ties with Cameroon increase because of the Chadian–Cameroonian oil pipeline.

CONCLUSION

While the economy of each of the six Sahelian states has diversified since independence, primarily through the introduction of mining and the expansion of manufacturing and, in the case of Senegal, information technology, the countries remain predominantly agricultural. Although standards of living have improved since the 1970s, many ordinary people remain poor, sometimes underfed and malnourished. Today, however, there are more options to improve their well-being. Over the last 30 years of prolonged drought, people have shown themselves to be adaptive and resilient in the face of environmental change and uncertainty. Many of the institutional constraints that limited opportunities for change in the past have been removed. To provide the Sahelian people with adequate food and housing, higher incomes, and greater economic opportunity, there must be widespread rural reform and changes in the means of production. Agricultural development and privatization should be given top priority in all national development plans. This does not necessarily mean large-scale irrigation schemes to produce cash crops, but rather improved farming techniques, more effective marketing, better storage facilities, new varieties of seed and breeding cattle, and the use of small storage dams and irrigation works. Privatization of the land market is still in its infancy. Movement in this direction, which has proved successful elsewhere and is central to attracting foreign investment, should continue. Greater regional cooperation, for example in power production, should be encouraged. Many positive changes could be implemented without great financial strain or social disorder, and with relatively little technological expertise, particularly if such change is led by private interests rather than government, or perhaps in some kind of transparent partnership. Without immediate and effective management of population growth, the success of these development programs may only be marginal. Experience elsewhere indicates that greater investment in women should be the first step in slowing the population growth rate.

BIBLIOGRAPHY

Adedeji, A. 1970. Prospects of regional economic cooperation in West Africa. *Journal of Modern African Studies,* 8(2): 213–231.

Baier, S., and P. Lovejoy. 1975. The desert-side economy of the Central Sudan. *International Journal of African Historical Studies*, 8(4): 551–581.

Bermúdez-Lugo, O. 2003. *The Mineral Industry of Mauritania. U.S. Geological Survey Mineral Report.* Reston, Va.: USGS. http://minerals.usgs.gov/minerals/pubs/country/.

CIEL. 2000. Letter to the president of the World Bank regarding the Chad/Cameroon pipeline. Center for Environmental Law, Washington, D.C. http://www.ciel.org/Ifi/chadcameroonproject.html.

Coe, M., and J. A. Foley. 2001. Human and natural impacts on the water resources of the Lake Chad Basin. *Journal of Geophysical Research*, 106(D4), 3349–3356.

Cole, R. 1991. Changes in drought-coping strategies in the Ségou Region of Mali. Ph.D. dissertation, Michigan State University.

Copans, J., ed. 1975. *Sécheresses et Famines au Sahel: Ecologie, Dénutrition, Assistance.* Paris: François Maspero.

Dalby, D., and R. J. Harrison-Church, eds. 1973. *Drought in Africa.* London: University of London.

Davis, R., and R. Hirji, eds. 2002. *Water Resources and Environment Technical Note Environmental Flows: Flood Flows.* Washington, DC: World Bank.

USEIA. 2003. Country analysis brief: Economic Community of West African States (ECOWAS). Washington, DC: U.S. Department of Energy, Energy Information Administration.

FAO. 2002. FAOSTAT. Food and Agricultural Organization of the United Nation. Rome: FAO.

Giannini, A., R. Saravanan, and P. Chang. 2003. Oceanic forcing of Sahel rainfall on interannual to interdecadal time scale. *Science*, 302 (5643): 1–10. www. sciencexpress. org.

Glantz, M. H. 1975. *Politics of Natural Disaster: The Case of the Sahel Drought.* New York: Praeger.

Grove, A. T. 1974. Desertification in the African environment. *African Affairs*, 73(291): 137–151.

Hewitt, K. 1983. *Interpretations of Calamity: From the Viewpoint of Human Ecology.* Boston: Allen & Unwin.

Hulme, M., R. Doherty, T. Ngara, M. New, and D. Lister. 2001. African climate change: 1900–2100. *Climate Research*, 17(2) Special 8.

Mbendi. 2003. Profile: Burkina Faso. Industrial Development Corporation of South Africa. Capetown, South Africa: Mbendi.

Kirk, W. 2003. *Iron Ore.* U.S. Geological Survey Mineral Report. Reston, Va.: USGS. http://minerals.usgs.gov/minerals/pubs/commodity/iron_ore/.

Magistro, J., and M. Lo. 2001. Historical and human dimensions of climate variability and water resource constraint in the Sénégal River Valley. *Climate Research*, 19(2): 133–147.

McCarthy, J. J., O. F. Canziani, N. A. Leary, D. J. Dokken, and K. S.White. 2001. *Climate Change 2001: Impacts, Adaptation and Vulnerability*. Intergovernmental Panel on Climate Change. Rome: United Nations Environment Program and the World Meteorological Organization. http://www.grida.no/climate/ipcc_tar/wg2/001.htm.

McMillan, D. 1995. *Sahel Visions: Planned Settlement and River Blindness Control in Burkina Faso*. Arizona Studies in Human Ecology. Tucson: University of Arizona Press.

Meillassoux, C. 1974. Development or exploitation: Is the Sahel famine good business? *Review of African Political Economy*, 1: 27–33.

Miles, W. F. 1994. *Hausaland Divided: Colonialism and Independence in Nigeria and Niger*. Ithaca, N.Y.: Cornell University Press.

Mobbs, P. 2004. *The Mineral Industry of Chad*. U.S. Geological Survey Mineral Reports. Reston, Va.: USGS.

Mobbs, P. 2001. *The Mineral Industry of Chad*. U.S. Geological Survey, Mineral Report. Reston, Va.: USGS. http://minerals.usgs.gov/minerals/pubs/country/.

NGO Working Group on the Export Development Corporation. 2001. Halifax Initiative. *Reckless Lending, Vol. II: How Canada's Export Development Corporation Puts People and the Environment at Risk*. Halifax, Nova Scotia: NGO Working Group.

Raynault, C. 1975. Le cas de la Région de Maradi (Niger). In *Sécheresses et famines au Sahel: Ecologie, dénutrition, assistance*, J. Copans, ed. Paris: François Maspero.

Roncoli, C., K. Ingram, and P. Kirshen. 2001. The costs and risks of coping with drought: Livelihood impacts and farmers' responses in Burkina Faso. *Climate Research*, 19: 119–132.

Rosenthal, J. E. 1974. Survival in the Sahel. *War on Hunger*, 8(8): 1–40.

Sen, A. 1982. *Poverty and Famines: An essay on Entitlement and Deprivation*. Oxford: Clarendon Press.

Sheets, H., and R. Morris. 1974. Disaster in the desert: Failures of international relief in the West African drought. Washington, D.C.: Carnegie Endowment for International Peace.

Szczesniak, P. A. 2002. *The Mineral Industry of Burkina Faso, Mali, Mauritania, and Niger*. U.S. Geological Survey Mineral Report. Reston, Va.: USGS. http://minerals.usgs.gov/minerals/pubs/country/.

UNCTAD. 2005. Foreign Direct Investment Database. New York: United Nations Conference on Trade and Development. http://stats.unctad.org.

U.S. Library of Congress. no date. *Mauritania: A Country Study*. Washington, D.C.: Library of Congress. http://countrystudies.us/mauritania/.

USGS. 2001. *Earthshots*, 8th ed. Reston, Va.: U.S. Geological Survey. www.usgs.gov/Earthshots.

Watts, M. 1987. Drought, environment, and food security: Some reflections on peasants, pastoralists, and commoditization in dryland West Africa. In *Drought and Hunger in Africa: Denying Famine a Future*, M. Glantz, ed. Cambridge, U.K. Cambridge University Press, pp. 171–212.

Webb, J. 1995. *Desert Frontier: Ecological and Economic Change Along the Western Sahel, 1600–1850*. Madison: University of Wisconsin Press.

CENTRAL AFRICA

Central Africa, sometimes called equatorial Africa, is one of the continent's most distinct geographic regions. Composed of seven states—Cameroon, the Central African Republic (CAR), the Congo, Equatorial Guinea, Gabon, the Democratic Republic of the Congo (DRC), and São Tomé and Príncipe—the region is dominated physiographically by Africa's greatest river, the Congo (Figure 1) and ecologically by humid, lowland forest. Central Africa has great ecological diversity, possessing vast humid forests, savannas, semi-deserts, lakes, swamps, mangrove forests, and coral reefs. Much of the region, especially central DRC, is plateaulike and lies below 3,000 feet (900 m). In the periphery there is great topographical variety. The Ruwenzori Mountains adjoining the state of Uganda rise above 13,000 feet (3,900 m), while the Mitumba Mountains, which form the eastern flank together with the lakes of the Western Rift Valley (Lakes Albert, Edward, Kivu, and Tanganyika), generally stand above 5,000 feet (1,500 m). In Cameroon, beyond the Congo Basin proper, the Adamawa Massif rises to similar heights. The central lowlands of the DRC, once one of Africa's great interior lakes, are flat, heavily forested, and often swampy. Lakes Mai-Ndombe and Tumba are remnants of this ancient feature.

Most of the people who live in central Africa speak languages of the Bantu subfamily of Niger–Congo, although many in the Central African Republic speak languages of the Adamawa–Ubangi subfamily of Niger–Congo. In the mountains of northern Cameroon live many speakers of the Chadic branch of Afroasiatic, while on the plains of northern Cameroon, Fulfulde speakers (Atlantic subfamily of Niger–Congo), whose ancestors came from the Senegal River valley several centuries ago, raise their livestock.

The peoples of central Africa are hardworking; the region is well watered and rich in natural resources, and it has great potential for development. But the near-term prospects for development are not encouraging. At independence over 40 years ago, the peoples of the central African countries had great hopes and ambitions for a bright future but, with few exceptions, the ensuing decades have been disappointing. The vast region is compact and focused on the Congo Basin but, despite a geographic shape that should favor communications, the countries of the region and the provinces of the countries remain disarticulated (Figure 2) and transportation infrastructure

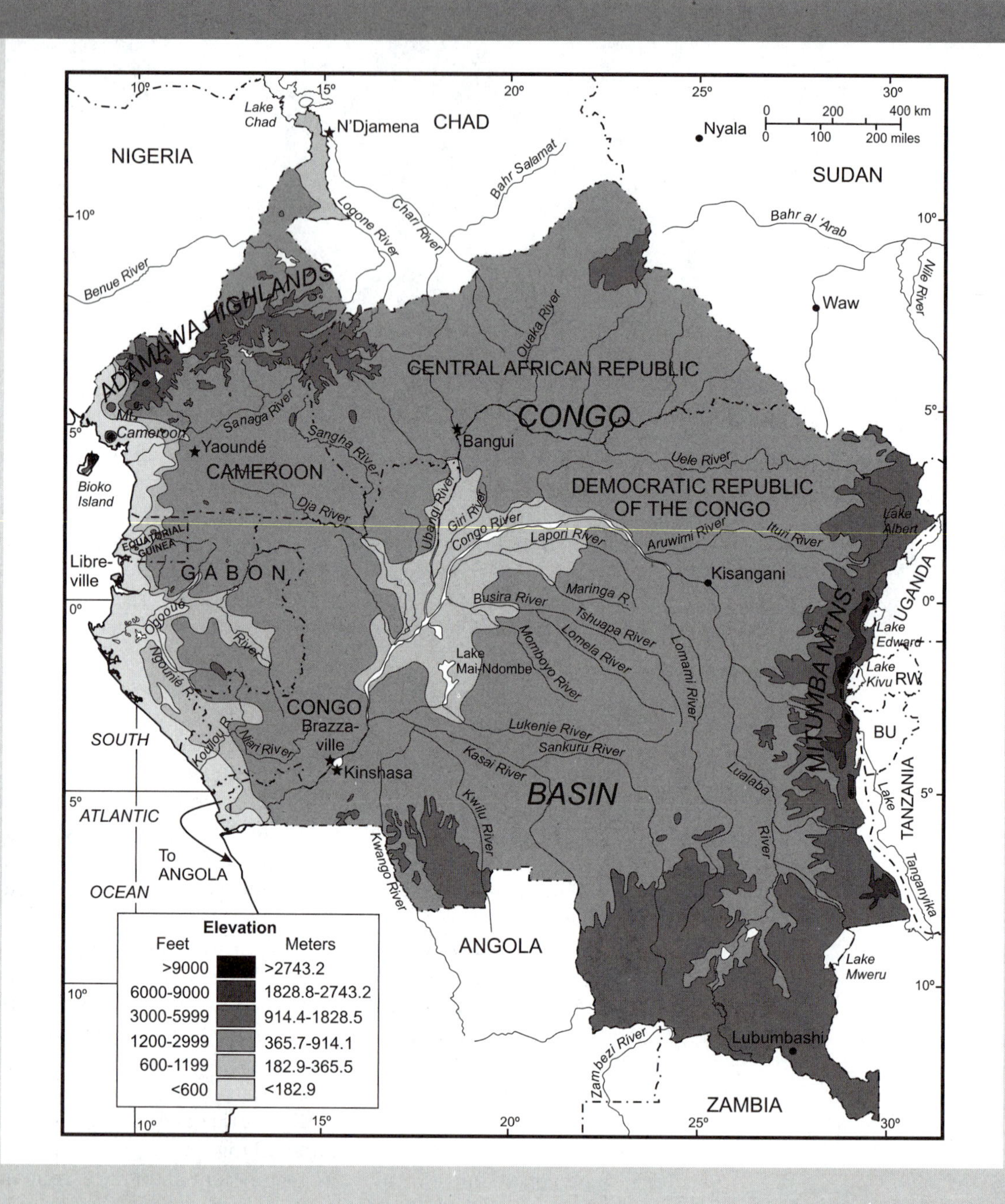

FIGURE 1 Physical features of central Africa.

is in a deplorable state. After the departure of the colonialists, public goods such as infrastructure (roads, rails, telecommunications) were little invested in and have deteriorated. Questions are being raised about whether the anchor of central Africa, the Democratic Republic of the Congo, will remain intact over the next decades.

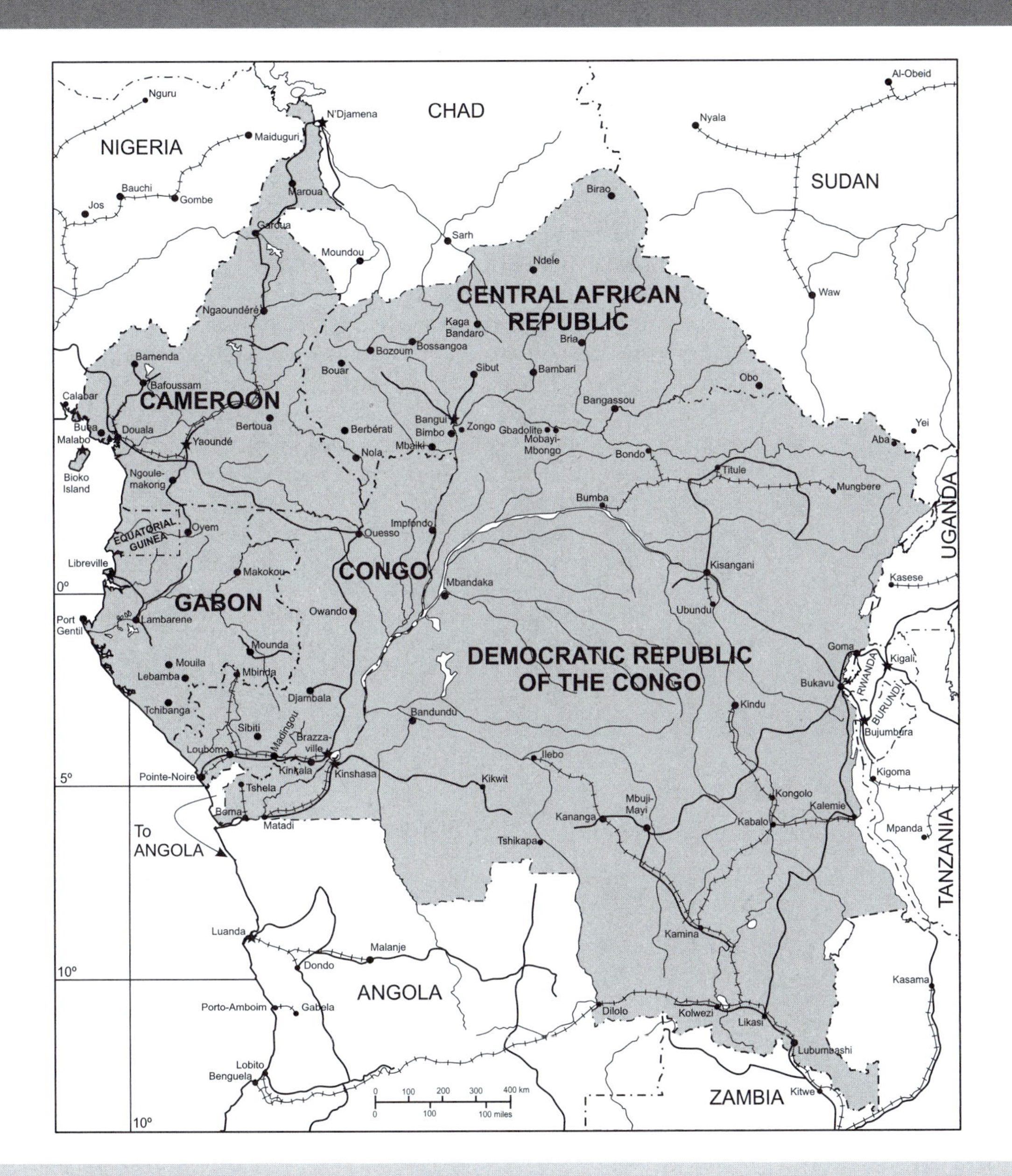

FIGURE 2 Countries, capitals, and central places; rivers, rail lines, and surfaced roads in central Africa.

In one sense the Cold War distracted the governments of the competing superpowers, governments of the more developed countries, and the intellectuals of the world from the important task of human development in the less developed parts of the world. In another sense, the Cold War warped intra- and international relations to such an extent that all issues were seen through the lens of superpower competition—even local conflicts tended to inflate themselves into the ideology and jargon of the Cold War. Governments came to power and lost it by violent means and often gained the supported of a superpower simply by paying lip service to the approved ideology of that superpower. At the same time, there was little international and no local accountability, as the central African governments mismanaged the wealth and resources of their respective countries and the superpower patrons and allies turned a blind eye to the mismanagement, corruption, and human rights abuses perpetrated by their client states.

Fortunately, the Cold War has ended, the world has changed as a consequence, and for the foreseeable future we will live in a unipolar world with the United States as the sole superpower. Cynical leaders of less developed countries can no longer play one superpower off another in a bid to attract financial and military support from Russia or the United States. Central Africa is a good region of Africa and the world in which to examine the local impacts of the Cold War and, more importantly, to monitor the regional implications of the end of superpower competition for the hearts and minds of Africans. Since the end of the Cold War, regional issues in central Africa are being framed in a more local context, and the geographies of some central African countries are changing. The ultimate political ramifications of these local movements remain to be seen. The focus of this section is on the historical geography of the region and the potentials and problems for development.

CHAPTER 15

Spatial and Cultural Influences in the Growth of the Democratic Republic of the Congo

The Democratic Republic of the Congo is central Africa's giant in a number of ways. It is the third largest country in Africa, with an area of 905,365 square miles (2,345,000 km^2). It is also the wettest country in Africa: as a consequence of its equatorial location and its basinlike characteristics, the DRC possesses 25% of the average annual internal renewable water resources of the entire continent (UNEP 2002). After the Amazon forest, the forest of the Congo Basin is the second largest contiguous forest in the world. The DRC is also richly endowed with minerals.

The DRC is a land of great distances, difficult communications, and severe environmental obstacles to development. The rich copper fields and economic core of Katanga lie some 1,700 miles (2,700 km) from Boma, the DRC's coastal port in the narrow proruption west of Kinshasa. Coffee grown in the foothills of the Ruwenzori Mountains, adjoining the state of Uganda, must travel some 1,800 miles (2,900 km) by road, rail, and riverboat to the Atlantic Ocean prior to export. The Congo River, Africa's largest, loops through the country, plunging over numerous rapids and falls making uninterrupted navigation impossible. Between Kinshasa and its mouth, in a distance of only 215 miles (345 km), the Congo drops almost 900 feet (270 m) as it traverses the Crystal Mountains, and goods from the interior must be transferred to trains for the journey to the coast. From Kinshasa to Kisangani and from Kinshasa to Ilebo, the Congo and its major tributary, the Kasai, are navigable for 1,085 miles (1,740 km) and 513 miles (820 km), respectively. The Ubangi can be used by modern riverboats as far as Bangui, capital of the Central African Republic.

Whatever the interior produces must be priced high enough to permit several transshipments. The nonnavigable sections of the Congo have been circumvented by railroads, but each break of bulk increases the cost of the product. It was Henry Morton Stanley, the first European to know the area well, who observed that without railroads the territory was not worth a penny. The Belgians in fact built over 80,000 miles (128,000 km) of roads, but even now many parts of the country remain isolated from modern transportation. During the rainy season, roads may be impassable for weeks; a very small mileage, mostly near the urban centers, is asphalted and usable throughout the year. The country's vastness and the huge costs of building communications between the various areas of population concentration scattered around the "empty heart" have perpetuated a high degree of internal fragmentation and have delayed the evolution of a Congolese people that might be referred to as a nation. Yet the Democratic Republic of the Congo has tremendous economic potential. Its mineral wealth, hydroelectric potential, and forestry resources are among the greatest in Africa, if not the world. Certainly greater development would have occurred during modern times had the country not been bedeviled by a series of political crises.

PHASES LEADING TO TODAY'S DRC

Four distinct phases can be recognized from the early 1880s to the present, and each produced obstacles to human development above and beyond those of physiography and environment. These phases fuse into a history that the historian Robert Edgerton has identified as been one in which "ordinary people refused to surrender their compassion for one another, their quest for happiness, and their hope for a better future" in the face of despotic, brutal leadership (Edgerton 2002:vi). During each phase, the country had a different name. From 1885 it was the Congo Free State; from 1908 to independence in 1960 it was the Belgian Congo; from 1960 to 1971 it was the Democratic Republic of the Congo. The name of the country was then changed to "Zaïre" as part of the government's "Africanization" or "authenticity" drive. Also changed were the names of cities, regions, geographical features, streets, and buildings. The most important place name changes are given in Table 15-1. After the change in government in 1997, the name of the country was returned to "Democratic Republic of the Congo." Throughout this chapter the name used to refer to the Democratic Republic of the Congo will be that used in the period under discussion: Congo Free State under King Leopold, Belgian Congo after the Free State was taken over by Belgium, Zaïre during the reign of former general Joseph D. Mobutu, and DRC thereafter.

TABLE 15-1 PLACE NAME CHANGES IN THE DEMOCRATIC REPUBLIC OF THE CONGO.

Places	Colonial Name	Name Change Under Mobutu	Current Name
Towns	Albertville	Kalemi	Same
	Bakwanga	Mbuji-Mayi	Same
	Baningville	Bandundu	Same
	Baudoinville	Moba	Same
	Coquilhatville	Mbandaka	Same
	Costermansville	Bukavu	Same
	Elisabethville	Lubumbashi	Same
	Jadotville	Likasi	Same
	Leopoldville	Kinshasa	Same
	Luluabourg	Kananga	Same
	Paulis	Isiro	Same
	Ponthierville	Ubundu	Same
	Port Francqui	Ilebo	Same
	Stanleyville	Kisangani	Same
	Thysville	Mbanza Ngungu	Same
Natural features	Congo River	Zaïre	Congo
	Lake Leopold II	Mai-Ndombe	Same
	Lake Rete	Ishangelélé	Same
	Lake Delcommune	N'Zilo	Same
	Stanley Falls	NgaKéma	Boyoma Falls
	Stanley Pool	Malebo	Pool Malebo
Districts	Katanga	Shaba	Katanga
	Leopoldville	Bandundu	Same
	Orientale	Upper Zaïre	Orientale

Precolonial states will be referred to by their customary names, for example, the Mangbetu Kingdom or the Kingdom of Kongo.

Long before the modern period, the principal kingdom that existed in the Congo Basin was the Kingdom of Kongo, which included parts of today's Congo Republic, western DRC, and northern Angola. The Kingdom of Kongo probably emerged in the fourteenth century and possessed a hierarchy of territorial and departmental officials. The Kongo state was the only one in central Africa to have a national currency, and it was the most centralized of the central African states at the time (Vansina 1990). Beginning in 1483, when the first Portuguese ships arrived at the mouth of the Congo River, peoples, products, and economies were drawn into the Atlantic trading system, a precursor of the global economy. Portugal and Kongo exchanged diplomatic missions.

Vansina (1990) suggests that the Atlantic trade was an economic stimulus that had the same relative importance for the Congo region as the Industrial Revolution had for Europe. Foreign values and ideas challenged the peoples of the Congo Basin, while new wealth strengthened the kingdoms of the Congo. Ultimately, however, the old order was overwhelmed as new trading and administrative structures were developed. The quest for territory, slaves, and ivory drew Arabs from Zanzibar and Egyptians from the Sudan into the region during the second half of the nineteenth century. After the arrival of H. M. Stanley on the Congo River in 1879, conflict between Europeans from the west and Arabs from the east seemed inevitable. The Europeans conquered and co-opted the Arabs during the last decade of the nineteenth century, but it took 40 more years for Europeans to destroy the last vestiges of local resistance.

Congo Free State

The person whose name became closely connected with the Congo was Henry M. Stanley, a British-born, American journalist. Having succeeded in his earlier search for Livingstone (whom he had found at Ujiji, located on the east bank of Lake Tanganiyka, on November 10, 1871), Stanley for the first time entered the Congo Basin. After his initial contact with the unknown interior, he went to Europe, trying first to interest the British in establishing their influence in this part of Africa. The "scramble for Africa" began to gain momentum, and Stanley's work became known to King Leopold II of Belgium. Leopold, who had visions of an empire for Belgium, saw his country without colonies at a time when Britain, France, Portugal, and the Netherlands already possessed vast overseas realms. In 1876 the king convened the Brussels Geographic Conference, with the purpose of supporting exploration and research in equatorial Africa. In 1878, Leopold and Stanley met and established an organization that aided Stanley on a number of subsequent journeys through the Congo. In effect, the explorer was now working for the Belgian king, founding settlements and signing treaties on behalf of Leopold with Congolese chiefs.

Thus Belgium entered the colonial scene in Africa and became involved in the disputes that marked that period of expansion. In 1884 German chancellor Otto von Bismarck called the Berlin Conference, at which European nations, the United States, and Turkey attempted to settle these colonial conflicts by arbitrarily defining some boundaries. Belgium, relative latecomer in Africa, found itself backed in its claims to rights in the Congo region by Germany, which wished to impede the expansionist efforts of its chief rivals, Britain and France. Hence, the International Congo Association, an outgrowth of the original committee supporting Stanley's explorations, achieved recognition in Berlin, and a document called the Congo Treaty was signed. The sovereign state thus created was named the Congo Free State, and later in 1885 the Belgian parliament authorized King Leopold II to become king of this country also. In this manner, the Congo Free State was initially the personal possession of King Leopold, not a colony of Belgium.

The Congo Treaty gave Leopold the tasks of suppressing the slave trade in this area, creating effective control over the country, and improving living conditions for the Africans who had become his wards. Slaving was widespread in the Congo Basin, but the trade was anchored in Zanzibar and locally controlled from the town of Nyangwe, on the Lualaba River on the east side of the Congo Basin. One of the most powerful slavers was Hamid bin Muhammed, also known as Tibbu Tip, who came from a long line of East African slavers. Vast areas of the Congo Basin had been depopulated by slavers who sometimes captured over a thousand people in a single raid.

Leopold was very wealthy, but he visualized the Congo experiment as a money-making venture, as did the Arabs and Afro-Arabs that his forces eventually ousted from the region. Moreover, he expected quick returns for his initial investments. He sent a large number of agents into the Congo whose duty it was to gather specified quantities of ivory and rubber. While the agents forced the local population to collect these products, the empty lands were parceled out to concessionaires who paid levies on their profits. During this early phase, therefore, the products of the Congo came from the low-lying basin rather than the rim of the northern, eastern, and southern periphery. Barges were floated on the rivers, and produce-collecting stations were established at various points along their courses. Wild rubber was by far the most important product, and since labor costs were low and the demand great, profits were considerable.

The people of the Congo benefited but little from these arrangements. In fact, the change in the producing areas was largely detrimental: the king's agents used barbaric methods to enforce the gathering of quotas, and African opposition was met with devastating force. Entire villages were burned and their inhabitants massacred. It has been estimated that the African population during Leopold's reign was reduced by 3 million to 8 million. The African's fear of the slaver was replaced by a fear of the Belgian.

Meanwhile, Leopold long failed to adhere to one of the major conditions of his assumption of control over the Congo Free State, namely, suppressing the slave trade. Instead slavery remained rampant in the east until near the turn of the century, and the long-awaited campaign against it came only after the most powerful of the slavers began to pose an economic threat to the king. By 1885, King Leopold had established the "Force Publique" or national guard, to keep peace in the Congo, and in 1892 his army came into direct confrontation with the Arab slavers' armies for the control of the Congo. The first war over which the independence of the peoples of the Congo Basin would be decided was not fought between Congolese and Europeans but between Europeans and Arabs and Afro-Arabs. The second war over the Congo was to be fought 68 years later, just after Congolese independence from the European victors of the first war.

An effect of Leopold's acquisition of unoccupied land was that in a region where shifting, "patch" agriculture, hunting, and gathering are necessary adjustments to environmental conditions, the peoples were confined to lands that, at the moment the king acquired them, were only temporarily occupied; the designated territories were too small to accommodate the populations specified, and in any event unfit for permanent habitation. Thus, the local economies, such as they were, were totally disrupted. Forced labor destroyed the traditional social organization and broke up families. Local famines resulted. A country's development involves more than the extraction and sale of minerals and the gathering and exporting of forest products. It involves, also, such activities as the betterment of the people through the improvement of farming methods; the provision of social amenities, education, and health facilities; research on the eradication of local diseases; and the introduction of better types of crops. During its Free State period, the Congo produced a revenue that dwarfed the profits made in many other African territories. But except where the export-oriented communication system was concerned, the improvement of local conditions—one of the prime requirements stated in 1885 in the agreement produced at the end of the Berlin Conference—was minimal. For instance, from 1898 to 1903, the value of exports was $46 million, while imports totaled a mere $20 million.

The major impacts of the working relationship between Leopold and Stanley were the expansion of the area of the Congo, especially in the southeast, and the organization of a transport system that could efficiently handle the products of the interior. Leopold realized that the plateau surrounding the Congo Basin proper might contain valuable resources, and he managed to appropriate the area today known as Katanga at a time when the African chiefs there were planning to request "protection" from the British. The early colonialists also linked the rubber- and ivory-producing interior to the western Congo River port at Matadi. A railroad connecting Matadi and Kinshasa, about 200 miles (320 km) to the northeast, was begun in 1890 and completed in 1898. Construction of the short line between Kisangani and Ubundu did not commence until 1903, with completion as late as 1906. Thus the Stanley Falls (Boyoma Falls) had been bypassed, but another rail link was needed to connect Kindu and

Kongolo (see Figure 2, p. 405). It was built immediately afterward so that southeastern Katanga was linked to the western port of Matadi, but no less than five transshipments were necessary after the Kisangani–Ubundu leg. Katanga did not assume real significance in the economic sense until after the end of the Congo Free State.

The Belgian Congo

The First Four Decades

By 1904 the international outcry regarding the "Congo atrocities" had forced Leopold to appoint an official commission of inquiry. The king was forced to abandon his Free State, and in 1908 the Belgian parliament assumed responsibility for the territory. From 1908 until 1960, Belgium administered the Congo as a colony, and it was early in this period that the first real steps were taken in development toward the modern state. Among the legacies of Leopold were numerous concession companies, to which the king had granted long-term rights prior to the termination of the Free State. As early as 1891, such concessions had been granted to the Comité Spécial du Katanga, and in 1906 to the Union Minière du Haut Katanga. The Belgian government continued to recognize these concessions, and, as a result, these big companies exerted considerable influence over Belgian decisions concerning the colony. Shortly before 1960, the Union Minière's assets stood at just under $2 billion, and the mining group was contributing almost half of the Congo's annual tax revenue. Great changes came to the economy of the Congo between 1910 and 1920, for it was in 1913 that the export of rubber, long the major product, fell by 50%. The full impact of the mineral wealth of Katanga then began to be felt; and whereas the economic heart of the Congo had long been in the forests, the minerals of Katanga now came to form the country's economic backbone.

The whole history of the Congo during the colonial period was one of increasing economic dependence upon the wealth of Katanga; transportation routes were built to serve it, investments were made to accelerate development there, internal communications, hydroelectric projects, housing for the labor force, schools, and many other public works and facilities were constructed. Taxes collected from the products of other provinces were used to improve conditions in Katanga. During the colonial years, Belgium's investment in Katanga was considerable, but the returns were less than desired.

Four transport routes converged on Katanga during the colonial era. Mention has been made of the route from Kisangani to Matadi requiring five transshipments. A second route lay southward, and, by 1909, the Rhodesian railroad had reached Sakania on the Katanga–Copper Belt boundary. The Belgians immediately began the construction of a link to Bukama (about 75 miles/120 km southeast of Kamina) and had proceeded as far as Elisabethville (Lubumbashi), near the southern tip of the colony, when construction was interrupted by World War I. Although Sakania is only about 100 miles (160 km), as the crow flies, from Elisabethville (Lumumbashi), the rail line did not reach Sakania until 1918.

During the prewar period, while the Rhodesian connection was still under construction, Dar es Salaam handled some of the copper traffic (via Kigoma on Lake Tanganyika); after 1918, however, much of the traffic went southward via Victoria Falls to Beira, itself still a lengthy and expensive journey. Between 1923 and 1928, the route was built that has since carried a large share of Katanga's produce, linking Ilebo to Bukama on the River Congo. Requiring only two transshipments, this 1,724-mile route (2,758 km) carried almost half the Union Minière's copper shipments before the upheaval of independence.

Very soon after this route was completed, the shortest link to the coast was constructed, namely, the Benguela Railroad through Angola, directly westward over a distance of 1,312 mi (2,100 km). Finished in 1931, the Benguela Railroad was built partly with Belgian capital, and of course, no transshipments were required. Its terminal is the ocean port of Lobito, and its initial effect was the almost complete diversion of Katanga's copper traffic from the east coast (Beira) route. Later, development within Angola and rising production in Katanga put strain upon the west coast routes, and Beira, 1,624 miles (2,600 km) from Lubumbashi by rail, again began to handle Katanga's exports.

The last railroad to be constructed to handle Katanga and Copper Belt products was that linking Maputo (formerly Lourenço Marques) to the Zimbabwean (formerly Rhodesian) system, but it was never of major importance to Katanga. Just before independence, 48% of Union Minière's copper production traveled via Matadi, 30% went through Beira, and 22% via

Lobito. What Katanga produced, therefore, was sufficiently valuable to overcome the region's unfavorable location. This very factor led to some industrial development in the towns of Lubumbashi, Likasi, and Kolwezi.

The mining industry required supporting undertakings such as chemicals, explosives, cement, coal, and food processing. The copper ore itself, although among the richest in the world (5–6% purity) had to be reduced to copper matte (99% metal) before shipment to the coast, and this required large quantities of power. At first, coking coal was supplied from Europe, Rhodesia, South Africa, and even the United States. But the cost was high, the railways often congested, and the supply not always guaranteed, so the Congo's own deposits at Kalemi and Luena were brought into production. This low-grade coal proved unsatisfactory in the smelting operations, but it was used by the railways and by Katanga's industry. Eventually several hydroelectric facilities were developed in the region, and Belgium invested in Rhodesia's Kariba Dam, one of Africa's biggest hydroelectric schemes.

The mines were developed initially with labor-intensive methods, but from the end of the 1920s the projects became increasingly capital intensive. Migrant workers came not only from Katanga itself, but from the present-day countries of Zambia, Angola, Rwanda, and Burundi, as well as distant parts of the Congo. In 1927 Union Minière instituted a recruitment program that was designed to improve productivity by providing a more permanent workforce than had previously existed. Mine workers were recruited for a minimum of 3 years; housing and social amenities were provided; and miners were permitted to bring their families. To some extent the ethnic units intermixed, and in Katanga the process of acculturation was more complete than elsewhere.

Mining was not confined to copper and to Katanga. Uranium was mined until 1961 at Shinkolobwe just west of Likasi, while lead and manganese mines opened along the line of rail near Lubumbashi. Five hundred miles (800 km) to the northwest around Tshikapa, Kananga, and Mbuji-Mayi (then Bakwanga), industrial and gem diamonds were found in large quantitities, and, by World War II, the Congo had become the world's leading exporter of the industrial variety. Gold was found in the Kilo–Moto region in the remote northeast, while tin mines opened at Manono in Katanga. The mining concessionaires grew into major financial empires that exerted tremendous influence in the Congo's administration.

While the Belgian Congo's economic core developed in the extreme southeast, its administrative core emerged in the west. The estuary port of Boma was the colony's first capital, but in 1927 Belgium moved the colony's administration to Leopoldville (Kinshasa), then a small river town of only 23,000 inhabitants located across the Congo River from Brazzaville. It quickly became the Congo's largest city (400,000 at the time of independence) and Africa's most important river port. Light industry was introduced, but Leopoldville (Kinshasa) remained first and foremost an administrative center and the headquarters of the colony's major extractive industries and financial institutions. Thousands of Congolese came to Leopoldville in search of work and new opportunities, and from this immigrant group grew a small but important class who would later form the vanguard of the nationalist movement, and the elite that came to power at independence. Reflecting the condescending attitude of Belgians toward Africans of the time, these enterprising people were called *évolués*.

Between the economic and administrative cores lay the vast and sparsely populated Congo Basin, whose contribution to the export economy was limited to the production of rubber, palm products, coffee, cotton, and timber, most of which was derived from European-owned plantations that enjoyed strong government backing. These large-scale operations were not concentrated in any one area, but were widely scattered, making collection and shipment to Boma and Matadi both costly and time-consuming. Most products, and especially palm oil and palm kernels from the lower Kasai, Kwilu, and Congo rivers, and rubber from the Kisangani and Mbandaka areas relied heavily on river transport. In contrast, cotton, grown primarily in the drier regions along the northern border and in the south central regions of Kasai, was more dependent on road and rail service. Most of the timber cut for export came from the Mayumbe forests between Kinshasa and Boma, while the forests of the Lake Mai-Ndombe region were cut primarily for domestic needs. High transport costs incurred by road, rail, and riverboat kept the timber industry depressed.

While many Africans resisted the imposition of strange crops whose cultivation seemed to contribute little to the household food supply, others saw in them a means of acquiring wealth. Whereas chiefs in particular gained from the forced cotton cultivation during the colonial years, labor constraints were often felt in monogamous households. Except for western Katanga, where a free-market system for cotton prevailed, prices were strictly controlled by the cotton companies, with the result that producers sometimes were paid less than it cost to bring the crop to market (Likaka 1997). Price control was a strategy that was picked up by the postcolonial state and used to "tax" a variety of crops, much to the detriment of the development of agriculture.

In the remote highland areas of the east, where the climate was more attractive for European settlement than other parts of the territory, a different type of agriculture emerged. There, privately owned and operated estates were established by European settlers who emphasized high-value, low-bulk export crops that could withstand high transport costs. Arabica coffee was the principal commodity, but tea, pyrethrum, tobacco, and livestock were also important. Agricultural colonization also occurred in Katanga, but there emphasis was placed on producing cattle, hogs, and grains for local markets rather than for export. Taxes from plantation and estate agriculture were applied more to mining and industrial expansion than to further agricultural development.

The Belgian Congo on the Verge of Independence

Most Belgians believed that they were civilizing the Congolese and that their relations with the Congo would continue for a long time into the future—and most felt that the Congolese were not ready to govern themselves. Few Congolese had any formal education at all, and in such critical institutions as the Force Publique and Gendarmerie, only a small minority were even literate. In fact, Belgian paternalism meant that few Congolese had been educated beyond the primary level; the positions that required higher education were occupied by Europeans. The Belgian Congo's Force Publique, the national army, was made up of 23,000 soldiers, almost all of whom were illiterate, and 1,000 Belgian officers. At independence in 1960 there were no Congolese physicians, engineers, or even veterinarians.

The Belgians themselves were as unprepared for independence as the Congolese. The most farsighted of the Belgian academics caused a shock wave in 1955 when he recommended that the Belgian parliament adopt a 30-year program for the independence of the Congo that would prepare the Congolese to govern themselves (Edgerton 2002). Had such a plan been implemented in 1955, perhaps many of the problems that arose with the hasty departure of the Belgians in 1960 would have been averted. In any case, Belgian colonialism was really never about Congolese development but about Belgium's. From the days of Leopold until the end of the colonial era, the course of development and change was controlled by the Belgian government, the Roman Catholic Church, and big business. Little attempt was made to draw the colony's various regions and peoples together under the flag of a single and unified state.

The Belgian government, naturally the major force, saw its colonial task as a paternal one. Almost all important decisions were made in Brussels—not in the colony, where in any event none of the Congolese, European representatives of the Belgian government, or European settlers possessed any political rights or voice in the fate of the territory. Europeans and Congolese were socially segregated, and even the relatively educated Congolese, humiliatingly still called *évolués*, were kept at arm's length by most Europeans in the Congo. The Roman Catholic Church, while accomplishing much in primary education, strengthened the image of the Belgian as the father of the Congolese masses. Big business, part-owned by the Belgian state, did little to advance the average African farmer, although it provided both employment and revenues that accelerated the transition from a subsistence to an exchange economy.

Out of this triumvirate of government, church, and business arose the mutual-interest policy of paternalism, to the detriment of national development and unity. Never having participated, in the Western sense, in the making of decisions on policy that affected the Congo as a political entity, the Congolese seemed unready to accept a Western-type pluralistic democracy when independence was granted. Not only had the Congolese been deprived of the chance to negotiate with the colonial leadership, more importantly, they had done little negotiating with one another. There was no educated elite and no civil society, the prerequisites for a national dialogue and nation building. Political parties really did not have a united Congo as their major goal but were instead narrowly centered on peoples and regions. Of the few leaders who were able to see beyond ethnic limits, several also saw that Belgian administrative fragmentation of the Congo provided opportunities for the secession of areas in which they dominated.

Although a very gradual plan for independence had been rejected as being inconceivable when it was proposed to the Belgian parliament in 1955, the decolonization of the Belgian Congo was the most rapid and radical in Africa. In 1958 French president Charles de Gaulle set the stage for Congolese independence when he spoke in Brazzaville, just across the Congo River from Leopoldville (later Kinshasa), suggesting that French colonies in Africa could opt for immediate independence or join in a community of French territories (Communauté des Territoires d'Outre-Mer). Six months later, in January 1959, there was rioting in the Belgian Congo; places of business were looted, Europeans raped and beaten, and much physical property was destroyed. The riot was "the most decisive single event in the surge to independence and singularly prophetic of the revolution without revolutionaries which followed in 1960" (Young 1965: 290, in Edgerton 2002: 183). In August 1959 the Belgian government legalized political parties in the Congo for the first time; independence came quickly thereafter.

INDEPENDENCE

The Early Days

After roundtable negotiations with Congolese in Belgium in January and February 1960, the Belgian government agreed to complete independence in less than 6 months.

> Totally unprepared, political parties mushroomed throughout the country: 120 different parties contested the parliamentary elections for 137 seats in May 1960. So long oppressed by poverty, forced labor, heavy taxation and racial discrimination, the Congolese electorate were swept away by the wildest of expectations for the coming independence, expectations that could not possibly be met. (Shillington 1989: 392–393)

Political activist, Patrice Lumumba, a former postal clerk, was elected prime minister. He appointed as president his political rival, Joseph Kasavubu, a former priest.

The Congo became independent on June 30, 1960, and almost immediately began to teeter on the brink of breakup. The second Congo War caused the deaths of thousands as Congolese fought Congolese in an attempt to dismember the colonial state through secession. The United Nations was called in to prevent the breakup of the country, and periodic civil wars occurred thereafter. First the Force Publique and the Gendarmerie mutinied against their Belgian officers, and later the resource-rich provinces of Katanga (July 11) and Kasai (July 29) declared their secession, each attempting to go its own way independently of any central government in the capital.

Europeans fled the Congo in the face of public humiliation, beatings, rape, and murder. Of the roughly 30,000 Europeans who lived in the Congo's three largest cities prior to independence, only 3,000 remained 10 days later (Edgerton 2002).

Belgian paratroopers, flown in to restore order, were replaced at the insistence of Lumumba by United Nations troops, arriving at the same time were Soviet materiél, advisors, and economic aid, which Lumumba had solicited. Although the crisis of the Congo had been prominent in world headlines up to that point, the involvement of the Soviet Union raised the stakes of violent Congolese secession to the level of a Cold War superpower conflict. Wishing to see an independent, democratic, and market-oriented Congo that would be friendly to business and help the American government in its attempt to contain the spread of communism, the United States would have been happy to see Lumumba removed from power. The Soviet Union, on the other hand, saw in the African leader an ally who would promote communism in Africa. In August 1960 Prime Minister Lumumba and President Kasavubu began to fight for power, and instability appeared to be growing. The United States sent warships to the mouth of the Congo River and demanded that the Soviet advisors leave the country. A pro-Western officer, Colonel Joseph Désiré Mobutu, who had been given promises of support by the United States, seized power in Leopoldville on September 14, 1960, just 75 days after independence.

It would take 5 years for Mobutu to consolidate all power over the vast country into his hands, and even so, violent turmoil continued until 1967. In 1960, however, Stanleyville was in the control of Patrice Lumumba, Katanga was controlled by Moïse Tshombe, and Albert Kalonji was leading the Kasai secession. One of Mobutu's first acts was to close down the embassy of the USSR and order the Soviet advisors out of the country. Lumumba was captured by the Congolese Army (formerly the Force Publique, renamed and controlled by Mobutu) and executed by firing squad in Elisabethville on January 17, 1961. Although it is unclear who gave the order for execution, Lumumba had been repeatedly beaten by Mobutu's soldiers, and the riflemen had acted with "the agreement of Mobutu, Kasavubu, Tshombe, and their Belgian advisors" (Edgerton 2002: 195). The Katangan rebellion ended in 1963 after two and one-half years of fighting, and Moïse Tshombe was appointed prime minister by Mobutu, who then carried the rank of general.

The political crises that beset the newly independent state in 1960 were symptomatic of several spatial and structural imbalances in the economy, and in the control of political power and the state's resources. Since independence, Congolese governments have attempted to change these imbalances through specific economic and political policies. Of the various changes instituted to provide political stability and socioeconomic development, the most significant were the restructuring of the national, provincial, and local governments, nationalization of most sectors of the economy, a reordering of priorities in mining, agriculture, and industry, and the development of the transport system. In 1963, the half-dozen provinces of the former colony were fragmented into 21 units, each of which became a province with substantially greater powers. In addition, the city of Kinshasa and its immediate surroundings became a federal district. Two major factors played a role in this experiment: pronounced ethnoregionalist sentiments in some areas (like that of the Bakongo people in Kinshasa Province), and dissatisfaction with the status quo in areas located far from the provincial capitals (Young 1963). Katanga was divided into three provinces, and Kasai into four. The scheme was criticized as a reversal to tribalism, but actual ethnic homogeneity existed in less than a third of the 21 provinces. But "divide and rule" it was not.

Rebellions in 1964 and 1965 showed weaknesses in the system and the failure of both parliamentary institutions and federal government. In 1964, for example, Orientale Province erupted in a bloody rebellion that continued into 1965, involving the entire northeast of the country. The Simba Rebellion, as it was called, was led by Pierre Mulele, a guerrilla fighter, trained in China, who intended to create a Marxist–Leninist state in the Congo. His supporters were a mixture of ideologues including many Tutsi, whose ancestors had long lived in the Congo. Other Tutsi who joined the Simba Rebellion had settled in the Kivu provinces as refugees from Hutu extremism during the Hutu revolution of 1959–1962, which ousted the Tutsi monarchy from Rwanda. The Tutsi called themselves Banyamulenge.

The Congolese Army made little progress against the rebels until Tshombe brought in hundreds of mercenaries. In 1964 Belgian paracommandos at the head of the mercenaries and the Congolese Army began to systematically destroy the Simbas, and by 1965 the rebels were a spent force. Tshombe's tenure as prime minister was short, however. Mobutu seized all power in November 1965 in a bloodless coup d'état and found general support for the restoration of a unitary state and the end of parliamentary government. Tshombe fled the country for his life.

In 1967 a new constitution establishing centralized presidential rule was ratified, and all federal elements of the earlier constitutions were eliminated. The 21 provinces were reduced to eight, the old colonial boundaries being adopted, except that Leopoldville (Kinshasa) Province was divided into two parts. Elected assemblies at the provincial level were abolished. For administrative purposes, each region was divided into subregions and territories. At the local level, authority was held by chiefs and village headmen appointed by the government. At the regional level, governors were appointed by the president, and they were rotated from one region to another to assure better control from Kinshasa.

Authenticity and Zaïrization

Redistricting was the prelude to Mobutu's "authenticity" drive, a movement designed to create greater national identity—an effort to create a sense of nationhood from the plethora of regional, tribal, class, and ideological identities that had clashed so bloodily during the first few years of independence (Wrong 2001). Basic to this was the replacement of European names with traditional names, which brought

Mobutu into direct conflict with the Catholic Church. The Republic of the Congo became the Republic of Zaïre, President Joseph Désiré Mobutu became Mobutu Sese Seku Kuku Ngbendu Wa Za Banga, and the capital Leopoldville was renamed Kinshasa (see Table 15-1). People were also required to refer to one another as "citoyen" or "citoyenne" ("citizen") in place of "Monsieur" or "Madame." Fashion changed from the suit and tie for businessmen and bureaucrats to the abacost, a loose, collarless shirt (the word is a contraction of the French expression for "down with the suit"). This new philosophy brought changes to foreign policy and national economic planning: Zaïre sought closer ties with neighboring Congo and Guinea, formerly considered too radical, and it gave ideological support to Tanzania and Zambia in their stance against white-dominated southern Africa.

"Authenticity" also meant increased "Zaïrization" of the economy. In 1973, on the eighth anniversary of his accession to power, Mobutu ordered the immediate nationalization of all plantation companies and the Belgian mining company MIBA. Later the state took control of all large enterprises in the import–export business, services, agriculture, transport, industry, and construction. Union Minière, Zaïre's largest mining company, was nationalized in 1966 and renamed Gécamines.

"Zaïrization" was introduced not only to add "authenticity" but to revitalize the economy, which had suffered numerous setbacks since independence. It was not until 1966 that the GDP reached its 1958 level, and agricultural output in 1972 was below that of 1959. In 1974, following the nationalization of most agricultural schemes, Mobutu unfolded a plan to address the food deficit and make Zaïre self-sufficient in food by 1980. The food deficit was principally caused by the post-independence civil wars that had disrupted the economy but had been exacerbated by Mobutu's nationalization of agricultural enterprises. He promised to give massive support for the improvement of roads in rural areas, for the development of cooperative and credit facilities, and for the establishment of a national fertilizer industry. In Katanga, the mining companies were instructed to create 12,350-acre (5,000 ha) farms to reduce the region's dependence on maize imports from Zambia and Tanzania. Unemployed town dwellers were forced to relocate to these and other rural projects. Rice production, expanded with Chinese technical assistance, was expected to become Zaïre's third most important staple after cassava and maize during the 1970s.

After Mobutu

Although hopes for self-sufficiency and even prosperity were high during the early years of independence, the government of Joseph Désiré Mobutu, who had fostered the authenticity movement, came to represent the worst that could go wrong in postcolonial Africa. Throughout the 1980s and up until 1997 when President Mobutu would flee the country, Zaïre was "a parody of a functioning state [where] the anarchy and absurdity that simmered in so many other Subsaharan nations were taken to their logical extremes" (Wrong 2001: 10).

The movement that was to topple Mobutu in 7 short months of vigorous campaigning emerged in the distant east of the country. It had roots in the precolonial ethnic geography and in the postcolonial interethnic war and genocide in Rwanda, which had spilled over into the Kivu provinces of Zaïre periodically over the last 45 years (Lemarchand 2000). The movement's leader, Laurent Kabila, was originally from Katanga and had been a hard-line socialist supporter of Lumumba. He was elected to the North Katanga legislature in 1962, but Kasavubu closed down parliament and Kabila and others fled to Brazzaville. Along with Pierre Mulele, Kabila became a leader in the Simba Rebellion against the Mobutu and Tshombe government but retreated into the mountains of South Kivu Province when the rebellion was put down.

Kabila later became the head of the Alliance of Democratic Forces for the Liberation of the Congo (ADFL), comprising Banyamulenge who by 1996 may have numbered 500,000 people. These economically successful, long-time residents of eastern Zaïre wanted to obtain the rights of citizenship in Zaïre, political participation, and the ability to buy land, but were prevented from doing so by an anti-Tutsi immigration law enacted in 1981. Anti-Tutsi violence in eastern Zaïre in 1993 served to agitate the ADFL. Then, in 1996 they clashed with Hutu militia during the massive Hutu refugee immigration that had resulted from the triumph of the Tutsi-led Rwanda Patriotic Front (RPF) in Rwanda. Paul Kagame, who in 2000 would become the first Tutsi president of

Rwanda, then orchestrated a march against Kinshasa by the ADLF and allies from Uganda, Rwanda, and Angola.

Kabila took power in 1997 and found a treasury that had been bankrupted, a dysfunctional and predatory administration, a physical infrastructure in almost complete deterioration, and human and capital flight (U.S. Department of Commerce 2001). At the time of Mobutu's departure, the state was unable to provide the most basic public services to the people of Zaïre. Ninety percent of the labor force was engaged in subsistence activities in a country with vast, untapped potential for almost any sort of economic activity.

One year after Kabila's seizure of power, it became apparent that his regime would continue the undemocratic leadership and predatory rent-seeking of the preceding government. His opposition to political reform, his corruption, his continued dependence on Rwandan and Banyamulenge armed forces, and his subservience to Kagame in Rwanda caused his popularity to plummet soon after he came to power.

Kabila chose to yield to the anti-Tutsi elements in the Congo and expelled his Rwandan and Banyamulenge advisors and soldiers in July 1998. A month later, a rebellion was under way in eastern Congo led by a new group, the Congolese Rally for Democracy (CRD), backed by Kagame and President Yoweri Museveni of Uganda. Many Congolese army garrisons switched sides; the rebels took key towns and cities across the country and even had control of the huge hydroelectric plant at Inga, the principal source of power for nearby Kinshasa. At that point, the Angolan army intervened from Cabinda, dislodging the rebels from Bas Congo Province. By the end of the year, however, the rebels had consolidated their control over the eastern third of the country.

The conflict escalated further, and in addition to Angola the neighboring countries of Chad, Namibia, Sudan, and Zimbabwe intervened on the side of Kabila.

Kabila was assassinated in 2001 and was replaced as head of state by his son, Joseph, but major fighting continued to the end of 2002, and in the eastern part of the country there were skirmishes into 2003. The foreign troops left later in 2003, after Joseph Kabila negotiated a peace in which rebel forces would be integrated into the government and military. Three out of four vice presidencies and over 40 ministries were given to the rebels, and all have pledged to hold democratic elections within three years (*Economist* 2003).

In what has been called Africa's Great War, rogue traders, multinational corporations, and unscrupulous neighbors helped fuel the hostilities by buying "conflict" diamonds, gold, and other natural resources brought to market illegally by regional rebels and criminals with weapons. In a particularly egregious example, the president of Zimbabwe, Robert Mugabe, who was himself mismanaging his country's economy, agreed to help Laurent Kabila in return for access to mineral wealth. The civil war in the Congo, fought along preexisting regional cultural fracture lines and exacerbated by the interests of neighboring countries, was mainly about the control of natural resources and the hearts and minds of people. It caused the deaths of between 3 million and 5 million people in the eastern part of the country alone. Only 10% of those deaths were violent, the vast majority being caused by starvation and disease in what some have called the greatest humanitarian disaster since World War II (IRC 2003). In such a resource-rich country it is cruelly ironic that today the overwhelming majority of Congolese live in poverty and per capita income is not as high as it was during colonialism, in the years just prior to independence.

ECONOMIC ACTIVITIES

Agriculture is the principal occupation of the population in the DRC, but the contribution of agriculture to the GDP is only 55%; industry contributes 11%, and services constitute the remaining 34%. In the many years of economic instability and civil strife, the farm calendar has been continually interrupted in many areas of the DRC, and there has been growing hunger in rural areas. The U.S. Department of State (2001) estimates that in 2000 fully 90% of the population of the DRC was engaged in subsistence activities rather than the cash economy. As a consequence, any description of the economic activities of the DRC is likely to be in error until the economy recovers. Furthermore, much of the economic activity that should normally contribute to GDP consists of illegal trading and goes unrecorded.

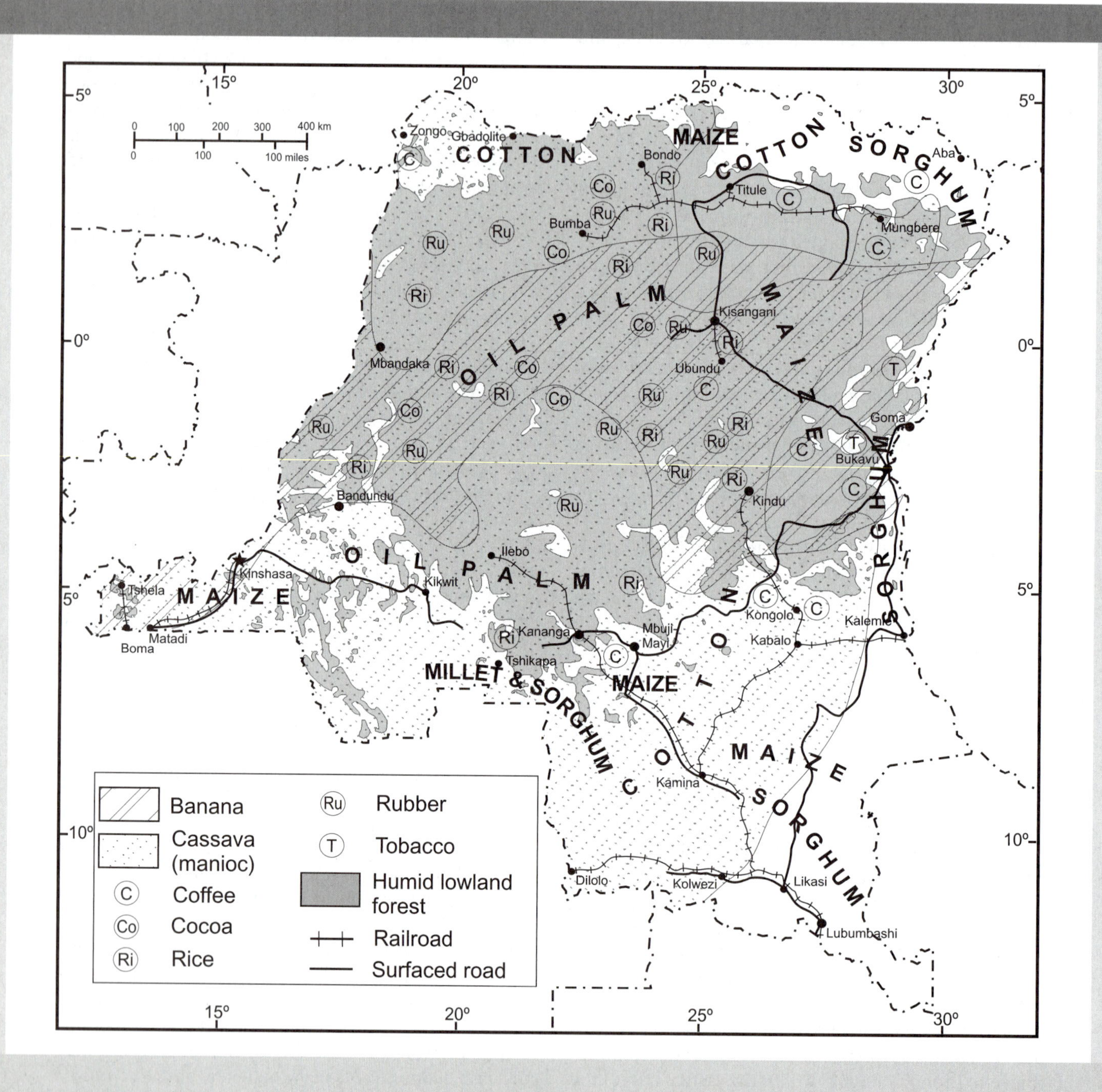

FIGURE 15-1 Location map of selected cash and food crops for the Democratic Republic of the Congo.

The general geographic pattern of farming in the DRC follows that of many other African countries with similar geographic variability. In the drier periphery of the country the cereals millet and sorghum are grown, while the more humid regions produce tubers, maize, bananas and plantains and other fruit, rice, and cash crops such as the oil palm, rubber, coffee, and tobacco (Figure 15-1).

It is difficult to believe that in such a well-watered country only 10% of the land is used for agriculture, with 3% planted to field and tree crops and another 7% used as permanent pasture; 73% of the country is forest. Since independence, agriculture has experienced some overall successes and recent setbacks, but great potential remains for expansion. With the exception of rubber, which has been declining for over 30 years, the production of many food and cash crops has decreased since the late 1990s when civil war broke out (Figures 15-2 to 15-4). Up until that time, agriculture had been increasing—principally as a result of expansion. Among the major food crops, the output of cassava more than doubled, while that of maize and rice more than tripled. Inflation, a big problem in Zaïre under Mobutu, has been so since his ouster, as well. Many farmers, unable to make a living in traditional markets in an environment of uncontrolled prices, have reverted to subsistence farming or have turned to growing foodstuffs for the burgeoning urban markets. The most intensively cultivated areas adjoin the towns. Even there, however, land has been abandoned because of poor transport and civil strife, and because imported foods may be purchased at lower prices in the urban centers.

Mining has long been the real source of the nation's wealth (Figure 15-5), accounting for three-quarters of all exports, and historically has contributed 25% to the GDP. Over recent years, the contribution to GDP has declined to 6% (Coakley 2001). In the past, copper alone has generally accounted for 60% of all exports, followed by cobalt, industrial and gem diamonds, cadmium, petroleum, zinc, manganese, tin, germanium, uranium, radium, bauxite, iron ore,

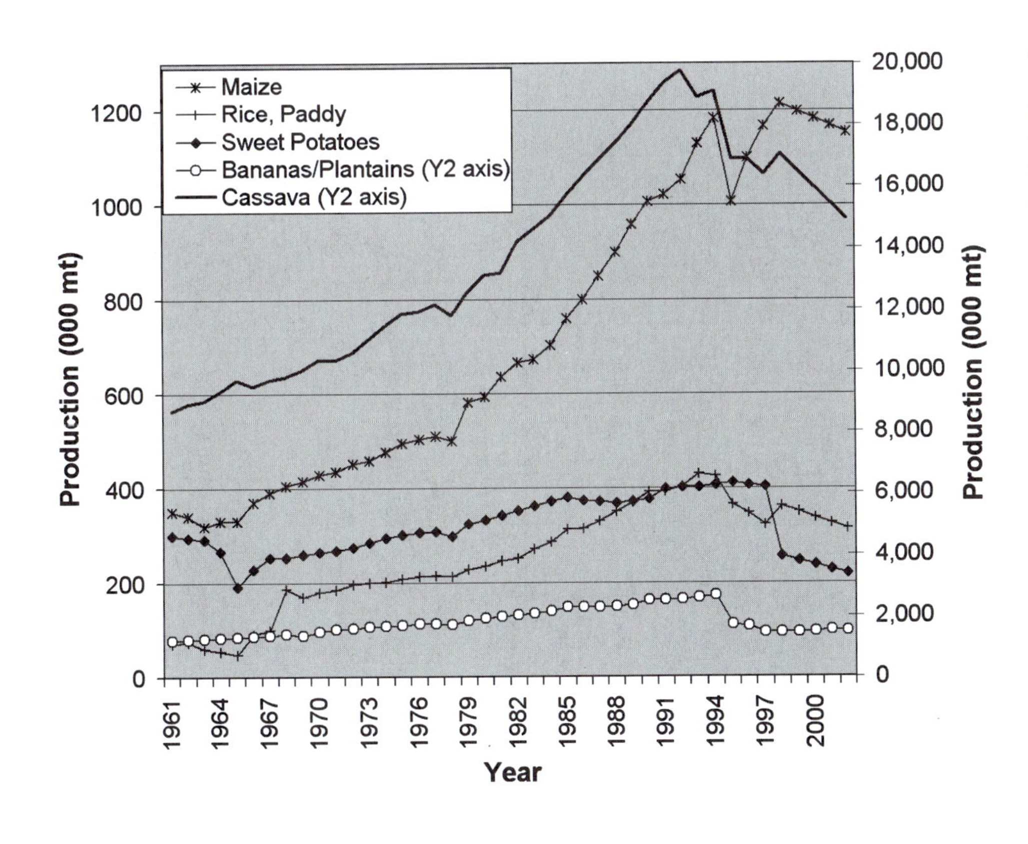

FIGURE 15-2 Major food crop production in the Democratic Republic of the Congo, 1961 to 2002.

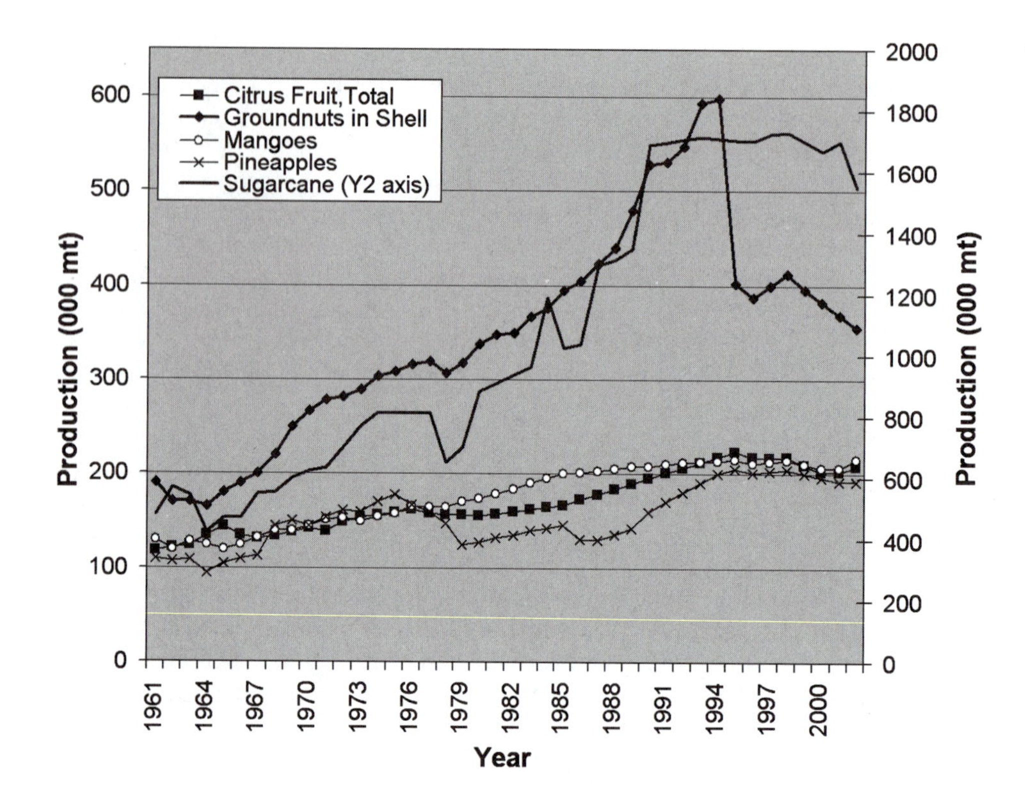

FIGURE 15-3 Selected export crops, DRC, 1961 to 2002.

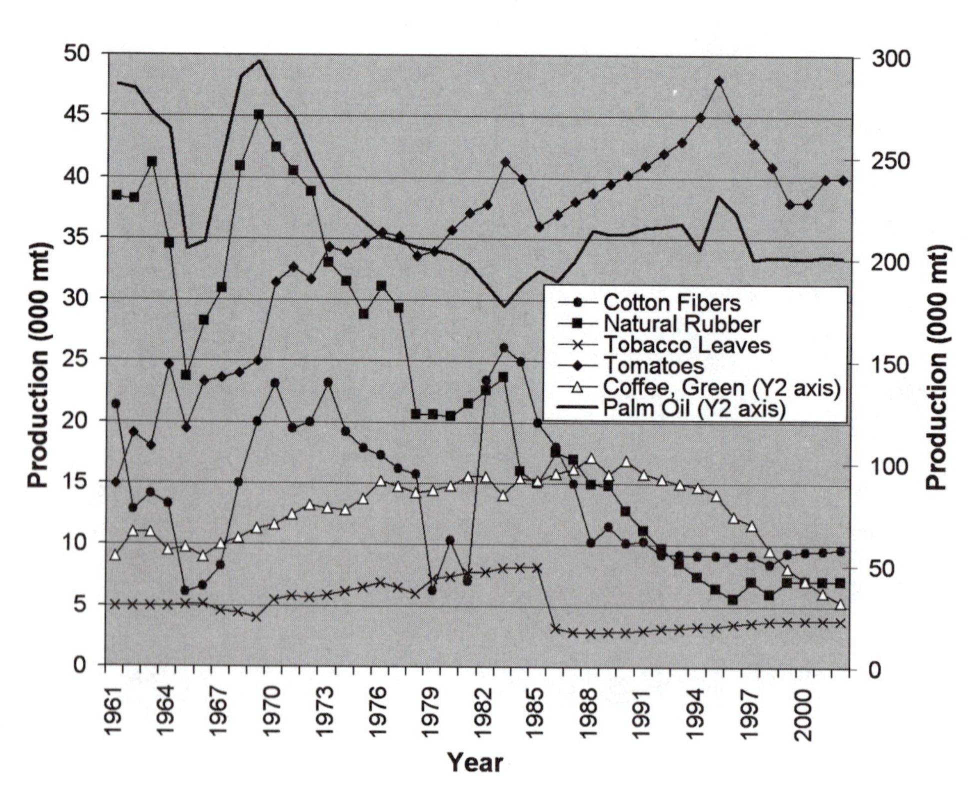

FIGURE 15-4 Selected export crops, DRC, 1961 to 2002.

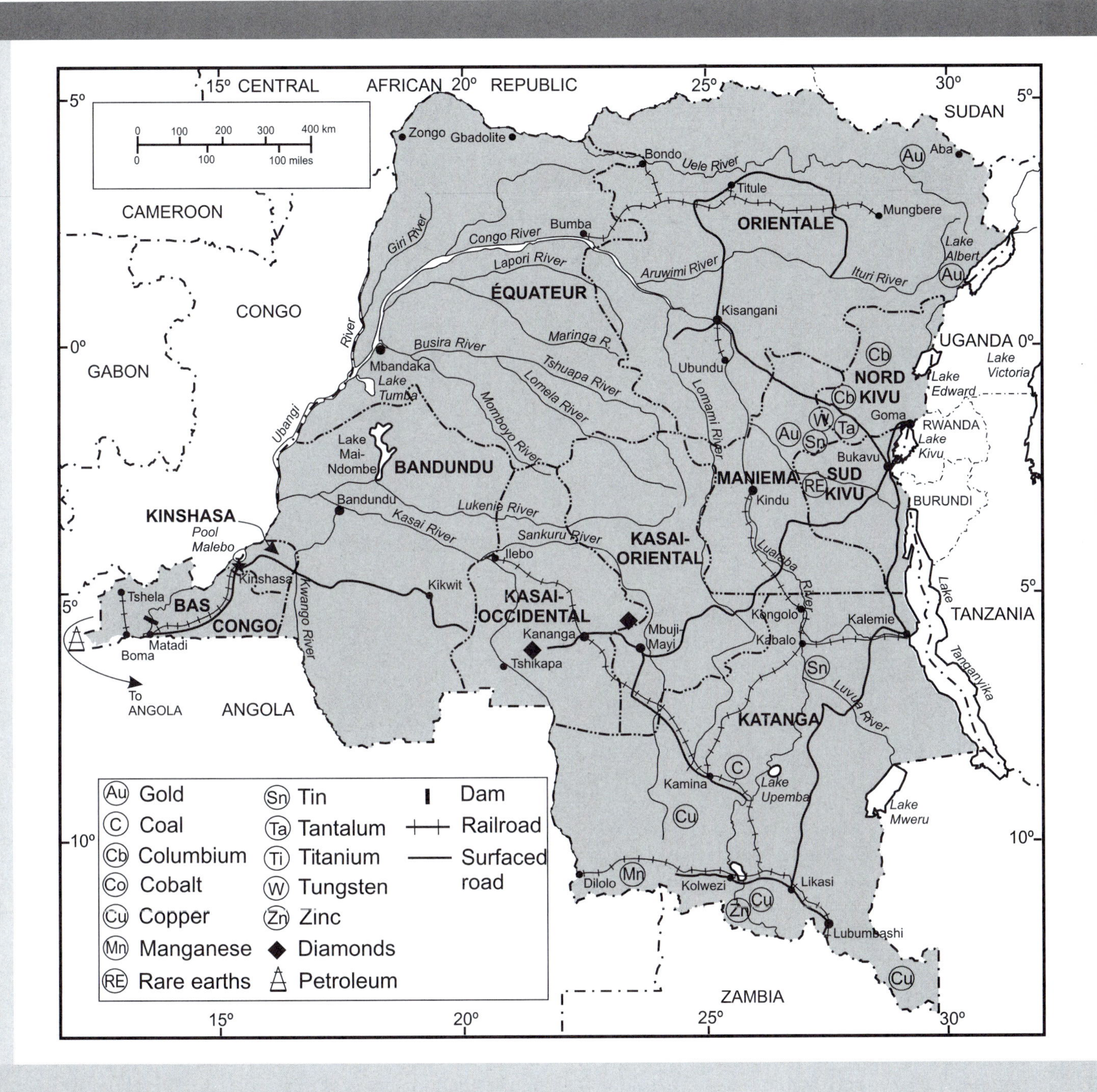

FIGURE 15-5 Economic map of the Democratic Republic of the Congo.

TABLE 15-2 MINERAL PRODUCTION BETWEEN 1958 AND 2001, IN THE DEMOCRATIC REPUBLIC OF THE CONGO.

	Production (000 metric tons unless otherwise specified)										
Commodity	**1958**	**1964**	**1970**	**1973**	**1990**	**1992**	**1994**	**1996**	**1998**	**2000**	**2001**
Cobalt	6	7	15	15	39.5	13	0.8	2	5	7	4.7
Columbite-tantalite ore (coltan) (mt)					19	15.5	2			0.2	0.1
Copper	237	276	387	500	509	275	30	40	40	21	21
Gold (mt)					9.3	7	0.8	1.3	0.1	0.05	0.05
Manganese	319		329	334							
Silver (mt)					84	29.5	0.9	0.5	0.5		
Tin	11	5	1	1	2	1	0.1				
Zinc	114	105	109	82	115	46	0.5	3	1	0.2	1
Diamond (millions of carets)					19.4	13.5	16.3	20.6	26.1	17.7	18.2
Coal, bituminous	294	106			100	61	11	10	5		1
Petroleum, crude (millions of 42-gallon barrels)					10.6	8.7	9.1	10	9.4	10.3	11.5

Sources. Peemans (1975), Rapports Banque du Zaïre (various issues), Quarterly Economic Review *(various issues), U.S. Geological Survey Mineral Reports (1994, 1998, 2001).*

and coal. A half-million metric tons of copper was produced in 1973 and 1974, and the state-owned copper complex, Gécamines, had investment plans for the period 1975–1980 costing about $500 million, with a view of expanding productive capacity to 600,000 metric tons. It was expected that an additional 3,000 metric tons of cobalt would be produced as well.

Other major mineral projects under development have included a uranium enrichment plant at the Inga hydroelectric scheme on the Congo River west of Kinshasa, and offshore oil production near the Cabinda border. Contrary to the plans and aspirations of the 1970s, however, total mining production declined dramatically since 1958 (Table 15-2 and Figure 15-6) with the exception of diamonds. Copper production peaked in 1990 and has declined precipitously. Gécamines was also in decline, its poor condition attributed to "a combination of aging equipment; lack of domestic and international investment; lack of spare parts; shortages of fuel, lubricants, and sulfuric acid; problems with transporting ore and finished products; theft of finished products; debts owed to the state electrical company and Société Interrégionale Zaïroise de Rail (Sizarail); flooding of open pit mines; and the inability to retain professional and other personnel because of disruptions caused by the war and other factors" (Coakley 2001: 2).

Cobalt, a by-product of copper production, peaked in 1990 as well, but the DRC remains the world's leading producer of cobalt and is the fifth largest producer of copper. The planned uranium enrichment plant was never completed, although offshore oil production became a reality, and the country's yield is about 10 million barrels a year. The production of diamonds appears to be stable over the years, but the figures are deceptive because much of the diamond production was marketed illegally by Mobutu and his cronies. In any event, after years of neglect, underinvestment, hyperinflation, and civil war, formal mining activities have almost ceased. It is difficult to quantify the export of informally mined and illegally exported gold, diamonds, coltan, and other minerals, but such exports were a significant spur to foreign intervention in the civil war in the DRC in 1998 (United Nations

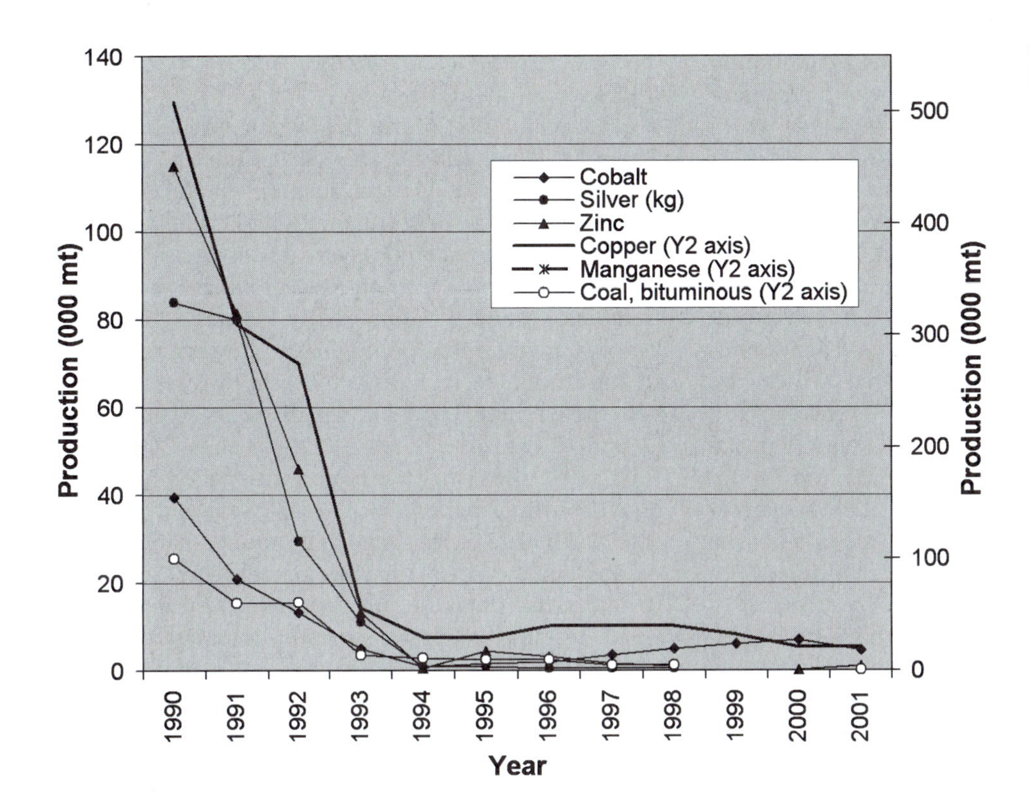

FIGURE 15-6 Decline in an industry: mining in Zaïre/DRC, 1990 to 2001.

Security Council 2002). With peace, mineral production should regain its former share of the GDP and provide some capital for investment. Without political stability, the formerly productive mining sector will not be redeveloped because there will be no investors.

Although the outlook for hydrocarbons is promising in terms of reserves—the DRC possesses 187 million barrels of petroleum reserves, one billion cubic meters of natural gas, and 88 million tons of coal reserves (EIA 2002)—the greatest energy asset of the DRC is its abundant water. The DRC has the greatest hydroelectric potential in Africa (150,000 MW), and 13% of the world's potential. Ironically, only 7% of the population has access to electricity, and electricity production has been stalled at less than 10% of the gross potential that had been installed by 1976.

Energy development has been closely linked with the development of mining and metallurgical industries that, before the collapse of the mining sector in the 1990s, consumed about 80% of the country's total electricity output. The largest hydroelectric power generation scheme is on the lower Congo at Inga, where the river falls 315 feet (96 m) in 9 miles (14.5 km) and the theoretical potential is 40,000 megawatts. Total power generation today is only 1,775 megawatts. Phase I of the Inga project (351 MW) was opened in 1972. Phase II added a further 1,424 megawatts in 1982. Inga sends 500 kilovolts of electricity to Kolwezi, and 200 kilovolts is exported to neighboring Zambia, Zimbabwe, and South Africa. The Congo obtains one-third of its energy needs from the Inga Dam.

Energy production has the potential to make a significant contribution to the national economy and has attracted the interest of foreign investors. According to the DRC Energy Ministry (EIA 2002), the country earned $500,000 a month from energy exports in 2001, and earnings of $5 million a month were projected for the near future, given appropriate and sufficient investments.

Two large expansion projects and the rehabilitation of Inga I and II are under discussion. The German Siemens corporation became involved in the Inga facility in 2001, investing $1 billion in its rehabilitation and in the power lines to Kinshasa. The two megaprojects consist of "Inga 3" and "Grand Inga." Inga 3 consists of a 3,500-megawatt expansion and the creation of a regional electricity export project connecting Inga to Lagos, Nigeria, on what is called the Western Energy Highway (WEH). It is unlikely that Inga 3 will come to fruition because Grand Inga, a more ambitious scheme, seems to be preferred by investors. Grand Inga is proposed to be the largest generating facility in Africa, producing 39,000 megawatts. The projects which the African Development Bank believes to be feasible, will consist of an export link, the Northern Energy Highway (NEH), through the Congo, the Central African Republic, and Sudan, to Egypt. Grand Inga will probably cost $4 billion and will be implemented in four stages; the 6,000-megawatt stage 1 is to be completed by 2010. The NEH is estimated to add $5 billion to project costs.

Large power generation schemes have been constructed for Africa in the past, generally through loans and aid to governments. In the present economic context (a liberalized economy with a stock market), shares in the Inga power company will be a freely traded and open to any investor in the world. It is possible that a multiregional, multicontinental, profit-oriented facility will yield economic returns attractive enough to draw serious long-term investment in addition to loans.

Escom, a South African company, is playing a leading role in the development of the Inga site and in the development of the Pan-African electric grid (Figure 15-7). Escom is the principal architect of an ambitious 50-year vision for African development to be powered by electricity. The company argues that electricity prices in southern Europe and the Middle East are high enough to warrant linking the African grid to Europe and Asia to meet demand for cheaper electricity.

The Escom plan is not so far-fetched. Algeria has been exporting natural gas via pipeline to Europe for years. In 2001 Morocco exported electricity to Europe for the first time. Critics counter that there will be substantial transmission losses across such a vast area and that use from continent to continent differs in the voltage (110 vs 220 V) and in type of wall plug (2 prong vs 3 prong). Advocates argue that such regional divergences in other parts of the world were resolved through standardization and can be resolved in the same way in Africa.

The population of the Democratic Republic of the Congo has increased from 30 million in 1984 to 57 million in 2003. The population is unevenly distributed across the country, the Congo Basin being the most sparsely populated (Figure 15-8). Dense clusters of population exist in both Kivu provinces in the extreme east along the Ruwenzori Mountains (156 and 214 per square mile; 83 and 60/km^2, respectively) and in the provinces of Bas Congo and Kinshasa in the extreme west (183 and 2,038 persons per square mile; 71 and 783/km^2).

In 1977 the Congolese nation then called Zaïre had 10 cities with a population over 100,000, but in 2003 the DRC had 25 cities above that number (Table 15-3). As was true during the first two decades after independence, most Congolese cities are not industrially diversified. Matadi, the nation's seaport, is situated on one of Africa's largest natural harbors, 80 miles (130 km) from the Atlantic Ocean; Kananga specializes in cotton, diamonds, and palm oil; Lubumbashi's major industries are mine related; Mbuji-Mayi produces little but diamonds. Kisangani produces metal goods and furniture; it is the country's major textile center as well. Kisangani, which possesses an international airport and one of the campuses of the University of the Congo, is the terminus for river shipping from Kinshasa. Kolwezi is a center for copper and cobalt mining and processing. Likasi is a major transportation and industrial center specializing in mining, chemicals, and copper refining.

Kinshasa, located on Pool Malebo, is the DRC's largest and most diversified industrial city: engineering, auto assembly, building and repairing of ships, clothing and other textile-related enterprises, tanning, chemicals, and foods, are among the industries represented. It is the terminus for the railroad from the port of Matadi and is itself a port on the Congo River for shipping as far as Kisangani. Kinshasa is an educational and cultural center (Box 15-1), with the Lovanium University of Kinshasa, the National School of Law and Administration, and several schools and research centers. Like all other Congolese cities, Kinshasa is short of qualified personnel, yet unemployment among unskilled workers is a serious problem, exacerbated by almost a decade of civil war and economic collapse.

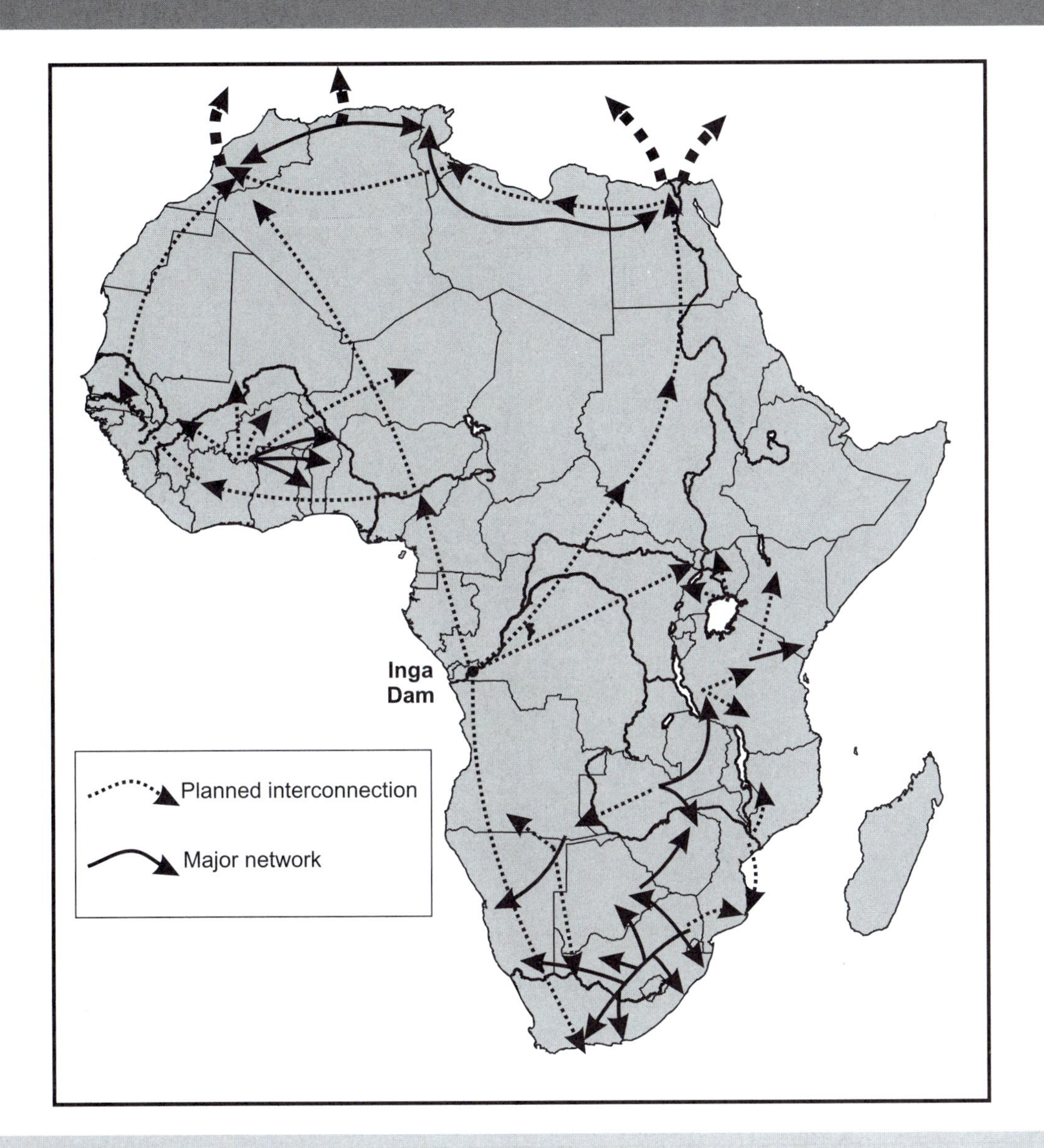

FIGURE 15-7 The proposed electric grid for Africa and connections to Europe and Asia. ***Adapted from Hale (2002).***

With the exception of Lubumbashi, the provincial capitals are not supported by lower-order towns and cities, and their trade areas are small. Interprovincial economic linkages are weak, and are restricted by inadequate communications and a generally poor population. Vast areas of unproductive land still separate these towns and resource regions.

Size has always been both an asset and a liability to the DRC. Its almost 1 million square miles (2.3 million km^2) encompasses great mineral wealth,

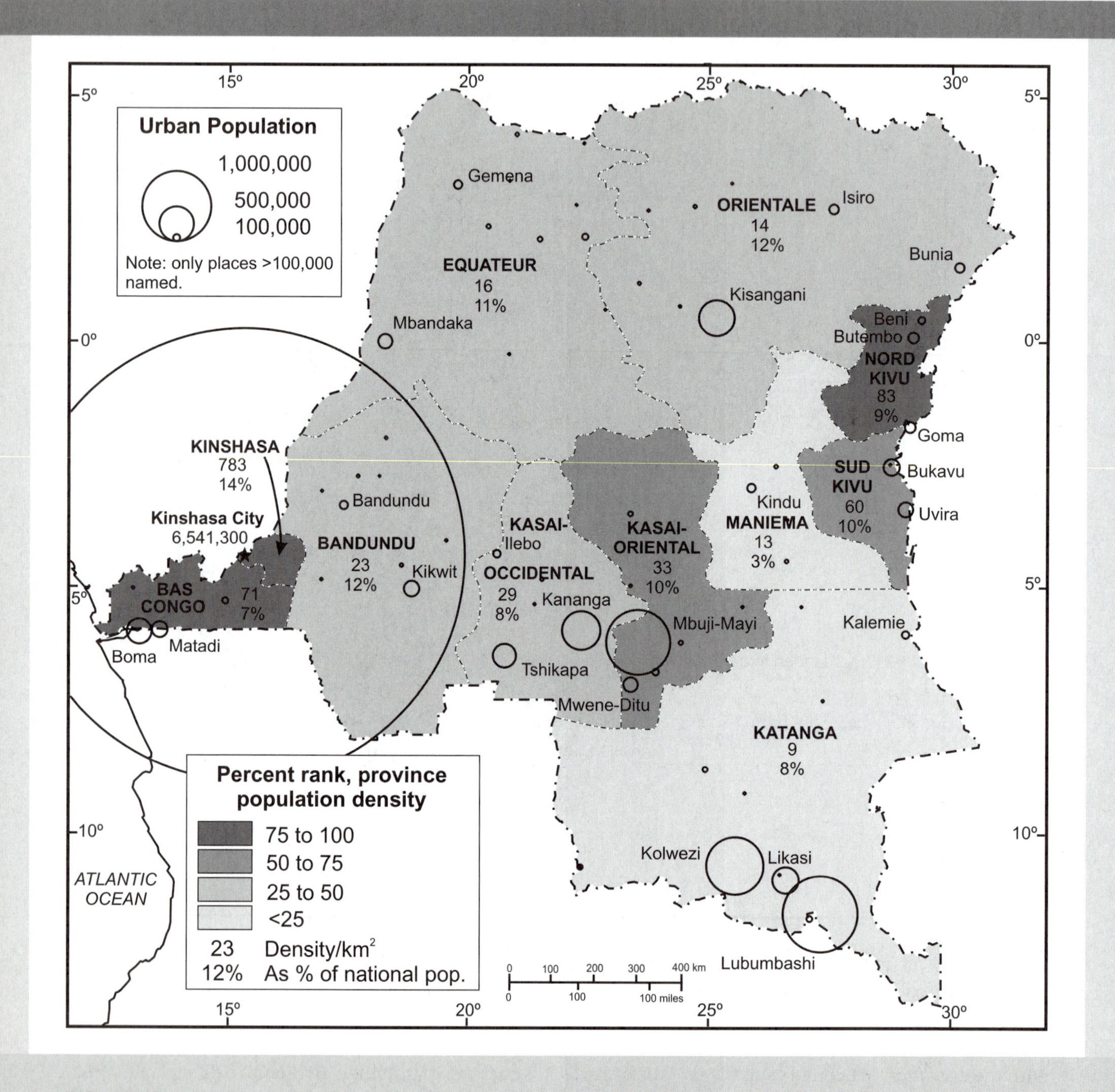

FIGURE 15-8 Provincial and urban populations of the DRC.

TABLE 15-3 URBAN POPULATION IN THE DEMOCRATIC REPUBLIC OF THE CONGO FOR 1984 AND 2003.

		Population	
Urban Place	**Province**	**1984**	**2003**
Kinshasa	Kinshasa	2,664,300	6,541,300
Lubumbashi	Katanga	564,800	1,105,900
Mbuji-Mayi	Kasai-Oriental	486,200	938,000
Kolwezi	Katanga	416,100	832,400
Kananga	Kasai-Occidental	290,900	557,800
Kisangani	Orientale	317,600	523,000
Likasi	Katanga	213,900	385,700
Boma	Bas Congo	197,600	363,700
Tshikapa	Kasai-Occidental	116,000	337,800
Bukavu	Sud Kivu	168,000	241,100
Matadi	Bas Congo	138,800	230,000
Kikwit	Bandundu	149,300	228,000
Mbandaka	Équateur	137,300	211,800
Mwene-Ditu	Kasai-Oriental	94,600	208,100
Uvira	Sud Kivu	74,400	207,500
Butembo	Nord Kivu	73,300	154,600
Goma	Nord Kivu	77,900	154,200
Isiro	Orientale	78,300	139,700
Bunia	Orientale	59,600	134,500
Gemena	Équateur	63,100	124,300
Kindu	Maniema	66,800	122,800
Bandundu	Bandundu	63,600	120,600
Kalemie	Katanga	73,500	109,100
Ilebo	Kasai-Occidental	53,900	108,600
Beni	Nord Kivu	44,100	100,700

Source: World Gazetteer (2003).

about 10,000 miles (16,000 km) of navigable waterways, and sufficient topographical variety to provide unrivaled hydroelectric potential. On the other hand, this vastness encompasses great ethnic and linguistic diversity that made political unity difficult long before the days of King Leopold and continues to impose obstacles in the present. Furthermore, it may be argued that such diversity has facilitated the conquest of the region by outsiders (first Arabs, then north Europeans) and that the DRC is a country that cannot survive in its present, impossibly large form.

Extreme size also has made transport both costly and time-consuming, and if problems of distance are to be overcome, great expenditures on the road and rail networks will be required. The uneven distribution of natural resources across the Congo Basin has underlain, reinforced, and exacerbated periodic rebellions aimed at secession along regional and subregional lines. The problems of size and diversity will not be easy to overcome in a developing country whose population is predominantly rural and widely distributed, and in need of improved health and education facilities, higher incomes, and better housing. Moreover, development funds are limited and dependent on exports whose prices are subject to severe fluctuations.

BOX 15-1

DIFFUSION OF THE "KINSHASA SOUND"

Traditional music, found in the rural areas of the Congo, focuses on the marimba, thumb piano, and drum. In urban places traditional music has been crowded out by the "Kinshasa sound," or "soukous," an electric, steely music originally belted out in bistros and backstreet dive bars throughout the capital in the early years of independence. But the origins of soukous go back much farther, according to ethnomusicologist Dr. Kazidi wa Mukuna (1993), who states that early soukous music developed among the work crews of European companies between 1900 and 1930. The electric guitar came to Kinshasa in 1954 and was incorporated into local jazz bands such as OK Jazz and Dr. Nico, which later enjoyed some international renown.

During the 1970s, the rumba brass component of the jazz bands began to be edged out in favor of electric guitars, sometimes as many as five, and a different sound began to emerge. Zaiko Langa Langa ("land of our ancestors"), one of the first bands to achieve international stardom with this supercharged electric sound, was so successful that dozens of look-alike bands sprung up during the 1970s. The leader of Zaiko was Papa Wemba, who was famous (or infamous depending on one's point of view) for his flamboyant dress and elegant lifestyle. He inspired a generation of urban youth with the superficialities of flashy dress and image of urban life in the same way that Western rock-and-roll musicians do to many young people—to the consternation of their parents. Soukous lyrics are generally in Lingala, the lingua franca of central Africa, and are apolitical in nature, focusing on relationships and love. During the 1970s, many soukous musicians moved to other parts of Africa and also to Paris to take advantage of greater opportunities, and soukous became an international big business.

BIBLIOGRAPHY

Coakley, G. 2001. *The Mineral Industry of the Congo (Kinshasa)*. U.S. Geological Survey Mineral Report. Reston, Va.: USGS. http://minerals.usgs.gov/minerals/pubs/country/.

Economist. 2003. Congo: Not as bad as dad. *Economist*, November 15.

Edgerton, R. 2002. *The Troubled Heart of Africa: A History of the Congo*. New York: St. Martin's Press.

EIA. 2001a. Southern African Development Community. Washington, D.C.: U.S. Energy Administration Information. http://www.eia.doe.gov/emeu/cabs/sadc.html.

EIA. 2002. Inga Hydroelectric Facility. Washington, D.C.: U.S. Energy Information Agency. http://www.eia.doe.gov/emeu/cabs/inga.html.

Escom Corporation. No date. Escom envisions Pan-African electricity grid centered on Inga in the DRC: http://www.eskomenterprises.co.za/.

Hale, B. 2002. Africa's grand power exporting plans. BBC News online, October 17. http://news.bbc.co.uk/1/hi/business/2307057.stm.

IRC. 2003. *Mortality in the Democratic Republic of Congo: Results from a Nationwide Survey*. New York: International Rescue Committee.

Lemarchand, R. 2000. The crisis in the Great Lakes. In *Africa in World Politics: The African State System in Flux*, 3rd ed., J. W. Harbeson and D. Rothschild, eds. Boulder, Colo.: Westview Press, pp. 324–352.

Likaka, O. 1997. *Rural Society and Cotton in Colonial Zaïre*. Madison: University of Wisconsin Press.

Peemans, J. P. 1975. The social and economic development of Zaïre since independence: An historical outline. *African Affairs*, 74(295): 148–179.

Shillington, K. 1989. *History of Africa*. London: Macmillan.

Thomson, A. 2000. *An Introduction to African Politics*. London: Routledge.

UNEP. 2002. *Africa Environmental Outlook: Past, Present, and Future Perspectives*. Nairobi, Kenya: United Nations Environment Program.

United Nations Security Council. 2002. *Final Report of the Panel of Experts on the Illegal Exploitation of Natural Resources and Other Forms of Wealth of DR Congo*. New York: United Nations.

U.S. Department of State. 2001. *Country Commercial Guide: Democratic Republic of the Congo*. Washington, D.C.: Department of State.

Vansina, J. 1990. *Paths in the Rainforest: Toward a History of Political Tradition in Equatorial Africa*. Madison: University of Wisconsin Press.

wa Mukuna, K. 1993. L'évolution de la musique urbaine au Zaïre durant les dix premières années de la Deuxième République (1965–1975). *L'Aquarium*, 11–12: 65–71.

Wrong, M. 2001. *In the Footsteps of Mr. Kurtz: Living on the Brink of Disaster in Mobutu's Congo*. New York: HarperCollins.

Young, M. C. 1963. The Congo's six provinces become 21. *Africa Report*, 8(9): 12–13.

Young, M. C. 1965. *Politics in the Congo: Decolonization and Independence*. Princeton, N.J.: Princeton University Press.

CHAPTER 16

Development Contrasts in Central Lowland Africa

The smaller and northern portion of central lowland Africa (the Democratic Republic of the Congo being the larger), forms a corridor from the Bight of Biafra and the mouth of the Congo River as far east as Sudan, and as far north as the margins of the Sahara (Figure 16-1). It comprises three republics derived from former French Equatorial Africa, the Central African Republic (CAR), the Congo Republic, and Gabon; Equatorial Guinea, once a Spanish colony; and the United Republic of Cameroon, consisting of the former French trusteeship and part of the former British trusteeship. Sometimes called "Middle Africa," this sparsely populated region (only 21 million inhabitants in an area four times the size of California) is predominantly forested, and rich in minerals. There has been considerable progress in development—yet much remains to be done. Four of these countries are listed as possessing a "medium" level of development by the United Nations Development Program (2002) while the fifth, the Central African Republic, ranks near the bottom. Several natural and institutional constraints to economic progress and political stability are shared by all these states, but each has its own particular development prospects and priorities. These are reviewed against the background of the territories' colonial heritages and in the context of change in contemporary Africa.

THE COLONIAL EXPERIENCE AND THE DEVELOPMENT OF THE POLITICAL UNITS

French Equatorial Africa before the Second World War possessed many of the characteristics of the Congo Free State prior to World War I, but the human rights abuses continued much longer. From the early 1500s, the major product of the region had been captive human beings, taken westward from the coasts to the Americas and eastward across the Sudan to the Arab world. As many as 150,000 captives may have been taken annually as late as 1840. The period of European exploration, starting in earnest in the 1820s, saw Britons such as Dixon Denham and Germans like Heinrich Barth and Robert Flegel travel across the region in the interests of their respective homelands. French interests were stimulated primarily by Pierre de Brazza, who founded the city of Brazzaville (1880) and explored parts of present-day Congo and Gabon.

French exploration continued between 1880 and 1890, when the upper reaches of the Ubangi River were explored and treaties signed. The object of France's push into this interior region was the establishment of an axis across Africa as far as the borders of what was then Abyssinia, and French flags were eventually planted that far east. But British power forced the French to withdraw from the entire Nile Basin. German influence, meanwhile, was established in Cameroon, and the British had taken most of present-day Nigeria, while the Congo Free State had been created to the south. French penetration was thus confined to a northerly path, and Chad became a part of the corridor of France's realm in Central Africa.

After the period of exploration, conquest, and consolidation came the first attempts at administration. The Berlin Conference of 1884–1885 had recognized a free-trade zone in the Congo Basin, a move that affected the spheres of influence of Belgium as well as France. Although it was felt that direct French government efforts to administer the region might lead to international protests, the Belgian system of permitting

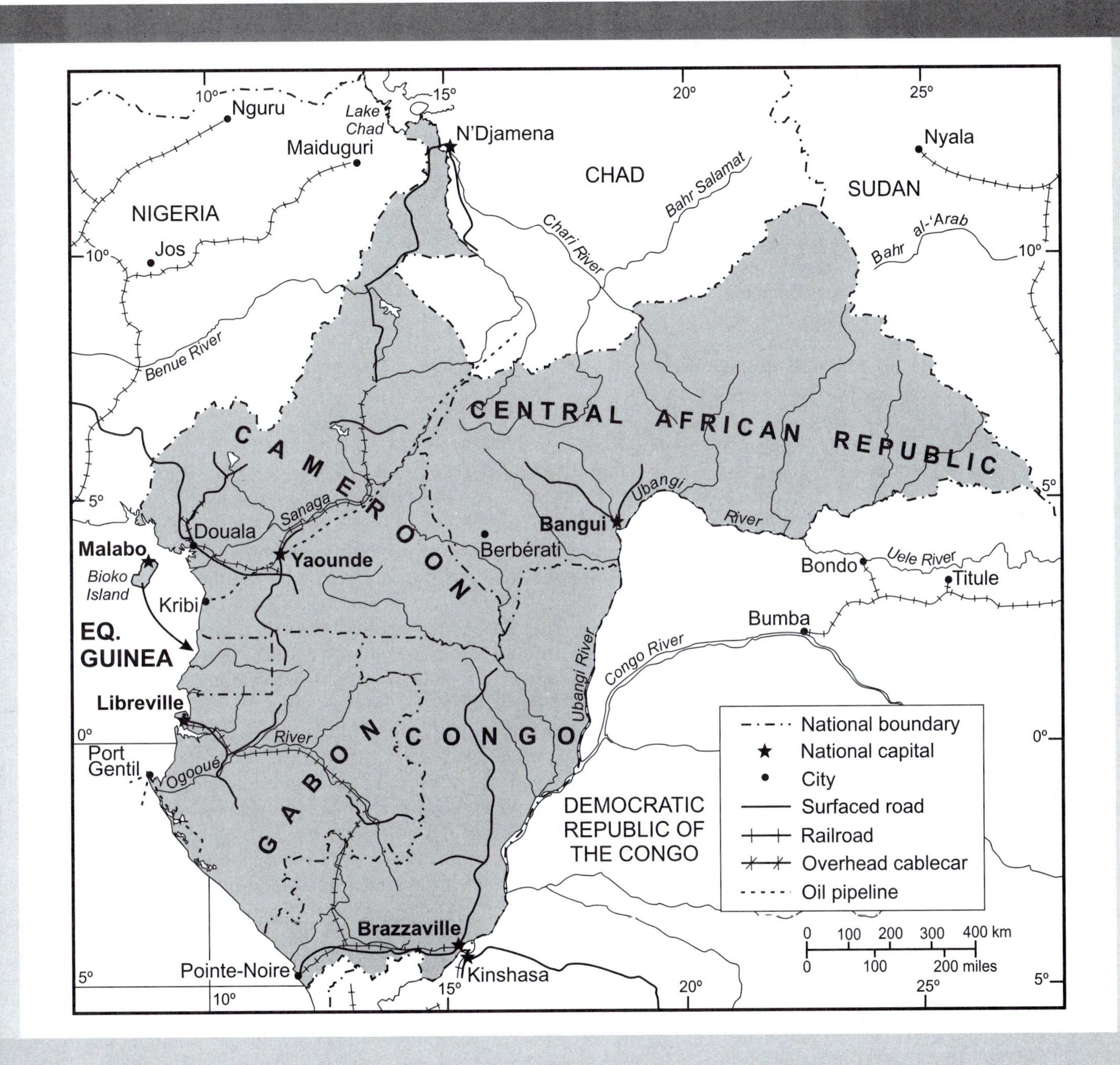

FIGURE 16-1 Central lowland Africa: Cameroon, Central African Republic, Congo, Equatorial Guinea, and Gabon.

private companies to exploit the Congo Free State was deemed successful. Thus a similar approach was used in the French areas, with the result that the period of exploration was followed by a phase during which the concessionaires, which possessed a monopoly of trade and a measure of sovereignty, literally ruled the region. The consequences of unfettered reign by organizations accountable only to their shareholders resembled those that led to the outcry over "Congo atrocities."

During the early years of European rule, it was thought that boundless wealth lay in this part of Africa. After the concession companies' early rush to exploit wild rubber and ivory, however, the reality of the situation began to penetrate. Central lowland Africa was, and still is, hot, humid, and disease ridden. The tsetse fly prevents the use of oxen or horses for transportation of goods overland, road-building materials are scarce, the forest dense; the population is sparse, and the labor supply limited at best. Thus, human portage, along tedious and narrow pathways was the only way of carrying supplies to the interior settlements and exports to the coast. The companies accepted the tremendous cost of this system until as late as the mid-1930s, when the railway between Brazzaville and Pointe-Noire, begun in 1922, was finally completed.

Just as Stanley had emphasized the need for railroads in the Congo, so Brazza had envisaged a railroad system for French Equatorial Africa, of which the Pointe-Noire–Brazzaville line was the priority project. But Brazza wanted the line to lie entirely in French territory, not, as Leopold suggested, partly in Belgian and partly in French territory as a joint enterprise. Thus, the Matadi–Leopoldville line was built by the Belgians alone, and materials too heavy for human portage sometimes reached the interior of French Equatorial Africa via that connection. But French Africa benefited little from the Belgian link: since the port of Matadi was congested, and the railroad operated near saturation point, there was little space for French freight.

Further north, a railroad was built from the port of Douala in Cameroon for the export of bananas, cocoa, and palm oil from plantations established during the German colonial era. In Gabon, the Ogooué and Ngounié rivers were virtually the only communications links until after World War II. No railroad was built there during the colonial era, and very few roads connected Libreville, the capital, with its forested interior.

When, after the end of World War II, France was confronted with a need for stimulating development in this grossly underdeveloped part of its colonial realm, it faced several major obstacles, not the least of which was the detrimental effect of decades of neglect. Distances are great—as great, indeed, as those in the Democratic Republic of the Congo. The known mineral resources were few, agricultural diversification had barely begun, and communications were rudimentary. As Belgium had done earlier, France initiated a Ten-Year Plan with an emphasis on transportation, both external and internal. Begun in 1946, the program bogged down in several areas simply because local conditions were not sufficiently well known or understood. Ultimately, its effect was at least to create a structure upon which further development could be based in 1960, when independence came to French Equatorial Africa. Since independence, the countries embarked on their national development with the same statist and sometimes Marxist–Leninist solutions that have become emblematic of much of what went wrong with Africa. Since the 1990s, the governments of the region have been relinquishing their control over the economy and the political process, encouraging private ownership of the means of production and multiparty democracy. Whether these strategies will overcome the obstacles to development that each country faces remains to be seen.

It was only after independence that oil and natural gas were discovered within the territory of five of the six countries of central Africa, and development of these resources has been the main focus of postindependence growth in most. In recent years high oil prices have meant windfall income to governments. As is evident from Table 16-1, human development indicators such as literacy rate are relatively high except for the Central African Republic, which lacks petroleum. Mere possession of petroleum or other sought-after natural resource is not any guarantee of high levels of development, however, and in many countries in Africa and elsewhere, lack of political stability and even civil war have undercut potential gains from natural resource wealth. In central Africa itself, the Democratic Republic of the Congo is an example of extreme poverty despite enormous wealth in natural resources. In addition, Gabon still possesses millions of square miles of pristine rain forest, while the eastern Central African Republic seems not to have recovered from the chaos

TABLE 16-1 DEMOGRAPHIC AND ECONOMIC INDICATORS FOR CAMEROON, THE CENTRAL AFRICAN REPUBLIC, THE CONGO REPUBLIC, EQUATORIAL GUINEA, AND GABON.

Country	Population (millions)	Infant Mortality Rate (%)	Literacy Rate (%)	Urbanization (%)	Per Capita GDP ($PPP)	Petroleum Reserves (millions of 42-gallon barrels)	Natural Gas Reserves (trillions of cubic feet)
Cameroon	16.8	77	76	48.9	1,703	400	3.9
Central African Republic	4.1	98	47	41.2	1,172	0	0
Congo Republic	3.4	84	81	65.4	825	1,500	17.5
Equatorial Guinea	0.49	95	83	48.2	15,073	1,300	2.4
Gabon	1.4	57	63	81.4	6,237	2,500	2.3

Sources: Population Reference Bureau (2003), Yager, Coakley, and Mobbs (2004), U.S. Geological Survey (2004), Helders (2004). Mobbs (2004a)

and depopulation of nineteenth century slave raiding. In terms of the proportion of the national population living in cities and towns, Gabon's population is the most urbanized of the five countries discussed in this chapter (Figure 16-2), but Cameroon has a better developed urban structure; one could call it a hierarchy of urban places. This, and the twin millionaire cities of Douala and Yaoundé, port and capital respectively, seem to belong more to the dense urban and rural population clusters bordering the Gulf of Guinea than to central Africa. But when one considers the urban population as a percentage of national population, it is tiny Gabon that seems to have made the rural-to-urban transition: over 80% of Gabonese live in urban places, far higher than the overwhelming majority of African states and much higher than the world average. Overall, however, the countries of central Africa are among the least densely populated on the continent.

GABON: RICHNESS IN RESOURCES

Gabon is a small compact country, rich in mineral and forest resources: dense tropical forests extend from the coast to the Congo border and into the interior highlands, where there are rich deposits of uranium, manganese, and iron ore (Figure 16-3). The climate is moist and hot, the soils are heavily leached, and malaria and other tropical diseases are endemic. Gabon's population is 1.4 million, and its annual population growth at 2.2%, is just slightly lower than the African average (2.4%) but over twice that of AIDS-afflicted southern Africa (1.0%).

If central Africa has one renewable resource in abundance, it is the forest; three-quarters of Gabon is covered by humid lowland forest. Timber was the mainstay of Gabon's economy during the early years of independence and still is the second most important export. Timber industries there, however, have suffered from a series of conditions detrimental to their growth. First, industries based on the exploitation of certain tree varieties are hampered by the likelihood that any specific type is going to be widely dispersed, given the multitude of trees make up the rain forest. Thus, downed trees may have to be moved over considerable distances, even to the waterways along which they can be floated down to the coast, and this presents its own problems. Second, the labor supply is limited, and since internal consumption of lumber is small, most of it must be exported. Third, loading facilities at the small coastal ports are inadequate, and wood, being bulky and heavy, is an expensive product to ship over long distances. Thus, only the most valuable woods, commanding a high enough price on the world market to withstand the cost of production and transportation, can form the basis for a successful timber industry.

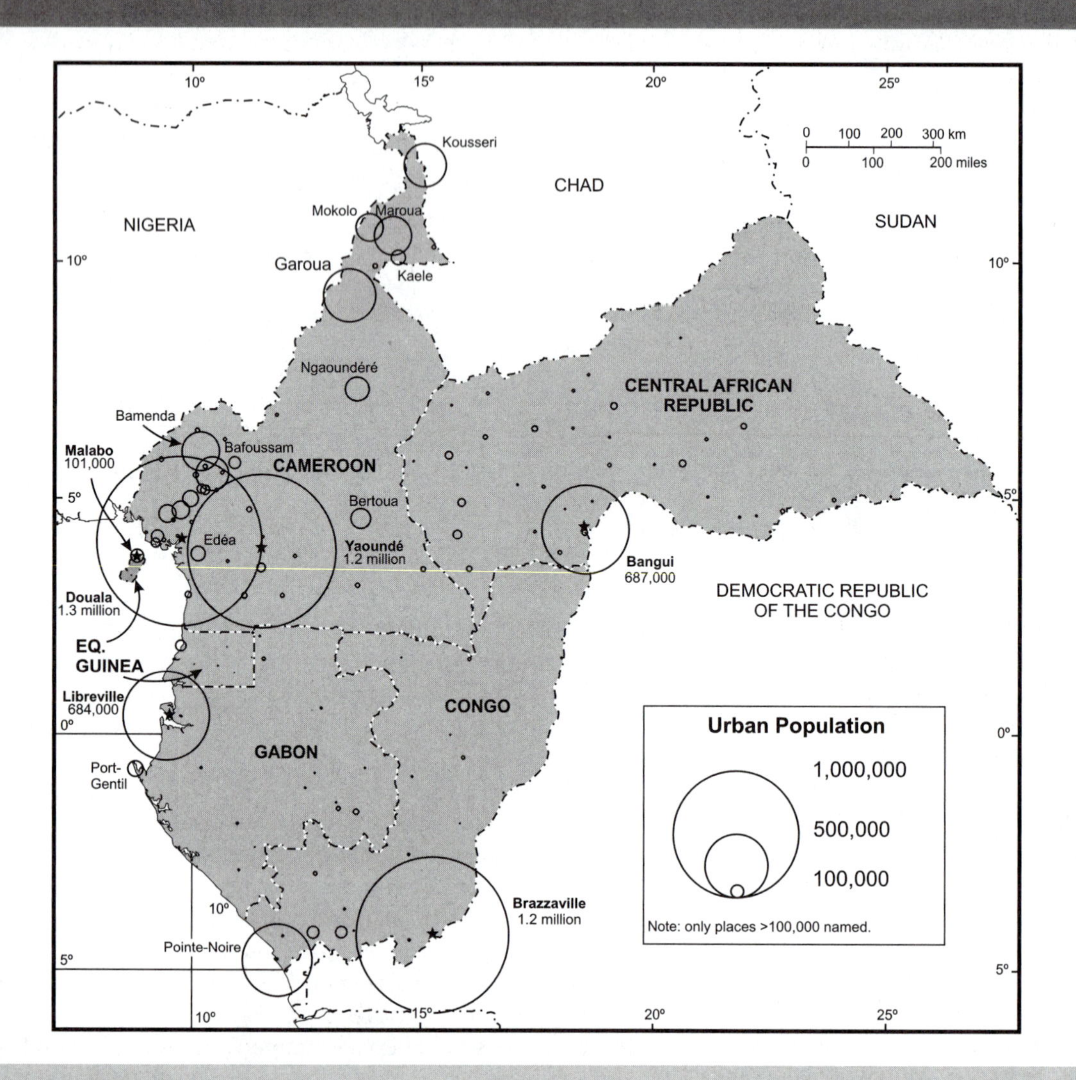

FIGURE 16-2 Urban population of Cameroon, Central African Republic, Congo, Equatorial Guinea, and Gabon.

Gabon is fortunate in that its forests contain some of the most valuable woods produced anywhere in the world. Indeed, the development of this country has taken place not in spite of the forest but because of it, and because a system of navigable waterways, focusing upon the Ogooué River, traverses it. More than the scarcity or wide dispersion of tree types, what has limited the timber industry (in addition, of course, to fluctuations in the world prices) has been the perennial shortage of labor. Gabon, along with Congo and Equatorial Guinea, (Box 16-1) has the world monopoly on the production of okoumé, a light wood used in the manufacture of plywood. The forest also yields ebony and mahogany, and for many years these woods produced 90% of Gabon's export revenues.

Inevitably some of the mismanagement characteristic of the rubber- and ivory-exploiting concession companies also has marked the timber industry. The

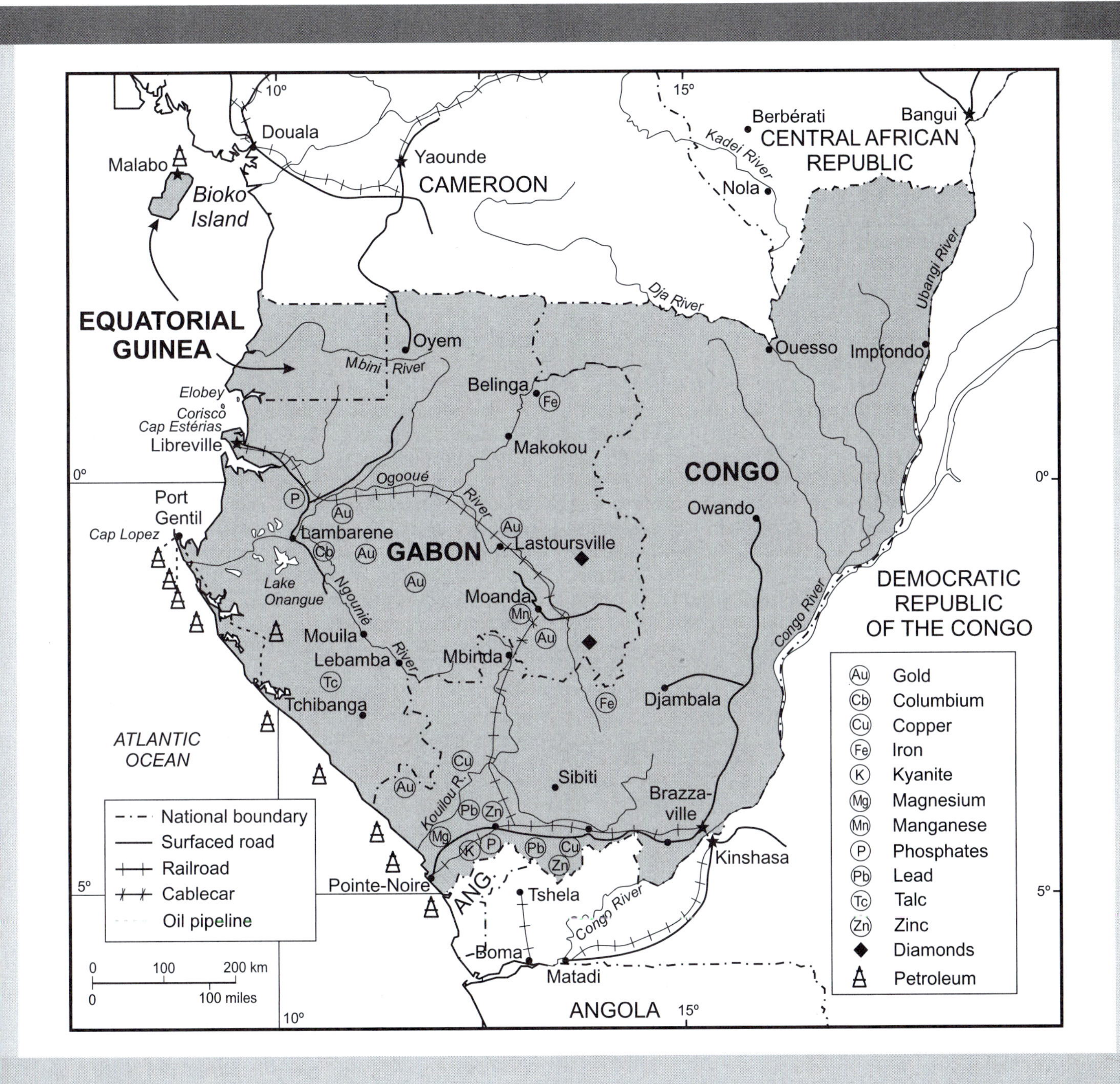

FIGURE 16-3 Economic map of Congo, Equatorial Guinea, and Gabon.

BOX 16-1

EQUATORIAL GUINEA

Equatorial Guinea (10,820 mi^2; 28,025 km^2), one of Africa's ministates, is divided for administrative purposes into two provinces: Bioko (formerly Fernando Póo) and Río Muni. Bioko is a small volcanic island 30 miles (50 km) off the Cameroon coast; its soils are rich, and rainfall is high. Río Muni, the larger of the two provinces, is bordered by Cameroon on the north and by Gabon on the east and south. Its narrow and densely forested coastal plain, cut by the fast-flowing and unnavigable Mbini River, gives way to a succession of valleys and spurs of the Crystal Mountains.

After almost 200 years of Spanish rule, independence was granted in 1968. A year later, following internal political disorders and antisettler riots in Río Muni, most Spanish residents fled the country, and the fragile economy—based on cocoa, coffee, and timber—almost collapsed. Cocoa is the major export crop, 90% of which is grown on African-owned plantations on Bioko and worked by Nigerian contract labor. The rest is grown by small farmers in both provinces. Coffee is grown chiefly by Fangs in northern Río Muni, while the timber industry is dominated by highly capitalized European-owned companies operating on the mainland. Okoumé is the chief wood. Other less important exports are bananas and palm oil. Until recently the industry of Equatorial Guinea was virtually nonexistent with no mineral production. All this has changed with recent discoveries of large offshore petroleum and natural gas deposits. Petroleum exports jumped 34% in 2004 making Equatorial Guinea the sixth largest petroleum producer in Africa. Oil exports were worth $4.6 billion in 2004 but only $1.1 billion in 2000 (Mobbs 2004b) oil accounted for 91% of government revenue in 2004. A liquified natural gas facility is planned. Spain is the principal trading partner and source of financial and technical assistance.

Almost one-third of the approximately 500,000 Guineans live on Bioko, and of these, 101,000 live in Malabo (formerly Santa Isabel), the republic's capital and main economic, educational, and religious center. Despite its size, Equatorial Guinea has great ethnic variety, with the Fang being the largest and most influential group in Río Muni and immigrant and refugee Nigerians being dominant on Bioko. Although some Nigerian groups have advocated the island's annexation or purchase, the Nigerian government has never espoused this claim. The governments of both Cameroon and Gabon have refrained from seeking the annexation of Río Muni, despite strong regional ethnic affinities, territorial contiguity, and potential economic gain.

coastal belt is almost worked out, and reforestation and other conservation practices have become imperative. Cutting has moved inland to what is known as Zone II, where highly capitalized French firms have dominated the industry. Here there is great potential, although river transport is less practical than along the coast and in Zone I.

In 2002, in an effort to protect the diverse natural environments of Gabon and to stimulate conservation and ecotourism, President Omar Bongo announced that Gabon would create 13 new national parks, protecting over 10% of the nation's land (Figure 16-4 and Table 16-2). The parks were created after several years of study to represent the variety of flora, fauna, and natural environments in the country and was praised by environmentalists (Gabon 2004) and will protect 11,294 square miles (29,251 km^2) of land. It is unlikely that such parks could ever have been created without the petroleum riches that have been the strength of the Gabonese economy since the 1970s.

Whereas forestry dominated Gabon's money economy for more than six decades, minerals have accounted for most of the country's economic growth since the mid-1960s. Rapid growth in the mining sector has been led by petroleum, Gabon's principal nonrenewable resource. Oil exports now provide 80% of

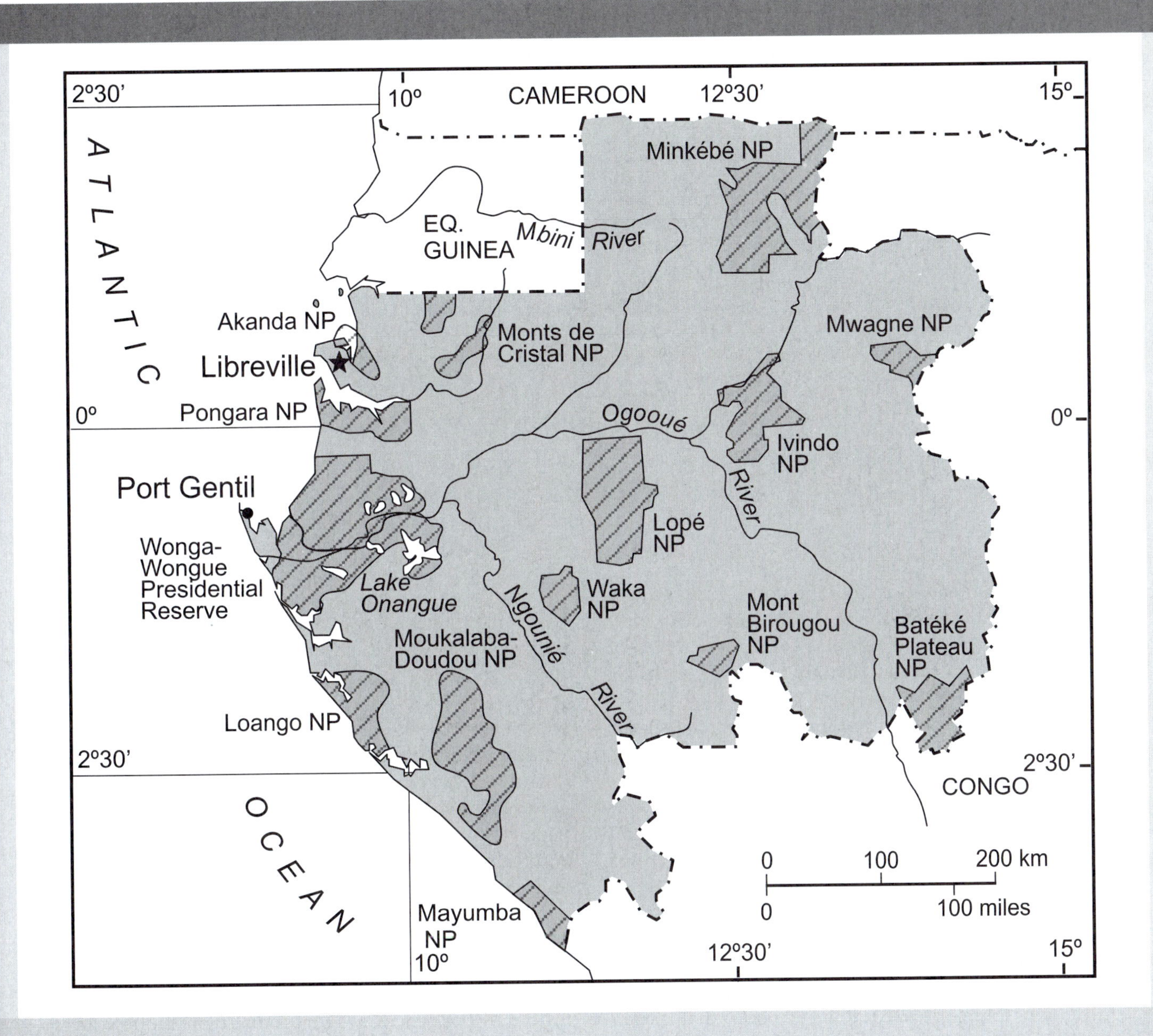

FIGURE 16-4 New national parks (NP) in Gabon.

foreign revenues and 33% of GDP; production peaked in 1997, but high prices and new discoveries should allow oil to dominate exports for many years to come. Exploitation began in 1957 near Port Gentil and witnessed spectacular growth in the 1970s with the opening of the Gamba–Ivinga and Anguille fields. In 1975 production exceeded 73 million barrels, two-thirds of it coming from the newly discovered offshore fields. During the 1990s and the early years of the twenty-first century, Gabon consistently produced

TABLE 16-2 GABON'S NEW NATIONAL PARKS.

Park	Description	Location
Akanda	Extensive mangrove swamps, tidal flats, migratory birds	Northwest coast near Cap Estérias
Batéké Plateau	Savanna, forest, and gorillas	Southeast extremity
Ivindo	Elephant wallow, Langoué Bai	Central Gabon
Loango	Elephants, hippopotami, gorillas, leopards, whales	Southwest coast
Lopé	Gorillas, chimpanzees, mandrills	Central Gabon
Mayumba	Nesting leatherneck turtles	Extreme southern coast
Minkébé	Elephants	North
Mont Birougou	Mountain environments and diverse flora	South
Monts de Cristal	Great diversity of flora in highland cloud forests	Northwest
Moukalaba-Doudou	Elephants and papyrus marshes	Southwest
Mwagne	Largest animal wallow in Gabon; bongos, otters, elephants	Northeast
Pongara	Elephants and forest buffaloes	West coast
Waka	Ancient village sites, diverse vegetation	Central

Source: Adapted from Quammen (2003).

about 300,000 barrels a day. The Port Gentil refinery meets domestic demands and supplies most of the new requirements of Cameroon and Congo. Over 85% of the crude oil is exported, mainly to France, Curaçao, Senegal, and Côte d'Ivoire.

Gabon's other important minerals are manganese, iron ore, columbite, and talc. Over 200 million tons of manganese ore is known to exist in the highlands near Moanda (Figure 16-3), and Gabon is the world's fifth largest producer after Russia, Brazil, Australia, and South Africa. Until the new trans-Gabon railway was completed, production was limited to 2 million tons per year, the ores being transported by overhead cableway to Mbinda and then by rail to Pointe-Noire, Congo. In the remote northeast near Belinga (Figure 16-3), there is a 1 billion-ton deposit of high-grade iron ore. During the 1970s there was hope that the rail system would be extended to Belinda to make the mining of the iron ore feasible. Unfortunately, this leg of the national system has still not been completed. Iron ore of inferior quality is located closer to the coast and port of Tchibanga, but there are no plans for its development. Finally, at Lastoursville on the upper Ogooué, small quantities of gold are mined, mainly by artisanal methods; lead, columbium, gold, diamonds, zinc, and phosphate are known to exist in profitable amounts.

For decades, Gabon's export economy has been hampered by poor transport and restricted geographically to the lower reaches of the Ogooué River. The trans-Gabon railway is beginning to alter that. Work started on the 350-mile, (560 km) line in 1974 and was completed in 1986. It was built not only to serve the mining interests and the once all-important forest industry but also to stimulate the rural economy. Roads are to be built linking the railway with the small and widely dispersed agricultural regions such as the Woleu and Ntem valleys in the extreme north, where cocoa is an important cash crop.

Gabon does not meet its own food requirements, and the government has never placed much emphasis on agricultural development. Farming is difficult in the rain forest areas, and agricultural output is hampered further by the general awareness that more money can be made in the mines and lumber camps. According to the FAO (2004), Gabon's top three agricultural exports (rubber, palm oil, and cocoa) brought in only $2.1 million in 2002, a tiny part of the national income.

Gabon made a move toward multiparty democracy in the 1990s, but power has remained in the hands of President Bongo. In 2002 the constitution of Gabon was amended to allow him to be elected president for an unlimited number of terms. He has held the office of president since 1967 and is the longest-serving head of state in Africa. Voters returned Mr. Bongo to office in Presidential elections in 2005. The most important issue was the unequal distribution of wealth in Gabon. Close to half the population lives below the poverty line—remarkable for a country with such a high GDP—and this reflects the fact that a small elite wields power and commands the resources of the country.

CONGO: THE ASSETS OF LOCATION

One of Congo's major assets is its geographical location (Figure 16-1). It has been a gateway to the interior ever since the French began their administration of their "Equatorial" Africa. Stretching as it does from the narrow Atlantic coastline south of the equator along the Congo and Ubangi rivers to nearly latitude 4° north, this region became France's principal access route to the Central African Republic (then Ubangi-Chari). Brazzaville became the major city, administrative capital, and dominant service center for the French equatorial realm, while Pointe-Noire became the region's busiest and best-equipped port.

The transit of goods to and from the interior became the major source of revenue for Congo. Brazzaville itself developed into a modern city, thanks to considerable French investment in government infrastructure, cultural centers, and, to a lesser extent, in industry. Its hinterland could not compete with that of Kinshasa, but it performed a break-in-bulk function between the coastal railroad and the Ubangi River. The region surrounding it saw little development. Brazzaville's raison d'être was its governmental functions, and with the dissolution of the federation, it was faced with stagnation and decay. Pointe-Noire, more favorably situated for future development, retains the monopoly over the external trade of much of interior Central Africa and has developed into a large (150,000 in 1975, 560,000 in 2004) industrial port with a petroleum refinery, chemical plants, sawmills, ship repair yards, and numerous import substitution industries.

Forests cover over 50% of the Congo Republic, and the focus of the forest industry, as in Gabon, is on stands of okoumé, limba, and mahogany; vast areas are being cut, and vast areas are being replanted. At present, lumber is trucked on the Ouesso-to-Brazzaville highway or floated down-river to Brazzaville and then railed to Pointe-Noire. Until the 1980s when petroleum began to make a significant contribution to the economy, timber and wood products generally constituted almost half the exports, and there was an overdependence on forestry products. At the time, Congo was considered poor in resources and, although it had much variety in its mineral and agricultural production, there was very little quantity.

Petroleum first began to be produced in 1957, and production has increased substantially since that time. Petroleum has enabled Congo to emulate Gabon in diversifying its economy. High oil prices in the 1980s jump-started the Congolese economy, and high prices in 2003–2005 have been helping the country rebuild after civil war. The country produces about 260,000 barrels of petroleum per day, accounting for 95% of export earnings, 50% of GDP, and 60% of the government's income (EIA 2004). Although Congo has been Africa's fourth largest petroleum producer, it appears likely that Equatorial Guinea will soon occupy that position as the smaller nation begins to pump its substantial offshore deposits. Congolese production has declined over the last few years but is expected to increase as new fields come into operation.

Like the majority of African countries, Congo is not self-sufficient in staple foods, yet it produces a number of export crops. The long list of crops actually harvested might give the impression that this is an area with great possibilities: bananas, cassava, sugar, rice, coffee, cocoa, peanuts, and citrus fruits are but a sample. Indeed, this may be a country with real potential. At present, however, the rural population is simply too small, too widely dispersed in tiny villages, and too poor to produce any of these crops in quantity. What can be done was indicated after World War II, when a group of French colonists settled in the valley of the Niari River, a tributary of the Kouliou River in western Congo (Figure 16-3), began to cultivate rice, peanuts, and tobacco for export, in addition to market-garden produce for Brazzaville, Lonbomo, and Pointe-Noire. Today the region supplies the entire country with its sugar needs, and sugar is the leading food export. The scheme has been an example for other similar ones, but they are still experimental and make only a small contribution to the export economy.

More beneficial to Congolese agriculture would be entrepreneurial investment in agriculture for profit. Nevertheless, agriculture is severely handicapped by unfavorable natural conditions. The soils are sandy and deeply leached, and dense forest extends over much of the north, reducing the cultivable land to pockets scattered across the country's southern half. Temperatures and humidity are perpetually high, and rainfall is heavy except along the coast. All but the higher areas around Brazzaville and the extreme southwest are plagued by tsetse flies, which preclude most forms of animal husbandry.

In the area of mineral exploitation, also, Congo can boast wide variety but little quantity. As recently as 1969, mineral exports accounted for less than 5% of total exports. Copper, lead, gold, zinc, and diamonds have been produced, but only lead has regularly ranked among the exports, largely because of the proximity of the mines to the Congo–Ocean railway. Deposits of phosphate, kyanite, and iron are also mined. Substantial magnesium deposits at Makola and Youbi in Kouilou Region, in proximity to the railway, will provide 60,000 metric tons of ore per year for production of the metal.

Industrial development was bleak after independence owing to the smallness of the local market, the poverty of the population, the small scale and limited technology of agriculture, and the limited mineral production. Industrial output has been concerned mainly with the processing of agricultural and forest products and is concentrated in the capital, the Niari Valley, and Pointe-Noire. The inflow of foreign capital into manufacturing virtually ceased following the government's nationalization of industry in 1965, when the People's Republic of the Congo was declared and ties with the socialist bloc were increased. Marxism–Leninism was abandoned in 1990, but the country's larger industrial concerns, such as the cement, textile, and printing plants, are still state owned. The government has plans to privatize these organizations soon, in the wake of the privatization of an oil parastatal in 2003.

In 1992 the Congo had its first multiparty election, and disputes over that election and the following one in 1997 led to brief civil wars. Violence after the 2002 election followed the earlier pattern, and peace was not reestablished until 2003. Although Gabon and the Congo Republic hold many things in common, they are quite different in their postindependence political experience. Gabon has enjoyed relative stability, while the Congo has had serious problems from which it is still trying to recover.

CENTRAL AFRICAN REPUBLIC: ELUSIVE POLITICAL STABILITY

The Central African Republic lies across an upland that forms the transition zone between the forested Congo Basin and the steppe-and-desert basin of Chad, and it occupies part of the Sudan–Chad divide as well. Thus, it straddles several climate and vegetation belts. The southwest is covered by dense forests, but northward an increasingly dry savanna replaces this growth. Since the country lies higher than either Gabon or Congo, the climate is somewhat cooler in the central parts, and perhaps the best evidence for this amelioration is the cattle population of 3.1 million (against a mere 90,000 in Congo) (Table 16-3). The most important consequence of the republic's location with reference to the major physiographic features of central Africa is that a

TABLE 16-3 LIVESTOCK POPULATIONS FOR CAMEROON, THE CENTRAL AFRICAN REPUBLIC, THE CONGO REPUBLIC, EQUATORIAL GUINEA, AND GABON IN 2001.

Country	Cattle	Sheep	Goats	Pigs
Cameroon	5,900,000	3,800,000	4,400,000	1,350,000
Central African Republic	3,100,000	220,000	2,600,000	680,000
Congo Republic	90,000	96,000	280,000	46,000
Equatorial Guinea	5,000	37,600	9,000	6,100
Gabon	36,000	198,000	91,000	213,000

Source: United Nations Food and Agriculture Organization. (2003).

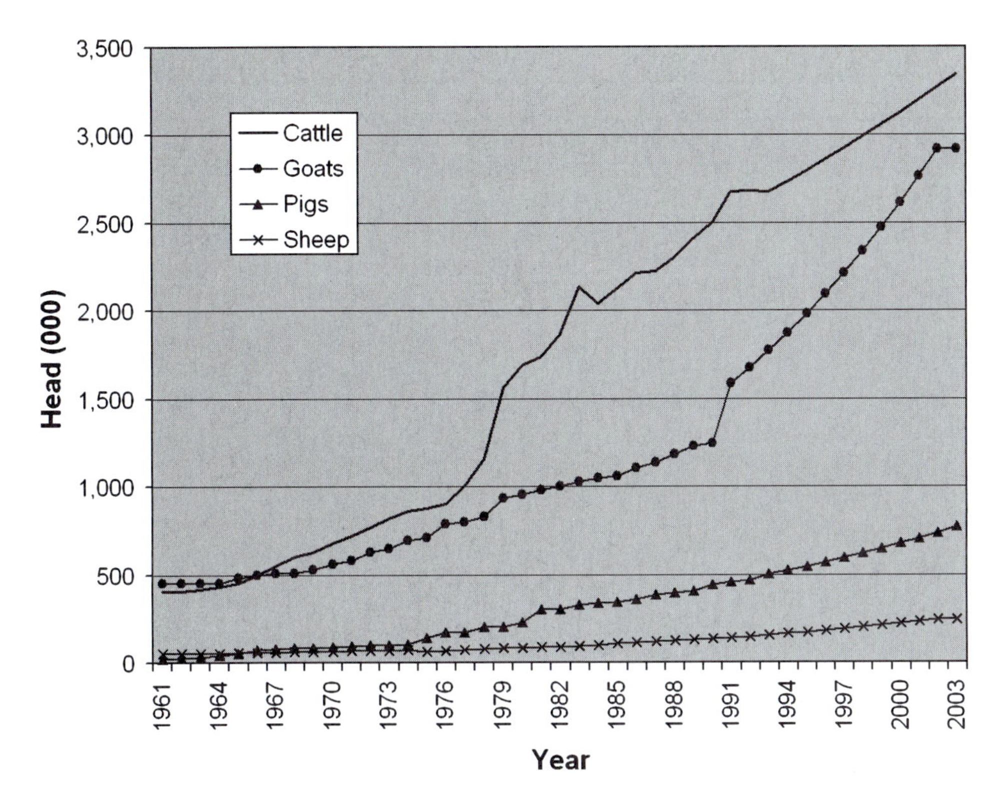

FIGURE 16-5 Cattle, goat, pig, and sheep populations for the Central African Republic, 1961 to 2003.

certain variety is brought to the economic picture. While historically there has not been the exchange of goods that has developed in West Africa between the peoples of the dry north and the forested south, the resource base is more diversified, for instance, than that of its southern neighbor, Congo. Prolonged drought in the Sahel beginning in the 1970s has demonstrated that latitudinal specialization and exchange relationships do exist between the CAR and its drier, northern neighbors. Figure 16-5 shows the influx of cattle from the north, which began in the 1970s. Judging from the data presented in Figure 16-5, it appears that the herders remained. It is likely that animal husbandry and market farming in the CAR will benefit from the rising incomes in Chad brought on by the oil boom.

Despite the CAR's resource diversity, in measures of human development the country does not compare very favorably with the Congo (Table 16-1). Although beneficial in some respects, the location of the republic also has some disadvantages. Its capital and major river port, Bangui, is more than 900 miles (1440 km) from the sea. Toward the east and northeast, the country gets very dry, and there is a rainfall deficiency over much of the nonforest zone. Soil erosion is extensive in some parts of the country as a result of careless cutting for plantation development. As elsewhere in central Africa, there are labor problems, and the country shares with Gabon and Congo the problem of urban unemployment, especially in Bangui, with its population of 687,000.

To some extent, the human problems of the Central African Republic are the result of the abuses of the recent past. There, the rule by colonial concession companies led to the worst conditions in all of French Equatorial Africa, and the opposition was bitter and desperate. When the period of exploitation came to an end, there remained an enmity between black and white that continues to affect relations today. The past hangs heavily over the Central African Republic, for in a sense the forced labor on the land, imposed by the concessionaires, alienated the farmer from the soil and led to the exodus to the cities after World War II. The government's major effort to stimulate the cotton industry has been impeded by these factors, even though a number of areas in the central and southern regions are suitable for the cultivation of this crop.

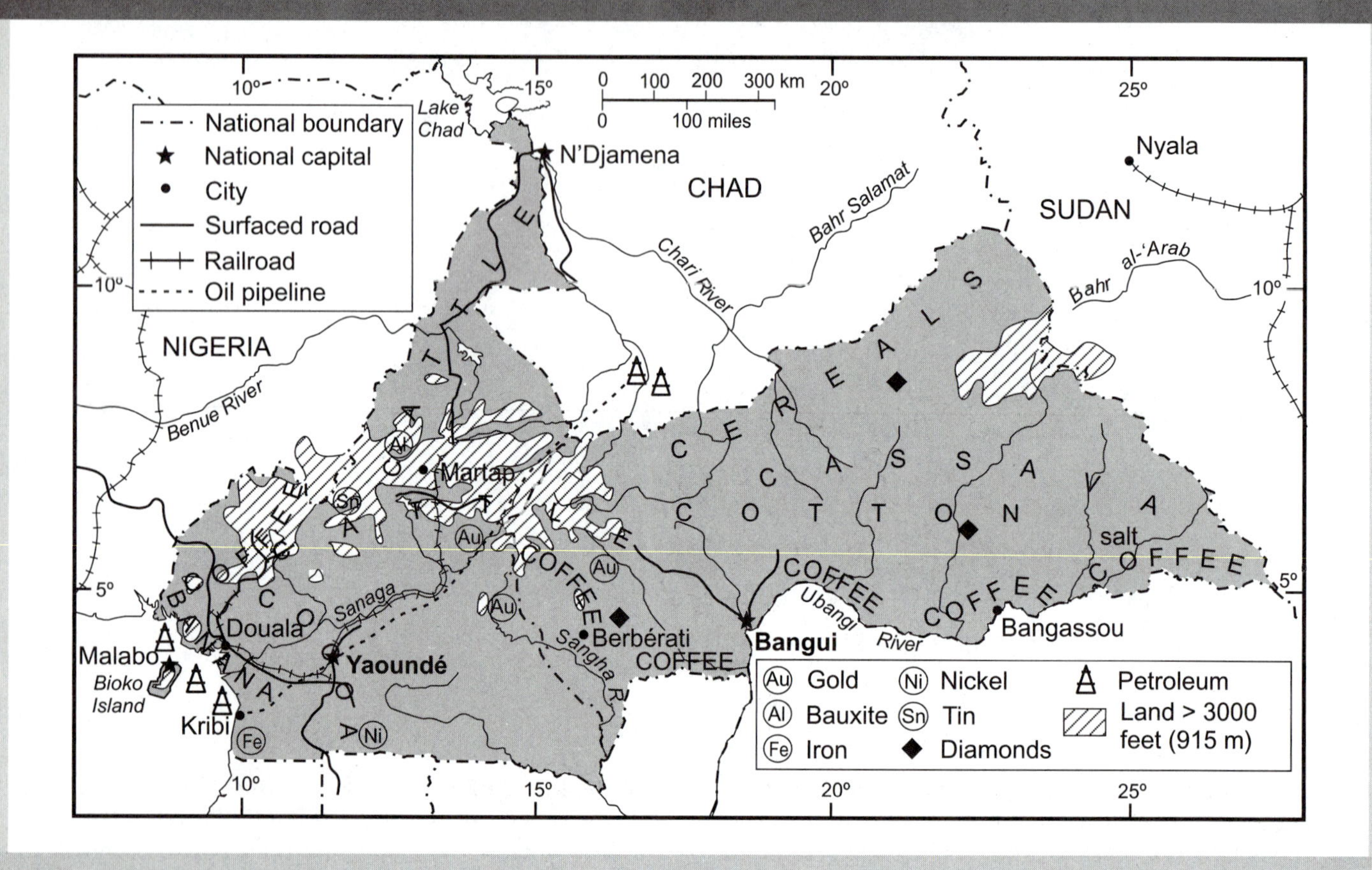

FIGURE 16-6 Economic map for Cameroon, with its capital, Yaoundé, and the Central African Republic (Bangui).

Another source of African resentment involved coffee. Some two-thirds of the acreage under coffee, the republic's second largest before independence export after diamonds, was in European hands. Coffee does well along the edge of the forests in the Upper Sangha Valley and the Upper Ubangi Valley, and it is an easier crop to grow than cotton. Africans accused Europeans of attempting to divert African coffee growers' efforts toward cotton, leaving the easier, more profitable crop to the Europeans. Such difficulties beset the country throughout the period of contact with Europe, and they continued after independence. Nevertheless, time may solve the human problem, if the economic opportunities remain.

Coffee production and exports peaked in 1988 and 1989 respectively. In addition to cotton and coffee, such crops as sisal, tobacco, cocoa, and rice can be cultivated and are grown in small quantities. Cassava and maize constitute the principal food crops in the CAR, as in many parts of Africa. In general, the areas offering the best opportunities lie in the southwest (Figure 16-6), in the region of the Upper Sangha, and in the southeast, along the northern slopes of the Upper Ubangi, broadly in the vicinity of Bangassou. Shortly after independence there was a brief period of general prosperity in the late 1960s which was attributed to the successful implementation of "Operation Bokassa." Under this program, supported by the European Development Fund and various French agencies, increased production in cotton, coffee, livestock, and staple foods was achieved. Insecticides, fungicides, and fertilizers were made available; villages were regrouped to provide

BOX 16-2

THE SUCCESSION OF CENTRAL AFRICAN GOVERNMENTS SINCE INDEPENDENCE

The Central African Republic became independent from France in 1960, and its first president, Barthèlémy Boganda, was killed in an accident just prior to taking office. His cousin, David Dacko, took his place as president and served from 1960 to 1966, when he was deposed in a coup d'état led by Colonel Jean Bedel Bokassa. Bokassa became increasingly autocratic, presiding over a corrupt regime in which human rights abuses were rife. He abolished representative government and made himself emperor of the Central African Empire. He is most famous for his extravagant coronation, modeled on that of the English monarchy. David Dacko returned in 1979 as the leader of a coup, and Bokassa went into exile. Exactly two years after taking power, Dacko was himself overthrown by General André Kolingba.

During the middle 1980s, it seemed as if the Central African Republic was moving toward a more representative model of government. A constitution was drafted and approved through a national referendum, and a "House of Commons" (Assemblée Nationale) was created. After the fall of the Soviet Union there was increasing pressure for multiparty democracy. Amid some irregularities, Ange Félix Patassé was elected president in 1993. In 1999 he was reelected. His administration was challenged by frequent mutiny in the army and civil unrest. On March 15, 2003, the Patassé government was overthrown in yet another military coup d'état, this time led by General François Bozize. The constitution was suspended and the Assemblée Nationale dismissed. In early 2004, however, a process was put in place leading to presidential and legislative elections in 2005 and General Bozize won the presidential race.

poles of development; and rural electrification was introduced to help small farm-based industries located away from major towns. But inefficient administration, domestic politics, a deterioration of relations with France, and prolonged unfavorable weather conditions brought this short period of economic growth to a close. President Bokassa, himself the leader of a coup in 1966, was deposed in a French-led coup in 1979 (Box 16-2). The 1980s and 1990s were years of decline for the cultivation of some crops, and the agricultural picture today is mixed but appears to have potential. Some crops have increased significantly (rice, maize, sesame), while other crops have declined (Table 16-4).

The availability of road materials and the presence of an intensive river system have produced a network of communications which, for central Africa, is good, although it does little to reduce the high cost of the lengthy journey of all exports to the sea and their overseas markets. At present, 90% of the Central African Republic's exports travel 700 miles (1,120 km) of the Ubangi and Congo rivers to Brazzaville, and another 300 miles (480 km) of rail to Pointe-Noire. The Ubangi River is not navigable for half the year. Although the mineral exports (diamonds and some gold) are not bulky, heavy freight charges on the agricultural exports and almost all imports have led to a search for alternative routes to the sea. Several railway proposals have been under consideration, the most feasible route being from Bangui via Berbérati to the trans-Cameroon railway that terminates in Douala (Figure 16-6). Not only is the port of Douala nearer to the Central African Republic, but the journey would be much faster than the river–rail route to Pointe-Noire and would require no transshipment. Such a link could stimulate development in the western region but would, of course, be detrimental to Congolese interests. Less feasible routes, but nevertheless under consideration, are from Bangui to the Sudan, and from Bangui to the new trans-Gabon line.

Although agriculture accounts for more than half the gross domestic product, it is diamonds and timber that generate the greatest export earnings. The little manufacturing that exists addresses local consumption rather than the export market. Industrial development

TABLE 16-4 AGRICULTURAL PRODUCTION OF SELECTED CROPS IN THE CENTRAL AFRICAN REPUBLIC, 1961-2003.

Crop	Averages (mt) 1961-1965	Averages (mt) 1999-2003	Difference (mt)	Difference (%)
Bananas	45,400	115,200	69,800	154
Cassava	672,000	561,420	-110,580	-16
Coffee, green	9,120	11,392	2,272	25
Groundnuts in shell	61,500	120,680	59,180	96
Maize	32,000	105,800	73,800	231
Millet	9,800	11,320	1,520	16
Plantains	48,400	82,000	33,600	69
Rice, paddy	4,620	24,840	20,220	438
Seed cotton	28,020	22,157	-5,863	-21
Sesame seed	7,660	38,020	30,360	396
Sisal	240	329	89	37
Sorghum	35,800	47,780	11,980	33

Source: FAO (2004). United Nations Food and Agriculture Organization (2004).

has barely begun; manufacturing contributes little to GDP and is focused on the processing of agricultural and forest products. Most manufacturing concerns are located in and around Bangui, with the textile and leather industries constituting the chief industrial sector. Light industry (making of beer, soap, paint, bricks, footwear, and eating implements; assembly of bicycles and motorcycles) is also found around the capital. A major asset is the Boali hydroelectric scheme about 50 miles (80 km) from the capital, which provides power for these enterprises. The power is insufficient, however, and frequently interrupted. French influence is still dominant, despite moves by several governments to reduce it, but French political and military intervention has probably given the country a degree of political stability that it would otherwise have almost completely lacked.

Despite some advances, the Central African Republic remains a poor and underdeveloped country whose major constraints to development may be traced to decades of misrule by unelected (mostly military) governments. It is certainly true that France did little to provide its colony with the essentials for political stability and economic progress—but in the face of nearly a half-century of independence and our greater understanding today of the issues that confronted the newly independent African states, we must realistically reassess the task that was before France. In retrospect, the food crop versus cash crop debates of the 1970s and 1980s distracted attention from the real issues in Africa: raising rural incomes and providing both rural and urban employment.

The CAR faces unenviable challenges, some of which are common to other African countries while others are not. Geographically, CAR is landlocked, and access to the world's markets is severely constrained. A first-rate transportation infrastructure would do much to alleviate the disadvantages of location, but CAR produces little that is worth transporting a long distance. The citizens of the CAR are mainly subsistence farmers who need encouragement to engage in commercial production to raise their incomes. But years of civil unrest have discouraged farmers and have pushed them deeper into production simply for consumption, a recipe for continued poverty.

If economic growth and human development are to be achieved, there must be civilian rule, along with sustained international technical and financial assistance, within a framework of domestic economic and administrative reforms. There are hopes that petroleum will be found in the north, near the large Chadian deposits. It is likely that the CAR will face the same challenges as Chad in the exploitation of this resource. Politically, the CAR has to get its house in order before investors will return, and the government has been taking steps in that direction. In 2004, encouraged by the serious progress toward representative government, the International Monetary Fund approved $8.2 million to help rebuild the country. Realistically, politically, economically, as well as geographically, the CAR has a long road to go.

CAMEROON: PROGRESS AND COOPERATION

The wedge-shaped state of Cameroon, extending from Lake Chad in the north and from the Sangha River in the east to a narrow but important coast line in the west, never was a part of French Equatorial Africa. Influenced in turn by the British (until 1884), the Germans (until World War I), and the French (who held a mandate over the bulk of the territory from 1922 to 1946 and a trusteeship until 1960), Cameroon forged ahead of the remainder of central Africa at an early stage and has remained in the lead ever since. Reunited with part of the British sector of the original mandates, this is the most populous country in former French-administered central Africa (16.9 million people: more than 2.5 times its size in 1975), though at 183,568 square miles (475,439 km^2), it is by no means the largest.

Cameroon's internal physiographic variety is considerable, and as a result a wide range of crops can be cultivated. Its north–south extent is no less than 10 degrees of latitude, so that conditions in the south and along the rather narrow coastal belt, are those of the tropical rain forest, whereas in the north, dry savanna and steppe lands occur. Topographically, the country rises in a series of escarpments toward an interior plateau covering the bulk of its area, but a number of prominent features mark the landscape: Mount Cameroon (13,354 ft, 4,006 m) in the extreme west, and the Adamawa Highlands farther north, rise from the coastal belt and the interior plateau, respectively. In the extreme north, Cameroon forms part of the central part of the Chad Basin, Lake Chad, and the land also drops in the extreme southeast, where the rivers are tributaries of the Congo (Figure 16-6).

Most of the development has taken place to the west of a line drawn north–south, slightly east of Yaoundé. Included in this western region are the important highlands of the former British section of Cameroon, which are among the more densely populated parts of the country and the scene of early settlement by Europeans, as well as most of the major towns and communications. There also lie the major port, Douala (1.3 million), with a capacity of the same order as that of Pointe-Noire, the agricultural region of the Sanaga Valley; the capital, Yaoundé (1.2 million); and the growing industrial center of Edéa, near a major hydroelectric project on the Sanaga River. Both Douala and Yaoundé have increased in size by over a million people since the mid-1970s. Two railroad arteries lead to Douala: one connects the port to Nkongsamba, the railhead for the highlands, and the other leads to Ngaoundéré, serving the northern areas, Yaoundé, the Sanaga Valley, and Edéa.

Cameroon's potential for development is by no means limited to this western area. The initial extraction of the usual products from the coastal and southern rain forest, the proximity to the coast and a good harbor, and the western location of the highlands, where much of the early development took place, have led to the early and continuing preeminence of the west in terms of economic geography. Along the western coast lie extensive petroleum deposits. Oil is the most significant sector of the economy, followed by aluminum smelting and processing. But the most important resource in Cameroon since independence has been stability, which, has permitted the growth of many productive endeavors. Agriculture has grown, transportation and communications infrastructure has been built and expanded, and industry has been encouraged. Although agriculture today provides a smaller share of the GDP than in the past, one of the country's major assets lies in the variety of crops that can be grown, a variety that is impressively large because of the environmental diversification of the republic.

Apart from the typical unfortunate consequences of military conquest, the peoples of Cameroon were more fortunate than many others in Africa as far as the nature of their colonial administration was concerned, and this has much to do with the present potential for progress. The German administration at a very early stage embarked upon a program of crop research and drew up an economic blueprint calling for the construction of an embryonic transportation system, limitations to land alienation, the perpetuation of certain

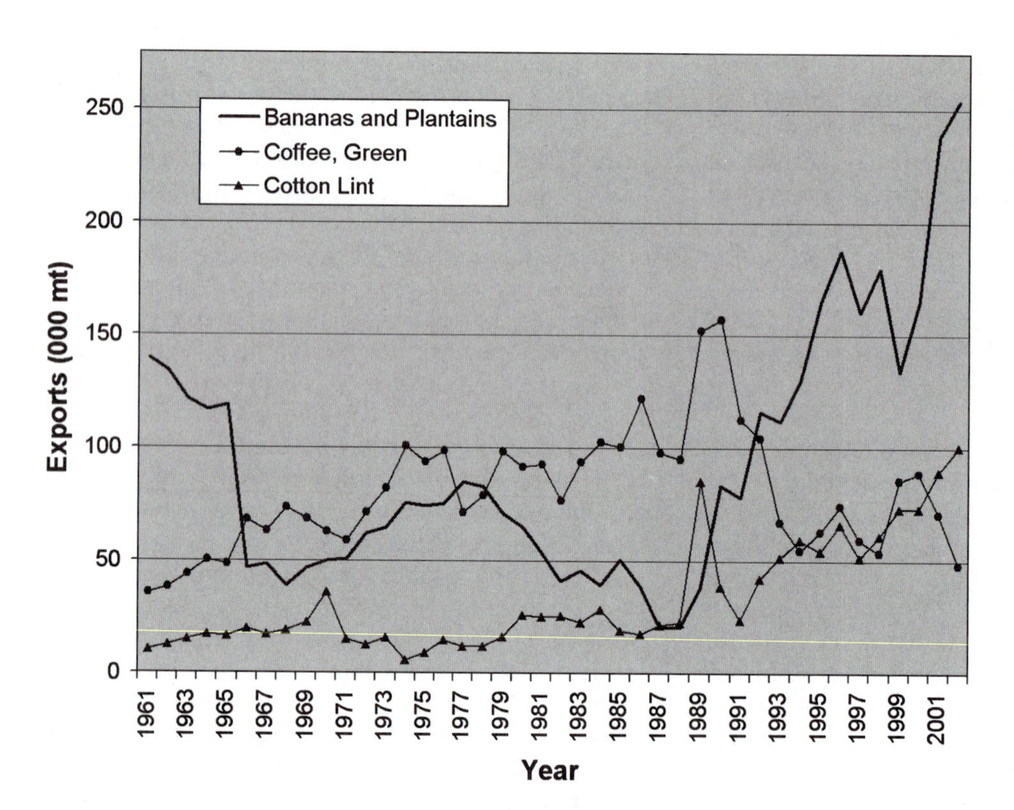

FIGURE 16-7 Cameroon's banana, coffee, and cotton exports, 1961 to 2002.

facets of African traditional authority, and the participation of Africans in agricultural development. Despite the brevity of their period of tenure, the Germans had had some notable success by the time the French took over the responsibility for Cameroon. Although certain aspects of French administration in the mandate were not designed to reap the maximum possible harvest from the base that had been laid, Cameroon was and remained ahead of French Equatorial Africa proper. Especially important in Cameroon's economic development was the early introduction of cash crops and the encouragement given to African farmers at an early time. Thus cocoa made its appearance among the country's exports before 1914; and by World War II, African producers were exporting all the Cameroonian cotton, cocoa, peanuts, and corn, and the bulk of oil palm products. The pattern persists today, with small farmers dominating agricultural export production with the exception of rubber, bananas, and coffee (which are still produced on large plantations begun during the German era). The "three Cs" of Cameroon, cocoa, coffee, and cotton, accounted for 69% of all Cameroonian agricultural exports in 2002. Export earnings from agricultural products have declined with lowered world prices, although exported volumes have increased (Figure 16-7) and cocoa processing has developed to add value. Processed cocoa products now account for about one-quarter of cocoa-related exports (Figure 16-8).

This early participation of African growers greatly facilitated the introduction of other crops found suitable for the region. Tea is grown along the highland slopes, and rice, cotton, and peanuts are cultivated in the far north. The Adamawa Highlands and the high grassland areas in the former British Cameroon territory support a sizable cattle population (Figure 16-6), and although the region is subject to recurrent drought, stock losses have not been as severe as in the Sahelian countries. Domestic consumption of meat is low, so that surplus cattle are exported to Nigeria, and meat is shipped to Equatorial Guinea, Congo, and the Democratic Republic of the Congo.

The years since independence have seen important growth in the Cameroon economy and structural change; in 1997, under the auspices of the International Monetary Fund, Cameroon began to restructure its economy to encourage the greater growth of private

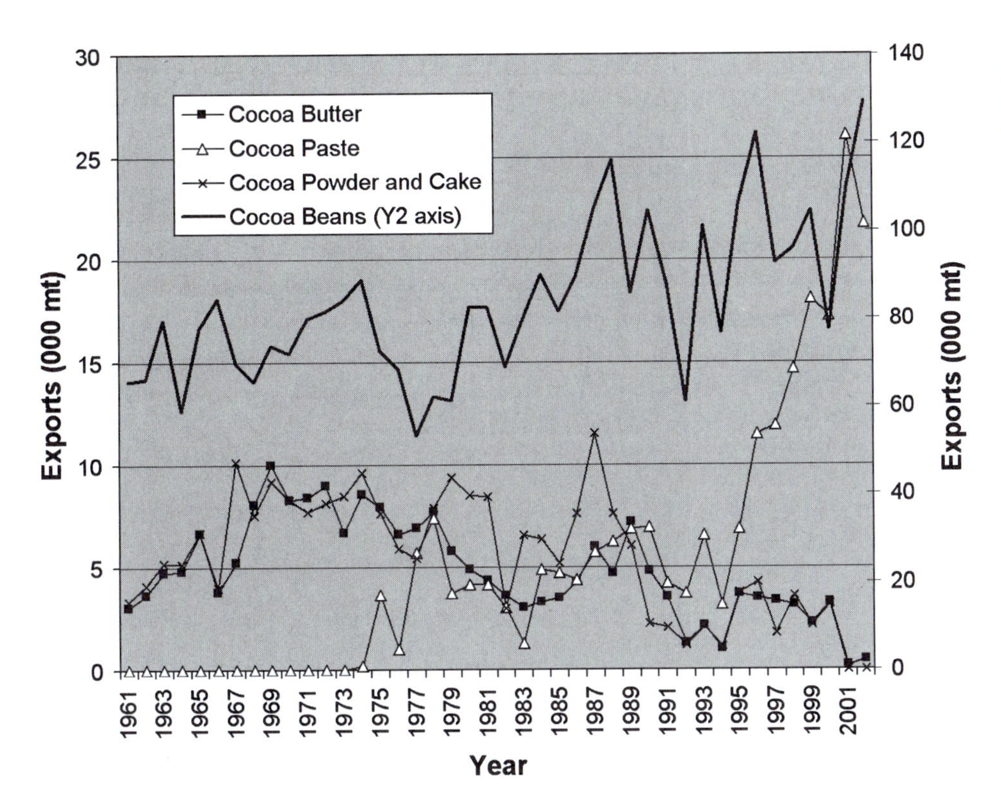

FIGURE 16-8 Cocoa products exports for Cameroon, 1961 to 2002.

enterprise, responsible government, and transparency (U.S. Department of State 2000). State corporations were privatized, although the state's role in the economy was nothing like that in the avowedly socialist or Marxist–Leninist countries of Africa, and taxes and other legal disincentives to investment were revised or scrapped.

Cameroon remains heavily dependent on primary exports and on overseas investment and technology to sustain growth. Its considerable resource diversity, past performances in agricultural and manufacturing production, and the relatively centrist and experimentalist political economic policy (compared with other francophone states) have helped create a measure of economic viability and potential for expansion uncommon in tropical Africa. Having cultivated a diversified and viable rural economy, the government is now promoting and privatizing the manufacturing sector. Cameroon's manufacturing output ranks third in francophone Africa, after Côte d'Ivoire and Senegal; but only a small percentage of the labor force is engaged in manufacturing, and manufactured goods account for only 5% of exports. Oil dominates Cameroonian industry, accounting for 35% of exports. Cameroon is the sixth-largest oil producing country in Subsaharan Africa. Its reserves of petroleum are estimated at 400 million barrels and it has substantial and unexploited reserves of natural gas (Yager, Coakley, and Mobbs 2004). Cameroon produced 35 million barrels in 2004 and will exhaust the resource in little more than a decade without new discoveries. Early industries such as petroleum refining, paper and pulp, textiles, cement, fertilizer, sugar, and aluminum utilize locally produced raw materials. The metals industry is dominated by the Edéa aluminum smelting complex and rolling mill, located east of Douala, which utilizes imported and local ores and locally generated hydroelectric power. There are significant deposits of iron, tin, bauxite, cobalt, nickel, rutile, diamond, and gold. The oil pipeline from Doba oil fields in Chad to the Cameroonian port of Kribi completed in 2003 will provide jobs and income for years and is expected to provide the government with $20 million a year for 25 years.

Cameroon is actively seeking to strengthen its regional economic links through membership in the

CEMAC (Communauté Économique et Monetaire de l'Afrique Centrale), and through developing routes to the sea and sources of manufactures for its landlocked neighbors, especially Chad. Its share of regional trade has been increasing, and its export partners have become more diverse. Italy and Spain account for over 30% of Cameroon's exports and France, the former colonial power, only 13%. Nevertheless it is France that provides most of the technical assistance and financial loans. The rail extension from Yaoundé to Ngaoundéré, completed in 1974, and funded by the United States and the EU countries, has opened up the northern regions, providing southern Chad with a new outlet. The line made exploitation of the northern bauxite deposits near Martap feasible (Figure 16-6), and linkages with Chad are growing with the new pipeline. Plans to extend the railway to Moundou (Chad) and to build a branch line to Berbérati (Central African Republic), under consideration for decades, may yet go forward in anticipation of oil-led development in the interior.

The relative success in the development of Cameroon's mineral and agrarian economy must be measured against the results of similar efforts in other parts of lowland central Africa. Major problems still confront the country, whose political geography has always been complex. The extreme northern region suffers from remoteness from the country's capital and southern core area, and its natural outlet has been through Nigeria, with which there are also strong social and cultural affinities. The long-term division of the country into British and French territories has presented problems upon the reunion of the southernmost British section with the French zone. Lying across the Bantu ethnolinguistic transition zone in the southwest, and penetrating the Muslim African world in the north, Cameroon's ethnic variety is as great as that of its physiography, and this has been an obstacle in the struggle for national unity.

Under the federal constitution adopted in 1961, Cameroon moved to increasing political, economic, and social integration, while pronounced regional variations in the economy, dating back to earlier administrations, persisted. In 1972 Cameroonians overwhelmingly approved a new constitution creating a unitary state. A completely centralized administrative system then prevailed until the structural adjustment programs of the 1990s.

BIBLIOGRAPHY

EIA. 2004. *Congo-Brazzaville.* Washington, D.C.: U.S. Energy Information Agency. http://www.eia.doe.gov/emeu/cabs/congo.html.

Gabon, government of. 2004. *Gabon National Parks.* http://www.gabonnationalparks.com/gnp-nationalparks.

Helders, S. 2002. *World Gazetteer.* http://www.world-gazetteer.com.

Mobbs, P. 2004a. *The Mineral Industry of Congo (Brazzaville).* Reston, Va.: U.S. Geological Survey.

Mobbs, P. 2004b. *The Mineral Industry of Equatorial Guinea.* Reston, Va.: U.S. Geological Survey.

Population Reference Bureau. 2003. *World Population Data Sheet.* Washington, D.C.: Population Reference Bureau.

Quammen, D. 2003. Saving Africa's Eden. *National Geographic,* 204(3): 50–75.

U.S. Department of State. 2000. *FY 2001 Country Commercial Guide: Cameroon.* Washington, D.C.: Bureau of Economy and Business, U.S. Department of State.

United Nations Food and Agriculture Organization. 2004. FAOSTAT. Rome: United Nations.

United Nations Food and Agriculture Organization. 2003. FAOSTAT. Rome: United Nations.

UNDP. 2002. *Human Development Report, 2002: Deeping Democracy in a Fragmented World.* United Nations Development Program. New York: Oxford University Press.

Yager, T. R., G. J. Coakley, and P. M. Mobbs. 2004. *The Mineral Industries of Benin, Cape Verde, the Central African Republic, Côte d'Ivoire, and Togo.* U.S. Geological Survey Mineral Report. Reston, Va.: USGS. http://minerals.usgs.gov/minerals/pubs/country/.

EAST AFRICA, THE HORN, AND SUDAN

East Africa, the Horn of Africa, and Sudan is a region of large cultural and physical contrasts. Location explains much of the difference: the region is defined by, focused on, and anchored by the African Rift Valley. Rifting and associated volcanic activity created a wide variety of physiogeographic and biogeographic regions that provide for a diversity of plant and animal life and land-use systems. With the exception of the major cities of the region, the densest human populations are found there, as well as the greatest regional ecological diversity. The Horn and Sudan are about half the size of the United States, and East Africa is about the size of Mexico. The total population is close to and will soon exceed that of the United States (Table 1). The sheer size of the region, in addition to the depth of time that humans have occupied and moved into and out of it, have strongly influenced the development of complex human languages, identities, politics, cultures and societies, and relations between peoples. Burundi, Kenya, Rwanda, Tanzania, and Uganda, make up East Africa. Ethiopia, Eritrea, Somalia, tiny Djibouti, and Sudan along the western flank constitute the Horn of Africa (Figure 1). The countries of the region are internally diverse and straddle cultural as well as environmental transition zones similar to most countries of Africa.

Both Arab and European civilization and culture have exerted profound influences on the people of the region: Arab influence, which has been strongest in the north and along the coast, developed along with the ancient Indian Ocean trade. Sudan's north is culturally more akin to Arab/Islamic North Africa, while its southern part has a closer cultural fit with the peoples of central Africa, East Africa, and the Horn. Perhaps more than any other country in Africa, Sudan, the continent's largest country, shows the dilemmas of the rather arbitrary division of the African landmass. Arab colonization penetrated deep into the Sudan along the Nile Valley into Uganda and other southerly interior locations. From the eastern shores of Tanzania, Arabs and Swahili extended their control west all the way into the Congo Basin, where Europeans, expanding from the western side of the continent during the latter part of the nineteenth century, contested their hegemony. Today the Arab cultural legacy is reflected principally in Arab and Islamic identities and, in many cases, Islamist politics and conflict with indigenous peoples. The

TABLE 1 DEMOGRAPHIC AND GEOGRAPHIC DATA FOR EAST AFRICA AND THE HORN AND SUDAN.

Region	Country	Population (2004)	Area		Population Density
			Km²	Miles²	
East Africa	Burundi	7,802,000	27,834	10,747	779
	Kenya	33,520,700	581,787	224,629	149
	Rwanda	8,594,100	26,338	10,169	845
	Tanzania	36,581,300	945,088	364,901	107
	Uganda	26,219,200	242,554	93,651	344
	Total or average	112,717,300	1,823,601	704,096	170
Horn and Sudan	Djibouti	765,300	23,200	8,958	85
	Eritrea	4,067,000	121,100	46,757	87
	Ethiopia	72,035,400	1,127,127	435,186	166
	Somalia	11,555,300	637,657	246,210	47
	Sudan	39,162,100	2,505,810	967,499	41
	Total or average	127,585,100	4,369,516	1,704,600	76
Total		240,302,400	6,238,495	2,408,696	241

Source: Helders (2004).

relatively recent European impacts profoundly permeate the formal institutions of governance today, in addition to education, land use, economies of the region, and Christian identities. East Africa's centuries-old trading towns, however, were at the same time a product of and instrumental in trade across the Indian Ocean. The Swahili stone towns are located along a thousand-mile stretch of coast between Somalia and Mozambique, but most are centered in today's Kenya and Tanzania. The Swahili culture, long in decline, has been significant in the development of African identity over a vast region, and while the Swahili themselves are few, their language is widely used, uniting many Africans. Although Omanis had been involved in East Africa for many centuries, their involvement become more aggressive in the seventeenth century and East Africa became part of the Omani Empire, first based in Muscat (Musqat), Oman, and later (1837) on the island of Zanzibar.

Later still, during the European colonial period, British-ruled East Africa constituted a nascent East African "community." The independent governments that succeeded the British neither maintained the relationships that had been built over time nor sought to become closer to a union, principally because they pursued paths to development that entailed fundamental conflicts between their political and economic assumptions and their expectations. The East African Community was neglected for over 20 years, but lessons learned during the first development decades after independence and ensuing changes have reduced—even reversed—contradictory political and economic policies.

During the last decade of the twentieth century and during the first years of the twenty-first, the East African countries are coming to resemble one

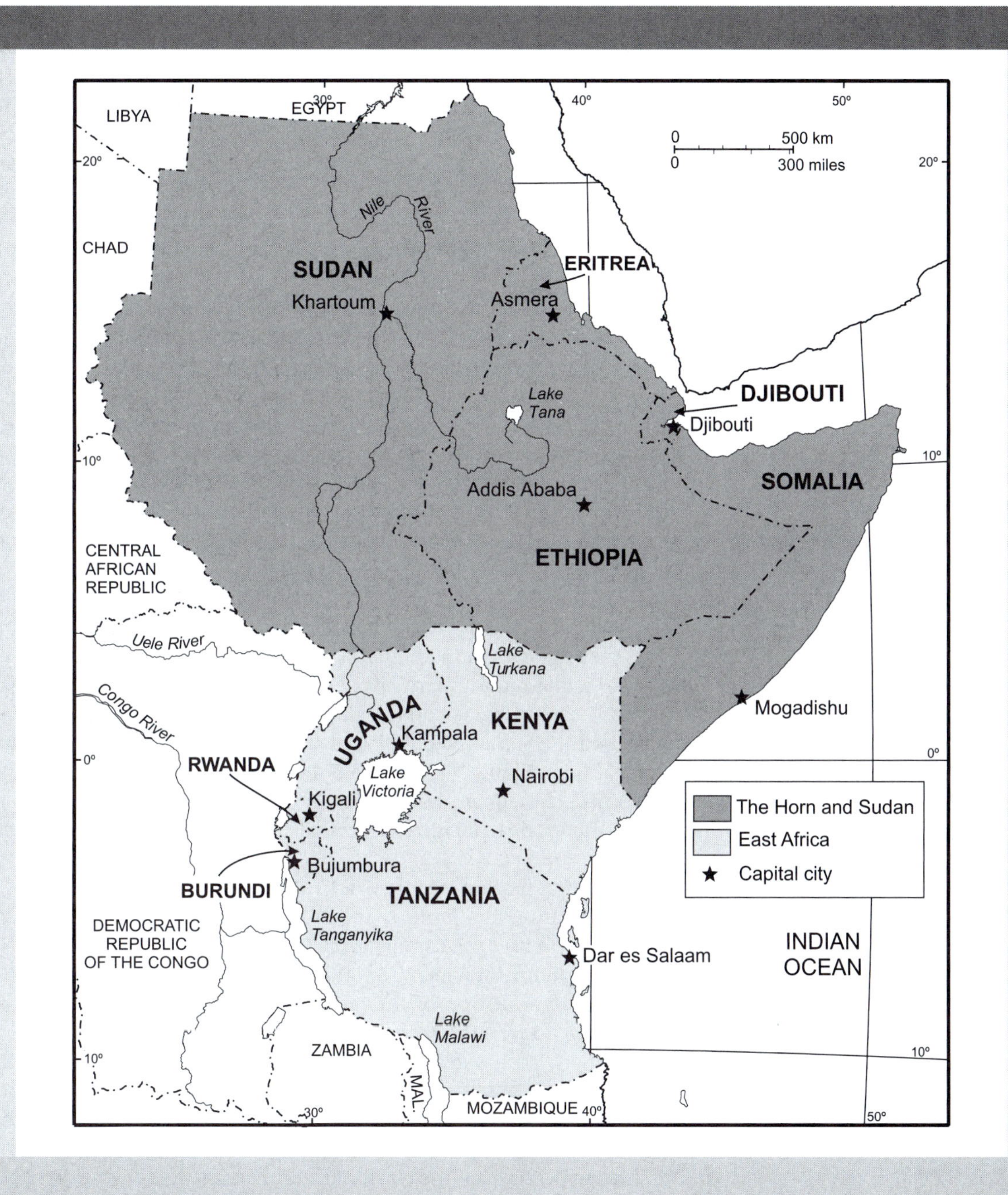

FIGURE 1 Region of East Africa, the Horn, and Sudan.

another more and more—at least in their economic planning. In other dimensions they are as fractured as ever. Ethnic conflict has been ongoing in Uganda for close to 20 years, while decades of poor relations between Hutus and Tutsis in Rwanda and Burundi have led to conflict and genocide. Almost all East African countries participated in Africa's "Great War" in the Democratic Republic of the Congo, a war in which Rwandan Tutsis played a significant military role and ultimately became kingmakers.

The "Horn" of Africa, including the lands north of Uganda and Kenya and east of the White Nile, which arises in Lake Victoria, is an area of immense physiographic diversity. Its physical core is constituted by the vast Ethiopian Plateau, rising to over 13,000 feet (3,900 m) in many places, rent by great rift valleys and cut elsewhere by spectacular declivities. The plateau still effectively prevents modern communications from reaching all of the heart of the northeast. At the source of the Blue Nile (at Lake Tana), the plateau wrings from the air more moisture than any surrounding territory. It possesses excellent soils and good climates, as well as barren wastes and inhospitable environments. Desert and steppe lands bound the plateau in all directions, separating it from the coasts of the Red Sea and Indian Ocean, from the valley of the Nile, and from the good lands of Kenya and Uganda.

The northeast is as diversified politicogeographically as it is physiographically. There, Arab and Bantu, Islamic and Christian, local and foreign empires have met. A modern political framework has been superimposed upon an area that retains many of the characteristics of its initial feudal condition. Today, this framework fragments the Horn into Ethiopia, Eritrea, Djibouti, and the Somali Republic (Somalia), consisting of the former Italian and British Somalilands. The present boundaries of the Horn are indeed superimposed, subsequent boundaries, and they are not, in several areas, approved by the local population, as several successful secessionist movements bear witness.

Close to 88.5 million people live in the Horn (34 million in 1975), and the great majority (72 million) reside in Ethiopia. Djibouti has a population of about 765,000, and the coastal Somali Republic about 11.5 million. There are many more Somali people (from the ethnic, religious, and other points of view) than the population of the republic might suggest. Indeed, there are Somali people under Ethiopian rule and also under the administration of Kenya. Formerly, the Somalis were further fragmented by the Italian and British division of Somaliland. Neither does the population of Ethiopia justify the term "nation." Apart from the Somalis living in (mainly eastern) Ethiopia, there are numerous peoples of other ethnicities within the confines of the state, which is itself the result of the amalgamation of a number of rival kingdoms, their consolidation under powerful leadership, and its subsequent expansionist policies. It is this expansionism that carried Ethiopian rule into Somali territory, and the ensuing definition of the boundaries that created one of the major, perpetual conflicts in this part of Africa.

During the Cold War, Somalia and Ethiopia switched ideological camps, changing allegiance from one superpower to its competitor. Somalia later descended into years of anarchy and competition for control by subnational, regional, and tribal warlords. Sudan has endured decades of civil war between the Arabo-Islamic north and the Christian–animist south, and more recently with the former kingdom of Dar Fur in the west. After a military coup in 1990 against a democratically elected government, Sudan instituted repressive Islamic law (Sharia) and prosecuted the war (many called it jihad) against the southern regions with a renewed vigor. Osama bin Laden briefly called Sudan home during the early 1990s.

Peace still is not at hand. In 1991 Eritrea defeated Ethiopia after over 30 years of bloody strife, but despite achieving independence two years later, Eritrea continues to be plagued by unresolved land disputes and military conflict. Ethiopia broke with its feudal past in 1974, embarking on a Soviet-style state that involved bloody purges, mass deportations, and a "Red Terror" that lasted until 1991, when the regime was overthrown by northern rebels piloting Sudanese tanks. Djibouti, capitalizing on its strategic location at the Bab el Mandeb, has acted as a transit point for goods into the Horn, especially during the Eritrea–Ethiopia war over the Badme region (1998–2000) when Ethiopia shifted its cargo from the rail line running through Eritrea to the Ethio–Djibouti line.

With the exception of Kenya and Tanzania, every country of the region has been wracked by civil war in recent years. In northern Uganda over a million refugees live in camps in fear of rebel armies. Neither Kenya nor Tanzania, formerly in opposing ideological camps, has experienced civil war since independence, although the Tanzanian army was involved in ousting the dictator Idi Amin from Uganda in 1979.

Wealth, natural resources, and power are distributed unequally throughout the region, which like most of Africa experienced great changes after the collapse of the Soviet Union and the end of the Cold War. Only recently have elections replaced one-party rule and dictators. Chapters Seventeen through Twenty-three will explore these and other issues in detail.

The contrasts between lowland equatorial Africa, as exemplified by the Congo Basin, and highland East Africa, which comprises Tanzania, Kenya, Uganda, Rwanda, and Burundi, are sharp and significant. For the most part, East Africa is higher, cooler, drier, and less forested. The vast majority of the people are semisedentary pastoralists and farmers who depend directly on the land for their livelihood; yet much of the region has only marginally useful agricultural and pastoral land. Only 4% of the five-state region may expect to receive more than 50 inches (1,270 mm) of rainfall in four years out of five, while half the region may expect less than 30 inches (760 mm). Droughts can strike with devastating force and frequency in the northern areas and in central Tanzania.

Much of East Africa is a gently undulating plateau that lies more than 3,500 feet (1,050 m) above sea level (Figure 2). The highest areas adjoin the rift valleys along the region's western boundary and east of Lake Victoria: the Ruwenzoris of Uganda and the Aberdares of Kenya rise above 13,000 feet (4,550 m), while several volcanoes tower even higher above the expansive savanna plains. Snow-capped Mount Kilimanjaro exceeds 19,300 feet (5,880 m), and on the equator, glacier-fringed Mount Kenya rises to 17,065 feet (5,200 m). On the lower slopes of these and other mountain

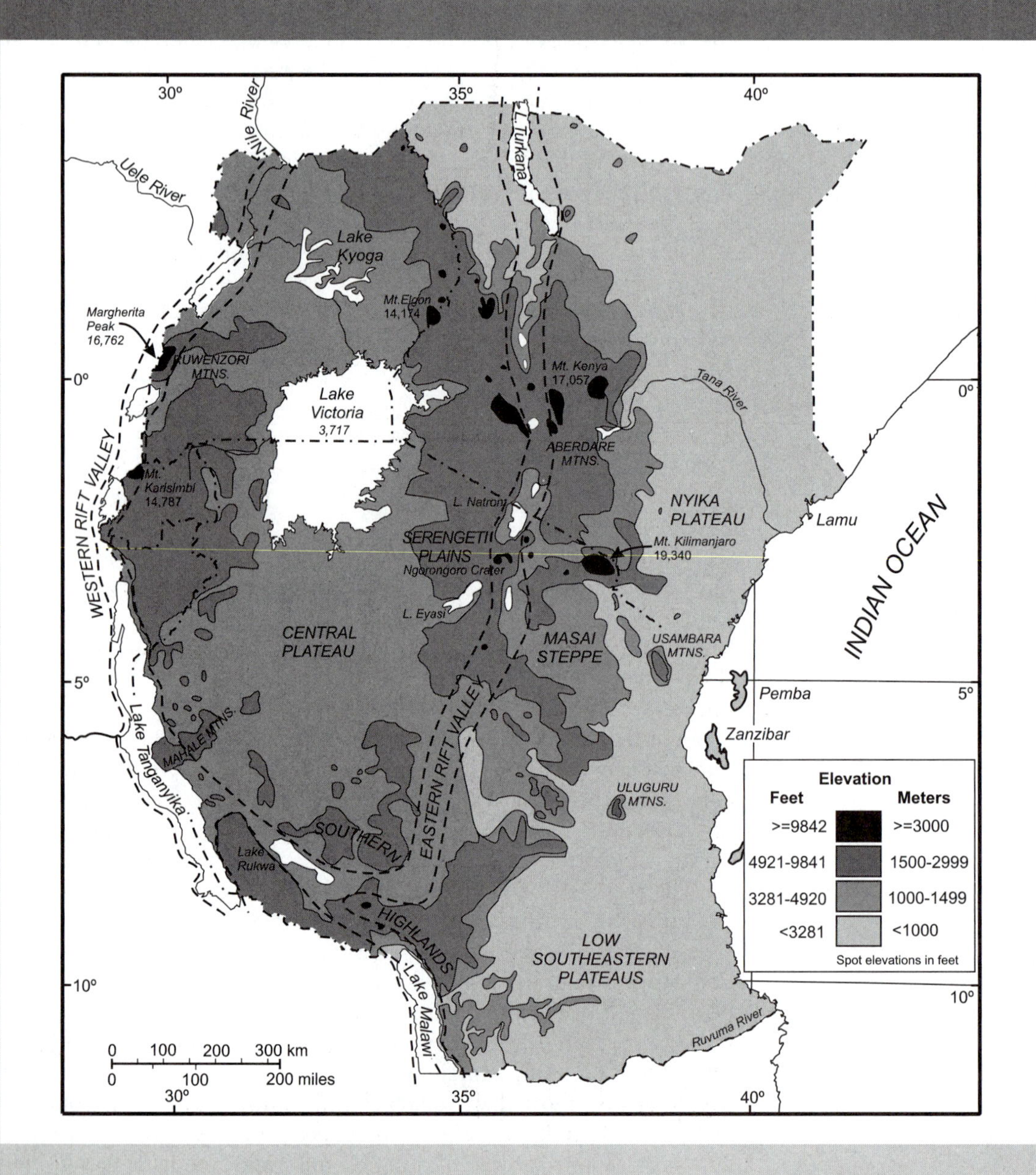

FIGURE 2 Physiography of East Africa.

zones, where soils are rich and rainfall is generally high, population densities are among the highest in Africa, and agricultural land use is intense. Population in the region clusters in the highlands along the Rift Valley, along parts of the coastal strip, and is dense on the larger coastal islands (Figure 3).

The region contains two of the least urbanized countries in Africa and two, Kenya and Tanzania, which are experiencing overurbanization as their

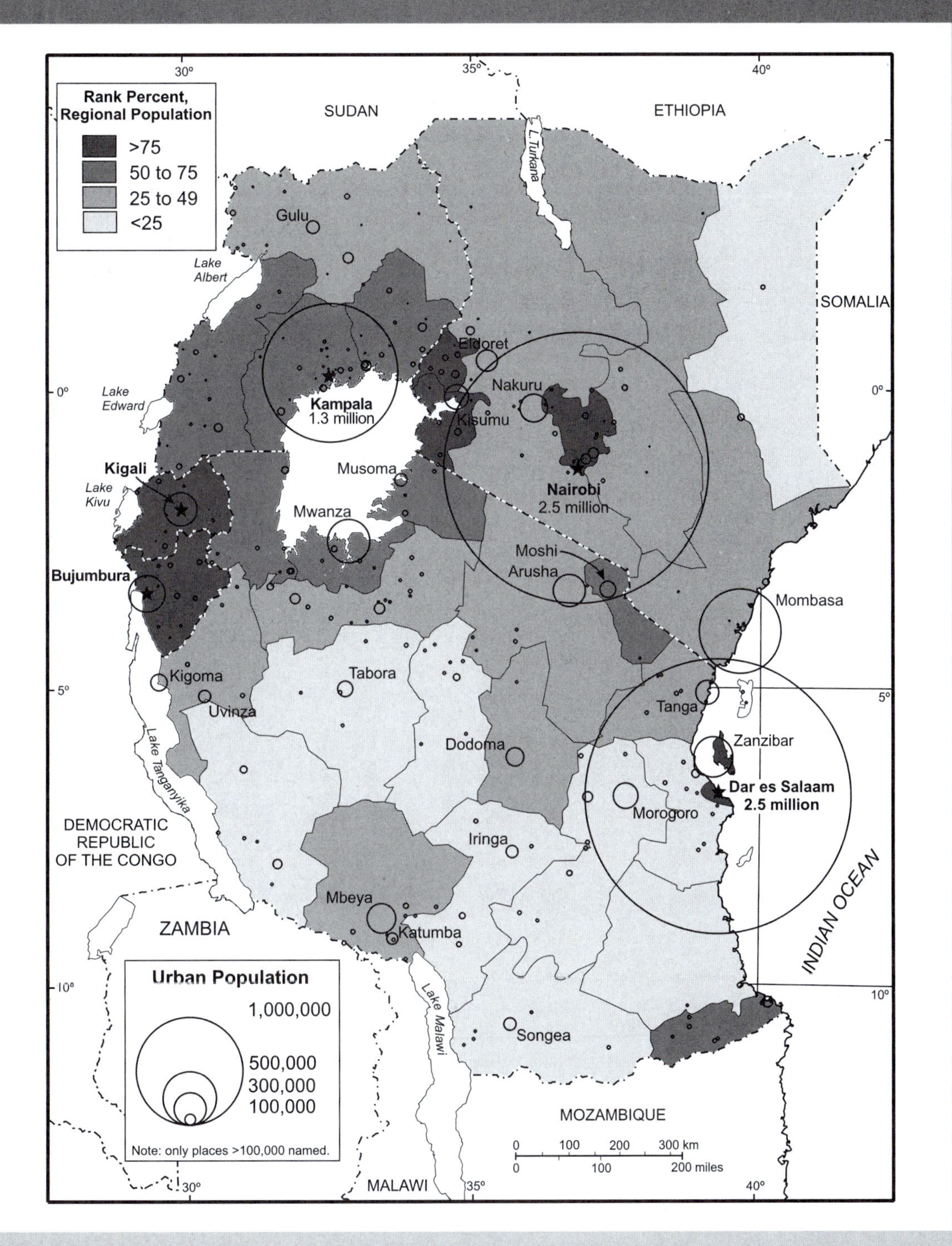

FIGURE 3 Provincial (district) population of East Africa in quartiles as a rank percentile of regional population and urban population.

primate cities have grown remarkably fast over the last 30 years. The two largest cities are the capital of Kenya, Nairobi, and Dar es Salaam, Tanzania; each has a population of 2.5 million. Kampala, Uganda's capital, is over a million, while Kenya's second city and major port, Mombasa, is over 750,000. The huge difference between the size of these cities and the rest is noteworthy. A lack of connectivity in the urban–urban, urban–rural networks is problematic and has impeded economic development and national integration. Unlike central Africa to the west or southern Africa, highland East Africa has few known mineral resources, and mining plays only a minor role in each of the national economies. Rwanda and Burundi, two tiny countries of the region perched high in the Rift Valley mountains, have been wracked by ethnic tension and conflict over the last few decades that reached the status of genocide in the early 1990s. The conflict spilled out of Rwanda to encompass neighboring countries, causing political and economic disruption and loss of life.

Of the winds of change that have been sweeping this part of Africa, democracy, capitalism, and the market economy, market economics is not new to the region. People in the region have been involved with the long-distance Indian Ocean trade for centuries, and it has mainly been postindependence governments (Tanzania in particular) that attempted abolish free enterprise and the market and to manage the economy from their capital city bureaus. Over the past decade or so Tanzania has attempted to change the developmental framework it established soon after independence and today, after a "false start" (Dumont 1969), may be said to be converging with its neighbor Kenya.

CHAPTER 17

False Start in Tanzania

Of the five East African states, the United Republic of Tanzania is by far the largest (364,900 mi^2; 945,087 km^2), larger in fact than the others combined. It is also the most populous (16.2 million in 1977; 36 million in 2004). Tanzania comprises mainland Tanganyika, formerly a UN trust territory held by Britain, and the islands of Zanzibar and Pemba, formerly a single British protectorate. In April 1964, following several months of open conflict between Zanzibar's African majority and its Arab and Asian minorities, these two independent territories merged under the name Tanzania and embarked on a unique program of social and economic development that sought greater economic self-sufficiency. A "cooperative approach" to development was adopted which, according to its proponents, was based in part on "traditional values," socialist theory (particularly Maoist and dependency theory), and on the strong intellectual and charismatic leadership of the country's first president, Julius Nyerere. The development of self-reliance through a program of radical rural reform led by state bureaucrats was the major thrust on the ground. That this program failed is testament to its impracticality.

RESOURCE LIMITATIONS

Although Tanzania is a tropical country, rainfall totals are remarkably low and variability is high (Figure 17-1). A broad belt stretching from northeast to southwest across the heart of the country has less than 20 inches (500 mm) of annual rainfall, and only on the highest topographic prominences do recorded totals exceed 70 inches (1,780 mm). Only one-fifth of the area can expect (with 80% probability) more than 30 inches (762 mm) of rainfall in a given year, and less than 5% can expect over 50 inches (1,270 mm). Even where the annual rainfall is higher, it may come during a relatively short wet season, limiting cultivation and rendering pastoralism a very precarious business during the dry season. An associated feature of the highly concentrated rainfall is its arrival in intense storms, leading to excessive runoff and erosion. The high temperatures and great amount of evaporation further limit the usefulness of the precipitation. The best-watered areas lie spread about the dry heart of the country and include the coastal belt, the Kilimanjaro and Meru slopes and surroundings, the Lake Victoria region, the western section beyond Tabora, the Southern Highlands (around Lake Malawi), and the belt extending from there northeast to Morogoro. In total, perhaps about one-fifth of the country may be described as adequately watered.

Tanzania, which forms part of the divide between three of Africa's greatest rivers (the Zambezi to the Indian Ocean in the south, the Nile to the Mediterranean Sea in the north, and the Congo to the Atlantic Ocean in the west), does not itself possess any such large watercourses. Indeed, the central part of the country has two major basins of internal drainage, Lake Rukwa in the south and the Rift Valley trough in the north. Thus the local drainage systems are of but minor significance on a continental scale. The eastern part is drained largely by the Rufiji River and its tributaries, and, in the west, the Malagarasi River follows a brief course into Lake Tanganyika. Many streams are ephemeral or intermittent and are of no permanent importance, and during the dry season people often have to undertake long treks to obtain water. The Rufiji, which drains to the Indian Ocean, has major potential for irrigation and hydroelectric power development, while the Pangani to the north already supplies energy to Arusha, Moshi, Tanga, and Dar es Salaam.

Soil is another important resource, yet much remains to be learned concerning the country's pedology. Soil analysis and mapping are slow and expensive,

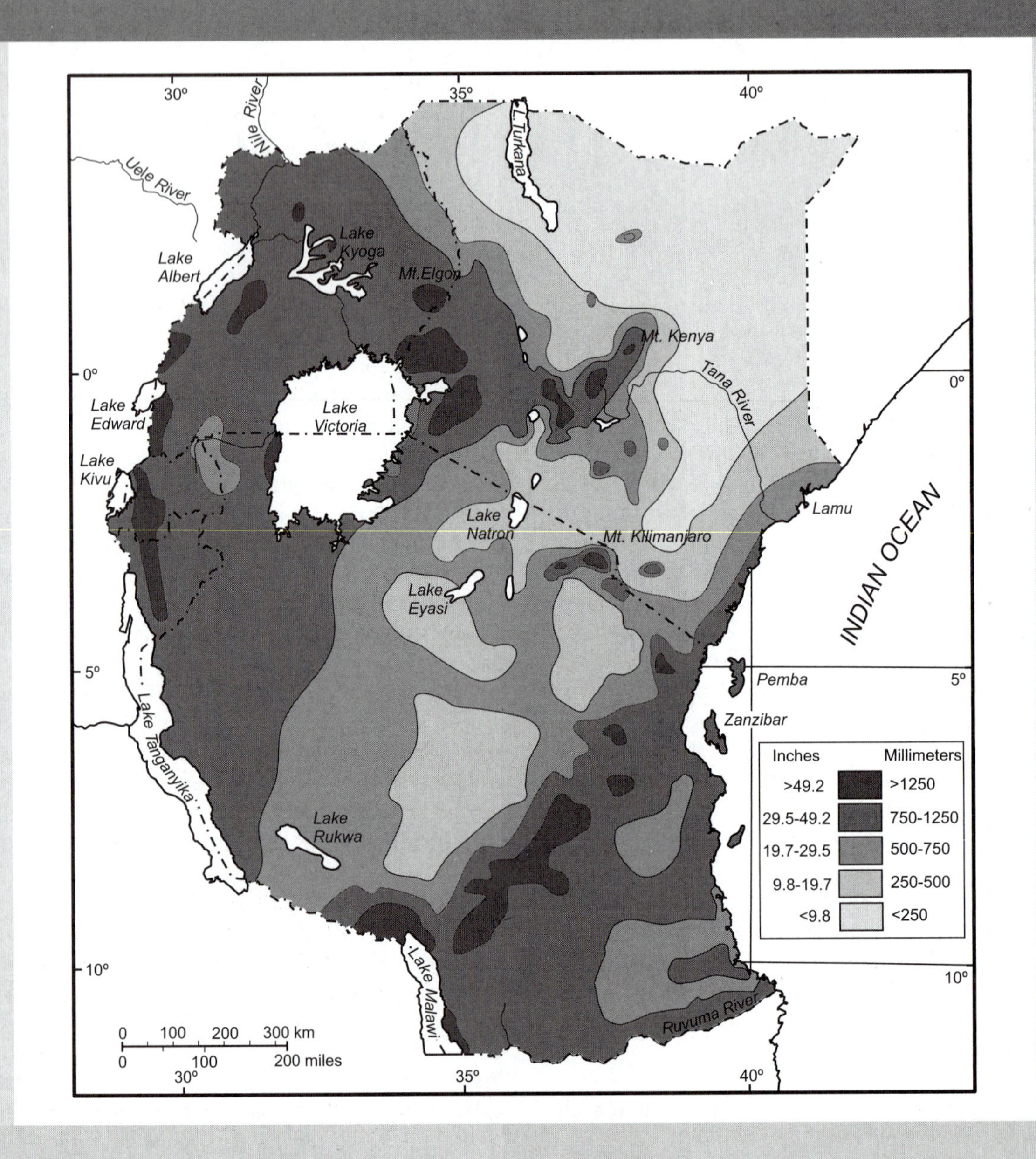

FIGURE 17-1 Precipitation levels in East Africa (80% probability).

and except in the important agricultural areas, the soils are known only in general terms. It is clear that the most fertile are the reddish-brown soils derived from volcanic materials in the highlands of the north (Kilimanjaro, Meru) and the south and southwest (around the north end of Lake Malawi). Good soils, although variable in certain qualities, also exist in the alluvial valleys where they await irrigation (Berry 1972). The remainder of the country is largely covered by the familiar, deep-weathered, reddish soils that develop on the crystalline rock formations of the African plateau, with their limited productivity. Under proper

BOX 17-1

A MAASAI SENSE OF PLACE

The distant harsh mountains, Ildonyo Ogol, are stranded, as if forsaken by nature. They are composed of granite and are covered with thorny shrubs and acacia trees. They are the habitat of baboons and klipspringers, which my people call stone goats. To the left, facing the rift, stands another majestic mountain called Makarot, which commands a breathtaking view of a stretch of open country known as Serenget. This, God's country, is my home. My old ancestors won it from the ferocious Iltatua, a people now pushed to the shore of Lake Eyasi. Wells dug by them are reminders of their past history. This country, known as the Korongoro (Ngorongoro), is so lovely that I do not regret the banishment of the Iltatua.

I grew up here unconscious of the beauty of the landscape but aware of the abundance of wildlife. When I was young, I would chase zebras with my friends until we were swallowed by the dust and had to shout so the zebras wouldn't trample us as they stampeded by. Once when I was tending lambs with one of my sisters, I saw my father and another man running at full speed toward us, and from their faces I could tell there was something seriously the matter. They grabbed us by the hands, lifted us to their shoulders and put us in a tree nearby. Then they pointed to a pair of huge animals that appeared to me like moving rocks. We watched as warriors from a nearby kraal caught up with them. The two beasts, which I came to learn were rhinoceroses, galloped at full speed for their lives. Engilusui, Oloongojoo, and Naibor Soit were small, isolated hills facing my home, to one side of which Oldopai (Olduvai) appeared like a dark rift. Its bold tree line cut the plain in two halves and continued far south, as if to an unknown destination. All those places were so familiar, expansive pastures for our herds, lush and green during heavy rains.

Source: Saitoti (1986), p. 6.

farming practices, including the application of fertilizers, these soils can produce good annual yields in the moister areas.

Although Tanzania does not possess the dense forests of lowland equatorial Africa, the majority of the country lies under bush, parkland savanna, and patches of denser scrub. The natural vegetation has been affected by human activities, the original growth having been replaced over wide areas by a tree-poor grassland savanna, home to the pastoral Maasai (Box 17-1). Twenty-five percent of the country remains infested by the dreaded tsetse fly, which thrives under the prevalent conditions and limits human occupation; only small parts of the country are free from malaria. In low-challenge areas where the tsetse infestation is less severe, rural people have successfully controlled trypanosomiasis (sleeping sickness). In regions of greater tsetse populations, high transmission rates and drug resistance hobble progress in fighting the disease (Chizyuka 1998). The cattle population is denser in areas where tsetse is less prevalent: in the highlands along the eastern and western Rift Valleys and around Lake Victoria.

Mineral deposits are widely distributed over Tanzania, with several of economic significance. Nowhere in Tanzania do minerals lie juxtaposed with a region of agricultural importance, the whole supporting an urban–industrial core; there is no Witwatersrand, Copper Belt, or Katanga, nor is there a highland core such as that of Kenya. Rather, mineral development has been isolated, with most production coming from peripheral regions (Figure 17-2). Since independence, mining has generally accounted for less than 3% of the GDP, but as Tanzania opened its doors to the world after years of austere "self-reliance," natural resource exports have been increasing. Since 1996 the mining sector of the economy has grown at 16% per year (Yager 2001), mainly because of greater gold and diamond production. While mineral exports amounted to only about 5 to 10% of all exports during the 1970s,

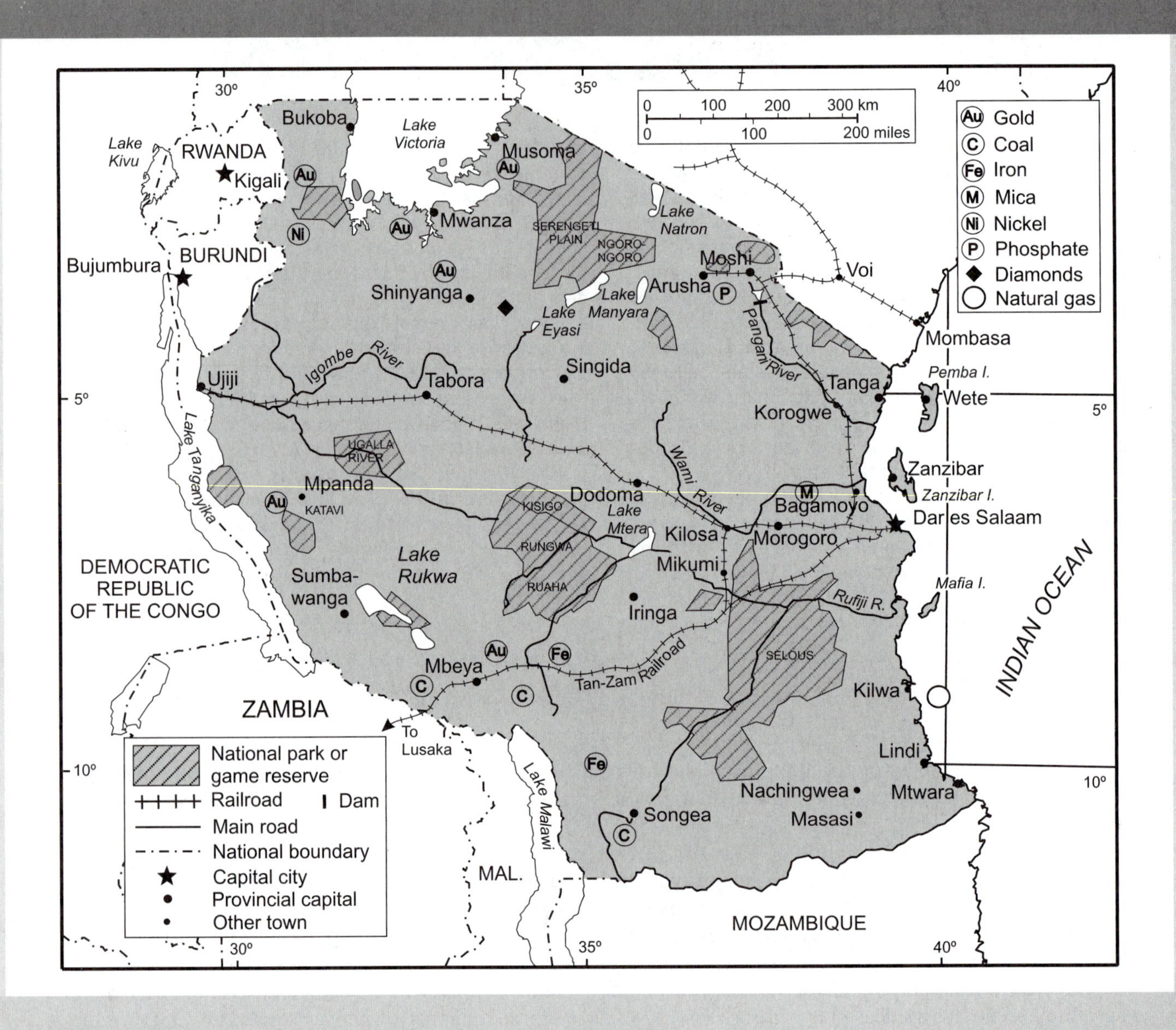

FIGURE 17-2 Towns, transportation, and natural resources in Tanzania.

by 2001 they accounted for 40%. Of that 40%, gold exports represented nearly a third. Gold exports earned over $260 million in 2001 compared with $3.4 million in 1998, while foreign exchange earned through diamond exports doubled, to $30 million. In that same year Tanzania became Africa's third largest gold exporter.

Until recently, gold was found principally in the southern highlands in the Mpanda and Mbeya areas and on the central plateau near Lake Victoria from the Kiabakari and Geita fields. Until the mid-1960s, gold was mined in substantial quantities (up to 1,100 lb; 500 kg annually), but high production costs, obsolete equipment, low world market prices for gold, and apparently dwindling reserves forced the closure of most operations. In the 1990s, under a new economic framework, gold became a major contributor to the Tanzanian economy. In 2001 the Bulyanhulu gold mine opened in northern Tanzania near Mwanza, adding significantly to gold production.

The single most important deposit of diamonds is located at Shinyanga, where a kimberlite pipe, similar to those of South Africa and Botswana, was discovered in 1940, and production began soon after. Production reached a peak in 1967 (988,000 carats), when diamond sales comprised nearly 90% of total mineral sales, and diamonds were Tanzania's third most important export after coffee and sisal. Diamond exports steadily declined to a low of 17,000 carets in 1994. Privatization during the middle 1990s brought renewed interest from international investors in diamond mining and exports picked up again but are still hard pressed to equal a third of their 1967 peak.

Rich phosphate deposits occur south of Arusha and Moshi, and pyrochlore has been discovered at Mbeya. Substantial deposits of high-grade iron ore, noncoking bituminous coal, and limestone are located in the Ruhuhu Valley and associated highlands north of Lake Malawi, where there is also a great potential for hydroelectric power. Before the Tan-Zam Railway was completed, these resources were considered too remote to be exploited. Today the transportation infrastructure needs rehabilitation and extension.

Tanzania has some state-of-the-art metal manufacturing and processing factories, as well as less modern facilities. The privatization of the economy has led to foreign investment and revitalization of industry. Exploration continues, and prospects seem promising for the exploitation of platinum-group metals; deposits of the ores of titanium, vanadium, and zirconium have been found in the heavy mineral sands near Bagamoyo, located near the coast north of Dar es Salaam, and Msimbati along the southeastern coast. Offshore exploration for petroleum is under way, and deposits of natural gas have been located at Songo Songo Island but are not yet producing. The government intends to pipe the gas to fuel electricity generation on the mainland. Although it has been true that overall, Tanzania's natural resources, especially those required by industry and for export markets, are limited in quality, quantity, and by accessibility, increased exploration may help provide the infrastructure to reach less accessible deposits.

THE IMPRESS OF THE PAST: ARAB, GERMAN, AND BRITISH COLONIALISMS

The Tanzanian space economy (spatial organization of economic activity and how that organization changes over time with development) has developed in several distinct phases and under different raisons d'être. The region has been long linked to the Indian Ocean coastal states through long-distance trade. Before the Portuguese took control of much of the African side of the Indian Ocean trading system in 1505, trade was carried on by Swahili living in the relatively independent mercantile towns along the coast. Trade goods were obtained from inland peoples generally living beyond the "nyika" ("coastal desert" in Swahili), and the Swahili acted as middlemen in this far-flung network. The Portuguese were ousted by Omani Arabs in 1699, after which the sultanate of Oman began to organize the space economy of the region. From Zanzibar, the Omani Arabs subdued the Swahili trading towns along the coast and organized a trading empire to serve their interests. In 1837 the sultan of Oman transferred his capital to Zanzibar.

Until the arrival of the German Colonization Society in 1884, the region's main contact with the outside world was through Omani Arab trading and slave-raiding caravans entering from Bagamoyo (Glassman 1995). Except for the first efforts to build some coastal stations (including the settlement of Dar es Salaam) and some self-serving alliances with inland tribes (Box 17-2), the Arabs took no measures to introduce any form of territorial organization effective enough to encourage the formation of nation states. Indeed, that was not their intention, but they achieved economic and political control of the coast nevertheless. During the chaos of the slave-raiding years, the sultan of Zanzibar and his allies among the Swahili traders extended their influence into the interior of the region through treaties and alliances with local chiefs. This was the same tactic that would be used to great effect by the Germans and British years later (and for a time, Arab and European competed in the treaty-making process). The Europeans, however,

BOX 17-2

THE KAGURU AND THE CARAVANS

Ukaguru is a land which has had a long and difficult contact with outsiders. The Kaguru live in an area between several very different cultural groups which were traditionally at odds with one another. Although a people who place a low value on warfare, they have been forced to deal for generations with some of what are reputed to be the most aggressive peoples in East Africa, especially the Maasai and Hehe. In general, the Kaguru preserved themselves by defensive combat from their mountain areas and sometimes making temporary alliances with some of their enemies in order, in turn, to raid others, sometimes even other Kaguru.

Kaguruland lies astride what was once one of the major caravan routes which led from the Arab-dominated Indian Ocean ports to the great inland lakes of Central Africa. In the 19th century most of the Arab and African caravans traversing most of Tanzania brought goods eventually to the island port of Zanzibar and the ports facing Zanzibar on the Tanzanian coast. Nearly all such caravans passed through the general vicinity of Kaguruland, which formed a relatively safe corridor between areas to the north and south which were arid, mountainous, and dominated by warlike peoples. It is estimated that shortly before Europeans colonized the area, nearly 100,000 persons in caravans passed annually through Kaguruland, mainly involved in trade in ivory and slaves (Beidelman 1962). By the late 19th century areas near the coast had been hunted out of ivory and reduced of their easily captured inhabitants. Then, areas such as Kaguruland served simply as sites for caravan stations where traders could rest, take supplies and water, and organize locals for various subsidiary services. Caravaneers sought slaves and looted for supples. Arabs readily sold arms to Africans and sought to encourage intertribal warfare in the hope of purchasing the defeated from the victors. In addition, the intrusion of aliens from the coast and also down the Rift Valley appears to have brought human and livestock epidemics, including smallpox, rinderpest, cholera, East Coast fever, and meningitis. All of these factors led to serious upheavals and movements of local peoples. Those who had lost livestock often sought to replace these from their neighbors' herds. Aggressive or ambitious persons, previously held in check by their neighbors with whom they were militarily equal, would begin trading for arms before their neighbors did so and would then exploit their new advantage while they could.

Arab and African traders benefitted from the upheavals they sowed, and yet, out of self-interest, they also sought to establish spheres of stability along which their own affairs of trade and travel might proceed. In the late 1870s, the Arabs recognized and supported a local Kaguru leader in Mamboya named Senyagwa Chimola. They gave him arms, cloth, and beads, and he provided them with local labor, building materials, a building site, and food; he also allowed African strangers associated with the caravan trade to establish settlements along the caravan route. The Arabs recognized Senyagwa as the sultan, encouraged him to take an Arab name, Saidi, and helped him extend his political influence.

Source: Beidelman (1971), pp. 11–12.

were interested in terminating the profitable East African slave trade, and this in turn meant disrupting the accords between the Omani Arabs and the Swahili traders through which the latter had acquired labor and weapons to help perpetuate this traffic, and their way of life. Indeed, since the end of slavery Swahili culture has been in decline. Nourished by food crops produced on mainland slave plantations and enriched by the export of spices and slaves, the Omani and Swahili elites lost their economic base after the emancipation of the slaves (Middleton 1992). During the second half the nineteenth century, the Omani sultanate became a wealthy state that was known around the world. Slave raiding and trading was ended by treaty with the British in 1873, although the local practice of slaveholding was not; as a result, plantation slavery expanded as did the production of cloves and other sought-after spices on the islands along the coast, especially Pemba and Zanzibar islands.

Several factors led to the decline of Omani hegemony over the coast: the overexploitation and destruction of the natural resources on which trade depended, Swahili resentment of Omani control, the abolition movement, and the coming of European and American traders who, paradoxically, were allied to the sultans but at the same time worked in competition with the Omani rulers' interests. The Zanzibari system broke down in the last decade of the nineteenth century and was replaced by a time of social and political chaos and autarky.

> The slave trade had virtually ceased, leaving vast numbers of local African slaves, war captives, and refugees to be exploited as local laborers and ivory carriers to the coast, with great cruelty and loss of life. This was the other side of the Swahili coin: their brief but brutal exploitation of the disorganized and demoralized interior peoples. It is little wonder that at the same time the advent of European colonial rulers and missionaries appeared preferable to Arab and Swahili brutality, or that even today the names "Arab" and "Swahili" are abhorred in parts of the interior. (Middleton 1992: 49–50)

The 30 years after 1884 mark the German period of occupation, and several of the tangible assets of Tanganyika were created during this time. Having initially used the settlement at Bagamoyo as the administrative headquarters, the Germans shifted the capital to Dar es Salaam and began to develop the site. Karl Peters, founder of the German Colonization Society and governor of German East Africa, recognized the need for railway links with the interior, in the interests of both effective control over the rebellious African population and economic development. Thus, the railroad from Dar es Salaam to Kigoma (near Ujiji, on Lake Tanganyika) was completed as early as 1914, having been begun in 1905. The line to Moshi, the first railroad to be laid in a German colony, was begun in 1893 and completed by 1911. The first German settlements were on the coast, but the Tanga–Moshi railroad opened progressively distant areas for agriculture, and it was thought that the railroad to the Great Lakes would capture trade in that region. This proved to be a miscalculation, the only sizable volume of such trade coming during a brief period of early exploitation in Katanga in the Belgian Congo.

In addition to their work on the transportation system, the Germans, as in Cameroon, did much research in the field of agriculture. Unlike their practice in Cameroon, however, the Germans made virtually no effort to stimulate African interest in cash cropping. The Germans established European-owned plantations and introduced, among other crops, sisal (which became the country's leading export product for a time), coffee, cotton, and tea. The plantations were situated on alienated land near the railroads in the hinterlands of Tanga and Dar es Salaam, but the period of expansion under German rule was too brief for the volume of exports to attain significance. The real contribution of the Germans was the successful introduction of several crops that in the future were to become mainstays of development.

World War I terminated German rule, and Tanganyika became British under the mandate system of the League of Nations. Britain encouraged African participation in cash cropping despite objections from local European farmers. The Chagga people of the slopes of Mount Kilimanjaro responded to Britain's inducements, planting coffee as a cash crop where formerly only subsistence crops had been grown (Figure 17-3). The Kilimanjaro Native Planters Association was formed, and crops were sold communally. As production attained significance, European planters, from whose estates the coffee trees had spread to the Chagga, expressed concern over the competition that would result and the dangers of inferior-quality crops and inadequate protection from disease. Eventually, the region was to become an outstanding example of cooperation among African and European farmers, occasional friction notwithstanding. By the outbreak of World War II, the new economic system was firmly entrenched.

The innovations were successful for at least three reasons. First, against much local resistance, administrative control had been established through the expropriation of land, imposition of taxes, maintenance of low wages, construction of roads and railways from strategic control centers on the coast, and the strict enforcement of British colonial values. Second, market-oriented production under the direct control of a colonial elite had been established through the introduction of plantations, essential processing industries, and monopolistic trading companies. Third, traditional economies were being transformed by the more capital-intensive, export-oriented, colonial economy. Thus African labor

FIGURE 17-3 Coffee beans on the bush. *Christine Drake.*

had been drawn into the money economies of the mines, plantations, towns, and cities, and traditional economies were being transformed.

Dar es Salaam, the capital, had developed into the main center of collection, distribution, and opportunity during the time of the British. In this developing primate city were centered sources of British capital, as well as administrative services and large numbers of people from other parts of the territory. Its railways to Kigoma, Mwanza, and Arusha carried cotton, coffee, tea, pyrethrum, tobacco, and other crops grown on European estates and African lands for foreign markets. In the immediate vicinity of Dar es Salaam itself, extensive sisal and coconut plantations had been established, and smaller acreages of cashew nuts, soy beans, and rice were in production (Figure 17-4). Further north, Tanga had developed into an important port serving the richly productive slopes of Mounts Kilimanjaro and Meru and the immediate hinterland, which accounted for much of the colony's sisal output. Zanzibar, a protectorate separate from the mainland and not united with it politically until 1964, was already the world's largest producer of cloves and an exporter of spices.

Around the shores of Lake Victoria, commercial agriculture, especially cotton production, was largely in African hands (Figure 17-4 and Table 17-1). But considerable acreages of once-productive land were destroyed through poor farming methods, lack of adequate equipment, capital-limited economic incentives, poor managerial ability, and overgrazing, a practice common to so many areas of Africa. Cash cropping spread southward as more and more Africans participated in the colonial economy, assisted by the extension of the network of roads to, for example, the towns of Mwanza, Musoma, and Shinyanga, which were later to become significant market centers. In the extreme southeast, Mtwara became the port for the disastrous

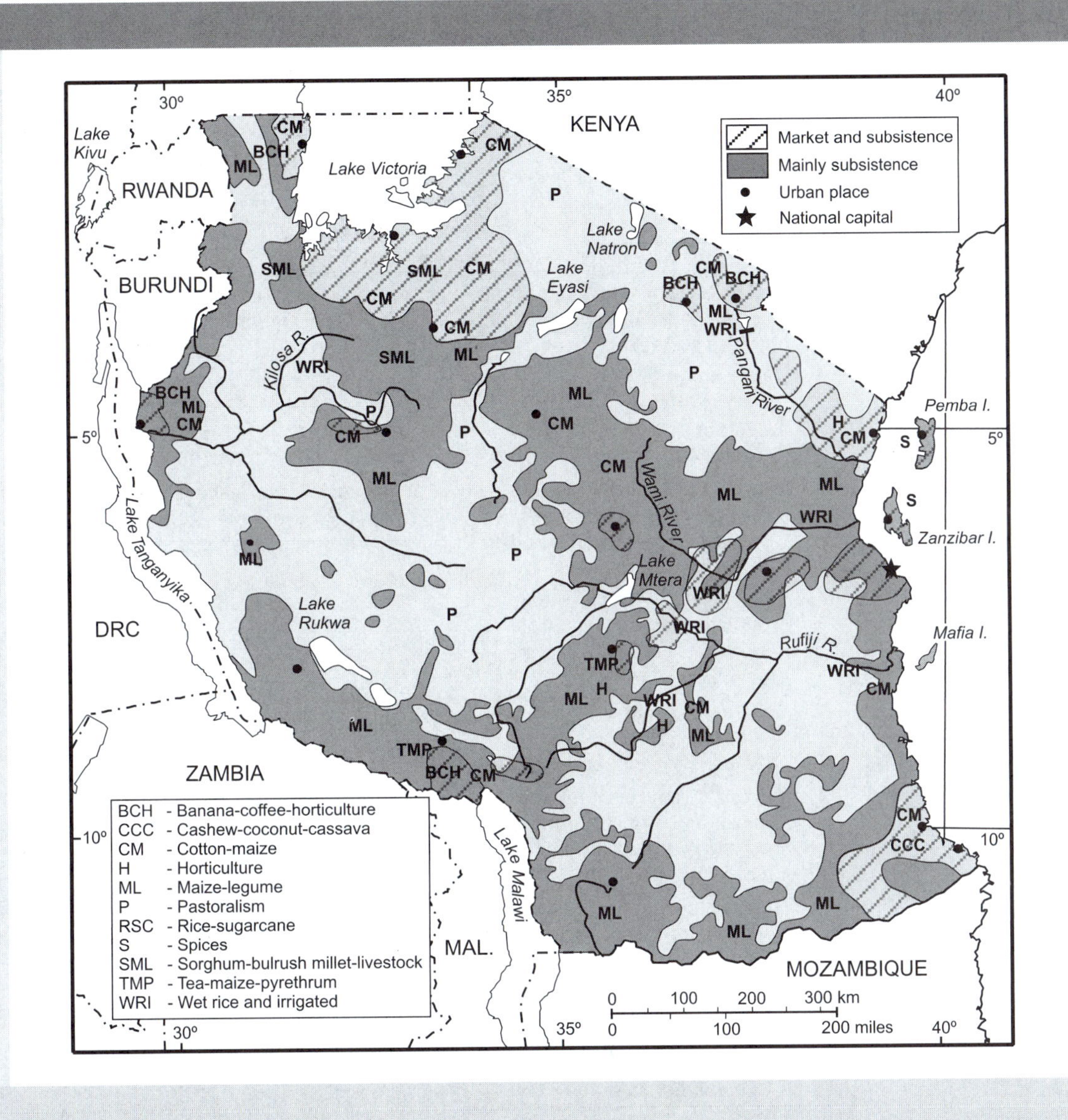

FIGURE 17-4 Agriculture regions in Tanzania.

groundnut scheme (Box 17-3) in the late 1940s, and for a while part of a secondary core region that functioned independently of Dar es Salaam. Elsewhere, economic activity focused on mining mica near Morogoro, diamonds at Shinyanga, lead and gold at Mpanda, tin at Karagwe, and salt at Nyanza.

Agriculture is the foundation of the Tanzanian economy and supports 80% of the population. It is overwhelmingly a smallholder system in which farmers work, on average, from about 2 to 7 acres (1–3 ha). Eighty-five percent of the farmed area in Tanzania is used to produce food crops. Seventy-five percent of the farmed area is cultivated with the hoe, 20% by the ox-drawn plow, and 10% by tractors (Tanzania 2004). A complex agricultural geography emerged during the colonial period in Tanganyika and Zanzibar. It was focused on a variety of food and cash crops, built on a long traditions of African land-use management, and

TABLE 17-1 FARMING SYSTEMS IN TANZANIA, THEIR CHARACTERISTICS, AND PRINCIPAL LOCATIONS.

Farming System	Characteristics	Location
Banana-coffee-horticulture	Tree crop Highly intensive land use Volcanic soils with high fertility Suitable land scarce	Kagera, Kilimanjaro, Arusha, Kigoma, and Mbeya regions
Maize-legume	Land not scarce Shifting cultivation Maize, legumes, beans, and groundnuts intercropped Arabica coffee	Rukwa, Ruvuma, Arusha, Kagera, Shinyanga, Iringa, Mbeya, Kigoma, Tabora, Tanga, Morogoro, Kahama, Biharamulo
Cashew-coconut-cassava	Low rainfall Low soil fertility Land not scarce Shifting cultivation	Coast region; eastern Lindi and Mtwara
Rice-sugarcane	Larger scale modern rice operations; sugarcane is grown on plantations and by many out growers	Alluvial river valleys sugarcane is grown on 4 large plantations at Kilombero, Mtibwa, Moshie, and Kagera
Sorghum-bulrush millet-livestock	Sorghum, millet, maize and cotton, oilseeds and rice Intense population pressure Declining soil fertility	Sukumaland; Shinyanga and rural Mwanza
Tea-maize-pyrethrum	Tea, maize, Irish potatoes, beans, wheat, pyrethrum, wattle trees, and sunflower	Njombe and Mufindi districts in Iringa region
Cotton-maize	Cotton, sweet potatoes, maize, sorghum, and groundnuts Intensive cultivation Livestock kept	Mwanza, Shinyanga Kagera, Mara, Singida, Tabora and Kigoma, Morogoro, coast, Mbeya, Tanga, Kilimanjaro, and Arusha
Horticulture based	Vegetables (cabbages, tomatoes, sweet pepper, cauliflower, lettuce, and indigenous vegetables) and fruits, (pears, apples, plums, passion fruits, and avocado) Maize, coffee, Irish potatoes, tea and beans	Lushoto district; Tanga region, rural Morogoro; Morogoro region and rural Iringa
Wet rice and irrigated	Mainly small-scale and rainfed	River valleys and alluvial plains: Kilombero, Wami valleys, Kilosa, Lower Kilimanjaro, Ulanga, Kyela, Usangu, and Rufiji
Pastoralists, agropastoralists	Deep attachment to livestock and simple cropping system Shifting cultivation of sorghum millet Moderate population density (30 people/km^2) Limited resource base; poor and variable rainfall	Semiarid areas: Dodoma, Singida, parts of Mara and Arusha; Chunya district, Mbeya, and Igunga district in Tabora

Source: Tanzania (2004).

BOX 17-3

THE EAST AFRICAN GROUNDNUT SCHEME

The ill-fated East African groundnut scheme illustrates the problems of attempting large-scale, top-down commercial agriculture in a tropical environment without adequate pilot testing and without proper understanding of the region's human and environmental potentials and limitations. Following World War II, Britain's margarine, food, and industrial oils were in short supply, so Britain decided to invest heavily in a vast groundnut-producing scheme in East Africa, mainly in southern Tanzania. The plan called for the eventual clearing of about 2.4 million acres (972,000 ha) in Tanzania alone, and another 810,000 acres (328,000 ha) in Kenya and Zambia. Estimated at $70 million, the expenditure was unlike anything ever before experienced in East Africa, and the expectations were unrealistic: the south would be transformed into a thriving agricultural-industrial region within a matter of years, and thousands of Africans would be brought into the modern sector of the economy.

After only superficial surveys of the soil, moisture conditions, traditional farm practices, and labor conditions, British engineers cleared thousands of acres of dense bush with secondhand bulldozing machinery, including tanks made obsolete during World War II. The port of Mtwara was improved, and a railway was started toward Nachingwea. But the soils of the region soon became leached of their nutrients, packed by the heavy equipment, and baked by the tropical sun. There was widespread labor unrest, serious breakdowns of machinery, and general mismanagement of operations. The first crop was minimal compared with expectations, and as early as 1949 it had become clear that the region was not suited to the type of agriculture being imposed on it. The funds supporting the scheme were nearly exhausted, and the entire project had to be revised several times before being officially abandoned in 1951. Today, small quantities of cashew nuts, sisal, grains, and even some groundnuts are produced on these lands, but southern Tanzania is not the rich agricultural region it was intended to be.

has continued to change to this day (Börjeson 2004). There are islands of intensive agriculture, vast areas of extensive use of the land, and plantations. "Systems" rather than "simplicity" seems be a more appropriate description of the diverse enterprise of Tanzanian farming. The government of Tanzania recognizes 10 farming systems in Tanzania (Table 17-1 and Figure 17-4).

SOCIALIST PRINCIPLES AND PRIORITIES DURING PRESIDENT NYERERE'S TENURE

During the colonial era, socialists in Africa and in the metropole argued that the capitalist economy was geared toward the exploitation of the territory's people and resources for the profit of British investors rather than local residents. According to this critique, very little expenditure of time, effort, and money was given to improving the welfare of the African; and little understanding of, and respect for, traditional ways of life was shown. In reaction to these perceived economic, social, and political realities, the Tanzanian leadership and various political organizations, especially the Tanzania African National Union (TANU), formulated a program of development based on their perception of traditional values and socialist philosophy. The principal architect of this special brand of African socialism was Tanzania's first president, Julius Nyerere, who died in 1999.

In the years immediately following independence (1961), Nyerere issued several statements on the strengths of traditional African values, the spirit of cooperation built around the extended family as the basis of socialism, the need for greater self-reliance and egalitarianism, and the urgent need for rural reform. The most important of these ideas were stated in a

BOX 17-4

ZANZIBAR

Zanzibar, united with Tanganyika since April 1964 in the United Republic of Tanzania, comprises two islands—Zanzibar and Pemba. The islands are inhabited by a great variety of people of whom the Shirazis form the majority. The Arab, Asian, and Somali minorities have traditionally dominated commerce and trade. Arabic is spoken on the islands, but the vast majority of people speak KiSwahili, related to the languages of the Mijikenda, Pokomo, and Comorians but unique in the high number of borrowed words from Arabic, Portuguese, and Hindi. For several centuries, Zanzibar, called Ungudya by the local Swahilis, was an Arab stronghold whose domain extended onto the African mainland. Until it became a British protectorate in 1890, Zanzibar had been a center of a large and prosperous slave trade, and an entrepôt for much of East Africa. However, as the mainland economics developed under German and British rule, and the ports of Dar es Salaam, Mombasa, and Tanga prospered, Zanzibar's commercial functions and hinterland decreased. Although only 25 miles (40 km) from the mainland, Zanzibar's economic links with the mainland are weak.

Zanzibar is the third largest producer of cloves in the world. The spice-bearing trees were introduced to the island in 1818, and by 1859 Zanzibar was the world's largest producer. Today the island accounts for about 10% of the world's output, down from almost 50% of the world's cloves in the early 1960s. Pemba has most of the clove trees, but Zanzibar Town is the chief processing, marketing, and distribution center. Two-thirds of the cloves and their extracts are exported to Southeast Asia (especially Indonesia); the rest goes mainly to Europe and America. The industry has long been threatened by disease, tropical storms, fluctuating world prices, and, in the last few years, by synthetic substitutes for oil extracts.

Coconuts, coconut oil, cinnamon, cumin, ginger, cardamom, pepper, copra, ropes, and matting are also exported, but further economic diversification is required. The islands could produce a great variety of tropical crops for both local and mainland markets.

With its mixture of peoples and cultures, fascinating history, open-air markets, and its own special urban landscape, Zanzibar Town is an unforgettable tourist destination. Tourism has replaced cloves as Zanzibar's major foreign exchange earner, and the tourist dollar has largely transformed the islands. There are over 100 hotels and guest houses on the island, and impacts on the local ecology are bound to be felt. Water is scarce on the island and problems are already visible: even now, some hotels pipe their water from the mainland.

document known as the Arusha Declaration (1967), later to be amplified in a series of papers authored by the president. The declaration was a TANU party statement of intent to build socialism in Tanzania with a view to promoting self-reliance, rural development, and good leadership.

In the words of Nyerere, Tanzania was committed "to build a society in which members have equal rights and equal opportunities; in which all can live at peace with their neighbours without suffering or imposing injustice, being exploited, or exploiting; and in which all have a gradually increasing basic level of material welfare before any individual lives in luxury" (Nyerere 1968). To create this kind of society, Nyerere advocated building upon the basic principles of *ujamaa*, a Swahili word meaning "familyhood," or "togetherness." These fundamentals—respect of others, sharing of basic goods, and the obligation to work—had characterized the way of life in Tanzania long before the colonial intrusion. Nyerere incorporated into his concept of ujamaa the instruments necessary to defeat the poverty and hunger that existed in traditional African society. In other words, the shortcomings of the traditional systems had to be corrected, and new inputs from the more technologically developed societies had to be added.

In the Tanzania of ujamaa socialism, land and labor, not money, were considered Tanzania's major assets; foreign involvement ("domination") in industry, technical assistance, and employment were replaced by Tanzanian government–owned concerns (parastatals); commercial banks and other financial institutions were nationalized; mass education supplemented elite education; private investment in the countryside by small urban entrepreneurs was deliberately checked; government spending was meant to encourage rural development rather than urban industrialization; and rural development was planned to occur in ujamaa villages, organized along socialist lines, not through large estates and plantations, whether owned by foreigners or by indigenous Tanzanians.

UJAMAA VILLAGIZATION AND RURAL DEVELOPMENT

Once the Tanzanian government had stated its intention to reduce its dependence on external sources of development (foreign capital, labor, technology), self-reliance was an automatic corollary; and since Tanzanian society was basically rural, government theorists decided that self-reliance must be rurally based. Hence the Arusha Declaration's emphasis upon rural reform. But collective living, village resettlement, and rural reform predated the Arusha Declaration by several years. In Nyerere's first Five-Year Development Plan (1964–1969), two approaches to rural development were outlined, and one was adopted. Development could be achieved through the gradual "improvement" of agricultural methods using existing extension services and without resettlement, or by the radical "transformation" of production methods through the concentration of investment and trained manpower in a few selected areas. The latter method was adopted.

The plan called for the establishment of 74 village settlements, each composed of 250 families, who were to be removed from high-density areas such as Kilimanjaro, the Pare and Usambara mountains, and the southern highlands, and resettled in less populous areas with presumed agricultural potential. Each settlement was to grow one major cash crop using new methods, fertilizers, and a high degree of mechanization. It was hoped these settlements would stimulate neighboring farmers to improve their methods of production, and industrialization would follow. Although 22 pilot schemes had been established by the end of 1965, it soon became clear that failure was inevitable. Most schemes were overcapitalized, mechanization proved to be uneconomic, planning was inadequate, and there was too much central government participation in the operations and far too little enthusiasm and involvement on the part of the farmers (Berry 1972). Recognizing these shortcomings, the government modified its approach and gradually phased out the program.

Despite the lack of success in planned settlement, the principal method to socialize agriculture and increase domestic farm production came through the growth of ujamaa villages, the rationale of which was outlined in Nyerere's *Socialism and Rural Development* (1967). Communally owned land was formed by people who lived and worked as a community. Removed from their farmsteads and resettled in a nucleated village, these Tanzanians farmed together, marketed together, and undertook the provisions of local services and small local requirements, such as digging boreholes and storing water. Farm equipment, livestock, storage sheds, and other units of production belonged to the village, not to any individual, and the intention was to create an economy of scale in which such things as farm machinery, ordinarily not affordable nor practicable on the small East African farms, could be purchased and shared by the members of the cooperative. That this intent was not realized is illustrated in Figure 17-5, which shows a steep dip in the number of tractors in Tanzania beginning in 1970.

The unintended consequences of forced collectivization included the suppression of the strongly individualistic and competitive attitudes that had been fostered under colonialism and were present in most rural areas. Decisions were now reached collectively, and each village operated as an economically independent unit (Omari 1974). In theory, each village was to develop according to the social customs and experiences of its people, and according to local environmental and economic conditions. Thus a certain degree of local autonomy was supposed to prevail, which was considered essential for the program's success. The reality was far different than the state planners had anticipated. People were forced to move in a hurried fashion, and their former homes were bulldozed and burnt to prevent the displaced families from returning.

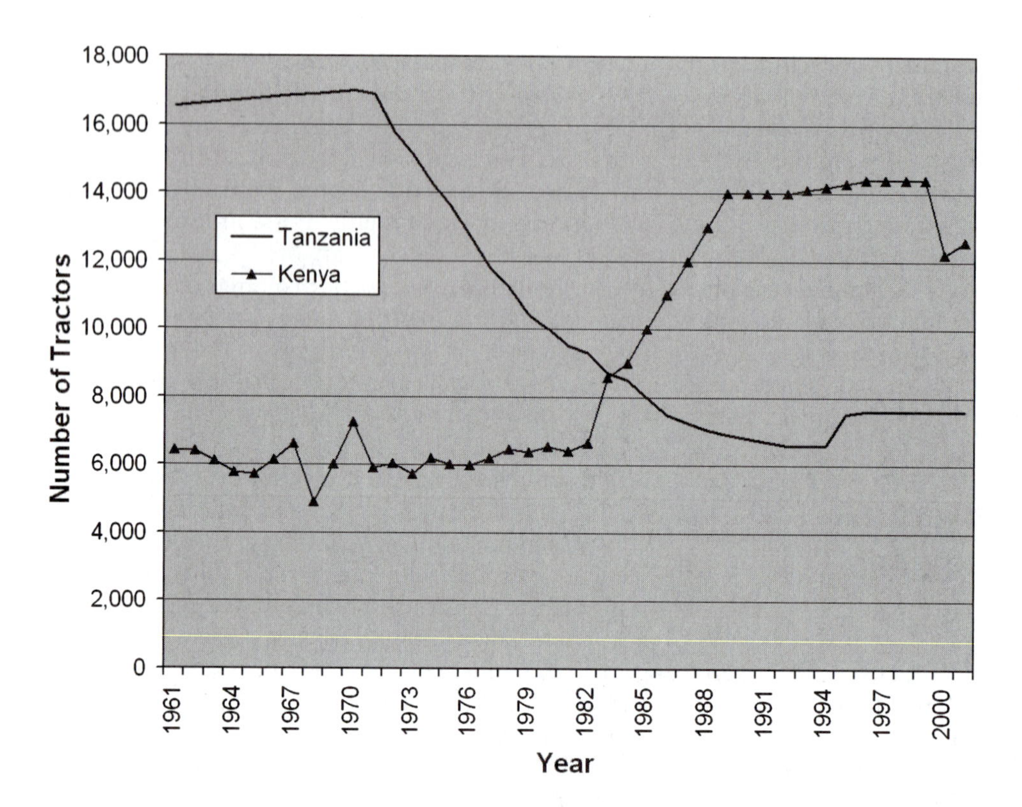

FIGURE 17-5 Mechanization in Tanzania and Kenya, 1961 to 2001.

By 1976 some 3 million Tanzanians (18% of the total) lived in over 7,000 ujamaa villages. Voluntary resettlement was slow, except in the poorer, more arid regions, where wells, power, seed, and livestock, provided by the central government, were major incentives. Villagization was intended to be voluntary, for, in the words of Nyerere, "Viable socialist communities can only be established with willing members; the tasks of leadership and of government is not to try and force this kind of development, but to explain, encourage and participate" (Nyerere 1967). Since 1974, and perhaps contrary to Nyerere's personal intentions, resettlement became compulsory in several densely populated areas in Mara, Mwanza, Kigoma, and Shinyanga, and in the Ufipa, Iranga, and coastal regions. There the population was resettled in "development villages," whose economic base was not necessarily farming. Little thought seems to have been given to new means of livelihood, and the system was reminiscent of the African resettlement schemes in apartheid South Africa. The villages were uniformly planned, the houses laid out in rows along roads and around schools and administrative offices; frequently, however, the people were without adequate water, sewerage, or power facilities.

The cooperative spirit was absent in these new schemes. Although such forced regrouping into villages lowered the government's costs of providing schools, clinics, and other social services, the program provoked resistance, ultimately proving counterproductive to the socialist goals of the government and generating antigovernment sentiment and action. Widespread food shortages and near-famine conditions in many rural areas in the middle 1970s have been attributed to rural dissatisfactions. Greatest resistance to ujamaa villagization was from the Chagga coffee farmers, who live on the slopes of Mount Kilimanjaro. They were the most prosperous and well-educated group in Tanzania, and rather than being wrested from their traditional homes and farms and placed into communal villages by outsiders during the socialist experiment, they threatened to destroy their lucrative estates. Despite the lack of popular enthusiasm, ultimately over half the Tanzanian population was resettled in ujamaa villages (Reed and Kulindwa 2001).

Agriculture has changed much over the last 25 years in Tanzania. Some crops have been almost completely abandoned, while new crops have emerged. Sisal exports declined dramatically beginning in 1970 and today are 94% below what they were in the early 1960s. The export of pyrethrum flowers declined 62% over the same period. Other export crops, for example sesame, declined around the same time, but production has picked up over the last 15 years. Although raw cotton exports have declined, the export of processed cotton has increased, as have exports of refined sugar, shelled peanuts, cocoa beans, peas, palm oil, and rice. The latter four products are relatively new: their export began during the middle 1990s after the lifting of government price controls on most agricultural commodities.

INDUSTRIAL POTENTIALS AND PROBLEMS

Tanzania's industrial potentials and problems must be seen in the context of the country's colonial past and the socialist philosophy that came to replace it until it too was discarded. According to socialist thinking, the combination of colonialism and capitalism does not develop manufacturing for internal markets, nor does it contribute to the creation of towns and cities other than the capital, a few ports, and essential administrative centers. According to socialist doctrine, these towns were first and foremost geared to meet the needs of the metropole, not those of Tanzania. Most resources produced were for export and shipped in their raw state, not converted into manufactured goods for internal markets. The transport network reflected this orientation according to TANU; it is deficient in links between regions (Figure 17-2). On the other hand, because the quantities involved were often small, it was uneconomic to process on site most of the goods that were produced in the colony and ultimately exported in a raw state. Certain products, diamonds, for example, are cut by a few highly paid specialists in just a few places in the world—yet diamonds are produced in hundreds of localities.

In addition, the internal market was so poorly developed that Tanzania was unable to absorb many of these manufactured goods. In villages where the lowest level of vending is found, vendors typically sell kerosene by the cupful and matches and cigarettes one or two at a time. Tanzania was and still is a poor country, and there is little purchasing power in the rural areas. Aside from certain light and inexpensive manufactures such as matches, textiles, footwear, soap, flour milling, beer brewing, and sugar refining, which were set up during the colonial period, it seems unrealistic to have expected the colonial governments to somehow "fully develop" Tanzania into a consumer society. Nonetheless, the colonial governments could have done more to encourage manufacturing in the colonies. Perhaps things would have been different had there been a longer formal political association between Tanzania and Britain.

Manufacturing as a percentage of the gross domestic product has been increasing; today it accounts for 16% of GDP. Food processing, together with the beverage, textile, and cigarette industries, dominate the manufacturing sector, but there are several metal processing plants. Current priorities, as in most African states, include greater outputs of refined oil (from imported crude oil and, it is hoped, from domestic production), as well as cement, fertilizer, paper products, metal goods, and chemicals, all from industries that are capital—rather than labor—intensive. Paradoxically, during the socialist period of self-reliance, foreign participation was still important and foreign financing actually increased, although the source of funds changed. Specifically, Tanzania became more dependent on socialist sources of investment and developed closer economic ties with the People's Republic of China in the interest of acquiring a major railroad.

The British colonialist Cecil Rhodes had envisaged a Cape-to-Cairo railway that would consolidate British interests throughout the length of Africa and traverse present day Tanzania and Zambia. But for many political and economic reasons, that imperial dream was never realized. In the early 1960s, Presidents Kaunda of Zambia and Nyerere of Tanzania became interested in a railway link between their two countries that would provide landlocked Zambia with an alternative route to the sea that skirted white-controlled southern Africa: Rhodesia, South Africa, Angola, and Mozambique. Several feasibility studies were conducted, and requests for financial aid were submitted to Britain, the United States, the Soviet Union, France, West Germany, Canada, the African Development Bank, and the World Bank. The project was rejected as being uneconomic.

Tanzania and Zambia then turned to the People's Republic of China. After brief studies of the Tanzanian section of the proposed line, a tripartite agreement among Tanzania, Zambia, and China was signed in Beijing in 1967, under which the Chinese agreed to finance and build the railway from Dar es Salaam to the Zambian Copper Belt. Known as the Tan-Zam Railway—or as the Uhuru (Freedom) Railway in East Africa—and built at an estimated cost of $400 million, the line represented China's largest aid project thus far in the less developed world and the third most costly in Africa, after the Aswan and Volta dams (Graham 1974).

Construction began in April 1970, and the 1,100-mile (1,770 km) line was completed in late 1975. Up to 15,000 Chinese and 45,000 African workers were engaged on the project during the peak construction period; at the time this represented 20% of Tanzania's labor force in paid employment. The railway was financed with an interest-free loan, repayable over a period of 30 years from 1983. Approximately 70% of the cost was for the Tanzanian section, mainly because of its greater length and more difficult terrain, but Zambia shared equally in the repayment. Local costs, mainly for labor and materials, were met by a commodity credit arrangement facilitating the purchase of Chinese products, such as clothing and household utensils.

Although built primarily for political reasons, the Tan-Zam Railway was intended to assist economic development in both Tanzania and Zambia. In Tanzania, the railway traverses rice and sugar estates in the Kilombero Valley and the southern highlands, where tea, coffee, pyrethrum, wheat, maize, and cattle have great potential. For Zambia, the railway meant a more secure outlet for its copper, possible new agricultural and forestry development in the northeast, and easier access to oil and other imports. Over the years, there has been a lack of investment in the Tan-Zam Railway. The capacity of the system is underutilized because there is a scarcity of engines, and the Zambia copper industry has declined as a result of mismanagement. Even so, the government is looking for investors willing to assume the risk of rehabilitation.

In recent years India has become Tanzania's greatest export partner (15%), followed by Japan (12%) and several north European countries and Kenya. South Africa is its major import partner followed by China, Kenya, and India. With the further development of India, China, and Southeast Asia, there should be some knock-on effects for Tanzania in today's Indian Ocean trade.

Industrial activity has been concentrated in Dar es Salaam (2.5 million) for reasons common to most African capitals—economies of scale, abundant and willing labor supply, and links to the rest of the world. The main industries of the capital include oil refining, food processing, and cigarette manufacturing, as well as production of goods for local and national consumption. Most importantly, Dar es Salaam handles Zambian copper exports.

Industrial and population growth quickened in Dar es Salaam in the early 1970s as a result of real and perceived opportunities stemming from the construction of the Tan-Zam Railway. Indeed, immigration increased to alarming proportions before the state intervened to control population movements by adopting "influx-control measures" reminiscent of the South African apartheid state.

To control the growth of Dar still further, and to lessen the regional inequalities of industrial output, the socialist government of Nyerere instituted a policy of industrial decentralization. For the period 1970–1975, places other than the capital were scheduled to receive some 81% of planned industrial investment and 76% of planned employment. Industrial decentralization was a logical and necessary part of the dogma of socialist planning, and it is no surprise that it appeared as a central part of Tanzania's program of self-reliance. In socialist thought, "class" distinctions such as the private ownership of property and inequality in the possession of wealth and assets are to be eradicated, along with distinctions between town and country. For socialist planning, economic geography was irrelevant, and the prevailing doctrine was that the purchase price of a good should be the same ("equal") at any location in the national space, regardless of where the item was produced and the costs of transportation.

Tanzania also pursued an industrial policy focused on substituting local manufactures for imports (Cliffe and Saul 1973). Consonant with socialist dogma regarding product pricing, Tanzanian socialists thought that industry should be located, wherever possible, close to markets. But much of the market in Tanzania was (and continues to be) rural and inadequately served by road and rail, which presented enormous logistical problems to the country's industrial planners. The problems were not only where to encourage industry and where to concentrate investment but also what types of industry should and could be established. Industries with strong forward and backward linkages, such as petrochemicals, heavy engineering, and vehicle manufacture, while theoretically desirable in almost any country's development, are quite unrealistic in Tanzania except for one or two places, such as Dar es Salaam and Tanga. Industries more likely to appear in the smaller and more peripheral towns are those utilizing local resources; but textile plants, fruit and vegetable canneries, and other such establishments cannot provide the motive force necessary for rapid industrialization at a national level. Furthermore, industrialization could have been integrated (at both the regional and national level) with the villagization program in some way, and perhaps such linkages would have led to success—although the difficulties in planning an economy from top to bottom would have become very apparent had such linkages been tried.

Contrary to socialist theory, geography does matter, and the costs of overcoming geographic space cannot be theorized away. In addition, a glance at any economic geography text used in any market economy at the time would have shown the Tanzanian planners that some goods are best produced close to the market, for example, those that gain weight through processing: bottled drinks, prefabricated construction elements, automobiles. On the other hand, products that lose weight during processing are often best located near the resource that is processed. Thus smelters, pelletizing plants, lumber mills, and the like are all situated near the resource because the processed product costs less to transport than the heavier raw material.

In 1986 Tanzania began to abandon socialism and to liberalize the economy, inviting investment from the capitalist world, welcoming joint ventures, and embracing export-led growth. Since that time, and especially since the middle 1990s, the Tanzanian economy has been diversifying and growing. Tanga (221,000), located on the bay of the same name, is Tanzania's second largest port and an important outlet for the agriculturally prosperous Mount Kilimanjaro region, including Arusha (300,000). Among its industries are textiles, food processing, and rope making, but these and others have suffered from the strong competition of Mombasa located across the border in Kenya, with which the Kilimanjaro region is linked by rail.

Dodoma (168,000), Tanzania's legislative capital, lies about 300 miles (480 km) west of Dar es Salaam (Figure 17-2). Situated in a sparsely populated agricultural region, Dodoma is connected by road and rail with several other designated growth points together with Dar es Salaam. Politically, the creation of Dodoma was meant to unite the country and shift the focus from the Swahili-dominated coast to more neutral ground. The Nyerere government hoped it would provide both industrial and ideological stimulus to what an underdeveloped region, but neither goal was accomplished; the Tan-Zam Railway does provide some employment, though. There is little industry in Dodoma, although there is economic activity when the legislature is in session. Most government bureaus remain in Dar.

TOURISM AND THE FUTURE

Among Tanzania's greatest assets are magnificent game reserves, national parks, and scenic attractions (Figure 17-6); island ecotourism is taking hold as well. From Mount Kilimanjaro in the east to Lake Victoria in the west lies a zone of incomparable scenery and wildlife. The Serengeti and Selous wildlife sanctuaries, Mounts Kilimanjaro and Meru, the Ngorongoro Crater, Lake Manyara, and the coastal regions of the lakes and Indian Ocean, all rank among Africa's finest tourist attractions. Although neglected as "antisocialist" for 20 years, the tourist trade is growing (there were over 223,000 tourist arrivals in 1993 and half a million in 2001), but Tanzania is not favorably located with reference to the main lines of communication. Nairobi has a far more frequented airport than Dar es Salaam, and internal communications, as well as hotels and other facilities, are better elsewhere in East Africa. Tourist arrivals in Kenya have remained stable at between 825,000 to 925,000 persons from 1993 to the present (World Bank 2003). Ironically, many tourist trips to Tanzania's game reserves begin and end in Nairobi, hence Tanzania has not reaped the maximum possible returns from these assets. With the encouragement of private initiative, many individuals have organized facilities and activities for tourists and attractive Web sites are online to lure the tourist dollar.

FIGURE 17-6 Zebras graze in the Ngorongoro Crater. *Christine Drake.*

Tanzania's postindependence programs to industrialize, to diversify the economy, and to become more self-reliant were abandoned in the wake of continued economic decline. Strong economic and ideological ties with the China and a commitment to the eradication of poverty and ignorance distinguished Tanzania from its neighbor to the north, Kenya, and from most other African states, but in the end did not work. Although the "African socialism" of Tanzania was not as strident and as militant as the more doctrinaire Marxism–Leninism of Mozambique or Angola, it nevertheless discouraged foreign investment, did great harm to people and the economy, and fostered a legacy of distrust for government in the general population. A comparison with its neighbor, Kenya, beset by a variety of economic and social problems itself over the same time period, illustrates this point. Kenya never adopted a socialist approach to its development but instead encouraged investment and foreign aid as a way to modernize. Today Kenya is considered by the United Nations to possess a medium level of human development, while Tanzania is given a low rating (UNDP 2002). Both countries have about the same amount of income or consumption inequality: the share of income or consumption controlled by the richest 20% of Tanzanians is just slightly less than that of Kenya—surprising for a country that was focused for so many years on erasing inequality between "classes" and between town and country. For the 32 African countries for which there are data, the mean is 51.8% (standard deviation 7.66) and the median 49.7%. Both Kenya and Tanzania are below the mean—slightly less inequality than average for the group—but are not significantly different from each other (Table 17-2).

Although Nyerere's experiment in socialism provided the country with a measure of political stability (Swantz and Tripp 1996) that has been rare in Africa, economic stability was fleeting and ironically dependent on Western, market-oriented donor nations for financial aid and support. Even today Tanzania receives over $1 billion in aid each year. Nevertheless, a different picture emerges after a glance at the literacy rates for African countries at the low end of human development. The literacy rate for Tanzania (Table 17-2) is much higher than for other countries in the low-development class. This, then, is the bright side of Nyerere's legacy. On the other hand, the long-time president's experiment with socialist planning represents a false start in Africa, something from which the country is still recovering. In the socialist struggle against capitalism, "workers" were pitted against "capitalists" (or bourgeoisie), "progressives" against "reactionaries," and all identities except a Marxist "class" identity became meaningless. One of the objectives of socialism was to develop "class consciousness" on the part of the so-called working class so that the workers could recognize and subsequently fight the common enemy—capitalism.

Although Nyerere's socialist ideology in practice was anathema to sustained growth and development, one benefit may have been the fostering of a national, rather than ethnic, identity. Julius Nyerere did two

TABLE 17-2 DEVELOPMENT INDICATORS FOR KENYA AND TANZANIA AS OF 2001.

	Country	
Measure of Development (2001)	**Kenya**	**Tanzania**
Crude birthrate (births per 1000 pop.)	35	40
Crude deathrate (deaths per 1000 pop.)	20.1	21.2
Literacy rate (%)	83.3	76
Infant mortality rate (deaths per 1000 pop.)	78	104
Life expectancy (years)	46.2	43.7
Telephone lines per thousand population	10	4
Agriculture (% of GDP)	19	45
Industry (% of GDP)	18	16
Share of income or consumption of the richest 20% (%)	51	45
Trade (% of GDP)	61	40
Official development assistance	$0.5 billion	$1.2 billion
Per capita GDP (US$ PPP)	970	520

Sources: World Bank (2003), UNDP (2002).

very important things after he came to power: first, he removed tribal membership from the list of questions on the national census. References to other than Tanzanian identity became "incorrect" thereafter. Although such change surely was superficial outside the elite circles, other steps were taken to build national identity. Second, and more importantly, he had Swahili designated as the national language: a clear effort to build a national identity. There is some irony in the choice of Swahili as a national language, given the manner in which the Swahili language diffused, especially in the nineteenth century.

In the end, the socialist experiment in Tanzania failed: it is difficult for government, especially poor government, to manage an economy in an efficient way. The heavy-handed behavior of the Nyerere administration in promoting socialist development increased suspicion and fear of the government among the people and worked to decrease the government's legitimacy. Unlike the former Soviet Union, Tanzania had neither the military force, nor the wealth in natural resources, nor even the human capital to force Nyerere's model of development to fruition. Ironically, the government programs and policy ultimately undermined the very concept of national self-reliance that Nyerere was trying to promote: it undermined the economy and forced the government to become dependent on foreign aid; it alienated the population and made people less inclined to cooperate; and it removed all elements of democracy. After taking power, Nyerere created a one-party state and vested all political and economic power in himself and his political party. There were no venues for discussion and dialog—dissent was not tolerated—and thus there was no way to stop or change the implementation of misguided policies until Nyerere left office.

BIBLIOGRAPHY

Allen, J. de V. 1993. *Swahili Origins*. London: James Curry.

Beidelman, T. O. 1962. A history of Ukaguru, Kilosa District: 1857–1916. *Tanganyika Notes and Records*, 58 and 59: 11–39.

Beidelman, T. O. 1971. *The Kaguru: A Matrilinial People of East Africa*. New York: Holt, Rinehart, & Winston.

Berry, L. ed. 1972. *Tanzania in Maps*. New York: Africana Publishing.

Börjeson, L. 2004. The history of Iraqw intensive agriculture, Tanzania. In *Islands of Intensive Agriculture in Eastern Africa*, M. Widren, and J. E. G. Sutton, eds. Oxford: James Curry, pp. 68–104.

Chizyuka, G. 1998. FAO liaison officer's summary report on trypanosomiasis. Rome: Food and Agriculture Organization of the United Nations.

Cliffe, L., and J. Saul. 1973. *Socialism in Tanzania*. Dar es Salaam: East African Publishing House.

Dumont, R. 1969. *False Start in Africa*. New York: Praeger.

Glassman, J. 1995. *Feasts and Riot: Revelry, Rebellion, and Popular Consciousness on the Swahili Coast, 1856–1888*. Portsmouth, N.H.: Heinemann.

Graham, J. D. 1974. The Tan-Zam Railway. *Africa Today*, 21(3): 27–41.

Helders, S. 2004. World Gazetteer. http://www.world-gazetter.com

Kusimba, C. M. 1999. *The Rise and Fall of the Swahili States*. Walnut Creek, Calif.: Altamira Press.

Legum, C., and G. Mmari, eds. 1995. *Mwalimu: The Influence of Nyerere*. London: James Curry.

Middleton, J. 1992. *The World of the Swahili: An African Mercantile Civilization*. New Haven, Conn.: Yale University Press.

Nyerere, J. K. 1967. *Socialism and Rural Development*, Dar es Salaam: Government Printer.

Nyerere, J. K. 1974. From Uhuru to Ujamaa. *Africa Today*, 21(3): 3–8.

Nyerere, J. K. 1990. *Freedom and Socialism: Uhuru Na Ujamaa. A Selection from Writings and Speeches, 1965–1967*. Oxford: Oxford University Press.

Omari, C. K. 1974. Tanzania's emerging rural development policy. *Africa Today*, 21(3): 9–14.

Reed, D., and K. Kulindwa. 2001. Tanzania. In *Economic Change, Governance, and Natural Resource Wealth: The Political Economy of Change in Southern Africa*, D. Reed, ed. London: Earthscan Publications, pp. 41–68.

Saitoti, Tepilit Ole. 1986. *The Worlds of a Maasai warrior: An Autobiography*. Berkeley: University of California Press.

Swantz, M. L., and A. M. Tripp (1996) *What Went Right in Tanzania: People's Response to Directed Development*. Dar es Salaam: Dar es Salaam University Press.

Tanzania, government of. 2004. *Participatory Agricultural Development and Empowerment Project (PADEP): Environmental and Social Framework Report*. Dar es Salaam: Government of Tanzania. http://www.tanzania.go.tz.

UNDP. 2002. *Human Development Report 2002: Deepening Democracy in a Fragmented World*. United Nations Development Program. Oxford: Oxford University Press.

World Bank. 2003. *World Development Indicators*. New York: World Bank.

Yager, T. R. 2001. *The Mineral Industry of Tanzania*. U.S. Geological Survey Mineral Report. Reston, Va.: USGS. http://minerals.usgs.gov/minerals/pubs/country/.

CHAPTER 18

Kenya

The Problem of Land and Governance

South of Ethiopia, east of Uganda, and north of Tanzania lies Kenya, East Africa's heartland. A revolution brought Kenya independence in 1963; the leader who was thrust into national prominence during that period, Jomo Kenyatta, three times elected to the presidency, headed a government that was stable and conservative for 15 years. That stability and conservatism brought Kenya substantial progress and comparative prosperity, but it also generated criticism among other African countries more strongly committed to socialist objectives. Throughout Kenyatta's tenure in office and to the present, the country's development issues focused on landownership and land reform. Indeed, for a very long time, land has figured prominently in Kenya's domestic politics. Although the issues of the day during the rule of Kenyatta's immediate successor, former vice president Daniel arap Moi, came to focus on corruption, crony capitalism, and autocratic rule, the land issue was ever present. In the 2002 presidential election, Moi, after 24 years in office, was defeated by Mwai Kibaki, who ran on a reformist, anticorruption platform that advocated economic liberalization.

TERRITORIAL DEFINITION

A geometric boundary was drawn to separate German East Africa from a territory known until 1920 as the British East Africa Protectorate. Although the product of convenient agreement rather than a carefully considered adjustment to local conditions, as is true of so many other African boundaries, this line ran through dry, relatively empty lands; the British part extended from Lake Victoria along the northern slopes of Mount Kilimanjaro to the coast. The people to be affected were the nomadic Maasai cattle herders of the region.

One consequence of the definition of the German-British boundary in East Africa was the channeling of British control toward what was then called the Lake (Victoria) Region. What remained of the British sphere of influence in East Africa included only one port, Mombasa. Like the Germans, the British recognized that effective communications with the hinterland were a prime requirement for effective government and economic progress, and railroad construction from Mombasa toward the densely populated Lake Region began in 1895, nearly a decade earlier than the Germans' first steps in connecting Dar es Salaam with the interior by rail.

Although the Germans and the British had agreed on the location of the border between their respective claims lying between Lake Victoria and the coast, the hinterland around and beyond the lake remained in dispute. In 1888 the British East Africa Company was created, which campaigned in Britain to rouse government and public interest in spreading British influence in this part of Africa. Company officials realized that the people of the kingdom of the Baganda, along the northwestern shores of Lake Victoria, were likely to play a leading role in a developing East Africa. In 1893, largely owing to the efforts of the company, Uganda—of which the Baganda Kingdom forms a part—became a part of the British sphere of influence. For years, the claim made by Karl Peters to this region for Germany had been disputed, and the abrupt termination of the role played by Emin Pasha, the German agent Eduard Schnitzer, against the spread of British influence led to German recognition of British hegemony in Buganda.

FIGURE 18-1 Towns, transportation, and resources of Kenya.

Having thus secured the entire northern Lake Region, the British, less troubled than the Germans by rebellions, began to exercise control. Work on the railroad begun at Mombasa continued, and a number of railway workshops were built where the city of Nairobi stands today (Figure 18-1). The British East Africa Company, unlike the British South Africa Company and the Royal Niger Company, did not thrive as a result of mineral concessions and agricultural development. Burdened with administrative functions, London was reluctant to take over; and hampered by a lack of direction in policy, the company was in frequent financial difficulties. Eventually, Britain assumed the responsibilities of a protectorate over the entire territory first opened by the company, including also Zanzibar and Uganda.

THE KENYAN HIGHLANDS

East of Lake Victoria the British East African domain included much land that could at best be described as savanna (Figure 18-2). In general terms, only the southwestern quarter of the territory could be considered an asset; the administration of the remainder was a liability. The railroad that was under construction from Mombasa toward the Lake Region reached the steep ascent to the higher areas of the territory in 1900. The initial object, to connect Mombasa with Kisumu on Lake Victoria, was achieved in 1903. The railroad had been built in view of future potentialities rather than existing opportunities, and at the time it was completely uneconomic. In addition to its role in extending British control over Kenya, the railway had other anticipated benefits. Often ignored by analysts is the humanitarian element to British imperialism: the desire to eliminate slavery, disease, and poverty in the colonized territories. To reduce the drain upon the home exchequer in financing such expensive undertakings, British administrators sought means of generating income within the overseas possessions themselves.

Connecting Mombasa and Kisumu by the shortest possible route, the railroad crossed southwestern Kenya in its entirety, including the region that has come to be known as the Kenya Highlands. This is Kenya's most diversified area, and tropical Africa's most extensive highland zone, as the bulk of it lies over 5,000 feet (1,500 m). The soils are among Africa's best, derived in large part from volcanic rocks. There are extensive areas of rather flat land, and the climate is cool and sufficiently moist to permit specialized

FIGURE 18-2 An acacia on the Kenyan savanna. *Gritty.org.*

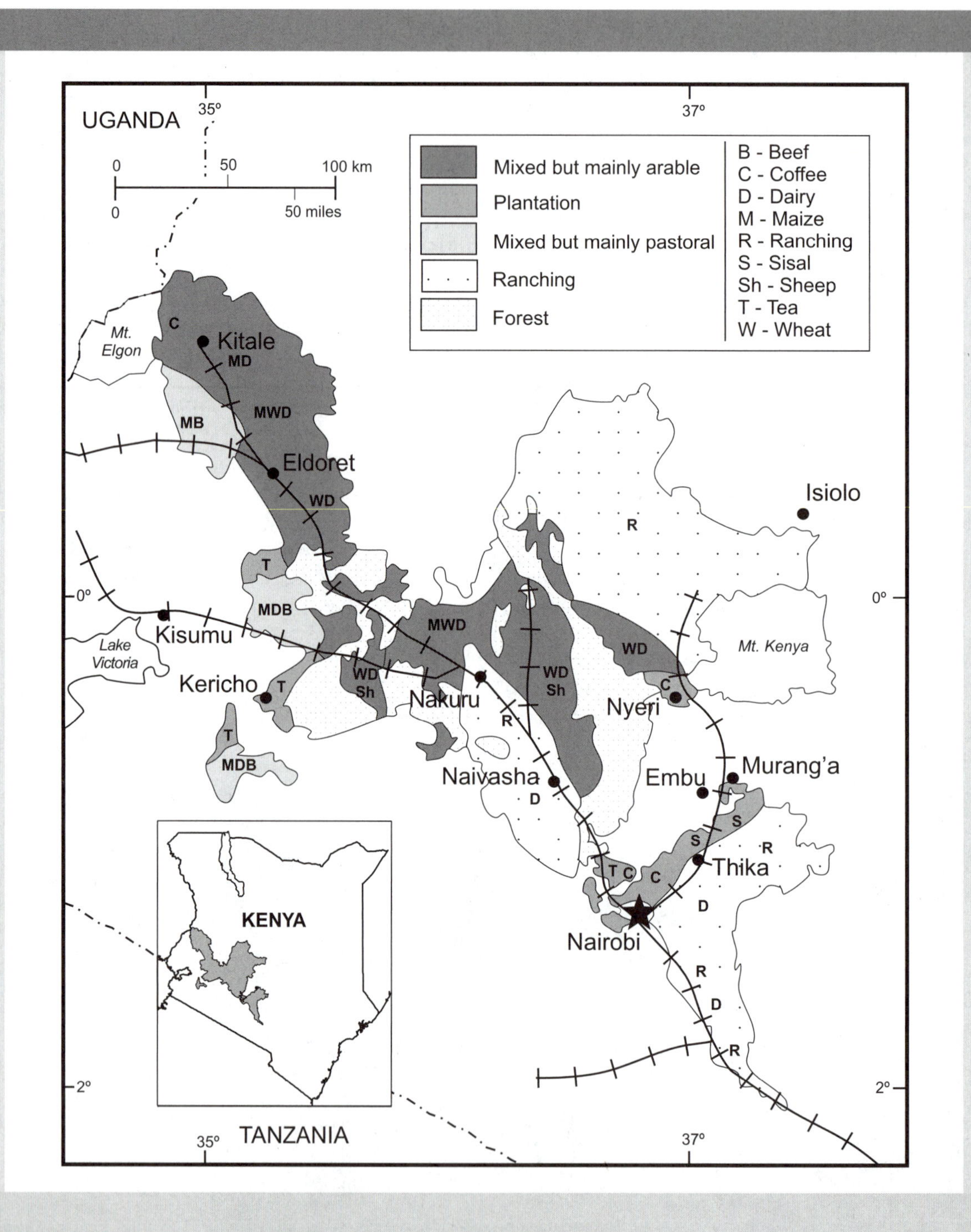

FIGURE 18-3 Land use in the Highlands Kenya.

cropping. Depending somewhat upon the contour boundary selected, the Kenya Highlands cover some 60,000 square miles (96,600 km^2). This includes several extremely high sections, such as the Aberdare Mountains and Mount Kenya, as well as the lower lands in the Eastern Rift Valley, whose floor nevertheless lies several thousand feet above sea level. Included also are steep escarpments and dense forests, but in general the land is of excellent quality for intensive agriculture and large plantations (Figure 18-3).

None of this potential was evident when the railroad was being laid across the Highlands, which are divided into two sections by the Rift Valley. There was no dense African population on the fertile lands and, of course, there were no plantations. The lands had been occupied by many people in the past, as subsequent research has proved, but they were vacant just at the time Charles Eliot, first commissioner of British East Africa, saw them. Envisaging a thriving European farming community in the Highlands, economic prosperity, civilization—and a railway that could pay for itself—Eliot advised the British government to encourage European immigration into the protectorate. The first few settlers arrived before the turn of the century, but white immigration into Kenya really began in 1902. Initially occupying the land around Nairobi, the white settlers soon spread out to other parts of the Highlands. With this, the seeds of friction had been sown, to reach fruition a half-century later.

LAND AND THE AFRICANS

Kenya's resources are concentrated in the southwestern sector of the country, and most of its people occupy this region. This was true in Kenya before the coming of the British settlers, as it has been true after their arrival. As noted earlier, however, at the exact moment in history marking the first organized settlement by Europeans, Charles Eliot saw British East Africa in a unique condition: the majority of the African people had vacated the lands they normally occupied and cultivated. Accordingly, land was granted to British settlers under the Crown Lands Ordinance, which stipulated that land not in beneficial occupation at the time was at the disposal of the Crown. Areas of sparse or haphazard cultivation might also be considered for settlement, as indeed they were.

It was under these conditions that the East African area placed under European freehold or leasehold increased rapidly, in fact, out of all proportion to the numerical growth of the white population. From the beginning, it must have been clear that the settlers could never expect to farm the huge areas placed under their control. Eventually, no less than 16,700 square miles (43,250 km^2) was thus alienated.

What had caused the African population to vacate the Highlands? The Kikuyu people, who occupied much of the land between the Aberdares, Mount Kenya, and the lands of the Maasai to the south, had cleared great tracts of the forest land for cultivation. Having displaced the indigenous hunters and gatherers toward the end of the nineteenth century, the Kikuyu are said to have compensated the original occupants with livestock for the lands relinquished. Prior to the arrival of the Europeans, the Kikuyu lived in an uneasy balance with their enemies, the raiding, cattle-herding, nomadic Maasai and the Kamba (Figure 18-4). In the south, some land may have been relatively empty because of the aggressiveness of the Maasai, but the Maasai were already declining in strength when the first Europeans arrived there.

In 1898 and 1899 a sequence of events occurred that changed the situation drastically. In 1898 a great smallpox epidemic ravaged the population, followed directly by an outbreak of rinderpest that decimated the livestock. Beginning during this period, a drought persisted for many months, ruining the crops that might have saved many people, and while it was breaking, an unprecedented invasion of locusts followed. These four disasters reduced the African population of the region, and the survivors turned northward and fled back in the direction of Murang'a (then Fort Hall) near Embu. The general effect was the temporary depopulation of the eastern sector of the Highlands—and it was just at this time that the British commissioner first saw the lands as "empty."

The catastrophe also affected the Maasai seriously and dealt them a series of blows from which they were never fully to recover. In 1904, to accommodate the new railroad, which was to run through the lands of the Maasai before reaching the Highlands proper, the beleaguered cattle herders were confined to two reserves, one to the south of the line, and the other in Laikipia, which is to the north. This move completely emptied the Rift Valley from Naivasha to Nakuru, as

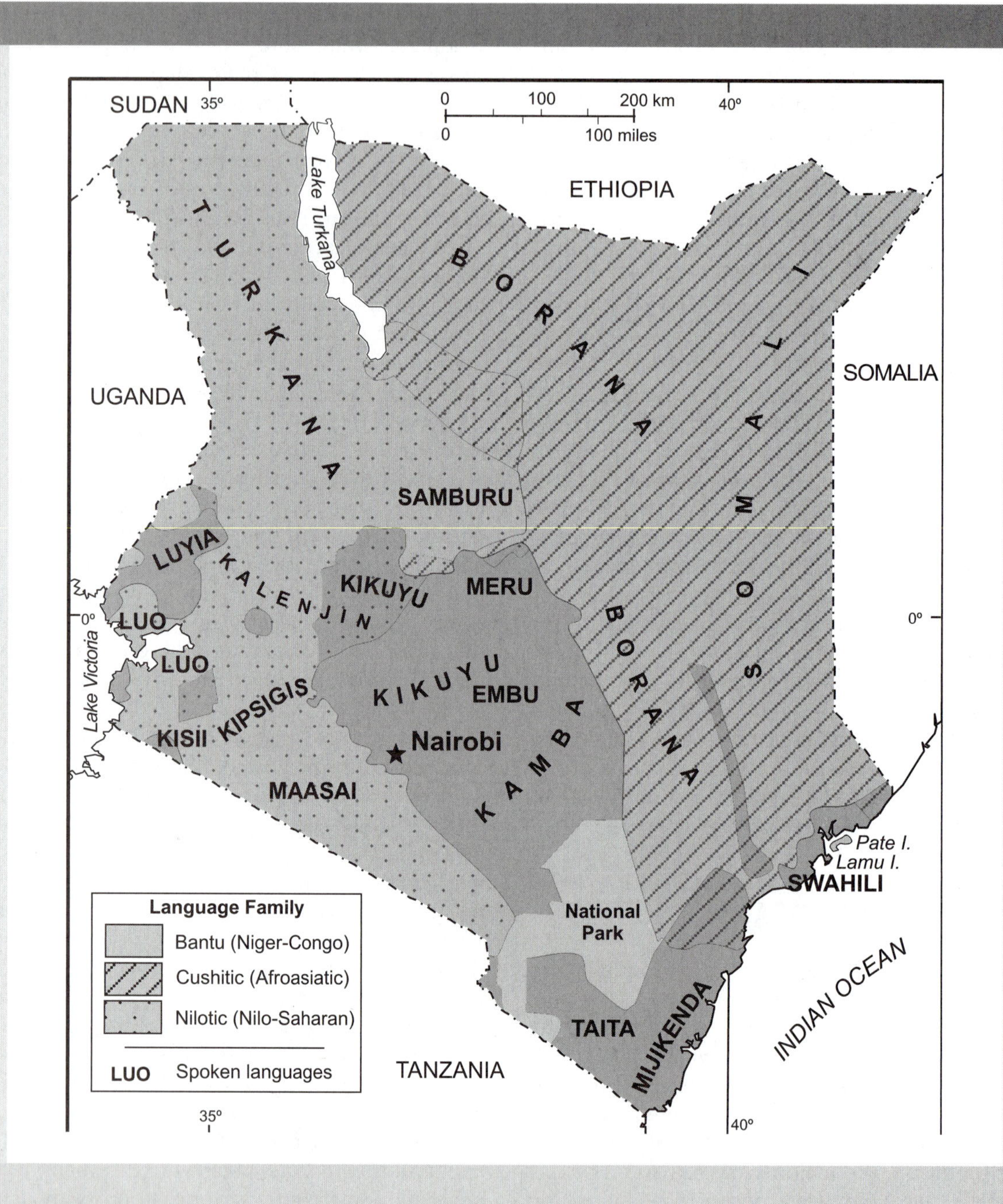

FIGURE 18-4 Ethnolinguistic groups of Kenya.

well as the lands east of Nairobi. In 1911 the northern Maasai reserve was closed, and the people moved across the railroad to the southern area, described as ample for their needs in official documents. It was never made clear just what the criteria were for the determination of the land requirements of a nomadic pastoral people dependent upon the vagaries of a variable climate, but the effects upon the Maasai became obvious.

It is true that the lands available to the various African peoples in the first decade of the present century were sufficient for their immediate needs, which had been reduced as a result of the disasters of 1898 and 1899. Of course, confinement in demarcated areas did not immediately alter the Africans' farming methods, and as time went on, the practice of patch agriculture began to reduce the capacity of the soil. Lands that were initially adequate eventually failed to provide the required yields, a result the British administration wrongly ascribed to "wasteful and harmful methods" of African agriculture. Meanwhile, Africans who had left their section of the Highlands in 1899 began to return, either to overcrowd the lands set aside for African occupation or to find that the lands they left under extraordinary circumstances had been reserved for European settlers. As the reserves became areas of severe population pressure, many Africans made their way to Nairobi and to the European settlers' farms in search of wage labor. Twelve thousand were thus engaged outside their home areas by 1912; this number had increased to 152,000 in 1927, and in 1939 to as many as 200,000. As the number of settlers and their prosperity increased, Nairobi became the thriving center of a wealthy farming region. The town grew rapidly and attracted many unskilled laborers, for whom jobs were insufficient and housing grossly inadequate. In contrast, there were fewer than 10,000 European settlers in the entire country in 1914, and under 30,000 in 1948. During the short but intense period of British rule, the European population of Kenya never exceeded 70,000.

Since the early 1920s, land has been the central issue in Kenya. After the end of World War I, when a number of Kikuyu returned from service in other parts of the world, greater political organization among the Africans began to emerge, and the oft-expressed aim was to effect a change in the government's land policies. Pressure rose, in response to which the secretary of state for colonies in 1932 appointed the Kenya Land Commission (the Carter Commission). This body was to investigate the land needs of the African population and to consider every claim made by the local people. As a result of the commission's lengthy deliberations, some hundreds of square miles changed hands, but basic policies (such as the reservation of certain areas on the basis of skin color) were not altered.

When, after World War II, African veterans received work permits and European settlers obtained land concessions, the land crisis entered its crucial stages. African grievances were rife. For example, restrictions had been placed on the cultivation by Africans of certain cash crops such as coffee and tea on the grounds of the need to maintain good quality and that African production would cause a deterioration of the level of Kenyan exports. In addition, the European planters did not especially want the competition of African farmers, and, finally, it was thought that African land was best put to use under food crops. These practices and policies, as well as the differences between wage scales for Africans and non-Africans, unemployment in Nairobi, segregation regulations, and the requirement that Africans (only) carry an employment registration certificate, created increasing tensions. The return to Kenya of Jomo Kenyatta late in 1946 from Britain was followed by much political and illegal activity, and the Mau Mau crisis broke in earnest in 1950. The "stolen lands" issue plunged Kenya into an abyss of division as the Africans initiated an unprecedented campaign of rebellion, murder, and destruction.

Underlying the entire matter was the contrast between the African and European approach to the ownership and use of Kenya's land. The Africans, and especially the Kikuyu, viewed land as the only real security, something sacred, eternally the possession of the people who once occupied it and depended upon it—but something that had always been in abundance. The temporary abandonment of such communally owned, land did not mean surrender of ownership, and when the Kikuyu returned to find the lands of their people occupied by settlers it was, indeed, from their point of view, a case of stealing. The settlers, of course, applied European concepts of landownership to the Highlands they had found nearly empty: the acreage was parceled out, purchased, fenced, and partly cultivated. Many African squatters were permitted to live on unused portions of such land, but they were compelled to work 180 days a year in exchange. Thus reduced to servitude by the fact of their presence on lands they considered theirs, the Africans within the Highlands declared to be under white ownership had a grievance, as had those in the African reserves.

The traditional system of land tenure and land use of the Kikuyu is known as the *githaka* system. *Githaka* means land, and the concept includes all land owned by *mbari*, an extended family or subclan, each member of which was entitled to cultivate part of the *githaka*. The plot, *ngundu,* could not be sold or leased without permission of the *mbari's* leaders, and on it the user was free to grow whatever he chose, usually millet, maize, sweet potatoes, and beans. On the death of the user, the *ngundu* was passed to his son. Part of each *githaka* was used for communal grazing, and the average *ngundu* was probably less than 4 acres (1.6 ha) (Taylor 1969). Men were responsible for the raising and milking of livestock, breaking of previously uncultivated land, and the care of certain crops. Women did most of the cultivating, and communal grazing of the whole *mbari* was supervised by one or two men. A strong sense of identity and belonging existed between the individual and his land.

LAND AND THE EUROPEAN SETTLER

At their maximum extent (excluding forest reserves), European land in the Kenya highlands comprised 11,571 square miles (29,970 km^2), or about 5% of Kenya's total land. Approximately 7 million acres (2.8 million ha) was divided among fewer than 4,000 farms and estates, which yielded more than 80% of Kenya's agricultural exports and much of its domestic requirements (McMaster 1975). Just before independence in 1963, these holdings provided employment for 6,900 full- and part-time European workers and for almost 300,000 Africans (42% of the total wage labor force). They also contained about 1 million landless Africans. To the European mind, land represented wealth, security, and power; it was the personal property of those who owned it. As the region developed through the infusion of capital and technology, the so-called white Highlands became the richest and most extensive European-controlled agricultural area in East Africa, and the economic heart of Kenya.

The settlers introduced such crops as coffee (almost exclusively the arabica variety), tea, pyrethrum, sisal, and a host of cereals, vegetables, and fruits. Whereas European settlement in Africa has been attracted mostly by mineral finds (Witwatersrand, the Copper Belt, Katanga), Kenya's settlers came to cultivate the soil, and in this respect the Kenya situation, from the large-scale point of view, was unique. The achievements of the settlers should not be underrated. Their number was never large (all of Kenya contained at the height of the colonial period fewer Europeans than the Copper Belt), but their impact upon the country was at least as great. In the Highlands grew an agrarian economic core yielding such rich harvests that Kenya could develop in step with other parts of colonial Africa without a single mineral figuring significantly among the exports. The European settlers brought progress in the spheres of education and health and built towns and a network of roads and railways, but they also planted the seeds of racial unrest. The landless majority was reduced to a state of servitude. Its population grew rapidly, its geographic mobility was severely restricted, and its economic opportunities were limited. Demands for more land were made, but even had these been met, the land problem would not have been solved. African leaders exploited the land issue to ensure their own legitimacy, knowing it would provoke strong reactions among the people, and in 1950 the Mau Mau movement was born.

During the first two and a half years of "the emergency" more than 10,000 people died (fewer than 100 were European), race relations were worse than ever before, and the economy was dealt an almost irreparable blow. But in 1953 a Royal Commission came to Kenya to study means of improvement, and in 1955 it recommended that the policy of reserving land on the basis of race should be terminated.

LAND REFORM AND RESETTLEMENT

The first phase of land reform began with the implementation of the Plan to Intensify the Development of African Agriculture in Kenya (The Swynnerton Plan) in 1955, the result of the Royal Commission of 1953 (Odingo 1971). The plan called for a change in African ownership from customary tenure to individual freehold in order to prevent the negative effects of population pressure on limited land. This meant the enclosure and registration of existing rights, the consolidation of land fragments, the introduction of lucrative cash crops, high-yielding livestock, and marketing facilities, access to agricultural credit, and the provision of rural water supplies and technical assistance. The plan aimed at providing Africans with the means to transform their system from subsistence agriculture to modern planned cash cropping. Economic farming on a large number of fragmented parcels of land was clearly impossible, thus consolidation into freehold units was the cornerstone of the plan. Table 18-1 illustrates this process as envisioned in the 1960s.

TABLE 18-1 EVOLUTION OF THE FARMING SYSTEM IN THE KIKUYU AREA OF KENYA.

Year	Type of Farming	Cropping Pattern	Livestock Economy	Land Rights and Tenure
About 1860	Shifting cultivation	Maize, beans, mixed cropping	Ample grazing, Zebu cattle, goats	Ample land. Rights of land use: communal
About 1920	Fallow system	Maize, beans, sweet potatoes, mixed cropping	Limited grazing, Zebu cattle, goats	Limited land. Rights of land use: communal
About 1950	Permanent cultivation	Maize, beans, sweet potatoes, banana, wattle (building material)	Roadside grazing, mainly goats, some Zebu cattle	Rights of land use: most communal; grazing turned into individually cropped plots
About 1960	Permanent crops + permanent cultivation + some ley farming	Coffee, maize, beans, sweet potatoes, banana, potatoes, vegetables, leys	Roadside grazing and some improved breeds of cattle	Private property rights: land can be leased and mortgaged
Estimated expansion path for 1980	Permanent crops and permanent cultivation	Coffee, tea, beans, hybrid maize, sweet potatoes, use of mineral fertilizer and manure	Improved breeds of cattle, poultry, pigs	Private property rights: land can be leased and mortgaged; an active land market

Source: Ruthenberg (1976).

Although the logic of the traditional methods of dryland cultivation under a rather variable climate was lost on most colonial officials, it has since been acknowledged that farming small scattered plots reduces the risk of loss from uneven rainfall, pest infestations, and the like. Cultivating different crops in a variety of ecological niches seemed like good management to African farmers, who knew how variable and uncompromising the climate could be from one year to the next. Ruthenberg, commenting on the process in the 1970s, indicated that improved breeds of cattle and hybrid maize were being adopted by more and more farmers, but population growth that exceeded expectations and demands to meet subsistence needs threatened to slow down further diversification of crops and greater commercialization.

At first, many Africans treated the Swynnerton Plan with suspicion. The Luo around the shores of Lake Victoria and the Abaluhya further north resisted all land reform efforts for 6 years, while soil erosion, poor farming methods, and population increases impoverished them still further (Jones 1965). In Kikuyuland, where land pressure was greatest, agricultural production rose by about 15% per year as a result of reorganization, land use intensification, the use of fertilizers, and hybrids. By 1974, about 12 million acres (4.8 million ha) had been reorganized into over 650,000 individual holdings. Coffee, pyrethrum, pineapples, tea, and tobacco were introduced, and the quality of livestock was greatly improved. Vast areas of the Kenya Highlands were literally relandscaped, and new settlement patterns and areas of production emerged. But the plan was not without its problems. Even in 1955, there was insufficient land to go around in Kikuyuland, so that consolidation and resettlement into viable economic units actually intensified the problems of landlessness. In places illegal subdivision occurred, and land once devoted to high-value cash crops reverted to staple foods. Migration to Nairobi, Nakuru, and other urban centers accelerated as the population swelled, and the number of landless increased.

The second phase of land reform was far more ambitious and initially more problematic than the first. It centered on the decision to open up the Highlands for farming by all races. The first step, beginning in 1961, was the purchase, by the Kenyan government, of a little over 1 million acres (405,000 ha) of European-owned farms and estates, out of which 36,000 African-owned farms were created. Next, the government instituted a program of instruction in agricultural techniques and marketing and encouraged the formation of cooperative societies. Since commercial production was essential to provide cash to repay the loans, careful planning of crop types, intensity of land use, and methods of production had to be ensured. Three types of settlement were planned: high-density smallholder settlements, intended for Africans with limited capital and agricultural expertise; low-density smallholder settlements for more experienced farmers with some capital; and a scheme for large-scale and cooperative farms and ranches.

Farms in the low-density schemes were designed to yield a net income of about $500 a year, the average farm being only 37 acres (15 ha). Normally a farmer had to be a member of the tribe in whose area the land was situated. Farms in the high-density schemes averaged 27 acres (11 ha) and were expected to yield a net annual income of $140 (Whetham 1968). Because most farmers prior to being resettled were either unemployed or landless, they lacked both capital and experience and were in debt almost to the full value of their land and stock. Once selected and installed, they were supervised by government officials and offered technical assistance wherever possible. For a while, group interests threatened to jeopardize the scheme before it had a chance to get started, but by 1974 some 36,000 families had been resettled on about 1.2 million acres (486,000 ha). The average cost of establishing each small-scale farm was more than $2,000, and financing came from loans and grants from the United Kingdom, West Germany, and the World Bank.

The success or failure of resettlement should be assessed from the point of view of its purpose. This attempt at land reform was not economically motivated. Rather, it had a political goal: to relieve an explosive situation and help stabilize a moderate nationalist government by creating a new landed middle class that acted as a buffer against agitation by the rural masses. But resettlement did not solve the problem of landless Kikuyu nor the problem of urban unemployment to which landlessness contributed. The economic consequences are well documented. Over 5 years, the Kenyan government spent about $28 million to buy out European settlers, and $25 million of this amount left the country; another $30 million went to converting the estates that had been purchased into African-owned farms. During the first 5 years of resettlement, production fell in almost every type of operation except for the low-density schemes. There, traditionally high outputs were maintained because of the intensity of supervision and technical advice, because of the large amounts of development capital, and because the settlers had been selected for their farming abilities. In 1967, for the first time, Kenya's small farmers contributed more than half the total output of marketed agricultural products.

Before decolonization of the Kenya Highlands, Europeans accounted for almost all of Kenya's coffee exports, and coffee was and remains the leading export. Today, almost half the output is produced by several hundred thousand smallholders. Yields on the large estates (numbering just a few hundred) are more than double those on small farms, and the quality is higher. In recent years, coffee (predominantly high-grade arabica) has contributed about one-fourth of the gross farm revenues and earned Kenya $276 million per year on average from 1995 to 1999 (although world prices fell thereafter). The major producing areas are immediately north of Nairobi, the Western Highlands, and the slopes of Mounts Kenya and Elgon (Figure 18-3).

The land tenure system in Kenya continues to be bifurcated. Eighty percent of the land is organized under communal tenure and areas are tribally demarcated—and not without controversy (Box 18-1). The other 20% is under the private tenure system, which is based on individual ownership and title. Transforming communal into individually owned land, the stated goal of the government of Kenya, involves adjudication, consolidation, and registration of individual title (Konyimbih 2002). Land adjudication has been continuous in most provinces but especially in Nyanza, Western, Rift Valley, Eastern, Northern, and Coast provinces. By 1991, over 17 million acres (about

BOX 18-1

LAND CONFLICT, ETHNIC IDENTITY, AND THE PARLIAMENTARY SYSTEM IN KENYA

MP WARNS OVER LAND IN POKOT

Kapenguria Member of Parliament Samwel Moroto yesterday said the Pokot community would use force to reclaim their land in Turkana and Trans Nzoia districts. Pokot land, he claimed, stretched 41 kilometres inside Turkana in the Kalim'ngorok area and he asked their neighbours to prepare for "war" if they cannot surrender their land. The MP, who was elected following fiery cabinet minister Francis Lotodo's death, said that since he had been sworn in as MP "if they (the non-Pokot) are not ready to surrender the land peacefully my kinsmen should not worry because I'm going to protect their interests."

The MP was sworn in on Wednesday together with his South Mugirango counterpart after winning in the January 12 by-elections. Students from Chewoyet joined the residents in giving Moroto a warm welcome as thousands of Kapenguria residents stormed towards the stadium to hear his speech.

Source: The Nation [*Nairobi newspaper*], *March 25, 2001.*

7 million ha) had been adjudicated and registered to individuals. Between 1994 and 1996, 24 million (9.7 million ha) additional acres was registered. But by the beginning of the twenty-first century only 6% of the total land area of Kenya had been registered under individual titles (Konyimbih 2002).

It is clear that the land reform process has stalled in Kenya, and the interests of two groups are opposed on the land issue. There are those who wish to abandon the tribal land tenure system in Kenya and universally adopt a Western private property model. Against this model are arrayed a variety of groups who fear that their own people will not be able to compete in a nontribal society based on individual landownership (King 1977). Experience in successful land reform from around the world suggests that real progress comes from the suspension of the normal political process in which there is direct intervention by a powerful figure (the king, for example), the military, a socialist revolution, or an overwhelming explosion of rural unrest that threatens the political and social order (Konyimbih 2002). It is uncertain what will happen in Kenya. Change may be brought about by rural unrest powered by a rapidly growing, increasingly landless, population (Hunt 1984), or the problem may be resolved in another way.

POPULATION PRESSURE, POLICY, AND THE LAND QUESTION

In 1966 Kenya became the first country in Africa south of the Sahara to adopt an official population policy. The program was viewed as an integral part of social and economic development. The objective was to reduce the population growth rate by 1% in over 10 years. The program was successful in its education efforts and in the improvement of the health of mothers, and although it failed to rapidly curb population growth (Hungu 1987), it probably was instrumental in reducing the desire for large families. The eventual impact of Kenya's population growth on the land question is uncertain at this time, although in the 1970s it was thought that Kenya was on a short road to disaster. Kenya's population was once the fastest growing in Africa. The 1969 census put the Kenyan population at 10,942,705. In 1977 it was estimated to be 14,500,000 and increasing at 3.3% per year. It reached 32 million in 2003—but it appears to be slowing. Given that only 17% of the total area is suitable for cultivation and pastoralism under available technology and practices, and that 75% of the people live on less than 20% of the land, Kenya appears to face unenviable population

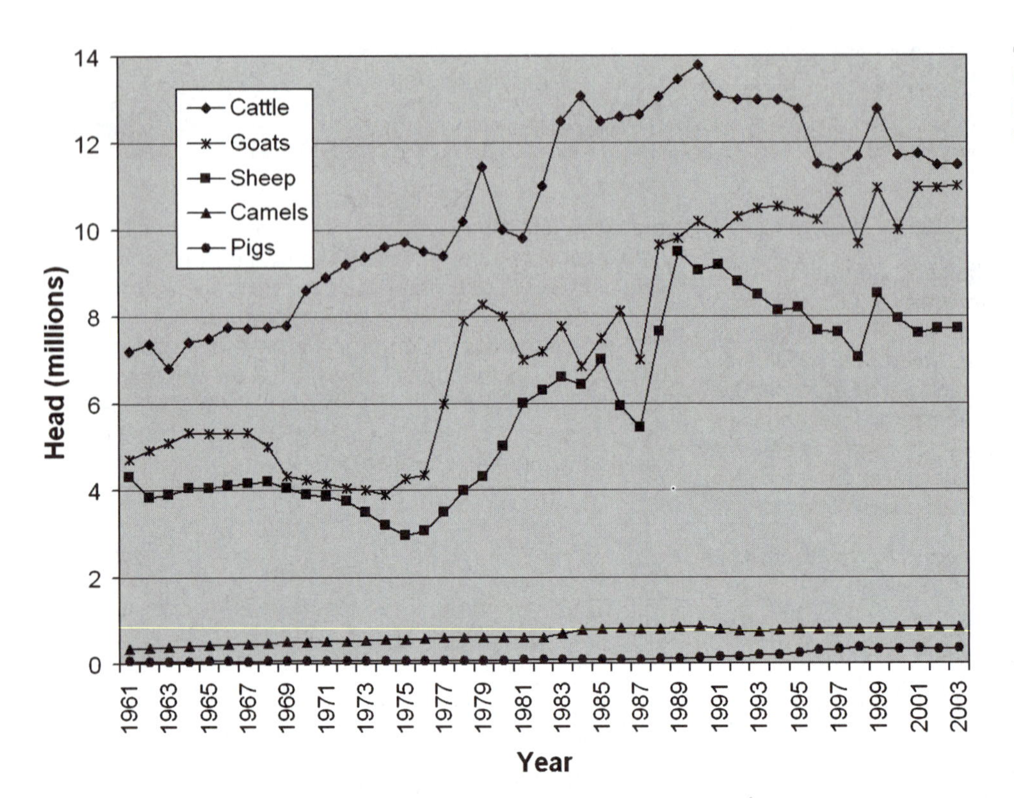

FIGURE 18-5 Livestock populations in Kenya, 1961 to 2003.

problems. If to these already alarming figures are added the numbers of the large (and growing) livestock populations (Figure 18-5), the situation seems even more ominous, especially when it is realized that much of the land in the northern and eastern cattle areas is subhumid and subject to prolonged and devastating droughts.

On the other hand, increased livestock populations mean a greater source of manure to maintain agriculture yields, and one of the goals of the Swynnerton Plan of 1955 was to encourage the integration of livestock raising with farming for this very purpose. In any case, livestock populations have declined by 2 million Tropical Livestock Units (TLUs) since their peak in the 1980s, although this change has been differentially distributed by livestock type: the most demanding stock, cattle and sheep, have declined the most, and less demanding stock have remained the same or increased slightly.

With respect to human populations, in *Fertility Decline in Kenya: Levels, Trends, and Differentials*, the most comprehensive study of Kenyan fertility, researchers found that Kenya is going through a demographic transition and, although population momentum is high, reproductive behavior is changing. The study from the African Population Policy Research Center found that the government's reproductive education programs that began in the 1960s were having an effect: both men and women interviewed in the study expressed the desire for smaller families (Fapohunda and Pokouta, no date). Their desires are borne out by these facts: the total fertility rate of women has declined from 8 per woman in 1983 to 4.9 in 2005 (Population Reference Bureau 1983, 2005). Perhaps the demographic transition will resolve the land issue in Kenya. Furthermore, the percentage of the population that is rural and dependent on agriculture is declining; from 90% in the 1970s to 75% today. As the country urbanizes, the proportion of rural agriculturalists is expected to decline even further. Nevertheless, it is important to point out that 75% of over 30 million people today represents many more rural residents than in 1975. Rural densities exceed 1,200 persons per square mile (460/km^2) in parts of Kakarnega District (Western Province), Kisil District (Nyanza Province), and several areas of Kikuyuland. Resettlement and land reform in these areas have done little if anything to alleviate the pressure, evidence of which still abounds: rural–urban migration, soil erosion, the cultivation of marginal land, malnutrition, and social distress.

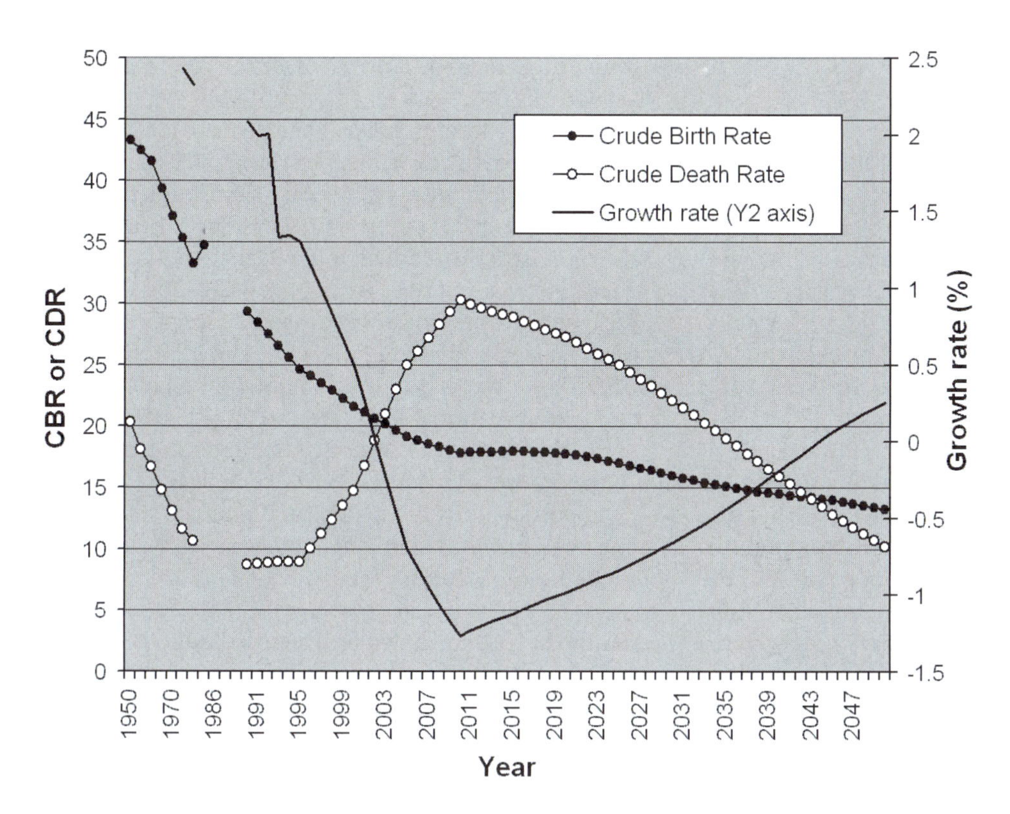

FIGURE 18-6 Kenya's crude birth and death rates (CBR, CDR) and projected population growth rate, 1969 to 2050.

Complicating Kenya's population problems is HIV/AIDS. According to projections by the U.S. Census Bureau, Kenya is entering a period of very slow population growth brought on by the spread of HIV/AIDS. The crude birthrate (CBR) is expected to steadily decline over the next 50 years, but the death rate is projected to rise from a low of under 10 per thousand in 1985 to 18 in 2017 (Figure 18-6). This trend will reduce the population growth rate to less than zero and, in the process, provoke social crises in the country. Average life expectancy prior to the AIDS epidemic was close to 60 years in Kenya. Today it is 44 years and likely to fall further before beginning to rise again. Unfortunately for agricultural production, those in their most productive years are the most likely to be struck down by AIDS.

CHARACTERISTICS OF TODAY'S AGRICULTURE

Kenya has one of the most diversified agricultural economies in Africa, and exports bring in about one billion dollars of foreign exchange a year. A wide variety of crops are grown on large estates, plantations, and by smallholders. African smallholders have made spectacular inroads in Kenya's tea production, especially in the high-density settlement areas. Prior to independence, tea (Kenya's second most important export) was almost exclusively a plantation crop, with the largest concentrations being around Kericho and Limuru. In 1975 about one-third of the total production came from smallholdings, but by 2001 smallholders produced 61% of Kenya's tea.

Another crop successfully produced on small farms is pyrethrum. Over 90% of this daisy like plant used in the manufacture of insecticides comes from small farms, many of them located in the higher altitudes of Central and Nyanza provinces. Small farms produce significant and increasing quantities of sugar, tobacco, rice, fruits, and vegetables. Opportunities for irrigation in Kenya are limited. The Mwea Irrigation Scheme (11,250 acres; 4,556 ha), one of the most well known, provides about 70% of Kenya's rice supplies together with smaller amounts of cotton, maize, and other crops (Hornby 1973). There, rice is produced on small tenant holdings, each consisting of about 4 acres (1.6 ha) of paddy. Small farms also account for the bulk of the country's principal staple, corn, while most of the sisal and wheat is grown on extensive farms and estates.

NAIROBI AND MOMBASA

Kenya's capital, Nairobi, has a population of 2.5 million. It is linked to Mombasa (780,000), the principal port, by railway (Figure 18-1). The two cities are separated from each other by an extensive area of low rainfall, relatively light population density (Figures 2 and 3, pp. 454 and 455) and low economic productivity. Each city has its own distinct hinterland, but the two are functionally interdependent and together form part of Kenya's emerging national core.

Nairobi started as a break-in-slope railway camp at the edge of the Kenya Highlands a century ago when the Mombasa line was laid to Lake Victoria. Over the years it has developed into Kenya's principal administrative, commercial, and industrial center, and thus the country's prime generator, transformer, and distributor of the forces of change and modernization. Its growth has been part of the general economic growth that has characterized not only Kenya but also Uganda. The 1948 census put Nairobi's population at 118,976, and that of 1969 at 509,286. On average about 57,000 new people were added to Nairobi's population each year from 1969 to 2004. Its current growth is estimated at 8 to 10% per year, but since housing, jobs, industry, and basic social services are not growing enough to keep pace, shantytowns are expanding, and unemployment is rising. The government hopes that increasing privatization will provide more opportunities for employment, and experience elsewhere fuels this hope.

Nairobi has more than half the country's industrial establishments and an equally impressive share of the industrial output and employment. Its processing industries include grain milling, clothing, tobacco, beverages, printing, pharmaceuticals, publishing, communications and transportation equipment, and a host of construction-related industries, many of which depend on locally derived agricultural and forestry resources. The government has been attempting to arrest this primacy and to encourage decentralization to Kisumu, Naivasha, Nakuru, and Kitale, where adequate transport, resources, markets, and labor exist. The railway yards and engineering works are the largest in East Africa, however, and the capital forms the hub of a dense transport network that extends into Uganda and Tanzania. Nairobi is strategically located with respect to several of Africa's popular game reserves (Amboseli Tsavo and Serengeti), and its airport is the busiest in East Africa.

Increasing attention has been focused on Nairobi since 1996, when the leaders of Kenya, Tanzania, and Uganda revived the East African Community (EAC) that had broken down after independence. The city itself, once strictly planned on racial lines, is one of Africa's most attractive, with broad tree-lined streets and boulevards, modern high-rise buildings, and spacious residential areas (Figure 18-7). The parliamentary buildings, university, mosques, and marketplaces, and the Nairobi National Park are all tourist attractions, which, together with the game reserves, bring almost a million overseas visitors to Kenya each year.

Important to Nairobi's cultural and economic life are the Asians (mainly Hindis), who have always faced much discrimination. They have historically dominated Nairobi's retail trade, and throughout the country have held skilled and semiskilled jobs, generating resentment among many African Kenyans. The cultural impact of Asians on East Africa is considerable. Apart from the tangible influences such as the architectural peculiarities of Nairobi and parts of Mombasa, the Asians brought with them and retain their home religions, languages, and living habits. Like the Africans, they were restricted in their ability to occupy land during the colonial period, and their representation in Kenya's governmental affairs has always been disproportionately small. Since independence, discrimination against them has intensified, resulting in several hundred expulsions from the country and many cancellations of licenses to trade. These measures were partly due to the cultural isolation in which many Asians continue to live and partly to their envied economic status. The 1969 Asian population of Kenya was estimated to be 139,000, but the most recent estimates indicate that there are only 64,000 Indians, comprising 50,000 Gujaratis, 10,000 Eastern Punjabis, and 4,000 Goans in Kenya (SIL International 2004).

About 250 miles (400 km) southeast of Nairobi lies the port of Mombasa–Kilindini. Its hinterland comprises not only Kenya but also Uganda and the Moshi–Arusha region of northeastern Tanzania. It is the best equipped harbor in East Africa, and whatever development takes place in the interior will benefit the city, already the second largest in Kenya and fourth in East Africa. Unlike Nairobi and Dar es Salaam,

FIGURE 18-7 Nairobi city center.
David Moorehead, USDA.

Mombasa is an old city, established by Arab traders as part of the Indian Ocean trade, and then occupied for centuries by Arab slave traders. It fell to Portuguese invaders in 1505, and came under the control of the Zanzibar Sultanate in 1840. Still later in 1885, British overlordship sanctioned this merger by creating the Kenya Protectorate under the administration of Zanzibar. The Kenya Protectorate, a strip of the mainland 10 miles (16 km) wide and 52 miles (84 km) in length, was formally attached to the sovereign Republic of Kenya in 1963.

Throughout its history, Mombasa developed as a part of the Indian Ocean littoral, with its Arab and Asian influences, rather than as an African settlement. It had many competitors—among them Zanzibar, Kilwa, Malindi, Mogadishu, and Lamu—some overshadowing Mombasa during certain periods. Today, Mombasa eclipses them all. One hundred and twenty-five years ago, slaves and ivory from the interior lakes region comprised Mombasa's major exports, and its orientation was oceanward toward Zanzibar, other Arab strongholds, and South Asia. The coming of British administration in the late nineteenth century was attended by many momentous changes: slavery was abolished; Africans crowded the island in search of work; Europeans came in large numbers, terminating old practices and initiating new; and Indian labor was imported to build the mainland railway. Each immigrant group left its mark in the landscape, and today Mombasa has several distinct ethnic enclaves. From 1900 to 1905, when the headquarters were shifted to Nairobi, it was the capital of mainland British East Africa. Since then, Mombasa has been the pulse of its hinterland (de Blij 1968).

Today, Mombasa has thriving industries. The principal industry is tourism. Others include refining of imported crude oil, metal fabricating, chemicals, plastics, textiles, food processing, beer brewing, cement, and glass manufacturing. Despite this apparent industrial emphasis, Mombasa's areas of specialization lie in storage, assembly, packing, sorting, and distribution. Its immediate hinterland is not rich in resources other than sisal and copra, and its role in the modernization process of coastal Kenya has been small. Mombasa and Nairobi represent two nodes in the Kenya nation, each with well-defined spheres of influence and dependence (Soja 1968). They have played an important role in the development of Kenya and in national integration.

In the Northeastern Province especially among the Somali and Oromo (Borana) peoples, there is little government control or identification with the Kenyan state. There lawlessness has increased since the collapse of order in Somalia in 1991, guerrilla warfare has erupted periodically on land and piracy along the Somalia Coast, and the land is poor and the opportunities very limited.

The recent free and fair elections in 2002 presented Kenya with another opportunity to set a new course for itself, having learned from the costly mistakes of the past. Whether change comes gradually, if at all, or by some unforeseen suspension of the political process, one cannot predict with certainty.

BIBLIOGRAPHY

African Population Policy Research Center. 1998. *Fertility Decline in Kenya: Levels, Trends, and Differentials.* Nairobi: Population Council.

de Blij, H. J. 1968. *Mombasa: An African City.* Evanston, Ill.: Northwestern University Press.

Fapohunda, B. M., and P. V. Poukouta. no date. *Trends and Differentials in Desired Family Size in Kenya.* Nairobi, Kenya: African Population Policy Research Center. http://www.uaps.org/journal/journal12v1/TrendsandDifferentials.htm.

Hornby, W. F. 1973. The Mwea Irrigation Scheme. *Geography,* 58: 255–259.

Hungu, J. 1987. Kenya's population policy. *Populi,* 14(2): 36–40.

Hunt, D. 1984. *The Impending Crisis in Kenya: The Case for Land Reform.* London: Gower.

Jones, N. S. C. 1965. The decolonization of the White Highlands of Kenya. *Geographical Journal,* 131: 186–201.

King, R. 1977. Land Reform: A World Survey. London: Bell.

Konyimbih, T. M. 2002. Major issues of smallholder land policy: Past trends and current practices in Kenya. *Land Reform, Land Settlement, and Cooperatives,* 2001/02.

McMaster, D. N. 1975. Rural development and economic growth: Kenya. *Focus,* 25(5): 8–15.

Odingo, R. S. 1971. *The Kenya Highlands: Land-Use and Development.* Nairobi: East African Publishing House.

Population Reference Bureau. 2005. World Population Data Sheet. Washington, D.C.: Population Reference Bureau.

Population Reference Bureau. 1983. World Population Data Sheet. Washington, D.C.: Population Reference Bureau.

Ruthenberg, H. 1976. *Farming Systems in the Tropics.* Oxford: Clarendon Press.

SIL International. 2004. *Ethnologue.* 14th ed. Dallas: SIL International.

Soja, E. 1968. *The Geography of Modernization in Kenya.* Syracuse, N.Y.: Syracuse University Press.

Taylor, D. R. F. 1969. Agricultural change in Kikuyuland. In *Environment and Land Use in Africa,* M. F. Thomas and A. W. Whittington, eds. London: Methuen.

Whetham, E. 1968. Land reform and resettlement in Kenya. *East African Journal of Rural Development,* 1: 18–29.

CHAPTER 19

Uganda

East African Emerging Economy?

The northwestern sector of former British East Africa is occupied by the state of Uganda, smallest (91,134 mi^2; 236,026 km^2) of the three major units and, with a population of over 25 million, by far the most densely populated. Landlocked Uganda shares the waters of Lakes Victoria, Edward, and Albert to the extent of about 14,000 square miles (36,300 km^2), and in addition there is much territory under swamps and marshes. The country lies largely on a plateau just under 4,000 feet (1,200 m) above sea level, dropping to lower elevations toward the Sudd Basin in the north, and diversified by some great mountains west and east such as Ruwenzori and Elgon. The southern part of the country, especially, enjoys the ameliorating effects of elevation upon the tropical temperatures, and almost all of it receives over 30 inches (762 mm) of rainfall a year. As a consequence, broadly speaking, the natural vegetation of the north is savanna, while that of the south is forest. Most of the southern forests have been cut for agriculture.

On closer inspection, the compact territory of Uganda lies in a number of transition zones. By virtue of its location in the east-central part of equatorial Africa, its natural vegetation includes the savanna lands of the east and the forests of the Congo margins in the west. The swampy lake regions of the south give way to the dryness and rain-deficient conditions that characterize the Sudan Basin northward. Uganda also is situated astride linguistic and ethnic transition zones: south of Lake Kyoga people speak languages from the Bantu subfamily of the Niger–Congo superfamily, in the north they use mainly the languages in the Nilotic superfamily, and in the northeast and northwest, Sudanic languages (Figure 19-1). The south is predominantly agricultural, much of the north pastoral. From the east have come Arabs, Europeans, and Asians, all of whom have made their impact in the country. And from the north came many of the political ideas out of which the strong traditionalism of present-day Uganda emerged.

UNITY AND FRAGMENTATION

Uganda became an independent state in 1962, after a lengthy sequence of political difficulties was in some measure resolved. In spite of its compact shape and relatively small areal extent, the internal variety of this country, in terms of ethnic groupings, traditions, and degree of economic development, is very great. This is true of many other African countries, but in Uganda the spatial arrangements so strongly favor the southern part of the country, mainly Buganda Province, that the final political unification of the territory, long administered by Britain as a protectorate, presented deeply rooted problems. These involved two major centrifugal forces which, even when the desire for independence was the strong centripetal force, tended to dominate the country's internal political geography.

The favored southern part of Uganda is home in large part to the Baganda people, who number about one-sixth of the country's total population, around 4.5 million. These Baganda, along with over 2 million non-Baganda, live mainly in the region known as Buganda, where the traditional authority was the king, or kabaka. One major centrifugal force in Uganda's political geography was the reluctance of Buganda to lose its privileges by being merged into a larger Uganda, the modern state. A second was the reticence of many of the non-Baganda peoples of the

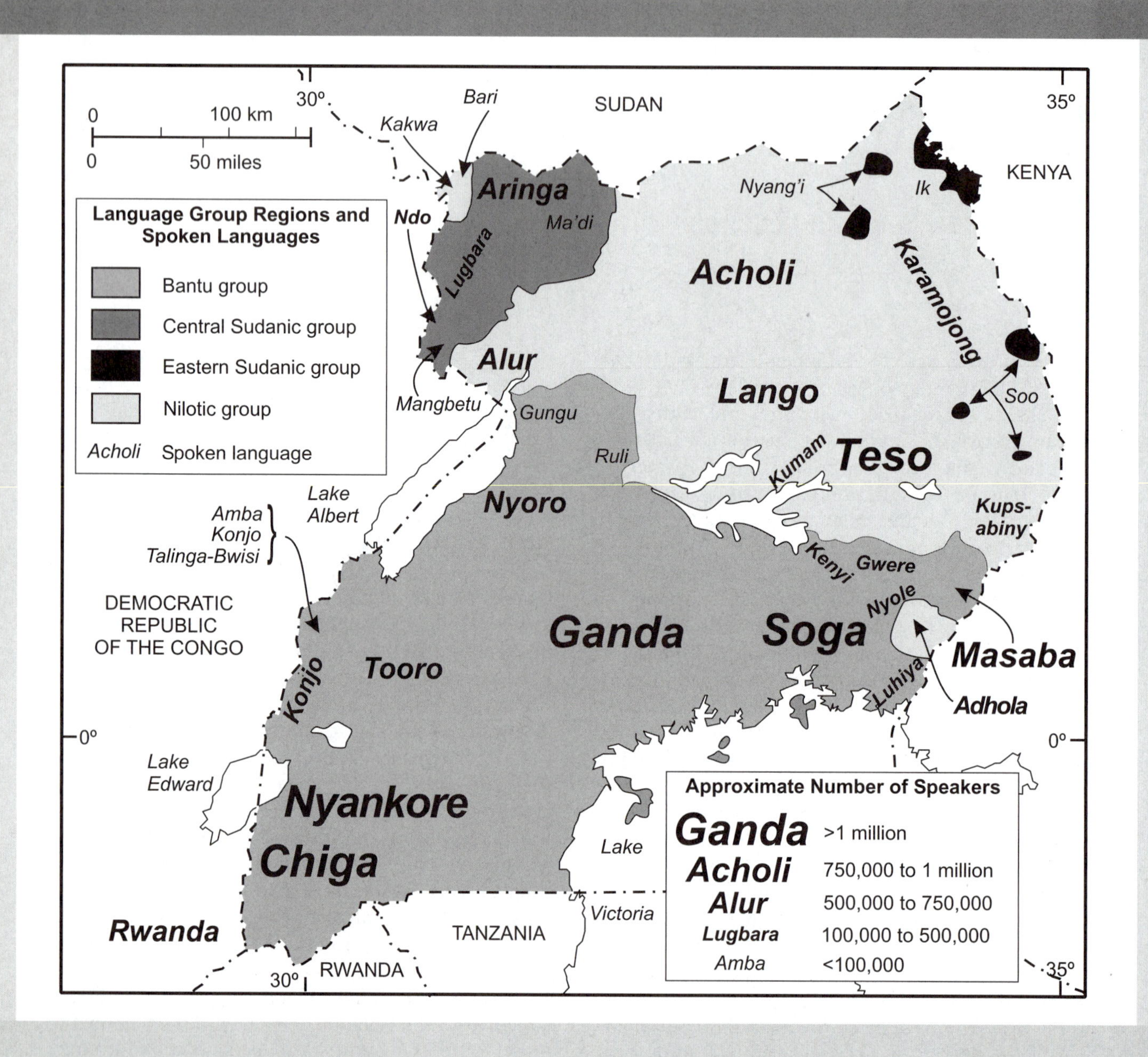

FIGURE 19-1 Language groups, spoken languages, and number of speakers in Uganda.

country to support an independent state in which most of the power would lie in the favored south. In brief, the problem was how to fit Buganda into a larger Uganda, and the centrifugal forces remained sufficiently strong to prevent the evolution of a unitary state, the most common type of state elsewhere in Africa. A complicated federal arrangement was the product of centripetal forces seeking to end colonial administration while retaining certain amounts of autonomy within this multicultural microstate.

In many respects, Uganda is almost the complete opposite of Tanzania, and the comparison yields valuable insights. Uganda, although much smaller and far more densely populated, is primarily a country of smallholder agriculture with plantations making a very small contribution to export revenue. In recent years Uganda has exported over twice the high technology products as Kenya and over six times that of Tanzania as a percent of GDP (World Bank 2004). The major contrast with Tanzania emerges when the spatial organization of the countries is considered: while Tanzania's development can still be best described as low, and a national core area, if recognizable at all, is in the initial stages of development in the northeast, Uganda has a core area that is as well defined as any. All factors seem to have worked together to make this so. When European colonial administration began, it naturally was focused on the region that was at that time best organized. This, of course, was the kingdom of the Baganda, lying on the shores of Lake Victoria (Figure 19-2). Had the outlet of the country been north or west, it would have been necessary to construct transportation lines in that direction, with the result that Buganda would have been less isolated from the rest of the country. But, as it happened, the natural exit was southeastward, through Mombasa, and the modern transport routes, focusing upon Buganda, came from that direction. That meant that they failed to cut through any part of the former protectorate except Buganda and the southern part of the Eastern Region. Thus, Buganda's political eminence was supplemented and reinforced by several additional advantages: the British set up their administrative headquarters in Buganda, and when cash cropping began, the Baganda reaped all the advantages because the most suitable areas in terms of climate and pedology were also those near the lake and railroad.

The Factor of Historical Geography

The political entity of Uganda was a creation of European colonialism, and the area prior to the first arrival of the Nyamwezi traders and the Arabs from the east in the late 1700s probably did not possess any elements of unity. Neither did it have the kind of contact with the outside world that marked the kingdoms of West Africa; for centuries no caravans reached the peoples of Uganda, and there was no organized exporting of products.

The internal heterogeneity of the area of present-day Uganda, due in large part to its character as an ethnic transition area, was expressed in political ways long before the first European explorers reached the headquarters of Buganda and was exemplified by the sophisticated state organization of the kingdom of the Baganda visited by the explorers Speke and Grant in 1862 (see Chapter Four). This empire was by no means the only organized political area in the region, nor was it the first to have occupied a dominant position. The kingdoms of Bunyoro, Ankole, and Toro, whose areas also were incorporated in the Protectorate of Uganda, were similar in their organization and had dominated the region centuries before.

Buganda was by far the most important of these politically organized units when the European invasion began, and in the period of colonial administration that followed, the kingdom played a leading role (Apter 1961). But in precolonial Uganda, when there was no force binding the larger territory together, each of the kingdoms, and the tribal peoples of the more loosely organized areas elsewhere, had existed separately. Each had a physiographically rather well-demarcated territory (Figure 19-2): Buganda between Lakes Victoria and Kyoga; Bunyoro between Lakes Kyoga and Albert, the Victoria Nile, and the Kafu River; Toro on the eastern slopes of the Ruwenzori Mountains north of the Katonga River; and Ankole west of Lake Victoria and south of the Katonga. These kingdoms had their times of greatness and decline; during the course of history, they had expanded at the expense of the less well-organized peoples around them and had encroached upon each other. The frontiers between them really were frontiers in the technical sense of the word: either they were undesirable lands, with swamps or marshes, or they were areas of conflict and attempted expansion.

When, during the middle 1800s, European contact was made, the Buganda kingdom was the largest, best organized, and most powerful in the region. Situated in the southeast of the area beyond Lake Victoria, nearest to the European bridgeheads on the coast, Buganda presented to the early explorers a fertile field for missionary and trade activities. But in the less organized parts of the region, the slave trade was active, and where the Europeans were not exercising effective power and propagating Christianity, Arab traders were converting the people to Islam. Uganda, having been a

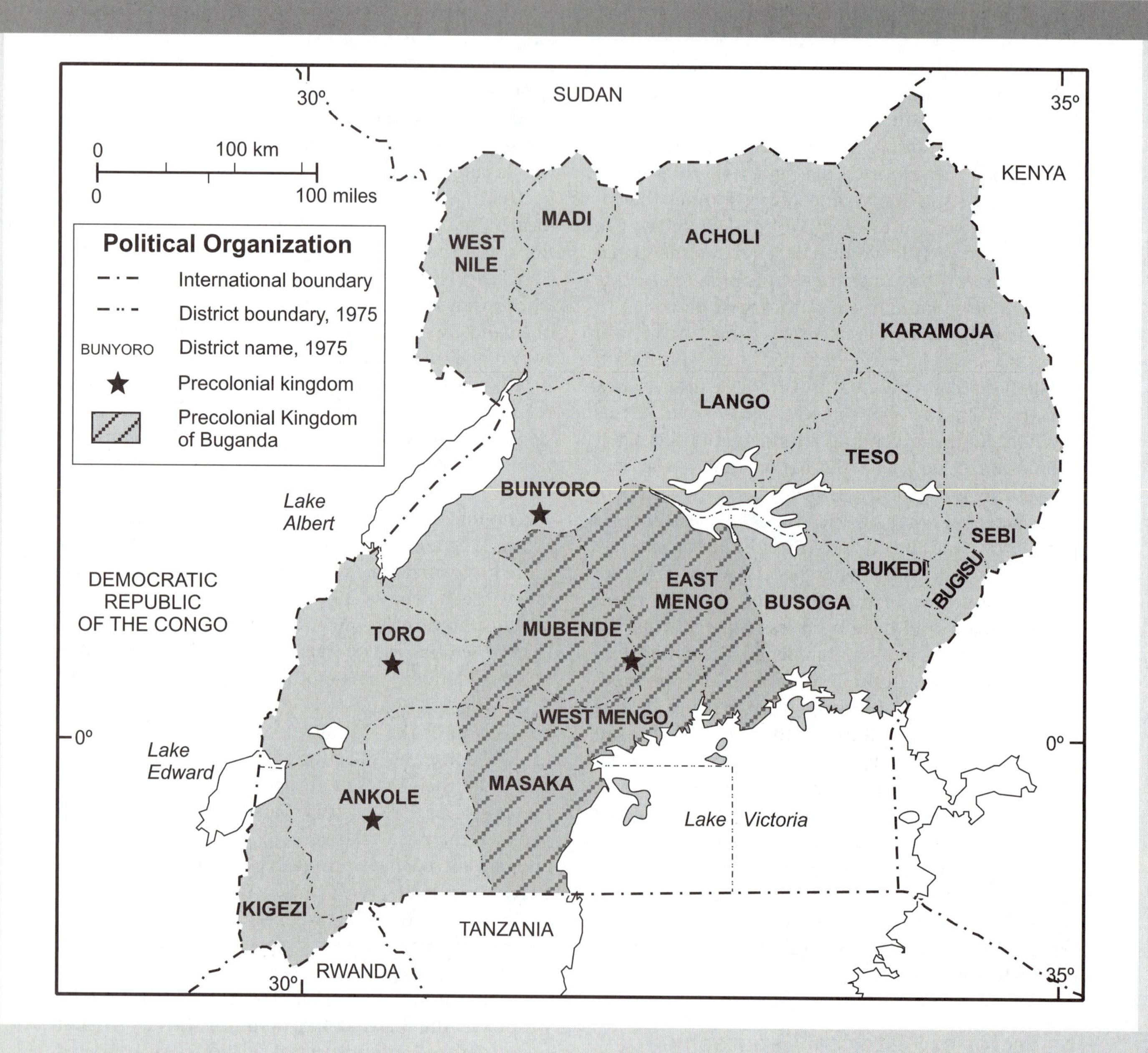

FIGURE 19-2 Administrative districts of Uganda circa 1975 and in precolonial kingdoms.

meeting place of indigenous peoples, now became an area of competition between the proponents of these religions. Egyptian influences were felt in the north, European in the south. This conflict, and a real conflict it was, became superimposed upon the regional ethnic, political, and economic contrasts that already existed. In effect, the area was in a state of instability until after the establishment of a British protectorate in 1894 over the Kingdom of Buganda and its extension in 1896 over Bunyoro, Ankole, and Toro.

Thus, there was little to warrant the incorporation of so much diversity into a single political entity. It was natural that Buganda should be selected by the British as the headquarters of administration, for in the Uganda region, Buganda was the most powerful unit, and effective control there was the prime requirement for the establishment of order. Indeed, the Baganda revolted against British overlordship not long after the establishment of the protectorate, and when the uprising had been put down, the kingdom was given special status in the Uganda Protectorate according to the Buganda Agreement of 1900. But then the British found themselves confronted by the task of forging a political whole out of the great variety in a country of nearly 30 distinct peoples, some of whom were being ruled on the basis of strong local customs and traditions, while others were socially fragmented and politically disorganized. Among the existing local authorities, moreover, there were considerable differences in terms of strength, competence, and desire to cooperate.

The Factor of Political Geography

For a state to emerge and organize disorganized peoples in a geographic area, the centrifugal forces that can tear a state apart—or prevent one from forming—must be overcome by centripetal forces. The most fundamental centripetal force is the very idea of the state itself. Other centripetal forces could be a common language or dialect, a common religion, a shared historical experience, a common cultural identity, or linkages such as road or railroad connections that bring people together.

The evolution of the state of Uganda provides an interesting illustration of Stephen B. Jones' classic unified field theory in political geography. The theory places "idea" and "state" at opposing ends of a continuum, which links together five phases Jones identified as political idea, decision, movement, field, and political area (Jones 1954). In this theory "field" refers to the variable geographic areas or regions in which some part has the potential to come under the influence of the political idea and form a politically organized area (political area). The political idea is put into practice through individual and group decisions that are manifested in movement (circulation or spread across space) of the political idea and other centripetal forces.

> There are no upper or lower limits on the magnitude of an idea. There are upper limits on decisions, movements, fields, and political areas, though these limits change with events (often, but not necessarily upwards). Such ideas as the great religions, nationalism, liberalism, and communism have created fields. Ideas, fields of exploration, in some cases political areas have expanded. (Jones 1954: 121)

In the case of Uganda, the political idea, existing first in the minds of British administrators and later developing among African residents of the region, was that eventually colonies and protectorates in general, and Uganda in particular, should become independent. The specific concept arising from this idea was the establishment of a united Uganda out of the diversity within the protectorate's borders. In the case of Uganda, the final decision determining the nature of the political entity was preceded by a series of earlier decisions, including the 1894 and 1896 protectorates over Buganda and the other kingdoms, and the 1900 Buganda Agreement. Although the new state did not emerge until more than a half-century later, the decision to foster self-determination in a future national state was evidently made during the period of the area's protectorate status. Subsequent decisions were adjustments to the developments taking place within Uganda as the third phase—"movement,"—took place.

In the application of the field theory model to Uganda, movement, or circulation, is seen to have taken several forms: political power projection, economic control, control of cultural icons. Britain ruled Uganda from Buganda territory, and the administrative headquarters was located in Entebbe (whereas the traditional seat of the Kabaka has always been within a few miles of Kampala). In areas with only a loose political organization, it was necessary virtually to create and superimpose responsible local authority and to project power across the national political space. Baganda personnel were often used to staff administrative offices in such non-Buganda territories, perhaps the most striking single example of movement (communication) in the context of this model. But there were other instances. Efforts were made to imprint the Buganda pattern of administration on the other kingdoms. The nature of landownership and occupancy was changed, communal ownership being replaced by a form of individual holding by chiefs and headmen. This facilitated the hierarchical introduction of a number of cash crops and accelerated the change from a subsistence to an exchange economy, an essential element of the field phase of the model.

Movement in Uganda has been a slow process. It should be remembered that the initial idea was that of a united Uganda, requiring, of course, the partial submergence of tribal and local loyalties in favor of allegiance to a larger state. Movement, in the Jones model, is, among other things, the spread of the idea of a state. In several other countries (such as Tanzania with its Tanzania African National Union at independence), strong national political parties have grown which have proved capable of fostering a national loyalty in addition to local and tribal attachments. This process has been less effective in Uganda, a factor that has inevitably made itself felt in the nature of the resultant political area.

The study of field phenomena in the model should be preceded, in this particular case, by a reference to the field characteristics already clearly defined at the time of the initial (idea) phase. Uganda's territorial extent was delimited, with only minor subsequent adjustments, at about the time of the establishment of the protectorate. Therefore, it is not possible here to speak of a "field" from which a defined political area eventually emerged. The character of the field changed significantly, but there was no phase of territorial consolidation through war and expansion, as for instance in the case of Israel, the product of the idea of Zionism.

Furthermore, Uganda, even before it was the Uganda Protectorate, possessed a core area, even though this area, Buganda, did not actually serve as such. In Buganda the protectorate possessed an area of advanced politicoterritorial organization, considerable concentration of power, the beginnings of an exchange economy, and a degree of urbanization. But if a map were drawn of Uganda at that time, there would be no network of communications focusing the rest of the region on this core area, no integrating movement and circulation, no spread and adoption of ideas originating there. The chain described by Jones had to be started before a functioning core area for a larger Uganda could emerge.

The special position of Buganda in Uganda was recognized by the Buganda Agreement of 1900 and was emphasized by the appointment in this province of an official called a resident rather than a provincial commissioner. Uganda was divided administratively into four provinces of which Buganda was one. The Western Province included the kingdoms of Toro, Ankole, and Bunyoro, and the Busoga "kingdom" was the main political entity in the densely populated Eastern Province. The Northern Province included some of the least developed parts of the country, with its many small village and clan communities. Buganda Province so far outstripped the rest of the country as the decades of the protectorate wore on, and its individuality so intensified, that its function as the core area of a larger Uganda was actually impaired (Bakwesegha 1974). The services performed by the kabaka's government were far more complete than those of any of the other local governments in Uganda.

By independence, much of the country's productive capacity lay in Buganda. Its per capita income, level of education, and degree of urbanization were higher than anywhere else. Buganda was at the center of the country's road and rail network with virtually all roads leading to Kampala, the capital (Figure 19-3). The railway from Kasese to the eastern border (and beyond to Mombasa) crossed the entire province.

The individualism, singularity, and separatism of Buganda has been a major centrifugal force in the political geography of this country. Uganda possessed many of the elements required for a healthy transition to independence: the success of cash cropping has ensured economic viability, and a series of maps showing the urban centers, resources, communication grid, and core area would seem to support the assertion that the field phenomena are prerequisite for a state. But the old forces of fragmentation had not been submerged sufficiently by the idea of the state, even in the independence year 1962, to permit the organization of the political area—the final phase of the model—as a unitary state, for which its shape, size, and communications network seem to be so suitable. While Buganda Province possessed far more cohesion than any other part of Uganda, and political sophistication and levels of education were more advanced in this region, the total population of 2 million represented only about 30% of that of Uganda. In an independent state with a government based on the universal franchise, therefore, Buganda could not expect to dominate politically as it dominated economically and socially.

Thus, Uganda was faced with demands for secession and independence from the very core area upon which its future as a state was to depend. The British government, in an effort to find a solution to this problem, in 1960 established an investigative body known as the Relationships Commission. In 1961

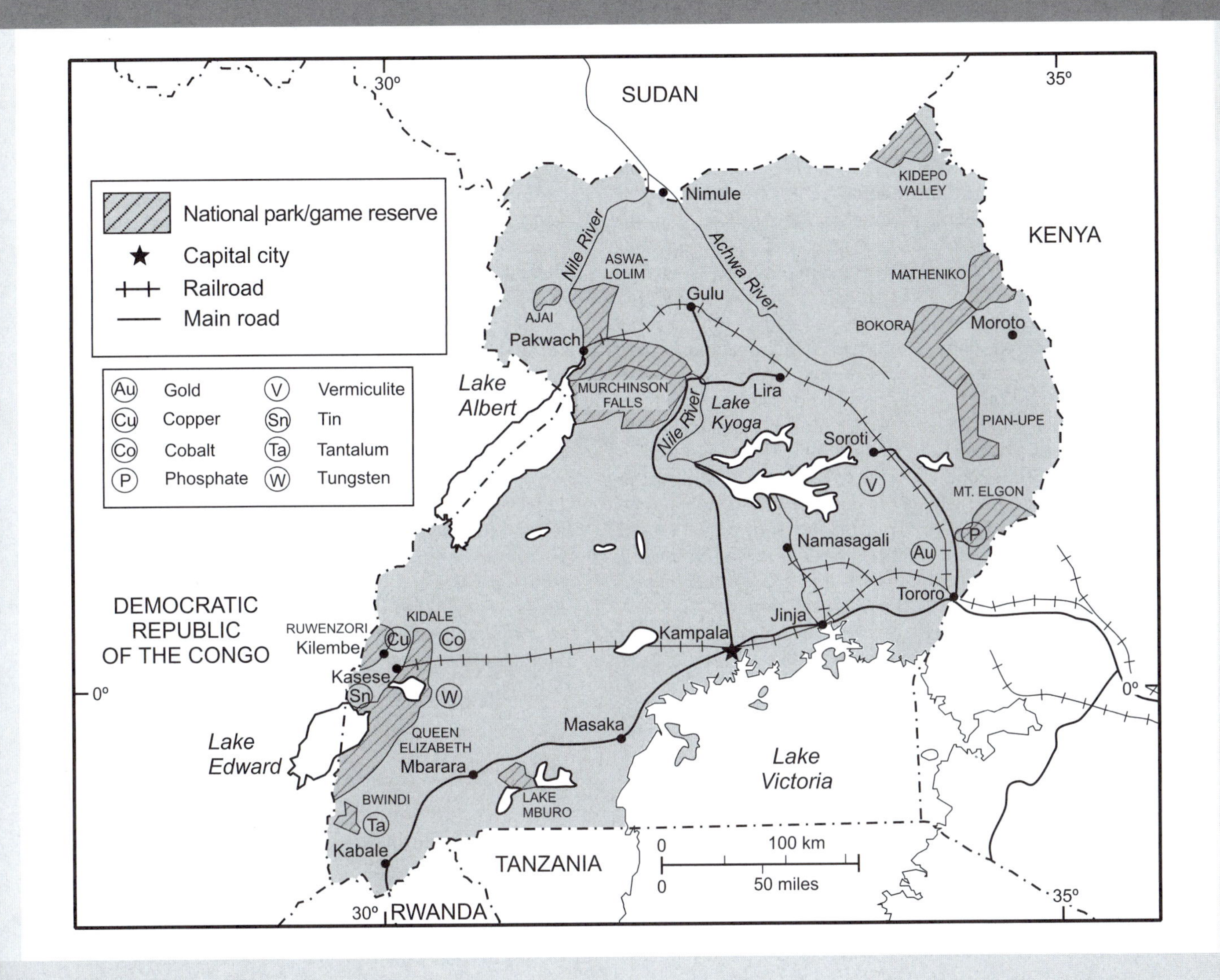

FIGURE 19-3 Towns transportation, and resources of Uganda.

these experts recommended that Uganda be served by a strong central democratic government, with which Buganda would be a federal relationship and Toro, Ankole, and Bunyoro in a semifederal relationship. Broadly on this basis, the political area became the state of Uganda late in 1962.

It soon became clear that Uganda's strong regional loyalties, fostered by the federal arrangement, were a hindrance to coherent development planning. Friction increased between the "kingdoms" as old land disputes and ideologies began to conflict. Events rapidly moved to a crisis, and following a series of uprisings along the borders with Sudan and the then Zaïre, Prime Minister Milton Obote, a northerner of the Lango ethnic group, suspended the constitution. In 1966 the federal status of Buganda and the other kingdoms was abolished, and a unitary state was created. Violence erupted in Buganda, the kabaka fled to England, and Buganda was

BOX 19-1

THE ASIAN EXODUS

In August 1972 President Idi Amin ordered all noncitizen Asians to leave Uganda within 3 months. He accused this population of being isolated and corrupt, of deliberately sabotaging African competition in commerce and industry, of being disloyal to the government, and of hoarding goods and money. By November, all but a few hundred of Uganda's estimated 75,000 Asians had been evacuated, leaving behind assets worth $400 million. Most held British passports, but perhaps as many as 15,000 Kenya nationals also fled rather than face persecution and possible extermination. The majority entered Britain against considerable public opposition, while others were resettled in Canada, India, and elsewhere. Tanzania, Zambia, and Malawi vociferously criticized the expulsion, while Kenya (with 140,000 Asians of its own) remained silent. Almost all the Asians' property was appropriated by government officials and the armed forces, and the economy was thrown into complete disarray. Amin, who was from a minority Muslim ethnic group in Uganda, sent a delegation to Pakistan in 1974 to specifically recruit Muslim doctors, lawyers, mechanics, and teachers to alleviate the shortage of (predominantly Hindi) skilled labor he himself had created.

The Asians (mainly of Indian origin and of diverse linguistic and religious backgrounds) first came to Uganda as indentured laborers to build the Kenya-Uganda railway. Once given their freedom, few opted to return but stayed to become traders and artisans, and gradually to occupy skilled and semiskilled jobs in industry and the civil service. By 1920, it was colonial policy that trade and commerce be reserved for Asians, who were nevertheless restricted from owning land. Because of this, most Asians were urban dwellers, and they bought and sold local produce and acted as a source of rural credit. They once controlled 74% of Uganda's wholesale trade and much of the sugar industry. Before the exodus, Uganda's largest private industrial organization (20,000 employees) was Asian owned. At no time, however, was the economic power of the Asians convertible into political influence.

parceled into four districts: Masaka, Mubende, West Mengo, and East Mengo.

Idi Amin, Obote's illiterate protégé and hatchet man—and later his bitter rival—took control of Uganda in January 1971. He immediately began purging the military of potential Obote supporters by executing northern troops (mainly Acholi and Langi). Amin's military coup was not immediately accepted in the region: the Organization of African Unity was hesitant, and Kenya, Tanzania, and Zambia rejected it. Amin affected a pro-West, pro-Israeli orientation, and his economic ideology was objectionable to Julius Nyerere and Kenneth Kaunda, the socialist presidents of Tanzania and Zambia respectively. Tanzania welcomed Obote and helped him raise a military force that, along with the Tanzanian army, would invade Uganda and depose Amin.

Amin's military administration was brutal and murderous; although no exact figures are available, it is estimated that about 300,000 people were murdered during his time in office (Byrnes 1990). Amin expelled the Asians (mainly Hindis) in 1972 (Box 19-1) and seized their property, ostensibly for the benefit of the "common Ugandan"; fact, however, he distributed the expropriated assets to his army cronies (Kabmegyere 1972). Amin, a nominal Muslim who found his religion in the process of seeking financial assistance from the Islamic states of Libya and Saudi Arabia, fanned the flames of religious communalism in Uganda and pitted majority Christians against Muslims. The army virtually looted the country, with senior military officers running the parastatals for their personal enrichment. As the internal political situation became increasingly unstable, Amin attacked Tanzania in 1978 under the pretext that Nyerere was the cause of his problems. Tanzania mobilized its army and, with the support of Obote's exiles, entered Kampala in merely five months. Amin fled to Libya in 1979, then to Saudi Arabia, where he died in 2003.

In a highly flawed election held in 1980, Obote's party, the Ugandan People's Congress (UPC), took over the government. Obote's second period in power had begun, and it was to be even more difficult, traumatic, and murderous for the Ugandan people than the tyrannical reign of Amin. Civil war began shortly after Obote resumed power. The rebel army, led by a southerner, Yoweri Museveni from Ankole, called itself the National Resistance Army (NRA). Obote's army, comprising mainly of Acholi troops, herded southerners into concentration camps and pillaged the south. As Museveni's army progressively occupied more and more territory and it became apparent that the tide of battle would ultimately go in his favor, Obote fled. The battle raged on for one more year, with Sudanese mercenaries fighting with the (mainly) Acholi troops on the government side. In early January 1986 the NRA marched on Kampala and were enthusiastically welcomed by the population. The mercenaries and the government army fled north.

Museveni inherited a shattered economy and a traumatized population in 1986, but his government has acted in forthright ways to rehabilitate the economy, the country, and the people. He has liberalized the economy, removed price controls, brought down inflation, and encouraged foreign investment and entrepreneurialism. Uganda was one of the earliest countries to be hit by HIV/AIDS and, during the height of the deadly epidemic, the Ugandan government—in stark contrast to many other African countries—publicly discussed the problem and adopted an aggressive program of AIDS education that helped reduce the incidence of the disease in the country. Asians expelled during Amin's time have been invited to return, reclaim their property, and be compensated for their losses. Many have accepted the invitation. The country experienced solid economic growth throughout the late 1980s to the present. Poverty, although still high, has declined throughout the country (Bigsten and Kayizzi-Mugerwa 2001). Museveni put away his fatigues and was elected president in 1996 and reelected in 2003 and 2006. Up until the 2006 elections political parties were forbidden to field candidates for political office; that is, candidates had to run for office as individuals. According to Museveni, the National Resistance Movement (heir to the National Resistance Army) was a mass movement rather than a political party.

Political instability continues in the northern areas caused by the formerly Sudan-supported Lord's Resistance Army (LRA), which is predominantly an Acholi rebellion. The LRA surfaced in 1987 but has been very active since 1995. Hundreds of thousands of people moved from their villages to towns and camps to protect themselves. The objective of the LRA is messianic; its leader, Joseph Kony, wishes to overthrow the present government and rule by the biblical Ten Commandments. LRA activity consists of looting, rape, murdering villagers, and kidnaping children to become child soldiers. It has recently targeted Catholic clerics. The Sudanese government cut its support for the rebels in 2002 and allowed the Ugandan army to pursue the LRA into Sudanese territory. Nevertheless, at the end of 2003 there were over 900,000 internally displaced persons in northern Uganda.

In an apparent attempt to forge a national fabric out of the disparate pieces of Ugandan cloth, Museveni reorganized the political units of the country at two different levels. At the lowest level the 18 ethnically based, districts were divided into 56, and gone were the names of the former kingdoms (including Buganda), as well as any ethnically associated designation (compare Figures 19-2 and 19-4). Each new district took the name of its new capital. In addition, four new regions were created. Except for the Central Region, which corresponds to the former Buganda Kingdom, the regions cut across ethnic boundaries.

The Factor of Economic Geography

Uganda is one of three African counties (Botswana and Mauritius are the other two) that have experienced sustained economic growth over the past 15 years—the result of sound economic policy by the Museveni government. The national budget, as in so many African countries, is heavily dependent on foreign aid, and there are the familiar, if less widespread, problems of corruption and mismanagement that must be addressed if Uganda is to continue to grow.

Geographically, perched high on the mountains of the African Rift, Uganda has an enviable climate, but the country has always suffered from its landlocked location. The railroad from Mombasa reached Kisumu on Lake Victoria as early as 1901, at which time goods were transported to the railhead by steamer, but later the railhead was extended into Uganda to eliminate the water link. Development has closely been tied to the expansion of the transport network, and a number of feeder roads to the central railroad, a branch line northward to Gulu and Pakwach, another to Kampala and on to Kasese, near the southwestern border, and

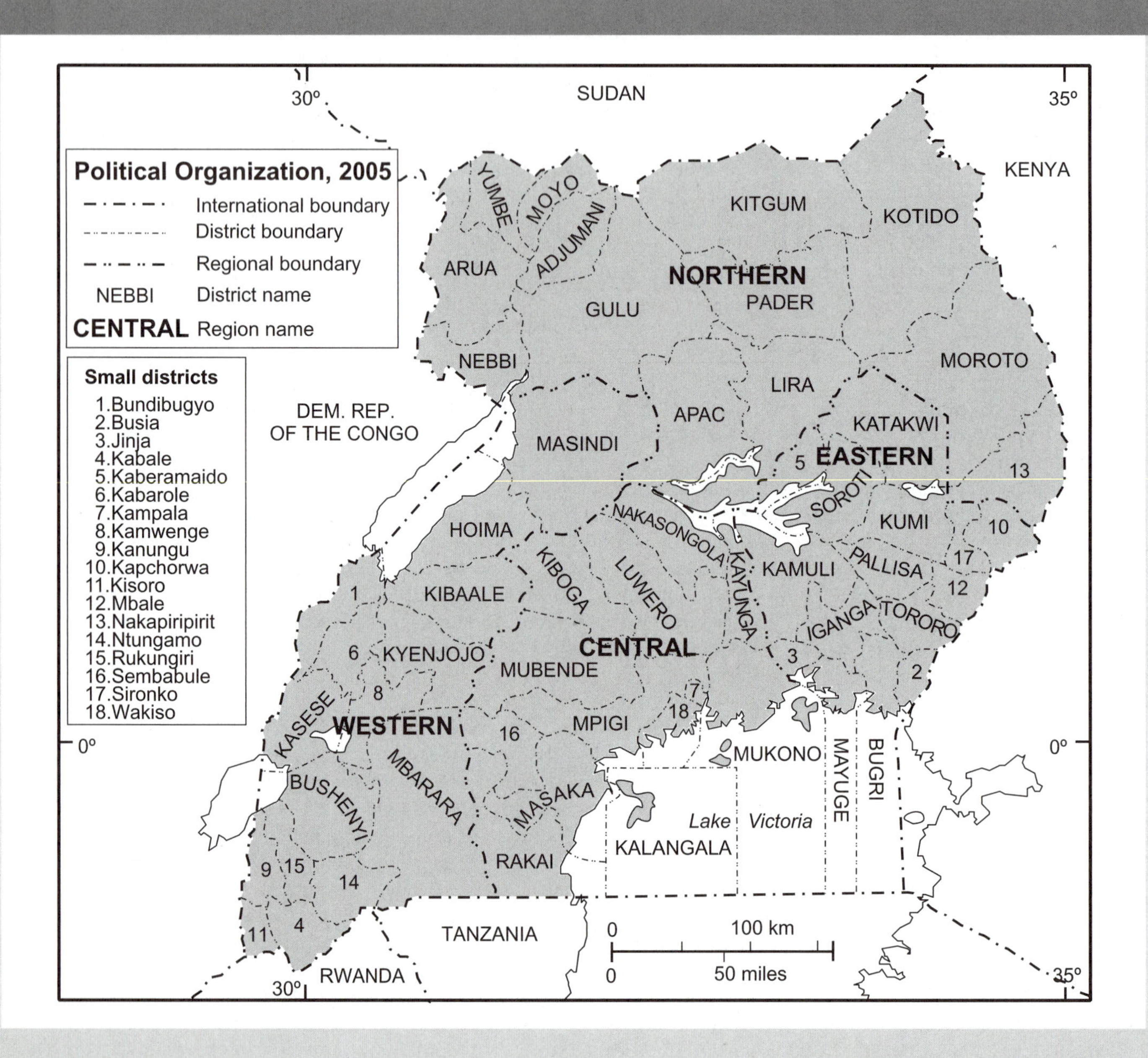

FIGURE 19-4 Current administrative districts and regions of Uganda.

an increase in inland water transport have been important factors. Cotton was introduced during the first decade of the twentieth century and proved to be a success, although the industry was severely affected by fluctuating prices on the world market. After World War II, cotton cultivation rose sharply, partly as a result of the establishment of African cooperative unions and their participation in the cotton ginning industry. Until then, this industry had been largely a non-African enterprise. Close to 3 million acres (1.2 million ha) of

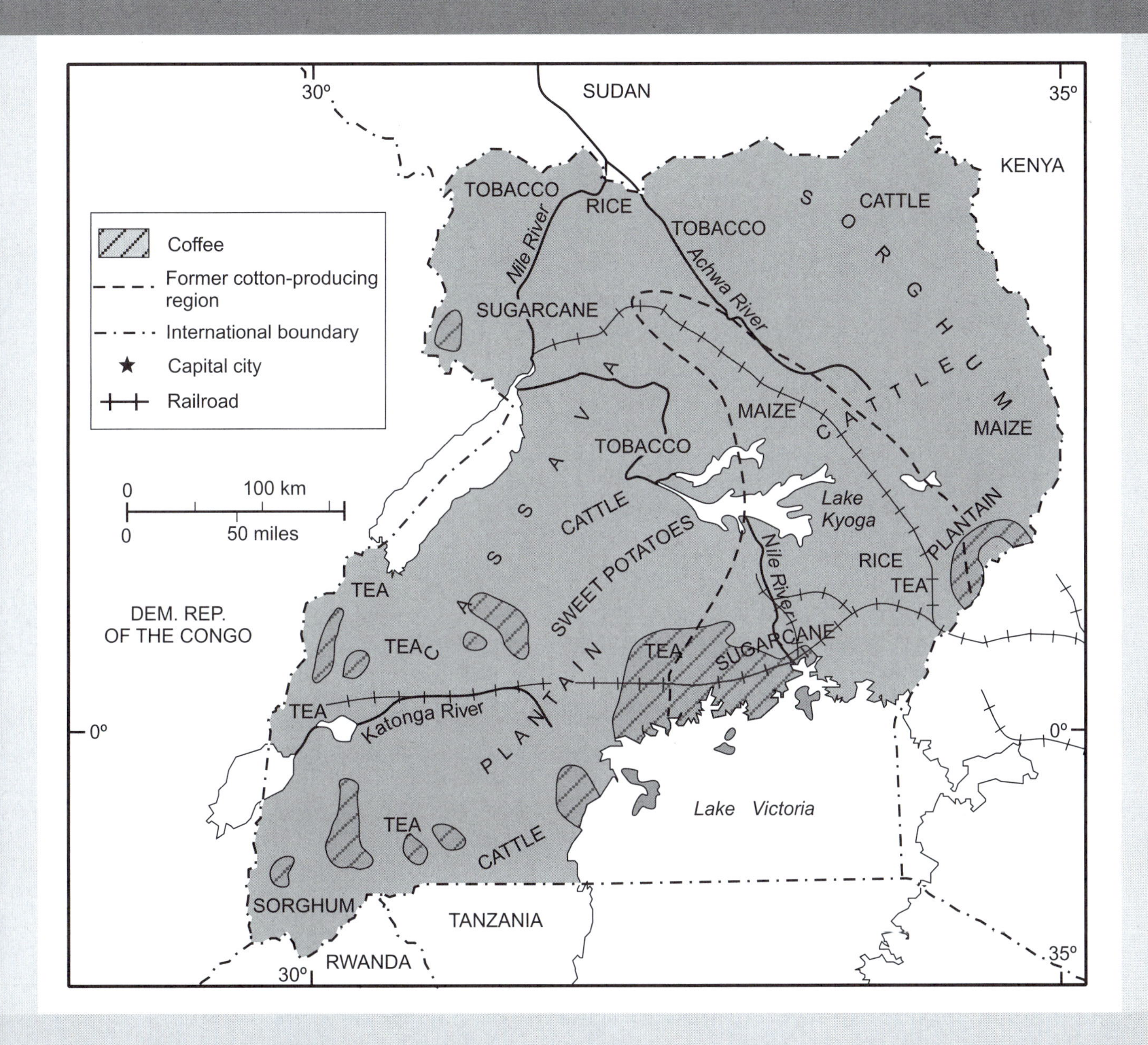

FIGURE 19-5 Agricultural regions of Uganda.

cotton was harvested at the peak of cotton production in 1972. As indicated in Figure 19-5, the highest intensity was in the former Busoga, Teso, and Bukedi districts of the southeast, where climatic conditions were good and distances to the railway were not great. Formerly, cotton was grown in many parts of Uganda and was expanding along the southern foothills of the Ruwenzori and north of Lake Kyoga. Most commonly it was grown as a rain-fed crop, without irrigation, on plots averaging less than one acre (0.5 ha). Cotton

TABLE 19-1 CHANGE IN PRODUCTION FOR SELECTED CROPS, 1961-1965 TO 1999-2003.

Crop Type	Crop	Average Production (mt) 1961-1965	Average Production (mt) 1999-2003	1999-2003 as Percent of 1961-1965 Change (%)
Food crop	Bananas	200,000	610,000	305
	Plantains	3,900,000	9,599,000	246
	Cassava	1,120,000	5,275,800	462
	Sweet potatoes	572,000	2,491,800	436
	Maize	215,200	1,148,000	533
	Millet	444,200	579,600	130
	Sorghum	275,600	403,800	147
Cash crop	Coffee, green	139,160	193,553	139
	Peanuts in shell	127,400	144,000	113
	Rice, paddy	3,180	111,600	3,509
	Seed cotton	203,221	66,800	33
	Sesame seed	32,600	101,600	312
	Soybeans	880	139,400	15,841
	Tea	6,714	31,133	464
	Tobacco leaves	4,511	26,717	592

Source: FAO (2004).

declined with the chaos of Amin's governance. One of the principal reasons for the failure of the crop to recover was widespread looting of cattle, which undermined the practice of animal traction. Although cotton is still grown and exported, the cotton area harvested today is 30% of what it was before Amin came to power (Table 19-1).

The cultivation of most other crops declined under Amin's rule as well, especially the cash crops (Figure 19-6). There was a massive increase in the growing of subsistence crops, bananas/plantains, sweet potatoes, and cassava as, under conditions of uncertainty and danger, people focused on their immediate needs. Later, however, as civil war spread and people fled, even these essentials were affected (Figure 19-7).

Uganda's production of coffee has increased in comparison to the early 1960s (Table 19-1). Uganda produces over 20% of the coffee exported from Africa on average, but its share of the world market has been declining. This is because new producers are coming on line (Vietnam, for example, which is moving toward a market economy), and others have been increasing their output (Brazil); yet others (Angola and Côte d'Ivoire) have been declining. Initially, coffee was grown almost exclusively by the few European plantation owners and Asians. Later, Africans took an interest in this cash crop, and by the time of independence, about 30,000 acres (12,150 ha) was in plantations and some 600,000 (243,000 ha) in individual African farms. After World War II, coffee cultivation rose dramatically in Buganda, to the extent that coffee is now Uganda's leading export.

The Baganda grow mainly the robusta variety, which is not as prized as arabica coffee but is especially useful in the preparation of instant coffee, hence the ready market for this product. Arabica coffee is grown on the slopes of Mount Elgon and in Kigezi, but its acreage is far less than that of robusta. Coffee is economical in its land requirements and is more popular than cotton was in its heyday because it can be harvested after only about 80 full days of work per acre versus 140 days for cotton. Coffee and tea are by far the most important cash crops and regularly yield about 80% of the total export earnings. Tea is produced on both estates and smallholdings, the major regions being Fort Portal in the western highlands, Jinja, and Mount Elgon. It was once exclusively an estate crop, but since the mid-1960s the government has encouraged smallholders (especially in Buganda) to produce more. Most of the estates were once owned

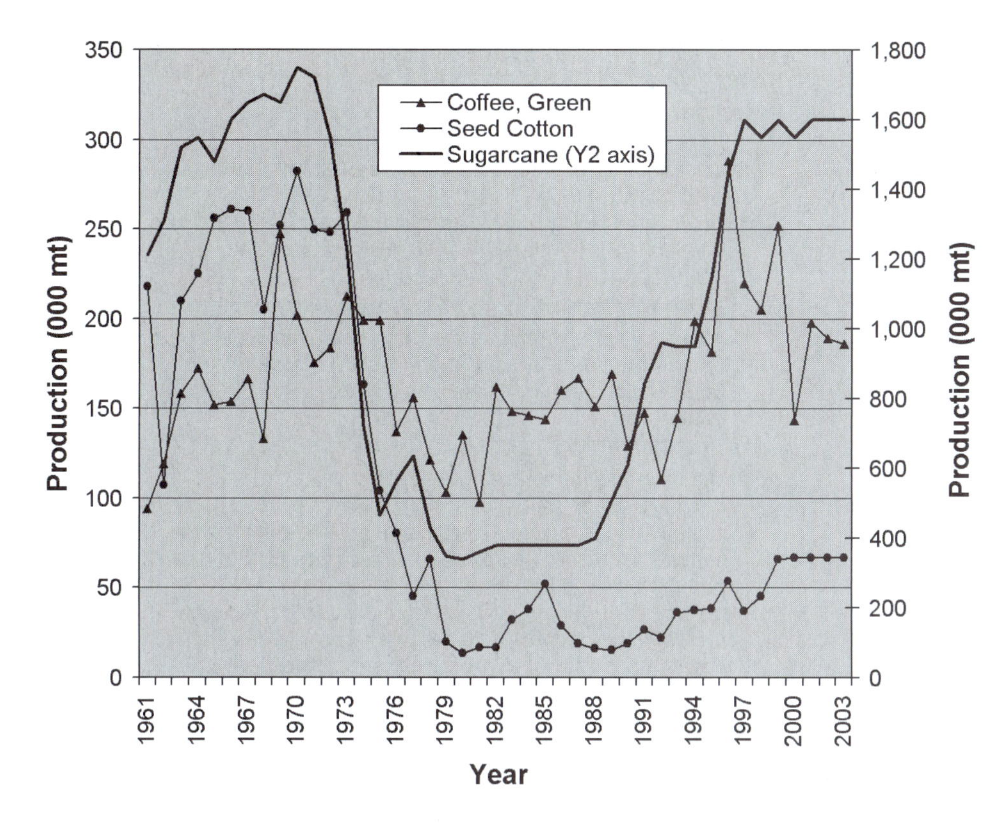

FIGURE 19-6 Production of the cash crops coffee, cotton, and sugarcane in Uganda, 1961 to 2003.

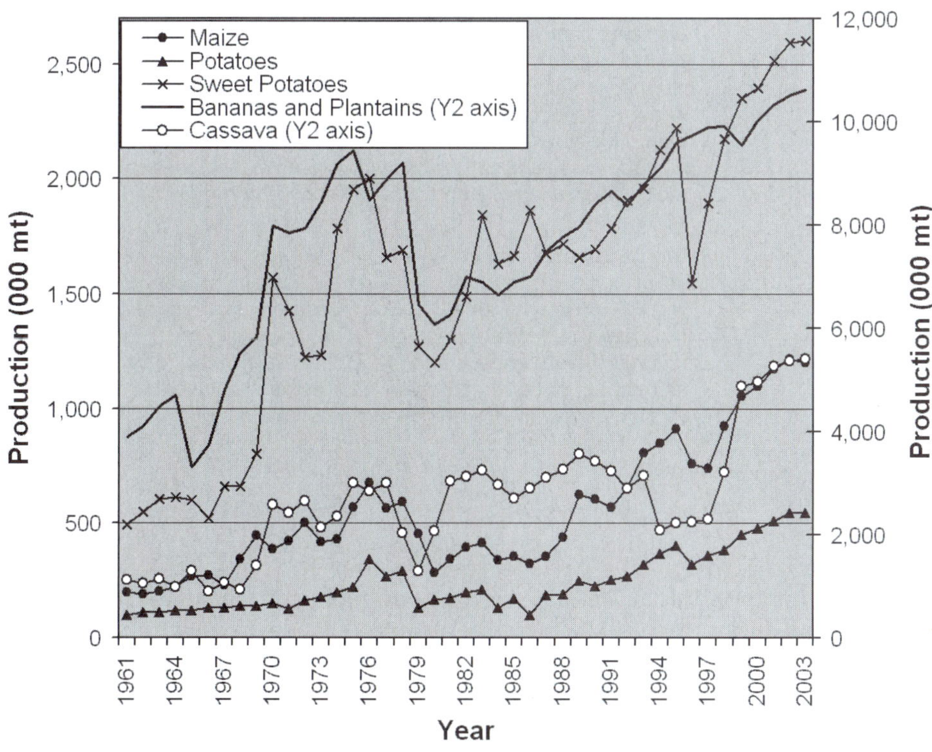

FIGURE 19-7 Production of food crops in Uganda, 1961 to 2003.

TABLE 19-2 INDICATORS OF AN EMERGING ECONOMY AND UGANDA'S RELATIVE PERFORMANCE.

Area	Variables	Uganda's Case
Macroeconomy	Macroeconomic stability	Stabilization efforts have been successful. Inflation is under control.
	International competitiveness	External competitiveness is improving slowly. Horticulture is internationally competitive but there are only a few manufacturing firms that are standouts in regional trade. Tariffs still need to be brought down.
Microeconomy	Competitive domestic markets	Less government interference in the domestic economy than in the past. End of Coffee Marketing Board's monopoly. Competition is still weak in other areas, particularly utilities.
	Stable financial system	An institutional constraint that the country still must deal with.
Human resources and infrastructure	Human capital for competitive investment	Some progress in the area of education with the introduction of universal primary education. But technical skills are scarce. Healthcare services are limited with consequent lower health status indicators for the Ugandan population.
	An effective physical infrastructure	Major roads are good in the south but poor in the north. The telecommunications and energy infrastructure need development but both are being restructured with international investment.
Governance and policies	Unbiased institutions	Some improvements but the process has been slow and uneven. Private banks have made nonperforming loans to politicians and their relatives.
	Good governance	Despite public sector streamlining and significant improvements in the 1990s there is still rent seeking and corruption by officials. Decentralization efforts may help development spread outside of the southern core area.
	A broad-based development pattern	Solid gains in poverty reduction have been made but in northern regions, especially where civil unrest has occurred, there is cause for concern.
	Political maturity	The main constraint to Uganda's future growth is political. Too much corruption or political interference in the economy may stifle growth.
Self-reliance	Reduced aid dependency	There has been a reduction in aid to GDP.
	Controlled level of foreign debt	Debt pressure on the national budget has been relieved by the HIPC (Heavily-Indebted Poor Country) program of the International Monetary Fund.
	Domestic savings and foreign private investment	Domestic private and foreign private investment have been increasing, but a large share of investment still comes from foreign governments of international agencies.

Source: Adapted from Bigsten and Kayizzi-Mugerwa (2001)

by resident Asians and expatriate British, but they were taken over by the Ugandan government during the Amin era and "managed" by the Uganda Development Corporation. Uganda's sugar estates were likewise expropriated from Asians, and the production of both tea and sugar was unstable until the mid-1980s, when Museveni took control. To further diversify the agricultural economy, the government is encouraging the increased production of such nontraditional cash crops as soybeans (Table 19-1), cocoa, avocado, mango, papaya, jackfruit, passion fruit, pineapple, cut flowers, and vanilla beans.

The export crop economy is concentrated in the south between Lakes Victoria and Kyoga. In the northeast, southwest, and in the northern areas, food crops predominate. The northeast is the least productive of all regions, and its peoples, the pastoral Karamojong, have resisted government efforts to increase food production and improve their livestock. Spatially, the peripheral areas belong to the subsistence sector (McMaster 1962) and are special problem areas in the economy, while Buganda forms the economic core, its population politically and socially dominant in national affairs.

Compared with its agricultural output, Uganda's mineral production is small, unlikely to ever contribute more than a small percentage to the GDP. New mineral regulations that encourage exploration and investment have attracted international attention, and in recent years exploration for petroleum has been undertaken and hydrocarbons discovered. Whether these resources are present in any economic quantity is not yet clear. Uganda's copper has long been mined at Kilembe (Figure 19-3). Mining began in 1948, but production was held down until the Mombasa-to-Kampala railroad had been extended west to Kasese (in 1956), and a smelter was completed at Jinja. The smelter is supplied with low-cost power from the Owen Falls Dam (close to where the Nile leaves Lake Victoria). Iron is mined at six different locations in Uganda and is processed at the steel mills at Jinja. Rock phosphate is mined at Sukulu in the east near the Kenyan border. Uganda mines some gold and columbium–tantalite but has been reexporting production from the Democratic Republic of the Congo, particularly during the DRC civil war.

Jinja, with a population of 92,000, has developed into Uganda's main manufacturing town; its major industries include copper and iron smelting, steel rolling, grain milling, textiles, and food processing. There are two steel mills and five steel processing plants in Jinja. Kampala (1.3 million), like most capitals and primate cities, has attracted more labor from its surroundings than can be employed. It has several light industries dependent on electricity from Jinja, but, on the, whole, Ugandan industry has been struggling to compete with imported goods. Industrial expansion was slower than projected during the Amin and Obote years, but the liberalization policies and development priorities of the Museveni government have encouraged investment.

Impressive growth rates, favorable investment climate, improvements with political stability and other factors have prompted analysts to ask whether Uganda has become an "emerging economy" (Bigsten and Kayizzi-Mugerwa 2001), one capable of sustainable economic growth. The answer is "no" although great progress has been made in many areas. Table 19-2 indicates areas in which progress has been made and where more needs to be done. The evidence indicates that the emergence of Uganda as a sustainable economy is a work in progress linked intimately to political concepts identified by Jones (1954) decades ago.

THE EAST AFRICAN COMMUNITY

Several frameworks for regional integration have united Kenya, Tanzania, and Uganda, and the countries have long cooperated in a wide range of nonpolitical activities. In 1917 Kenya and Uganda formed a customs union that was later joined by Tanganyika. Interterritorial cooperation was first formalized in 1948 by the East African High Commission. This agency provided a customs union, established a common external tariff, issued currency and postage, and dealt with common services in transport and communications, research, and education. Following independence, these integrated activities were reconstituted, and the High Commission was replaced by the East African Common Services Organization (1961–1967), which many observers believed would lead to the political federation of the three member states. But the new organization ran into difficulties because of lack of joint planning and fiscal policy, separate political policies followed in each state, and Kenya's dominant economic position. Kenya had the lion's share, especially in manufacturing, and was exporting more manufactured goods within the organization than its member states. In 1964 an agreement was signed in Kampala whose aim was to reduce Kenya's industrial primacy by redistributing some of the industry to less prosperous areas. But because of Nairobi's natural advantages and its access to resources, capital, and technology, Kenya's lead went unchecked.

In 1967 the East African Common Services Organization was superseded by the East African Community

(EAC). The EAC aimed to strengthen the ties between the member states through a common market, a common customs tariff, and a range of public services; in this way, the members hoped to achieve balanced economic growth, "the benefits of which shall be equitably shared." The treaty that enabled the organization had several innovative provisions that were designed to permit greater decentralization than had previously been possible. The East African Harbour Authority moved to Dar es Salaam, the post and telegraph offices moved to Kampala, and the EAC's administrative headquarters were located in Arusha, Tanzania. A transfer tax system was instituted that permited Uganda and Tanzania, under certain conditions, to tax imports of manufactures from Kenya for the purpose of protecting their own infant industries. The treaty also established the East African Development Bank to provide financial and technical assistance to promote industry. The bank was required to devote 77.5% of its funds equally to Tanzania and Uganda and the balance to Kenya.

Despite these controls, Kenya's industry and exports grew more rapidly than either Uganda's or Tanzania's. The EAC was disbanded in 1977, and the three countries went their separate ways until the Commission for East African Cooperation (1993–1999), which provided the framework for the emergence of the East African Community in 2000. The objective of the new EAC is to eventually move to political and economic union of the three countries based on free-market principles, with emphasis on the private sector and trade liberalization. The EAC is an example of African economic and perhaps political unity. Other East African countries have formally applied to join. Rwanda applied for membership in 1996 and Burundi in 1999, but they were never approved.

There is much merit to a broader association, but there are many problems to be resolved. The lack of effective road and rail links between Tanzania and its northern partners is a major impediment to the community's stated objectives. Geographically, Kenya and Uganda are united, but Tanzania is an outsider. Each country has internal problems but perhaps, in a manner similar to what has happened in parts of the European Union, East African supranationalism will permit a degree of devolution and local autonomy that will peacefully quell ethnic nationalism and secessionist movements within each country. In any event, the next step forward is to encourage the acceptance of the political idea of the East African Union among the populations of Kenya, Tanzania, and Uganda.

BIBLIOGRAPHY

Apter, D. E. 1961. *The Political Kingdom in Uganda*. Oxford: Oxford University Press.

Bakwesegha, C. J. 1974. Patterns and processes of spatial development: The case of Uganda. *East African Geographical Review*, 12: 46–64.

Bigsten, A., and S. Kayizzi-Mugerwa. 2001. *Is Uganda an Emerging Economy*? Uppsala, Sweden: Nordiska Afrikainstitutet.

Byrnes, R. M. 1990. *Uganda: A Country Study*. Washington, D.C.: U.S. Library of Congress.

FAO. (2004). FAOSTAT. United Nations Food and Agriculture Organization. Rome: U.N. Food and Agriculture Organization.

Ibingira, G. S. K. 1973. *The Forging of an African Nation*. New York: Viking Press.

Jones, S. B. 1954. A unified field theory of political geography. *Annals of the Association of American Geographers*, 44: 111–123.

Kabwegyere, T. 1972. The Asian question in Uganda. *East African Journal*, 9: 10–13.

McMaster, D. N. 1962. *A Subsistence Crop Geography of Uganda*. London: Geographical Publications.

Uganda, government of. 1967. *Atlas of Uganda*. Entebbe: Lands and Surveys Department.

World Bank. 2004. World Development Indicators. New York: World Bank.

CHAPTER 20

Rwanda and Burundi

Reaping the Whirlwind

Rwanda and Burundi, land of a thousand hills, enchanting scenery, beautiful climate, killing fields, genocide. These two very similar states (Table 20-1) epitomize to many observers both outside and inside Africa the fundamental intractability of Africa's social problems and the utter depths of injustice and inhumanity to which those problems can lead. Although it is difficult to explain such events, such brutal treatment of one group by another is not uncommon in history as the examples of Bosnia, Nazi Germany, the destruction of the native peoples of North America, the horrors of the African slave trade, and the callous brutality of the Islamic jihadists easily demonstrate. In this chapter we will briefly look at land use, economy, and demography as these factors have influenced the origin of the Rwanda and Burundi states, and the genesis of the conflicts.

LIVELIHOOD AND POPULATION

The republics of Rwanda and Burundi lie in one of Africa's least accessible regions. Physiographically, they form part of highland equatorial Africa: volcanic mountain masses reach over 14,000 feet (4,260 m) in the Virunga Range of the north, the Western Rift Valley forms the western boundary, and the Nile rises in the east. Both countries are dominated by the backbone of mountains, which trends north–south and forms the eastern edge of the great rift occupied by Lakes Kivu and Tanganyika. Most areas lie between 3,000 and 8,000 feet (900–2,400 m), and the higher elevations are the favored parts (Figure 20-1). There, the dreaded tsetse fly is absent, soils are often volcanic and rich, and precipitation exceeds 50 inches (1,270 mm) annually. By contrast, many of the valleys in the peripheral areas are low, hot, and disease infested. Thus, although these countries possess rich agricultural areas that could support high population densities, the actual pattern of occupation of Rwanda and Burundi's 21,000 square miles (54,000 km^2) shows that many prefer to live in the highlands (Figure 20-2).

Vertically, three main crop regions can be recognized, of which the lowest (2,500–4,500 ft; 760–1,370 m) is favorable mainly to subsistence crops such as corn, beans, and bananas, although some coffee and cotton are grown. The middle zone (4,500–6,500 ft; 1,370–1,980 m), covering the greatest part of both countries, is the most densely populated, and in addition to the subsistence crops normally grown at these elevations in equatorial regions, coffee, tobacco and irish potatoes are cultivated. Finally, in the highest zone (6,500 ft; 1,980 m and over), cash crops such as wheat, barley, tea, tobacco, flowers, and pyrethrum can be grown, and some European settlers, who at the time of independence numbered over 8,000, have established plantations. In each of the three zones, subsistence crops such as corn, beans, sweet potatoes, cassava, and a variety of cereals are grown, occupying the vast majority of the acres of cultivated land. Although it can rain anytime in the higher mountains, there are two rainy seasons, one from September to December, and one from February to May. June, July, and August are the months with the lowest precipitation.

Arabica coffee is the dominant export of each country (80–85% of the total in Burundi, 40–45% in Rwanda), most of it grown in small stands scattered among basic food crops. Ninety percent of Burundi's foreign exchange is earned through the export of coffee and tea. Annual production of coffee has averaged

TABLE 20-1 DATA TABLE FOR RWANDA AND BURUNDI.

Variable	Burundi		Rwanda	
	1975	2003	1975	2003
Area	10,745 mi^2 (27,830 km^2)		10,170 mi^2 (26,340 km^2)	
Population (millions)	3.8	7.8	4.2	8.6
Population density	354/mi^2 (137/km^2)	726/mi^2 (301/km^2)	413/mi^2 (160/km^2)	845/mi^2 (326/km^2)
Urban population (%)	2	8	3	5
Population engaged in agriculture (%)	96	–	92	–
Crude birthrate (per 1000)	48	40	50	40
Crude death rate (per 1000)	25	19	24	21
People ages 15–49 with HIV/AIDS in 2001 (%)	0	8.3	0	8.9
Infant mortality rate (per 1000)	150	75	133	107
Ethnic groups (% of population)	Hutu (84) Tutsi (15) Twa (1)	Hutu (84) Tutsi (15) Twa (1)	Hutu (88) Tutsi (11) Twa (1)	Hutu (89) Tutsi (10) Twa (1)

Sources: Best and de Blij (1977), Population Reference Bureau (1975, 2003), SIL International (2004).

about 20,000 tons in each country since 1976, the bulk of which is sold to the United States and the EU, and shipped through Dar es Salaam and Mombasa. Rwanda normally has produced twice as much tea as Burundi, averaging 13,300 and 6,600 metric tons respectively from 1995 to 2004 (FAO 2005). Both countries are attempting to diversify their export base by increasing production of tea, irish potatoes, vegetables, cotton, pyrethrum, forestry products, and minerals. Rwanda exports small quantities of tin concentrates and wolframite, gold, tantalum, and beryllium. Burundi is even less well endowed with minerals (Figure 20-3).

Rwanda and Burundi are the most densely populated countries of Africa and are among the least urbanized and poorest (Table 20-1). Practically everyone lives directly off the land, of which about 40% is considered to be cultivable, and between 30 and 40% is under pasture. The remainder is in forest, game reserve, and lakes, or is excessively steep and rocky. In Rwanda the population density is 845 per square mile (326/km^2), and in Burundi 726 per square mile (301/km^2). In 1900, the population density was estimated at 22 and 21 per square mile (58 and 55/km^2) for Rwanda and Burundi, respectively (Chrétian 2003).

Widespread poverty and overcrowding are major obstacles to development, but the need for population control has been strongly resisted by the Roman Catholic Church (du Bois 1973). Close to 90% of the people are rural and widely dispersed, and not concentrated in villages, thus making the provision of schools and hospitals more costly. Bujumbura, the capital of Burundi and formerly of the whole trusteeship, is the only large town; its population of 340,000 is up from 80,000 in 1975. Kigali, capital and largest town in Rwanda, has a population of 305,000 (only 30,000 in 1975). Virtually all nonagricultural activities are concentrated in these two centers.

Since most of the opportunities for farming and pastoralism lie in the higher areas, the central plateau, running north and south along the eastern side of the Nile–Congo divide, is by far the most densely populated part of both countries (Figures 20-1 and 20-2). Toward the lower western slopes, population densities decrease considerably, and they are lowest in the disease-ridden valleys of the lower east. As part of their 10-year plan of the 1950s, the Belgians designed a program for the relocation of people, with the object of reducing population pressure where it had become most serious. But the task was a difficult one, for it meant a considerable change in habitat, crop possibilities, and climate for those who were transferred. The only empty areas in Rwanda and Burundi were the lower valleys, where tsetse and other pests had kept the population totals low. Having cleared some of these areas for resettlement, the Belgians made a start with the alleviation of population pressure on the plateau. By 1960, a third of the holdings were

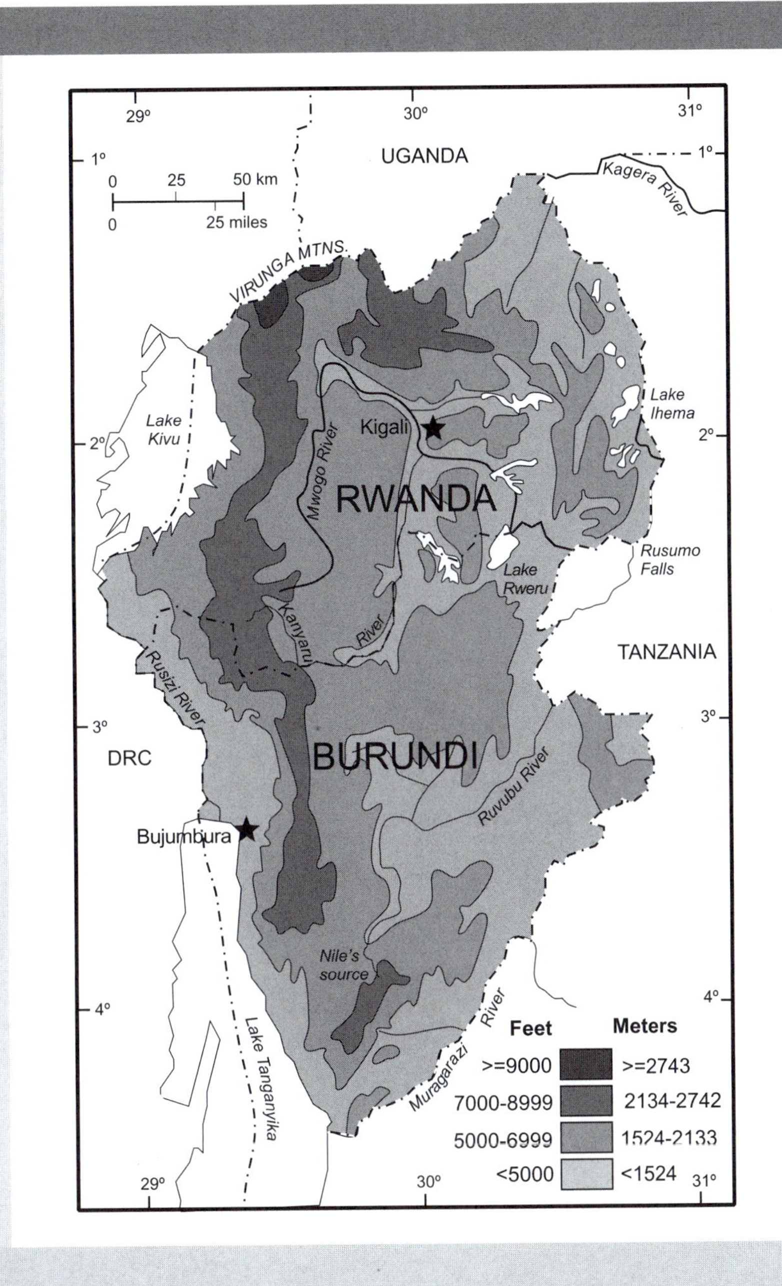

FIGURE 20-1 Relief and drainage map of Rwanda and Burundi.

abandoned because of resistance to the scheme, poor management, and flooding. In 1969 the Burundi government reinstituted the program through the *paysannat* system, which had been developed in the old Belgian Congo. A *paysannat* was a large-scale farm divided into smallholdings of about 4.8 contoured acres (2.0 ha) that have access to roads and on which cash crops must be grown. Communal grazing land and watering points were provided, together with social services including a school, dispensary, veterinary office, and

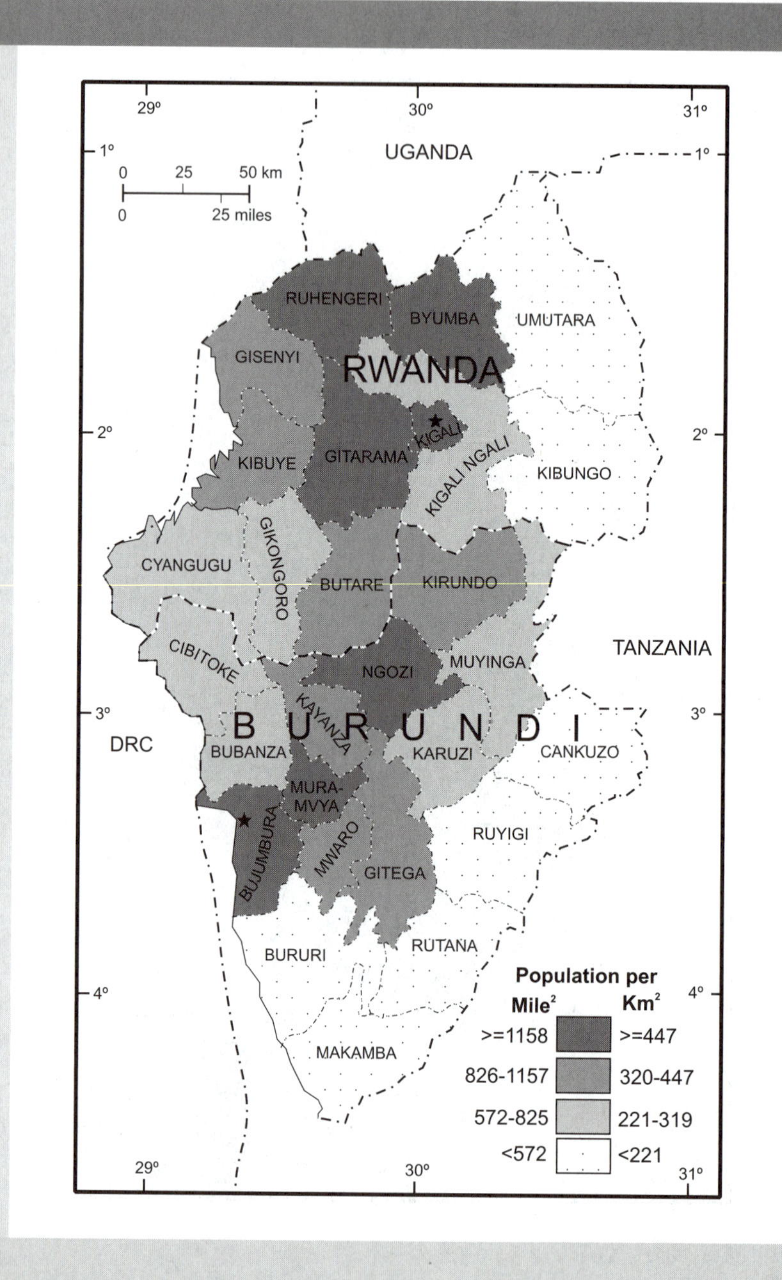

FIGURE 20-2 Population density of Rwanda and Burundi by district.

nutrition center. The people were thus grouped into villages in an attempt to bring greater social as well as economic benefits to the farmer (Baker 1973).

None of these measures, of course, permanently overcame the problem, for the population has continued to grow, and the opportunities for additional land availability constantly dwindle. Population control seems like an unlikely solution, since rural people need large families for farm labor. Off-farm employment could be a solution, but a study on rural

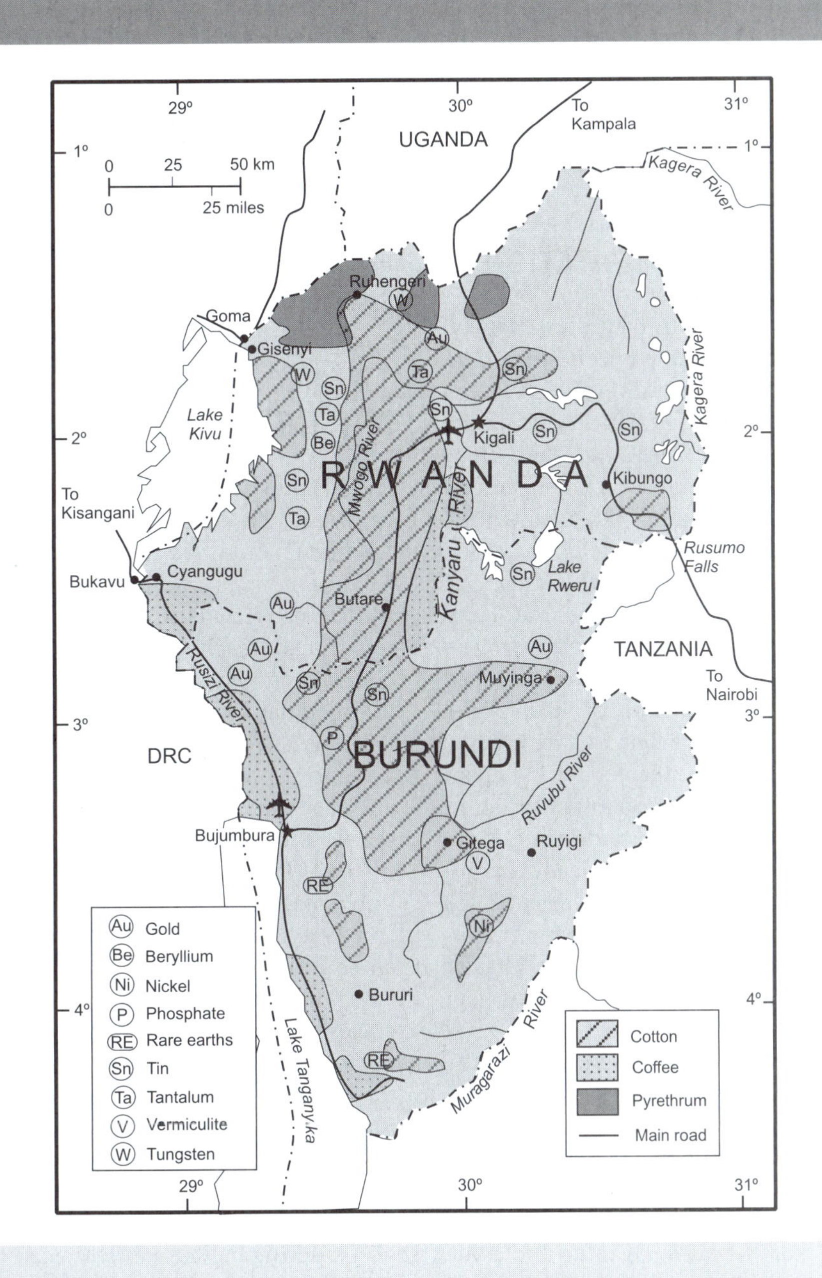

FIGURE 20-3 Agriculture and natural resources of Rwanda and Burundi.

inequality undertaken for the U.S. Agency for International Development (USAID) found that members of well-off rural households, rather than the rural poor, were more likely to take off-farm employment (Clay et al. no date). Perhaps the pastoral industry in Rwanda and Burundi will one day become a major source of revenue. Presently, the pastoralists are not market oriented, but it is unclear how long a subsistence orientation to cattle can persist: the resources of Rwanda and Burundi alike appear to offer very limited possibilities for a rapidly growing population. An expanded production of cash crops could find ready markets, for the range of possibilities is considerable. But any increased cash crop production requires a major increase in the acreage allotted to such crops, and that acreage, under present systems of land occupancy is not available.

POLITICOECONOMIC REORIENTATIONS

Since independence in 1962, both countries have experienced important changes in their economic orientations and political leadership, and both have seen genocide on a scale unsurpassed in Africa. During the Belgian era they were part of the Congo Free Trade Area, which meant that goods moved freely from one Belgian territory to another, and that Ruanda–Urundi was essentially an appendage to the Belgian Congo. High tariffs virtually eliminated trade with Tanzania, Uganda, and Kenya, and most resources and labor moved westward. Adequate communications were never developed: only the poorest roads linked Ruanda–Urundi with the border town of Bukavu and the river town of Kindu, in the Belgian Congo, while Bujumbura was served by lake steamer from Kalemi (formerly Albertville), the Belgian Congo's railhead on Lake Tanganyika. But since Rwanda and Burundi are closer to the Indian Ocean and Dar es Salaam than to the Atlantic and Matadi, transport to the east was faster and cheaper. Thus Belgium secured extraterritorial rights from the British at the port of Dar es Salaam, but these were transit rights only and trade was kept firmly in Belgian hands.

Following the independence of Zaïre and the collapse of its economic and political institutions, Rwanda and Burundi turned toward Uganda and Tanzania for their import requirements and for access to their overseas markets (Baker 1973). But communication routes were almost nonexistent, there being only two dirt tracks from Kigali to Uganda, and none from Burundi to Tanzania. Burundi's only link with its eastern neighbor was via lake steamer to Kigoma, a route also used by the Rwandans. But when relations between Burundi and Rwanda themselves deteriorated in 1963, Rwanda found itself almost entirely dependent on its grossly inadequate northern outlet, the goods being trucked to Kasese and Kampala, and then railed to Mombasa.

Transport has since been improved between Rwanda and Uganda, and an all-weather highway was built to Kampala. Tanzania has joined with Rwanda in developing hydroelectric power on the Kagema River near the Rusumo Falls, and is cooperating with Burundi in mapping the water and mineral resources of the Ruvubu River. Thus, both Rwanda and Burundi are now oriented to the east and have applied for membership in the East African Community—although their membership has been delayed for years.

THE BELGIAN ADMINISTRATION AND INDEPENDENCE

Ruanda and Urundi, prior to World War I, were among the most densely populated districts of German East Africa. When armed hostilities erupted in East Africa, Belgium participated, using Congolese troops to attack the Germans from the west. By September 1916, Belgian forces had reached Tabora, in the central part of German territory, and one battalion actually penetrated to the Indian Ocean at Lindi. At the end of the war, the Belgians had captured not only Ruanda and Urundi, but also most of the District of Kigoma and parts beyond, and were thus in effective control of nearly a third of the people and territory of the former German sphere.

In 1919 the Supreme Council of the League of Nations, meeting at Versailles to consider the question of the dissolution of Germany's colonial empire, placed all of former German East Africa under British mandate. This decision aroused great resentment in Belgian circles, in view of the effort Belgium had expended in the East African war theater and because the area captured by the Belgian forces contained land suitable for European settlement and for livestock raising, opportunities that were rather scarce in the adjacent Congo. Thus, Belgium expressed its desire to retain for occupation, at least, a section of the territory in question, and the matter was reopened for discussion. The League of Nations granted Belgium the right to negotiate directly with Great Britain, and an agreement was reached whereby Belgium received, under mandate, almost the entire area of the territories of Ruanda and Urundi. Boundary definition, delimitation, and demarcation (in part) took place during 1923 and 1924. In 1925 "Ruanda–Urundi" became an integral part of the Belgian Congo. As a result, Belgium came to rule

BOX 20-1

THE GREAT LAKES TRADITION OF HIERARCHICAL ORGANIZATION AND DIVINE KINGSHIP

On the western slopes all along the mountains bordering the uplands . . . . in the area west of Lakes Kivu and Edward, the forests harbor tiny principalities ruled by a *mwámi*. In the 19th century, going north to south, they belonged to Pere, Nyanga, Tembo, and Nyindu speakers. These languages belong to the great lakes group. They strike us as theater-like states, because their minuscule size was matched by elaborate rules for succession, accession to the throne, and royal burial, by a complex titulature surrounding the royal office, and by intricate royal rituals and a plethora of emblems, as if these district-sized kingdoms were the equals of the great kingdoms that lay beyond the rift itself. These societies are in part involuted reminiscences of what others farther east had been before large states appeared in the great lakes area, in part imitations of these kingdoms, and in part quite original elaborations. Divine kingship, a rainmaking ritual as the legitimation par excellence of the monarchy, a stress on divination, and patrilineal clans were all ultimately part of the old great lakes heritage, although some details associated with them may, in fact, be relatively recent borrowings from farther east.

Vansina (1990), p. 185.

the most densely populated region of former German East Africa, including over one-third of the total population of that dependency.

A primary reason for Belgian interest in Ruanda and Urundi was the relationship of this area to Katanga, which was developing and in need of labor. The Belgian government initiated large-scale transfers of the Ruanda–Urundi population to the mining area and soon found itself subjected to criticism, which also had accompanied the 1925 incorporation of Ruanda–Urundi into the Congo. Thus, the labor policy was altered somewhat, but the territories continued to provide large quantities of labor for the mines of Katanga until much later.

In securing the administration of Ruanda–Urundi, however, Belgium had inherited problems as well as assets. The area of the two territories combined is just under 21,000 square miles (54,000 km^2), and their total population, which today numbers over 16 million (Table 20-1), already exceeded 3 million in the mid-1920s. Population pressure increased constantly, while the variability of rainfall (highest in the Congo–Nile watershed, lowest in the western lowlands) caused periodic famines in certain areas.

The Belgians faced an especially difficult problem of administration in Ruanda–Urundi, where an effort was made to impose a democratic form of government upon one of the clearest examples of what may be called feudal Africa. In substance, these problems were not entirely different from those faced by the British in Uganda, where access was easier and more living space available. The regions possessed long traditions of social hierarchy and monarchy. These kingdoms were part of swath of principalities and kingdoms that had developed in a similar fashion and extended from the eastern Congo Basin up into the Rift Valley (Box 20-1).

The Belgians found the territory, occupied by three distinct groups—Hutu, Tutsi, and Twa—and the social system was based on that situation. Today perhaps 13% of the total are Tutsi, 86% are Hutu, and about 1% Twa (Batutsi, Bahutu, and Batwa, respectively). A feudal class system known as *ububake* developed whereby the Hutu were permitted the use of Tutsi cattle and land, and in exchange rendered personal and military service to the Tutsi (Lemarchand 1970). In this organization, the omnipotent ruler was the Tutsi king (*mwámi*), who delegated power to the chiefs of his people. There was, of course, no semblance of democracy, and without education the imposition of democracy had little meaning.

The origin of this system and, especially, the ruling Tutsi, is a subject of some dispute. One view holds that the Tutsi, who are predominantly pastoralists, migrated into the area from the Nile Basin. According to this theory, the tall, proud pastoralists of Nilotic origin arrived in the area about five centuries ago and gradually established social and political dominance over all other groups. It is not rare for a pastoral group in Africa to dominate agriculturalists and even create a social hierarchy that could be called feudal. The Fulani have done it in West Africa; elsewhere the ruling caste of the Beni Amer in the Sudan and Eritrea, the Nabtab, claim to have come from the Nile Valley and established their feudal rule over "Tigre." The alternative theory is that the Tutsi dominance over the Hutu is not a tribal or an ethnic difference but a social or class difference that developed over time within one society and one people. "Tutsi" originally meant "a person rich in cattle" but later came to mean "noble." Hutu, originally referred to a subordinate or follower of a more powerful person and came to be applied to the vast majority of the people (Des Forges 1999).

Surprisingly, there are no remnants of any Nilotic language in Kinyarwanda, the northeast Bantu language that Tutsi speak today. Kinyarwanda is really only slightly different from Rundi, which the Hutu, speak, and the languages are mutually intelligible. On the other hand, oral history in the Great Lakes area supports the migration theory of the Tutsi (d'Hertefelt 1965). There is no evidence that the Hutu and Tutsi differ genetically, but known Nilotic populations have not been studied much. Moreover, since the blood sampling across Africa that exists today is not fine enough to permit comparisons of Hutu and Tutsi with respect to genetic differences, both are grouped together as "Northeast Bantu" (Cavalli-Sforza et al. 1994).

The French historian Jean-Pierre Chrétian believes that the social differentiation between Tutsi and Hutu is rooted in "cultural cleavages between agriculturalists and cattle keepers in the Great Lakes [which were] a function of the region's ancient human settlement and complex economic and social structuring" (Chrétian 2003: 281). Chrétian suggests that European colonialists reinforced and exacerbated the social distance between Tutsi and Hutu and bear some responsibility for the result. Aristocratic Tutsi were few in number, and in Rwanda, the Tutsi monarchy was strongest in the central region of the country; in the north, Hutu clans and kings were dominant. In the Belgian system of indirect rule, the reservation of official posts for Tutsi permitted the central Tutsi to extend their control over the entire country; in the process, "Tutsi" became confounded with "chief" and a national "chiefly caste" was created (d'Hertefelt 1965). During the colonial period identity cards were issued on which the holder had to indicate his or her ethnic identity.

TABLE 20-2 YEARS OF SCHOOLING IN ASTRIDA TOWNSHIP DURING COLONIAL TIMES.

Location	Tutsi (%)	Hutu (%)
Township		
≤5 years	37	29
>5 years	46	13
Illiterate	17	58
Rural area		
≤5 years	21	16
>5 years	10	2
Illiterate	69	82

Source: d'Hertefelt (1965).

During the colonial time the Tutsi were more likely to get an education than Hutu (Table 20-2) thus reinforcing distinctions between them. Only about 20% of the children of Ruanda–Urundi received any elementary education from the Belgians, according to d'Hertefelt's estimation. After independence, however, about 75% of school-aged children attended primary school.

Ruanda–Urundi, therefore, was a country of vested interests (Tutsi) and smoldering resentment (Hutu), which frequently found expression in open conflict. Although making some improvements in traditional agriculture and pastoralism within its mandate, and toward the end appointing Hutu to positions of authority, Belgium found the obstacles in the path of educational, social, and political progress insuperable. The feudal relationship was progressively abolished beginning in 1954, but the land tenure system was left intact. In 1957 Hutu intellectuals began to indicate their dissatisfaction with the inequality and commented that the Belgian administration was "as accountable for the persisting caste inequality as was the traditional sociopolitical system" (d'Hertefelt 1965: 436).

In the late 1950s two Hutu political parties were formed, the reformist, Aprosoma, founded by Joseph Gitera, which worked primarily in the south, and the socialist–revolutionary mass party, Parmehutu (Party of the Movement of Emancipation of the Bahutu), founded by Grégoire Kayibanda, which was strong in the northern districts and rapidly spread across the country. The Tutsi monarchy formed the Rwanda National Union (UNR) as a counterweight to Hutu nationalism. The "racist" nature of Parmehutu was evident to all, and in 1959 Parmehutu provoked a series of atrocities against Tutsi in the north. Thousands of houses were put to the torch and several hundred Tutsi murdered. In retaliation, many Hutu leaders were assassinated. As if bowing to Hutu pressure, the Belgian administration replaced half the Tutsi chiefs and 60% of the subchiefs with Hutu officials—the Hutu revolution. In 1960 in local elections the Hutu parties (especially Parmehutu) won landslide victories. In 1961 the Parmehutu government declared a republic and abolished the monarchy, an action that was confirmed in a later general election. The king, some of his councilors, and an estimated 150,000 Tutsi went into exile. When independence was granted in 1962, the territory was divided into two new states, Rwanda and Burundi.

The central event that shaped the political scene until 1990 was the Hutu revolution of 1959–1961, which institutionalized anti-Tutsi activities as accepted behavior.

> The Batutsi became scapegoats for every bad memory of the past, including colonial oppression, memories that were continually and vindictively recalled. Every time a Hutu regime encountered a problem, the Tutsi in the country were a priori considered suspect. They thus were obliged to remain discreet about their lot. For a generation, all harassment by the local authorities toward them was treated as an understandable and spontaneous manifestation of popular resentment, and not a single murder was punished. (Chrétian 2003: 305)

More Tutsi refugees poured out of Rwanda in 1964 and 1973 as tens of thousands were murdered by Hutu. By the late 1980s Tutsi refugees totaled about 700,000 people in Uganda, Zaïre, and Tanzania. Parmehutu remained in power until it was toppled in a coup d'état in 1973, at which time the National Revolutionary Movement for Development (MRND) took power under General Juvénal Habyarimana. The Hutu military officers who were behind the coup were discontented with the lack of economic progress and by the favoritism shown toward Hutu in the president's home area. The new military government took a far more moderate stand on the Hutu–Tutsi issue than had its predecessor, attempting to restore and strengthen diplomatic and economic ties with its neighbors, especially Tanzania and Burundi.

In Burundi the Hutu–Tutsi conflict was not an issue in the runup to independence nor immediately after—the government comprised both Tutsi and Hutu—and the main issue seemed to be whether the new country would have a pro-Western government or one that was revolutionary as in the Congo and Tanzania. Events in Rwanda spilled over into Burundi in 1965, however, and a Hutu coup d'état was carried out. The Tutsi king fled, and massacres of Tutsi began. The army, controlled by the Tutsi, took power and exacted a terrible retribution against Hutus, particularly those in government. In 1972 Hutu exiles began a rebellion in which thousands of Tutsi were murdered. In retaliation, an estimated 150,000 Hutu were massacred. Tens of thousands of innocent women and children were dragged from their homes and schools and systematically clubbed to death, while Tutsi soldiers and civil servants murdered thousands of Hutu farmers, teachers, and others with automatic weapons, machetes, and spears. Approximately 3.5% of the country's total population was liquidated in a matter of weeks, yet the OAU and the United Nations took no action, and raised little effective protest.

After the events of 1965 and 1972, the gulf between Tutsi and Hutu in Burundi became as great as in Rwanda. Since independence, close to 2 million people have been slaughtered in civil wars in both countries (Figure 20-4). In just a few months in 1994, close to a million Tutsi men, women, and children, and some moderate Hutus were massacred in systematic attacks orchestrated by the Hutu-led government (Gourevitch 1999). Hutu radio broadcast encouragement to the murderers with such exhortations as "The graves are not yet full!" (Figure 20-5). The bodies of the victims were dumped into rivers and latrines, or buried in mass graves.

A Tutsi army, the Rwandan Patriotic Front (RPF) raised in the diaspora led by Paul Kagame—and supported by Yoweri Museveni in Uganda—invaded Rwanda and put an end to the slaughter. One and one half million Hutu fled, mainly to the then Zaïre. In 1996 trials of Hutu accused of genocide began. In the same year RPF and Ugandan soldiers crossed into Zaïre to strike against the Hutu extremists who were reorganizing and rearming in the refugee camps in eastern Zaïre. The armies, drawn westward in the pursuit of the Hutu extremists, ultimately marched on Kinshasa,

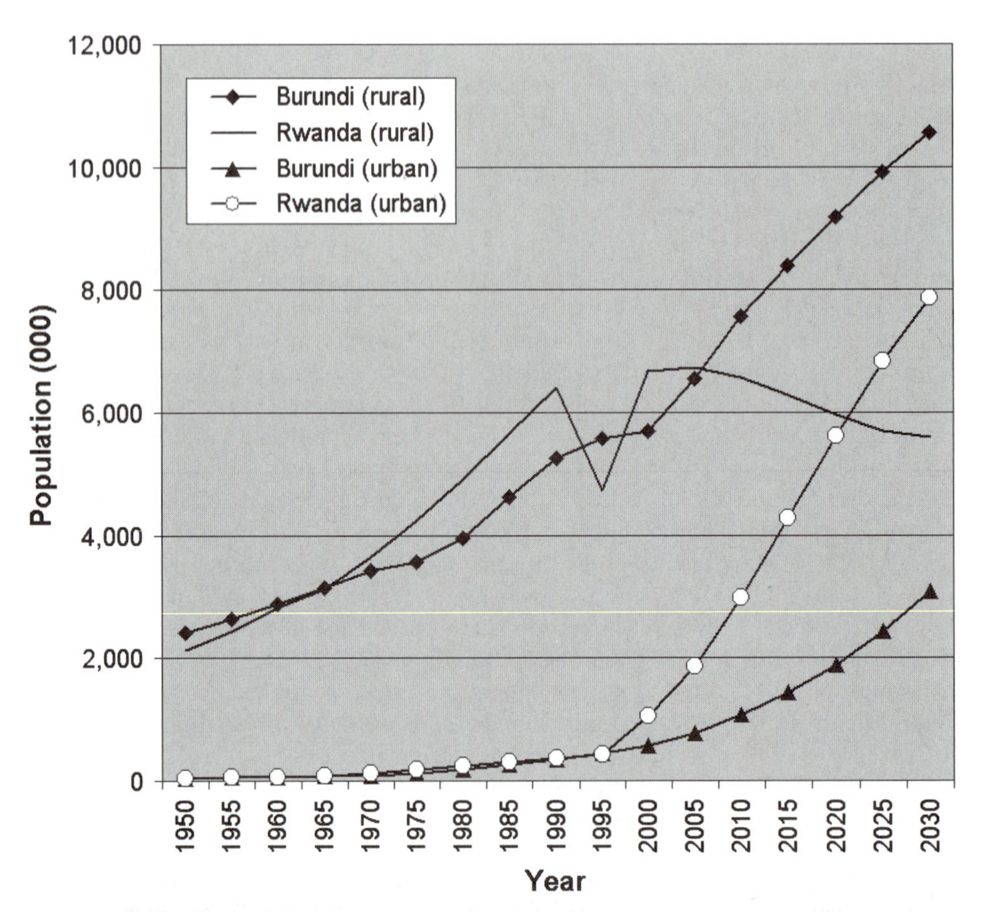

FIGURE 20-4 Rural and urban population, Rwanda and Burundi, 1950 to 2030.
Source: Adapted from FAO (2005).

FIGURE 20-5 Memorial and museum of the 1994 Hutu-led Tutsi genocide.
Steve Montgomery.

ousting Zaïrean president Mobutu Sese Seko and installing Laurent Desiré Kabila in his place. This war drew in from neighboring countries regular or irregular troops interested in obtaining control over some of the rich resources of the Congo Basin. In 1997 and 1998 Hutu extremists from the Congo crossed into northwest Rwanda and attempted to organize a rebellion. The RPF put down the rebellion, but overall tens of thousands of people, mostly civilians, died (Des Forges 2004).

SOWING THE WIND

The reasons underlying the atrocities in Rwanda and Burundi are not easy to understand. It is not solely a question of the majority seeking political power and a greater share of the resources against a despotic ruling minority. The two groups are involved in a conflict whose roots lie deep in the history of their relations of overlord and serf, and in the process of social transformation set in motion before, during, and after the colonial interlude (Des Forges 1999; Verwimp 2004; Mamdani 2001). The Hutu revolution is Rwanda in 1959–1961 had a decisive psychological impact on ethnic self-perceptions in Burundi, and in a way the revolution sowed the wind, setting the stage for what was to come; the accession to power of Hutu politicians in Rwanda led many of their kinsmen in Burundi to share their political objectives, in turn intensifying fears of ethnic domination among the Tutsi of Burundi. The exhortations of racist politicians desirous of maintaining their hold on power, shortages of land, widespread poverty and hunger, and the exploding populations (André and Platteau 1998) are all contributors to these disorders. Rwanda and Burundi, once dependent on and divided by Belgium, remain internally fractured.

BIBLIOGRAPHY

André, C., and J. P. Platteau. 1998. Land relations under unbearable stress: Rwanda caught in the Malthusian trap. *Journal of Economic Behavior and Organization*, 34 (1): 1–47.

Baker, R. 1973. Rwanda. *Focus*, 23(10).

Best, A., and H. J. de Blij. 1977. *African Survey*. New York: Wiley.

Clay, D. C., T. Kampayana, and J. Kayitsinga. no date. *Inequality and the Emergence of Non-Farm Employment in Rwanda*. Agricultural Surveys and Policy Analysis Project (ASPAP). Washington, D.C.: U.S. Agency for International Development.

Cavalli-Sforza, L. L., P. Menozzi, and A. Piazza. 1994. *The History and Geography of Human Genes*. Princeton, N.J.: Princeton University Press.

Chrétian, J.-P. 2003. *The Great Lakes of Africa: Two Thousand Years of History*. New York: Zone Books.

Des Forges, A. 1999. *Leave None to Tell the Story: Genocide in Rwanda*. New York: Human Rights Watch. http://www.hrw.org/reports/1999/rwanda/.

Des Forges, A. 2004. *Leave None to Tell the Story: Genocide in Rwanda—Ten Years Later*. New York: Human Rights Watch. http://www.hrw.org/reports/1999/rwanda/.

d'Hertefelt, M. 1965. The Rwanda of Rwanda. In *Peoples of Africa: Cultures of Africa South of the Sahara*, J. Gibbs, ed. New York: Holt, Rinehart, & Winston, pp. 403–440.

Du Bois, V. D. 1973. *Rwanda: Population Problems, Perception, and Policy*. Hanover, Va.: American Universities Field Staff.

Food and Agriculture Organization. 2005. FAOSTAT. Rome: United Nations Food and Agriculture Organization.

Gourevitch, P. 1999. *We Wish to Inform You That Tomorrow We Will Be Killed with Our Families: Stories from Rwanda*. New York: Picador.

Lemarchand, R. 1970. *Rwanda and Burundi*. New York: Praeger.

Mamdani, M. 2001. *When Victims Become Killers: Colonialism, Nativism, and the Genocide in Rwanda*. Princeton, N.J.: Princeton University Press.

Population Reference Bureau. 1975, 2003. *World Population Data Sheet*. Washington, D.C.: Population Reference Bureau.

SIL International. 2004. *Ethnologue*, 14th edition. Dallas: SIL International.

Vansina, J. 1990. *Paths in the Rainforest: Toward a History of Political Tradition in Equatorial Africa*. Madison: University of Wisconsin Press.

Verwimp, P. 2004. Peasant ideology and genocide in Rwanda under Habyarimana. In *Genocide in Cambodia and Rwanda*, S. E. Cook, ed. New Haven, Conn.: Yale University Genocide Studies Program Monograph Number 1. http://www.yale.edu/gsp/rwanda/.

CHAPTER 21

Ethiopia

A Legacy of Feudalism, Imperialism, and Great Power Geopolitics

Ethiopia often is described as one of the oldest independent states in Africa. It is true that the territory of Ethiopia was not successfully claimed by any of the colonial powers in the "scramble" of the 1880s, although attempts at annexation were indeed made, and parts of the Horn that might have become Ethiopian territory did fall to the Europeans. Ethiopia lay within the sphere of influence mainly of Italy, but the period of effective control over the entire country by the Italians was limited to a few years preceding and during World War II, 1936–1941.

These historical conditions notwithstanding, to describe Ethiopia as having been a sovereign nation state since before the arrival of the European intruders is misleading. Less even than Uganda had the country progressed toward unification and internal consolidation. Being much larger, and affected by severe physiographic obstacles to circulation, Ethiopia presents great restrictions to movement of all kinds. Until the late nineteenth century, the plateau and its periphery were occupied by a number of feudal kingdoms (whose essential structure was not very different from that of Buganda), sultanates (in the eastern margins), and politically fragmented tribal peoples. Only then did the first steps toward the modern state take place, with hundreds of years of almost total isolation finally coming to an end.

Several geographical factors have contributed to the development of present-day Ethiopia, with its great complexity and heterogeneity. The area's relative location and physiography have played major roles (Figure 21-1). Mediterranean peoples made contact with the shores of the Horn when the Red Sea was the only sea route to the Indian Ocean; Greco-Egyptian and Roman excursions led to some landings and intermittent associations. The center of territorial organization in Ethiopia lay in the region of Aksum (modern Axum) and covered a part of Eritrea, the Ethiopian Province of Tigray, and a section of the Arabian Peninsula opposite. This Aksum Kingdom, itself a successor of Nile Valley empires (Nubia, Kush), was supreme from the first to the seventh centuries of the common era. It was in this period that Christianity was introduced to the region, as missionaries from Egypt (the Coptic Church) settled among the people of Aksum.

In the early years of the common era, therefore, the main center of organization in the Horn was not as isolated from the outside world as some of the later kingdoms were to be. Aksum was located against the northern extremity of the Ethiopian Massif and extended to the Red Sea coasts. Nowhere else in the region was there the organization, architecture, art, literature, or prosperity of this kingdom. But as Islam rose in the east, Aksum declined, and the period of contact with the Mediterranean (limited as it was) and India ended. But Christianity had taken hold among the rulers of the empire in the fourth century, probably within a decade of its becoming the state religion of Rome (Phillipson 2000), and when the remnants of Aksumite power withdrew into the protective interior plateaus, it was the beginning of a permanent strife between Coptic Christian kings and the proponents of Islam. Coptic orthodox Christianity or, as it is called in Ethiopia, the Ethiopian Orthodox Church, differs from Roman Catholicism in a variety of ways, most importantly on the nature of Christ, divine or human. Copts have their own Pope in Alexandria, Egypt. The Coptic Pope has traditionally appointed the Ethiopian orthodox bishop (abuna) but Ethiopians began appointing their own bishop in 1959.

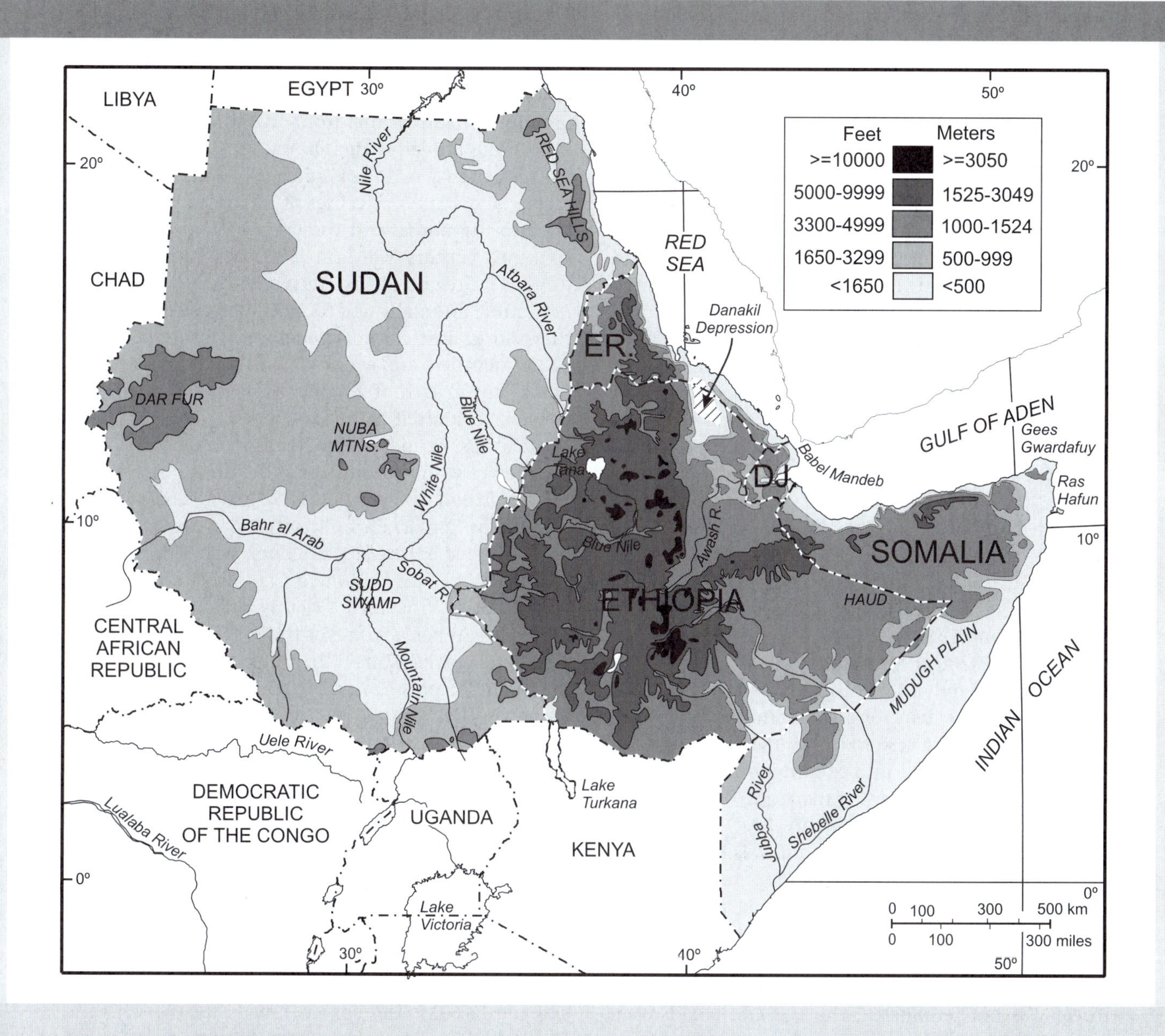

FIGURE 21-1 Elevations and major physical features of the Horn and Sudan.

As Islam spread into the marginal areas, the Horn went through its dark ages. Not until the twelfth century did the Zagwé dynasty (c. 1137–1270) arise in the interior, its Christian rulers, in addition to building many monolithic churches, engaged in ceaseless battle with their Muslim neighbors. Except for the contact between the Muslims and the center of Islam on the Arabian Peninsula, there was total isolation; the kings in the highlands were no more in touch with the outside world than were the early kings of Uganda. They waged their wars against the Muslims with varying fortunes. At times, the Islamic forces penetrated deep into the plateau, while at other times the Christian kings expanded their area of hegemony. They used the military assets of the highlands to good advantage, but by the sixteenth century, the tide began to turn decisively against them. The Muslims of the peripheral areas had begun to become politically organized, and a strong sultanate, Adal, had arisen centered upon Harar. Before the middle of the century, a charismatic Islamic warrior, Ahmad ibn Ibrahim, supported with rifles and cannon from the Turks, was able to unite the lowland Muslims and deliver to the much less well-armed Christians a crushing defeat. It appeared that Christianity in this region was doomed as, with few exceptions, Christians were forced to convert or die (Pankhurst 2001).

The apparently final defeat of the Ethiopian emperor, Galawdéwos, happened to coincide with the Turkish defeat of the Mamluk regime in Egypt, so that the entire Red Sea seemed likely to fall under the domination of Islam. But then another European power with Mediterranean interests became involved, the Portuguese, who saw one of their routes to the Indies seriously threatened and decided to take action. In 1535, having suffered defeat after defeat, Ethiopian Emperor Lebna Dengel sent a Portuguese resident of his court to Portugal to summon aid. The small Portuguese army landed in Mitsiwa (Massawa) in 1541, one year after the emperor's death and the succession of his son, Galawdéwos, and, equipped with cannon and musket, advanced into the interior. There, allied with the remnants of the Christian kings' armies, they defeated the Muslims on February 21, 1541, not far from Lake Tana. Ahmad ibn Ibrahim was killed in battle, and Muslim hegemony evaporated almost overnight. The victory was the turning point in the Christians' fight for survival against the encroachments of Islam. Those who had been forced to convert to Islam returned to Christianity. Although the Islamic threat was not terminated, the war effort had taken such a toll from both sides that neither was able to deal a final blow to the other.

One side might eventually have prevailed, were it not for the migration, beginning in the early sixteenth century, of the Oromo people from the southeast. Armed conflict between Oromo and the Ethiopian Empire continued for two centuries as hundreds of thousands of Oromo flooded onto the plateau and, in the process, separated the empire from the state of Adal. The origin of these Cushtic-speaking pastoralists appears to have been southern Ethiopia or northern Kenya, and they were drawn into the north by the power vacuum created by the collapse of Ahmad Ibn Ibrahim's jihadist armies from Adal in 1541 (Pankhurst 2001). The arrival of the Oromo on the plateau gave rise to a lengthy period of upheaval, political disorganization, and wars. The center of Ethiopian power withdrew to the north, and the Oromo, who by the mid sixteenth century were on horseback, spread northward to the region of Lake Tana and beyond. They were not united internally and fought among themselves, as they did against common enemies. Thus, a period of chaos prevailed, and eventually no less than six "kingdoms" arose, each of which was ruled by a man who considered himself to be the emperor of all Ethiopia. During this period, several groups of Oromo allied themselves with the Emperor at the time.

Confusion also reigned in religious circles during this era, as Roman Catholic missionaries of the Jesuit order entered Ethiopia's kingdoms and attempted to convert the people to Roman Catholicism. Ethiopians were Coptic Christians, a heresy according to Rome. Ethiopian Emperor Susneyos (ruled 1607–1632) sought an alliance with Jesuits to help them parry the Oromo onslaught. Repression and bitter strife over Emperor Susneyos' pro-Catholic policies, which included an official proclamation repudiating the Coptic belief in the nature of Christ, the central issue in the schism with Roman Catholicism, further divided the hard-pressed leadership and people. Coptic Christians considered the Emperor's proclamation as apostasy. In 1632 Susneyos called a state council to consider the restoration of the former faith; it voted to reestablish Coptic Christianity in place of the divisive Roman Catholicism. Susneyos died at the end of 1632 and was succeeded by his son Fasiladas, who expelled the troublesome missionaries (1633). The emperor made treaties with the Muslim rulers of the ports of Mitsiwa and Suakin (Sudan) to prevent Europeans from entering the empire—and for the ensuing 150 years the country was almost entirely isolated from European contact.

Fasiladas moved the capital of the empire permanently to Gondar in the north in 1636. Increased weapons imports during the eighteenth century helped turn the tide against the Oromos as the empire took the battle to Oromo territory. Yet paradoxically, involvement of some Oromo groups in the affairs of the empire increased: some provided soldiers for the emperor, others were given high positions in the imperial bureaucracy. Emperor Iyasu married an Oromo woman of the Yajju group. Iyasu's half-Oromo son, Iyo'as, took the throne in 1755, and much of the land occupied (and claimed) by the Oromo became integrated into the Ethiopian state. But the state was unable to successfully deal with separatist tendencies, and there was no central authority until the later nineteenth century. Three separate states existed: Tegray in the north, Amhara, and Shawa in the south. In 1849 Tegray sought military assistance from Queen Victoria of England to defend itself against advancing Muslim armies but was politely refused.

Not until 1855 did change begin once more, and in this year the phase that led directly to modern Ethiopia may be said to have begun. Kasa Haylu, distantly related to the royal family, ascended to the throne after having defeated several of the feudal rulers on the plateau, in effect accomplishing the first step in the direction of consolidation. As emperor, Kasa assumed the name Téwodros II (Theodore II), and he initiated a series of administrative, social, and religious reforms. Although he was not always successful in imposing them, his effort in this direction was the first to have been made in Ethiopia. Meanwhile, hostilities against rebel chiefs, Muslims, and all who displeased him continued. In addition, Theodore had to deal with a new factor on the Ethiopian scene: expressions of political interest in the Horn on the part of Europe. Such connections could have been advantageous to Ethiopia, but Emperor Theodore mishandled diplomatic relations badly.

In 1862, desiring closer relations with Britain, Theodore asked not only that Queen Victoria receive an Ethiopian embassy in London but that the British arrange for the free passage of Ethiopians through Ottoman territory, assistance in the purchase of modern firearms, and the services of a road engineer and artisans. The British, allied with the Ottoman Empire against Russia to parry the Russian threat to British Imperial India and not wishing to jeopardize the Crown's growing economic dependency on Egyptian cotton, ignored Theodore's request.

When he had not heard anything from the queen in two years, Theodore concluded that he was being abandoned to his Egyptian and Ottoman enemies and imprisoned the British consul and his staff. The British consul in Aden, then a British protectorate, was sent to meet with Theodore, while preparations were being made to send him an engineer and some artisans, who arrived in Mitsiwa in 1866. Theodore continued to blunder diplomatically, bickering about releasing the British prisoners before or after the arrival in Ethiopia of the engineer and artisans. An angry Britain dispatched a rescue force that landed at Mitsiwa in 1867. The force was joined by tribesmen who had suffered oppression under Theodore, and in 1868 the emperor's army was defeated. Theodore killed himself on the battlefield.

Britain's first incursion was not permanent. Having achieved its objective, the force withdrew, leaving Ethiopia in renewed disarray. This coincided with the opening of the Suez Canal and unprecedented European activity along the entire east coast of Africa. Pressures upon Ethopia increased from several sides. Egypt briefly entered the stage by taking the Eritrean coast and southeastern Ethiopia as far west as Harar, and in 1869 an Italian concern purchased the Red Sea port of Assab. The Mahdist rebellion in Sudan brought invasions into western Ethiopia.

Ethiopia's own internal division was a major factor endangering its survival. Emperor John IV emerged as the dominant figure out of the feudal chaos, and he repelled the Italian advances from the port of Mitsiwa. But he had a rival, whose power in the south and west was on the increase while John was occupied with the war in the east. This feudal king, Menelik, formerly imprisoned by Theodore, was encouraged by the Italians to open hostilities against John, and to this end he was given arms. In return, he was promised the throne. John, however, was killed in the war against the Mahdists, and king Menelik became Emperor Menelik II in 1889.

Menelik immediately faced the aggressive forces of colonial imperialism. During the period of his accession to the throne, Italy occupied Eritrea and proclaimed a colony in that country. Immediately after becoming emperor, Menelik signed the Treaty of Ucciali, which was to become the first serious source of conflict between his regime and the Italians. The Amharic text of the treaty, which was the only document that was actually signed, states that, should he so desire, Menelik could make use of Italian diplomatic channels for his business with other powers and

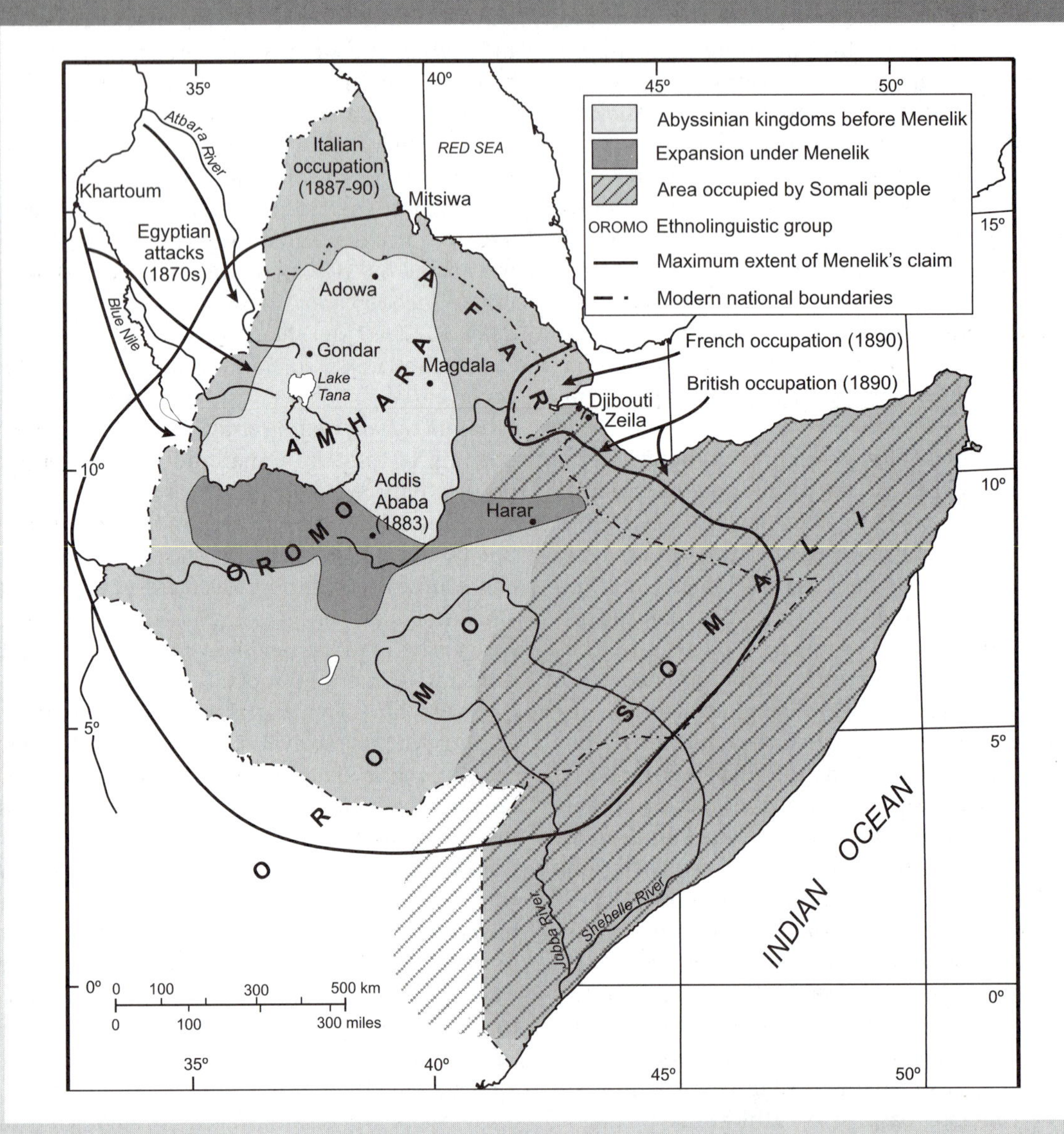

FIGURE 21-2 Political and ethnolinguistic geography of the Horn during the last decades of the nineteenth century.

governments. The Italian translation, on the other hand, states that Menelik consented to make use of such channels, and in these terms Ethiopia was virtually a protectorate of Italy. This, indeed, is what the Italians professed to believe, and when Menelik violated the terms by making direct contact with Queen Victoria, the inevitable crisis arose. War ensued, and in 1896 the Ethiopian forces triumphed in Battle of Adowa, extracting from Italy a new treaty recognizing Ethiopia's sovereignty.

Menelik now embarked upon his own imperialist campaign and used the European powers to his advantage. He expanded the territory under his sway far to the southeast, south, and west, and signed treaties with the colonial powers defining the boundaries of Ethiopia much as they are today (Figure 21-2). Having founded the modern capital, Addis Ababa (Adis Abeba: "New Flower"), in 1883 he renewed Theodore's efforts to initiate reforms by taking major steps to modernize the country. He negotiated with the French, who had occupied the port of Djibouti, for the building of a railroad from this port to the capital. He also began a road-building program, and established schools, postal services, public utilities, and other modern amenities.

Menelik II consolidated Ethiopia and withstood European intervention at a time when most of Africa was being parceled out by the colonial powers. The emperor's death, in 1913, again deprived Ethiopia of strong central leadership at a crucial time. Even before his death, France, Britain, and Italy mapped out a division of the country into their own desired spheres of influence, to take effect in case of the disintegration of the Ethiopian state. Italy planned to connect its two possessions of Eritrea and Somaliland across Ethiopian territory; France wished to safeguard its interest in the Addis–Djibouti railway; and the British wanted to protect the source of the Blue Nile and the region around Lake Tana. It seemed as though this division would indeed take effect, since for many years after the death of Menelik, the country was without a leader of his stature. Menelik's grandson and proper successor was youthful, irresponsible, and leaned toward Islam; his reign was predictably short. One of Menelik's daughters, Zauditu, then became empress, with the young Ras Tafari (Haile Selassie) designated as heir to the throne. A struggle ensued, and the divisive forces of feudalism again were strongly felt in Ethiopia as various chiefs sought and achieved individual power.

Naturally, the situation was extremely detrimental to Ethiopia, and when Haile Selassie was crowned in 1930, the effects of two decades of stagnation were evident everywhere. But Haile Selassie had shown signs, even before his coronation, of desiring the end of Ethiopia's isolation. He had successfully applied for the country's admission to the League of Nations in 1923, and had engaged in treaties and cooperative projects with a number of European states. Under his rule, the country's first constitution was written, a first parliament assembled, and social reforms initiated. In the 1930s a new law against slavery was passed—it is a reflection upon the Ethiopian situation that the practice has yet to be eliminated altogether.

In his attempts to unify and consolidate Ethiopia, Haile Selassie, like other Ethiopian rulers before him, faced an insuperable task. Ethiopia's location has made it a battleground between Christian and Muslim. Its coastal fringes lie on one of the world's most important maritime avenues, and while its high interior mountains afforded a haven for an island of Christianity amid an ocean of Islam, its coasts attracted Muslim and European alike. But when modern times came, and boundaries were drawn around Ethiopia, the strife of ages had left a legacy of deep and fundamental division, too strong to be overcome in a matter of decades. The very physiography that once had helped ensure the survival of the Christian kings now became a major obstacle in the effort to build a nation state there in the Horn.

Viewing Ethiopia's politicogeographical characteristics of shape, the country would appear to have several assets. It is large, but compact. No lengthy protrusions extend from the country's main area (as in the Democratic Republic of the Congo and Namibia). The asset of proximity to the coast was enhanced by several opportunities for port development, even though the colonial holdings of Italy, Britain, and France long prevented direct access to the sea. The capital is located in a central position, which would appear to help bind the state together.

But apart from these politicogeographical features, other conditions seem to perpetuate internal division and fragmentation in Ethiopia. The same physiography that once protected the kingdoms separates the peoples, languages, and religions within the state. Having once hindered the invasion of the plateaus, physiography now makes communications difficult, slows the spread of ideas from the capital, isolates communities from each other, and raises the cost of importing and exporting goods. In a region far from seaports, railway lines, and paved roads, goods are transported as they were over 2,000 years ago (Figure 21-3). True, Addis Ababa is situated in the middle of a radiating network of communications, but the network thins out rapidly, roads become tracks and eventually mere paths (if they continue at all), and the greater the distance from the capital, the less effective the contact with it (Figure 21-4). And the less effective the contact, the less integrated are the outlying parts with the heart of the state.

If the result of these conditions were merely a limitation of movement, Ethiopia's problems would be like those of other underdeveloped countries requiring

FIGURE 21-3 A riverbed serves as a dry-season road in a mountainous environment.
Steven Blakeway.

improved communications systems to stimulate development and foster a national spirit among peoples, some of whom are located in remote areas. But in Ethiopia the consequences are much more serious, and their solution will require more than a better transport network. Strong actual and latent centrifugal forces exist within the state: Muslims and people who practice traditional religions comprise about 45 and 12% of the population, respectively. The Islamic center of Harar and the capital of Addis Ababa seem worlds apart (Box 21-1), but they are within 250 miles (400 km) of each other.

The period of comparative stability and progress under Haile Selassie was to be interrupted in a violent manner in 1935. In the previous year, during efforts to demarcate the boundaries in the region where Italian, British, and Ethiopian territory met, Italian Somali forces had clashed at Walwal with Ethiopian troops trying to control the area. Ethiopia took the matter to the League of Nations, but Italy, having massed armies on the Eritrean and Somali borders, invaded Ethiopia before the international body could act. The campaign ended before the middle of 1936, and Haile Selassie fled to Europe. Italy's half-decade of rule in Ethiopia was marked by cruel repression of the local population, as Italy experienced the troubles of any authority attempting to establish effective control over the vast country. A vigorous program of road building and economic development was initiated, which still has favorable effects today. Having interrupted Haile Selassie's reform program in 1936, the Italians were themselves unable to complete their plans, for in 1941 British troops invaded Ethiopia and, with the aid of patriotic forces, defeated the Italians. The emperor returned to his capital in the same year, and the sovereignty of the state was restored.

Since World War II, Ethiopia has been struggling to modernize its economy and break with the past. But little has been accomplished. The capital, Addis Ababa (2.8 million), has more than tripled in size since the 1970s, but has provided few impulses of change and modernization. In addition, it has grown at the expense of its surroundings. Forty-three percent of the people are still illiterate, the rural majority remains impoverished by archaic systems of land tenure and methods of production; almost half the population still lives more than 10 miles (16 km) from all-weather highways. Nowhere has it been easy to break with tradition and overcome the limitations imposed by the prevailing feudal system. But in 1974, following years of dissension in Eritrea, widespread drought and famine, and general dissatisfaction with the lack of progress in the social and economic spheres, Haile Selassie was dethroned. The coup that unseated the emperor was mounted by army officers calling themselves the Dergue. Their leader, Colonel Mengistu Haile Mariam, imposed military rule. But this did not mean the end of feudalism, for the culture is engrained in the region, and the people remain deeply divided by religious, social, and ethnic prejudices.

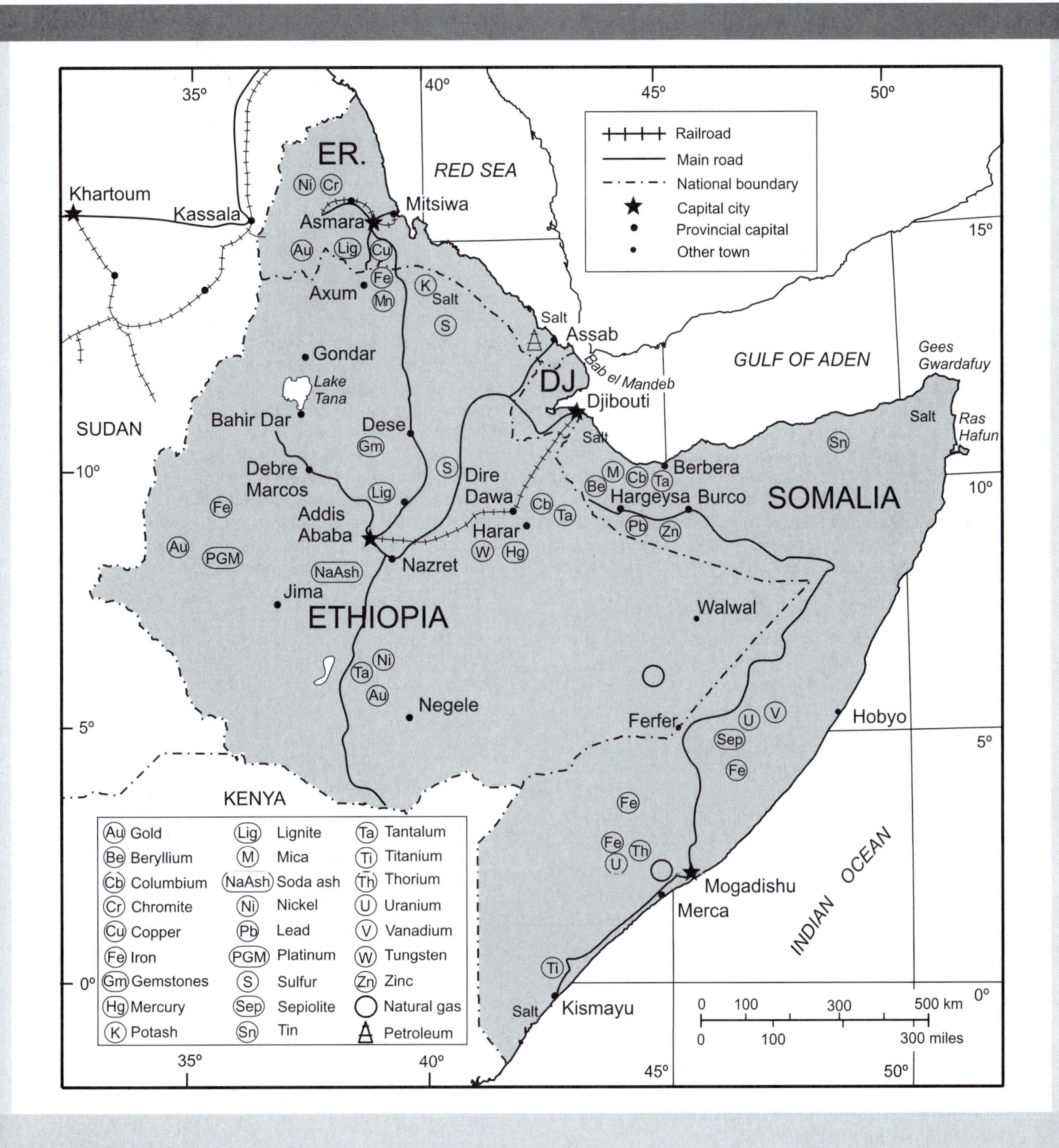

FIGURE 21-4 Towns, transport, and resources in the Horn of Africa.

BOX 21-1

ON THE CULTURAL FRONTIER IN EASTERN, LOWLAND ETHIOPIA

Jijiga is a divided city—by common, through unspoken, consent. The northern half is inhabited by Christians, mostly Amharas, and the southern by Muslims, mainly Somalis. The Somali man is easy to identify: he is almost always dressed in a *sherit*, a long multi-coloured garment stitched like a sack, which is tied about the waist, hanging loose at the bottom. Most carry a cane in one hand, and a piece of twig in their mouth, with which they brush their teeth if they happen to have nothing better to do. The nomads who venture into town may wear huge daggers at their waists. They are usually barefoot, though they may carry sandals to wear in town—balanced carefully over the shoulder, hanging from the end of their canes.

The women wear colourful, loose-fitting dresses that fall to their feet. As the woman walks, she holds her dress with one hand so it doesn't drag on the ground. She might drape a vibrantly coloured shawl over her shoulder and head, and wear sandals on her feet. I do not recall once seeing a Somali woman outdoors in the company of her husband or a male friend, even on major social occasions, like Ramadan, or at weddings or funerals.

The northern half of town appears to have consulted with a different fashion designer. The men dress in various forms of Western jackets and pants. The women signify the different phases in their lives by the colour and make of the dresses they wear. A girl invariably picks colourful outfits, cut according to the style of the time—though always hanging well below the knee. She might even indulge herself and wear a T-shirt and a pair of jeans. After marriage, however, her wardrobe will undergo a drastic transformation. A woman who is wife and mother wears a traditional white dress and head-wear, and a *netela*—a white, lightweight shawl with colourful trim.

Although I was born and raised in the northern part of town, the excitement and intrigue of the other half was not lost on me. A family friend, Mustafa, brought home everything I needed to know.

Source: Mezekia (2000), pp. 7–8.

LAND AND LIVELIHOOD

Ethiopia's problems in the economic sphere are as severe as those due to its political geography. Once again, the initial impression is favorable: this is primarily an agricultural country, and it is comparatively well endowed with good climates and adequate soils. Estimates of the amount of arable land vary, but perhaps 11% of the land area may be so classified, and no less than 20% is capable of carrying livestock (FAO 2004). But an agricultural economy, to rise above the subsistence level, requires much more than just land; it needs adequate transport facilities and other forms of organization. It is because of the paucity of these necessities that current development falls far short of the country's potential.

The isolation of the agricultural areas is as great as that of its various peoples. Much of Ethiopia is so rugged, and slope incidence so high, that it is hardly possible to describe the obstacles the topography puts in the way of communication. The areas of arable land are for similar reasons widely scattered. Areas of cultivable land, whether in the valleys, on gentler slopes, or on upland surfaces, often are separated by impassable declivities—and more important, they are also separated from the few routes to internal and external markets (Figures 21-4 and 21-5).

Three distinct environmental zones, which are actually altitudinal belts, are recognized in Ethiopia. In the hot lowlands is the *kwola*, which reaches up against the valley and plateau slopes as high as about 5,000 feet (1,500 m). This is tropical Ethiopia, and it includes also

FIGURE 21-5 Terraced slopes and farmhouse. *Steven Blakeway.*

the deserts and steppe stretches around the foot of the highlands (Figure 21-6). There, bananas, dates, and other fruits thrive, as well as coffee in the higher parts. Above the *kwola* lies the *woina dega,* or temperate belt, extending up to between 8,000 and 9,000 feet (2,400–2,700 m) Only about 13% of Ethiopia is forested, and most of the forest areas lie in this belt, which also sustains a wide variety of crops. Cereals, fruits such as the fig and orange, grapevines, and other Mediterranean plants thrive there. There is much pastureland, and thus a large cattle and sheep population in this zone. Highest is the *dega,* extending to the mountain areas of the country and including more pastureland and areas suitable for cereals such as wheat and barley.

This wide variety of conditions permits the cultivation of a large number of crops, and it has been said that with proper care and cultivation, Ethiopia's soils and climates make it possible to raise successfully almost any type of crop. The limitations imposed by lack of communications, education, agricultural organization, modern implements, and incentive have slowed the development of a healthy agrarian economy. Furthermore, since World War II, population growth has reduced and virtually eliminated fallows in the most productive highland areas of the country (McCann 1995). As elsewhere in plateau Africa, erosion is severe, and conservation practices are in their initial stages. Most of Ethiopia's farmers are mired in a life of small-scale cultivation with some small local sales for cash. The main staple crop is a cereal, teff, varieties of which have been bred to grow almost everywhere, but the country can produce far more than the population requires of almost every crop grown. Indeed, Ethiopia is often described as a future breadbasket of the north and Middle East.

Coffee, the most valuable export product of Ethiopia, earned $160 million in foreign exchange in 2002. The manner in which coffee is cultivated and harvested typifies much of what is problematic in Ethiopia's internal conditions. The country is extremely well suited for coffee of the arabica variety; indeed, this plant can be left untended and still produce well. There are veritable wild coffee forests in Ethiopia from which the beans are simply gathered, and if coffee seedlings are planted, they are often left without any form of care. The total harvest comes in large part from the wild forests, in addition to the production from the small plots of the local small farmers and a relatively minor contribution from a few large plantations (Figure 21-6). Tea production has been expanding, and the output of cotton and sugar can't keep up

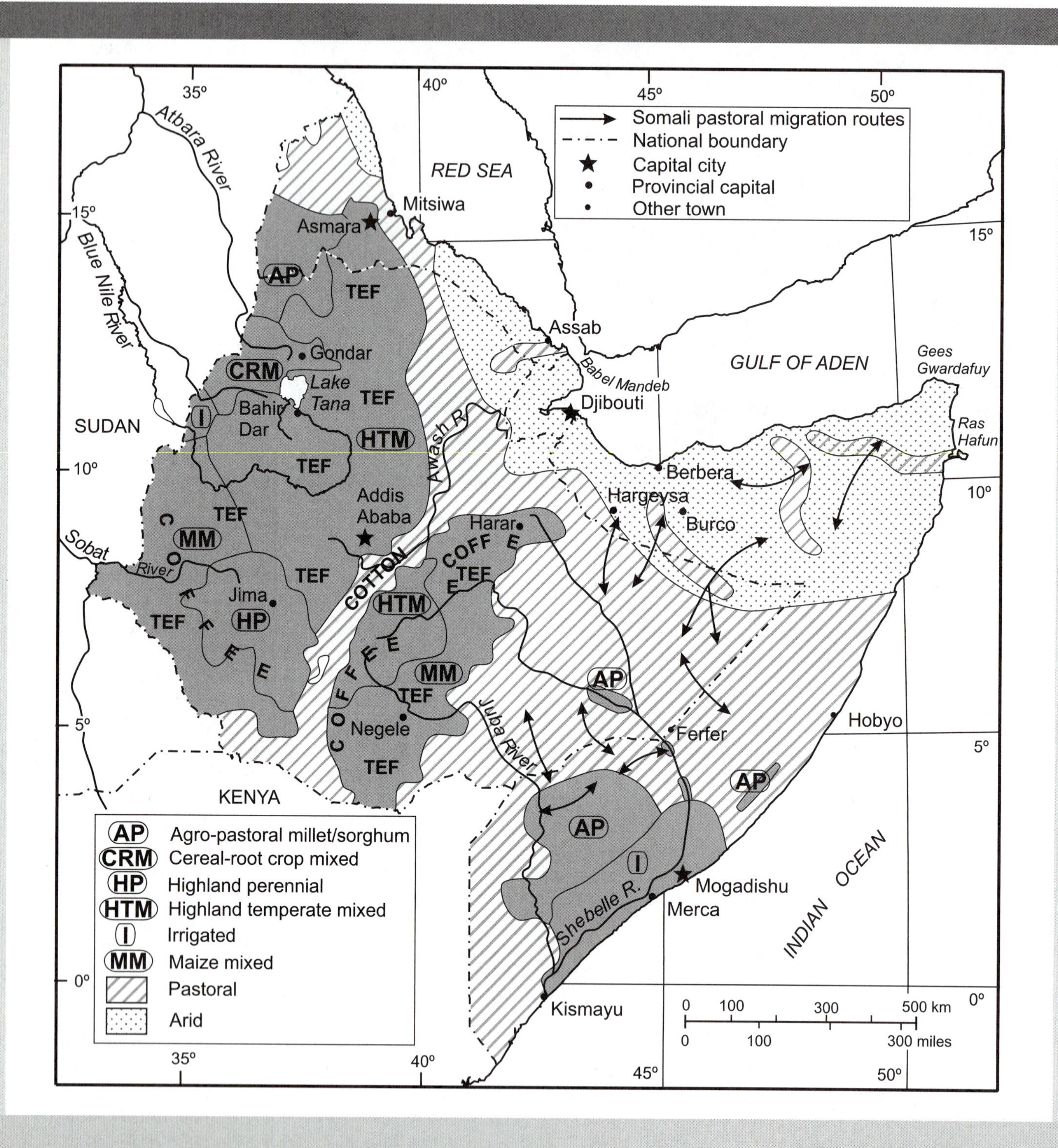

FIGURE 21-6 Agricultural regions in the Horn of Africa.

FIGURE 21-7 A Camel train is led through "Farewell Mother" Pass in Eritrea.
Steven Blakeway.

with demand from local industry (U.S. Department of Commerce 2000).

Some export revenue is obtained from hides and skins. There are over 35 million cattle in the country: 11.5 million sheep, 10 million goats, and an extraordinary number of asses—3.5 million, 38% of the Subsaharan African total. The livestock is maldistributed in relation to the available pastureland but is reflective of specialization by certain groups of people and the social conditions in the territory; especially large concentrations occur in the lands of the nomadic and seminomadic peoples of the lowlands. In the south, not surprisingly, as throughout Subsaharan Africa, cattle are a source of wealth and are kept as a kind of investment rather than for the market. Although in much of the country cattle are used as beasts of burden, slaughtered, and sold for meat and hides, they serve Ethiopians mainly as an investment. In the areas where precipitation totals are low, environmental variability is great and the dry season lengthy and intense (Figure 21-7). As a result, many cattle in these overpopulated, overgrazed regions are underfed and weakened, and disease is rampant.

Agriculture earns 60% of national exports, employs 80% of the 72 million people, and contributes half of the GDP. But its potential is nowhere near fully developed and agricultural output is growing at less than 2% per year. The most important element in agricultural improvement in Ethiopia, as elsewhere, is education, to promote an appreciation of the need for reform in land tenure and ownership, soil and water conservation practices, maintenance of irrigation and other structures and equipment, and improvements in transport, communications, marketing, and credit facilities. Of these, the need for land reform is paramount, now just as much as before the "revolution" of 1974.

Before the emperor was deposed by the Dergue, about 80% of the land was owned by four groups: the imperial family (15%), the Coptic Church (20%), the traditional aristocracy (20%), and the government (25%). The distinction between land owned by the government and by the emperor was not always clear. The church, the emperor, and the government paid no taxes, while taxation of large landowners was nominal. Half of Ethiopia's farmers were tenant farmers, who, on the basis of either custom or law, were required to pay tithes to their landlords of between half and three-fourths of their total harvest; in some cases the feudal landlord obliged them to provide labor and other services. Many different forms of land tenure and land tax existed, making reform measures difficult to conceive and even more difficult to implement (Hoben 1973). During the late 1960s, attempts were made to modernize the land tax system as a prerequisite to land reform. Cultivated land was taxed at higher rates than fallow land, but in parliament it was argued that the reverse should be in effect (Harris 1974). This was to force the aristocracy to use the land more efficiently, boosting production for both domestic and foreign markets or, failing that, to sell land to the tenants. It was also proposed that landholdings be limited to 988 acres (400 ha)—some estates in the south measured

BOX 21-2

MENGISTU HAILE MARIAM, LEADER OF THE DERGUE

Colonel Mengistu Haile Mariam headed the junta which in 1974 overthrew the government of Emperor Haile Selassie in a bloody coup. Known as the Dergue, the junta consisted of about 100 junior officers drawn from all regions of Ethiopia. Proclaiming a revolutionary agenda for the country, the Dergue inaugurated its rule by sending some 60 senior officials of the emperor's government to the firing squad. The emperor and the patriarch of the dominant Coptic Church were both secretly killed in the months that followed. Colonel Mengistu emerged as the Dergue's undisputed leader after orchestrating the physical elimination of rivals from within.

In 1976 Col. Mengistu gave a dramatic send-off to a campaign of terror that he officially dubbed the "Red Terror." He threw to the ground before a huge crowd in the capital bottles filled with a red substance representing the blood of enemies of the revolution: the "imperialists," and the "counter-revolutionaries," as members of rival leftist groups were labeled. In particular, the campaign targeted students and young people suspected of membership in the Ethiopian People's Revolutionary Party (EPRP). Thousands of young men and women turned up dead in the streets of the capital and other cities in the following two years. They were systematically eliminated mainly by militia attached to the "Kebeles," the neighborhood watch committees which served during the Dergue period as the lowest level of local government and security surveillance units. The Kebeles required families to reimburse the administration for the price of bullets used to kill victims when they reclaimed their bodies for burial.

The process of eliminating the "counter-revolutionaries" was quite organized. Each neighborhood committee would meet to discuss how to eliminate individual suspects, and each member would sign on documents to confirm the decision reached at the meeting. Copies of the document would be sent to different levels of the administrations and the party apparatus. The centralized killing enterprise thus left mountains of documentary evidence of its crimes.

Cold War rivalries helped the Dergue to flourish and tighten its hold on power. It became the main client of the Soviet bloc in Africa, and received massive shipments of arms to help it counter serious challenges from several armed insurgencies by ethnic and regional liberation movements seeking to break away from centuries of centralized hegemony by Ethiopia's ruling elite.

Source: Human Rights Watch (1999).

hundreds of thousands of acres—and that sharecropping tithes be limited to 33% of the produce. But these and other reform measures were successfully blocked by parliaments composed primarily of the landed aristocracy.

Hardships caused by the archaic system of land tenure were aggravated by the severe droughts of 1969–1975, the worst since 1916. Drought first struck the sparsely populated northern areas of the country, Wollo, Tigray, and Eritrea, where approximately 200,000 persons died in 1971–1973 alone, and stock losses were up to 90%. It then spread to the southeastern provinces, where an equal number of deaths were reported and an estimated 2 million persons, mainly nomads, were short of food and water. For political reasons, the military government, in power since 1974, refused to allow international relief agencies to provide either food or medicine for the starving pastoralists in the dissident province of Eritrea. And the Eritreans accused the Dergue of using food as a military weapon. In the southern region, the Dergue publicized the drought but failed to attract the required amount of international assistance.

An estimated 300,000 persons from Wollo and Tigray, in the northeast, unable to pay their tithes or even harvest enough food for themselves, moved off the land and into government relief camps, or migrated to unsettled areas in Gojam and Welega, west of Addis Ababa.

Commonly they resettled on vacant government-owned land and on large commercial farms and estates where food and water were available, but in most cases they were evicted and taken to refugee camps. Land these Ethiopians abandoned, sold, or leased was taken over by the aristocracy and consolidated into large holdings. In late 1973 Haile Selassie had decreed that all such land be redistributed to the original owners and occupants, but the decree was not honored by the Dergue, which later nationalized all land. Today the state still owns all land and leases it to tenants, hardly an improvement over the ancient feudal system.

The Dergue transformed cotton plantations into state farms on the Soviet model. For cash crops not directly under the control of government cadres, however, production was difficult to maintain at the low prices offered by the government. Farmers preferred to focus on their food crops. A widespread famine in 1984–1985 caused the death of hundreds of thousands, mostly children and old people. The extent of the famine was exacerbated by a governmental villagization campaign and by secessionist uprisings in the provinces. The Dergue was ruthless in its attempts to stamp out the dissent, which had begun almost as soon Mengistu's Junta took power (Box 21-2).

The drought of 2000 was the worst since the early 1980s, but rain was well distributed and sufficient in 2001 to result in a bumper harvest. According to the World Food Program of the United Nations (WFP 2004) there has been drought in Ethiopia almost every other year in one region or another, and in any given year 2 million Ethiopians live on food aid.

Scores of government-to-government and non-governmental organizations work in Ethiopia providing assistance to the destitute. Others bring technical assistance to all sectors of the economy in an attempt to improve processing, production, and the level of development. The World Bank has also played an important role in developing cash crops for export and import substitution (sugar and cotton) in the Awash Valley. Industry has been started utilizing these resources and locally generated hydroelectricity. But industry is still negligible in terms of its contributions to the GDP (about 5%), employment, and annual capital expenditure. Most factories were established with Italian, Greek, and French capital, and are concentrated in Dire Dawa (338,000), Harar (176,000), and the Addis Ababa–Nazret development corridor along the line of rail to Djibouti. Addis Ababa (2.8 million) has the greatest concentration of industry, mainly light manufacturing, but it is primarily an administrative center. Described as a mask behind which the rest of the country is hidden, Addis serves as the headquarters of the African Union, a focal point for communications, and is centrally situated within the country, in one of the most densely populated regions. The poor showing of manufacturing results from the small purchasing power of the domestic market, the lack of mineral exploitation, a shortage of skilled labor and technology, and inadequately developed hydroelectric power potential.

THE ERITREAN ISSUE, SECESSION, AND THE END OF THE DERGUE

Eritrea became a politicoterritorial entity in 1890, when Italy proclaimed it a colony. By the Treaty of Ucciali of 1889, Menelik II had recognized the Italian possessions on the Red Sea, and in subsequent years the Italians used bases in Eritrea for attacks on the emperor's lands. This, and controversy over the terms of the treaty, resulted in the hostilities of 1896, when Italian forces were annihilated near Adowa. Later that year, Menelik and the Italians redefined the boundary between their respective realms. In 1935 Eritrea again became a base for Italian military activity, in preparation for the invasion of Ethiopia. Roads, bridges, port facilities, and airfields were built, improved, and expanded, and thousands of Europeans entered the colony. The campaign was successful but short-lived, and by 1941 Eritrea, as well as Ethiopia, had been wrested from Italian control.

After World War II, Italy renounced its rights to Eritrea, and the United Nations attempted to decide upon the territory's future. Several commissions were given the task of determining the most acceptable course of action, but none managed to produce a suitable blueprint. Eventually, the General Assembly itself recommended the federation of Eritrea with Ethiopia, with the former colony to retain a considerable degree of autonomy.

This recommendation was put into effect in 1952, but a decade later Eritrea was unilaterally absorbed into Ethiopia as a province. A long war of secession then commenced, led by the Eritrean Liberation Front (ELF), a nationalist movement of Muslims and

Christians, and the Marxist Eritrean Popular Liberation Front (PLF). In the early years both groups had support from Libya, Somalia, Iraq, and other Arab states, and between them they controlled most of the territory after 1975, the year of an uprising against Ethiopian rule. In 1977 Mengistu broke with Washington, dissatisfied at what he perceived to be undeserved criticism and unhappy at the amount of military assistance provided by the United States. Ethiopia then allied itself with the Soviet Union, a more natural ideological fit, whereupon Western aid began to trickle in more rapidly to Eritrea, which by the 1990s was receiving substantial support from the West.

The thousands of deaths that resulted from the mishandling of the 1984–1985 drought and famine, the inappropriate resettlement schemes, the failed villagization programs, and the failed state farms set the stage for the end of the Dergue. But it was not until the fall of communism in Europe that the Dergue was finally overthrown. Allied Eritrean and Tigray armies, along with Amhara and Oromo opposition groups, marched on Addis Ababa in 1991 and brought down Mengistu and his murderous regime. In 1993 Eritreans almost unanimously voted for independence from Ethiopia. Over the years of war, hundreds of thousands of people died in conflict with government troops. The years since Eritrean independence have not been without conflict. In 1998 war between Eritrea and Ethiopia broke out again concerning the border area around the town of Badme. Hostilities ceased in 2000, and a peace plan was brokered, but Ethiopia has raised several objections to the plan and never approved it. Both countries suffered terribly as a result of the war, focusing on conflict rather than their own development.

Eritrea is not a rich country, and its prospects are limited by its environment and resource endowment. It consists of two main physiographic regions, the coastal plain and the interior plateau. The former varies in width from 10 to 50 miles (16–80 km), and the latter is an extension of the Ethiopian Plateau. The descent from the highlands, which are mountainous in parts and reach 6,000 to 8,000 feet (1,800–2,400 m), is partly abrupt, as in the north, and elsewhere steplike. The lowlands are hot and dry, and in the highlands, rainfall may exceed 20 inches (500 mm) but is highly seasonal. Thus Eritrea is much less well endowed with agricultural possibilities than Ethiopia, and in the past the latter has supplied most of the food consumed in Eritrea.

Overgrazing is a serious problem, irrigation a frequent necessity, and the available acreage capable of sustaining sedentary agriculture is small. Industrial development is limited to the processing of the country's small food production and the treatment of hides and skins. Asmara (408,000), located on the cooler plateau and on the railway to Mitsiwa (32,000), is the capital and main manufacturing center.

Eritrea's population is as heterogeneous as that of the rest of Ethiopia. In the highlands live Coptic Eritreans, in the coastal plain are several nomadic Muslim peoples (possibly 55% of the population), and in the southwest are Nilotic farmers. In addition there is an urbanized minority and a remnant of the once-large European population. Tigrinya and Tigray, two related South Semitic languages of the Afroasiatic family, are spoken over most of the country, while Cushitic languages prevail in the northwest and the Nilotic language, Kunama, is spoken in the southwest. Apart from the few European expatriates, Eritreans generally share a low standard of living, although the literacy rate has been rising (men 70%, women 47%). The country is about equally divided between Christians and Muslims and there is considerable linguistic diversity. But during the time of Ethiopian rule, Muslims and Christians were united in their resentment and opposition of the imposition of Amharic (the language of Ethiopia's politically dominant Amhara group) as the official language, and the elimination of local languages from schools. They also resented the influx of Amhara officials and the erosion of their own culture. The future of Eritrea depends on how well it overcomes the disadvantages it faces: subsistence-based agriculture, a semiarid climate, illiteracy, high unemployment, and a population with few skills. A more market-oriented economy might draw in millions of dollars of investment from expatriate Eritreans who have done well in their foster countries.

It is true that the loss of Eritrea in 1993 was a blow to Ethiopia. It meant the loss of Ethiopia's only coastline, the ports of Assab and Mitsiwa, about 190 miles (300 km) of railroad, one-fourth of its industrial output, and the country's only oil refinery. All of this has been a shock to Ethiopia, but until the Badme conflict, Eritrea and the new Ethiopia seemed to have been working well together. Indeed, it appeared that the transit function would provide much-needed income to Eritrea and at the same time permit Ethiopia access to the global market. Since the war, Ethiopia has been using the Djibouti line for most of its imports and exports and has been exploring transport options with the government of "Somaliland," the self-proclaimed republic in northern Somalia.

It may very well be that the long-term realities of independence will be grim for Eritrea—as they have been up to this point. Eritreans may one day realize that the price of self-determination—high enough in blood, instability, and missed opportunities after over 30 years of bitter struggle—was economic dependence on the international financial community. Perhaps a looser, federal, union with Ethiopia would have given the Eritreans the autonomy they desired and the economic security they needed in their marginal environment. One wonders whether the Eritreans have, perhaps, behaved in a way reminiscent of old Téwodros II, emperor of Ethiopia, who, convinced of the importance of his royal person and his empire, acted rashly, repeatedly burning bridges with a potential ally and benefactor and bringing about his own downfall.

BIBLIOGRAPHY

Harris, J. E., ed. 1974. *Pillars in Ethiopian History.* Washington, D.C.: Howard University Press.

Hoben, A. 1973. *Land Tenure Among the Amhara of Ethiopia.* Chicago: University of Chicago Press.

Human Rights Watch. 1999. *Mengistu Haile Mariam.* Washington, D.C.: Human Rights Watch. http://www.hrw.org/press/1999/nov/mengistu.htm.

McCann, J. C. 1995. *People of the Plow: An Agricultural History of Ethiopia, 1800–1990.* Madison: University of Wisconsin Press.

Mezekia, N. 2000. *Notes from the Hyena's Belly: An Ethiopian Boyhood.* New York: Picador.

Pankhurst, R. 2001. *The Ethiopians: A History.* Oxford: Blackwell.

Phillipson, D. W. 2000. Axsumite urbanism. In *Africa's Urban Past,* D. M. Anderson and R. Rathbone, eds. Oxford: James Curry, pp. 52–66.

United Nations Food and Agriculture Organization. 2004. FAOSTAT. Rome: United Nations Food and Agriculture Organization.

U.S. Department of Commerce. 2000. *Country Commercial Guide FY2002: Ethiopia.* Washington, D.C.: U.S. Department of Commerce.

WFP. 2004. *Ethiopia: More than 10 Million Lives at Risk.* United Nations World Food Program. http://www.wfp.org/newsroom/in_depth/Ethiopia.html.

CHAPTER 22

Somalia

Irredentism and Fragmentation

The Somali Republic, commonly known as Somalia, received its independence in 1960 when the British Somaliland Protectorate was united with the Italian-administered United Nations Trust Territory of Somalia. It forms the Horn of Africa, extending along the south shore of the Gulf of Aden to Gees Gwardafuy (Cape Gardafui), and then south to Ras Chiamboni (Ras Jumbo) beyond the Juba (Giuba) River. It is a poor dry land with few known mineral resources, limited grazing, thin acidic soils, and only two rivers with a regular flow of water, both in the more humid southern region. The Juba River rises in the high Ethiopian Massif and flows south across the broad low-lying coastal plain before entering the Indian Ocean at Kismayu (Figure 21-1). The Shebelle, or "Leopard," River also rises in the Ethiopian Massif but fails to reach the sea except during periods of exceptionally heavy rains. Otherwise, it is frequently dry during December and January and disappears in a series of sandy depressions, having paralleled the coast from Mogadishu to Jamaame (Giamama). Both rivers provide for irrigated agriculture, permit permanent habitation in their lower courses, and form Somalia's agricultural core.

Northern Somalia, essentially the former British protectorate, is topographically more diverse than the south and is dominated by the rugged Ogo and Mijurtein (Medjourtine) mountains in the north paralleling the coast. In places rainfall is as high as 20 inches (500 mm), and perennial wells provide winter grazing. South of Hargeysa (Hargeisa) lies a vast wilderness of thorn-bush and tall grasses but no permanent water: the Haud (Hawd) and Ogo plateaus. These are the traditional grazing lands of the Somali herders, but today much of the region lies in Ethiopia.

Further south still is the vast, low-lying, almost featureless plateau that comprises the bulk of the territory. Rainfall is sparse and erratic, and desert conditions predominate. The heaviest rains fall between March and June and between September and December and are associated with the monsoons, but nowhere does rainfall exceed 20 inches (500 mm).

The predominantly pastoral Somali are homogeneous in language, religion (Muslim), and culture. They are bound together by a strong sense of nationalism but it is a nationalism that has been tempered by internal forces of fragmentation. Indeed, although Somali nationalism extends beyond the legal state, and for years the Somali have sought to include all their people in a single nation-state, over recent decades internally divisive forces have riven the state and overshadowed the quest for a Greater Somalia.

Of the regional ethnic Somali population, some 30% live in eastern Ethiopia, 3% in Kenya, and 2% in Djibouti, and the remainder, in Somalia (SIL International 2004). The number of Somali living abroad is not known. If Somalia's boundaries were extended to include all these people, its area would be increased by 63% to 400,000 square miles (>1 million km^2), and traditional migratory pastoralism could proceed unimpaired by the artificially drawn international boundaries imposed by colonial Europe and Ethiopia. While boundary readjustments of this magnitude have never been likely, Somali irredentism has remained strong since independence and has been a concern of the Kenyan and Ethiopian governments.

For over a decade, however, Somalis have been busy fighting among themselves rather than with their neighbors, although along with the refugees, instability has crossed borders as well. A model of democracy and good governance for its first 9 years after independence, Somalia today is a failed state (Metz 1992) and has become a harbor for international Islamic terrorists and other criminals. Since the fall in 1991 of Muhammad Siad Barre, who as military dictator held the country together for 21 years (1969–1991) by cynically playing one Somali clan off another, centrifugal forces of place (tribe and family) have fractured the Somali state, probably irrevocably, into two pieces, with a third still in anarchy. The Somali conflicts drew in the United States and the United Nations during the Clinton administration but neither potential moderator seriously tried to put an end to the conflict: the latter because it was not in the UN's brief and the former probably out of fear of negative poll results that might have jeopardized his reelection in 1996. In this chapter we will trace out the origins of the Somali state and its breakup, and explore a not-uncontroversial hypothesis regarding the cause of political and geographical instabilities in Somali society.

THE SOMALIS AND PARTITION

Although it was once thought that the Somali originated in the north and migrated to the south over the last thousand years, displacing Oromo pastoralists and Bantu famers (Lewis 1965), most scholarly opinion today holds that the ancestors of the Somali were from southwest Ethiopia and northern Kenya. Linguistic analysis indicates that the Somali language is most closely related to several languages spoken around Lake Turkana, for example Oromo. Somali belongs to the Omo–Tana branch of Eastern Cushitic. Cushitic itself is one of the several branches of the Afroasiatic superfamily of languages that comprises the Semitic, Berber, Ancient Egyptian, Chadic, and Omotic branches (Hayward 2000) in addition to Cushitic. Proto-Somali speakers, the ancestors of today's Somalis, split from the main nucleus of Cushitic language speakers about 2,500 years ago and migrated from the Lake Turkana region to pastureland in northern Kenya. They probably followed the Tana River valley (Ehret 1995) arriving at the Indian Ocean several hundred years later. Later a group moved north into present-day southern Somali and developed a mixed pastoral and agricultural economy (Metz 1992). Somali expansion, spurred by population growth perhaps, or a desire for greener pastures and economic benefits to be gained from controlling trade routes to the interior, followed two major routes: the valley of the Shebelle and its tributaries, and the coastal plain of the Indian Ocean littoral. By about 100 CE they probably occupied the Horn and Ogaden, an area roughly corresponding to their present distribution. Along the banks of the Juba and Shebelle and in fertile pockets between them, they found Bantu settlements of iron-using farmers.

Centuries later, Yemenite Arabs set up coastal city states such as Zeila, Berbera, Muqdisho (modern Mogadishu), and Brava. Like other coastal towns of East Africa, they were largely dependent for their prosperity upon the entrepôt trade between the interior (in this case Ethiopia), Arabia, and the markets of Asia. By the tenth century, a ring of coastal emporia had been established through which Muslim expansion was to follow. The towns were conquered by the Portuguese in the sixteenth century but recaptured by various Arab groups in the next. Then came the Turks, the Egyptians, the sultan of Zanzibar, and, by the end of the nineteenth century, France, Britain, and Italy. But during the fourteenth and fifteenth centuries, Somali expansion was temporarily checked as wars broke out between the Coptic Christian kingdom of Abyssinia and the Muslim states. These holy wars saw the penetration of Arab and Somali forces deep into the heart of Abyssinia, only to meet defeat in the sixteenth century. Later massive Oromo invasions from the southwest checked further Somali expansion and effectively divided the Somali and Abyssinian forces.

The first Western contacts with the Somali coast occurred in the early sixteenth century when Portugal sacked the northern port of Berbera. But Portuguese occupation was short-lived, being replaced in the seventeenth century by a non-European ruler, the imam of Muscat. British, French, and Dutch merchants visited the coast en route to and from India and the Far East, but it was not until the opening of the Suez Canal in 1869 that the Red Sea became a major avenue of European trade with Asia and the object of colonial aspirations.

At the time, the Red Sea coast was subject to the nominal suzerainty of Turkey, but governed by local authorities. In 1870 an Egyptian governor was appointed over the whole coast from Suez to Gees Gwardafuy and Egyptian garrisons were established at

Zeila, Berbera, and later at Harar. But in 1885, because of troubles stemming from the Mahdist revolt in the Sudan, Egypt withdrew, and the "scramble" for the Horn of East Africa began.

Britain sought additional territory to provide fresh meat and cattle for its garrison colony of Aden, which was considered vital for the security of British interests in the Indian Ocean. Thus the British busied themselves making treaties with the Somali east of Zeila, which led in 1887 to the establishment of the British protectorate of Somaliland. French and Italian interests were consistent with traditional reasons for colonial expansion. In 1862 France had acquired from the Afar the port of Obock and the adjoining coast north of the Gulf of Tajura in present-day Djibouti. But for almost two decades Obock lay forgotten by France, and it was not until war broke out in Madagascar, and Anglo-French rivalry caused the British to close the port of Aden to French shipping, that the French government saw the necessity of establishing its own naval base and coaling station (Lewis 1965). Thus France secured the port of Djibouti from the Issa Somali in 1885, and 3 years later, following negotiations with Britain, defined the boundaries of French Somaliland.

Italian entry into Somali proper began in 1889 with a concession midway along the east coast granted by the sultan of Obbia. Soon after, Italy acquired control over the coastal towns of Brava, Merca, Mogadishu, and Warsheikh and appointed a commercial firm as its administrative representative. Through the company, Italian interests spread into the interior until they clashed with those of Britain and Ethiopia. Britain and Italy signed treaties in 1891 and 1894, defining their respective spheres of influence, but neither the Somali or the Ethiopians were partners to the negotiations. Somali herders living in the interior Ogaden region became subject to Italian rule, while those in the Haud region came under the British. It was not until 1908 that the several loosely administered Italian "protectorates" were united as Italian Somaliland, stretching from Bender Ziada on the Gulf of Aden south to the Kenya border (Figure 22-1).

Somali conquest was not confined to the coast, nor was it solely at the hands of Europe. It was also effected in the interior by the Empire of Ethiopia, the only African state capable of contesting the Europeans and sustaining its own territorial ambitions. Emperor Menelik's expansionism had carried Ethiopian rule far eastward, beyond the limits of the Ethiopian Massif and to the lower sections of the Horn, leading to the coastal plain. These were, and are today, the grazing areas of the nomadic Somali herders, and although boundary demarcation did not take place (other than at a few fixed points great distances apart), intrusions by Ethiopian forces were common.

Menelik asserted that the Somalis had from time immemorial been the cattle keepers of the Ethiopians and had paid annual tribute to their masters until the Muslim invasion. It was on these grounds that he justified his claim to hegemony over at least a part of the Somalis' lands. The British, throughout the period of friction, showed an awareness of the problems facing the nomadic, pastoral people, who can survive the rigors of the local environment only by practicing a form of transhumance from the low plains to the foothill slopes and back. Menelik's actions and boundary demands interrupted the migratory pattern, and the effect upon the Somalis was serious.

In 1897 further boundary definition (although very little delimitation and demarcation) took place, as Britain and Ethiopia signed an agreement that did not lead to any exchange of territory but did terminate the nominal protection Britain had extended over certain Somali groups. Italy had likewise extended the territory nominally under its control well west of the defined boundary between Italian Somaliland and Ethiopia, and for several decades there was no real opposition. Then the Walwal incident occurred (1934) when Britain and Ethiopia finally attempted to demarcate the boundaries defined decades earlier. Haile Selassie was on the Ethiopian throne, and he recognized the encroachment of Italian rule over what Menelik had considered Ethiopian territory. The Walwal incident began when an Anglo-Ethiopian boundary-surveying team that had been marking the boundary between Ethiopian and British Somaliland arrived at Walwal on November 23, 1934. Although Walwal was located over 60 miles (100 km) on the Ethiopian side of the border, the Anglo-Ethiopian surveyors were met by an armed Italian force. The British surveyors withdrew from the area to avoid an international incident. Nevertheless, there was a border clash on December 5, 1934; no one was hurt and only one shot was fired, probably from the Italian side (Pankhurst 2001). The Walwal incident

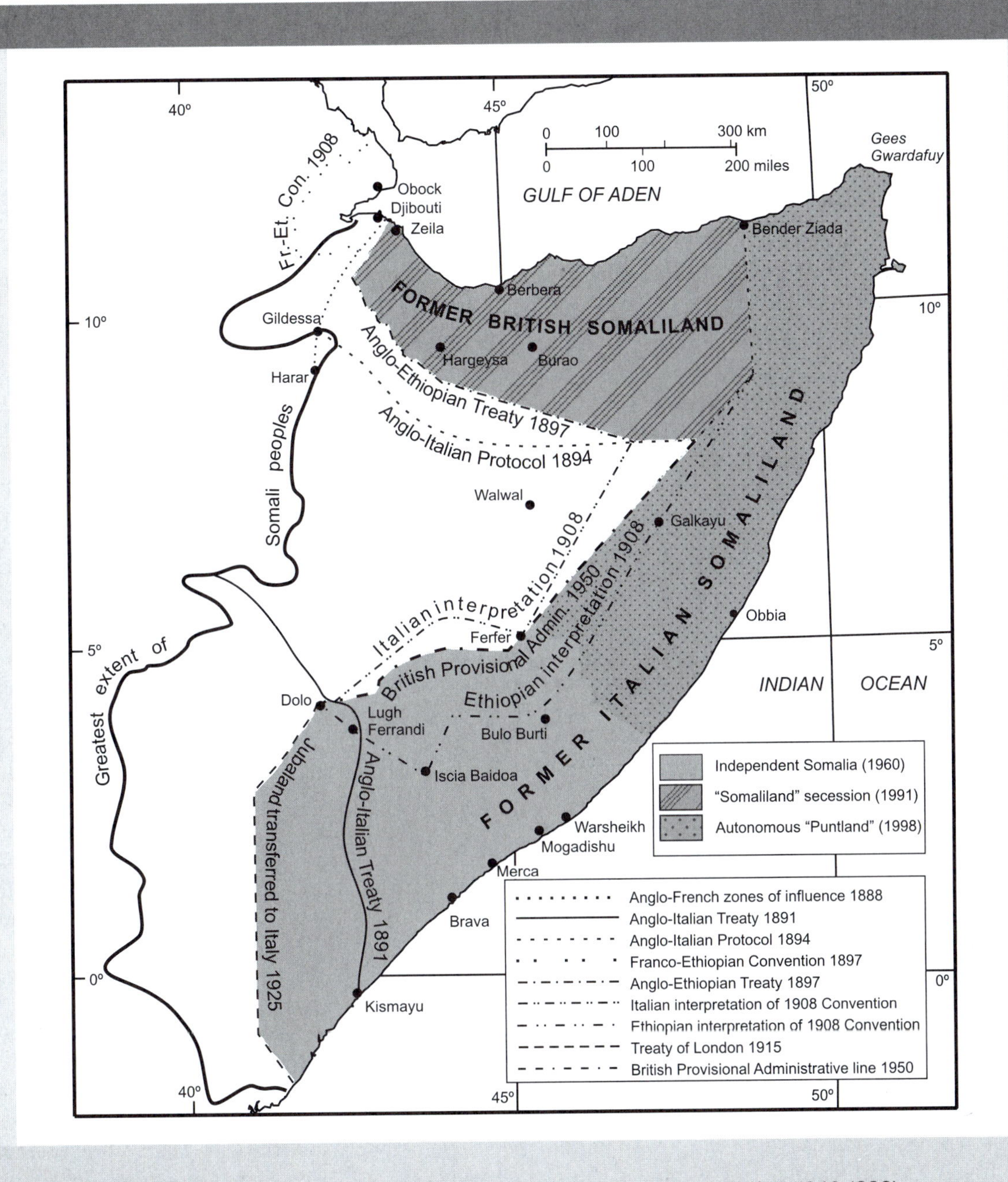

FIGURE 22-1 Defining the Somali state (1888–1960) and its internal fragmentation (1960–1998).

provided the pretext for the Italian invasion of the empire in 1935.

Between 1941 and 1947 the majority of the Somali people were under British military administration. This arrangement, highly desirable for the Somali, was the result of war conditions and was terminated when the British withdrew from the Ogaden and returned it to Ethiopia in 1948 and finally when the

United Nations took charge of the fragmented Horn in 1955. Ethiopia made vigorous efforts to persuade the United Nations that the Horn should be unified under Ethiopian rule, while Somali representatives insisted that the Somali people in the various sectors desired unification under Somali government. In the event, the boundaries as established previously (the Anglo-Ethiopian treaty of 1897, the Jubaland border of 1925, and the British Provisional Administrative line of 1950) were confirmed, and the only Somali unification that took place involved the merger of British and Italian possessions in the Horn under one flag as the Somali Republic (1960) after 10 years of United Nations Trusteeship.

SOMALI IRREDENTISM

Pre- and postindependence Somali leaders argued that the Somali people were bound together by a strong sense of nationalism and the desire for a Greater Somalia. This ideal was symbolically represented by a five-pointed star on the first independent Somali flag, each point representing Somali-occupied territory: eastern Ethiopia, northeastern Kenya, Djibouti, and the former Italian and British sectors that were united in 1960 as the Somali Republic. Pan-Somaliism has dominated government affairs since independence, and Somali irredentism has been strong ever since the imposition of the colonial boundaries. At times irredentism in Kenya and Ethiopia has been accompanied by much violence.

Somalia's boundary with Kenya first followed the Juba River, but in 1925 it was shifted west (the Jubaland transfer) to 41° east, adding about 12,000 square miles (31,000 km^2) to Italian Somalia in accordance with the secret Treaty of London of 1915 in which Britain promised to compensate Italy for its participation in World War I. But it left within Kenya an almost equal area whose population was predominantly Somali. To prevent further intertribal warfare in this remote and neglected region of Kenya, then known as the Northern Frontier District (N.F.D.), and today as the Northeastern District, British authorities set up a line beyond which Somalis were not permitted to travel (Mariam 1964). They also established administrative laws and techniques different from those applied elsewhere in Kenya, all of which reinforced the Somali sense of exclusiveness and exceptionalism. For instance, the Somali were taxed at the rate assessed on the Arab and Indian populations rather than at the lower tax set for the Africans. As a result, entire Somali communities crossed the border into Ethiopia or Somalia, where British administration was lax or nonexistent. Modern progressive programs in education and pastoralism applied in the south were not developed in the N.F.D., causing the gap in development between northern and southern regions to widen with time.

As Kenya's independence approached, Somali leaders in Kenya's N.F.D. demanded separation from Kenya and union with Somalia, and a British-appointed commission confirmed that this was the wish of the Somali majority. The British government opposed secession by the N.F.D. from Kenya, fearing it would encourage similar movements elsewhere in Africa, but believed Somali individualism would be protected by Kenya's federal constitution, then being drafted. When the Kenya African National Union (KANU) came to power, the modicum of federalism disappeared and Somali nationalists resorted to guerrilla warfare, which took thousands of lives and brought suffering and repression to the Somali civilian population. After 1967 these *shifta* wars subsided, and the Kenyan Somali have looked to Mogadishu, the capital of Somalia, for ideological and, reportedly, material support.

Somali irredentism in Ethiopia has been even more problematic, since far more Somalis are involved in a much greater area, and the international boundary line is still in dispute. The cartographic agreement of 1897 is lost, and reconstruction of the line from secondary sources has led to much confusion. Italian claims as of 1908 put the boundary 180 miles (288 km) parallel to the east coast, while the Menelik line is much farther to the east (Figure 21-1). Similarly, there are several interpretations of the 1908 Italo-Ethiopian agreement defining the boundary between the Shebelle River and the northeast corner of Kenya (Touval 1972). At issue are the seasonal grazing rights of the Somali in the Haud and elsewhere, now within Ethiopia. The British did not withdraw from the Haud until 1955 and stressed at the time its importance to Somalia by requiring the Ethiopians to guarantee the Somali free access to the grazing lands. The Somali government refused to recognize any boundary treaties to which it was not a signatory and thus has never discouraged Somali nomads from entering Ethiopian-claimed pastures. Soon after independence, border clashes became more frequent, and by 1964 those had developed into military conflict extending the entire length of the disputed boundary. Although open hostilities subsided that same year following a cease-fire arranged under the auspices of the then OAU, also in 1964 Ethiopia and Kenya signed a mutual defense pact against Somali aggression.

Somalia appeared to have abandoned its expansionist vision at that time, seeing its neighbors and the world arrayed against it. Shortly after the military takeover of the democratically elected government of Somalia in 1969, however, Somalia began a slow military buildup that was supported by its patron state, the Soviet Union. By 1974, Somalia had MiG aircraft, armored personnel carriers, and the largest tank force in Subsaharan Africa. Ethiopia, crumbling after years of drought and political instability, seemed unprepared to respond decisively. In 1976 Somalia began actively supporting a Somali secessionist guerilla movement in the Ogaden. By September 1977, the Somali army controlled almost 90% of the Ogaden and, although it suffered some defeats on the field of battle, appeared poised to hold the Ogaden. Unexpectedly, the Soviet Union shifted its support from Somalia to Ethiopia, sending in over 10,000 Cuban troops, 1,500 Soviet advisors, and tanks and other war materiél to bolster the Ethiopian war effort. This action, in support of the apparently more doctrinaire Marxist–Leninist regime that had come into power in Ethiopia in 1974, led directly to Somalia's humiliating defeat and withdrawal. The critical battle was in Jijiga, a small town located southeast of Harar, where the Somali army was outmaneuvered and over 3,000 Somali troops were killed. The battle of Jijiga occurred in February 1978 and the war ended one month later.

After abandonment of Somalia by the U.S.S.R., the African country's relationship with the United States warmed. A new agreement gave the U.S. Navy the right to use the strategic deepwater port at Berbera, which previously had been the domain of the Soviets. In return Somalia was given military and humanitarian aid. Within Somalia, centrifugal forces began to rend the state. The Somali Salvation Democratic Front emerged in the early 1980s in southern Somalia and was associated with the clans of the Majeerteen (Daarood Clan Family). In the north, armed resistance to the Siad Barre government were manifested in the armed movement of the Isaaq clans.

The Somali government's response was a brutal, scorched-earth policy in the north: the second largest city in Somalia, Hargeysa (328,000 in 2004), was destroyed in 1988, as well as many other northern towns and cities; tens of thousands of refugees streamed into Ethiopia, Djibouti, and across the Gulf of Aden into Yemen. In 1989, in response to growing concerns about human rights violations by the Barre government, the United States suspended all but humanitarian aid to that regime. Anarchy broke out in Mogadishu in 1989 as armed clan-based organizations, in this instance, the United Somali Congress (Hawiye clans), took to the streets. Instability and violence spread across the country, and the Barre government fled in 1991. Former British Somaliland declared independence as "Somaliland" in that same year.

Civil war continued as different interests competed to choose Barre's successor as president of Somalia, spreading further as groups of mobile fighting units associated with various Somali clans fought one another for supremacy. Clan-based organization engaged in a protracted struggle for power: the Somali National Movement (SNM) of the Isaaq clan, the Somali Salvation Democratic Front (SSDF) affiliated with the Majeerteen, the Somali Patriotic Movement (SPM) representing the Somali from the Ogaden region, the Somali Democratic Alliance (SDA) representing the Gadabursi from northwest Somalia, and the Somali Democratic Movement (SDM) associated with the Rahanwiin. Crisis loomed as prolonged drought jeopardized the lives and livelihood of hundreds of thousands of Somalis. The United Nations sent in humanitarian assistance and a multinational peacekeeping force in 1993 which remained until 1995. The attack on UN Peacekeepers in which 24 Pakistani soldiers were killed by militia associated with Muhammad Farah Aideed (Hawiye clan) led to a manhunt by United Nations and U.S. forces to capture him. But continued killing of UN and U.S. soldiers and lack of success in the Aideed manhunt led to a pullout of both the UN and U.S. military forces and humanitarian assistance. In 1998 the region of central Somalia declared itself autonomous "Puntland" (Figure 22-1) but did not declare independence. Somaliland and Puntland have irredentist disputes about their mutual boundaries, and within Puntland itself fighting has broken out several times with regard to territorial disputes between Puntlanders. Over the years of conflict in Somalia, refugees have fled to neighboring countries, Europe, and North America.

SOMALI FRAGMENTATION

Somalia is often cited as a classic, even quintessential, example of the post–Cold War failed African state (Harbeson and Rothschild 2000). External geopolitical factors are generally discussed with reference to the failure of such states, as well as internal shocks, for example drought. But without the strong hand of a Somali leader, or an outside threat against which the fractious Somalis could unite, the unitary Somali Republic was a state that had surprisingly little internal cohesion. Somalia comprises two core areas located on either extreme of the country, each with its own major city, Hargeysa and Mogadishu, separated by a poorly connected, vast expanse of sparsely inhabited desert. This geographical reality is exacerbated by the Somali sociopolitical realities.

The quest to unify the Somali people within a nation-state united Somalis for many years, but Somali society is based on lineage segmentation, which works against the emergence and persistence of a stable, centralized state (Metz 1992). A segmentary lineage system is one in which descent is traced to a common ancestor, generally through the male line. Over many generations such a descent structure resembles a branching hierarchy of descent from the founding father. According to anthropologist I. M. Lewis (1965, 1961), there are six agnatic (tracing descent through the father) groups of Somali in Somalia that are called clan families: the Daarood, Dir, Digil, Hawiye, Isaaq, and Rahanwiin. Each of these clan families identifies its own clan founder, from whom all subsequent descent branches. The clan families can be up to 30 named generations deep, although some of the clan genealogies may contain fictitious elements.

Clan families are broad and geographically dispersed (Figure 22-2), and although affiliation has value in modern party politics in Somalia, it is normally not as significant as lower levels of agnatic relationship. Three lower levels have been identified by Lewis: the clan, the primary group, and the *mag*-paying group. The clans, each led by a clan "sultan," may vary in size from roughly 10,000 persons to 100,000, and clan pastoral movements are localized to the extent that such migrations follow generally similar pathways from year to year. At a lower segmental level are the primary lineages, perhaps tracing 6 to 10 generations, but without a territorial ascription. At the base of the primary lineage is the *mag*-paying group. This group is the "fundamental political and legal unit in Somalia" and can be held responsible for the conduct of lineage group members. For example, if a member of one *mag*-paying lineage is killed by a member of another, blood revenge or blood money is taken from the offending group. Because individual culpability is not generally recognized as the basis of *mag*-paying, any member of a clan may end up paying for the offense of another. Thus in the identity politics of clan relations, if a man from Hawiye clan murders someone from Daarood, either Hawiye pays compensation in money to Daarood or *any* member of Hawiye may be murdered by Daarood in retribution for the crime.

Clan units join or split in opportunistic, and sometimes fleeting, alliances serving temporary interests. Such alliances are discussed, elaborated, and concluded at ad hoc councils of adult men.

> In a system of lineage segmentation, one does not have a permanent enemy or a permanent friend—only a permanent context. Depending on the context, a man, a group of men, or even a state may be one's friends or foes. This fact partially explains why opposition Somalis [never] hesitated to cross over to Ethiopia, the supposed quintessential foe of Somalis. Ethiopia was being treated by the Somali opposition as another clan for purposes of temporary alliance in the interminable shifting coalitions of Somali pastoral clan politics. (Metz 1992)

The inherently fractious lineage segmentation system works against the development of a stable modern state. Once bounded in such an entity, the traditional sociopolitical relationships are revealed to be inherently unstable. It is noteworthy that the two political units to emerge out of former Somalia are clan-based and, on the whole, have been relatively peaceful. The following anthropologist's description reflects the ad hoc, fleeting, and fragmentary nature of clan-based politics in the years after the emergence of the Republic of Somaliland in 1991: "While the Isaaq had made peace with their non-Isaaq neighbors (the Gadabuusi, Dulbahante, and Warsangeli), they were themselves more likely to fall apart, along segmentary lines, into their constituent clans (the Habar Awal, Harbar Yuunis, Habar Tolja'alo, etc.)" (Lewis 1994: 232).

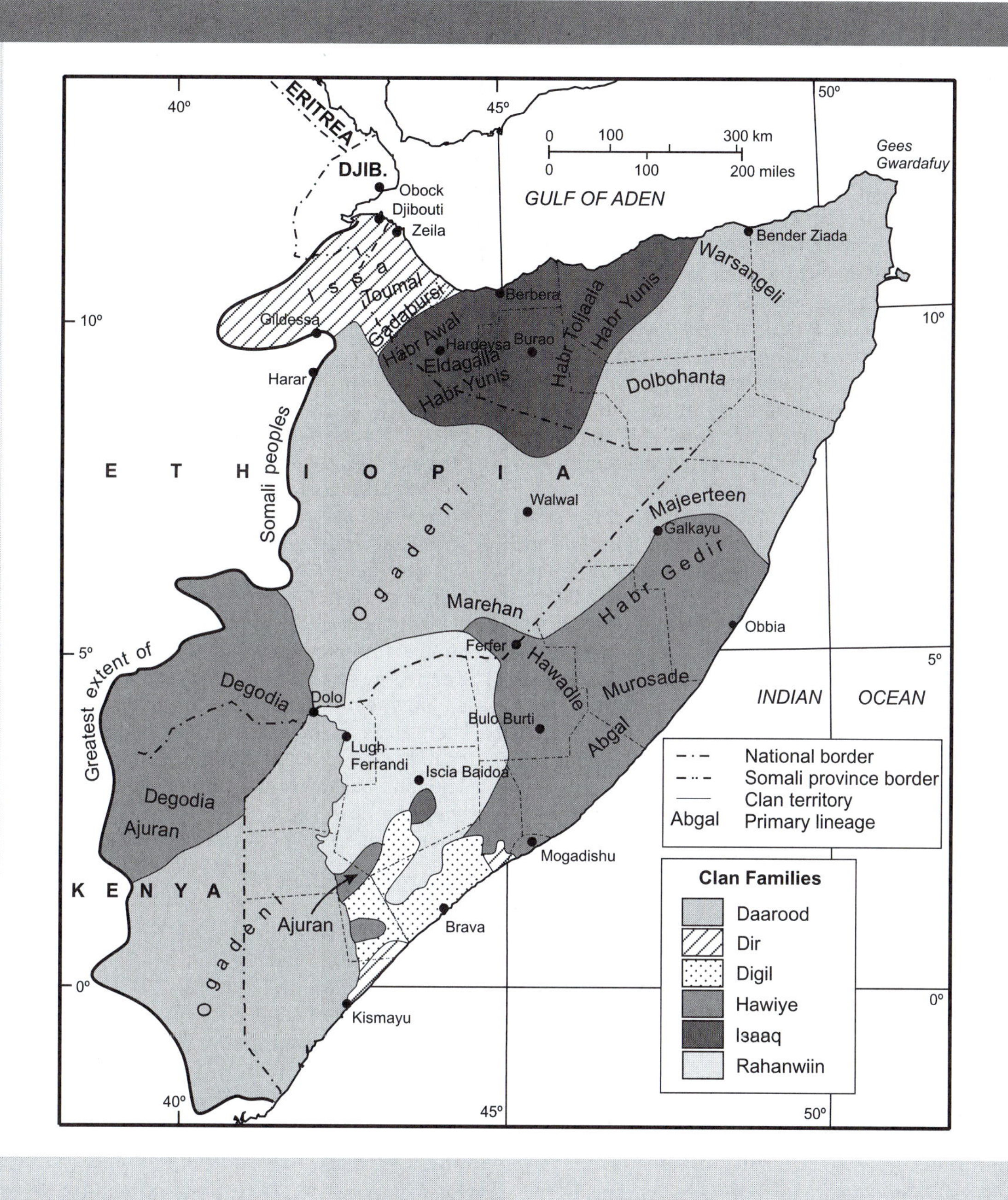

FIGURE 22-2 Borders, clan families, and primary lineage group names of Somalia.

ECONOMIC PROSPECTS AND PROBLEMS

Somalia's national economy has disintegrated, and few reliable statistics are available. There is no national currency, local businesses print their own money, and day-to-day life for the ordinary Somali is extremely difficult (Box 22-1). Some generalizations about economic prospects and problems can nevertheless be made. The likelihood that Somalia will ever have a diversified economy, rather than one that is overwhelmingly pastoral, is extremely low.

BOX 22-1

SOMALI SINGLE-PARENT FAMILY SURVIVAL

When mama came back from visiting Grandmama, she said she had made a decision to stop selling bread and to start selling milk. She had met some other women who were selling milk and they had told her how much money they were making—more than the little bit she had been making selling bread. She decided to buy a milk container, because milk you can sell every day, all day long. People need it all the time—morning, lunch, and evening.

Now she left the house early in the morning, at about sunrise, after she had prayed, and she would come home after sundown. She would sell four or five containers full of milk a day. When she had sold a whole containerful, she would go back to the market and buy another one. She didn't come home during the day to eat. When she did get home in the evening, she would eat a little fruit, and inside the milk container would be lots of food. Some days, though, she was so tired that she fell into bed and went to sleep as soon as she came home, and sometimes her feet would get swollen because she had to walk so far with such heavy loads. We had such respect for her . . . and I used to rub her feet in the evening when she was lying down.

When it came time for her to come home in the evening, we were all excited—especially me, because I was younger and I loved bananas and she used to bring some for me every day. So I would be outside looking for Mama, and everybody else would come out too, because she always brought them a little cake or something sweet to eat before dinner was ready. Then we would cook a big meal to eat at nightfall. We used to eat rice a lot, and pancakes and flat bread, with meat and vegetables. Hassan began to work too—he would go to the market with Mama in the morning and carry things for people, so he was bringing home a little money. We stayed in Mogadishu for eight months. By then the two pregnant cows that Grandmama was looking after had had their calves, so we had six cows, and Mama had enough money saved for us to go back home. When we got back to Mango Village, [near Jawhar in the Shebelle Valley] Mama started doing a little business again. This time we kept our cows in town; there were many other people who did the same thing. They hired someone by the day to take the cows outside of town for grazing in the morning, watch them all day, and bring them back in the evening. She had to pay money for that, but people put their animals together and hired one man to watch all of them. She did that, and she sold the cows' milk, and she was selling eggs too. Once again she had started something new.

Source: Barnes and Boddy (1995), pp. 49–51.

It lacks all the requisites of a modern economy. Even its predominantly pastoral economy is handicapped by low and erratic rainfall, periodic droughts, overgrazing, inferior stock, frequent invasions of locusts, outbreaks of livestock disease, and a reluctance on the part of pastoralists to transform livestock raising from a way of life into an industry. Despite these limitations, livestock and livestock products make the greatest contribution to GDP each year. Live goats, sheep, and camels form the bulk of its exports, most being shipped to Saudi Arabia and Kuwait. Bananas, introduced during the colonial period in Italian Somaliland, are exported as well.

Close to three-quarters of Somali are nomadic and seminomadic pastoralists. In the northern regions they occupy the lowlands during the winter months and trek with their livestock (predominantly sheep and goats) to the highland pastures in summer. Further south, cattle are more numerous; camels are highly valued in all regions, and economic transactions of the pastoralists operate on a camel standard (Lewis 1965). The FAO has estimated that if controlled grazing were enforced, the existing range could support two or even three times the number of livestock, but most attempts to manage the pasture have failed.

Agriculture is also severely restricted by the harsh environment (Figure 21-6) and has been affected by

BOX 22-2

LOCAL ECONOMIC DIMENSIONS OF CLAN CONFLICT

SOMALIA: "BANANA WAR" LEAVES EIGHT DEAD.

NAIROBI At least eight people were killed and over 10 wounded in heavy fighting which broke out in the southern coastal town of Marka, 100 km south of the capital Mogadishu [in November 2003], according to local sources.

The fighting—between the Sa'd and Ayr subclans of the Habar Gedir—was concentrated in and around Marka and the town of Shalaanbood, 17 km northwest of Marka, the sources told IRIN on Monday. According to a local aid worker, who requested anonymity, the clashes were triggered when Yusuf Muhammad Indha'ade, the commander of an Ayr subclan pro-Transitional National Government militia, imposed taxes on goods imported or exported through Marka port. Sa'd businessmen, opposed to the move, sent in their militia and took over a checkpoint in Shalaanbood manned by Indha'ade's militia "and then proceeded to attack Marka," he said.

"They were planning to export huge quantities of bananas to Arab countries, and therefore wanted to take control of the port. This is all about bananas and money. You can call it a banana war. But neither side is in control of the port," the aid worker added.

The fighting, which broke out on Saturday, had now subsided, but Marka was said to be "very tense." The Sa'd militia had reportedly lost two battlewagons (technicals) and a Landcruiser. Meanwhile, the two sides had withdrawn from the town and were said to be waiting for reinforcements, he said. All businesses were closed in Marka on Monday, with residents expecting more trouble, a local businessman told IRIN: "We are waiting for the inevitable resumption of the fighting."

The violence also arose from disagreements over the formation of a new administration for the town, with the Sa'd opposing "any administration involving Indha'ade," the businessman added. Meanwhile, Habar Gedir clan elders in Mogadishu have reportedly begun serious mediation efforts towards ending the violence in Marka, local sources said.

Source: IRIN (2003).

civil war (Box 22-2). Less than 1% of the area is cultivated, and only about 13% is cultivable, most of which is in the Juba and Shebelle valleys. There the emphasis is on food crops (sorghum, maize, and cassava), although irrigated cash crops such as sugar and bananas are also important. Bananas were once Somalia's leading export, the plantations being established by Italian settlers, with the industry protected and subsidized by the Italian market. Although this industry is in Somali hands, clan conflict has undercut the banana industry. The Hawiye clan controls much of the banana crop, and clan leaders received a return on exports. Beginning in 1991, rival clans prevented the docking of banana boats to diminish the income of the Hawiye (Norfolk Education and Action for Development Centre 1995).

Nonagricultural and nonpastoral economic activities are of minor importance. Without capital, technology, resources, markets, functioning communication network, and an integrated transport system, there is little likelihood of development. There are no railways, and few good roads except in the vicinity of Mogadishu (1.2 million), the capital and primate city. With U.S. aid, the southern port of Kismayu (183,300) has been expanded to accommodate the projected growth in Somalia's agricultural exports from the Juba region, and roads between it and Mogadishu, destroyed during the civil war have been improved.

Somalia has experienced many droughts over the last 30 years and the toll has been high: at least a third of the goats and sheep, more than a quarter of the cattle, and perhaps a tenth of the camels have perished. Vast areas of grazing in the north and northeast have been devastated and cannot be used for years to come. In mid-1975, possibly a million pastoralists were encamped in relief centers where they received food, water, and medical supplies. The impact of drought in the 1980s was exacerbated by civil unrest, and hundreds of thousands of people were put at risk. Clan competition for the control of humanitarian assistance and corruption in the distribution of assistance contributed to human suffering and ruined what might have been successful international assistance. Ultimately the experience contributed to the reluctance of the United States and other countries to become involved in another humanitarian crisis that emerged in 1994, the genocide in Rwanda.

Gross overstocking of sheep, encouraged by rising prices in the 1910s, led to the installation of thousands of cement tank reservoirs. This, in turn, meant more overgrazing. Installing the tanks became a lucrative business until the government banned it. We know now that the drylands have limits, and that traditional pastoralism is effective as an extensive land-use strategy only when human and livestock populations are relatively low. A successful pastoralist is one who increases the size of the herd. Such logic is valid under environmental constraints but inappropriate today because boreholes, reservoirs, and the like have removed the principal constraint to livestock overpopulation and environmental overuse in the drylands (Helland 1998).

There has been one positive side effect of the pre–civil war droughts: an increase in literacy and education. The government, agreed in 1972 on a Somali script (there being no written Somali language before then), closed most urban, formal schools for a year and had the teachers and students go to the camps of the pastoralists. Under normal conditions this would not have been feasible, but during the drought the pastoralists were concentrated in relief camps and around water holes. In this way the revolutionary government's official doctrine of "scientific socialism" was introduced at the grassroots level. Somali has replaced Italian, English, and Arabic as the official written medium and it was thought that in time, the regional differences resulting from the colonial era would lessen (Best and de Blij 1977). This has not been the case, as the preceding discussion clearly indicates. Somaliland has been developing its political and economic relationships with Ethiopia, however, and by virtue of its possession of the deepwater port of Berbera, may benefit from the Ethiopian export trade.

THE CHIMERICAL SOMALI NATION

The history of Somali political geography concerns the creation of an African state through negotiation and treaty by Ethiopia, an African empire, the British and French empires, and Italy. Creating the new state also entailed Somali ethnic nationalism and the idea of a Greater Somalia based on the putatively common culture of the Somali. The ultimate breakup of the state was due to the inherent instability of the Somali sociopolitical fabric. United against a common foe, whether European colonialism, Ethiopia, or Western imperialism, the Somali clans fought for and were joined in a seemingly culturally homogeneous state that used military force to push irredentist claims to the center of national ideology. Later, shifting official state ideology in the Horn led to changing superpower interest in Somalia, a decline in military support, and a rather rapid collapse of effective irredentist momentum. The homogeneity of the Somali nation, always more apparent than real, proved to be ephemeral.

Ethnic nationalism is contrary to the philosophy of socialism, but in Somalia, state-sponsored ethnic nationalism dovetailed temporally with the Soviet interest in acquiring strategic geographical assets to counter the United States and its Western allies. The Soviets ostensibly supported Somalia and the half-hearted socialism of the Siad Barre government, but the real value of Somalia to the USSR was the deepwater port of Berbera, which was made into a naval base. In the early 1970s, during the Siad Barre regime, the Soviet Union established a state-of-the-art naval port and missile base in this ancient port, which had been occupied by the Egyptians in 1875 in an era of different ideologies. From the 1960s to the end of the 1980s, Soviets and Americans competed to influence and control the governments of the countries situated at the southern end of the Red Sea close to the Bab el Mandeb strait and the Gulf of Aden, through which much of the world's trade passes. At that time, most of Western Europe's petroleum passed through the Suez Canal. Over the years, thousands of lives in the Horn and across the Gulf of Aden in Yemen were lost in the pursuit of chimerical political objectives, and perhaps it is fitting that the geographic choke point that it was focused on, the Bab el Mandeb, means in Arabic "the gate of lamentations."

One hypothetical but very important question on ethnic diversity has been implicitly raised by the lineage segmentation issue. Generally, ethnic diversity works as a centrifugal force in a country, as happened in Sudan. Some countries seem to manage multiple cultures in a way that does not cause secession—but not without a price. Bitter conflict led to political reconfigurations and accommodation policies in federalist Nigeria. If Somalia had been part of a larger state, perhaps federal rather than unitary, that included a large proportions of non-Somalis—it would have held together. In such a federal system is it too far-fetched to see the Somali clans joining together as one or two or three political blocs against political groupings of Amhara, Oromo, or others for example? Would a larger, federal system have preserved the autonomy of territorially based clans and subclans while at the same time providing, by virtue of size, a viable geographical space comprising a variety of resources, ecological regions, and opportunities? The answer to this question cannot be known, but what is becoming clear is that nationalism, the ethnic group and "natural" territory as state—a concept almost universally held to be desirable—is no longer so.

DJIBOUTI

Djibouti, formerly known as the French Territory of Afars and Issas—and before that French Somaliland—was the last French possession on the African mainland. French occupation dates back to 1862, when Afar (Danakil) chiefs ceded the port of Obock and its environs. As Italian and British interests in the Horn and Red Sea expanded, France acquired additional land to the south from the Issa Somali, and subsequently built a coaling station and naval base at Djibouti, today the capital and principal city of the small republic.

The large, well-sheltered harbor of Djibouti (465,000) is the obvious seaport for Ethiopia. In 1896 Menelik formally recognized French Somaliland and signed agreements with the French initiating the construction of the railroad to Addis Ababa. The first rails were laid in 1897, and the line was completed in 1917. For several decades, Djibouti was virtually the sole export harbor for Ethiopia's small external trade. Following the incorporation of Eritrea into Ethiopia, more and more trade began to pass through Mitsiwa (Massawa) and Assab, but beginning with the Badme war between Eritrea and Ethiopia in 1998, relations cooled between Asmara and Addis Ababa and traffic increased on the Djibouti line. Economically, Djibouti is heavily dependent on revenues earned through railroad transit. Djibouti has little agriculture and generates few exports other than leather and skins to France, and it must import all its industrial raw materials, machinery and oil. Retail trade is controlled by Arabs and Asians. Inflation, generally a problem for such poor countries, is not an issue as the Djibouti currency, the Djibouti franc, is tied to the U.S. dollar.

The Republic (8,500 mi^2; 22,000 km^2) has a population of approximately 765,000 that is evenly divided between the Afar who have ties with Ethiopia, and the Issa who have ties with Somalia. Both groups are Muslim, speak related Cushitic languages, and are nomadic beyond the towns. The Issa are more urbanized than the Afar and make up two-thirds of Djibouti City's population. Although both Ethiopia and Somalia threatened to take over to Djibouti if France granted it independence (Shilling 1975), in the event nothing happened. At one point Somalia saw the acquisition of Djibouti as a necessity in its quest for a Greater Somalia. Ethiopia based its claims upon a historical presence in the area and on its obvious economic interests.

In 1967, when the future of the French territory was put to a vote, nearly 60% of the voters opted for the continuance of French rule. The Somalis contested the outcome, and many suspected agitators and illegal residents were deported. The Afar remained strongly committed to French rule as the only means to retain their ethnic independence, while the Somali were equally committed to independence and union with Somalia. Meanwhile, French rule continued until 1977, when Djibouti gained its independence. The first president of the Republic of Djibouti, the Somali Hassan Gouled Aptidon, created a one-party authoritarian state and ruled until 1995. In 1991, the Afar minority took up arms against the government and fought a guerrilla war until 2001. Multiparty elections were held in 1999. In the United States' war on terror, Djibouti has allowed the stationing of hundreds of Marines.

Regionally, Somali–Ethiopian issues become complicated in Djibouti. The Issa Somali form the majority in the capital of Djibouti, and they are the largest ethnic group along the line of rail from Djibouti to the Ethiopian border. If the Djibouti Issa should one day choose to put pressure on Ethiopia, they could cut Ethiopia off from its major port. And, if Somalia were to annex Djibouti, the Somali government might acquire the leverage it needs to force Ethiopia to relax its control over Somali-populated areas elsewhere. But none of this is likely to happen in the near future for three reasons. First, Somalia has no patron superpower to support it politically and militarily and no superpower conflict to exploit to its own advantage. Although a developed nation could probably conquer tiny Djibouti if French military forces left, holding it in the face of international opposition and with the large expenses of occupation would be another story. Finally, with the departure of military strong man Siad Bane, the glue that had held Somalia together for over 20 years is gone; internal centrifugal forces are dominant, and Somalia itself has broken into three pieces.

BIBLIOGRAPHY

Ahmed, A. J. 1995. *The Invention of Somalia*. Asmara, Eritrea: Red Sea Press.

Barnes, V. E., and J. Boddy. 1995. *Aman: The Story of a Somali Girl*. New York: Vintage Press.

Best, A., and H. J. de Blij. 1977. *African Survey*. New York: Wiley.

Clarke, W., and J. Herbst. 1997. *Learning from Somalia: The lessons of Armed Humanitarian Intervention*. Boulder Colo.: Westview Press.

Ehret, C. 1995. The Eastern Horn of Africa, 1,000 to 1,400 AD: The Historical Roots. In Ali Jimale Ahmed, ed. The Inventions of Somalia. Pages 233–262.

Harbeson, J. W., and D. Rothschild. 2000. *Africa in World Politics: The African State System in Flux*, 3rd ed. Boulder Colo.: Westview Press.

Hayward, R. 2000. Afroasiatic. In *African Languages: An Introduction*, B. Heine and D. Nurse, eds. Cambridge, U.K.: Cambridge University Press, pp.74–98.

Helland, J. 1998. Institutional erosion in the drylands: The case of the Borana pastoralists. *Eastern Africa Social Science Research Review*, June 1998. Addis Ababa: Organization for Social Science Research in Eastern and Southern Africa (OSSREA). http://www.ossrea.net/.

IRIN. 2003. "Banana war" leaves eight dead. Integrated Regional Information Networks, Office for the Coordination of Humanitarian Affairs (OCHA). New York: United Nations. http://www.irinnews.org.

Lewis, I. M. 1961. *A Pastoral Democracy*. Oxford: Oxford University Press.

Lewis, I. M. 1965. *The Modern History of Somaliland: From Nation to State*. New York: Praeger.

Lewis, I. M. 1994. *Blood and Bone: The Call of Kinship in Somali Society*. Asmara, Eritrea: Red Sea Press.

Little, P. D. 2003. *Somalia: Economy Without State*. Bloomington: University of Indiana Press.

Mariam, M. W. 1964. The background of the Ethio-Somalian boundary dispute. *Journal of Modern African Studies*, 2: 189–219.

Metz, H. C. 1992. *Somalia: A Country Study*. Washington, D.C.: U.S. Library of Congress.

Norfolk Education and Action for Development Centre. 1995. Banana wars in Somalia. *Review of African Political Economy*, 64: 274–275.

Pankhurst, R. 2001. *The Ethiopians: A History*. Oxford: Blackwell.

Shilling, N. A. 1973. Problems of political development in a mini-state: The French Territory of the Afars and Issas. *Journal of Developing Areas*, 7: 613–634.

Shilling, N. A. 1975. Problems of political development: The French Territory of the Afars and the Issas, *Focus*, 26(2): 14–16.

SIL International. 2004. *Ethnologue*, 14th ed. Dallas: SIL International.

Touval, S. 1972. *The Boundary Politics of Independent Africa*. Cambridge, Mass.: Harvard University Press.

CHAPTER 23

Sudan

Bridge Between African and Arab?

The Republic of Sudan is Africa's largest state (967,494 mi^2; 2,505,798 km^2). Sparsely populated by about 18 million people in 1975 and 38 million in 2003, the Sudan, extending from the southern border of Egypt to the northern border of Uganda, shares the peoples and cultures of Subsaharan Africa, as well as those of the Middle East and Mediterranean Africa. Indeed, the Muslim north is the dominant region of the country in terms of size as well as population, influence in government as well as economic development. But the southern states form a large part of the state (Figure 23-1) and are of growing economic importance. The zone of greatest transition between the Arab north and African south lies along the 12th parallel, below which physical conditions change also: emerging from the dry heart of the central basin, the country becomes moister and the vegetation denser, as the land rises to the North Equatorial Divide separating the Nile Basin from that of the Congo.

Extending latitudinally over 1,200 miles (1,900 km), the Sudan appears to constitute a geographical link between cultural regions, perhaps to a greater degree than any other African state. No other transitional state is quite as sizable, and none incorporates as large and as diverse a population, although Ethiopia's total population is much greater. Neither can Ethiopia really be called a bridge between Arab and African from the administrative point of view. Furthermore, the physiographic fragmentation and ethnic, religious, and linguistic heterogeneity of Ethiopia create a far more complex picture than that of the Sudan, with its low relief, unending plains, and basic cultural geographical division.

Although a bridge in the territorial sense, the Sudan remains a deeply divided country, where the south was resisting the implementation of administrative decisions taken in the north even before independence. In part, this has been the result of the tenuous communications between the government and people of the north and those of the south. Just as the Somali of eastern Ethiopia look east rather than toward Addis Ababa, many southern Sudanese look south rather than to Khartoum or the Middle East. The matter has been compounded by religious differences, for in the south many Africans have adopted Christianity and find themselves in a Muslim state. Although this apparent dichotomy may not by itself constitute cause for the conflicts that have dogged Sudan since independence, it has created a certain tension, at times exacerbated by outside forces (Johnson 2003).

THE POLITICOGEOGRAPHICAL ENTITY

The Sudan is one of Africa's large, compact political units. All of its northern and northwestern boundaries (with Egypt, Libya, and northern Chad) are geometrical, and they lie in desolate, dry terrain for the greater part. In the east, the boundary corresponds more or less to the western limit of the Ethiopian Massif. The southern border shows some adjustment to physiographic conditions, running along divide areas between major drainage lines. The Sudan has its own seaport, Port Sudan, on the Red Sea; but although the state is not landlocked, parts of it are a thousand miles (1,600 km) from that exit. The southern provinces are much farther from the railroads connecting the central Sudan with Port Sudan than they are from the Uganda railhead at Gulu, which is linked to Mombasa.

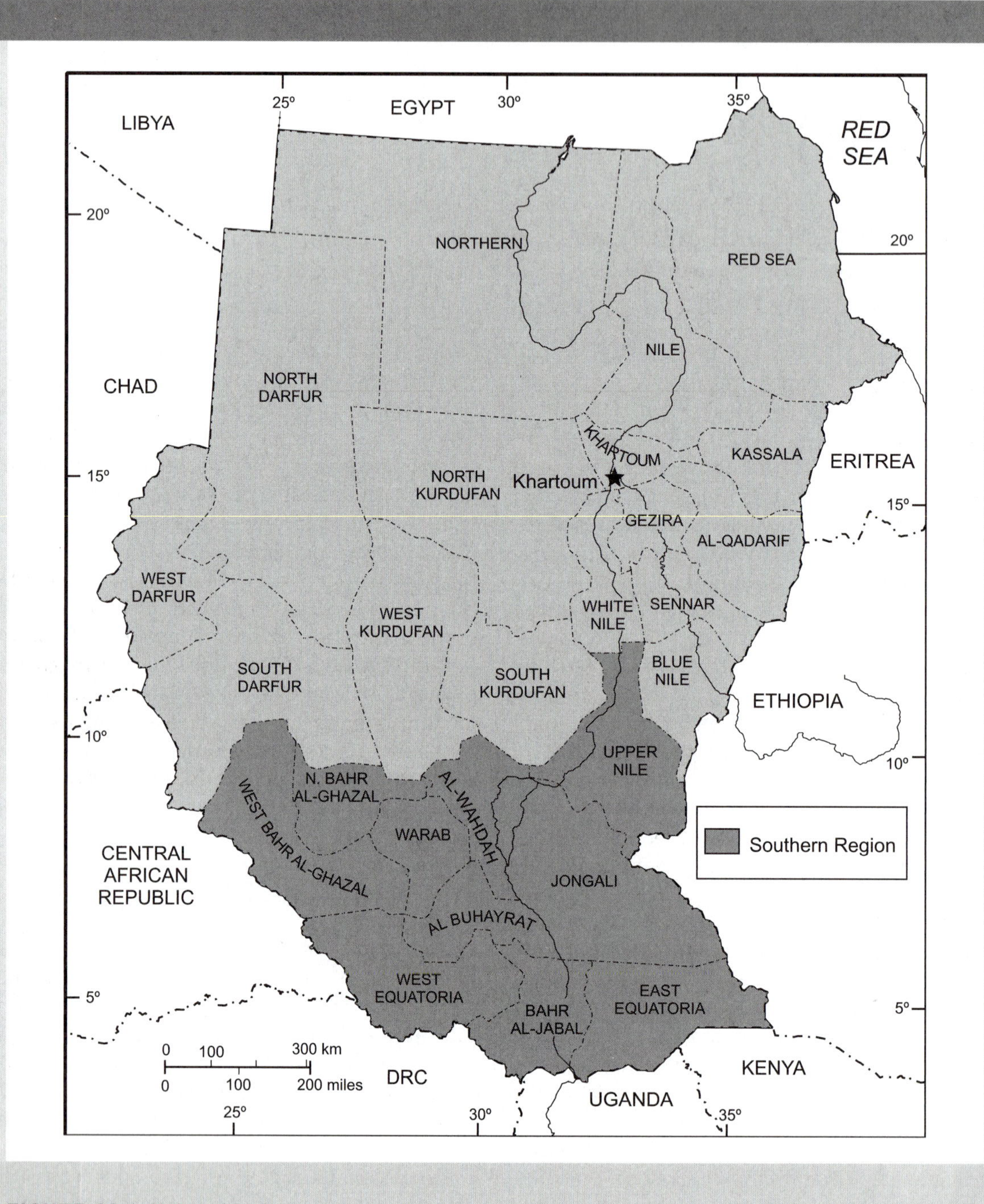

FIGURE 23-1 States of Sudan.

The orientation of the Sudan's physiographic belts is latitudinal, and a series of climatic zones can be recognized from the desert north to the moist savanna of the extreme south (Figure 23-2). With the exception of some places in the southern Red Sea Hills (Figure 23-3), the vegetative cover is practically nonexistent north of the latitude of Al-Fasher, and the average annual rainfall is below 10 inches (254 mm). Across the Kurdufan Plateau the total has reached between 25 and 30 inches (635–762 mm), and in the southwestern margins it exceeds 50 inches (1,270 mm). This east–west alignment of environmental zones has not promoted contact between the north and south of the country, especially since the physiography of the south renders the building of modern transport routes difficult and expensive, and in places almost impossible. Hence, the factor of physiography is added to that of distance in the separation of the south from the northern regions.

The great longitudinal unifier, of course, is the White Nile. In the extreme south, the great river enters the Sudan via a series of rapids, all of which lie upstream from Juba. Beyond Juba, the Nile is navigable to Khartoum, and over considerable stretches it forms the only possible means of communication. The gradient is very low as the river braids its way through the Sudd, reentering a well-defined valley in the area of Malakal. In the latitude of the Sudd, the Sudan consists of swampland and marshes, and the road from Juba to Khartoum, which traverses this region, is usable only part of the year. But the river is navigable all year round and continues to carry most of the traffic between the southern provinces and the northern parts of the country.

It is along the Nile that the "bridge" character of the Sudan between black and Arab Africa can be observed to some small extent. Not only have people of the south begun to adopt modes of dress that are normal in the north, but in the settlements, the square mud huts of the desert north and the round, thatched huts of the south stand side by side. Influence and control of the Khartoum government are greater along and near the Nile than in the peripheral areas. But the river and the areas immediately adjacent to it form only a small part of the vastness of the Sudan, and interregional contact is much less evident in the distant areas. In fact, the Sudan has also played a role in the east–west contact between African and Arab, involving peoples and ideas from beyond the boundaries of the present political entity. The savanna belt has been an avenue of penetration for Islam and a zone of transit by Africans on pilgrimages to Mecca through the Red Sea port at Suakin (Swakin). Some of the people involved in these movements settled in the region of the Sudan and made their impact upon local modes of living. The modern state incorporates an area in which north–south diversification is greater than that from east to west, and if Sudan today were to form a real bridge between Arab and African, it would be through increased north–south contact and cooperation.

The republic has its origins in a turbulent past. As in the Horn, local as well as European imperialism played a part in the course of events, and the boundaries are in several places the result of the consolidation of colonial claims. The local imperialism is that of Egypt, whose interest in the lands to the south is as old as the state itself. The first exports of the Sudan included slaves, ivory, and ostrich feathers, and the Egyptians intermittently sought to extend their power over the source of these products. The invasion of Islam up the Nile Valley was halted by the Christian kingdom of Makurra located near the First Cataract (Figure 23-2) but after 1317, when the Dongola (Dunqulah) Cathedral was converted into a mosque, Islamization proceeded more rapidly. After the fall of Dongola, only isolated pockets of Christianity, which had formed during Roman days, remained as Arab Muslims penetrated south, west into the Dar Fur region, and much later to the Ethiopian Massif.

After various small Muslim states had formed and failed, a large state, later called the Fung Sultanate, arose centered upon Sinnar (modern Sennar) on the Blue Nile. At the height of the Fung in the sixteenth century, it extended up to the Third Cataract. Although the origins of the Fung Sultanate are obscure, the people of the Fung probably emerged from the pastoral groups who lived in the grasslands between the upper Nile and the foothills of western Ethiopia. By around 1600 they were probably mostly Muslim. The state thrived during the 1600s, taxing sedentary farmers who lived in the valley, and the region between the White and Blue Nile became a core area for the entire Sudan, a position it continued to hold until it was toppled by the Turks in 1821. The Fung Sultanate held the center of the stage for well over two centuries, asserting its rule over far-flung tribal peoples, and in a sense it foreshadowed the rise of the modern Sudan.

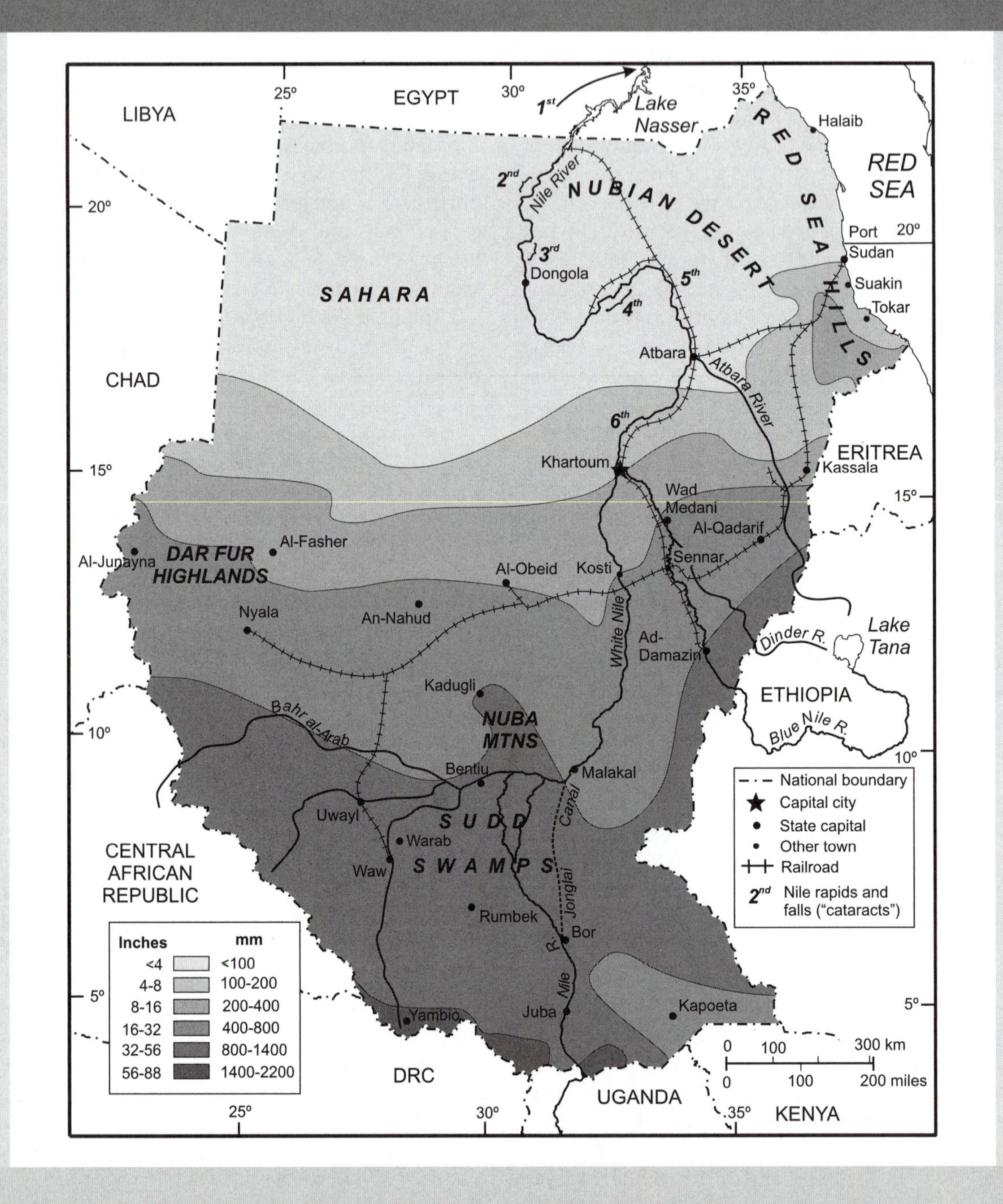

FIGURE 23-2 Precipitation, major physical features, towns, and railways of Sudan.

FIGURE 23-3 The mist oasis of Erkowit, Red Sea Hills. *Roy Cole.*

In 1821 the empire fell as a result of an Egyptian invasion led by Muhammad Ali, Pasha of Egypt, and an Egyptian government was established at Khartoum in 1830. The period of Egyptian rule is called the Turkiyya and it lasted until 1881. The entire eastern Sudan had fallen to the Egyptians, and the connection of Sudanese territory with the ports of Suakin and Mitsiwa (the latter subsequently lost to the Italians) dates from this period. British influence was now beginning to be felt, and the slave trade, prohibited in 1877, was actively suppressed in 1884 in Egypt and 1889 in Sudan (Spaulding and Beswick 2000). Economic chaos developed as a result, for slavery had been one of the economic mainstays of the tribes in the transition zone between black and Arab Africa. Under these conditions, the country was ripe for revolt, which came in 1881, when Muhammad Ahmad proclaimed himself *al-mahdi al-muntathar*, "the divine leader chosen by God at the end of time to fill the earth with justice and equity" (Holt and Daly 1979: 86). Muhammad Ahmad persuaded many southerners to join him in the revolt, but to their dismay the intentions of the "Mahdi" were much more mundane: the revolt was really about the continuance and control of the lucrative slave trade not rebellion against harsh Turkiyya, as many have thought (Holt and Daly 1979). The revolt slowly spread, and within a few years, the Mahdi's followers, the Sufi Dervishes (Darwish), controlled large parts of the country. For 13 years the Sudan was reduced to a state of disorder. Economic development was at a standstill, the livestock population was greatly reduced through the lack of care and willful destruction, disease was rampant, southerners continued to be enslaved, and general disorganization prevailed. It has been estimated that the population itself was reduced from 8.5 million to 3.5 million in the period between the defeat of one British general, Charles Gordon by Mahdist forces in Khartoum (1885), and General Horatio Kitchener's victory at Omdurman in 1898.

THE CONDOMINIUM

After the successful Anglo-Egyptian effort to quell the Mahdist uprising in the Sudan, an administration was set up whereby both Britain and Egypt had a share of the government. The Egyptian khedive (ruler) would appoint a British governor general, in whom the supreme powers in the country were vested. The governor general could legislate by proclamation and was the military and civil commander. Lord Kitchener, who had led the conquering forces, was the first governor general, and he established an administrative framework in which the policy-making positions were held by British and the majority of the civil servants were Egyptian. Egypt paid a large share of the costs of administration and control, and the country became known as the Anglo-Egyptian Sudan, although in effect it remained under British control. Egypt itself remained nominally a part of the Turkish Ottoman Empire, and when World War I erupted, it found itself involved with both sides: with Britain in the administration of the Sudan, and with Turkey (which had joined with Germany and Austro-Hungary) in the war itself. Britain declared a protectorate over Egypt, solving the immediate problem but creating the basis for a new one. Egyptian nationalism arose against British overlordship, not only in Egypt, but also in the Sudan, over which the Egyptians wished to extend their sovereignty.

During the entire colonial period, the focus of activity, political as well as economic, was in the region north of latitude 12° north. The southern part of the Sudan, while under British administration, was almost completely separated and isolated from such developments taking place elsewhere, and was little involved in the continuing administrative struggle among the Egyptians, northern Sudanese, and British. To the linguistic, ethnic, physiographic, and religious individuality of the south was added a form of administrative separation that had the character of a protectorate in all but name. Thus, rather than submerging the large-scale regional differences of the Sudan, British administration actually helped foster them, leaving the problem of actual integration and adjustment to the independent state declared in 1956. Under British administration, therefore, the Sudan did incorporate Arab and African under a single government, but the country did not really serve as a meeting ground between the two. The British deepened the division that already existed by supporting the spread of Christianity in the south while discouraging efforts to propagate Islam.

Until the Egyptian evacuation from the Sudan in 1924, administration of the condominium was based on the principle of "direct rule." The subsequent introduction of "indirect rule," or "native administration" through tribal sheikhs and chiefs, tended to accentuate the politicogeographical contrasts between north and south. A "southern policy" was implemented to prevent the spirit of nationalism from spreading southward and to separate the then three southern provinces from the rest of the country, with a view of their eventual assimilation in a broader British East African Federation (with Uganda, Kenya, and Tanganyika). Some British officials saw a southern Christian Sudan as a buffer against the Arabization of East Africa (Albine 1970). Thus Muslim and Arabic-speaking people in the south were evicted, or prevented from entering in the first place. A pass system was introduced to control interregional movement, and northern merchants were replaced by Greeks and Christian Syrians, whereas southerners were discouraged from practicing any northern customs they had acquired. Christian missionaries (mostly Roman Catholic Italians), although excluded from the north, became responsible for education in the south.

Little effort was made to provide the requisites for widespread economic development and social betterment in either north or south, although it was decided at an early stage that the Sudan's economic future must be based on long-staple cotton from the Jezira Scheme near the confluence of the two Niles. The scheme limited the amount of funds available for development elsewhere, and once it was fully operative, the tenants and administrative board successfully prevented the profits from being spent outside the region. Thus the regional economic inequalities were accentuated, and the south formed part of the condominium's economic backwater.

As early as 1944 an advisory council was constituted in Khartoum, with representation from and responsibilities for the six northern provinces. The peoples of the south, far removed from the mainstream of Sudanese development, were not identified with the Sudanese struggle for independence, and they were not represented in the council. Neither did they feel the impact of Egyptian involvement in Sudanese affairs: they were far removed from Egypt itself and from Khartoum, the scene of much of the Anglo–Egyptian–Sudanese friction. The days of temporary solidarity between parts of the south and north, resulting from the Egyptian thrust southward in the early 1800s, were long forgotten. Deeper divisions, involving race and religion, and going back also to the days when the northerners were slavers and the southerners actual and potential slaves, remained. Indeed, it may be said that the southern peoples desired either British or Egyptian rule in the Sudan, but not Sudanese rule. In the decades before independence, the Egyptian

Crown did much to retard Sudanese nationalism and delay independence, partly because of concern over the future use of the waters of the Nile. While the rift between the Sudanese and Egypt widened, the delay in the coming of independence was welcome in the south, where the replacement of British administrators by northern Sudanese officials was an unpleasant prospect. Thus, on the question of the attitude to Egypt, southern and northern Sudanese held entirely different opinions.

The internal fragmentation of the Sudan was long minimized by the uniform and high quality of British rule, even though the British could do little to wipe out regional differences. In the south, the British were unquestionably quite popular, since they were seen as security against the imposition of northern rule in the form of northern Sudanese officials. When independence came, and even before the actual date of its achievement, violence accompanied the withdrawal of British government and military personnel from the south and their replacement by Sudanese (of course, northern Muslims). In August 1955, the Equatoria Corps of the Sudan Defense Force mutinied, and for some weeks held most of the south, excluding the garrison town of Juba, which remained in the hands of forces loyal to the Khartoum government. The refugees who streamed over the Uganda and Congo borders included both civilians escaping from Khartoum rule and government officials seeking sanctuary. Belgium closed its Congo border, but among some Acholi leaders in northern Uganda there was talk of uniting the southern extreme of Sudan with Uganda. The British administration in Uganda negated these suggestions, which, however, were undeniably popular with a segment of the population of the southern Sudan.

Thus, Egyptian and British activities in this part of Africa have resulted in the establishment of a unique state, which is part of the Subsaharan African cultural realm as well as of the Middle East, internally divided in such a manner that people in the south consider themselves occupied by the north. And the north has done nothing to disabuse the southerners of this point of view, far from it. Democratic government, predictably, faltered and failed, and military–Islamic leadership has replaced it. Efforts to develop the south, to educate and integrate the peoples in the framework of the country, were never seriously part of the strategy of the government in Khartoum.

THE NORTH AND ECONOMIC DEVELOPMENT

Sudan is primarily an agricultural country, and the rainfall limits to cultivation are offset by some of Africa's best opportunities for irrigation. The agricultural capacity of the country is concentrated in the region that once was the heart of the Fung Sultanate, between the White and the Blue Nile rivers. To be sure, this is not the only region capable of development, for possibilities also exist along the Nile River north of Khartoum, along the Gash River (Wadi Qash), near Kassala, the Baraka River (Khor Baraka) in the extreme east at Tokar, and along the Atbara River, a tributary of the Nile. The south, also, can sustain sedentary cash cropping in certain areas, and there development has barely begun in the face of such major obstacles as poor communications, lack of education and incentive, and years of civil war.

Agriculture in the Sudan is dominated by cereals for food crops (Figure 23-4) and a variety of other cash crops. Production has increased tremendously for some crops (Table 23-1). Depending mainly upon its agricultural industries, the Sudan is thus not without bright prospects. In spite of the country's reliance upon farming, it has enjoyed financial solvency practically continuously from 1913 to the early years of independence. This was first achieved through the export of gum arabic, of which the Sudan is the world's major producer.

South central Sudan has been the main source area of this product, making an early and major economic contribution to the country. Al-Obeid, chief auction center for the harvest, was connected by rail (via Sennar and Kassala) to Port Sudan (Figure 23-5). But although this railroad was later extended to Nyala in Darfur Province, no southward line was constructed until the 1960s. Gum arabic, which is used principally in medicine, confectionery, and textile manufacture, is tapped from *Acacia senegal* (hashab in Sudanese Arabic), *Acacia seyal* (talh) and *Boswellia papyrifera* (gafal). Ninety percent of gum comes from *A. senegal* (El Wasila 1993), which produces the only type of food-grade gum permitted for use in the United States. Indiscriminate tapping, resulting from the ready market and strong demand, has endangered the industry, which is now supported by the planting of trees. The

FIGURE 23-4 Varieties of sorghum found in southern Sudan. *Steven Blakeway, Vetwork, U.K.*

TABLE 23-1 COMPARATIVE PRODUCTION FOR SELECTED CROPS, 1961-1965 TO 1999-2003 IN SUDAN.

Crop	Average Production (mt) 1961-1965	Average Production (mt) 1999-2003	Difference Between 1961-1965 and 1999-2003 (mt)	Change%
Bananas	67,000	72,700	5700	9
Castor beans	6,356	1,000	−5356	−84
Cassava	200,000	10,100	−189,900	−95
Chickpeas	1,537	26,800	25,263	1,644
Citrus fruit, total	95,800	146,192	50,392	53
Dates	57,800	312,928	255,128	441
Fruit, total				
Fruit excluding melons, total	526,480	1,114,100	587,620	112
Groundnuts in shell	328,700	1,090,200	761,500	232
Maize	17,291	52,600	35,309	204
Mangoes	50,000	192,800	142,800	286
Millet	302,560	587,600	285,040	94
Rice, paddy	1,209	12,150	10,941	905
Seed cotton	436,322	206,400	−229,922	−53
Sesame seed	178,338	252,800	74,462	42
Sorghum	1,256,079	3,520,400	2,264,321	180
Sugarcane	171,800	5,361,152	5,189,352	3,021
Tomatoes	98,000	693,504	595,504	608
Watermelons	67,200	142,800	75,600	113
Wheat	35,723	259,800	224,077	627

Source: FAO (2004).

FIGURE 23-5 Port Sudan's central square. *Roy Cole.*

region of production extends from Kassala Province through Blue Nile and Kurdufan to Darfur, all "northern" provinces.

In the mid-1920s a cash crop, cotton, became the Sudan's leading export product and has remained so ever since (Figure 23-6). Cotton, which contributes as much as three-quarters of the total annual export revenue, is cultivated throughout the country, under irrigation along both Niles and without it in the wetter southern provinces (Figure 23-7). The contribution of

FIGURE 23-6 Warehouses for cotton export in Port Sudan. *Roy Cole.*

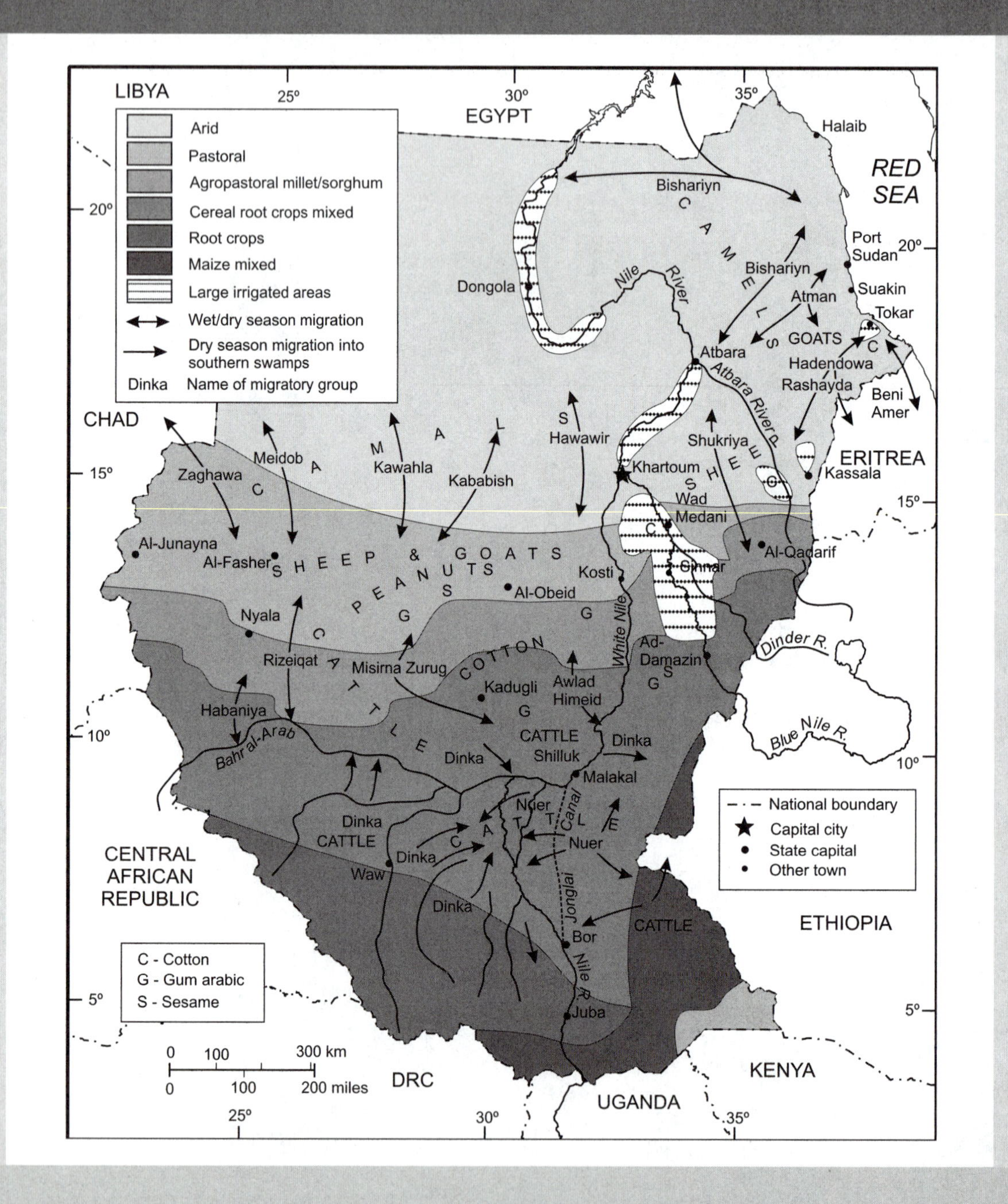

FIGURE 23-7 Land use in Sudan.

the south toward the annual cotton crop should not be overestimated, however, since only the northernmost Nilotic peoples (Shilluk and Dinka) were really involved in the cash crop economy. The south is capable of greatly increased cotton production nevertheless, and its share in the industry is likely to increase if there is real peace in the country and southerners can return to their homes. Sudan has several textile mills and a developing textile industry.

The great cotton-producing area in the Sudan is the Jezira ("island") between the Blue and the White Niles. There, the crop is grown in one of the greatest irrigation projects in the world. The Blue Nile is dammed at Sennar, a project begun in 1913, delayed by World War I, and completed in 1925. The area was connected to the coast by rail, and the plan was initially a partnership between the government, two concession companies (Sudan Plantations Syndicate and Kassala Cotton Company), and the individual tenant farmers. In 1950 the concession companies' leases expired, and the Sudan Jezira Board replaced them.

The triple partnership worked extremely well. Of the two companies, the syndicate played the dominant role, and a complete reorganization of local agriculture was achieved. The arrangement was such that the profits were divided between the tenant farmers, the government (which had provided the capital and the cost of maintaining the irrigation works), and the companies, these parties receiving 40, 40, and 20%, respectively. In 1980, following a 1965 World Bank recommendation that had initially been rejected by the Sudanese government, the Jezira Board gave full value of the marketed crops to the tenant, requiring only that the tenant pay for board services for inputs such as irrigation, fertilizer, and seeds, (Wallach 1988).

In 1950, when the role of the private firms was taken over by the Sudan Jezira Board and the scheme was therefore in effect nationalized, the profit-sharing system was altered somewhat. The tenant farmer after that time received 42% of the profits directly and another 6% indirectly in the form of social services, while the government got 42% and the Sudan Jezira Board, in charge of the entire operation, was allotted 10%.

The project continues to function successfully, with nearly 80,000 tenant farmers participating, each cultivating a piece of land averaging about 40 acres (16 ha). Careful land husbandry practices have been enforced, protecting the soil. A rotation system has been developed whereby cotton is planted on a certain plot every third year. Following the cotton year, the soil carries a crop of millet or durra, and is then left fallow the next season. Nevertheless, yields per acre are very variable, and the total harvest is great mainly because of the enormous extent of the scheme. Average yields are, for instance, substantially less than those of Egypt, although the increased use of fertilizers is expected to improve this situation.

With the completion of the Managil extension, the Jezira–Managil Scheme accounts for about 60% of Sudan's cotton output and covers about 1.9 million acres (770,000 ha) of cultivable land stretching 186 miles (300 km) from north to south and 62 miles (100 km) from west to east. Each year over a half-million acres (202,000 ha) is devoted to cotton, but this amount flucturates from year to year depending on international prices, drought, and food crop priorities. Although much of the cotton is exported (to China, Japan, and India), some is used in the textile mills of Khartoum North, Wad Medani, Hasaheisa, and other local towns. East of the Blue Nile, the government has developed the 300,000-acre (121,500 ha) Rahad Irrigation Scheme (completed in 1978), which utilizes waters diverted from Ar-Roseires, a dam on the Blue Nile, and adds substantially to Sudan's production of medium-staple cotton, groundnuts, and sorghum.

A third major irrigation scheme is situated further east on the Atbara River below the Khashm el-Girba Dam. This scheme, completed in 1966, was for the resettlement of 53,000 Nubians from Wadi Halfa and environs displaced by the waters backed up behind the Aswan High Dam. Twenty-five villages of 250 to 300 houses each were built around a new service center called Halfa (35,000), and each displaced farmer was given 2 feddans (1 feddan = 1.038 acres) for each feddan he previously had farmed (Fahim 1973). There are plans for over 500,000 feddans of irrigated land on which the settlers must cultivate cotton, wheat, and groundnuts in a three-crop rotation system. The tenants and management equally share the net returns from cotton, while all profits from other crops go to the farmer. Sugarcane, long cultivated in the Sudan, was introduced to the scheme in the 1970s and is processed locally.

Each of these irrigation schemes is situated in the central north, which makes the greatest contribution to the country's economy, far in excess of the dry northern fringe and the dominantly subsistence south. Here in the economic heart together lies the major population center of 5 million people: the capital, Khartoum (1.5 million), and its twin cities of Omdurman (2.2 million) and Khartoum North (1.3 million). This is the most densely populated area of Sudan, producing by far the bulk of the exported products and consumer-oriented manufacturing. Most of the industrial establishments are located in Khartoum North. Long the focus of administrative, religious, and educational activity, the region was the core of the Fung Sultanate of old. Lying 490 miles (785 km) by rail from Port Sudan (468,000), the principal port, it forms the major domestic market for the country.

Petroleum was discovered in southern Sudan in 1978, but exploitation has been disrupted by civil war. In 1999 Sudan completed a 930-mile (1500 km) pipeline to Port Sudan, began oil exports, and had its first budget surplus in decades. The pipeline' capacity of 250,000 barrels a day could be increased to close to 500,000 bbd with additional pumping stations. A new refinery has been built at Port Sudan, and one is planned for Al-Obeid. Positive growth has been augmented by an IMF-led economic restructuring program. Recent growth has been over 6%, and it appears that such growth will continue into the future. Sudan has a dual economy: the growing industrial economy is located in the north, mainly along the Nile Valley, and is controlled by northerners. Outside the Nile Valley, fluctuating world prices for farm products have kept standards of living at low levels. In the south, development problems have been exacerbated by decades of civil war and drought as well.

CONFLICT IN THE SOUTH AND RECONSTRUCTION

In contrast with the northern provinces, the southern provinces have not received large-scale development expenditures in agricultural projects, nor have they contributed much to the national economy. Indeed, for much of its existence as an independent country since 1956 Sudan has been riven by civil strife. From 1955–1972 the region was embroiled in a civil war that saw over a million war-inflicted deaths and an equal number of refugees fleeing across the borders and into the bush (El-Bushira 1975). Government forces retained control of the main towns, while the rebel *Anya-nya* organization established a basic administration elsewhere, although many regions and perhaps a million persons were largely ungoverned. The military regime that seized power in Khartoum in 1958 attempted a policy of enforced Arabization and Islamization of the south. Christian missionaries were expelled, the education process collapsed, transport and communications routes were disrupted or poorly maintained, and general disorder prevailed. The population, facing punitive raids by both the Sudan army and the *Anya-nya,* was forced to supply the combatants with food and materials. Cash cropping was replaced by basic subsistence production. Famine crops such as yams and cassava, which are easily planted and harvested, replaced the more nutritious millet. Food shortages, deteriorating diet, periodic drought, and the general absence of medical facilities and relief supplies brought a rising death toll (Roden 1974). Survivors either drifted to refugee camps in the south or sought work in northern agricultural schemes, construction sites, and factories. By 1971, about 800,000 southerners were living in the north. The population of Juba, which had dropped from 20,000 to 7,000 in 1965 following a Sudan Army massacre, rose to 120,000 by 1971, while Waw grew from 10,000 to 60,000 over the same period. During that time little employment was available, and the population was dependent on relief supplies.

In May 1969 General Muhammad Jaafar al-Nimeiry seized power and soon after declared his government's policy of solving the age-old north–south dispute by granting regional autonomy to the three southern provinces within a secular and socialist Sudan. But stability was not achieved until the signing, in 1972, of the Addis Ababa Agreement between the Sudanese government and the *Anya-nya* southern rebels. Under the agreement, the Sudan government recognized that all of Sudan's citizens were equal regardless of their race, religion, or color, and measures to assure the decentralization of power were adopted (Warburg 1991). In this framework, the south, having gained regional autonomy within a federal structure of government, dropped its aim of secession. It was granted a Regional People's Assembly with a High Executive Council, and with representation in the National People's Assembly located in Khartoum. The head of the High Executive

BOX 23-1

THE JONGLAI CANAL: DEVELOPMENT OR DISASTER?

The potential of a canal to bypass the Sudd swamps and deliver water to the north was recognized as early as 1901 by the British administration of the Sudan (Johnson 2003). Work on a canal did not get started until 1978, when a 175-mile (280 km) trench, the Jonglai Canal, was begun at Bor; it was to bypass the Sudd and connect to the main Nile near Malakal. Engineers predicted that the Jonglai Canal would cut evaporation losses by 50% and provide about 3.7 million acres (1.5 million ha) of reclaimed land that could be put under irrigation (Waterbury 1979). Mixed farming, pastoralism, fishing, and specialized estate cultivation were all to be potential activities. It was thought that papyrus from the swamps could be processed at Malakal to make paper and cardboard, and further south expansion of forestry. A road was to parallel the canal and reduce the distance between Juba and the north, and feeder roads were to create a network, linking Juba with the surrounding region. Construction was halted in 1984 because of the rebellion in the south, has not started again, and is unlikely to begin again.

However, enough of the canal was dug so that southerners could see that "the development of southern Sudan was not an equal priority with the expansion of existing irrigation schemes downstream. As with oil, so with water: Khartoum proved itself to be more concerned with the extraction of the South's resources with the minimum return for the region itself" (Johnson 2003: 48).

Construction of the canal would have meant the destruction of southern Dinka and Nuer cultures. The Dinka, Nuer, and other groups live on high ground within and surrounding the Sudd. They migrate with their livestock (principally cattle) into the swamps during the dry season when grazing is plentiful and later, as the swamps fill with floodwaters from the Nile, retreat to their home areas for farming. The draining of the Sudd and the removal of the Dinka, Nuer, and others appeared to the southerners as one more way in which the Khartoum government was planning to depopulate and take over the south.

Council was also a vice president of the republic. The agreement also provided for the return and rehabilitation of southern Sudanese refugees, the reintegration of *Anya-nya* rebels into the Sudanese armed forces, and the recognition of Christianity and the English language in the south, together with Islam and Arabic in the north. The Constitution contained some language that southerners rejected: particularly Article 9, which established Islamic law (Sharia) and custom as the main sources of legislation and naming Arabic as the official language of Sudan.

Oil was discovered in the south near Bentiu in 1978, and the government attempted to redefine the boundary to annex the oil-producing region to the closest northern province. Furthermore, southerners began to suspect a power play to control the natural resources (Box 23-1) when the government announced that oil would be refined only in the north (Spaulding and Beswick 2000).

In 1980 President Nimeiry unconstitutionally dissolved the Southern Regional People's Assembly. Full-scale civil war broke out when he subsequently declared that Islamic law would be implemented throughout the entire country. The Nimeiry government was toppled in a coup in 1985 and, after a one-year transitional period, elections were held and Sadiq al-Mahdi, leader of the Ansar Party and a descendant of Muhammad Ahmad, was elected prime minister. As long ago as 1966, Mahdi, an Islamist, had declared that "the failure of Islam in southern Sudan would be the failure of Sudanese Muslims to the international Islamic cause. Islam has a holy mission in Africa and southern Sudan is the beginning of that mission" (Malawal 1981: 41). His government began arming tribal Arab militias, who were charged with depopulating areas in the south. Events did not go as well as he had planned.

The new rebellion in the south was spearheaded by a new organization led by Colonel John Garang, who held a Ph.D. in economics from Iowa State University. The Southern People's Liberation Movement (SPLM) and its military wing, the Southern People's Liberation Army (SPLA), has engaged in almost continuous warfare with successive northern governments ever since (Burr and Collins 1995). Up to 2005 there were approximately 30 other rebel groups in the south; some of them made up of Muslims, and a few were led by Muslims. Refugees flowed in every direction, particularly north, where over a million live in Khartoum alone. During the 1980s the SPLA won a series of stunning victories against the northern army, taking towns throughout the south. When Kurmuk, in the northern Blue Nile Province, fell in 1989, shock waves were sent through the north. Kurmuk, on the Ethiopian border, is located just a few miles south of Ad-Damazin and the dam at Ar-Roseires. It appeared to many that the SPLA was taking the war to the enemy. The northern population was uneasy, the economy was declining, and inflation was draining the hard-won savings of the citizens. In this climate of uncertainty, the government of Sadiq al-Mahdi was ousted in a military coup d'état on June 30, 1989, and a new government, led by General Omar al-Bashir, took power to resolve the issue by force. It soon became evident that the general was in alliance with radical Sudanese Islamists led by Hassan Al Turabi, who appeared to have a major influence on policy in the country. Indeed, three northern businessmen were executed for trading currency, a serious violation of Sharia law. The government instituted price controls, took control of the nation's food supply, and began gearing up for military confrontation in the south. In 1991 the radical Islamist and terrorist Osama bin Laden moved to Sudan, along with other terrorists, causing concern in the United States that Sudan had become a sponsor of international terrorism. The Sudan government continued to pursue its policy of arming Arab tribal militias charged with depopulating areas of the south (and most recently Darfur in the west).

Over the years of the rebellion, the military fortunes of the southerners have waxed and waned, each new victory giving them more bargaining power, and each loss strengthening the government's resolve. Cracks appeared in the SPLM in the 1990s, and the SPLM lost some of its gains over the north, but since that time there has been rapprochement within the movement of those with differing viewpoints. Life in the south has been completely disrupted by northern armies and militias, by drought, and by the activities of the rebels themselves. International relief efforts to help southern Sudan have been hobbled by the country's government, which has "requisitioned" relief food trains on the way to Waw and prevented the movement of relief food and other humanitarian aid to that region. In several interviews with former U.S. ambassador Donald Petterson, General al-Bashir declared that the International Red Cross was working for the SPLA (Petterson 1999). In 1999 the general declared a state of emergency and dissolved the National Assembly after a power struggle. In 2001 Hassan Al Turabi, the architect of political Islam in Sudan, was arrested, imprisoned for 3 years, released, and rearrested in 2004 for alleged involvement in plotting a coup against the al-Bashir government.

Internationally mediated talks between the rebels and the Khartoum government took place in Kenya in 2002, and in that same year General al-Bashir met with Colonel Garang in Kampala, Uganda. It was agreed that the south would be exempted from Islamic law and that 6 years after the implementation of peace, the south would be able to vote on secession. The most divisive of all is oil. The oil is located in the south—but just over the border (Figure 23-8). Questions still remain regarding whether the Khartoum government is negotiating in good faith or merely buying time (Figure 23-9), but in late 2004 a peace agreement was signed in Naivasha, Kenya. The provisions of the agreement, finalized in January 2005, are as follows:

1. The south will have autonomy for 6 years, at which time there will be a referendum on secession in the southern provinces.
2. Both the northern and southern military forces are to be unified into one force if, after 6 years, the southerners do not opt for secession.
3. Oil wealth is to be shared 50:50 between the southern and the northern states.
4. Northerners will occupy 70% of the jobs in Sudan's central administration, while 30% will go to southerners. In the Blue Nile state and in the Nuba Hills area of South Kurdufan and in the Abyei district of West Kurdufan, northerners will take 55% of the jobs and southerners 45%. (Abyei represents the land of 9 Ngok Dinka chiefdoms that was transferred to Kurdufan in 1905 by the British.)

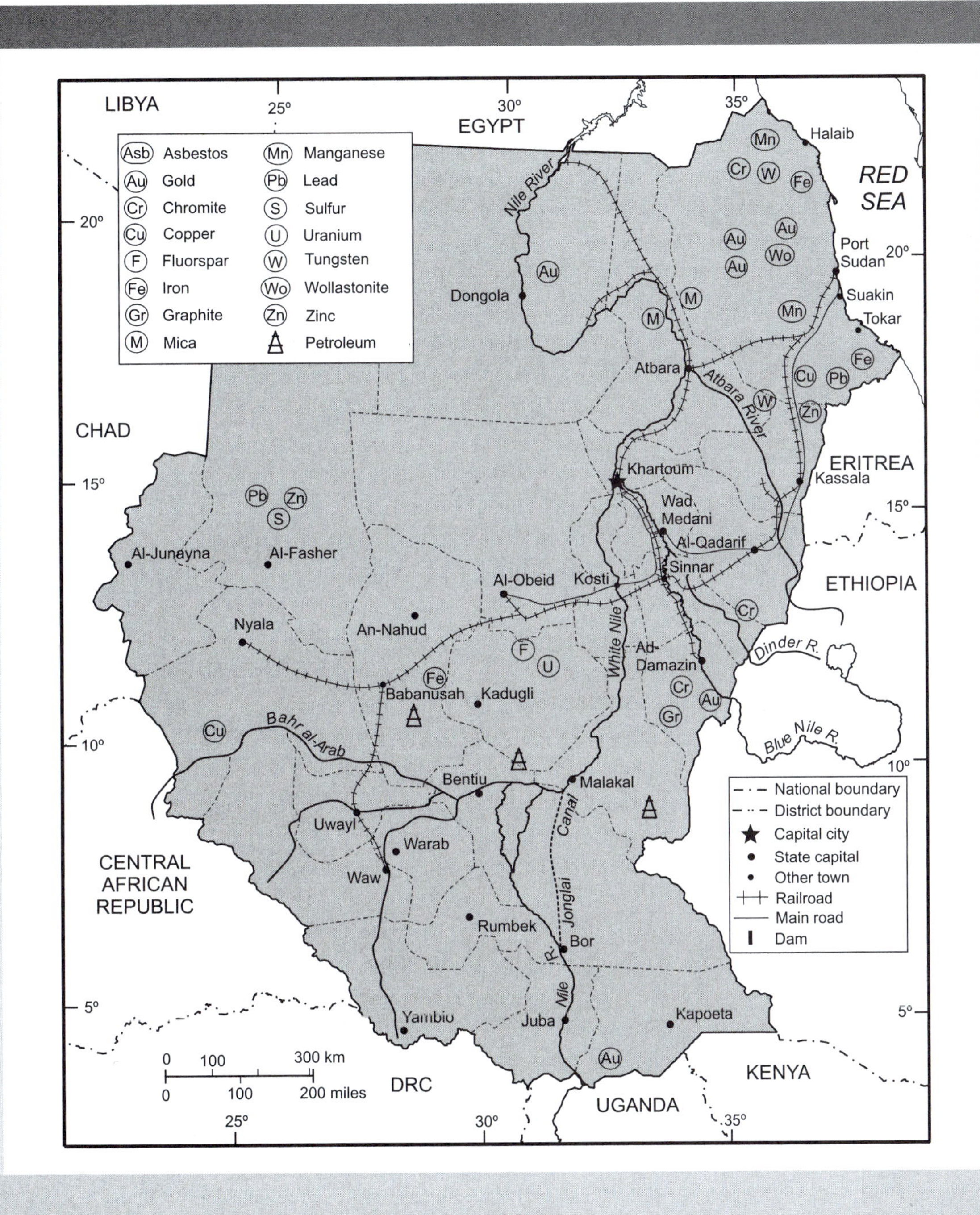

FIGURE 23-8 Towns, transportation, and resources of Sudan.

FIGURE 23-9 AK-47 in southern Sudan cattle camp. *Steven Blakeway, Vetwork, U.K.*

5. Sharia will only apply in the northern states. The application of Islamic law in Khartoum, where a large population of southerners resides, is to be decided by an elected assembly.

While peace was looming in the south, conflict broke out 2003 in western Sudan in the Darfur region, and hundreds of thousands of refugees streamed into Chad. Outside observers charge that the Sudan government has, as they did in the south, armed local militias who are depopulating large areas of their inhabitants and burning villages. U.S. Secretary of Defense Colin Powell visited the region in 2004, as did U.N. Secretary General Kofi Annan. The Khartoum government has pledged that it is not involved in the ethnic cleansing, but government forces and aircraft have been repeatedly observed working with the Arab militias. Another national minority, the Beja, took up arms against the government in 2004 in Red Sea province.

BRIDGE AND BARRIER

Before the recommencement of war in the early 1980s, there were many who thought that Sudan could well become the breadbasket for the Near East: it has 40% of the potentially cultivable land of the Arab world and is strategically situated to supply Arab neighbors with food. Most of this potential lies along the Nile, especially in the south. Arab oil money has financed in part, the Rahad irrigation project, the aborted Jonglai Canal, mechanized agricultural production of cereals near Al-Qadarif, a petroleum products pipeline from a new refinery in Port Sudan to Khartoum, a textile mill at Sennar, and numerous smaller agricultural processing plants. There is also the need to upgrade and expand the presently inadequate transport network, one of the greatest impediments to rapid development. But nothing can be done until the problems in the south and west are resolved completely.

Sudan's peculiar internal politicogeographic realities have made it difficult for successive leaders to adopt a single foreign policy toward neighboring states. Policies that have attempted to bridge the Arab and African worlds have foundered on the region's politicogeographic problems. As dissent increased, the Khartoum government seems to have opted to destabilize its neighbors by supporting their rebel movements. Relations with Uganda have been strained as a result of Sudan's support of the Lord's Resistance Army, and in the past Sudan has supported Ethiopian rebel movements as well. Since the oil boom of the 1970s, Sudan has courted oil-rich Arabian peninsular states. Relations with Egypt have often been strained, most recently in 1995, by the Sudan-sponsored attempt on the life of Egyptian president Hosni Mubarek while he attended a conference in Addis Ababa. Fighting in western Sudan spilled into Chad in 2004.

The transition from the heart of the Arab world to Africa along the Nile River, begun in Egypt, ends in Uganda. Indeed, the most obvious zone of cultural transition lies across the Sudan, but it would be a mistake to assume that the peoples of the northern and southern parts of the country are part of homogeneous Arab and African populations living beyond. The idea of a two-way bridge between Arab and African worlds is chimerical. In the former "Western Sudan," from Senegal to Niger, the bridges that developed between North Africans and Subsaharan Africans were mediated by trading groups in desertside towns (Baier and Lovejoy 1975). Transportation was so slow, however, and distances between the population centers in North Africa and the savanna and forest peoples in the south were so great that both sides were relatively autonomous partners in long-distance trade. Cultural momentum was with the universalizing and expansionist Islam which, for the most part, spread gradually and peacefully.

With the decline of the trans-Saharan slave trade and the development of ocean-going transportation, the desertside "ports" of the Sahel declined. Unlike any of the other desertside developments south of the Sahara in historical times (Gombe, Timbuku, Gao, Agadez, for example), Sudanese Arabo-Islamic culture has developed a political, administrative, economic, and cultural monopoly and momentum that never emerged along the Sahel Corridor in the Western Sudan. Part of the reason for the Sudanese Arab-Islamic cultural juggernaut is explained by the geography of the Sudan itself, and the institutions and infrastructure set up by the British during their administration. Unlike in West and North Africa, a population cluster has developed in the Sahara itself, along the banks of the Nile, halfway between the Mediterranean and the southern savanna lands. Second, this population rather than the southern one, has relatively easy access to oceangoing transportation. Third, the British administration, which was later reinforced by the Khartoum governments, built only road and rail links that drained the south and converged on Khartoum, ignoring the natural corridor for the southern half of the country through Uganda and Kenya. Economically, the transportation monopoly of the north means that products must go to northern markets and exits. To break the northern monopoly, southern Sudan would have to develop rail and road links with both Uganda and Kenya.

Short of secession, the south cannot resist the northern monopoly. British rule really only delayed the inevitable effort to integrate the south with the north. Although it could be argued that not fostering the integration of south with north was the only major British error in the modern administration of the Sudan as a dependency, what would integration have meant? The disdain of northerner for southerner and of Arab Sudanese for non-Arab is profound and longstanding and is rooted in Muslim religious exclusivity and racism (Abdel Salam and de Waal 2001). What integration would be possible under such circumstances?

Spaulding (2000) believes that Sudan's problems between the north and the south since independence result from the unequal geographic distribution of natural resources and the expansionist character of the Arabo-Islamic culture of the north. The exploitative relationship between the resource-poor Arabo-Islamic north and the resource-rich but politically disorganized south has involved slavery, rape as an instrument of terror, and mass murder with genocidal intent as the northern governments have tried to depopulate the south so that northerners could take control of the land and resources. During the time that the British indirectly administered Sudan, the British reinforced northern power.

Geographically, southern Sudan will always be the hinterland of someone else; nevertheless, the independence of southern Sudan and a reorientation toward the Southern tier of countries (Uganda, Kenya, Ethiopia) with whom there is much in common, would go far to rectify the northern bias that threatens to destroy the south.

But perhaps nothing so symbolizes Sudan's failure as a cultural "bridge" as does the conflict in Darfur, where African Muslims (the Fur people and other minorities) are victimized by Arab militias aided and abetted by the Khartoum regime. But this particular conflict, like that in southern Sudan, is not new: a glance at Figure 3-3 on page 45 demonstrates the fragmented geography of the Nilotic languages. The isolation of the Fur language speakers, surrounded in a sea of Arabic, is apparent. The ancestors of the Arabic speakers expanded into Darfur (and elsewhere) over the last few hundred years. Many of those inhabiting the Darfur region have only converted to Islam in the nineteenth century. But such conversions are evidence of which way the tide of war will turn. Unlike southern Sudan, the Fur peoples have no outlet to the world.

Although Darfur does not possess the petroleum resources or agricultural potential represented by the vast, well-watered plains of southern Sudan, Spaulding's (2000) description of north-south violence in Sudan as conflict over the control of natural resources is apt for Darfur as well. Darfur is centered on highlands where water is relatively abundant: precipitation is higher than on the surrounding plains; and rivers and streams provide water for orchards and other forms of intensive agriculture. The highlands of Darfur extend for about 200 miles (322 km) from north to south from the Meidob Hills to the foothills of Jebel Marra. High points in these rugged mountains are Jebel Marra (10,131 ft; 3,088 m), Jebel Gurgei (7,713 ft; 2,351 m), and Jebel Teljo (6,410 ft; 1,954 m). The potential of these mountains has drawn Arab invaders to this region; however, the ruggedness of the terrain has enabled the Fur, the Meidob (Tiddi), the Berti, and others to successfully defend their homelands.

Geographically, historically, and culturally these relict peoples are similar to peoples of the Nuba Hills in south central Sudan, the Tedagu (Tubu) who live in the Tibesti Mountains in northern Chad, the Tuareg of the Ahaggar Mountains in southern Algeria, and other peoples who have been encapsulated and isolated by expansionist Arabs (and other militarized Muslim groups). But the governments of Chad and Algeria are no longer interested in their distant mountain minorities. In Sudan, the quest for land and resources has been powered by Arab nationalism and Islamic jihadist ideology, fueling ethnic cleansing in all regions of the Sudan: southern Sudan in the world spotlight yesterday; Darfur, today. Unlike southern Sudan, which may reorient itself through its southern outlets to the world, encircled Darfur has no such option. The fact that virtually all of the people living in the Darfur highlands are Muslim today demonstrates their encapsulation. Years of armed resistance of southern Sudanese peoples to Arabo-Islamic expansion and international pressure made Sudan a pariah state and paved the way for peace and possible independence for its southern states. Such an option for Darfur is impossible.

Long before independence and continuing to the present there has been a persistently growing Arab-Islamic occupation of Sudan. Migrating into Sudan hundreds of years ago, Arab Muslims used the Nile Valley as a staging area from which to subdue and occupy much of the rest of Sudan. That process continues to this day. Sudan is neither a bridge nor a barrier between the Arab peoples of the north and black Africa. The most appropriate metaphor to describe the relationship is a military one: Sudan is a bridgehead.

BIBLIOGRAPHY

Abdel Salam, A. H., and A. de Waal, eds. 2001. *The Phoenix State: Civil Society and the Future of Sudan.* Asmara, Eritrea: Red Sea Press.

Albine, O. 1970. *The Sudan: A Southern Viewpoint.* Oxford: Oxford University Press.

Baier, S., and P. Lovejoy. 1975. The desert-side economy of the central Sudan. *International Journal of African Historical Studies,* 8(4): 551–581.

Burr, J. M., and R. O. Collins. 1995. *Requiem for the Sudan: War, Drought and Disaster Relief on the Nile.* Boulder, Colo.: Westview Press.

El-Bushira, El-Sayed. 1975. Regional inequalities in the Sudan. *Focus,* 26(1): 1–8.

Fahim, H. M. 1973. Nubian resettlement in the Sudan. *Ekistics,* 36(212): 42–49.

El Wasila, Omer El Karim. 1993. *Gum arabic—An essential non–wood based products in Sudan.* Paper presented in Regional Expert Consultation on Non-wood Forest Products (NWFP) for English-Speaking African Countries, Arusha, Tanzania, October 17–22, 1993. http://www.fao.org/docrep/X5325e/x5325e0e.htm.

Holt, P. M., and M. W. Daly. 1979. *The History of the Sudan: From the Coming of Islam to the Present Day.* London: Weidenfeld & Nicolson.

Malawal, B. 1981. *People and Power in Sudan.* London: Ithaca Press.

Johnson, D. H. 2003. *The Root Causes of Sudan's Civil Wars.* Oxford: James Curry.

Petterson, D. 1999. *Inside Sudan: Political Islam, Conflict, and Catastrophe.* Boulder, Colo.: Westview Press.

Roden, D. 1972. Sudan after the conflict. *Geographical Magazine,* 44: 593–598.

Spaulding, J., and S. Beswick. 2000. *White Nile, Black Blood: War, Leadership, and Ethnicity from Khartoum to Kampala.* Lawrenceville, N.J.: Red Sea Press.

Warburg, G. 1991. The *Sharia* in Sudan: Implementation and repercussions. In *Sudan: State and Society in Crisis,* J. Voll, ed. Bloomington: University of Indiana Press.

Waterbury, J. 1979. *Hydropolitics of the Nile Valley.* Syracuse, N.Y.: Syracuse University Press.

Wallach, B. 1988. Irrigation in Sudan since independence. *Geographical Review,* 78(4): 417–434.

SOUTHERN AFRICA

More than 130 million people live south of the Congo–Tanzania border in a region of immense environmental and cultural contrasts. Bantu-speaking agriculturalists comprise the vast majority of the population, while there are small populations of Khoisan speakers who predate the arrival, over a thousand years ago, of the iron-using farmers. The island nations of Madagascar, Comoros, Mauritius, and Seychelles have long been dominated by a plethora of outsiders. People from the Indonesian islands came to Madagascar over a thousand years ago, and their distinctive cultures contrast sharply with those on the continent; Arab and Swahili have economically and politically dominated the East African coast and the coastal archipelagos since the tenth century of the common era. Most recently, southern Africa has been politically and economically dominated by European resident minorities and by Europe: the Portuguese presence in Angola and Mozambique dates back to the 1480s; Dutch merchants and farmers settled the Cape over 350 years ago; the French established trading posts on the Madagascar coast in the late seventeenth century; and British settlement, which began in South Africa in the early 1800s, had crossed the Zambezi by the beginning of the twentieth century. From these early beginnings and coastal strongholds, the European sphere of influence spread across the subcontinent, bringing to southern Africa a unique set of politicoeconomic conditions, constraints, and problems, many of which are still with us today.

Nowhere else in Africa has the European presence been as strong as in southern Africa. Principally as a consequence of the moderate and healthful climates in southern Africa, hundreds of thousands of Europeans settled in the highlands and along some of the breezy coasts; over time they took the levers of power, consigning native peoples to environmentally marginal "homelands" and "native reserves," similar in all but name to the Indian reservations of North America. South Africa, Zimbabwe, and Kenya are probably the best-known examples of European settler colonies in Subsaharan Africa, although Algeria was once a significant colony of the French in North Africa. In each of these well-known cases, the European colonial power partitioned the territory and appropriated many of the best lands for its own nationals to settle.

Europeans tended to prefer more temperate, salubrious environments, and settler colonies grew in the Mediterranean climates in both northern and southern Africa and in highlands. In the middle 1950s, there were about 1 million European settlers in Algeria, the second largest settler colony in Africa at the time. In 1962, after a long, bloody war, Algeria achieved independence. The overwhelming majority of European settlers left Algeria during the war or shortly before independence, relinquishing their control over the land.

In contrast, the European settler population in southern Africa in the mid-1950s was around 5 million out of a continental population of around 240 million (Fage 1988), and the proportion of settlers to indigenous persons was variable from country to country and from region to region, often increasing in the years prior to independence where independence was unexpected. Table 1 presents the settler population (those who regarded Africa as home) at the height of the settler population: after 1950 and before independence. For Angola, Mozambique, and Zimbabwe, the greatest number of settlers was in the early 1970s, while for North Africa it was in the early 1950s.

The decolonization of Africa is notable for its overall lack of violence (Legum 1999), perhaps due in large part to the rapidity with which it occurred. In settler colonies, however, the level of violence was much greater, particularly when the colonial power resisted granting independence. The overwhelming majority of European settlers in North Africa left at or shortly before independence (Morocco, 1956; Algeria, 1962). In Algeria, independence was granted after a long, bloody civil war, fought mainly in the 1950s, that divided families on both sides of the Mediterranean and perhaps cost a

TABLE 1 ESTIMATES OF MAJOR EUROPEAN SETTLER POPULATIONS IN AFRICAN COUNTRIES AFTER 1950 AT THEIR HIGHEST POINT PRIOR TO INDEPENDENCE.

Region	Country[1]	Colonial Power	Settler Population	Percent of Total Preindependence Population
North Africa	Algeria	France	1,000,000	10
	Libya	Italy	50,000	5
	Morocco	France	300,000	4
	Tunisia	France	300,000	8
Southern Africa	Angola	Portugal	480,000	8
	Mozambique	Portugal	90,000	3
	South-West Africa	Germany, South Africa	91,000	12
	Northern Rhodesia	Britain	70,000	3
	Rhodesia	Britain	225,000	7
	South Africa	Britain	3,000,000	20
Central and East Africa	Kenya	Britain	60,000	9
	Belgian Congo	Belgium, France	75,000	<1

Source: Adapted from Fage (1988), Hance (1975), U.S. Census International Database (2002).

[1] Europeans in the rest of Africa constituted less than 1% of the total population in any given country.

million lives. Southern Africa, where the greatest numbers of European colonists had settled on the continent, was different from North Africa. In all of the settler colonies of southern Africa, the post–World War II years saw increasing European settlement. Despite increasing anticolonial violence, the migration of Europeans continued into the mid-1970s in the Portuguese colonies of Angola and Mozambique, but rapid and massive decolonization occurred after political changes in Portugal brought an independence-minded government to power.

Although many Europeans have left since independence, the former English colonies of southern Africa have had no mass exodus similar to Algeria, Angola, or Mozambique. In these former colonies the settlers have, with the recent exception of Zimbabwe which may be an anomaly, retained control over their estates, factories, and shops, and have continued to play a significant role in the economy and government. European settlers also established large estates in Malawi, Namibia, Swaziland, and Zambia although the settler populations were relatively small.

Settler-colonialism was experienced differently in each of these southern African countries, but all countries share similar circumstances of conquest, land partition, the occupation of prime lands by European settlers, and the monopolization of power. The colonial power encouraged the migration of people from other parts of the empire to satisfy the demand for labor in the African colonies. Thus, for example, Indians migrated to South Africa and ultimately took an important role in trade and commerce. In every colony the settlers were interested in making a home and in wresting a living from the land and through commerce. In many settler colonies the vast majority of settlers lived in urban settings very similar to those in Europe. On the farm, European agriculture often contrasted with the African smallholder, who was focused on producing food for the family rather than on profit and commodities for trade. The European enterprises were ambitious and drew in African laborers, generally attracted by wages but sometimes obliged to participate to pay colonial taxes and sometimes simply forced.

In most settler colonies, the colonial philosophy was one of paternalism in which the African needed European guidance to assimilate to European culture and enter as a full member into modern society. In some colonies barriers were erected against such contact and Africans were merely regarded as a supply of cheap labor. In any event, the cultural and social landscapes created by large numbers of European settlers is a legacy to which southern Africans are still trying to adjust. The land shortage that African farmers face in Namibia, South Africa, and until recently, Zimbabwe has been artificially created and can be remedied only by redistribution. Many of the issues related to such adjustment will be discussed in this section of the text.

SOUTHERN AFRICA AS A REGION

The region of southern Africa begins at the southern edge of the Congo Basin and extends to the southernmost tip of the continent. Out of this area of 2.5 million square miles (6.5 million km^2), six landlocked countries (all once

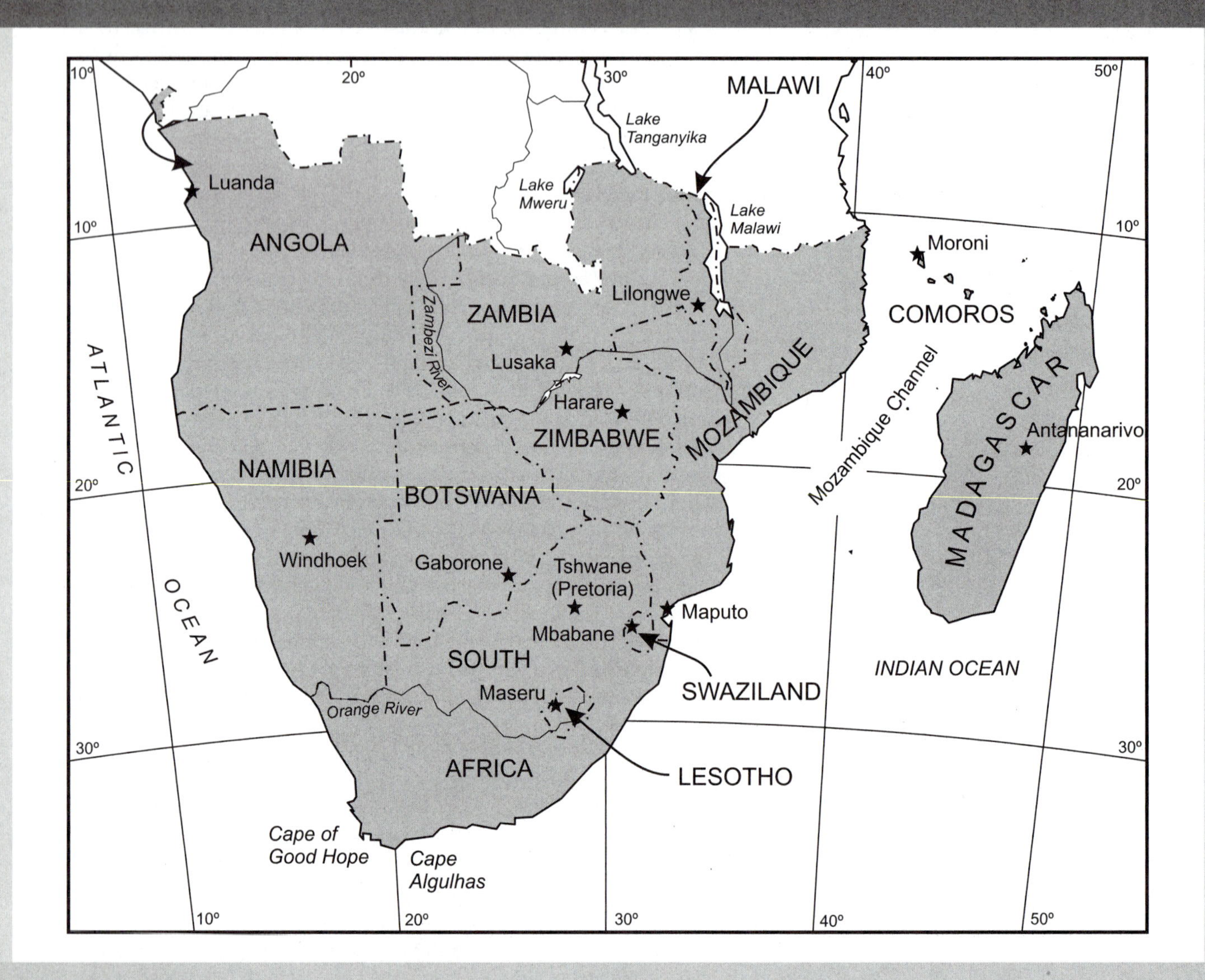

FIGURE 1 Southern Africa: states and capitals.

British) and six coastal states (of which two are island nations) have been carved (Figure 1). Although the process of regionalization is sometimes more art than science, in several ways southern Africa is one of the most integrated regions on the continent. Physically, much of southern Africa is a high plateau, bounded by narrow coastal plains and the Great Escarpment, and drained by some of Africa's largest rivers: the Zambezi, Limpopo, and Orange, among others. The economic, political, and cultural focus of southern Africa is on the economic powerhouse, South Africa. Whether its neighbors were working for or against it during the apartheid years, South Africa has been the glue that has bound the countries of the region together in a variety of ways. South Africa has the most developed agriculture, manufacturing, and consumer economies and possesses a well-integrated road and rail network linked to its neighbors. In a very real sense, all (rail) roads in southern Africa lead to South Africa. The region is also united by its common experience of late decolonization: Botswana and Lesotho in 1966; Swaziland in 1968; 1974 and 1975 for the former Portuguese possessions of Mozambique and Angola; 1980 for Zimbabwe; and 1990 for Namibia. And although South Africa has been independent from Great Britain since 1910, white minority rule there ended just in 1994. States surrounding South Africa are similar in their dependence on South Africa for employment in mining and manufacturing; seasonal labor migration from the surrounding states has been historically very high. During the apartheid era, the surrounding states sought to reduce their dependence on their rich neighbor, South Africa, and there is some unease today—even in Africa—regarding the dominant economic, political, and military roles South Africa seems to be taking for itself in the region.

South Africa is fabulously rich and has the resources to initiate broad development-related investments in profit-making ventures—South African corporations are heavily involved in the economies of their neighbors. The countries of southern Africa have had a long history of cooperation as well in the Southern Africa Development Conference and the Southern African Customs Union.

The peoples of the region are among the most diverse in Africa and in this regard the region can be considered unique. The majority of southern Africans are of ancient African ancestry, speaking closely related languages of the Bantu subfamily of Niger–Congo. Minorities in the region include the small populations of Khoisan speakers located mainly in the southwest of the region, Arabs, North Europeans and immigrants from the Indian subcontinent of more recent provenance, and descendants of ancient migrants from Indonesia. The region is characterized by rich and diverse natural resources, varied agriculture, a wide range of environments, and in almost all southern African countries a distinctive legacy of past inequality.

A more ominous commonality that links the vast majority of the countries of the region is the extraordinarily high prevalence of HIV/AIDS: rates are the highest in the world. Southern Africa has the greatest potential for development of any region in Africa. Yet despite this promise, the reality of AIDS has created a grim future for millions of southern Africans in the first decades of the twenty-first century, with potential consequences that could compromise the stability of the region.

CHAPTER 24

South Africa

Challenges of Racism, Rebellion, and Reconciliation

In terms of resources, South Africa is the richest country in Africa. The diversified resource base includes gold and diamonds, iron ore and good-quality coal, alloys, and copper, as well as fertile soils and adequate water supplies in several areas. South Africa is the continent's most developed country in economic terms, possessing a sizable iron and steel industry, plants for the conversion of coal into oil, and more than half the industrial establishments in Africa. It supplies a higher percentage of its own needs for manufactured goods than any other African country, while it leads the continent in the production of gold, platinum, coal, chrome, sugarcane, and corn. With only 5.5% of the continent's population, South Africa's GDP has hovered around one-fifth of the African total.

South Africa's share of *Subsaharan* Africa's total GDP has declined from 41% in 1998 to 33% in 2002 (World Bank 2003), reflecting both a decline in the value of the rand against the dollar and increasing development in other areas of Subsaharan Africa. South Africa's share of Subsaharan manufactured exports has remained over 60% since the last years of apartheid, and South Africa is Africa's largest energy consumer and second largest producer. South Africa's infrastructure is second to none in Africa. An integrated system of roads, rail lines, and airways links the country's cities, which are among the largest and most modern in Africa.

South Africa's human resources, too, are varied. Every continent has contributed to the complex demography of this large (472,500 mi^2; 1.2 million km^2) country (Figure 24-1). The arrival of the various component groups was not simultaneous, but each has made its specific contribution to the development of the state. Today, the country displays an incredible patchwork of people, languages, religions, customs, and modes of living. The spectacular developments made in the economic sphere in certain regions lie in sharp contrast to the low human development in others. Despite the apparent success in its economic development, South Africa's approach to human development has been much less successful, earning the country years of international opprobrium and civil strife because of its long history of legally sanctioned separation of the "races." Although separation of the "races" has ended as official policy, the rigid racial social geography established by the white South African government is a legacy that today's South Africans are struggling to overcome—and must overcome if the country is to maintain the developmental momentum it already possesses and move forward in extending such development to all South Africans.

From 1948 to 1994, South Africa was engaged in a unique experiment in multiracial development—an experiment that was imposed by the white minority upon the African majority and the Asian and mixed-race communities. Officially termed "separate development" or "multinational development" by the ruling Nationalist Party, but more commonly known as *apartheid,* the program envisaged a number of independent racial states carved out of the existing country, called "independent homelands" by their supporters and "Bantustans" by their detractors (Box 24-1). Boundaries were superimposed on a preexisting cultural landscape in an effort to create 10 independent African states and a single white majority state. To achieve this, the Nationalist government designed and implemented plans to change regional economic conditions, relocate industries, build new towns, resettle populations (often involuntarily), develop new political institutions, and devise new legal systems. The

FIGURE 24-1 Provinces, some cities, and major rivers of South Africa.

costs, human and economic, were incalculable and in many ways insurmountable—nevertheless, the government embarked on a racially divisive and violent policy that ultimately failed. To an increasing degree, up to the collapse of apartheid in 1994, South Africa was a laboratory for the political and economic geographer seeking to predict the consequences of this singular attempt to solve the local version of a worldwide problem, a quintessentially human problem: how to live with diversity and difference. In this chapter, an effort is made to present these developments and their legacy in a geographical perspective without losing sight of their roots in a combination of unusual historical circumstances.

BOX 24-1

WHAT WAS APARTHEID?

Apartheid is an Afrikaans word meaning "apartness" or "separateness." South Africa's white minority government was committed to the political, social, and territorial separation of what the Nationalists described as four "races": White, Black, Asian, and Coloured. Apartheid was based on the idea that integration would result in the loss of cultural identity, and a brutal, Orwellian police state was developed to enforce racial policy. Apartheid was implemented at three levels: "grand apartheid" was concerned with dividing the country to ensure white control; "urban apartheid" focused on creating separate living and business areas in the cities; and "personal" or "petty" apartheid minimized personal contact between people of different races (Christopher 2001). Only whites were represented in the national parliament. The Asians and Coloureds, individuals of mixed racial parentage, had separate representative councils with limited legislative and administrative powers, while the African majority was reorganized politically and spatially into quasi-independent homelands or "bantustans."

Although the white government recognized four races, the South Africans were more commonly labeled either "white" or "nonwhite," and much of the apartheid legislation reflected these designations. Entrances to all public facilities such as railway stations, post offices, and libraries were marked "Whites Only" and "Nonwhites." Park benches, elevators, beaches, hotels, and sports facilities were similarly marked, while each race had its own schools, universities, and residential areas.

In apartheid-era South Africa, mixed marriages and sex across the color line were criminal offenses. The Group Areas Act reserved for each race a separate residential area in the urban regions. Often these areas were separated from each other by buffer zones such as cemeteries, highways, and vacant land. The influx of rural Africans to the cities was restricted by law to the number of jobs available, and certain skilled jobs (e.g., mine engineering) were reserved for whites. Every South African was issued an identity card (pass) that stated the holder's race and legal address. Africans were required to carry them at all times, and failure to produce one upon police demand meant a fine, imprisonment, or even banishment to the designated homeland. Each day almost 4,000 Africans were arrested for pass violations. Africans had to regularly prove to the municipal authorities and police that they were gainfully employed and legally resident in designated urban areas. At Sharpeville in May 1960, police shot indiscriminately with automatic weapons into an unarmed crowd of women and children demonstrating against the pass laws, killing 67 and wounding 186. The riots of June 1976, in Soweto and other African townships are further examples of black opposition that existed to South Africa's race laws.

Many of apartheid's more vocal and influential opponents were detained without trial, often for months. Intellectuals were put under house arrest and "banned." All "nonwhite" opposition was considered illegal and brutally repressed. The apartheid government routinely tortured and murdered its opponents.

THE PHYSICAL SETTING

South Africa is a land of diverse natural landscapes that have both helped and hindered human settlement and development. Adverse relief and rainfall conditions have seriously restricted the extent and nature of agriculture, while the rich and diverse mineral resources have provided the basis for Africa's most important mining economy and mining-related industries. The 1,800-mile (2,900 km) coastline from Mozambique to Namibia has few natural harbors, and the adjoining coastal lowlands are narrow, varying from desert in the west to marshland in northern KwaZulu-Natal. In

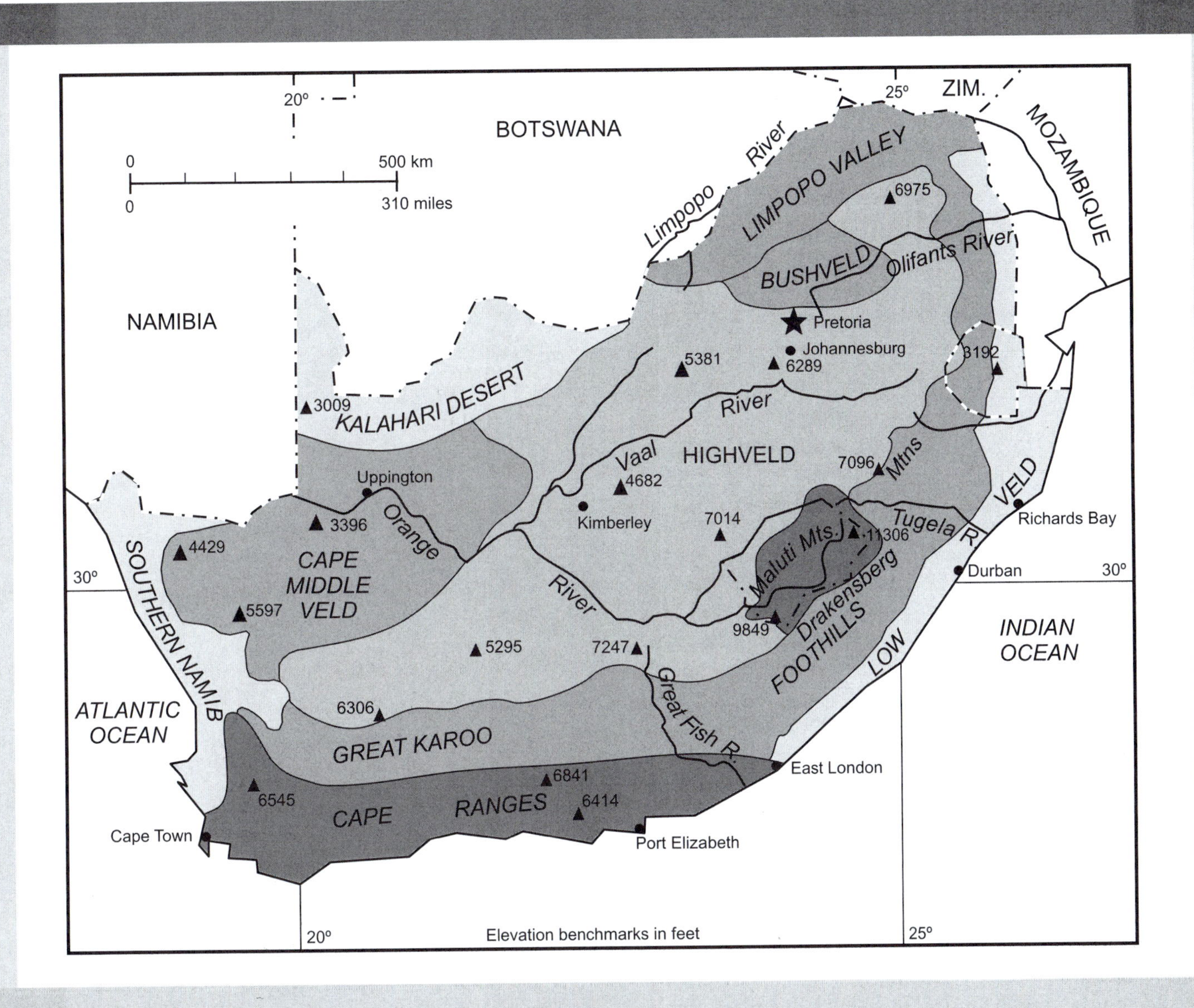

FIGURE 24-2 Physiographic regions of South Africa.

the southwestern Cape, the east–west-trending fold mountains of the Cape Ranges rise to heights of 6,000 to 7,500 feet (1,800–2,250 m) and were effective barriers to early colonial expansion into the interior (Figure 24-2). Behind them lies the vast Karoo, a sparsely settled dry-land region that merges with the Kalahari and Namib deserts of Botswana and Namibia, respectively. Behind the dissected coastal perimeter further east, the spectacular Drakensberg Mountains reach 11,000 feet (3,300 m) near the Lesotho border and stretch from Mpumalanga Province southwest into the Eastern Cape Province. The Drakensberg Escarpment rises, in places,

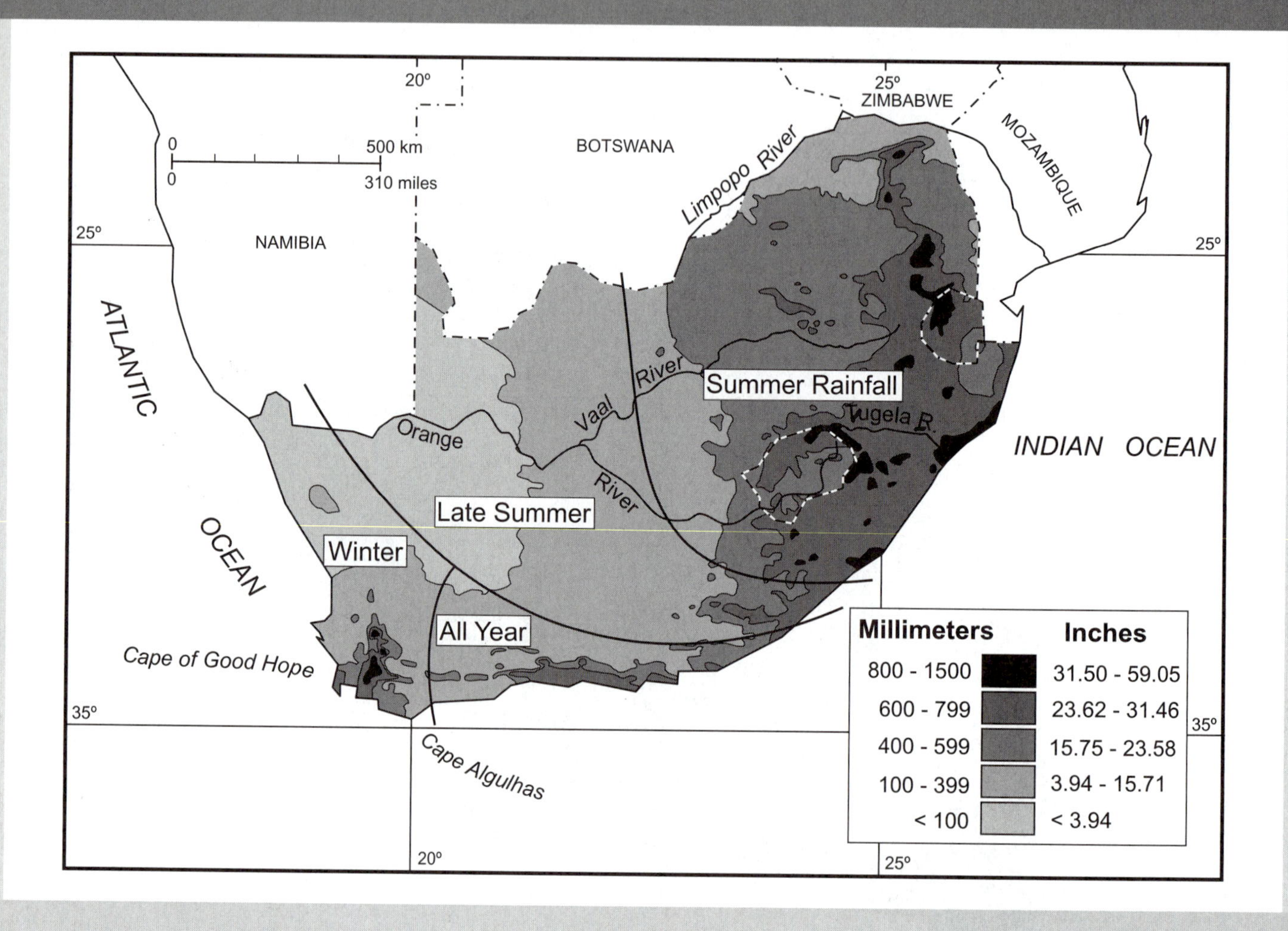

FIGURE 24-3 Mean annual precipitation for South Africa.

almost 6,000 feet (1,800 m) above the adjoining lowlands, where it forms a major barrier to communications between the coast and the interior plateau.

The interior plateau, or highveld, stands between 3,000 and 6,000 feet (900–1,800 m) and is centered on the northeastern third of the country and includes the Free State, Gauteng, and Mpumalanga in their entirety and parts of Limpopo and North-West provinces. Here lies South Africa's underground wealth—its gold, coal, copper, diamonds, platinum, asbestos, and other minerals. The Witwatersrand, a low range of hills centered on Gauteng Province, forming the watershed between the Vaal and Olifants rivers, is Africa's richest gold-producing zone.

In general the highveld landscape is gently undulating, broken in places by isolated hills and highland areas such as the Zoutpansberg and Waterberg mountains located along the southeastern side of the Limpopo Valley. Flowing westward across the highveld from its headwaters in Lesotho, is South Africa's major river, the Orange, and its principal tributary, the Vaal. Both provide the basis of irrigated agriculture and supply the interior urban centers with their industrial water requirements, while water from the Orange River is also being diverted southward toward the Cape and several south coast towns. The Limpopo forms South Africa's northern boundary and drains much of the northern provinces before emptying into the Indian

Ocean. Hundreds of smaller rivers such as the Tugela, Umzimvubu, and Great Fish pour out of the Drakensberg, cutting deep valleys across the provinces of KwaZulu-Natal and Eastern Cape. Few have been dammed to control erosion or to provide much-needed irrigation water and hydroelectric power.

South Africa has great climatic diversity tempered predominantly by elevation and latitude: a summer-dry "Mediterranean" climate prevails around Cape Town; a humid subtropical climate occurs along the KwaZulu-Natal coast; highland savanna and steppe characterize much of the highveld; and semidesert and desert conditions cover much of the Cape interior. Rainfall is highest in the Drakensberg and decreases in general from east to west (Figure 24-3). Drier parts of Northern Cape Province receive less than 8 inches (200 mm), while coastal KwaZulu-Natal, the Drakensberg, and parts of the Cape Ranges receive over 56 inches (1,400 mm) annually. Only 10% of the area receives over 30 inches (760 mm), the amount generally required for grain farming. Both evaporation rates and rainfall variability are high, and in the interior droughts are common and occasionally prolonged. Winter snowfalls are common in the Drakensberg and are not unknown in Johannesburg, while overnight frosts occur from June through August in the higher elevations.

The diversity of climates, soils, and topography makes possible a diversity of agriculture. It is unfortunate that the heaviest rainfall occurs either in the higher elevations or on land too steep and broken for cultivation. The combination of steep slopes and high rainfall—especially in the Drakensberg foothills of KwaZulu-Natal and Eastern Cape has produced much soil erosion and reduced the amount of land available for farming despite serious attempts to promote conservation practices. Since the majority of South Africans are farmers, this is a serious constraint to local development.

PEOPLES AND REGIONS: SOUTH AFRICA'S MAJOR POPULATION GROUPS

According to the current official classification of people in South Africa, South Africa's population (44.8 million in 2001) consists of four major groups: black Africans, whites, Coloureds, and Indian-Asians (Table 24-1). There are 11 official languages. These population groups of South Africa are differentially distributed across the nation. People of mixed ancestry ("Coloured," in the South African terminology) are in a majority in the Northern and Western Cape provinces (Figure 24-4). Elsewhere Africans form a clear majority at the provincial level. People of European ancestry are mainly urban and form roughly a quarter of the urban population of both Gauteng and KwaZulu-Natal provinces. People of Asian ancestry are most numerous in urban centers of KwaZulu-Natal Province.

Almost four-fifths of the population of South Africa today is of African ancestry, and the vast majority are Bantu-speaking peoples. They are culturally and ethnically diverse and comprise four major language groups. The largest linguistic group is the Nguni, which includes the Zulu, Xhosa, Ndebele, and Swazi. The Sotho-speaking South Africans include the Tswana, Pedi, and South Sotho or Basuto, while the two smallest linguistic groups are the Venda and the Shangana-Tsonga (Figure 24-5). Although the pre-European distribution of Africans was general, it is not surprising that the highest pre-European population densities in South Africa were achieved in the disease-free uplands of the north and along the humid east coast.

Along the foot of the Great Escarpment, the coastal belt afforded easier migration routes for the iron-using agropastoralists, and there the ancestors to today's Xhosa and Zulu made the greatest penetrations to the south. The Xhosa were the first to confront the advancing Europeans in the eastern Cape during the eighteenth century, and the Great Fish River became the scene of a series of wars between these two pastoralist groups. The Zulu rose to power when they found strong leadership in the then Natal, and the distribution pattern results largely from their influence. The Zulu might was broken by the European invaders over a century ago, but the Zulu nation survives today. The present distribution of Bantu-speaking Africans is very similar to the past, except for the urban centers, which have drawn hundreds of thousands to city life.

People of European ancestry form the second largest group (4.4 million), and like the Africans they are culturally and ethnically diverse but principally of Dutch and English extraction (4.3 million). Although comprising only 10% of the population today (Table 24-1), they continue to dominate many aspects of life in South Africa, particularly in upper levels of management and in the urban financial sector of

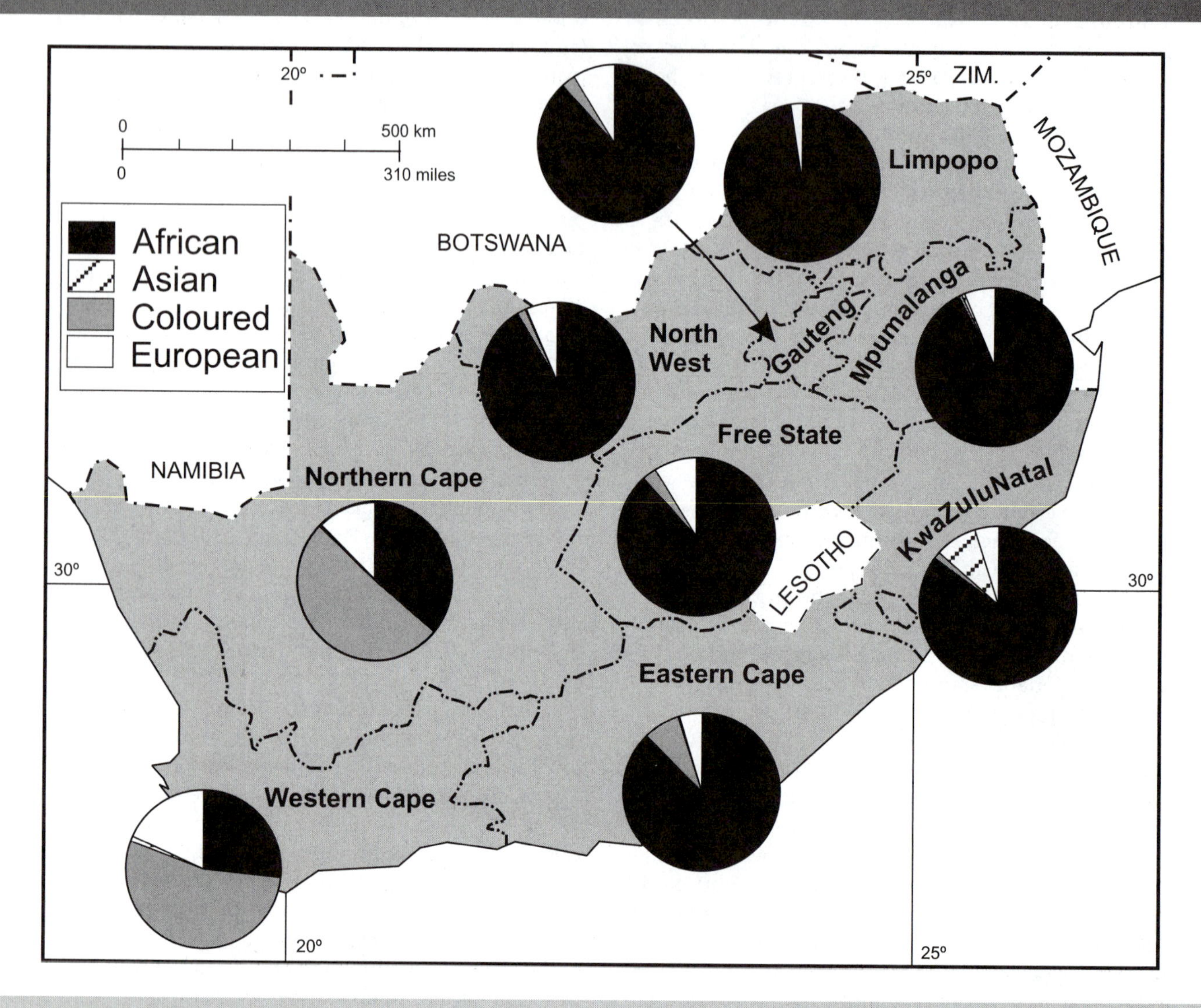

FIGURE 24-4 Proportions of people of African, Asian, Coloured, and European ancestry in South Africa's provinces in 2001.

economy. Permanent white settlement dates back to 1652, when Jan van Riebeeck established a supply station at Table Bay (Cape Town's natural harbor) for ships of the Dutch East India Company sailing between the Netherlands and Southeast Asia. Large-scale settlement was officially discouraged, but the population grew and was augmented by French Huguenot and German immigrants, and slaves imported from Malaysia, Mozambique, and East and West Africa (Wilson and Thompson 1969). The immigrant groups brought with them their languages, values, and essentials of life, including their staple crops (wheat, vines, and citrus) and technology. Gradually they spread eastward in search of new land beyond official Dutch control, and by the eighteenth century had clashed with the southward-migrating Xhosas. In 1806 Britain gained control of the Cape, and for the first time colonialism was followed in earnest: English was declared

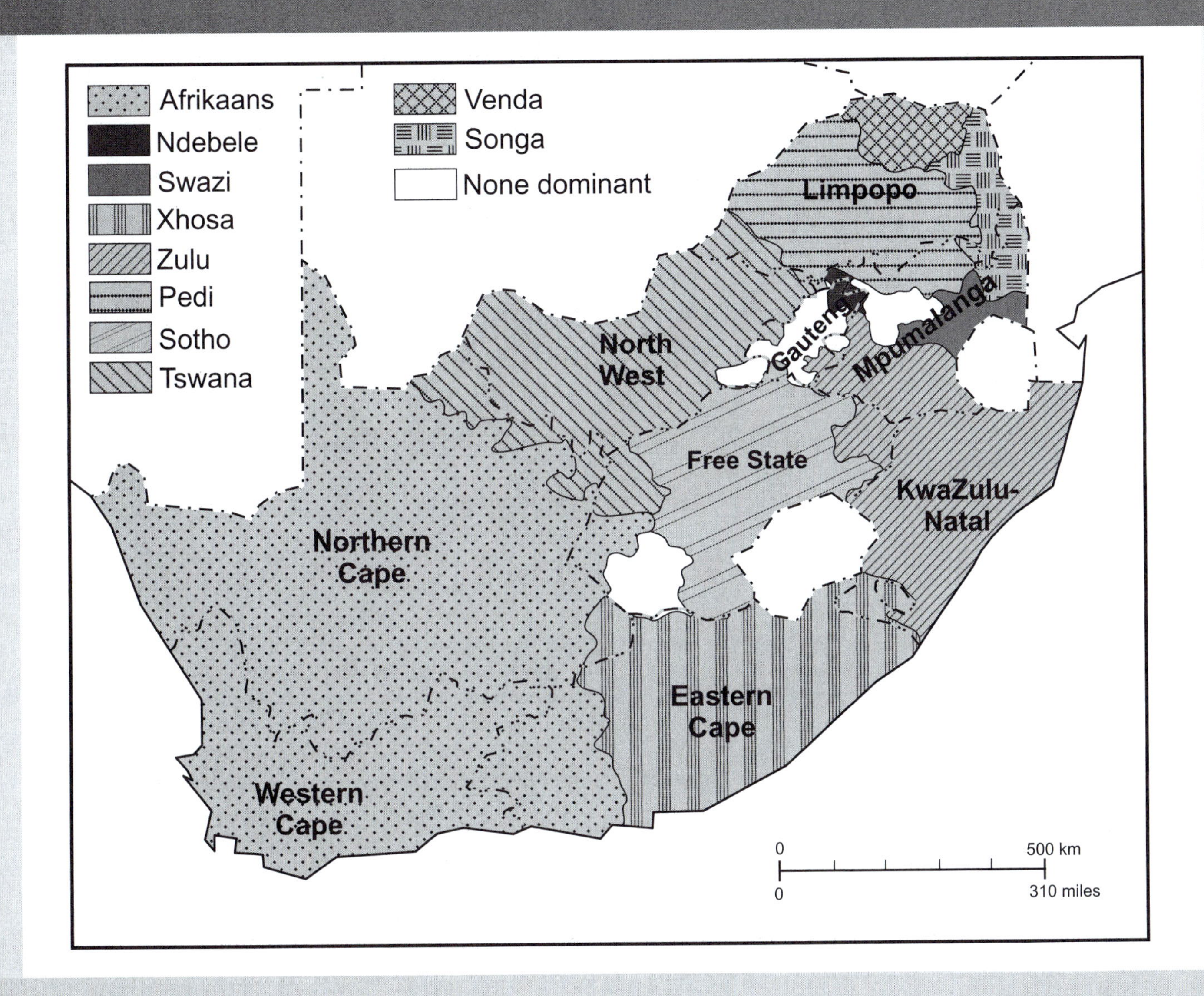

FIGURE 24-5 Dominant languages of South Africa.

the sole official language, the courts and education systems were anglicized, British settlement was encouraged, and slavery was abolished in 1834. British colonialism spread to Natal in the 1840s, while Afrikaners, the descendants of the original Dutch immigrants, pushed into the interior (1836), followed soon after by the British and other European groups.

Among the white immigrants, the Afrikaners have separated themselves most completely from their European heritage. They identify themselves as intensely with Africa as do black Africans; they are fiercely proud of their language, which has evolved from Dutch and has a unique flavor today, and many do not think of any part of Europe as their ancestral or cultural homeland. The English-speaking South Africans, on the other hand, have not relinquished their ties with Britain and consider themselves Europeans first and foremost. There are exceptions,

TABLE 24-1 SOUTH AFRICANS 1900–2001, BY ANCESTRY.

Year	African Ancestry Population (thousands)	African Ancestry Percent	European Ancestry Population (thousands)	European Ancestry Percent	Mixed Ancestry Population (thousands)	Mixed Ancestry Percent	Indian-Asian Ancestry Population (thousands)	Indian-Asian Ancestry Percent	Total Population (thousands)
1900	3,400	68	1,100	22	400	8	100	2	5,000
1910	4,000	68	1,250	21	500	9	124	2	5,874
1911	4,019	67	1,276	21	523	9	152	3	5,970
1921	4,697	70	1,276	19	545	8	164	2	6,682
1930	6,000	70	1,750	21	575	7	200	2	8,525
1936	6,597	69	2,003	21	769	8	220	2	9,589
1940	6,200	66	2,100	22	800	9	250	3	9,350
1946	7,830	69	2,372	21	928	8	285	2	11,415
1950	8,000	65	2,600	21	1,200	10	450	4	12,250
1960	10,908	68	3,088	19	1,509	9	477	3	15,982
1970	15,058	70	3,751	17	2,018	9	620	3	21,447
1980	21,000	72	4,400	15	2,800	10	800	3	29,000
1990	26,500	74	5,100	14	3,250	9	1,000	3	35,850
2000	34,300	78	4,300	10	4,000	9	1,100	3	43,700
2001	35,416	79	4,294	10	3,995	9	1,116	2	44,820

Sources: Best and de Blij (1977), Christopher (2001), Statistics South Africa (2001).

naturally, but the sentiments among whites are mostly divided along these lines. An appreciation of this is indispensable to the analysis of the Afrikaners' efforts to consolidate their position during the apartheid years and in their efforts to maintain their cultural identity afterward.

Ironically, it was not the Afrikaners who brought to the interior of South Africa the modern ways of life which they view with such pride today. The nineteenth-century Dutch wished to escape from British overlordship at the Cape, and they entered the interior with the intent of establishing pastoral republics, ranching cattle and sheep. Thus, they became known as *Boers* (a Dutch word literally meaning "farmer" but used in a pejorative sense by the English), and their economic mode of life resembled more the Africans' than that which was developing in Europe.

The discovery of diamonds near the confluence of the Orange and the Vaal rivers (1867), and the subsequent discovery of gold along the Witwatersrand (1886), drew thousands of European immigrants who brought with them capital, technical skills, and the knowledge of Europe's progress in the Industrial Revolution. Opposed bitterly by the Boers, these people infiltrated the interior republics and founded cities such as Kimberley and Johannesburg, exploiting resources by means unavailable to pastoralists. Against the Boers' will, the isolation of the interior came to an end, as railroads penetrated the highlands, roads were built, and urbanization gained momentum. After their defeat in the Anglo-Boer War, the Afrikaners were to participate in the economic development of their country—a transformation that was initiated by more recent immigrants from Europe. Their level of participation rose constantly, and Afrikaners in due course began to attain positions of power. The ratio of rural to urban dwellers among Afrikaners changed continuously in favor of the latter. In the early days of the union of South Africa, the majority of Afrikaners still lived on the land, and the bulk of the English-speaking people and other immigrants resided in the growing cities. Today, the Afrikaner is as much an urban dweller as is his or her English-speaking contemporary. Over 90% of South Africa's whites are urban dwellers (Table 24-2). They enjoy the highest standard of living of any group; they control the vast majority of the country's wealth, industrial resources, and land; and, until the dismantling of apartheid in the early 1990s, they had absolute political power.

TABLE 24-2 PERCENT OF SOUTH AFRICA'S POPULATION THAT IS URBAN BY GROUP, 1911-1991.

Year	African	White	Coloured	Indian-Asian	Total
1911	13	53	50	53	25
1921	14	60	52	60	28
1936	19	68	58	70	32
1946	24	76	63	73	36
1960	32	84	68	83	47
1970	35	87	74	86	48
1991	43	91	83	96	57

Sources: Republic of South Africa, (1974) p. 75; official census reports (various years); U.S. Census International Data Base (2002).

The Coloured population (3.9 million official 2001 census), represents a mix principally between Khoikhoi and European and constitutes the majority in two provinces. It is primarily Afrikaans speaking, and forms the largest population group in Cape Town and in almost all other towns in the Cape Province. The Coloured population initially resulted from sexual relations between Dutch men and Khoikhoi women during the early years of European colonization, when the western Cape was the only region of Southern Africa that was permanently settled and organized by Europeans. Cape Town was the only real town, and the Coloureds worked there and on the surrounding farms as slaves.

The Khoikhoi were pastoralists who raised sheep and cattle in the Cape region. They were related to the San people, some of whom still hunt and gather in South Africa and Namibia today. Livestock raising probably diffused to the hunters and gatherers of Southern Africa over 2,000 years ago from East Africa. Some groups of San adopted the practice and differentiated themselves from their groups of origin over the ensuing centuries. When the Dutch came to the Cape in 1652, they found Khoikhoi pastoralists, whom they called "Hottentots." And their San hunter–gatherer cousins they called "Bushmen." Although the Coloured population is mainly a product of intermarriage between Khoikhoi and Europeans, cultural and biological differences among the Coloureds are great; the group includes not only persons of mixed white and African blood (Figure 24-6), but also Malay people who have remained remarkably distinct racially for many generations. Slaves were emancipated from 1834 to 1838, and, although their official status changed, the Coloured continued to work on the farms of their former masters. There was a distinct racial geography—the farther away from Cape Town (Figure 24-7), the greater the proportion of Khoikhoi in the Coloured population and the more abject their conditions of servitude.

Throughout the 1960s the Coloured community grew at 3% per year, but since 1970 the growth rate has dropped to about 2.4% as birthrates declined more rapidly than death rates. Like their white contemporaries, the Coloured peoples speak either

FIGURE 24-6 Township children, Cape Town. *Sara Fisher.*

FIGURE 24-7 Cape Town from Table Mountain. *Sara Fisher.*

Afrikaans or English, but Afrikaans predominates. This fact attests to their early association with the Cape Dutch settlers, even though the Dutch community never integrated the Coloured community with its own. About one-third of the Coloured church-going population attends the Dutch Reform churches, another third attends English churches, and 5% are Muslims. The number of Coloureds adhering to the Dutch churches, in view of the segregationist philosophies of these institutions is remarkable. Islam is most prevalent in Cape Town, where it is practiced by the Malay peoples as it has been for centuries.

Like other nonwhites, the Coloureds faced discrimination in jobs, wages, housing, and general amenities during apartheid, and they were denied representation in the national parliament. Various avenues of employment were closed to them by law, and a sizable portion of the community has always lived in poverty in both the urban and rural areas (Figure 24-8). During apartheid, Coloured workers could not compete successfully with skilled Europeans, and they lost jobs in the unskilled field to Africans who worked for lower wages. In Cape Town they have maintained themselves in only a few industries, including stevedoring, textile work, building, and food processing. Still, there are more Coloureds in manufacturing occupations (mostly general labor) than in agriculture and other primary industries, and the difference is growing despite the problems they face in the cities (Figure 24-9).

The smallest of the racial groups is the Asian, at 1.1 million (Table 24-1). In 1860 the British brought the first group of indentured Indian laborers to work the newly founded sugar estates of Natal. The practice was to continue for a half-century; but in 1913, when the Asian population numbered about 155,000, the South African government stopped it, having become alarmed at size of the influx. Most Indians, whose period of indentured labor had expired, declined the government's offer of free passage back to India and remained to develop market gardening near the cities and to enter the general urban labor pool. Wealthy Asian merchants immigrated at the turn of the century and established successful commercial enterprises in many South African cities and especially in Durban. Today their stamp is unmistakable there, and over one-third of this city's population is Asian.

The great majority of South Africa's Asians are Indians, and about 70% of these are Hindu. They speak a variety of Indian languages (although English is the lingua franca) and separate themselves from the 20% who are Muslims, many of whom have moved to the interior provinces. Since 1913 the Asian population

FIGURE 24-8 Cape Town township dwellings in 2002. *Sara Fisher.*

has increased between 2 and 3% per year. While most of the economically active Asians were employed almost exclusively in agriculture before the turn of the century, today about two-thirds of them work in manufacturing, commerce, and finance. Large numbers, especially in metropolitan Durban, are in textile work, general manufacturing, retail trade, construction, and the transport business. Some have achieved considerable wealth and property and a university education; a growing number are entering the professions.

Seventy-two percent of all Indian-Asians in South Africa live in KwaZulu-Natal (Figure 24-4), the greatest concentration being metropolitan Durban, and 20% live in Gauteng Province. During apartheid, they were not permitted to live in the Orange Free State, nor were they able to move freely from one province to another. Under the Group Areas Act, which reserved for each race a specific and separate residential area, tens of thousands of Asians were removed from city centers to more peripheral locations, where business opportunities were reduced and major life style changes had to be made. From central Durban, for example, more than 230,000 Asians were moved to Chatsworth and other towns some 10 to 15 miles (16–24 km) away. An even greater number of Coloureds from Cape Town's District Six were brutally removed under the Group Areas Act during the 1960s (Western 1997).

If demography is destiny in a cultural sense, then South Africa's demographic future is no puzzle to geographers. In an absolute sense, the African, Coloured, and Indian-Asian populations have increased 10-fold since 1900, while the white population has grown only fourfold. As a result, the proportions of people of African, European, mixed, and Indian-Asian ancestry have changed relative to one another over time (Table 24-1). People of African ancestry have increased their overall majority from 68% to 79% of the total population from 1900 to 2001, while the relative

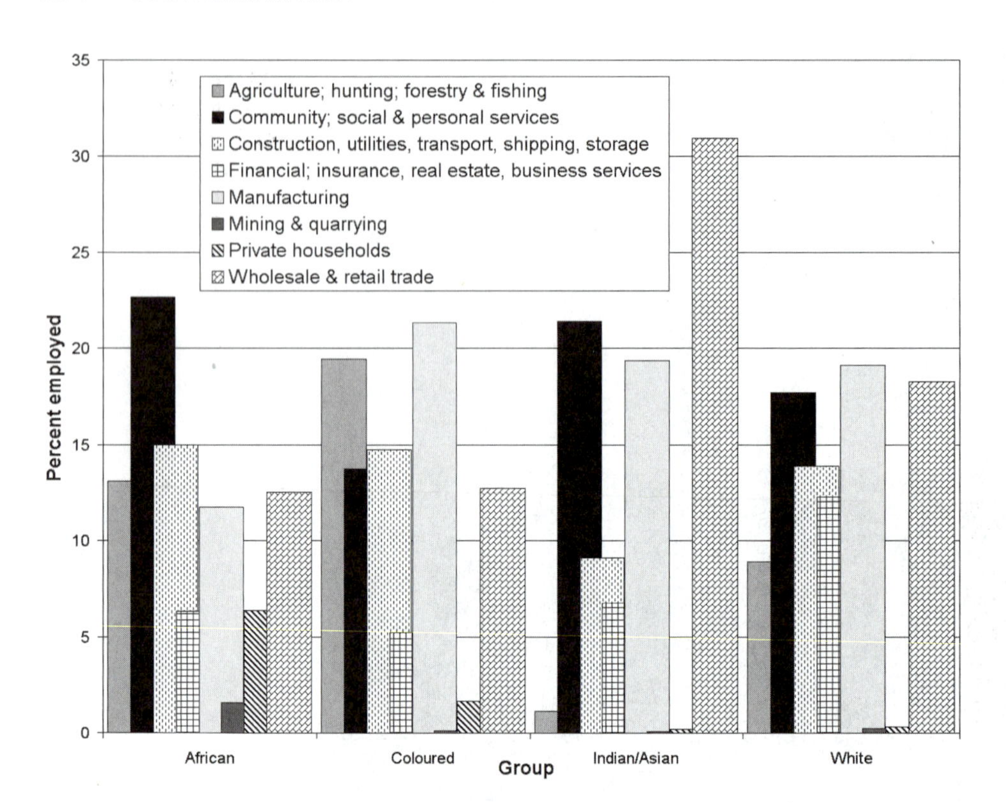

FIGURE 24-9 Employment by population group in South Africa in 2001.

proportion of people of European extraction has decreased from 22% to 10% from 1900 to 2001. The Coloured and Indian-Asian populations have remained about the same relative to the national population. The dramatic decline in the white population can be ascribed to a lower birthrate and emigration. Many left because of the moral reprehensibility of South African policy during white minority rule, while others may have preferred not to live under black majority rule. The policy of apartheid, while similar to segregation in the southern states of the United States after the Civil War, went far beyond it and, because of its anachronistic social model and totalitarian social engineering, was doomed to failure.

A NATION DIVIDED: THE POLICY OF SEPARATE DEVELOPMENT

During the apartheid era few would have asserted that South Africa's peoples were indeed a nation in any sense of the word and, although apartheid has officially ended, the question of nationhood and national identity remains to be answered today. The question is how a nation thus fragmented can develop along desirable lines. The Brazilian answer has been integration, and although not completely successful, the South American country is often cited as one that has averted failure. The United States, over 100 years after a bitter civil war and after decades of segregation, adopted a melting-pot model as a framework to bring diverse peoples together in a cultural sense. South Africa is similar to the United States in that it is a country in which the colonial immigrants chose to remain rather than return to their native countries. Independence in 1776 for the United States and in 1910 for South Africa meant something far different for each immigrant population and its descendants than it did for the rest of Africa during the 1960s. South Africa is different because the colonial immigrants were always a minority, while in the United States, most native peoples and their identities were removed through disease, war, intermarriage, and the reservation system; the rest became a tiny, peripheralized minority. Unlike most African colonies, in South Africa the colonizer stayed on, and the South African response to diversity was along completely opposite lines to recent American models, although in South Africa the segregation it practiced in the 19th century was very similar to that imposed by the Jim Crow laws in the southern United States.

Asserting that integration leads to cultural and moral decay as well as to racial pollution, the white ruling minority produced a blueprint of separate living that had no rival anywhere in the world and was considered a unique and odious solution to a unique set of circumstances. Although admittedly intensifying the divisive factors between South Africa's peoples, it was, according to its architects, intended to lead to political independence for each people while promoting economic interdependence. The South African state, therefore, was envisaged as developing along lines similar to that of the British Commonwealth. There was not to be a federation and central government, but a group of adjacent, racially based units, each internally independent in political matters. Counter to the official South African view, critics asked how could there be economic interdependence when the whites had almost all the valuable land, as well as possession of industry and control of the levers of state and power that relegated people of nonwhite ancestry to secondary and subservient roles? According to its defenders, the goal of separate development was the preservation of Western culture in South Africa. The mingling of European, African, and Asian cultures, although it had gone on for centuries, was deemed undesirable, and it was the policy of the South African government to provide each racial group its own "homeland" in which the different cultures, their purity unthreatened, were to develop. This involved the relocation of hundreds of thousands of people who had for generations been thrown together by the forces of economic development, urbanization, and mutual need. A complete reorientation of the South African economy, massive relocation of industry, and alterations to the transport network were among the major problems facing the state in consequence of the regional form of apartheid.

A glance at South Africa's history shows that race has played an important role from the early days. In 1814, to make way for English, Welsh, Scots, and Irish settlers, the British military and their Khoikhoi allies removed Xhosa farmers from their land west of the Fish River in today's Eastern Cape Province. In a pattern remarkably similar to that of the United States formal urban residential segregation began with the increase of African rural-to-urban migration that occurred after the abolition of slavery in 1834 (Christopher 2001). Afrikaners began more formal separation of the races in 1857, when the Dutch Reformed Church of South Africa mandated the creation of separate (and inferior) churches for Coloured worshipers. Afrikaners responded negatively to the British emancipation of the slaves from 1834–1838, viewing it as contrary to God's plan.

In the 1840s about one-tenth of the Cape Afrikaner population (about 6,000 people) migrated to the highveld on the north side of the Orange River and took with them about the same number of Coloured "servants" to establish their own state—what was to be called the "Orange Free State." Referring to the Coloured servants, an observer close to one of the leaders of this migration, the so-called Great Trek, observed that "It is not so much [the slaves'] freedom that drove us to such lengths, as their being placed on an equal footing with Christians, contrary to the laws of God and the natural distinction of race and religion" (Bird 1888: 459).

During the apartheid era (1948–1994), the policy of separate development manifested itself in three ways. Grand apartheid was concerned with the removal of Africans to "homelands" and their political independence from South Africa. Urban apartheid was concerned with separating people of different races in cities and towns and in providing separate facilities for each race (separate post offices for white and nonwhites, for example), while personal apartheid, or petty apartheid, was focused on the day-to-day regulation of people of different races (for example, prohibiting sexual relations between people of different races). Although the first legislation controlling the behavior of people of difference races was passed in the early nineteenth century, the legislative foundation of the modern apartheid state was based on laws passed in independent South Africa in 1913, 1923, and 1927 (Table 24-3) and augmented in a flurry of additional legislation mainly in the 1950s.

The totalitarian ideology of separate development became policy after the Nationalist Party came to power in 1948. Separation, the Nationalists argued, had to be effected between the races in all spheres of life. Serving as the legal framework regulating South Africans' behavior and contact between the "races" was a mass of legislation affecting labor, housing,

TABLE 24-3 LEGISLATION PASSED IN SOUTH AFRICA IN SUPPORT OF APARTHEID BETWEEN 1913 AND 1970.

Date	Act	Description
1913	Natives Land Act	Purchase or lease of land by blacks was made illegal except in reserves, restricting black occupancy to less than 8% of South Africa
1923	Natives (Urban Areas) Act	The legal foundation of urban residential segregation
1927	Natives Administration Act	Made the African areas subject to separate political administration from the rest of South Africa and subject to rule by decree rather than by parliament
1949	Prohibition of Mixed Marriages Act	Prohibited mixed-race marriages
1950	Immorality Act	Prohibited sex between white and black people
1950	Population Registration Act	Created a national race registry
1950	Group Areas Act	Forced urban residential racial segregation
1951	Bantu Authorities Act	Established black homelands
1952	Natives Laws Amendment Act	Restricted black residency in urban areas
1952	Natives (Abolition of Passes and Co-ordination of Documents) Act	The "pass law": forced blacks to carry identity documents at all times; pass, containing photographic identification and information on place of origin, employment, and police records, was necessary for travel outside one's designated homeland and for employment in urban areas
1953	Native Labour Act	Prohibited strikes by blacks
1953	Bantu Education Act	Legal framework for separate and unequal education for black South Africans
1953	Reservation of Separate Amenities Act	Forced the separation of whites and nonwhites in all public places
1954	Natives Resettlement Act	Legal mechanism for forced removals to homelands
1956	Natives (Prohibition of Interdicts) Act	Denied black people court appeal for forced removals
1959	Bantu Investment Corporation Act	Provided for the creation of financial, commercial, and industrial development projects in black areas
1959	Extension of University Education Act	Mandated separate university education for Africans, Asians, Coloureds, and Europeans
1959	Promotion of Bantu Self-Government Act	Grouped black South Africans into eight "ethnic" groups each assigned to a independent homeland
1967	Terrorism Act	Permitted indefinite detention without trial
1970	Bantu Homelands Citizens Act	Forced every black South African to become a citizen of his or her homeland and took away blacks' South African citizenship

rural–urban migration, and other important matters. One of the government's first undertakings within this framework was to order an inventory of the human and environmental resources of the African reserves. The Tomlinson Commission, tasked with this job, presented its findings in 1955. In addition to inventories of the natural resources and the socioeconomic conditions of the widely diverse and highly fragmented reserves, the report made recommendations for their improvement. The Tomlinson Report became the government's blueprint for territorial separation and so remained, reinforced by subsequent legislation, until the demise of the apartheid state. Of all the unusual characteristics of the apartheid model, one of the most unusual was that it advocated the breakup of the South African state. State fragmentation was rare (Yugoslavia, the USSR) in the twentieth century and was almost always accompanied by violence. All the more peculiar that a state would want to break itself up, as South Africa was proposing during the apartheid regime.

The Nationalist government saw South Africa not as one nation with a common citizenship and rights, but as many "nations," each entitled to self-determination within its prescribed geographical area. These "nations" were defined in terms of language, culture, and tradition, and each, according to the government's racist ideology, was to retain its identity and determine its own future. Territorial separation, which was supposed to make that possible, in practice so deepened and perpetuated colonial era segregation that it could be labeled a "crime against humanity." The areas that were set aside for the African majority were called "homelands" or "bantustans." The apartheid government of the Nationalist Party intended that they become fully independent sovereign states, free to apply for membership in the United Nations and other international organizations.

Ten homelands were ultimately designated: Transkei, Ciskei, KwaZulu, Bophuthatswana, Lebowa, Venda, Gazankulu, (Basotho) Qwaqwa, KwaNdebele, and the Swazi Territory, KaNgwane. They varied considerably in area, resources, *de jure* and *de facto* populations, and territorial makeup (Table 24-4). All but Qwaqwa and KwaNdebele were territorially fragmented. Extreme fragmentation occurred in KwaZulu, homeland of the Zulu nation. Collectively the homelands comprised only 13% of the republic's total land area, and they formed a discontinuous arc from the Great Fish River of the eastern Cape, north through Natal and the Transvaal, and west to the Botswana border (Figure 24-10). The bulk of South Africa (87% of the area) was reserved for the whites; Coloureds and Asians were allowed to continue to reside in segregated locations in the white area.

In 1951 over 60% of the African population lived in areas that were to be designated white and in virtually all the "homelands" the population was ethnically diverse. Separating the peoples of South Africa amounted to a monumental undertaking that caused great human suffering. Despite the forced movement of population to create racially and ethnically "pure" homelands, the United Nations did not recognize these racial and ethnic entities, nor did other countries. Furthermore, during apartheid many black urban dwellers had never seen their homelands with which they are supposed to identify. The results of a survey in Soweto, the country's largest African township, located southwest of Johannesburg, demonstrated the dangers and dilemmas of the homeland policy. Seventy percent of the survey respondents said that they would prefer to live in South Africa under multiracial government rather than under a white government or in the homelands under tribal government; 61% said they would not accept a good job in the homelands if it were offered them, and 88% preferred black South Africans to form one nation irrespective of tribal origins. It was the urban African who presented the greatest challenge to the government in its implementation of multinational development.

TABLE 24-4 AREAS AND POPULATIONS OF THE APARTHEID-ERA HOMELANDS.

National Homeland	Designated Group	Capital	Area (mi^2)	Number of Territorial Fragments (1975)	Population in 1970	
					De facto	De jure
(Basotho) Qwaqwa	South Sotho	Phuthaditjhaba	198	1	24,000	1,254,000
Bophuthatswana	Tswana	Heystekrand	12,589	19	884,000	1,658,000
Ciskei	Xhosa	Debe Nek	3,000	19	524,000	924,000
Gazankulu	Shangana	Giyani	3,395	4	267,000	650,000
Transkei	Xhosa	Umtata	18,000	2	1,734,000	3,005,000
KwaZulu	Zulu	Ulundi	12,119	29	2,097,000	4,026,000
Lebowa	North Sotho	Leboa Kgomo	6,753	9	1,083,000	2,019,000
KaNgwane	Swazi	Schoemansdal	1,235	2	118,000	460,000
Venda	Venda	Makwarela	3,062	3	264,000	358,000
KwaNdebele	Ndebele		580	1	178,000	234,000
Total					7,173,000	14,588,000

Source: Best and de Blij (1977).

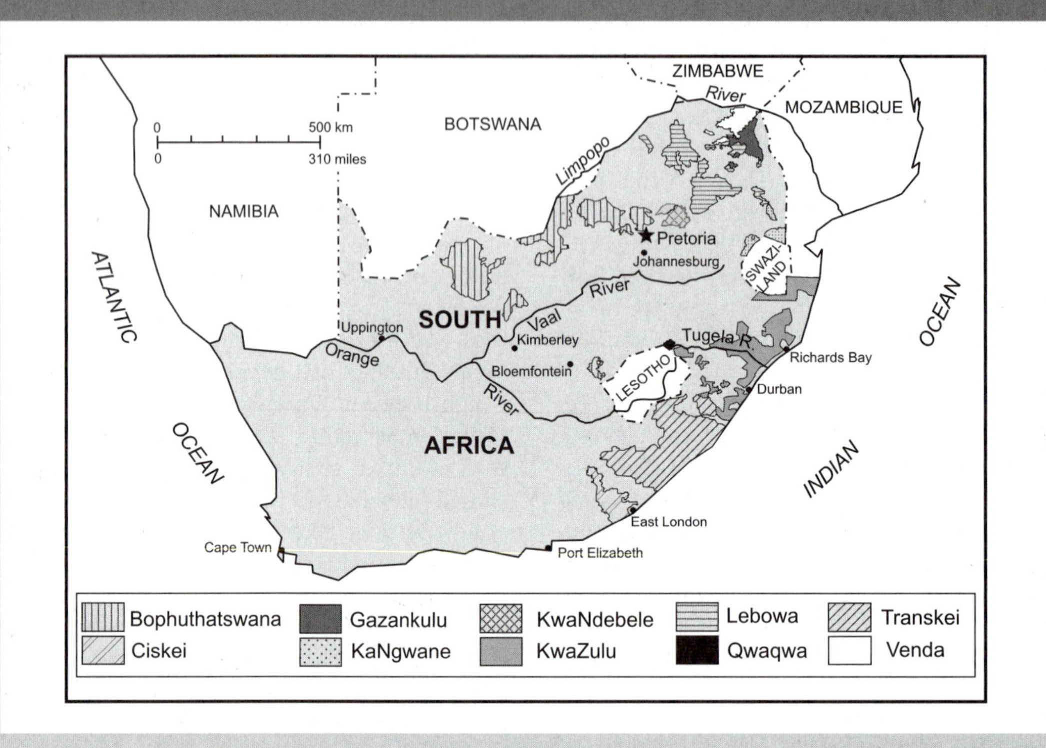

FIGURE 24-10 Homelands in apartheid South Africa.

Although white industrialists were encouraged to decentralize their operations to the homelands and peripheral regions to relieve some of the congestion of the metropolitan areas and to stimulate the homeland economies, cooperation was virtually nil. A few towns were built in the homelands, but few had industry and a diversified economic base, and most were merely residential appendages of the richer and economically more productive white areas. To varying degrees the physical infrastructures and educational facilities were improved, and each homeland was provided with rudimentary political and administrative systems, albeit closely controlled by officials in Pretoria, the administrative capital.

Agriculture, the mainstay of the homeland economies, was supposed to be reorganized under apartheid but in actual fact the homelands were located in the most agriculturally marginal areas of South Africa. The African homelands were never part of South Africa's generally prosperous agricultural economy. Within the South African space economy, they formed the overcrowded, economically depressed periphery. East of the Drakensberg, the homelands were located in predominantly mountainous, broken, or hilly terrain, with only small areas of flat or gently undulating land where cultivation is possible. In all, one-third of the total land area of the homelands was mountainous, 20% was hilly to broken, and 46% was gently rolling or flat. Much of the latter lay in dry Bophuthatswana and Lebowa, where about 79% of the total area was used for grazing, and only 15% was cultivated. In 1955 the Tomlinson Commission estimated that if the homelands were properly planned, they could support a farming population of about 2.1 million, while an additional 258,000 could earn a living in forestry and mining (South Africa 1955). During the mid-1970s there were about 8 million Africans in the homelands, and several million more were supposed to be resettled there from white areas. This plan would have required rural planning, resettlement, and general development on a scale not seen before. Just a few years after the publication of the Tomlinson report, de Blij observed that the "land allotment to Africans on the basis of the Tomlinson Commission's recommendations would send the Africans into a battle not for advancement without limits but for bare survival which would hold back their changes of improvement . . . as never before" (de Blij 1962: 252).

In addition to their agricultural marginality, the former homelands were resource poor: most mineralized regions were beyond their borders. The absence of large mineral deposits within the homelands was a natural consequence of the European desire to control and exploit the country's resources. The homelands were regions in which the Europeans, mainly from their point of view, had little or no interest. Lebowa and Bophuthatswana were the most favorably endowed, having asbestos and chrome mines and deposits of manganese, platinum, and iron ore. In contrast, Transkei, Ciskei, and KwaZulu had no significant mining activities and few known minerals. In 1974, when South Africa's mineral production exceeded $6 billion, only $60 million came from minerals produced in the homelands.

Reorientation toward separate development proceeded very slowly and was never completely implemented. Even so, the postapartheid governments have found it difficult to reverse the centuries of underdevelopment of the majority population. Opposition to apartheid emerged early but resolutely and first manifested itself in peaceful and later sometimes violent protest. The African National Congress (founded in 1912) began passive resistance in the early 1950s during the early years of Nationalist Party control, but as the implementation of apartheid progressed, resistance increased. The more radical Pan Africanist Congress (PAC) was formed in 1959 and later adopted the slogan "one settler, one bullet." The police massacre of pass-law demonstrators at Sharpeville in 1960 crystallized opposition to apartheid around the world. Civil unrest, arrest without warrant, torture, and murder became commonplace during late apartheid.

With the independence of the former Portuguese colonies of Angola and Mozambique in 1975 and later Zimbabwe (1980), the writing was on the wall for the white minority government in Pretoria, although there was resistance right up to the end. In 1976 and 1977 hundreds of black South Africans died in conflict with the police in Soweto and in other black townships. In 1977 the first of many embargoes and sanctions were applied to pressure South Africa to change its policies. After violence in the townships from 1984 to 1986, the South African government established a state of emergency during which wide powers of arrest and detention were given to the police, who perpetrated great abuse on the nonwhite populations.

In the mid-1980s the government took the first steps toward majority rule by making contacts with representatives of the African National Congress. In 1986 the hated pass laws were repealed. In 1989 Frederik de Klerk became the leader of the National Party and later president. De Klerk ushered in a new era in race relations in South Africa. He released Nelson Mandela and other political prisoners from prison and began dismantling the institutions of apartheid, beginning with racist legislation. The land acts were repealed, the Group Areas Act was undone, and the state of emergency was lifted. The so-called independent homelands of Bophuthatswana and Ciskei, ruled in an autocratic manner and subsidized by Pretoria, were absorbed into South Africa. Nine new provinces were carved out of the state. In 1994 the first democratic election that included all South Africans was held, and Nelson Mandela was the winner and first black president. In 1999 elections were again held and the African National Congress was again the winner. Thabo Mbeki was sworn in as president.

Apartheid was an impossible policy, bound to fail; it could never have succeeded, and there are five reasons why. We cite first the increasing demographic imbalance between the whites and blacks. The old fear of the early white minority in South Africa was well founded—but the racist policies to address it were not. Although whites constituted over 20% of the South African population up until the 1950s, the higher African reproduction rate and white emigration virtually assured that the whites would become an increasingly smaller segment of the population.

Apartheid was also condemned because of the high financial and political costs of implementing and administering it. Thompson calls apartheid a burdensome and cumbersome financial "extravagance" and points out that it involved "three parliamentary chambers, fourteen departments of education, health, and welfare (one for each "race" at the national level, one for each province, and one for each Homeland), large

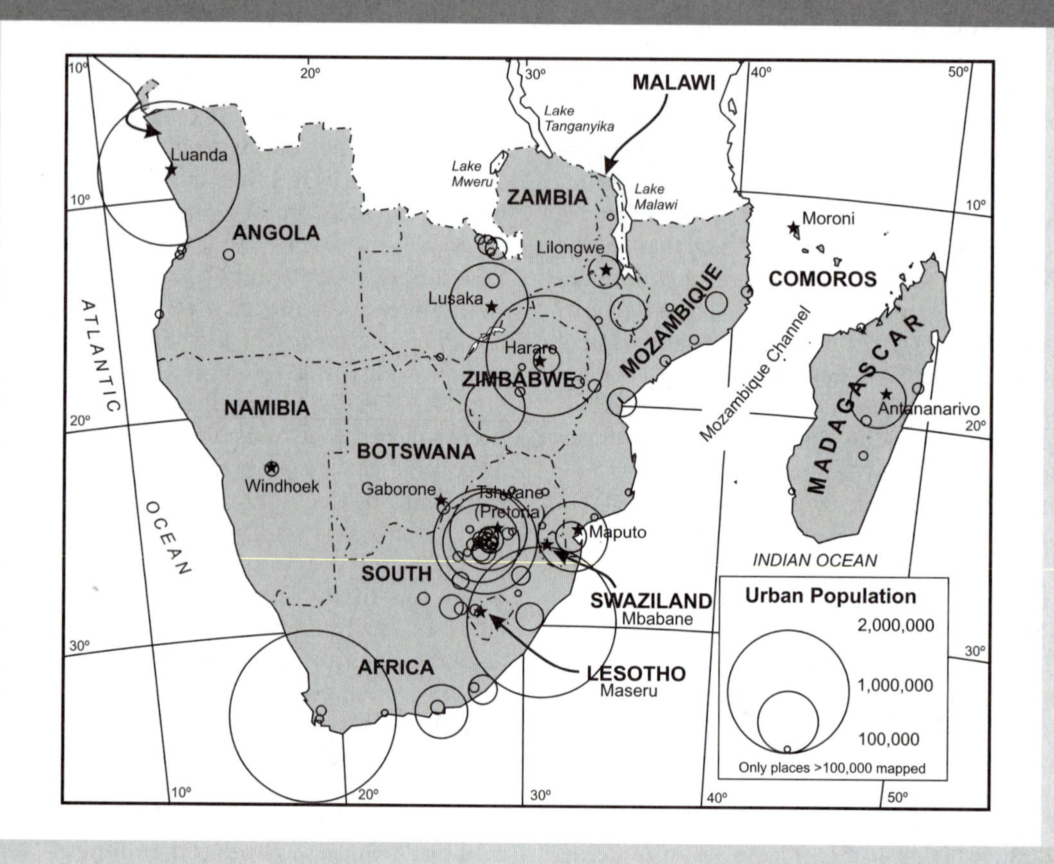

FIGURE 24-11 Urbanization in Southern Africa, 2003.

military and security establishments, and financially dependent Homelands" (Thompson 2001: 242).

The third reason for apartheid's failure was economic: South African exports, when they were not being boycotted, bore a higher cost because of this burden and were not competitive on the world market. A black middle class was emerging in South Africa, and the black contribution to the national economy was increasing. Blacks were no longer mainly farmers who could be consigned to rural reservations, and black professionals, skilled workers, and their white counterparts were becoming increasingly interdependent.

The fourth reason entails the collapse of the Soviet Union and the Communist bloc, ending the capitalist–communist rivalry for the hearts and minds of people around the world and marking the emergence of a unipolar world. South Africa lost its importance as a bastion of capitalism in a socialistic and sometimes hostile Africa (that was itself moving toward more market oriented solutions to economic organization and development). And finally, apartheid never really worked—even during the height of apartheid, only 47% of the African population ever resided in areas reserved exclusively for them, and 35% were urban dwellers.

SOUTH AFRICAN CITIES DURING AND AFTER APARTHEID

Cities and towns, geographers have often said, are a response to a region's needs. As might be expected from the vigorous economic development of South Africa, there has been much urbanization. Moreover, unlike almost all other African countries, South Africa possesses a pyramid-like hierarchy of urban places: there are a small number of very large cities at the top, a greater numbers of towns, and even more small towns and villages at the base (Figure 24-11). In contrast to South Africa's well-developed urban hierarchy, most African countries are dominated by a primate city, usually the capital. In part, heavy urbanization of South Africa is due to the large settler population; other reasons include the location along the Witswatersrand of the mining industry and large manufacturing complexes.

Urbanization in South Africa was first stimulated by mining, but the contribution of mining to the national income has declined in favor of manufacturing and even agriculture. Yet the cities continued to grow, and Johannesburg, once based entirely on mining, had 60% of its white, 50% of its Indian-Asian, and 20% of its Coloured population engaged in professional and administrative activities in 2001. The figure for Africans, 15%, is an increase over the past but shows what a long way there is to go for the majority. Ten years after the dismantling of apartheid, the Indian-Asian and Coloured participation in administration and professional occupation was encouragingly high.

South Africa is more urbanized than in any other part of Africa, and Johannesburg (2.1 million; greater metropolitan area, 8 million), Cape Town (3.6 million), and Durban (3.2 million) have no rivals in terms of urban development and industrial output on the continent. As break-in-bulk places, the cities on the coasts benefited from the progress of mining and industry on the plateau, and Port Elizabeth (1.1 million) and East London (423,500) vie with Cape Town and Durban for the trade of the interior. On the highveld, the Witwatersrand became the core of the entire country. Further gold discoveries were made, and towns sprang up east and west from Johannesburg, so that a lengthy urbanized belt developed. North of Johannesburg, Pretoria (1.47 million) prospered as the country's administrative center, and southward, Vereeniging (462,800) on the Vaal River benefited from ample water supply and the proximity of coal deposits. Gauteng Province thus became the hub of all South Africa, with an east–west manufacturing and mining axis and a north–south administrative and transportation axis. Johannesburg alone has over 2 million inhabitants today (8 million in the metro area), but the two axes of Gauteng Province, representing only 1% of the area of South Africa, included over 20% of the population of the entire country in 2004. A megalopolis is developing there.

South Africa's cities are dynamic and viable, and places of modernization and social change. Here lie the country's major industries, financial institutions, universities, and technological might. In short, the cities represent South Africa's wealth and prosperity. Johannesburg is South Africa's largest city, and its problems are symptomatic if not entirely representative of the country's multiracial centers. Its African population (like that of Pretoria, Germiston, and other cities) has long increased faster than housing could be provided, and conditions in the city's shanty towns have been among the worst forms of urban squalor in the world. To be sure, the cities have their share of lower-class white suburbs, but the major solution for the prevalent urban decay was seen as removing the Africans, not providing more housing.

After the establishment of majority rule in 1994 the government embarked on a program to build housing for urban poor that has met with mixed success, perhaps in part due to the high expectations for rapid change once the white minority government was removed from power. In the urban centers of South Africa, and especially on the Witwatersrand, the diverse peoples of South Africa have long been thrown together in a common economic effort. In spite of past residential segregation and job restrictions based on race, the fortunes of black and white in South Africa have become inextricably mixed.

The urban centers, from Cape Town (the first city) and Kimberley (the first mining town) to Johannesburg and Durban, have attracted the capital investment and skills of foreigners and labor of the African. Together, they achieved South Africa's economic progress, most dramatically reflected in the urban sprawl and high-rise core of the Golden City. Up until the 1980s when sanctions were applied against South Africa and it became a pariah state, Johannesburg was the continent's financial capital, overseas investors displayed much confidence in South Africa's future, and every sector of the economy blossomed.

The process of urbanization in South Africa was not the same under apartheid as it has been since 1994

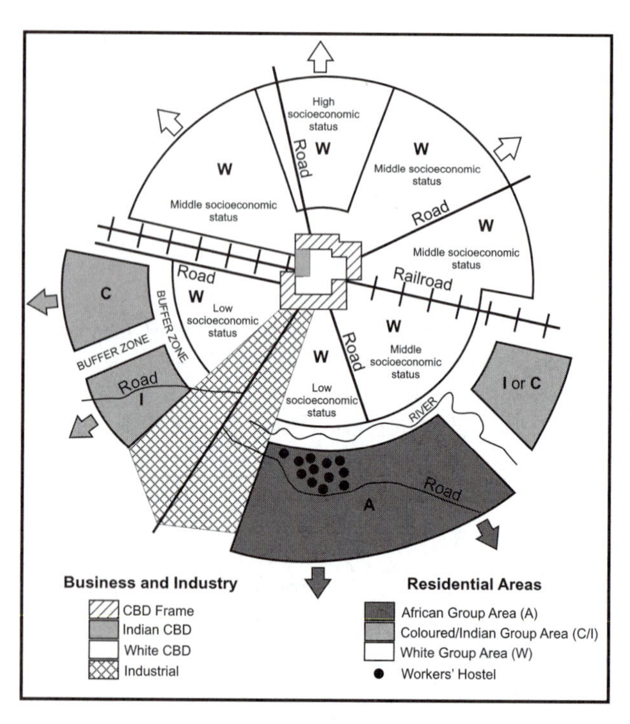

FIGURE 24-12 The model Apartheid city, showing residential and industrial areas and central districts (CBD). *Adapted from Davies (1981).*

when apartheid officially ended. Apartheid generated a distorted urbanization: all the major industrial towns and urban centers were situated in the white areas; the homelands existed only as dormitory townships that housed the labor required across their borders. For example, the KwaZulu homeland's largest "towns" during the height of apartheid, Umlazi (250,000) and KwaMashu (230,000), were merely residential appendages of Durban, having no manufacturing and few tertiary industries of their own because the competition in Durban was too strong.

In many ways, Johannesburg is unique, being not only the largest South African city but also the heart of the country's most intensely industrialized area and the financial capital. East and west of Johannesburg, along the outcrop of the gold-bearing reefs of the Witwatersrand, a string of mining towns sprang up, and their urban areas are merging into a South African megalopolis. Racial segregation was strictly enforced under the Group Areas Act during apartheid (Figure 24-12). Buffer zones such as expressways, railways, cemeteries, and mining and industrial areas separated the various residential group areas (Figure 24-13). Johannesburg's northern boundary is within sight of expanding Pretoria, the country's administrative capital and most rapidly growing city. Southward lies the lifeblood of the industrialized Witwatersrand—the Vaal River and the water supplies of Vaal Dam. There also lie Vereeniging, the iron and steel complex, and the mines and industries in the Free State (Formerly Orange Free State). The Johannesburg area, therefore, has been the focal point of mining, industry, communications, commerce, financial affairs, and administration in South Africa. The city has received a succession of stimuli that have kept it in the forefront of rapidly urbanizing South Africa.

Johannesburg exemplified several other characteristics of the South African urban scene, including the distortions of the apartheid era (Beal et al. 2002). Its areal extent is vast, covering perhaps four times as much territory as a European city of comparable population. Its core and central business district are congested and small compared with the entire urban region. Vertical development is considerable and comparable to that of any American city of similar population. The city's functional zones, as a result of the decline of mining and the growth of industry, are interdigitated, although under apartheid the residential areas were strictly segregated according to race. Unlike the cities of former French and Portuguese Africa, the civic and administrative buildings of South Africa's cities were dispersed throughout the central city. This is one of the points of contrast between the urban places of British-influenced Africa and those of the remainder of the continent.

During the late 1980s and especially after 1991 when the Group Areas Act was abolished, urban segregation declined across the country (Figure 24-14). Middle-class Africans moved into white group areas in larger urban areas, and class rather than race was becoming the main influence on residential choice, particularly in the inner suburbs of the city (Tomlinson et al. 2003). Small towns and outer suburbs have remained very segregated (Christopher 2001). Some low- and medium-density residential areas near the city center were rezoned for high-density settlement. The land-use specialization and residential segregation in South African city is beginning to take on a North American, multinodal morphology (Figure 24-15).

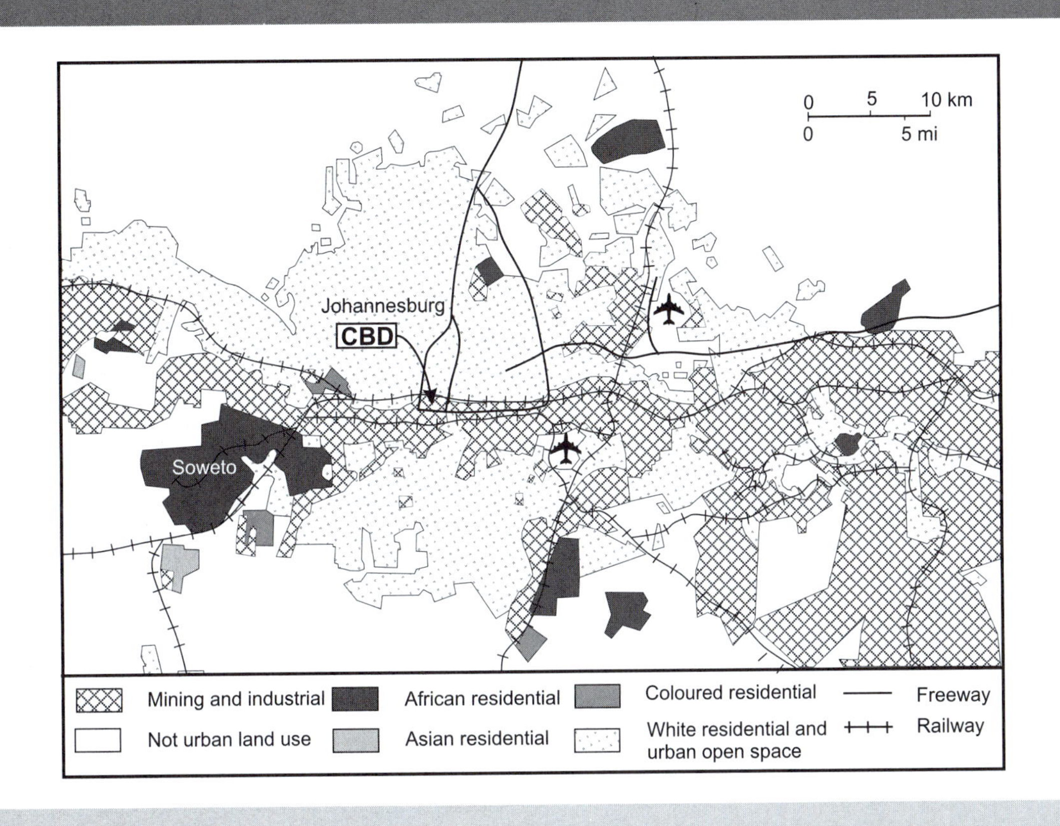

FIGURE 24-13 Group areas in Johannesburg in 1973 (during the last decades of apartheid).

AGRICULTURE AND MINING

After an early phase dominated by agriculture, the South African economy between 1929 and 1943 was sustained mainly by mining. Since 1943, manufacturing has taken the lead, now contributing almost three times as much as mining and agriculture combined. These manufacturing industries have not relied only upon overseas markets; indeed, their success is largely a reflection of the increased capacity of the South African market. Even though their respective shares were disproportionate, Africans, Coloureds, and Asians benefited from the economic progress of the country as well as the whites. Thus, a multiracial market with constantly increasing purchasing power developed even during apartheid and has developed to a much greater degree since its demise.

The four industrialized regions of South Africa are the Witwatersrand, the Durban-Pinetown area, the region around Cape Town (Figure 24-16), and the area of Port Elizabeth. Mining stimulated these industries in large measure. The manufacture of explosives became a major industry on the Witwatersrand, and the need for mining boots stimulated the manufacture of footwear in Port Elizabeth and Durban. The need for coal and electricity increased constantly. An iron and steel industry was established at Pretoria, with a subsidiary plant near Vereeniging. A further step toward self-sufficiency was taken during the apartheid period, with the construction of the world's largest oil-from-coal plant at Sasolburg in the northern part of the former Orange Free State.

As the urban centers grew, so did the market for agricultural products. No longer a subsistence pastoralism,

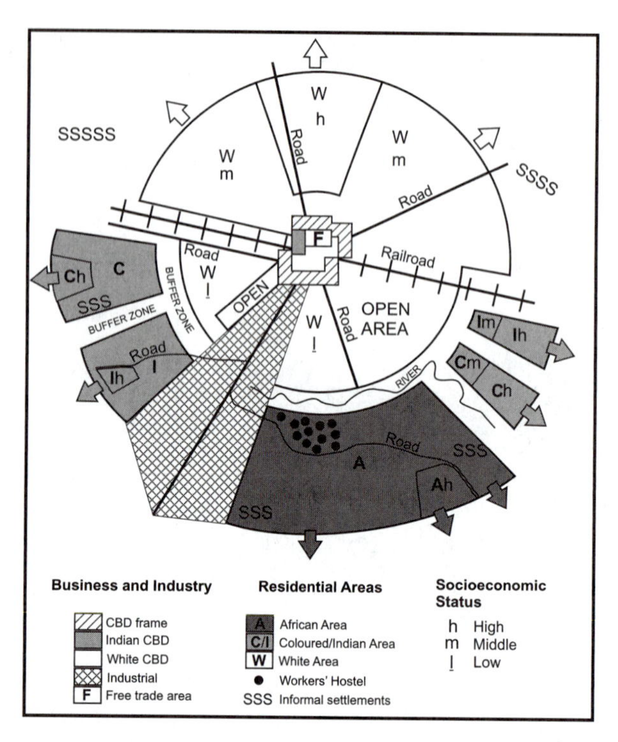

FIGURE 24-14 The late apartheid city.
Adapted from Simon (1989).

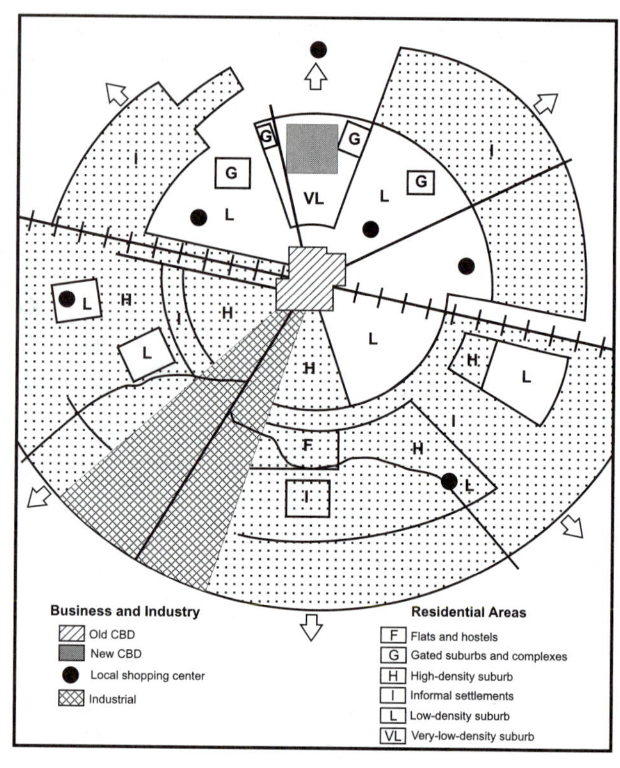

FIGURE 24-15 The postapartheid city.
Adapted from Christopher (2001).

FIGURE 24-16 Cape Town train station.
Sara Fisher.

FIGURE 24-17 Viticulture in Cape Province, near Stellenbosch. *Sara Fisher.*

South African agriculture produced a vast variety of specialized fruits and vegetables (Figure 24-17), a higher corn yield per acre than anywhere else on the continent, wheat, a variety of other cereals, and sugar. Near the cities, market gardening and intensive agriculture developed. But whether around the cities or in the vast interior, the bulk of the labor was carried on by Africans under the direction of white landowners.

Africans were for decades migratory laborers, leaving homeland for city, mine, and farm. In the cities a permanently settled African population exists that is as solidly urban—though poorer—as its white, Coloured, and Asian counterparts. The homelands of South Africa during the apartheid period—for example, Kwazulu, in today's KwaZulu-Natal province, and the Transkei, in the Eastern Cape—although affected by the steady stream of emigrants, changed little as South Africa developed. Traditional agriculture, rocky soils, steep slopes, poor soil management, and underdevelopment characterized these and other reserves, and the exodus was the result of local conditions as much as of the attraction of the cities and mines of the modern, industrial sector.

Although environmental conditions are problematic and only 15% of the total land area is arable, South Africa has a strong and diversified agricultural and pastoral economy. Sharp contrasts in farming techniques, types of farming, and yields occur from place to place, but especially between the white and African farmers. In the former, specialized commercial agriculture is emphasized (Figure 24-18), while subsistence production prevails in the latter (Figure 24-19). White farms produce over 90% of the country's agricultural output in the monetary sector, including much of the corn (maize) and sorghum consumed by the African population. Cattle and sheep have traditionally had higher earnings

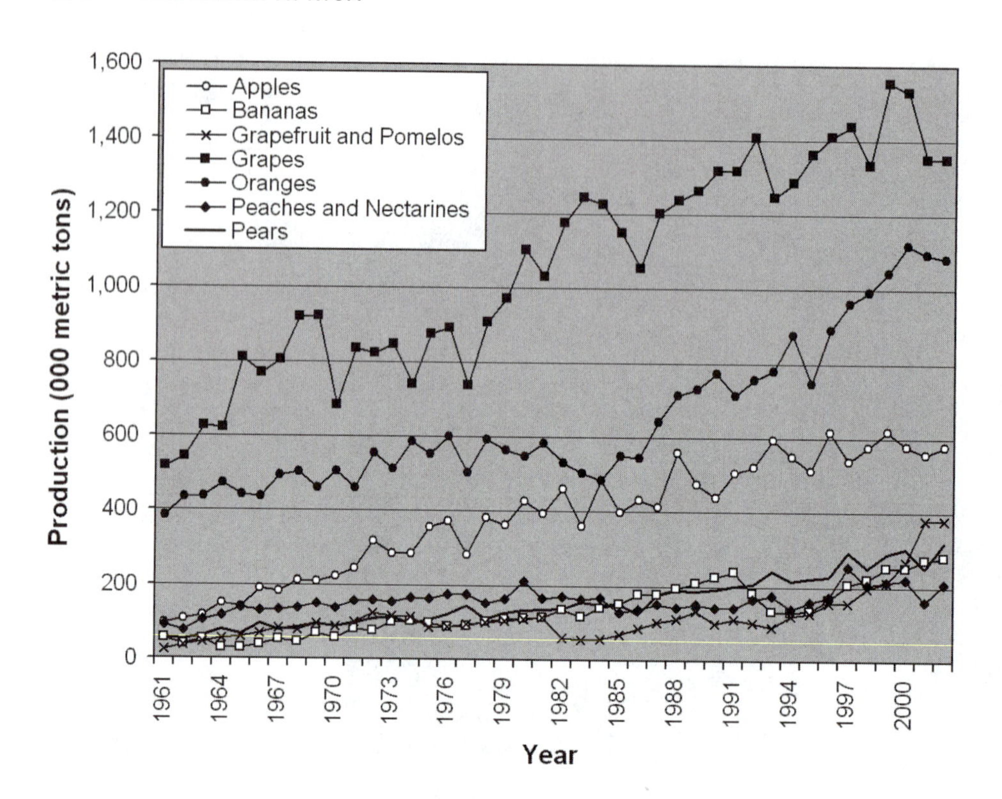

FIGURE 24-18 Fruit production, 1961 to 2002, South Africa.

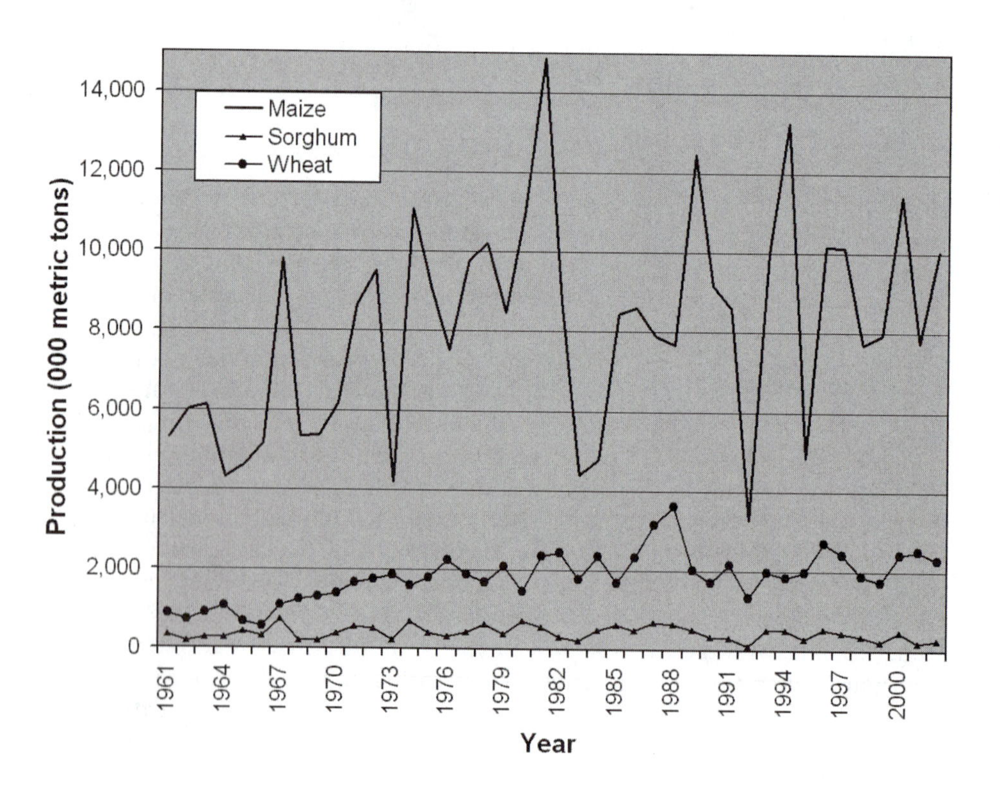

FIGURE 24-19 Production of three major cereals (maize, wheat, and sorghum) in South Africa between 1961 and 2002, South Africa.

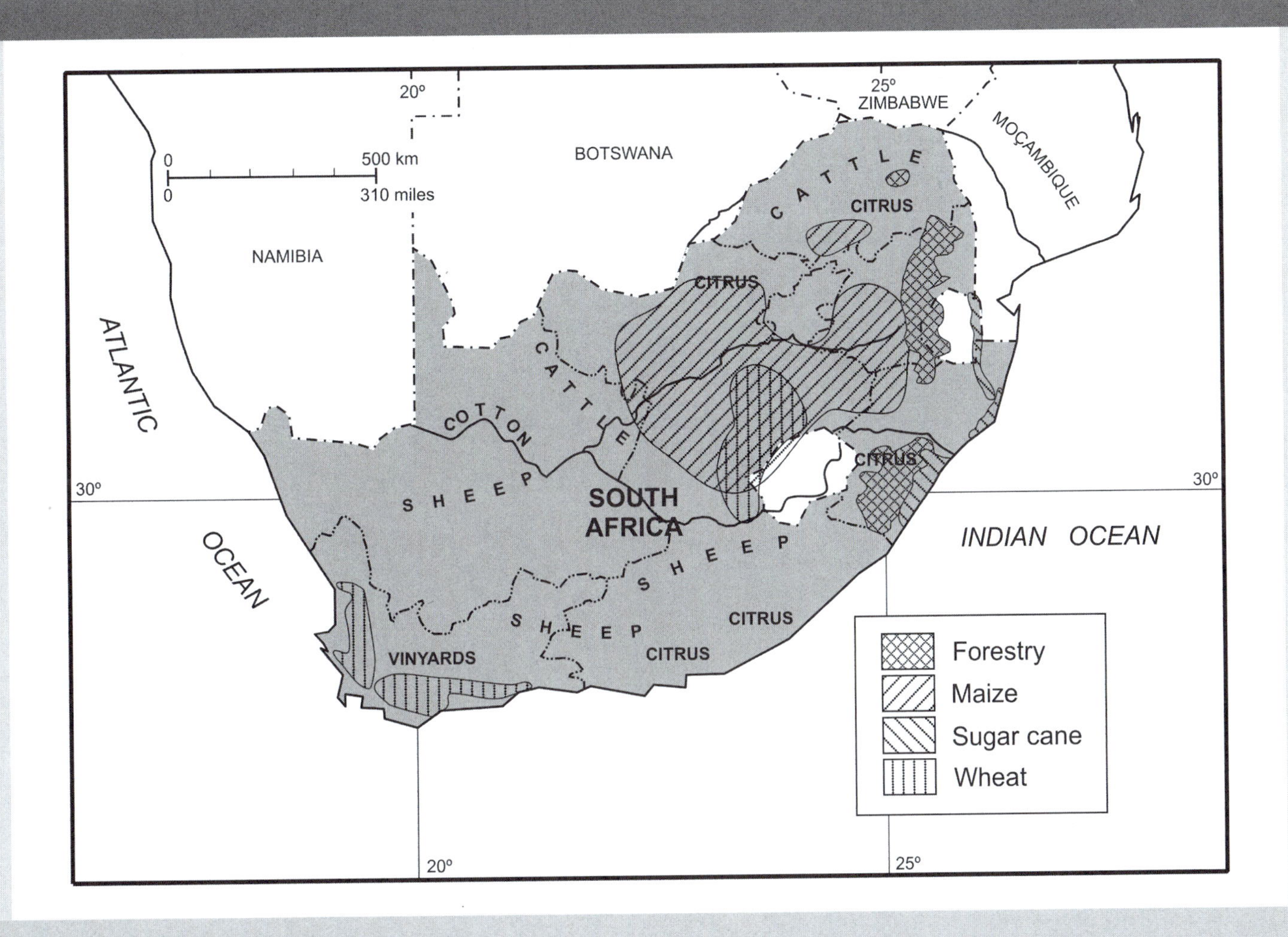

FIGURE 24-20 Land use in South Africa.

than crops and are raised extensively by commercial enterprises for market as well as by small-scale producers. Most of South Africa's wool is produced in the Karoo and in the Free State. Livestock is an important source of cash for many villagers.

Several agricultural regions may be defined (Figure 24-20). Intensive agriculture occurs in the southwestern Cape, where viticulture, horticulture, and deciduous fruit growing predominates and where the Coloured population still provides much of the farm labor. Sugarcane is grown on large white-owned estates along the KwaZulu-Natal coastal plains, and in a few of the adjoining areas. Sharp contrasts in land use exist between communal and commercial farming. On the highveld, maize and wheat are the major crops, while citrus, tobacco, dairying, and ranching are locally important. Although South African agriculture is more mechanized than in most African countries, labor is very important; African and Coloured workers provide most of the farm labor, sometimes on a seasonal and contractual basis. Employment in agriculture varies from province to province as well as by race and by sex (Figures 24-21 and 24-22). Small areas of sisal and cotton are grown in the northern regions, while tea, sugar, and subtropical fruits are produced on a small scale in KwaZulu-Natal and in Eastern Cape Province.

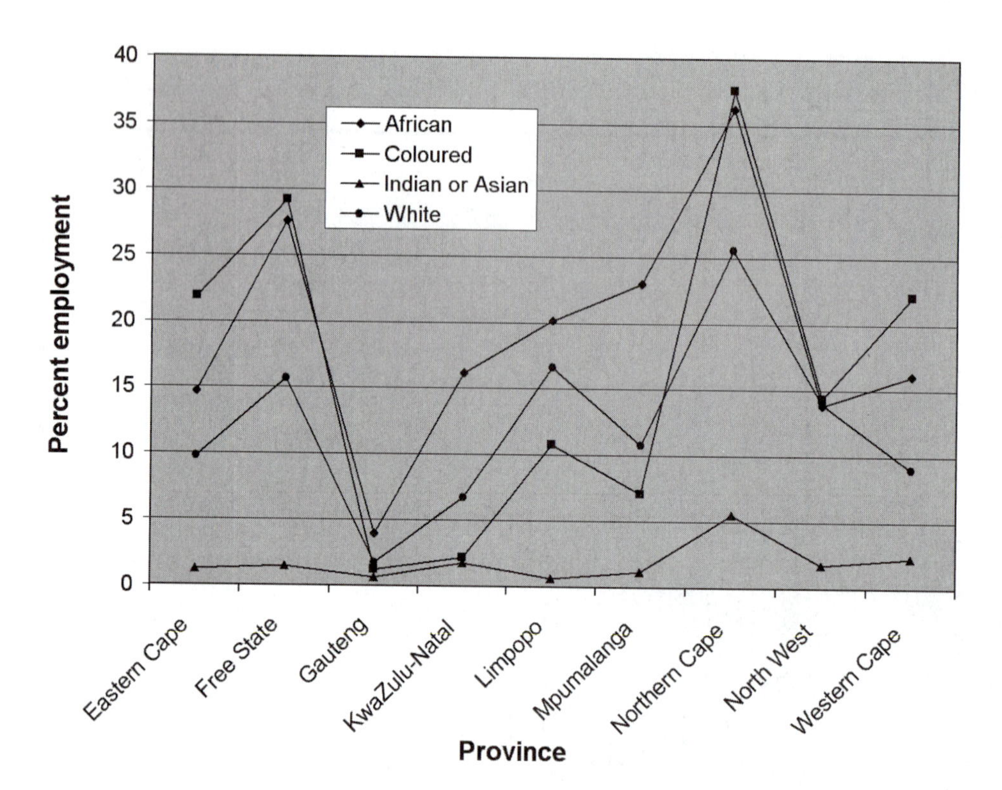

FIGURE 24-21 Male employment in agriculture, forestry, hunting, and fishing by province and group for South Africa in 2001.

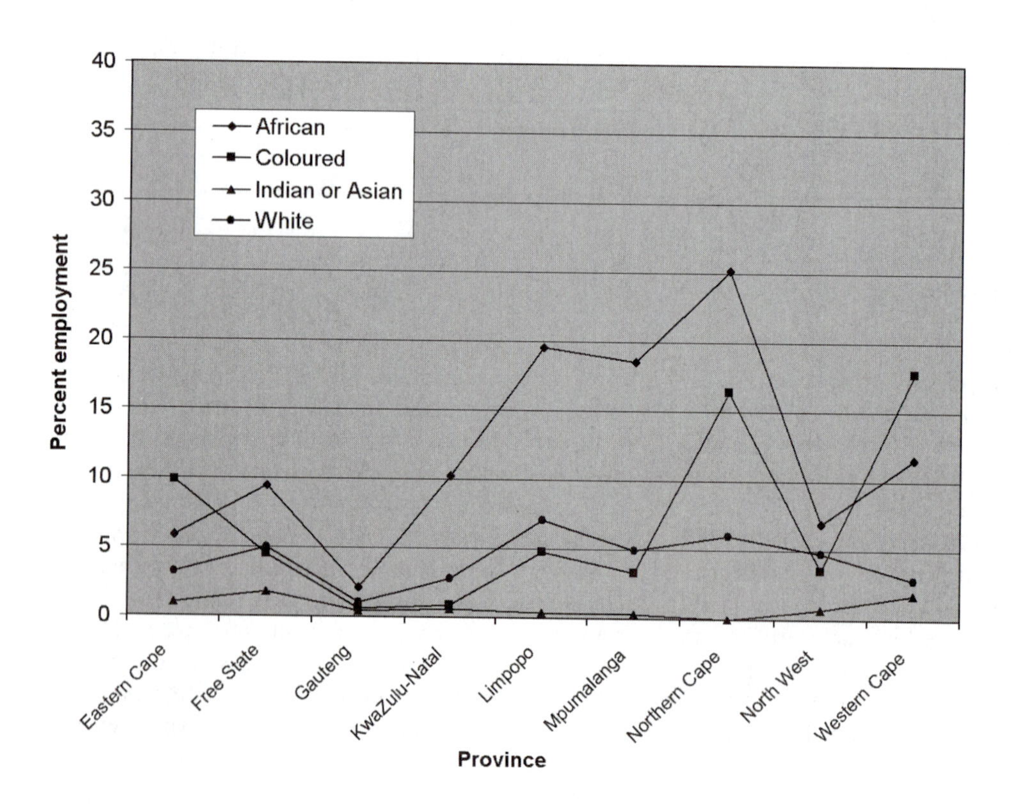

FIGURE 24-22 Female employment in agriculture, forestry, hunting, and fishing by group and province for South Africa in 2001.

Maize, wheat, and sorghum are the most important cereals grown in South Africa, although small quantities of barley, millet, rice, and rye are grown as well. Although production is trending upward on average, the fluctuations in production depicted in Figure 24-19 illustrate the impact of periodic drought. Commercial agriculture has been growing and diversifying. The irrigated area in South Africa has almost doubled since 1961 to 3,669,500 acres (1,485,000 ha), only 9% of the arable land. Twenty-one percent of the irrigated area is devoted to the production of sugar, which has tripled in output over 1961; production in 2002 was 23 million metric tons. Unless technology changes significantly, it is unlikely that the irrigated area can be greatly increased in the future.

The type of fruit produced in South Africa reflects its varied environment. Citrus is widely grown in the more Mediterranean areas of the country, while in cooler uplands apples and pears are popular. South Africa is the largest producer of apples and pears in all of Africa.

In contrast to the trends elsewhere in Africa, particularly in the more marginal regions just south of the Sahara, livestock populations in South Africa have remained relatively constant over the last 30 years (Table 24-5). The severe impact on livestock mortality of the cycles of periodic drought such as that experienced in the Sahelian countries over the last 30 years simply does not exist in South Africa. Although the causes are not yet clear, one hypothesis cites the wetter, cooler climate; another, the role of scientific management.

TABLE 24-5 LIVESTOCK POPULATIONS IN SOUTH AFRICA: CHANGES IN TYPES FOR 1961-1965 AND 1998-2002.

Type of Stock	Averages 1961-1965	1998-2002	Change (%)
Cattle	12,565,400	13,655,077	8.7
Goats	5,236,400	6,624,121	26.5
Pigs	1,462,400	1,573,509	7.6
Sheep	37,514,403	28,893,189	-23.0

Source: FAO (2004).

During the apartheid era, the homelands were divided by the state into farm units varying in size according to the environment and prevailing agricultural practices. Dams, irrigation canals, and contour banks were constructed, and the farms were divided into grazing camps, arable plots, and residential units. Despite all the efforts of the state planners, overgrazing was common, yields were low, and soil erosion was widespread. Rural poverty was high. Since the demise of apartheid there has been land redistribution. In 1994 the Restitution of Land Rights Act was passed by the South African legislature and a commission set up to investigate claims of lost land since the passage of the Natives Land Act of 1913. Almost 64,000 land claims were filed by 1998, the deadline the commission had set. In addition, land was purchased by the government for settlement schemes for African farmers. The government had the goal of redistributing 30% of the arable land of South Africa by 1998, 5 years after the passage of the Provision of Land and Assistance Act in 1993 that enabled it.

Land reform pilot projects were undertaken in all provinces, but the greatest area was along the border between the Northern Cape and North West provinces. Both the restitution of land and land redistribution programs made slow progress and gross inequalities remain. In 1998 there were 1.2 million African small farmers cultivating only 42 million acres (17 million ha) of the former homelands, while a mere 55,000 white farmers owned 252 million acres (102 million ha) (Christopher 2001). To reduce the rural poverty and malnutrition, not only must the land worked by Africans be increased but also traditional farm practices must be changed and nonagricultural jobs created. Yields are kept low by the unavailability of credit and consequent lack of investment, archaic tenurial systems, and overstocking. It is also true that because men migrate to work in mines and manufacturing, there is excessive dependency on the work of women, children, and aged people.

MINING AND MANUFACTURING

Mining accounts for about two-thirds of all exports and 12% of the GDP. The value of South Africa's principal metal products was over $13 billion in 2001. Processed metals were worth $2.9 billion, while hydrocarbons contributed $250 million. About $3 billion in wages was paid, and close to $600 million in taxes. The importance of mining is much greater than these figures suggest. Mining provided the catalyst for industrial development, railway construction to the interior, and the transportation of labor; in addition, it permitted the subsidization of a large part of South African agriculture. About 407,000 people work directly in mining, and 55,000 in metals processing. For the entire country, African men comprise 73% of mine workers, although participation in mining activities by group varies from province to province. Many mine workers migrate from neighboring Botswana, Lesotho, Swaziland, Mozambique, and Malawi. These migrants work long hours for low wages and live separated from their families in the mine compounds for the length of their contract, normally 9 to 12 months.

South Africa's mineral output is prodigious; an astounding 59 different mineral commodities are produced. In 2003 South Africa had 920 operating mines, of which 116 were diamond mines, 59 were coal mines, 41 were gold mines, and 21 mined platinum group metals (PGM). It is the major world producer of andalusite, chromite, ferrochrome, gold, PGM, vanadium, and vermiculite and also a major world producer of antimony, ferromanganese, manganese, titanium, and zirconium (Coakley 2003) (Table 24-6).

TABLE 24-6 PRINCIPAL MINERAL PRODUCTS OF SOUTH AFRICA AS A PERCENTAGE OF THE WORLD'S RESERVES, PRODUCTION, AND EXPORTS.

	Proportion in World (%)		
Product	**Reserves**	**Production**	**Exports**
Aluminum	No data	3	3
Andalusite	No data	36	46
Antimony	8	3	26
Chromium	76	45	28
Coal	11	6	12
Diamonds	No data	10	No data
Gold	52	17	No data
Iron ore	1	4	4
Manganese	80	20	25
Phosphate rock	7	2	2
Platinum group metals (PGM)[1]	56	46	No data
Titanium	20	23	6
Uranium	9	3	No data
Vanadium	44	57	65
Vermiculite	40	45	95
Zinc	4	<1	<1
Zirconium	22	28	39

Source: Chamber of Mines, South Africa (2002).

[1] Platinum, palladium, rhodium, ruthenium, iridium, and osmium.

Most of South Africa's gold mines are located on the Witwatersrand near Johannesburg (Figure 24-23) and are still active and profitable after over a century of working, although most of the republic's gold now comes from fields of the Western Rand, and from Welkom in the Free State. South Africa possesses just over half of all the known gold reserves in the world (Table 24-6). Most of South Africa's gold is located in thin beds about a meter thick in conglomerate layers of rock located in an area about the size of Virginia: the Witwatersrand Basin in central South Africa. Conglomerate is a type of rock that is made up of smaller pieces of rock that have been cemented together over time. Witwatersrand gold, like most of the gold in South Africa, is of alluvial origin, called placer gold. The gold itself originated about 3.5 billion years ago in ancient crust (greenstone belts) lying the east, west, and northeast of the Witwatersrand Basin. Over 500 million years the crust was eroded by ancient river flow, carried, and concentrated in the Witwatersrand Basin, which was once a large lake. Then, about 2.7 billion years ago, the gold and other transported materials were covered by several miles of lava and later cemented together under heat and high pressure. The sedimentary layers that contain the gold of the Witwatersrand were tilted when a large meteorite, on the same order of magnitude as the famous one that blasted into the Yucatan Peninsula, slammed into the Witwatersrand about 2 billion years ago, bringing material from the upper mantle to the surface as well as tilting the sedimentary layers of the Witwatersrand radically upward, preserving them from erosion (Kirk et al. 2003). The northern edge of this tilted basin, where the layers were exposed at the surface, was where gold was first discovered in 1886. Because the reefs of gold are thin in most places, South African gold mining has been very labor intensive. Rising labor costs have made mechanization economic, and a hydraulic gold drill has been increasingly put to use. Mining employment has declined from about 700,000 people in the 1970s to just over 400,000 today.

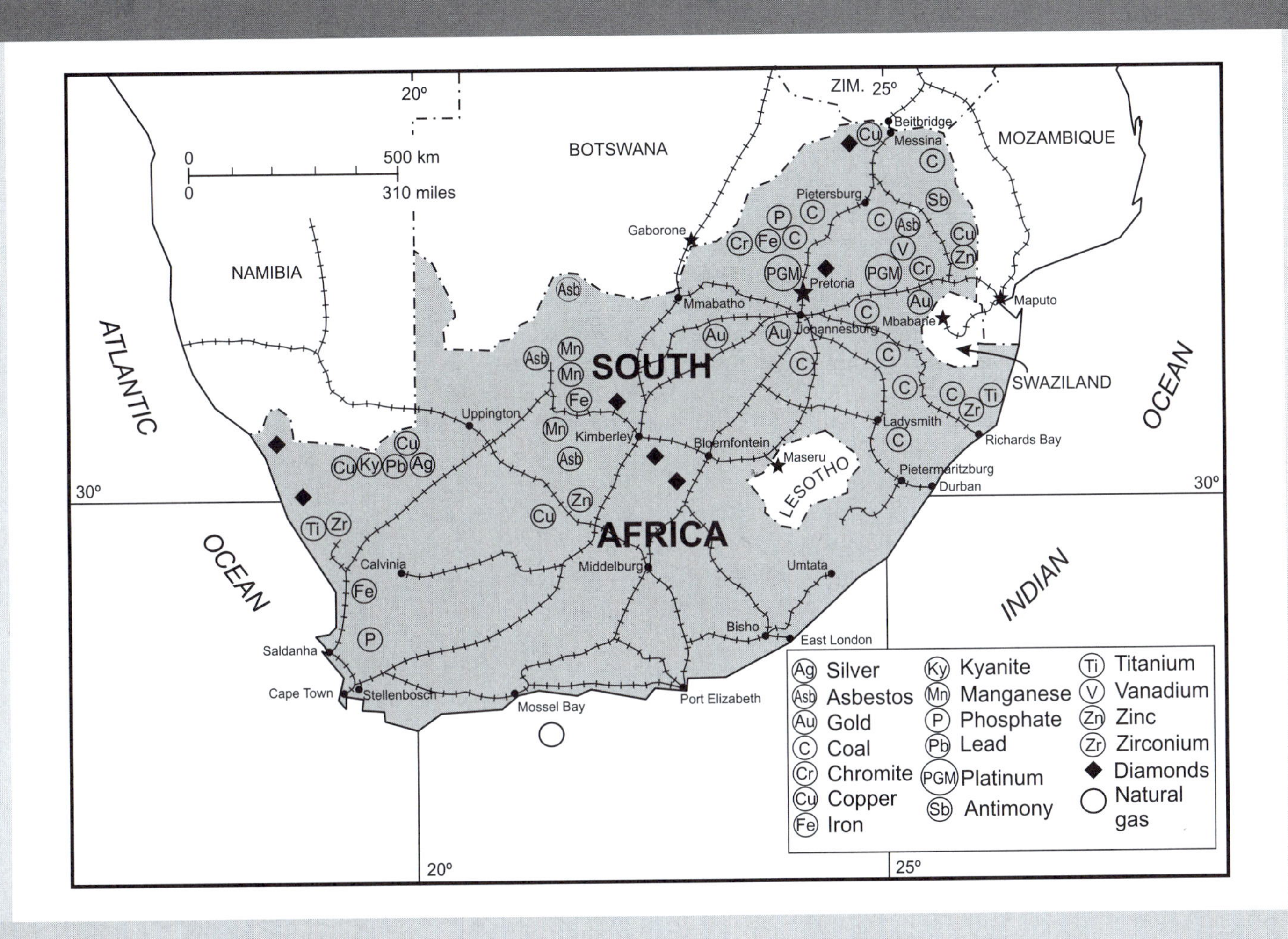

FIGURE 24-23 Minerals and railroads in South Africa.

Although gold was the top metal export up to 2001, platinum group metals have overtaken it because of declining gold prices. In terms of the contribution to the economy, PGM earned \$3.9 billion in 2001, while gold, coal, and diamonds earned \$3.4, \$3.1, and \$1 billion, respectively. The platinum group metals are present in South Africa in such abundance because of another geological accident. An upwelling of magma about 60 miles (100 km) northeast of the Witwatersrand Basin that occurred about 2 billion years ago contains the PGM (as well as chromium) and constitutes the world's largest deposit (Kirk et al. 2003). This feature, called the Bushveld Igneous Complex (BIC), extends 250 miles (400 km) into Limpopo Province and is associated with the mineral rich "Great Dyke" of Zimbabwe.

Diamonds, the first mineral to be mined on a considerable scale, are found in Kimberley, Pretoria, and at Jagersfontein in the southern Free State. Diamonds originate in the upper mantle about 90 miles (150 km) beneath the surface of the earth and are formed under great heat and pressure. They are brought to the surface in volcanic pipes, called kimberlite pipes after the city of Kimberley, South Africa, where they were first noticed by science. Over 90% of South Africa's diamonds are mined from kimberlite pipes; the rest are of alluvial origin. Alluvial diamonds have been eroded away from their volcanic sources by the action of water, transported, and concentrated along water courses and beaches.

East and south of the Witwatersrand are Africa's largest known reserves of bituminous coal (55 billion metric tons total), a particularly valuable asset because South Africa, with no significant petroleum reserves has a high energy-consumptive economy. Coal deposits extend over an area that runs 430 miles (700 km) north to south and 300 miles (500 km) east to west. The coal in Limpopo and Mpumalanga provinces is mainly bituminous and is found in thick seams. The deposits in KwaZulu-Natal, only about 2% of all the coal in South Africa, are relatively thin and comprise the harder anthracite coal. South Africa produces almost 90% of its electricity from coal and has the cheapest electricity in the world. South Africa possesses the fifth largest coal reserves in the world and is the second largest exporter of coal in the world. Coal is used in a sophisticated synthetic fuels program that is the largest in the world, developed during the time of apartheid under the rubric of national security. South African production increased from 66 million metric tons in 1974 to 224 million metric tons in 2000. Exports have remained relatively consistent in relation to production: 2.3 million metric tons of coal was exported in 1974 and 69 million metric tons in 2000, accounting for 34 and 30% of total production for each year, respectively.

Elsewhere titanium, copper, iron ore, manganese, vanadium, nickel, zirconium, and many more that have contributed to the country's mining–industrial economy. Sixty-three percent of the titanium is exported to the United States, to be used mainly in the production of paints, plastics, and paper. Only 6% of titanium is used to produce lightweight metal. South Africa is the world's leading producer of vermiculite, used in insulation, and andalusite, used in glassmaking and optics.

In 2001 South Africa introduced the Minerals Development Bill, to give the state more control over the mining industry. By this bill, the state has exclusive control over mineral rights long held by private individuals, national companies, and multinational corporations. According to the South African government, the purpose of the bill is to redress past discrimination by freeing up unexploited mining rights held by large companies so that black South Africans can participate, protect the environment, and promote sustainable development across the country. Private industry has been concerned about the bill, particularly with regard to respecting the rights of mining concerns, disrupting current mining and exploration, and disturbing the competitive mining environment.

INDUSTRIAL GROWTH

In South Africa, industrial development has taken place at a rate that may be described overall as phenomenal. Until after 1930, the country depended mainly on the export of gold, diamonds, wool, and corn, the bulk of its manufactured goods was purchased from abroad. Then, government action, tariff protection, the emergence of a home market, and isolation during the war all combined to stimulate industrial development. During the height of the apartheid era, South Africa was forced to "go it alone" because of international sanctions and opprobrium. The establishment of an iron and steel industry and the development of automobile assembly, engineering, and metalworking industries created South Africa's largest and most important industrial complex. More persons are employed in this type of industry than in any other and the value of the output is highest. Textiles are next in importance, followed by the chemical industry, fertilizers, foods, and clothing manufacture. Initially, these industries specialized heavily to supply a local demand, the chemical industry depending on its sales of explosives to the mining industry, and the engineering industries being likewise tied to mining. The location of the South African industries has depended on the presence of raw materials, the existence of markets, and availability of power and water supplies as well as labor. Since the economic core of the country has long been the Witwatersrand, the development of an impressive industrial complex in association with the mines is no surprise. Markets appear to have been the dominating factor, and thus the highly diversified Witwatersrand complex dwarfs that of Durban, Cape Town, and Port Elizabeth.

Most of the multinational corporations that pulled out of South African during the 1980s have now returned. Most of the major automobile manufacturers are again producing in South Africa, using it as a platform to produce vehicles for domestic consumption and export. Volkswagen and Toyota produce 21 and 20% of the vehicles in South Africa, respectively. Other companies that produce cars and trucks in South Africa

include Daimler Chrysler (10%) and Delta (9%). South Africa's domestic vehicle market is the second largest (after Brazil) in Africa, Latin America, and the Middle East. Despite its unfavorable location, far from the markets of the northern industrialized states, it is poised to increase its vehicle exports there and within Africa because the cost of South African labor is competitively low. In 1995 South Africa exported fewer than 10,000 vehicles, but in 2003 it exported 126,000 vehicles and 140,000 in 2005.

Industrial development is being encouraged by national and international government policy. Two policies are worth mentioning here. Under the U.S. African Growth and Opportunity Act of 2000, the United States has given African manufactures tariff-free access to the U.S. market. Since the passage of this act, sales of South African car parts, for example, have increased dramatically in the United States. General Motors entered the South African vehicle manufacturing scene in 2004 by purchasing a controlling interest in a domestic company, Delta Motors. Possibilities for growth and expansion exist on the African continent and, although Japanese car manufacturers began to dominate the car markets across Africa in the 1970s and 1980s, auto assembly in South Africa can give GM a share of that growing market. If Mercedes South Africa can competitively produce right-hand-drive vehicles for the British market, Americans may soon see some South African General Motors products with left hand drive in the United States. In a program similar to the North American Free Trade Agreement (NAFTA), the South African government has the Motor Industry Development Program (MIDP), which is designed to encourage growth in the vehicle industry by allowing exporters to import vehicle components duty-free (Economist 2004).

In a regional sense the South African government has taken the initiative to promote southern African integration in four major ways since the end of apartheid. The South African government is stabilizing the water supply in the region through the Lesotho Highlands Water Project, which is intended to impound water through a series of large dams. The purchase of water from Lesotho by South Africa will provide a significant revenue to Lesotho. In addition, South Africa has embarked on the reconstruction of the electrical transmission lines of Mozambique's Cahora Bassa Dam. This will promote greater regional integration in the use of electricity. The third major project, the Trans-Kalahari highway linking Namibia and Botswana to South Africa's Gauteng Province, was completed in 1998. The fourth initiative is the Gauteng–Maputo Spatial Development Initiative (SDI). Through the SDI, the South African government has taken an active role in the development of industry around the country and has identified a number of regions for increased industrial investment.

South Africa's Spatial Development Initiative and Industrial Development Zones (IDZ) are attempts to attract investment to specific areas in which the unemployment is greater than 20%. The SDI is focused on 10 areas of South Africa with the intention of creating "corridors of development" through the improvement of transportation infrastructure and the development of a variety of initiatives depending on the location; tourism, agriculture, and light industry. The Maputo Development Corridor, the most important of the 10 areas, extends from Gauteng Province across Mpumalanga Province to Maputo, the capital of Mozambique. It promises greater integration with the deepsea port facilities at Maputo, South Africa's most important southern African trading partner (Nel 2000). Other SDI projects are focused on the southeast coast and on Cape Town.

The Industrial Development Zones are intended to be state-of-the-art development centers similar to the export processing zones, which are a familiar sight in many parts of the world. Multinational and domestic companies have been attracted to the IDZs because of investment incentives, a tax holiday, and support for research. Most of the IDZs are located on the coast (Port Elizabeth, Saldanha Bay, East London, Durban, Richards Bay) or near reliable rail transportation (Johannesburg, Pietersburg, Uppington).

During the apartheid era one of the major issues facing the state was the control of labor. The peculiar ethnic social geography of South Africa has always been manifested in its manufacturing employment. Employment in manufacturing varies by province and by race/ethnic group with white, Indian-Asian or Coloured men and women being more likely to be employed in manufacturing than Africans (Figure 24-24 and 24-25). In a very similar sense, labor is still the issue today. In the twenty-first century one of the major problems has been finding gainful employment for an increasing national population and migrants. The labor force in South Africa in 2001 was 16.4 million people, an increase of 3.5 million over the previous

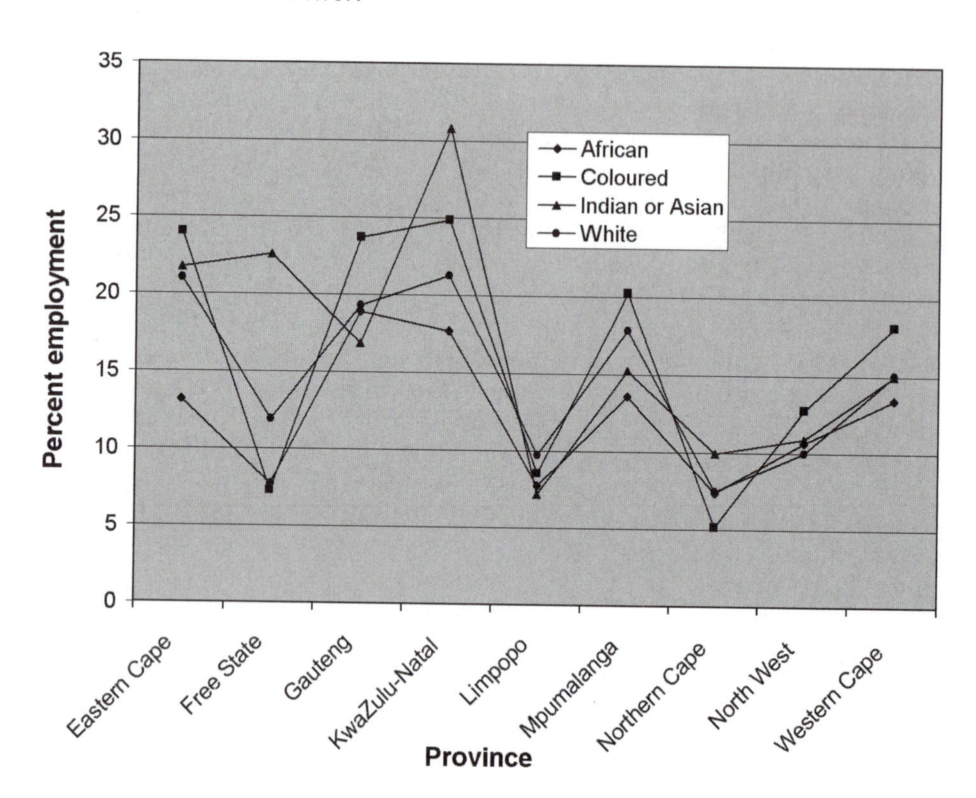

FIGURE 24-24 Manufacturing employment for men by group and province in South Africa, 2001.

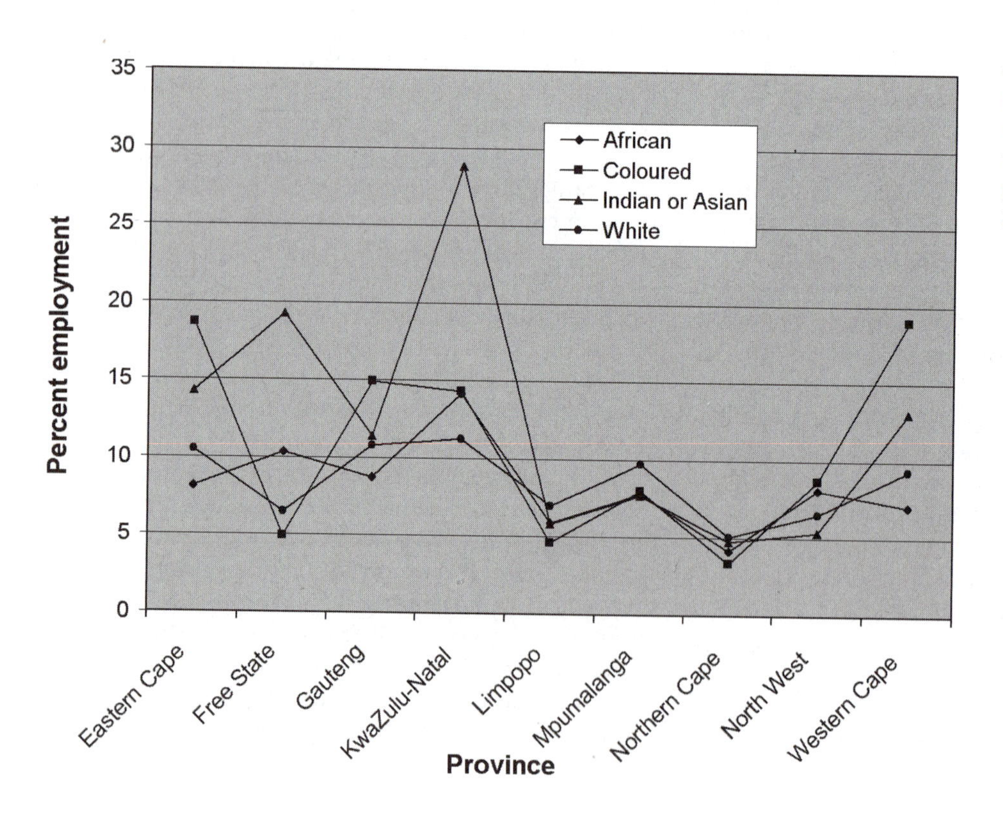

FIGURE 24-25 Employment in manufacturing for women by group and province in South Africa 2001.

decade (Economist 2004). Added to that are perhaps 2 million illegal economic and political migrants from Zimbabwe who are looking for work.

In the near term, South Africa's labor problem is similar to many of its less developed neighbors in Africa: it has too many unskilled laborers and too few skilled. It could solve this problem in the short term by liberalizing its rigid immigration laws to allow more skilled immigrants to enter the country. There would likely be beneficial effects on the employment of less-skilled South Africans from such economic stimulation. In the medium to long term, economic growth and wise investment of resources by the government in education and training for the majority will help impart skills to the unskilled South African.

South Africa is rich in resources, possesses a variety of climates, and is inhabited by a hardworking population that represents a diverse human mosaic. Despite these gifts, the road ahead will not be easy for South Africa. The country is being challenged by globalization after decades of statism and protectionism. Its possession of numerous strategic resources will help it become more competitive but will not solve every problem. The country has had a terrible legacy of inequality and discrimination that, if the experience of the United States is any indication, will be very difficult to overcome.

Perhaps the biggest challenge to the country and the region is the spread of human immunodeficiency virus (HIV), which may precipitate regional social and economic instability in as much as South Africa's neighbors in southern Africa are experiencing similar problems with HIV/AIDS. The HIV infection rate among 15- to 49-year-olds in South Africa is 20% (higher in Gauteng and KwaZulu-Natal and lower in Limpopo, Northern Cape, and Western Cape provinces). Infection rates for the other countries of the region are generally high as well: Botswana, 39%; Lesotho, 31%; Malawi, 16%; Mozambique, 13%; Namibia, 23%; Swaziland, 33%; Zambia, 22%; Zimbabwe, 34%. Madagascar has had little HIV/AIDS.

The spread of HIV/AIDS is having an increasingly significant impact on the South African labor force, and companies have had to develop plans to deal with increasing numbers of sick and dying workers. Estimates from the U.S. Census (Figure 24-26) indicate

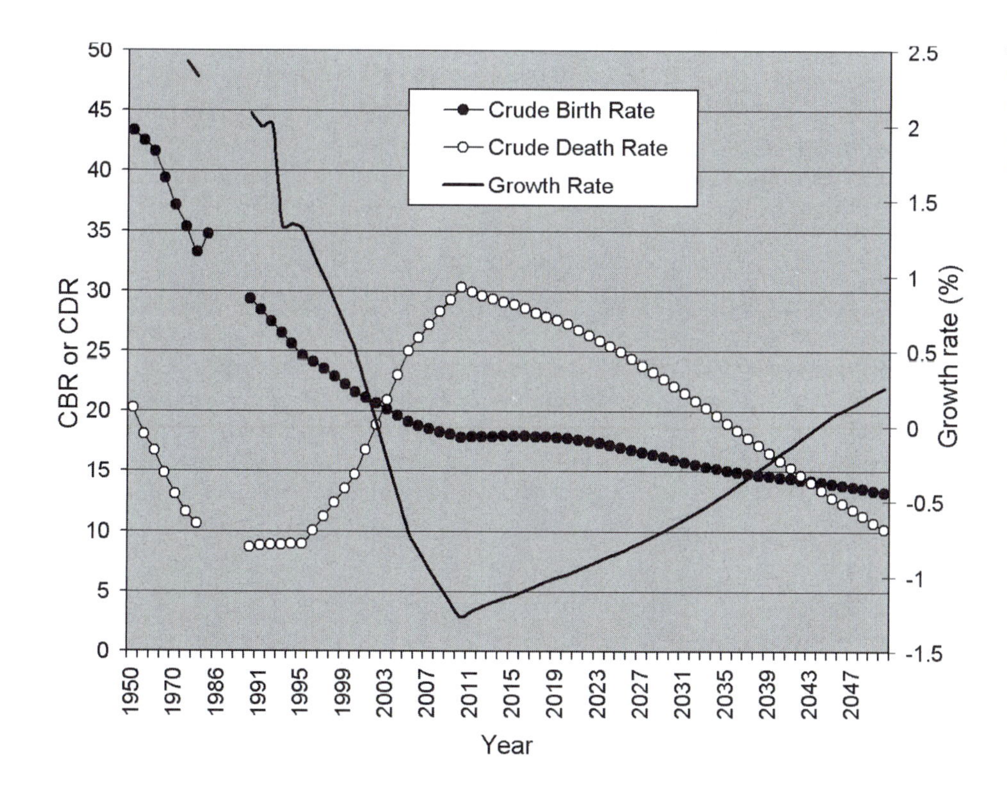

FIGURE 24-26 Crude birth, crude death, and growth rates for South Africa, 1950 to 2050.

that mortality from AIDS will peak in 2010 and decline thereafter. The disease could reach up to 25% of the labor force, and the average life expectancy could drop from 54 years to 38 years as a result. Fighting HIV/AIDS in an effective way will take a massive investment of resources (Tshabalala-Msimang 1999), and the principal challenge will be in finding and mobilizing such resources in a timely way.

BIBLIOGRAPHY

Beal, J., O. Crankshaw, and S. Parnell. 2002. *Uniting a Divided City: Governance and Social Exclusion in Johannesburg.* London: Earthscan Publications.

Best, A., and H. J. de Blij. 1977. *African Survey.* New York: Wiley.

Bird, J. 1888. The annals of Natal: 1495 to 1845. In *A History of South Africa*, L. Thompson, ed. New Haven, Conn.: Yale University Press, 2001, p. 88.

Christopher, A. J. 2001. *The Atlas of Changing South Africa.* London: Routledge.

Coakley, G. 2003. *The Mineral Industry of South Africa.* U.S. Geological Survey Mineral Report. Reston, Va.: USGS. http://minerals.usgs.gov/minerals/pubs/country/

Davies, R. J. 1981. The spatial formation of the South African city. *Geojournal*, supplementary issue 2: 59–72.

de Blij, H. J. 1962. *Africa South.* Evanston, Ill. Northwestern University Press.

Economist, 2004. South Africa: Running to stand still. *Economist*, February 14, 2004: 43–44.

Fage, J. D. 1988. *A History of Africa.* London: Unwin Hyman.

F.A.O. 2004. FAOSTAT. Rome: United Nations Food and Agriculture Organization. http://faostat.fao.org.

Hance, W. A. 1965. *The Geography of Modern Africa.* New York: Columbia University Press.

Helders, S. 2006. World Gazeteer. www.world-gazeteer.com.

Kirk, J. J. Ruiz, J. Chesley, and S. Titley. 2003. The origin of gold in South Africa. *American Scientist*, 91(6): 534–541.

Legum, C. 1999. *Africa Since Independence.* Bloomington, In.: Indiana University Press.

Nel, E. 2000. Economies and economic development. In *The Geography of Southern Africa in a Changing World*, R. Fox, and K. Rowntree, eds. Oxford: Oxford University Press, pp. 114–137.

Simon, D. 1989. Crises and change in South Africa: Implications for the apartheid city. *Transactions of the Institute of British Geographers*, 14: 189–206.

South Africa. 1955. Summary of the report of the commission for the socioeconomic development of the Bantu areas within the Union of South Africa. (The Tomlinson Report). Pretoria, South Africa: Government Printer.

South Africa. 1974. South Africa, 1974. Johannesburg: Perskor Printers.

South Africa. 2001. *South Africa's Mineral Industry, 2000–2001.* Department of Minerals and Energy, Republic of South Africa. http://www.dme.gov.za.

Tomlinson, R., R. Beauregard, L. Bremmer, and X. Mangcu, eds. 2003. *Emerging Johannesburg: Perspectives on the Postapartheid City.* London: Routledge.

Thompson, L. 2001. *A History of South Africa.* New Haven, Conn.: Yale University Press.

Tshabalala-Msimang, M. 1999. *HIV/AIDS and STD strategic plan for South Africa: 2000–2005.* United Nations program on the Acquired Immune Deficiency Syndrome. http://www.unaids.org/EN/other.

U.S. Census. 2002. International Database. http://www.census.gov/ipc/www/.

Western, J. 1997. *Outcast Cape Town.* Berkeley: University of California Press.

World Bank. 2003. *World Development Indicators.* http://web.worldbank.org.

Wilson, M., and T. Thompson, eds. 1969. *The Oxford History of South Africa.* New York: Oxford University Press.

CHAPTER 25

Development Contrasts in Three Landlocked Southern African States

Botswana, Lesotho, and Swaziland

Although Botswana, Lesotho, and Swaziland are separate political entities, they share several important politicogeographical characteristics. First, each is landlocked, and each bounds the Republic of South Africa: Lesotho lies entirely within the republic; Swaziland, although nearly enclosed by the provinces of Mpumalanga and KwaZulu-Natal, has a short but crucially important eastern boundary with Mozambique; and Botswana shares most of its borders with South Africa and Namibia but is also bounded by Zimbabwe. Of the various neighboring countries, South Africa wields by far the most influence in each of the three states.

Traditional leadership has always been strong in these territories: two of them are constitutional monarchies, and Swaziland is the last absolute monarchy in Subsaharan Africa. Moreover, the three countries are ethnically homogeneous; ethnic homogeneity (and strong leadership) helped ensure the survival of traditional values under colonialism and have contributed to the strong sense of nationhood that currently prevails.

In addition, until their independence (Botswana and Lesotho in 1966, and Swaziland in 1968), the three territories were British dependencies collectively known as the British High Commission Territories. As such, they depended on Britain for administration, services, and financial assistance for whatever development projects were contemplated. Britain's investments in their infrastructures and social services were minimal, and for more than 70 years Britain failed to meet its colonial responsibilities. This resulted in part from the belief, shared by both Britain and South Africa, that the High Commission Territories would be incorporated someday into South Africa. It was not until that possibility was removed following World War II that Britain accepted its responsibilities and provided the means for development and modernization.

Another common factor, in view of their proximity to South Africa, is the ongoing domination of the three territories by the republic in the political and economic arenas. South Africa provides manufactured goods, employment for their surplus labor, markets for their resources, and technical assistance and capital for their mining and agricultural projects. At one time all three countries used the South African rand, but today each has a separate currency that is, ironically, at par with the rand. Botswana uses the pula; Swaziland utilizes the lilangeni, adopted in 1974; while Lesotho uses the loti, adopted in 1980. The South African rand is legal tender in all three countries, but at a discounted rate. The three territories form part of the periphery to the South African core, and thus a relationship of interdependence exists between them and South Africa. Although politically independent sovereign states, they are integral and inseparable parts of the South African sphere of influence.

The final similarity we recognize here is that, in each territory, economic development is based on agricultural or mineral resources. Botswana's development, based on the export of minerals, began to exceed the others in the middle 1970s and can be considered an economic success story. Lesotho and Swaziland have made considerable progress in their human development but continue to face economic and, for Swaziland, environmental obstacles.

During the apartheid period, Swaziland was the most conservative in its relationships with South Africa, while Botswana attempted to follow (within practical limits) a measure of independence, and Lesotho oscillated between general compliance with and open opposition to South Africa's policies. Thus, in several ways, Swaziland, Lesotho, and Botswana were South Africa's hostages.

On the other hand, important differences between these countries, especially in their areas, natural environments, resources, and development potentials (Table 25-1) set them apart from one another. Although Botswana is enormous in area compared with tiny Lesotho and Swaziland, much of its territory is dry and sparsely settled. The small, mountainous kingdom of Lesotho, perched high in the Drakensberg Mountains, contrasts sharply with the large, arid, and plateaulike country of Botswana, which occupies much of the Kalahari Basin and is vastly richer. Swaziland, the smallest of the three and having the largest expatriate population and foreign ownership of its land, lies along the Great Escarpment itself and has great topographical, climatic, and pedological diversity. Its sparsely populated western highveld, where humid, temperate conditions prevail, differs significantly from the densely settled subtropical middleveld, which in turn differs from the hot, semiarid, and less populous eastern lowveld.

Each of the three countries has its own distinct development problems and prospects that arise from its own unique environment and history, and from its external relationships with Britain and South Africa. In Lesotho, two major problems are intensifying traditional agriculture and increasing manufacturing development; Swaziland has difficulties associated with overgrazing, maintaining soil fertility, droughts, and landownership; Botswana is trying hard to maintain the remarkable developmental successes that have been achieved since independence. Botswana has had

TABLE 25-1 COMPARATIVE DATA: BOTSWANA, LESOTHO, AND SWAZILAND.

	Botswana		Lesotho		Swaziland	
Variable	**1977**	**2003**	**1977**	**2003**	**1977**	**2003**
Area [mi^2 (km^2)]	224,606 (581,727)		11,718 (30,350)		6,703 (17,361)	
Total population (millions)	0.8	1.6	1.2	1.8	0.6	1.2
Population growth rate (%)	2.2	0.3	2.0	1.1	2.7	1.5
Life expectancy (years)	44[1]	37	46[1]	37	44[1]	45
Infant mortality rate per thousand (%)	97[1]	60	114[1]	89	149[1]	65
HIV infection rate (% of 15- to 49-year olds)	0	38.8	0	31	0	33.4
Urban population (%)	38	54	2	17	5	25
Literacy rate (%)	20	79	59	83	36	81
Per capita GNP (U.S. dollars)	270[1]	2,980[2]	120[1]	470[2]	400[1]	1,180[2]
Annual GDP growth (%)	na	6.1[3]	na	1.7[3]	na	2.5[3]
Net secondary enrollment (%)	na	70	na	21	na	44
Number of people per physician	15,300	4,167	24,900	20,000	8,300	6,667

Sources: Population Reference Bureau (1976, 2003), World Bank (2003).

1. 1976.
2. 2002.
3. 1998–2002.

one of the fastest growing economies in the world since independence and is one of the few economic success stories in Africa. All three countries have been burdened by infection with HIV, which has erased all gains made since the 1970s in increasing life expectancy. Botswana has the highest infection rate in the world, while Swaziland is second, and Lesotho is fourth (Table 25-1).

BOTSWANA: AFRICAN SUCCESS STORY

Botswana, the largest of the three former High Commission Territories, occupies much of the Kalahari Basin, a vast and gently undulating plateau that lies between 3,000 and 4,000 feet (1,000–1,300 m) above sea level. Although there is little topographical variation except in the extreme east, Botswana is a land of great surface contrasts, the most striking features being the Kalahari Desert, the Okavango Delta, and the Makgadikgadi Pans (Figure 25-1). The Okavango Delta is a 6,500 square mile (16,835 km^2) swampland on the northern edge of the Kalahari Desert, where the Okavango River terminates, having flowed across southeastern Angola and the Caprivi Strip. Summer rainfall in Angola seasonally augments the Okavango River flow, which takes about 6 months to traverse the delta. Only around 3% of the flow ever makes it across the entire delta. Although it is peripheral to the major concentration of population in Botswana, the area is very important in attracting tourists to Botswana. In unusually rainy years, waters from the delta flow into the adjoining Mbabe and Chobe depressions and occasionally into Lake Ngami and the salt pans of Makgadikgadi. The only other permanent rivers are the Chobe and the Limpopo, which form parts of the northern and southeastern boundaries, respectively.

Before the mining of diamonds in Botswana, aridity was the single most important natural impediment to development. As the economy has developed, however, more and more people have moved into the city and away from the natural environment. Rainfall is light and erratic and decreases from northeast (Kasane, 27 in.; 687 mm) to southwest (Tshabong, 10 in.; 249 mm) (Figure 25-1). Only 20% of the country has an average annual rainfall above 20 inches (500 mm). Variability is high throughout, with the steepest gradient occurring in the southeast, where population densities are highest. All regions experience a summer maximum precipitation, with the heaviest falls occurring during brief and highly localized thunderstorms.

Temperatures are high (especially during the summer), and strong winds sweep across the open country, causing much moisture to be lost through evapotranspiration. The porous sandy soils reduce still further the amount of moisture available for pastoral and agricultural use. Because of these conditions, and because of the sparse distribution of surface water, Botswana is very dependent on groundwater for its daily water requirements. About three-fourths of the country is covered with porous Kalahari sands. In the extreme southwest there are small stretches of open sand dunes, but elsewhere the sands are covered with grasses and acacias typical of steppe and savanna environments. The soils are frequently highly alkaline and low in humus and thus of low agricultural value. In all regions the vegetation and soil are delicately balanced, and if abused by poor agricultural and pastoral practices, they quickly degenerate, and recovery is slow.

Of all the countries of southern Africa, Botswana has experienced the greatest problems with regard to over-use of the land. The depletion of vegetation, rangeland composition changes, and soil erosion have all been associated with overstocking and the impact of livestock, particularly near watering points (Darkoh 2000). Given the environmental limitations of Botswana, and generations of abuse from people and their livestock, the amount of land suitable for farming and pastoralism is extremely limited. Only 6% of the area, all of it in the east, can sustain semi-intensive mixed farming, and agriculture accounts for less than 5% of GDP. The best areas are around Gaborone, Kanye, Francistown, and Mahalapye, and they are already farmed to capacity, and ecological overstress is evident. In the adjoining areas, and extending for about 80 miles (130 km) west of the line of rail, semi-intensive pastoralism prevails, cultivation of mainly sorghum is widely scattered, and crop yields are consistently low. Then to the west, desert conditions take over, and neither pastoralism nor agriculture is practiced. In the northern region of Chobe, millet and to a more limited extent, maize, are grown. Scattered around the towns and villages, sometimes to a distance of 40 miles (64 km), are the communal grazing areas, agricultural "lands" (farms) and "cattle posts." With

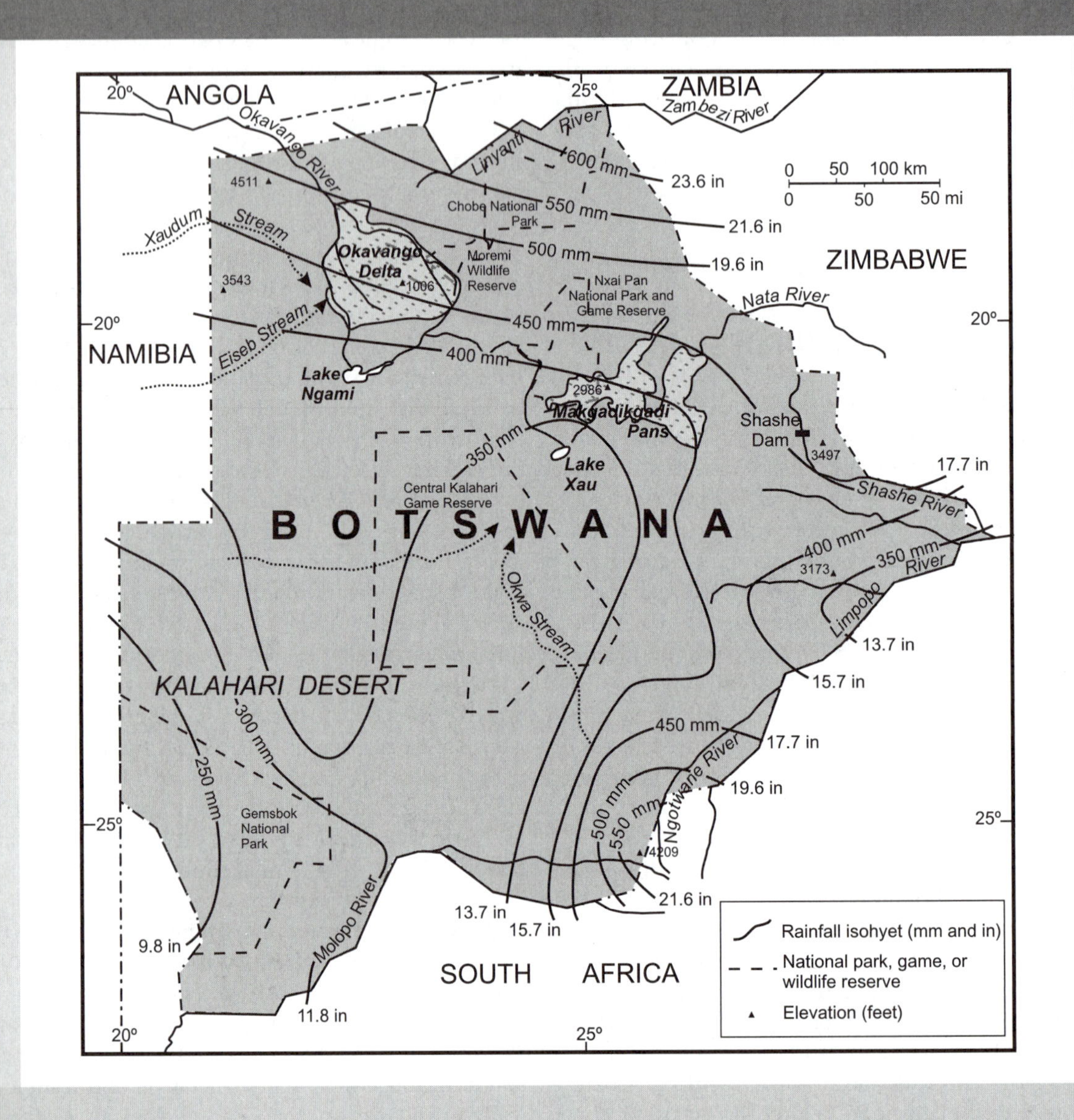

FIGURE 25-1 Major physical features of Botswana, with parks and rainfall.

the onset of the summer rains, women and children leave their villages for the farm, returning after the harvest. Men and boys are frequently absent tending their cattle in the more distant cattle posts, so that up to one-third of the village populations may be absent at any one time. Traditionally the chief authorized the commencement of plowing and harvesting, and he allocated the farms and cattle posts.

Of the three states discussed in this chapter, only Botswana has developed a specialty in raising livestock, particularly cattle; Botswana *is* cattle country, similar in environment to Texas. Cattle in Botswana are grazed on communal and freehold rangelands that in the past were rarely fenced or scientifically managed. But there lurks controversy. Fencing has had a negative impact on wildlife, the basis of tourism; and scientific management benefits the rich. With the lure of export income from the European Union and in tandem with the growth in mineral exports, there has been greater investment in the livestock sector, including veterinary fences, the maintenance of livestock routes, research on breeding and management, subsidized purchase of improved bulls and artificial insemination (Hubbard 1986), and an interest in transforming pastoralists into ranchers.

But there are obstacles to increasing and stabilizing production. The tenure system hinders herd improvement because grazing control has been virtually impossible, and stock of good and poor quality freely intermix (Smit, 1970). Most areas are overgrazed, and few cattle posts have sufficient water. In the more humid regions, each animal requires about 20 acres (8 ha) of grazing, while in the drier areas up to 50 acres (20 ha) is required. The highest concentrations occur in the southeast, in the Shashe–Motloutse drainage areas, tributaries of the Limpopo River, and east of the Okavango Delta along the Botietle River. The lowest concentrations are found in the Kalahari Desert, the Makgadikgadi Pans, and the Okavango Delta, where the range is poor and tsetse fly is common. Cattle raising in a drought-prone environment like Botswana is a risky enterprise. Those with the greatest assets survive over the long term—and even then there are no guarantees.

About three-quarters of the national herd is owned by 15% of the population (Mulale 2002). The largest herds belong to the chiefs, European ranchers, and those who have acquired income from working in the mines and cities of South Africa. Since livestock and livestock products have, until the mid-1970s, accounted for 75 to 95% of Botswana's exports and have provided the principal source of cash income in the rural sector, this ownership pattern has been particularly significant. Today there are more options for people. During the severe droughts of 1965–1966, when the national herd was reduced by about one-third, it was the small cattle owner who suffered most (Figure 25-2). The cattle industry suffers not only from the vagaries of climate but also from a variety of diseases (nagana, foot and mouth disease, and anthrax) and the rigors of transportation. During the 1990s the government slaughtered several hundred thousand cattle in the northwest District to reduce the spread of disease. Eradicating disease is a difficult enterprise with range cattle, since infection from wildlife is possible at almost any juncture.

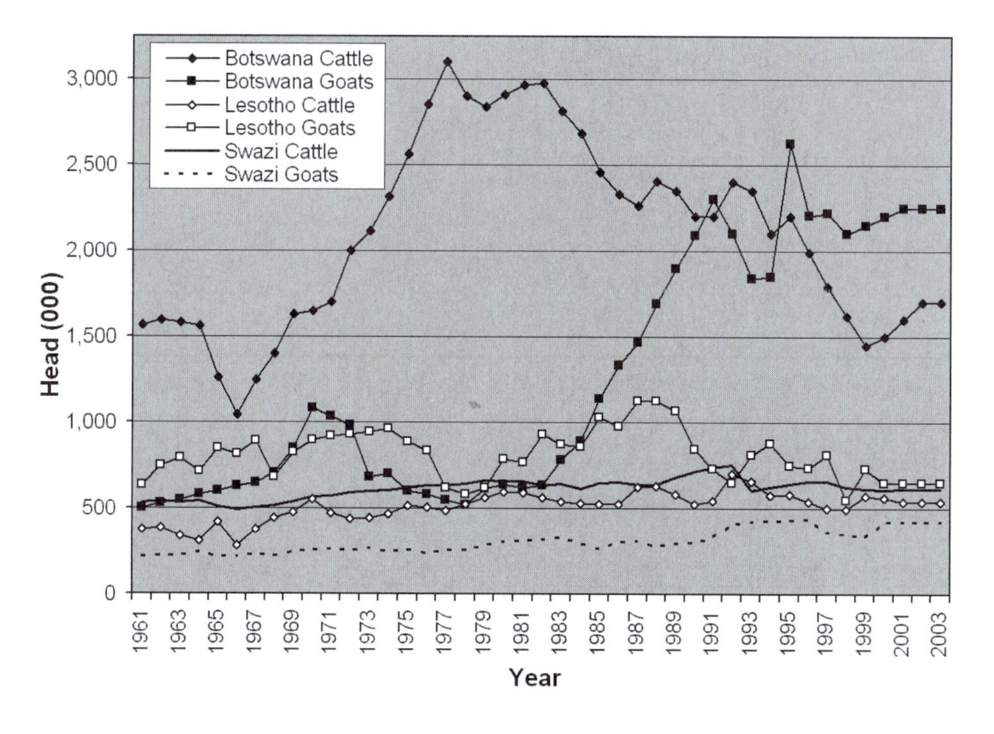

FIGURE 25-2 Head of cattle and goats in Botswana, Lesotho, and Swaziland, 1961 to 2003.

Botswana's livestock industry has undergone considerable transformation since the early 1970s. The number of cattle has been cut in half since peaking in the mid-1970s and is currently just above the early 1960s level (Figure 25-2). The number of families without cattle has been increasing (from 32% to 49% of farm families from 1980 to 1996), and most farm families have significant nonfarm income (Whiteside 1997). As cattle population went down, goat numbers went up and have increased by about 1.5 million head.

Up until the early 1970s, live cattle were trekked from the north directly to Zambia and Zimbabwe and south to South African markets. Since that time live cattle exports have plummeted to almost nothing and processed meat exports have burgeoned enormously (Figure 25-3). Processing the meat, rather than simply exporting the animal on the hoof, adds value to the product and commands more money in the market. The principal slaughterhouse and meatpacking plant is at Lobatse, near Gabarone. Cattle are trekked or trucked from around the country to the slaughterhouse and canning factory. Trekking is hazardous because of limited and irregular water supplies and grazing, and many producers sell their livestock to local intermediaries, who sell them in turn for a profit to big buyers. At present, chilled carcasses comprise the major livestock exports.

The increase in the goat population may be due to the hardiness of the goat in comparison to cattle. Goats will eat almost anything and can subsist on the most impoverished of ranges. In addition, they are attractive because they are cheaper, more drought resistant, and less demanding of labor than cattle or sheep. The raising of goats has been increasing among the poor and cattleless in Botswana (Whiteside 1997). Cattle possess much prestige in addition to economic value in Botswana, and it is unlikely that numbers will decline in the immediate future—on the contrary.

Almost 70% of Botswana's meat exports are sent to the European Union, which gives Botswana greatly reduced import tariffs (IRIN 2004). The rest of the exports go to other countries where prices are higher than the EU, particularly Norway and Réunion. The remainder is marketed elsewhere. Health-conscious Europeans prefer Botswana's beef because it is range fed and much less likely to be infected with bovine spongiform encephalopathy (BSE, or "mad cow" disease). Although over 40% of Botswanans make their living in agriculture, agriculture is necessarily limited and constrained by the environment. The income

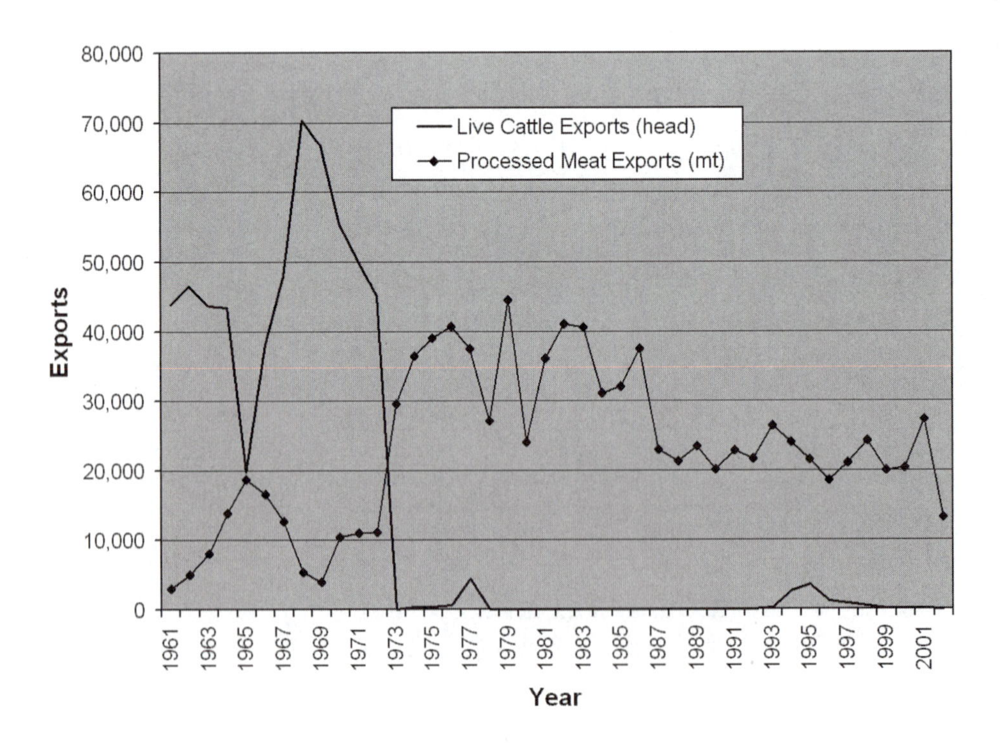

FIGURE 25-3 Live cattle and processed meat exports in Botswana, 1961 to 2002.

from mineral production more than assures imports of a wide variety of foodstuffs, but the Botswanans need rural employment.

One region of possible agricultural expansion is the Okavango Delta, where preliminary soil and water surveys indicate there are about 1.4 million acres (567,000 ha) of potentially irrigable land capable of producing maize, rice, millet, and numerous industrial crops. But such development is a political and environmental issue and in the event would call for great inputs of capital, the construction of roads and canals, and the bringing of farmers. Development there is controversial and unlikely for many years, if ever, but over the last 30 years the Southern Okavango Integrated Water Development Project (SOIWDP) has been proposed and much debated. The purpose of the project is to provide water to the town of Maun and irrigate 25,000 acres (10,000 ha) for agriculture. Tourism, a significant contributor to the economy, will likely be the principal land use in the Okavango area, rather than large-scale agricultural development.

The population of Botswana is concentrated in the eastern zone, especially in the southeast around Gaborone, and is almost absent (except for small bands of San hunters and gatherers) from the Kalahari Desert and the Okavango Delta. The population is relatively homogeneous and is composed of eight major Setswana-speaking groups (about 60%), and several smaller non-Batswana groups including San, Hereros, Europeans, and Asians. Setswana is the language of the Tswana. The Batswana comprise eight subgroups: Bamangwato, Bakwena, Batawana, Bangwaketse, Bakgatla, Bamalete, Barolong, and Batlokwa. Unlike all of its neighbors in southern Africa except South Africa, over half of Botswanans live in towns. This is a manifestation of the wealth of minerals development that has enriched the country, and its good management.

The largest town in Botswana is the capital, Gaborone (195,000). Francistown, the next largest town, has a population of 87,000, less than half the size of Gaborone. Gaborone was established as the capital in 1965 just before independence. Until then the administration was run from Mafikeng, South Africa. Like Brasilia and Lilongwe, Gaborone is a planned, forward-looking capital whose site was selected with the hopes of stimulating growth in a frontier region (Best, 1970). It is principally an administrative center and in addition to government ministries and offices, the University of Botswana (1976) and the Botswana Agricultural College (1967) are located there. Botswana is a nation of small towns: there are 23 with a population between 10,000 and 87,000, and 79 with fewer than 10,000 people each. Botswana is administered through 10 provinces or "districts" and four "town councils." Gaborone, in addition to being the national capital, has its own local government. Other town councils are in Selebi-Phikwe, Francistown, and Lobatse (Figure 25-4).

Industry and Minerals

Botswana has a variety of service-oriented and consumer products plants, and light manufacturing: banks, hotels, breweries, furniture, processed foods, leather goods, clothing, and vehicle assembly. These are located in mainly in Lobatse (31,000), Francistown (87,000), and Gaborone (195,000), all situated on the line-of-rail (Figure 25-5). The Botswana Meat Commission, located in Lobatse, is the largest meat processing industry in the country and controls the export of beef, formerly the major export earner of the country.

Botswana's economy entered a new phase in 1971 with the opening of the De Beers (Anglo-America Group) diamond mine at Orapa, 233 miles (375 km) north of Gaborone. The kimberlite pipe is worked by open-cast methods. It was thought that production could hold at about 4 million carats a year, but in fact production has vastly exceeded all early expectations (Figure 25-6). Fifteen percent of the diamonds are of gemstone quality. A second mine is being developed at Letihakane and a third at Jwaneng, 350 miles (563 km) north and 115 miles (185 km) west of Gaborone, respectively.

The largest mining operation centers at Selebi Phikwe—where a 44 million ton deposit of low-grade copper–nickel ore was brought into production in 1973 and is still producing. Financed initially by West German banks, the Industrial Development Corporation of South Africa, Canada, the World Bank, and USAID, this multimillion dollar project has run into numerous unexpected difficulties with the smelting operations, the low-quality coal being used from nearby Morupule, and water supplies from the Shashe River. The mine required a new railway spur, additional roads, a thermal power station, and has given rise to Botswana's largest township (30,000 in the mid-1970s and 52,300 in 2003).

FIGURE 25-4 Towns, transportation, and political organization in Botswana.

Copper and nickel are also mined about 12 miles (20 km) east of Francistown at Selkirk Mine. Mining provides the government with much-needed revenues for the improvement of roads and health and education facilities, and for investment in the livestock industry. Diamonds today contribute 35% of the GDP, 45% of government revenues, and 85% of export revenue (Coakley 2001). The gemstones brought in almost $2.25 billion in 2001. Copper, nickel, and cobalt production was worth $230 million in the same year. It is expected that diamond mining will remain economically viable for 25 to 30 years and nickel–copper mining for about 15 years.

Although the Botswana government is usually a partner in all mining enterprises in the country, the mining industry operates on free-market, private

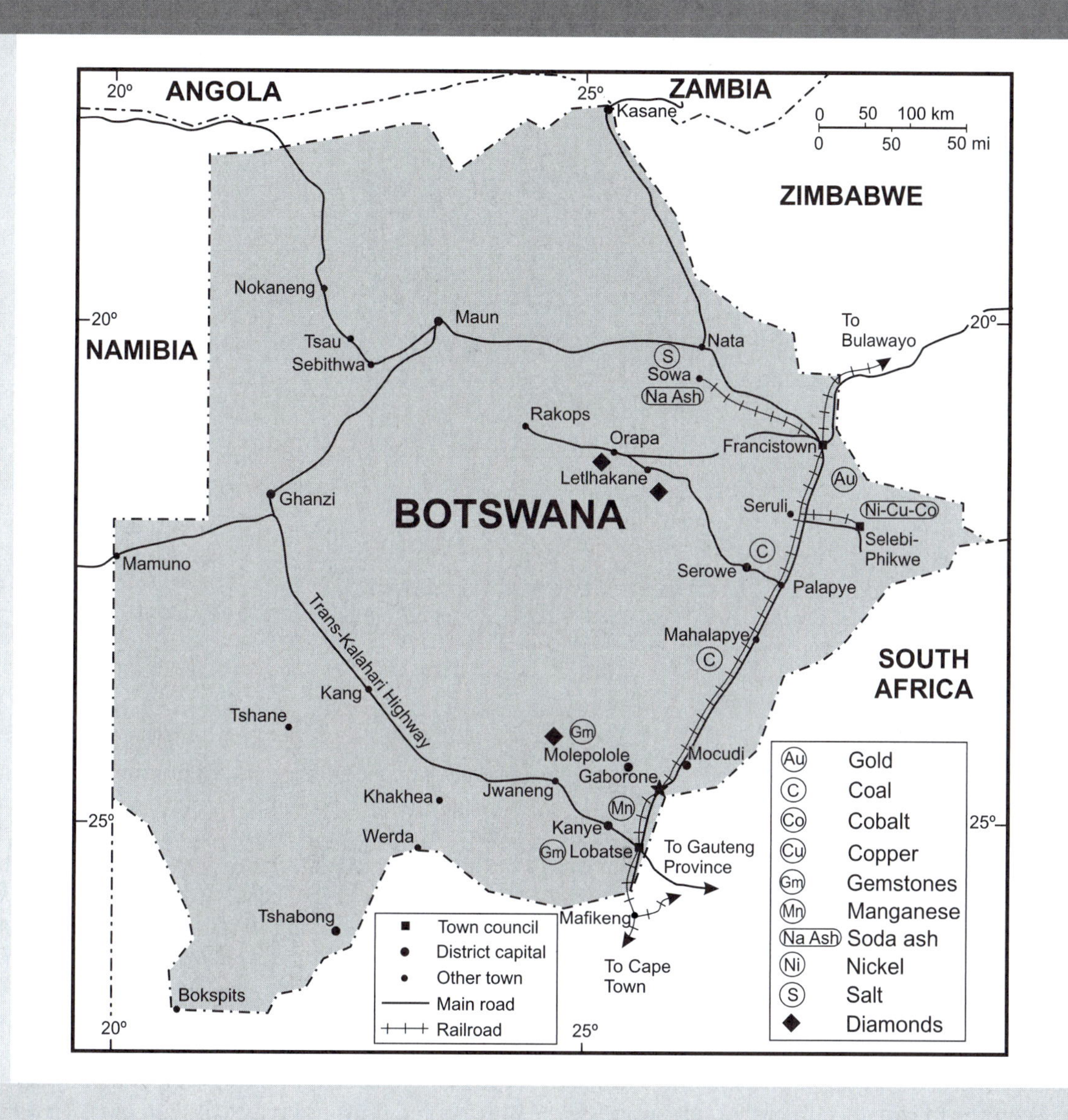

FIGURE 25-5 Botswana mining.

ownership principles and government share is normally well below 50%. The advantages of political stability, favorable investment climate, and low taxes will continue to bring foreign direct investment (FDI) into Botswana's mining sector. The most significant challenge for Botswana is to diversify the economy away from dependence on diamonds and metals for the vast majority of its earnings. Botswana's environment is a natural attraction for tourists from all over the world, and there has been much investment in tourist support infrastructure, part of the government and private sector initiative to diversify the economy. Tourism is one of the most rapidly growing industries in Botswana. Regular air service and paved roads link the major attractions of the Okavango Swamp and the Kalahari Desert with Gaborone and with neighboring countries.

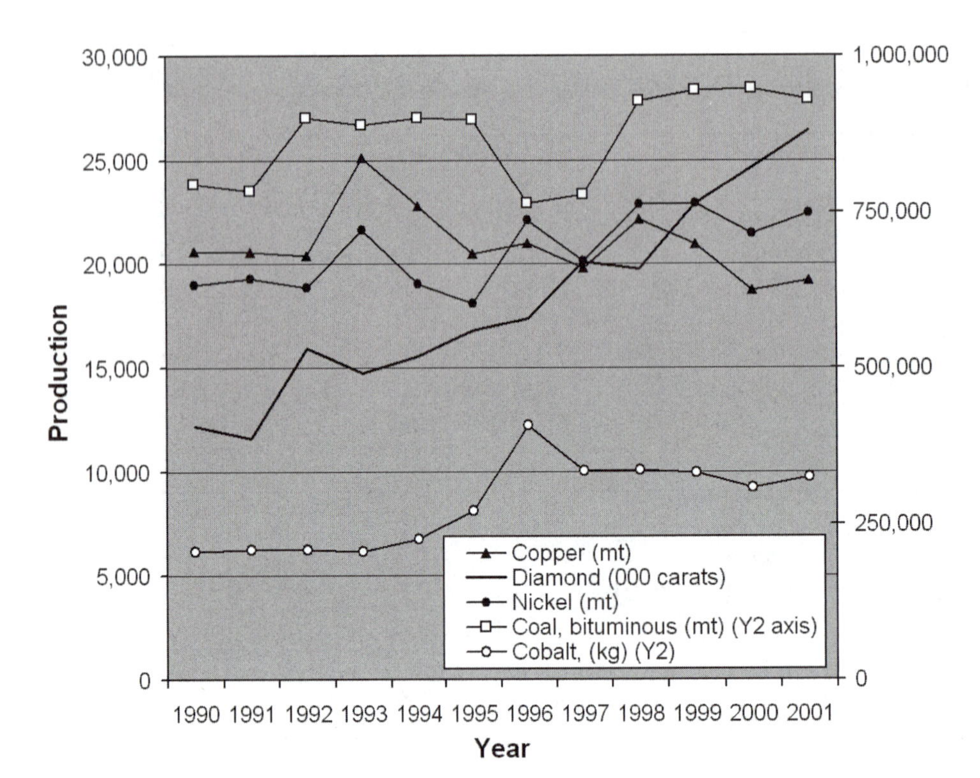

FIGURE 25-6 Mineral production in Botswana, 1990 to 2001.

Botswana Today

Botswana is obviously limited by its remote and landlocked situation, but in today's information economy, where good communications infrastructure can overcome distance, location may not be very important. The government is encouraging the development of a state-of-the-art international financial services center in Botswana to assist in economic diversification and to help integrate Botswana to a greater degree into the world economy. One of the purposes of the financial services center is to convince international businesses to relocate to Botswana.

The Globalization Index of *Foreign Policy* magazine has ranked Botswana number 30 (just below Japan) of the 62 countries studied. Botswana is the most globalized country in Africa and the third most globalized developing country in the world, well ahead of the newly industrializing countries of Asia (Kearney 2004). Botswana is becoming more integrated regionally as well as globally. Although Botswana has always had relatively good road and rail access to South Africa, the Trans-Kalahari Highway, which opened in 1998, links Walvis Bay, Namibia, with Gaborone, Botswana. One of the major purposes for building the highway was to relieve the intense import–export pressure on southeastern South African ports such as Durban, which is 5 days sailing time away from the port of Walvis Bay. A road and rail link connecting Botswana with Zimbabwe and Zambia is accessible via paved road.

Although Botswana was described in the 1970s as "a classical underdeveloped open economy, exporting primary products and importing manufactures, primarily from South Africa" (Best and de Blij 1977), this description must be revised in light of new evidence of the great changes that have taken place in the human and economic development of Botswana since that time. Since the early 1970s, Botswana's economy has been the fastest growing in the world (Acemoglu et al. 2001) and, with continued sound leadership, appears poised to continue its successes.

There are several reasons for its past successes that provide a solid basis for future growth. Like many countries in Africa, Botswana is rich in natural resources. Many countries in Africa sadly attest to the fact that possessing wealth in resources is not a sufficient cause for development. But in addition to riches in natural resources, Botswana has had responsible, committed government; the people of Botswana have been cooperative and patient rather than militant or revolutionary; the government has invested wisely in infrastructure, health care, and education and other public institutions rather than consume the national income on prestige projects or rent-seeking; and debt has been avoided, with Botswana continuously running an economic surplus (Samatar 1999). Furthermore, despite the heavy HIV/AIDS burden borne by the people of Botswana, the government, unlike its big neighbor to the south, recognized the nature of the problem early on and is implementing a medical solution. At present, Botswana can be considered an African success story.

LESOTHO: A CONSTITUTIONAL MONARCHY

The mountain kingdom of Lesotho may be divided into three physiographic regions running northeast to southwest. Each has its own distinct climatic and pedologic characteristics and agroeconomic potentials and problems. In the far west lie the borderlands and the "lowlands," really a misnomer, since much of the land is higher (5,000–6,000 ft; 1,525–1,630 m) than the bulk of the South African highveld (Figure 25-7). Here the gently undulating plateau is studded with mesas and buttes, the sandy soils are impoverished and severely eroded, and rainfall is light (under 30 in.; 760 mm) and erratic. Despite these adverse conditions, the region, which represents only 17% of the total area, supports about 40% of the population. Here arable densities approximate 600 persons per square mile (230/km^2), and the Basuto live in both nucleated towns and villages and dispersed farmsteads. The lowlands contain the capital, Maseru (178,300), and the country's largest villages. Here also is the densest communications network, which centers on the only paved road stretching from Hlotse (Leribe) to Mafeteng (Figure 25-7), about 675 miles (1,087 km).

A smaller but also densely populated lowland follows the Orange (Senqu) River, which cuts through the southern mountain ranges. In both regions, maize and millet form the staple crops, although winter wheat is also widely grown in the Orange River Valley. In the center of the country lies a belt of foothills whose elevations and slopes become steeper and more barren toward the east. Its soils are generally more fertile and less eroded than to the west, but agriculture is not so fruitful. Communications become difficult, few roads connect the region, and pastoralism dominates.

Finally, the eastern section of Lesotho (58% of the area) is composed of the scenically magnificent Maluti–Drakensberg Mountains, where slopes are steep and soils are thin and badly eroded. Here rainfall is heavy (40–70 in.; 1,016–1,778 mm), much of it falling as summer thundershowers. The growing season is short, and winter snows may last several weeks. Cultivation is limited but expanding as population pressure mounts elsewhere.

Pedologically, Lesotho is one of the poorest countries in Africa. Unfavorable parent material, unreliable rainfall, sparse natural vegetation, and strong relief have combined to hamper soil development. Overgrazing and mismanagement have destroyed large portions of the areas that were best endowed, and one of the characteristics of present-day Lesotho is the large number of gullies and heavily eroded fields, fast reducing the available usable soil acreage. Indeed, soil erosion is the country's greatest obstacle in the face of agricultural self-sufficiency (Smit 1967). Almost 40% of the total cultivated surface is really unsuitable for cultivation, with the greatest incidence of soil erosion occurring in the extreme west and in the Orange River Valley. Although the largest proportion of the fields in the mountains are on steep slopes, soil erosion is less common than in the foothills and the lowlands. This is because of the lower population densities, because the soil is not so exhausted after a shorter period of cultivation, and because soils here are less sandy than those in the lower-lying areas.

Lesotho constitutes one of the very few countries in Africa that may be called a nation-state because the great majority of its people are of one nationality—the Basuto. It is also one of the few monarchies in Africa. The first Basuto leader, Moshesh (c. 1780–1870), welded the nation out of the peoples scattering (mfecane) before the marauding Zulu, consolidated them in the rugged mountains of the Drakensberg, and then ruled them with wisdom and cunning such as no other leader displayed. He successfully withstood the early European imperialists, both Boer and Briton, and guided his people into a position of strength that

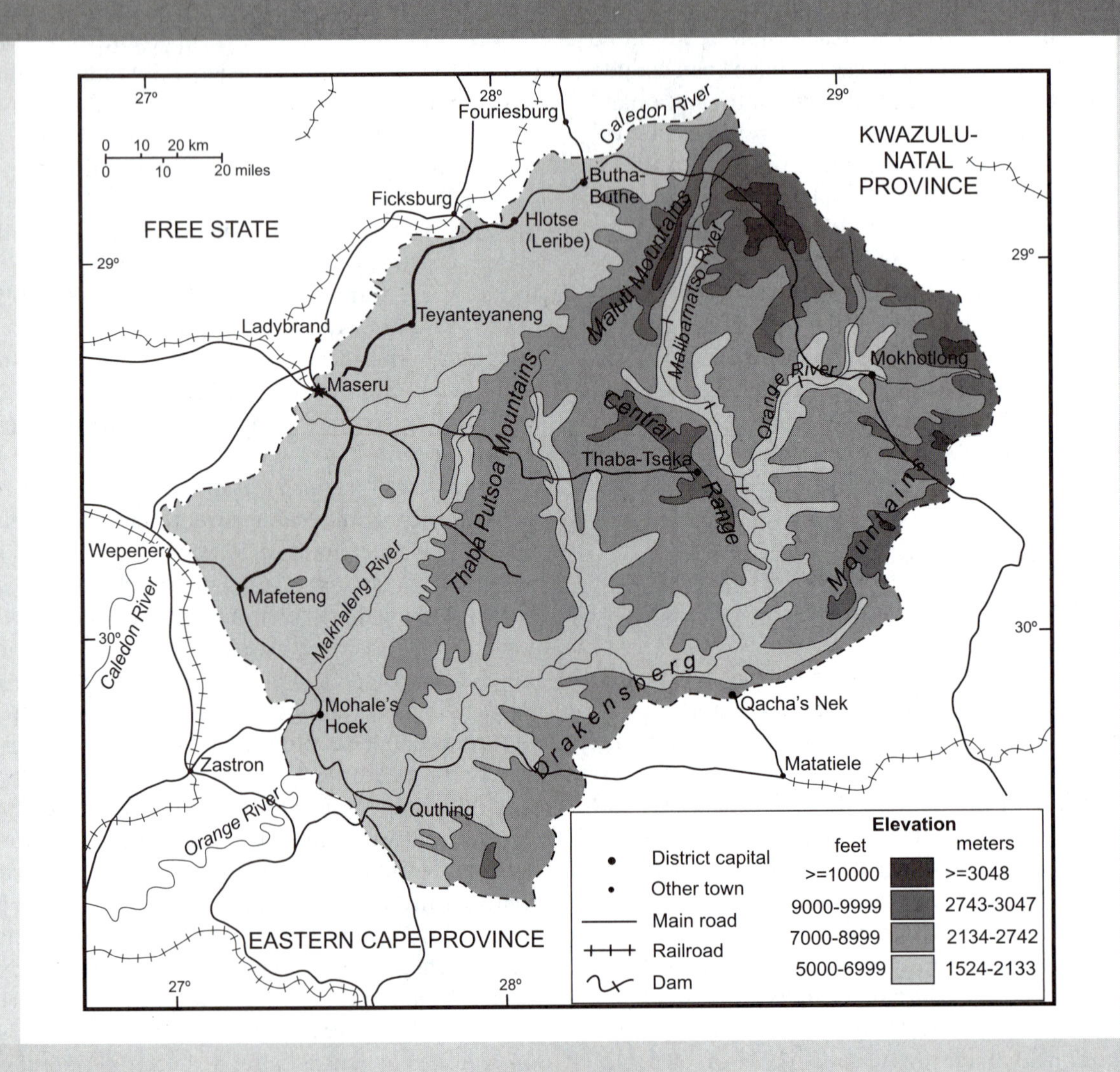

FIGURE 25-7 Relief, infrastructure, and principal towns of Lesotho.

ensured their survival long after his death. He was unable to prevent the isolation of his nation, and the very isolation that once ensured Basuto survival has become a great liability.

One of Moshesh's aims was the establishment of a British protectorate over his country at a time when both the British Cape Colony and the Boer's Orange Free State Republic sought its annexation. He eventually succeeded, and when the Union of South Africa was created, the Basuto people were permitted to remain a separate political entity. They gained their independence from Britain in 1966. Although the legal system in Lesotho reflects English common law from the time of the British protectorate and Romano-Dutch law borrowed from South Africa, the political system that has developed is one of traditional African political organization (chiefdoms) and African monarchy in a framework of European parliamentary democracy.

The government of Lesotho is a constitutional democracy; its chief of state is King Letsie III, and the head of government is Prime Minister Pakalitha Mosisili. The king has no executive or legislative power and serves only as a symbol of national unity. The real power in Lesotho is held by parliament and the prime minister. Lesotho has a bicameral parliament with a Senate composed of 33 unelected members. Twenty-two of the senators are chiefs of the principal tribal groups of Lesotho. Eleven members are appointed by the ruling party. The National Assembly is made up of 120 members; 80 are elected by a direct popular vote, while 40 obtain their seats via proportional vote by party. In a parliamentary system, the leader of the party that wins the majority becomes the prime minister and forms the government.

Since independence, Lesotho's path toward democracy has not been as straightforward as Botswana's and has been punctuated by several stumbles. Instead of relinquishing power in 1970 after losing the first postindependence elections, the prime minister of the ruling Basotho National Party (BNP) annulled the election and imprisoned the leadership of the winning party, the Basotho Congress Party (BCP). For 16 years the BNP ruled by decree until forced from power by a military coup d'état in 1986. At that time the king, Moshoeshoe II, formerly only a ceremonial head of state, was given executive powers. A year later he was deposed by the military and his son was installed as King Letsie III. In 1993 constitutional government was reestablished with the election of the BCP. In 1997 a faction broke away from the BCP and under the leadership of the former prime minister, Ntsu Mokhehle, formed the Lesotho Congress for Democracy (LCD). The LCD (later led by Pakalitha Mosisili), won the general election of 1998. The BCP, now in opposition, rejected the results and violence erupted. After a rebellion by junior military officers, the government requested military assistance from the South African Development Community (SADC) to restore stability. The SADC force withdrew in May 1999, having restored calm.

To address the structural political imbalance in the National Assembly, Lesotho added the 40 seats that are filled by party on a proportional basis, to ensure that there would always be an opposition in the National Assembly. In the 2002 election, the LCD won the general election again but significant opposition exists in the National Assembly where although the LCD has 79 of the 80 seats that are elected by constituency, 9 opposition parties control all of the proportionally elected 40 seats. The next parliamentary election will be in 2007. In 2005 the first ever nationwide local government elections were held and the ruling Lesotho Congress for Democracy (LCD) won.

Eighty-five percent of the resident population of Lesotho is engaged in farming, mainly for subsistence. According to tradition, all land belongs to the paramount chief. Acting as a trustee for the Basuto, he administers the land through the local chiefs. Every Mosuto (Basuto male) householder has a traditional right to three fields—one for maize, one for wheat, and one for sorghum. Usually these fields are scattered so that everyone gets both good and inferior land. Today, because of population growth and the limited amount of land available, not everyone has the full complement. Indeed, some people have no land at all. Land not used for two consecutive years can be granted to someone else. The same is true of land owned by a Mosuto who leaves the area permanently. Migrant workers retain their right of possession as long as the land is cultivated for them. The local chief has the authority to expropriate and reassign land as he chooses.

According to the communal land tenure system, land may not be enclosed because that would encourage individual occupation and prevent communal grazing, which traditionally follows harvesting. Not all land is cultivated each year. Indeed, approximately one-fourth remains idle for the lack of a plow or seed. Plowing and harvesting begin only after authorization from the chief, and one man's poor judgment can spell disaster for many. A poor harvest means not only food shortages that year but also a lack of seed to plant in the following year. Thus, unless money can be raised with which to buy seed, the shortage is perpetuated.

The continuous demand for food, seed, and cash forces the Basuto, especially the men, to seek work in South Africa. Plowing must begin as soon as the first rains fall; but if the rains are late, crops run the risk of frost damage before they mature. Traditionally, all land was cultivated with the hoe, but now the plow is widely used, permitting the cultivation of larger areas. The practice of plowing across the contour has promoted soil erosion. Very little crop rotation is practiced, and as maize is planted year after year in the same field, much of the soil is exhausted. In the past, fertilizer was rarely applied, and as the soil deteriorated it was abandoned. Today, shifting cultivation is no longer possible as virtually all cultivable land is used, and little manure is applied since, in the absence of firewood, cattle dung is used for fuel. Commercial fertilizer is beyond the means of most farmers.

Despite these challenges to farming, there is no lingering land question, as there has been in other countries of the southern Africa region: Swaziland, Zimbabwe, Namibia, and, of course, South Africa. British policy restricted European settlement in the territory to administration and commerce; all land remained in the hands of the Basuto.

Industry

After independence, animal products, primarily cattle, sheep, wool, and mohair, constituted up to three-fourths of Lesotho's exports, while diamonds made up the balance. Today, in contrast, manufactures constitute 75% of the exports, while wool and mohair and live animals comprise the rest. The vast majority of the country's exports are purchased by the United States. Lesotho has few known natural resources other than water and diamonds, and it lacks the capital, technology, markets, and communications required for major manufacturing. Its most valuable natural asset is water, especially the Orange River.

In 1986 the Lesotho Highlands Water Project (LHWP), financed by the World Bank, began the first phase of a 30-year water resources development program. In 1998 it began to sell water to the Witswatersrand in South Africa, bringing in needed capital. As a result of the project, Lesotho has become self-sufficient in the production of electricity and, if it manages its supply judiciously, will be able to power light industry for domestic consumption and regional export. The production of textiles, mainly for export to the United States, has rapidly grown recently with the assistance of Asian (mainly Chinese) investors, who are taking advantage of a growing market in southern Africa as well as finding a tariff-free point of entry into the clothing markets of the United States. Lesotho has become Subsaharan Africa's largest exporter of clothing to the United States.

Lesotho and the other countries of the Southern Africa Customs Union (SACU) stand to gain much from the African Growth and Opportunity Act passed by the United States Congress in 2000. This law allows 6,000 products of southern Africa to be exported tariff-free to the United States. Clothing exports from Lesotho in 2002 totaled more than $320 million and, for the first time, the number of employees in manufacturing was greater than that for government employment. Alluvial diamonds have been mined for many years in Lesotho, mainly by artisanal methods, but Lesotho is known to possess 33 kimberlite pipes in the Liqhobong Kimberlites Complex, located about 125 miles (200 km) northeast of Maseru. Lesotho also has deposits of coal and uranium but it is not known whether they are significant enough for commercial mining.

Lesotho Today

Although during the apartheid years Lesotho seemed to be an economic and political hostage to South Africa, since the fall of the apartheid regime in the early 1990s Lesotho has begun to diversify its economy, often with South African investment. Employment opportunities in South Africa, once looked upon as unwelcome economic dependence, now are sought after. Work in the mines, factories, and other industries in South Africa is bringing wealth back to Lesotho as well as providing employment for Lesotho nationals. Basuto men began to seek work on the mines of the Witwatersrand and in the Free State during the early years of the twentieth century as the mining sector developed. A permanent system of migration developed, and with it the dependence of Lesotho on the South African economy grew. Basuto families depend on wages earned by their kin working in South Africa for the payment of taxes and for the purchase of imported foods and merchandise.

In reality, labor is Lesotho's principal export. Each year about 35% of the active adult males are employed in South Africa's mines, cities, and farms. Migrant remittances and deferred pay generally exceed the value of exports and constitute the principal source of expendable income within Lesotho. Less than 10% of the Basuto are wage earners within the country, and 85% of the population is engaged in subsistence agriculture.

Lesotho is an ethnically homogeneous country that has faced obstacles in creating representative government but has succeeded in developing a unique blend of European and customary political organization as manifested in its constitutional monarchy. Most citizens of Lesotho have experienced a rise in their standard of living since independence, but such gains may increase with greater involvement in the world economy. In the short term, perhaps the greatest immediate hope for growth can be achieved through tourism. Lesotho's spectacular scenery, winter sports potential, and gambling facilities are all within a few hours drive from the Witwatersrand, Durban, and the Free State. If tourism is to be developed to its fullest, roads within Lesotho will need upgrading and extending. At present, the most scenic region has but a few third-class roads, which can be blocked by snow in winter and washed out by rain in summer.

SWAZILAND: THE LAST MONARCH

The Kingdom of Swaziland is Africa's last remaining absolute monarchy and one of the few in the world today. Although once a major form of governance around the world, monarchy has an anachronistic, and even paradoxical, air about it in the twenty-first century. In the early 1800s, the Swazi monarchy helped protect Swaziland. Later in the same century it perversely ignored the interests of the Swazi people and ceded the country to Europeans. Today some view the king as the keeper of tradition while, at the same time, others see him as the guardian of hereditary privilege and an impediment to real political progress for the majority of Swazis. There have been efforts to reduce the power of the king in the past; Swaziland adopted a constitution at independence in 1968 but it was suspended by the king in 1973. Political parties have been banned. Executive, legislative, and judicial powers are concentrated in the hands of King Mswati III.

Swaziland, smallest of the former High Commission Territories, appears to have the greatest potential for growth. It possesses a proven resource base, there is a small but growing industry, and there is some climatic and pedologic diversification. Moreover, the Swazi people, although not entirely homogeneous ethnically, share a common history (including successful resistance to Zulu aggression), an attachment to and identification with Swazi territory, and a rapidly growing degree of political consciousness and desire for progress. There are about 1.2 million Swazi in Swaziland proper, and hundreds of thousands live and work beyond the borders in an area of South Africa that formerly was part of the kingdom.

The distribution of the Swazi nation over an area far larger than Swaziland itself is a reflection of one of the major problems facing Swaziland today: the problem of land. The Swazi, although not conquered by the Zulu, were unable to prevent their powerful southern neighbors from claiming large portions of their territory. Subsequently, when contact was made with the first European invaders, Swazi chiefs carelessly ceded away vast tracts of land. Most guilty was Chief Mbandzeni (Umbandine), who by the time of his death in 1889 had assigned the entire country of the Swazi to Europeans. Despite an impressive military organization (the Swazi helped the British crush the Zulu in 1879), they offered little resistance to the Boers, who took their lands along the lower Komati River the following year. Nor did the Swazi oppose the Portuguese annexation of their land east of the Lebombo Plateau near the growing colony of Mozambique. By 1887 the Swazi had lost most of their winter pasture to Boers, and by the following year they had lost most of their mineral wealth to English prospectors (Davenport and Saunders 2000). Thus, the Swazi by the end of the nineteenth century had lost all they had successfully defended against the Zulu.

Having ceded their land and aware of their predicament as squatters on what was once Swazi territory, the people appealed to Britain for protection. The British, realizing the problems involved, refused to establish a protectorate over the Swazi when the appeal was made in 1893. Indeed, Britain was concerned over events on the South African Plateau at this time and was prepared to allow Swaziland to become a part of the South African Republic. Doubtless this would have happened, had the matter not been under review at the time of the outbreak of the Boer War; when the war was over, the British agreed (in 1902) to establish a protectorate over the Swazi. Thus, the Swaziland Protectorate came into being through a series of the kind of accidents that make history.

The establishment of the protectorate did not end the land problems of the Swazi people. With or without protectorate status, Swaziland had been ceded to European settlers. With the Swazi indicating their desire to remain outside a unified South Africa, it became necessary to define the international boundaries of the country. This was completed in 1907, while a specially

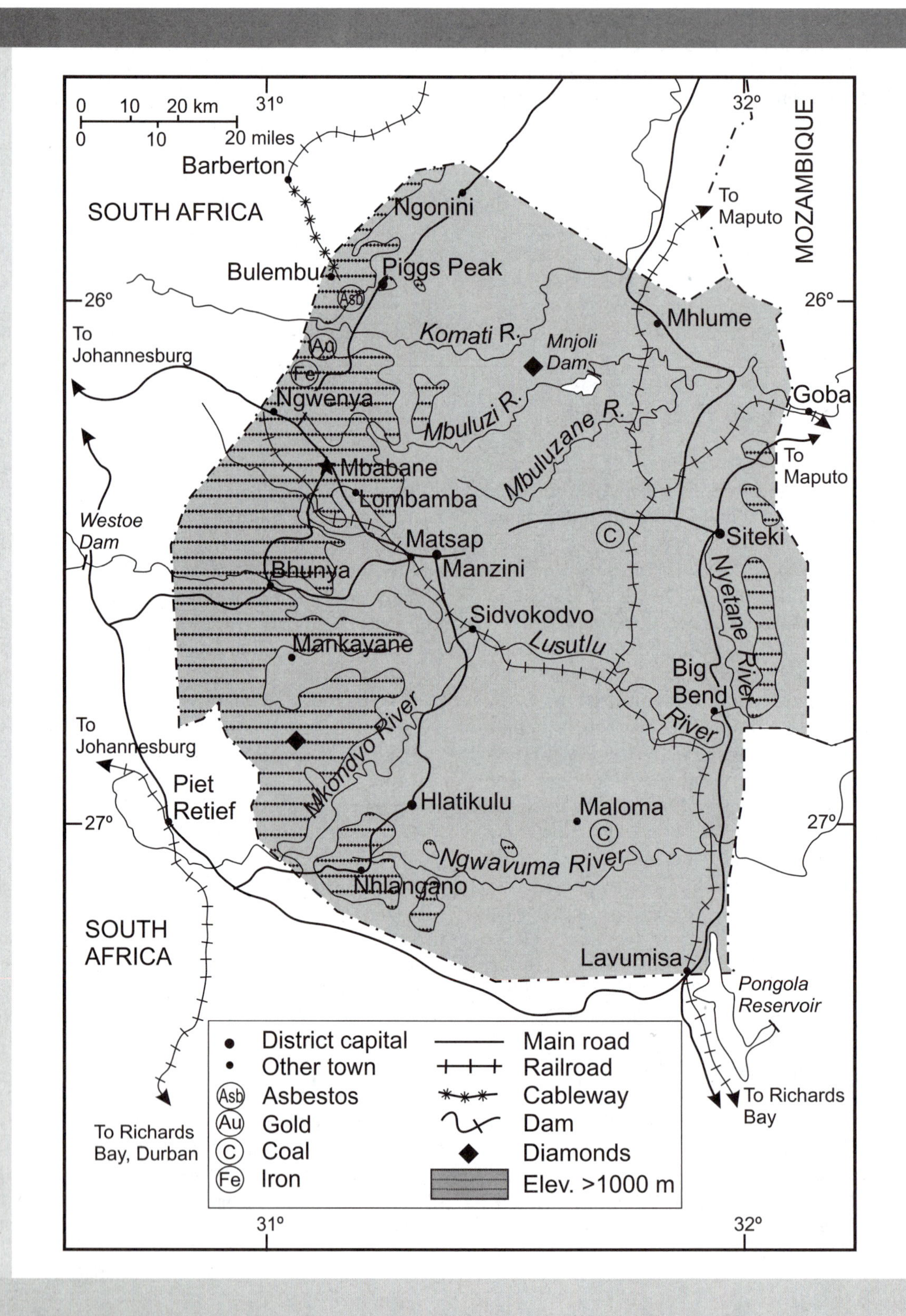

FIGURE 25-8 Principal towns, infrastructure, minerals, and relief of Swaziland.

appointed commission was attempting to unravel the complex of adjacent and overlapping concessions with the aim of returning to the Swazi people some of the land they once possessed. The commission in 1907 published a plan for the partition of the territory along lines that essentially form the basis of present-day land division in Swaziland. After the delimitation of its international boundaries, the area of Swaziland was found to be 6,704 square miles (17,364 km^2), all under concessions. The commission proposed that exactly one-third of the total land of each concession be returned to the Swazi nation, that the pieces of land thus deducted be consolidated as much as possible, and that these stretches of territory form the home of the Swazi people.

The Swazi, in return for this allotment of land, were given from July 1, 1909, to June 30, 1914, to leave the land remaining under concession and move to their own areas. Any Swazi able to make private arrangements with the European landowners to remain on the ceded lands could do so. In fact, most Swazi were able to stay on the ceded land, and no major migration of Swazi areas resulted. The important achievement of the commission, nevertheless, was the arrangement by which the Swazi nation obtained tribal lands with which the people could identify themselves, on which the chief and elders could live, and where Swazi practices of landownership, cattle and goat herding, and so on, could be maintained.

In 1914 the population of Swaziland was perhaps somewhat over 100,000, with about 1,000 European settlers, and, at that time, population pressure on the land was much less than it is today. The commission dealing with the territory's partition, in recognition of a future greater need for land, developed a mechanism by which land could be purchased from the white landowners for inclusion into the Swazi nation's areas, or for Crown land which at a later stage might be vested in the Swazi nation. The growth of the Swazi people outstripped the increase in available land. In 2005, with the population of Swaziland estimated to be 1.2 million, about 60% of the total area comprises Swazi nation land, the remainder being held on a freehold or leasehold basis. The patchwork quilt pattern persists, and on the Swazi nation land lives 71% of the total population. Population densities are highest in the cooler, relatively more humid western third of the country where elevations are above 3,280 feet (1,000 m) (Figure 25-8).

In the mid-1970s the Swazi population was estimated to be about 500,000 people and total head of livestock were estimated at about 950,000. The cattle population was estimated to be 200,000 head above what could be supported by the land. By 2002 the human population had more than doubled, but livestock populations remained relatively constant. There has been continuity and change with regard to livestock—a principal subsistence base and export of Swaziland—with regard to total tropical livestock units, but larger stock are becoming a smaller share of the livestock population (Table 25-2).

TABLE 25-2 PRINCIPAL LIVESTOCK IN SWAZILAND IN 1975 AND 2001.

	1975		2001	
Stock	**Frequency**	**Percent**	**Frequency**	**Percent**
Cattle	622,000	65	615,000	54
Sheep	35,000	4	32,000	3
Goats	261,000	27	445,000	39
Pigs	18,000	2	34,000	3
Horses/Asses	17,000	2	16,000	1
Total head	953,000	100	1,142,000	100
Total Tropical Livestock Units	670,400		682,100	

Source: FAO (2004).

The government of Swaziland has made considerable effort to improve agricultural practices but has been faced with challenges: a feudal land tenure system, soil erosion, and prolonged and recurrent drought. Security of land tenure is a problem faced in many parts of Africa where private property in land is forbidden or discouraged. Private ownership of land is impossible in many areas of Africa where land tenure rights are restricted to "use" rights (or usufruct) and in some socialist or formerly socialist African states where private property is viewed as counterproductive to building communism. Constraints on landownership in Swaziland are of the former kind: Swazi nation land (SNL) is owned by the monarch in trust for the Swazi nation and cannot be owned outright by individuals. Individuals can have use rights to cultivate it. Each adult male has the right to land under the *khonta* system (Mushala 2000). The khonta system is very pro-family: land is allocated to married couples or sometimes to a single female through her son. A man cannot get khonta land without a wife and a woman cannot get land without a husband (or a son, if single). Those who receive khonta land are required to pledge allegance to the chief who controls the land. Land is managed by chiefs in the "interests" of their Swazi community and if, for example, a farmer dies or moves away, the use rights to the land farmed by that person will be transferred to another member of his family who remains in the village or to another family in the village. Thus there is no security of land tenure for the individual, and any improvements a farmer might make on a piece of land may be taken away by a chief. With population growth, the khonta system has some perverse consequences: the plots of land allocated get smaller and more fragmented.

In addition, there are tribal customs that reduce the number of cattle taken each year for slaughtering; these include their use as a form of currency, the prestige associated with numbers (rather than quality), and their inclusion in the customary bride price. Nevertheless, some Swazi have accepted such principles as contour farming, fallowing, and other forms of soil conservation, and are farming for profit rather than subsistence. Some model farms run by Swazi earn as much money as do many farms owned by the Europeans. Since, however, the land initially given to the Swazi nation through the partition of the concessions was by no means Swaziland's best land, opportunities for this sort of development have been few. It is considered vital that other Swazi adopt the necessary practices to prevent the degree of soil erosion from which large parts of Lesotho has suffered, and local technicians are constantly working to effect these changes. Another mark of progress is the virtual elimination of the dreaded cattle disease, East Coast fever, by the enforcement of a requirement for regular dipping of all cattle. Perhaps the greatest challenge to Swazi farmers has been drought. The year 2003 was the fifth consecutive year of drought in the country; especially hard hit have been the eastern lowlands. The World Food Program of the United Nations has assisted up to 25% of the Swazi population in the hardest of the last few years.

The division of land toward the end of the nineteenth century marked the beginnings of Swaziland's social and economic dualisms, which have persisted to the present. Indeed, over the years they have intensified and their spatial manifestations are most pronounced (Maasdorp 1976). Today there are islands of modernization and development surrounded by broader areas of traditionalism and poverty. The modern economy largely represents manufacturing, mining, agricultural, and commercial interests, while the traditional economy is characterized by subsistence cultivation and pastoralism. Real national viability can be achieved only with the complete integration of these two sectors through the structural transformation of the social and economic systems. Little attempt was made to integrate them until the 1960s, and progress accelerated during the 1980s. Nevertheless, over 80% of Swazis are subsistence farmers and herders (80% own no land); many of the rest work in the growing urban formal economy and in government.

During the colonial period, and especially in the earlier years, Swaziland was an impoverished dependency lacking both the infrastructure and government encouragement needed for development. Britain refrained from providing development capital since there was a possibility that Swaziland would be incorporated into the Union of South Africa; the African state, in turn, would not undertake large-scale investment without some guarantee of incorporation. The Swazi themselves were totally deficient in the skills and capital required for modernization, while the revenues derived from the small quantities of gold, tin, tobacco, cotton, and cattle produced by the few

European settlers were also insufficient to transform the economy.

After World War II, Swaziland experienced considerable economic diversity and growth when Britain belatedly accepted its responsibilities for the promotion of economic development, and it became clearer that incorporation with South Africa would not materialize and independence would be achieved. The British Commonwealth Development Corporation, in partnership with private enterprise, established timber plantations of exotic softwoods and eucalyptus in the highveld, erected a pulp mill at Bhunya, and established irrigation schemes and a sugar mill in the lowveld at Mhlume (Figure 25-8). This government initiative encouraged additional investment of private South African and British capital in the citrus, sugar, forestry, and cattle industries, and saw the improvement of roads, education, health facilities, and a greater participation of Swazi in the market economy. In 1964 a railway was opened between the new iron ore mine at Ngwenya and Maputo in Mozambique, which provided Swaziland with an outlet for its exports and permits easier access to petroleum products and other bulky imports. Today industry accounts for almost 45% of the Swazi GDP, but mining contributes less than 2% and fewer than a thousand Swazi jobs (Coakley 2000). Diamond mining ceased in 1996, and asbestos mining stopped in 2000. The government has liberalized the taxation of mining to attract foreign investment. The contribution of agriculture to GDP is now less than 20%.

Swaziland has benefited from its proximity to South Africa, which serves both as a source of investment capital and as a market for its products. As a result, Swaziland's economy is more diversified than those of Lesotho and Botswana. The food processing and small manufacturing industries have grown since independence. The Swazi government has encouraged the development of export-led growth by investing in export industries and by attracting foreign investment (mainly South African), and exports account for 65% of GDP (U.S. Department of Commerce 2002). Other exports are wood pulp (Figure 25-9), timber, and wood products, followed by sugar and minerals (mainly iron ore and asbestos). Meat, fruits, rice, cotton, and tobacco are also exported.

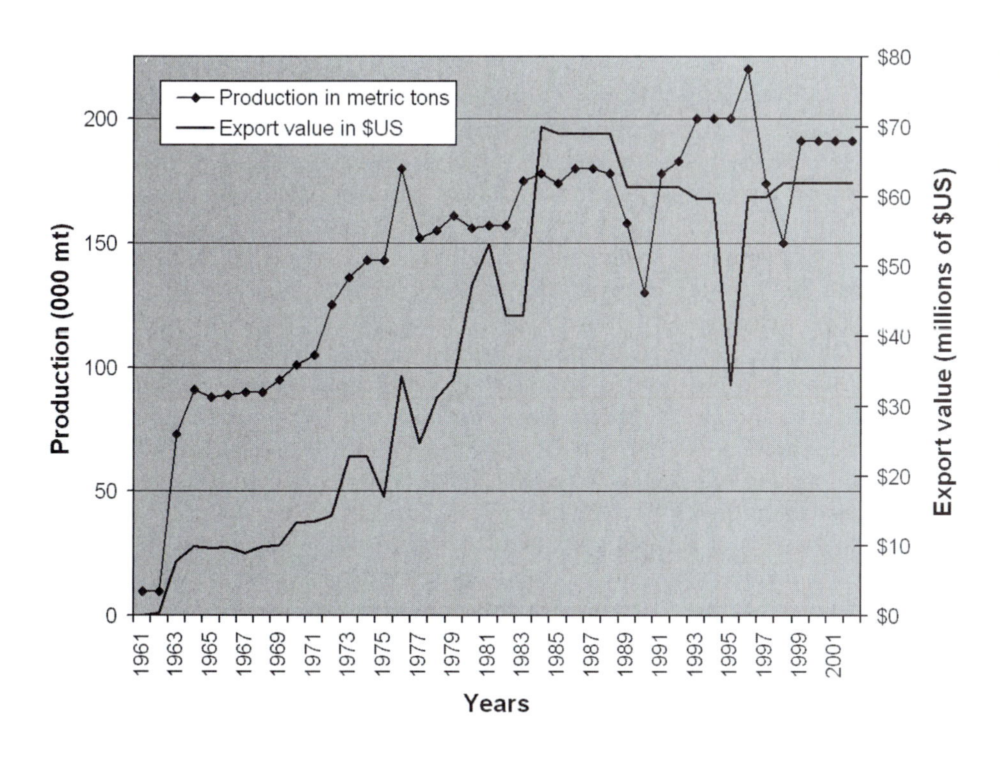

FIGURE 25-9 Production and export value of wood pulp in Swaziland, 1961 to 2002.

There is only one major export partner, South Africa, which imports close to 75% of Swaziland's products (soft drink concentrates, textiles and apparel, coal, wood pulp for paper manufacture). The European Union and Mozambique are minor export partners. South Africa provides about 90% of all Swazi imports (mainly machinery, transport equipment, clothing, and general merchandise). It is no exaggeration to say that the Swazi economy is just a small part of the South African economy.

Swaziland has an unusually good transportation infrastructure and is well linked to its South African neighbors and to the world. Its railroads are well run and profitable and link the industrial and urban areas of Swaziland to the South African ports of Richards Bay and Durban, and Maputo in Mozambique. Through Beitbridge, in Zimbabwe, Swaziland is connected to Zambia, Malawi, and the Democratic Republic of the Congo.

The economy is concentrated in four core regions. The national core is the most diverse and includes the Usutu Forest, the Bhunya pulp mill, the Malkerns Irrigation Scheme, the Ngwenya iron ore mine, Manzini (75,000) and its industrial satellite township, Matsap, and the administrative capital, Mbabane (69,000). Matsap has several light industries including a brewery, a cotton gin, and an abattoir, and a one-stop dry port and international airport. The national core accounts for almost half of Swaziland's primary and secondary output, and contains the legislative capital, Lobamba, and more than half of the territory's European population (35,000). Subsidiary resource core regions are located at Big Bend and Mhlume, where sugar, citrus, and rice are produced on commercial farms, and around Piggs Peak, where a vast private afforestation scheme is located (Figure 25-8). Each region has become an important source of local employment and each is linked to external markets. Surrounding these regions of growth lies traditional Swaziland, encompassing 85% of the total area, possessing three-fourths of the total population, and producing only 10% of the primary and secondary output. Its functional linkages with the cores are low, and it represents the subsistence sector of the dual economy.

Swaziland has several ingredients required for national economic viability but is challenged by the lack of political pluralism anchored in the anachronistic institution of monarchy, a feudalistic land tenure system, a dry environment, and a high HIV infection rate. To maintain the level of development it has already achieved, the country must continue to attract foreign direct investment for export-led manufacturing. In many ways, the interest of the international community hinges on the political climate of the country. Lavish new palaces built for the many wives of the king do not encourage donor countries to provide food aid during drought nor do they attract investors. Irredentist claims made by the Swazi royal family for land in Mpumalanga and KwaZulu-Natal provinces of South Africa may serve only to alienate their powerful neighbor. Mounting internal and external pressures strongly suggest that King Mswati III will be the last monarch of Swaziland. That will only be for the better of the people and the country.

SOUTH AFRICAN INFLUENCES AND CONTROLS OVER THREE SMALL NEIGHBORS

South Africa's economic interests in and influences upon Botswana, Lesotho, and Swaziland have been strong for over a century, and they are likely to remain so because of the combined forces of location, history, politics, and economics. The political, economic, and cultural geographies of the four countries have been inextricably interwoven into one cloth. Over the years the three smaller countries have become inseparable economic dependencies within the South African sphere of influence. They depend on South Africa for manufactured goods, employment for their surplus labor, and markets for many of their exports produced with South African capital and technology; they are tied to the South African economy by the South African Customs Union and by the South African rand. When the three territories chose to remain outside the projected Union of South Africa in 1909, an economic arrangement (the customs union) was agreed upon, which was to tie them almost inextricably into the South African economic framework. It meant that all goods leaving the three territories had to travel through South Africa and be transported by South African carriers. Thus transport links between South Africa and the territories developed, and South African quality control standards were introduced in marketing. Furthermore, the customs revenues were shared according to a formula that favored South Africa and prejudiced the junior members. In 1969, and during the 1980s, the agreement was renegotiated on a more equitable basis, but the same linkages remain. A free South Africa has been a transforming force on the countries that were once called its "hostages." There appear to be greater economic benefits, at least for Lesotho and Swaziland, in closer political integration with rich neighbor South Africa that could help these small countries transcend the boom-and-bust vulnerability of their small economies. Botswana, an African success story, has capitalized on its location close to the South African core and its location midway between South Africa and the deepwater ports of Namibia. Its wealth is natural resources, political stability, cultural cohesion, and responsible leadership should continue to help prospel Botswanans to higher levels of human development in the future.

BIBLIOGRAPHY

Acemoglu, D., S. Johnson, and J. Robinson. 2001. *An African Success Story: Botswana.*

Best, A. 1970. Gaborone: Problems and prospects of a new capital. *Geographical Review*, 60(1): 1–14.

Best, A., and H. J. de Blij. 1977. *African Survey*. New York: Wiley.

Centre for Economic Policy Research, Discussion Paper Number 3,219. London: CEPR.

Coakley, G. 2000. *The Mineral Industry of Swaziland.* U.S. Geological Survey Mineral Report. Reston, Va.: http://minerals.usgs.gov/minerals/pubs/country/.

Coakley, G. 2001. *The Mineral Industry of Botswana.* Reston, Virginia: U.S. Geological Survey Mineral Report. Reston, Va.: USGS. http://minerals.usgs.gov/minerals/pubs/country/.

Darkoh, M. 2000. *Desertification in Botswana.* RALA Report Number 200. Reykjavik, Iceland: Agricultural Research Institute.

Davenport, R., and C. Saunders. 2000. *South Africa: A Modern History*. New York: St. Martin's Press.

FAO. 2004. FAOSTAT. Rome: Food and Agricultural organization of the United Nations.

Hubbard, M. 1986. *Agricultural Exports and Economic Growth; A Study of the Botswana Beef Industry*. London: Kegan Paul.

IRIN. 2004. Botswana: Beef industry woes expected to continue. *IRIN News*, January 16, 2004.

Integrated Regional Information Networks, United Nations Office for the Coordination of Humanitarian Affairs. Nairobi, Kenya: United Nations. http://www.irinnews.org.

Kearney, A. J. 2004. Measuring globalization: Economic reversals, forward momentum. *Foreign Policy*, March/April. http://www.foreignpolicy.com/.

Maasdorp, G. 1976. Modernization in Swaziland. In *Contemporary Africa: Geography and Change*, C. G. Knight and J. L. Newman, eds. Englewood Cliffs, N.J.: Prentice-Hall, pp. 408–422.

Mulale, K. 2002. The challenges to sustainable beef production in Botswana: Implications on rangeland management. Paper presented at the 17th Symposium of the International Farming Systems Association, Lake Buena Vista, Fla., November 17–20, 2002. http://www.conference.ifas.ufl.edu/ifsa/posters/Mulale.doc.

Mushala, H. M. 2000. Some perspectives on poverty and land degradation on Swazi National Land. In *Perspectives on Poverty in Swaziland.* Workshop organized by the Organization for Social Science Research in Eastern and Southern Africa (OSSREA), December 7–9, 2000, Mbabane, Swaziland. http://www.ossrea.net/nw/swaziland/Swasiland-dec00.htm#P13_127

Population Reference Bureau. 1976. *World Population Data Sheet.* Washington, D.C.: Population Reference Bureau.

Population Reference Bureau. 2003. *World Population Data Sheet.* Washington, D.C.: Population Reference Bureau.

Samatar, A. I. 1999. *An African Miracle: State and Class Leadership and Colonial Legacy in Botswana Development.* Portsmouth, N.H.: Heinemann.

Smit, P. 1967. *Lesotho: A Geographical Study*. Pretoria, South Africa: Africa Institute.

Smit, P. 1970. *Botswana: Resources and Development.* Pretoria, South Africa: Africa Institute.

U.S. Department of Commerce. 2002. *Country Commercial Guide FY2002: Swaziland.* Washington, D.C.: U.S. Department of Commerce.

Whiteside, M. 1997. *Encouraging Sustainable Family Sector Agriculture in Botswana.* Gloucester, U.K.: Environmental and Development Consultancy.

CHAPTER 26

Zimbabwe

A Failure in Partnership

Although the government of Zimbabwe since independence in 1980 has been socialist in ideology, it has governed a capitalist economy with an export-oriented commercial agriculture and well-developed mining and manufacturing sectors (Binns 1994). Until recently, a vibrant, export-oriented economy has contributed to a relatively high GDP and a general level of prosperity not found in most African countries. Yet Zimbabwe, once a moderately developed country, has experienced a decline in its human development over the last several years, and chances of near-term recovery do not appear good. The economy has been contracting, and there has been civil strife, with widespread food shortages and refugees fleeing from city and from countryside.

At the crux of the current problems in Zimbabwe were the unequal distribution of land between the European settlers and black Zimbabweans, the militant Marxist roots that the ruling party had not entirely escaped when it achieved legitimacy in public office, and President Robert Mugabe's desire to hang onto power by any means possible. Since Mugabe came to power in 1980, he patently failed to tackle the land reform issue except in the most perfunctory way until his hold on power was questioned. In 2000, in a hasty move to undercut political opposition after multiparty elections indicated that its 20-year reign would probably end, the Mugabe government chose to completely expropriate the settler-farmers rather than compensate them, using intimidation and violence against the rest of the political opposition to "win" the 2002 elections. The economy came to a halt.

But Mr. Mugabe did not have to cut off his nose to spite his face; addressing inequality in Zimbabwe did not have to become a model of what is to be avoided in African land reform. Several other African countries with similar land problems are watching Zimbabwe very closely. Experience with land reform in Kenya has shown that settler estates can be transferred successfully to African ownership in a largely peaceful and productive manner. Ironically, before the Mugabe government took preemptive and violent action, the opposition party had set forth an attractive program to peaceably address the land reform issue. We will examine the sociogeographic roots of this problem in this chapter.

Zimbabwe, once a stronghold of white supremacy and a theater of bitter racial war, occupies a geographically well-defined area in southern Africa. In many ways it is a natural fortress, bounded on the north and south, respectively, by the Zambezi and Limpopo rivers, and on the east by the Great Escarpment and the Inyanga Mountains. To the west lie the semiarid plains and desert wastelands of Botswana, themselves separated from the Atlantic by the tablelands and deserts of Namibia. Within this compact and landlocked plateau country of 150,803 square miles (390,580 km^2), European settlers and indigenous Africans have failed to produce an integrated society where there is equality in the social and economic spheres.

The Zambezi has played an important role in the settlement and development of Zimbabwe (Figure 26-1). The river itself has dangerous rapids and waterfalls, and its steep valley sides are hot, disease ridden, and densely vegetated. In the precolonial past, it hindered the southward expansion of the Shona and the northward expansion of the Ndebele and Sotho peoples, and it

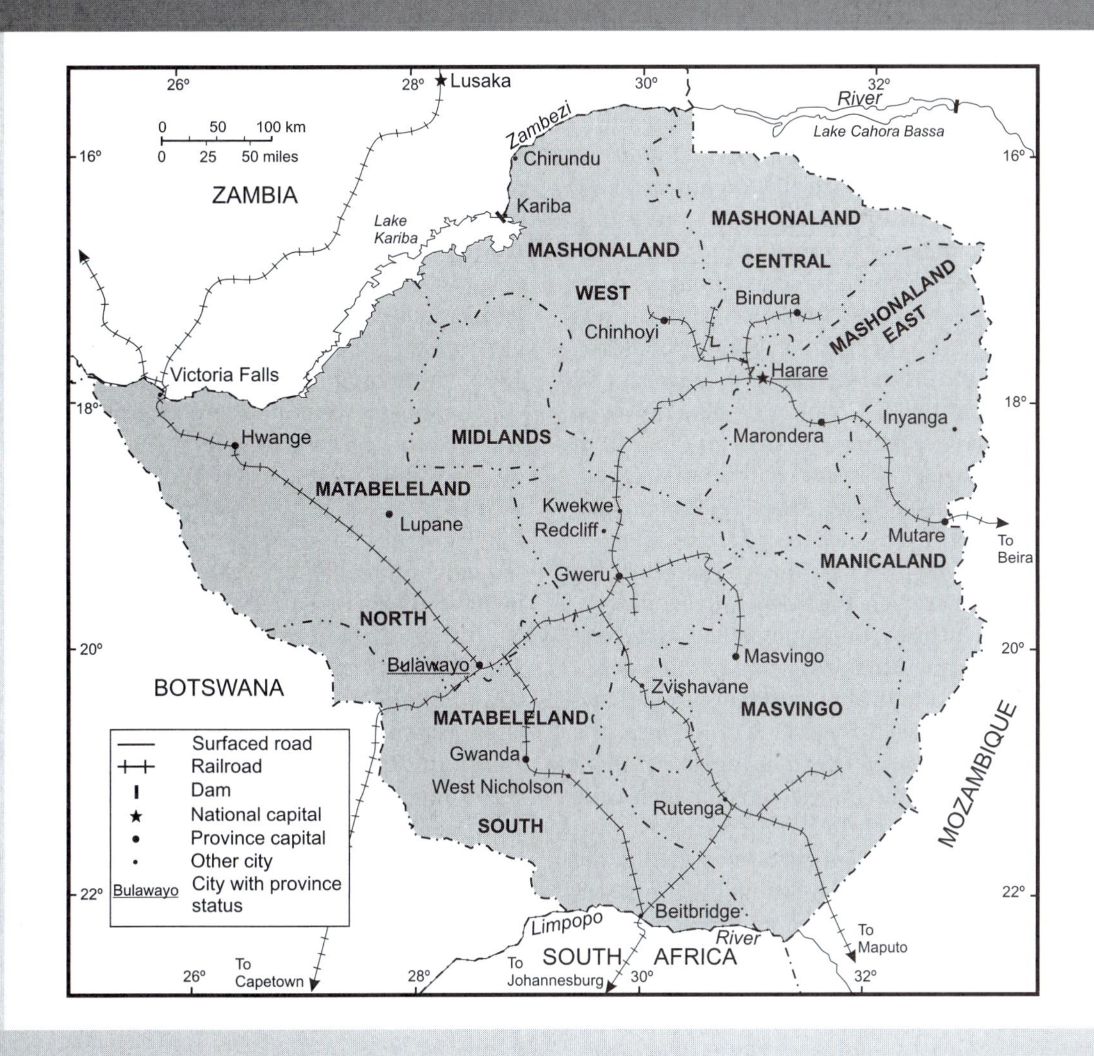

FIGURE 26-1 General identification map for Zimbabwe.

marked the southern limit of the slave trade (Thomas 1997). During the colonial era, when Zimbabwe was known as Rhodesia, the Zambezi was the boundary between British protectorate administration and limited European settlement in Zambia and Malawi and more direct colonial rule and extensive settlement in the south. The river was also a divisive factor in the former Central African Federation. Since the 1960s, the Kariba Dam on the Zambezi has produced electricity for Zambia and Zimbabwe.

FROM COLONY TO MULTIPARTY STATE TO DE FACTO ONE-PARTY RULE

From 1890 until 1980 European settlers had control over the export economy and strategic resources, dominated the parliament, and struggled to preserve their privileged way of life in what was then known as Rhodesia. With independence in 1980, a single African party has monopolized power in Zimbabwe, named after Great Zimbabwe, a state founded in the twelfth century by Shona peoples who traded with the Swahili and, through them, in the Indian Ocean long-distance trading network. The Shona-speaking peoples comprised a loose federation, each group occupying a geographically distinct area. In the early nineteenth century, they were dominated by Nguni-speaking peoples, the Ndebele, who migrated from the south. Controversy surrounds the reason for the migration. The conventional explanation has cited drought, competition for scarce resources, violence, and distress migration for a plethora of Nguni-speaking peoples in southeast Africa, the Zulu, Ndebele, Sotho, Swazi, and Xhosa (Shillington 1987). Cobbing (1988), however, suggests that slave raiding of the interior from Mozambique's Delagoa Bay set the Nguni-speaking peoples to flight. Later, the Ndebele, led by Chief Lobengula, were driven by northward-migrating Dutch colonists called Boers (see Chapter Twenty-Four) to seek refuge on the Zimbabwe plateau where the Ndebele live today.

In 1859 the Ndebele, sometimes called Matabele, permitted the London Missionary Society to open a mission at Inyati, and soon afterward European adventurers and hunters entered the region. Following the accession to the throne of king Lobengula in 1868, European activity was discouraged in Matabeleland, although it continued further north and east in Mashonaland (Figure 26-1). Gold was discovered in the vicinity of Gweru, and by the 1880s many Europeans in South Africa believed that the area north of the Limpopo was rich in precious minerals, possibly a "second Witwatersrand."

International interest in the region quickened as the scramble by European powers for African territory gained momentum. German imperialists competed with the Portuguese in their search for a continuous band of territory across southern Africa, while the British government was persuaded to maintain and extend its hold on Bechuanaland (now Botswana) to provide a corridor north between the South African Boers and the Germans. Cecil Rhodes saw in the region its strategic qualities and persuaded the British High Commissioner for South Africa to obtain Lobengula's promise that the Ndebele would not enter into agreement with any foreign power without British approval. Lobengula agreed, and by placing his territory in British hands, he prepared the way for British expansion. The treaty of 1888 granted the British complete and exclusive charge of all metals and minerals in Lobengula's kingdom, but not the right to possess land. In the years that followed, land was expropriated and its division between colonizer and colonized became one of Rhodesia's major political issues.

The British South Africa Company was formed in 1889, and under its charter Rhodes made preparations to occupy the territory north of the Limpopo River, in particular, Mashonaland. At the head of a small group of white settlers and adventurers called the Pioneer Column, Rhodes pushed through the Bechuanaland Protectorate and onto the Zimbabwean plateau (highveld) east of the Ndebele stronghold. By September 1890 the column had established Fort Salisbury (Harare today), where the settlers disbanded, and each was free to select 3,000 acres (1,210 ha) of farmland provided under the charter. Additional land was later secured by treaty between Rhodes and Lobengula and, following the chief's defeat in battle in 1893, European settlement spread into Matabeleland, especially around Bulawayo, Lobengula's abandoned capital.

In 1895 the territory was named Southern Rhodesia and was administered by the British South Africa Company. By 1900, Bulawayo had become Rhodesia's largest town and chief railway center. The line from South Africa through Bechuanaland had reached it by 1897, and 5 years later the line reached Salisbury, already linked by rail with Beira in the then Portuguese Mozambique (Moçambique). The line was extended north from Bulawayo to Hwange in 1903 and crossed the Zambezi at the Victoria Falls early the following year. Thus the basic physical infrastructure of the colony was completed within 15 years of the settlers' arrival, and the stage was set for both colonial consolidation and expansion.

But economic prosperity and progress were slow to materialize as droughts, outbreaks of rinderpest, and East Coast fever wrought havoc on ranching and other enterprises. Recurrent uprisings and rebellions by both the Ndebele and the Shona brought death and destruction to many areas, and it was not until the Shona were violently suppressed in 1897 that widespread European occupation was assured. Discontent with the British company's power and administration was common, and the company itself soon realized that Rhodesia's gold resources were not nearly as extensive as had been supposed. Knowing that gold would not fill its coffers, the company decided to protect and improve its assets by trying to build up a larger European population through the encouragement of land settlement and farming, especially along the line of rail. It also relaxed the law requiring that only enterprises in which it had an interest could undertake mining operations. The result was a surge in prospecting, the opening of many mines and mineral-related industries, and steady increase in registered farms.

In 1923 a referendum was held among the European electorate on whether to seek "responsible government" as a self-governing colony of Britain or to ask for union with South Africa. By a vote of 8,774 to 5,989, Rhodesia became a self-governing colony under whose constitution Britain retained a right of veto on discriminatory and constitutional matters (Bowman 1973). Although this prevented the removal from Africans of their largely theoretical right to the franchise, it did not prevent the establishment of a racially stratified and segregated society.

After the Second World War, Rhodesia prospered as minerals were exploited and industry developed. In 1953 it entered into political union with Northern Rhodesia and Nyasaland (Zambia and Malawi today). This federation, lasting 10 years, brought great economic benefits to Rhodesia, especially to the European population. The demand for textiles, automobiles, and other consumer goods greatly expanded; tobacco production surged; the communication network around the national core developed; and major undertakings, such as the Kariba hydroelectric project, were begun. Rhodesia profited from the copper revenues of Zambia, the labor from Malawi, and markets throughout the federation. Its contributions to the development of Zambia, Malawi, and its own African peoples were far less evident. But federation produced a politically conscious African working class in the urban areas, and African nationalist movements gained momentum, spurred on by independence elsewhere on the continent.

Yet federation failed. The reasons were many, but especially important were the lack of effective spatial and economic integration and the refusal by Rhodesia's whites to share with their African fellow citizens the profits of their labor and the power of government. In 1962 a white supremacist political party, the Rhodesian Front Party (RFP), came into power led by Ian Smith. Following the dissolution of federation, the Rhodesian government resumed the powers that had been transferred to the federal government in 1953 and fully expected to achieve independence from Britain, as Zambia and Malawi had in 1964. But Britain was not prepared to give independence before guarantees were incorporated into the constitution for majority rule, elimination of racial discrimination, and for immediate improvement in the political status of black Rhodesians. Under these new political circumstances, Rhodesia's white minority government refused to accept the British conditions and on November 11, 1965, Ian Smith's government unilaterally declared independence and a state of emergency. A new phase of Rhodesian history was born.

The period of white minority "independence" lasted 15 years—a period of political uncertainty and guerrilla warfare, particularly after 1972. International sanctions and a trade embargo imposed on Rhodesia by the United Nations increased the economic hardship faced by all Rhodesians. The change in government in Portugal in 1974 and the coming of majority rule in its colonies probably was the last straw. Rail links, except through South Africa, were closed and black nationalist guerrillas used Zambia and Mozambique as staging areas for strikes deep into Rhodesia. The white minority government agreed to majority rule and began meeting in Geneva with formerly exiled black nationalist leaders such as Bishop Abel Muzorewa, Robert Mugabe, leader of the Marxist Zimbabwe African National Union (ZANU), whose constituency was mainly among the Shona, and Joshua Nkomo, leader of the Zimbabwe People's Union (ZAPU), which drew its supporters mainly from the Ndebele people.

In 1979 an accord known as the Lancaster House Agreement called for a cease-fire, preindependence elections, a period of transition under British rule, and a democratic constitution providing for majority rule that would protect minority rights. Of particular importance was the stipulation (to be effective for 10 years—to 1990) that landowners would not have their land confiscated by the government and that any transfers that did take place would be on a "willing seller–willing buyer" basis (Human Rights Watch 2002). During the negotiations that resulted in the Lancaster Agreement, it was agreed that the country would be called Zimbabwe. Sanctions were lifted in late 1979. Robert Mugabe and the Zimbabwe African National Union (ZANU) were elected to form the first government. Zimbabwe was formally granted independence from Britain on April 18, 1980.

Tensions that surfaced after independence between the supporters of Joshua Nkomo (ZAPU) and Robert Mugabe (ZANU) culminated in several years of fighting in Matabeleland, Nkomo's base. Nkomo was dismissed from his post by Mugabe, who accused the ZAPU leader of conspiring to overthrow the government. Mugabe's government maintained the state of emergency declared under the Smith government and sent in the army to put down the rebellion in Matebeleland, the traditional area of support for ZAPU, at a cost of thousands of Ndebele lives. His main rival out of the way, Mr. Mugabe further consolidated his power in 1987 with the merger of ZANU and ZAPU as ZANU-PF and the amendment of the constitution to permit himself to become president of Zimbabwe.

In 1990 ZANU-PF won 117 of 120 seats in parliament—a de facto one-party state. Ten years later, Mugabe became embroiled in a foreign adventure by sending Zimbabwean troops to the Congo in support of Laurent Kabila. There have been charges that the cash-strapped Mugabe expected to be rewarded with mineral riches, but his involvement has drained the Zimbabwean economy (Walker and Clark 2000). Popular opposition to ZANU-PF rule grew as the economy declined. The party won parliamentary elections held on June 24 and 25, 2000, a victory attributed to well-documented use of violence and voter intimidation by Mugabe's forces. The Movement for Democratic Change (MDC), a new, prodemocracy party led by Morgan Tsvangirai, however, had seriously threatened Mugabe's hold on power. The MDC has widespread black and white support, particularly among the educated, and it made significant gains, threatening ZANU-PF's hold on power. Specifically, the MDC won 57 of the 120 seats contested in the election of 2000, while the ZANU-PF seats dropped from 117 to 62. Clearly, Mugabe was losing his grip. But it was over the land issue that the conflict finally crystallized, and Mugabe was able to defeat the opposition. ZANU had won by securing the most votes in rural areas (Johnson 2001), but the cost, ironically, was deepening economic decline.

LAND APPORTIONMENT AND AGRICULTURE

Prior to European colonization of Zimbabwe, land was owned and occupied according to customary law, which generally meant communal ownership and grazing rights and the extensive production of food crops such as maize, sorghum, and beans. There was adequate land for all, the ratio of people to land was low, and a particular plot of land was simply abandoned for another plot elsewhere once the soil became exhausted. The major natural limitations to people and livestock were the broken terrain and excessive rainfall in the Chimanimani and Inyanga mountains, high temperatures, and low annual rainfall (<30 in.; 762 mm) in the Zambezi Valley, even lower rainfall (<16 in.; 406 mm) in the southeast lowveld, and recurrent and widespread disease such as East Coast fever and sleeping sickness (Figure 26-2). The most productive soils, reliable rainfall, and nutritious grasslands occurred in a broad belt stretching along the highveld northeast to southwest through the central regions.

These customary practices and areas of production were dramatically altered by Cecil Rhodes's Pioneer Column and the British South Africa Company. Each pioneer (settler) claimed up to 3,000 acres (1,210 ha), while the company both sold and leased large areas of land to incoming settlers and companies, and rewarded each member of the victorious columns of the Matabele campaigns (1894) with 6,000 acres (2,420 ha) of land (Kay 1970). Such actions prompted additional African uprisings and eventually led to the establishment of so-called native reserves, which were supposed to give the African in Rhodesia some security and protection from the settlers. Beyond these reserves, land was held by the company or designated for European use, although a small amount was available for African purchase.

In practice, the native reserves allowed the white administrators to shunt Africans to poorer, drier lands in the same way that Indian reservations did in North America or South African "homelands" did in South Africa under apartheid. By 1899, 16 million acres (6.5 million ha), or 15.6% of the territory, had been allocated to the white settlers, although much of it remained unoccupied and undeveloped (Griffiths 1985). In the early years of the twentieth century, the Native Reserves Commission took more and more land from Africans to reduce the competition white farmers, eager to supply the growing towns and mining centers, faced from African farmers. Furthermore, by the 1930s the government purchasing board for cereals had established regulations that discriminated against African-produced maize.

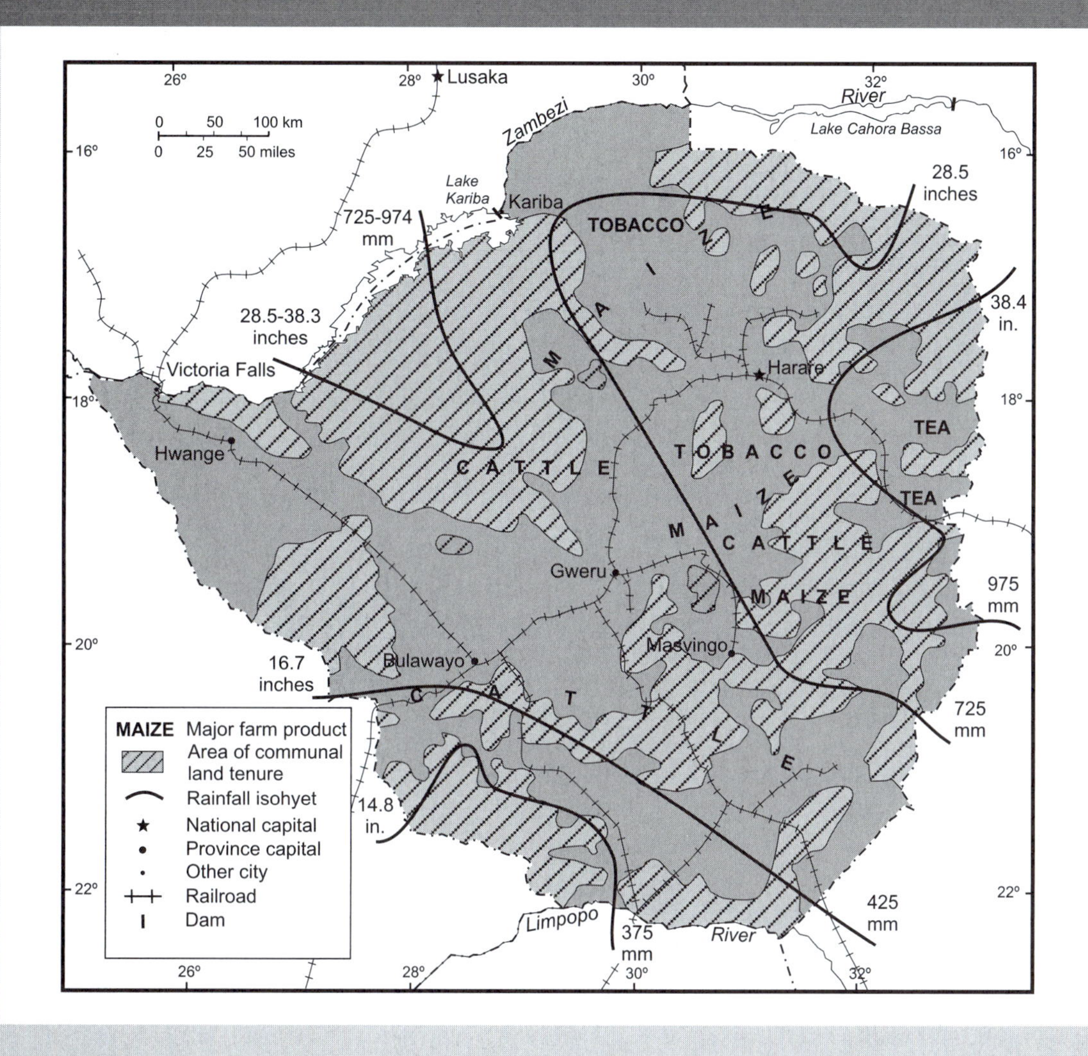

FIGURE 26-2 Population of Zimbabwe by district, with overlay of communal areas' wards.

In 1930 Rhodesia passed the Land Apportionment Act, which, with its various amendments, was the cornerstone of Rhodesia's residential segregation policy. The original act reserved half the colony for Europeans (48 million acres; 19 million ha) and one-third for African occupation (32 million acres; 12.9 million ha); the balance went unassigned. The white lands in general occupied the core of the plateau around the major cities, where the soils are fertile, rainfall generally reliable, and mineral wealth exceptional. The fringing African lands were of poorer quality, less accessible to urban centers, and spatially more fragmented. The act also prohibited Africans from growing certain cash crops, most notably tobacco. After World War II, higher prices for tobacco encouraged European immigration and settler farming. From 1945 to 1950, for example, 45,000 Europeans immigrated to Rhodesia, bringing the white population from 89,000 to 125,000 (Binns 1994).

During the 1950s and 1960s the amount of African land was increased to 46%, while white-owned land

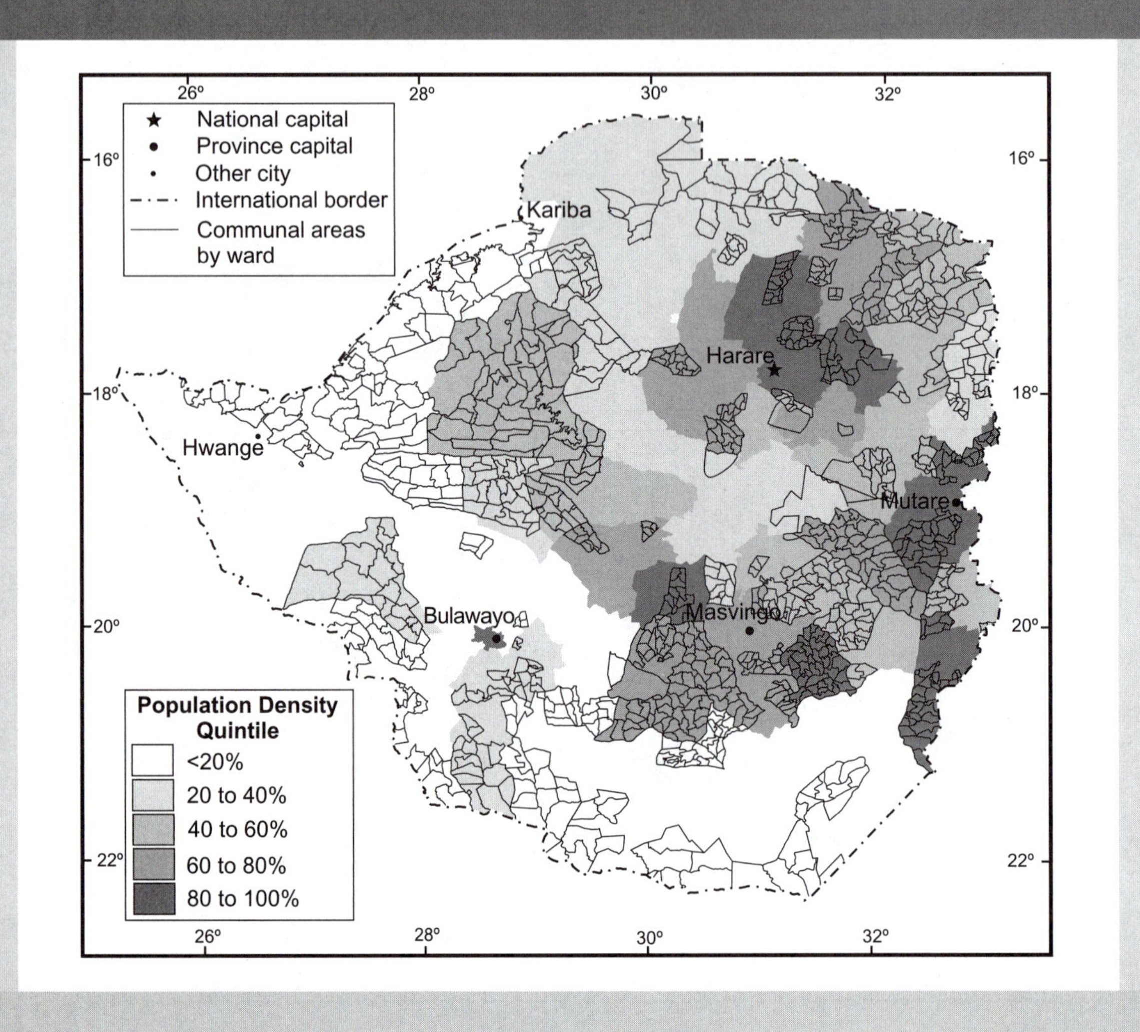

FIGURE 26-3 Land use, infrastructure, and precipitation in Zimbabwe.

was reduced accordingly. In 1969 the Land Apportionment Act was replaced by the Land Tenure Act, which gave equal areas (45 million acres; 18.2 million ha) to Africans and whites, while approximately 8% of the total area was designated "national" land, to be used as game reserves, national parks, and special-purpose areas. This had the effect of consolidating the smaller fragments into more viable units while retaining the basic pattern of a European core—that included the most productive part of the country and virtually all the cities—and a dry, rural, disconnected African periphery (Figure 26-3).

Land apportionment reserved for the white minority the most suitable farmlands, where tobacco, sugar, cotton, and other cash crops could be grown with adequate space for expansion and economies of scale. In contrast, the Africans faced overcrowding and mounting population pressure, and their opportunities for commercial production were limited (see Table 26-1). These conditions were aggravated still further by the Native Land Husbandry Act of 1951. This act attempted both to change communal landowning in the reserves to permanent individual title and to destock the African herds. The aim was to improve agricultural

TABLE 26-1 CHARACTERISTICS OF FOUR FARMING TYPES IN ZIMBABWE IN 1995 (1 HECTARE = 2.471 ACRES).

1. The large-scale commercial farming sector
   - 4,700 farming units
   - 12.8 million hectares of land
   - 33% of the total land area
   - Size range from 200 to 8,000 hectares (average 3,000 ha)
   - Freehold tenure
   - Mostly white Zimbabweans
   - Some wealthy black Zimbabweans
2. Small-scale commercial farming sector
   - 8,500 farming units
   - 1.4 million hectares
   - 3.6% of total land area
   - Generally less than 150 hectares in size
   - Land was set aside during the colonial period for the few blacks who could afford to purchase land under freehold tenure
3. Communal farming sector
   - Supports over 60% of the national population
   - Occupies 16.4 million hectares
   - 42% of total land area
   - Average size of holdings is 2 hectares
   - Usufructuary tenure
4. Resettlement areas
   - Postindependence development
   - 8% of the national land area
   - Most created out of former large-scale commercial lands
   - Some vacant land used

Source: Human Rights Watch (2002).

productivity, but it was seen by many as the racial elite's way of destroying the tribal system and keeping a permanent workforce of landless Africans in the cities, since farmers who violated the conservation measures were deprived of all rights to land. So strong was African reaction to the Land Husbandry Act that the government was forced to suspend it, but not before local nationalist movements had made political capital of its injustices.

By the early 1970s, Africans outnumbered Europeans by 21 to 1 (5.3 million to 250,000), and a larger proportion of Africans than Europeans depended on the land (Griffiths 1985). After independence there was much white emigration. Binns (1994) estimates that by 1985, five years after the commencement of black majority rule, there were only 120,000 whites left in Zimbabwe. Independence has solved neither the land problem nor the population problem. In a comparative study of commercial and communal (African) land use in Zimbabwe, Prince (2001) found that communal lands were overcrowded and degraded in relation to the commercial, principally white-owned, areas. The majority of the African population resides on either communal lands (51.4%) or in urban areas (31.6%) (Prince 2001).

There were several attempts to address the land tenure problem in Zimbabwe after independence using the willing seller–willing buyer model. These early efforts were financed by contributions mainly by Great Britain which, as the former colonial power, probably felt some moral responsibility for the problem. Great Britain also played an active role in the Kenyan land reform of the 1950s and 1960s.

In 1990 the Lancaster House Agreement expired. In 1992 the Zimbabwean Parliament passed the Land Acquisition Act, which provided a legal framework for the purchase and transfer of European land to Africans that went beyond the willing seller–willing buyer model and allowed the government to take land regardless of the wishes of the landowner. Under this legislation, a fair price was to be set by a six-person committee. President Mugabe had difficulty in obtaining financing for land reform. He was widely accused of using the land reform issue to reward party loyalists and dispense political patronage. In 1996 with £30 million in British aid, 73,500 landless black families were resettled on 110 formerly white farms. In 1998 Mr. Mugabe sought an additional $2 billion in donor aid to resolve the landownership issue. But talks broke down between the government of Zimbabwe and donors because of concerns about transparency and the respect for the legal process.

The Movement for Democratic Change (MDC), which emerged in Zimbabwe in 1999, was truly national in its constituency. The party's members were both black and white Zimbabweans from across the country. In addition to developing positions on multiparty democracy and responsible government, the party had developed a "people-centered" land reform program that it said it would implement once in power. The focus of the plan would be on purchasing idle land, not working farms. In addition, the MDC advocated a tax on idle land to give landowners an economic incentive to free themselves of unproductive resources.

Early in 2000, more land reform legislation was pushed through by the president, and a law permitting the seizure of white land without compensation (except for infrastructural improvement) was written into the books, giving Mugabe "fast-track" authority to seize farms. The president declared that compensation for the land would be paid only for infrastructural improvements such as buildings, not for the land itself. If any compensation was to be paid, then the former colonizer, Britain, was to do it. In the event, no compensation was paid by the Mugabe government for anything.

The disorderly land seizures threatened to become a destabilizing force in the region, and the leaders of neighboring countries were concerned. International discussions aimed at systematizing the land reform process in Zimbabwe, held in Abuja, Nigeria, in September 2001, led to an agreement whereby Britain would finance subsequent land reform with the proviso that Mr. Mugabe and his government abide by the rule of law. The meeting recognized that as a result of historical injustices, the patterns of landownership and distribution in Zimbabwe needed to be rectified in a transparent and equitable manner. After the signing of the agreement, the occupation of farms, the murder of farmers, land seizures, and using expropriated land as political patronage continued for 2 years until almost all white-owned land was taken. Great Britain was reluctant to finance land purchase under such circumstances, and Mr. Mugabe found no other takers.

A Human Rights Watch Report (2002) indicated that there were serious human rights abuses and questioned whether the landless poor were indeed benefiting at all from the land seizures. Those getting land were required to demonstrate support for the ruling party, ZANU-PF.

As many predicted (Prince 2001), production shortfalls did occur when the European farmers' land was expropriated after 2000. It is uncertain whether the export orientation of agriculture will continue now that most of the European farms have been seized or whether the former estates will be reorganized for smallholder production of subsistence crops to feed the growing population. The government has nationalized all land, and land may only be used with a lease from the government. All the available evidence indicates continuing steep decline in Zimbabwe's farming sector despite the governments promises of "bumper" harvests (Figures 26-5 and 26-6; Box 26-1).

MINERAL RESOURCES AND INDUSTRY

Zimbabwe is an agricultural country. Tobacco exceeded gold in its contribution to the economy in 1945 for the first time (Rubert 1998) and continues to be the greatest foreign exchange earner with $437 million in revenues, much higher than minerals (gold, $164 million; ferroalloys, $113 million; nickel, $72 million). But Zimbabwe is richly endowed with minerals, and mining makes a significant contribution to the economy. The greatest concentration occurs in and adjoining a narrow geological structure known as the Great Dyke, which extends from north of Harare 280 miles (450 km) to the south through the central plateau (see Figures 26-4 and 26-7). The Great Dyke is 6.8 miles (11 km) wide. Note the evidence of past east–west movement along the Great Dyke that is visible a Figure 26-7. Gold, chrome, platinum, nickel, palladium, asbestos, cobalt, rhodium, and iron ore are all found in this zone.

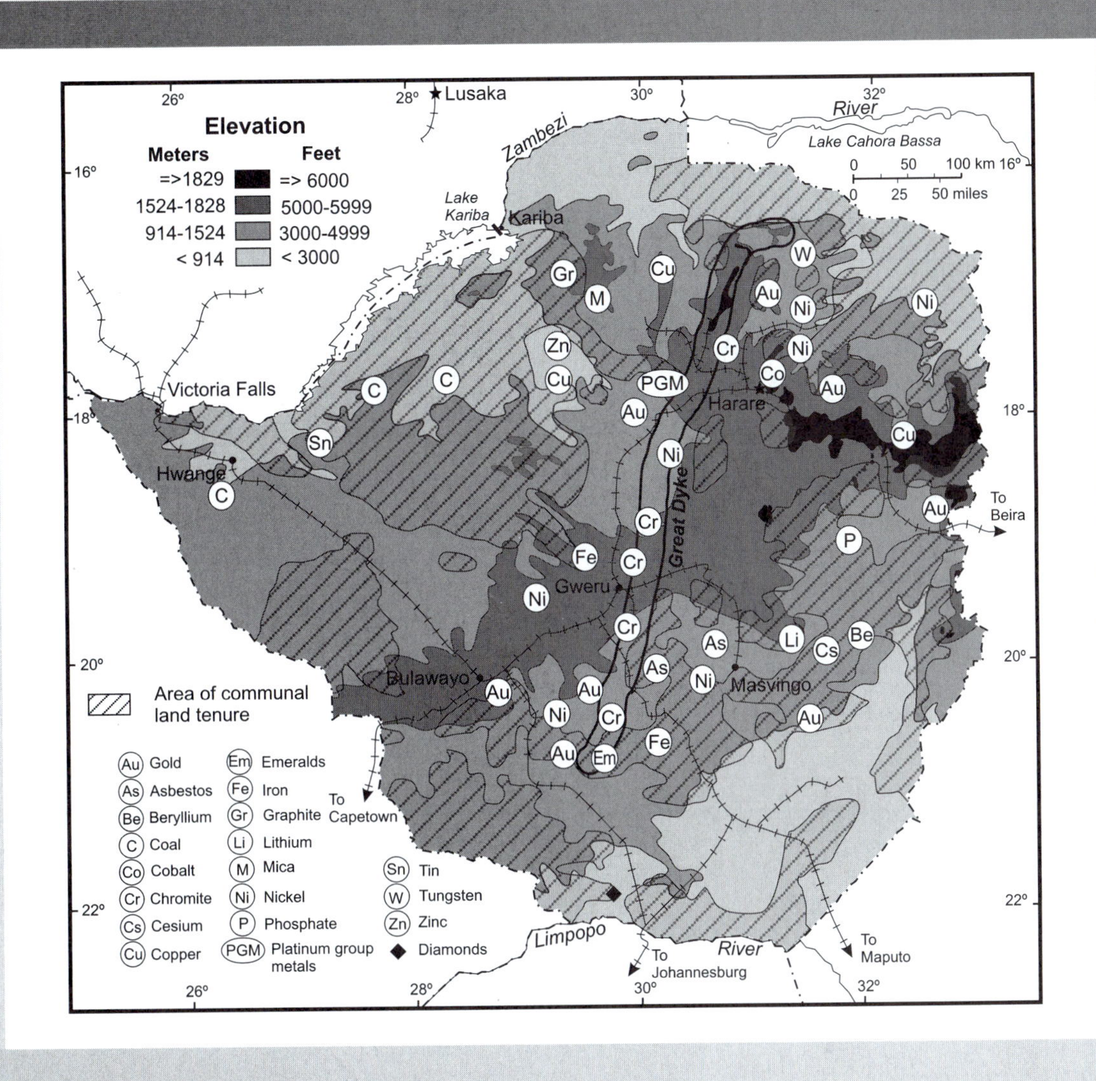

FIGURE 26-4 Mining, elevation, and communal areas of Zimbabwe.

Mining's share of the GDP is gradually declining, but minerals figure prominently in the list of Zimbabwe's exports. For more than 50 years before independence, gold was Rhodesia's leading mineral export and the main reason Europeans occupied the area—they thought that it would be rich like the Witwatersrand in South Africa. Many of the smaller and older workings closed after World War II when gold still accounted for more than a fourth of all mineral revenues. Sanctions imposed following Rhodesia's unilateral declaration of independence in 1965 provided an incentive to increase gold output to improve the balance of trade. Today's major revenue-producing minerals are copper, platinum, nickel, tin, and chrome. Coal is also mined at Hwange. Coal supplies the country's iron and steel works at Kwekwe and Redcliff, where local iron ores and limestones are also mined. Although Zimbabwe has no petroleum deposits, it could produce oil from coal. Of all exports in 1997, gold and ferroalloys represented 14 and 7%, respectively (CIA 2001).

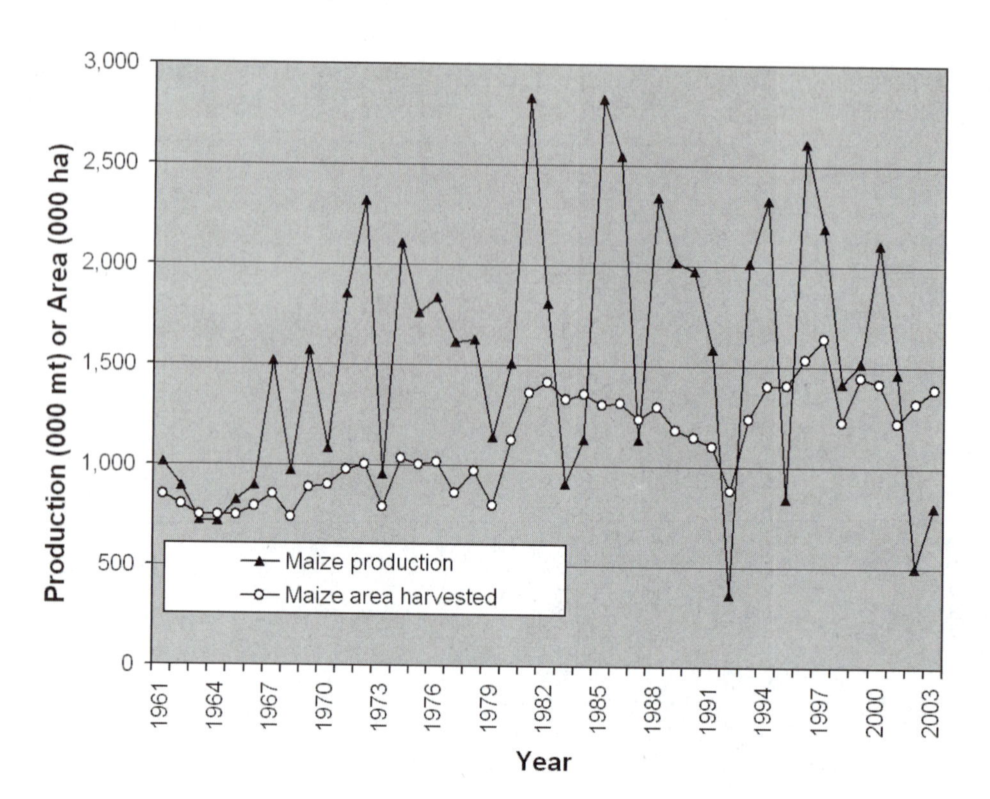

FIGURE 26-5 Maize production and area harvested in Zimbabwe, 1961 to 2003.

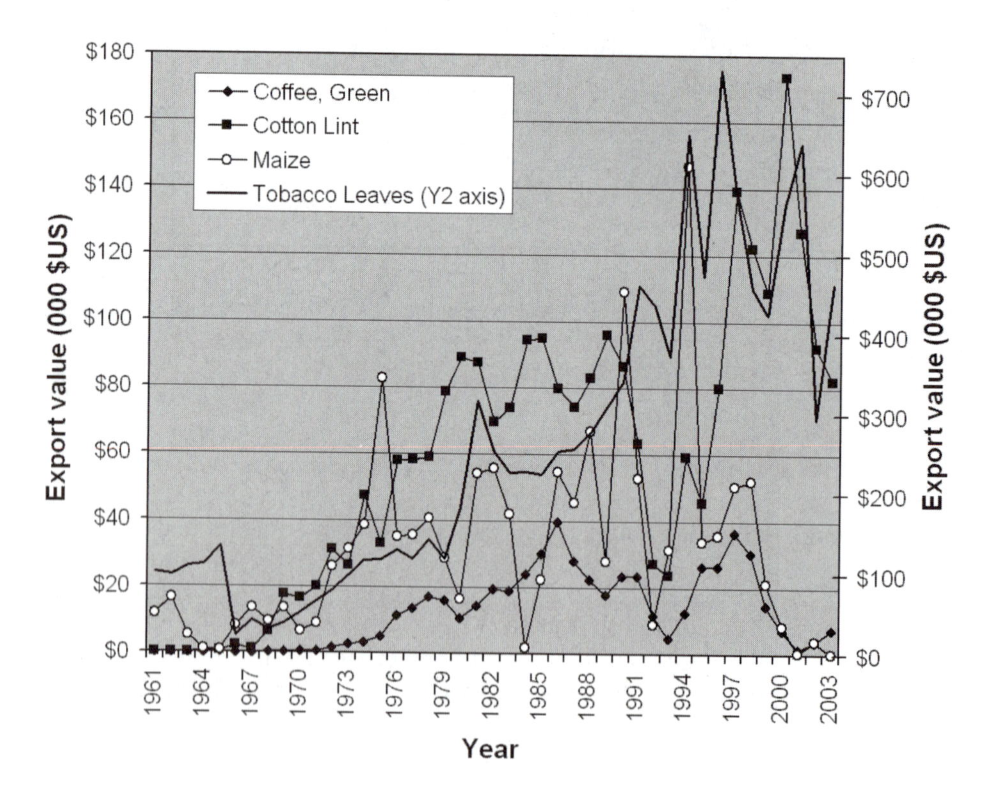

FIGURE 26-6 Value of four of Zimbabwe's export crops, 1961 to 2003.

BOX 26-1

ORWELLIAN NEWSPEAK IN ZIMBABWE IN 2004

President Mugabe should be applauded for economically empowering black Zimbabweans, the real owners of the country by giving them back their land which had been stolen from their forefathers by the unrepentant Rhodesians and their kith and kin. The last three years have seen the newly resettled farmers doing wonders on their pieces of land, contrary to what was being said by prophets of doom. This year [2004] the country is going to have a bumper harvest as it expects 2.3 million tonnes of maize. Thus, the country will not need to import maize, neither will it ask for any donations.

The sanctions and other moves taken by our detractors have failed to stop the land reform programme. This historical achievement has however not gone down well with some of our neighbours who are controlled by the imperialists. This is the reason why they are harassing and slashing our people. Some of these countries are still under bondage as they do not own their means of production. Even their civil service is not run by themselves. When they see the emancipation and total independence of the Zimbabwean people there is bound to be jealous [sic], hence the ill-treatment.

In Zimbabwe we are lucky to have an able and dynamic team led by President Robert Mugabe which strongly believes in home grown solutions for the country's economic problems. We have the land reform programme around which the economy is set to revolve, grow and survive. What we now need is to continue fine tuning our home grown solutions and be cautious in our dealings with the IMF and the World Bank.

Source: Land reform bearing fruits, by ZANU-PF correspondent. http://www.zanupfpub.co.zw/.

Zimbabwe has one of the largest, diversified, and developed manufacturing sectors in Africa. The principal industries are centered on metals and metal products, food and beverages, tobacco, textiles, and chemicals (Binns 1994). Sanctions against the white minority government imposed during the period of the unilateral declaration of independence encouraged self-reliance in production and fostered the growth of local industry. Local manufacturing was protected from international competition for 15 years, and the value of manufacturing grew at 5.4% per year from 1965 to 1980, when it accounted for 25% of GDP. Multinational corporations, which control mining, were forced to reinvest rather than repatriate profits (Potts 1992).

Industry is powered by coal-fired power plants (78%) and hydroelectric power generated by the Kariba Dam (22%). The coal is mined near Hwange. The Kariba Dam was built across the Zambezi River before independence, when Rhodesia was federated with Zambia and Malawi and was a major centripetal force in the union. Located midway between the Zambian Copper Belt and Rhodesia's Great Dyke, it supplied both regions with power (starting in 1960) from generators located on the south bank. This 420-foot-high (128 m) dam, 1,900 feet (579 m) across the top, was an economic necessity, not merely an object of national prestige. Its power supplemented that derived from the Hwange collieries and contributed to the rapid expansion of mining and industry in both Zambia and Rhodesia. In the 1970s it provided 80% of Rhodesia's total consumed electricity. Today it provides only 22%.

The south bank generators served both Rhodesia and Zambia until the completion of the plant on the Zambian bank. A two-lane highway across the dam wall, opened when the project was completed, lessened considerably the road distance between Harare and Lusaka. Behind the dam wall spreads Lake Kariba, encompassing about 2,000 square miles (5,180 km^2). The lake was stocked with fish to provide the basis for a fishing industry in a region low in protein consumption, and it also has become an important international tourist attraction. About 50,000 persons were relocated to make way for the lake.

FIGURE 26-7 The Great Dyke from space.
NASA.

Although manufacturing declined in Zimbabwe in the 1990s and especially since 2000, compared with most African states, the country's development of manufacturing is far advanced. Most industry is concentrated along the line of rail between Harare and Bulawayo, and virtually all of it lies within the former white-owned areas (Figure 26-4). Harare (656,000 in 1982; 1.9 million in 2003) and Bulawayo (413,000 in 1982, 965,000 in 2003) account for more than three-fourths of all manufacturing establishments, of the net output of manufacturing, and of the total labor force in manufacturing industries. Like most primate cities, Harare has the greatest variety of industry, and it is the seat of government. Bulawayo is the chief regional center and manufacturing city in the south. Its industries include metal refining, food processing, textiles and clothing, auto assembly, and tobacco processing. It is also an important railway junction.

Zimbabwe's heavy industry is located in Kwekwe and Redcliff, where steel tubes, rails, and other metal products are manufactured. Other important industrial towns include Mutare, the chief regional center on the border with Mozambique and the site of a strategic oil refinery; Gweru, home of Zimbabwe's largest ferroalloy plant and footware industry; and Kadoma, Zimbabwe's largest textile town, centrally located southwest of Harare.

A FAILURE IN PARTNERSHIP

The unequal land distribution between black and white Zimbabweans was an issue that required resolution, but the manner in which the Mugabe government addressed it was neither well thought out nor in the interests of the country. Poverty has increased and the economy has collapsed. What price land reform? Before 2000, 26% of Zimbabweans lived in poverty, 31% of the poor lived in rural areas, and the vast majority of the poor were black. And yet in 2003, the Central Intelligence Agency reported that 70% of Zimbabweans lived below the poverty line, with inflation fluctuating from 300% to 500%. What has Mugabe's land reform really accomplished?

The mining sector, long dominated by foreign companies and white Zimbabweans, is also experiencing difficulties: declining foreign investment and capital flight. From 1995 to 1999, investment in industry declined 1.2% on average. President Robert Mugabe then declared that the assets of British and other foreign mining companies in Zimbabwe would be seized. "After land, now we must look at the mining sector" (Mining Journal 2000).

After ruling for over 20 years in a mode that some have called "open plunder" (Johnson 2000) and an openly rigged presidential election in 2002, the octogenarian Mr. Mugabe is still clinging to power. The political opposition is on the run, its leader tried for treason. The ruling party appears intent on running opposing political parties completely out of government. Mr. Mugabe uses the tools of state at his disposal including state funds, organized violence, and intimidation to stay in power. It does not seem likely that he can continue much longer, and once he is gone, his unpopular ZANU-PF party will not be far behind. But the damage to the economy has already been done. Zimbabwe, once a moderately developed African country with a vibrant, although inequitable, farming sector and growing industry has, over the past few years, experienced catastrophic economic shocks, widespread food shortages, unemployment, economic uncertainty, and outmigration of economic refugees—and international opprobrium. The opposition party, the Movement for Democratic Change, appears to be a party of unification; with strong support among both blacks and whites, it seems to have bridged the political and racial gap between Zimbabweans and may have the potential to address the inequalities of Zimbabwean society and competently lead the country. Although Zimbabwe has taken a radically different course from multiparty democracy and has begun to resemble the colony that it once was, it is certain that the days of autocratic leaders with one-party rule in Zimbabwe are numbered.

BIBLIOGRAPHY

Binns, T. 1994. *Tropical Africa*. London: Routledge.

Bowman, L. 1973. *Politics in Rhodesia: White Power in an African State*. Cambridge, Mass.: Harvard University Press.

CIA. 2003. *World Factbook. Zimbabwe*. http://www.odci.gov/cia/publications/factbook/.

Cobbing, J. 1988. The Mfecane as alibi: Thoughts on Dithakong and Mbolompo. *Journal of African History*, 29: 487–519.

Griffiths, I. 1985. *An Atlas of African Affairs*. New York: Methuen.

Human Rights Watch. 2002. *Zimbabwe: Fast track land reform*. Human Rights Report, Number 15(1): A. New York: Human Rights Watch. http://www.hrw.org/reports/2002/zimbabwe/.

Johnson, R. 2001. Mugabe's thugs set up urban terror squads. *The Times* (London), April 8, 2001.

Mining Journal. 2000. Zimbabwe mines threat. *Mining Journal*, 334(8588), June 23, 2000. http://www.mining-journal.com/MJ/23jun2000.htm.

Potts, D. 1992. The changing geography of southern Africa. In *The Changing Geography of Africa and the Middle East*, G. P. Chapman and K. M. Baker, eds. London: Routledge, pp. 12–51.

Prince, S. 2001. *Current Land Status in Zimbabwe*. College Park, Md.: Biogeography Group, University of Maryland. http://www.inform.umd.edu/geog/LGRSS/Projects/degradation.html.

Rubert, S. C. 1998. *A Most Promising Weed: A History of Tobacco Farming and Labor in Colonial Zimbabwe: 1890–1945*. Athens: University of Ohio Press.

Thomas, H. 1997. *The Slave Trade*. New York: Simon & Schuster.

Walker, T., and J. Clark. 2000. Congo diamond link to be probed. *The Times* (London), June 4, 2000.

CHAPTER 27

Unfinished Business

Addressing Inequality in Landownership in Namibia

Namibia, once known as South-West Africa, occupies a unique position in the annals of African colonial history. Once a German colony, then a mandate of the League of Nations, Namibia lacked a clearly defined and universally recognized legal administration until 1990. For 55 years, it was the subject of heated debate within the United Nations and at the International Court of Justice; it had been a prize coveted by South Africa since 1919. Confusion over its legal status resulted in the imposition of South African laws and institutions, and the material advancement of the colonial minority at the expense of the African majority. In 1990 Namibia became independent, dismantled the institutions of apartheid imposed by South Africa, and sought an identity and a place in Africa and the world. The country has great wealth but has been challenged by many problems, for example, the unpredictability of the environment and HIV.

One critical issue that the country has not addressed is the issue of the ownership of the land. Most of the best land in Namibia is owned by descendants of settlers who took the land through conquest, and local people have little access to good land. There are two models before the Namibian people with which to address this problem, each based on the experience of a country whose best lands were dominated by European settlers. The first, and most recent example, is that of Zimbabwe, where President Robert Mugabe's government evicted the settlers from 2001 to 2003 and confiscated their land, houses, machinery, and even harvests without any form of compensation. The alternative example is Kenya, where in the former "White Highlands" around Nairobi, the transition in landownership took place over period of years and settlers were paid for their land and other assets. There was no shortfall in production in Kenya, but Zimbabwe is currently in a crisis from which it will not recover for many years. In which direction will Namibia go?

PHYSICAL ENVIRONMENT

Namibia is a land of diverse, but relatively dry, environments (Figure 27-1). The country has four major physical and environmental regions: the Namib Desert, the Great Escarpment, the Interior Plateau, the Kalahari Desert, and the Caprivi Strip. The Namib Desert extends along the Atlantic coast of Namibia and is approximately 60 miles wide (100 km) at any given point. It is rocky and cobbled in the north and south but extremely sandy in the central section. The Namibian segment of the Great Escarpment, which rises to heights of 6,500 feet (2,000 m), is located east of the Namib Desert and extends the entire length of the country in a north–south direction. The escarpment rises abruptly from the coastal desert. Most of Namibia comprises a high plateau with elevations between 3,500 and 5,600 feet (1,100–1,700 m) (Figure 27-2). The densest populations occur on the plateau, especially near the northern border with Angola, where rainfall is at its heaviest. On the eastern side of the Interior Plateau is the Kalahari Desert. The Caprivi Strip receives greater than average rainfall and constitutes a relatively well-vegetated environmental region of Namibia.

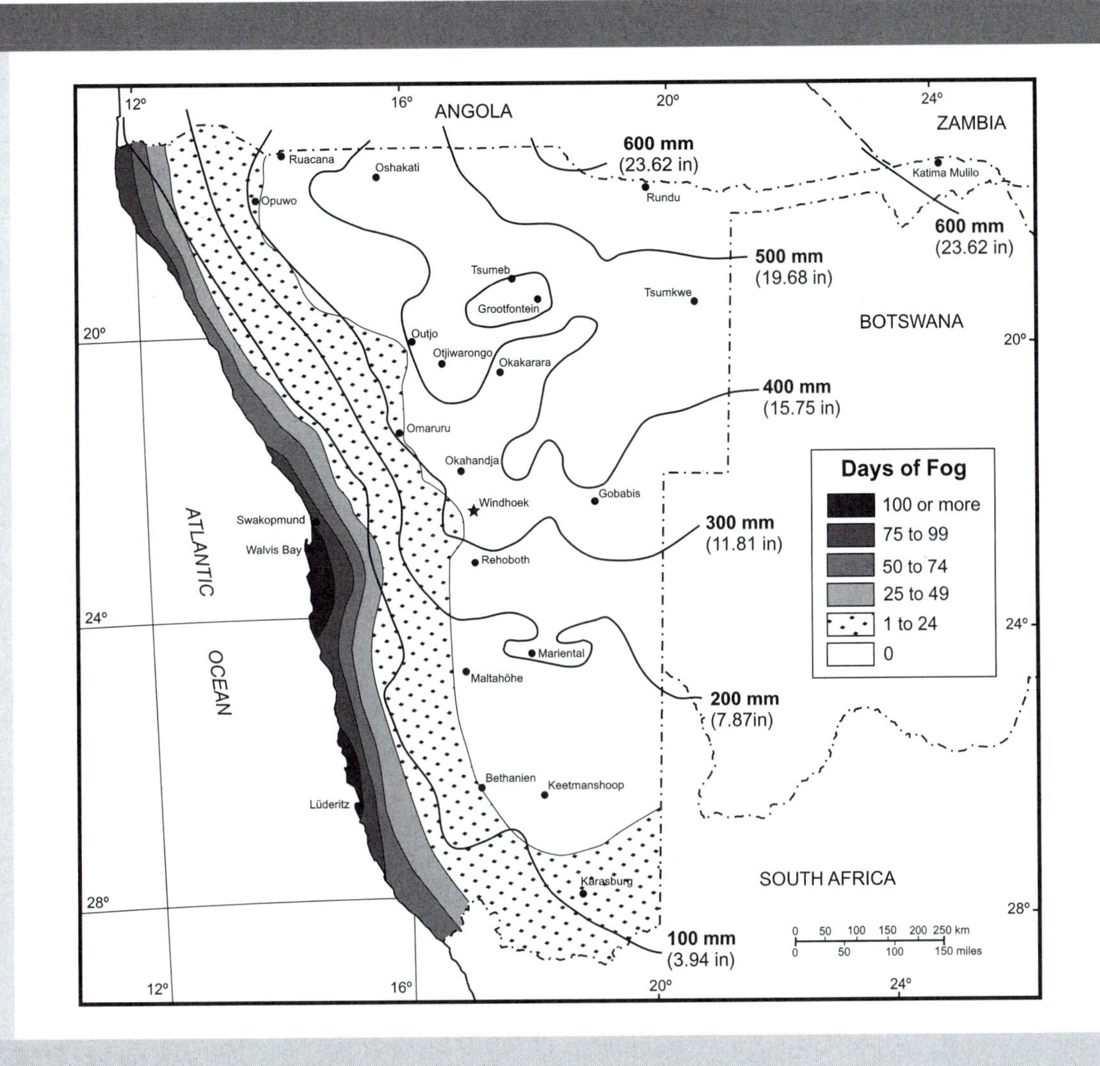

FIGURE 27-1 Days of fog and isohyets of rainfall in Namibia.

COLONIAL LEGACY

European exploration of the Namibian coast dates back to the fifteenth century when Portuguese navigators landed at Cape Cross (1484) and Walvis Bay (1487). Like the Dutch who followed them in the seventeenth and eighteenth centuries, the Portuguese made no attempt to colonize or even establish permanent settlements along the desert coast. That phase of Namibian history was left to Germany and Britain in the late nineteenth century, following a brief period of missionary activity by Cape Dutch, British, and German church organizations.

In 1878 Britain annexed Walvis Bay on behalf of the Cape Colony, and the area was incorporated into the Cape of Good Hope in 1884. Germany laid claim to the remainder of the coast from the Orange River to the Cunene and inland to the twentieth degree of longitude, while a corridor was claimed northeast to the Zambezi. By 1883 the southern, drier part of Namibia that was occupied by pastoral communities

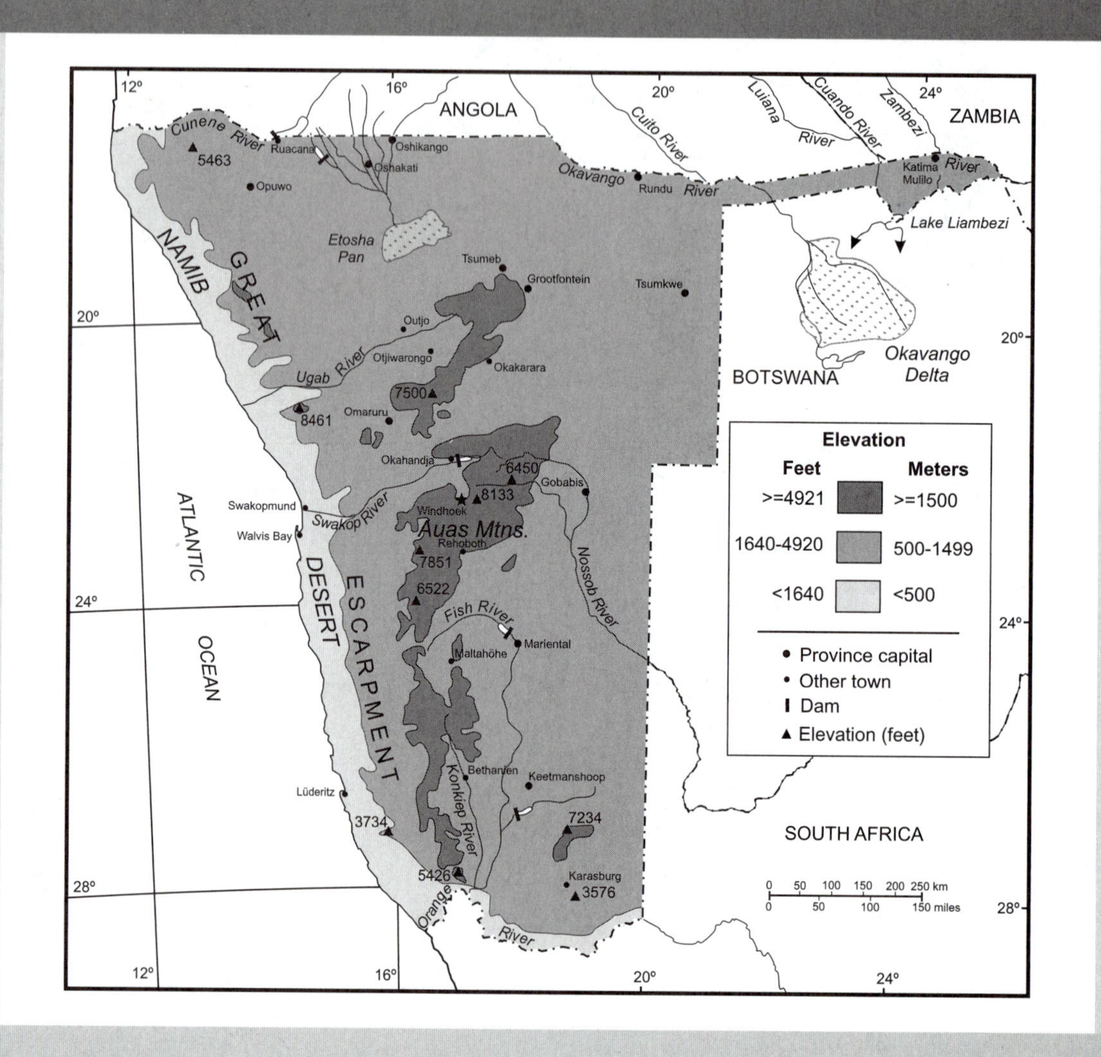

FIGURE 27-2 Major physical features of Namibia.

had been divided up by eight concession companies (Amoo 2001). German occupation began after 1897, focusing first on Lüderitz on the coast and then on Windhoek in the interior highlands. There the settler community confiscated land and cattle from the Nama and the Herero and forced the local peoples into labor units to work their newly acquired farms. Five years after German occupation began, only 38% of the 322,395 square miles (83.5 million ha) that had been parceled out remained in African hands: European settlers had acquired 14,286 square miles (3.7 million ha), concession companies had acquired 112,742 square miles (29.2 million ha), and the colonial administration had acquired 74,132 square miles (19.2 million ha). In 1903 the Bondelswarts Nama revolted against German rule but were quickly suppressed along with the Herero, whose numbers, on the extermination orders of General von Trotha, were reduced from 80,000 to 15,000. The German administration appropriated all tribal land and required any black Namibian who wished to obtain land to petition the colonial governor. Following these and other uprisings, European settlement spread to Otjiwarongo, Tsumeb, and Keetmanshoop aided by railway construction and government-sponsored land schemes.

In 1906, the German authorities set up a "Police Zone" over central and southern Namibia where they exercised direct rule (Figure 27-3). The small German armed forces were unable to extend their military control over the Ovambo kingdoms in the more densely populated north, and formal German authority was never exercised there. As discoveries of minerals were made and European settlement increased, the northern border of the Police Zone, the "Red Line," was pushed further north. Diamond discoveries near Lüderitz in 1908 brought more German and white South African settlers to the colony, and by 1914 the European population totaled 15,000.

When World War I began, South African troops occupied the German colony in the name of the Allied cause, and the territory was placed under South African military rule. Following the war, as provided for by the Treaty of Versailles, Germany ceded its colonies to the principal allied and associated powers. The former colony became a class C mandate of the League of Nations, and according to the mandate, the territory was officially transferred to "His Britannic Majesty" to be governed on his behalf by the Union of South Africa, which then paid allegiance to the British Crown. Under the mandate, South Africa was given full powers of administration and legislation over Namibia as an integral part of the union, and the union was obliged to "promote to the utmost the material and moral well-being and the social progress of the inhabitants of the territory." For South Africa's General Jan Smuts, this relationship amounted to annexation in all but name.

In the interwar years, the economic and administrative links between Namibia and South Africa were continuously strengthened: European settlement was encouraged; South African mining companies acquired all diamond rights from German-owned operations; a European legislative assembly was established in Windhoek, the capital; communications were expanded and linked with the Cape; German settlers were encouraged to become British citizens; and the African reserve system was introduced and modeled after that in South Africa. In 1920 indigenous Namibians' ties to their ancestral land beyond the Red Line were cut when the South African government formally declared that land inhabited or owned by tribal groups in South-West Africa (Namibia) would become "Crown land," effectively giving control of such land to the state. After this declaration, Crown land was classified according to identifiable tribes. Fourteen native reserves were set up, and customary land allocation functions were transferred to colonial authorities. In 1933, following several years of severe drought and worldwide economic depression that brought near disaster to Namibia's farming and mining sectors, the South African government pressed for the incorporation of Namibia into the union. This request was rejected by the Mandates Commission of the League of Nations and opposed by the remaining German settlers, who preferred a return to German rule, or at least closer association with their former metropole.

THE UNITED NATIONS CONTROVERSY

With the demise of the League of Nations in 1946, South Africa refused to place its mandate under the new United Nations trusteeship system, but immediately and unsuccessfully sought the territory's incorporation. That same year the government produced a "petition" signed by 208,850 Africans, in addition to a majority of the whites in the territory, appealing to the United Nations to permit the final inclusion of the territory in South Africa. The petition was rejected by the United Nations as fraudulent when it was realized that the literacy rate of the African population (then about 430,000) was so low that the number of signatures could not possibly reflect the desires of people who were aware of the meaning of the document they had signed. Meanwhile, the South African government, led by the victorious Afrikaner Nationalist majority of 1948, began to perpetrate in Namibia the practices of *apartheid* that were being legalized in South Africa. In 1949 the UN General Assembly placed the matter before the International Court of Justice at The Hague, requesting an opinion on several aspects of the issue. The court judged that South Africa was at fault for not having submitted petitions and reports on Namibia to the United Nations. In addition, it stated that South Africa could not unilaterally change the political status of the territory. Finally, the court ruled by eight votes

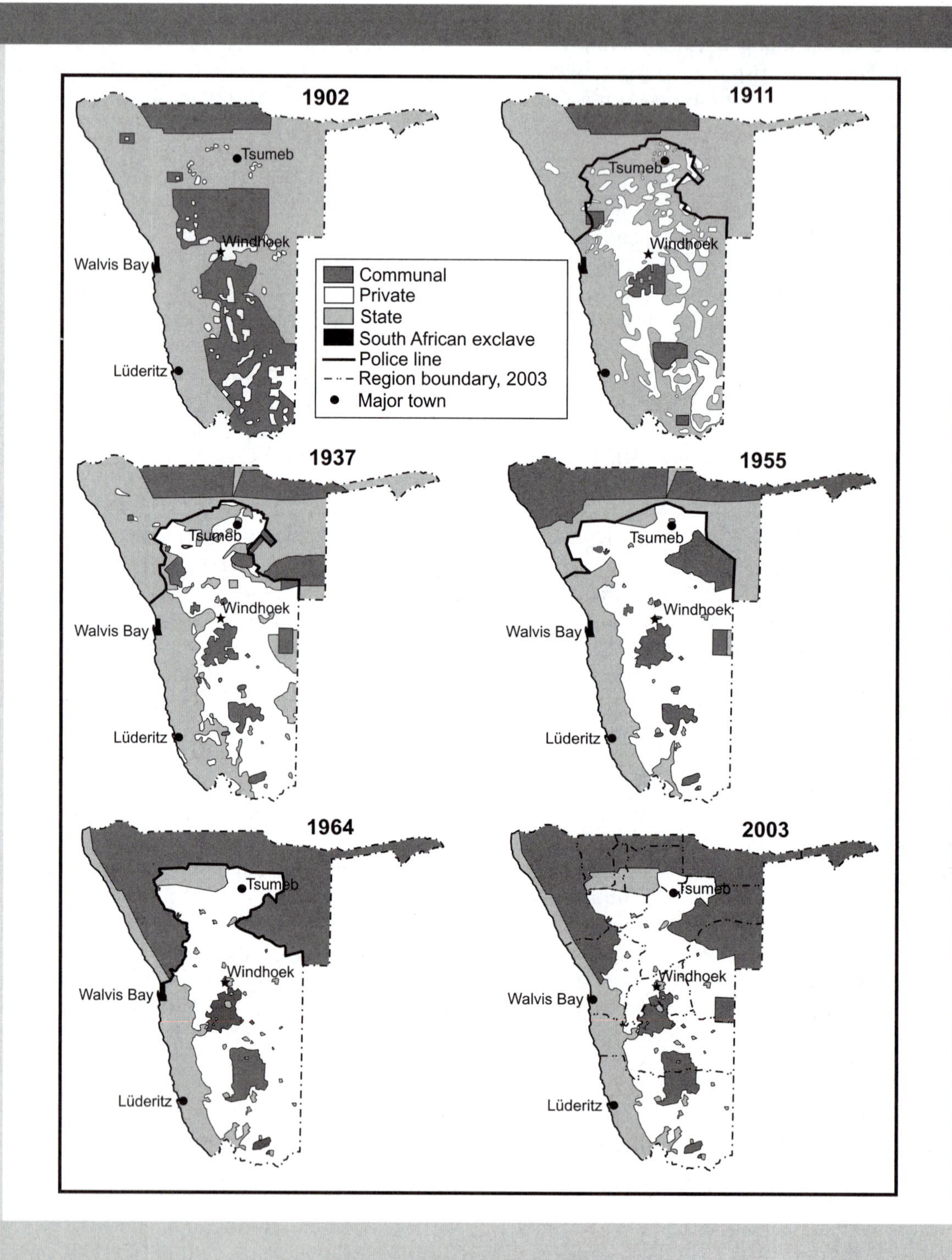

FIGURE 27-3 Changing geography of land tenure in Namibia, 1902 to 2003.

to six that South Africa was at no time legally required to place Namibia under the trusteeship system.

The decision rendered at The Hague unquestionably strengthened South Africa's hand in the debate. In the first place, no decisions of the International Court of Justice are binding, and the major question of the relations of Namibia with the United Nations had, for all intents and purposes, been decided in favor of South Africa. The union, therefore, continued to intensify its policies of racial segregation within the territory and its political integration with the South African state. Appeals from within the region went unreported and unheeded.

South Africa's continued implementation of apartheid in Namibia and its refusal to recognize UN decisions prompted Ethiopia and Liberia in 1960 to charge that South Africa had violated both the terms and spirit of the mandate and its obligations to the United Nations. After 6 years of deliberations, the International Court of Justice rejected the claims on grounds that Ethiopia and Liberia had no standing before the court. Despite the inconclusiveness of the decision, the UN General Assembly resolved that South Africa should be stripped of its mandate and that responsibility for the territory should be assumed by an 11-nation council appointed by the United Nations.

What South Africa had acquired in 1919 was beginning to appear to be a real liability: it was vast, largely desert and steppe, sparsely populated, and expensive to organize and govern even after the attempts made by the Germans to develop it. There were those who referred to the acquisition as "Smuts's folly." And it did seem that, apart from some revenues from diamond production, the Namibian territory could contribute little to a fast-developing, wealthy state. The tenacity with which South Africa defended its position in Namibia was indicative of the potential usefulness of the land as an integral part of the South African state. The period since World War II has seen the evolution of a new Africa and the confrontation of ideological opposites, black and white nationalism. To the latter, fighting for a permanent place on a black continent, the possession of territory was a matter of vital interest. South Africa's pressures on Britain for the transfer of the former High Commission Territories (now Botswana, Lesotho, and Swaziland) and its refusal to yield in the Namibian question are reflections of this geopolitical situation. By continuing to rule Namibia, South Africa was extending its power from the Cape to the Zambezi River, and the South African government found this psychologically satisfying, politically desirable, and strategically reassuring.

Furthermore, the economic importance of Namibia to the South Africans had changed greatly since 1919. Mineral production, particularly diamonds, had given the country an importance disproportionate to its low population and its arid and semiarid environment; in addition to diamonds, Namibia has uranium, gold, copper, lead, zinc, tantalum, and other minerals. But South Africa coveted Namibia not only because of its mineral and agricultural resources or because the white minority there sought its protection. The South African state believed that it deserved the fruits of its considerable investment in the region, as well as its efforts to organize the economy and to develop the transportation network. In addition, the South African government argued that the precarious agricultural sector of the Namibian economy would make the territory permanently dependent upon South Africa, and that political independence was impossible under such circumstances of economic dependence. These factors, coupled with the desires of the resident white minority and the security requirements of the South African state, created in Namibia a virtual fifth province of the South African Republic. The United Nations was completely unable to seriously influence the unfolding of events to South Africa's advantage.

THE ODENDAAL COMMISSION

As a result of UN criticism of South Africa's administration of Namibia, South Africa appointed the Odendaal Commission in 1962 to inquire into the moral and material welfare of Namibia's African population. In particular, the commission was directed to investigate the agricultural, mining, and industrial potentials of the reserves and to determine the need for additional health and educational facilities. The commission concluded that Namibia's black populations differed in customs, language, levels of economic development, and social and political systems and "would prefer to have their own homelands and communities where they would retain residential rights, political say and their own language to the exclusion of all other groups" (South Africa 1964). However, the commission rejected any system of representative democracy, fearing the Ovambos (who comprised almost one-half the African population) would dominate all groups and lower the standards of administration and government, which in turn would hamper the European settlers "to whom the Territory mainly owed its economic progress." The commission further suggested that the South African government take over most branches of administration then in control of the territory's administration and take initiative in demarcating and developing African homelands.

Not unexpectedly, the Odendaal proposals came under scathing attack from most sectors of the United Nations. They were denounced by the Special Committee on Ending Colonialism on grounds they would intensify apartheid, legalize racial discrimination, dismember the territory, and exacerbate tribal antagonisms. In view of this criticism, South Africa temporarily refrained from implementing its homeland policy but pushed on with selected development projects that were designed to appease public opinion as much as to promote the local economies. In 1968 the South African government enacted legislation to create "independent" nations similar to the Bantustans in South Africa. In 1969 the UN Security Council called on South Africa again to withdraw its administration from Namibia, but the republic ignored the injunction. By a vote of 13 to 2, the International Court of Justice upheld the UN position and found that "the continued presence of South Africa being illegal, South Africa is under obligation to withdraw its administration from Namibia immediately and thus put an end to its occupation of the Territory." The opinion was subsequently endorsed by the Security Council but was again ignored by South Africa (Dugard 1973). Soon after, the republic outlined its policy of "self-determination" within an apartheid framework, which called for the territorial dismemberment of Namibia into one white and 10 African homelands. This intensified world opinion against South Africa and provoked widespread unrest within Namibia, led by nationalistic organizations such as the South West African People's Organization (SWAPO), which sought a territorially unified, independent Namibian state. Nevertheless, the homeland policy was implemented, and despite much opposition from those being affected, Ovamboland became Namibia's first self-governing homeland in 1972.

AFRICAN HOMELANDS

South Africa's homeland policy in Namibia was very similar to that in the republic in both rationale and execution. South Africa maintained that there were ten African groups distinct from one another in language, culture, and history, whose separate identities must be preserved in separate geographic territories, each with its own political and social institutions. The homelands were located in two separate regions: a northern zone composed of Kaokoland, Damaraland, Ovamboland, Kavangoland, East Caprivi, Bushmanland, and Hereroland; and three spatially separate southern homelands, Namaland, Tswanaland, and Rehoboth (Figure 27-4). Collectively the 10 homelands comprised 130,000 square miles (336,700 km^2), or 41% of the total land area. Unlike the South African homelands, each Namibian homeland was a single territorial block, and the seven northern units formed an unbroken buffer between the white-controlled south and the newly independent, militant, Angola to the north. These 10 artificially defined, landlocked areas had much in common: they were peripherally located with respect to the settler-controlled economy on which they depended for employment and consumer goods; they were lacking in almost all the resources, technology, capital, skills, and physical infrastructures necessary for economic prosperity; local political and administrative expertise, long suppressed under the apartheid system, was scarce; mining, manufacturing, and commercial agriculture were almost nonexistent; the predominantly subsistence agricultural–pastoral economies were severely restricted by light and erratic rainfall, poor groundwater resources, and poor pasture land; there were no opportunities for acquiring the skills in most occupations in the adjoining modern sector of the economy; and there was little prospect of material progress, since there was almost no capacity for saving or for the acquisition and application of modern techniques of production. Indeed, the homelands were neither politically nor economically viable and were merely poor dependencies of apartheid South Africa. The northernmost tier of homelands corresponded with the old German "treaty lands," later the "native reserves," and on account of their climate and tribal histories were areas the German colonists had avoided (Figure 27-3).

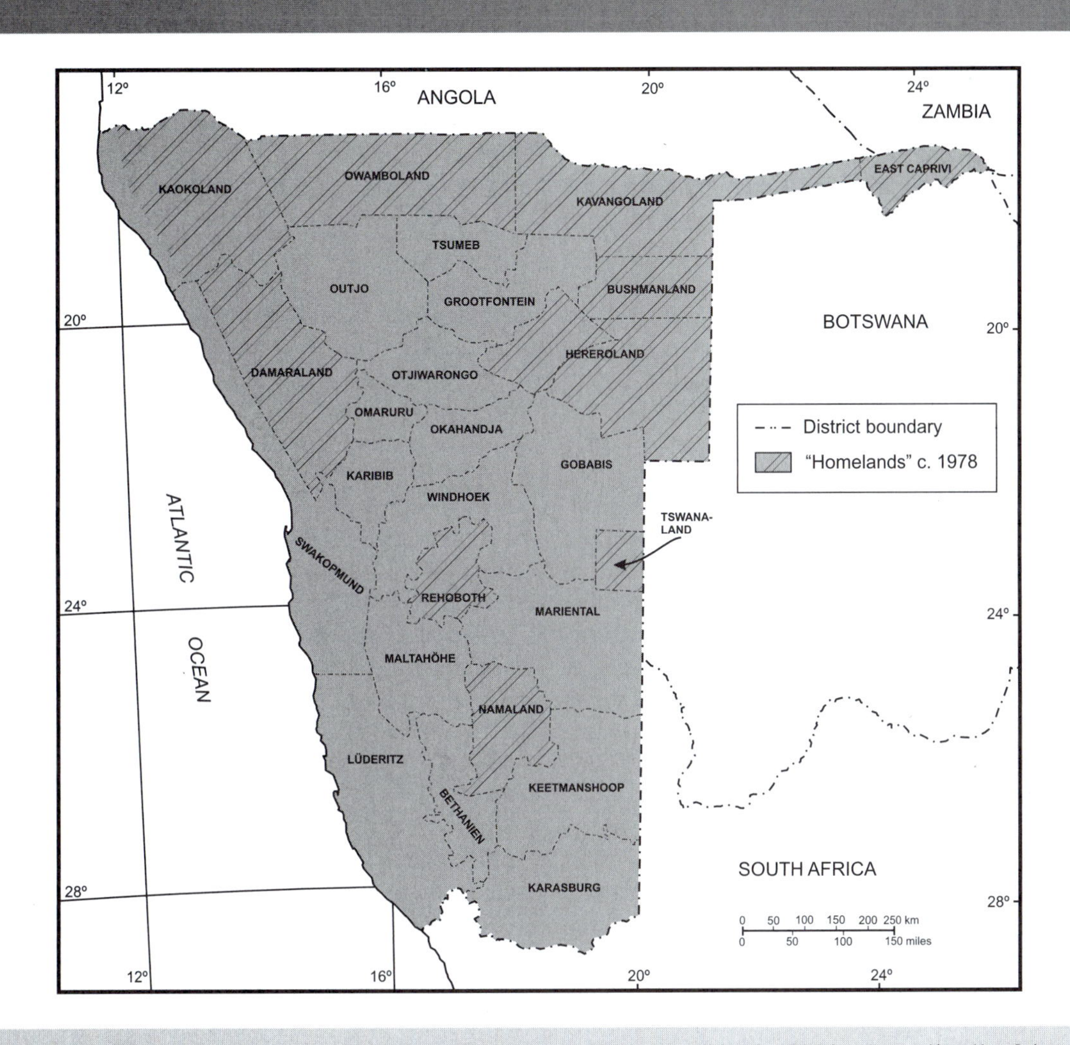

FIGURE 27-4 Provinces and "homelands" of South-West Africa (Namibia), in 1978 implementing the Odendaal recommendations of 1964.

Annual rainfall decreases from 24 inches (610 mm) in East Caprivi (Figure 27-1) to only 2 inches (50 mm) in the west, averaging only 5 inches (127 mm) in Damaraland and 20 inches (508 mm) in densely populated Ovamboland.

Before the arrival of European colonialists, indigenous populations were accustomed to migrating from one grazing area to another, with the result that most regions became ethnically diverse and not as homogeneous as the South African government wanted, or even publicly maintained, them to be. As was the case in South Africa during apartheid, many thousands of Africans resided within the white areas. If the Odendaal plan and subsequent homeland legislation had been fully implemented in the interests of producing ethnically homogeneous units, 95% of the Damaras, 66% of the Bushmen, and 74% of the Hereros would have had to be uprooted from their locations and resettled in areas officially designated as theirs. In contrast to most other areas of Namibia during the period of South African control, about 95% of the Ovambos resided in their homeland, the remainder being temporarily absent as contract workers in the mines and towns in the southern Police Zone.

The three separate southern homelands were equally poor in resources, although Namaland and Rehoboth were more favorably situated with respect to employment opportunities and transport facilities. In any case, given their institutional and environmental limitations, the homelands had little chance of being anything but poverty-stricken dependencies of white-controlled South Africa.

AFRICAN OPPOSITION TO SOUTH AFRICAN COLONIALISM

Opposition has taken various forms over the last century of colonial rule. Since World War II, the Namibians have appealed to the United Nations for redress of their grievances, greatest of which have been the alienation of their land, the contract labor system, restrictions of freedom of movement, and the state of poverty in the African areas that stems from the colonial system. For two decades these appeals went unheeded. The UN's ineffectiveness in the matter provoked protest actions by SWAPO, local leaders, the church, and the public. In December 1971, for example, a prolonged and effective strike by contract workers shut down the Tsumeb mine and smelter complex, together with other mines, and brought the railways, fishing, and construction industries to a standstill. The South African government countered by extending the powers of the police force and repatriating all strikers and illegal residents to the reserves. Protest was also manifested in the form of noncompliance of government regulations and the boycotting of elections of government-appointed candidates in the homelands.

Toward the end of 1974, following the internal disorders in adjoining Angola which triggered renewed guerrilla warfare in Ovamboland and the Caprivi Strip, the South African government adopted a new strategy to appease African opposition. It proposed an independent Ovamboland, free to amalgamate with the Ovambos across the border if the Angolan government would consent, and a loose confederation of the rest of the population groups dominated by the economy and the political system run by European settlers and their descendants. This too was rejected, since it would not provide for an economically sound and politically stable state. Meanwhile, the U.N. Security Council called on South Africa to grant Namibia independence on the basis of territorial integrity and to withdraw its administration by May 1975. South Africa failed to comply, but in September 1975 it convened a constitutional conference in Windhoek attended by representatives of all of Namibia's population groups. The conference adopted a resolution calling for greater equity in the administration of the territory as a step toward independent statehood. After the independence of Angola from Portugal, the military wing of SWAPO based in southern Angola launched an armed struggle for liberation. Other attacks occurred in the remote Caprivi Strip adjoining Zambia and Botswana.

In 1977 a joint diplomatic effort to win independence for Namibia was started by five members of the United Nations Security Council, Canada, France, West Germany, Britain, and the United States, in negotiation with South Africa, and the neighboring states of Angola, Botswana, Mozambique, Tanzania, Zambia, and Zimbabwe. A result of these talks, to which SWAPO was also a party, was UN Security Council Resolution 435, calling for the end to hostilities, the restriction of South African and Namibian military activities, and elections in Namibia under UN control. Although South Africa agreed to Resolution 435 early in 1978, it undermined its own credibility by unilaterally holding elections in Namibia that same December. Discussions continued for 10 more years. Perhaps significant in bringing the South Africans to the negotiating table were the military upsets that the South African army suffered in southern Angola in 1987 and 1988. From October 1987 to June 1988, the South African army suffered humiliating defeats in infantry and tank battles with the Angolan army, reinforced by Cuban troops (Davenport and Saunders 2000). In the latter part of 1988, during 7 months of negotiations in London under U.S. mediation, representatives from Angola, Cuba, South Africa, along with Soviet observers, met and agreed on the withdrawal of Cuban troops from Angola. These negotiators also drafted a joint proposal for the implementation of UN Resolution 435.

Implementation of Resolution 435 began on April 1, 1989, under a United Nations and South African transition team. In their political negotiations with South Africa behind closed doors and in their negotiations "by other means" in the bush, the

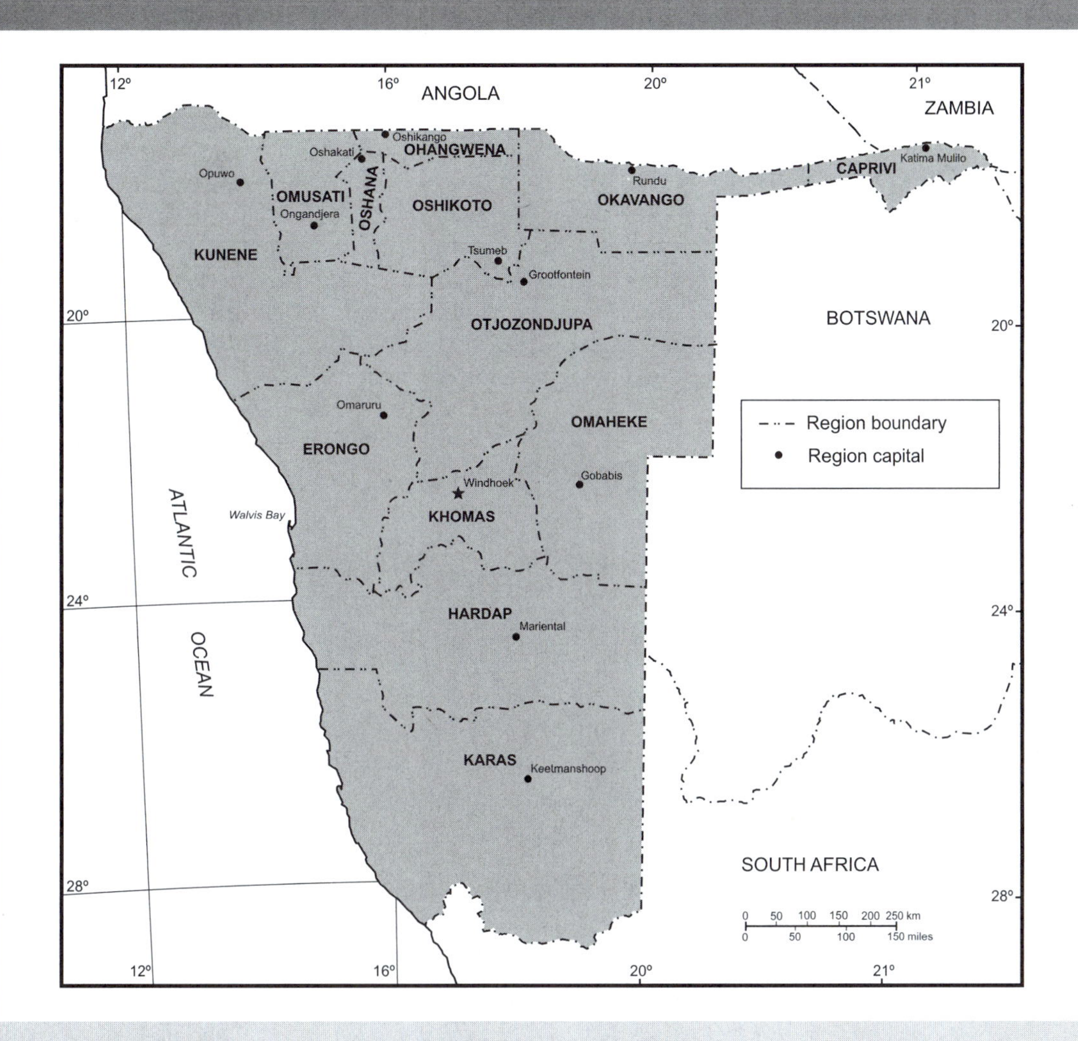

FIGURE 27-5 The internal political organization of Namibia in 2004.

Namibians demonstrated their intention to preserve the territorial integrity of their country and to secure their rights to self-determination, national unity, and independence. Having long been considered part of the prize, the Namibians wanted only an equal share of the wealth to which they contributed.

Elections were held in November 1989 to elect the Constituent Assembly, which would create the legislative framework for independence, including a constitution and its own transformation into the National Assembly. SWAPO took close to 60% of the vote. The following year a constitution had been drafted and on March 21, 1990, Namibia achieved independence. Provisions were made for a bicameral legislature with a lower and an upper house, the National Assembly and National Council, respectively. The internal political geography of newly independent Namibia underwent a transformation with majority rule. Homelands were abolished and 13 regions took the place of the 26 apartheid-era districts (Figure 27-5).

Elections were held in 1992 to elect the governments for the 13 newly created provinces. Four years later

Walvis Bay and adjacent islands, formerly South African exclaves, were ceded to Namibia by South Africa as part of the UN agreement. Although the political geography of Namibia changed dramatically after independence, the official land designations did not. The threefold nomenclature of communal, private, and state land remained. Yet although the independence of Namibia was largely about access to resources for all Namibians, and the government since independence has considered inequality to be a primary concern, a radical redistribution of land similar to that which has taken place in Zimbabwe has been avoided. We can expect this problem to be addressed more aggressively in the near future.

ECONOMIC ACTIVITIES

Namibia has a high per capita GDP compared with many African countries; this is due, in large measure, to international trade. Namibia's small export sector depends on good transportation for its growth. The country has one of the most well-developed road and rail systems in Africa and it is well connected to other members of Southern Africa, especially South Africa. There are over 41,000 miles (66,000 km) of road, of which barely 14% is paved. The well-maintained rail system consists of almost 1,500 miles (2,400 km) of rail. A well developed fiber-optic cable network links urban areas to each other and to the world, and the more isolated rural areas are to be linked through wireless connections. Harnessing the hydroelectric power of the Cunene River project in the north and the use of Namibia's natural gas deposits in the production of electricity will help encourage economic growth and attract further investment. Further integration with the Southern Africa Regional Electric Grid will provide more power for economic and human development. Walvis Bay, one of the best deepwater ports in Africa, is likely to become a significant port for South African and other SADC countries' imports and exports via the Trans-Kalahari highway and the Trans-Caprivi highway.

Three export processing zones (EPZ) have been designated: Walvis Bay, Oshikango, and Katima Mulilo. Oshikango is located on the northern border with Angola, and Katima Mulilo is located in the Caprivi Strip along the Trans-Caprivi highway that links Zambia, Zimbabwe, and Namibia. The growing manufacturing sector contributes about 15% to the GDP. If Angola is tardy in rehabilitating and extending its transportation network to more tightly bind southern Angola to the rest of the country, Walvis Bay will probably take over as the principal port serving the southern region of Angola. The role of Walvis Bay in Botswana's export economy is also very important with its easier access to Europe and the Americas.

Mining accounts for 20% of Namibia's GDP. In 2001 total exports earned over $1 billion and mineral exports accounted for 70% of all exports, with the United States, South Africa, the Germany, Britain, and Japan being the major markets. Of the mineral exports, diamonds are the most important earner, but Namibia is the world's fourth largest exporter of uranium and the second largest producer of salt in Africa. Of the 2001 mineral exports, diamonds accounted for $530 million, uranium for $155 million, gold for $24 million, and zinc for $16 million, with the remainder being made up of a variety of other minerals. All mineral rights are held by the Namibian government, as is customary in many countries, but most mining is done by private companies.

Although the existence of diamonds was known at an early stage, the real potential of the resource has only recently come to light: diamonds are the most important mineral produced and the main revenue earner. Since 1970, output has averaged between 1.6 and 2.0 million carats annually (Table 27-1). About 98% of the diamonds are gemstones, most of which are extracted from raised coastal beaches and offshore between Oranjemund, along the extreme southwest coast, and Conception Bay, located north of Lüderitz, a region formerly closed to the general public (Figure 27-6). Most of the diamonds are produced by private mining companies both domestic or foreign. The center of mining in the interior is Tsumeb, a rapidly growing town where some industrial development related to mining has been taking place.

Copper is smelted from locally produced metal as well as from imported copper concentrate. Three large copper–lead–silver mines operate in the country: Khusib Springs, Kombat, and Otjihase (near Windhoek). The uranium mine at Rössing has been producing for nearly 30 years. Additional mining operations include zinc and lead near Rosh Pinah in the extreme south of the country, fluorspar at Okorusu (north of Otjiwarongo), and salt at Swakopmund.

TABLE 27-1 MINING PRODUCTION (METRIC TONS UNLESS OTHERWISE SPECIFIED), IN NAMIBIA, 1991 TO 2001.

	Year					
Product	**1991**	**1993**	**1995**	**1997**	**1999**	**2001**
Antimony (Sb content)	10	8	–	–	–	–
Arsenic, white	1,800	2,290	1,661	1,297	–	914
Copper (Cu content)	31,700	29,500	22,530	17,879	–	12,392
Gold (kg)	1,860	1,953	2,394	2,417	2,005	2,851
Lead (Pb content)	15,000	11,600	16,084	13,577	9,885	13,025
Silver (kg) (Ag content)	91,300	72,000	69,000	41,000	9,670	12,679
Uranium (U content)	2,890	1,980	2,366	3,775	3,171	2,640
Zinc (Zn content)	33,200	17,624	30,209	39,658	35,140	31,803
Diamond (000 carets)	1,190	1,141	1,382	1,416	1,633	1,490
Fluorspar (content)	34,600	43,466	36,889	23,160	71,011	81,245
Lithium minerals	1,190	739	2,611	669	–	–
Salt	141,000	132,585	303,986	492,780	503,479	523,000
Semiprecious stones	545	779	785	1,993	524	620
Sulfur (S content)	65,000	56,900	51,330	46,476	–	28,606
Wollastonite	305	824	967	194	347	440

Source: Mineral Reports of the U.S. Geological Survey, 1992-2002.

Reserves of natural gas are 2.2 trillion cubic feet (62 billion m^3), and exploration for more offshore natural gas is under way, led by Shell Oil. Currently the country meets its energy needs through the production of electricity from turbines at the Ruacana Dam on the Cunene River in northern Namibia (250 MW) and from some coal-fired plants, and by importing electricity from South Africa (600 MW). Since there are no perennially flowing rivers in Namibia with the exception of the Cunene along the northern border and the Orange along the southern border, cheap hydroelectric power is simply not an option.

When apartheid South Africa was in control of Namibia, it set up the same system of racially organized industry that was operating at home. Mining was dependent on cheap migratory labor from the homelands that South Africa had imposed upon Namibia. Most of the approximately 40,000 laborers (of whom 70% were Ovambos) contracted to work at a fixed rate for periods of between 6 and 18 months in the mines and mine-related industries. As in South Africa at the time, prospective workers were recruited through government-run labor bureaus and were graded according to physical fitness and age for work in the mines, factories, and farms. Minimum wages were laid down for each class of laborer, ranging from about $17 a month for an unskilled mineworker to $8 for a farm laborer less than 18 years of age. The workers were not free to choose their employers or the type of work they were to do. Trade unions and strikes were illegal, and workers could not break contracts before the agreed date of expiration.

In 1990 the newly independent government of Namibia authored legislation to encourage international investment. The Foreign Investment Act of 1990 provides for a variety of incentives to investors such as the repatriation of profits, security of land title and tenure, the availability of foreign currencies (foreign exchange), international arbitration in the event of dispute, and just compensation in the event of expropriation (Coakley 1996). In 1992 the Namibian National Assembly passed legislation to replace the South African legislation that had controlled the mining sector during South Africa's occupation of Namibia. Mining and Exploration Act 33 of December 1992 provided for standard licensing and incentives to encourage foreign and domestic investment in mineral development and the payment of royalties to the Namibian government. In 1998 royalties were further reduced through new legislation in an effort to spur

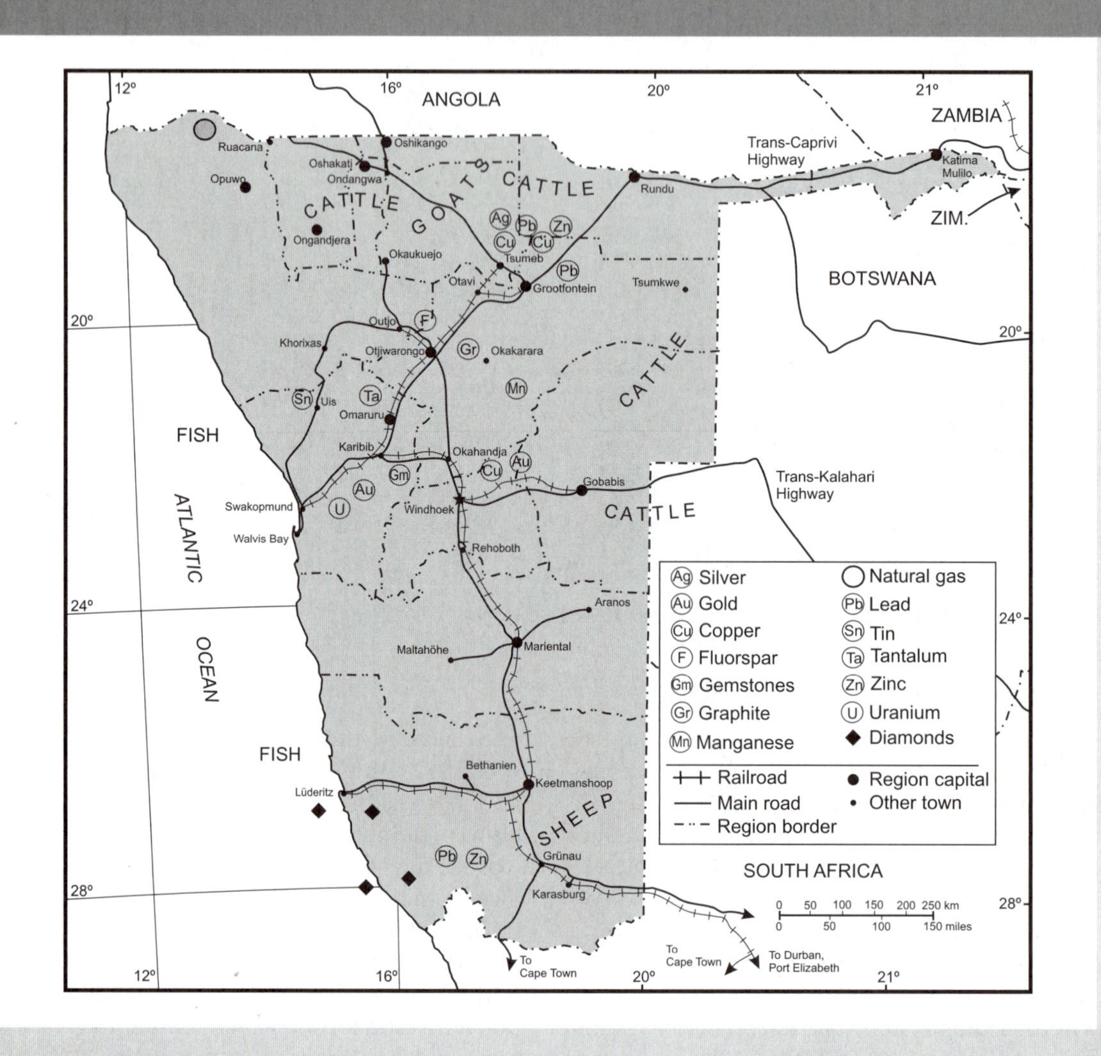

FIGURE 27-6 Extraction and livestock activities in Namibia.

investment. The mining sector in Namibia is expected to grow substantially for the foreseeable future. New mines in Tsumeb opened in 2000 and offshore diamond mining is experiencing continued success. Exploration for metals and additional deposits of diamonds has been encouraged by previous success. The potential for the development of value-adding industries, principally manufacturing, metal processing, and gemstone cutting and polishing, to process mineral resources is great (Coakley 2001).

The marine fishery is also important in Namibia (Figure 27-7). The cold Benguela Current has helped produce one of the world's most bountiful fishing areas off the Namibian coast. The industry was originally based on the fishing of lobster and whitefish for local consumption (although the United States was an important market for lobster), but it is now based on the large-scale processing of sardines and pilchards. Fishmeal, canning, and oil extraction account for 90% of the total fishing revenues, with South Africa and the

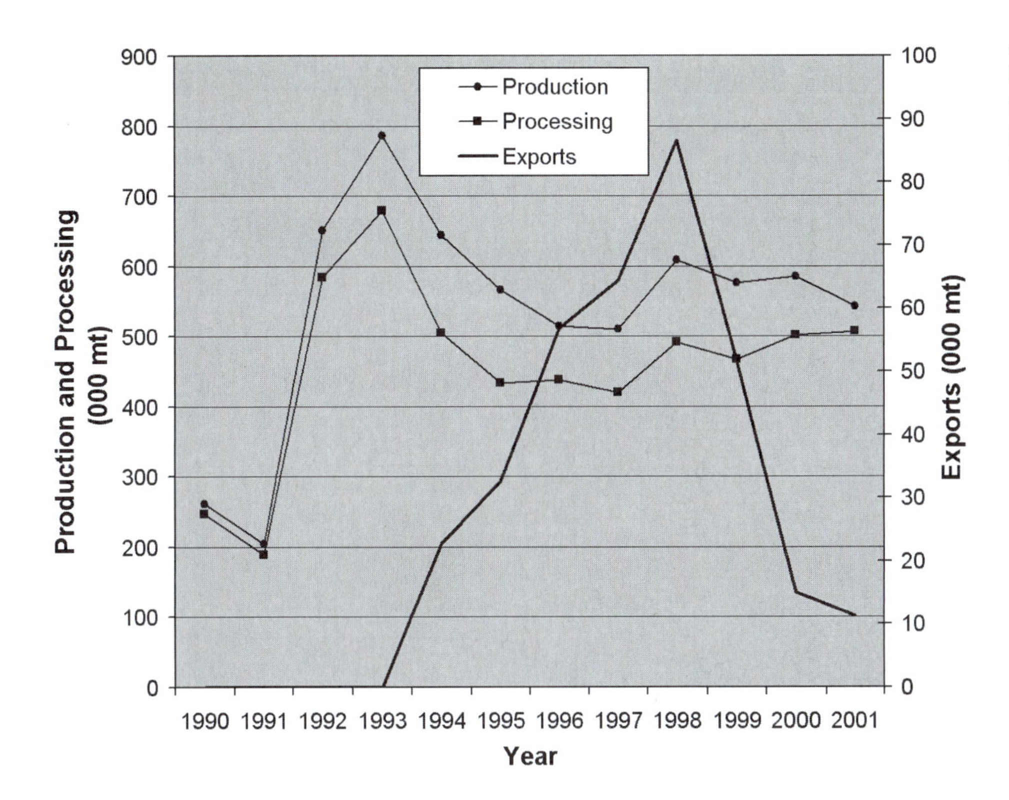

FIGURE 27-7 Namibian fishing production, processing, and exports, 1990 to 2001.

United States being the principal markets. The industry is based at Walvis Bay and Lüderitz. Fish production increased after the newly independent government declared its exclusive rights to a 200-mile (320 km) fishing zone along the coast. The fishing industry is largely controlled by settlers, and the government has been working to address the disparity in the ownership of this means of production. The Namibian government has been accused of pursuing the Namibianization of the fishing industry with excessive zeal, and some foreign fishing companies with Namibian fishing licences have alleged that to renew their licenses, they have been forced to partner with Namibian individuals or companies selected by the government.

AGRICULTURE

Although Namibia's fishing and mineral resources are almost unrivaled in southern Africa, its agricultural potential is quite limited. Namibia is the driest country south of the Sahara. Evapotranspiration rates are high throughout the region, and the soils are generally salty. Perennial streams are absent except along the northern, southern, and Caprivi borders, and close to 75% of the water used comes from boreholes. Twelve major dams (and hundreds of small ones) in Namibia supply water for irrigation and domestic consumption. Irrigated agriculture and livestock raising use about 55 and 23% of the water consumed, respectively (NamWater 2004). In the northern part of Namibia where most of the nonsettler population lives, almost all of the water comes from Angola. Water from the Cunene River's Calueque Dam, some 25 miles (40 km) across the Angolan border, is diverted south into Namibia, where it is used for agriculture and domestic consumption. Millet, maize, and cassava are the main food crops, while rice, sorghum, groundnuts, and jute are the major cash crops in the area.

Agricultural land in Namibia is divided into two types, communal and private (Figure 27-8 and Box 27-1). Land tenure in communal areas is restricted to use (usufruct) rather than ownership rights, and production is oriented toward food production for

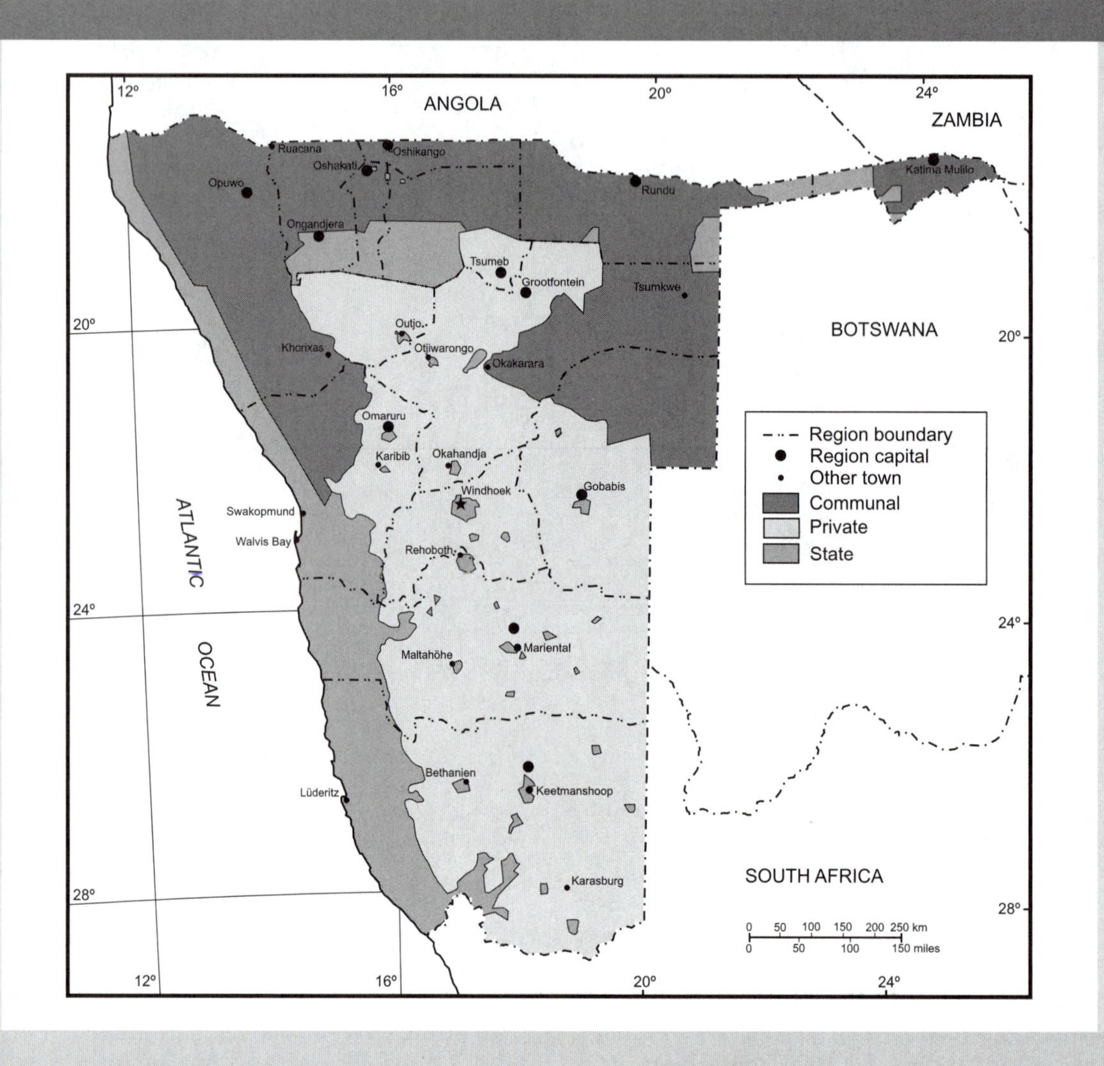

FIGURE 27-8 Land tenure, towns, and postindependence provinces of Namibia.

the family. Root crops, millet, and some maize are grown in the communal areas, and livestock are kept as well. The commercial farming areas, dominated and even jealously guarded by European settlers, are mainly located in the northern third of the country. The principal areas of settler-owned farms receive less than 12 inches (505 mm) of rain each year. Only the northeastern extremity, including the Caprivi Strip and the Ovambo area, has an annual rainfall of over 20 inches (508 mm), still subhumid given the high evapotranspiration rates. Eastern Namibia merges into the Kalahari Desert, although latosols and a thorn tree-studded grassland make the term "steppe" more applicable. In general, soils are poor, and the grasslands are capable of supporting only extensive pastoralism. Irrigated agriculture is carried on in a few valleys.

The European and multinational farms are large (up to 40,000 acres; 16,200 ha) and primarily devoted

BOX 27-1

LAND TENURE IN NAMIBIA BEFORE EUROPEAN COLONIALISM

Prior to the colonial era, and during the early period of settlement, the southern part of the territory now known as Namibia consisted of Great Namaqualand or Namaland. The central parts consisted of Hereroland and Damaraland, and later the coloured groups. The northern part consisted of Kaokoland, Ovamboland, and the Okavango. The far north east belonged to the middle Zambezi Bantus, who consisted of the Masubya (Bekuhane), Yei (Koba), Mabukushu (ha-Mbukushu) as major groups and later a tribe known as the Mafwe (Fwe). These independent indigenous tribes occupied and ruled the territory. The use of land by these indigenous peoples during the pre-colonial era was largely determined by two distinct production systems: pastoral and agricultural. Communities in southern and central Namibia such as the Nama, Herero, Damara, and Basters led a predominantly pastoral existence. The scarcity and unpredictability of pastures required these communities to disperse widely over the territory in small groups in order to utilise existing resources efficiently. Consequently, no fixed boundaries existed among different communities, although loosely defined areas of jurisdiction by small chiefs were generally recognised. Members belonging to these defined areas of jurisdiction recognised and practised communal ownership of land. In the northern regions, the indigenous population combined settled agriculture with animal husbandry, and land was owned by the community as a whole. Permanent usufruct was granted to arable plots, but the allodial title vested in the community.

Pre-colonial South-West Africa, therefore, cannot be described as terra nullius, i.e., devoid of any land tenure system. However, the land policies of the various colonial powers not only deprived the indigenous people of their land, but also had an influence on their traditional communal land tenure structures.

Source: Amoo (2001), pp. 88–89.

to cattle ranching in the north and central plateau, and to the raising of karakul sheep in the south (Figures 27-6 and 27-9). Karakul pelts (Persian lamb) earn tens of millions of dollars in foreign revenues each year, with Germany, France, and the United States being the major markets. Namibia is the world's leading exporter of karakul sheep products. Karakul sheep originate in Asia and comprise one of the oldest sheep breeds in the world. The first karakul were brought to Namibia 100 years ago by European settlers. Cattle, meat, and butter are sold in South Africa almost exclusively. As with mining, the commercial agricultural sector depends on cheap and plentiful African labor today, just as it did under South African rule.

Namibia was in a peculiar politicogeographic position while under the control of South Africa. Although it was vulnerable to the whims of outside markets and the maintenance of communications depended on an outside power, Namibia internally faced the perpetual threat of crippling droughts, which had the potential to devastate the agricultural economy, requiring imports of food grains from South Africa. At such times, which were relatively frequent owing to the high variability of precipitation, the territory's dependence on South Africa was clearly a liability. Returns from the karakul sheep industry were drastically reduced, cattle died in great numbers, meat sales to the republic dwindled, and costly imports were necessary. South Africa truly had life and death power over Namibia. From the politicogeographic point of view, this aspect of the dependence of Namibia upon South Africa was the most serious: when the economy prospered, it did so on the basis of South African markets. When it failed, it required South African support.

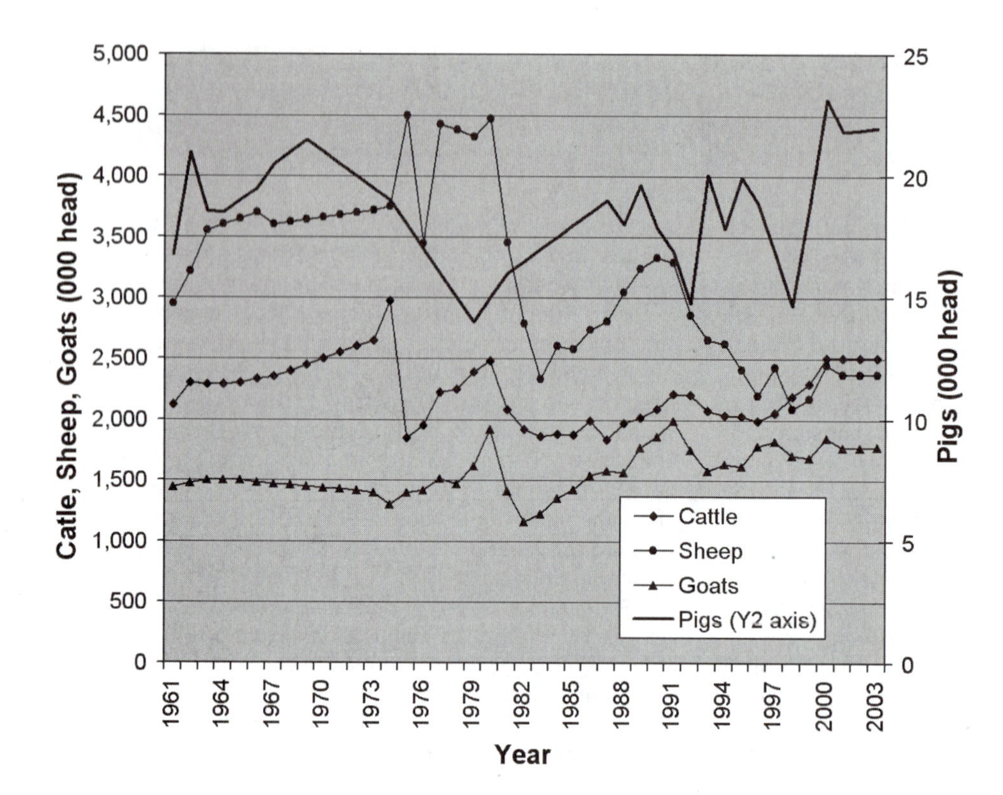

FIGURE 27-9 Namibian livestock production, 1961 to 2003.

UNFINISHED BUSINESS?

The private/communal land tenure problem Namibia has been experiencing is similar to that experienced by many other African countries (Moyo 2003). The problem has been compounded in Namibia by foreign, often multinational, control of much productive land. In 1991, just after independence, Namibia had 6,292 farms of which 6,123 were owned by white settlers and foreigners; constituting 97% of the commercial farms and 95% of the 132,819 square miles (34.4 million ha^2) of commercially classified land in Namibia at the time (Geingob 1991). Although the government states that it wishes to create a unitary land tenure system in Namibia and has taken some steps in that direction, the fact of the matter is that the tripartite system of state, private commercial, and communal lands developed during colonialism is still in place and, as a consequence, Namibia's population still has unequal access to the land and resources. Namibia's settler population, representing only a small minority of the total population of the country, controls its land and industry. The government has taken steps to help remedy the land question in Namibia, but since independence the "willing seller–willing buyer" mechanism has been operating very slowly: there are political pressures to expropriate the settler-farmers; after all, SWAPO's roots lie in leftist liberation ideology undertaken on "behalf of the people" according to the official slogans. Sam Nujoma, Namibia's first president, has expressed support for the forced expropriation of land occupied by settler-farmers spearheaded by Zimbabwe's president, Robert Mugabe. Nujoma's successor, Hifikepunye Pohamba (elected 2004), former Land Minister in Nujoma's government, is an advocate of expropriation. Industry and international investment have fled Zimbabwe; the economy has been shrinking; inflation exceeds 500%; and black as well as white Zimbabweans have been emigrating in the years since President Mugabe initiated seizures of white-owned land, and later movable assets and industry. President Mugabe has been roundly condemned by the international community—although he has been applauded by some African leaders. If the same social and economic chaos experienced by Zimbabwe is to be avoided, Namibia must resist the temptation to rashly expropriate land held by the settler population and multinational corporations (Boyle 2001). Although President Nujoma had applauded Mugabe's policies, he resisted acting precipitously to satisfy insistent political pressure from his left. His successor may not be so prudent.

Robin Sherbourne, an economist and researcher at the Namibian Institute of Public Policy and Research, observed (2004) that from 1991 to 2003, 11% of commercial land had been redistributed through purchase to black Namibians and that the current rate of redistribution is 1% a year. His research indicates that of the two methods of land redistribution employed in Namibia, market-based purchases of commercial land using government loans extended to black small, but commercially oriented, farmers has been much more successful than the government program to purchase land from willing sellers to resettle landless Namibians. The latter scheme replicates the communal land tenure arrangements used in the subsistence farm sector and does not give farmers title to the land, as under the loan program.

Sherbourne suggests that several ways in which the land reform process can become more responsive. First, the commercial nature of Namibia's commercial farmland should be recognized and maintained. Development policy should focus on creating a successful and fully commercial black farming sector rather than on resettling many landless people on commercial farmland. Commercial farming is a skilled occupation using scarce resources; it is not for everybody and should not be considered as a "way out" for those who have nothing. Over the short to medium term, redistribution should take precedence over production, but radical (and corrupt) redistribution following the Zimbabwe model must be avoided. The government has enacted a land tax to put economic pressure on commercial landowners who have land not in production. In the end, policy makers must ask themselves what kind of agriculture Namibia wants and needs. Does it want to encourage small-scale, subsistence agriculture with communal land tenure? Large-scale, export-oriented, industrialized farming on privately owned property? Or something in between?

BIBLIOGRAPHY

Amoo, S. K. 2001. Towards comprehensive land tenure systems and land reform in Namibia. *South African Journal of Human Rights,* 19: 87–108.

Boyle, H. 2001. The land problem: What does the future hold for South Africa's land program? A comparative analysis with Zimbabwe's land reform program: A lesson on what *not* to do. *Indiana International and Comparative Law Review*, 11(3): 665–696.

Coakley, G. J. 1996. *The Mineral Industry of Namibia.* U.S. Geological Survey Mineral Report. Reston, Va.: USGS. http://minerals.usgs.gov/minerals/pubs/country/.

Coakley, G. J. 2001. *The Mineral Industry of Namibia.* U.S. Geological Survey Mineral Report. Reston, Va.: USGS. http://minerals.usgs.gov/minerals/pubs/country/.

Davenport, R., and C. Saunders. 2000. *South Africa: A Modern History*. New York: St. Martin's Press.

Dugard, J. 1973. *The South-West Africa–Namibia Dispute.* Berkeley: University of California Press.

Geingob, H. 1991. Opening address at the Land Conference on Land Reform in Namibia. Quoted in Amoo (2001).

Moyo, S. 2003. The land question in Africa: Research perspectives and questions. Paper presented at Codesria Conferences on Land Reform, the Agrarian Question and Nationalism, in Gaborone, Botswana, October 18–19, 2003. http://www.codesria.org/Links/conferences.dakar/moyo.pdf.

NamWater. 2004. Windhoek: Namibia Water Corporation. http://www.namwater.com.na/en/Home.asp.

Sherbourne, R. 2004. Rethinking land reform in Namibia: Any room for economics? Windhoek, Namibia: Institute of Public Policy and Research.

South Africa. 1964. Report on the commission of enquiry into South West African affairs, 1962–1963. Pretoria: Government Press.

CHAPTER 28

Zambia and Malawi

On the Tightrope

Zambia and Malawi are two landlocked states that once were part of the Central African Federation, together with Rhodesia (then Southern Rhodesia). While having very different natural environments, resources, economies, and political leadership, Zambia and Malawi share similar problems in view of their landlocked position north of the Zambezi, and their former politicoeconomic orientations toward Britain and the south. Both countries have experienced political and economic instabilities caused by war in neighboring countries. Additionally, both countries have depended on one or two products whose exports were expected to meet the needs of government and population alike. In Zambia the product was a mineral, while Malawi depended on tobacco and human labor. Finally, both countries endured decades of single-party rule and long-term presidencies. In Zambia after independence, government and economics were put on a collision course, while in Malawi limited economic opportunity and realpolitik led to a policy of cooperation with apartheid South Africa that most independent black African countries found repugnant. In the 1990s both countries adopted economic liberalism, multiparty democracy, and export-led development.

EARLY CHIEFDOMS, COLONIAL BEGINNINGS, AND ORIENTATION

The areas now known as Zambia and Malawi were originally settled by Bantu-speaking peoples from the north. Archaeological evidence indicates that a single iron-using, farming, and cattle-raising culture existed in the region of central Zambia to Lake Malawi about a thousand years ago and persisted for two centuries (1000–1200). These agricultural people used iron axes to clear forests for cultivation. During the 1400s the first chieftaincies in the region arose. The early chiefdoms and their peoples were known as "Maravi," or "peoples of the fire," probably referring to the groups' ceremonial use of fire. It is likely that one of the products local groups exported for the Indian Ocean trade was ivory. In the fifteenth century the Maravi pushed east to Lake Malawi, and the Luba–Lunda kingdoms extended south to the Kafue River and east to Lake Bangwuelu. In the early nineteenth century, small numbers of Nguni-speaking peoples (Sotho, Zulu for example), arrived from the south. By the mid-nineteenth century the various peoples of Zambia and Malawi were largely established in the areas they occupy today, the dominant groups being the Nyanja (Chewa), Bemba, Lomwe, Yao, Tonga, and Tumbuka. These groups speak remarkably similar Bantu languages; so close that some linguists consider them dialects of the same language (Table 28-1). A rich diversity of agricultural and pastoral practices developed and prevailed alongside highly competitive political systems and long-distance trade, especially in captives for the Arab slave trade.

The impacts of the Arabs and Swahili slavers and traders on local groups was profound—devastating for those who were plundered and enslaved; but for those who were allies, there was cultural assimilation. The impact of can be seen in Table 28-1 for the Nyiha language, spoken mainly in the Northern Province of Zambia. A glance at the table will show that in Nyiha the numbers 6 through 9 are not similar to the other languages listed in the table; obviously they do not fit the Bantu language pattern. These words were

TABLE 28-1 COMPARATIVE TABLE OF THE NUMBERS 1 TO 10 FOR EIGHT CENTRAL BANTU LANGUAGES SPOKEN IN MALAWI AND ZAMBIA.

	Malawi				Zambia			
Language	**Tumbuka**	**Nyanja**	**Yao**	**Lomwe**	**Bemba**	**Nyiha**	**Nsenga**	**Tonga**
Province or Region	**Northern**	**Central, Southern**	**Southern**	**Southeast**	**Northern, Copper Belt, Luapula**	**Northern**	**Eastern and Central**	**Southern**
1	mo	chimodzi	mo	mosa	cimo	yooka	mo	mwi
2	uiri	ziwiri	wili	pili	fibili	vavili	wiri	bili
3	tatu	zitatu	tatu	taru	fitatu	zitatu	tatu	tatu
4	nayi	zinai	mcheche	nai	fine	zine	ne	ne
5	nkonde	zisanu	msano	tanu	fisano	zisano	sano	sanu
6	nkonde na chimoza	zisanu ndi chimodzi	msano na mo	tanu na moha	mutanda	sita	sano na mo	musanu mwi
7	nkonde na viuiri	zisanu ndi ziwiri	msano na wili	tanu na pili	cine lubali	saba	sano na wiri	musanu bili
8	nkonde na vivatu	zisanu ndi zitatu	msano na tatu	tanu na taru	cine konse	umunana	sano na tatu	musanu tatu
9	nkonde na vinayi	zisanu ndi zinai	msano na mcheche	tanu na nai	pabula konse	tisa	sano na nai	musanu ne
10	khumi	khumi	likumi	li-kumi	ikumi limo	ishumi	kumi	ikumi

Source: Rosenfelder (2004).

borrowed from the Arabic (through Swahili probably), sitta, sab'a, thamaniya, and tis'a (ستة, سبعة, ثمانية, and تسعة.) for 6, 7, 8, and 9, respectively. Sounds in the Arabic words that were unfamiliar to the Nyiha speakers were simplified to fit their language. What would account for such borrowing?

It is likely that the Nyiha, or some groups of Nyiha, worked so closely with the Arab and Swahili slavers and traders that the Arabic numbers simply replaced the original Nyiha ones. An example for which we know more about relates to French and Arabic in Algeria. As a result of their long colonial relationship, the French words for counting were used in Algeria in place of Arabic words up until recently. Cultures that borrow in this manner may be in a subordinate relationship with a dominant culture in which the language of the dominant culture becomes more important for certain activities. In addition, more than just numbers diffused to peoples in northern Zambia and Malawi. Muslims are more numerous in the Northern Province of both countries than elsewhere in the country.

As the region was plundered by slavers, and the delicately balanced subsistence economies were destroyed, Christian missionaries, traders, and explorers penetrated from the south and east. David Livingstone reached the Victoria Falls in 1855 and spent the next 18 years mapping out the terrain, spreading the Gospel, and promoting legitimate trade. His travels took him up and down the Zambezi and Shire rivers, the length of Lake Malawi, and into northern Zambia. His vivid descriptions of the land and peoples provided the impetus for further missionary work and whetted the appetites of those with colonial ambitions, among them Cecil Rhodes.

Cecil Rhodes, who had made his fortune from diamonds and gold in South Africa (between 1870 and 1890), wished to see the highveld of central and eastern Africa under British rule and settlement. He also wanted to annex to his personal mining empire the copper deposits of present-day Katanga. The British government, concerned at the time with maintaining British supremacy at the Cape and anxious to prevent German, Portuguese, and Boer settlement north of the

Limpopo, granted a charter to Rhodes's newly formed British South Africa Company in 1889, giving it powers to make treaties and conduct administration north of this river. Thus, in 1890, the Pioneer Column, a group of about 190 whites, left Rhodes's farm near Kimberley and trekked north through eastern Botswana and onto the highveld of what is now Zimbabwe. There the company concentrated its activities, subjugated the Africans, and expropriated their land.

Rhodes's personal ambitions north of the Zambezi were never fully realized (the rich copper deposits of Katanga falling to King Leopold II of Belgium) but, through treaty and coercion, most of what became Northern Rhodesia (Zambia) came firmly within the British sphere of influence. The company accepted administrative responsibility for the territory in return for the mineral rights, but its profits were few. Ores were discovered at Kansanshi, Kabwe (formerly Broken Hill), and in the area of the present Copper Belt, but transport costs inhibited small-scale operations, while large investments went to the Congo and elsewhere. By 1909 the British South Africa Company had built a railway across the territory to Katanga, and copper began to move south to the Indian Ocean port of Beira. Coal moved north from Hwange (formerly Wankie), Rhodesia, and the small settler community along the line of rail provided the mines with maize and cattle. Further east, following successful Church of Scotland missionary activity and opposition to the slave trade, the British established the Nyasaland Protectorate (Malawi) in 1891.

Colonial rule meant a new economic order. In both territories European settlers obtained land at nominal prices and, with the introduction of hut and poll taxes, thousands of Africans were forced to work several months of each year on European estates growing coffee and tobacco in Nyasaland and tobacco and maize in Northern Rhodesia. Still others were forced to meet their needs by working in the mines and on the farms of Southern Rhodesia and South Africa. By 1903, some 6,000 migrants had left Nyasaland for work in the south. Land alienation meant that many Africans were made tenants-at-will, with little or no legal right to the land they cultivated, and the result was a typically colonial dual economy. But real economic progress was slow in coming. Distances were great, communications inadequate, and markets small. The Europeans of Northern Rhodesia resented the British South Africa Company's restrictive policies on land and mineral rights, and the imposition of an income tax in 1920. The company itself was in financial difficulty, so in 1924 it transferred its administrative responsibilities for Northern Rhodesia to the British Colonial Office, and the territory became a protectorate.

In the late 1920s, huge deposits of copper ore were discovered south of the Congolese border, and the Copper Belt was born. By 1930 there were nearly 30,000 Africans employed in the region, and the European population for the territory as a whole reached almost 14,000, up from 3,600 a decade earlier. But the Great Depression delayed development: thousands of employees were dismissed, and many Europeans left the territory. With economic recovery, the mines were brought into full production; mining-related industries were started in Mufulira, Chingola, and Ndola; and the region became the protectorate's economic heart (Figure 28-1). European immigration accelerated, many of the new arrivals settling not only in the Copper Belt but also in towns and on farms along the line of rail.

In neighboring Nyasaland, events took a different course. The railway from Nsaje to Blantyre-Limbe was built between 1903 and 1908 and was extended north to Salima, near Lilongwe, in 1935. Between 1900 and 1920 the number of Europeans increased as tea and coffee plantations and other agricultural enterprises were developed in the Shire Highlands. Few settlers went to the northern areas. In Nyasaland too, economic dualism prevailed, and the Africans became increasingly dependent on work opportunities in Southern Rhodesia and South Africa. Contact with the south meant rising expectations, firsthand experience in social discrimination, and rural discontent.

Following World War II, Britain proposed that Northern Rhodesia, Nyasaland, and Southern Rhodesia be united in a federation, and in 1953 this became a reality. Most Africans opposed closer links with Southern Rhodesia, fearing stronger settler control north of the Zambezi. In particular, they feared land alienation like that in the south, and they feared an erosion of political rights guaranteed under the protectorate system. Despite their protests, the new arrangement was imposed and the Central African Federation was established.

Britain argued that federation would benefit each component from an economic point of view,

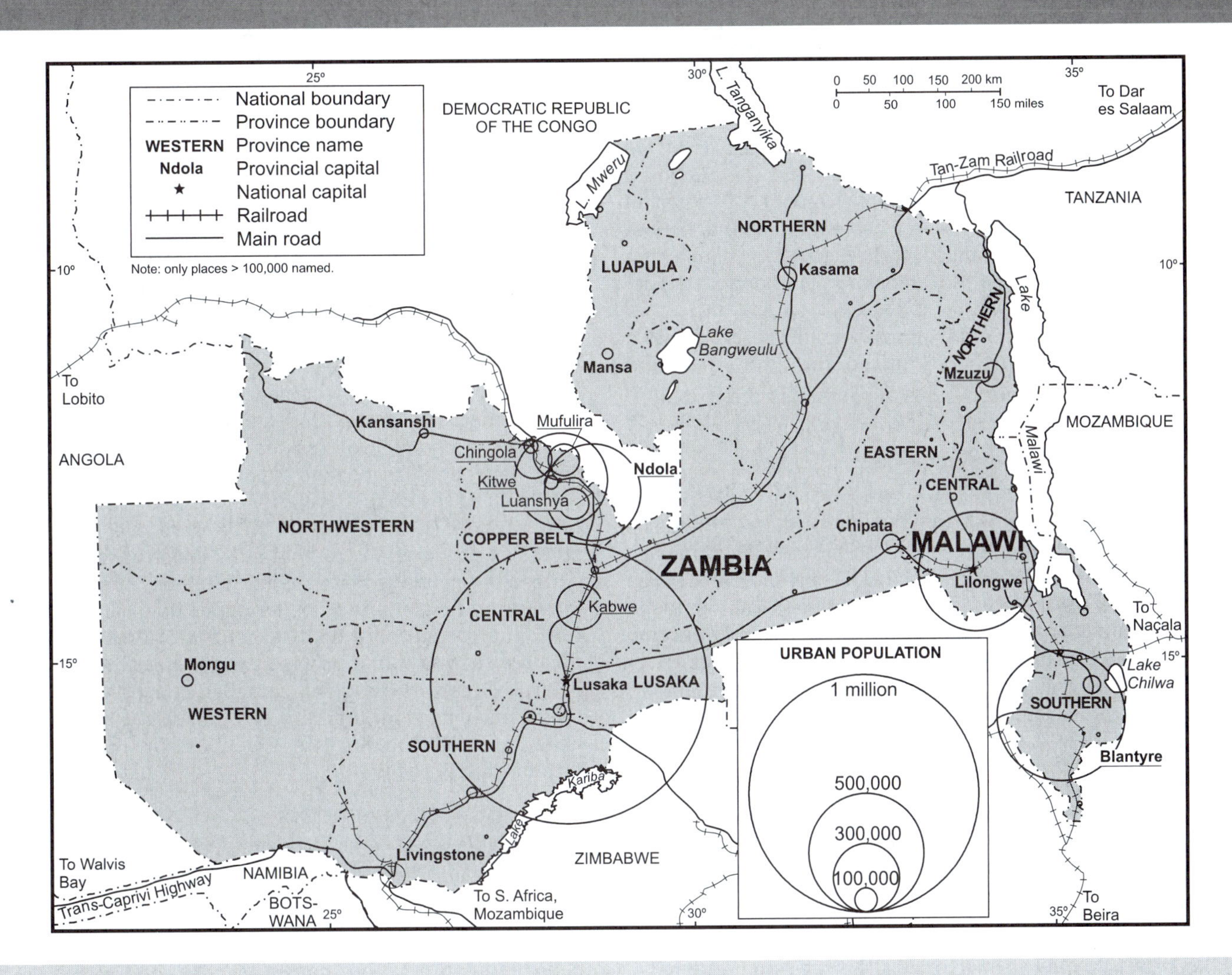

FIGURE 28-1 Provinces, urban places and populations, infrastructure, and international connections for Zambia and Malawi.

providing a better balance, greater diversification, and a wider resource base for the economic efforts of the new state. Federation worked primarily to the advantage of the European settlers in Southern Rhodesia, who dominated the federal parliament, seated in Harare (formerly Salisbury). Southern Rhodesia diversified and prospered under the arrangement, but the northern partners did not. Much of the revenue generated in the Copper Belt and along the line of rail was used to fund development projects south of the Zambezi, not in Northern Rhodesia where the needs were greater. More importantly, federation meant an increased dependence of the northern partners on Southern Rhodesia for jobs, manufactures, and technicians. Consequently, outside the mining sector in Northern Rhodesia, and agriculture in Nyasaland, development was extremely limited.

Federation, however, produced closer economic ties between the two Rhodesias than between either of the Rhodesias and Nyasaland (Sills 1974). The railway

BOX 28-1

DR. KENNETH KAUNDA, ZAMBIA'S FIRST PRESIDENT

Dr. Kenneth David Kaunda, born on April 28, 1924, led Zambia to independence in 1964 and was Zambia's first president. He held that office until 1991 when, in the first elections to ever be held in Zambia, he lost to former trade union chief Frederick Chiluba, the candidate of the Movement for Multiparty Democracy (MMD).

The son of David Kaunda, a minister and missionary of the Church of Scotland, the future president attended Fort Hare University in South Africa. He prepared for a career in education, and after having served as a teacher at Lubwa School, he became headmaster there. In 1950 he became secretary of the Northern Rhodesia African Congress Party. He opposed the formation of the Federation of Rhodesia and Nyasaland in the early 1950s. In 1958 he formed the Zambia African National Congress (ZANC), was banned in 1959, and was arrested and imprisoned in 1960. After his release from prison Kaunda became president of the United National Independence Party (UNIP), and in a 1962 election he won a seat in parliament, became minister of local government and social welfare in the federation he had so strongly opposed, and set in motion the process that would quickly result in the dissolution of the federation in 1963.

Kaunda banned all political parties except his own after violence broke out during the elections of 1968, and 4 years later he made Zambia a one-party state. He went on to win five elections but was forced by internal opposition to restore multiparty democracy. His administration was plagued with charges of corruption, and after the decline in the copper industry, Kaunda and UNIP lost much public support. Nevertheless, he was able to maintain his position because he and UNIP controlled the levers of power in the one-party state. Opposition was repressed. Kaunda's economic policies made Zambia increasingly dependent on copper exports for income. He nationalized the copper industry in 1969. By the end of his tenure in office (1991) there were over 300 state-run corporations (parastatals). Most were privatized in the mid- to late 1990s. Voted out of office in 1991, he became head of an opposition party in 1995 and carried on a political feud with President Chiluba. Fearing Kaunda's presidential ambitions, in 1996 Chiluba's government passed a constitutional amendment barring Kaunda's election to the presidency. In 2000, in his late seventies, Dr. Kaunda retired from public life.

connection over the Victoria Falls Bridge strengthened these ties: coal moved freely from Hwange northward, together with manufactured goods from Southern Rhodesia and South Africa, while this became the Copper Belt's principal outlet to its overseas markets. Nyasaland, in contrast, separated from Southern Rhodesia by a wedge of Mozambique territory, was linked by rail with Beira. Road connections between the two northern partners were poor, and they had little to exchange with each other. The two Rhodesias were drawn even closer together as partners in the massive Kariba Dam project. Mounting African opposition to federation, especially from the northern members, led to a British reconsideration of it and finally, in 1963, to dissolution. In July 1964, Nyasaland became independent under the name Malawi, and three months later Northern Rhodesia became Zambia. Independence meant new raisons d'être for each state, and in the case of Zambia, significant reorientation in its politicoeconomic alignments.

ZAMBIA

Reorientations After Independence

Zambia's independence was led by Dr. Kenneth Kaunda (Box 28-1), staunch opponent of the Federation of Rhodesias and Nyasaland, and by the United National Independence Party (UNIP). In 1966, two years after independence, Dr. Kaunda, an avowed democrat and defender of the rights of man, declared that "We must think and think and think again about how best we shall serve and not about how important we are as leaders of our people, or how we can safeguard our own positions as leaders. To my fellow leaders on the continent of Africa I would venture to send this message—that our task and challenge is to try and help establish governments of the people, by the people and for the people" (Kaunda 1966). It is ironic that during his tenure as president of Zambia, Dr. Kaunda seemed more interested in staying in power than in establishing and promoting democracy. We will examine some of his achievements in this chapter.

With independence, Zambia attempted to reduce its dependence on the industrial complex of white-ruled Southern Africa. An unexpected impetus to this was provided for in November 1965 when Rhodesia unilaterally declared its independence of Great Britain. To force the Ian Smith regime to back down, the United Nations imposed sanctions against Rhodesia, and since Zambia was so dependent on Rhodesian technology, transit, and manufactures, there was good reason to fear economic (and political) chaos in Zambia. The immediate problems centered on the export of copper and the import of oil and other essential goods. Britain, Canada, and the United States provided military planes to handle these materials, but not in sufficient quantity to meet both local and overseas needs. The shortage of coal and transport dislocations retarded growth of copper production, and it was not until 1969 that the 1965 output levels were surpassed. Oil became a scarce commodity, and the cost of food and most imported goods soared.

Zambia met the challenge by concentrating its efforts (with considerable sacrifice) on developing its own industrial and energy resources, increasing its industrial output of both consumer and capital goods, and most importantly by utilizing Tanzanian and Angolan outlets for its copper. A coal mine was brought into production at Maamba to replace imported supplies from Hwange, and electrical output was increased from the Victoria Falls, while construction of a new power station on the Zambian side of the Kariba Dam was started. A new hydroelectric project was started on the lower Kafue, and later another on the middle Kafue. In 1972 the Kafue River project was completed and a 600-megawatt power plant went into operation. The Kafue project was specifically developed during white minority rule in Rhodesia (Zimbabwe) to reduce Zambia's dependence on Kariba, and especially on the output from the Rhodesian turbines. The Kafue River rises in the Copper Belt and joins the Zambezi near Chirundu, about 40 miles (64 km) downstream from the Kariba Dam (Figure 28-3). In its upper reaches it provides water to the Copper Belt's towns and mines. It then flows through the tsetse-infested Lukanga Swamp south to the Itezhi-Tezhi Gorge, where a 900-megawatt capacity dam was later built as well. There the river turns sharply east and flows through the Kafue Flats, where cotton and maize are important cash crops, and fishing is an important industry. At the small industrial town of Kafue, water is diverted to Lusaka, Namwala, and other towns along the line of rail. East of the railway the Kafue plunges some 1,900 feet (580 km) in 12.5 miles (20 km) into the Zambezi Gorge. The copper industry uses about 70% of total output. The supply of water is regulated from Itezhi-Tezhi. Together with the Kariba Dam and power installations at Victoria Falls, the Kafue project made Zambia self-sufficient in its electricity needs. Today, Zambia obtains almost 100% of its electricity from hydropower, and the export of electricity contributes significantly to the economy.

By 1971 the main road between Zambia and Tanzania was paved, a 1,060-mile (1,705 km) pipeline had been laid between Ndola and Dar es Salaam by 1968, and a refinery opened at the latter site soon after. In addition, plans pushed ahead for the Tan-Zam Railway between Dar es Salaam and the Copper Belt. At the same time the Zambian government acquired a 51% holding in mining and strategic industries in an effort to control the state's industrial requirements. Zambianization of the economy and administration was accelerated as well—with the state at the helm. Efforts to check rural–urban migration and to increase production of maize, meat, and other foods to offset Rhodesian imports were far less successful.

The situation worsened in January 1973 when Rhodesia closed its borders with Zambia, ostensibly in retaliation for guerrilla incursions from across the Zambezi. A month later, Rhodesia reopened the border to the copper traffic (since Rhodesia stood to lose valuable transit dues), but the measure backfired as Zambia decided to reroute all its exports to the north. Many states came to Zambia's aid by subsidizing copper exports by road through Tanzania and Kenya,

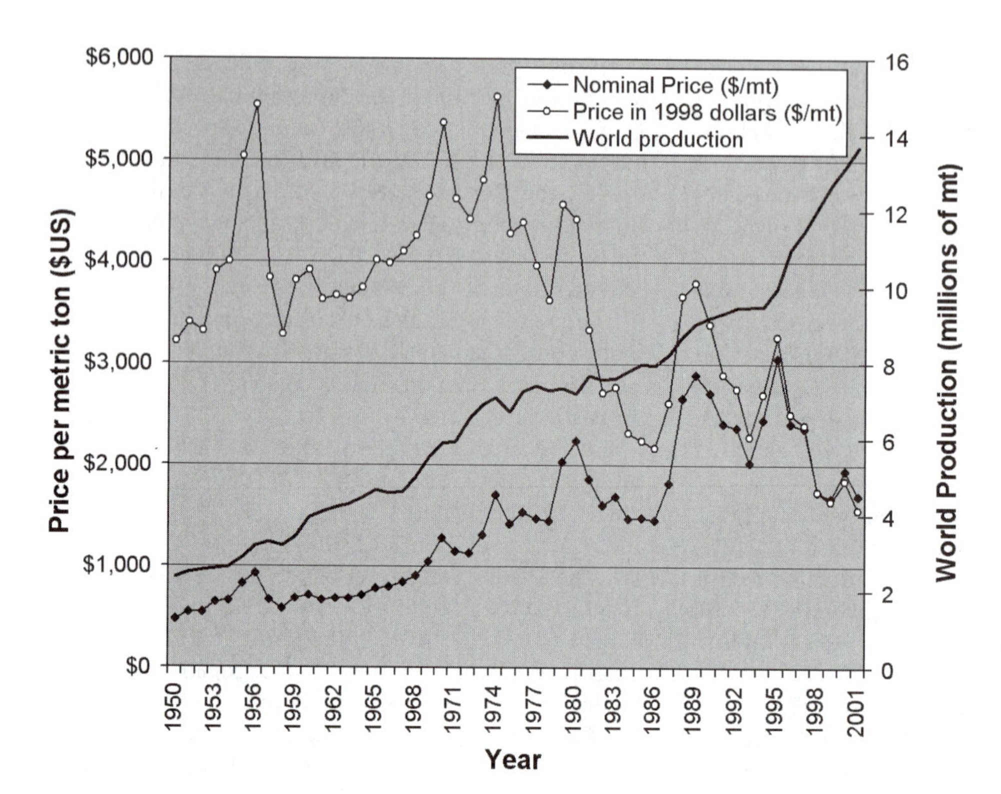

FIGURE 28-2 World copper production and prices, 1950 to 2002.

while essential mining supplies and oil were airlifted from South Africa and the Mozambican port of Beira, respectively. The original rerouting plan placed most of the burden on Dar es Salaam and Lobito, Angola, which were the best equipped to handle the 61% of imports and 50% of exports that previously had traveled via Rhodesian Railways. Dar was to take 43,000 tons of imports and 20,000 tons of exports each month, while Lobito's monthly quotas were 35,000 and 30,000 tons, respectively. Zambia's projected monthly imports and exports were 117,000 and 67,000 tons, but because of fluctuating copper prices, the flow of copper was irregular, and so too was the inflow of merchandise.

Zambia is a landlocked country, and its imports and exports must be received and expedited by neighboring countries under any circumstance. Rail distances are tremendous, and transportation charges can add up to 30% to the cost of a good (U.S. Department of State 2000). Four ports handle most of Zambia's foreign trade: Dar es Salaam, Beira, Durban, and Walvis Bay. Each has its particular assets and problems. Zambia's natural and closest outlet (just over 1,000 miles; 1,600 km) to the west is Lobito, a modernized port with refrigerated and open air storage theoretically capable of taking all of Zambia's exports. This port closed in 1975 and remained effectively closed for 28 years as a consequence of the Angolan war for independence and subsequent civil war. Even before war broke out, increasing port surcharges, port inefficiencies, a shortage of rolling stock, and political uncertainties in Angola (and even Zaïre), made Lobito far from ideal. Just prior to that closure, Lobito was handling 40% of Zambia's imports and 30,000 tons of copper each month. The Benguela Railway normally operated at capacity, serving both Zambia and Zaïre, but even at the time of the war in Angola it needed realigning and upgrading. Today, much more investment is needed to rehabilitate the track and rolling stock if Lobito is to become competitive with Zambia's options on the east side of the continent. The distance from Ndola in the Zambian Copper Belt to either Dar es Salaam, Tanzania, or Beira is almost the same: 1,250 miles (2,000 km). Durban, South Africa (via Zimbabwe), is over 1,550 miles (2,500 km) away.

Dar es Salaam, Tanzania's principal port and former capital, has been the main outlet for Zambia's copper and the country's major port of entry for manufactures. Following the unilateral declaration of independence by the white minority government in Zimbabwe, copper was hauled to Dar over the Great North Road—once known as the "Hell Run" because of its deplorable conditions. Before the road was upgraded in 1971, some sections were impassable during the rainy seasons, others in the mountains were too narrow for the large trucks, and maintenance facilities en route were grossly inadequate. Costs were high, the service unreliable, and Dar itself had insufficient ore-loading and storage facilities. By mid-1976 the new Tan-Zam Railway (or Tazara), built with Chinese financial and technical assistance, was operating near to capacity, but Dar's port facilities were unable to adequately handle the higher volume of traffic. The three remaining ports, Beira, Durban, and most recently, Walvis Bay, handle the remainder of the traffic. Although there is no direct connection between the port of Beira and the Copper Belt, and goods must travel through Zimbabwe, Beira is important to Zambia; and facilities in the Mozambican port have been modernized. A direct link either between Ndola and Beira or Ndola and Naçala would be about 1,100 miles (1,770 km) long and the cheapest route to the sea from the Copper Belt. Zambia and Malawi have agreed in principle to a rail connection to feed into the Naçala system, and serious discussions are under way with Mozambique regarding the extension of the Beira line northward beyond Tete and into Zambia. The low price of copper since the 1970s may discourage such investments (Figure 28-2). Zambia's transportation infrastructure (roads and rails) is reasonably adequate in general; 20% of the main roads are paved and 20% are of gravel. In 1990 the road link to Walvis Bay, Namibia, began to be used, and since that time the Trans-Caprivi Highway has made road transportation via Walvis Bay more attractive. In addition, Zambia has one of the best telephone systems in Subsaharan Africa.

Aid and Reform in Zambia

Before independence and as late as 1979, Zambia was viewed as a middle-income developing country with a per capita GDP three times Kenya's, twice Egypt's, above such developing countries as South Korea, Malaysia, Turkey, and Brazil, and fast approaching that of some of the poorer European countries like Portugal and Spain (Ferguson 1999). Zambia had what appeared to be a solid industrial base, reliable markets, and a relatively educated, capable, and urbanizing population—in short, a model for successful development after independence. All of this changed during the late 1970s, and Zambia experienced "reverse development" from which it has yet to emerge. By almost every measure, Zambia has lost ground in its human and economic development.

Why did it happen and, more importantly, why has economic stagnation, and the concomitant human misery, persisted for so long? It is well known that the world market price for copper plummeted in the late 1970s (Figure 28-2). Is Zambia a victim of the world market and globalization? A careful investigation reveals that the problems are much more complex, rooted locally as well as internationally. Several variables were involved, and the price of copper is a proximate cause rather than an ultimate cause of Zambia's decline. Let's look at some of the other variables.

After independence Zambia became a one-party state that monopolized the extraction, transportation, marketing, and profit from copper, Zambia's principal resource. This dual monopoly over both political and economic power had profound, far-reaching, and ultimately pernicious consequences for both the political life and the economy of the country. In the national equation there were three variables, each of which was erroneously believed to be a constant by the Zambian government and by Zambians who reflected on their future. The first "constant" was the state monopoly over economic activity, the second was the official party monopoly over political expression, while the third was world demand (hence price) for copper.

The late-1970s decline in demand for copper due to the economic downturn and instability in world economies had been induced by the rise in petroleum prices beginning earlier in the decade. As customers substituted cheaper plastics for copper in such items as tubing, the Zambian economy was plunged into crisis, revealing the inappropriateness of the government's economic policies.

It wasn't that the Zambian government smothered free enterprise, it simply did not permit it. The state owned and managed the economy as a part of the normal responsibilities of sovereignty. The state was the principal employer, teacher, planner, welfare provider, and economic risk taker. The state took revenue generated from mining and employed it elsewhere in the socialist economy, neglecting to invest in modernizing the mining industry and rail transportation. Over time, production became expensive, and just how expensive was revealed when the days of high prices ended. In a free-market democracy, on the other hand, adherence to disastrous, short-sighted economic policies would rapidly lead to a new government with more progressive economic policies.

But not only did the Zambian government control the economic activity in the country, it also controlled political expression and paternalistically believed its policies were in the best interests of the people—after all hadn't the government led the people to independence? Thus by clinging to the first two "constants," Zambia was poorly positioned to respond quickly in economic terms when world demand for copper dropped, invalidating the third presumed constant.

In the 1980s continued policy disasters and economic deterioration forced the UNIP government of Kenneth Kaunda to adopt a structural adjustment program (SAP) developed by the World Bank and International Monetary Fund to address the crisis. But President Kaunda did not fully implement the changes because political protest in the towns and cities threatened to undermine his traditional base of political support—and after all, free-market economics as espoused by the World Bank and IMF was antisocialist and would pull the government safety net out from under the people.

Political monopolies are almost invariably corrupting, but economic monopolies are not always bad. Research has shown that where economic competition exists, a monopolistic company will invest in research and development and innovate to maintain its position (Etro 2004). Where a single entity has the power to keep out competitors, however, monopoly stifles change and becomes fat and lazy. The problem for Zambia was the unwise vesting of the one-party, paternalistic state as the national "firm." There was little investment in the copper industry as the government siphoned off revenue generated from copper exports in the good days for unproductive national projects, and much was lost through rent-seeking (corrupt practices) by officials. When prices declined, rather than innovate, the government allowed the state, the economy, and human development initiatives to decline; in just over a decade Zambia went from "middle income" to "basket case." In the late 1980s and early 1990s, economic decline became the rallying cry of the political opposition and, allied to Western donor nations, the structural adjustment program was revived. But it was too late for UNIP and Kenneth Kaunda. In 1991 the Movement for Multiparty Democracy (MMD) ended Kaunda's 17-year tenure as president of Zambia. The MMD made economic reform and structural adjustment part of the party platform, and as a result the World Bank and Western donor countries expected Zambia to became a model of reform and adherence to structural adjustment (Rakner et al. 1999). Paradoxically, the financial assistance that flowed into Zambia after the change in government acted as a disincentive to real change in the early years of the MMD regime (1991–1994). After the MMD's victory, corruption continued under Kaunda's successor, Frederick Chiluba. Presiding over the privatization of many of the state corporations, Chiluba channeled money to himself and his friends. After his unsuccessful bid to change the constitution so that he could run again for president, his hand-picked successor, Levy Mwanawasa, won the presidency. Soon after leaving office, Mr. Chiluba was involved in a long legal process for corruption, apparently on a far greater scale than was practiced during Kaunda's time. Chiluba was never indicted because the prosecutors were unable to have his immunity from prosecution lifted.

Revival of the Zambian copper industry depends on how cheaply copper can be produced and marketed. Zambian production costs, from the pit right up to the port, are higher than those its competitors face. For Zambia to become competitive in all of its mineral exports, especially copper, it must move from state to private ownership of the means of production. The state is an inefficient employer, often corrupt when it is too involved in the economy, and tending to discourage incentives for individual initiative when it is the employer of first resort.

Privatization of the Zambia copper industry (ZCCM) began in 1995, stalled for several years, and was completed in 2000. Growth of the mining sector then began to pick up, contributing to the 3 and 5.2% increases in GDP in 2000 and 2001, respectively. Many of the over 300 Zambian parastatals had been privatized along with copper, and the bloated civil service was reduced somewhat, as well. Now the copper industry needs capital investment, modernization, and streamlining, as do the transportation infrastructure and management. Transformation of the insolvent Zambia Railways would make mineral and agricultural exports cheaper and reduce the costs of imports; in addition, a

BOX 28-2

PRESIDENT HASTINGS KAMUZU BANDA, MALAWI'S FIRST PRESIDENT

Dr. Hastings Kamuzu Banda, Malawi's first president, ruled Malawi for 30 years. After receiving his early education in rural missionary schools, he obtained a medical degree in the United States. While practicing medicine in London, he made contact with other African expatriates and became involved in the early independence movement. He returned to Nyasaland (today's Malawi) to work against the Federation of Rhodesia and Nyasaland. He rose in the ranks of his political party, the Malawi Congress Party, which enjoyed large successes during the time of the federation.

In 1964 Nyasaland became independent Malawi, with Dr. Banda as its prime minister. He became president in 1966 and had himself appointed president-for-life in 1971. Banda's administration became increasingly corrupt and autocratic; the life president led a life of luxury while the political opposition was jailed or killed. Increasing unrest and pressure from the Western donor countries in the early 1990s motivated Banda to abandon the one-party-state concept and hold multiparty elections. He lost the presidency in 1994 to Bakili Muluzi, a Muslim from northern Malawi from the United Democratic Front (UDF) Party. Dr. Banda died in 1997. Muluzi was reelected in the 1999 election. In 2004 Bingu wa Mutharika of the UDF was elected president in a vote that was marred by anomalies including charges of election fraud, demonstrations, and killings.

well-functioning, efficient rail system would encourage tourists to return to Zambia. The government has made halting steps toward the privatization of Zambia Railways. According to the World Bank, employment must shrink from 6,000 to about 1,000 employees. During the Kaunda years and even later, despite the rhetoric of reform by the MDP, employment with Zambian Railways was a gift of patronage given to reward loyalty.

MALAWIAN ORIENTATIONS

Malawi's relationships with its neighbors are, and have always been, different from Zambia's. This is because Malawi is bounded by three rather than eight other territories, Mozambique, Tanzania, and Zambia. Geographically Malawi is dominated by Mozambique (and not Zimbabwe), and its political leadership has adopted and adhered to a "southward-looking" policy of economic and political alignment in the past. South Africa, and not Zimbabwe, has been its principal southern ally and trading partner. When independence was granted in 1964, it was generally assumed that Malawi would seek closer political and economic ties with its independent colleagues to the north. This included support of the OAU in its determination to liberate the white-controlled southern territories, condemnation of Rhodesia's unilateral declaration of independence, and membership in the East African Community. Malawi's president, Hastings Kamuzu Banda, steadfastly resisted such measures, and Malawi relied heavily upon South Africa for its economic needs during the apartheid years and upon Portuguese Mozambique for access to the sea. Banda (Box 28-2) warned other African states of the dangers of accepting aid from Communist countries, supported dialogue and détente with South Africa, forbade Mozambique's Frelimo "freedom fighters" from using Malawian territory, and established diplomatic relations with South Africa. Contrary to what most African leaders were doing during the apartheid years, Banda repeatedly stated that if Africa's leaders were genuinely interested in promoting the welfare of blacks in white-ruled southern Africa, they should talk with, and not fight, the political leadership in South Africa and Rhodesia (Reltsma 1974). Such a position led some African states to call for Malawi's expulsion from the OAU.

Banda's policy, unlike Kaunda's, was based on realism and was much more cooperative than confrontational than the policy adopted by Zambia. More than 100,000 Malawians were employed each year during the apartheid years in South Africa's mines, and an additional 200,000 worked in Rhodesia. More than twice as many Malawians worked outside Malawi than had

paid employment in it during those years. Even today, although reduced, international employment occupies a fifth to a third of Malawians. During the period of white minority rule in South Africa and Rhodesia, remittances from Malawian workers constituted Malawi's third most important source of foreign revenue. During the 1980s and 1990s South African mining became less labor intensive, and employment opportunities for Malawians have decreased. Close relations with South Africa during Banda's tenure in office extended into other fields as well. South Africa helped Malawi build a new capital city at Lilongwe and an international airport to serve it. It provided both technical assistance and financial aid for the national highway between Lilongwe and Zomba (the old capital), and the rail extension east of the Lichinga (formerly Vila Cabral) to Naçala line in Mozambique.

Malawi's first capital was established toward the end of the nineteenth century at Zomba, about 40 miles (64 km) northeast of the country's twin cities, Blantyre-Limbe. Zomba never shared in the development of the south and was almost destroyed by a storm in 1946. A commission was established to select a new site, but interest waned during the years of federation with the Rhodesias. In 1964 H. K. Banda, who was then prime minister, proposed to parliament that the capital be moved to Lilongwe, and this was agreed to in 1968. Lilongwe was selected partly because of its central location in Malawi and partly to encourage economic development north of the economically dominant southern region. The city sits in the largest and one of the most productive agricultural plains in the country, where tobacco is the principal cash crop. The new capital was added to old Lilongwe, historically the chief service center for much of the north. The city itself has four major areas: old Lilongwe, Capital Hill (the site of government buildings and main shopping areas), and two industrial zones and their associated residential areas. Population in 1975 was estimated at 55,000, and by the end of the century it had reached 500,000. Built with South African assistance, Lilongwe is joined by modern highway with Chipata (in Zambia), Blantyre-Limbe, and Zomba, and the railway was extended to it from the central Malawi port of Salima. Special concessions have been offered to entrepreneurs wishing to establish industry there.

Landlocked Malawi is totally dependent on Mozambique for access to the sea. Beira, due south of Malawi, is its closest outlet, with which it has been linked by rail since 1908. Until the opening of the Liwonde–Cuamba–Naçala line to the west in 1970, the Beira route carried almost 100% of Malawi's tea, tobacco, cotton, and coffee exports, together with most of its imports. From 1970 to 1975, much of the traffic originating in central and northern Malawi (along with Zambian copper) was exported through Naçala, one of the best deepwater ports in Africa. Both routes faced disruption during Mozambique's anticolonial wars of the late 1960s and early 1970s. While Mozambique was still Portuguese, and Malawi was part of the Central African Federation, a limited amount of traffic passed between the two territories. But during Mozambique's war of independence, this was severed completely as Frelimo units mounted their attacks on the Cahora Bassa Dam and the town of Tete, midway between Rhodesia and Malawi. The Naçala corridor has still has not recovered the traffic it carried during the 1970s; the track and port need to be renovated. In 2003 Malawi, Mozambique, and Zambia signed a memo of understanding regarding the rehabilitation of the port of Naçala and the transport corridor through Malawi to Zambia. The work will be done by two companies from the United States.

Malawi's transport links with Zimbabwe have never been good, and thus the amount of interstate traffic has been small. Malawi's transport links with Tanzania are almost nonexistent. This is because Malawi's core region lies in the south (Figure 28-1), while its essentially underdeveloped and remote northern areas adjoin equally underdeveloped and remote areas in Tanzania. When the line is extended from Chipata to the Tan-Zam Railway (< 200 miles; 322 km), Malawi will, in fact, have a route (albeit circuitous) through Tanzania.

From independence to 1979, Malawi experienced high economic growth rates and had a relatively favorable balance of trade. Savings as a percentage of GDP increased to 20% in 1979, industrial production expanded at 10% a year, and the average economic growth was 6%. From 1970 to 1980, exports increased in value by $237 million from $48 million to $285 million (Chinsinga 2002). In the early 1980s, the economy began to contract and has never recovered. According the Malawian social scientist Blessings Chinsinga (2002), the causes of the crisis were principally external: the OPEC oil price increases in the 1970s

exacerbated by serious drought in 1980–1981, a sharp decline in the terms of trade, a rising interest rate for international borrowing, the closure of the Beira and Naçala rail corridors, the influx of an estimated 300,000 Mozambican war refugees, and declining aid levels. In contrast, the Malawian economist Chilowa (2000) identified five areas of the Malawian economy that had impediments to growth and development: low growth in smallholder agricultural exports, a very narrow export base dominated by tobacco, dependence on imported fuel and declining stocks of fuelwood, deteriorating finances of parastatal corporations, and govern- ment control of prices and wages.

ZAMBIA AND MALAWI: SOME SIMILARITIES AND CONTRASTS

Besides being landlocked, Zambia and Malawi share a number of important geographical, cultural, and developmental characteristics. They also have their differences. Malawi is a small but environmentally diverse elongated state less than one-sixth the size of Zambia, which is essentially almost compact, with much greater environmental uniformity. Despite their great size differential, their populations are similar (Table 28-2), but Zambia has had a larger expatriate element in the past (50,000 vs 7,000). Malawi's considerable variations in altitude and location within the Rift Valley system produce a wider range of climatic, soil, and vegetation conditions than in Zambia. It possesses some of the most fertile soils in Africa: in the lakeshore plains, the Lake Chilwa–Palombe Plain, the lower Shire Valley, the Lilongwe–Kasungu high plains, and in the tea-producing areas of Chole, Mulanje, and Nkhata Bay. Good soils and humid conditions mean that half the country is cultivable, but less than that is currently in use, with the greatest intensity of land use and the highest population densities occurring in the south.

In contrast, Zambia is essentially a vast, gently undulating plateau with wide shallow basins, lakes, swamps, and grassy plains. It forms the divide between the Zambezi and the Congo drainage systems, and its

TABLE 28-2 PHYSICAL, DEMOGRAPHIC, AND ECONOMIC COMPARATIVE DATA, MALAWI AND ZAMBIA. 1975 (OR CLOSEST DATE) AND 2003.

Variable	Malawi		Zambia	
	1975	2003	1975	2003
Area [mi^2 (km^2)]	45,725 (73,572)		290,583 (467,548)	
Population (millions)	4.9	11.7	5	10.9
Population density [mi^2(km^2)]	107 (41)	256 (99)	17 (6.6)	38 (14.7)
Life expectancy (years)	41	38	44	35
Infant mortality rate per thousand (%)	148	108	157	95
Total fertility rate (%)	7	6.5	6.9	5.9
Population growth rate (%)	2.4	2.2	3.1	1.5
HIV/AIDS infection rate (%)		15		22
GNP per capita (U.S. dollars)	100	160	380	231
Urban dwellers (%)	7.6	15	34.8	40
Literacy rate:		63.7		81.8
Percent of population in agriculture		86		85
GDP by sector: (%) Agriculture		55		15
Industry		19		29
Services		26		56
Inflation rate (%)		27		21
Percent of population below poverty line		55		86

Source: Population Reference Bureau (various dates), Population Data Sheet; World Bank (2003) World Development Indicators.

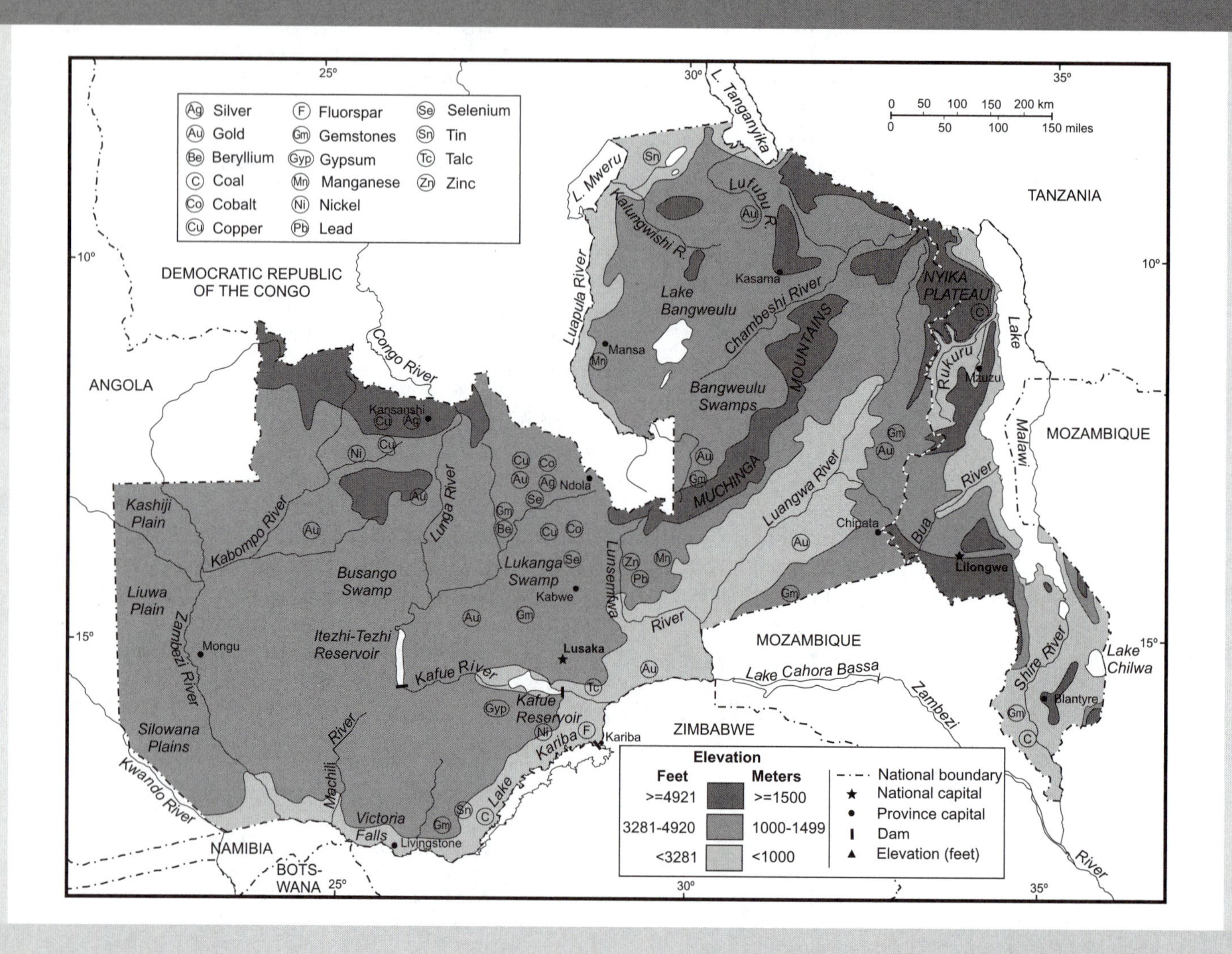

FIGURE 28-3 Physical geography and minerals of Zambia and Malawi.

major topographical features are the Zambezi River and the Muchinga Mountains (Figure 28-3). Its soils are widely infertile, lacking humus and depth on much of the plateau, and badly eroded, shallow, and infertile in the escarpment zones. Population densities vary considerably but are heaviest along the line of rail and around the provincial and district capitals such as Ndola, Kasama, and Mongu and, of course, the national capital, Lusaka (Figure 28-1).

Good environmental conditions in Malawi have contributed to the country's agricultural economy. Tea, cotton, groundnuts, and some tobacco are grown around Blantyre, Mulanje, and Chiromo, while the major tobacco area centers on Lilongwe. Small pockets of cash cropping occur in the northern highlands and along the lakeshore. In Zambia, agriculture plays a negligible role in the export economy—but in a diverse economy it should. Productivity is low, surpluses are few, and the main crops are maize, groundnuts, sorghum, and millet. In Zambia, where the aims are to improve income and nutritional standards, and lower imports, agricultural development is a top priority, neglected during the copper boom years and difficult to afford thereafter. Agriculture in both countries has suffered from periodic drought since the 1990s.

Zambia's golden years of copper-based growth in the 1960s and early 1970s, in contrast to Malawi's agrarian orientation, are reflected in several ways in Table 28-2. Industry's contribution to Zambia's GDP is much higher than that of the predominantly agricultural Malawi. Second, Zambia's relatively high urbanization rate for both years of data reflects industrial cities and towns that grew up along the Copper Belt (and the line of rail). In addition, the literacy rate in Zambia is very high, reflecting urban opportunity—and a government that had the wherewithal to invest in education. The high percentage of Zambians below the poverty line reflects the decline of the copper industry, the deterioration of the social welfare system that depended on it, and a decline in state involvement in the economy. It is expected that a private market economy will generate enough employment for the people, taxes for the government, and wealth in general that the level in human development that was experienced in the past can be regained.

In both countries there are regional and structural imbalances in the economy. Economic dualism is greatest in Zambia. Large disparities of income and living conditions exist between mining and other workers, between Zambians and expatriates, and between persons on the line of rail and those in the periphery. Minerals dominate domestic exports, with copper alone bringing in 90% of the foreign exchange. Copper has financed much of Zambia's manufacturing, most of which is centered on the Copper Belt, and in Lusaka, Kabwe, and Livingstone (Maramba). Since the late 1960s numerous capital-intensive industries (auto and truck assembly) and resource-oriented industries (explosives, chemicals, oil refining, and wire manufacturing) have been established.

In contrast, industry in Malawi is primarily based on agriculture. Although greater emphasis is being given to import substitution in Malawi, over 90% of its foreign exchange is earned through agricultural exports and 40% of that is from tobacco alone (Figure 28-4). According to the Census of Malawi (2003),

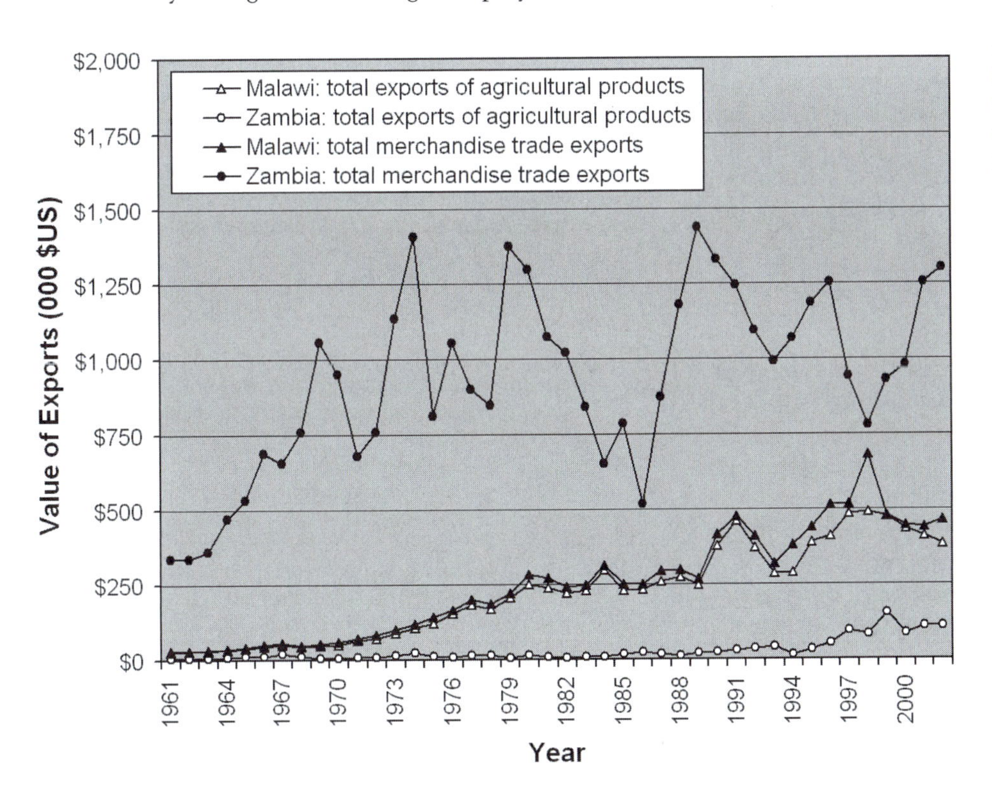

FIGURE 28-4 Agricultural and total merchandise exports for Malawi and Zambia, 1961 to 2003.

tobacco exports accounted for 47% of all export revenue for 1996 to 2000 on average, while tea represented 7% and sugar 5%. For tobacco exports, Malawi stands to gain from the collapse of the Zimbabwean economy since the late 1990s and from the decline in Zimbabwean tobacco exports. The fact remains, however, that Malawi has an undiversified economy and will continue to suffer the ups and downs of demand for the few products it does export. It would be wise for Malawi to diversify its agricultural production because Malawi's situation with regard to tobacco is analogous to Zambia's problem with copper: it is dangerous to be a single-commodity producer and exporter. Disaster awaits. Moreover, tobacco and sugar are crops that have lost some of their attractiveness over recent decades. The dangers of smoking are becoming apparent to more and more people. Sugar is overproduced around the world and its share of the sweetener market has been reduced by artificial sweeteners.

Both countries are being increasingly drawn into an integrated southern African region, and this is manifested in a variety of ways, trade, transportation, and power being three important ones. South Africa is the principal import partner of both Malawi and Zambia, and increasing connectivity is evident from Figure 28-1. Since majority rule in Zimbabwe and South Africa in 1980 and 1994, respectively, power production in southern Africa has been converging rather than diverging. The Southern Africa Power Pool (SAPP) proposed by South Africa in 1995 is intended to put the countries of southern Africa on a common electrical grid so that power can be shared between countries. A key element of this international plan is to upgrade existing physical plant and to built new. Zambia is upgrading its Kafue project to 750 megawatts, upgrading its connection to Inga in the Democratic Republic of the Congo, and connecting its national grid to Tanzania, while Malawi is doubling its power output to 128 megawatts. Zambia intends to sell most of its surplus electricity to Zimbabwe. Ironically less than 20% of Zambia's and Malawi's population have access to electricity.

During apartheid, Malawi, resource poor and sandwiched between resource-rich countries, adopted the role of labor pool for South African industry. This pragmatic policy and political astuteness permitted the one-party rule of President Hastings Banda to continue until he was basically forced out of office. Even though Zambia chose confrontation and Malawi chose cooperation, both countries experienced the negative impacts of their own internal political and economic monopolies: one-party rule and long-term presidencies and state control of the economy. After the establishment of multiparty democracy and economic liberalism, both countries experienced the ambiguity, uncertainty, and vulnerabilities of democracy and the free market. There is no way back to the old days, and with South African aid, investment, and influence, both countries may prosper.

BIBLIOGRAPHY

Chilowa, W. 2000. Adjustment impact on social policy implementation in Malawi. Paper presented at the Annual Colloquium of the South African Research Institute for Policy Studies, Regional Integration in Southern Africa: Past, Present and Future, September 24–28, 2000. Harare, Zimbabwe: SARIPS.

Chinsinga, B. 2002. The politics of poverty alleviation in Malawi: A critical review. In *A Democracy of Chameleons: Politics and Culture in the New Malawi*, H. Englund, ed. Stockholm: Nordiska Afrikainstitutet, pp. 25–42.

Etro, F. 2004. Innovation by leaders. *Economic Journal*, 114(495): 281–303.

Ferguson, J. 1999. *Expectations of Modernity: Myths and Meaning of Urban Life on the Zambian Copperbelt.* Berkeley: University of California Press.

Kaunda, K. D. 1966. Speech delivered at the opening ceremony of the University of Zambia, March 18, 1966. Lusaka, Zambia. Reproduced in the *Internet Modern History Sourcebook*. http://www.fordham.edu/halsall/mod/1966Kaunda-africadev1.html.

Malawi, government of. 2003. *Census of Malawi*. Zomba, Malawi: National Statistical Office of Malawi. http://www.nso.malawi.net/.

Rakner, L., N. van de Walle, and D. Mulaisho. 1999. *Aid and Reform in Zambia: Country Case Study*. New York: World Bank. http://www.worldbank.org/research/aid/africa/zambia2.pdf.

Reitsma, H. J. 1974. Malawi's problems of allegiance. *Tidjschrift voor Economische en Socialegeografie*, 65(6): 421–429.

Rosenfelder, M. 2004. Numbers from 1 to 10 in over 4,500 languages. http://www.zompist.com/numbers.shtml.

Sills, H. D. 1974. The break-up of the Central African Federation. *African Affairs*, 73(290): 50–62.

U.S. Department of State. 2000. *Country commercial guide: Zambia, FY 2001*. Washington, D.C.: U.S. Department of State.

U.S. Geological Survey. 2003. *Copper Statistical Compendium*. Reston Va.: USGS. http://www/minerals.er.usgs.gov/minerals/pubs/commodity/copper/stat/.

CHAPTER 29

Mozambique and Angola

A Legacy of Lusotropicalism and Marxism

Before the independence of Angola and Mozambique Portugal had had a relationship with both areas for about 500 years. A philosophy, called "lusotropicalism," that explained and justified the colonial relationship developed over the years. "Lus" is a word that refers to Portugal. Lusotropicalists argued that God entrusted Portugal with the mission of civilizing the non-European peoples of the world. In this view, Portuguese colonialism was held to be very different from that of other European colonizers. Portuguese colonialism, in contrast to that of Britain, France, Italy and the rest, fostered the development of multi-racial societies in which the life of the colonized was improved. The "theory" was developed by the authoritarian Salazar dictatorship (1932–1968) in Portugal to justify its continued possession of colonies in the face of the independence movements that developed mainly after World War II. In the event, Portuguese development of Angola and Mozambique was minimal and favored Portuguese colonists. The delay of independence by over 15 years compared to the majority of African countries was the result of the lusotropicalism theory and what might have been a smooth transition into independence in 1960 ended up being a catapault into Cold War power politics that had catastrophic consequences for the peoples of both countries. After armed struggle and a chaotic independence, both Angola and Mozambique adopted Marxism–Leninism as guiding philosophy and framework of national development planning. Practically speaking, both countries had abandoned it about 10 years after independence and had renounced it formally just a few years later. Decision making was too centralized; 5-year plans developed in the capital were too far removed from the reality in the countryside to be workable; and rural populations, for which the family had been the unit of production and consumption, were resistant to collectivization. The "big push" ideology that was so focused on the development of heavy industry seemed inappropriate to rural, agricultural Africa.

Both countries had become involved in costly, chaotic civil wars in which civilians were the principal losers. Angola is perhaps the classic case of a country rent by a civil war that was fueled by ethnicity, mineral wealth, ideology, and foreign intervention. Insurgency in Angola began before independence. But while Mozambique's civil war lasted for 17 years, the civil war in Angola lasted 10 years longer—far longer than the usual conflicts that emerged in Africa after independence.

Angola's experience with civil war marks a sharp contrast to its relatively resource-poor cousin on the Indian Ocean side of southern Africa, for Mozambique has enjoyed peace since the early 1990s. In 1992 when Mozambican rebels and government signed UN-brokered peace accords, Angola, after almost 30 years of war, held its first multiparty election. The results were contested by a former rebel group, and the country was plunged into 10 more years of civil war. The cultural and physical geography of Angola contributed enormously to the basic fault lines between the main warring factions: the northern MPLA (People's Movement for the Liberation of Angola) and the southern UNITA (National Union for the Total Independence of Angola). The MPLA government controlled the national oil wells, located in the north, and fueled their military campaigns with petrodollars. UNITA, on the other hand, controlled the diamond fields, which were located in their domain in the south. Revenue from both sources enabled what might have been a short

TABLE 29-1 DEMOGRAPHIC AND ECONOMIC INDICATORS FOR ANGOLA AND MOZAMBIQUE, VARIOUS DATES.

Variable	Year	Angola	Mozambique
Total population (millions)	1975	6.2	10.3
	2000	13.1	18.3
	2003	25.2	17.5
Area [mi^2 (km^2)]		481,351 (1,246,693)	309,496 (801,591)
Population density per square mile (km)	2003	27(11)	59(23)
Annual population growth rate (%)	1975-2000	3	2.3
	2000-2015	3.1	1.7
HIV infection rate (%)	2003	5.5	13
Urban population (as % of total)	1975	17.8	8.7
	2000	34.2	32.1
	2015	44.1	48.2
Population under age 15 (as % of total)	2000	48.2	43.9
	2015	48.5	41.8
Population aged 65 and above (as % of total)	2000	2.8	3.2
	2015	2.6	3.4
Total fertility rate (per woman)	1970-1975	6.6	6.6
	1995-2000	7.2	6.3
GDP per capita PPP (U.S. dollar)	1985	1,510	380
	1995	1,840	730
	2001	2,040	1,140
Average annual GDP growth (%)	1998-2002	6.7	9.1
Manufacturing value added (% of GDP)	1997-2001	4	12
Industry value added (% of GDP)	1997-2001	66	24
Agriculture value added (% of GDP)	1997-2001	8	29
Fuel exports (as % of total exports)	1991	95	9

Sources: World Bank (2003), Population Reference Bureau (2003).

but bloody conflict to escalate into continual warfare, drawing in, under the aegis of the Soviet Union, Cuban soldiers on the MPLA side, as well as mercenaries sponsored by the U.S. Central Intelligence Agency on the UNITA side—a classic, Cold War proxy war!

The end of the Cold War in the late 1980s led to the diminution and ultimate disappearance of the ideological edge to the civil war and the arms and money associated with the global struggle for hearts and minds. Angola, unlike Mozambique, had sufficient resources to prosecute the war long after the superpowers lost interest in the conflict and would probably have prosecuted it indefinitely but not for the death of the leader of UNITA, Jonas Savimbi, in 2002. In Mozambique, on the other hand, the principal resistance movement, Renamo (Mozambican National Resistance), was not homegrown. Rather, it was started by the Rhodesian security services in 1976 to combat guerrilla groups based in Mozambique that had been attacking minority-ruled and racist Rhodesia. South Africa supported these black guerrilla fighters during the apartheid era in an effort to wear down and destabilize its antiapartheid neighbor. South Africa supported UNITA in Angola for the same purpose.

The conflicts were complex, long lasting, and ultimately exhausting for both countries. Today, Mozambique has essentially recovered, but Angola, the richer of the two, has not yet emerged from the economic crisis that almost uninterrupted conflict provoked. Prolonged conflict in Angola, starting about 10 years prior to independence, has had several areas of impact. There have been massive numbers of internally displaced people, unprecedented rates of urbanization, especially in the capital, increasing poverty, the breakdown of families, transformations of ethnicity and ethnic identity, and most recently, widening inequality (Hodges 2004). In this chapter we examine the historical, political, economic, cultural, and environmental geographies of these very similar (Table 29-1) but paradoxically sometimes very different countries.

MOZAMBIQUE

From the southern boundary of Tanzania, along 1,700 miles (2,700 km) of southeast African coastline to the borders of South Africa, lies Mozambique. Mozambique is a large, but not a populous or very productive country. Its area of 309,496 square miles (801,591 km^2) extends, elongated and Y-shaped, along the eastern flank of the southern African plateau (Figure 29-1). In 1975 the population was estimated at just over 10 million (Table 29-1) and by 2003 it had almost doubled. In the middle 1970s about 60% of Mozambique's population lived in the area lying to the north of the Zambezi River, the country's great dividing line, which roughly splits the country into two equal parts. By 2004, the northern provinces contained just over half (52.8%) of the total population.

The Portuguese were influential in the early development of Mozambique's economic geography, and during their time in Mozambique, it was the southern part of the territory that generated the bulk of the revenues. The country's capital and leading port, Maputo (formerly Lourenço Marques) served as a transit for the rich South African mining–industrial complex centered on Johannesburg. Labor from southern Mozambique has been employed in South Africa's mines and industries for many decades. By comparison, the northern regions have always been more remote and much less productive.

The demographic momentum is swinging toward the southern half of the country, particularly the extreme, and rapidly urbanizing, south around the capital. Strategically located in relation to the South Africa's economic core, Maputo has become one of South Africa's major ports and has been integrated into the South African economic core region centered on Johannesburg.

Mozambique's spatial morphology is complex. In the south, the country is only 55 miles (88 km) wide; in the middle, about 200 miles (320 km), and in the north, some 500 miles (800 km). A wide wedge of territory extends westward along the Zambezi River, separating Zimbabwe and Zambia. Malawi, in turn, extends deep into the heart of northern Mozambique. The country's pronounced territorial attenuation has always had political, economic, and social significance. During the decade of insurgency, for example, the populous north proved fertile ground for the Tanzania-based Frelimo (Front for the Liberation of Mozambique) forces. Similarly, Mozambique extends deeply into southern Africa, and during the time of white minority rule in Zimbabwe (to 1980) and South Africa (to 1994) the former Portuguese colony fractured the buffer zone that long protected the south against the march of black African nationalism. Today, the geographical extension of the country has implications for national integration and unity, but potential problems seem to be diminishing in importance with time.

Physical Environments

Most of Mozambique lies on the seaward side of southern Africa's Great Escarpment, occupying the region's most extensive coastal plain. Nearly half the country's total area lies less than 750 feet (229 m) above sea level. This coastal lowland is widest in southern Mozambique, where it covers virtually the entire territory; but northward it becomes narrower, and higher country approaches closer to the coast. In interior central and northern Mozambique, elevations rise to a plateau level ranging from 500 to 2,000 feet (150–600 m), and near the western boundaries with Malawi, Zambia, and Zimbabwe, there is mountainous terrain reaching as high as 7,000 feet (2,100 m). Hence Mozambique's topography generally declines from west to east so that, from the Rovuma River in the north to the Limpopo in the south, streams have an eastward orientation.

The pivotal geographical feature of the country is the Zambezi River. The Zambezi is a physiographic as well as a historical divide: almost all of Mozambique's higher and more rugged terrain lies to the north, while most of the south is gently undulating, coastal lowland. The country's colonial imprint was stronger to the south of the river: both the capital and the second city, Beira, and all the really effective communications with the interior, developed in southern Mozambique.

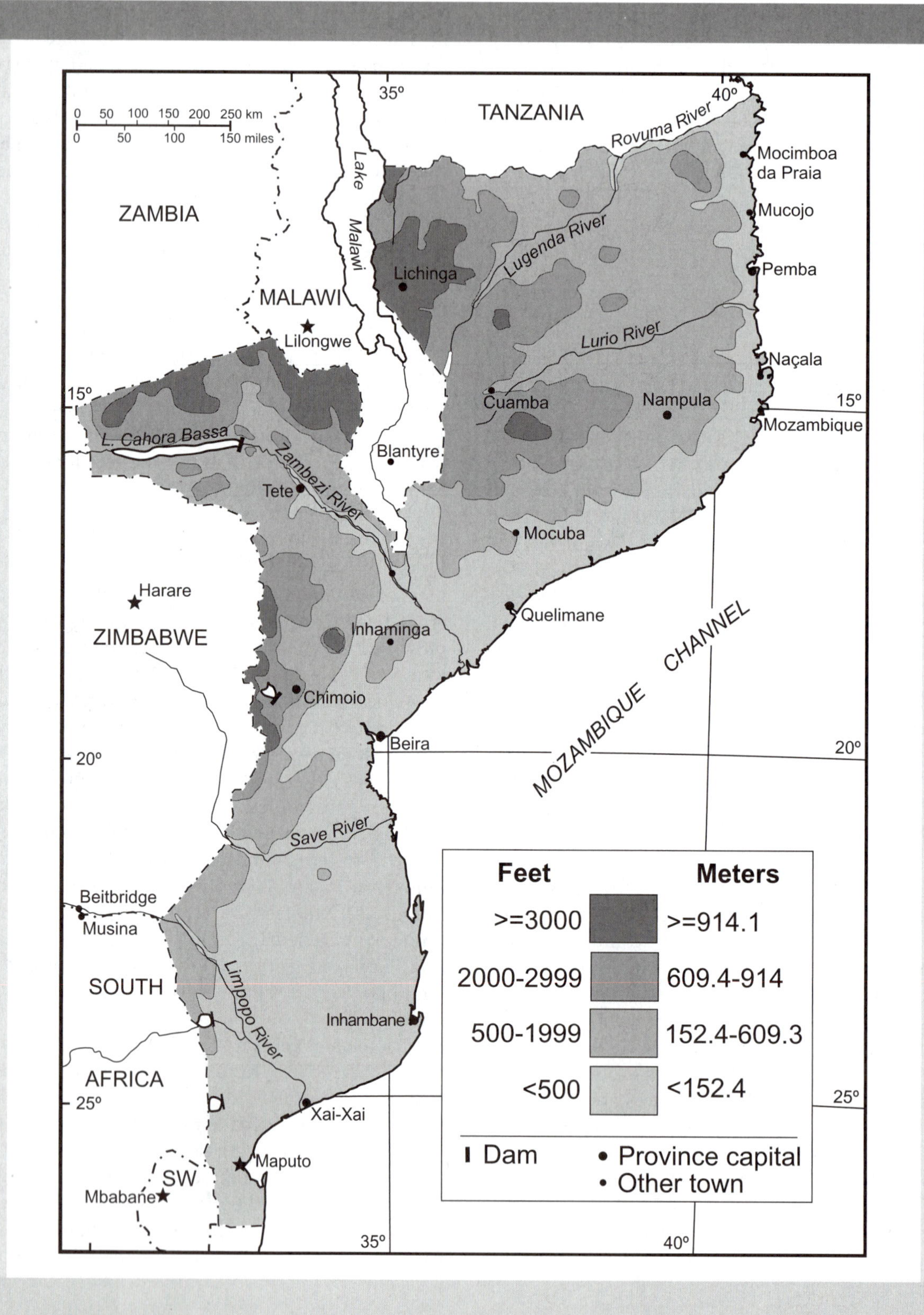

FIGURE 29-1 Physiography of Mozambique.

The Zambezi Valley itself is the scene of the huge Cahora Bassa hydroelectric project. Compared with the south, there is little modern development in northern Mozambique. But it was the savanna-clad, village-dotted, subsistence-crop north that was most directly exposed to the impact of African liberation forces, and there the insurgents achieved their first major successes against Portuguese power. For a time there was talk of partition in Mozambique, and the only feasible dividing line seemed to be the Zambezi. Events in Portugal overtook such designs.

Most of Mozambique is tropical savanna country, and heat and high humidity prevail over all but the high-elevation sectors. Although the coastline lies in the path of the southeast trade winds, and the warm Mozambique Current flows southward offshore, Mozambique receives rather low annual rainfall, totals averaging 30 inches (760 mm) along the coast and increasing significantly only against the high slopes of the far western interior. Low elevation interior areas share the marginal rainfall that is characteristic of coastal stations: Tete, far up the Zambezi, records only 23 inches (580 mm). The rainfall comes mainly during the high-sun season (October or November to April), a rainy season that is accompanied by such high temperatures that much of the country is permanently moisture deficient. Only north central and northwestern Mozambique and the high slopes along the Zimbabwean border generally maintain favorable moisture balances. This condition is reflected in the natural vegetation, for large areas of the savanna are sparse and erosion prone, and bush country covers much of the gently rolling south. In common with much of savanna Africa, soils are generally rather infertile, with the exception of more productive alluvial soils in sections of the river valleys and the *machongas,* the humus-rich, moist patches of lowland between interior dunes of the coastal plain.

Precolonial Spatial Patterns

The major events that shaped the historical geography of Mozambique had their origins and foci beyond the country's present borders. The great majority of the ancestors of the present population arrived here during the great Bantu migrations from the north and west during the first millennium of the common era; but later a major readjustment occurred, resulting from the emergence of the Zulu Empire in South Africa. The peoples of Mozambique are part of the Central Bantu cluster, but the Zambezi River again performs its divisive function: descent rules to the north are in accordance with the matrilineal principle, while patrilineal societies prevail to the south. The peoples south of the Zambezi are related to those in Zimbabwe, including the Thonga and Shona; northern Mozambique is occupied principally by the Makua, Makonde, and Yao peoples, whose domains extend into adjacent Tanzania and Malawi.

The first non-Africans to arrive on Mozambique's coasts were not Europeans, but the Persians and Arabs who had created an Islamic sphere of influence along the East African coast that extended as far south as Sofala. Eastern Africa's most powerful Swahili city-states were Malindi, Mombasa, and Kilwa; but along the coast of what is today Mozambique the Muslim traders had also established themselves in substantial settlements at Mozambique Island and Sofala, below present-day Beira. These coastal towns were the foci for a trading hinterland where Swahili traders were predominant (Box 29-1). When the Portuguese first arrived in the area, Sofala and Mozambique Island (Mozambique) were under the sway of more powerful Kilwa, and Mozambique was part of the Indian Ocean trading region. From India and the Middle East came metal tools, cotton textiles, and leather goods; in return, the Swahili traders were able to send ivory and, importantly, gold from the plateau interior. Sofala was known for its gold export. And, of course, Mozambique provided the Arab traders with a steady supply of human captives bound for Middle Eastern markets.

The European Encounter

Portuguese ships first reached Mozambique during the last years of the fifteenth century, and almost immediately Portugal was embroiled in disputes and conflicts with the Arabs. But Lisbon's principal interest was in the Far East, and colonization of East Africa was not a high priority. Portuguese efforts to secure footholds along the East African coast were designed to protect and reinforce the Indian Ocean trade routes. Even when Mozambique's boundaries were defined in the 1890s, Portugal still did not control its colonial domain effectively. In terms of spatial organization and territorial integration, Mozambique in 1890 was

BOX 29-1

THE SWAHILI

"Swahili" is an Arabic word meaning coastal dweller and refers to the distinctive Afro-Arab culture that developed along the East African coast from Mozambique to Somalia over centuries of interaction between Africans and Asians, mainly Arabs and Persians.

The Swahili language (Kiswahili) is a lingua franca in eastern and central Africa and is spoken by millions as a first and over 30 million people as a second language. The language developed as a pidgin language of trade. Its grammar and syntax are Bantu rather than Arabic, although the Arabic loan words are fundamental and substantial. The Arabic content varies with distance from the coast: the greater the distance, the less the Arabic influence. There are a variety of Swahili dialects, each reflecting very local Bantu influences.

little different from Portuguese East Africa three centuries earlier. As late as 1894 Maputo was attacked and partially destroyed by a powerful African force. Not until 1902 was control finally established over the long-dissident Barué region. North of the Zambezi River, colonial consolidation came still later. Indeed, the Portuguese "pacification campaign" in Yao country was not over even when World War I commenced and German East African troops entered northern Mozambique.

The colonial spatial system of Mozambique, therefore, was substantially the product of the twentieth century. During the 1890s various colonial administrators wrote treatises proposing new systems of organization. One significant product of this period was the circumscription, a civil administrative unit designed to strengthen colonial control over the population and to mobilize their labor in the economic interest of both colonial Mozambique and metropolitan Portugal. But ultimately the developments that were occurring in southern Africa's highveld interior—in South Africa and in Rhodesia—had greater impact on Mozambique than the new Portuguese policies had. While Portugal's administrators were directing the pacification campaign and reorganizing Mozambique's provinces and circumscriptions, the full dimensions of the Witwatersrand gold fields were being realized, Johannesburg was growing into a major mining center, European settlement in Rhodesia (Zimbabwe) was expanding, the inevitable Anglo-Boer War took place, and the Union of South Africa was established in 1910. The beginning of the twentieth century signaled the start of unprecedented economic and urban development in southern Africa, and its shock waves strongly affected the course of events in Mozambique.

In 1975, following a decade of war between Frelimo and Portuguese armed forces, nearly five centuries of Portuguese influence and control in the region came to an end. Independence also brought major changes in the relationships between Mozambique and its neighbors, as economic ties forged during the colonial period were broken and new ties with the Soviet bloc were fostered. The war for independence became civil war and continued into the early 1990s. Since that time, even though there are serious development problems in every region of the country, Mozambique has been one of the fastest growing economies in the world, profiting from its location next to South Africa.

Colonial Legacies

South of the Zambezi River, Mozambique lies interposed between ocean and highveld, astride the most direct routes from the plateau to the outside world. North of the river, Mozambique was never in that position. Communication lines to colonial Northern Rhodesia (Zambia) were always oriented south and westward; the eastern route did not materialize until the construction of the Tan-Zam railroad—which traverses Tanzania, not Mozambique. And while

Mozambique forms the obvious outlet for Malawi, development in colonial Nyasaland never matched that of South Africa and the Rhodesias, so that Mozambique gained little advantage from its monopoly. Portugal's colonial presence, therefore, focused strongly on southern Mozambique, for the economic geography south of the Zambezi River soon revealed its potentials and opportunities. Lourenço Marques (Maputo), the territory's southernmost town, had been made the capital of Mozambique in 1898; before the first decade of the twentieth century was over, its port had been connected by rail to Johannesburg. The shortest overland distance from Johannesburg to the coast was through Lourenço Marques, and the port facilities required expansion to handle the growing tonnage there. At about the same time the locational advantages of Beira, opposite Southern Rhodesia (Zimbabwe), led to the construction of a railroad to link its port to Salisbury (later Harare) and other areas of that developing colony's heartland.

Mozambique would have benefited from its break-in-bulk situation even under normal circumstances, but in southern Africa during the early decades of the twentieth century, things were far from normal. The feverish growth of the Witwatersrand generated not only an expanding volume of trade, but also a demand for a commodity Mozambique could supply: labor. The Portuguese consolidated their locational advantages by negotiating the Mozambique Convention with South Africa. According to the terms of this agreement, Mozambique would supply to South Africa a certain number of laborers annually, and in return South Africa would guarantee the passage through Lourenço Marques of at least 40% of the trade (by volume) generated in the industrializing region around Johannesburg. This unique arrangement secured for Lourenço Marques' a competitive position against Durban, the South African port most favorably positioned to serve the Witwatersrand.

The impact of these circumstances upon urban development in Mozambique can still be observed in the townscapes of the country's first and second cities. Anglo-Portuguese investment transformed these long-dormant towns into modern, spacious, skyscrapered urban centers whose central business districts and suburbs alike reflect three-quarters of a century of primacy. In 1955, Lourenço Marques was linked by rail to Rhodesia, so that both Beira and the capital could serve the Rhodesian hinterland. Up until the 1990s, the major roads of Mozambique paralleled the railroads in their orientation to the interior. Internal communications in Mozambique were never adequate but have improved. The road link between southern and northern Mozambique, separating Beira and Quelimane on either side of the Zambezi Delta, however, is still insufficient. There are vast swampy stretches between the Limpopo and Save deltas, the Zambezi Delta, and most of the northern provinces away from the coast that are poorly served, if at all, by roads.

The colonial spatial framework of Mozambique as it emerged during the first half of the twentieth century was insular and linear: insular in that the largest cities and towns were islands of wealth and growth in a sea of poverty, and linear in the sense that the modern transport lines, in Mozambique as in other parts of the world, stimulated some growth poles along their routes. The Portuguese located their capital city, Lourenço Marques, in one of the most sheltered ports along the coast. As geography would have it, only 55 miles (88 km) separated Lourenço Marques from the South African border at Ressano Garcia and a further 250 miles (402 km) from the South African mining and industrial heartland, the Witswatersrand. During the colonial period several substantial towns (including Moamba) and extensive farms serving the nearby urban market developed in this short corridor. Lourenço Marques in 1970 had a population of over 200,000 and, shortly before the European exodus, perhaps as many as 50,000 Europeans—the largest and wealthiest domestic market of Mozambique.

The Portuguese developed a second corridor from Beira to the Rhodesian border at Vila de Manica (Manica), and still another linear area of development extended along the Zambezi River with Tete as its focus. That region received a major boost when the decision was made to proceed with the Cahora Bassa hydroelectric project above Tete on the Zambezi, but the Portuguese never enjoyed the benefits to be derived from this scheme. A fourth, incipient corridor of development emerged during colonial times in the hinterland around the coastal town of Mozambique, from where an internal railroad was constructed via Nampula to Vila Cabral (today's Linchinga). In the comparatively densely populated triangle between Mozambique, Vila Cabral, and Quelimane, there developed the country's most effective internal transport network.

In southern Mozambique, the subdivision of the districts into circumscriptions paid off well. Combined with a very severe labor law that was tantamount to involuntary servitude to the colonial state, the administrative arrangement of the circumscription in the rural areas produced a steady stream of "volunteer" labor for the South African mines, men who recognized such work as the only alternative to serfdom. In each circumscription, the South African mining companies had their recruitment centers, and they were aided in their search for manpower by the Portuguese administration at local as well as provincial levels, for the rewards lay in the terms of the Mozambique Convention. This system became a cornerstone in the economic policy of the Salazar regime in Portugal (1932–1968), when the demands placed by Lisbon upon the overseas empire grew markedly. And long before the attention of the world in general was turned to the excesses of colonial oppression, there was widespread criticism of the practices prevailing in Mozambique.

Southern Mozambique had the twin advantages of location and labor that could be mobilized; Mozambique north of the Zambezi had neither. Northern Mozambique's large peasant population was brought forcibly into the modern economic sphere through the practice of compulsory cropping, mainly of cotton. During Salazar's *Estado Novo,* villagers were compelled to plant areas of cash crops in proportion to their numbers and to the area of their food crop plots. To add more than 1 acre (0.4 ha) per person each year to the already marginal food crops spelled disaster for many villagers, for the cotton fields came to replace, not supplement, the subsistence crops. Local famines occurred, but the colonial administration used force to sustain production. The cultivators did not fare well when they sold their meager production on markets at prices controlled by the Portuguese buyers. In 1956, when the system was in full operation, over 500,000 sellers received an average of $11.17 per person for their year's labor in the cotton fields (Harris 1958).

In addition to the revenues gained from its transit functions, labor provision, and forced cropping, Mozambique during the middle of the twentieth century also witnessed the expansion of European agriculture in favored areas of the country. This development is reflected by the rapid increase in the European population of the country, the principal architects and beneficiaries of such development. As recently as 1930, the European population of Mozambique was only 16,000, and during the decade of the 1930s it grew by only about 10,000. But during the 1940s, more than 20,000 Europeans, mostly Portuguese, arrived in Mozambique to settle, and during the 1950s, Mozambique's European population nearly doubled, from under 48,000 to over 90,000. Urban population distribution as of 2003 is shown in Figure 29-2.

Notwithstanding the accelerating pace of change in Africa during the 1960s, Portuguese settlers continued to arrive in large numbers in Mozambique even after the first indications of the coming African rebellion could be perceived. In part this was the result of farm settlement programs set in motion before political problems engulfed Mozambique (and then Portugal itself); until 1970 Lisbon believed that its centuries-old presence in the African empire would be sustained and that the insurgents could be accommodated (if not defeated outright). Thus even while the war was waged, work on the Cahora Bassa hydroelectric project continued, farmlands were laid out, and farm settlements were built. One of the earlier and larger farm projects was the Lower Limpopo Valley Scheme, where a quarter-million acres (101,175 ha) was designated for irrigation and occupation by 10,000 farm families. In the hinterland of Beira another major scheme, involving 75,000 acres, (30,350 ha), was laid out. By allocating land to African as well as European families, the colonial administration hoped to create the sort of nonracial community that would prove multiracialism possible in Portuguese Africa, while encouraging European immigration and strengthening the white sector. But that was not to be.

The agricultural development program eventually fell victim to the rising tide of African nationalism in Mozambique. At the time of independence almost 100,000 acres was being irrigated (40,000 ha) throughout the country. Despite the potential for opportunity, the settlers finally saw the writing on the wall. Thousands of farmers left for Portugal as independence approached, and the Lower Limpopo Valley Scheme reached only about one-quarter of its projected dimensions. Other schemes were similarly abrogated. Nevertheless, Mozambique's agricultural potentials were substantially proved by what was achieved.

Today over 250,000 acres (101,175 ha) is under irrigation. In the hinterland of Maputo approximately to the latitude of Inhambane, sugar grows in

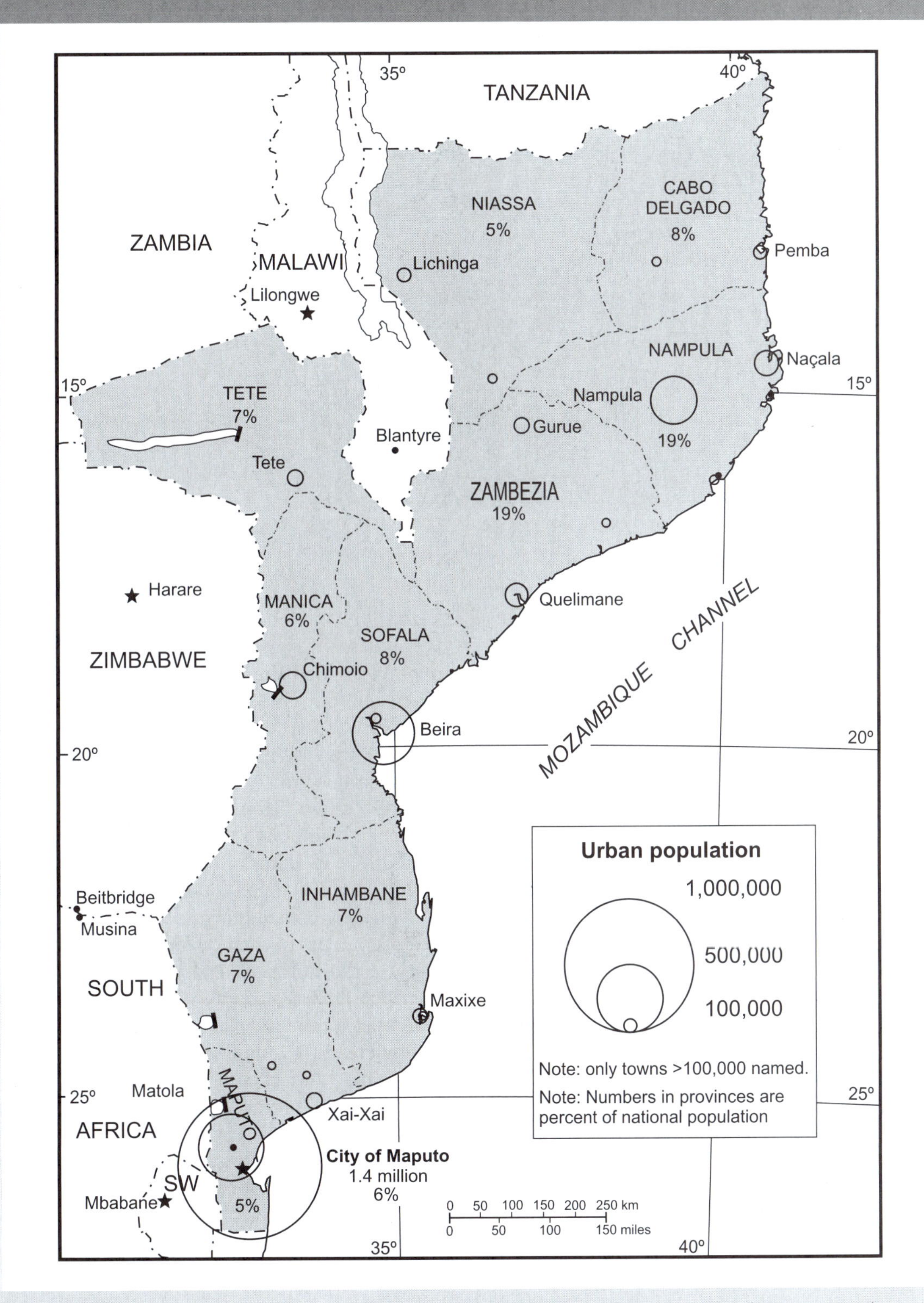

FIGURE 29-2 Mozambique's urban (graduated circles) and provincial population (province as percentage of total population) as of 2003.

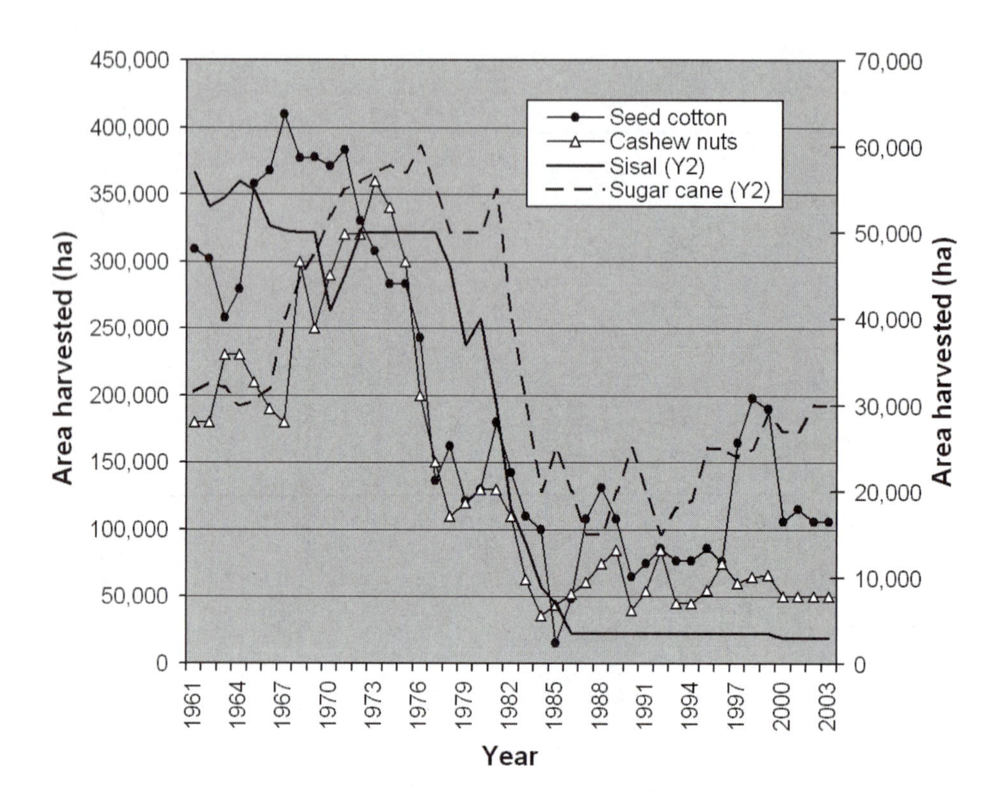

FIGURE 29-3 Area harvested from 1961 to 2003 for four cash crops grown in Mozambique that were virtually abandoned after independence.

exportable quantities, as does rice; a variety of fruits including citrus, pineapples, and bananas also do well in the southern part of the country. Sugar estates also developed along the Zambezi River and in the hinterland of Quelimane, and remnants of the cotton plantings still exist in many areas of the country, especially in the north. Along the coast, especially between 17° and 20° south, the coconut palm abounds, and copra (dried coconut meat, the source of coconut oil), coconut oil, tea, and related products have accounted for over a significant portion of exports in the past. The coconut plantation at Quelimane, with 50,000 acres (20,350 ha) and about 4 million palms, is often identified as the largest of its kind in the world. Cashew nuts, sisal, tea, tobacco, and groundnuts are also produced in relatively small quantities in various areas of Mozambique (sisal and groundnuts in drier upland areas, tea and tobacco along better-watered slopes). In recent years timber exports, mainly from Manica and Sofala districts, generated about 3% by value of annual exports.

Mozambique's agricultural industry, of course, was oriented during colonial times toward Portugal: cotton, sugar, tea, and other farm produce went to Portuguese markets. In the 10 or so years prior to independence when the settlement of Portuguese farmers in Mozambique was growing, the export orientation that agriculture took was clearly evident. Of the four crops plotted in Figure 29-3, cotton, cashew, and sisal were quintessentially colonial cash crops. Having been forced to grow cotton for the metropole, Mozambican farmers abandoned its cultivation when the Portuguese left Mozambique and concentrated their efforts on food crops. Aside from cotton, cashew, sisal, and sugarcane were grown mainly by the Portuguese, and the impact of their departure on production is evident in the numbers. The postindependence government of Frelimo adopted a communist policy and reorganized the rural geography of the country in light of Soviet and Chinese models (Box 29-2). In recent years, there has been a move toward increasing the cultivation of cotton and sugarcane—two very useful crops under any political framework—and Mozambicans have begun to reexamine other crops that had been introduced by the Portuguese.

The African farmers' subsistence crop, as elsewhere in southern and East Africa, is maize (corn), which is

BOX 29-2

THE MARXIST-LENINIST, ONE-PARTY STATE OF MOZAMBIQUE AND AGRICULTURAL DEVELOPMENT POLICY

On 25 December 1974, when Portuguese army officers established a multi party democracy in Lisbon, they entrusted the destiny of Mozambique to one party, the Frente de Libertação do Moçambique (Mozambique National Liberation Front), or Frelimo. The Front, founded in June 1962 under the leadership of Eduardo Chivambo Mondlane, managed to win the sympathy of the international community and had the military support of both China and the Soviet Union. Unlike in Angola, Frelimo managed on the eve of the Portuguese revolution of 25 April 1974 to cause serious problems for the colonial troops, most of whom were African in origin. By 1974, it was clear that Marxism-Leninism predominated among Frelimo's leadership. After Frelimo's second congress in 1968, the significance of the anti-imperialist struggle, as formulated by Samora Machel in accordance with the Chinese notion of "liberated zones," gradually took the shape proposed by Mondlane shortly before his death in 1969: "I conclude today that Frelimo is more socialist, revolutionary, and progressive than ever, and that our line is now firmly oriented toward Marxist-Leninist Socialism."

In the belief that the essentially rural Mozambique could take shape only as a one party state, [the Frelimo leadership] sought to control the country through a process called "villagization." This policy was first implemented in the early 1970s in the "liberated zones . . .". Frelimo decided to extend it throughout the territory. All "peasants" (80% of the population) were expected to abandon their traditional homes and to regroup in new villages. In the initial enthusiasm of independence, the population responded quite favorably to the government's requests, creating collective farms and sometimes cooperating in the construction of communal buildings, although they generally refused to inhabit them and soon abandoned the communal fields. On paper it appeared that the country was under the careful control of a hierarchical administration through a network of Communist cells.

In 1977 the Frelimo leaders had openly proclaimed their allegiance to the Bolshevik ideal, calling for extended collectivization and closer links with the international Communist movement. Various treaties were signed with the countries of the Soviet bloc, which provided arms and military instructors in exchange for close support of the Rhodesian nationalists of the Zimbabwe African National Union (ZANU).

The intelligentsia rapidly became disenchanted with the movement. . . . After Frelimo's Fourth Party Congress in 1983, the organization . . . put a halt to the policy of collectivization that had had such disastrous consequences. Every time the government militia had burned another haystack to try to ensure the villagization quota, it had increased [the armed opposition], Renamo's support. The severe damage done to traditional systems of agriculture, together with the wildly erratic exchange rates for consumer goods versus foodstuffs, had led to severe problems with the food supply [to the extent that], according to . . . UNICEF, . . . 600,000 died of hunger during this period.

Source: Courtois et al. (1999), pp. 702–704.

not part of the overseas trade. Mozambican farmers and their new government treated food crops (Figure 29-4) differently from the export, or cash, crops depicted in Figure 29-3. The area devoted to maize has more than doubled since independence and is the major crop by area produced in Mozambique today. Livestock holdings have always been limited because of the prevailing diseases. The new political situation in Mozambique after independence caused an antiexport reorientation of agricultural production. The triumph of the Marxist–Leninist independence movement and the significant Portuguese exodus that followed contributed to the

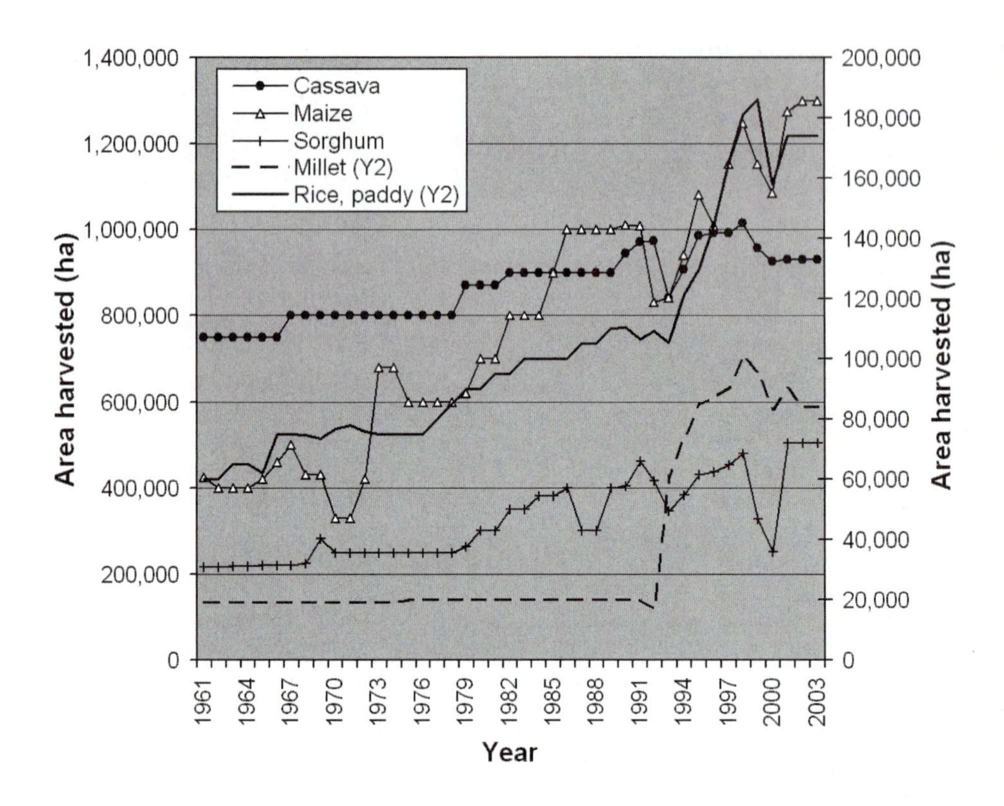

FIGURE 29-4 Area harvested for five food crops grown in Mozambique, 1961 to 2003.

dramatic decline in the export of some of these products (Figures 29-5 to 29-9), but there were other contributing factors: the antibusiness atmosphere created by the government, villagization programs (Courtois et al. 1999), and mismanagement of the economy.

Not all exports declined, however. For example, over the 15 years prior to independence the Portuguese had developed the fishing industry, which was becoming a significant contributor to the economy at the time of independence. Unlike other export-oriented activities, production of fish did not decline as the state-owned corporations took over the functions of the biggest, formerly privately owned, fishing companies. After Marxism ceased to be the dominant theory of government, the antibusiness atmosphere evaporated and fish, as well as most other, exports have increased. The government has become very interested in fostering capitalist agriculture and increasing incomes in the rural areas. Research has shown that cash-cropping "can have a positive effect on smallholder food production, particularly of maize" (MAP/MSU Research Team 1997: 1).

Industry

Although Mozambique is well mineralized, the mining industry has never been a major contributor to the economy (< 1%), but there is much small-scale, local mining (artisanal mining) that goes unrecorded. The government would like to raise the contribution of mining to GDP to 10% over the next few years. Over recent years great outside interest has been shown toward Mozambique's mineral potential, and the sector has been growing. By 1994 the major ports, railroads, and roads had been substantially rehabilitated after decades of conflict and neglect (although much work remains to be done), and in 2002 the government of Mozambique passed a law that gave mining concession owners guarantees over their concessions, granted exclusive rights to exploitation to small and artisanal miners, and allocated state funds for geological survey and mapping (Yager 2002). Precious gemstones, tantalum, and graphite are found in the northern half of the country associated with a Precambrian granitic formation (Figure 29-10). Coal is mined at Moatize, near

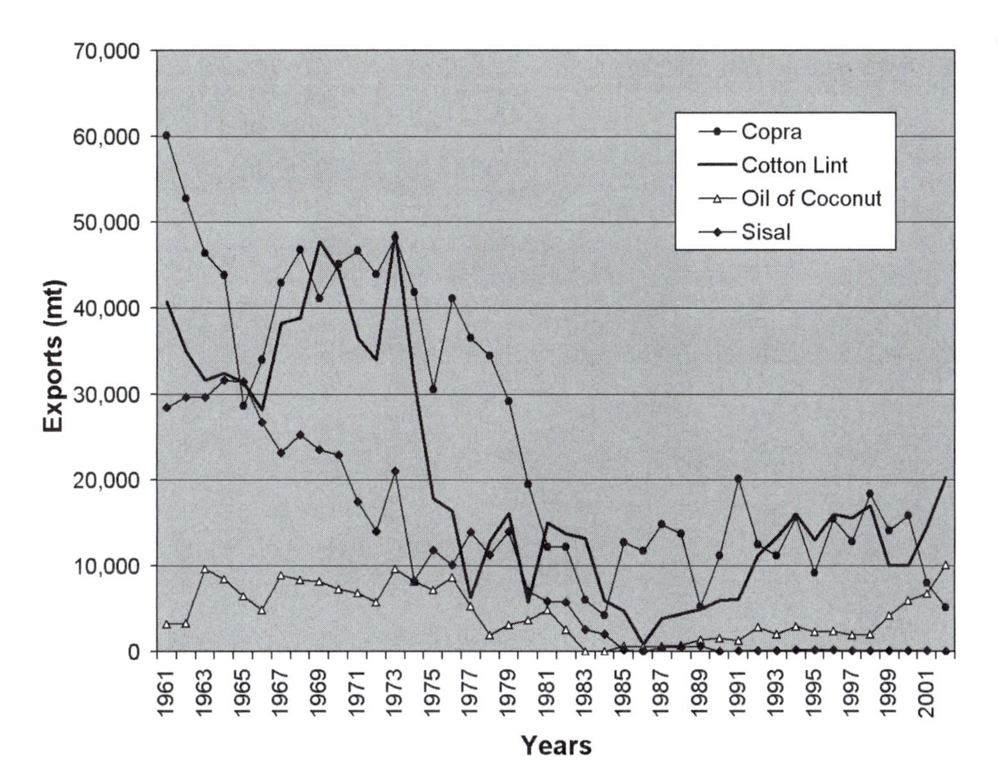

FIGURE 29-5 Exports from Mozambique of the cash crops copra, cotton lint, oil of coconut, and sisal, 1961 to 2002.

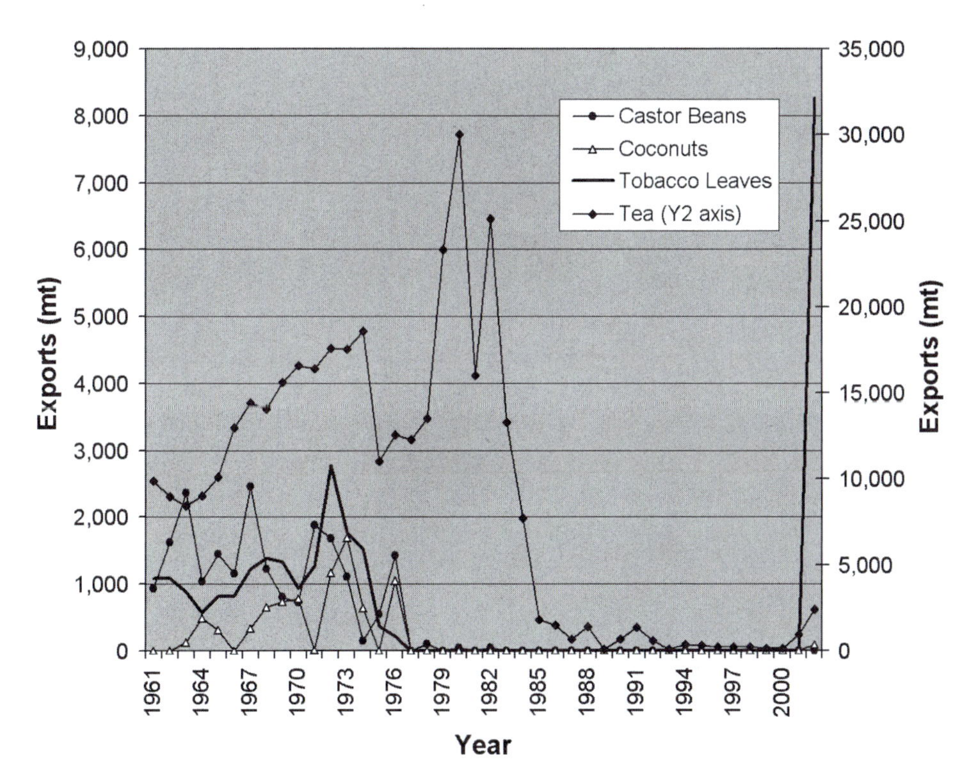

FIGURE 29-6 Exports from Mozambique of the cash crops castor beans, coconuts, tobacco leaves, and tea, 1961 to 2002.

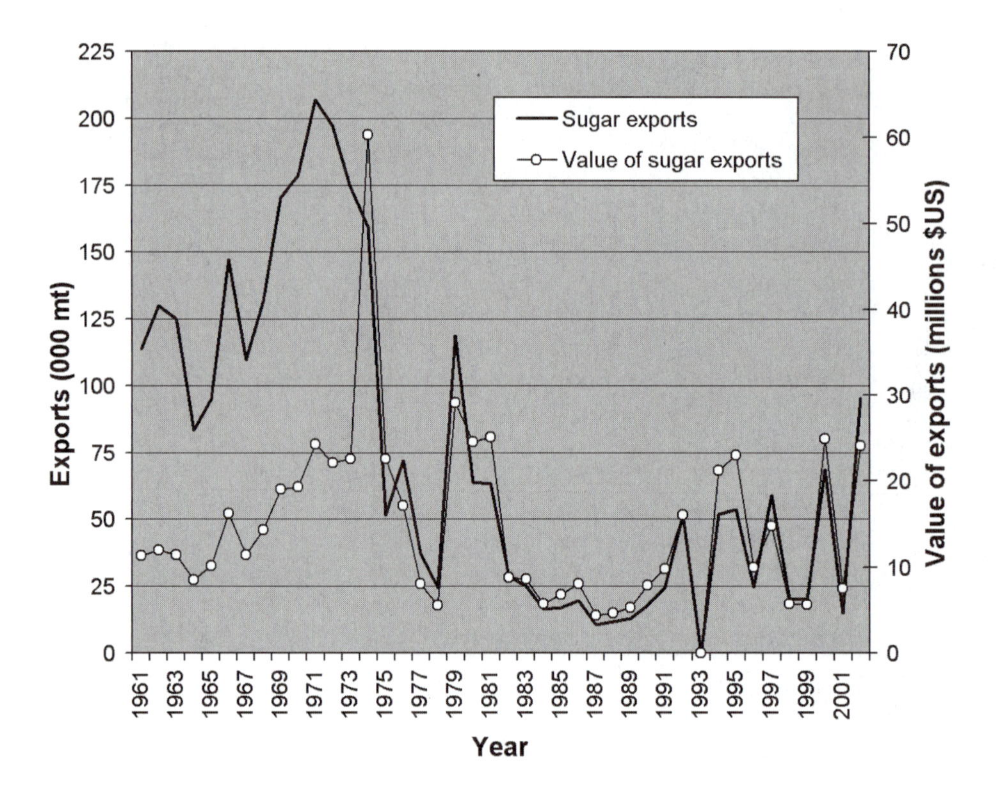

FIGURE 29-7 Sugar exports from Mozambique and their value, 1961 to 2002.

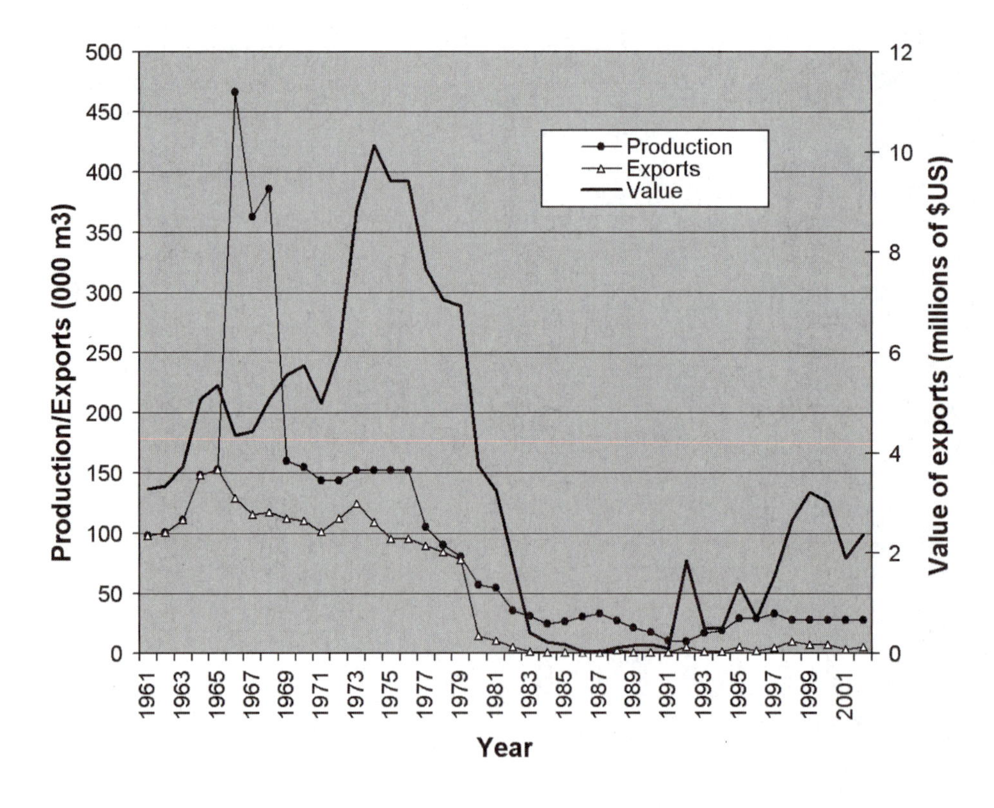

FIGURE 29-8 Mozambican timber production, exports, and value, 1961 to 2002.

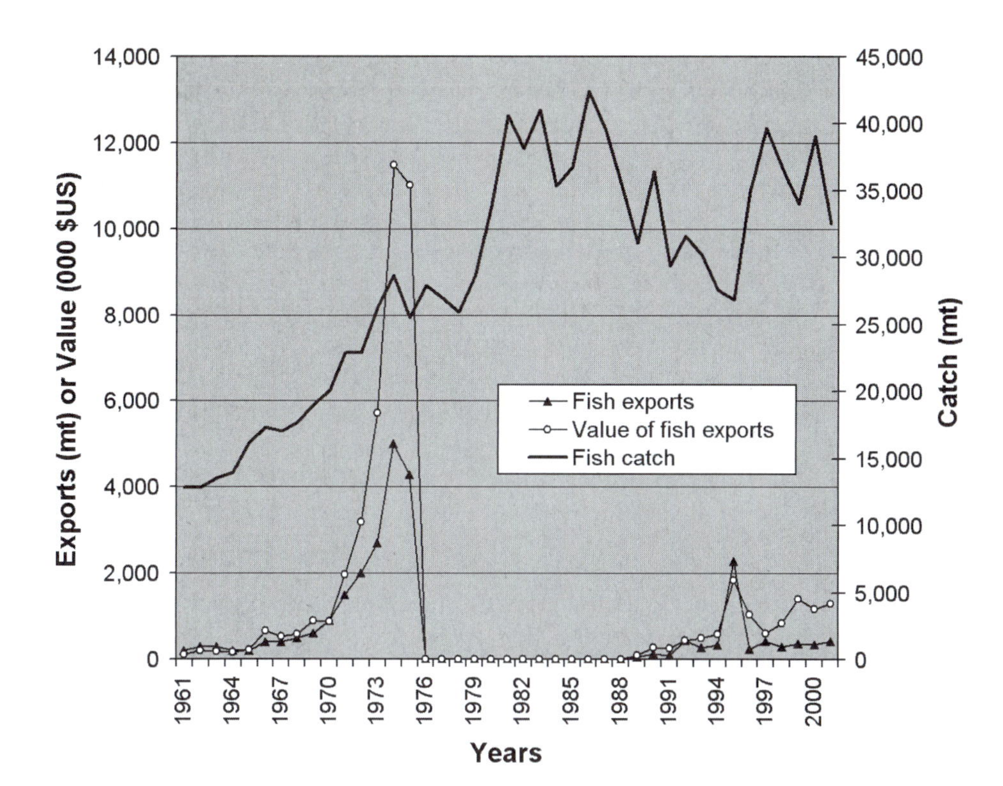

FIGURE 29-9 Mozambican fish catch, exports, and value of exports 1961 to 2001.

Tete, and has long been the only major operation of its kind in the country; other minerals have been mined at small workings elsewhere.

Before 1970, some potentially significant discoveries were made. Near Namapa, a promising iron ore deposit was located that attracted Japanese capital for development, but it has not been producing in years; and deposits of manganese and asbestos, found near the Swaziland border, have not been exploited. Along the Tanzanian border near Lake Malawi is a little explored Precambrian belt of greenstone from which gold has been mined on an artisanal basis for many years. It is hoped that once exploration gets under way there will be exploitable quantities of minerals there. The natural gas discovery near Moamba, northwest of Maputo, stimulated hopes that a search for oil on the coastal plain or on the adjacent continental shelf might succeed, but no petroleum has been discovered. Nevertheless, the natural gas reserves are estimated at 2.2 trillion cubic feet (63 billion m^3). The sands located along the Indian Ocean littoral contain ilmenite and rutile, sources of titanium, used mainly in the production of paints.

In 2000 Mozambique began smelting Australian bauxite ore to produce aluminum (Table 29-2). A second smelter has been built. Mozambique became Africa's second largest aluminium producer in 2003 (Yager 2003).

Mozambique never developed a major industrial base during colonial times. Industries were oriented either to the small local market or to distant Portugal: at Maputo the oil refinery, small steel plant, and fertilizer and chemical factories supplied local demands, while other industries processed raw materials (cotton, sisal, tea, etc.) prior to their shipment to Portugal. Some Mozambican products found their way to South African and Rhodesian markets, but the role of these countries as trading partners was always far overshadowed by their transportation requirements.

During the 1960s hopes were expressed by the colonial administration that the Beira–Zambezi Valley region might become Mozambique's mining and industrial heartland. Geographically there is much sense to this idea. The coalfields near Tete, connected by rail to Beira's port, served the Beira–Rhodesia railroad (some coal was also shipped coastwise to local power stations). The Cahora Bassa hydroelectric project was believed to be the power source that would stimulate industrial as well as agricultural development. Jute and cotton mills at Chimoio (formerly Vila Pery), sawmills

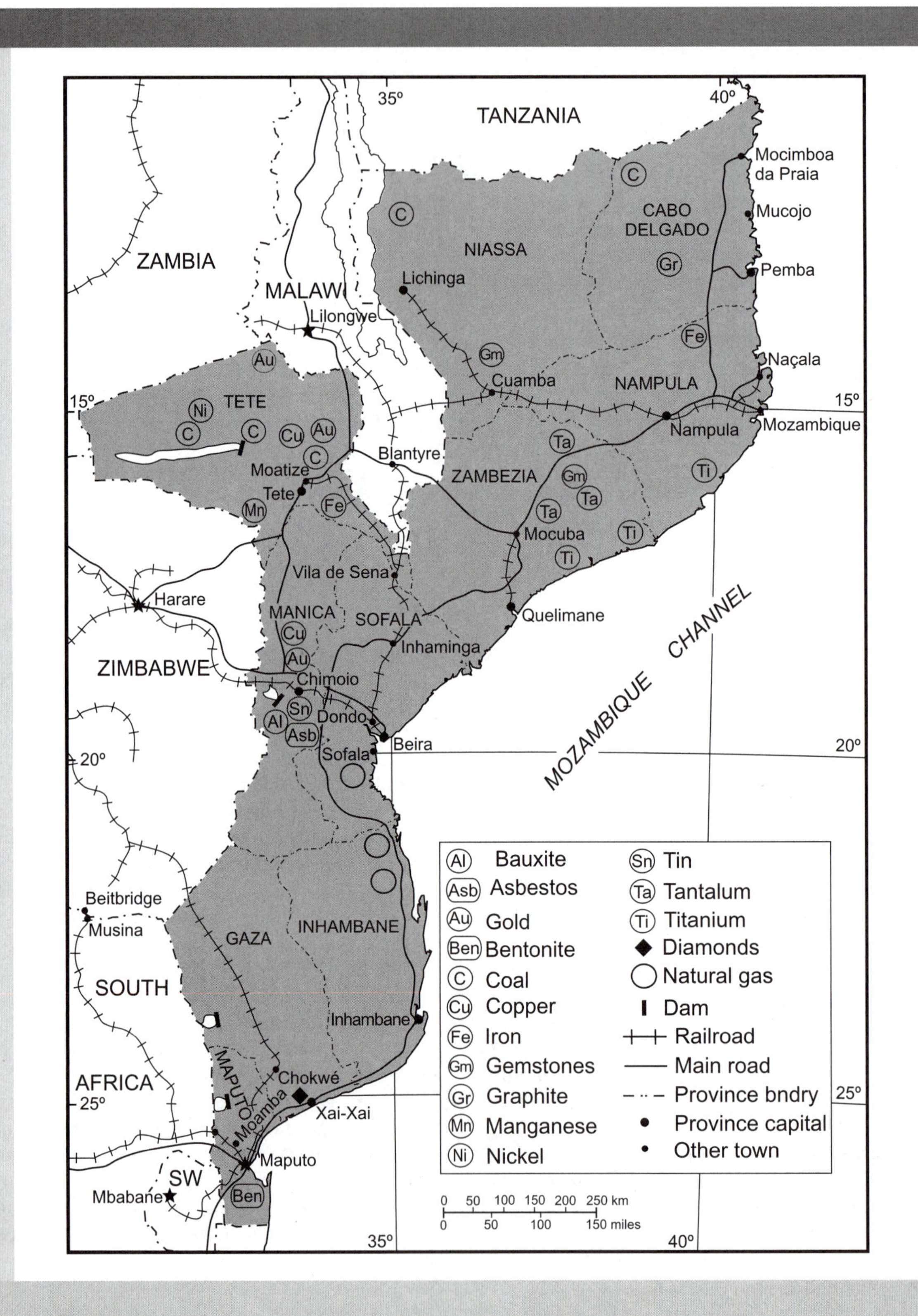

FIGURE 29-10 Minerals, infrastructure, provinces, and principal towns of Mozambique.

TABLE 29-2 MAJOR MINERAL PRODUCTION IN MOZAMBIQUE, 1990 TO 2002.

	Aluminum (000 mt)									
Year	As Bauxite	Refined	Cement (000 mt)	Coal (000 mt)	Copper (Cu content) (000 mt)	Gemstones (kg)	Gold (kg)	Graphite (000 mt)	Natural Gas (million m^3)	Tantalite (kg)
1990	7,190		79		28	2,560	63			
1991	7,760		80			1,280	394			266
1992	8,340		30	48,000		750	296			
1993	6,000		20	48,400		560	149	10		
1994	9,620		60	58,190	259	7,000	6,804			
1995	10,700		60	40,000		8,000	6,800	3,019		
1996	11,459		180	40,000		1,862	67	3,283		
1997	8,218		220			1,091	6	5,125		
1998	6,130		212			1,465	17	5,889	1	
1999	7,883		216	8,573		1,400	19	4,007	1	
2000	8,130	53,800	270	16,115		1,000	23		1	
2001	8,592	266,000	265	27,600		115	22		1	27,000
2002	9,119	273,200	274	43,512		1,326	17		2	46,900

Source: U.S. Geological Survey Mineral Reports (various years).

along the rail link to Malawi, and the discovery of gold in Manica and Sofala provinces and more iron ore in the Zambezi Valley all contributed to the anticipation that the long-divisive Zambezi might become the country's center of gravity. Although the hopes of the Portuguese were to be denied by events, the geographic attraction of this area remains. But as the Maputo industrial corridor merges with the South African industrial core of Johannesburg, it is likely that the Zambezi Valley will be a second-tier industrial area.

Although the amount of electricity generated at the underused Cahora Bassa facility is second in all of southern Africa, it is tiny compared to the enormous power generated by South Africa (Table 29-3). All countries in Africa are dwarfed in their electrical production by South Africa. Mozambique is fortunate to share a border, and that particular stretch of border, with such an economically powerful country.

A third-tier urban–industrial area is emerging in the north, focused on Nampula (population 372,000), located at the center of mining operations in the north along the line of rail from Lichinga to the coast. Although the future course of its development is largely dependent on mining prospects in this largely unexplored region, Nampula occupies a strategic location and seems poised to grow, drawing in its presently loosely connected periphery more tightly in the process. What Mozambique might have been without its South African connection and its transport functions lies revealed in coastal towns such as Inhambane (64,000), Quelimane (181,000), and the less populous Mocimboa da Praia. Although these towns have grown in recent years, they are still small ports that participate mainly in coastal shipping, possessing limited communications with mainly agricultural hinterlands, and industries restricted to the processing of agricultural products. These small urban places reflect the modest development of the interior—and in their contrast to Maputo and Beira, they reveal the magnitude of the impact made by Mozambique's transit function.

Problems of Development

Mozambique has one of the fastest growing economies in the world today, although it started from a low base level. Industries include food and drinks, chemicals (fertilizer, paints, soap), aluminum smelting, petroleum products, textiles, cement, asbestos, glass, and tobacco processing. Exports are focused on aluminum, prawns (shrimp), cashews, cotton, sugar, citrus, timber, and electricity. Nevertheless, Mozambique is predominantly an agricultural country, and one of the principal foci of development has to be on creating market-oriented farming systems that produce large surpluses.

TABLE 29-3 ELECTRICITY OVERVIEW: 2000, SOUTHERN AFRICA (BILLION KILOWATT-HOURS EXCEPT WHERE NOTED).

Country	Consumption	Generation	Installed Capacity (gigawatts)	Exports	Imports
Angola	1.11	1.12	0.586	0	0
Botswana	1.45	0.5	0.217	0	0.99
Comoros	0.02	0.02	0.005	0	0
Lesotho	0.1	0	0	0	0.1
Madagascar	0.76	0.82	0.285	0	0
Malawi	0.77	0.83	0.308	0	0
Mauritius	1.2	1.29	0.365	0	0
Mozambique	0.93	7.02	2.388	5.7	0.1
Namibia	0.89	0.03	0	0	0.86
Seychelles	0.15	0.16	0.028	0	0
South Africa	181.52	194.38	43.11	4.55	5.29
Swaziland	0.9	0.36	0.131	0	0.56
Zambia	5.84	7.82	1.786	1.54	0.1
Zimbabwe	10.48	6.43	1.881	0	4.5
Regional total	206.12	220.78	51.09	11.79	12.5

Source: USEIA (2004).

Mozambique's population distribution is characterized by a concentration along the coast, a strong rural agglomeration, especially in the Mozambique and Zambezia districts and in the Maputo–Inhambane region, and very low densities in the upper Limpopo, upper Zambezi, and upper Rovuma–Lugenda areas (each of these areas adjoins the national boundary). The great majority of the people, of course, live in dispersed homesteads, small hamlets, or villages characteristic of the eastern and southern African rural scene, with their surrounding, communally worked fields of maize, the principal subsistence crop. The growing urban economy is not growing fast enough to absorb the influx of population from the countryside (Figure 29-11), therefore development emphasis must be placed on rural employment. At the same time the urban economy must grow: continued investment must be made in urban manufacturing capacity and services not only to provide employment but to diversify the economy.

It is axiomatic that healthy and well-educated people make the best employees, wherever they may be employed. The tasks that lie ahead in the fields of education and health are enormous but much progress has been made since independence. When the Portuguese left in 1975, there were few educated Mozambicans. At independence, perhaps 6% of Mozambique's nearly 10 million inhabitants were literate. Reliable birth and death rates were not available at the time, but various estimates suggested that births averaged 43 per thousand annually and deaths 23 per thousand. Officially, infant mortality was given as between 90 and 100 per thousand, but some observers believed it was double that. Whole regions of Mozambique never had medical facilities to speak of; tens of thousands of children never saw a school. It is not surprising that, shortly before independence, president-designate Samora Machel appealed to Portugal for $150 million in aid to survive the first six months of independence. Economic uncertainties and staggering postwar needs rendered Mozambique's future problematic.

According to the most recent census (1997), there have been accomplishments in the human development of Mozambique since independence. One measure of human development, illiteracy, has declined from 90% in 1980 to about 66% of the population in 1997. Although 66% is still alarmingly high, the rate for women is higher yet, but declining. The regional (north–south) and rural–urban dichotomies in human development persist (Table 29-4) but are not unbridgeable.

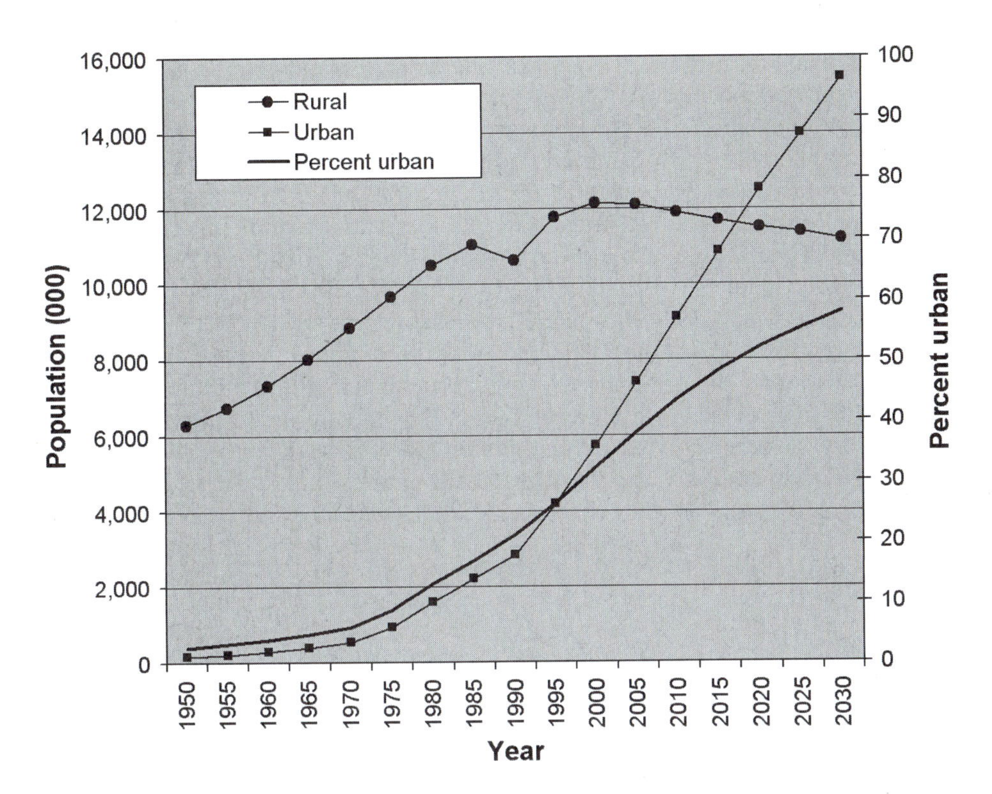

FIGURE 29-11 Rural and urban populations and percentage urban population for Mozambique, 1950 to 2030.

TABLE 29-4 DEMOGRAPHIC AND DEVELOPMENT DATA FOR 10 MOZAMBICAN PROVINCES, MAPUTO CITY, AND COUNTRY OF MOZAMBIQUE, 1997.

Province/City	Area (mi²)	Population[1]	Percent of Total Population	Population Density (mi²)	Illiteracy Rate (%)		Percent Rural
					Men	Women	
Maputo city	232	966,837	6	4,160	7.1	22.6	0
Maputo	9,944	806,179	5	81	20.2	45.9	37
Gaza	29,231	1,062,380	7	36	35.8	63.0	75
Inhambane	26,492	1,123,079	7	42	35.1	66.4	80
Sofala	26,262	1,289,390	8	49	35.9	74.8	59
Manica	23,807	974,208	6	41	38.5	73.9	72
Tete	38,890	1,144,604	7	29	50.0	81.0	85
Zambezia	40,544	2,891,809	19	71	53.2	85.2	87
Nampula	31,508	2,975,747	19	94	56.7	85.9	75
Cabo Delgado	31,902	1,287,814	8	40	60.0	88.5	83
Niassa	49,829	756,287	5	15	52.2	84.2	77
Mozambique	308,641	15,278,334	100	50	44.6	74.1	71

Source: Mozambique Bureau of the Census (2004).

[1] Total for Mozambique does not agree exactly with Mozambique's Census Bureau.

To increase the human development of all Mozambicans, there must be increased investment in agriculture and redevelopment of the commercial and export orientation that was lost with abrupt and violent decolonization and years of internecine warfare. There must be intensification of exploration and mining, reorientation of transport systems to promote national integration, allocation of development investment to diminish the regional disparity between the deprived north and the far more developed south, and stimulation of Mozambique's industrial base. Most of the investment must come from outside. Greater investment from and integration with the South African economy are to be welcomed and will have positive effects throughout Mozambique.

ANGOLA: MARX AND OIL

Angola presents a paradox. Tony Hodges has observed that although Angola has one of the best natural resource bases in Africa, its wealth has not underpinned development and prosperity but has instead been associated with "years of conflict, economic decline, and human misery on a massive scale" (Hodges 2004: 1). The trauma of over 30 years of conflict presents a terrible contrast between the potential of Angola at independence and the current state of the population. Few would have imagined at independence that Angola's elites would lead the country to ruin so quickly; after all, there was so much natural resource wealth. Exacerbating regional politics to control such wealth—and the hearts and minds of people—were Cold War superpower competition, manipulation of superpower rivalry by regional Angolan interests, ethnic and ideological competition, corruption, and greed. We shall examine the background to the protracted conflict and prospects for the future.

Five hundred years of Portuguese rule in Africa came to a convulsive end in November 1975 when Angola, Lisbon's most profitable colonial possession, achieved independence amid chaos and civil war. The conflict among rival nationalist movements, each with its cultural and regional core areas, ravaged Angola for months; it bitterly divided the Organization of African Unity, damaged U.S.–Soviet détente, brought yet another foreign army to Africa, and opened a new chapter in the breakdown of the buffer zone (Namibia) that isolated South Africa from its independent African adversaries.

Angola is one of those creations of European colonial competition that threw together peoples of diverse cultures and histories and separated others with strong common traditions and other bonds. The vast country (481,351 mi^2; 1,246,693 km^2) extends from the forests of the Congo Basin to the margins of the Kalahari Desert, and from a lengthy Atlantic coastline to the heart of south-central Africa. Among its over 13 million people, the Bakongo in the north have ties with the Bakongo of western Democratic Republic of the Congo; in the central plateau the Ovimbundu are part of a greater nation that extends into Namibia. Angola's core area, including the capital, Luanda, and its hinterland, is the domain of the Mbundu. In the early 1970s the European population of Angola reached nearly 500,000, strongly concentrated in Luanda, the other urban centers, and on the better agricultural lands.

Physical Environments

Angola is a large and essentially rectangular country lying wholly within the tropics. A narrow coastal plain, arid in the south and becoming gradually more humid and broader in the north, gives way in a series of steps to a highland interior plateau that averages between 3,000 and 5,000 feet (1,000–1,500 m). Central Angola, source area of the Cunene, Cubango (Okavango), Cuando, Cuanza, Cuango, and other major rivers, rises to heights above 8,500 feet (2,600 m) in the Bié Plateau (Figure 29-12). Because the plateau drops precipitously to the coast in a series of large escarpments, none of the rivers (other than the Cuanza) is navigable for any great distance, and none figured prominently in Portugal's early exploration into the interior. Only small valley stretches, most notably along the Cunene (which forms the southwestern boundary with Namibia) are irrigated, but most have great and largely untapped hydroelectric potential. Several northeastern rivers, among them the Chicapa, Cassai (Kasai), and Cuango, all of which flow into the Democratic Republic of the Congo, have cut deep valleys into the hard African tableland, exposing rich diamondiferous gravels. In its northwestern corner Angola shares with the Democratic Republic of the Congo the wide and navigable estuary of the Congo River, beyond which lies the oil-rich enclave of Cabinda.

Angola has tropical climates tempered by altitude and cool maritime breezes. Along the coast, the northward-flowing Benguela Current and prevailing

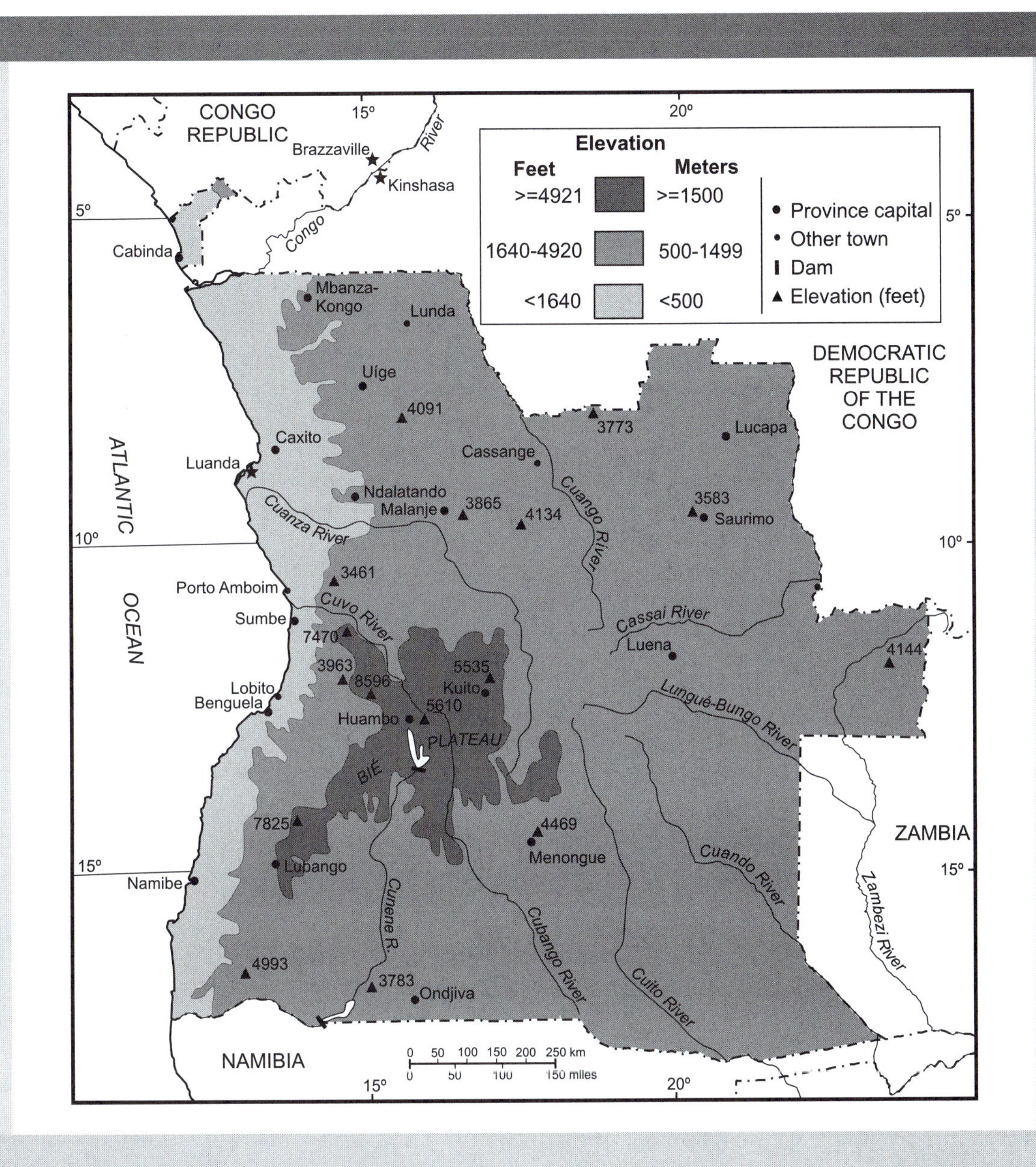

FIGURE 29-12 Physiography of Angola.

atmospheric conditions of high pressure combine to inhibit precipitation, so that much of the Atlantic lowland is either arid or semiarid. Namibe (formerly Moçamedes), for example, has an average annual rainfall less than 2 inches (51 mm), Lobito 9 inches (229 mm), and Luanda only 16 inches (405 mm). This littoral is noted for its frequent fogs and relative coolness as cold water from the ocean bottom wells up close to shore. Over 70 inches (1,780 mm) of rainfall in the interior forests of Cabinda, in the northern plateau around Uíge (formerly Carmona), and along the eastern border with the Democratic Republic of the Congo. There precipitation is decidedly seasonal with one or two peak periods, the heaviest rains coming between September and April. In general, average annual temperatures decrease latitudinally and altitudinally, so that on the Bié Plateau almost temperate conditions prevail, and winter frosts are not unknown.

Northern Angola has patches of tropical rain forest, and in many valleys and sheltered slopes of the northern subplateau zone, a rain-and-cloud forest thrives because of the condensation of moisture brought in by the westerly sea breezes. There the agricultural potential is similar to most humid tropical areas, and coffee and oil palm have proven successful. Central and eastern Angola's natural vegetation is woodland savanna, patches of which have been cleared to make way for maize, coffee, and a host of other crops. The Bié Plateau supports the densest human populations. Southern and coastal Angola are the least favorably endowed regions in terms of soil, rainfall, and vegetative cover for human settlement. Dry savanna and steppeland conditions prevail in the southern interior, and in the more humid parts, pastoralism is possible.

Colonial Beginnings and Policies

Portuguese presence in Angola dates back to 1482 when Diogo Cão landed at the mouth of the Congo River and established contact with the Manikongo people. Initial contacts were brief and cordial but led to the establishment of a slave-trading port at Mpinda. Permanent Portuguese settlement was delayed until 1575, when Paulo Dias de Novais, the newly appointed governor, founded the garrison of Luanda. For the next 250 years, Portuguese interests were confined to the profitable slave trade and to the establishment of military garrisons and religious missions in the immediate vicinity of Luanda, Benguela, and other coastal ports. Loose administration was organized through these points, but Angola itself was economically controlled from Brazil (Duffy 1962). An estimated 3 million Angolan slaves were taken to Brazil's plantations. Angola derived virtually no social or political gains from this administrative arrangement, while food crops including maize, cassava, and rice were introduced to the region from the New World and elsewhere.

The abolition of the Atlantic slave trade in 1807 by the British, emancipation of slaves in British territories in 1836, and the suppression of illegal slaving by the British led to the abolition of the slave trade in 1836. Slavery itself, however, was only truly abolished in the colonies in 1878. The abolition undermined the colonial economy of Angola, for slaves had been the greatest source of revenues. Thus, in the second quarter of the nineteenth century, Portuguese policy was motivated by the need for new revenues and new sources of production. Between 1836 and 1861 the area of Portuguese control doubled, and by the late 1870s military garrisons and mission stations were firmly implanted on the Bié Plateau and as far east as the Cuango River. Expansion met with fierce resistance, but the Angolan peoples were no match for the militarily superior Portuguese.

Portuguese expansionism was eventually thwarted by British, Belgian, French, and German interests in central Africa. Portugal had hoped to link its territorial claims in Mozambique with those in Angola in a cross-African axis, but Henry Stanley had begun acquiring land in the name of the International Association of the Congo, and both Germans and Belgians had staked claims in the Lunda and Cassange areas (Figure 29-12). Portuguese aspirations were further negated at the Berlin Conference of 1884–1885, and after much diplomatic maneuvering between Portugal and London following Cecil Rhodes's treaties with Lobengula in Matabeleland, Portugal and Britain negotiated a treaty (1891) that set the present boundaries between Mozambique, Malawi, Zambia, Rhodesia, and Angola. But the implied colonial "right of occupation" far from guaranteed effective control with the newly defined boundaries. This was achieved only after 30 years of military "pacification" and the introduction of an administrative system. Portuguese colonial rule did not begin in any systematic way until the second decade of the twentieth century.

The course of Angolan development was determined primarily by politicoeconomic conditions in, and government policy emanating from, Portugal. Throughout its tenure in Africa, Portugal was a poor country: its per capita income and literacy rates were the lowest in western Europe; a considerable proportion of its people lived at the subsistence level; infant mortality rates were among the highest in Europe, while birthrates were among the highest in the world; the majority of Portugal's inhabitants made their living from agriculture, while manufacturing contributed little to the national economy; freedom of thought and action were not tolerated by dictatorships; and the Portuguese remained largely uninformed about changing world conditions and very conservative in their attitudes. Little wonder that "development" was so painfully slow in Angola (and Mozambique). Portugal looked upon its colonies (officially referred to as "provinces" after 1951) as reservoirs of human labor and natural resources to be exploited for the benefit of the metropole first, Portuguese settlers second, and the Africans last. Angola was to finance its own development, produce financially profitable development schemes, but not compete with Lisbon.

Policy was paternalistic; the African was treated like a child with little or no culture or civilization worthy of recognition. Beginning in 1890, Portuguese policy toward the Angolans was one of "tendential assimilation," officially respecting local institutions and customs while gradually attempting to bring the Africans into contemporary life (Abshire and Samuels, 1969). From this developed the *assimilado* system, an *assimilado* being an African whose fulfillment of certain educational, financial, and social requirements entitled the individual to equal "citizenship" alongside the Portuguese. Only the *assimilado* had the right of unrestricted movement within Angola (and Mozambique); all others required permission of the local administrator to leave their *circumscricao* (district). The *assimilado* represented Angola's elite, and before the system was abandoned (1961), they numbered only 40,000, under 1% of the total population. As a group, the *assimilados* overlapped with mixed-race Angolans, possibly numbering 150,000 just before independence (Hance 1975).

Such investments as were made in Angola went mainly to impressive visible projects, particularly infrastructure designed to support the Portuguese colonists. These included dams, port facilities, railway improvements, and settlement schemes, from which black Angolans derived little material benefit. Until the 1960s, funds were not generally available for African schools, hospitals, clinics, housing, and farm improvement. As in Mozambique, a steady supply of cheap labor had to be assured for the plantations, mines, and other economic projects. Every African male had to show that he was productively employed 6 months of the year or face conscription as a laborer for the government or a private employer. Contract labor became the most obvious form of human injustice and provided the focus abroad for the condemnation of Portuguese rule (Duffy 1962).

Unlike Mozambique, Angola became a large settler colony. Following World War II, Angola was seen as a place affording relief from population pressure in the Portuguese metropole and from which resources should be extracted to bolster Portugal's faltering economy. In 1940 Angola's Portuguese population numbered only 44,083 (1.2% of the total), but by 1960 it numbered 172,529, and before independence it had reached nearly a half million. Most of this phenomenal increase resulted from the settlement of Portuguese small farmers in government-sponsored *colonatos* or settlement schemes such as Cela (north of Huambo), and the far more successful Matala Scheme on the Cunene River. Migrant families were usually provided with free transportation, a house, farm holding and basic farm tools, seed, and a cash subsidy for 2 years. These schemes, first established exclusively for Portuguese but later extended to Africans, provided Portugal with a means to control possible discontent in selected rural areas, besides stimulating the local economy and producing foods for Portuguese markets. Economic costs were generally high, but the schemes were justified politically. In addition to small farmers, Portuguese colonists established large-scale commercial farms, and a network of small- to medium-scale "bush traders" emerged linking the African and European producers in the countryside in a series of steps to the urban marketplace (Hodges 2004).

Colonial Angola had the second largest European population in Subsaharan Africa (after South Africa), but its colonists never had high educational standards. In 1950, for example, half the colonists had no formal education at all; only 17% had attended school for 5 or more years. Fewer than 2% of the immigrants arriving between 1953 and 1964 had more than 4 years of education (Bender and Yoder 1974). In Angola the colonists were the elite, but in Portugal, ironically, they had represented the poor majority. Although Portugal gave great emphasis to its settlement projects, the overwhelming majority of the Portuguese were employed in commerce, not farming. Portuguese monopolized positions as waiters, bus drivers, bank clerks, and taxi drivers, jobs commonly held by Africans elsewhere in colonial Africa. Two-thirds of the professionals in 1968 lived in only 3 of the 16 districts—Luanda,

Benguela, and Huambo; 42.4% lived in Luanda, the capital. Thus, spatially, the European presence was very concentrated.

Independence, Civil War, and Ethnopolitical Strife

Colonial rebellion, Cold War superpower competition, and ethnic conflict all played a role in Angola's decades of internal conflict since independence. Angolan independence was preceded by years of local rebellion and urban violence, and the rise and fall of aspiring African nationalist groups whose only apparent common objective was the end of colonialism. Regional and not necessarily national interests frequently took precedence in the struggles against Portugal, and these regional concerns emerged arrayed against one another after independence.

Prior to independence, three major divergent movements emerged in the struggle against the Portuguese: the Angola National Liberation Front (FNLA), founded in 1962, headed by Holden Roberto, and backed by Mobutu's Zaïre, France, and the United States; the Soviet-backed, Marxist-Leninist, Popular Movement for the Liberation of Angola (MPLA), founded in 1956, and headed by Agustinho Neto; and the moderately socialist National Union for the Total Independence of Angola (UNITA), founded in 1966, headed by Jonas Savimbi, and backed by certain Portuguese business interests, China, and South Africa. Ideological, personal, ethnic, and regional differences split the groups and kept them feuding to the advantage of Portugal. The MPLA's strongest and most consistent support came from the Mbundu people, situated in the general vicinity of Luanda and its hinterland. The FNLA's strongest support came from the Bakongo people further north, especially around Uíge. UNITA's support was probably the broadest, including all areas south of the Benguela Railway, and focused upon the Ovimbundu on the Bié Plateau.

Competition between the liberation groups intensified following a series of events in Portugal. In April 1974 the 42-year-old Portuguese dictatorship was overthrown in a bloodless coup, led by army officers disenchanted with the slow rate of decolonization in Africa and convinced that the revolutionary wars could be settled only by a political solution. Years of fighting had strained the Portuguese economy and demoralized its people. Thus, facing military defeat in the colonies and civil war at home, war-exhausted Portugal granted Angola and Mozambique their independence in 1975.

When the Portuguese withdrew from Angola on November 11, 1975, they conferred independence and sovereignty on the "Angolan people" without recognizing any particular movement as representing Angola's legitimate government. All three nationalist movements held independence ceremonies in the areas they controlled. What followed was one of the bloodiest and costliest civil wars ever fought in Africa. The MPLA, whose strategic position at the heart of the country and in control of Luanda gave it an advantage that overcame its numerical inferiority, expanded its sphere of domination with the aid of Soviet arms. Later they were also supported by 13,000 Cuban troops. The FNLA and UNITA announced that they had formed a coalition government in Huambo in the central provinces, but that was not to last. A northward thrust drove the FNLA forces back toward the Zaïre border, and a group of mercenaries hastily recruited in Europe by the CIA in support of FNLA suffered heavy casualties without having any real effect upon the course of the conflict (Stockwell 1978).

As the FNLA retreated and the northern front failed, the MPLA, now bolstered by Cuban troops, accelerated its southward push on the divided armies of UNITA. The headquarters of UNITA, Huambo, fell to the MPLA–Cuban forces in January 1976, shortly after the withdrawal of South African forces began. South African troops had entered Angola to protect installations along the Cunene River that were deemed vital to irrigation projects in South-West Africa (Namibia), which was under the control of South Africa at the time. The MPLA's southward push stopped short of a confrontation with remaining South African forces, and there were reports that the MPLA's leaders had allayed South African fears concerning the Cunene dams. As the MPLA's military successes increased, the Organization of African Unity, its member states, and a growing number of countries around the world recognized its leaders as the legitimate government of Angola.

The impacts of Angolan independence and civil war on the displacement of population were varied but significant. The first big movement concerned the Portuguese and has already been mentioned. The Ovimbundu, who had migrated from the central plateau (mainly the provinces of Huambo and Bié) to work on Portuguese coffee plantations in Uíge Province, fled at the outbreak of civil war and returned to their homeland in the central provinces. Many Bakongo who had fled to Zaïre during the anti-Portuguese violence, from 1961 to independence, returned to Angola. The third major population displacement has been out of the central provinces, as the UNITA insurgency grew stronger in the mid-1980s. Most of these refugees or their descendants live precariously in shantytowns and high-rise, "vertical" slums.

At the regional level, in white South Africa and Rhodesia (Zimbabwe), the anticolonial, Soviet-backed Cuban troops had an immediately destabilizing effect. The presence of Cuban forces in southern Africa gave a new dimension to Soviet influence on the continent, and it altered the balance of strength in the rapidly disintegrating buffer zone (Namibia) north of the South African republic. Most immediately, Rhodesia felt the impact of its new, greater vulnerability: Mozambique, Angola's ideological ally, closed its Rhodesian border and announced a state of war in March 1976. The white minority regime in Salisbury (Harare) faced greater isolation than ever before, as its troops fought battles against insurgents along its borders, negotiations with black leaders failed, and South Africa remained aloof.

Pretoria faced its own challenge in South-West Africa, where the course of events appeared also to be overtaking the slow and unproductive negotiations whose objective was social and political reform acceptable to the South African government. The South West African People's Organization (SWAPO), viewed by many at the time as a legitimate representative of Namibian African aspirations, was excluded from the talks: the pattern was all too familiar. Throughout the 1980s, the United States supported UNITA and called for the removal of Cuban troops from Angola and an end to Soviet assistance. Negotiations between Angola, Cuba, South Africa, and the United States led to the withdrawal of Cuban troops in 1989. The Angolan government and UNITA signed a cease-fire in 1991, with the proviso that multiparty elections be held the following year. These elections were won by the MPLA candidates and UNITA returned to the bush. Then, in 1994 the Angolan government and UNITA signed the Lusaka Accords, which sketched out a United Nations–approved plan to end the conflict. The Lusaka Accords provided for the integration of UNITA troops into the Angolan army and the demobilization of both government and UNITA fighters. Conflict resumed in 1998; the MPLA-dominated government renounced unity government, expelled UNITA members, and declared Savimbi persona non grata. Fighting continued into 2002, ceasing with the death, in that year, of the UNITA leader.

The Angolan civil war had local and regional as well as international contexts. At the local level were the three main competing "parties," each drawing its members mainly from a single ethnic group. At this level the struggle was about controlling the national levers of power and gaining access to land, jobs, resources, and wealth. At the regional level, anticolonial, communist boots on the ground in Angola (Cuban and Angolan soldiers) showed the racist, anticommunist, white minority governments in southern Africa that the unthinkable alternative of black majority rule—and even Communist rule—could be forced upon them. Superpower proxy involvement in the conflict raised the stakes of the game and elevated combat to far deadlier levels with modern training, modern automatic weapons, tanks, and land mines.

Indisputably, the human costs of the Angolan civil war have been incalculable and far higher than they would have been without outside intervention. Locally the conflict persisted far longer than the superpower rivalry that gave it such ideological impetus at the beginning. Certain larger questions may not have answers but deserve careful thought. At the regional level, for example, did the presence of Cuban troops in Angola prevent South Africa from invading and installing a friendlier government? Did the superpower proxy war in Angola contribute to the demise of apartheid or did it increase the intransigence of white South Africa and ultimately prolong the trauma of apartheid? On the other hand, the system was beginning to break down in the 1980s. Did Angola hasten its end?

Zimbabwe achieved majority rule in 1980 after years of international opprobrium and armed struggle by fighters based mainly in Marxist Mozambique and supported by the Soviet Union. But by the time South-West Africa became independent Namibia in 1990, with SWAPO in power, Soviet-sponsored international communism was a spent force and there was only one superpower, the United States. South Africa, an international pariah, began moving toward a more pragmatic solution to relations between the races in the late 1980s and held its first all-racial elections in 1994.

The Economy: Agriculture, Mining and Manufacturing

As long ago as 1893 a Portuguese government official remarked of Angola: "We have good land and labor to work it; we lack only capital and initiative." Such an observation could have been made at independence in the mid-1970s, and even today, if one excludes the mining sector. Although it is true that land has been cleared and preempted for export crops (coffee, cotton, sisal, and others) and mines have opened and processing plants have been built, much of Angola's vast agricultural potential and mineral wealth still awaits the input of capital, the development of domestic markets and transport systems, and a political system conducive to economic investment. The first half of the twentieth century saw virtually no development of Angola's mineral resources other than manganese ore from the highlands east of Luanda (Figure 29-12). Exports were dominated by coffee and cotton grown by Angolan farmers under strict colonial supervision, and for which the colonial government paid prices below free-market levels.

Between 1970 and 1975, Angola's economy showed buoyancy and confidence. The GDP (held at constant prices) grew at approximately 11% per year. Most of the stimulus was derived from price and quantity increases in exports that flowed in part from Portugal's revised and more liberal economic policies. Exports and manufacturing diversified, while cotton and coffee producers received higher prices than ever before. Coffee was the major crop in terms of export sales, and during the 1950s Angola was the largest coffee producer in Africa and consistently the fourth largest coffee producer in the world. Up to independence, cotton was Angola's second agricultural export, and for decades it was grown by local farmers supervised by European concessionaires who were given a buying monopoly at fixed prices. As in Mozambique, the system was abolished in 1961 in favor of free cultivation and sales. Output did not materially change as a result: yields per acre increased, while the area under cultivation decreased. Other important cash crops included sisal, sugar, maize, and oil palm products, most of which in the colonial era were produced on large corporate plantations and *colonatos* rather than on traditional African holdings.

Up to independence, agriculture regularly contributed from 20 to 25% of GNP, engaged about 88% of the population and was regionally specialized and highly localized. The exodus of the Portuguese after independence and ensuing civil war changed the agricultural and industrial economies in profound ways. First of all, the production of virtually all of the plantation crops introduced by the Portuguese plummeted at independence (Figure 29-13). Coffee, formerly a principal export of Angola, provides a dramatic illustration. From 1961 to 1975 Angola and Côte d'Ivoire competed for the top spot in coffee production. In 1974, the year before independence, Angola was the number-one African producer of coffee. By 1976, it had slipped to eighth place; by 1990 to eighteenth place; and by 2003, almost to the bottom, well below tiny countries like Malawi, Equatorial Guinea, and even Liberia, itself engaged in civil strife.

The decline in agriculture was not limited to cash crops but affected almost all food crops as well. Food crop production declined slowly for maize and rice after independence and just slightly for tubers (cassava, sweet potatoes) and millet. With the exception of rice, cultivation of each of these crops has recovered and now exceeds pre–civil war levels (Figure 29-14). With the exception of cassava, per capita food crop production is just a fraction of what it was before independence. The World Food Program of the United Nations indicates (FAO 2004) that Angola has received over 1 million metric tons of food aid from 1998 to 2002 and 2.3 million metric tons of food relief over the 10 years from 1993 to 2002. Food aid imports are a necessity for Angola and will be for the foreseeable future until farming recovers.

Close to the same proportion of the Angolan population was engaged in agriculture at the close of the twentieth century (85%) as at independence (88%), but the nature of the involvement of many Angolans in agriculture has changed. They have moved from a "colonial semi-poverty to autonomous self-subsistence of a very low kind" (Birmingham 1995: 93). The government means to change the agriculture sector and is investing in crops such as coffee, coconut palm, sugar, and sisal, formerly grown on large Portuguese estates and by Angolan smallholders. Fortunately, the nation

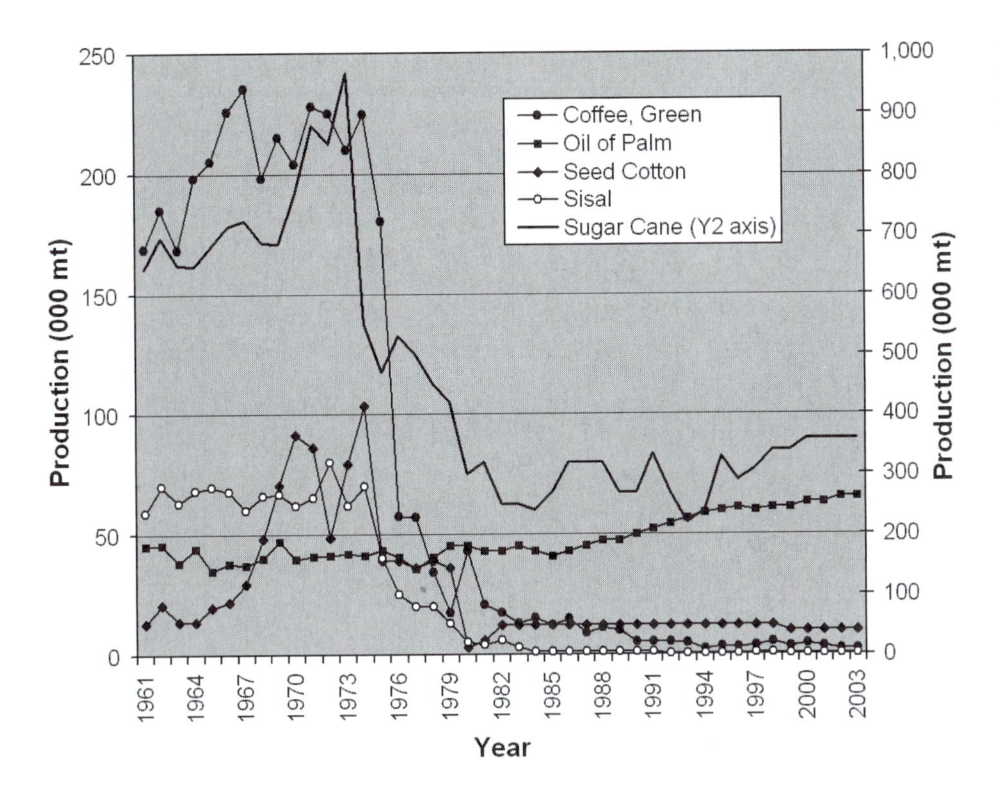

FIGURE 29-13 Angolan cash crop production, 1961 to 2003.

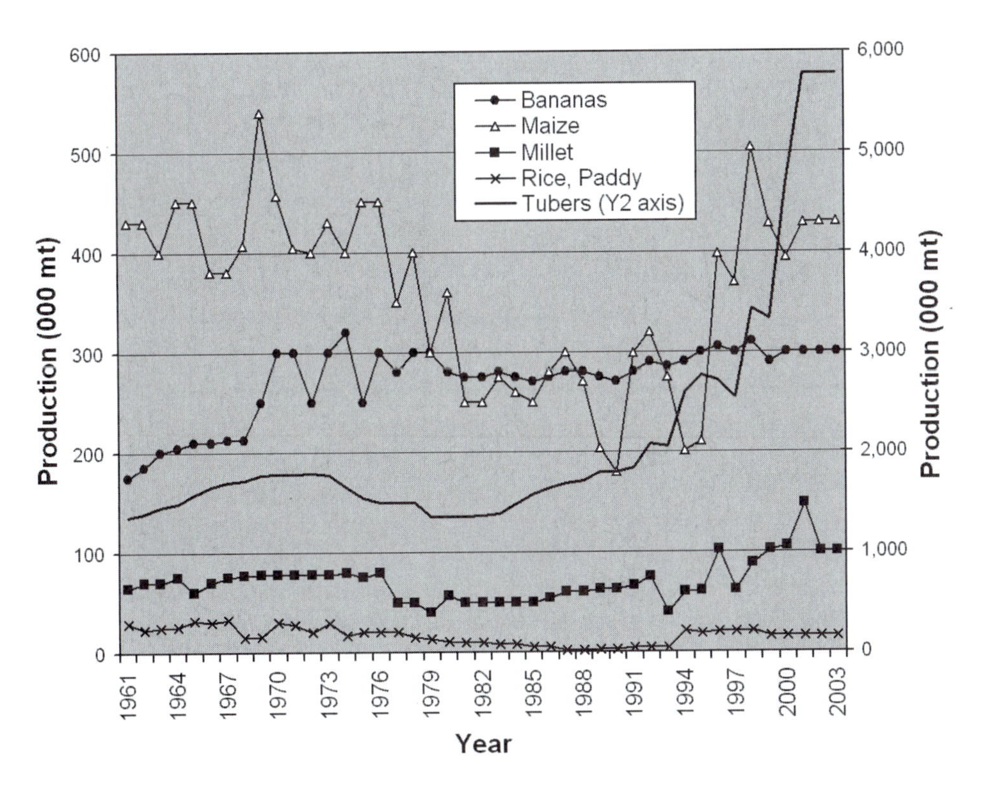

FIGURE 29-14 Angolan food crop production, 1961 to 2003.

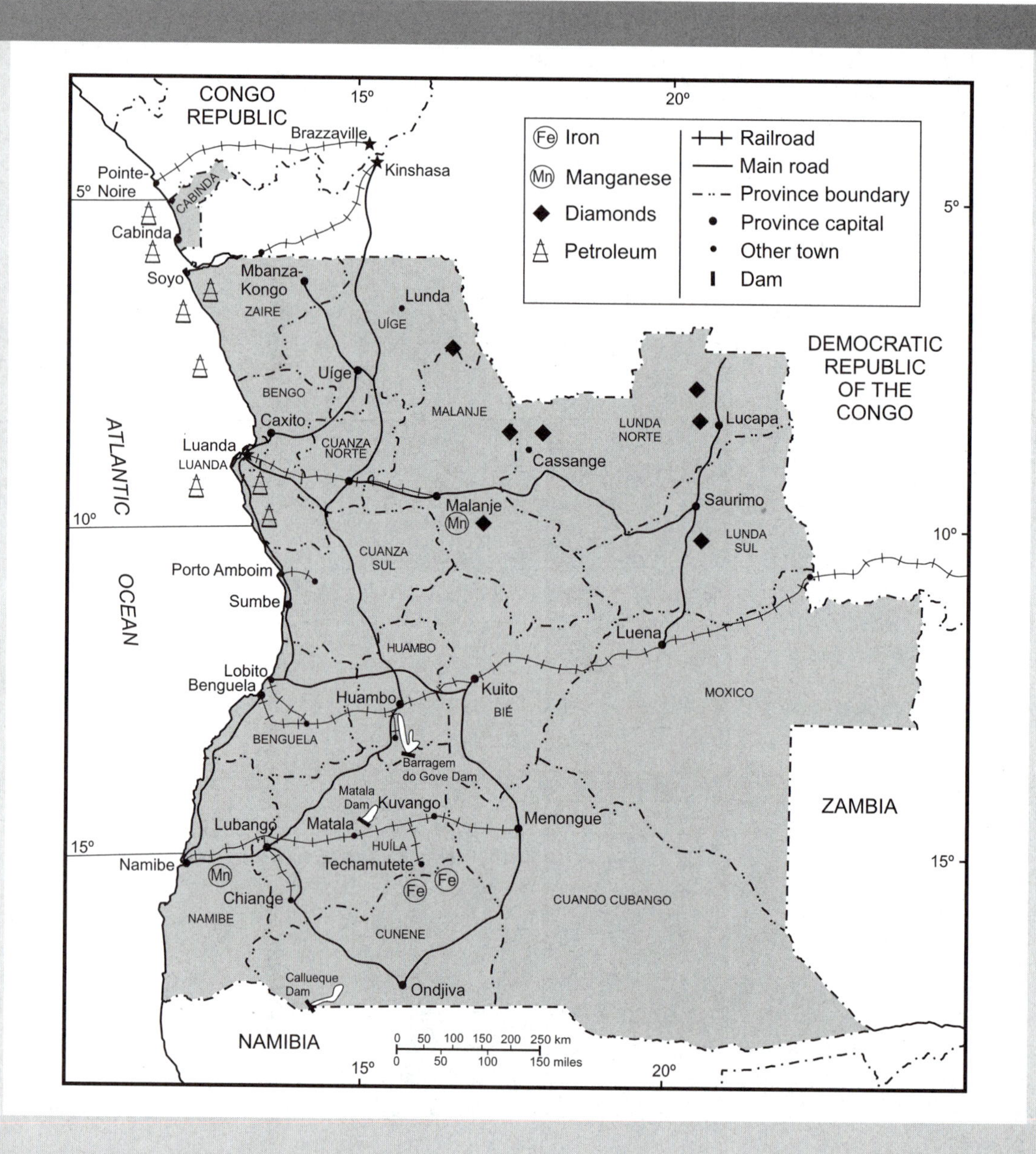

FIGURE 29-15 Infrastructure and minerals in Angola.

has natural resource wealth that could be invested in developing agriculture and manufacturing. There are lessons to be learned from Nigeria's love affair with oil and neglect of agriculture.

Angola has a rich and diversified mineral base, possibly one of the richest (and least explored) in Africa. Until the 1950s little was known of this mineral potential, and mining was confined to diamonds and manganese, both in the northern region. By the mid-1970s, mineral production also included petroleum, iron ore, copper, manganese, and sulfur (Figure 29-15). There are phosphates, bauxite, uranium, zinc, lead, and gold as well. Petroleum has headed the list of exports since 1973 and promises to be Angola's

BOX 29-3 CABINDA

Cabinda is a 2,800 square mile (7,252 km^2) territory on the Atlantic coast sandwiched between the Republic of the Congo and the Democratic Republic of the Congo and separated from Angola proper by the Democratic Republic of the Congo's only outlet to the sea (Figure 15-17).

Until the discovery of oil in 1966, the Angolan enclave was neither a major asset nor a liability to Angola or to the Portuguese. Timber, coffee, and cocoa were its major exports. In the weeks immediately preceding Angolan independence, however, Gulf Oil was pumping 160,000 barrels of oil per day from what is possibly one of the largest oil fields in Africa. That yielded the Angolan treasury about $40 million a month and represented 85% of Angola's oil production. Since 1973, oil has been Angola's leading export.

Cabinda's oil and location were critical geopolitical issues in the postindependence Angolan civil war. Each of the major liberation movements (MPLA, UNITA, and FNLA), was committed to keeping Cabinda a part of the Angolan state because of its assets. However, a separatist movement–the Front for the Liberation of Cabinda, formed in 1963–sought sovereign status for the enclave and declared its independence in 1975. The pronouncement went unheeded, and Cabinda remains a part of Angola. During the civil war, Cabinda was occupied and controlled by the Soviet-backed MPLA. About half of the territory's 80,000 inhabitants reside in the city of Cabinda.

The people of Cabinda face a similar situation to the people who live in the Niger Delta of Nigeria. Their resources are the major contribution to the national GDP, but local people have little or no control over the resources and realize little benefit. As in the Niger Delta, separatists have kidnapped oil workers. The Angolan government has rejected complete independence for Cabinda but appears willing to discuss autonomy. Thus far, however, no progress has been made.

principal source of foreign revenues for years to come. Annual production during the early 1980s was about 125,000 barrels a day, increasing to about 500,000 barrels a day from 1989 to 1995. Since 1996, exploration has been increasing: 25 significant deepwater petroleum fields have been discovered, and production was 890,000 barrels a day in 2002–03 and 1.25 million barrels a day in 2005, and it is expected to increase to 2 million barrels a day by 2008 (Coakley 2003). Oil accounts for about 90% of Angola's exports, 80% of government revenue, and 45% of GDP (Coakley and Szczesniak 2001) and has done so for some time. Although the government has a considerable war debt, it has had a favorable balance of trade over the last few years. The Cabinda field has been the most productive (Box 29-3), accounting for 70% of Angolan production. Petroleum reserves have been estimated at 12 billion barrels, while natural gas reserves are 279 billion cubic meters (Coakley and Szczesniak 2001).

Angola, long Subsaharan Africa's number-two producer of oil, is expected to take Libya's place as the second largest African producer after Nigeria. The Angolan government, Angolan entrepreneurs, and a wide variety of international interests are investing heavily in petroleum production: $20 billion was scheduled to be invested between 2002, when a lasting peace seemed to be in the offing, and 2007. A liquefied natural gas (LNG) plant in Soyo, located at the mouth of the Congo river, is scheduled to begin operation in 2007 (USEIA 2004).

During the long years of civil war, much of Angola was off-limits to mineral exploration. The oil sector, however, remained unaffected and under the control of the Angolan government, which used oil revenue to purchase weapons. In addition, an international accounting firm has stated that almost a quarter ($4.2 billion) of Angola's estimated oil revenue earned between 1997 and 2002 could not be accounted for by the Angolan government (Human Rights Watch

2004). It is alleged to have disappeared through "rent-seeking," as such corruption is euphemistically termed, but the Angolan government hotly denies any wrongdoing.

Angola possesses both alluvial and kimberlite deposits of diamonds. The largest mine, located about 35 miles (56 km) south of Saurimo in Lunda Sul Province, is situated over a large kimberlite pipe and is operated by the Sociedade Miniera de Catoca, a joint venture of Angolan, Russian, Israeli, and Brazilian interests. Australians, Canadians, Portuguese, and South Africans are involved in other joint ventures. Alluvial diamonds are mined principally in the Cuango Valley in Lunda Norte Province, and on a smaller scale elsewhere. It is estimated that Angola possesses over 300 million carets of diamonds, 90% of gem quality. The U.S. Geological Survey reports that during the last years of the civil war diamonds worth $1 billion were annually exported from Angola. About half were produced by artisanal operations not under the control of the government and smuggled out of the country; it is believed that UNITA, operating in the northeastern provinces, routinely accounted for about $250 million worth of smuggled diamonds each year. Despite the best efforts of the United Nations to combat the sale of conflict diamonds, they entered the market nevertheless and, in the case of Angola, helped UNITA rebels purchase weapons and land mines, maintain troops, and prolong the civil war. Exploration for nickel, platinum group metals, and other minerals is under way, hampered in part by the presence of land mines.

During the colonial era, manufacturing never received high priority in the scheme of development. But increased economic autonomy during Portugal's final years of rule resulted in expansion of Angola's raw material processing industries and growth in import substitution industries. Food-processing industries, led by sugar refining, fish product preparation, and vegetable oil extraction, were the most developed. Beverages, tobacco, and textiles were important growth industries, with slowly expanding domestic markets.

Angola faces many of the difficulties encountered elsewhere in Africa in its struggle toward industrial development. The two major handicaps have been scarcity of skilled labor and limited purchasing power, added to which have been the destructive effects of war. Angola imports almost all of its finished goods (and most of its food). Viewed spatially, Angola's commercial economy is linear and insular. Primary development has occurred in and around the major towns and resource regions that are linked together by three railway arteries running parallel to each other and perpendicular to the coast (Figure 29-15). Secondary development has occurred along the coast and highways that connect these arteries. Between these development axes and associated growth centers lies the periphery: subsistence farming, empty but potentially productive agricultural lands, and untapped mineral resources.

Little attempt has been made to integrate the various regions into viable units. Luanda, the capital, for example, has poor communications with all but the northern region and like Maputo, is a primate city. Like Mozambique, Angola has several distinct development corridors, but unlike Mozambique, only one is dependent on the transit traffic from the mineral-rich interior beyond its borders. At present, that rail link to the Democratic Republic of the Congo is in need of rebuilding. An Italian company has expressed interest in its renovation (U.S. Department of Commerce 2001). The three ports in Angola are functioning but need to be upgraded. The country is working toward rebuilding destroyed industry and infrastructure and to link the three isolated centers of the country together on a southern Africa electric grid.

Angola has possessed an important fishing industry centered in Luanda, Lobito, Benguela, and Namibe. The cool, nutrient-rich Benguela Current supports large numbers of fish. Angola's fish industry was massively larger than Mozambique's prior to independence but in both countries fish exports virtually stopped at independence. Whereas production increased in Mozambique after the departure of the Portuguese, Angola's fish industry has not recovered completely. The Angolan government has created an investment climate attractive to private business and production, however, and exports have been rising steadily since the mid-1990s (Figure 29-16).

Northern Angola contains the greatest diversity of natural resources and economic activities in the country. Its principal city and port is Luanda (population 475,300 in 1970, but 2.7 million today), on which the inadequate northern transport network focuses. The Luanda railway is Angola's oldest, and currently most degraded, dating back to the 1880s and originally intended as a trans-Africa line. It terminates 264 miles

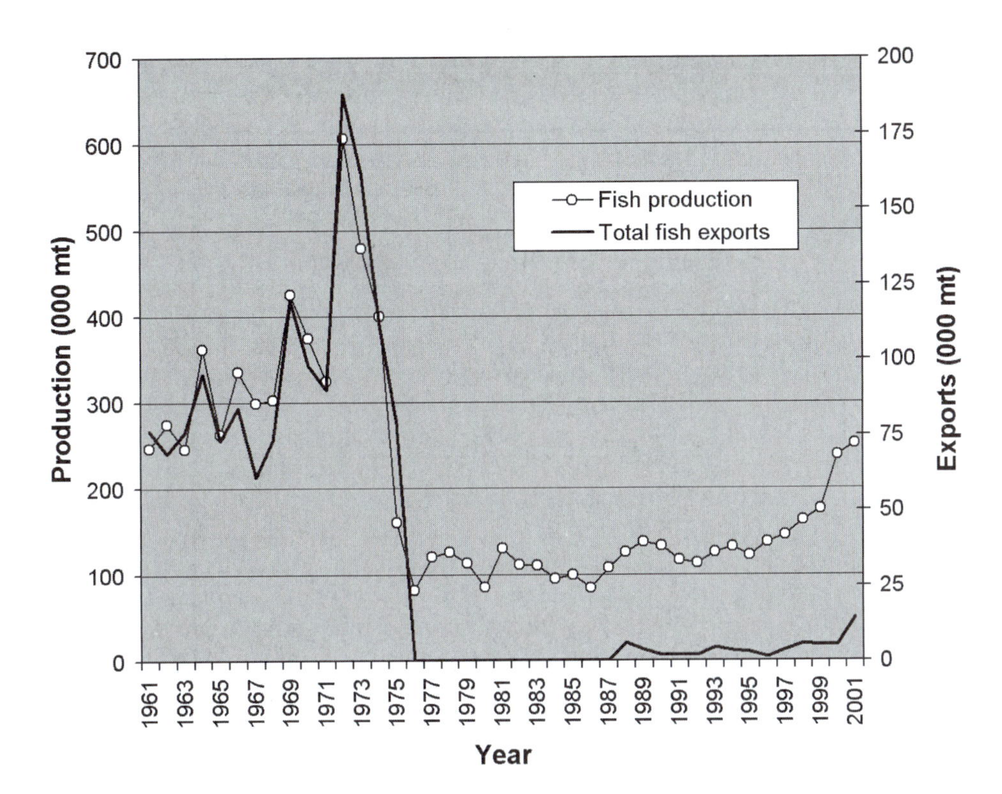

FIGURE 29-16 Angolan fish production and exports, 1961 to 2002.

(423 km) inland at Malanje (71,600), a formerly important agricultural, transshipment, and administrative center on the plateau (Figure 29-15) that lost its luster during the civil war. The Malanje area is a principal producer of maize. The region was formerly a large center for the cultivation of cotton, tobacco, and sugar, but these crops have declined in importance since decolonization. To the east were the profitable cotton areas of the Cuango Valley, cultivated during the time of the Portuguese but abandoned at independence. West of Malanje are once-productive manganese mines.

During the colonial period the region's economic mainstay was coffee, almost all of it being *robusta*. The major coffee-producing area centers on the Uíge Plateau. Prior to the uprisings of 1961, one-third of the regional harvest came from Angolan farms. The insurrection of March 1961, marking the onset of organized opposition to Portugal's rule, in one night took the lives of about a thousand coffee growers and threatened to set aflame all of northern Angola. Damage to the coffee plantations was only moderate: equipment was destroyed but the plants were left untouched, since the rebels hoped to inherit the abandoned European holdings—which they did (Abshire and Samuels 1969). The largely Ovimbundu labor force fled.

Along the moister coastal plains north of Luanda, sugar and cotton are grown, and in the lower Cuanza Vailey southeast of Luanda, market gardening is being extended using by waters diverted from the Cambambe Dam. This is to meet the food requirements of the capital, which had depended heavily on Malanje, and other areas.

Luanda itself remains Angola's largest port and chief manufacturing city. It experienced a phenomenal rate of industrial and urban growth in the 1960s and early 1970s following the discovery of oil both locally and in Cabinda (44,600). Its population doubled during the 1960s, and the whole townscape was changed by a building boom producing modern avenues, skyscrapers, theaters, schools, and hospitals. Its industries include oil refining, chemicals, textiles, food processing, car and truck assembly, and a host of others, many of which were geared to the relatively affluent Portuguese population concentrated in the city prior to independence. Luanda drew to it countless thousands of Angolans in search of jobs and a better way of life, only to find disappointment and deplorable living

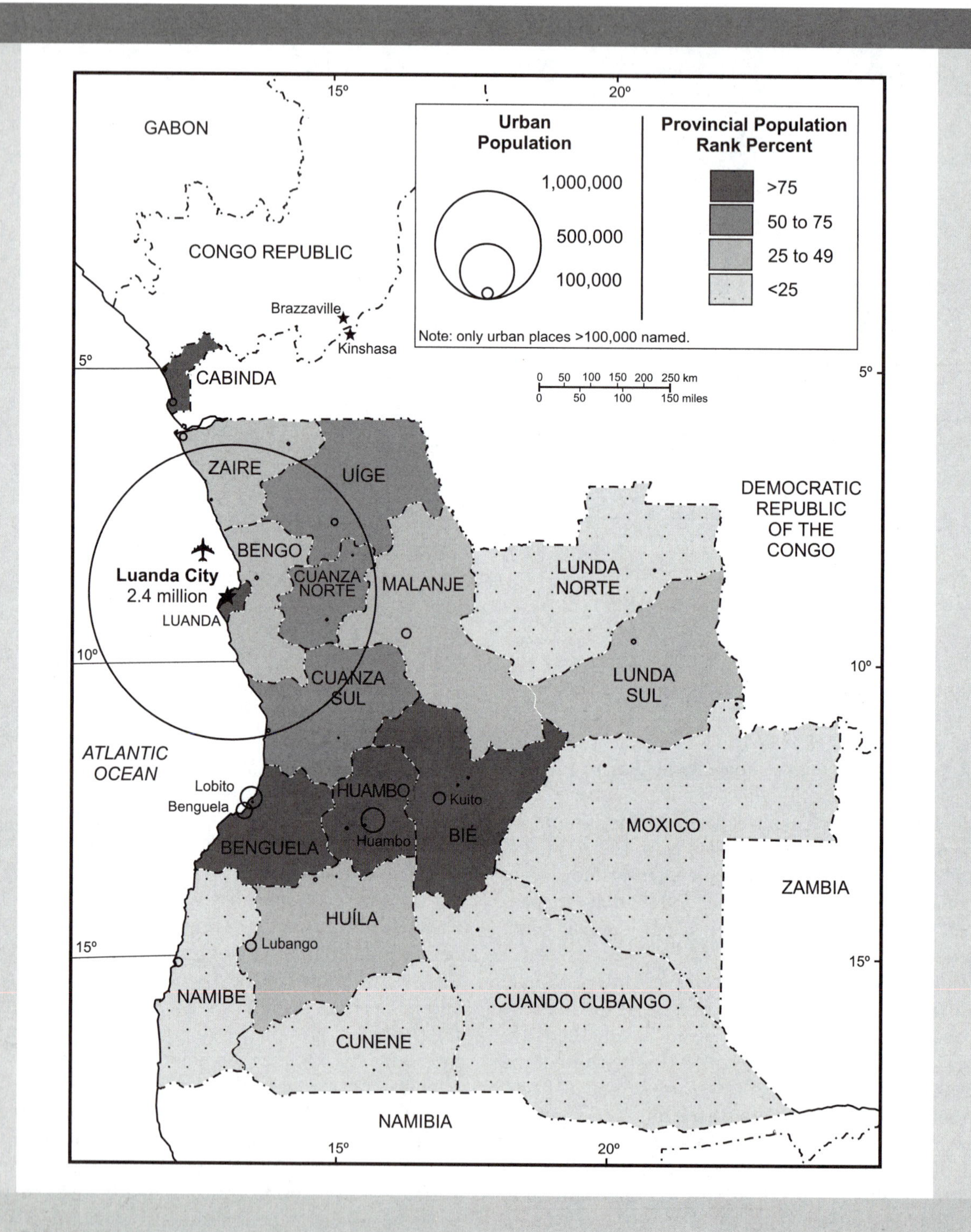

FIGURE 29-17 Urban population and percentile provincial population densities, Angola 2004.

conditions in the *muceques* (slums) that ring the city proper and contain one-fourth of Luanda's metropolitan population. Total Portuguese withdrawal and fighting near the city in late 1975 brought industry to a halt, and even greater economic hardship to its people. Over the decades of civil war, hundreds of thousands of Angolans have left their home areas and migrated to Luanda in search of peace and stability. The population of the greater metropolitan area was close to 3 million people in 2004 (Figure 29-17), an anomalously high population density for such an overwhelmingly rural country.

Central Angola focuses on the Bié Plateau and the Benguela Railway, which traverses it from the port of Lobito to connect with the interior rail system at the border with the Democratic Republic of the Congo (Figure 29-15). The Benguela Railway, which follows an old slave route across Angola, was started in 1903 from the port of Benguela just south of Lobito to connect with the recently discovered copper riches of Katanga. The original concession was held by a British company headed by a personal acquaintance of Cecil Rhodes. The company experienced numerous setbacks in building the line, which reached the eastern border only in 1929, and the copper mines in 1932. Since then the line has been beset by difficulties stemming from the control of alternative transport routes from Zambia and the Democratic Republic of the Congo to the coast: at times the DRC, then Zaïre, preferred to ship its copper and manganese through Kinshasa and Matadi; Zambian exports during and after the Rhodesian blockade and prior to completion of the Tan-Zam Railway put severe strains on the already over used route; and during the Angolan civil war, Zambian and local traffic was again disrupted. Most of the internally displaced people who have not migrated to Luanda are located on the fringes of the towns of central Angola.

The two major economic nodes along this central route were Huambo (formerly Novo Lisboa) and Lobito–Benguela. Huambo (population estimated at 95,000 in 1970 and 173,600 today) is the primary regional center for the rich and agriculturally productive Bié Plateau. This densely settled region produced much of Angola's maize and important quantities of *arabica* coffee. Between Huambo and the coast, in the lower and drier areas around Cubal, east of Lobito, vast plantations of sisal, often Angola's third most valuable agricultural export existed during the colonial period.

Lobito (70,000 in 1970 and 137,300 today), which handles a more specialized cargo than Luanda, but approximately an equal volume of traffic, has one of the finest natural harbors in Africa. Protected by a 3-mile-long (5 km) sandspit parallel to the shore, the harbor is deep, but the facilities need renovation. Lobito developed into Angola's second most important industrial city after the Benguela Railway was extended to it from the port city of Benguela in 1920. Today Lobito's main industries include cement works, food processing, and general metal fabricating.

Southern Angola has lagged behind all but the remote eastern regions in economic prosperity, a condition that has become worse in the years since independence. The main foci during the time of the Portuguese were the Rocadas and Matala settlement schemes along the Cunene River, the rail terminus and fishing port of Namibe (formerly Moçamedes), and more recently the Cassinga iron ore mines near Techamutete, which are connected by rail to the Namibe lines east of Matala (Figure 29-15). The mines came into operation in 1957 and by 1970 were producing over 6 million tons of high-grade hematite ore each year. Before the civil war, iron ore was Angola's fourth ranking export by value: the government of Angola expects to reestablish iron mining in the near future. The southern region is less favorably endowed with rain than the rest of the country, thus its agriculture is extensive rather than intensive, and cattle are the mainstay of the rural economy.

CONCLUSION

Mozambique and Angola have common colonial and postcolonial experience but differ in at least one important respect. The peoples of both countries had been involved with the Portuguese for centuries; both adopted militant versions of Marxism–Leninism at independence; and both experienced years of civil strife. Civil war in Angola was longer and more devastating, however. Angola faced the reality of independence amid a staggering array of problems and obstacles and did not meet the challenge. Apart from the lack of preparedness so familiar upon colonial withdrawal, Angola's bridges, railroads, and other facilities were heavily damaged; population had been severely dislocated, and regional animosities intensified.

The totalitarian political ideology borrowed from the Soviet Union was grafted on the Portuguese legacy of authoritarian governance and incorporated into the winning faction's framework for "development." Making matters worse, the insurgency continued until 2002. The human impact of the prolonged conflict has been perhaps close to a million dead, 4 million displaced, and tens of thousands maimed (mainly from land mines). Angola has accumulated an enormous war debt, and it will take years to rebuild the infrastructure necessary for development. Fortunately Angola has the natural resources to finance its recovery. The question remains whether it will use them wisely or squander them.

During the civil war, the great opportunities presented by oil and diamond wealth constituted a motive to fight and a means to do so (Hodges 2004). In the end, the overwhelming petroleum wealth controlled by the state was the factor that tipped the balance in its favor. It is ironic that Mozambique, relatively poor in resources in comparison to Angola, has fared better over the last 20 years, profiting from its location near the industrial giant, South Africa. If Angola manages its vast wealth moderately well, it should be able to do as well as Mozambique.

If the "natural resource wealth" hypothesis of Hodges (2004) and the United Nations Human Development Report (2003) is correct, the vast natural resources possessed by Angola constitute an inherently corrupting force that will continue to be difficult to manage for the good of the nation rather than narrow interest groups. For Angola, this hypothesis can only be tested by time. It is a country of considerable potential, but it will be some years before its peoples can fully reap its benefits.

BIBLIOGRAPHY

Abshire, D. M., and M. A. Samuels, eds. 1969. *Portuguese Africa: A Handbook*. New York: Praeger.

Bender, G. J., and P. S. Yoder. 1974. Whites in Angola on the eve of independence. *Africa Today*, 21(4): 23–37.

Birmingham, D. 1995. Language is power: Regional politics in Angola. In *Why Angola Matters*, K. Hart and J. Lewis, eds. London: James Curry, pp. 91–95.

Coakley, G. J., and P. A. Szczesniak. 2001. *The Mineral Industry of Angola*. U.S. Geological Survey Mineral Report. Reston, Va.: USGS. http://minerals.usgs.gov/minerals/pubs/country/.

Coakley, G. J. 2003. The Mineral Industry of Angola. U.S. Geological Survey Mineral Report. Reston, Va.: USGS.

Courtois, S., N. Werth, J. L. Panné, A. Paczkowski, K. Bartošek, and J. L. Margolin. 1999. *The Black Book of Communism: Crimes, Terror, Repression*. Cambridge, Mass.: Harvard University Press.

Duffy, J. 1962. *Portuguese Africa*. Cambridge, Mass.: Harvard University Press.

FAO. 2004. FAOSTAT. Rome: Food and Agriculture Organization of the United Nations. http://apps.fao.org.

Mozambique, government of. 1987. *Census of Population*. Maputo: National Institute of Statistics. www.ine.gov.mz/.

Mozambique, government of. 1997. *Census of Population*. Maputo: National Institute of Statistics. www.ine.gov.mz/.

Hance, W. A. 1975. *The Geography of Modern Africa*. New York: Columbia University Press.

Harris, M. 1958. *Portugal's African Wards*. New York: American Committee on Africa.

Hodges, T. 2004. *Angola: Anatomy of an Oil State*, 2nd ed. Bloomington: Indiana University Press.

Human Rights Watch. 2004. Some transparency, no accountability: The use of oil revenue in Angola and its impact on human rights. *Human Rights Watch Report*, 16:1(A). New York.

MAP/MSU Research Team. 1997. Smallholder cash-cropping, food-cropping and food security in northern Mozambique: Summary, conclusions, and policy recommendations. Working Paper Number 25. Maputo: Government of Mozambique.

Stockwell, J. 1978. *In Search of Enemies: A CIA Story*. New York: Norton.

United Nations. 2003. *Human Development Report, 2003*. New York: United Nations.

U.S. Department of Commerce. 2001. *Angola Country Commercial Guide FY 2001*. Washington, D.C.: U.S. Department of Commerce.

USEIA. 2004. *Angola Report, 2004*. U.S. Energy Information Agency. http://www.eia.doe.gov/.

Yager, T. R. 2002. *The Mineral Industry of Mozambique*. U.S. Geological Survey Mineral Report. Reston, Va.: USGS. http://minerals.usgs.gov/minerals/pubs/country/.

Yager, T. R. 2003. *The Mineral Industry of Mozambique*. U.S. Geological Survey Mineral Report. Reston, Va.: USGS.

CHAPTER 30

Madagascar

Ecological Diversity and Sustainable Development

The large island country of Madagascar, population 18.5 million (2006), traces its history not only to Africa but also to Asia. Madagascar's African connections were forged substantially by France, the country's colonizer, for the French administered the island as part of their African possessions. But Madagascar's cultural heritage is largely Indonesian, and despite its membership in the African Union and its participation in African diplomatic matters, there has been very little active contact between the island and eastern Africa, its nearest neighbor. Along some parts of the coast of Madagascar, there are linguistic ties with people of the Comoro Islands (Box 30-1) and the mainland. In certain spheres the links with francophone West Africa have been stronger, though not as strong even as the ties with France itself. Madagascar's insular isolation is symbolic of its aloofness.

Madagascar's physiography is unmistakably African. Its remarkably straight eastern coastline lies at the foot of a high escarpment that rises steeply, sometimes in a series of steps, to elevations near 10,000 feet (3,000 m) that are sustained by crystalline and volcanic rocks; westward, this high plateau surface declines to lower elevations, and the basement rocks are covered by sedimentaries similar to those that fill Africa's great basins. Like Africa, Madagascar has few good natural harbors. The east coast has few adequate inlets, and dangerous offshore coral reefs form a hazard to navigation; the country's leading port, Toamasina, occupies one of the few harbor sites available (Figure 30-1). Coastal communication takes place between Toamasina and Farafangana along the Pangalanes Canal, a valuable intracoastal waterway at the foot of the escarpment but behind a raised shoreline. The mangrove-studded west coast tends to be inhospitable to port development; only the northwestern sector of the island provides some good natural harbors, including Antsiranana (formerly Diégo-Suarez) on Madagascar's northern tip—not the most favorable position in relation to the country's core area. In terms of relative location, Toamasina (formerly Tamatave) has the greatest advantage, but the surface connection between Toamasina and the capital, Antananarivo, involves negotiation of the great eastern escarpment.

Climatic conditions are as diverse as the island's physiography. They vary from tropical rainy (monsoon) conditions along the entire east coast, where rainfall exceeds 100 inches (2,450 mm) and hurricanes strike each year, to low-latitude steppe conditions in the extreme southwest. Most of the highland interior and north coast areas are savanna. The island lies in the path of the moist southeast trades, so that much moisture is wrung out of the atmosphere before reaching the lower west side, which suffers from high rainfall variability and strong seasonal regimes. Temperate conditions, similar to those in the highlands of Kenya and Zimbabwe, prevail in the Ankaratra Mountains near Antananarivo. Here the indigenous forest has been cut, soil erosion is among the worst in Africa, and population pressure is high. Less than one-fifth of the forests that once covered Madagascar remain. The highland forest was removed through the expansion of agriculture before the arrival of the French, while European timber concessions removed much of the lowland forests during the colonial period. The loss of the forests is exacerbated for Madagascar because its isolation from the continental landmass prevents natural regeneration of the forest stock. Making matters worse has been the severe erosion that has denuded formerly forested areas of their soils.

BOX 30-1

THE UNION OF THE COMOROS

Between northern Madagascar and the African mainland lies a group of four volcanic islands called the Comoros. The total area of these islands is about 840 square miles (2,170 km^2), and the population was estimated at 350,000 in 1977, 450,000 in 1991, and 630,000 in 2003. In 1960 the French-administered Comoros achieved internal autonomy under the terms of de Gaulle's Fifth Republic. A political crisis arose in 1975 when the voters on one of the islands, Mayotte, indicated a preference for continued ties with France, a position that clashed with the independence-minded peoples of Grande Comore (the largest and most populous island—363,000), Mohéli (the smallest—31,000), and Anjouan (252,000). Mayotte's population (194,000) is mainly Christian, but Islam prevails elsewhere; Mayotte became French as early as 1843, but the other islands fell under French control after 1885. France responded favorably to Mayotte's expressed intentions, but the government on Grande Comore unilaterally declared the Comoros independent in 1975. In retaliation, France cut off aid to the islands.

The Comoros has suffered through 19 coups d'état since independence and much political instability, including a takeover by South African mercenaries. In 1997 Anjouan and Mohéli declared independence from the Comoros, favoring instead a return to French rule. In 1998 they voted to break away from the Comoros, but the situation remained unresolved until December 2001, with the approval by voters of Grande Comore, Mohéli, and Anjouan of a new constitution and a federal system. Under the new constitution, each of the three islands elects its own president, and the federation is headed by the president of the Union of the Comoros. The union presidency rotates every 4 years among the three presidents. Arabic and French are official languages, but Swahili is spoken widely. The legal system is a mix of French and Islamic law that was instituted as part of the new constitution, and it remains to be seen if such a mix between secular and religious law will be contentious.

The Comoros are overpopulated and poverty stricken. The islands have poor transportation links, the land is excessively fragmented, there are few natural resources, the population is rapidly increasing, and educational opportunities are limited. Spices and perfume products are exported, as well as some tropical fruits and tobacco. But there is no manufacturing to speak of, not even in the capital, Moroni (60,000), on Grande Comore. Corn in the lower areas and rice on the higher slopes are staple foods, and there is heavy dependence on the coco palm. The union is a net importer of rice, the staple food. Opportunities are severely limited, the government is still internally divided and dependent on foreign aid, and there is a steady, but telling, stream of emigration, especially to Madagascar. Tourism development may be the answer for the Comoros: the islands have been a favorite destination for tourists, especially South Africans.

Floral and faunal biodiversity on Madagascar is high—and endangered. The diversity of Madagascar's vegetation reflects the wide range of environments that exist on the island. For well over a century, biogeographers and botanists have recognized Madagascar as a treasure of floristic diversity (Lowry et al. 1997), which they attribute to the island's geological history and its geographic position in the western Indian Ocean. Madagascar separated from the African landmass about 165 million years ago, and most of the island has remained above water throughout the evolution of flowering plants, which has facilitated numerous plant colonizations and much evolutionary diversification.

Madagascar has a wide range of endemic fauna as well. There are five primate families comprising 32 distinct species, all of which are native to Madagascar. Most animal groups on Madagascar are relict populations that exist nowhere else on earth—at least outside of the fossil record—and in this respect Madagascar can be thought of almost as a museum. The study of such populations is very important in reconstructing the

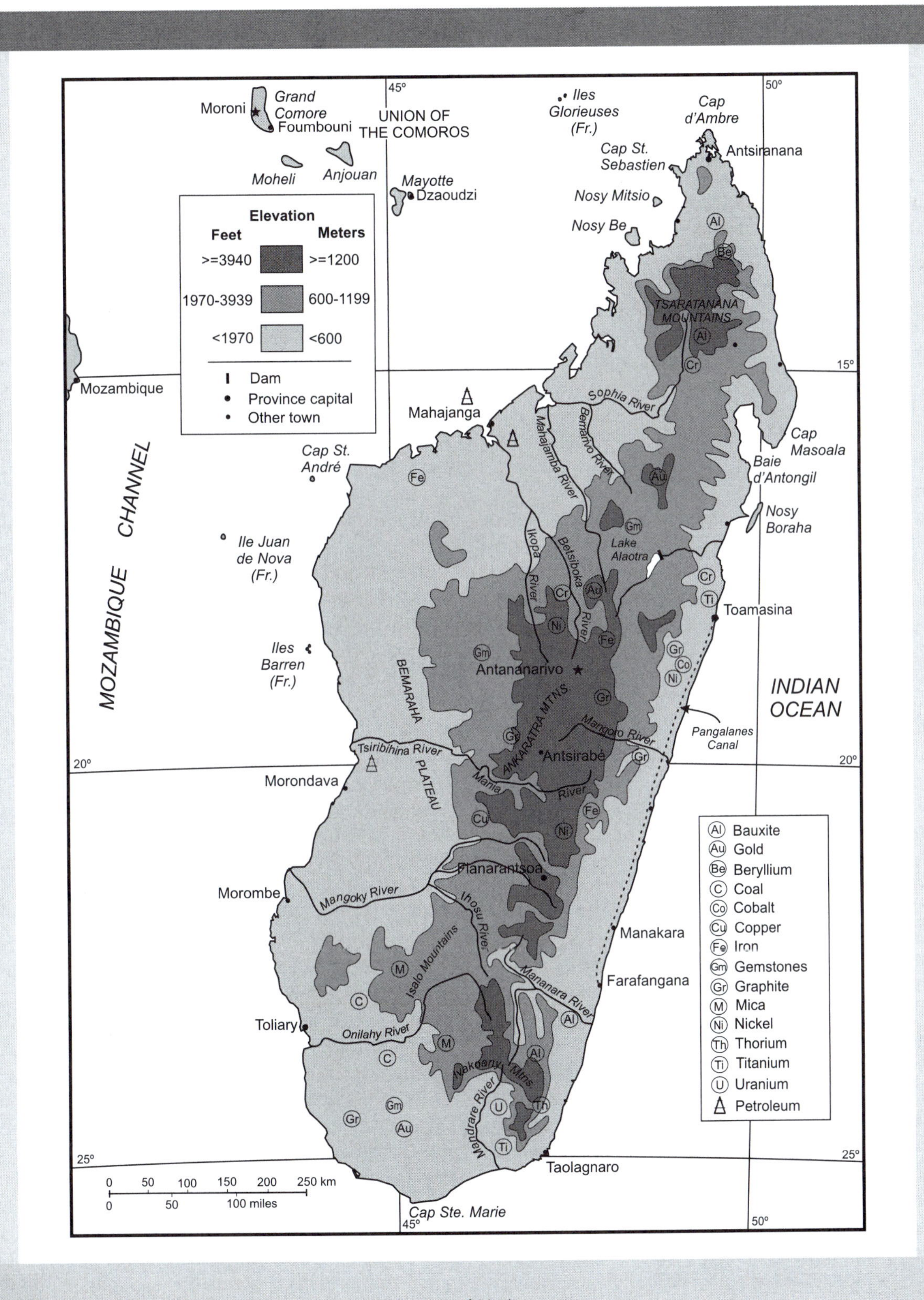

FIGURE 30-1 Major physical features and minerals of Madagascar.

geographies prior to the breakup of Gondwana and in understanding the linkages between modern taxa and those found on Madagascar (Wright 1997). In addition, recent research on global winds has confirmed the long-distance dispersal (LDD) hypothesis: that the Asian character of the flora on the island is due principally to the prevailing winds and westward-flowing currents from Southeast Asia (Muñoz et al. 2004), which aid in the transportation of seeds and plant material.

If names such as Antananarivo, Fianarantsoa (the southern plateau city), and Toamasina appear to differ from those on the African mainland, the contrast confirms Madagascar's linguistic distinctiveness. Madagascar's peoples speak an Austronesian language that is closely related to languages spoken in Southeast Asia, with some Bantu loan words (Box 30-2). The living language that is closest to Malagasy is Ma'anyan, spoken in south Borneo.

Just how Madagascar came to be a Southeast Asian outpost remains a matter for debate. One theory holds that Southeast Asians first reached the African coast, mixed to some extent with the local peoples, and then emigrated to Madagascar. In coastal Kenya, cultural phenomena such as the use of the long house suggest that an early contact of this kind may have occurred. This would account for the fact that most Malagasy have some black African characteristics. The "dual inheritance" of the Malagasy has a distinct geographic distribution, however: Malagasy with more African characteristics are located on the south and west sides of the island, closer to the continent, while in the highlands, particularly among the ruling groups, distinctly Indonesian faces are apparent. In any case, the evidence suggests that Madagascar's modern population first reached the island probably around 1,500 years ago. The individuals who made the journey sailed either with the Northwest Monsoon Current from Southeast Asia to Sri Lanka and then on to Madagascar or directly across the Indian Ocean with the South Equatorial Current. It may be that Indonesians came in groups, possibly at different times, and that the cultural similarities observable today are a result of the cultural convergence of different peoples. In any case, it seems remarkable that the island of Madagascar was uninhabited until the arrival, from over 4,000 miles (6,435 km) away, of these colonists. Large, square-rigged, outrigger boats, a common mode of transportation between the islands of Southeast Asia and a common sight around Madagascar today, were used to make the trip across the Indian Ocean.

Madagascar is one of the few African countries characterized by monolingualism (Wolff 2000): over 90% of the population speaks Malagasy. Others monolingual countries in Africa are Botswana (Setswana), Burundi (Kirundi), Lesotho (Sotho), Rwanda (Kinyarwanda), Somalia (Somali), and Swaziland (Seswati). There are four closely related Malagasy languages spoken on the island. Standard Malagasy, used by the vast majority of people, particularly in the highlands, is expressed in over 20 dialects. The most widely spoken dialect is Merina (35% of the population), but others are Betsimisaraka (19%), Betsileo (15%), Antandroy (7%), Tanala (5%), and Antaimoro (4%). Merina became the literary dialect after the arrival of the British, who first transliterated this dialect into the Latin alphabet. Antankarana is a related Malagasy language, spoken in the extreme north of the country in Antsiranana Province, that is 70% lexically similar to Merina, while Tsimehety and "Southern" are, respectively, 68 and 61% similar to Merina. The Tsimehety ethnic group live in the northern part of Mahajanga Province. Over 3 million people speak Southern Malagasy, just over 1 million speak Tsimehety, and less than 100,000 speak Antankarana (Grimes 2000). The Merina and Betsileo are concentrated on the central plateau, while the other groups live in proximity to the coast. Betisimaraka live mainly along the eastern coast between Maroantsetra, on Cop Masoala, and Mahanoro, some 110 miles (175 km) below Toamasina. Atandroy live in the extreme south of the island.

THE COLONIAL PERIOD

Little is known about Madagascar's territorial organization prior to about 1500. The earliest information dates from the time of the Arab and Persian traders, who established trading posts and slave stations in northern Madagascar during the fourteenth and fifteenth centuries as they had along the East African coast. During that period the Merina (or Hova) peoples probably were settling in the plateau heartland around present-day Antananarivo, and eventually a powerful kingdom arose there. King Andrianampoinimerina, who reigned from 1787 to 1810, united the Merina people. His son and successor, King Radama I (1810–1828), worked with the British to end the slave trade and was

BOX 30-2

LINGUISTIC EVIDENCE FROM AGRICULTURE FOR THE DUAL INDONESIAN-AFRICAN INHERITANCE

The dual inheritance of the Malagasy can be seen in the farming and livestock vocabularies. The ancient Indonesian colonists brought with them across the Indian Ocean many of the crops they depend upon today. The words that they use to refer to these crops seem closely related to words used in Indonesian. Livestock, however, were introduced much later from the African continent. Compare the similarities and differences in these all word lists.

DOMESTICATED PLANTS

English	Malagasy	Indonesian	Swahili
Banana	fontsy	punty	ndizi
Coconut	nio	niu	nazi
Ginger	tamutamu	tamu	mtangawizi
Gourd	voatavo	tavu	dundu
Pacific yam	ovy	ubi	mtubwi
Sugarcane	fary	pari	muwa
Taro	taho	taro	jimbi

DOMESTICATED ANIMALS

English	Malagasy	Indonesian	Swahili
Cattle	omby	termak	ng'ombi
Chickens	akoho	ayam	akoko
Goats	osi	kambing	mbuzi
Guinea fowl	akanga	–	akanga
Sheep	ondri	domba	kondoo

Source: Adapted from Gade (1996).

rewarded for his cooperation with assistance in modernizing his army. As a consequence of military modernization, the Merina were able to extend their control in the highlands. Radama's wife, Ranavalona I, succeeded him (1828–1861) but was an isolationist; she stopped trade with Europeans and forbid the practice of Christianity in 1835. The reign of Ranavalona I was riven by internecine wars, which weakened the kingdom. The succeeding administration was welcoming to European traders and missionaries.

By the last quarter of the nineteenth century, Merina control had been extended to the entire island except the south and the west. Queen Ranavalona II (1868–1883) was baptized in a public ceremony. In 1883 the French occupied Toamasina and 2 years later declared a "protectorate" over Madagascar. Rainilaiarivony, prime minister during the reigns of Ranavalona II and her successor, Ranavalona III (1883–1896), organized resistance to the French, but the last Merina resistance was broken in 1895 and the monarchy abolished in 1896.

Madagascar never ranked high in France's order of colonial priorities (Metz 1994). The first decade of French control witnessed the suppression of two insurrections, but during the 1920s a movement arose in favor of assimilation and *département* (administrative district of France) status for Madagascar. France rejected this effort, and World War II temporarily defused the rising tide of Malagasy nationalism. But a major revolt erupted in 1947, evidence that while the formula of the Fourth Republic might be appropriate for conditions in West and Equatorial Africa, it was inadequate for Madagascar. Tens of thousands of lives were lost as France put down the nationalist rebellion, and the country was still recovering in 1958, when de Gaulle gave France's African dependencies the opportunity to choose independence under the terms of the Fifth Republic. Madagascar chose autonomy within the French Community, and in 1960 the independent Malagasy Republic (*République Malgache*) was proclaimed.

EARLY DAYS OF INDEPENDENCE

The first president of the Malagasy Republic, Phililbert Tsiranana, a Tsimihety, continued a close relationship with France and developed economic relations with apartheid South Africa. Tsiranana, reelected in 1965 and in 1972, was an autocratic leader who favored the coastal peoples over the Merina. Dissatisfaction with his administration grew, and was exacerbated by Madagascar's economic dive after independence. After months of demonstrations, he handed power to the military. Military rule lasted in one form or another until free elections were held in 1993. Economic decline continued and was hastened by the country's turn to the left in 1975, with the election in a referendum of Lieutenant Colonel Didier Ratsiraka as president. Relations with France and other West European countries were severed, and the government proceeded with the implementation of "socialism from above." Closer political and economic relations were established with socialist countries such as the Soviet Union, China, and North Korea.

In conformance with socialist ideology, the Supreme Revolutionary Council supported by the military developed institutions to implement "progressive" ideals. The socialization of the economy, termed "economic decolonization," included extensive nationalization of key areas, as well as agricultural reform. Agricultural reform, in turn, was an attempt to mobilize the "peasants" and to defuse the ethnic rivalries that underlay the political disorders of the early 1970s. Rural reform was based on the traditional *fokonolona* communes and distribution cooperatives. "Fokonolona" is the Merina name for a village or hamlet organization traditionally composed of the heads of all village households. It was the basic unit of local government under the Merina kings, but during the colonial era It was largely suppressed. In 1962 the system was reactivated but under the socialist Revolutionary Council, the fokonolona became the vehicle for rural socialist transformation.

Fokonolona were viewed as a means of making the rural areas more democratic and narrowing the income differentials of the various sections of the population and at the height of the program, there were about 10,000 fokonolona communes. The program was a costly failure. Mentioning the fokonolona program by name, the International Labor Office (ILO) of the United Nations tartly stated that "the usefulness and feasibility of cooperatives in agrarian reform is questionable" (ILO 2001).

President Ratsiraka borrowed heavily to finance his development programs, and the country became heavily indebted. Nationalization of the economy had a negative impact on production and led to food shortages, price increases, and popular discontent, manifestations of which were repressed. Kept in power by authoritarian rule backed up by the military, President Ratsiraka began to abandon some of the unworkable socialist policy only a few years after taking power in 1975, adopting a structural adjustment program of the World Bank. In 1989 President Ratsiraka was reelected in an electoral process that outside observers have called fraudulent. Popular unhappiness with the president led to marches on the capital by hundreds of thousands of citizens in 1991. Amid growing violence, Ratsiraka agreed to implement a transition process to democracy, including a constitution and multiparty elections. In 1993 Albert Zafy, a Tsimihety, won the presidental election. The new president was faced with calls for the development of a federal system to give more autonomy to the regions and diffuse regional tensions. In 1996 President Zafy was impeached for abuse of power, and former president Ratsiraka was reelected for a 5-year term. In 2001 Ratsiraka lost his bid for reelection to Marc Ravalomanana, a wealthy businessman and free trade advocate. Ratsiraka refused to recognize the winner and would not leave the presidential palace. After a military struggle lasting several months, it became clear that his forces would not prevail, and Ratsiraka fled the country.

DEVELOPMENT PROBLEMS AND PROSPECTS

During the colonial era, development was limited by Madagascar's geographic isolation, by the conservatism and ethnocentricity of the Malagasy, and by an undistinguished colonial policy. During the years of French rule, France's response to Madagascar's economic needs and potentials confirmed the island's lowly position among the priorities of Paris. The surface communication system developed slowly; the Antananarivo–Toamasina rail line was completed in 1913 (branches extend northward to Ambohidava and Morarano on Lake Alaotra and southward to Antsirabé), and the Fianarantsoa–Manakara line was laid by 1935; but railway development under French occupation never went farther (Figure 30-2). No rail was laid after independence. Only about 2,000 miles (3,200 km) of road was paved by independence, and since that time, although over a thousand more miles of road has been paved, both road and rail infrastructure have deteriorated. During the wet season whole regions of Madagascar are still effectively isolated. On balance, what the French did accomplish during the colonial period in the development of infrastructure, agriculture, sanitation, and in raising the standards of living was significant, particularly when viewed in the context of demands made from other parts of the French empire and especially in light of what little has been accomplished since their departure.

Since independence, this insular state has experienced a slower rate of economic growth than most African states but growth has been picking up. Madagascar's export economy, based mainly on coffee, has not paralleled that of Côte d'Ivoire or Kenya, and the provision of essential social services and industries has been slow. With the brief exception of the "socialism from above" period in the middle 1970s, like much of francophone Africa, Madagascar has relied heavily on French aid, technical assistance, and markets.

In 1971 Madagascar broadened its donor base and sought financial and technical assistance from China and the oil-rich Arab countries (notably Libya and Kuwait). Expatriates still either run or manage most of the export-oriented agricultural estates, mining operations, and industries, although the state made an attempt during the years of the socialist Supreme Revolutionary Council, to take active control of banking and insurance, energy, transport, minerals, and foreign trade. Ethnic rivalries, especially between the Merina and various coastal groups, or *côtiers,* have compounded the economic ills and have contributed to political strife that has occurred since independence.

Agriculture is the mainstay of the economy, providing employment for close to three-fourths of the population. The contribution of primary activities to exports has been declining. In 1990 primary economic activities accounted for 85% of merchandise exports. But by 2000, according to the United Nations (UNDP 2002), this figure had fallen to 48% and manufactured exports had increased from 14% to 50% of merchandise exports. Agriculture is hampered because the red lateritic soils that cover much of the island are very infertile, severely eroded, and require expensive treatment to enable viable production of food and cash crops. Where volcanic materials penetrate the surface, especially in the central highlands around Antananarivo and Fianarantsoa, better pedologic conditions have resulted. Some of the richest soils are found in the savanna areas and along the valley floors. In most places erosion is due primarily to systematic firing of the bush and forest, and to subsequent overgrazing. In the highland environment, increasing degradation of the soil and the replacement of forests by native grasslands has meant that the land is now used mainly for livestock raising.

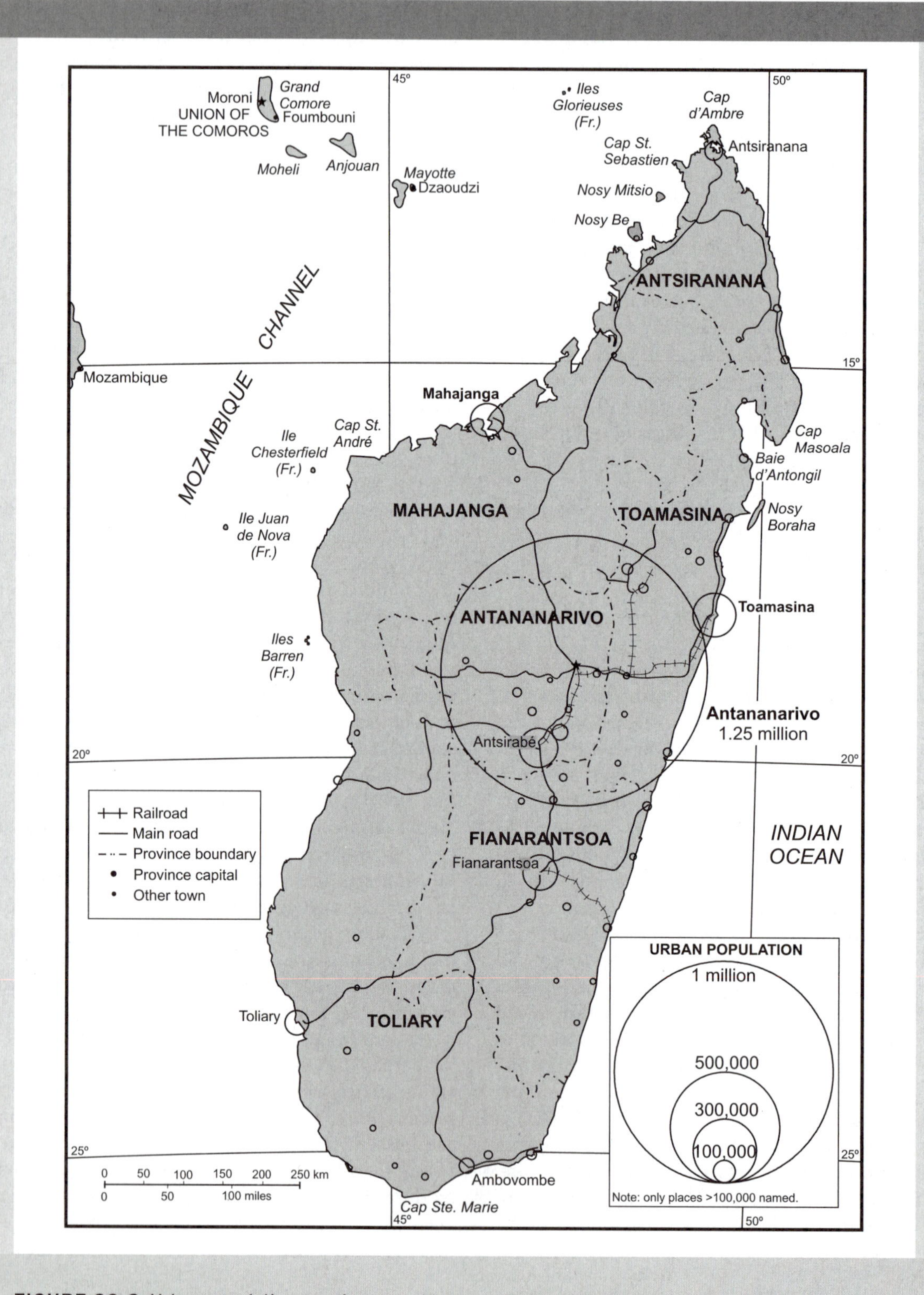

FIGURE 30-2 Urban population, provinces, and transportation infrastructure of Madagascar.

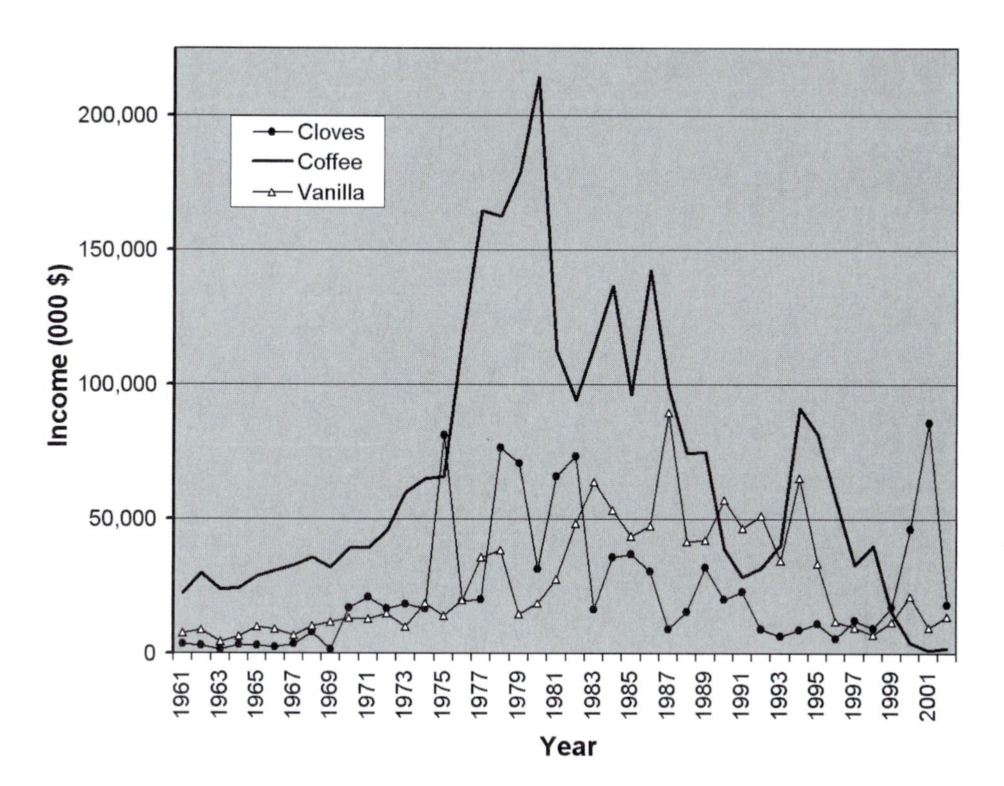

FIGURE 30-3 Value of principal Malagasy agricultural exports, 1961 to 2002.

A wide range of cash crops removes much of the instability experienced by many primary producers. There are six types of spices and condiments grown on the island, a wide range of fruits and vegetables, several oil crops, and sugarcane. Robusta coffee historically has been the leading export, but production and exports of coffee have declined since the peak in 1980. During the late 1990s and early twenty-first century, coffee production skidded into insignificance, and cloves became the leading agricultural export in value (Figure 30-3). Most coffee is produced on small east coast farms badly in need of replanting and reorganization.

From 1990 to 1992 vanilla became the leading agricultural export in terms of value, and up until 1995 Madagascar was the world's leading producer of the crop. Vanilla is a high value-for-weight crop that became popular as an additive to soft drinks. The addition of flavoring agents to drinks increasingly became fashionable in rich countries during the 1990s, and as a result demand for vanilla rose markedly. Incomes of vanilla producers rose dramatically as well. The crop is difficult to grow outside Mexico, the area in which it evolved, where certain insects pollinate the flowers. On Madagascar no insects or birds are attracted to the vanilla flower, and so pollination has to be done by hand. As vanilla became more and more expensive, artificial substitutes were developed. The substitutes do not taste as good as natural vanilla, but the increased supply of the ersatz flavoring caused natural vanilla prices to fall. Both cloves and vanilla are grown along the narrow east coast lowlands (Figure 30-4). These industries periodically suffer from cyclone damage, and recent years have seen severe competition from other clove producers (Zanzibar and recently Indonesia), fluctuating prices in the world market, and competition from synthetic flavorings derived from coal tar.

Migration is being encouraged from overpopulated rural areas, such as the Betsimitatatra Plain around Antananarivo, the borders of Lake Alaotra, and sections of the east coast, to underpopulated regions with an agricultural potential. These include the Betsiboka Valley of the north, and the Morandava and Mangoky basins in the west, where there is great potential for rice, raffia, sugar, tobacco, and livestock. A lack of transport and capital, rather than adverse environmental conditions, have limited development in these regions. Such agricultural colonization, necessary under Madagascar's high population growth rate, is having adverse effects on the environment.

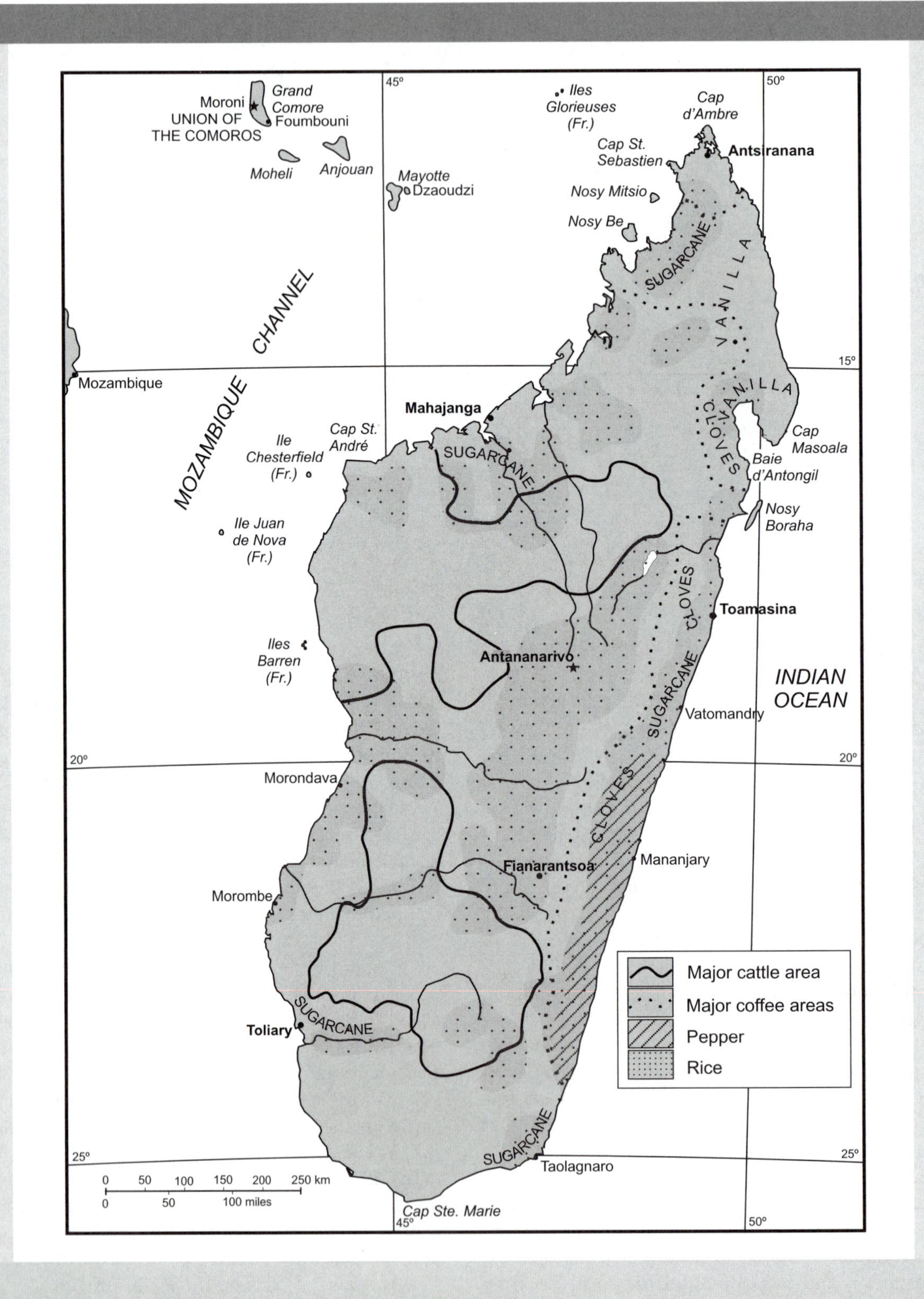

FIGURE 30-4 Principal agricultural regions of Madagascar.

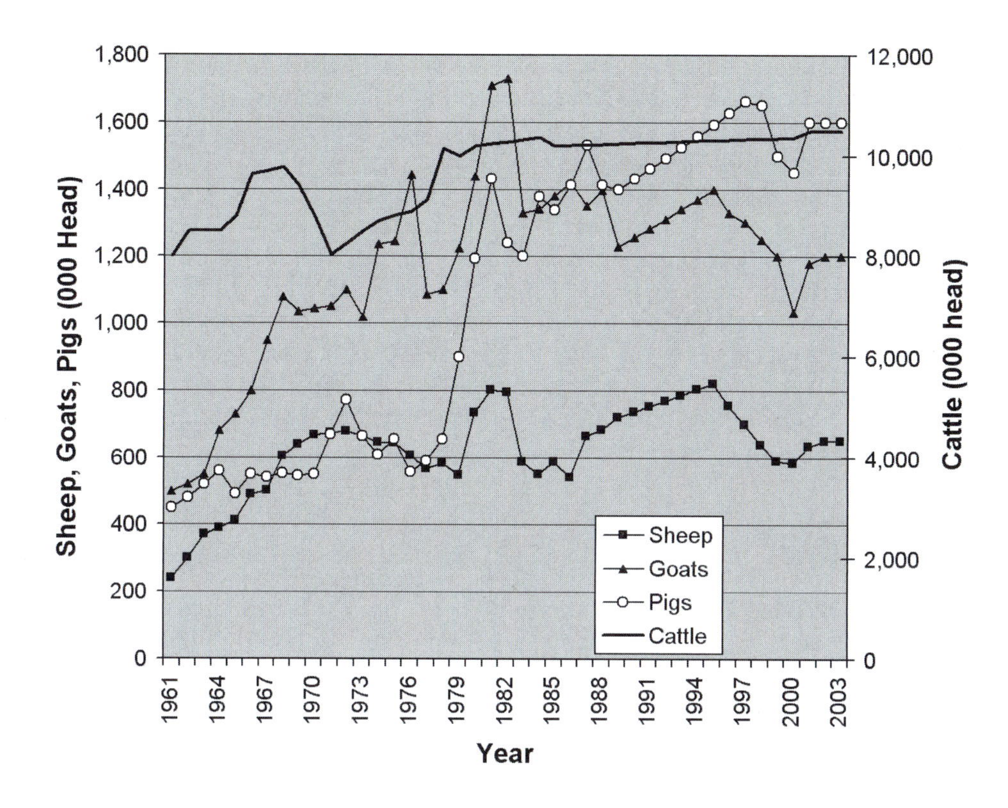

FIGURE 30-5 Madagascar's populations of cattle, goats, sheep, and pigs, 1961 to 2003.

Madagascar's traditional means of livelihood have characteristics of both mainland Africa and Southeast Asia. The similarities with mainland Africa lie more with the pastoral economy than with agriculture, as reflected in the language used to refer to livestock (Box 30-2). Over three-fourths of the island is used for grazing by the 11 million cattle and 3.5 million pigs, goats, and sheep. Unlike most other regions of Africa, cattle, goat, and sheep populations have remained rather steady over the last 30 years, but the numbers of pigs have increased by over a million head and have become the second most kept form of stock after cattle (Figure 30-5).

The livestock densities are highest in the southern and western parts of the island, where almost every person owns some cattle, and the average per capita meat consumption is one of the highest in the less developed world. Cattle are kept as a form of reproducible capital and as a mark of prestige. In addition, they are slaughtered and eaten on ceremonial occasions, which occur rather often in comparison to other African pastoralist cultures, whose people drink milk but seldom eat meat. Selective breeding and controlled range management are rarely practiced. Rather, most of the cattle are allowed to roam and graze at will, with little provision taken against the popular custom of cattle stealing. If more scientific methods were adopted, more meat could be sold within Madagascar, especially in the protein-deficient and densely settled highland areas. There, and on the eastern seaboard, cattle are used primarily for trampling the rice paddies prior to planting and for providing traction.

In several ways, subsistence farming in Madagascar (especially among the Betsileo and Merina) is strongly reminiscent of Southeast Asia: the emphasis is on paddy rice; the farms are small (2–3 acres—about 1 hectare), fragmented, and intensively cultivated; two rice crops are grown each year if climate permits; and terracing of the hillsides is a common feature. Yields are generally low but have been increasing in recent years: average rice yields of 1.8 metric tons per hectare in 1961 had increased to 2.3 metric tons in 2003. In contrast, Egyptian intensive production of rice yields over 9 metric tons per hectare, and Japan's rice fields yield on average over 6 metric tons per hectare. Rice is the main subsistence crop, accounting for 94% of all cereals grown, almost half the total value of agricultural output, and one-third of all cultivated land.

Per capita rice consumption is one of the highest in the world—most Malagasy eat rice three times a day. If greater applications of fertilizer were used, improved rice strains were adopted, and the fields were consolidated, Madagascar could become one of the world's leading rice exporters. East Africa is a logical market, but first Madagascar must feed its own population. Since the late 1960s the gap between supply and demand has widened, and the island is now an importer of rice. As Malthus pointed out over two centuries ago, importing food crops is not necessarily a bad thing, provided the domestic population is able to pay for the imports somehow. Singapore, for example, no longer produces its own rice but Singaporeans, most of whom now work in the industrial and service sectors of the economy, are able to purchase rice on the international market.

Cultivation of rice and other crops destined for domestic consumption, such as cassava and sweet potatoes, has grown rapidly since independence. Madagascar produces close to 3 million metric tons of rice every year, almost 3.5 million metric tons of tubers (70% of which is cassava), and 2.2 million metric tons of sugar cane. Export of Malagasy agricultural products was facilitated in 1994 when the Malagasy franc, which had been pegged to the French franc, was allowed to float. This made Malagasy exports more attractive on the international markets because they were cheaper. On the other side of the coin, imports became more expensive; perhaps benefiting Malagasy domestic industrial production.

During the 1970s, industry was limited to the manufacture of a limited number of consumer goods and the processing of agricultural/pastoral products in Antananarivo, Fianarantsoa, and the three largest ports. Local-born Indians, Chinese, and Comorians controlled much of the industry, and commerce and had not integrated socially with the Malagasy. Years of ineffective government planning and policy implementation, including the nationalization of vital sectors of the economy during the Ratsiraka era, discouraged private investment and encouraged the flight of capital, skilled (mostly expatriate) labor, and managerial expertise. Development today is still spatially discontinuous, there being only small isolated nodes of production separated by larger areas of subsistence, poverty, and outmigration.

Since 1990, however, Madagascar has made some profound changes in its development philosophy and has become a business-friendly environment. The floating of the Malagasy franc helped make Madagascar competitive on the world markets. Also to attract business, the government provided tax incentives and regularized the tax schedule. Although only 15% of the hundreds of new businesses that have blossomed since 1990 are owned by Malagasy, tens of thousands of new jobs have been created. Over half of the businesses that have developed on Madagascar are French, with about 20% owned by Mauritian and South African concerns. Major industries include meat processing, soap making, brewing, tanning, sugar refining, textiles, glass, cement, vehicle assembly, paper, and oil refining. Tourism is growing, as well. Like most developing countries, Madagascar suffers from a grossly inadequate transport network, a small internal market, a shortage of labor and skills, and a high rate of inflation. But with greater regional integration and freer trade, opportunities for Malagasy to improve their lives are increasing.

Mining in Madagascar has been noted for its high-grade chromite ores, graphite, and mica. The country also mines beryllium, gold, rare earths, and gemstones. Madagascar has significant deposits of gold, but most of the gold is mined artisanally and actual output is unknown. The gold was deposited in lode type in mainly Archean formations. Recoverable deposits of bauxite, coal, copper, lead, manganese, nickel, petroleum, thorium, tin, titanium, zinc, and zirconium exist as well (Figure 30-1). Madagascar's poor transportation network (mainly roads) has prevented the exploitation of many minerals (Yager 2003).

In a country where less than 10% of the population lived in urban areas in 1950 (330,000 out of a total population of 4.23 million), close to 30% do so today. The rural–urban ratio is expected to rise to 50% by 2030, when the urban and rural populations will be 16.9 and 16.5 million, respectively. Madagascar's urban structure is characterized by the extreme primacy of Antananarivo, which has a population of 1.25 million (Figure 30-2). With the exception of Antsirabé (181,000), which is located in the same region as the capital city, only the cities of Toliary (114,400), Mahajanga (153,000), Fianarantsoa (162,000), and Toamasina (202,000), regional anchors in the urban hierarchy as province capitals, possess populations over 100,000. Of the remaining top 50 towns in population, the median size is 30,000. The province of Antananarivo, home of the mega-primate city of Antananarivo, is the most urbanized of all provinces. Population density is highest in Antananarivo Province as well (229 persons/mi^2; 88/km^2), twice that of the next ranking province, Toamasina (105/mi^2; 41/km^2). The least dense population is in the lower and drier western side of the island in Toliary and Mahajanga provinces (40 and 34 persons/mi^2; 16 and 13/km^2).

Unique Madagascar has taken some positive steps in creating a climate more favorable for human development than that which existed in the past. Its products are more competitive internationally because its currency is no longer artificially pegged to the French franc. It has made the transition from the one-party state to multiparty democracy. It has opened its borders to international investment and trade. The question remains of how to protect its unique biodiversity while at the same time developing its resources and sustaining large and rapid demographic growth.

BIBLIOGRAPHY

Gade, D. 1996. *Madagascar*. American Geographical Society. Blacksburg, Va.: MacDonald and Woodward.

Grimes, B. ed. 2000. *Ethnologue*, 14th ed. Dallas: SIL International.

ILO. 2001. *Promotion of Cooperatives*. ILO Report V(1), International Labor Conference, 89th session, June 5–21, 2001. Geneva: International Labor Office of the United Nations.

Lowry, P. P., G. E. Schatz, and P. Phillipson. 1997. The classification of natural and anthropogenic vegetation in Madagascar. In *Natural Change and Human Impact in Madagascar*, S. M. Goodman and B. D. Patterson, eds. Washington, D.C.: Smithsonian Institution Press, pp. 93–123.

Metz, H. C. 1994. *A country study: Madagascar.* Washington, D.C.: U.S. Library of Congress. http://lcweb2.loc.gov/frd/cs/mgtoc.html#mg0002.

Muñoz, J., A. Felicisimo, F. Cabezas, A. Burgaz, and I. Martinez. 2004. Wind as a long-distance dispersal vehicle in the Southern Hemisphere. *Science*, 304 (5674): 1144–1147.

UNDP. 2002. *Human Development Report, 2002*. United Nations Development Program. New York: United Nations.

Wolff, H. E. 2000. Language and society. In *African Languages: An Introduction*, B. Heine, and D. Nurse, eds. Cambridge, U.K.: Cambridge University Press, pp. 298–347.

Wright, P. C. 1997. The future of biodiversity in Madagascar: A view from Ranomafana National Park. In *Natural Change and Human Impact in Madagascar*, S. M. Goodman, and B. D. Patterson, eds., Washington, D.C.: Smithsonian Institution Press, pp. 381–405.

Yager, T. (2003). *The Mineral Industry of Madagascar.* U.S. Geological Survey Mineral Report. Reston, Va.: USGS. http://minerals.usgs.gov/minerals/pubs/country/.

INDEX